Lecture Notes in Civil Engineering

Volume 268

Series Editors

Marco di Prisco, Politecnico di Milano, Milano, Italy

Sheng-Hong Chen, School of Water Resources and Hydropower Engineering, Wuhan University, Wuhan, China

Ioannis Vayas, Institute of Steel Structures, National Technical University of Athens, Athens, Greece

Sanjay Kumar Shukla, School of Engineering, Edith Cowan University, Joondalup, WA, Australia

Anuj Sharma, Iowa State University, Ames, IA, USA

Nagesh Kumar, Department of Civil Engineering, Indian Institute of Science Bangalore, Bengaluru, Karnataka, India

Chien Ming Wang, School of Civil Engineering, The University of Queensland, Brisbane, QLD, Australia

Lecture Notes in Civil Engineering (LNCE) publishes the latest developments in Civil Engineering - quickly, informally and in top quality. Though original research reported in proceedings and post-proceedings represents the core of LNCE, edited volumes of exceptionally high quality and interest may also be considered for publication. Volumes published in LNCE embrace all aspects and subfields of, as well as new challenges in, Civil Engineering. Topics in the series include:

- Construction and Structural Mechanics
- Building Materials
- Concrete, Steel and Timber Structures
- Geotechnical Engineering
- Earthquake Engineering
- Coastal Engineering
- Ocean and Offshore Engineering; Ships and Floating Structures
- Hydraulics, Hydrology and Water Resources Engineering
- Environmental Engineering and Sustainability
- Structural Health and Monitoring
- Surveying and Geographical Information Systems
- Indoor Environments
- Transportation and Traffic
- Risk Analysis
- Safety and Security

To submit a proposal or request further information, please contact the appropriate Springer Editor:

- Pierpaolo Riva at pierpaolo.riva@springer.com (Europe and Americas);
- Swati Meherishi at swati.meherishi@springer.com (Asia - except China, and Australia, New Zealand);
- Wayne Hu at wayne.hu@springer.com (China).

All books in the series now indexed by Scopus and EI Compendex database!

J. N. Reddy · Chien Ming Wang · Van Hai Luong ·
Anh Tuan Le
Editors

ICSCEA 2021

Proceedings of the Second International
Conference on Sustainable Civil Engineering
and Architecture

 Springer

Editors
J. N. Reddy
Department of Mechanical Engineering
Texas A&M University
College Station, TX, USA

Chien Ming Wang
School of Civil Engineering
The University of Queensland
St. Lucia, QLD, Australia

Van Hai Luong
Faculty of Civil Engineering, Ho Chi Minh
City University of Technology (HCMUT)
Vietnam National University Ho Chi Minh
City (VNU-HCM)
Ho Chi Minh City, Vietnam

Anh Tuan Le
Faculty of Civil Engineering, Ho Chi Minh
City University of Technology (HCMUT)
Vietnam National University Ho Chi Minh
City (VNU-HCM)
Ho Chi Minh City, Vietnam

ISSN 2366-2557   ISSN 2366-2565 (electronic)
Lecture Notes in Civil Engineering
ISBN 978-981-19-3305-9   ISBN 978-981-19-3303-5 (eBook)
https://doi.org/10.1007/978-981-19-3303-5

This Springer imprint is published by the registered company Springer Nature Singapore Pte Ltd.
The registered company address is: 152 Beach Road, #21-01/04 Gateway East, Singapore 189721,
Singapore

Preface

On behalf of the organizing committee, it is our pleasure and great honor to welcome all of you to the Second International Conference on Sustainable Civil Engineering and Architecture 2021 (2nd ICSCEA 2021). This conference is organized by the Faculty of Civil Engineering, Ho Chi Minh City University of Technology (HCMUT), and Vietnam National University-Ho Chi Minh City (VNU-HCM), Vietnam.

The conference aims to strengthen collaboration and cooperation between Vietnamese and international civil engineering and architecture scholars. The organizers intend to hold this conference biennially in order to provide an innovative environment for scholars in civil engineering and architecture to share, exchange, connect, and work together to build a better world. Vietnam, as a developing country, views civil engineering and architecture as critical fields for the country's rapid transformation into a developed and modern state.

Following the success of the first ICSCEA in 2019, the second ICSCEA 2021 featured two plenary lectures and over a hundred paper presentations in the fields of civil engineering and architecture across two sets of nine parallel sessions. We are extremely appreciative of all of the plenary speakers' and participants' support and hope that the papers included in this volume of proceedings will inspire you to pursue new directions in your professional work and research.

We would like to express our gratitude to all authors for their participation in this conference. Additionally, we would like to express our gratitude to the staff members of the HCMUT and VNU-HCM for their assistance during the conference's preparation stage. We would like to express our heartfelt appreciation to the dedicated reviewers for their time and efforts in enhancing the scientific quality of the manuscripts. Finally, but certainly not least, we wish to express our gratitude to the sponsors. Without this invaluable assistance, it is difficult to envision this conference succeeding.

Ho Chi Minh City, Vietnam

J. N. Reddy
Editor of the 2nd ICSCEA 2021
Proceedings

C. M. Wang
Editor of the 2nd ICSCEA 2021
Proceedings

V. H. Luong
Editor of the 2nd ICSCEA 2021
Proceedings

A. T. Le
Editor of the 2nd ICSCEA 2021
Proceedings

Organization

International Scientific Committee

Chien Ming Wang
Chung Bang Yun
Dai Jian
Hiroshi Katsuchi
Hitoshi Tanaka
Hoang Nguyen
Hong Hao
Jaroon Rungamornrat
Jin-Ho Park
Junuthula N. Reddy
Keh Chyuan Tsai
Kenichiro Nakarai
Kenji Kawai
Kok Keng Ang
Lanh Si Ho
Lindung Zalbuin Mase
Massimo Menenti
Mike Xie
Nigel K. Downes
Ong Wee Keong
Pruettha Nanakorn
Roderik Lindenbergh
Ryunosuke Kido
Sachie Sato
Salman Azhar
Sritawat Kitipornchai
Sugimoto Mitsutaka
Takeshi Satoh

Teerapong Senjuntichai
Tomoaki Utsunomiya
Yeong Bin Yang
Yong Han Ahn
Yuko Ogawa
Yun-Tae Kim

Local Scientific Committee

Anh-Thang Le
Cao Thanh Ngoc Tran
Chau Lan Nguyen
Chau Ngoc Dang
Dao Nguyen Khoi
Dinh Nhan Dao
Duy Liem Nguyen
Hai Chien Pham
Hien Vu Phan
Hong Tham Duong
Hung Dang Thanh
Huu Loc Ho
Mai Anh Duc
Minh Tam Nguyen
Minh Tuan Ha
Nam Tran Tuan
Nghia Nguyen Hoai
Ngoc Huyen Trang Tran
Ngoc Loi Dang
Ninh Thuy Nguyen
Phu-Cuong Nguyen
Quang Hung Tran
Quang Minh Nguyen
Quang Trung Nguyen
Quoc Hoang Vu
Quoc Viet Dang
Quoc Y. Nguyen
Tan Ngoc Than Cao
Thanh Canh Huynh
Thanh Danh Tran
Thanh Linh Trinh
Thanh Phong Nguyen

Thanh Son Nguyen
Thanh Viet Nguyen
Thi Bich Thuy Nguyen
Thi Thao Nguyen Nguyen
Thi Truc Lieu Tran
Trong-Phuoc Huynh
Truc Thi Minh Huynh
Truong Ngoc Son
Tuan Anh Nguyen
Van Chung Nguyen
Van Hieu Nguyen
Van Phuc Le
Van Quang Le
Van Tuan Tran
Viet Huy Nguyen

HCMUT's Scientific Committee

Anh Thu Nguyen
Anh Tuan Le
Anh Tuan Nguyen
Ba Vinh Le
Cong Hoai Huynh
Danh Thao Nguyen
Duc Duy Ho
Duc Hoc Tran
Duc Long Luong
Hai Yen Tran
Hoai Long Le
Hoang Linh Tran
Hoang-Hung Tran Nguyen
Minh Long Nguyen
Minh Phuong Le
Minh Thi Tran
Ngoc Thi Huynh
Nguyen Ngoc Cuong Tran
Phuong Trinh Bui
Quy Phuoc Dao
Quynh Nga Tra Nguyen
Song Giang Le
Tan Phat Huynh

Thai Binh Nguyen
Thanh Hai Do
Thanh Long Tran
Thanh Phong Mai
Thi Bao Thu Le
Thi Bay Nguyen
Thi Hong Na Le
Thu Ha Nguyen
Tien Sy Do
Tuan Duc Ho
Van Hai Luong
Van Qui Lai
Viet Hai Vo
Vu Hong Son Pham
Xuan Loc Luu
Xuan Long Nguyen
Xuan Qui Lieu

Local Organizing Committee

Chairmans

Anh Tuan Le
Van Hai Luong

Members

Thai Binh Nguyen
Phuong Trinh Bui
Ngoc Thi Huynh
Bich Phuong Thi Nguyen

Sponsors

Contents

Plenary Lectures

Floating Breakwaters: Sustainable Solution for Creating Sheltered Sea Space

C. M. Wang and H. P. Nguyen

Abstract This paper is concerned with floating breakwaters which provide a sustainable solution for creating a sheltered sea space from strong waves. We shall first present motivations for using floating breakwaters, their advantages and some example applications. This is followed by recent advances in materials, modelling, analysis and design of floating breakwaters, and recommendations for future research and developments.

Keywords Floating breakwater · Wave attenuation performance · Hydrodynamic analysis · Modelling

1 Introduction

Oceans cover more than 70% of the Earth's surface, and play a vital role in the global socio-economic development. Oceans have contributed about US$ 3 trillion to the global GDP by 2016, created opportunities for tourism for almost 200 countries, been responsible for nearly 10% of the global trade, and provided humans with enormous supply of seafood, clean water and energy [1]. Ocean developments involve the construction of very large floating structures such as floating roads and bridges, floating energy plants, floating aquaculture platforms and floating buildings and cities. The safety of these structures under wave action is of utmost importance in order to avoid significant economic losses, and negative environmental and social impacts. A conventional solution for protecting marine structures and fragile shorelines is to use bottom-founded breakwaters which can be constructed by using quarried rocks or concrete caissons resting on a foundation at the seabed. While the bottom-founded breakwater solution is effective in blocking ocean waves, it may be not economical for water depths larger than 6 m or for soft seabed due to the enormous foundation costs [2].

C. M. Wang (✉) · H. P. Nguyen
School of Civil Engineering, The University of Queensland, St Lucia, QLD 4072, Australia
e-mail: cm.wang@uq.edu.au

© The Author(s), under exclusive license to Springer Nature Singapore Pte Ltd. 2023
J. N. Reddy et al. (eds.), *ICSCEA 2021*, Lecture Notes in Civil Engineering 268,
https://doi.org/10.1007/978-981-19-3303-5_1

3

For sea sites with soft seabed and deep waters, floating breakwaters may be a more promising solution because this type of breakwaters does not require a huge amount of filled materials and a long construction time for seabed foundations like the traditional bottom-founded ones. When compared to bottom-founded breakwaters, floating breakwaters also possess the following advantages:

- they allow for water circulation through the gap between the seabed and the lower hull of the breakwater, which can result in improving water quality and minimizing impacts on marine ecosystems in the lee side of the breakwater;
- they have a low profile relative to the water surface for all tidal periods, which minimizes their presence on the horizon;
- they can be rearranged, removed, relocated, expanded and downsized more easily.

These advantages of floating breakwaters have stimulated extensive research and developments in this space in recent decades, and floating breakwaters have been used for a variety of purposes including:

- protecting harbors and acting as a berth for vessels. Examples for such applications are the Holy Loch breakwater in Scotland and the floating breakwater in Fezzano, SP, Italy (see Figs. 1 and 2).
- protecting offshore fish farms. For example, a 450 m steel floating breakwater was used to protect an offshore fish farm that is exposed to wave heights up to 11 m in Uwajima, Japan [3].
- providing space for congested coastal cities. The 350 m long floating breakwater in Monaco has been used for not only to attenuate ocean waves but also to provide space for car park and shopping center [4].
- forming a storm barrier for fragile coastlines. Figure 3 presents a design concept of a floating forest which is a mega floating breakwater and a windbreak to protect vulnerable shorelines and coastal infrastructures from waves and winds in extreme storm events [5].
- providing tourism attractions. For example, the weather side of a V-shaped floating breakwater with high waves is suitable for surfing, the lee side having sheltered

Fig. 1 Holy Loch breakwater in Scotland <https://www.maritimejournal.com/news101/marine-civ ils/port,-harbour-and-marine-construction/breakwater_beats_the_weather_at_holy_loch>

Fig. 2 Floating breakwater at Fezzano, SP, Italy, courtesy of Ingemar srl

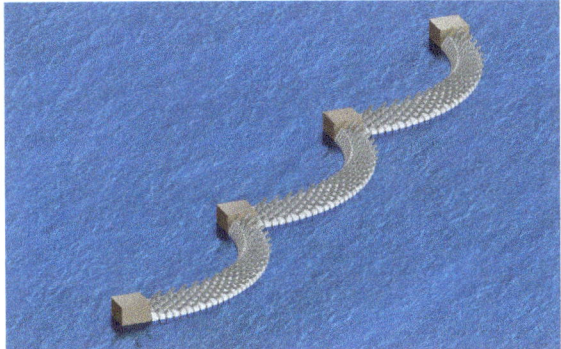

Fig. 3 Design concept of floating forest

waters can be used for swimming, and the lower part of the breakwater can be designed as an artificial reef to attract divers [6].

- creating a calm patch of ocean space for a floating city, as shown in Fig. 4.

Fig. 4 Floating breakwater for protection of floating city <https://www.seasteading.org/>

2 Materials for Floating Breakwaters

Floating breakwaters can be made of timber, plastic, reinforced or prestressed concrete, or steel. Timber was a preferred material due to its large availability in the 1800s when floating breakwater concepts were first proposed, e.g. for use in the Plymouth Sound [7]. Timber floating breakwaters are subject to severe degradation due to large mechanical loads from ocean waves and wood-degrading organisms in coastal zones [8]. Nowadays, timber is usually used in combination with plastic and steel. A common design for such a combination comprises: (i) plastic tubes or boxes that play a role as buoyant structures and are free from corrosion caused by sea environments and organisms; (ii) a timber platform that is used to provide safe sidewalks; and (iii) a galvanized steel frame that is used to provide additional strength to the breakwaters to withstand wave loads and improve abrasion resistance [9].

Another common material for floating breakwater is plastic such as high-density polyethylene (HDPE) [10]. Plastics have a high resistance to biofouling and corrosion [11]. Plastic floating breakwaters are usually light, and can be formed into various configurations. They can also be easily constructed in-land, towed to deployment sites by small vessels, assembled and anchored to the seabed. The setbacks of plastic floating breakwaters are that they suffer from large deformation under wave action due to their flexibility and shallow drafts because of light weights, and also their limited effectiveness in attenuating in long waves where sufficient drafts are required for blocking water particle motions.

Steel has also been used for floating breakwaters (e.g. see Fig. 5). Its main advantage lies in their high tensile strength that allows floating breakwaters to be used in high energetic sea environments. For example, a steel floating breakwater was used at an offshore site (with wave heights up to 11 m) in Uwajima, Japan [3]. Steel breakwaters have another advantage that they can be designed and constructed by utilizing the extensive knowledge, experience, available technologies and supply chains for

Fig. 5 Floating steel breakwaters in Burlington marina, Canada <http://www.kropfindustrial.com/marine/floating-breakwaters>

offshore steel structures and ships. However, steel structures require regular mainte-nance to reduce the rate of degradation caused by corrosion and fatigue cracking [12], leading to high maintenance costs. Experience in the maritime sector shows that main-tenance costs may account for about 20%–40% of the total operating expenditure [12].

An alternative material for floating breakwaters is concrete. Floating concrete breakwaters are usually heavier than their steel counterpart. The heavy selfweight of concrete breakwaters results in a larger draft that increases wave attenuation perfor-mance. Concrete breakwaters also require less maintenance, and have a longer design life than steel breakwaters because well designed concrete exhibits better corro-sion resistance and higher durability. Some estimates (see for e.g. [13]) showed that maintenance costs can be reduced by 10%, and the design life be increased by 20 to 40 years by using concrete for marine structures. These enormous bene-fits have resulted in common use of concrete for floating breakwaters in practice. Two examples of massive floating breakwaters using concrete include the Monaco breakwater and the Kan-on breakwater (Fig. 6) [14]. A main drawback of concrete is with its low tensile strength, which may cause considerable cracks in the structure leading to corrosion of steel bars and the loss of buoyancy. To overcome this draw-back, prestressed tendons and/or steel plates should be used for reinforced concrete floating breakwaters (e.g. see the design of Kan-on floating breakwater in Ujina, Japan [14]).

3   Design of Floating Breakwaters

In designing floating breakwaters, the following issues need to be considered [3, 4]:

- Attenuation performance in long waves (e.g. relative to the breakwater width). Floating breakwaters allow for more transmitted wave energy when compared to bottom-founded breakwaters. Their applications are usually limited to wave

Fig. 6 Kan-on floating prestressed concrete breakwater in Ujina, Japan [14]

periods up to 5 s (e.g. [3]). Such wave conditions are found in nearshore and in relatively calm areas. For more exposed sites, breakwater dimensions have to be increased to achieve a required level of wave attenuation performance. This translates to an increase in costs for materials, mooring systems, manufacturing, and installation.

- Safety and durability of the structure. To maintain buoyancy, the structure must be designed to not only meet the strength requirements against environmental forces and accidental ship collision, but also to minimize corrosion and cracking. In addition, stress concentrations at connections between floating modules, and between the modules and mooring systems may affect the cost effectiveness and safety of the connector design.
- Robustness of mooring systems. To increase the wave attenuation performance, the structure needs to be designed to: (i) increase the portion of incident waves being reflected; and/or (ii) increase frictional and turbulent dissipation [3]. The former case directly leads to an increase in the horizontal forces acting on the structure and hence on the mooring system.

Addressing these issues in a cost-effective manner requires innovations in the design of breakwater's cross-sectional and plan shapes, connector systems and mooring systems. An alternative solution is to integrate floating breakwaters with other purpose marine structures in order to share costs for mooring, operation and maintenance. Next sub-sections will briefly present advances in designs of concrete and steel breakwaters, and integrated systems to be used for wave attenuation and other purposes. We focus on concrete and steel floating breakwaters as they have greater potential to be used in exposed sites to support future developments for ocean space creation and blue economy activities.

3.1 Breakwater Cross-Sectional Shapes/Configurations

A traditional cross-sectional shape of a concrete or steel floating breakwater is a box shape as shown in Fig. 7a. Analysis results for this traditional design showed that the width-to-wavelength ratio may need to be larger than 0.35 to attenuate up to 50% of incident wave energy [2]. In order to increase the breakwater performance in wave attenuation, a simple modification is to extend the outer side walls as shown in Fig. 7b. This modified design allows for larger wave reflection and turbulent energy dissipation due to the extended side walls, while minimizing the increase in the material volume [15]. Another solution is to have horizontal 'wings' (or bilge keels) at the bottom of floating breakwaters as shown in Fig. 7c. These attachments help reduce the roll motion of floating breakwaters, and the amplitude of radiated waves [16]. The wave attenuation performance may also be increased by using two pontoons placed side by side (i.e. dual-pontoon floating breakwaters) as shown in Fig. 7d [7, 17]. A rather similar solution is to have a floating breakwater comprising two pontoons, but with a confined air chamber (see Fig. 7e). The air pressure inside the chamber can be controlled to maximize energy dissipation [18]. Recently, attention is given to floating breakwaters comprising porous elements (e.g. Fig. 7f). The porous elements can provide increased turbulence, reduce breakwater motions and forces on mooring systems as compared to impermeable elements. Example of physical models of porous floating breakwaters can be found in [19, 20].

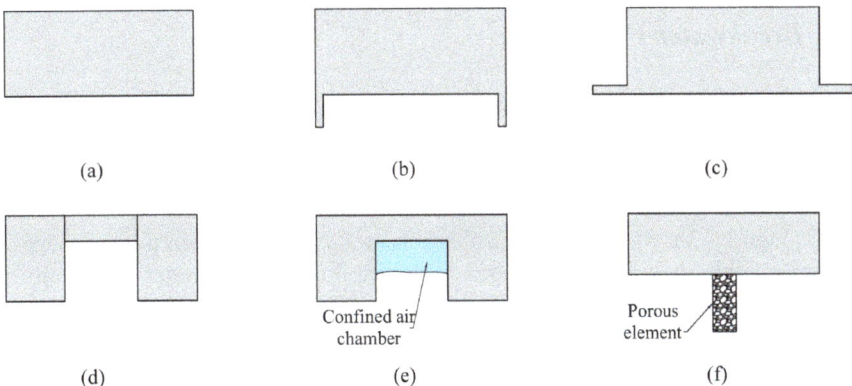

Fig. 7 Illustrations of cross-section designs for: **a** box-type breakwaters, **b** breakwaters with extended side walls, **c** breakwater with horizontal 'wings', **d** dual-pontoon breakwaters, **e** breakwaters with confined air chambers, **f** breakwaters with porous elements

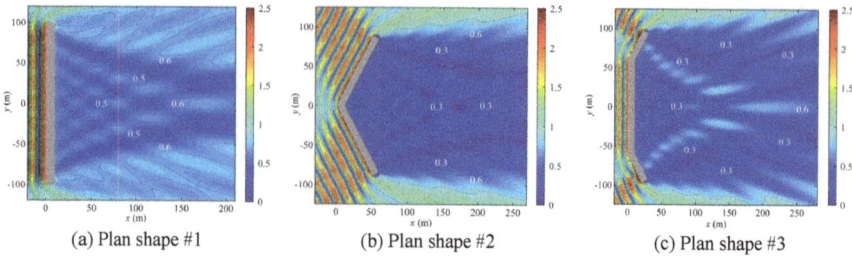

(a) Plan shape #1   (b) Plan shape #2   (c) Plan shape #3

Fig. 8 Contours of wave elevation amplitude normalized with respect to incident wave amplitude for wave period 4 s, water depth 10 m, draft 2 m

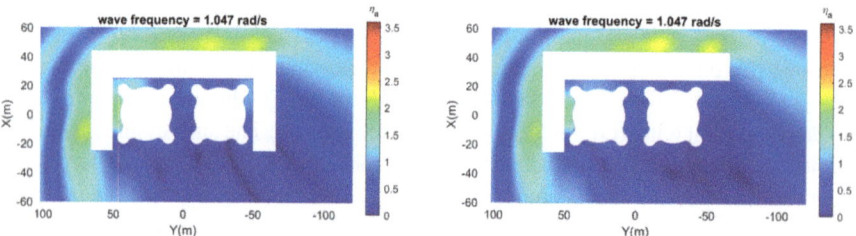

Fig. 9 Contours of wave elevation amplitude normalized with respect to incident wave amplitude for L-shaped and U-shaped breakwaters protecting hydrocarbon storage facility [21]

3.2 Breakwater Plan Shape

In addition to cross-sectional shapes, the plan shape of floating breakwaters is of concern in breakwater design. In order to examine the effect of the breakwater plan shape on wave attenuation, we conducted a numerical study on heave-only box-type floating breakwaters of the same material volume, but with different plan shapes. Figure 8 shows the wave transmission coefficients for different plan shapes. The wave fields show that the use of a V-shaped floating breakwater can result in increasing the wave attenuation performance by 60%, as compared to the conventional straight breakwater. Zhang and Magee [21] recently numerically investigated floating breakwaters with different plan shapes for protecting a floating hydrocarbon storage facility (see Fig. 9). They found that the plan shape of floating breakwater have considerable effects on the transmitted wave field and hydrodynamic motions of floating storage tanks [21].

3.3 Module Connections

Another important task in designing floating breakwaters is to engineer durable and cost-effective connection solutions for floating breakwater modules. The length of floating breakwaters may stretch over a few hundreds or even kilometers depending on the sea space to be sheltered. Such a huge breakwater causes significant difficulties for manufacturing and installation if it is made in one piece. To facilitate the construction process, a modularity approach has been widely adopted in practice. For example, floating concrete breakwaters can be assembled from modules with lengths of about 20 m [22]. The connector system for adjacent modules can be rigid or flexible.

- Flexible connections between breakwater modules can be made by using post-stressed tendons/wires, chains or rubber-bolt systems to connect the top parts of modules. The use of chains has been adopted in early designs of floating breakwaters as given in a design guideline [3], but it seems to be less common nowadays possibly to avoid collision between breakwater modules from relative motions. Post-stressed tendons/wires, or rubber-bolt systems [22–24] can be used to tighten the top parts of adjacent modules, and better control the relative motions. In order to avoid collision, breakwater modules need to be separated at a sufficient distance (e.g. by a rubber cylinder, and by trimming the bottom part of the modules [22, 24]).
- Rigid connections can be made by connecting both the top and bottom portions of adjacent modules; using joining bolts as in Fig. 10, or by welding of the modules as in the case of the Mega-Float [25], or using corner and side connectors as for the Marina Bay performance stage [26]. Floating breakwaters with rigid connections are free from floating module collisions, and do not have spacing between modules where waves can be transmitted. However, a long breakwater assembled from many rigidly connected modules may experience large bending moments causing considerable elastic deformations (as seen for very large floating structures [27]). In addition, installation of rigid connections may be difficult, e.g. divers may be required for installation of connections as shown in Fig. 10b.

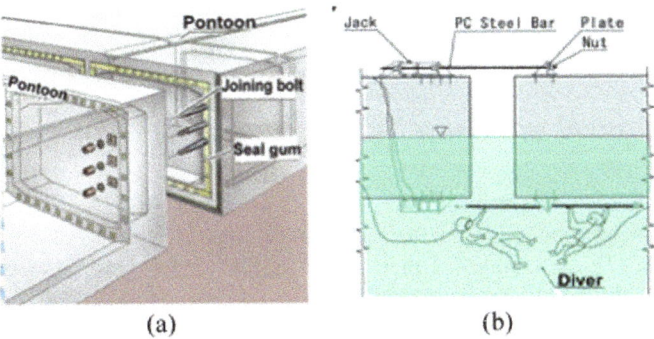

(a)   (b)

Fig. 10 Rigid connection between modules of Kan-on floating breakwater [14]: **a** jointing bolts; **b** work of divers to assemble breakwater

3.4 Mooring Systems

Common mooring systems for floating breakwaters include: piles, mooring dolphin systems and mooring lines.

- Pile and mooring dolphin systems completely restrain horizontal motions of floating breakwaters, but they allow for free motions perpendicular to the water surface [14, 28]. When compared to piles, mooring dolphin-rubber fender systems are more robust and applicable for large floating breakwaters (e.g. see [14]). These solutions are preferred for shallow waters, usually up to 10 m depth [2]. For deeper waters, large wave forces and water depths may cause enormous horizontal forces and bending moments on the piles and mooring dolphin systems, which can significantly increase the costs of the mooring design.
- Mooring lines are normally used for deeper waters [2, 3]. They allow floating breakwaters to move in all directions to a certain extent, which results in reducing wave forces acting on the structure and mooring system. Mooring lines can be catenary or taut [29]. Catenary mooring lines are easier for installation, and less affected by tidal variations [30]. However, they are longer (about 3 times of water depth [29]), and thus may have larger impacts on underwater environments and seabed. In addition, catenary mooring lines provide lower effectiveness in mitigating breakwater motions. Taut and catenary mooring lines are usually made of chains, wires and synthetic ropes. Chains provide catenary effects due to their heavy weights. They also possess good abrasion properties, and are usually preferred for seabed and fairlead segments [29, 31]. Wires and synthetic ropes can provide greater elasticity which is of importance for taut mooring lines. Combination of chains, wires and synthetic ropes may potentially be an optimal solution in terms of costs and safety [29]. An example of such combination can be seen in designs of floating aquaculture platforms [32], where chains are used for seabed and fairlead segments, and synthetic ropes are used for the middle segment that is under certain pretension.

3.5 Integration with Other Purpose Marine Structures

Innovations in ocean engineering promise to provide human with solutions to increase renewable energy production, create more tourism attractions, and deal with the negative effect of climate change on marine ecosystems. Integrating these sustainable climate-resilient developments with floating breakwaters has potential to significantly reduce capital and operational costs, through cost sharing for construction, installation, mooring, foundation, and maintenance. Such an integration approach also enables an efficient use of ocean space, as a given area can be used for multiple purposes. An example of integrated systems is between floating breakwaters and wave energy converters (WECs) [33]. Such an integrated system may help wave energy industry unlock its significant potential that is estimated to be around 2GW only for the region within 50 km from the shore [34]. Another example of integrated systems is to use breakwaters for not only attenuating ocean waves, but also for regenerating marine life or as tourism attractions (as for the proposal of the Webber reefs [6]).

4 Modelling of Floating Breakwaters

A floating breakwater comprises two main components: a floating structure and a mooring system. These components are subject to wave loads which result from fluid motions of undisturbed waves, and the fluid–structure interaction. Modelling the structure, mooring system, fluid motions, and fluid–structure interaction is required for accurate estimates of wave loads acting on the structure, stresses in the structure, mooring forces, and wave attenuation performance.

4.1 Modelling of Structure

Most studies have been carried out by modelling floating breakwaters as infinitely long and rigid bodies in a two-dimensional domain. The hydrodynamic problem of two-dimensional breakwaters can be solved analytically, and the wave attenuation performance can be easily estimated by some approximation formulas (e.g. see [15]) which provide great simplicity for practical engineers. However, two-dimensional modelling may lead to an overprediction of wave attenuation performance; and hence an under-design of floating breakwaters because it does not account for diffraction and radiation effects at the breakwater ends [35, 36]. The two-dimensional modelling approach should only be adopted for a large ratio between the breakwater length and wavelength. Numerical examples in our previous study [35] indicate that such a ratio should exceed 10.

In order to improve the accuracy of the estimates of breakwater hydrodynamic performances, recent efforts have been given to develop three-dimensional modelling of floating breakwaters, e.g. by Diamantoulaki et al. [28]. This modelling approach is also able to account for non-straight floating breakwaters, and three-dimensional interaction between floating breakwaters and the protected marine structure. For example, Wang et al. [37] examined an arch-shaped floating breakwater (see Fig. 3). Tay [38] investigated a floating breakwater with an L-shape in plan. Zhang and Magee [17] studied the wave attenuation performance of breakwaters in consideration of their interaction with the protected floating hydrocarbon storage tanks.

Modelling of floating breakwaters has been developed to account for the flexibility of floating breakwaters that may result from the following sources:

- flexible connections between breakwater modules. As floating breakwaters are frequently built in form of a modular structure with flexible connections between

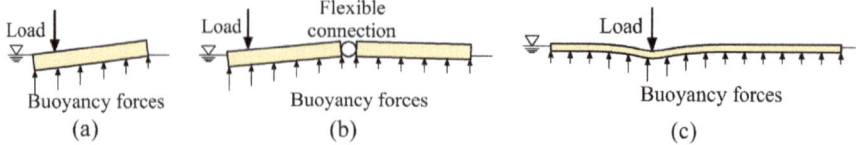

Fig. 11 Global response of: **a** a conventional rigid breakwater, **b** a breakwater with a flexible module connection, and **c** a flexible breakwater under a concentrated static load [18]

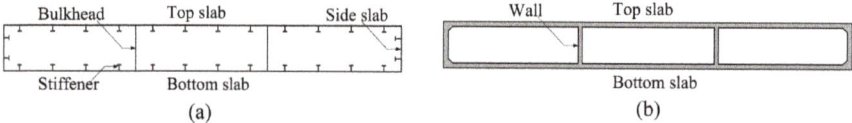

Fig. 12 Example cross sections of: **a** steel pontoon and **b** concrete pontoon

floating modules, their behaviors cannot be regarded as a single rigid body, especially under oblique wave action (see Figs. 11a and b).

- large horizontal dimensions of floating breakwaters with shallow depths, and/or materials with low elastic modulus such as high-density polyethylene. An example is that the Mega-float [39] with 1 km in length, 60 to 120 m in width and 3 m depth, and comprising rigidly connected steel modules may be considered as a flexible structure whose responses under applied loads are dominated by elastic deformations, instead of rigid-body motions (see Fig. 11c).

The flexibility of floating breakwaters can be accounted for by using various modelling approaches available in structural engineering, such as those based on plate or shell theories [40, 41]. The shell theory is more general, and is able to account for detailed structural components including side walls, top/bottom slabs, and stiffeners of the floating structure (e.g. see Fig. 12). However, the complexity of the shell theory cause difficulties for ocean engineers and researchers in the modelling process. In addition, enormous computational resources required for the associated simulations make the shell theory to be resorted to at a detailed design stage. At a preliminary design stage where the main aims are to estimate the wave attenuation performance and hydrodynamic properties, floating flexible breakwaters may be modelled as equivalent plates according to the Kirchhoff or Mindlin plate theories [40]. The equivalent plate has almost the same vibration properties (i.e. natural frequencies and mode shapes) as the realistic flexible breakwater [42]. When compared to the Kirchhoff plate theory, the Mindlin plate theory is able to account for the effects of transverse shear deformation and rotary inertia which are more important for thicker plates. The Mindlin plate theory also provides more accurate stress-resultants as its formulation involves only first derivatives of the transverse displacement and bending rotations.

4.2 Modelling of Mooring Systems

Mooring systems affect motions of floating breakwaters, and they need to be treated in floating breakwater models. Floating breakwaters restrained by piles or mooring dolphin systems usually have only heave motions (i.e. the motions perpendicular to the water surface). However, some designs of connections between piles (or mooring dolphin systems) and the breakwater platform may allow for pitch and roll motions of the breakwater (e.g. see [39, 43]). A common modelling approach for piles and mooring dolphin systems is to impose boundary conditions on the plate model of freely oscillating floating breakwaters to reflect the constraints on the breakwater motions, whereas the interaction of piles and mooring dolphin systems with the fluid is often neglected for simplicity.

If mooring lines are adopted for the station keeping purpose, they can be modelled as equivalent linear springs. The spring stiffnesses for a given configuration of mooring lines can be obtained by using some formulas given in [44, 45]. The linear spring model is simple, but it is unable to account for dynamic behaviors of mooring lines, e.g. time-dependent changes in mooring configurations due to the breakwater motion, and the interaction between mooring lines and the fluid. A more advanced modelling approach was proposed by Loukogeorgaki and Angelides [46], where a procedure was established to integrate a hydrodynamic model of freely oscillating floating breakwaters with a mooring line model for predicting mooring line forces and drag damping. The established procedure requires multiple iterations to accurately account for the changes in breakwater motions due to the mooring line forces and drag damping, and vice versa. However, this modelling approach [46] does not account for inertia of mooring lines which is important for catenary mooring lines. The inertia may be considered by using a lumped-mass model [47] or a finite element (FE) model [29] of mooring lines.

4.3 Modelling of Fluid Motions

Fluid motions are commonly described by using the linear potential wave theory where the velocity potential exists and satisfies the Laplace equation and boundary conditions on the free surface, seabed, at infinity, and the wetted surface of the breakwater (e.g. see [35]). This theory is able to give reasonable accuracy for relatively small wave steepness (i.e. the ratio between wave height and wavelength). A study [48] for wave steepness of about 0.04 and for 2D box-type floating breakwaters showed a good agreement between measured and estimated transmission coefficients. Another study [49] for a larger wave steepness of 0.07 showed that the transmission coefficients were overestimated by up to 35% by using the linear wave theory, and the lowest accuracy of the estimated transmission coefficients was seen when the resonance occurs leading to large motions of the breakwater. The inaccuracy of the linear wave theory for such a case results from its limitation that it cannot account for turbulence and viscosity. This can be overcome by using the Navier–Stokes equations (e.g. see [50]). However, the use of the Navier–Stokes equations requires more computational resources when compared to the adoption of the linear wave equations; and thus, the trade-off between accuracy and computational resources for different wave theories need to be considered carefully.

4.4 *Modelling of Fluid–structure Interaction*

For floating breakwaters without porous elements, the interaction between fluid and the structure is commonly represented by the boundary condition that on the wetted surface of the structure, and the fluid velocity is the same as the structure velocity (e.g. see [35]). For porous floating breakwaters, a special attention needs to be given on how to illustrate the fluid flow through the porous medium. Such an illustration is usually based on semi-empirical formulas that show the relation between the fluid motions on the two sides of the porous medium and the medium properties. Some modelling approaches for floating structures with porous elements can be found in [51].

5 Analysis of Floating Breakwaters

Designing floating breakwaters involves hydrodynamic, structural, and mooring analyses. The hydrodynamic analysis focuses on the wave attenuation performance of floating breakwaters and their motions under wave action and applied loads. Mooring analysis is carried out to examine forces and moments acting on the mooring system, which is needed for designing a cost-effective and robust station keeping system. Structural analysis aims to study the stresses in the floating structure including at module connections.

Traditionally, hydrodynamic analysis of floating breakwaters is performed in two-dimensional regular waves [3]. The transmission coefficient, defined as the ratio between the transmitted wave height and the incident wave height, is used to show the wave attenuation performance of floating breakwaters. For a two-dimensional floating breakwater, the transmitted wave height is almost constant from a certain distance behind the breakwater. This is, however, not the case for 3D breakwaters where transmitted wave heights within a predefined surface area in the lee side may vary significantly (see Fig. 7). Thus, for 3D breakwaters, the transmission coefficient corresponding to the transmitted wave height at a single surface point cannot reflect the overall wave attenuation performance. This results in the need to redefine an index showing the overall wave attenuation performance of floating breakwaters (which may be called a representative transmission coefficient). The representative transmission coefficient can facilitate comparison between different designs of floating breakwaters, and hence allowing one to achieve an optimal breakwater design for maximizing the effectiveness in breaking harsh ocean waves. Michailides and Angelides [52], and Wang et al. [37] made initial efforts by using the mean transmission coefficient over a predefined surface area. Recently, Nguyen et al. [35] proposed an alternative definition of the representative transmission coefficient where within the predefined surface area, the probability for having transmission coefficients smaller than the representative transmission coefficient is equal to a user-defined percentage, say 90%.

Owing to recent advances in modelling of floating breakwaters, a more rigorous hydrodynamic analysis can be performed with consideration of three-dimensional effects, realistic irregular waves, mooring line configurations and module connection properties. The need to perform such an advanced hydrodynamic analysis has been highlighted in the literature. For example, Nguyen et al. [35] showed that using a regular analysis may result in significantly over-estimating the wave attenuation performance, as compared to that in realistic irregular waves. Results from hydrodynamic analysis of cable-moored floating breakwaters [53] revealed that the number of mooring lines and hinge connections between breakwater modules has a direct effect on the breakwater performances.

6 Concluding Remarks

This paper presents the advantages of floating breakwaters, some practical examples, and recent advances in materials, designs, modelling and analysis of floating breakwaters. Some key advances include:

- the use of reinforced or prestressed concrete floating breakwaters for higher structural integrity than those made of wood and plastics, and for less maintenance costs than steel breakwaters;
- the modifications to cross-sectional and plan shapes, module connections and mooring systems of floating breakwater for cost reductions, greater safety, and better wave attenuation (especially in long waves);
- the shift from two-dimensional modelling and analysis approaches in simplified regular waves towards three-dimensional approaches in irregular waves with the ability to account for nonlinear fluid–structure interaction, interaction between fluid and mooring lines, porous elements, breakwater rigidity, and flexible module connections;
- the evaluation of wave attenuation performance by using a newly-defined index called a representative transmission coefficient, and in consideration of the interaction with the protected marine structure.

Moving to offshore sites is identified as a trend in the on-going developments in ocean space to avoid congestion in nearshore environments, for larger space and better water quality. Floating breakwaters have potential to be used to protect marine structures in such sites, but their safety, durability, and cost-effectiveness are of main concern due to the lack of experience for their deployment in high energetic environments. Future research and developments are needed to enable commercialization of floating breakwaters in such environments. Some recommendations for future studies include:

- conducting a feasibility study including cost estimates for construction of floating breakwaters in exposed sites. This will allow to determine economic viability of floating breakwaters, and identify key cost components of floating breakwaters that are needed to be reduced;
- proposing innovative breakwater designs for cost reductions when compared to existing breakwater designs, while meeting the strength, serviceability and environmental requirements;
- developing appropriate modelling approaches for some recently proposed breakwater designs that have potential for practical applications in exposed sites, such as three-dimensional breakwaters with porous elements as in [19];
- conducting more physical model tests for three-dimensional floating breakwaters with different configurations to provide reliable data to validate modelling and numerical techniques, and to evaluate the hydrodynamic and structural performances of different breakwater designs.

Acknowledgements This research was supported by the Australian Government through the Australian Research Council's Discovery Projects funding scheme (project DP170104546), with additional funding provided by the Hyundai Engineering and Construction and ARC NanoCOMM Hub. The views expressed herein are those of the authors and are not necessarily those of the Australian Government or Australian Research Council.

References

1. Toward a Blue Economy: A promise for sustainable growth in the Caribbean. https://openkn owledge.worldbank.org/handle/10986/25061. Accessed 1 July 2021
2. McCartney BL (1985) Floating breakwater design. J Waterw Port Coastal Ocean Eng 111(2):304–318
3. PIANC (1994) Floating breakwaters: a practical guide for design and construction. The World Association for Waterborne Transport Infrastructure
4. Wang CM, Wang BT (2014) Large floating structures. Springer, Singapore
5. Wang CM, Han MM, Lyu JW, Duan WH, Jung KH (2021) Floating forest: a novel breakwater windbreak structure against wind and wave hazards. Front Struct Civ Eng (accepted)
6. V-reef. https://webber-reefs.com/v-reef/. Accessed 27 July 2021
7. Dai J, Wang CM, Utsunomiya T, Duan W (2018) Review of recent research and developments on floating breakwaters. Ocean Eng 158:132–151
8. Treu A, Zimmer K, Brischke C, Larnoy E, Gobakken LR, Aloui F, Cragg SM, Flæte PO, Humar M, Westin M, Borges L, Williams J (2019) Durability and protection of timber structures in marine environments in Europe: an overview. Bio Resources 14:10161–10184
9. Wave Attenuators. https://thedockdoctors.com/wave-attenuators. Accessed 25 July 2021

10. Wave Eater. https://waveeater.com/. Accessed 27 July 2021
11. Beveridge MCM (2004) Cage aquaculture, 3rd edn. Blackwell Publishing, Oxford
12. Abbas M, Shafiee M (2020) An overview of maintenance management strategies for corroded steel structures in extreme marine environments. Mar Struct 71:102718
13. Fernández RP, Pardob ML (2013) Offshore concrete structures. Ocean Eng 58:304–316
14. Kusaka T, Ueda S (2015) Ujina floating Ferry Pier and Kan-On floating breakwater, Japan. In: Wang CM, Wang BT (eds) Large float. struct., Springer, Singapore
15. Ruol P, Martinelli L, Pezzutto P (2013) Formula to predict transmission for π-type floating breakwaters. J Waterw Port Coastal Ocean Eng 139(1):1–8
16. Christensen ED, Bingham HB, Skou Friis AP, Larsen AK, Jensen KL (2018) An experimental and numerical study of floating breakwaters. Coast Eng 137:43–58
17. Williams AN, Abul-Azm AG (1997) Dual pontoon floating breakwater. Ocean Eng 24:465–78
18. Nguyen HP, Wang CM, Tay ZY, Luong VH (2020) Wave energy converter and large floating platform integration: a review. Ocean Eng 213:107768
19. Ji CY, Guo YC, Cui J, Yuan ZM, Ma XJ (2016) 3D experimental study on a cylindrical floating breakwater system. Ocean Eng 125:38–50
20. Han MM, Wang CM (2021) Modelling wide perforated breakwater with horizontal slits using Hybrid-BEM method. Ocean Eng 222:108630
21. Zhang C, Magee AR (2021) Effectiveness of floating breakwater in special configurations for protecting nearshore infrastructures. J Mar Sci Eng 9(7):785
22. Concrete Breakwater (2021). https://www.sfmarina.com/wp-content/uploads/2021/01/SFB W1000-2020.pdf. Accessed 14 July 2021
23. Jiang D, Tan KH, Wang CM, Dai J (2021) Research and development in connector systems for Very Large Floating Structures. Ocean Eng 232:109150
24. Breakwater (2021). https://marinetek.net/products/pontoons/breakwater/. Accessed 17 July 2021
25. Fujikubo M, Suzuki H (2015) Mega-float. In: Wang CM, Wang BT (eds) Large float. struct. Springer, Singapore
26. Koh HS, Lim YB (2014) Floating performance stage at the Marina bay, Singapore. In: Wang CM, Wang BT (eds) Large float. struct. Springer, Singapore
27. Watanabe E, Utsunomiya T, Wang CM (2004) Hydroelastic analysis of pontoon-type VLFS: a literature survey. Eng Struct 26:245–256
28. Diamantoulaki I, Angelides DC, Manolis GD (2008) Performance of pile-restrained flexible floating breakwaters. Appl Ocean Res 30(4):243–255
29. Barltrop NDP (1988) Floating structures: a guide for design and analysis, vol 2. Oilfield Publications, Houston
30. Müller K, Matha D, Tiedemann S, Proskovics R, Lemmer F (2017) Qualification of innovative floating substructures for 10MW wind turbines and water depths greater than 50m. https://cor dis.europa.eu/project/id/640741. Accessed 29 July 2021
31. Xu S, Wang S, Guedes Soares C (2019) Review of mooring design for floating wave energy converters. Renew Sustain Energy Rev 111:595–621
32. Shen Y, Greco M, Faltinsen OM, Nygaard I (2018) Numerical and experimental investigations on mooring loads of a marine fish farm in waves and current. J Fluids Struct 79:115–136
33. Zhao XL, Ning DZ, Zou QP, Qiao DS, Cai SQ (2019) Hybrid floating breakwater-WEC system: a review. Ocean Eng 186:106126
34. Gunn K, Stock-Williams C (2012) Quantifying the global wave power resource. Renew Energy 44:296–304
35. Nguyen HP, Park JC, Han M, Wang CM, Abdussamie N, Penesis I, Howe D (2021) Representative transmission coefficient for evaluating the wave attenuation performance of 3D floating breakwaters in regular and irregular waves. J Mar Sci Eng 9:388
36. Penney WG, Price AT (1952) Part I. The diffraction theory of sea waves and the shelter afforded by breakwaters. Philos Trans R Soc London Ser A Math Phys Sci 244(882):236–253
37. Wang CM, Han MM, Lyu J, Duan WH, Jung KH, Kang An S (2020) Floating forest: a novel concept of floating breakwater-windbreak structure. Lect Notes Civ Eng 41:219–34

38. Tay ZY (2020) Performance and wave impact of an integrated multi-raft wave energy converter with floating breakwater for tropical climate. Ocean Eng 218:108136
39. Suzuki H (2005) Overview of Megafloat: concept, design criteria, analysis, and design. Mar Struct 18(2):111–132
40. Liew KM, Wang CM, Xiang Y, Kitipornchai S (1998) Vibration of mindlin plates - programming the p-version Ritz method. Elsevier, Oxford
41. Liu GR, Quek S (2003) The finite element method: a practical course. Butterworth Heinemann, Oxford
42. Wang CM, Tay ZY (2011) Very large floating structures: applications, research and development. Procedia Eng 14:62–72
43. Cox R, Beach D (2006) Floating breakwater performance - wave transmission and reflection, energy dissipation, motions and restraining forces. In: Proceedings of the first international conference on the application of physical modelling to port and coastal protection, Porto, Portugal
44. Jain RK (1980) A simple method of calculating the equivalent stiffnesses in mooring cables. Appl Ocean Res 2:139–142
45. Al-Solihat MK, Nahon M (2016) Stiffness of slack and taut moorings. Ships Offshore Struct 11(8):890–904
46. Loukogeorgaki E, Angelides DC (2005) Stiffness of mooring lines and performance of floating breakwater in three dimensions. Appl Ocean Res 27(4–5):187–208
47. Ji C, Cheng Y, Yang K, Oleg G (2017) Numerical and experimental investigation of hydrodynamic performance of a cylindrical dual pontoon-net floating breakwater. Coast Eng 129:1–16
48. Sannasiraj SA, Sundar V, Sundaravadivelu R (1998) Mooring forces and motion responses of pontoon-type floating breakwaters. Ocean Eng 25(1):27–48
49. Zhang H, Zhou B, Vogel C, Willden R, Zang J, Zhang L (2019) Hydrodynamic performance of a floating breakwater as an oscillating-buoy type wave energy converter. Appl Energy 257:113996
50. Chen Q, Zang J, Birchall J, Ning D, Zhao X, Gao J (2020) On the hydrodynamic performance of a vertical pile-restrained WEC-type floating breakwater. Renew Energy 146:414–425
51. Mackay EBL, Johanning L (2020) A BEM model for wave forces on structures with thin porous elements. J Fluids Struct 102:103246
52. Michailides C, Angelides DC (2012) Modeling of energy extraction and behavior of a Flexible Floating Breakwater. Appl Ocean Res 35:77–94
53. Diamantoulaki I, Angelides DC (2011) Modeling of cable-moored floating breakwaters connected with hinges. Eng Struct 33:1536–1552

Parameter Identification for Linear System Using Multiple Model Estimation

Jixing Cao and Ser-Tong Quek

Abstract Kalman filter (KF) has gained wide adoption in system identification of engineering systems. It is a recursive estimation method under linear and Gaussian assumptions. In practice, a single model based on KF may not be able to capture the structural performance well for complex systems. To address this problem, KF estimation using multiple models is proposed. This method employs KF with different transition and measurement matrices, each of which can be assigned (if necessary) with different initial states, process and measurement noises to describe the system. The outputs of these models are then integrated to obtain the overall estimates through a weighted combination, where the weights are determined using the likelihood function. A numerical model is employed to illustrate the procedure and evaluate the accuracy of the proposed KF estimation with multiple models. The estimated results indicate that the proposed method is robust and reliable, with potential for system identification under a wider variety of situations.

Keywords System identification · Kalman filter · Multiple model estimation · Likelihood function

1 Introduction

Structural health monitoring (SHM) has been extensively used to evaluate the performance and health condition of civil infrastructure [1]. One approach in SHM focuses on the system identification from measured responses. The methods employed in system identification can be classified under frequency-domain or time-domain. Frequency-domain methods usually obtain the modal parameters of frequency, damping ratio and mode shape through the Fourier transform (FT) of the responses. The identified modal parameters can be further used to update the finite element model [2] and diagnose for structural damage [3]. To avoid the energy leakage in FT, time-domain methods are developed as alternatives. Time-domain methods not only

J. Cao · S.-T. Quek (✉)
Department of Civil and Environmental Engineering, National University of Singapore, Singapore 117576, Singapore
e-mail: st_quek@nus.edu.sg

© The Author(s), under exclusive license to Springer Nature Singapore Pte Ltd. 2023
J. N. Reddy et al. (eds.), *ICSCEA 2021*, Lecture Notes in Civil Engineering 268,
https://doi.org/10.1007/978-981-19-3303-5_2

can estimate the modal parameters from the time-history data, they can also reflect the time-variation of the physical parameters. Among the algorithms developed within time-domain methods, the Kalman filter (KF) [4] is one of the most effective estimators for linear dynamic systems, which has been employed for parameter estimation, model updating and damage identification. KF techniques have been extended for systems that are nonlinear and/or non-Gaussian. For example, the extended KF and unscented KF have been proposed for nonlinear systems, and the Particle filter is applicable for non-Gaussian cases. Although these techniques under the family of KF have enriched the adoption of system identification in engineering, the performance of state estimation may still be difficult and inaccurate for complex systems, such as the coupled translational and torsional responses in asymmetric structures, some of which can be difficult to uncouple. It is envisaged that for such systems, a single KF model may not be sufficient. Hence, the use of multiple models is explored.

The multiple model estimation (MME) technique employs multiple KFs, each of which represents peculiar patterns of dynamics, to better characterize the system. This concept was originally proposed by Blom and Shalom [5] for linear system with Markovian coefficients. Subsequently, the challenge of designing the appropriate set of models and how these sets can be integrated have been pursued by many researchers with the aim of improving the performance and robustness of the MME approach. To investigate the influence of the number and type of models in the set on the estimation results, the model classes based on extended KF, unscented KF, and Particle filter have been pursued and compared [6]. The results indicate that there is an appropriate number of models for optimal performance. To combine the multiple outputs from the models, Li et al. [7] put forward an exponential decay term to determine the weights of the filters. Kottath et al. [8] assumed that all the models had equal weights initially and updated the model weights based on the measured data, in which those with low weight factors were eliminated.

The afore-mentioned studies focused on the fields of maneuvering target, control system in fault tolerance, and radar system, and there seems to be no publicly published works in the field of structural engineering. In this work, the MME method is developed for coupled structures. The algorithm is extended to jointly estimate the model parameters and state vector from the measured response. A numerical model with torsional and translational coupled responses is employed to illustrate the estimation process.

2 Framework of Multiple Model Estimation (MME)

In Sect. 2.1, the Kalman filter (KF) for optimal state estimation of a linear input–output Gaussian system in time-domain is briefly introduced. Due to the presence of uncertainties, the estimation is characterized probabilistically, which for Gaussian processes, the mean and covariance of the estimates are obtained. Section 2.2 introduces the concept of decomposing a system into multiple KF models and how they can be combined using weights which changes with each time step. The weights at each time step are derived based on the likelihood of the measured data and the prior probabilities associated with the models. In Sect. 2.3, the MME technique is extended to jointly estimate the state vector and model parameters. The detailed procedure of MME for parameter estimation is then described.

2.1 Brief Review of Kalman Filter

A linear dynamic system can be expressed as a second-order governing differential equation, that is,

$$M\ddot{x}_k + C\dot{x}_k + Kx_k = f_k. \tag{1}$$

in which M, C and K are matrices of mass, damping and stiffness, respectively; x_k, $\dot{x}_k$ and $\ddot{x}_k$ are vectors of nodal displacement, velocity and acceleration responses at time step k; and f_k is the external dynamic force vector.

Let the state vector $x_k = (x_k, \dot{x}_k)^{\mathrm{T}}$. Equation (1) can be rewritten in the state space form as

$$\dot{x}_{k+1} = Ax_k + Bf_k. \tag{2}$$

in which the transition matrix A is given by

$$A = \begin{bmatrix} 0 & 1 \\ -M^{-1}K & M^{-1}C \end{bmatrix}. \tag{3}$$

where I is the identity matrix, and the external force matrix B is given by

$$B = \begin{bmatrix} 0 \\ M^{-1} \end{bmatrix}. \tag{4}$$

The measurement equation of the response (output), denoted by $y_k = \begin{bmatrix} x_k, \dot{x}_k, \ddot{x}_k \end{bmatrix}^T$, is formulated as

$$y_k = Hx_k. \tag{5}$$

in which the measurement matrix H is given by

$$H = \begin{bmatrix} I & 0 \\ 0 & I \\ -M^{-1}K & M^{-1}C \end{bmatrix}. \tag{6}$$

For linear systems with process and measurement noise, each following the Gaussian distribution with zero mean and respective covariance denoted by Q and R, the equations to predict and update the mean state vector and covariance are as follows:

$$\tilde{x}_{k+1|k} = A\hat{x}_{k|k} + Bu_k. \tag{7}$$

$$\tilde{P}_{k+1|k} = A\hat{P}_{k|k} A^T + Q. \tag{8}$$

$$K_{k+1} = \tilde{P}_{k+1|k} H^T \left(H \tilde{P}_{k+1|k} H^T + R \right)^{-1}. \tag{9}$$

$$\hat{x}_{k+1|k+1} = \tilde{x}_{k+1|k} + K_{k+1}\left(y_{k+1} - H\tilde{x}_{k+1|k} \right). \tag{10}$$

$$\hat{P}_{k+1|k+1} = (I - K_{k+1}H)\tilde{P}_{k+1|k}. \tag{11}$$

where $\hat{x}_{k|k}$ and $\hat{x}_{k+1|k+1}$ are the posterior state vector at time step k and $k + 1$ respectively; $\hat{P}_{k|k}$ and $\hat{P}_{k+1|k+1}$ are the posterior estimate covariance at time step k and $k+1$ respectively; $\tilde{x}_{k+1|k}$ and $\tilde{P}_{k+1|k}$ are the prior state vector and prior estimate covariance respectively at time step $(k + 1)$; K_{k+1} and y_{k+1} are the Kalman gain and measurement response respectively at time step $(k + 1)$; u_k is the external excitation at time step k.

The KF uses an innovation term (term in brackets in Eq. (10)) to incorporate the measured data and provide optimal estimates (given by Eqs. (10) and (11)) of the system states using the Kalman gain K_{k+1} in Eq. (9). In MME, the sub-models are linear systems and these equations are used as part of the solution procedure.

2.2   Multiple Model Estimation

Conceptually, KF, being a recursive estimator under both linear and Gaussian conditions, admits the use of the principle of linear superposition. Hence, a complex linear and Gaussian system may be decomposed into a set of simpler KF models, each of which can be first solved as described in Sect. 2.1. The output from each individual model can then be integrated to provide the overall system output. The quality of the output hinges strongly on the combination rule which is dependent on how the system is decomposed.

The simplest method of weighted average is adopted for each time step, with the weights (where their sum is 1) updated at each time step. The updating is done using the likelihood function of the innovations (which are the deviations of the prediction from the measured responses) at the current time step $(k + 1)$ through the equation

$$\mu_{k+1}^{j} = \frac{\left(\mu_k^j \mathcal{L}_{k+1}^j\right)}{\sum_{j=1}^{m}\left(\mu_k^j \mathcal{L}_{k+1}^j\right)}. \tag{12}$$

where μ_k^j and μ_{k+1}^j denote the weights for model j at time steps k and $k+1$, m is the number of models and the likelihood function is given by

$$\mathcal{L}_{k+1}^j = \frac{1}{\sqrt{2\pi \cdot \det(S)}} \exp\left[-\frac{1}{2} r^\mathsf{T} S^{-1} r\right]. \tag{13}$$

in which $S = H\tilde{P}_{k+1|k} H^T + R$ is the covariance matrix of the innovations and $r = y_{k+1} - H\tilde{x}_{k+1|k}$ is the innovation or residual vector associated with model j. An initial (time step $k = 0$) set of weights is assumed to be known, denoted by $\mu_0 = \left[\mu_0^a, \mu_0^b, \cdots, \mu_0^j, \cdots, \mu_0^m\right]$, with the weight vector at time step $k+1$ denoted by $\mu_{k+1} = \left[\mu_{k+1}^a, \mu_{k+1}^b, \cdots \mu_{k+1}^j, \cdots, \mu_{k+1}^m\right]$.

The system response statistics $\hat{x}_{k+1|k+1}^S$ and $\hat{P}_{k+1|k+1}^S$ at time step $(k+1)$ are obtained by integrating the response statistics $\hat{x}_{k+1|k+1}^j$ and $\hat{P}_{k+1|k+1}^j$ at time step $(k+1)$ using the weight μ_{k+1}^j of each model j to yield

$$\hat{x}_{k+1|k+1}^S = \sum_{j=1}^{m} \mu_{k+1}^j \hat{x}_{k+1|k+1}^j. \tag{14}$$

$$\hat{P}_{k+1|k+1}^S = \sum_{j=1}^{m} \mu_{k+1}^j \left[\hat{P}_{k+1|k+1}^j + \left(\hat{x}_{k+1|k+1}^j - \hat{x}_{k+1|k+1}^S\right)\left(\hat{x}_{k+1|k+1}^j - \hat{x}_{k+1|k+1}^S\right)^\mathsf{T}\right]. \tag{15}$$

2.3 Parameter Estimation Using Multiple Model Estimation

Besides predicting response (state vector), one may be interested in estimating some unknown parameters of the system concurrently. The formulation remains valid for this purpose by regarding the unknown system parameters as additional states that are augmented to the state vector, that is, $X_k = [x_k, \theta_k]$. If the model parameters are time-invariant, the corresponding first differential state is $\dot{X}_k = [\dot{X}_k, 0]$. The augmented state vector expands the state-space model to

$$\dot{X}_{k+1} = A^{\text{aug}} X_k + B^{\text{aug}} f_k. \tag{16}$$

$$Y_k = H^{\text{aug}} X_k. \tag{17}$$

in which the new transition matrix A^{aug}, external force matrix B^{aug} and output matrix H^{aug} are augmented as

$$A^{\text{aug}} = \begin{bmatrix} 0 & I & 0 \\ -(M)^{-1}K & (M)^{-1}C & 0 \end{bmatrix}. \tag{18}$$

$$B^{\text{aug}} = \begin{bmatrix} 0 \\ (M)^{-1} \\ 0 \end{bmatrix}. \tag{19}$$

$$H^{\text{aug}} = \begin{bmatrix} -(M)^{-1}K & (M)^{-1}C & 0 \end{bmatrix}. \tag{20}$$

Adding the unknown parameters to the state-space representation enlarges the state vector without changing the system property, which means the algorithm of KF is still valid for the augmented state vector estimation, including for MME systems. The solution process of MME is summarized as follows:

1. Build a set of sub-models with weights μ^j, $j = a, b, \cdots, m$, to represent the system of interest

$$\begin{aligned} \dot{X}^j_{k+1} &= A^{\text{aug},j} X^j_k + B^{\text{aug},j} f_k + v^j_k, & v^j_k &\sim \mathcal{N}(0, Q^j), \\ Y^j_k &= H^{\text{aug},j} X^j_k + w^j_k, & w^j_k &\sim \mathcal{N}(0, R^j). \end{aligned}$$

2. Initial values are given by.
    a. each filter weight $\mu^j_0 = 1/m$,
    b. MME state estimate $\hat{X}^S_0$,
    c. MME state covariance $\hat{P}^S_0$,
    d. each filter process noise Q^j,
    e. each filter measurement noise R^j.

3. For each model:

    a. Propagate the next state (prediction)

$$\begin{aligned} \tilde{X}^j_{k+1|k} &= A^{\text{aug},j} \hat{X}^S_{k|k} + B^{\text{aug},j} u_k, \\ \tilde{P}^j_{k+1|k} &= A^{\text{aug},j} \hat{P}^S_{k|k} \left(A^{\text{aug},j}\right)^{\text{T}} + Q^j. \end{aligned}$$

    b. Calculate the Kalman gain

$$K^j_{k+1} = \tilde{P}^j_{k+1|k} \left(H^{\text{aug},j}\right)^{\text{T}} \left(H^{\text{aug},j} \tilde{P}^j_{k+1|k} \left(H^{\text{aug},j}\right)^{\text{T}} + R^j\right)^{-1},$$

c.   Update the prediction

$$\hat{X}^{j}_{k+1|k+1} = \tilde{X}^{j}_{k+1|k} + K^{j}_{k+1}\left(y_{k+1} - H^{\mathrm{aug},j}\tilde{X}^{j}_{k+1|k}\right),$$

$$\hat{P}^{j}_{k+1|k+1} = \left(I - K^{j}_{k+1}H^{\mathrm{aug},j}\right)\tilde{P}^{j}_{k+1|k}.$$

d.   Calculate the likelihood

$$\mathcal{L}^{j}_{k+1} = \frac{1}{\sqrt{2\pi \cdot \det(S)}}\exp\left[-\frac{1}{2}\left(r^{j}\right)^{\mathrm{T}}\left(S^{j}\right)^{-1}\left(r^{j}\right)\right].$$

e.   Update the weights

$$\mu^{j}_{k+1} = \mu^{j}_{k}\mathcal{L}^{j}_{k+1}.$$

Repeat steps (a − e) for all the KF models.

4.   Normalize the weights

$$\hat{\mu}^{j}_{k+1} = \frac{\mu^{j}_{k+1}}{\sum^{m}_{j=1}\left(\mu^{j}_{k+1}\right)}.$$

5.   Obtain the estimate results

$$\hat{X}^{S}_{k+1|k+1} = \sum^{m}_{j=1}\hat{\mu}^{j}_{k+1}\hat{X}^{j}_{k+1|k+1},$$

$$\hat{P}^{S}_{k+1|k+1} = \sum^{m}_{j=1}\hat{\mu}^{j}_{k+1}\left[\hat{P}^{j}_{k+1|k+1} + \left(\hat{X}^{j}_{k+1|k+1} - \hat{X}^{S}_{k+1|k+1}\right)\left(\hat{X}^{j}_{k+1|k+1} - \hat{X}^{S}_{k+1|k+1}\right)^{\mathrm{T}}\right].$$

6.   Repeat steps (3–5) until all the measurements are depleted.

3   Numerical Example

3.1   Case Study

To demonstrate the effectiveness of MME algorithm, a two-floor building under bidirectional earthquake excitation is presented in Fig. 1a. The concentrated mass of floor 1 and 2 are $600 \times 1e3$ kg and $500 \times 1e3$ kg, respectively. The polar moments of inertia are $J_t = [J_1, J_2] = [7.2, 5.6] \times 1e7(\mathrm{kg} \cdot m^2)$. The lateral stiffness of floors 1 and 2 in x and y directions, respectively, are $k_x = [k_{x1}, k_{x2}] = [0.55, 0.43] \times 1e9(\mathrm{kN/m})$, and $k_y = [k_{y1}, k_{y2}] = [0.48, 0.39] \times 1e9(\mathrm{kN/m})$. It is assumed that Rayleigh damping is appropriate and that the centres of mass are located at the geometric centres of floors 1 and 2, while the resultant centre of stiffness of the

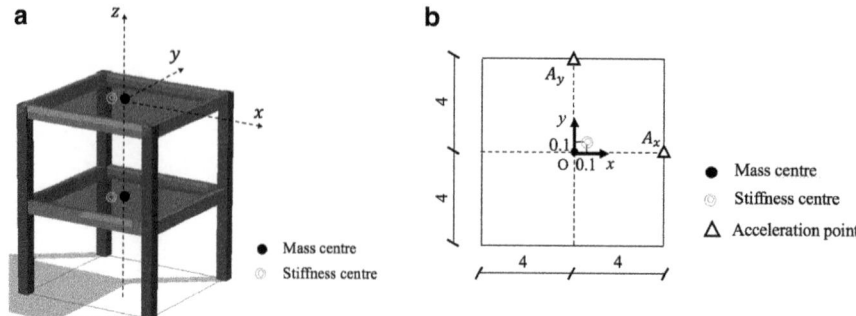

Fig. 1 Geometric information of the model, **a** diagram of 3D model, **b** plane of floors 1 and 2 (unit: m)

structure in plan is offset at 0.1 m from the geometric centre in both the x and y directions (see Fig. 1b). Because the centres of mass and stiffness do not coincide, the equation of motions will lead to coupling in the horizontal and torsional responses. The stiffness and mass matrices for a 3-DOFs system (x, y for translational and t for rotational DOFs) are modelled as

$$
\boldsymbol{K}_w = \begin{bmatrix} \boldsymbol{K}_{xx} & 0 & \boldsymbol{K}_{xt} \\ 0 & \boldsymbol{K}_{yy} & \boldsymbol{K}_{yt} \\ \boldsymbol{K}_{tx} & \boldsymbol{K}_{ty} & \boldsymbol{K}_{tt} \end{bmatrix}, \boldsymbol{M}_w = \begin{bmatrix} \boldsymbol{M}_{xx} & 0 & 0 \\ 0 & \boldsymbol{M}_{yy} & 0 \\ 0 & 0 & \boldsymbol{J}_{tt} \end{bmatrix}.
\tag{21}
$$

in which $\boldsymbol{K}_{xx}$ and $\boldsymbol{K}_{yy}$ are translational stiffness in x and y directions; $\boldsymbol{K}_{tt}$ is the torsional stiffness about an axis perpendicular to the x–y plane; $\boldsymbol{K}_{xt}$ and $\boldsymbol{K}_{yt}$ are cross translational-torsional terms; $\boldsymbol{M}_{xx}$ and $\boldsymbol{M}_{yy}$ are translational mass in the x and y directions, respectively; $\boldsymbol{J}_{tt}$ is polar moment of inertia. The damping matrix is formed using $\boldsymbol{C}_w = \alpha \boldsymbol{M}_w + \beta \boldsymbol{K}_w$, whose coefficients α and β are determined by

$$
\alpha = \frac{2\omega_1 \omega_2}{\omega_1 + \omega_2} \xi, \beta = \frac{2\xi}{\omega_1 + \omega_2},
\tag{22}
$$

here ω_1 and ω_2 are the first two natural frequencies; ξ is the damping ratio set at 3%.

Two horizontal ground acceleration records from the 1940 El Centro earthquake are selected as inputs for the base excitation, whose peak ground accelerations in x and y directions are scaled to 0.2 and 0.15 m/s^2. The time histories of structural response are calculated by the Runge–Kutta method. The accelerations in the x and y directions of each floor are measured at points A_x and A_y (see Fig. 1b), respectively. To simulate measurement noise, an artificial noise of level $N_L = 30\%$ is superimposed on the exact response as

$$
a_N = a_E + N_L \cdot N_S \cdot \sigma_E.
\tag{23}
$$

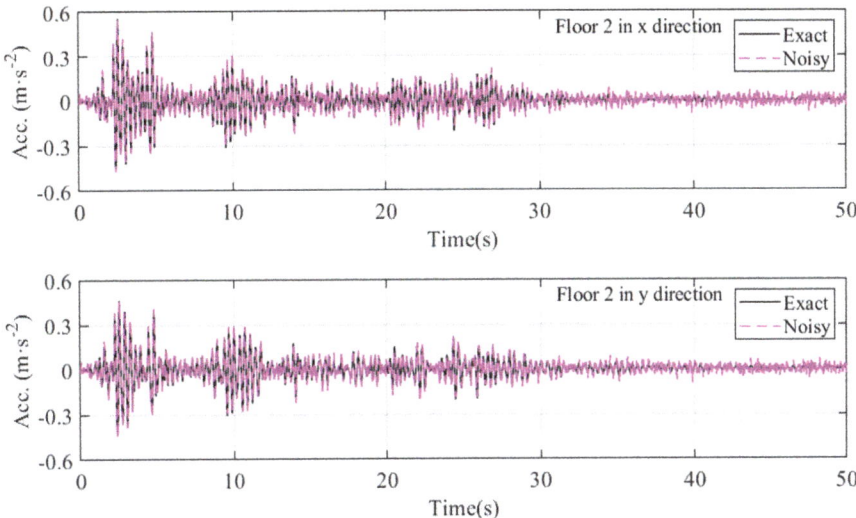

Fig. 2 Noisy acceleration responses of floor 2 in x and y directions

where a_N and a_E are the noisy and exact accelerations; σ_E is the standard deviation of the exact response; N_L denotes the noise level and N_S denotes the standard normal noise generated using the randn function in MATLAB. Figure 2 displays the noisy responses of floor 2 in both the x and y directions.

3.2 Identification Results

The goal is to estimate unknown system parameters from the noisy acceleration responses. Two unknown parameters $\boldsymbol{\theta} = \left[k_{x2}, k_{y1}\right]^T$, namely the x-direction stiffness in floor 2 and the y-direction stiffness in floor 1, are to be identified, in which the superscript T denotes matrix transpose. Hence, the augmented state vector of this system is written as $\boldsymbol{X}^S = \left[\boldsymbol{x}^x, \boldsymbol{x}^y, \boldsymbol{x}^t, \dot{\boldsymbol{x}}^x, \dot{\boldsymbol{x}}^y, \dot{\boldsymbol{x}}^t, k_{x2}, k_{y1}\right]^T$, in which $\boldsymbol{x}^x$ and $\boldsymbol{x}^y$ are horizontal displacements in the x and y directions, and $\boldsymbol{x}^t$ is the torsional displacement along the z direction; $\dot{\boldsymbol{x}}^x$ and $\dot{\boldsymbol{x}}^y$ are horizontal velocities in the x and y directions, and $\dot{\boldsymbol{x}}^t$ is the torsional velocity along the z direction. This system is decomposed into two linear sub-models. The first model (model a) includes all the structural responses related to the x direction, whose state vector is expressed as $\boldsymbol{X}^a = \left[\boldsymbol{x}^x, \boldsymbol{x}^t, \dot{\boldsymbol{x}}^x, \dot{\boldsymbol{x}}^t, k_{x2}, k_{y1}\right]^T = \boldsymbol{E}^a \boldsymbol{X}^S$, where the state vector position matrix $\boldsymbol{E}^a = [1, 0, 1, 1, 0, 1, 1, 1]$. Since the augmented state vector $\boldsymbol{X}^a$ contains unknown parameters, the state space equation is formulated as

$$\dot{\boldsymbol{X}}_k^a = \boldsymbol{A}^a \boldsymbol{X}_k^a + \boldsymbol{B}^a \boldsymbol{f}_k + \boldsymbol{v}_k^a. \tag{24}$$

in which v_k^a is the process noise (accounting for model error) that is assumed to have zero mean and covariance matrix Q_k^a; and the transition matrix A^a in Eq. (24) is augmented by

$$A^a = \begin{bmatrix} 0 & I & 0 \\ -(M_x)^{-1}K_x & (M_x)^{-1}C_x & 0 \end{bmatrix}. \tag{25}$$

here the matrices of stiffness K_x and mass M_x are given by

$$K_x = \begin{bmatrix} K_{xx} & K_{xt} \\ K_{tx} & K_{tt} \end{bmatrix}. \tag{26}$$

$$M_x = \begin{bmatrix} M_{xx} & 0 \\ 0 & J_{tt} \end{bmatrix}. \tag{27}$$

The external force matrix B^a in Eq. (24) is extended to

$$B^a = \begin{bmatrix} 0 \\ (M_x)^{-1} \\ 0 \end{bmatrix}. \tag{28}$$

When only the horizontal acceleration data in the x and y directions, denoted by $\ddot{x}^x$ and $\ddot{x}^y$, is recorded $Y = \begin{bmatrix} \ddot{x}^{ax}, \ddot{x}^{ay} \end{bmatrix}^T$. The measurement equation is formulated as

$$Y_k^a = H^a X_k^a + w_k^a. \tag{29}$$

in which $Y_k^a = F^a Y$, here the measurement position matrix $F^a = [1, 0]$; w_k^a is the measurement noise, assumed to have zero mean and covariance matrix R_k^a; H^a is given by

$$H^a = \begin{bmatrix} -(M_x)^{-1}K_x(M_x)^{-1}C_x 0 \end{bmatrix}. \tag{30}$$

In the second model (model b), all the structural responses related to the y direction are considered, whose state vector is given by $X^b = [x^y, x^t, \dot{x}^y, \dot{x}^t, k_{x2}, k_{y1}]^T = E^b X^S$, where the state vector position matrix $E^b = [0, 1, 1, 0, 1, 1, 1, 1]$. Since the models a and b are linear systems, the form of state equation (Eq. 24) is the same, that is,

$$\dot{X}_k^b = A^b X_k^b + B^b f_k + v_k^b. \tag{31}$$

where the values of the process noise v_k^b can be different from the values of v_k^a; and the augmented transition matrix A^b is given by

$$A^b = \begin{bmatrix} 0 & I & 0 \\ -(M_y)^{-1} K_y & (M_y)^{-1} C_y & 0 \end{bmatrix}. \tag{32}$$

where the matrices of stiffness K_y and mass M_y are given by

$$K_y = \begin{bmatrix} K_{yy} & K_{yt} \\ K_{ty} & K_{tt} \end{bmatrix}. \tag{33}$$

$$M_y = \begin{bmatrix} M_{yy} & 0 \\ 0 & J_{tt} \end{bmatrix}. \tag{34}$$

The corresponding measurement equation is expressed as

$$Y_k^b = H^b X_k^b + w_k^b. \tag{35}$$

in which $Y_k^b = F^b Y$, where the measurement position matrix $F^b = [0, 1]$; w_k^b is the measurement noise that can be different with w_k^a; H^b is given by

$$H^b = \begin{bmatrix} -(M_y)^{-1} K_y (M_y)^{-1} C_y 0 \end{bmatrix}. \tag{36}$$

The initial weights of each model are set as $\hat{\mu}_0 = [0.5, 0.5]^T$. The initial state estimate $\hat{X}_0^S = \begin{bmatrix} 0, \hat{\theta}_0 \end{bmatrix}^T$, where $\hat{\theta}_0 = \left(1.2 * k_{x2}^{\text{Exact}}, 0.85 * k_{y1}^{\text{Exact}}\right)$ is an initial guess of exact values of the parameters (alternatively, can use the nominal values of the parameters). The process noise of the two filters are $Q^a = Q^b = (1e-8) * I_{10\times10}$, in which $I_{10\times10}$ is a 10×10 identity diagonal matrix. The measurement noise of the two filters are $R^a = R^b = (1e-8) * I_{8\times8}$. According to the procedure of MME described in Sect. 2.3, the estimated accelerations are obtained presented in Fig. 3 (only the first 15 s shown). The estimated accelerations capture exact responses well. Figure 4 shows that the estimated time-history parameters converge to exact values within 3 s.

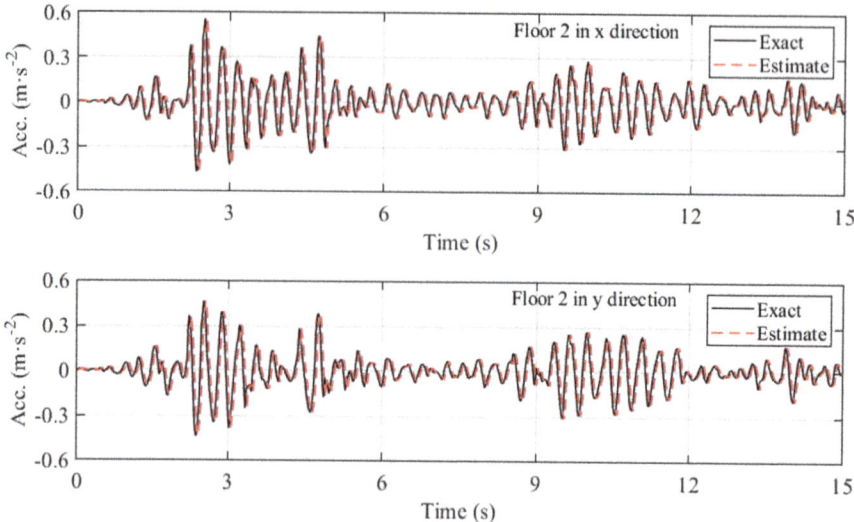

Fig. 3 Comparison of accelerations between exact and estimated results

3.3 Effect of Response Noise

To investigate the robustness of the MME algorithm, four cases with different noise levels are added to the exact acceleration responses, namely, cases A: $N_L = 20\%$; B: $N_L = 30\%$; C: $N_L = 40\%$; and D: $N_L = 50\%$. These noisy responses are used as the measured data from which the accuracy in determining the unknown system parameters are inferred. Since the system parameters are assumed to be time-invariant, the determination of the unknown parameters is affected by the process noise (see Eqs. 24 and 31), Q^a(and Q^b) = (1e−8) * $I_{10\times10}$. The process covariance is kept as the same for all the four cases, while the measurement covariance may change due to the different levels of measurement noise. For cases A and B, the measurement covariance are assumed to be R^a (and R^b) = (1e−8) * $I_{8\times8}$, and R^a (and R^b) = (1e−7) * $I_{8\times8}$ for cases C and D. The estimated acceleration responses for the four cases are compared in Fig. 5. All the four cases can track the exact accelerations well.

To further quantify the estimated time-history responses, the relative root mean square error (RRMSE) is employed, defined as

$$\text{RRMSE} = \frac{\sqrt{\frac{1}{N}\sum_{k=1}^{N}\left(R_k - \hat{R}_k\right)^2}}{\sqrt{\frac{1}{N}\sum_{k=1}^{N}(R_k)^2}} * 100\%. \qquad (37)$$

in which R_k and $\hat{R}_k$ are the exact and estimated structural response, respectively; N is the total number of time steps used in the structural responses. Table 1 reports the

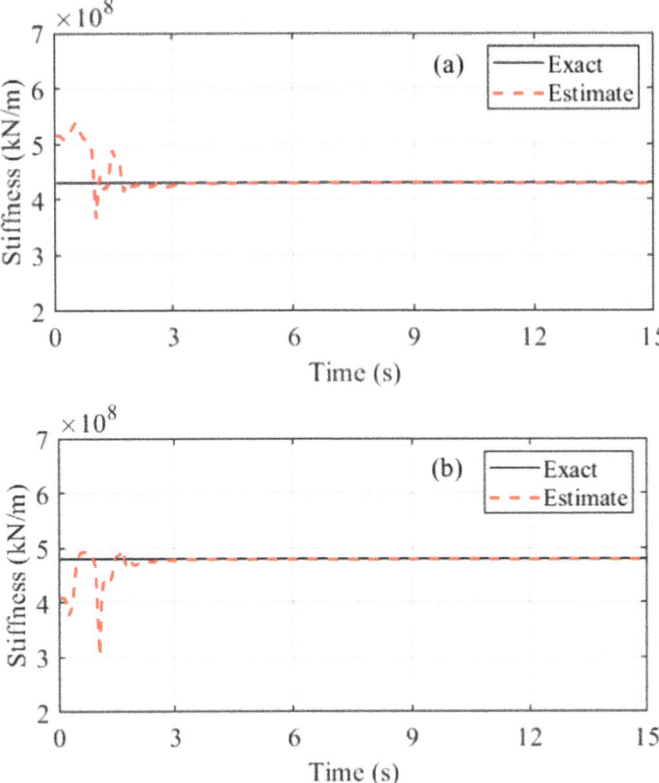

Fig. 4 Comparison of estimated parameters with exact values, **a** time history of k_{x2}, **b** time history of k_{y1}

RRMSE of the accelerations between the estimated and exact results. The RRMSE of all the four cases are less than 9%. As the noise level increases from cases A to D, the values of RRMSE increase as expected.

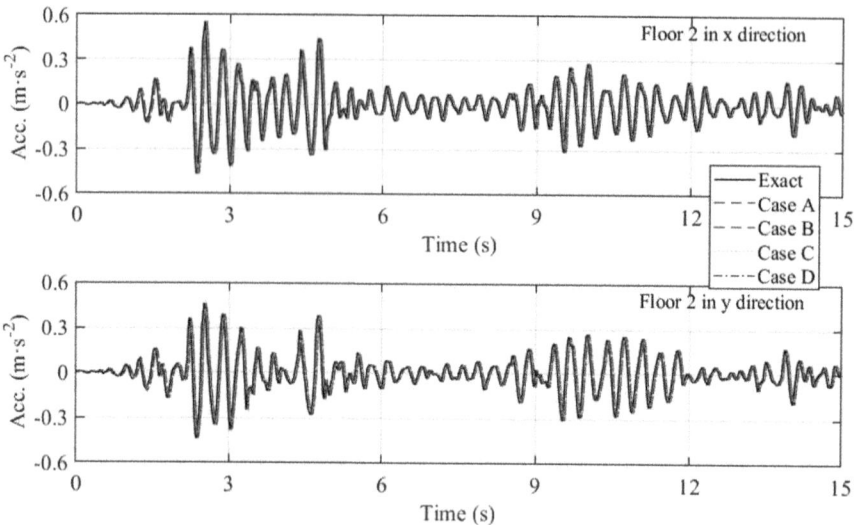

Fig. 5 Comparison of time-history accelerations between exact and estimated results for four cases

Table 1 RRMSE (%) of accelerations between exact and estimated results for four cases

Case	Case A	Case B	Case C	Case D
Floor 1 in x	0.78	1.85	4.29	6.18
Floor 2 in x	0.97	2.85	5.70	8.24
Floor 1 in y	0.90	2.12	5.06	6.80
Floor 2 in y	0.70	2.82	6.12	8.85

Figure 6 presents the time-history of the parameters estimated by the four cases. All the estimated parameters fluctuate initially but gradually converge to the exact values within 3 s. The temporal average over the last five seconds are taken as the final estimated values, as reported in Table 2. The relative error between the estimated and exact parameters are within 1%. Although this study uses a simple case to illustrate the procedure of MME, the estimated results indicate that the proposed method seems robust and reliable, with potential for system identification under a wider variety of situations.

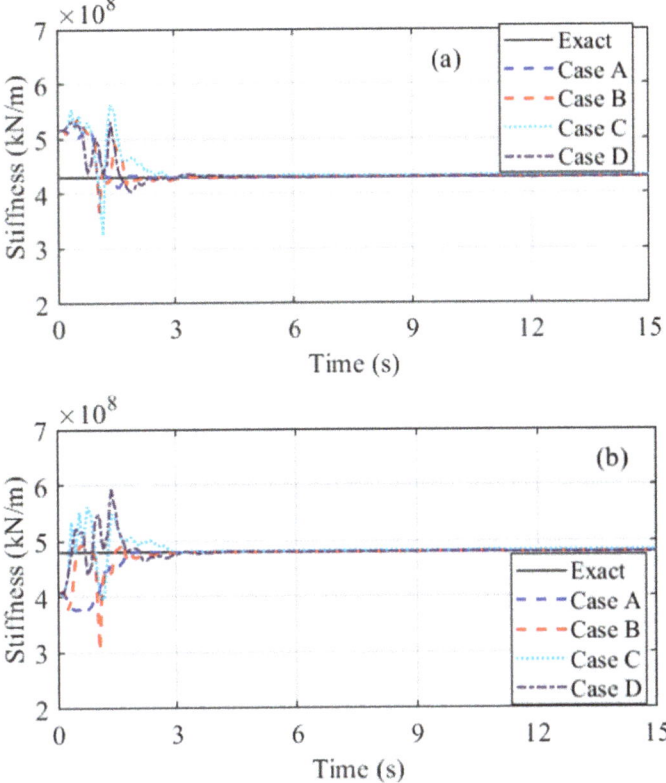

Fig. 6 Comparison of estimated parameters with exact values for four cases, **a** time history of k_{x2}, **b** time history of k_{y1}

Table 2 Relative error (RE) of parameters between exact and estimated results for four cases

Parameters	Case	Case A		Case B		Case C		Case D	
	Exact	Estimate	RE (%)	Estimate	RE (%)	Estimate	RE (%)	Estimate	RE (%)
k_{x2}(kN/m)	4.3e8	4.295e8	0.11	4.294e8	0.15	4.327e8	0.63	4.299e8	0.03
k_{y1}(kN/m	4.8e8	4.798e8	0.04	4.799e8	0.03	4.825e8	0.52	4.788e8	0.25

4 Concluding Remarks

This paper introduces the multiple model estimation (MME) method to jointly estimate the unknown parameters and the state vector for complex structures. This
method constructs a bank of sub-models, each of which uses different state equations to represent the different characteristics of the dynamical system. The multiple
outputs are fused to give an overall estimate through a linear combination of the

responses obtained from the sub-models using time-variant weights. These weights are updated probabilistically in time using the likelihood value of the innovations (that is, deviation of prediction from measurement values). Numerical simulated structural responses from a three-dimensional translation and torsion coupled two-storey building are used to illustrate the performance of the proposed MME algorithm. The estimated response and system parameters agree well with the exact responses and values, indicating that the MME algorithm has potential for applications in complex structures. The effect of measurement noise (up to 50%) on the estimated results is also investigated. Nevertheless, more work is currently ongoing before confidence can be gained for application to real complex systems in practice.

References

1. Sony S, Laventure S, Sadhu A (2019) A literature review of next-generation smart sensing technology in structural health monitoring. Struct Control Health Monit 26(3):e2321
2. Xiong HB, Cao JX, Zhang FL, Ou X, Chen CJ (2019) Investigation of the SHM-oriented model and dynamic characteristics of a super-tall building. Smart Struct Syst 23(3):295–306
3. Hou R, Xia Y, Zhou X (2018) Structural damage detection based on l1 regularization using natural frequencies and mode shapes. Struct Control Health Monit 25(3):e2107
4. Meinhold RJ, Singpurwalla ND (1983) Understanding the Kalman filter. Am Stat 37(2):123–127
5. Blom HAP, Bar-Shalom Y (1988) The interacting multiple model algorithm for systems with Markovian switching coefficients. IEEE Trans Autom Control 33(8):780–783
6. Akca A, Efe MÖ (2019) Multiple model Kalman and Particle filters and applications: a survey. IFAC-PapersOnLine 52(3):73–78
7. Li S, Jiang X, Liu Y (2014) Innovative Mars entry integrated navigation using modified multiple model adaptive estimation. Aerosp Sci Technol 39:403–413
8. Kottath R, Poddar S, Das A, Kumar V (2015) Improving multiple model adaptive estimation by filter stripping. In: 2015 IEEE recent advances in intelligent computational systems (RAICS). IEEE, pp 11–16

Architecture Session

An Adaptive Facade Configuration for Daylighting Toward Energy-Efficient: Case Study on High-Rise Office Building in HCMC

Van Tung Nguyen, Thi Hong Na Le, Hung Tien Le, and Phan Bao Long Nguyen

Abstract Natural daylight within buildings is one of the solutions to effectively reduce energy consumption in high-rise office buildings (HOB). The management of natural lighting depends largely on the characteristics of the building envelope (BE) in the building, especially the facade system. Adaptive facade (AF) is one of the solutions in the BE system of the building that helps to solve numerous problems in energy-efficiency and in particular, the balance of natural lighting. In this research, it is proposed that a kinematic AF system integrated onto the single-layer glass facade structure be implemented in HOBs in Ho Chi Minh City (HCMC), through the study of the typical case being the LIM Tower office building located in central HCMC. The kinetic AF system is integrated in order to improve the quality of natural lighting through 3 statistics: Annual Sunlight Exposure (ASE), spatial Daylight Autonomy (sDA), and Daylight Factor (DF). Results from simulations utilizing Rhinoceros-Grasshopper software (Computer-Aid Design) and Climate Studio plugin show that the AF phenotypes significantly reduce the luminance in the room—a reduction of appropriately 50% compared to the case without AF. In cases of the proposed AF phenotypes, the ASE index decreased below 10% compared to natural daylight conditions and achieved 3 points according to LEED V4.1. During the daily opening and closing cycle of the AF, the ASE and sDA indices don't observe many sudden fluctuations and remained stable within the allowed lighting range.

Keywords Adaptive facades · Daylighting · Energy-efficient · Office building

V. T. Nguyen (✉) · H. T. Le · P. B. L. Nguyen
Faculty of Mechanical - Electrical and Computer Engineering, Van Lang University, 69/68 Dang Thuy Tram Street, Ward 13, Binh Thanh District, Ho Chi Minh City, Vietnam
e-mail: tung.nv@vlu.edu.vn

T. H. N. Le
Faculty of Civil Engineering, Ho Chi Minh City University of Technology, 268 Ly Thuong Kiet Street, 14 Ward, District 10, Ho Chi Minh City, Vietnam

1   Introduction

Energy consumption in the industrial and building sectors as reported by IEA is the area with the highest energy consumption in the global economy [1]. In order to reduce energy consumption in buildings, the design forms of passive and active design are two main groups of solutions, in which passive design is the solution that is more often taken interest in because of its applicability. Passive design is a solution aimed at reducing energy consumption through the influence of the BE system such as the usage of shading devices.

The AF solution is one of the components of the building envelope system, whose function is interacting and responding to the natural environment in real-time [2]. AF is understood as a multifunctional highly adaptive system, the insulating system between the interior and exterior of a building, this facade is capable of changing features, functions, or behaviors by itself to meet the requirements of usage efficiency, with the aim of improving the energy-efficiency of the building. The AF helps to balance natural lighting and prevents radiation from transmitting into the building through mobile shading panels. The AF form is widely applied in many practical works and in researches, which shows the great potential that this system offers. Typical buildings with AF applied are presented in Fig. 1.

AF affects the issue of energy-efficiency through it is ability to adjust natural daylight to suit the internal occupational environment. Daylight indicators through simulation have the effect of establishing limits for lighting in accordance with design standards. Common daylight indicators include ASE, sDA, and DF. ASE being the proportional value of the locations with the number of hours receiving direct sunlight inside the room. In particular, ASE measures locations exposed to direct sunlight above 1000 lx and received over 250 h. sDA surveys locations achieved adequate sunlight exposure during standard working hours (8 am to 6 pm) in the workspace. To reach sDA requirement the surveyed positions must yield a minimum of 300 lx in half of the day's working hours (50% of the occupied period). According to the LEED V4.1 standard, buildings will be assigned 1 to 3 points for designs with appropriate ASE and sDA statistics (Table 1) [3]. Also according to LEED V4.1, the ASE index of below 10% needs to evaluate luminance inside the working plane. (iii)

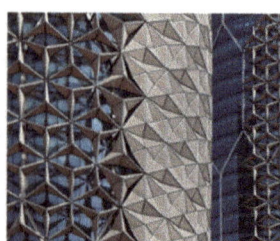

Fig. 1 Typical AF phenotypes in buildings, from left to right: House of Natural Resources in Zurich, Al Bahar Towers, Kiefer Technic Showroom (source: [8])

Table 1 Natural daylight rating scale in option 1 according to LEED V4.1 (source: [3])

New Construction, Core and Shell, Schools, Retail, Data Centers, Warehouses and Distribution Centers, Hospitality		Healthcare
Percentage of regularly occupied floor area	Points	Points
The average sDA300/50% at least 40%	1	1
The average sDA300/50% at least 50%	2	2
The average sDA300/50% at least 75%	3	Exemplary performance

Daylight Factor (DF) is the percentage between indoor and outdoor illuminance as determined under CIE Overcast (lux) sky conditions. According to the United States Green Building Council (USGBC), DF must achieve a minimum of 2% in 75% used space.

The diversity of AF phenotypes has been presented in many previous studies [2, 4], some of which are interested in natural lighting issues such as [5, 6]. Research by A. Tabadkani et al. [4], presented analysis on the kinetic AF and natural lighting comfort through the Useful Daylight Illuminance (UDI) index. Kinetic adaptive facade are defined as complex mechanical systems in which a certain kind of motions like displacing, sliding, expanding, folding or transforming, ensure variable geometries and mobility of the system [2]. Research by P. Bakmohammadi [7] studied lighting and energy in classrooms through UDI, DA, ASE indices, concerning user comfort. When existing effects change the size or angle of the window, the illuminance and energy consumption index (EUI) also change simultaneously. However, there are yet no studies on the change of motion of kinematic AF affecting the lighting indices ASE, sDA, and DF in high-rise offices located in HCMC. At the same time, there are no proposed AF solutions following the new evaluation method of the LEED V4.1.

In this study, the kinetic AF model is proposed to be equipped on the northeast facade of the HOB which is the building's main facade. The case study used is LIM Tower HOB located in HCMC as it is typical for all buildings with a single-layered glass facade. The research focuses on the impact of structural design and motion of the mobile AF (kinetic facade) towards the ASE, sDA, and DF indices in accordance with the criteria for rating green-buildings such as LEED V4.1. The implementation process is divided into 3 main stages to enhance the efficiency of natural daylight. Of which, phase 1 is to build a HOB model consist of a typical floor for natural lighting simulation. There on, the survey of the typical floor natural lighting coefficient in the above HOB is taken. Phase 2 presents the survey of ASE, sDA, and DF coefficients in the room when equipped with the simultaneous AF system. Finally, stage 3 alters the design variable of the AF to find a suitable solution for the lighting problem according to LEED V4 standards.

The study was carried out using Rhinoceros-Grasshopper software, the plugin used in the analysis of daylight factors is ClimateStudio of Solemma. Climate Studio is an improved and updated plugin from DIVA-for-Rhino. Many studies on the accuracy of software DIVA-for-Rhino and ClimateStudio have also been taken in prior researches [6, 7]. This study provides AF solutions to the problem of natural lighting for HCMC, at the same time, building a design process towards lighting comfort adhering to LEED V4.1 standard.

2 Setup Experiment

2.1 Daylight Modeling

In order to properly apply AF to HOBs, it is necessary to investigate the daylight coefficients inside the workspace. Surveys on daylight quality in HOB in HCMC were carried out on the 25th floor of LIM Tower (Fig. 2a). The total floor area is 920 m^2, the height between floors is 3.2 m. The daylight factor is calculated on a grid plane with an area of 48 m^2. Each grid module is spaced 0.6 m apart (conforming to the calculation grid proposed by LEED) and the grid system is located 0.76 m above the floor, which coincides with the standard working position proposed by LEED V4.1.

The kinetic AF system is integrated on the northeastern facade of the building (Fig. 2b). The integration of Kinetic AF into the building is intended to solve lighting

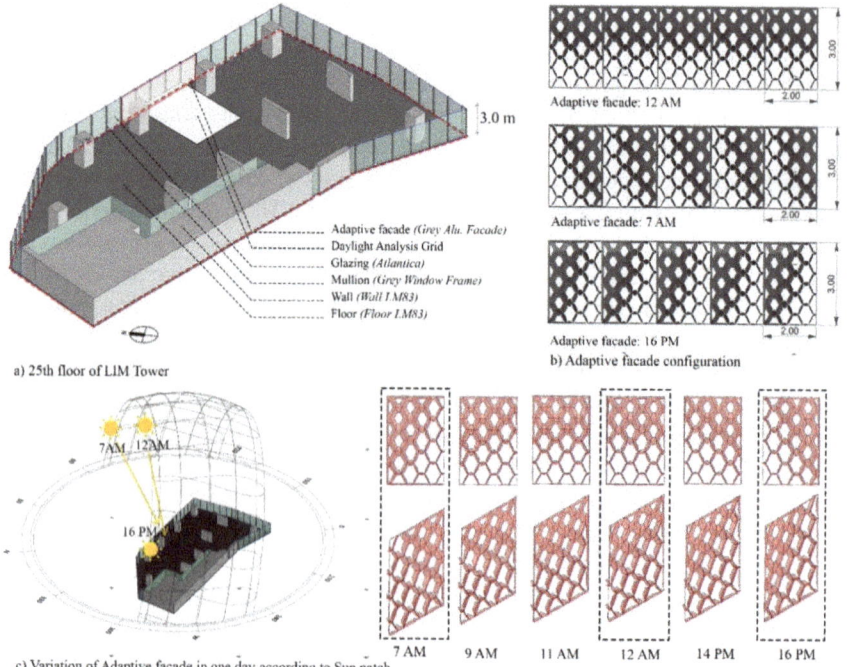

Fig. 2 Case study for daylight simulation and adaptive facade phenotypes (source: [8])

Table 2 Material system setup for daylight simulation (source: authors)

Construction	Type	Roughn ess	R_{vis} (tot)	R_{vis} (diff)	R_{vis} (spec)	T_{vis} (tot)	T_{vis} (diff)	T_{vis} (spec)
Ceiling LM83	Matte ceiling	0.10	70.0%	70.0%	0.0%	0.0%	0.0%	0.0%
Floor LM83	Matte floor	0.00	20.0%	20.0%	0.0%	0.0%	0.0%	0.0%
Grey Window Frame	Glossy others	0.20	18.1%	16.9%	1.2%	0.0%	0.0%	0.0%
Wall LM83	Matte wall	0.20	18.1%	16.9%	1.2%	0.0%	0.0%	0.0%
Grey Alu. Facade	Glossy ext. build	0.10	37.2%	34.7%	2.5%	0.0%	0.0%	0.0%

Glazing	Layers	T_{vis}	R_{vis} Front	R_{vis} Back	U value	SHGC
Atlantica	Single	66.3%	6.4%	6.4%	5.82	0.53
Atlantica-solarban 60 (3)	Double	45.9%	8.1%	8.3%	1.66	0.30
Atlantica-solarban 67 (3)	Double	40.2%	10.8%	18.1%	1.66	0.29

problems such as reducing the ASE index below 10% while keeping the sDA above the 50% threshold as announced by LEED shown in Table 1. The AF is composed of sheets of hexagon cells. In each AF there are 8 cells vertically and 5 cells horizontally. The inside of the hexagon cell sheets are then divided into small surfaces and are capable of rotating around the attraction structure from 0 to 80% according to the solar pattern. The orbit of the sun was determined on three marks: 7 am, 12 am, and 16 pm on June 15 (Fig. 2c). The simulation of the sun's orbit is done through the Ladybug plugin that operates on the Rhinoceros-Grasshopper platform [9].

Materials for daylight calculation are presented in Table 2. Accordingly, the materials used for ceiling, floor, and wall are Ceiling LM83, Floor LM83, and Wall LM83, respectively. The material for the AF uses a gray aluminum façade and the mullion uses a gray window frame material. Glazing materials were surveyed on three types of materials: Atlantica, Atlantica Solarban 60 (3), and Atlantica Solarban 67 (3) to evaluate daylight performance.

Table 2 presents the material properties used in the simulation. The material data is extracted from ClimateStudio v1.0. In which, materials are divided into two main groups of glazing materials and construction materials. For construction materials, the listed material parameters include the coefficient of surface roughness, Visual Reflectance (R_{vis}) which includes Specular Reflection (R_{vis}-spec), Diffuse reflection (R_{vis}-diff), and Total Reflectance (R_{vis}-tot). Indicators of construction materials are not capable of transmitting light (Visual transmittance-T_{vis}). For glazing materials, interfering indicators of lighting calculations include T_{vis}, front and back Visual reflectance, heat transfer coefficient (U-value), and Solar Heat Gain Coefficient (SHGC). The SHGC, U_{val} and R_{vis} indices of the three glazing materials is provided by the International glazing database (IGDB). The material parameters will directly interfere with the calculation of the ASE, sDA, and DF illuminance factors.

Table 3 Surveying the daylight of HOB LIM Tower through glass materials (source: authors)

	Atlantica single glazing	Atlantica-solarban 60 (3)	Atlantica-solarban 67 (3)
ASE	33.7%	33%	32.3%
sDA	100%	96.15%	87.69%
Mean DF	5.97%	4.31%	3.8%
LEED V4.1	3	3	3

3 Results and Discussion

3.1 Case Study

In hot and humid areas like HCMC, buildings using glazing often have inappropriate daylight intensity. In order to evaluate the daylight ability of the LIM Tower case study, the lighting indicators including ASE, sDA, and Mean DF (Average DF) were surveyed on many different types of glazing (Table 3). The results from Table 3 show that all glazing phenotypes achieved 3 points of LEED V4.1. A high sDA index from 87.5% to 100% indicates that the room receives a good amount of light for working processes. However, the ASE index is still marginally higher than the regulation of less than 10% of LEED. A high ASE number suggests that the luminance in the room is too high and needs to be managed. At the same time, a high DF index (from 3.8 to 5.97%) also shows that the room receives more natural light than allowed. From the results, the problem to be solved is to reduce the ASE index below 10% to ensure the balance of the glare control in the room while keeping the sDA at high levels to achieve 3 points of LEED V4.1.

3.2 Adaptive Façade Parameters

Based on the inadequacies in ASE index encountered by HOB LIM Tower as presented in Table 3. The kinetic AF in the form of the hexagon cells model is integrated to satisfy the natural daylight criteria. Surveys were performed on six different scales of the hexagon structure as shown in Fig. 3.

The resulting daylight indices are presented in Table 4. From those, it can be seen that the ASE index is significantly lower than the single glazing phenotypes of the

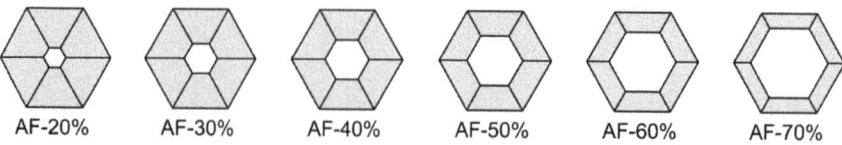

AF-20%   AF-30%   AF-40%   AF-50%   AF-60%   AF-70%

Fig. 3 Scale changes of hexagon cells (Source: authors)

Table 4 Investigate the change in the ratio of hexagon cells affecting natural daylight (source: authors)

	AF-20%	AF-30%	AF-40%	AF-50%	AF-60%	AF-70%
ASE	0.7%	3.8%	6.92%	11.54%	15.38%	15.38%
sDA	66.92%	70%	71.54%	72.51%	73.85%	74.62%
Mean DF	0.88%	0.9%	1.06%	1.2%	1.36%	1.64%
LEED V4.1	3	3	3	2	2	2

case study. Surveying on 6 scales of AF, the ASE index ranged from 0.7 to 15.38%, of which 3 cases met the ASE condition under 10% were AF-20%, AF-30%, and AF-40% and had ASE figures of 0.7, 3.8, and 6.92% respectively. The DF index was also significantly lower than the case study. However, with DF index of less than 2%, it is necessary to equip additional lighting equipment to enhance artificial lighting. As recommended by LEED V4.1, the ASE index should be less than 10% to ensure the quality of natural daylight, according to which three cases of AF are proposed to be suitable for lighting conditions are AF-20%, AF-30%, and AF-40%.

According to LEED V4.1 an sDA greater than 50% is counted as 2 points and greater than 75% is counted as 3 points for natural lighting (Table 1). In Table 4, the sDA index of all the AF cases is above 50% and below 75%. This shows that all the surveyed cases of AF got 2 points of LEED V4.1. However, all cases with ASE index under 10% are added 1 point.

3.3 Adaptive Facades Daylight Performance Throughout the day

Table 4 has shown 3 suitable AF phenotypes including AF-20%, AF-30%, and AF-40%. A study based on three AF phenotypes is proposed to investigate the daylight quality through the opening and closing movements of hexagon cell sheets (Fig. 2). The survey results show that there are no significant changes in daylight indicators in the three-time marks of 7 am, 12 am, and 16 pm (Table 5). All cases had ASE of less than 10%, the highest was 8.4% as in the case of AF-40% (12 am) and the lowest was 0.77% as in the case of AF-20% (7 am). The sDA index is always above 50% with the lowest case being 60.77% (AF-20%) and the highest case being 99.23% (AF-40%). From the simulation data, all movements in all day of the AF achieved the maximum score for natural lighting according to LEED V4.1. The survey on the DF index shows that all cases are below 2% and need to be equipped with additional lighting equipment to ensure the necessary amount of light. The AF phenotypes consistent with LEED V4.1 are shown in Fig. 4.

Table 5 Investigation of natural daylight indices in all day of selected AF phenotypes (source: authors)

	Adaptive façade scale								
	AF-20%			AF-30			AF-40%		
	7 AM	12 AM	16 PM	7 AM	12 AM	16 PM	7 AM	12 AM	16 PM
ASE	0.77	5.38	3.08	3.8	7.6	3.8	6.92	8.4	6.92
sDA	66.92	60.77	66.92	70	63.85	64.62	71.54	95.3	99.23
Mean DF	0.88	0.79	0.9	0.9	0.83	0.97	1.06	1.39	1.56
LEED V4.1	3	3	3	3	3	3	3	3	3

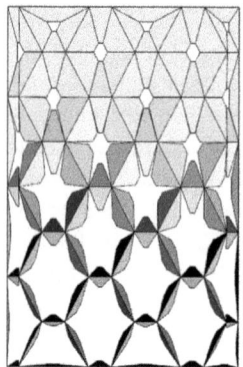 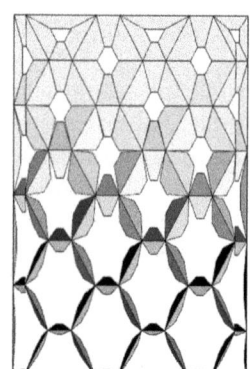 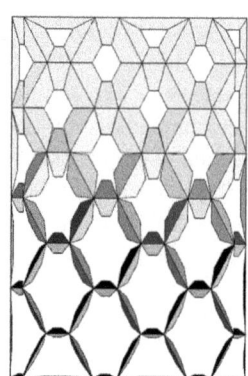

Fig. 4 The proposed AF phenotypes fit LEED V4.1 criteria, from left to right: AF-20%; AF-30%; AF-40% (source: authors)

4 Conclusion

The AF phenotype shows significant improvements in enhancing the efficiency of natural daylight energy-efficiency. Research on the AF phenotypes applied to HOB in HCMC has achieved positive results for natural daylight through the ASE, sDA, and DF indices. The proposed AF phenotypes have an ASE index of less than 10%, an sDA of more than 50%, and a score of 3 for LEED V4.1. The study's result opens up a system of solutions suitable for conditions in hot and humid climates such as HCMC and is a basis for designers and architects to apply in construction projects with an emphasis on energy efficiency and sustainable development.

Acknowledgements This paper is funded by Van Lang University and ATC Lab (Van Lang University) in the cost of implementation and simulation equipment. We acknowledge the support of time and facilities form Ho Chi Minh City University of Technology (HCMUT), VNU-HCM for this study.

References

1. International Energy Agency (2015) World energy outlook special report, world energy outlook Spec. Rep., Paris, France
2. Tabadkani A, Roetzel A, Li HX, Tsangrassoulis A (2021) Design approaches and typologies of adaptive facades: a review, vol 121. Automation in Construction, pp 103450
3. USGBC (2020) LEED V4.1: Building design and construction, us green build. Counc., Washington, D.C
4. Alkhatib H, Lemarchand P, Norton B, O'Sullivan DTJ (2020) Deployment and control of adaptive building facades for energy generation, thermal insulation, ventilation and daylighting: a review, vol 185. Automation in Construction, pp 116331
5. Tabadkani A, Valinejad Shoubi M, Soflaei F, Banihashemi S (2019) Integrated parametric design of adaptive facades for user's visual comfort, vol 106. Automation in Construction, pp 102857
6. Hosseini SM, Mohammadi M, Guerra-Santin O (2019) Interactive kinetic façade: improving visual comfort based on dynamic daylight and occupant's positions by 2D and 3D shape changes, vol 165. Building and Environment, pp 106396
7. Bakmohammadi P, Noorzai E (2020) Optimization of the design of the primary school classrooms in terms of energy and daylight performance considering occupants' thermal and visual comfort, vol 6. Building and Environment, pp 1590–1607
8. Tung NV (2020) Adaptive facade architecture solutions in office buildings in Ho Chi Minh City in the direction of energy efficiency, Master thesis, Ho Chi Minh University of Architecture
9. Roudsari MS, Pak M (2013) Ladybug: a parametric environmental plugin for grasshopper to help designers create an environmentally-conscious design. In: Conference of the international building performance simulation association, vol 13, pp 3128–3135

An Analysis of Architectural Features of Vietnamese Pagodas in an Giang Province

Thi Hong Na Le

Abstract An Giang province is a potential rich place of spiritual tourism 310 pagodas. An Giang (AG) is considered one of the two provinces with the most pagodas in the Southwest cultural sub-region of Vietnam. This paper focuses on identifying architecture features and values in Vietnamese temples of AG province. Today, there are six typical pagodas in AG province, including An Hoa, Hoa Thanh, Linh Son, Phi Lai, Tay An and Ancient Tay An Pagoda. Architectural characteristics of AG pagoda can be codified from the construction site, master plan, layout, materials to be used for decorative and sculptural art. From that point, the tangible and intangible values in architecture of pagodas in AG province will be clarified. The results will be regarded as the basis for preserving and promoting the architectural values of pagodas in AG province in order to enhance activities of spiritual tourism in the local sustainable development plan.

Keywords Pagoda architecture · An Giang pagoda · Heritage architecture

1 Introduction

Apart from the existing Khmer Buddhism, when transmitted to the South, Buddhism was influenced by the local culture, giving birth to many endogenous religions, enriching the spiritual life of the people. Buddhism appeared in AG before the establishment of the province, going through many ups and downs along the history. AG is the birthplace of the most endogenous religions in the South, so the religious life is very diverse. In AG province, the temple always occupies a special place. Apart from the communal house, Vietnamese pagoda has absorbed and shown all the spiritual culture of village community. AG is one of the two regions with the highest

T. H. N. Le (✉)
Faculty of Civil Engineering, Ho Chi Minh City University of Technology (HCMUT), 268 Ly Thuong Kiet Street, District 10, Ho Chi Minh City, Vietnam
e-mail: na.bmkt@hcmut.edu.vn

Vietnam National University Ho Chi Minh City, Linh Trung Ward, Thu Duc, Ho Chi Minh City, Vietnam

J. N. Reddy et al. (eds.), *ICSCEA 2021*, Lecture Notes in Civil Engineering 268,
https://doi.org/10.1007/978-981-19-3303-5_4

percentage of pagodas in the Southwestern provinces (67% of all other religious buildings) with many temples recognized as Historic Monuments at the national and provincial levels [2].

The published works focus more on culture and history than architecture. Those studies are generalized but have not gone into detail or systematized the typical values of Vietnamese pagoda architecture in AG province. Some typical works such as "Vietnamese Buddhist Architecture" by the late Nguyen Ba Lang (2015), "The temples of Southside" by Nguyen Quang Tuan (1994), "Southside Temples architecture" by Pham Anh Dung (2014), articles "Cultural acculturation in the near-modern Vietnamese pagoda in the South" (2003) and "Landscape and architecture of ancient Vietnamese temples in the South" (2005) by Phan Huu Toan. A Master's thesis in architecture by Tran Dang Khuynh (2004) entitled "Cultural symbiosis in the population of religious—belief architecture in the Sam mountain—An Giang region" has studied very comprehensively the types of architecture of pagodas and beliefs at Sam mountain and cultural symbiosis in AG but did not consider valuable temples in the remaining areas [3]. Within the scope of this paper, the architectural features of the pagodas in the whole AG province will be systematized and clarified. Six typical Vietnamese temples were analyzed in terms of site planning, overall layout, layout composition, construction materials and decoration. There are also the historical relics, artistic architectural monuments at the provincial and national level, belief centers of endogenous religions, excluding Theravada Buddhist temples.

2 Overview of Typical Vietnamese Temples in an Giang Province

The Vietnamese pagodas in AG show many features of the Southside pagoda's architecture. However, the architecture between the temples differs depending on the local cultural and religious point of view. Six typical works (ancient temple, famous, attracting thousands of believers) analyzed in this article are An Hoa Pagoda, Hoa Thanh Pagoda, Linh Son Pagoda, Phi Lai Pagoda, Tay An Pagoda and Tay An Ancient Pagoda (Table 1). Although these temples have been restored many times but still retain many material and spiritual values [3].

An Hoa pagoda in Hoa Hao village, Phu Tan district, facing South, overlooking the Vam Nao river. It was built in 1935, remodeled in 1952 and expanded in 2019. An Hoa Pagoda is the most important religious architectural work of Hoa Hao Buddhism with an area of up to 32,000 m². **Hoa Thanh Pagoda** is located in Nhon Hung commune, Tinh Bien district. It covers an area of 15,000 m², built in 1847, located in the middle of a dense forest, facing the North, overlooking Vinh Te canal. The inner frame system bears bold nineteenth-century Southern architecture and Western-style outside architecture. After over 170 years, on August 4, 1992, the pagoda was recognized as the national heritage site by Decision No. 983-VH/QD. **Linh Son Pagoda** was built in 1912 in an area of 10,000 m², on the slopes of Ba The

Table 1 Site planning and overall layout (source: author)

Pagoda	Location	Direction	Master plan
An Hoa	Near the Vam Nao river	South, overlooking the Vam Nao River	Four consecutive houses. The auxiliary works are arranged in a free style. There are two ports most important, symmetrical on both sides of the main shaft. A small shrine is built in the front
Hoa Thanh	In the middle of the secluded countryside, near Vinh Te canal	North direction, overlooking Vinh Te canal	Four consecutive houses, developing in depth, symmetry through the main axis. In the front is Lien Tri pond, worshiping Guanyin statue. Around the building there are many trees
Linh Son	Ba The mountain slopes, in the architectural complex of Phu Nam	Southeast direction, back against mountain	II-shaped house, with two simple parallel houses. The whole is on the symmetry axis. The gate of the most important form. Wide temple yard, with many trees
Tay An	Located at the foot of Sam mountain (Chau Doc)	Northwest direction, back leaning on Sam mountain	III-shaped house. There is a mezzanine, onion-shaped roof. The gate, flagpole and main-house are on a symmetrical axis. There are three gates (one main gate and two secondary gates). There is a large shrine and a complex of many tombs
Ancient Tay An	Located on the bank of Ong Chuong canal (Cho Moi)	Northeast direction, overlooking Ong Chuong canal	III-shaped house, with three-span-and-two-wing (Right house and Left house are on both sides). There is a symmetry axis. There are three gates, the shrine and the altar in front. Behind the pagoda there is a lotus pond to store water

(continued)

Table 1 (continued)

Pagoda	Location	Direction	Master plan
Phi Lai	Located at the foot of Tuong mountain (Tri Ton)	Southwest	The layout is "shrines in front, temple behind". Triple gate. There are two flagpoles. There are bell floors, empty floors. There is a natural courtyard inside. Behind there is a shrine to worship the two-ancestors. The front has two small shrines. The front and back are both the main sides

mountain, in the area of Oc Eo cultural relic, Nui Sap town, Thoai Son district. In 1913, it was rebuilt into a solid structure in accordance with the contemporary Southern pagoda architecture. It was built on the royal court of the late Phu Nam dynasty, in the seventh century AD. It was recognized as a special national relic under Decision No. 1419/QD-TT dated September 27, 2012. **Tay An pagoda** is located at the foot of Sam mountain, Vinh Te village, Tay Xuyen district, Chau Doc city. It was built in the area of 15,000 m², divided into many levels. The pagoda was built by the former governor Doan Uan in the 7th year of Thieu Tri (1847). In 1957, it was restored with the facade and until now this architecture remains intact. The main facade looks towards the province, the back is based on the canopy of the mountain. The interior is characterized with Southern architecture while the exterior is influenced by Western architecture. On July 10, 1980, Tay An pagoda was recognized as a national architectural monument, according to Decision No. 92/VH.QD, and the Vietnam Record Book Center officially recognized it as the first pagoda to have the architecture as a combination of the Indian art style and the ethnic ancient architecture in Vietnam. **Ancient Tay An Pagoda** is a holy place where important items of Buu Son Ky Huong religion are kept. It is located on the bank of Ong Chuong canal, near Xeo Mon canal, Cho Moi district. The facade has a northeast direction, looking towards the canal. Site area is about 10,000 m². It was built before 1856 on the old land plot of Coc Ong Dao Kien, restored for the first time in 1952–1953, then completed and kept the intact architecture up to now. Tay An Ancient pagoda reflects traditional architectural values of the South and is different from contemporary ancient temples. There is an ingenious combination of Western elements with traditional architecture. In addition, over 160 years of history, the temple contains many local historical and cultural values. **Phi Lai pagoda** is located at the foot of Statue mountain, in the vestige of complex of Ba Chuc Tombs, Tri Ton district. It covers an area of 9,000 m², built in 1877, rebuilt in 1968 and restored in 1980 after being destroyed. The pagoda has undergone more than 140 years with many historical events, including the massacre of hundreds of people in 1978. The architecture of reinforced concrete, though modern, still has traditional values and is different. The pagoda was recognized as a national historical and cultural relic on December 4, 1997.

3   Site Planning and Overall Layout

Temple construction has always been very important for the villages in AG [4]. The pagoda is often associated with the mysterious mountainous area, where the legend of miracles and sacredness is revealed. Hoa Thanh Pagoda is located in a secluded wilderness with dense trees. Linh Son Pagoda is located on a mountain slope. Tay An pagoda is located at the foot of the mountain. Ancient Tay An Pagoda, An Hoa Pagoda is located on the cup of a Buddhist man. The pagodas in AG almost do not follow any main direction but geographical position, wind direction and always face the rivers or main roads. If built near the mountain, the back of the pagoda will be leaning against the mountain and facing the road (Table 1). This is a factor that shows practicality in the culture of AG people. Maybe after opening the road, some pagoda gates were moved to facilitate the traffic. The pagoda is often associated with water elements, according to Feng-shui and can be influenced by the river culture of the AG people and is convenient for the travel of devotees. The temple is often built near rivers and canals. When not situated near the river, the pond can be dug in the front to store water for domestic use and create Feng-shui elements. In addition, AG pagoda is often close to communal houses, temples and shrines, forming a common belief system for villages. Tay An Ancient pagoda was built near Nguyen Trung Truc temple or Tay An pagoda located in the spiritual belief community at the foot of Sam mountain, including Ba Chua Xu shrine, Thoai Ngoc Hau mausoleum, Vinh Te communal house and Confucius temple.

4   Spatial Organization

The temples all follow the layout with the main hall at the front and the suffix at the back. Particularly, Phi Lai pagoda has a "shrines in front, temple behind" layout (Table 2). Except for Linh Son pagoda and Ancient Tay An pagoda, which is organized in a T-shape, all four remaining temples have a rectangular plan. An Hoa Pagoda and Phi Lai Pagoda have three main doors, Tay An pagoda has two main doors and two side doors, while the rest have two main doors. The houses in these six temples are linked across rooms. The corridor system is planned around the main-house. These pagodas are built in the style of "Sap Doi House" (adjacent houses).

An Hoa pagoda has two gates of East and West, some shrines, one flagpole, Eastern and Western houses, dining house etc. The main-house and the rear have the form of a double house, including two houses: three-span-and-two-wing house and a four-pillar house. The pagoda has two roof layers, the main-house has a mezzanine

Table 2 Site combination (source: author)

Pagoda	Facades and Roof Plans	Floor plan (▨ Main-house ▨ back-house)	
An Hoa			The house is about to be lined up, four consecutive houses
Hoa Thanh			five parallel house folds
Linh Son			two parallel house folds
Tay An			three parallel house folds
Ancient Tay An			three parallel house folds
Phi Lai			front-temple, back-agoda with inner yard

(roof wall). Roof angle is straight, not curled, no symbolic decoration. The main building is a combination of front-space, main-house, back-house, which are linked together. The back-house extends longer, with a door to the front porch. **Hoa Thanh Pagoda** has an overall layout of gate, pond and the main main-house planned on the symmetry axis. The main-house and the back-house have the form of a double house, consisting of a fold of the three-span-and-two-wing house with two wings linked to the four-pillar-style house. Dining house, back-house built in the style of a three-room and two-leaned house. Main hall, surrounded with verandas, with balustrades designed in cement hyacinth patterns. The roof is designed in a two-story structure. The main-house has a mezzanine. **Linh Son Pagoda** has symmetrical architecture across the axis, separated from the tomb tower and the roads are planned in a free-style structure. There are 25 steps from the outside to the temple. The gate is a single form, with two floors, above is the statue of Maitreya Buddha. Going inside, the pagoda has a large yard, cool space with dense old trees. The two sides of the main-house are the tombs of the abbot over the generations. In front of the main-house there is a small shrine. The main-house is connected to the back-house. The main-house has the form of a traditional four-pillar house and is not surrounded with verandas, the two sides are the dining house and the monastery. The back-house is a three-span-and-two-wing house, with two doors connecting to the main-house. The roof has two stories without mezzanine.

Tay An pagoda has the main gate in the form of three-gates, three-layers roof, yin and yang tile. The main door is decorated with Bodhisattva statue. The two sub-gates are single form with one single-story roof. Behind the gate is a 16 m high flagpole, below is a yellow lotus symbol. Next is the main-house, the classroom and the ancestor-house. In front is a tower with two floors. The lower floor is square, with the statue of Buddha Shakyamuni when he was a child, Buddha's statue in meditation, the two feet of the tower in front has the statue of Colossus. The upper floor is round, arranged standing Buddha Shakyamuni statue, onion-shaped dome in the Indian Buddhist architecture. The front space has a curved roof made of concrete, two square towers with two-layer roof. The main-house and back-house have the type of double house. **Ancient Tay An pagoda** is different from the contemporary ancient temples. The main-house, the back-house, the Buddhist school, left and right houses are linked together to form a large architecture. In the front, there is a small shrine and three-gates. Behind is the lotus lake, the grave area and the dining room. The back-house is the four-pillar house type, connected with the main-house, the Buddhist school, and left and right houses are connected to across room. Two left and right houses run long together along the two sides of the main-house and the back house (or rear house). Behind the back-house, there is a small courtyard, but now it is built to cover again. The back-house is long, including seven rooms with two wings. The main-house has a dual-house style, the three-span-and-two-wing house was linked to the four-pillar house. **Phi Lai pagoda** has a complex of premises "shrines in front, temple behind". The main gate is in the form of three-gates, built in 1965. In front of the pagoda, there are small shrines on both sides and the flagpole in the middle. The shrine is linked to the main hall through two corridors, the inside has a courtyard with surrounded veranda. Along the corridor, there are two bell and drum towers, including two floors (lower-square, upper-octagon), with the onion-shaped roof. The architecture of the front house is a three-span-and-two-wing house, linked to the temple in the form of a four-pillar house. The main-house is a double house, with three-span-and-two-wing house, linked with four-pillars, with a mezzanine. The four houses connected to each other in a cross-room style. The column system made of reinforced concrete, brick wall, with many windows. Two inner corridors have flat roofs and reinforced concrete columns. In the middle of the courtyard, there is a small celestial altar with a flagpole.

5   Construction and Decoration Materials

In general, Vietnamese pagodas in AG are built with concrete, reinforced concrete and brick walls. Some pagodas are incorporating traditional Southern wooden structures for the inner frame and roofing such as Tay An pagoda, Tay An Co pagoda and Phi Lai pagoda. The decorative colors are simple with gold column walls, trim, threads, and white details. Except for An Hoa pagoda, the majority of pagodas have many reliefs and decorative details on façades, roofs, columns and doorways. The content of reliefs and decorative details is diverse in each temple with the cultural characteristics of Buddhism, Ethnicity, and the West. Inside the main-house, the main altars are often decorated with intricately carved blue bags, painted with gold lacquer (Table 3).

An Hoa pagoda has boxes between two roof floors decorated with Buddha relics. The difference is that the back-house worship the Tay An Master and only worship the image to commemorate the founder of the religion. The pagoda was built in the typical Southern architecture, simple but containing many cultural values. Over 80 years of history, from a small temple, the pagoda has gradually become larger, regarded as an important mecca of millions of Hoa Hao Buddhists. **Hoa Thanh Pagoda** has a roof angle that is not curved, the roof bank is attached with many reliefs and ceramic statues. The facade decorated many Western reliefs. In the main-house, there are many jackfruit wood worship statues, painted with brilliant golden lacquer, with artistic value of the nineteenth century. The space inside is a bit dark because there are surrounding verandas, creating a sacred feeling in darkness. **Linh Son Pagoda** is decorated with many poems, distich, and stylized lotus-shaped reliefs. In front is the statue of four-hands Buddha, two stone steles and Amitabha Buddha, Shakyamuni Buddha, Samantabhadra Bodhisattva, Tam Tang and two Colossus statues. Behind is the altar of Guanyin Buddha. The main hall and the rear chamber connected to each other through three arched doors. The back-house arranged many altars to the ancestors. four-hands Buddha statue is a special sculpture, which is the image of the god Vishnu made of black stone, 3.3 m high, with a round-shaped hat, dating back to the fifth century. After that, it was painted and repaired into a seated Buddha statue to worship according to the customs of Tonkin Buddhism.

In **Tay An pagoda**, the exterior decoration (Fig. 1) is influenced by the Indian Buddhist culture with details such as the veranda column row, the roof frills, the small statues at the bottom of the roof etc. but the interior space is completely different (solitary, sacred). The pagoda has many altars, about 200 statues of many stages, representing Vietnamese sculpture art in the nineteenth century. In **Ancient Tay An pagoda**, the veranda surrounds the main-house, taking indirect sunlight inside, making the space somewhat solitary and sacred. The main-house with mezzanine is the place to worship the altar of Tran Dieu plate and not to worship Buddha statue according to the viewpoint of Buu Son Ky Huong religion. The main facade is elaborately decorated, combining traditional Southern and Western architecture (Fig. 1). The roof is divided into three floors, decorated with many sophisticated crockery reliefs and exquisitely glazed ceramic statues in the shape of dragons, phoenixes, and dragon carp. Between the rooftops, there are boxes of drawers painted with legends, mascots, and charming paint. Handrails are decorated with retro-shaped reliefs. **Phi Lai pagoda** has columns which further decorated with distiches. Columns and doors are not elaborately decorated. The compartments between the roofs have charming and picturesque paintings. The roof frill is decorated with many reliefs with wooden roof support system. The space inside is very high and airy but somewhat solitary due to the surrounding corridor. A total of 20 altars with paintings, no worshiping

Table 3 Decoration materials and details of six Vietnamese pagoda in AG Province (source: author)

Pagoda	Materials	Details			
		Color	Culture	Relief	Others
Common features	Concrete, reinforced concrete, brick walls	Gold column walls, trim, threads, and white details	Buddhist culture, Vietnamese tradition and Western culture		The altars are decorated with intricately carved blue bags, painted with gold lacquer
An Hoa	Concrete, reinforced concrete, brick walls	Gold column walls, trim, threads, and white details	Cultural characteristics of the South of Vietnam		The boxes between two roof floors decorated with Buddha relics
Hoa Thanh			Vietnamese tradition and Western culture. nineteenth century art	There are many reliefs at the root and facade	The space inside is a bit dark, creating a sacred. Jackfruit wood worship statues are painted with brilliant golden lacquer
Linh Son			Vietnamese tradition, fifth century art Tonkin Buddhism	Stylized lotus-shaped reliefs	There are Buddha statues and altars to the ancestors. The special sculpture is seated Buddha statue
Tay An	Concrete, reinforced concrete, brick walls, incorporating traditional Southern wooden structures for the inner frame and roofing		Vietnamese tradition, nineteenth century art. Indian Buddhist culture		There are veranda column rows, the roof frills, the small statues at the bottom of the roof. The interior space is solitary and sacred. There are many altars and around 200 statues

(continued)

Table 3 (continued)

Pagoda	Materials	Details			
		Color	Culture	Relief	Others
Ancient Tay An		Gold column walls, trim, threads, and white details	Vietnamese tradition and Western culture. Buu Son Ky Huong religion	There are many delicate and ornate decorative reliefs	The space is quiet and sacred because it receives indirect light through the veranda. There is an altar to Tran Dieu plate, not a Buddha statue
Phi Lai		Gold column walls, trim, threads, and white details. Hammock door has an inlaid gold and red lacquer style	Vietnamese tradition Tu An Hieu Nghia religion	There are many reliefs at the roof	The space inside is very high and airy. There are distiches at the columns. Columns, doors are simple. There are charming, picturesque paintings. There are 20 altars with paintings, no worshiping statues, and red fabric

Fig. 1 Left to right: Decorative details of Buddha-worshiping tower; Tay An pagoda column; Facade decoration of Ancient Tay An pagoda; Decorative handrail of Phi Lai pagoda [3]

statues, and red fabric according to Tu An Hieu Nghia religion can be found here. The hammock door has an inlaid gold and red lacquer style.

6 Conclusion

Vietnamese pagoda in AG province is a cultural product of AG people. It reflects tangible and intangible values. In which, the object value is transmitted through architectural solutions and contains many features of Vietnamese folk architecture. Spiritual values are reflected through traditional, historical and spiritual value systems hidden inside the building. These values are economically meaningful to the temple, especially when AG province decides to promote the development of spiritual tourism. AG pagoda architecture has similarities with Southside pagoda but also has its own characteristics. Typical is a unique four-pillar house, created based on a chest house. The forms of decoration show a folklore culture, close to people's life and integrate many cultural elements. The layout of the altars in the temple is also very diverse, depending on the individual views of each religion. All these factors have created a different characteristic for each temple.

Acknowledgements I would like to thank the University of Technology, VNU-HCM for the support with time, facilities and funding for this study.

References

1. Board of Trustees of the Vietnam Buddhist Church in An Giang province (2017) Problems and proposed solutions for the sustainable development of the Vietnamese Buddhist Church in the future, An Giang province
2. General Statistics Office (2018) Economic census results 2017, Publisher. Statistics, Hanoi, pp 308–341
3. Duc LA (2020) Temple architecture of an Giang province, Ho Chi Minh City University of Architecture, Master's thesis
4. Them TN (2018) Vietnamese culture in the southwestern region. Ho Chi Minh Culture and Art Publishing House

Asian Cultural Heritage Conservation: Cholon's Heritage Resources in Vietnam's Transformation

Ngo Minh Hung

Abstract Saigon was founded by the army commander Nguyen Huu Canh in the seventeen century. It became the capital of the French colony of Cochin-China in 1883, part of French Indochinese Union which lasted until 1945, and the capital of the Republic of Vietnam (South Vietnam) until 1975. It has diverse cultural characteristics and a unique urban form. It was described as 'the pearl of the Far East', 'the door to Far East Asia', 'the Oriental Paris' (Guillaume, 1985) during the nineteenth and twentieth centuries. After the unification of the North and South Vietnam in 1975, Saigon was renamed Ho Chi Minh City (HCMC). Within the context of the 'Asian Century', Asia's rapid growth and urbanization have transformed global commodity markets and Asian societies where millions of people have migrated from rural areas to the cities thereby increasing the urban population. The rapid transformation of HCMC's economy in the last two decades is one of the success stories of Asia. The transformation of this economic Centre brought about tangible and intangible cultural changes at central places. A conservation plan has been drawn up in the past, such as Cholon town (2012), but this has not led to the conservation project to protect 'the spirit' of Saigon and Cholon, which had been considerably as the original cities in the past of HCMC. This paper, therefore, aims to re-assess cultural significance of old Cholon thematically to support action plans towards sustainable conservation in globalization context. It pays more attention to the district No. 5 as an important part of the inner city and is the location of many ancient historic places such as Gia Dinh region and Cholon (Chinatown), which date from the seventeen century and display characteristic architectural and urban features of the period. To deal with this issue, it is urgent that action is taken immediately to protect Ho Chi Minh City's surviving built heritage.

Keywords The spirit of Cholon · Heritage conservation · Cultural significance

N. M. Hung (✉)
Institute of Cultural Heritage and Development Studies, Van Lang University, 69/68 Dang Thuy Tram, Ward 13, Binh Thanh District, Ho Chi Minh City, Vietnam
e-mail: hung.nm@vlu.edu.vn

1 Introduction

Ho Chi Minh City was once known as Saigon and is still called this by many people in Vietnam today. It is not an ancient city by world standards, having been founded by the army commander Nguyen Huu Canh in 1698 CE. Even so, Ho Chi Minh City has a rich heritage reflecting its complex history and the diverse cultural characteristics of its inhabitants and best seen in a unique urban form comprising both individual structures and their arrangement in city plans and streetscapes. In the seventeenth century the Kinh Viet people were spreading south from their historic core in the Red River delta around the royal capital, Hanoi. This was a period of internecine struggle for power between two sets of war lords—the Trinh who came to dominate the north and the Nguyen who led the move southwards taking lands from previous occupants such as the Cham and Khmer. It was Nguyen Phuc Tru whose orders led to Nguyen Huu Canh capturing the small Cham market town, Baigaur, and renaming it Gia Dinh, a name later changed again to Saigon.

During the last thirty years of the eighteenth-century Vietnam was effectively split between north and south following the so-called Tay Son rebellion. However, another Nguyen prince, Nguyen Anh, reunited the country, had himself crowned Gia Long in 1802—the first emperor in Vietnam's last royal, Nguyen dynasty—and shifted the Vietnamese capital from Hanoi to Hue. In order to assume power Nguyen Anh had called on French military support, France's first official state intervention in the territory of modern Vietnam. France's imperialist ambitions did not rest there for long, however, and 50 years later the French navy bombarded and captured Saigon and, in 1862, the city was proclaimed capital of the French colony of Cochin-China. Subsequently it was incorporated within the French Indochinese Union that lasted from 1883 to 1945. Saigon grew quickly during the late nineteenth and early twentieth centuries and acquired a charm that led to it being described as the 'Oriental Paris', the 'Pearl of the Far East' and 'the door to Far East Asia' [1]. These epithets dwindled mid-century during the Japanese War (1940–45) and then especially during the Vietnam War (1955–75) when it was the capital of the Republic of Vietnam (South Vietnam).

After the unification of North and South Vietnam in 1975 as the Socialist Republic of Vietnam under the Vietnamese Communist Party based in Hanoi, Saigon was renamed Ho Chi Minh City. Within a decade, however, it became apparent to national leaders that the moves to socialize business and industrial activities and to collectivize agriculture in the south were failing. This was a main factor behind the development of the economic reform process, known as *Doi Moi*, which officially starting with the Sixth Congress of the Vietnamese Communist Party in 1986, has transformed Vietnam from a highly centralized planned economy to a socialist-oriented market economy. Vietnam's cities have borne much of the impact of *Doi Moi*, undergoing both inner city redevelopment and new development on the periphery. Ho Chi Minh City confirmed its status as Vietnam's largest metropolis in area and population terms and as the country's economic power house.

Asia's rapid growth has already transformed global commodity markets and Asian societies have themselves been irretrievably altered with millions of people migrating from rural areas to swell city populations. In these terms the rapid transformation of Ho Chi Minh City's economy in the last two decades can be seen as one of Asia's success stories. But the transformation and has brought about tangible and intangible cultural changes and these changes have, however, both positive and negative consequences. For instance, the old urban fabric of streetscapes, buildings and sometimes even treasured open spaces, built heritage that represented the distinctive cultural and architectural identities of the city's Vietnamese, Khmer, Cham, Chinese, French and Indian inhabitants, are being replaced by larger modern glass skyscrapers. The local Vietnamese scholar, Le Quang Ninh[1], has provided clear evidence that Ho Chi Minh City has lost about 30%[2] of its 108 registered architectural heritage items since 1993 [2]. In 2014 alone, hundred-year-old factories along Tau Hu canal and Tran Van Kieu Street and at the old Binh Dong quay as well as ancient shop-houses in the Ben Thanh area were demolished for redevelopment projects.

Within the context of the 'Asian Century', such redevelopment can be expected to increase significantly. Given the speed of change, it is urgent that action is taken immediately to protect Ho Chi Minh City's surviving built heritage. Regrettably, the city is in a dilemma about how to preserve the spirit of Saigon. Two conservation plans have been drawn up in the past—for the central city's architecture in 1995 and for the traditional Chinese quarter known as Cholon in 2012—but these have not led to practical conservation projects. On one hand, the Ho Chi Minh City conservation program of architecture and landscape had completed with various activities, such as the workshops, public hearings, study visits in France, Singapore, Malaysia and involved expats. After all, a book on "Sài Gòn 1698–1998: Kiến trúc-Quy hoạch" (Saigon 1698–1998: Architecture-Urban Planning) had been officially published as such invaluable reference for late policies from the City. On the other hand, the conceptual research (area 64 ha) on preserving Cholon (district 5), prepared by DCU, Spain in 2012, was submitted to HCMC's People committee for another reference, which has no more mention later on.

In late 2012, discussions among local scholars Le Quang Ninh, Nguyen Thi Hau and Nguyen Ngoc Dung led to the conclusion that conservation of the city's built heritage[3] must begin with the oldest long-ways (đường thiên lý) and the townscape, with priority given to those areas best reflecting Ho Chi Minh City's unique cultural identity and memories of old Saigon life, its river (Saigon River) and market (Cho

---

[1] The former director of first Ho Chi Minh City conservation program of architecture and landscape (in 1993) leading to the later inventory of 108 relics. Internet: http://vnexpress.net/tin-tuc/thoi-su/tp-hcm-lap-danh-muc-bao-ton-kien-truc-do-thi-2421220.html (accessed at 9.00am on July 22, 2015)

[2] Internet: http://sgtt.vn/Kien-truc-doi-song/Chi-tiet/174850/TPHCM-30-cong-trinh-kien-truc-co-gia-tri-lich-su-da-bien-mat.html (accessed at 11.40am on May 8, 2013)

[3] Those concepts were raised up in the conference on 'Urban and Architecture heritage in Ho Chi Minh City', organized by Vietnam Architect's Association, Ho Chi Minh City Architect's Association and Ho Chi Minh PC, dated 14 December 2012, in Ho Chi Minh City, Vietnam. Internet: http://vietnamnet.vn/vn/van-hoa/101230/loay-hoay-giu--hon-via--do-thi-tp-hcm.html (2:28pm, 10 July 2015).

Ben Thanh). In this book we argue that these nostalgia-driven approaches are insufficient and too generalized to protect the city's heritage. A clearer, more detailed understanding the origins and evolution of Saigon and Cholon and an analysis of the significance of the urban heritage elements that contribute to the city's identity is needed.

Another key issue is the need to take more fully into account the views of the local community about what is and what is not significant in their living environment and to bring the community into the management of the urban heritage so defined.

In fact, there is some basis for optimism in the Ho Chi Minh City case given that the People's Committee—the city's municipal government—has already adopted in 2010 the objective of achieving sustainable urban development. This opens up potentialities both for conserving its cultural heritage alongside economic development and infrastructure modernization and for consolidating its status as a major Asian industrial, service and scientific centre.[4] Local, national and international scholars will, however, need to take maintain pressure to ensure that conservation is not seen as a dispensable or even minor part of the management equation. However, once considered as an impediment to modernization and progress, heritage in Vietnam is now more likely to be seen as a 'vector of growth' [3].

2   Historical Transformation

- ### *Pre-Gia Dinh period*

   The area, on one hand, had attracted more trader, mainly Vietnamese by 1620 [4] to carry out commerce of agricultural product at the ports. To foster the trading activity, Khmer King-Chay Chettha II [5] established two custom posts for Vietnamese located closely to Tan Binh River (old name of present Saigon River) and Saigon arroyo (previous name of current Thi Nghe arroyo) in the East of Prey Nokor. First post located in Ben Nghe region, called *Ben Nghe post* (now Thai Binh market). Second post placed in Saigon region, called *Saigon post* in Chinese area. A number of Vietnamese, with background of soldier-peasants, moved to the region as the land of refuge in the seventeenth century [1] and rapidly increased afterwards.

   Typologically, the Ben Nghe post was highlighted, by the local scholars, on higher-area and at central point between Ben Nghe port (the East)—a confluence of Tan Binh River and Saigon arroyo and Dieu Khien Market (the West) and Saigon arroyo (the South) as main point of the early city towards *first form of commerce-port city*. The Saigon post, located about nearly four km away compared to first post, in the west to shape another town centre of Tan Thuan region (the East). There were two old pagodas (Cay Mai and Phung Son) of Viet and Chinese built and existed until today. It was defined as a centre of pure commerce and

---

[4] Point 2.c, Decision No. 24/QD-TTg on Master Planning of Ho Chi Minh City to 2025, approved by Prime Minister, dated 6th January 2010.

religion with Chinese majority towards *second form of trading and religious town* in the regions. Both single forms had likely been important trading centre in the Mekong Delta region. In other words, economic activity was important in these towns' foundation [1]. In this period, French also marked their visits by building religious buildings, such as the first southern Bishop's Palace (1659) on higher-area (now district 1 of HCMC) (Figs. 2 and 4).

After this period, the region gradually became Saigon, which passed over up and down periods from feudal regime to Western interference towards the Reunion. Those, upon the Author's point of view, are dividable into four folds: (1) The Viet localized Gia Dinh; (2) The French colonized Saigon; (3) Americanised Saigon and; (4) Saigon under Socialism respectively.

- *Gia Dinh period (1698–1860)—The Viet localized Gia Dinh*

  In 1788, the Southern region was re-organized and centralized, under feudal lord Nguyen Anh, to become the southern administration. This land played a role

Fig. 1 Location of Prey Nokor in Cochin-China and Cambodia before 1145 Source: PADDI (2015)

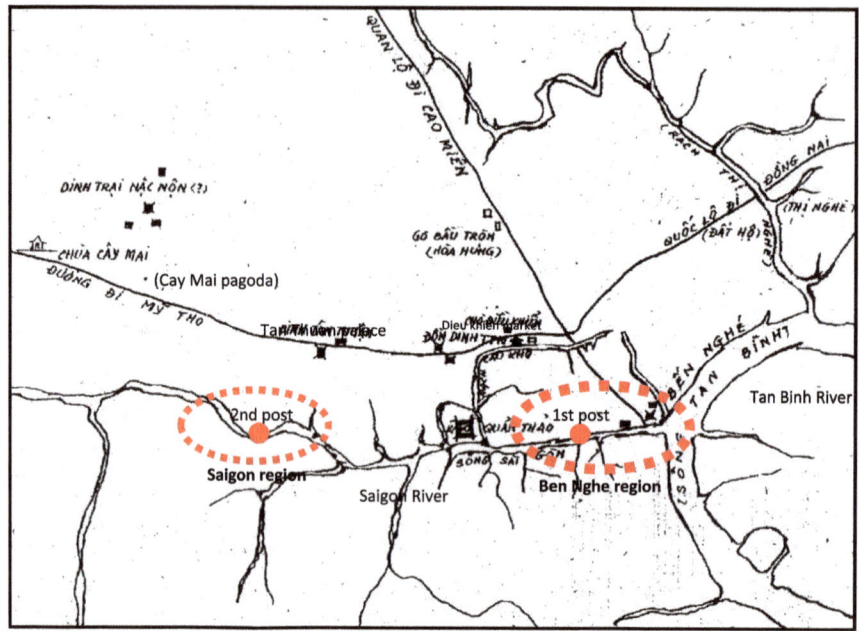

Fig. 2 The map of Saigon 1623–1679 (Source: General Science Library of HCMC 2014)

of main port in the South with the posts of Ca Tre and Vam Co along the Tan Binh River; together with Cau Kho sub-port on Saigon arroyo to control the goods that arrived and were shipped out (Fig. 2).

Later, the army commander Nguyen Huu Canh, in 1836, was sent to pacify land of Gia Dinh[5], which had previously been selected by Gia Long—first Emperor of Nguyen dynasty. In term of administration, Gia Dinh region covered about around 260.000 ha of land and was descriptive as ***a water and cool climate city***.

- **Gia Dinh land—the Imperial City**

The land (phủ Gia Định) covered entire the South, while the rural area belonged to two provinces Tran Bien and Phien Tran, being managed by the commander Canh of the Nguyen dynasty associated with the feudal institution, which applied over the administration, social-economic aspect and etc.

The Gia Dinh region was next naturally localized by Viet people after the market town Baigaur was taken over in 1698.[6] It was gradually coming to be

---

[5] Nghia M. Vo (2011) discussed: 'Gia Dinh' in Malaysia terms, orally meant *water* (ya); *cool and clear* (digin). Gia Dinh meant the City of clear and cool water.

[6] Justin Corfield (2013), in Historical Dictionary of HCMC, stated: Many ethnic Vietnamese soon moved to the market town, which grew in importance and in 1674, the Vietnamese from there launched an attack on Cambodia. In 1698, the town was taken over by Vietnamese and it was renamed Gia Dinh (p.vii).

called *Saigon*.[7] Since there, the Saigon is considerably initiative form of Vietnamese feudal city in the east. The region, based on the initiative commercial forms, grew and developed to become **Saigon** (mainly Vietnamese) and **Cholon** (predominantly Chinese). According to the map 1815, Saigon and Cholon were single towns, the First in the East and the Second in the West, with rice-field, cemetery and scattered hamlets in the centre. Therefore, the Gia Dinh had unostentatiously been taken such form *two towns in the region* in urbanist point of view.

The City was well protected by two fundamental systems: *firstly,* the Ban Bich rampart, built in 1772, ranging from the North to the West; *secondly,* Tan Binh River and Ben Nghe (Buffalo Quay) arroyo to separate the city to rural area. The region and cities being structured and well connected by two oldest long-ways (1) East–West way (now Nguyen Trai road), built in 1748 under Nguyen dynasty (Son, 2009), connecting Saigon to Hue and My Tho; (2) North-South road (now Cach Mang Thang Tam road) linking Saigon to Cao Mien (the map 1623–1679); and canal of horse bowel (Ruot Ngua) made in 1772.

Administratively, the boundary of the City had been changed sometimes. For instance, it was belonged to two provinces: Bien Hoa and Phien An in 1808 and again re-adjusted into Bien Hoa and Gia Dinh provinces after 1836 (Fig. 3).

- *Cholon Town*

   According to the map 1815 above, Cholon initiated from the old street (now Tan Da street) to Kim Lien Street. Its town was not large and similarly to Ben Nghe region but high density of Chinese and Vietnamese people. Cholon grew upon Ben Nghe arroyo and the banks of narrow canals. It was western commercial centre of the Gia Dinh land. The settlers, from China carry on trade, are the most part too poor to engage in occupations. In fact, there were two religious buildings built in Cholon quite early. The first building was the Giac Lam pagoda (1744) in the North of Cholon; next the Ba Thien Hau pagoda (1760) to respond basic demand of both Vietnamese and Chinese (mainly Cantonese). Those, on the one hand, indicated well socio-commercial development supporting the inhabitants, both physically and spiritually. On the other hand, it showed religious content within Cholon as representing *town of Chinese trade* and *harmoniously mixed religion.* In general, the Saigon-Cholon had not the Western sensed-cities until the French's occupation due to small grouped villages surrounded the citadel with principle trading activity of rice eventually.

- *The setbacks of Gia Dinh Imperial City, Saigon City and Cholon Town*

   A year later (1836), new Phung citadel had been built to follow *SE-NW* direction replacing the Bat Quai citadel. Its walls re-used the materials from the previous

---

[7] Kiem N.T (1958) interpreted: word 'Saigon' is translated from term 'Preikor' of Cambodia. Prei means Forest; Kor was Cotton; or Prei Nokor means Forest of the Capital (p.12). Nghia (2011) indicates 'Sai' in Chinese means firewood or twigs while 'gon' is equivalent to cotton stick or pole. 'Nokor' means City, Land from Sanskrit 'nagara'. The Chinese pronunciation is Tai Gan, Tai Ngon or Tai Gon (p.9).

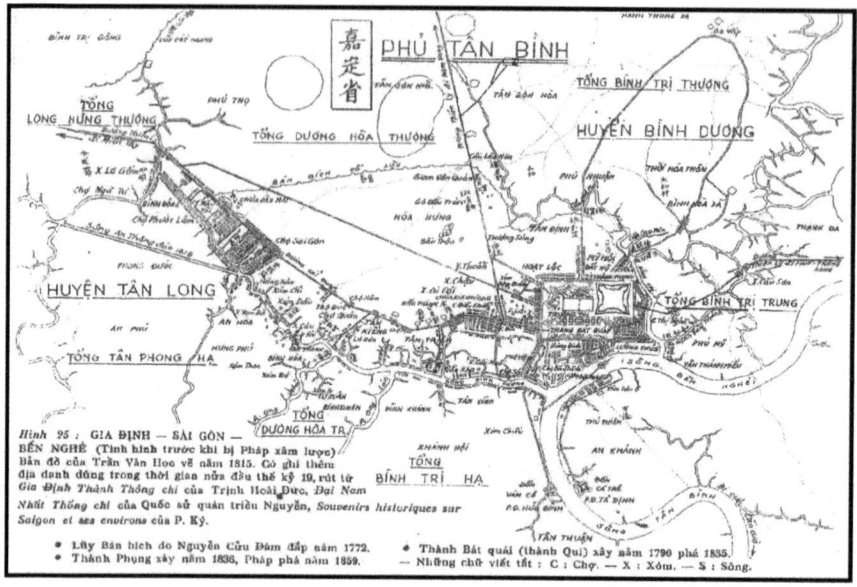

Fig. 3 The map of Gia Dinh-Saigon-Ben Nghe *(drawn by Tran Van Hoc in 1815)* Source: General Science Library of HCMC (2014), PADDI (2015)

one. It, also called Phuong or Gia Dinh citadel, was roughly about a half of the Bat Quai scale and located in the Northeast (now Nguyen Dinh Chieu, Nguyen Binh Khiem, Mac Dinh Chi and Nguyen Du streets of District 1, HCMC). In contrast to previous citadel, the Phung Citadel was such simplified Vauban form that designed in purpose and constructed smaller than the Hue Royal Citadel. The small scale presented a signal of more conflicts led to weaker connections of Nguyen dynasty to the South in contrast to increase of the central power in Hue.[8] Consequently, the downgraded Saigon seemed to be gradually taken into account of France.

This citadel was, however, surrounded by concentrated activities of the markets and buildings, such as the workshops of Chu Su and Voi, gunpowder storage, schools (now at the corner of Pham Ngu Lao and Nguyen Trai). Due to the French strategic demand on, at least, one of attractive sea-ports of Vietnam (in the report of France Southern Council, May 1857) in Cochin-China, naval General Commander Rigault De Genouilly associated with Spain soldier had attacked the Phung citadel causing its demolition later due to the French military's limitation[9] in maintaining the citadel. Later, French military base placed along the thousand

---

[8] Nghia (2011) indicated: To consolidation his central power around Hue, the political capital, Minh Mang tore apart and downgraded Saigon by destroying its economic center and shutting down its vision of modernity and riches (p.58).

[9] In the army and people Viet-Nam resisting Western invader (1847–1945), 1971, it presented: Date 19 Feb 1859, the French recognized Gia Dinh citadel was too large and could not spread out all soldiers to cover the citadel ... Rigault de Genouilly telegraphed back to France to get

miles roads (Thiên lý) linking Saigon with Cho Lon (Son, 1998). It clearly opened new period of Saigon colonized by the French. In other words, the Colonized Saigon started to be 'imposed cultural transition' of France. Lastly, the Phung (or Gia Dinh) citadel had shortly lasted about 23 years (1836–1859).

Alternatively, residential area was allocated along Tan Binh (Ben Nghe) River, bridges of Ong Lanh and Kho, arroyos of Ong Lon, Ong be and Tan Dinh (Son, 2009). The scholar Tran Van Hoc (1815) mapped residence concentrated in four areas, such as: Ben Nghe market (or the first Ben Thanh) and others (Dieu Khien, Tan Binh) and Saigon (now Cholon market).

Regarding the French maps for Cholon Town, it definitely allowed us to list 05 built constructions erected within the period. They are including (1) Central market (Cholon market); (2) Casino theatre; (3) Chinese theatre; (4) Monastery and; (5) Cholon church.

- ***Vernacular architecture characteristics***

    There were different prominent architecture, such as Vietnamese and Chinese corresponding to each area. At first, the vernacular architecture, rarely used brick and stone before French colony, was mainly made by precious wood structure and wall, double title roof for middle-class group; wood structure with low quality, earth wall, and cottage or nipa roof for common residents. Traditionally, the house had two separate spaces, one—for daily gathering or public activity; second—for the family's living, and worship place in the front.

    Moreover, Cholon seemed to allow commercial development itself, then its town structure was expanded accordingly to Chinese community's capacity. As a result, this commercial town, present district 5 in HCMC, did not well plan following linear form and strip of land in between the oldest long-way (Nguyen Trai road) and Saigon arroyo. In other word, Cholon is shaped commercial town (Phố Thị) or linear Big Market (Fig. 4).

- ***The French colonized Saigon (1861–1954)***

    According to French colonial policy in Cochin-China, their power, by issuing the land-code to secure rights to land for colonial ownership, practiced over *land, resource and people* [6]. For this reason, Saigon had been quickly re-planned to follow 'checker-board town' [1] or *strict grid-form* of military civil engineering style, practically and determinatively, for priority purpose of trade [7] and port; politic and administration and; military. The Admiral's decision, dated 11 April 1861, defined administrative limits of Saigon City (Ville de Saigon 1861), which regulated boundaries to Ben Nghe and Thi Nghe arroyos, Saigon River (Old Tan Binh River) and northern road connecting Cay Mai pagoda with the Chi Hoa defence line. The total City area occupied 25 km^2 for 5–600.000 inhabitants. By deploying the decision, general Saigon City was clearly separated into two space: *first*—Saigon and *second*—Cholon because of playing different roles functionally (Fig. 5).

---

permission destroying the citadel... Date 8 March 1859, he was permitted to destruct the citadel by the explosive...(p.78).

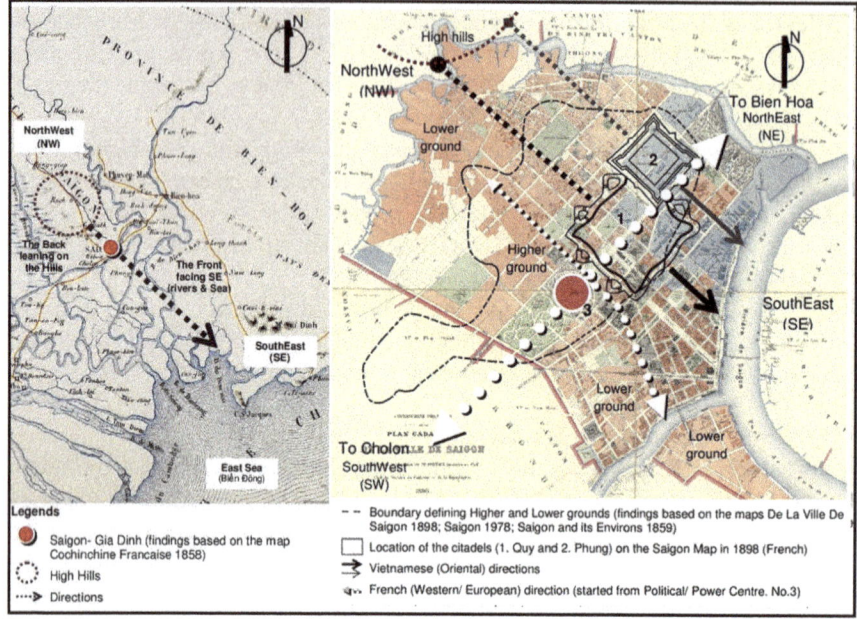

Fig. 4 Oriental and topological analysis of the Citadels and French development Source: the Author's findings (2021)

To concrete such colonial plans, the French government had detached Cholon from the City since 1864. Then, Cholon was considerably the town of Tan Long district of Tan Binh province. According to the administrative division, the City had been growing and recorded over historical transformations and the periodical maps.

The South was again separated into 20 areas in 1880. Saigon city and Cholon town were relocated into area No. 20 [8]. Moreover, the map 1882 indicated that there were rural villages between both cities, for example, Phu Thanh, Thai Binh, Nhon Hoa and Tan Thanh, Tan Hoa, Binh Yen and Tan Quang, Nhon Giang, Tan Kieng, Tan Chau and Hoa Binh and etc. Those located in two provinces[10] of Binh Chanh Thuong and Duong Minh. Hence, the City's boundary in 1882 did not much changes compared to itself in 1878.

Regrettably, Cholon town's limit seemed not to be changed due to no record on period 1867–1878. Base on the map 1816 and map 1882, Saigon City had expended widely to all directions. Eastern and Northern villages, such as An Binh, An Dong, Tan Thanh; Southern villages, Binh Dong, Phong Phu, Long Vinh, Un Long were emerged in 1881 to make the City larger approximately about 3 times in terms of population (39.806 people) [9].

---

[10] Corresponding to the decree of Cochin-China Governor, dated 13 December 1880.

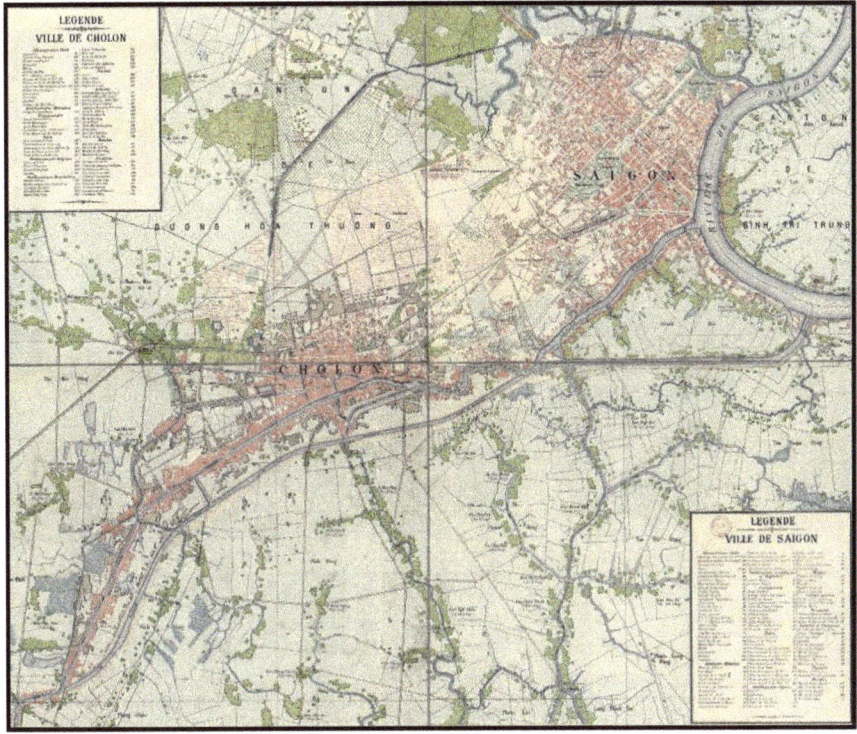

Fig. 5 The map of Saigon-Cholon in 1923 (Source: PADDI 2021)

Corresponding to those adjustments, the Saigon-Cholon twin-cities [10] under the French colony. It defined Saigon city was *major port, market, industrial and rice-milling centre* and Cholon became *rice capital* of French Indochina. The twin-cities were Indochina's most populous center [11]. From this time, the French rule applied entire city to accelerate Southern Vietnamese society in Saigon towards definition of new social categories.

Up to 1923, the twin-cities had such adjustment of administrative boundary by adding two villages (Tan Hoa and Phu Thanh)[11], which located in the middle and covered 344 ha. As a result, Saigon area was occupied about 1.317 ha. Next, the City kept expanding to Eglise road (now Tran Dinh Xu) and Route Strategique road (now Tran Phu), Cach Mang Thang Tam and Nancy road (now Nguyen Van Cu).

To the South, the City developed further, to Ong Doi and Bang arroyos, by compiling remained areas of Khanh Hoi village and Chanh Hung as well as added 447 ha. Total area of the City grew up to 1.764 ha. Therefore, Saigon City had 3 districts (district 1, 2 and 3) on *Plan de Saigon-Cholon 1923*.

---

[11] Tu (2007) showed that this expansion was probably made by the decree of Cochin-China's Governor, 30 December 1912

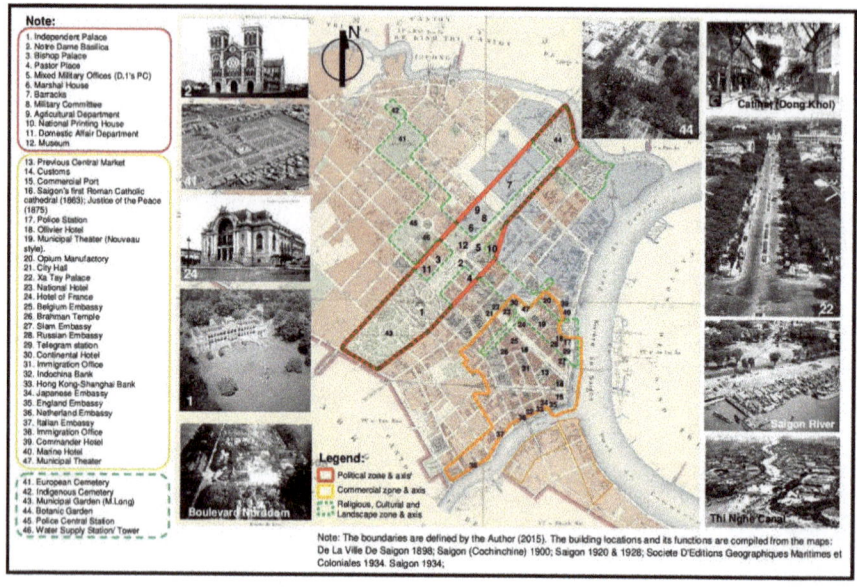

Fig. 6 Analysis of political and commercial zone/axis of Saigon City Source: the Author's findings (2021)

The map 1942 showed largely changed boundaries of Saigon-Cholon. It expanded to the West by combining Hoa Hung village; to the South (Bang Don arroyo to Te canal (now in district 7). Saigon consisted of 4 districts (district 1,2,3 and 6)[12]. District 6 comprised villages Khanh Hoi, Vinh Hoi of the City.

Since middle of twentieth century, after having occupied Ben Nghe-Saigon (1861), the French military Government had enacted policy on improving and constructing Ben Nghe area from the Nguyen dynasty's political-military centre to become colonial 'Capital city' (Thủ phủ) of Indochina. Saigon's development started following European urban form by the plans. Then, Saigon became political-cultural centre and Cholon was the centre manufacturing Chinese handicrafts and products. Up to beginning of twentieth century, Cholon was gradually urbanized but still keeping commercial identity of a 'China town' [12].

The French Master Plans towards Saigon City: The City had been planned and developed into 5 areas with separate functions that are: (1) political centre; (2) military area; (3) port-mixed business zone; (4) administrative/public service area and; (5) residential area (Fig. 6).

Spatial change of Cholon town: at this period, Cholon was an independent town of Chinese with more people migrated from China due to the unrest of China. It was organized by narrow-interlaced streets and shop-house in line of the streets. It appears that the town's structure and architecture has not been much physically impacted by

---

[12] According to the decree No. 2383-MI/DAA of the temporary Republic of Cochin-China, dated 10 May 1948.

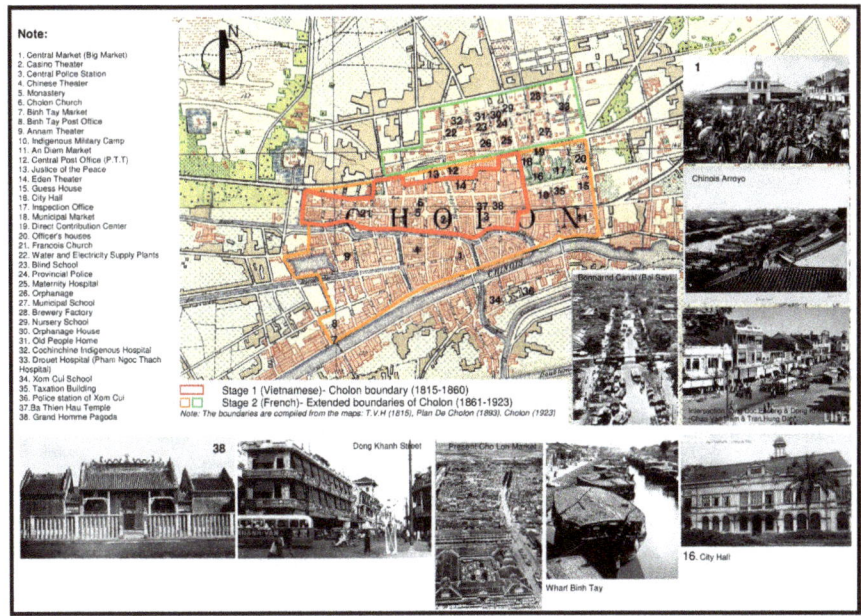

Fig. 7 Analysis of Cholon expansion—Stage 1 and 2 (Source: the Author's findings, 2021)

the political change because of two reasons: (1) the basic conflict was between the colonising French and the colonised Vietnamese; (2) Chinese was 'intermediaries' as essentially non-political before 1920 [10]. The Chinese (Hoa) community here acted linguistically and economically in the colonial areas. They commonly were self-employed merchants, factory owners or skilled craftsmen, manual workers or pretty traders. Hence, Cholon was 'a center for commerce and light industry' during 1867–1887 [33] and more flourishing because of new business methods carried out by the Chinese immigrants, who avoided the war between China and Japan during 1936–1937. Therefore, Cholon and Saigon were connected by Tran Hung Dao road that was filled up from the canals and arroyos [13].

In 1923, Cholon had some changes administratively. It expanded to cover two districts[13] (district 4 and 5) more to the Eastern villages (Tan Hoa, Phu Thanh, towards Nancy road); the Western and Southern villages (Tan Hoa Dong, Phu Dinh, An Lac, An Phu Tay, Phong Duoc, Binh Dang, Binh Dong, Chanh Hung, Tan Thoi Hoa, Chi Hoa).

Next stage, Cholon town continued expanding to the West by emerging 2 hamlets (Phu Thoi, My Trung), Phu Dinh village, Hoa Dong, Phu Lam of Tan Hoa Dong village); the North (covered Chi Hoa village and a part of Hoa Hung village); the South (included hamlet Cau Mat, Thai Phuc in Chanh Hung village). The rest of Cholon was remained appropriately (Fig. 7).

---

[13] Tu (2007) showed it was mentioned in the decree of the Governor, dated 5 January 1923.

Chinese in the twin-cities of Saigon-Cholon were traders of some sort followed by all kinds of craftsmen. Particular business in Cholon was mainly the trade and processing of rice, deerskin and other agricultural products [10].

Architectural transition and revolution: due to the limited construction materials after Saigon was colonized, the first French building, the Governor house, was the wooden assemble structure that had been carried from Singapore [14]. From high demand on solid buildings, the government allowed to import materials (steel, brick, cement and etc.) from France, Hong Kong; wooden panel, interior decorations from Singapore. As a result, there was a 'Western buildings' boom with brick, stone and red tile construction replaced traditional ones.

In Cholon, end of 19th and beginning of 20th, Chinese house mostly were single-story house, which made by wooden or brick wall, low double-tile roof, traditional door (main door and horizontal bar). Those located in Cholon street area and along present Tau Hu canal, Tran Van Kieu, district 5; wharf Binh Dong, district 8 and etc.

To 20th century, European architecture (Art Nouveau) gradually replaced vernacular architecture, especially in Saigon centre. Indigenous people had been familiar with huge and block buildings constructed by applying modern construction technique. It indicated that a mixed architecture between French and Chinese was appeared in following areas: Old market (Ho Tung Mau street, district 1), Cholon (Trieu Quang Phuc, Hai Thuong Lan Ong, Tran Van Kieu, Tran Hung Dao B, Chau Van Liem, Nguyen Trai… district 5), wharf Binh Dong (district 8) and etc.

- **Socialized Saigon (1975–1990)**

After the reunion in 1975, Vietnam played a new role. From 1975 to 1990, the economy was more concentrated and subsidised with the economic embargo seriously impacting the growth of Ho Chi Minh City (HCMC, old Saigon). Then, urban improvement program was not moving due to such consequence from the war and lack of motivation.

Saigon City was *firstly* re-organized into 18 districts, including 11 districts of Saigon City and 7 communes of Gia Dinh province. The population was about 3.498.120 inhabitants in 1975. Nevertheless, Saigon continued *secondly* converting 18 inner districts to 12 new districts (district 1, 3–6 and 8, 10, 11, Binh Thanh, Phu Nhuan, Go Vap and Tan Binh, Thu Thiem and Thu Duc commune). In addition, it was *thirdly* re-named Ho Chi Minh City (HCMC) on 02 July 1976. Hence, HCMC has 12 inner districts and 6 suburbs with 2.095 square kilometres remained until 1997. After 2003, HCMC boundary *next* had been adjusted to 24 districts (district 1–12, Binh Tan, Binh Thanh, Go Vap, Phu Nhuan, Tan Binh, Tan Phu and Thu Duc) and 5 suburbans (Binh Chanh, Can Gio, Cu Chi, Hoc Mon and Nha Be). Total area is about 44.213 k^2.

After the 1986 renovation policy, HCMC had huge changes by infrastructure, environmental and construction projects towards sustainable development. During this period, the administration area placed at the French building complex (previous Hotel De Ville or Dinh Xa Tay; Saigon municipal building) re-following direction of the Spiritual Axis of the Viet ancestor.

3 The City Development Impacting the Built Heritage

According to HCMC development plan over the periods, it showed that the misuse of urban planning in HCMC [20], which has failed to shape its urban development. From the plans 1993 and 1998, main development direction stressed East–West by implementing unrealistic infrastructure projects. For instance, project West–East expressway, an ODA project funded by JICA-Japan, built in 2005–2011, had completely removed the ancient trace, such as the wharfs Binh Dong, Ba Binh and Me Coc and Le Quang Liem [22], old Chinese group and row of shop-house along Tran Van Kieu street and the store buildings of the Saigon, and trading activity at the ports and along Thi Nghe canal. Moreover, the old commercial centre (Thuong Xa Tax—a 134 years old building-ended 30 Sep 2014) and rows of the heritage trees (63 trees for Thu Thiem bridge No.2 project; 51 trees in front of HCM Opera House and 57 trees in Ben Thanh Market for the Metro Project) [23], in the centre of District 1 just have been destroyed recently.

To other ancient buildings within residential zones, many of them have been occupied, encroached and demolished to build new functional constructions. People built everywhere without a thought to planning, aesthetics or zoning. Three to five story-buildings could be raised on 'a foundation of 16 square meters' [4]. Between the groups of buildings in the blocks is narrow alleys enough for people to walk by or for a bike to get through. Behind the front modern and nice buildings were more temporary and poor buildings that lodged more people—a growing population that Saigon needed to accommodate. Additionally, many 'ghost villages' [4] as slums and squatters, along Thi Nghe canal, have attached and surrounded old build heritage compounds everywhere in HCMC.

According to list of preserved relics, it showed that total relics of three district is about 41 buildings, covering 26 national relics and 15 municipal relics in term of levelling. In other words, it includes religious majority (20 relics), mainly in district 5. Most of revolutionary relics (6 of 11) next are located in district 1 while its historical relics is less than, about 7 relics. From this data, it may be argued that the Municipal recognition seemed to stress such place and site concerning revolutionary period (1954–1975) more than others within district 1.

Consequently, many architectural buildings under the Republic of Vietnam have not existed because its value related to old regime. To other buildings of Americanized Saigon, some of them have been well preserved in purpose of (1) marking victories of the revolution in 1975; (2) attracting tourist and cultural activities, such as Independent Palace for instance.

According to the conceptual partition on conservation areas, it showed that conserving HCMC means to implement preservation of those [24]. There are three architectural movements that are listed below [25]:

- *The Vernacular architecture:* it combined both national and local characteristics. For instance: pagoda Giac Vien, Giac Lam, Go, communal house Binh Tien, Nam Chon, tomb Ong, shrine Tran Hung Dao and so on. Its major features are traditional curved roof of Thang Long and Hue Citadel, double tiles, simple and rural architecture, the veranda (hàng hiên, cột tròn). Moreover, it also is architecture of Chinese, Indian and Burmese mixed to make more variations within the landscape.

- *Indochina architecture:* it presented on European and French civic buildings, such as hospital, school and office under French colony. This architecture could be seen over Roman columns, classical arches (vòm cuốn cổ điển), Gothic style with animal status, ceramic and colour glass and etc. These materials made Saigon become more unique and Indochina architecture.
- *Contemporary architecture*: this movement had not existed quite short and continued up to end of 20th. There are many buildings, such as the hotels: Caravelle, Rex, Palace on Nguyen Hue street, the banks: Vietcombank, BIDV; Vietnam National library, the hospitals: Unification, Cho Ray and high-rise buildings: Land Mark, New World, Sun Wah, Saigon tower and etc.

From renovation policy to now, newly unplanned architecture has widely developed, encroached and isolated ancient and old forms and broken Saigon's built heritage. Regarding the survey, there were only remaining 10 traditional houses over 100 years [26] with unique characteristic such as: 3 rooms—2 bays, double tiles and traditional structure and etc. Unfortunately, those have seriously been deteriorated and partly repaired and encroached as well as led to soon replacements by new and modern architecture. However, ranked architectural buildings, like pagodas Go (Phung Son), Giac Lam, Giac Vien, communal houses Binh Tien, Nam Chon, tomb Ong, shrine Tran Hung Dao and etc., are traditional and valuable architecture. All of these buildings are being seriously damaged and occupied by local residents for living, commercial purpose and waste bin. For instance, pagoda Go had been transgressed over the years by 132 households; a part of pagoda Giac Lam precinct was also encroached and illegally borrowed for non-religion use. Other constructions, such as communal house An Phu, Cay Sop and pagoda Giac Vien, have been left fallow and soon fallen down. HCMC, at present, still remain Hai Thuong Lan Ong Street as Ancient Street in Chinese area, and several French buildings, which have been isolated and surrounded by new development and modern/contemporary architecture [26].

4   Institutional System for Built Heritage Conservation

HCMPC had enacted the notice No. 46/TB-UB-QLDT, dated 17 May 1996, about inventory list of 108 architectural—urban items that had to be preserved. It indicated 4 zones, 8 streets, 9 landscape groups and 87 buildings, which included the Unification Palace and its surroundings, park 30/4, front park and Notre-Dame Basilica, City Port building, buildings of Shell and Esso company, the Cultural park (Tao Dan), Zoo park (Thao Cam Vien) and etc. [27].

However, the notice had not regulated building height within such conservation areas. As a result, the limited high-rise buildings, such as Saigon Tower, Diamond

Plaza, Plaza of Office Tower on Le Duan Street and Vincom Tower and others, had raised up at core of the preservation area in District 1. Those blocks and its modern architecture have seriously impacted 'spirit of Saigon'.

Up to May 2017, HCMC has 172 relics, which are listed and being protected by the Law on Cultural Heritage; as definitely as including 02 special national (historic) relics, 56 national relics (02 archaeological, 30 architectural arts, 24 historic), 114 city relics (48 historic, 66 architectural arts) [28]. After 16 years, only 36 relics had been added into the list. It means that protecting the built heritages moves quite slow and slower when the local authority and management system just were separated into Department of Culture and Sport and Department of Tourism of HCMC from 23 Oct 2014 [29].

For Cholon area in district 5, there was a decision no. 567/QD-UBND, dated 07 Dec 2007 of People's Committee of District 5 about adjusted detail plan approval, scale 1/2000. It defined ancient streets Trieu Quang Phuc, Hai Thuong Lan Ong, Cholon are existing residential area that need to be improved. However, conservation area and preservation regulation are unidentified [30]. Therefore, preserving the built heritage of district 5 might meet many difficulties eventually.

Regarding the approval on the Master Plan to year 2025, HCMC also defined conserving historic, cultural relics is one important goal after the factors of sustainable and economic development. It confirmed no new development within strictly conserving areas, which are concentrating unique and valuable architectural buildings, villas in current mixed—use centre of district 1, district 3 and a part of district 4, Ba Chieu area (Binh Thanh district); Cholon zone (district 5 and 6) with 120 ha. Those are clarified at point 6, page 5 of the decision. Unfortunately, there is no definition for such conservation area with clear boundary (or conservation boundary for built heritage of HCMC) and no particular sanction to protect the built heritage at all. However, this legal document highly focuses construction prohibition to natural conservation areas (Can Gio area); airport Tan Son Nhat; Dong Nai, Saigon and Nha Be river corridors; and rural areas. As a result, the remaining urban and architectural heritage will not fully be protected by phenomenon 'strategically isolating architectural buildings' of the local authority. Legal base for conserving the built heritage of HCMC [31, 32] are listed as follows:

1. Adjusted Law on Cultural Heritage No. 32/2009/QH12, dated 18 June 2009;
2. Decision No. 1706/2001/QD-BVHTT, dated 24 July 2001, of Ministry of Culture and Information, about approval on Comprehensive plan and development towards historic-cultural relic and landscape to 2020.
3. Notice No. 46/TB-UB-QLDT, dated 17 May 1996 about managerial implementation towards landscape conservation of HCMC.
4. HCMPC's policy on implementing an action plan, assessment research to the Wet-East Boulevard's landscape.
5. Decree 61/CP of the Government, dated 5 July 1994 about granting and selling old/ancient State-owned villas. This decree did not pay more attention on conservation matter.

6. Decision No. 188/1998/QD-TTg, dated 28 Oct 1998, of Prime Minister, about selling villas in Ho Chi Minh City and the Resolution No. 48/2007/NQ-CP, dated 30 Aug 2007 of Central Government about adjustment on purchasing policy to State-owned house. Unfortunately, those did not focus on conserving and protecting unique architecture of these buildings.
7. Document No. 3606/UB-QLDT, dated 19 Oct 1996, about dismounting (tháo dỡ) old villas before 1975. However, these policies only stopped doing researches on particular cases without such conservation regulations.
8. Decision No. 24/QD-TTg, dated 6 Jan 2010, of Prime Minister, about approval on adjusted Master Plan of Ho Chi Minh City to year 2025.

5 Re-Identification of HCMC's Built Heritage Settings and Its Building

The Gia Dinh region, initiated from the market towns of ancient Prey Nokor in 10th–11th century and identified Southern administration by the first emperor of Nguyen dynasty during 19th century, had conserved huge historical trace and value over the up and down periods. This research provides clear evidence of untouched worth of ancient Gia Dinh, the twin-cities Saigon-Cholon upon particular historical layers. Its result appears to confirm following findings:

- The Bat Quai citadel was the first largest Vauban citadel in Vietnam, especially in the South to play a complex role mixing multi-functions, military-political administration-residence, to develop trade and port activity towards the agricultural products. The citadel had definitely been considerable as such foundation in terms conceptual design, construction technology and western military art, which were absorbed for the Hue Capital Citadel and other citadels during the French colonial period in Indochina region.
- Under the feudal periods, Saigon and Cholon were single cities running different functions based on their own topographical—social—economical features. Therefore, Saigon was the administrative centre created relatively to military fort—commerce—port, while Cholon was known as the town of trade-religion and sub-ports. This reciprocal combination had caused Saigon-Cholon becoming one of the remarkable economic centres in Southeast Asia at that time.
- In the colonized period, new political axis (northeast-southwest) had been rotated to concrete clear the French economic strategy of exploiting the resources of the colonized countries, particularly Vietnam and Saigon-Cholon. The administrative center as gradually shifted to the West compared to the citadels of Nguyen periods as more importantly as its development changed perpendicular to the old Southeast-Northwest (SE-NW) as chosen by previous the Emperors. This directional change allowed French to open widely to three sides (towards Nam Vang, Hue and Western region, the map 1815) and easier connect to Cholon to enhance trading activities among French and Chinese in the City. Moreover, the City's characteristic, as demonstrated as the Paris of Asia, had been formed

through its grid-structure, five functional zones, European architectural buildings and landscape, which had been built for serving the French's long settlement in Cochin-China.

- According to the data analysis, the result illustrates first list of the buildings as assumingly as basic inventory list of the built heritage, such as Vietnamese, Chinese and French architecture, in Saigon-Cholon cities later on.

6 Concluding Remarks

Ho Chi Minh City (old Saigon) within the context of the 'Asian Century' is one of the success stories of Asia. Its transformation brought about tangible and intangible cultural changes at central places. To protect 'the spirit' of Cholon—one of the original cities in the past, development plans need to really consider cultural significance of old Cholon thematically to support conservation actions towards sustainable conservation in globalization context. The paper pays more attention to the district No. 5, which consists of many ancient historic places of Gia Dinh region and Cholon (Chinatown) with particular findings, which are able to be taken into account of future HCMC's general master plan readjustment to year 2040, vision 2060 and particular incentive policies towards its sustainable development.

References

1. Guillaume 1985 Saigon or the Failure of an Ambition (1858–1945). Colonial Cities. Springer, Dordrecht, pp 181–192
2. Son T (2013) Thành phố Hồ Chí Minh lập danh mục bảo tồn kiến trúc đô thị (VNExpress: vnexpress.net/thoi-su/tp-hcm-lap-danh-muc-bao-ton-kien-truc-do-thi-2421220.html)
3. Logan WS (2005) Cultural role of capital cities: Hanoi and Hue, Vietnam. In: Pacific affairs, vol. 78, no. 4. University of British Columbia, Canada, pp 559–576
4. Nghia MV (2011) Saigon: a history. McFarland & Company, USA
5. Justin C (2013) Historic dictionary of Ho Chi Minh City. Anthem Press, USA
6. Mark (2003) Land codes and the state in French Cochinchina c.1900–1940. J Hist Geogr 29(3):356–375 (2003). Elsevier Science Ltd.
7. Francois T (1999) Architecture and urban planning of Saigon under French colony. J Vietnam Archit 2/99
8. An (2003) Vietnam-Changes of Geographic name and Administrative boundary. Vietnam News Agency Publishing House, Vietnam
9. Thanh (2012) Administrative boundaries of Saigon- Ho Chi Minh City over the maps (period 1859-2005). The conference on Spatial development of Saigon-Ho Chi Minh City over the historical periods. HIDS, Vietnam
10. Thomas E (2010) Chinese politics in colonial Saigon (1919–1936): the case of the Guomindang. Chinese Southern Diaspora Studies, vol 4
11. Philippe P (2013) From the social to the political: 1920s colonial Saigon as a "Space of Possibilities" in Vietnamese Consciousness. Duke University Press, USA, pp 496–546
12. Hau NT (2007) A view on Saigon City (Van Chuong). www.vanchuongviet.org/index.php?comp=tacpham&action=detail&id=6431)

13. Son ND (2009) Saigon-Ho Chi Minh City-integration, modernity and identity. J Constr 12–2009:24–29
14. Tuyet (2000) Dwelling of old Saigon. Psychol J 3
15. Erik (2011) Saigon's edge: on the margins of Ho Chi Minh city. University of Minnesota Press, USA
16. Tuan (2009) Conservation and value improvement of architectural heritage in Ho Chi Minh City in development process. Ph.D. thesis. Ho Chi Minh University of Architecture, Vietnam
17. Kiem (1958) Saigon the past and present. In: Southern wind, No.1, dated 7 July 1958. Department of Culture, Criticism and Information, Vietnam
18. Natasha and Francois (2007) Urban planning and architecture of Saigon 1954–1975. In: Saigon past and present. Youth Publishing House, Vietnam, pp 105–108
19. SGDT and XD (2010) Remembering built Heritages of Saigon-1975. Saigon Invest Constr Mag 11–2010:20–23
20. Tat NV (2012) A valuable corner of Saigon Architecture. J Archit Live 68–69. http://tad.vn/news.php?id=229
21. Du H (2015) The misuse of urban planning in Ho Chi Minh City. Elsevier, Amsterdam, pp 11–19
22. Thuy X (2011) Restoring "commercial activity at the port" of Saigon in the Past. (Tien Phong: www.tienphong.vn/van-nghe/khoi-phuc-canh-tren-ben-duoi-thuyen-sai-gon-xua-529290.tpo)
23. Ngan T (2014) The row of old tree in Saigon is almost eliminated (Dan Tri: dantri.com.vn/Print-980767.htm)
24. Son LT (1998) Saigon-initial construction. J Vietnam Archit Assoc 6(44):23–24. 1993
25. Ninh (2000) Role of conserving architecture and landscape in HCMC's boom. J Archit Life 48/2000
26. Cuong, L (2011) Ho Chi Minh City: issues on preserving architectural heritage in development process. Vietnam Archit J. 199-11/2011
27. HCMC (1996) Thông báo 46/TB-UB-QLĐT. http://www.congbao.hochiminhcity.gov.vn
28. SVHTT (2017) Danh sách các công trình, địa điểm đã được quyết định xếp hạng di tích trên địa bàn Thành phố Hồ Chí Minh (đến hết tháng 5 năm 2017). http://svhtt.hochiminhcity.gov.vn
29. Khanh H (2014) The announcement on establishing HCMC Department of Tourist. VOV Online http://vov.vn/van-hoa/du-lich/cong-bo-quyet-dinh-thanh-lap-so-du-lich-tp-hcm-360014.vov. Accessed 24 Oct 2014
30. Quyen NH (2014) Conserving and Improving traditional environment of Chinese-Vietnamese in district 5, HCMC. Master thesis. Ho Chi Minh City University of Architecture. HCMC, Vietnam
31. Binh (2010) Preservation of architecture of old villages in district 3, Ho Chi Minh City. Master thesis. HCMC University of Architecture, Vietnam
32. Khanh (2011) Conservation and development of architectural space and landscape of Ben Nghe-Tau Hu Canal. Master thesis. HCMC University of Architecture, Vietnam
33. Thomas (2010) Chinese politics in colonial Saigon (1919–1936): the case of the Guomindang. In: Chinese Southern Diaspora studies, vol 4

Assessment Framework of Urban Spatial Adaptability to Drought—Flood Coexistence (DFC). A Case Study from Phan Rang-Thap Cham City, Ninh Thuan Province, Vietnam

Nguyen Quoc Vinh, Le Minh Ngoc, and Le Anh Duc

Abstract Measuring urban spatial adaptability to environmental stresses has recently been a popular topic in urban planning towards resilient cities. Urban Resilience has emerged over the past few decades across multiple dimensions from economics, society, environment, physics, and urban planning. Recently, many scholars in the planning have further investigated urban spatial Resilience to natural disasters and have stated that adaptive capacity is the critical factor to achieve Resilience. In Vietnam, the central provinces currently suffer from extreme climate events that cause heavy damage, such as droughts and floods. Ninh Thuan province, located in the South-Central Coast, at coordinates $11°18'14''$ to $12°09'15''$ Northern latitude, $108°09'08''$ to $109°14'25''$ Eastern longitude, is the typical place. The case study focuses on the capital city Phan Rang-Thap Cham City's context—one of the most developed and vulnerable communities in the province of Ninh Thuan—to build an assessment framework of urban spatial adaptability to drought-flood coexistence (DFC). The framework consists of rows of urban spatial components and columns of adaptive indicators and variables to D.F.C. Urban spatial adaptability of Phan Rang-Thap Cham is measured by two methods. The first one is a qualitative analysis of urban spatial components, such as spatial networks and their land covers, based on adaptive variables to D.F.C. The second one is a quantitative analysis of the examined urban spaces interacting with the Remote Sensing (R.S.) index such as Normalized Different Vegetation Index (NDVI) and Land Surface Temperature (LST). Both of the two methods were carried out thanks to the utilities of R.S. and Geographic Information System (G.I.S.). The content of the study is reflected in four parts: (i) Overview of the concepts of urban spatial Resilience and adaptability, (ii) Study methodology and steps, (iii) Assessment framework of urban spatial adaptability to D.F.C., (iv) Conclusions.

N. Q. Vinh (✉)
Faculty of Civil Engineering, Ho Chi Minh City University of Technology, VNU-HCM, Ho Chi Minh City 740128, Vietnam
e-mail: vinh.bmkt@hcmut.edu.vn

L. M. Ngoc
Landscape Architecture and Urbanism, Omgeving, Ho Chi Minh City, Vietnam

L. A. Duc
Faculty of Architecture, Van Lang University, Ho Chi Minh City, Vietnam

© The Author(s), under exclusive license to Springer Nature Singapore Pte Ltd. 2023
J. N. Reddy et al. (eds.), *ICSCEA 2021*, Lecture Notes in Civil Engineering 268,
https://doi.org/10.1007/978-981-19-3303-5_6

Keywords Drought-flood coexistence (D.F.C.) · Urban spatial resilience · Urban spatial adaptability · Adaptive indicator to D.F.C. · Adaptive variable to D.F.C

1 Introduction

From the second Industrial Revolution in the 1870s till now, on the pathway into the fourth giant leap known as the 4.0 era, the world has faced enormous challenges because of exponential urban growth and industrialization. One major challenge was climate change (CC) which is deemed an inevitable consequence. The Earth's systems have accordingly undergone severe changes with its hydrological systems, species, migrating patterns, etc., and intensified melting ice. In addition, extreme weather events are in spades: drought, inland and coastal floods, extreme precipitation, heat waves, landslides, cyclones, saline intrusion, etc., which happen more and more frequently [1]. Riverine, coastal and urban areas worldwide are unfortunately often victims of these natural hazards. Drought intensity usually follows above-normal temperatures and would escalate by surface warming, increasing greenhouse gases [2].

These environmental stresses on urban areas have attracted attention from many scholars worldwide to search for urban spatial Resilience and adaptability. These concepts are formed to consider that a city is a complex of Social-Ecological Systems (S.E.S.) [3]. To unlock the complexity of this S.E.S., the most effective and necessary tools and methods are R.S. and G.I.S. These, moreover, would help scholars and practitioners elaborate more scientific, systematic and integrated approaches to planning and urban design, notably in geospatial databases [4, 5]. Apart from the L.S.T.—central part of drought, the research will extract NDVI from R.S. images to analyze their inter-correlations [6, 7]. The outcome would clarify the positive and negative relationships between L.S.T. and NDVI at different surfaces. In the context of Phan Rang-Thap Cham in Ninh Thuan—the coastal province of the South—Central part of Viet Nam, at the end of 2020, it had to cope with severe floods caused by a tropical cyclone. But earlier in the same year, it had undergone extreme drought level of 3.[1] Drought and floods coexist in the provincial basins.

These concepts, utilities of R.S. and G.I.S. technologies, and findings in Phan Rang-Thap Cham City help define a measurement system of adaptive indicators and variables and assessment tools. The study content is composed of four parts: (i) Overview of the concepts of urban Resilience and urban spatial adaptability, (ii) Study methodology and steps, (iii) Assessment framework of urban spatial adaptability to D.F.C., and (iv) Conclusions.

---

[1] Drought level 3 according to Vietnam classification: (a) The shortage of next month rainfall is over 50%, lasting from 3 to 6 months and the shortage of water source in the drought area is over 70% compared with the average of many years; or (b) The shortage of next month rainfall is over 50%, lasting for more than six months and the shortage of water source in the drought area is over 50% to 70% compared with the average of many years.

2 Urban Spatial Resilience and Adaptability

2.1 *Urban Spatial Resilience and Adaptability*

The concept of *Resilience* was first introduced by ecologist Hollings (1973), specifying systems and their capability to deal with disturbances and external shocks while maintaining core structures and functions [8, 9]. The term was later referred to as the *adaptive cycle* [10–13], where dynamic systems are put into the central subject for research. These systems would show no static or equilibrium condition but a repeatedly four-phase adaptive cycle: (i) growth and exploitation; (ii) conservation; (iii) collapse or release; (iv) renewal and reorganization. In ecological and other complex systems, the concept of *"Panarchy"* ("The rule of nature" from the Greek god of Pan) has shown that there are connections among adaptive cycles at different levels. A sudden change occurs due to the interactions of slow and broad variables with smaller and faster variables [14]. Top-down control happens when slow and comprehensive features control the small and fast ones. A critical change in one adaptive cycle can cascade up to a vulnerable stage of a larger and slower cycle. In contrast, the larger and slower level could facilitate the smaller and faster ones, respectively, reorganizing the system.

In the 1990s, the notion of Resilience was widespread in planning and regarded a city as "S.E.S.", supported by ecosystem services: (i) provisioning, (ii) regulating, (iii) supporting and (iv) cultural services [15]. *Urban Resilience* was considered the ability of a social system to withstand disturbances and reorganize itself following disturbance-driven changes [16, 17]. Numerous studies conducted in the field of urban planning pointed out that spatial planning is strongly critical for enhancing urban Resilience [18–20]. In research to figure out an appropriate urban structure that diminishes multi-hazards, Fleischhauer (2008) presented various indicators for assessing resilience in spatial planning [19]. One of the effective strategies is to keep urban areas filled with open spaces, green or water surfaces instead of buildings. That would facilitate ecological diversity in the city to create "nurturing conditions for recovery and renewal after disturbance" [21]. Moreover, green spaces in urban areas would minimize urban heat islands [22], simultaneously improving biodiversity and increasing vegetated land cover, thereby showing greater Resilience to drought risk [23]. *Urban spatial Resilience* initially resulted from Graeme S. Cumming's study (2011) on coral reef, where he mentioned spatial Resilience as an emergent property of the spatial arrangement, differences and interactions among internal elements (amount, configuration, location) and external elements (context, connectivity and dynamics) [24]. It was later applied in planning by U. Hassler and N. Kohler [25] and Mohammad Reza Masnavi [26]. They considered urban spatial Resilience the spatial organization of physical components in a metropolitan area, where social spaces are necessarily arranged compatible with natural spaces.

Many researchers considered the meaning of *adaptability* as Resilience and close relationships with a group of relevant concepts, such as adaptability, coping ability, robustness, flexibility [27–30]. Pelling (2014) shared a notion of adaptability in three degrees: *Resilience, transition, transformation,* which is quite a coincidence with Resilience: *resist, absorb,* change [31], and with a four-phase adaptive cycle.

In short, the issue of change is a fundamental dimension of adaptability, both in terms of resistance to change and recovery from it. Therefore, urban spatial adaptability to D.F.C. is related to preparations to minimize outer attacks caused by droughts and floods, actions to cope with disturbances once they have occurred, and flexibility to change urban spaces. As such, adaptability represents an ongoing process [32], a time-scale of reshaping, reorganizing and developing new adaptive strategies. To measure *urban adaptability to D.F.C.* is essential to plan adaptive city strategies.

2.2 Urban Spatial Adaptability to Drought-Flood Coexistence (D.F.C.)

To put urban Resilience into practice, particularly to natural hazards or D.F.C. during the dramatic transformations, urban spatial adaptability has accordingly been discussed and studied [33]. The literature on Resilience in planning has emphasized preparation and mitigation, especially on the local scale [34]. This closely corresponds with a conventional approach of land-use planning to diminish external shocks. The realization that mitigation is often not sufficient to prevent disturbances from occurring, adaptation became an essential matter in many studies. S. Meerow (2017) defined it as the ability to preserve or quickly get back to proper functions, be adaptive, and modify systems that restrict adaptive capability [35]. Sendai Framework (2015) described Resilience as cities' ability to resist, absorb, accommodate and recover from the effects of a hazard [36]. Janssen et al. (2006) argued that a competent network would facilitate the mobility of human resources and information and associated human activities. Nelson et al. (2007) provided examples that the social ecosystems can contribute to more efficient management of the significant changes in the natural ecological system, such as drought [37]. Using the concept of the adaptive cycle, the notion of Resilience was interpreted by Lu et al. (2013) regarding a system's robustness (or mitigation) and rapidity (or flexibility or adaptation) resilient to floods in Rotterdam [38]. At different levels, variables have their control roles in keeping S.E.S. developed in a compatible way through adaptive cycles that can be applied for S.E.S. and natural hazards [39].

Mc Harg (1992), Godschalk (2003) and other scholars have also discussed the notion of adaptability concerning characteristics, that could be adaptive planning principles or *adaptive indicators*, such as compatibility [40], diversity [25, 41, 42], efficiency [34], connectivity [9, 41–44], robustness [9, 41, 45], redundancy [25, 41, 42], etc., or attributes of decision-making processes such as fluidity, reflexivity, contingency, multiplicity and polyvocality [18]. Among these indicators, *compatibility* and *connectivity* are the most adaptive to D.F.C. Therefore, they are selected to be the *core adaptive indicators*. Variables or variable spatial quantities could be tracked in most systems, including biological and physical, to understand the Resilience of systems. Ecological rehabilitation is considered an intermediate stage and is often lost due to the interaction between objects during operation at different levels of time and space [11, 46].

In the context of Phan Rang-Thap Cham City, urban spatial adaptability to D.F.C. is the critical ability of urban spaces to adaptively regulate floods and droughts and bounce back to another stable situation after facing uncertain events. To achieve that, an assessment framework is an essential tool to measure its current adaptability levels, from which adaptive strategies will be developed for the future.

3   Study Methodology

Based on the ecological approach, urban spatial adaptability to D.F.C. in Phan Rang-Thap Cham City could be measured by a matrix of urban spatial components and adaptive indicators to D.F.C. and assessment methods.

Urban spatial and D.F.C. Components. Urban space is a complex, heterogeneous and dynamic entity seen in the *water, vegetation, barren, and built-up.* These include spatial attributes such as structures (*amount, configuration, location* for internal interactions and *context, connectivity, dynamics* for external ones) [47], and functions (natural and social). Concerning structures, the quantitative entities such as spatial network or components' areas, 'amount' is measured through the number of accessibility and density of intersections for the first one [48] and is measured through figures extracted from LULC maps. *Location* is considered the *core attribute* of spatial components because of its dominant role in S.E.S. and is evaluated through qualitative analysis of ecological landscape patterns [49]. Focusing on natural ones only, functions are performed through ecological land covers and measured thanks to R.S. utilities. Therefore, *naturally, spatial networks* and *land covers* are concrete entities to measure urban spatial adaptability to D.F.C. D.F.C., on the other hand, is illustrated by R.S. indexes, for instance NDVI and L.S.T. [40, 50, 51].

Adaptive Indicators and Variables. Indicators are referred to, analyzed and learned from the previous studies and completed within the study context of D.F.C. These indicators are: compatibility, efficiency, diversity, connectivity, robustness and redundancy. On the macro scale, spatial variables are examined and analyzed at the entire catchment, including topography, slope, water resources, aquifers, maximum temperature area, etc. On the mesoscale, structures, functions and forms of natural spaces (water, vegetation) and artificial spaces (built-up such as transportation, urban cells or buildings) are investigated.

Assessment Tools of LULC, L.S.T. and NDVI. LULC maps of Phan Rang-Thap Cham, at three different points of time (1988, 2005 and 2020), are examined through R.S. images to explore land cover transformations (Fig. 2). Based on that, NDVI (Fig. 3) and L.S.T. (Fig. 4) maps are extracted from key figures of heat and green

Table 1 Detailed characteristics of the Landsat Satellite images used in the research

Path/row	Acq. date	Dataset	Producer	Attribute	Type
123,052	26/03/1988	TM	USGS	Ortho, GLS2000	GeoTIFF
123,052	29/03/2005	TM	USGS	Ortho, GLS2000	GeoTIFF
123,052	11/04/2020	OLI-TIRS	USGS	Ortho, GLS2000	GeoTIFF

(Source authors)

spaces, respectively, changed through the years obtained. Also, the interrelationships between different urban areas and R.S. indexes could be discovered.

LULC: The multispectral Landsat satellite images of the following table are used (Table 1). The Landsat images were registered to Universal Transverse Mercator projection, zone 49 N, WGS84 Datum with a spatial grid of 30-m resolution. The dates of the Landsat images are chosen to be as close as possible to the same vegetation season. The methodology used involves the following stages: image pre-processing, the design of classification scheme, image classification, accuracy assessment and analysis of the LULC changes.

- Image pre-processing: including cloud masking, layer stacking, sub-setting by administrative boundary;
- Classification scheme: utilizing four classes (Urban—Built-Up, Water, Barren soil, vegetation);
- Image classification: training data was used for Maximum Likelihood Classification (M.L.C.) and accuracy assessment. In this research, colour composites, ratio images (MNDWI, NDBI, NDVI), Google Earth and topographical map were used to generate training data.
- Accuracy assessment: such assessment was performed for the classified maps of all three steps. Error matrices were used to assess classification accuracy with two measures: overall and Kappa statistics.

NDVI. NDVI is used to quantify vegetation by measuring the difference between near-infrared (vegetation reflects) and red light (vegetation absorbs) [52]. The NDVI algorithm subtracts the red reflectance values from the near-infrared and divides them by the sum of near-infrared and red bands.

$$NDVI = \frac{(NIR - \text{Red})}{(NIR + \text{Red})} \tag{1}$$

The NDVI values are represented as a ratio ranging from -1 to $+1$. Negative values represent water. Around zero is bare soil, built-up areas, from 0.2 to 0.5 is for shrubs, grasslands, crops, and over 0.6 is dense green vegetation.

L.S.T. The final L.S.T. at the interface of the Earth's surface up to 2–3 m height of the air can be measured in the following steps: conversion of the Satellite Digital

Number (D.N.) to Radiance value, conversion of Radiance to At Sensor Temperature, Conversion of Satellite Brightness Temperature into L.S.T.

4   The Study Context and Results

4.1   The Context of Ninh Thuan Province and Phan Rang-Thap Cham City

Ninh Thuan Province. The coastal province in the South—Central part of Vietnam, located at the coordinates: 11°18′14″ to 12°09′15″ Northern latitude, 108°09′08″ to 109°14′25″ Eastern longitude (Fig. 1a).

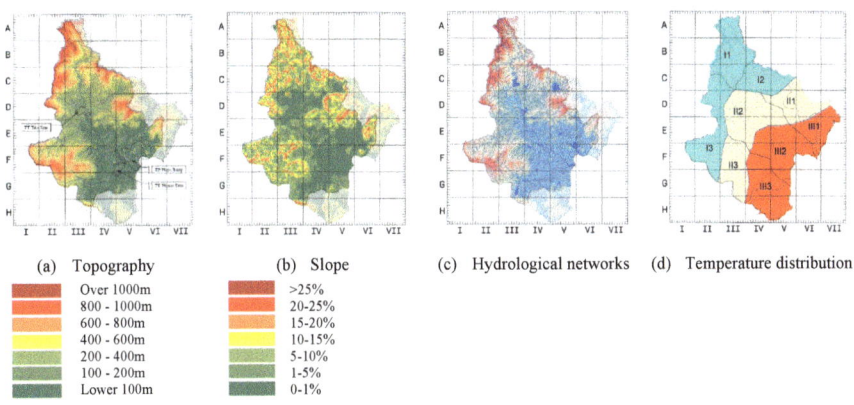

| (a) Topography | (b) Slope | (c) Hydrological networks | (d) Temperature distribution |

(a) Topography
- Over 1000m
- 800 - 1000m
- 600 - 800m
- 400 - 600m
- 200 - 400m
- 100 - 200m
- Lower 100m

(b) Slope
- >25%
- 20-25%
- 15-20%
- 10-15%
- 5-10%
- 1-5%
- 0-1%

Fig. 1 Natural features of river basins of Ninh Thuan province. *(Sources authors, background map from USGS)*

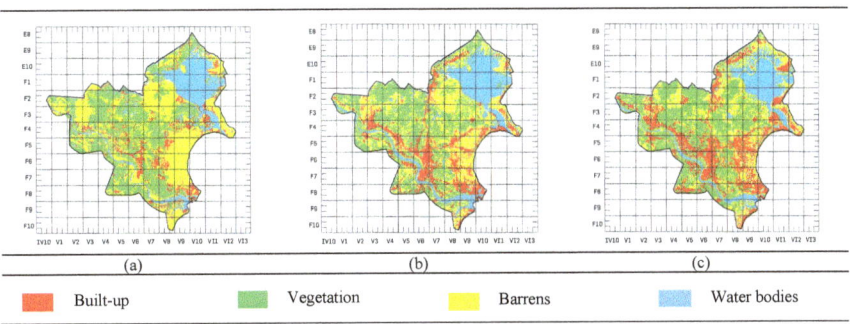

| Built-up | Vegetation | Barrens | Water bodies |

Fig. 2 Transformation of Phan Rang-Thap Cham urban spaces (1988–2020). **a** 1988, **b** 2005, **c** 2020 *(Sources authors, background map from USGS)*

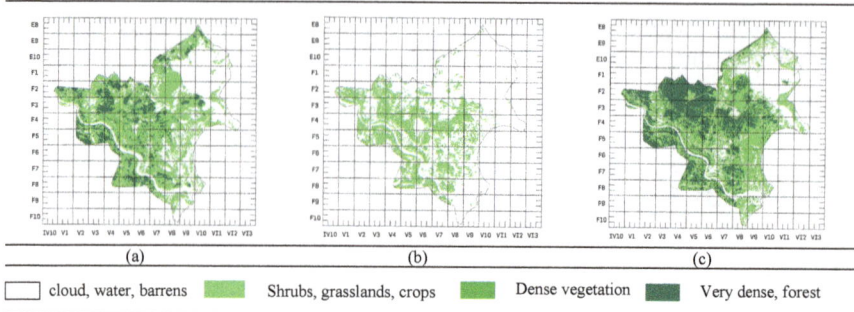

Fig. 3 Transformation of Phan Rang-Thap Cham's NDVI (1988–2020). **a** 1988, **b** 2005, **c** 2020 *(Sources authors, background map from USGS)*

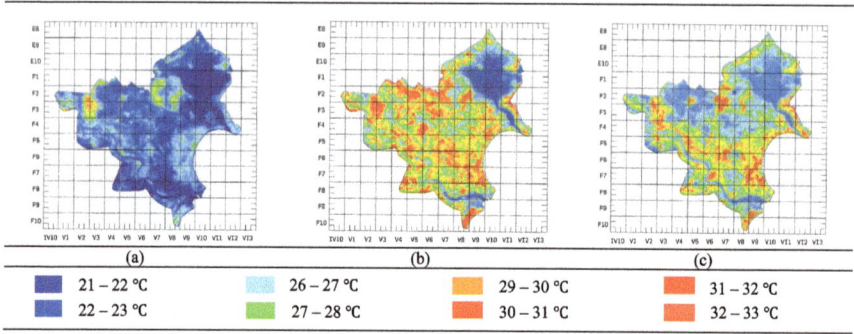

Fig. 4 Transformation of Phan Rang-Thap Cham's LST (1988–2020). **a** 1988, **b** 2005, **c** 2020. *(Source authors, background map from USGS)*

Topography and slope: three quarters of provincial extent surrounded by mountains up to over 1,000 m high (Fig. 1a). The terrain has quite a steep slope in the hills (>20%) and is relatively flat in the centre (<5%) (Fig. 1).

Hydrology: the total area of the basins is 3,092 km², including 46 rivers, mainstreams and four principal aquifers, of which the Dinh river basin covers the majority of the area with 3,000 km², and others with 92 km² (Fig. 1c).

Meteorology: dry (75–77%), hot (26–27 °C) and strong wind (2.3–5 m/s, the strongest is of 25 m/s) with high rate of vaporization. Three climate sub-regions can be distinguished. The coastal area (III) has the most severe droughts with an average rainfall of around 500–700 mm/year, where the studied site of Phan Rang-Thap Cham city is located. The plain area (II) also has drought with rainfall from 750 to 1,200 mm/year. The mountainous area (I) has a rainfall of 1,000–1,700 mm/year (Fig. 1d).

Extreme weather events: the climate alternates from droughts to floods [53–55].

Table 2 Changes in surface area (ha) and remote sensing index (LST and NDVI) in Phan rang-Thap Cham city (1988–2020)

Urban spaces		Study period (1988–2020)					
		1988	%	2005	%	2020	%
Nature	Water	1,703	17	2,198	22	1,536	15
	Vegetation	3,641	36	2,895	29	4,049	40
	Barren	3,532	35	2,831	28	1,891	19
Built	Imperviousness	1,226	12	2,188	22	2,635	26
Total area		10,102	100	10,112	100	10,111	100
Drought index	NDVI	0.25		0.12		0.30	
	LST (°C)	23.66		28.06		27.3	

(Sources authors, figures extracted from remote sensing images, USGS)

Phan Rang-Thap Cham City. It is located downstream of the Dinh River, and is the centre of politics, administration, economy, culture and science of the province. The natural area covers 10,111 ha, of which the built area is 2,519 ha.

Topography is relatively flat (<5%). Water bodies consist of the Dinh river, Nai lagoon and some canals.

Strongly influenced by socio-economic development and uncontrolled urbanization after the policy of "Open Economy" in 1986, spatial networks and land covers had been changed (Fig. 2 a–c) during the study period (1988–2020). Water surfaces had not been changed much, but barren vegetation and built-up spaces had been transformed (Table 2). The amount of bare decreased and was substituted by vegetation and built spaces. All these changes have impacted NDVI (Fig. 3 a–c) and LST index (Fig. 4 a–c).

4.2 Assessment Framework of Urban Spatial Adaptability to D.F.C.

From Phan Rang-Thap Cham city's available facts and figures, an assessment framework of urban spatial adaptability to D.F.C. was elaborately developed in detail. According to several referenced frameworks that assess the overall urban Resilience, many aspects are taken into account, which is generally institutional, economic, social and cognitive factors, apart from the information and physical characteristics [56–59]. A matrix for managing natural disasters was later referenced for this step [60]. It is an amalgam of 16 cells, organized into four rows and four columns. While the rows describe the four common areas of management of any complex system (*physical*, information, cognitive, social), the columns stand for the four phases of disaster management (planning/preparing, absorbing, recovering, and *adapting*).

5 Assessment Matrix

In the study context, a specific matrix for measuring urban spatial adaptability to DFC (Table 3) ranges from rows of urban spatial components and attributes (*location, amount, configuration*), columns of adaptive indicators (*compatibility, efficiency, diversity, connectivity*, redundancy, robustness, NDVI, LST), variables to DFC, and results. The *core adaptive variable* is the *location* and the *core adaptive indicators are compatibility, efficiency and connectivity*. Depending on the role of each component in SES, spatial attributes are selected to be examined, as shown in Table 3. For instance, water bodies are the dominant element for both droughts and floods, while built-up are the most impacted (being impacted) by water, location is a must concern in the SES. The x-marked cell on the table below stands for the urban spatial attributes relevant to adaptive indicators.

Indicators are learned from the previous research and accomplished for Phan Rang-Thap Cham's context. Variables were later comprehensively developed per the studied scale and context (Table 4).

Assessment Levels

The level of urban spatial adaptability to DFC is synthesized through levels of spatial component adaptability. Depending on components' ecological characteristics, adaptive indicators are selected to investigate their spatial attributes. Assessment of urban spatial adaptability is divided into three levels: low, average and high.

As described on Table 4, adaptability is measured by both qualitative and quantitative analysis. Regarding '*location*' and '*configuration*', these spatial attributes are qualitatively analyzed through ecologically spatial patterns and land covers. '*Amount*', '*NDVI*', '*LST*'—the quantitative facts and figures are extracted from LULC, NDVI, and LST. From the above theoretical bases, the main percentage is divided into the core indicators as follows: *compatibility* (approx. 30%), *connectivity* (approx. 30%), *efficiency* (approx. 20%) and 10% for the rest. The core adaptive variables are chosen as location, amount and configuration. In addition, according to ecosystem services [7], water and vegetation are the core natural spatial components to provide ecosystem services of providing water and regulating drought and flood; meanwhile, built up is the main component to consume the services.

The level of urban spatial adaptability is a fundamental base for further study on urban spatial planning resilient to DFC in particularly and to CC in general in the future.

Table 3 Assessment levels of urban spatial adaptability to DFC

| Scales | Spatial components | Spatial attributes | Adaptive indicators of urban spaces to D.FC | | | | | | | | Urban spatial adaptability to D.F.C. (Study area) |
			1. Compatability	2. Efficiency*	3. Diversity	4. Connectivity*	5. Redundancy	6. Robustness	7. NDVI	8. L.S.T	
Basin/Urban/Urban fragments	Water	– Location*	x						x	x	x
		– Amount		x							
		– Configuration			x	x	x	x			
	Vegetation	– Location*	x								
		– Amount		x							
		– Configuration			x	x	x	x			
	Barren	– Amount		x							
	Built-up	– Location*	x								
		– Amount		x							
		– Configuration			x	x	x	x			

* The core adaptive variables and indicators of urban spaces

(Source Author)

Table 4 Assessment description of the urban spatial adaptability to D.F.C

Urban spaces			Description of assessment methods of urban spatial adaptability to D.F.C. under adaptive indicators							
Scales	Spatial components	Spatial attributes	1	2	3	4	5	6	7	8
		Location	1. *Compatibility:* Natural and artificial water networks transformed at the appropriate DFC service locations							
		Amount	2. *Efficiency:* The water area changed through times ($\uparrow$, $\downarrow$, $\rightarrow$)							
	Water	*Configuration*	3. *Diversity of functions:* diversified with dynamic: river, stream/ static: lake, lagoon, etc 4. *Diversity of configurations:* evenly distributed, compatible with landscape across scales 5. *Connectivity:* Natural surfaces water connected with ecological networks 6. *Connectivity:* Natural surfaces water connected with external areas 7. *Connectivity:* Natural surfaces water connected with internal land covers 8. *Redundancy:* Spare spaces for regulating floods located at appropriate locations 9. *Robustness:* enough capacity to store and cope with floods							
	Vegetation	*Location*	10. *Compatibility:* Natural and artificial vegetation transformed at the appropriate DFC service locations.							
Basin/Urban/Urban fragments		*Amount*	11. *Efficiency:* The vegetation area changed through times ($\uparrow$, $\downarrow$, $\rightarrow$)							

(continued)

Table 4 (continued)

| Urban spaces | | | Description of assessment methods of urban spatial adaptability to D.F.C. under adaptive indicators | | | | | | | |
Scales	Spatial components	Spatial attributes	1	2	3	4	5	6	7	8
		Configuration	12. *Diversity of functions:* (nature: forest, agriculture, grasslands, etc.), artificial (artificial: parks, urban trees, etc.) 13. *Diversity of configurations:* evenly distributed, compatible with landscape across scales 14. *Connectivity:* Natural and social functions connected with ecological networks 15. *Connectivity:* Natural and social functions connected with external areas 16. *Connectivity:* Natural and social functions connected with internal land covers 17. *Robustness:* Enough capacity at appropriate locations to store, cope with droughts							
	Barren	*Location*	18. *Efficiencty:* The bare land area changed through time ($\uparrow, \downarrow, \rightarrow$)							
	Built-up	*Location*	19. *Compatibility*: Built-up transformed at the appropriate DFC services locations (Typopology high, slope from 5–10%)							
		Amount	20. *Efficiency:* The built-up area changed through time ($\uparrow, \downarrow, \rightarrow$)							
		Configuration	21. *Diversity of configurations:* (urban areas, urban blocks, small blocks in priority, small blocks in priority) appropriate with DFC service locations 22. *Connectivity:* Built-up (mainly transportation) connected with ecological networks 23. Connectivity: Built-up (particularly transportation) connected with external areas 24. Connectivity: Built-up (mainly transportation) connected with internal land covers 25. *Redundancy:* spare spaces for regulating droughts and floods at appropriate locations 26. *Robustness:* Roads, urban areas, blocks and buildings are strong enough to resist floods							

(Source Author)

6 Conclusions

As a complex system, a city is dynamic, self-organizing, and prepares itself with the ability to adjust to disturbances. The framework for assessing urban space's adaptability to natural disasters in general and D.F.C. in the same area, in particular, is an essential outcome in the context of global urban areas facing the impact of increasing CC.

The study introduces the methodology to develop a qualitative and quantitative assessment framework for the urban spatial adaptability to D.F.C., based on an ecological planning perspective. The case study of Phan Rang-Thap Cham city, Ninh Thuan province, towards a resilient and sustainable urban area is the typical case for provinces in the South-Central provinces of Vietnam and for other sufferings similar situation.

In conclusion, three remarks for urban spatial adaptability to D.F.C. are as follow: (i) The relevant concepts of urban spatial adaptability are pretty essential for Phan Rang-Thap Cham City and other similar context, (ii) Using R.S. and G.I.S. utilities could help much in separating urban spatial components to examine their adaptability, and (iii) Assessment framework of urban spatial adaptability to D.F.C. is a critical tool to further study on adaptive spatial planning.

Acknowledgements The authors wish to acknowledge the assistance and encouragement from the colleagues at the Department of Architecture, HCMC University of Technology, from Omgeving, Vietnam and the Faculty of Architecture, Van Lang University.

References

1. IPCC (2014): Climate change, adaptation, and vulnerability. Organ Environ 24:1–44. http://ipcc-wg2.gov/AR5/images/uploads/IPCC_WG2AR5_SPM_Approved.pdf
2. Dai A (2011) Drought under global warming: a review. Wiley Interdiscip Rev Clim Chang 2(1):45–65. https://doi.org/10.1002/wcc.81
3. McGrath B, Pickett STA, Cadenasso ML (2013) Ecology of the city as a Bridge to Urban Design. In: Resilience in ecology and urban design. Springer, pp 7–28
4. Jhawar M, Tyagi N, Dasgupta V (2012) Urban planning using remote sensing. Int J Innov Res Sci Eng Technol 1(1):42–57
5. Verma R, Kumari K, Tiwary R (2009) Application of Remote Sensing and GIS technique for efficient Urban planning in India. In: Geomatrix conference proceedings, October 2016, pp 1–23. http://www.csre.iitb.ac.in/~csre/conf/wp-content/uploads/fullpapers/OS4/OS4_13.pdf
6. Van TT (2011) Study of urban temperature changes under the urbanization impacts by remote sensing methods and GIS, A case study of Hochiminh City
7. Vinh NQ, Khanh NT, Anh PT (2019) The inter-relationships between LST, NDVI, NDBI in Remote sensing to achieve drought resilience in Ninh Thuan, Vietnam
8. Holling CS (1973) Resilience and stability of ecological systems. Ann Rev Ecol Syst
9. Eraydin A et al (2013) Resilience thinking in urban planning. Resil Think Urban Plan 106:39–51. https://doi.org/10.1007/978-94-007-5476-8
10. Folke C (2006) Resilience: the emergence of a perspective for social-ecological systems analyses. Glob Environ Chang 16(3):253–267. https://doi.org/10.1016/j.gloenvcha.2006.04.002

11. Holling CS, Gunderson LH (2002) Panarchy: understanding transformations in human and natural systems, pp 25–62. https://doi.org/10.1016/j.ecolecon.2004.01.010
12. Holling CS, Gunderson LH (2002) Resilience and adaptive cycles. Panarchy Underst Transform Hum Nat Syst
13. Walker B, Salt D (2006) Resilience thinking: sustaining ecosystems and people in a changing world
14. Gunderson L (2009) Comparing ecological and human community resilience. https://doi.org/10.1017/CBO9781107415324.004
15. Millenium Ecosystem Assessment (2005) Millenium Ecosystem Assessment: ecosystems and human well-being
16. Mileti D, Henry J (1999) Disasters by design: a reassessment of natural hazards in the United States. Nat Hazards 1–372. ISBN 0-309-51849-0
17. Walker B, et al (2002) Resilience management in social-ecological systems: a working hypothesis for a participatory approach. Ecol Soc 6(1). https://doi.org/10.5751/es-00356-060114
18. Davoudi S, Strange I (2009) Space and place in the twentieth century planning: an Analytical Framework and an historical review. Conceptions Sp Place Strateg Spat Plan 7–42
19. Fleischhauer M (2008) The role of spatial planning in strengthening urban resilience. In: Resil. cities to terror. other threat. proc. NATO adv. res. work. urban struct. resil. under multi-hazard threat. lessons 9/11 res. issues futur. work, pp 273–298. https://doi.org/10.1007/978-1-4020-8489-8_14
20. Hoa NT, Vinh NQ (2018) The notions of resilience in spatial planning for drought - flood coexistence (DFC) at regional scale. IOP Conf Ser Earth Environ Sci 143(1):20. https://doi.org/10.1088/1755-1315/143/1/012066
21. Ni'mah NM, Lenonb S (2017) Urban greenspace for resilient city in the future: case study of Yogyakarta City. IOP Conf Ser Earth Environ Sci 70(1):0–8. https://doi.org/10.1088/1755-1315/70/1/012058
22. Dissanayake D, Morimoto T, Ranagalage M, Murayama Y (2019) Land-use/land-cover changes and their impact on surface urban heat islands: case study of Kandy City, Sri Lanka. Climate 7(8):99. https://doi.org/10.3390/cli7080099
23. Crossman ND (2018) Drought resilience, adaptation and management policy (DRAMP) framework. Support Tech Guidel 17
24. Cumming GS (2011) Spatial resilience: integrating landscape ecology, resilience, and sustainability. Landsc Ecol 26(7):899–909. https://doi.org/10.1007/s10980-011-9623-1
25. Hassler U, Kohler N (2014) Resilience in the built environment. Build Res Inf 42(2):119–129. https://doi.org/10.1080/09613218.2014.873593
26. Godschalk DR (2003) Urban hazard mitigation: creating resilient cities. Nat Haz Rev
27. Wandel J, Smit B (2006) Adaptation, adaptive capacity and vulnerability. Global Environ Change 16(3):282–292. http://www.sciencedirect.com/science?_ob=ArticleURL&_udi=B6VFV-4KDBM15-1&_user=961305&_rdoc=1&_fmt=&_orig=search&_sort=d&view=c&_acct=C000049425&_version=1&_urlVersion=0&_userid=961305&md5=83ac11af855b7630cb2f47c45bb3152
28. Smithers J, Smit B (1997) Human adaptation to climatic variability and change. Glob Environ Chang 7(2):129–146. https://doi.org/10.1016/S0959-3780(97)00003-4
29. Gunderson LH (2000) Ecological resilience in theory and application. Ecol Syst 31:425–439. https://doi.org/10.1146/annurev.ecolsys.31.1.425
30. Adger WN, Kelly PM (1999) Social vulnerability to climate change and the architecture of entitlements. Mitig Adapt Strat Glob Change 4(3–4):253–266. https://doi.org/10.1023/a:1009601904210
31. Pelling M, Manuel-Navarrete D (2011) From resilience to transformation: the adaptive cycle in two Mexican urban centers. Ecol Soc 16(2):11
32. Carpenter S, Walker B, Anderies JM, Abel N (2001) From metaphor to measurement: resilience of what to what? Ecosystems 4(8):765–781. https://doi.org/10.1007/s10021-001-0045-9

33. Folke C, Carpenter S, Elmqvist T, Gunderson L, Walker B (2002) Resilience and sustainable development: building adaptive capacity in a world of transformations 31(5):437–440. https://doi.org/10.1579/0044-7447-31.5.437
34. Godschalk DR (2003) Urban hazard mitigation: creating resilient cities. Nat Hazard Rev. https://doi.org/10.1061/(ASCE)1527-6988(2003)4:3(136)
35. Meerow S, Newell JP, Stults M (2016) Defining urban resilience: a review. Landsc Urban Plan 147:38–49. https://doi.org/10.1016/j.landurbplan.2015.11.011
36. Sendai Framework for Disaster Risk Reduction (2015) Sendai framework for disaster risk reduction 2015–2030, pp 1–25. A/CONF.224/CRP.1
37. Nelson DR, Adger WN, Brown K (2007) Adaptation to environmental change: contributions of a resilience framework. Annu Rev Environ Resour 32:395–419. https://doi.org/10.1146/annurev.energy.32.051807.090348
38. Lu P, Stead D (2013) Understanding the notion of resilience in spatial planning: a case study of Rotterdam, The Netherlands. Cities 35:200–212. https://doi.org/10.1016/j.cities.2013.06.001
39. Elmqvist T et al (2004) The dynamics of social-ecological systems in urban landscapes: Stockholm and the National Urban Park, Sweden. Ann N Y Acad Sci 1023:308–322. https://doi.org/10.1196/annals.1319.017
40. McHarg I (1992) Design with nature. Wiley, New York
41. Sharifi A, Yamagata Y (2017) Urban resilience assessment: multiple dimensions, criteria, and indicators. Adv Sci Technol Secur Appl 2016:259–276
42. Suárez M, Gómez-Baggethun E, Benayas J, Tilbury D (2016) Towards an urban resilience index: a case study in 50 Spanish cities. Sustain 8(8). https://doi.org/10.3390/su8080774
43. Eraydin A, Taşan-Kok T (2013) Introduction: resilience thinking in urban planning
44. Allan P, Bryant M (2011) Resilience as a framework for urbanism and recovery. J Landsc Archit. https://doi.org/10.1080/18626033.2011.9723453
45. Bruneau M et al (2003) A framework to quantitatively assess and enhance the seismic resilience of communities. Earthquake Spectra. https://doi.org/10.1193/1.1623497
46. Holling CS (1985) Resilience of ecosystems: local surprise and global change. Glob Chang Proc ICSU Symp Ottawa 1984:228–269
47. Cadenasso ML, Pickett STA, McGrath B, Marshall V (2013) Ecological heterogeneity in urban ecosystems: reconceptualized land cover models as a bridge to urban design. In: Resilience in ecology and urban design, pp 107–129
48. Janssen M, Anderies JM, Elmqvist T, Mcallister RR (2006) Toward a network perspective of the study of resilience in social-ecological systems. Ecol Soc 11. http://www.ecologyandsociety.org/vol11/iss1/art15/
49. Ahern J (2013) Urban landscape sustainability and resilience: the promise and challenges of integrating ecology with urban planning and design. Landsc Ecol 28(6):1203–1212. https://doi.org/10.1007/s10980-012-9799-z
50. WMO and GWP (2016) Handbook of drought indicators and indices, no 1173
51. Bijaber N et al (2018) Developing a remotely sensed drought monitoring indicator for Morocco. Geosciences 8(2):55. https://doi.org/10.3390/geosciences8020055
52. Rouse W, Haas RH, Deering DW (1974) Monitoring vegetation systems in the Great Plains with ERTS, NASA SP-351
53. Annual Statistic of Ninh Thuan Province (2000). Annual Statistic 1995–1999
54. Annual Statistic of Ninh Thuan Province (2006) Annual Statistic 2005
55. Annual Statistic of Ninh Thuan Province (2020) Annual Statistic 2019
56. Da Silva J, Moench M (2014) City Resilience Framework. Arup. http://www.seachangecop.org/files/documents/URF_Bo. http://www.seachangecop.org/files/documents/URF_Booklet_Final_for_Bellagio.pdf%5Cnhttp://www.rockefellerfoundation.org/uploads/files/0bb537c0-d872-467f-9470-b20f57c32488.pdf%5Cnhttp://resilient-cities.iclei.org/fileadmin/sites/resilient-cities/files/Image
57. Figueiredo L, Honiden T, Schumann A (2018) Indicators for resilient cities, p 67
58. Tyler S, Moench M (2012) A framework for urban climate resilience. Clim Dev 4(4):311–326. https://doi.org/10.1080/17565529.2012.745389

59. Fox-Lent C, Bates ME, Linkov I (2015) A matrix approach to community resilience assessment: an illustrative case at Rockaway Peninsula. Environ Syst Decis 35(2):209–218. https://doi.org/10.1007/s10669-015-9555-4
60. Linkov I, Eisenberg DA, Plourde K, Seager TP, Allen J, Kott A (2013) Resilience metrics for cyber systems. Environ Syst Decis 33(4):471–476. https://doi.org/10.1007/s10669-013-9485-y

Assessment of Sustainable Development in Phu Quy Island, Binh Thuan Province Using Sustainable Development Index

Pham Viet Hai, Nguyen Thi Diem Thuy, Phan Thi Thanh Hang, and Dao Nguyen Khoi

Abstract The objective of this study were to develop the sustainable development index (SDI) at the local scale and to investigate the sustainable level of sustainable development for the Phu Quy Island District in Binh Thuan Province. The indicators for the SDI were selected based on the literature review and expert consultation. The results identified 52 indicators based on 17 sustainable development goals (SDGs) for three components of sustainable development (i.e., economy, society, and environment) to estimate the SDI. Furthermore, the result of SDI for the Phu Quy District indicated that the level of sustainability is good. Also, the results highly emphasized increases in sex ratio at birth (SDG5), ratios of employment and product in processing and manufacturing industries (SDG13), and forestland area (SDG15) to improve the level sustainable development of the Phu Quy District.

Keywords Indicator approach · Sustainable development · Phu Quy Island · Vietnam

1 Introduction

Historically, the concept of sustainable development was introduced in the Brundtland Report or Our Common Future in 1987 and the concept was defined as "the development that meets the needs of current generation without compromising the ability of future generations to meet their own needs" [1]. To apply this concept, three pillars, such as economy, society, and environment, were emerged to allow holistic approaches to issues of sustainable development.

P. V. Hai · N. T. D. Thuy · D. N. Khoi (✉)
Faculty of Environment, University of Science (HCMUS), 227 Nguyen Van Cu Street, District 5, Ho Chi Minh City, Vietnam
e-mail: dnkhoi@hcmus.edu.vn

Vietnam National University Ho Chi Minh City, Linh Trung Ward, Thu Duc District, Ho Chi Minh City, Vietnam

P. T. T. Hang
Institute of Geography, Vietnam Academy of Science and Technology, Hanoi, Vietnam

© The Author(s), under exclusive license to Springer Nature Singapore Pte Ltd. 2023
J. N. Reddy et al. (eds.), *ICSCEA 2021*, Lecture Notes in Civil Engineering 268,
https://doi.org/10.1007/978-981-19-3303-5_7

Sustainable development indicators (SDIs) have been developed to measure the progress of sustainable development and support decision-making process under consideration of its three dimensions. Regarding the SDIs, the Commission on Sustainable Development took the main responsibly in developing and implementing the national indicators of sustainable development [2], whose the last revised version containing 14 themes and 96 indicators was release in 2007. In regard to the Millennium Development Goals (MDGs), the UN Statistic Office identified 8 goals with 18 targets and 48 indicators and a deadline of 2015 for their achievement [3]. In the 2005, 4 new target and 10 indicators were added to the MDGs in the Millennium Summit. As the replacement of MDGs, the UN Sustainable Development Summit released the 2030 Agenda which consists of 17 Sustainable Development Goals (SDGs) with 169 targets and 232 indicators and a deadline of 2030 for their achievement [4].

Although the sustainable development issue appears in most development agenda at the international and national levels, this issue has become the concern of governments at the local and regional levels in recent years. Additionally, Rahma et al. indicated that there is a need to develop the SDI that can be easily implemented to monitor the progress of the sustainable development at the regional level [5]. These indicators are grouped into five topics, including people (93 indicators are related to the status and wellbeing of human beings), money (60 indicators examine different types of monetary flows), plans and policies (38 indicators examine how policies, regulations and laws are affecting the implementation and success of SDGs), production and consumption (20 indicators measure the flow of material and energy in the global economy), and planet (18 indicators measure the earth's physical systems, such as water levels, forest loss and agriculture).

The objective of this study was to develop the SDI for measuring the progress of sustainable level of regional development with a case study of the Phu Quy Island District in the Binh Thuan Province, Vietnam.

2 Study Area

Phu Quy is an island district of the Binh Thuan Province located in the South Coast of Vietnam (Fig. 1), far from 120 km southeast of the province. The district has a population of approximately 28,000 people and an area of approximately 16 km^2. The population density is approximately 10 times higher than the average density of the province. The main economy of the district is based on the fishery industry, tourism, and agricultural development. According to the master plan on socio-economic development of the Binh Thuan province up to 2020 (Decision no. 120/2009/QD-TTg on October 6th, 2009), the Phu Quy island district will be a center of fishing, aquatic product processing, fishing logistics services and sea transport and an eco-marine tourist site associated with national defense strengthening.

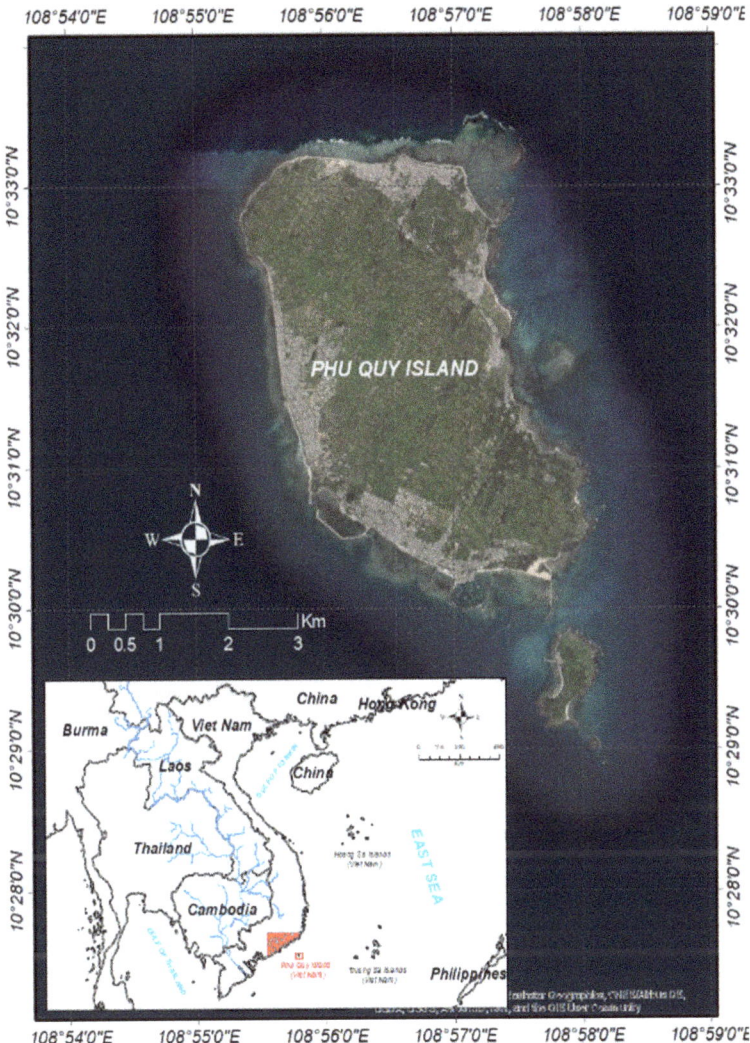

Fig. 1 Location of the Phu Quy island district

3   Methodology

3.1   Sustainable Development Index (SDI)

We developed a set of indicators based on the 17 sustainable development goals (SDGs) to measure the progress of the island districts of Vietnam towards the goals of SD. The indicators were categorized according to the three pillars of sustainable development, namely economy, environment, and society (Fig. 2). The sustainable

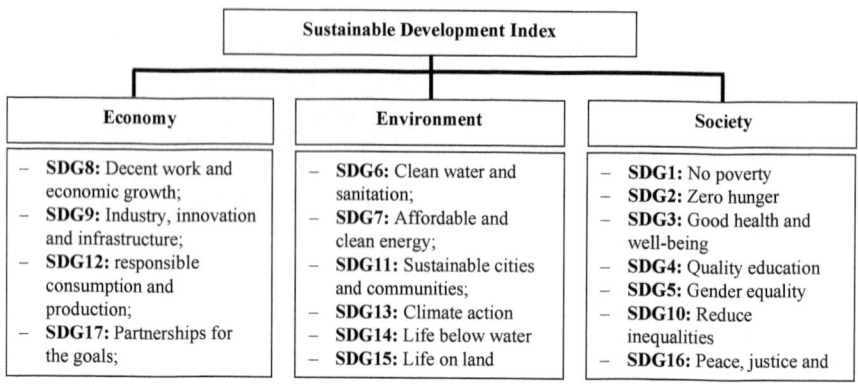

Fig. 2 Linkage between SDGs and the sustainable development index

development index (SDI) was estimated based on these three main dimensions, each dimensions included a different number of sub-dimensions or SDGs, and each sub-dimensions contained a different number of indicators. In the present study, the indicators for the SDI were selected based on the 17 United Nation SDGs with 169 targets and 232 indicators, Vietnam's set of sustainable development indicators with 158 indicators and 115 specific targets of Vietnam (circular No. 03/2019/TT-BKHDT), and the literature review of similar studies. Then, the selected indicators were re-evaluated for our case study based on expert interview and data availability. As a result, 52 indicators were been selected for 17 SDGs (Table 1).

3.2 Scoring and Weighting of the Indicators

As indicators were evaluated on diverse scales, they were first standardized. Standardization was based on the Human Development Index (HDI) of UNDP [6].

$$IndexI = \frac{I - I_{min}}{I_{max} - I_{min}} \tag{1}$$

where I is the original value of the indicator for the district, I_{min} and I_{max} are minimum and maximum values reflecting low and high sustainable of this indicator.

After standardizing indicators, the major index for each SDGs was calculated by the following equation:

$$S_j = \frac{\sum_{i=1}^{n} indexI_i}{n} \tag{2}$$

where n is the number of indicator in each SDGs and S_j is the value of SDG j for the district. The sustainable level of each main dimensions (i.e., economy, society, and

Table 1 List of 52 selected indicators based on 17 SDGs for calculating the SDI

	SDGs	Indicators		Unit	Explanation of indicators relative to SDIs
Environment	SDG6	6.1	Percentage of households supplied with clean water through a centralized water supply system (%)	%	A higher value indicates more sustainable
		6.2	Percentage of households using clean water sources (%)	%	A higher value indicates more sustainable
		6.3	Percentage of households using hygienic toilets (%)	%	A higher value indicates more sustainable
		6.4	Percentage of wastewater is collected and safety treated	%	A higher value indicates more sustainable
	SDG7	7.1	Percentage of households with access to electricity	%	A higher value indicates more sustainable
		7.2	Percentage of renewable energy in gross final electricity consumption	%	A higher value indicates more sustainable
		7.3	Diversification index of electricity source	–	Higher value indicates more sustainable
		7.4	Percentage of households using clean fuel	%	A higher value indicates more sustainable
	SDG11	11.1	Percentage of population living in permanent houses	%	A higher value indicates more sustainable
		11.2	Percentage of solid waste is collected and safety treated	%	A higher value indicates more sustainable
		11.3	Percentage of communes meeting the criteria for new rural standards	%	A higher value indicates more sustainable
	SDG13	13.1	Percentage of households having access to communication media (TV/radio, telephone)	%	A higher value indicates more sustainable

(continued)

Table 1 (continued)

	SDGs	Indicators		Unit	Explanation of indicators relative to SDIs
		13.2	Percentage of population to be disseminated of knowledge on typhoon prevention and mitigation	%	A higher value indicates more sustainable
		13.2	Economic losses due to natural disaster events	Mil. VND	A lower value indicates more sustainable
	SDG14	14.1	Rate of capture fisheries to regional fishery resources	%	Lower value indicates more sustainable
		14.2	Total area of marine protected areas	ha	A higher value indicates more sustainable
		14.3	Number of illegal fishing cases	–	A lower value indicates more sustainable
	SDG15	15.1	Percent change of forest land over the years	%	A lower value indicates more sustainable
		15.2	Percentage of forest cover	%	A higher value indicates more sustainable
Economy	SDG8	8.1	Ratio of revenue budget to total budget	–	A higher value indicates more sustainable
		8.2	Ratio of employment to population	–	A higher value indicates more sustainable
		8.3	Average production value per hectare of agricultural land	Mil. VND/kg	A higher value indicates more sustainable
		8.4	Tourism growth	%	A higher value indicates more sustainable
		8.5	Percentage of unemployment and underemployment	%	A lower value indicates more sustainable

(continued)

Table 1 (continued)

	SDGs	Indicators		Unit	Explanation of indicators relative to SDIs
	SDG9	9.1	Rate of waterway passenger transport to total passenger transport	%	A higher value indicates more sustainable
		9.2	Rate of waterway cargo transport to total cargo transport	%	A higher value indicates more sustainable
		9.3	Rate of processed and manufactured industrial products to the total products	%	A higher value indicates more sustainable
		9.4	Rate of employment in processing and manufacturing industries	%	A higher value indicates more sustainable
	SDG16	16.1	Rate of environmental pollution-causing establishments	%	A lower value indicates more sustainable
Society	SDG1	1.1	Percentage of poor households	%	A lower value indicates more sustainable
		1.2	Percentage of population having social insurance	%	A higher value indicates more sustainable
		1.3	Percentage of population living in households with access to basic living conditions	%	A higher value indicates more sustainable
	SDG2	2.1	Rate of malnutrition in under-5-year-old children	%	A lower value indicates more sustainable
		2.2	Percentage of agricultural land to be applied safety regulations or standards	%	A higher value indicates more sustainable
		2.3	Percentage of population struggling for foods	%	A lower value indicates more sustainable
	SDG3	3.1	Annual number of new HIV infections (per 1,000 uninfected population)	–	A lower value indicates more sustainable

(continued)

Table 1 (continued)

SDGs	Indicators		Unit	Explanation of indicators relative to SDIs
	3.2	Percentage of vaccinated children under 1-year old	%	A higher value indicates more sustainable
	3.3	Number of hospital beds (per 10,000 population)	–	A higher value indicates more sustainable
	3.4	Number of health workers (per 10,000 population)	–	A higher value indicates more sustainable
	3.5	Percentage of communes meeting the criteria for cultural standards	%	A higher value indicates more sustainable
SDG4	4.1	Primary school completion rate	%	A higher value indicates more sustainable
	4.2	Percentage of children attending primary school at the right age	%	A higher value indicates more sustainable
	4.3	High school completion rate	%	A higher value indicates more sustainable
	4.4	Percentage of schools meeting the national standards	%	A higher value indicates more sustainable
SDG5	5.1	Number of child marriage	–	A lower value indicates more sustainable
	5.2	Sex ratio at birth	–	A value closer to 1 indicates more sustainable
	5.3	Rate of female deputies in the People's Council	%	A higher value indicates more sustainable
SDG10	10.1	Expenditure growth of households	%	A higher value indicates more sustainable
	10.2	Income growth of households	%	A higher value indicates more sustainable

(continued)

Table 1 (continued)

SDGs	Indicators		Unit	Explanation of indicators relative to SDIs
SDG16	16.1	Percentage of citizen satisfaction with public services	%	A higher value indicates more sustainable
	16.2	Number of family violence incidents	–	A lower value indicates more sustainable
	16.3	Number of crime of social disorder and safety (per 10,000 population)	–	A lower value indicates more sustainable

environment) and the SDI were calculated using the following equation:

$$SDI = \frac{\sum_{i=1}^{m} S_i}{m} \tag{3}$$

where m is the number of sub-dimensions in each major dimensions.

After calculating the major dimensions and SDI, a radar chart was used to compare the sustainable level of each SDGs and dimensions. The SDI was scaled in the range from 0 (least sustainable) to 1 (most sustainable).

3.3 Data Collection

The present study used data from both primary and secondary sources. Secondary data on economy, society, and environment were collected from the Statistics Office of the Phu Quy District and local governments. Primary data were gathered from our household questionnaire surveys. A semi-structured questionnaire was designed in relation to SDI's dimensions and SDGs. A household survey was conducted with approximately 100 households in October 2019. The collected data were inputted and analyzed using Microsoft Excel version 2013.

4 Results and Discussion

The results shown in Fig. 3 indicate that the sustainability of the Phu Quy District estimated by the SDI is good based on the sustainable scale of 0 to 1. Specifically, the value of SDI is 0.80. In case of components of sustainable development, the values of SDI for society, economy, and environment are 0.85, 0.76, and 0.79, respectively.

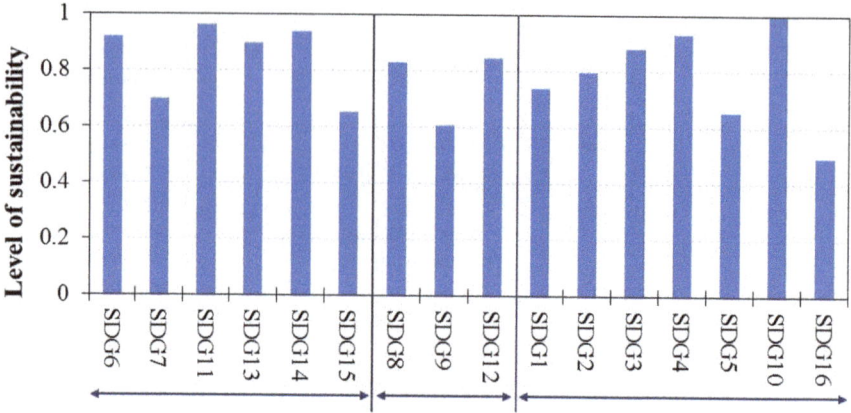

Fig. 3 SDI sub-dimensions for the Phu Quy District

Regarding the social aspect of sustainable, there are 7 SDGs with 23 indicators. Overall, the social component of SD is at the good level (0.79). In the Fig. 3, it is easy to recognize that SDG16 had lowest value of sustainable level (0.50). This is mainly attributed to that the number of crime of social disorder and safety are still high even though the number of crime had upward trend in the period 2016–2020. Furthermore, the sex ratio at birth was quite high (around 113 boys per 100 girls). The supporting policies to reduce the sex ratio at birth will increase the sustainable level of SDG5. For example, there are some supporting policies that can be considered, such as enhancing the public awareness to reduce practices of son preference and undervaluing of girls, as well as gender-based discrimination within the family.

In case of economic component of sustainable development, there are 3 SDGs with 10 indicators. Although the sustainable level of economic aspect is quite high (0.76), there are two indicators 8.2 and 8.3 related to SDG8 and indicators 9.3 and 9.4 related to SDG9 which slightly decreased the sustainable level of economic aspect. Thus, increased ratios of employment and product, especially in processing and manufacturing industries are emphasized to improve the sustainable level of economic aspect for the study area.

Environmental component of SD included 6 SDGs with 19 indicators. The sustainable level of environmental aspect is 0.85. There are two SDGs of 7 and 15 that decrease the sustainability for environmental component of the Phu Quy District. In the period 2016–2020, the forestland area decreased by approximately 62.81 ha due to increasing demands of agricultural and urban areas. This is the main reason decreased the sustainable level of SDG15. In case of SDG7, the low score is attributed to low contribution of renewable energy in gross final electricity consumption (indicator 7.2). In order to increase the score of SDG7, the development of renewable energy, including wind, tidal, and solar power should be emphasized to rise the contribution of renewable energy to the total electricity consumption in the island district.

5 Conclusion

In the present study, the SDI at the local scale was developed and applied to investigate the sustainable level of development for the Phu Quy Island District in the Binh Thuan Province. The main findings can be summarized as follows: (1) the 53 indicators based on 17 SDGs for three components of sustainable development (i.e., economy, society, and environment) were identified; (2) the result of SDI for the indicated that the Phu Quy District is at a good level of sustainable development (0.80); and (3) this study emphasized the decreased the sex ratio at birth (SDG5), increased ratios of employment and product in processing and manufacturing industries (SDG13), and increased forestland (SDG15) to improve the SD level of the Phu Quy District.

Acknowledgements The authors expresses their gratitude to Vietnam Academy of Science and Technology, Marine Science and Technology Program (KC 09/16-20), the project on "Scientific basic, orientation and solution for sustainable socio-economic development of Ly Son and Phu Quy Island Districts (KC.09.37/16-20)" in support of this research.

References

1. World Commission on Environment and Development (1987) Our common future. world commission on environment and development. Oxford University Press, Oxford
2. United Nations (2007) Indicators of sustainable development: guidelines and methodologies. UN, New York
3. United Nations (2000) Resolution 55/2 United National Millennium Declaration. UN, New York
4. United Nations (2015) Transforming our world: the 2030 Agenda for sustainable development. UN, New York
5. Rahma H, Fauzi A, Juanda B, Widjojanto B (2019) Development of a composite measure of regional sustainable development in Indonesia. Sustainability 11:5861
6. United Nations Development Programme (2007) Human development report 2007/2008. Fighting climate change: human solidarity in a divided world. UNDP, New York

Assessment on the Economic Damage Due to Urban Inundation in Ho Chi Minh City

Bui Viet Hung, Nguyen Ngoc Diep, and Nguyen Viet Hung

Abstract The most of urban objects living in coastal region as Ho Chi Minh City (HCMC) is adverse impacted by the inundation. Trading household is the important object of City's economic and serious damaged by the inundation. They make a lot of jobs and a large workforce cooperating with them such as the distribution, business services. The trading households are not only adversely affected as the households (houses, yard, asset) but also damaged more such as the goods loss, revenues reduced, trade delayed,.. and hard living. It throughs the investigation and analysis by the SPSS model, the study determines of damage level to all citizen including the trading households and the flood-damage correlative function. Based on the research results, it shows that the economic damage of trading household object is more serious with the others in HCMC. With the research's results (the tool on analysis and determination flood-damages), the district's and city's agencies will have a new tool to determine the damage level of objects including the trading household in their area. Through the assessment of flood-damage, the City assesses the effective of flood control methods.

Keywords Assessment flood-damage · Trading household object · Correlative flood-damage function · Urban inundation · Flood-damage distribution map

1 Introduction

The most of big cities of Viet Nam places at a coastal zone. So, they meet many adverse impacts form water with the urban inundation being the big problems. The urban inundation seriously affects the economy of all objects (citizen) living in

B. V. Hung (✉)
University of Science, Vietnam National University, Ho Chi Minh City, Vietnam
e-mail: bvhung@hcmus.edu.vn

N. N. Diep
University of Labor and Social Affairs, Secondary Branch, Ho Chi Minh City, Vietnam

N. V. Hung
Department of Construction, Ho Chi Minh City, Vietnam

these cities. The citizen including a poor and high-income are affected/impacted directly and indirectly by inundation. The citizen is facing the inundation impacts with increasing loss on the economic sector to make reduce their income. The Cities are facing the increasing financial deficits to adapt flood situation. Trading households are an important economic component due to their great contributions of the overall economic development and the job creation for the citizen and neighbor region. The trading households are diverse scale and different business. They are not only the households with small convenience store but also the product allocators or small producer working at their houses. So, they often stay and share the same infrastructure with other local residences. These problems lead a trading household to have much larger losses/damages and be more vulnerability to others in City.

The assessment of damages caused by urban inundation is the determination of damage level and the relationship between the flood factors and damage numbers by the corrilative function. The damage level analysis is seem as an activity of flood management. It is also used in analyzing the different damages of research's object affected. Therefore, this analysis is important when assessing the flood disaster mitigation options and make the economic optimization of possible measures [6].

There are many approaches to carry out the assessment of flood-damages due to urban inundation's impacts. The assessments of flood-damage [6] based on an approach of future risk maps and an financial appraisable for an insurance and compensation are used to determine a real economic damages when there are no specific relative information. When the flood-damage is converted into economics [7], the flood-damage values have identified two main ways with the directly and indirectly damages. The damages by costs of a natural hazard (the repairment) is the direct damages. And, the economic losses due to delays or opportunities lossed that is indirect damages. The costs (being a money number) are used for the restoration and improvement of inundation impacts. Thereto, citizen also contributes the cost to remain and repair an infrastructure damaged and urban services in their location. These citizen's payment are also considered as the direct damages [9]. In the case of lack of data, the mathematical model is applied to identify the inundation level with the flood depth and area. By the rate of damage on the land-use value and the its area, it determines the cost of flood-damage [5].

The importance in general assessment of disaster damage and specific urban inundation is the determination of damage groups according to local current economic-social conditions. The damage groups are often classified into both direct/tangible and indirect/intangible damages [6]. For trading household, the relative damage groups can be identified normally the direct damages [3] when costs used to repair/improve the problems occurred as a result of directly contacting with water (house, store, factories or facilities have low quality due to the maceration in water). Many relative studies have showed that a direct/tangible damage is the most important. Furthermore, this is the direct damage group identified easily; In fact, the disadvantageous affects/impacts of urban inundation is usually prolonged and concerned with the declines of citizen's health and income. So, these disadvantageous impacts of urban inundation are seem an indirect/intangible damages. The indirect/intangible damages [8] is the costs incurred as a consequences of affects/impacts caused by inundation in duration period or after happening. The indirect/intangible damages involve a slowing or stopping production and business, causing financial difficulties, as well as adversely affects for the trading households, tacking time to transport goods, transactions and demand production market lost due to the traffic jam or seriously damaged roads [7]. The tangible damage group [3] is when it could be objectively quantified, i.e. damage could be calculated directly monetary value as costs of repairment, overcoming consequences. The intangible damage group is when it is often related to the health and well-being of employees, loss of investment opportunities or profits [1].

The damage-depth correlative curve method is one of many solutions applied on the relative studies. The method shows the costs/damage's values as a correlative function (curvilinear regression) with flood depth. Urban inundation's depth is considered to be an important factor when considering physical damage [2, 5]. It also used the formula to calculate the tangible and intangible damages with the two flood main factors which are flood depth and duration (time) could cause a damage [8]. So, the flood-damage value is a total of tangible and intangible costs/damages [2, 8]. So that, the damage-depth curve method is also used to determine the indirect and intangible damages. The study have to consider the local characteristics about market conditions, income, labor rent cost by time (hour, day), which are applied to determine the indirect and intangible damages.

There are many formulas applied to identify the damage due to natural calamity and specifically urban flood. Such as:

$$DAMAGE = p(\%) * Area * Coeff.(h) * V [4] \tag{1}$$

$$DAMAGE = a_0 + a_1 H + a_2 L [8] \tag{2}$$

$$DAMAGE = D_2(0.06 + 1.42H - 0.61H^2) R (1 + ID) + DCLEAN [10] \tag{3}$$

$$DAMAGE = \sum_{i=1}^{n} Area_i \times f(H_i) [5] \tag{4}$$

where are: p(%) is a rate of construction on the resident area; V is the value per area usually as money. a_1, a_2, a_3 are correlative coefficients, H is a flood depth, L is flood inundation, R is mitigative factor (0, 85), ID is a rate of indirect damage (0, 2) and DCLEAN is cleaning fee.

Most of above formulas are used to qualify the direct/tangible damages caused by flood. The indirect/intangible damages relate usually the duration of flood. So, flood-damage function should include all flood's factors.

The main objective of study is a foundation to support to map the distribution inundation as well as flood-damage risk in HCMC. The damage's assessment and determination of trading households caused by urban inundation are initial surveying period of our study. The flood damages of trading households are determined through the cost's value to pay for the repairing and improving house, asset; the payment on health as well as the lost opportunities on increasing income. The study surveyed directly the economic damage caused by the inundation of trading households living and business at all districts of the HCMC. The study analyses and determines a correlative relationship between the damage and the flooding factors (flood depth and duration) and establish a flood-damage correlative function caused by urban inundation. Based on the flood-damage correlative function applied for each district of HCMC, the study used the flood-urban model (MIKEURBAN) to simulate the inundation situation and damage distribution maps in HCMC.

Therefore, in the study's scope as well as jounary's content, the study will only attend to assess the all direct and indirect/tangible and intangible damage groups of trading households caused by urban inundation in HCMC. After that, the study would like to present the part of results about the inundation and damage distribution maps at District 4 of HCMC on 2016–2019 period as an example.

2 Methodology

The study have analysed and determined the damage group of trading households living and business in HCMC. It is shows on the Table 1 below.

The sequence of research is done according to the steps below.

The Size of Survey and number of objects surveyed/interviewed: In the research, the direct damages (costs to repair a house, traffic vehicles; loss of goods, etc.) and indirect damages (take time to clean, selling delayed or be off to result the

Table 1 Classification of damages due to flood causes for a trading household

Iterm	The direct damages	Cost (10^6VND/year)				
B20	Repair a floor	☐< 5;	☐5–10;	☐10–30;	☐30–50;	☐>50
B21	Repair a fence, yard	☐< 5;	☐5–10;	☐10–30;	☐30–50;	☐>50
B22	Repair a electric assets	☐< 5;	☐5–10;	☐10–30;	☐30–50;	☐>50
B23	Repair a water pipé	☐< 5;	☐5–10;	☐10–30;	☐30–50;	☐>50
B24	Repair a vehicles	☐< 5;	☐5–10;	☐10–30;	☐30–50;	☐>50
B25	Rent a new tools	☐< 5;	☐5–10;	☐10–30;	☐30–50;	☐>50
B26	Good lose/damage	☐< 5;	☐5–10;	☐10–50;	☐50–100;	☐>100
B27	Business tools lose	☐< 5;	☐5–10;	☐10–50;	☐50–100;	☐>100
B28	Time for moving goods	☐<1;	☐1–2;	☐2–4;	☐4–8;	☐>=1 day
B29	Time for cleaning	☐<1;	☐1–2;	☐2–4;	☐4–8;	☐>=1 day
Iterm	The indirect damages	Cost (10^6VND/year)				
B30	Payment for healthy	☐<1;	☐1–2;	☐2–5;	☐5–10;	☐>10
B31	Traffic jam (hour/inundation)	☐<0.5;	☐0.5–1;	☐1–2;	☐>2	
B32	Working delays (hour/year)	☐<0.5;	☐0.5–1;	☐1–2;	☐>2	
B33	Number of jam (number/year)	☐1–2;	☐3–5;	☐5–10;	☐>10	
B34	Number of working out (number/year)	☐1–2;	☐3–5;	☐5–10;	☐>10	

worker's income decreased and seller's revenues decreased or loss) of trading households selling in HCMC, which are determined through the surveying directly objects affected by floods at 22/24 districts of HCMC.

Normally, the size of samples is determined by Yamane's formula:

$$n = \frac{N}{1 + N * e^2} \tag{5}$$

where are: n is number of samples; N is the total objects; e is a scope of choose's error.

It is not possible to determine the number of trading household affected and damaged by floods selling in the HCMC due to the absence of statistics from district authorities. Therefore, in the study, the number of implemented survey questionnaires is 2325 questionnaires for the period 2016–2018 and 1094 questionnaires for 2019 (the study started on November 2019).

Structure of survey questionnaires has 42 indicators and 162 different fields assigned to the indicators. The survey questionnaire consists of 3 different parts, in which, part A is general information, part B is the inundation situation in the research area and part C is the damage situation caused by the urban inundation. Part B is considered as the main content in determining direct damage, indirect damage with quantitative factors of lost time and delay of employees, stopping trading number and estimated revenue reduction or spending to overcome flooding.

The level of flood damage is determined follow: *The damage value of flood = The direct damage's value after flood + The indirect damage's value after flood* (Unit: 10^6 VND).

The direct damage's value is a summary of direct indicators (Unit: 10^6 VND); The direct damage's value is a summary of indirect indicators (Unit: 10^6 VND); The quantization of indirect damage's value is a wage/income of worker and revenue of trading households got if the inundation not happen (it is a multiple number of lost time due to inundation situation with labor/income cost).

The study applied a damage-depth correlative curve method (the flood-damage correlative function) to determine the damage by through the flooding duration and depth, which is widely used in domestic and foreign relative studies. Formula:

$$D = A_0 + A_1H + A_2T \tag{6}$$

where are: D is a damage of studied objects (resident area, industrial park, trading area); H is a flood depth (m); T is a duration (hour); A_0, A_1, A_2 are the parameters determined for the different land use.

To determine the correlation function's parameters, the study analysed the regression relationship between the total flood-damage values surveyed and the correspondent flood factors such as depth (h) and inundation time (T). The study uses the regression method to establish and evaluate the correlation degree (R^2) of above relationship.

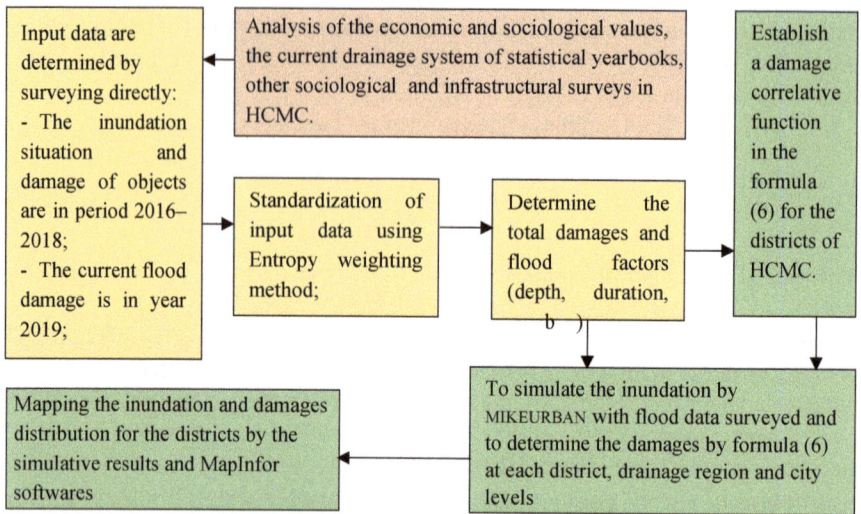

Fig. 1 The diagram of research's approach to determine the flood-damage corrilative function

The Flood and Damage Distributions in HCMC. The study applied the MIKEURBAN software to simulate the inundation in HCMC with the flood data surveyed on period 2016–2019 for the adjustment modelling results. From the simulating flood results, the study applied formula (6) to determine the correspondently damages. The research approach used in study is following the steps in Fig. 1.

3   Results and Discussions

The Survey Results: After the survey activity implemented at all districts of HCMC, the number of questionnaire papers (see Table 1) implemented and collected is about 3419 in period 2016–2019. The direct and tangible/indirect and intangible flood-damage of trading household affected are summarised on Tables 2 and 3. Also, the direct and tangible flood-damages are determined by the all costs to repair and improve all structures as house, yard,.. asset of trading households caused by floods occurred.

Based on the cost (including direct and indirect damage) listed on Tables 2 and 3, the inundation factor (depth and time) causing damages have occurred and be valuable when it has the depth about 20–45 cm and the duration about over 30 min.

The Statistical Analysis: The evaluation of the survey results on statistical significance as well as the square correlation level used the determining *P-Value* and R^2 values. These values are summarized on Table 4.

The square correlation coefficient R^2 between the damage level and the flooding factors including flood depth (cm) and duration (minute) shows that their correlation

Table 2 The flood-damages of trading householdes on period 2016–2019

Sign	Tangible damage (million VND)								Intangible damage (million VND)								Total damage
	House	Yard	Under construct	Assets	Vehicles	Equip	Goods	Total	Set up	Traffic jam	Delay	Not sell	Sick cost	Remade cost	Other	Total	
Q4-01	1.0	1.0	1.0	1.0	5.0	5.0		1.17	4.0	1.5	3.1		0.5			0.22	1.39
Q4-02	5.0			5.0	1.0	2.0	1.0	1.08	3.0	0.5	1.1	1.0	0.5	1.0	10.0	1.13	2.20
Q4-03	1.0			1.0	1.0	10.0		0.93	2.0	1.5	0.1					0.04	0.97
…	…																

Table 3 The flood factors of inundation causing damages on period 2016–2019

Sign	Short duration (minute)	Medium duration (minute)	Long dur. (minute)	Medium depth (cm)	Max. depth (cm)	Duration (minute)	Depth (cm)
Q4-01		75		15	30	75	22.5
Q4-02	30			45	45	30	45
Q4-03		75		20	30	75	25
...	...	......	...	...	...	...	....

Table 4 The square correlation coefficients R^2 between an inundation damage and factors and between two inundation factors

Objective: trading household		The square correlative coefficient (R^2)			
No	Location	Period 2016–2018		Current year 2019	
		Depth	Duration	Depth	Duration
1	Ho Chi Minh City	60%	69%	54%	62%

is tight level when R^2 are in the range of 50–80%. With the number of above square correlation coefficients, the flood-damage correlative functions will be nonlinear (the natural logarithmic function).

Determine the Coefficient Sets for Damage Correlative Functions Following: Based on the results of direct and indirect damage obtained and the statistical analysis method, the flood-damage correlative function applied to determine the flood-damages of trading household affected in the City, which are following the formula (1), has the coefficient sets such as (See Table 5).

Therefore, the correlative functions between inundation damage and two factors (depth and duration) as well as between damage and conversion depth are defined below.

$$D = -2.45 + 0.11\,h(cm) + 0.06\,T(minute) \tag{7}$$

Mapping the Inundation and Flood-Damage Distributions in HCMC: Application of MIKEURBAN model to simulate the inundation on period 2016–2019 with the results about flood factors (depth and duration). These flood factors simulated by model put in the flood-damage correlative function (3) to determine the correspondent

Table 5 The coefficient sets of inundation damage functions

No	Location	The flood-damage correlative function		
		Ao	Ah	AT
1	Ho Chi Minh City	−2.45	0.11	0.06

damages. For example, we have calculated and mapped the inundation and flood-damage distributions of trading household at District 4 (see below figures) (Figs. 2 and 3).

Thus, in the study, the flood-damage correlative function (7) is reasonable applied and according with the flood-damage value of trading household living and business in Ho Chi Minh City. After that, the study have mapped the flood-damage distribution for HCMC. Based on the calculation method, the results of the flood-damage correlative function with the flooding factors, some assessments are as follows:

– Using a damage curve method to identify the regression correlative function between the flood-damage and the two flooding factors (depth, duration) and the flood-damage value correspondently. The flood-damage correlative functions are set up with different parameters sets according with each district of HCMC.
– The correlation coefficient value (R^2) between the flood-damage and the flooding depth and duration have determined about from 50 to 80% to ensure that their

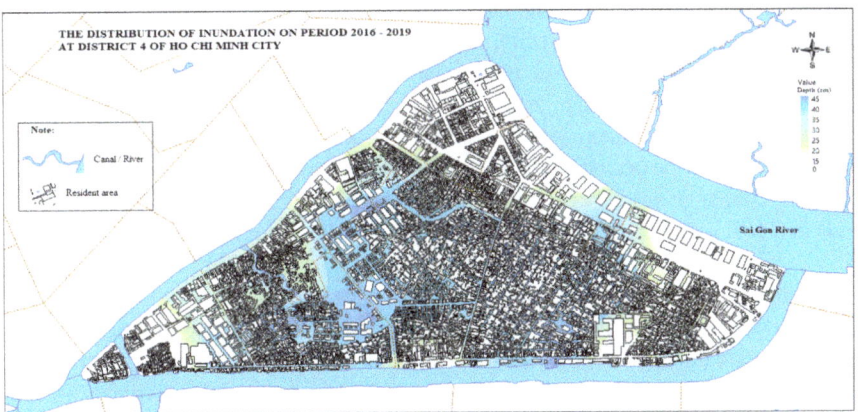

Fig. 2 The distribution maps of inundation at district 4

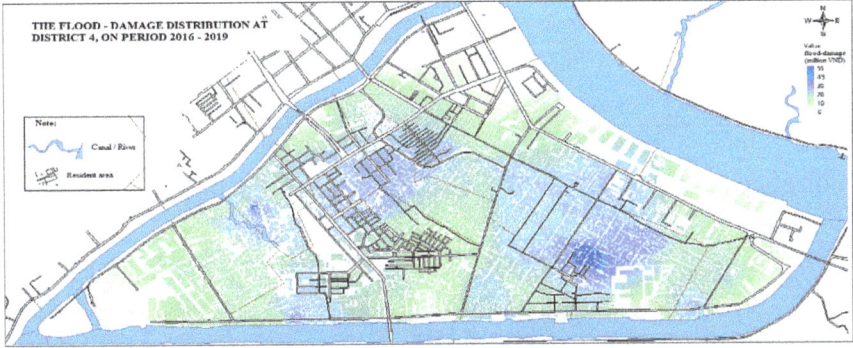

Fig. 3 The distribution maps of flood-damage of trading households at district 4

correlations are close. It means that the survey data series of damage, inundation depth and duration have same trend and their correlative functions are curvilinear regression form.
– With the above assessment, the results of determining the sets of parameters for the flood-damage correlative function ensure the scientific basis as well as meet the requirements of the state management of the impacts of natural disasters (urban inundation) in Ho Chi Minh City.

4 Conclusions

With the identification of direct/indirect damage types according to the flood-damages of trading household object, the study have founded the flood-damage corrilative function for Ho Chi Minh City on period 2016–2019. Based on the analysis results and application of MapInfor software to map the inundation and damage distribution. The analysis results show that the flood-damage correlative function meet the determination of the economic loss of a trading household selling in the City whom is adversely affected by inundation.

The study of flood damage assessment just establishing the form of correlative function between the damage with the flooding factors and initially determined the set of coefficients for the period from 2016 to 2019. The distribution maps of inundation and flood-damages are also established on period 2016–2019. The Climate change leads the local weather (in Southern East Region of Viet Nam including Ho Chi Minh City) being complicated and changing more extreme. It means that flooding situation be continues to evolve unpredictably and cause a lot of damage to the citizen in both extent and types. The study is implemented on period 2016–2019. So, it should make the formula of flood-damage corrilative function to be more stable and convenient with new conditions. The study will necessary continue to implement under the new ways. Such as, the relative surveys are implemented by the years (possibly with a narrower scale) in specific urban inundation management period.

Acknowledgements The article is implemented with results from the research "Surveying and assessment of economic damages caused by an inundation to socio-economic; mapping a current flood damage for the prevention of flooding, urban planning in Ho Chi Minh City".

References

1. Green C, Viavattene C, Thompson P (2011) Guidance for assessing flood losses. CONHAZ report
2. Handmer J, Reed C, Percovich O (2002) Disaster loss assesment guidelines. Department of EmergencyServices, State of Queensland, 90pp
3. Hammond MJ, Chen ASS, Djordjević D, Butler OM (2015) Urban flood impact assessment: a state-of-the-art review

4. Elisabetta Genovese (2006) A methodological approach to land use-based flood damage assessment in urban areas: Prague case study. The mission of the Institute for Environment and Sustainability is to provide scientific and technical support to the European Union's policies for protecting the environment and the EU Strategy for Sustainable Development

5. Bao LX, Van Cong M (2016) Economic risk assessment due to flooding, applying to the flood control project phase 1 in HCMC. J Sci Technol Irrig Environ 55:55–72

6. Merz B, Thieken, Kreibich H (2011) Multi-variate flood damage assessment: a tree-based data-mining approach. Nat Haz Earth Syst Sci 13(1):53–64

7. Olesen L, Löwe R, Arnbjerg-Nielsen K (2017) Flood damage assessment – literature review and recommended procedure. Cooperative Research Centre for Water Sensitive Cities. https://watersensitivecities.org.au/content/flood-damage-assessment-literature-review-recommended-procedure/

8. Price RK, Vojinovic Z (2008) Urban flood disaster management. Urban Water J 5:259–276

9. The Center for Neighborhood Technology (2014) The prevalence and cost of urban flooding: a case study of Cook County, IL, 32pp

10. Tenaga KTA (2003) Flood damage assessment: a review of flood stage–damage function curve

Establishing the Water Body Planning Structure for Traditional Craft Villages of Hanoi City, with Thematic Study on Traditional Weaving Craft Village in Phung Xa Commune, My Duc District

Tuyen Van Nguyen, Hanh Thi My Phung, Huong Thu Nguyen, and Thao Nguyen Phuong

Abstract Water bodies form a constituent part of the cultural values of traditional craft villages in Hanoi, Vietnam. Water bodies do not only improve the ecological environment and create landscape architecture but also preserve the traditional cultural values and support craft production activities in rural villages in Hanoi. Yet, urbanization and modernization of production has transformed villages and brought changes to the structure of water bodies there, as well as reduced their roles in the sustainable development of the craft villages. Through the survey in Phung Xa craft village (My Duc district, Hanoi), the article outlines the current situation of local water bodies and proposes solutions to establish the planning structure of water bodies in a way that will enhance their role and enable their potentials. The study results include viewpoints and principles to establish the planning structure, new functions, functional integration and structure of water bodies in craft villages.

Keywords Traditional craft villages · Water bodies · Planning structure

1 Introduction

Water is key to the uniqueness of craft products; its presence in the kneading process creates diverse ceramic products for different craft villages of the Red River Delta; water leaching improves bamboo elasticity which is input materials for craft products while water inclusion in ink making makes traditional paintings more appealing. Water bodies form typical village landscape such as village wells or water puppet theatre stage and space. Water bodies do not only represent the image of Vietnam but are also close to the development and evolvement of the spatial structure in traditional craft villages.

T. V. Nguyen · H. T. . M. Phung (✉) · H. T. Nguyen · T. N. Phuong
Hanoi University of Civil Engineering, No. 55 Giai Phong Street, Hai Ba Trung District, Hanoi, Vietnam
e-mail: hanhptm@huce.edu.vn

© The Author(s), under exclusive license to Springer Nature Singapore Pte Ltd. 2023
J. N. Reddy et al. (eds.), *ICSCEA 2021*, Lecture Notes in Civil Engineering 268,
https://doi.org/10.1007/978-981-19-3303-5_9

As of 2020, Hanoi authorities reviewed and recognized 276 traditional craft villages [3]. Among these, 17 traditional craft villages are encouraged to promote tourism-oriented traditional craft production [4]. Traditional craft villages that have been recently restored and developed such as Bat Trang pottery, Phu Vinh bamboo and rattan weaving, Van Phuc silk, etc. are outstanding models for integrated development of traditional crafts and tourism. The development model of traditional craft associated with tourism has become a tendency in Vietnam, as the parallel development of traditional crafts and tourism based on preservation and sustainable development on professional values—traditional villages [3]. Despite such potentials, the traditional craft villages remains underdeveloped and unsustainable due to increasingly serious environmental pollution. Water bodies are shrinking in terms of roles and sizes, becoming polluted and gradually replaced by concrete structures. The system of water bodies is disrupted and converted into domestic and industrial sewerage system in traditional craft villages. Therefore, it is necessary to restructure the water body system in order to preserve the existing water bodies, improve ecological environment and promote their roles in the economic development of traditional craft villages.

Spatial structure is still seen as "an abstract concept", covering the entire communal space in its development: from overall structure to components, from internal relations to external relations, etc. Water bodies are part of the communal space. Communal planning structure is one of the basic characteristics of the current practices in communal spatial organization, and it reflects the layout and interrelations between functional areas of a commune, ensuring that these areas are systematically linked in a robust structure and that the interactions between the most important areas of the commune in the development process are somehow maintained [2]. The planning structure of water bodies is one of the core elements of the green infrastructure system of a commune (part of the rural infrastructure system structure), reflecting the layout and the relationship between the constituent areas while ensuring that these areas are systematically linked [5]. Spatial structure of water bodies is consisted of three factors: Functional components; relations and connection forms between functioning areas; intended structure of water bodies.

2 Research Methodology and Subjects

The study applied two main methods: analysis and consolidation method and field surveys. While analysis and consolidation method refers to the reviews of research and legal documents produced in Vietnam and elsewhere, field surveys was conducted to collect additional information or to validate the aggregated data and information. Hanoi Peoples' Committee has set up the tourism orientations for seventeen traditional craft villages of: Bat Trang, Van Phuc, Ngo Ha, Du Du, Phu Vinh, Son Dong, Ha Thai, Kieu Ky, Thang Loi, Thiet Ung, Trach Xa, Dai Dong, Nhi Ke, Thuong Hiep, Phung Xa, Xuan La and Le Mat. The research selected seventeen traditional craft villages that are encouraged to promote tourism-oriented craft production following

Table 1 Impacts of weaving activities in Phung Xa commune on water bodies (*Source* Authors 2020)

No.	Steps	Products and direct impacts on water bodies	Pollution level (1 ÷ 5)
1	Spinning	Little impact	1
2	Weaving	Little impact	1
3	Chemical treatment	Washing liquids, waste textile dyes	5
4	Edge sewing	Little impact	1
5	Inkjet printing	Fabric detergent, print spills, screen cleaner, printing brush	2

the Hanoi's master plan on crafts and craft villages development [4] for the analysis on water usage for production, existing water body system, production impact on water body system. The research team selected these traditional craft villages where these aspects were most visible for further analysis. From this shortlist, the weaving craft village in Phung Xa commune, My Duc district was finally chosen.

3 Situation of Surface Water Pollution

According to the water resource analysis in 292 craft villages (out of a total of 1,350 craft villages in the city) in the 2017–2020 period by the Hanoi Department of Agriculture and Rural Development, 139 (47.6%) craft villages were seriously polluted, 95 (32.5%) craft villages were polluted, 58 (19.9%) craft villages were not infected; and only about 5.2% of wastewater discharged by craft villages was collected and treated. Most of the wastewater generated in craft villages in Hanoi is discharged directly into the environment with serious pollutant concentration. As reported [7], the concentration of pollutants such as COD, BOD5 or total coliforms in the wastewater discharged to the Day river section in the Phung Xa commune is dozens of times higher than the maximum level to be allowed in the current standards (Table 1).

4 Current Structure of Water Bodies in Phung Xa Commune

Phung Xa commune is located in the south of Hanoi belonged to the green belt zones of the master plan of Hanoi up to 2030 with visions to 2050 [8]. The spatial structure of water bodies in Phung Xa traditional craft village as followings:

Water Bodies: In general, water bodies in Phung Xa commune weaving craft village includes public and private water bodies such as rivers, canals, ditches, ponds, lakes and wells. With the traditional production and daily life activities of local people, water bodies are overused in a natural way regardless of their size. Ponds and lakes inside and outside of the village are losing their production role and culture as well as ecological values in the modernization and urbanization. Some ponds have been replaced by communal houses and flower gardens or converted and sold for residential purposes. Fishing ponds contribute very little to the villages' ecological balance and are also seriously polluted. Rivers and lakes are often the places of recreational and sports activities, such as cycling, fishing, boat racing, swimming, etc. These activities are no longer organized due to the impacts from polluted water bodies.

Water Body Connections: The traditional village ecosystem is independent and characterized with closed cycles. The energy and material cycles flow within villages and the commune, in which water bodies are both the connecting factor and the key component of the ecosystem. In Phung Xa commune traditional weaving craft village, the ecological links between lakes, ponds, irrigation canals and ditches with the Day river—the main irrigating river—have been seriously broken. Tourism services, entertainment and recreational activities was not properly developed in accordance with water body planning. These are important factors to be considered during the establishment of the structure for water bodies in Phung Xa commune traditional weaving craft village so as to facilitate tourism development following Hanoi city's orientations for craft village development (Figs. 1 and 2).

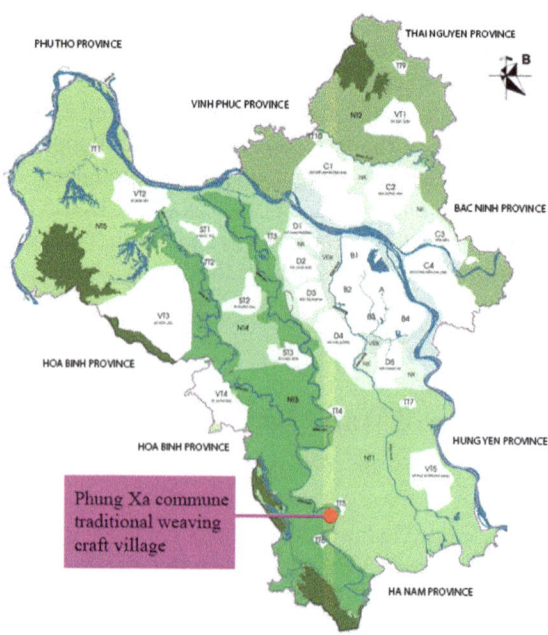

Fig. 1 Location of Phung Xa commune in the planning of Green Belt [8]

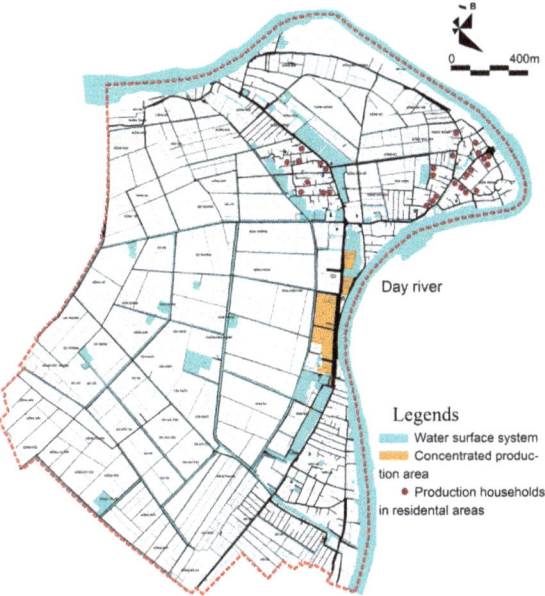

Fig. 2 Water body system and production (*Source* Authors 2021)

5 Proposed Solutions for Wastewater Treatment Technology in Phung Xa Commune

In Vietnam, large-scale textile and dyeing factories located in industrial zones are required to build wastewater treatment stations. Effluent must satisfy the discharge standards of the [6]. Textile and dyeing wastewater are characteristic with high levels in pH, temperature, chemical oxygen demand (COD), and color, or generally speaking, its composition is rather complex. Consequently, the wastewater treatment at large-scale textile and dyeing factories requires high-tech facilities. Wastewater must be treated mechanically, physiochemically and biologically (mostly in aero tanks) before going further with an advanced oxidation process. The advantage of this technology is that it can centrally treat a large volume of wastewater to deliver effluents that meet the current standards of the environmental technology. This solution is not economically appropriate for small-scale production areas, which are scattered in traditional craft villages. In addition, this solution does not take advantages of the existing pond system.

Plants of the subsurface flow wetland are spongy plants with cluster roots and suitable for wastewater environment, most often perennial aquatic plants which are wastewater-resistant, herbaceous with spongy stem and cluster roots, such as Typha Orientalis, Phragmites Communis, Cyberus Involucratus, Dracaena Fragrans, Scirpus spp or native plants which adapt and survive in wastewater conditions. These plants also show some more values in terms of economy and landscape. Plant biomass

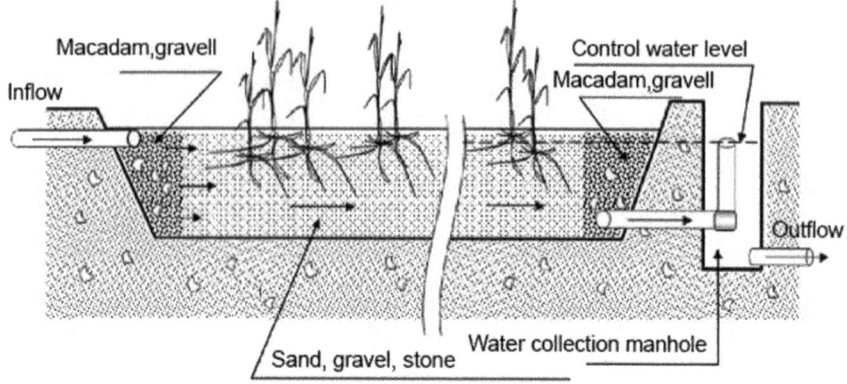

Fig. 3 Structure of a horizontal subsurface flow constructed wetland [1]

from the subsurface flow wetland can be used for various purposes: handicrafts (knitting products); paper (from source of cellulose fibers); commercial product (ornamental plants, flowers); or domestic animal feeding depending on the plant and market demand. Effluent of the subsurface flow wetland will be channeled to nearby ponds and lakes. For Phung Xa commune traditional weaving craft village, it is proposed herein to establish three decentralized wastewater treatment stations for the three main production areas. Treatment station 1—W1 (by upgrading the existing station and supplementing a subsurface flow wetland) will serve the printing and dyeing households in the commune center. Treatment station 2—W2 (new station) shall be built for craft households who are residing outside the dike in Thuong village. Treatment station 3—W3 (new station) shall be set up for craft households who are located inside the dike in Thuong village (Fig. 3).

6 Proposed Solutions to Establish the Planning Structure of the Water Bodies in Phung Xa Commune

6.1 Objectives and Principles of Establishing the Water Body Planning Structure

The planning structure of water bodies in Phung Xa commune traditional weaving craft village is based on the orientation to develop the textile and garment traditional craft villages into sustainable green supply chain and aims at the following objectives: Protect the existing water bodies, improve the environment, control the amount of wastewater discharged into the water bodies and promote ecological environmental balance in traditional craft villages; Preserve and promote the cultural values of traditional craft villages and traditional villages in consideration of water body formulation and development; Promote the production value of traditional craft activities and ensure water resources for agricultural production and domestic needs.

With these in mind, the paper proposes the following principles to establish the water body planning structure in Phung Xa commune traditional weaving craft village: Maintain existing water bodies; Create additional water bodies and greenery; Restore the connections of the water system; Control handicraft activities.

6.2 Proposed Model of Water Body Planning Structure

Functions of the Water Bodies: Water bodies should be functioned originally as natural and stable water bodies. About additional functions related to the promotion of water body development, these include historical and cultural relic areas such as communal house, pagoda, temple, etc.

They are supposed to be identified by the existence of tangible heritages (relics, banyan tree, etc.) or intangible heritages (festivals, beliefs, etc.), tourism and forestry service areas (including ecotourism, parks, sports, and outdoor recreation areas), low-density rural residential areas with orchards and ponds (such as garden houses and residential houses of low construction density i.e. less than 45%); agricultural land areas, (including land for perennial crops, vegetable planting, rice—fish, rice—fish—duck production, aquaculture and farms).

Connections Between Water Bodies: Ecology-and environment-based connections: It is necessary to clearly analyze the desired ecosystems for reasonable connections of greenery and water bodies and to pay attention to the principles of formulating clusters and regional connections for water bodies in order to minimize (or better avoid) fragmented development (Figs. 4 and 5).

Tourism-based connections: Thanks to the diverse functions of the water system, it is possible to organize numerous tourism programs, such as tourism-oriented craft activities; tours to learn about traditional communities, craft villages, communal cultural values, pagodas, traditional architecture; cultural and spiritual tourism activities associated with water-related relics, cultural heritage, festivals, and religious rituals; agricultural tourism to learn about the traditional "wet rice" culture; ecotourism associated with tourism and recreation activities along the Day river.

Culinary, recreation and sport-based connections: Water body can house a variety of tourism and recreation services. It is advised to organize sports cycling routes, site seeing tours, culinary service hubs in the village along the Day river and inter-field canals or provide countryside restaurants and culinary services along the Day river.

Production-based connections: Link waterway for the transport of raw materials and craft products to the market; connect different production activities in the commune, such as connecting cotton growing areas to production areas and to finished product gathering areas.

Shaping the Morphology of Water Bodies: It is suggested to develop water body system in Phung Xa commune in a network orientation: Water bodies should be

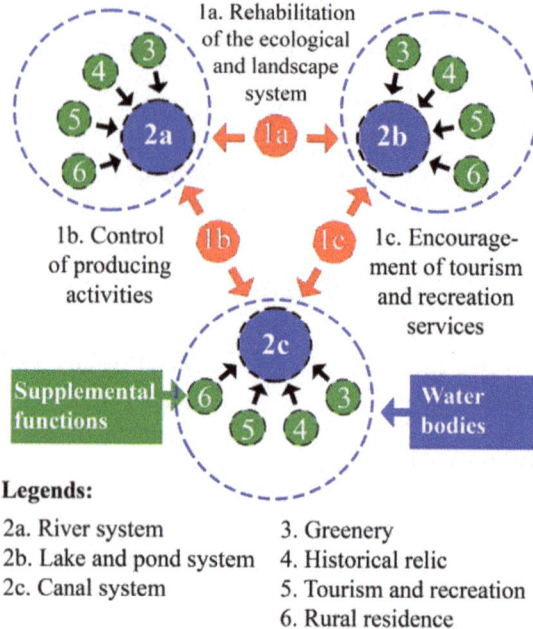

Fig. 4 Model of spatial structure for water bodies in traditional crafts (*Source* Authors 2021)

linked in an interconnected network using irrigation canals and transport routes. The Day River should be set as core and other components of the system should be mapped accordingly. Below are the expected water body directions: (1) The Day river direction: This section of the system focuses on greeneries along the river. Green space should be expanded with supplementary riverside cotton trees on available bare land. Village lakes and ponds: This section of the system is designed around the village ponds, lakes and wells. It will include irrigation canals to ensure smooth water circulation between ponds and lakes inside and outside of the villages, and between the village water bodies and Day river for tighter ecological connections. (2) Irrigation canal system: The irrigation canal serving agricultural production should be set at the core of this section. Additional trees and flowers should be planted on the canal banks to improve the diversity of the riverside eco-system.

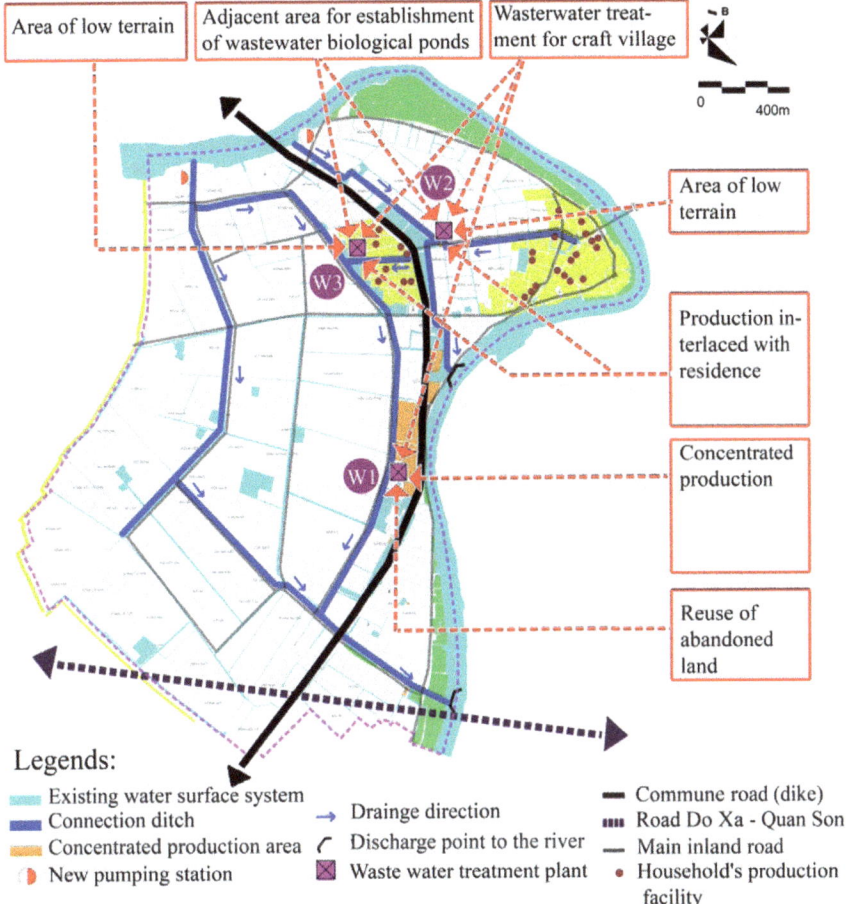

Fig. 5 Structure of water body system in Phung Xa commune traditional weaving craft village (*Source* Authors 2021)

7 Conclusions

Water bodies is an indispensable part of the natural ecosystem and cultural value values of traditional craft villages in Hanoi. Establishing water bodies structure is a practical work that contributes to the sustainable development, in which the focus is on improving the environment of craft villages and promoting their values. water bodies is currently narrowed down, disrupted and seriously polluted. In which, chemical processing (cooking, bleaching, dyeing) causes the most tremendous pollution to river and lake environment. Heritage spaces, tourist services and low-density rural residential spots associated with water surface systems should be recognized as constituting elements of the landscape.

With the analysis and assessment of the current situation of water bodies system of handicraft village of Phung Xa, the paper has systematized the theory of restructuring water bodies from the viewpoint, principles and establishment including the following elements: functional groups, structural linkages and spatial morphological orientation. The paper has applied principles craft villages, restructured the water surface space system on the basis of preserving the existing system of ponds, lakes and canals, adding new water bodies systems by revitalizing polluted ponds and lakes; extending new canals; creating connections with Day river; establishing three wastewater treatment stations to strictly control wastewater in three main production areas of the communes.

With the method of dividing into 03 pollution control areas associated with 03 treatment stations, wastewater from craft production activities in Phung Xa commune will be collected and thoroughly treated before being discharged into the common drainage canal system. This recommendation is suitable and utilizable with the natural terrain, meeting the current situation of scattered production activities and limited investment resources of the locality.

Acknowledgements This academic paper comes from a project coded B2018-XDA-17 and sponsored by Vietnam Ministry of Education and Training (MOET) and managed by Hanoi University of Civil Engineering (HUCE).

References

1. Vymazal J (2009) The use constructed wetlands with horizontal sub-surface flow for various types of wastewater. Ecol Eng 35(1):1–17
2. Cuong PH (2015) Urban Planning - course book for students in architecture. Construction Publishing House, Hanoi, p 62
3. Huong NT (2021) Model of planning for traditional craft-tourism villages in the Red River Delta, Ph.D. thesis, National University of Civil Engineering, Hanoi
4. Hanoi People's Committee (2013) Master plan for crafts and craft villages development in Hanoi to 2020 with a vision to 2030, issued attached to Decision No. 14/QD-UBND dated January 2nd 2013, Hanoi
5. Hanh PTM (2018) Establishment for the scheme of typical infrastructure of communes and rural settlements in the Red River Delta towards green infrastructure, coded B2018-XDA-17, Vietnam Ministry of Education and Training (MOET), Hanoi
6. Ministry of Natural Resource and Environment (2011) QCVN 40:2011 Industrial wastewater, associated with Decree No.47/2011/TT-BTNMT dated December 28th 2011, Hanoi
7. Ministry of Natural Resource and Environment (2018) Report on present status of National environment in 2018, Thematic paper on water environment of river basins, issued on July 31st 2019, Hanoi
8. Prime Minister (2011) Approval of the development master plan for Hanoi city to 2030, vision to 2050, Decision No. 1259/QD-TTg dated July 26th 2011, Hanoi

Linking Dynamic Urban Settlement Patterns to Environmental Infrastructure Needs: The Case of Da Nang City, Vietnam

Nigel K. Downes

Abstract This paper outlines an urban structure type approach to interpret the interlinkages between sustainable infrastructure provision and the existing and evolving urban fabric of the Central Vietnamese city of Da Nang. Urban structure type mapping has become a widely used tool for categorising cities into homogenous areas for both planning and monitoring actions. The work investigates the relationships between the built environment and the spatial patterns of supporting infrastructure. Spatially mapped urban structure types and infrastructures provide a clear picture of Da Nang's shortcomings and requirements towards strategizing effective cross-sectorial policy-making, planning and monitoring.

Keywords Urban structure types · Da Nang · Sustainable infrastructure provision · Urban environmental planning · Sustainable urbanization

1 Introduction

Cities require up-to date spatial information on both current land use patterns and functions in order to monitor, manage and plan the provision of basic public services, infrastructure and develop on a resilient and sustainable pathway. However, the local pace of transformation often makes conventional sources of information frequently inadequate for planning purposes. Data is often out of date, unreliable, in some cases simply unobtainable or not gathered and represented at a scale for effective planning [1, 3, 5]. Da Nang is seen as a driving force economic development of the entire central region. However, recent growth to renovate and establish new urban areas has resulted in rapid modernisation processes and the expansion of urban space. Yet this urban development has either been in accordance with short-term plans, or been somewhat unplanned. As a province level city, directly under the command of the central government continued urbanisation and infrastructure development,

N. K. Downes (✉)
College of Environment and Natural Resources, Can Tho University, Campus II, 3/2 Street, Ninh Kieu District, Can Tho, Vietnam
e-mail: nkdownes@ctu.edu.vn

© The Author(s), under exclusive license to Springer Nature Singapore Pte Ltd. 2023   133
J. N. Reddy et al. (eds.), *ICSCEA 2021*, Lecture Notes in Civil Engineering 268,
https://doi.org/10.1007/978-981-19-3303-5_10

either by incrementally expanding existing or creating new infrastructures is critical in addressing the socio-economic development goals and visons of the city.

Sustainable infrastructure planning typically makes use of a multi-scale approach, the local level represents the important spatial scale where the tangible connection between the supplier and service provider and that of the consumer or producer, where energy, food and water are consumed and refuse and wastewater are typically generated [5].

This study applies an urban structure (UST) classification scheme to understand the current elements for infrastructure planning and the improved understanding of current challenges and innovations. UST classifications have been previously used in a range of disciplines to facilitate an improved understanding of urban systems, detecting and delineating the built form and the make-up of cities.

2   The Urban Structure Types of Da Nang

Within this study the base map totalled 46,194 polygons. Each polygon was assigned a specific UST code following manual interpretation and classification of UST following the classification scheme developed by Downes et al. [4]. Free, publicly available images from the Google Earth tool were used for this research.

In total, thirty-nine different USTs, were mapped for Da Nang. Figure 1 shows the UST spatially mapped for the central and coastal districts of the city. As with the classification scheme proposed by Downes et al. [4], individual residential USTs were predominately classified based on the dominate building archetypes *i.e.,* buildings with similar or related attributes, the perceived level of planning, and density. The mapping results showed that are as varied in levels of planning, with high density around the urban centre, to more rural open settlement structures in the peri-urban and rural parts of Da Nang (Table 1).

3   Linking USTs to Urban Infrastructure

3.1   Water and Wastewater Sector

Water is essential for socio-economic development. As Da Nang has grown in population, so has the requirements to supply the city with usable water, as well as to treat generated wastewater. The city's water is supplied from the Vu Gia—Thu Bon River catchment and its distribution networks in Da Nang and neighbouring Quang Nam Province. Typically, greater piped distances translate as increased prices of connection, while concurrently, the nonexistence wastewater utilities translate to greater environmental degradation costs. Herewith, the longer-term repercussions of improper planning can be ultimately expensive.

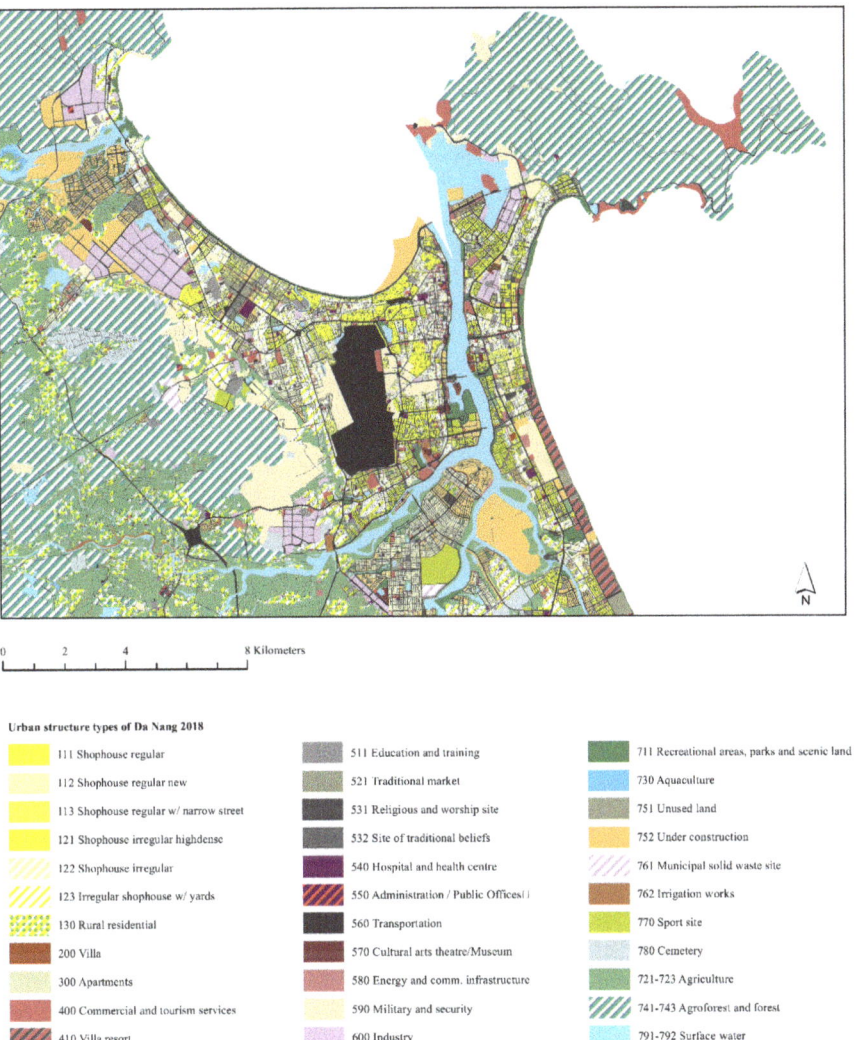

Urban structure types of Da Nang 2018

111 Shophouse regular	511 Education and training	711 Recreational areas, parks and scenic land
112 Shophouse regular new	521 Traditional market	730 Aquaculture
113 Shophouse regular w/ narrow street	531 Religious and worship site	751 Unused land
121 Shophouse irregular highdense	532 Site of traditional beliefs	752 Under construction
122 Shophouse irregular	540 Hospital and health centre	761 Municipal solid waste site
123 Irregular shophouse w/ yards	550 Administration / Public Offices( )	762 Irrigation works
130 Rural residential	560 Transportation	770 Sport site
200 Villa	570 Cultural arts theatre/Museum	780 Cemetery
300 Apartments	580 Energy and comm. infrastructure	721-723 Agriculture
400 Commercial and tourism services	590 Military and security	741-743 Agroforest and forest
410 Villa resort	600 Industry	791-792 Surface water

Fig. 1 The mapped urban structure types of Da Nang

Most residential urban structures types in Da Nang are connected to the water supply system. In the more peri–urban and rural areas, households may often have a small private well, as the main water source or to supplement water from the grid (Table 2). Moreover, in some more remote areas, bottled water is additionally used for drinking purposes or a single connection was seen shared among many households or communities. Bored wells in principle should be registered with the local government, but there are no restrictions or monitoring on withdrawn groundwater volumes.

Table 1 Overview of the urban structure type classification for Da Nang

Utilisation	No. urban structure types	No. of polygons	Surface area (ha)	Percentage utilisation category (%)	Percentage of total Da Nang administrative surface area (%)
Total urban structure types	39	46,194	103,758	–	100
Residential (4 classes)	9	19,209	6,773	100	6.5
Shophouse based-	6	13,492	3,754	70.2	3.6
Villa based-	1	44	22.9	0.2	0.02
Apartments based-	1	70	51.9	0.4	0.05
Rural residential	1	5,603	2,943	29.2	2.8
Commercial land and tourism services	2	775	2,182	100	2.1
Public	9	2,632	2,618	100	2.5
Industrial	1	208	887.7	100	0.8
Green and open space	15	21,417	83,266	100	80.2
Residual transport and surface water area	3	1,953	8,030	100	7.7

3.2 Solid Waste Management Sector

In Da Nang, solid waste management, including, both waste disposal and transfer facilities require space, yet the sectors integration into urban development planning has been secondary. Vetter-Gindele et al. [6] stated that the expansion of urban space, rapid population growth and high urbanisation rates have placed high pressures on the waste collection, transportation and disposal systems of Da Nang. The solid waste sector is outdated, overloaded and not kept pace with the population and spatial development of the city. This reactive planning has led to unstable collection and transportation facilities and mechanisms, and the reliance on outdated treatment technologies. This has led to pollution, site planning issues as well as social and public health issues. Waste generation volumes vary between areas and is highly influenced by the transformation of the city as a whole and the changing demographic of its inhabitants. Solid waste is collected using a verity of means, including hand cart,

Table 2 Connection and interlinkages between observed urban structures and the water and wastewater sectors in Da Nang. Adapted from Downes [5]

Urban structures type	Observed water and wastewater features
Shophouse regular	– Connected to the water grid – Septic tanks with tanks discharging effluent and greywater into existing combined sewage and drainage system
Shophouse regular new	– Connected to the water grid – Septic tanks – Separate or combined systems – Greywater maybe discharged directly to drainage system
Shophouse regular narrow streets	– Connected to the water grid – Septic tanks. Vary degrees of connection of septic tanks to discharge effluent overflow to back-alley sewers or drainage systems. Greywater discharged into drainage systems
Shophouse irregular high-dense	– Connected to the water grid – Septic tanks. Inner city location does not guarantee effluent connection to sewage system as accessibility issues due to building orientations and narrow alleyways. Greywater often discharged directly into drainage system
Shophouse irregular	– Connected to the water grid – Septic tanks located at back of house. Accessibility and connection issues relating to internal narrow alleyways. Greywater often discharged directly into drainage system
Shophouse irregular with yards	– Connected to the water grid – Some private wells – Individual septic tanks, with limited connection to sewage system. Greywater maybe discharged directly to drainage system, or adjacent environment
Rural residential	– Connected to the water grid – Private wells – Bottled water used for drinking – septic tanks systems, pit latrines or double vault composting toilets. Greywater maybe discharged directly to drainage system, or adjacent environment
Villa	– Connected to the water grid – Septic tanks located at back of house, sewers and drainage designed as site and service. Areas may contain separate or combined systems. Greywater maybe discharged directly to drainage system
Apartment	– Connected to the water grid – Large capacity storage tanks – Septic tank, discharging to the drainage system with large dimeter sewer connections. Limited small-scale sewage treatment plants. Greywater maybe discharged directly to drainage system

tricycle, fixed location waste bins, hourly collection services and collection by open
or compactor trucks.

Solid waste in Da Nang is typically collected and transported to transfer stations.
At these facilities, no further sorting or separation occurs, waste is simply compressed
and transported to the end facility, the city's solid waste treatment site. At the time
of writing, 100% of the collected solid waste is landfilled, buried and covered over
with soil (Table 3).

Compactor or older waste trucks are only able to collect waste in areas that have
large streets. Larger streets also allow for fixed collection bin placement. Here the
refuse bins are collected at least once a day. In areas with a lack of space for fixed
bins, a collection scheme is operated, involving waste materials being placed on the

Table 3 Connection and interlinkages between observed urban structures and the solid waste
management sector in Da Nang Adapted from [5]

Urban structures type	Observed solid waste management features
Shophouse regular	– Primary collection by truck – Transfer to treatment site – Space for fixed location collection bins
Shophouse regular new	– Primary collection by compactor or older trucks – Transfer to treatment site – Space for fixed location collection bins
Shophouse regular narrow streets	– Primary collection by hour scheme by trucks, tricycles (pedi-carts) or hand carts – Transfer to transfer facility and secondary transport to end treatment site
Shophouse irregular high-dense	– Limited primary collection access. Collection by hour by pedi-cart or hand cart only – Transfer to transfer facility and secondary transport to end treatment site
Shophouse irregular	– Primary collection by hour scheme by trucks, tricycles (pedi-carts) or hand carts – Transfer to transfer facility and secondary transport to end treatment site
Shophouse irregular with yards	– Primary collection by hand carts, pedi carts and trucks – Transfer to transfer facility or direct to end treatment site – Disposal of waste in adjacent environment
Rural residential	– Primary collection rates vary considerable depending on location – Direct transfer to end treatment site – Disposal of waste in adjacent environment
Villa	– Primary collection by compactor or older trucks – Direct transfer to end treatment site – Space for fixed collection bins
Apartment	– Primary collection by compactor or older trucks – Direct transfer to end treatment site – Space for fixed collection bins

street by residents in front of their homes and being collected within a short space of time, at a specific known time. Often this involves collection by pedi-carts or hand cart, due to narrow alleyways. In more rural and peri-urban areas, residents often bury waste, feed it directly to livestock, undertake composting of the organic elements, burn or dump illegally into the environment.

4 Summary

This paper investigated the spatial configuration of the built environment and patterns of water supply, wastewater and solid waste management in Da Nang, a rapidly developing provincial level city in central Vietnam. The work demonstrates the applicability of urban structure type mapping to gain greater understanding of the selected infrastructure sectors at an planning relevant scale. It is hoped the findings stimulate discourse among the key city stakeholders. Spatially mapped structure types and their main infrastructures provide a clear vision of Da Nang's shortcomings under rapid urbanisation and the requirements towards strategizing effective sustainability oriented cross-sectorial urban spatial development planning.

Acknowledgements This research was supported by the German Ministry of Education and Research (Bundesministerium für Bildung und Forschung, BMBF) within the research project Rapid Planning, grant number 01LG1301B.

References

1. Braun A, Warth G, Bachofer F, Quynh Bui TT, Tran H, Hochschild V (2020) Changes in the building stock of Da Nang between 2015 and 2017. Data 5(2):42
2. Downes NK, Storch H, Rujner H, Schmidt M (2011) Spatial indicators for assessing climate risks and opportunities within the urban environment of Ho Chi Minh City, Vietnam. In: Isocarp (ed) E-proceedings of 47th ISOCARP congress 2011 "Liveable Cities Urbanising World, Meeting the challenge". Wuhan China, Case study platform www.isocarp.net and Congress CD. ISOCARP, The Hague, pp 13
3. Downes NK, Storch H (2014) Current constraints and future directions for risk adapted land-use planning practices in the high-density Asian setting of Ho Chi Minh City. Plan Pract Res 29(3):220–237
4. Downes NK, Storch H, Schmidt M, Van Nguyen TC, Tran TN (2016) Understanding Ho Chi Minh City's urban structures for urban land-use monitoring and risk-adapted land-use planning. In: Sustainable Ho Chi Minh City: climate policies for emerging mega cities. Springer, pp 89–116
5. Downes NK (2019) Climate adaptation planning: an urban structure type approach for understanding the spatiotemporal dynamics of risks in Ho Chi Minh City, Vietnam (Doctoral dissertation, BTU Cottbus-Senftenberg)
6. Vetter-Gindele J, Braun A, Warth G, Bui TTQ, Bachofer F, Eltrop L (2019) Assessment of household solid waste generation and composition by building type in Da Nang, Vietnam. Resources 8(4):171

Model of Catlai Logistic Center—Ho Chi Minh City

Le Thi Bao Thu

Abstract Ho Chi Minh City aims to become the regional logistic center with the advantages of natural and social resources. According to the feasible study of logistic development, there are 07 logistic centers, distributed in the East–West-South-North of this city. In order to reach the goal that by 2025, it will trade 200 million tons of goods per year, Ho Chi Minh City needs to determine the appropriate paradigm for these logistic center and be ready to organize traffic corresponding to above mentioned volumes as well as consequent impacts to the spatial structure of the city.

Keywords e-commerce · Logistic center · Urban fulfilment center

Abbreviations

CBD   Centre business district
GRDP Gross Regional Domestic Product
HCM   Ho Chi Minh City
ICT   Information & Communication Technologies
VLI   Vietnam Logistics Research and Development Institute

1 Introduction

According to the Logistic Development Plan in Ho Chi Minh City, there are 7 Logistic centers in the city, with a total forecasted capacity of 200 million tons of goods per year. Among the 7 planned centers, there is a 2nd class center, which is the existing Cat Lai port area. As a specialized container port, Cat Lai Port accounts for 25%

L. T. B. Thu (✉)
Faculty of Civil Engineering, Ho Chi Minh City University of Technology, VNU-HCM, HCM City, Vietnam
e-mail: thu.bmkt@hcmut.edu.vn

© The Author(s), under exclusive license to Springer Nature Singapore Pte Ltd. 2023   141
J. N. Reddy et al. (eds.), *ICSCEA 2021*, Lecture Notes in Civil Engineering 268,
https://doi.org/10.1007/978-981-19-3303-5_11

market share of inland sea transport, nearly 50% market share of import and export containers of the country and 92% market share in Ho Chi Minh City. According to the Vietnam Maritime Administration, the output of goods through Cat Lai port continued to increase, in 2016 reached 47 million tons, in 2017 over 56 million tons, by 2018 more than 66 million tons. According to the HCMC Department of Transport, only in the first 6 months of 2019, the volume of goods arriving at this port has reached 58.8 million tons and is expected to deeply destroy the output in 2018 and exceed the plan by 2020 of 37 million tons per year.

Cat Lai port's cargo volume growth corresponds to the growth rate of industry in the Ho Chi Minh City area (including 7 provinces and cities) in the past 5 years. The port's growth outpaced the planning forecast, leading to disparities between the traffic volume and the capacity of the transport infrastructure, resulting in congestion from the port extending as far as Hanoi Highway out almost every day.

With the plan of becoming a Logistic Center by 2025 [3], Cat Lai is oriented not only as a logistics service center, but also as an international trade center. The issue on discussion is that the specific conditions of CatLai will be exploited efficiently in the model of Logistics service center or other model, especially in the context that CatLai now locates inside Thu Duc city, a leading part for developing of HCMC.

2   Research

2.1   Overview

2.1.1   Literature Review

a. Logistic Center
In the logistics industry, there are many terms when referring to the concept of logistics center. However, in general, these terms do not differ much in terms of interpretation, of which the most used are: logistics centers, logistics distribution centers, freight villages, distribution centres, freight distributon centres, logistics hub, logistics clusters, logistics park, logistics nodes, logistics zones, cargo shipment centers, gütervekehrszentrum-GVZ, distripark, central depot, depot base, distribution-storage center, interporti, platformes logistiques.

- M. Krzyzanowski gives a definition: logistics center is a multimodal transport terminal where operations take place exploiting cargo flows for different transport units, serving markets at regional; national and international level. In the view of M. Krzyzanowski, the main specific functions of a logistics center are: transportation, handling, warehousing, classification and labeling [6].
- André Langevin and Diane Riopel again believe that logistics center is a point element of the logistics network system, playing an important role in supporting the flow of goods, warehousing, handling, recycling; freight collection, handling shipments [1].

b. Urban Fulfillment Center

Urban consumers are increasingly using their phone, and consequently changing traditional retail. E-commerce emergence with innovation of technology in driving urge to think of urban node network, where fulfilment center can respond them in era of "see now-buy now-how fast I get it" [2].

2.1.2 Ho Chi Minh City in the Vision of an Important Logistic Center by 2025

Ho Chi Minh City is geographically located as the center of Southeast Asia, a gateway for international exchange with the advantages of river, sea, land, and air.

Ho Chi Minh City has the central location of the main production regions in the South: the Mekong Delta—agricultural production, the Southeast region of industrial production. Meanwhile, Ho Chi Minh City is also a major consumer market of the country with a population of about 13 million people, including more than 9 million local residents and 4 million non-residents. In the 2016–2019 period, the City GRDP increased by an average of 8.3% per year; the contribution to the country's GDP increased from 21.9% in 2015 to 22.3% in 2019.

According to the Chart [Fig. 1], service accounts for the largest proportion in the GRDP structure of the City and the service industry contributes more than 61%, of which transportation and storage accounts for 80.4% of the country.

According to the development orientation of Logistics services in Vietnam, among 18 (1st and 2nd class), and specialized aviation centers throughout Vietnam, there are 02 2nd class Logistic centers for the HCMC area and neighboring provinces. (According to the plan approved in Decision 1012 /QD-TTg) [5].

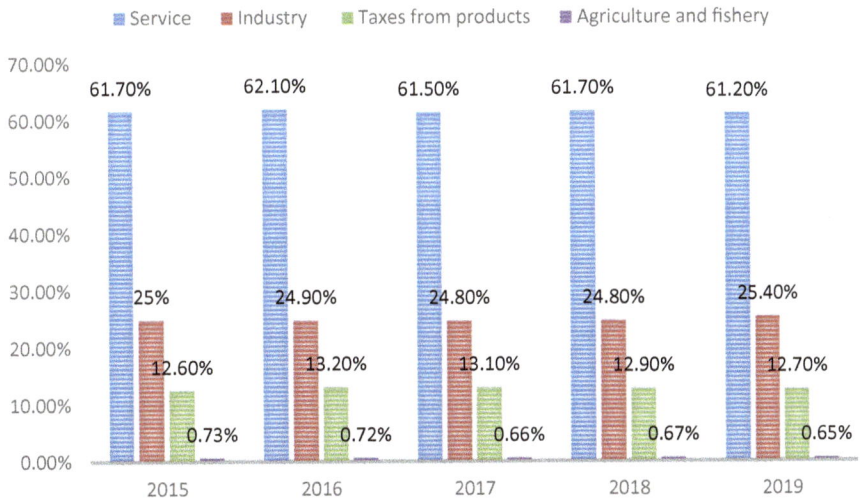

Fig. 1 Proportion of economic sectors of HCMC in the 2015–2019 period

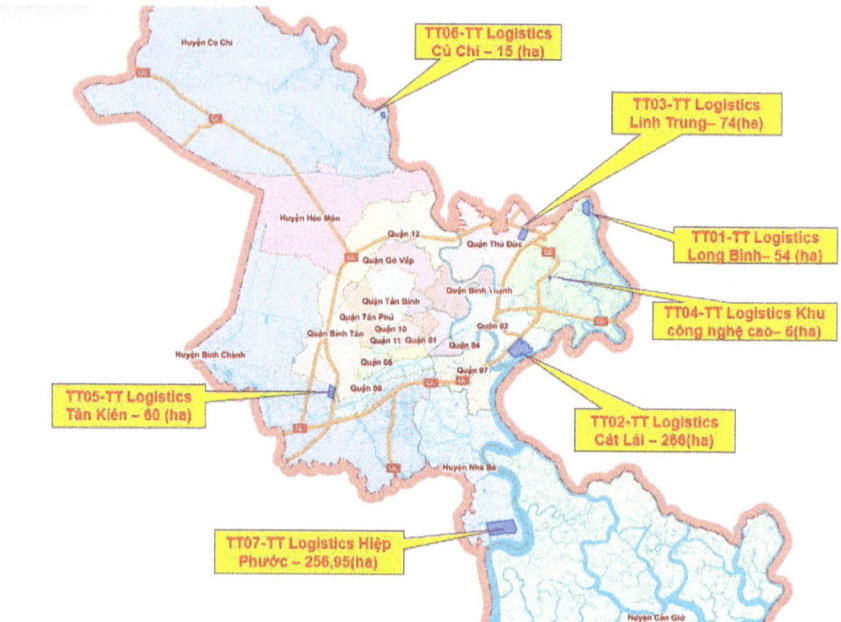

Fig. 2 Location of logistic centers of HoChiMinh City *Re: VietNam Logistics Institute Report (2019)*

According to the Logistic center development project of Ho Chi Minh City, there are 07 Logistic centers located in the East - West - South - North directions of Ho Chi Minh City, corresponding to production conditions and activities of each locality [Fig. 2].

In those, the existing Cat Lai port will develop as 266 ha 2nd class Logistic center, that serves the region of Ho Chi Minh City and its southward provinces.

2.1.3   Current Conditions of Cat Lai Port

Cat Lai port and Hiep Phuoc port are two transportation hubs receiving all domestic ports of Ho Chi Minh City relocated out of the inner city (District 4, District 1) since 2010.

From Cat Lai and Hiep Phuoc, 88% of import and export goods through the port are transported by road, 12% are transhipped by inland waterways.

According to statistics of Traffic Police at Cat Lai Station 2019, the average traffic volume of vehicles in and out of Cat Lai port is 19,000–20,000 turns per day and night, especially up to 26,000 vehicles in some days, as much as dual capacity of Nguyen Thi Dinh street- the sole route leading directly to the port. The traffic congestion extends from Cat Lai port to the end of Nguyen Thi Dinh street, cuts through the Mai Chi Tho axis-connecting to southward of the city and extends to Hanoi highway, causing congestion on the axis connecting to northern provinces of Ho Chi Minh City.

2.2 Research

2.2.1 Evaluate the Conditions to Build Logistic Center at the Location of Existing Cat Lai Port

– Adjacent entity: Soai Rap river; Nguyen Thi Dinh street, Vo Chi Cong street

Cat Lai port is the intersection point of key economic areas, with a channel connecting to CaiMep—ThiVai deep-water port cluster with a receiving capacity of 200,000 Teu ships. These are favorable conditions to form Logistic Center according to the model of an international trade center—logistics—a link in the global supply chain.

– Legal:

+ Decision 34 /QD-UBND dated 24/01/2002 on approving the detailed planning project of Cat Lai Industrial Port in Thanh My Loi Ward, District 2, in which approving the total land area of 66 ha for construction building a medium-sized general trading port including: bulk cargo, container cargo, grocery, agricultural and forestry products … with the task of directly serving import and export goods in Cat Lai Industrial Park and other industrial zone, in northeast of the city and from proximity provinces as Dong Nai and Binh Duong.

+ Decision 6707/QD-UBND dated December 29, 2012 approving the revision of general construction planning project of District 2 to 2020, scale of 1/5000: Since 2020, 20.52 ha for the function of local handicraft is shifting to trade—port services [7].

– Connection system between Cat Lai Logistics Center and main road areas:

Currently, the road system connecting to Cat Lai port can afford the requirements for container trucks to transport goods. However, the expected connection route from Binh Duong and Dong Nai still depends on the sole route Vo Chi Cong—Nguyen Thi Dinh, so congestion often occurs.

In addition to the simple functions such as warehouse, transportation, packaging … Cat Lai is intended to develop as commercial center (combining market and exhibitions or wholesale shopping centers) in the Southern economic zone of VietNam and the global scale. Therefore, the convenient location of Cat Lai when connecting to the inner city of Ho Chi Minh City is a premise for Cat Lai Logistics Center to develop in the future.

Proximity transportation of CatLai as follow:

Cat Lai locates along Vo Chi Cong Street, Vanh Dai 2, that is convenient to connect with sources of goods from adjacent areas (i.e. Industrial park with the role of the supply source) and with the inner city of Ho Chi Minh City (with the role of the consumer market) to develop Cat Lai as a trade-logistics center.

Cat Lai connects with Nation highway 1A via 40 km of Ringroad 2 road, then connects to Ringroad 3 in the future. The main road will include 10 lanes, 67 m planned width of road, is the infrastructure system capable of transporting goods.
− System connecting TT with economic regions:
+ Dong Nai, Binh Duong, Binh Phuoc: From Cat Lai to Dong Nai, Binh Duong via Mai Chi Tho Street and Hanoi Highway, are two roads with 10 lanes.
+ The Mekong Delta region: From Cat Lai to the Mekong Delta through Ringroad 2 and National Highway 1A.
− Waterway traffic connection system:
Ho Chi Minh City: Cat Lai is located adjacent to Soai Rap river system, convenient for exchanging goods with Dong Nai port, Hiep Phuoc Port (Ho Chi Minh City) or going further to Ba Ria—Vung Tau at Phu My—Cai Mep port cluster by route: Dong Nai river—Soai Rap river—Long Tau river—Thi Vai river.
The Mekong River Delta: Cat Lai connects with the Mekong Delta provinces by the route: Soai Rap River—Vam Co Dong River, Vam Co Tay …
− Connection system between the center and industrial parks: Convenient to connect with Cat Lai Industrial Park, Nhon Trach Industrial Park, Long Binh Industrial Park, Song Than Industrial Park [Fig. 3].

Situation of transportation in existing capacity of CatLai port:
Traffic from My Thuy intersection to Ba Cua bridge (on Vo Chi Cong street leading to Nguyen Duy Trinh street to Phu Huu port) frequently occurs traffic jams at the intersection of many traffic flows from the inner city to the Long Thanh—Dau Giay freeway. [Fig. 4].
Nguyen Duy Trinh road, leading to SP ICT port and Phu Huu new Port, has only 2 lanes, so traffic speed is slow.

2.2.2 Case Study: Logistics Thailand and Development Opportunities from Thailand 4.0 Program

According to the World Bank's LPI ranking in 2018, Thailand reached 32th rank in 2018 (from 45th in 2016), 2nd in Asean, ahead of Malaysia and ranked 7th in Asia.

Thailand has invested primarily in transport infrastructure as declared in the 12th five-year national socio-economic development plan (2017–2021) to cut logistics costs of the country to 12% of GDP by 2021 (This fee is 14% in 2016). According to this plan, Thailand not only develops logistics infrastructure in big cities but also but also improves logistics connectivity with neighboring countries.

The Thailand 4.0 program introduces a new economic model for this country to develop its strengths in the digital economy, thereby also creating a huge change in the logistics sector.

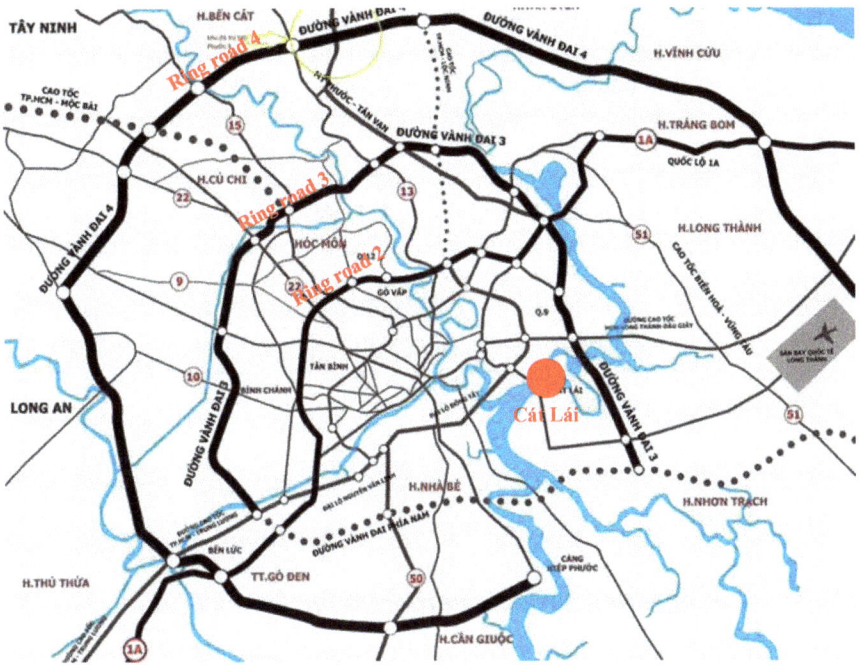

Fig. 3 Map of Ringroad system of HCMC *Re: Department of Transportation of HCMC*

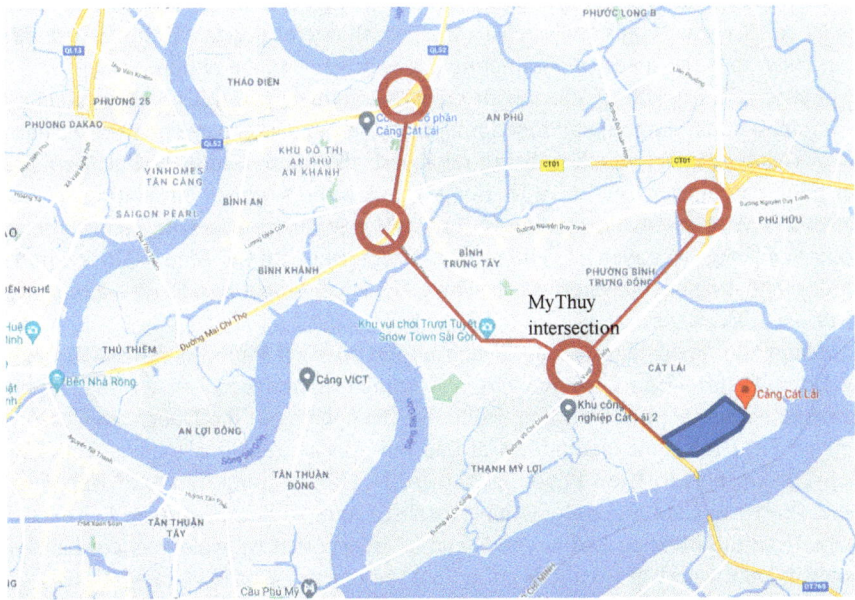

Fig. 4 Congested points/routes on CatLai's Proximity transportation system Re: Author's on resource of GoogleMap

The thriving e-commerce has created a demand for logistic space and has brought about significant changes in the supply chain and logistics operations in Thailand, especially in the warehouse and last-mile delivery segments. Logistic centers, all scale warehouse networks distributed throughout Thailand are gradually promoting efficiency.

The development of the ASEAN economic community (AEC), the location of a transport hub of Thailand in Greater Mekong Sub-region (GMS) is being strengthened. The Thailand 4.0 Initiative has increased Thailand's opportunities for cross-border transactions and imports and exports. [4]

HoChiMinh city with population of 13 millions, on the way to e-commerce lifestyle, should prepare infrastructure as a warehouse and last-mile delivery for a fulfilment center. Thailand experience is a further scenario for CatLai beside approved plan as a logistics center.

3   Discussion

3.1   Discussing about suitable paradigm for Cat Lai

(1) Scenario of Logistics center: Cat Lai needs to develop a complex of freight transport types, including roads for containers, railways, and waterways. With 200 million tons of cargo traffic per year, Cat Lai will receive about 75,000 container trucks per day. According to forecasts, the growth rate of goods will be 5–10% per year, faster than the speed of completing traffic projects in the region.

In case of a Logistics Center with its core activities, Cat Lai needs to connect with international outgoing and incoming cargo terminals, in particular the air hub, Long Thanh Airport, which is being deployed; the main seaport is Cai Mep-Thi Vai port. To connect Cat Lai with Long Thanh airport, goods circulating on the route East–West axis—the existing HCM City-Long Thanh-Dau Giay freeway; and Dong Van Cong—Nguyen Thi Dinh route. To connect with Cai Mep—Thi Vai port, goods circulating on the waterway are Dong Nai River—Soai Rap River—Long Tau River—Thi Vai River.

Meanwhile, goods from Long Thanh airport to Hiep Phuoc port pass through the Long Thanh—Dau Giay freeway and branch off the Ben Luc—Long Thanh expressway; not clogged by urban traffic like the opening of the Dong Van Cong—Nguyen Thi Dinh route leading to Cat Lai. For goods circulating by sea, from Cai Mep-Thi Vai port to Hiep Phuoc port, there is Thi Vai-Long Tau-Soai Rap river route, shorter to Cat Lai; or along the coastal sea route to Soai Rap river. For export goods from the Mekong Delta, goods can be transported by water via coastal sea routes, inland waterways without going into the inner city of Ho Chi Minh City or by expressway Can Tho-My Thuan-Trung Luong. -Ben Luc-Long Thanh [Fig. 5].

As a 2nd class Logistics center, CatLai would face some matters:

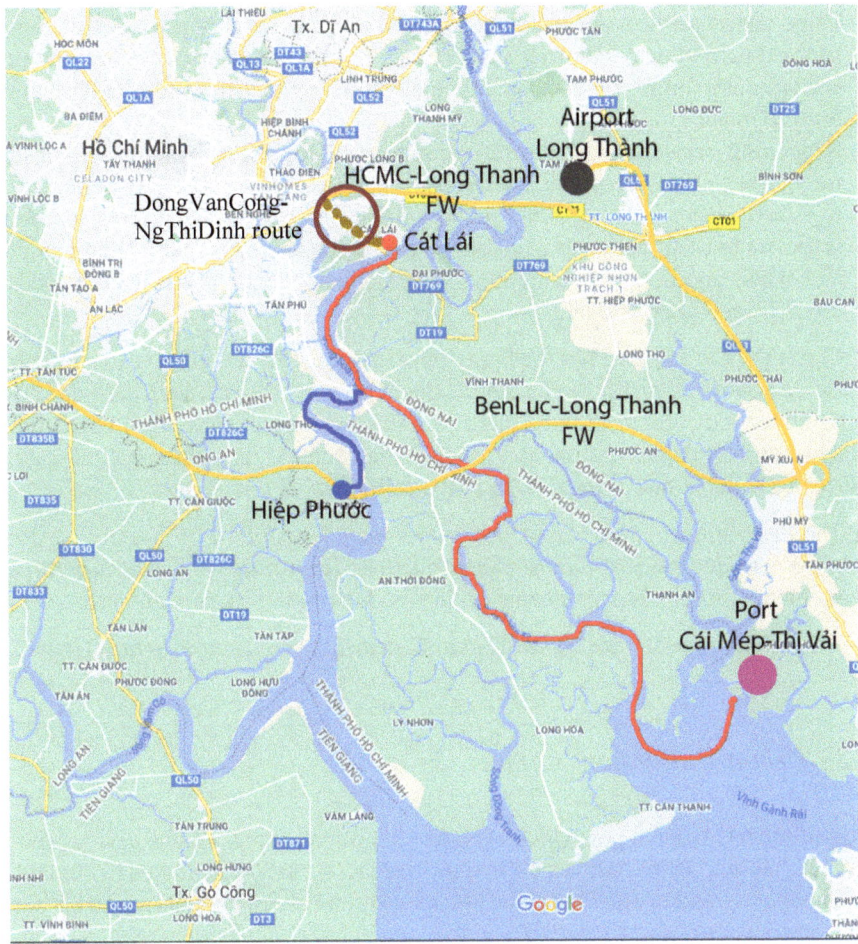

Fig. 5 Comparison between Cat Lai and HiepPhuoc port in connecting with seaport CaiMep and airport Long Thanh Re: Author's on resource of GoogleMap

– Requirement of railway for goods transport: cannot be provided in the time, because there is no plan of railway in the region of CatLai Logistics Center so far.
– Solution for traffic congestion: CatLai locates inside ThuDuc city, a form of innovative center of HCMC, where concentrating million of inhabitant to and from every day. Authorities ought to issue proper solution for the circulation that is somewhat serious now.

(2) Scenario of International trade center: Cat Lai needs relevant infrastructure and services of an industrial—international trade logistics center, specifically:

– The core services of logistics center include: forwarding, warehouse services, dock-yard services.

- Value added services include: organization of trade support activities such as national exhibitions, industrial support fairs, goods trade fairs; organize annual seminars and forums on the topic of trade, improving logistics and supply chains from Vietnam.
- Supporting services for industrial—commercial logistics: operating space of foreign businesses in the field of consumer goods, e-commerce (e-commerce logistics); commodity trading floor, entering into deals, receiving orders, processing orders, acting as a link in the global supply chain, distributing import and export goods, participating in global e-commerce with manufactured goods domestic and international.

Currently, Cat Lai area has the basic logistics infrastructure to serve exhibitions and trade of goods, which are industrial clusters adjacent to the port and a port capacity of 50 million tons/year. In the area, there are residential and urban areas with facilities suitable for international transactions, such as Pho Dong Villa, Citihome, CitiGrand. Proposing suitable model for Cat Lai port conditions.

- The international cargo transaction center includes: + Port; + Warehouse for goods introduction, inspection, implementation of transaction procedures, product promotion; Exhibition, conference complex, financial services, insurance, …

As an international trade center, CatLai has essential advantages to be effective.

3.2   Urban Fulfillment Center

Ho Chi Minh City has a population of about 13 million people. Mobile device penetration is close to 100%. Almost 90% of ICT businesses are concentrated in this city. Consumption trend of Ho Chi Minh City residents develops in the direction of "see now-buy now-How fast I get it". [2]

Location of Cat Lai has convenience in both sides: goods stock and delivery hub or Urban distribution hub: a commodity hub serving a market of 13 million people in e-commerce. This advantage of Cat Lai should be utilized in the context that Ho Chi Minh City is a city of a digital economy.

Cat Lai is located at both convenient access to the supply of goods and convenient delivery style to serve the needs of the market in e-commerce markets.

In particular, Cat Lai belongs to Thu Duc city, a newly formed institution with expectation of bringing an income of 19,000 billions VND to Ho Chi Minh City through digital-based economic activities. Citizens of Thu Duc City and CBD of Ho Chi Minh City are potential customers to make paradigm of Cat Lai fulfillment center possible.

4 Conclusion

Cat Lai is planned to be 2nd class Logistics center but it seems waste potential power of the area, that is resources of an innovative center of Ho Chi Minh City.

In terms comparing between Cat Lai and Hiep Phuoc—two port centers of Ho Chi Minh City, it is necessary to approach the concept of an international supply chain consisting of two parts. One is having core infrastructure for international freight forwarding; the second part has core infrastructure for commercial transactions. Therefore, the center of international supply chain in Cat Lai needs to separate the freight forwarding infrastructure with a capacity of 200 million tons per year by 2030 and transfer it to Hiep Phuoc. At that time, the problem of traffic jam caused by 75,000 container trucks to and from Cat Lai on the Dong Van Cong—Nguyen Thi Dinh axis will be completely solved. Cat Lai, meanwhile, with available infrastructure along with proximity to "Vung Tau—HCMC's CBD—Thu Duc City" has been upgraded to serve operations of an international trade center and an urban fulfillment center, consequently gain added value, high growth, in line with the direction of Thu Duc City—an institution being applied to properly exploit potential of the East area of Ho Chi Minh City.

With Cat Lai's location in Thu Duc city and Ho Chi Minh city, it is necessary to consider paradigm of an international trade center and urban fulfillment center to meet the demand of e-consumption.

References

1. Langevin A, Riopel D, Systems L (2005) Design and Optimization. Springer, Boston
2. Bimschleger C, Batel K (2019) Urban Fulfillment center - Helping to deliver on the expectation of same-day delivery, Deloitte Development LLC
3. Logistic Report (2019) https://moit.gov.vn/VietNam
4. https://www.mordorintelligence.com/industry-reports/thailand-freight-and-logistics-market
5. http://vlr.vn/logistics/news-1408.vlr
6. Meidute-Kavaliauskiene I (2010) Comparative analysis of the definitions of logistics centres
7. VLI (2109) Project to develop the logistics industry in the city to 2025, with a vision to 2030

Promote the Value of Planning and Landscape Architecture Under the French Colonial Period in Ba Vi National Park Towards Sustainable Tourism

Nguyen Viet Huy, Nguyen Quoc Thong, Nguyen Duc Vinh, and Pham Thuy Linh

Abstract Ba Vi Mountain offers a pleasant climate and a beautiful pristine landscape. Until the arrival of French colonists 90 years ago, the natural features and good environment of the Ba Vi mountain area were exploited for retreatment purpose. The environment and French architecture in the area have been devastated as a result of historical changes and neglect. Those ruins and landscapes must be repaired and developed in order to meet the necessities of current living. Later forest management legislation and tight land use restrictions, on the other hand, are limiting the growth of tourism services in the region. As an outcome, the research will actively contribute to our understanding of national culture, as well as educate young people about nature, social responsibility, and especially environmental conservation.

Keywords French colonial · Architecture · Ba Vi · Tourism · Sustainable

1 Introduction

Ba Vi is a magnificent mountainous place where the tale of Duc Thanh Tan Vien, one of Vietnam's "four immortals," was born. Aside from its historic and spiritual significance, Ba Vi Mountain offers a unique natural resource, a moderate temperature, and a diversified landscape. The French colonists arrived 90 years ago and explore the area's wonderful natural conditions. They discovered that it is an ideal location for recreational activities and resort development. As a result, they created

N. V. Huy (✉) · N. D. Vinh
Faculty of Architecture and Urban Planning, National University of Civil Engineering (NUCE), 55 Giai Phong Street, Hai Ba Trung District, Hanoi, Vietnam
e-mail: huynv@nuce.edu.vn

N. V. Huy · N. Q. Thong · N. D. Vinh
Association of Vietnam Architecture, 40 Tang Bat Ho Street, Hai Ba Trung District, Hanoi, Vietnam

P. T. Linh
Hanoi Architural University (HAU), Km 10, Nguyen Trai Street, Thanh Xuan District, Hanoi, Vietnam

J. N. Reddy et al. (eds.), *ICSCEA 2021*, Lecture Notes in Civil Engineering 268,
https://doi.org/10.1007/978-981-19-3303-5_12

a Ba Vi-based resort with five distinct elevations: 400, 600, 700, 800, and 1000 m. The design has served as a good model for urban planning and resort development in Vietnam. After the French colonialists left, the structures they left behind were abandoned, deteriorated, and overgrown with a variety of flora. The Master Plan for Conservation and Sustainable Development of Ba Vi National Park in 2010 as well as the Detailed Plan of the administrative and service subdivision I—Ba Vi National Park in 2014 was approved by the Ministry of Agriculture and Rural Development.

Currently, the area is located within the boundary of Ba Vi national forest and is managed directly by a technology business company to rebuild a high-end resort here. In the form of ruins, the works are identified as relics that need attention and special landscape highlights. The commencement of the construction of new resorts in the same area to harmonize and co-develop with the ruins should be carefully considered. The first constructions built in 2008, in very limited quantities and natural materials, are the right first steps. However, when the resort entered phase II with room demand increasing from 55 to 100 rooms, previously unresolved issues emerged and posed challenges to the area. Therefore, in order to simultaneously develop the develop historical, cultural and ecotourism site in Ba Vi National Park, the authors have researched and outlined a number of principles to preserve and promote the value of architectural ruins and kb in the area. of Ba Vi National Park.

2 Compare Construction Planning in the French Period with the Current Construction Planning.

2.1 French Construction Planning

On the basis of studying the master plan and detailed planning drawings of the Ba Vi tourist area during the French period, consulting archives, and surveying the remaining architectural ruins, it can be concluded that the French conducted a thorough study of the topography, climate, and landscape values of the Ba Vi tropical forest before designing the master plan. In proposing exploitation of specific sites and landscapes, the master plan of 400, 600, 700, 800, and 1,000 m heights clearly displays respect for topography and landscape. Simultaneously, the planning attached precise management requirements that must be followed by building works in the planned area. For example, the building must fulfill the art requirements and the regulations for public health protection set forth in the Governor of Tonkin's Decree on August 7, 1929; the major works must not exceed 1/5 of the land lot area; setback 6 m; the fence must be sparse, no more than 1.5 m high, and so on.

According to planning papers from the French period (1914–1951) relating to the Ba Vi region, National Archives Center I, the master plan area was 272.5 hectares until 1951, with the following components: Essence of 400 m was 196 ha, including 29 construction works; The height levels of 600–700 m, 55 ha area is relatively flat with 82 construction works; the height level of 800 m area is 3.5 ha, including 16 construction works; the height level 1000 m area is 18 ha planning with 34 projects (unconstructed) (Fig. 1).

Current Construction Planning

Two plans have been developed: Ba Vi National Park conservation and sustainable development plan (2010) and the administrative and service subdivision I - Ba Vi National Park detailed plan (2014). The initial works were planned and built on the levels of 400, 600, and 700 m based on these two plans (Fig. 2).

In general, the current plannings respect the master plan of the French period. Specifically, Ba Vi National Park is preserved; keep the main position is the resort tourism function with the main subdivisions and routes; conservation of architectural ruins and landscapes serving cultural and historical discovery tourism; construct limited and harmonious with architectural and landscape ruins for cultural tourism and relaxation. Especially, the first works were planned and built on the levels of 600 and 700 m of phase 1 of Melia Bavi Mountain Retreat for the purpose of resort

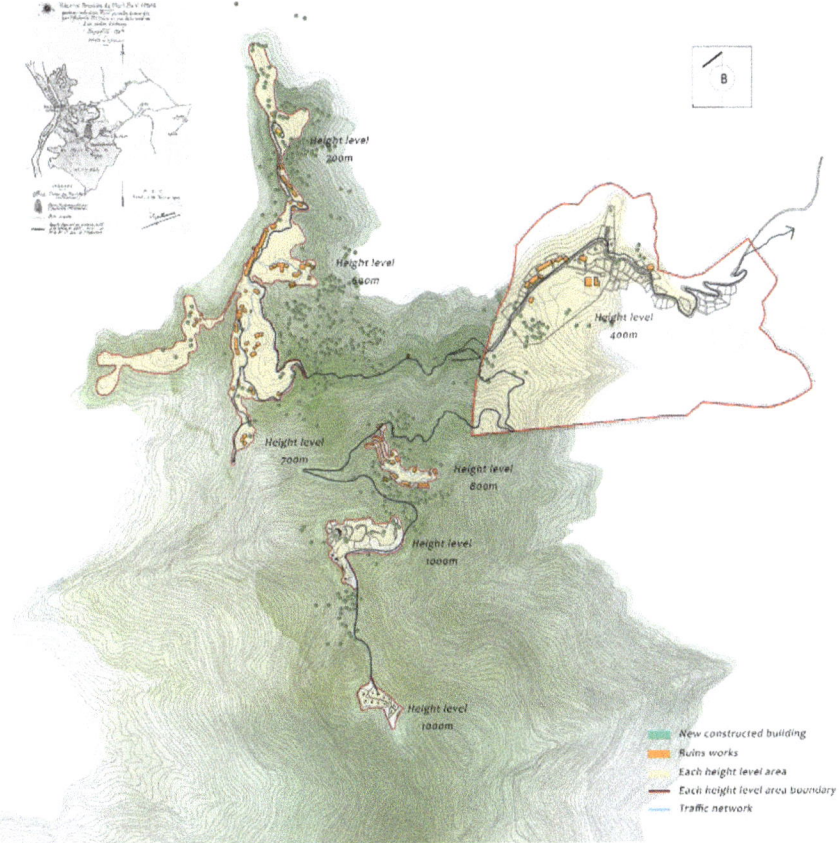

Fig. 1 Map of French planning in Ba Vi (1914–1951), redrawn based on French documents [2]

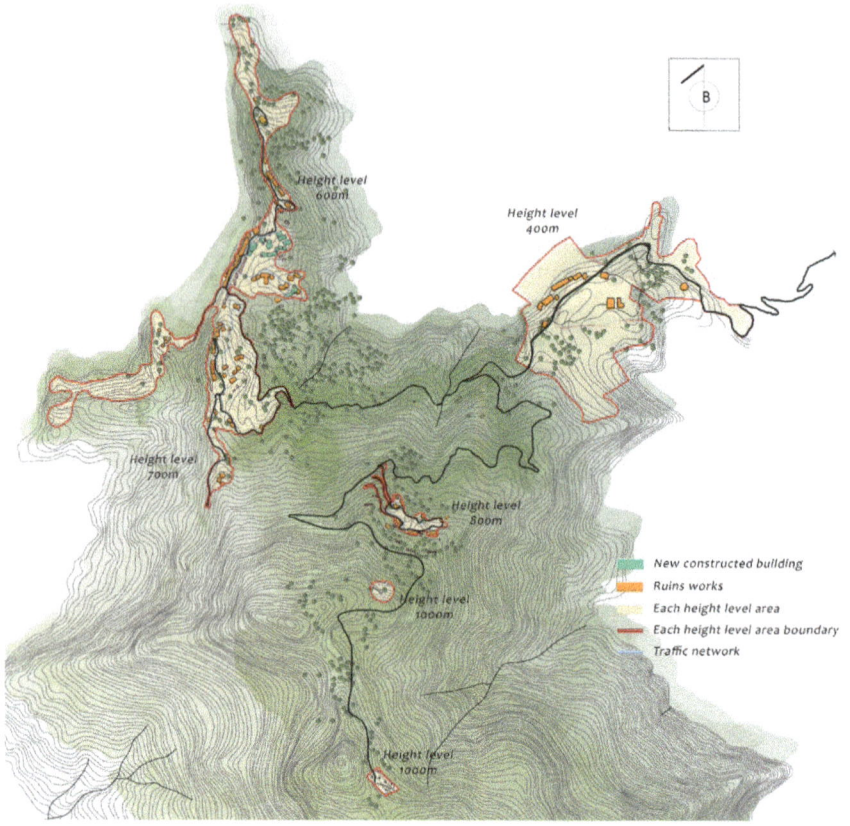

Fig. 2 Site master plan at pillars 600, 700, 800 m of administrative and service subdivision 1, Ba Vi National Park [4]

tourism and discovery that are considered successful because they have created. a new architectural complex with scale and architectural language in harmony with the landscape and on the basis of respecting and rationally exploiting the values of architectural ruins and landscapes (Fig. 4).

3   Current Status of Architectural and Landscape Ruins

Survey results of the current state of architectural and landscape analysis on the 400, 600, 700, 800 and 1,000 m levels in the Ba Vi National Park area show that there are still 127/161 works planned in the French period in the form of traces. Architectural monuments and ruins. On the 600 m level, there are architectural ruins of the main works, such as: 1-storey helipad with a ground area of 759 m²; The holiday home (girl's camp) with its ground and part intact on an area of 442 m².

On the 700 m level, there are remains of works: Colonel house, quite large scale in location with beautiful view, with intact ground size 26 × 46 m on area of 1196 m^2; The lieutenant colonel's house still has its foundation intact, the wall covering the size of 11 × 14 m.

On the 800 m level is the main work located in the center of the area is the Stone Church. The building still has a foundation with the size of 11 × 19 m and a wall of stone with the back wall still intact a large cross cut across the wall, clearly showing the shape of the old church building. Currently, the building is hidden in the forest with mossy stone walls, sunk in permanent fog, occasional or natural light shining through the cross, giving the church space a surreal, fanciful feeling full of charm (Fig. 3).

In the context of sustainable development, conservation is abiding by the law and serving the public good. It proposes solutions based on the rational application

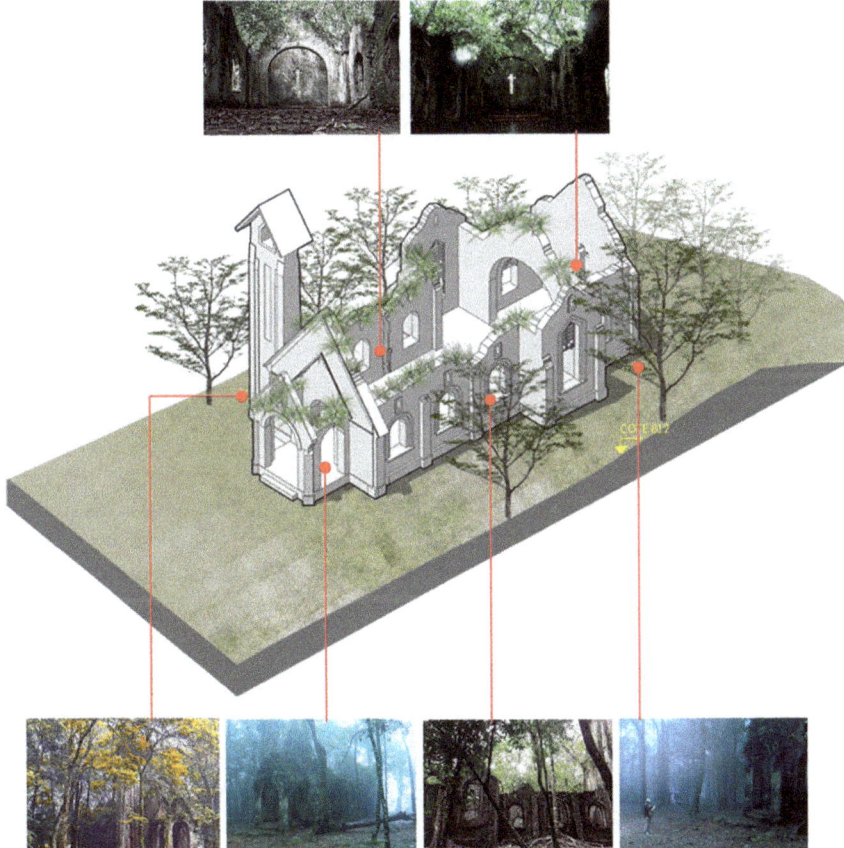

Fig. 3 3D perspective and current status of the building - Stone Church (*Source* Student from National Unviversity of Civil Engineering)

of intervention principles, which are suited for the location, features, significance, and present state of architectural ruins and landscapes in each conservation and development region. To promote the worth of architectural ruins, landscapes, and intangible cultural assets in the Ba Vi mountain area, in order to satisfy modern cultural and social demands while without disrupting or negatively impacting the ecology, natural scenery, or intrinsic history of Ba Vi's holy territory.

4   Principles of Preserving and Promoting the Value of Architectural Ruins and Landscapes

There are several intervening principles in the planning design and architecture of Ba Vi resort to conserve and develop the value of architectural ruins and landscapes for tourism requirements. The most essential intervention idea in the master plan is to link the ruins in order to enhance conservation efficiency, increase the value of architectural ruins and landscapes, and best satisfy tourism demands. On several levels, architecture and landscape create a system (Fig. 10). In terms of architectural design, with the goal of preserving and developing the value of architectural ruins, the landscape serving tourism basically has many distinct intervention principles, and it is possible to combine them depending on the individual architectural solution. The following are the primary ideas for conserving and developing the value of Ba Vi resort's architectural ruins and landscapes for tourism purposes:

Principle 1: Combination
With ruins in a vast area that are the area's attraction but have few traces remaining and historical architectural value that is not really typical, it's important to bring the ruins back to life, for example by catering to the demands of resort tourists. However, it is critical to limit the use of incompatible building with the natural landscape and to conserve the ruins' distinctive, identifiable elements in new architecture, resulting in a harmonious blend of new and old. In reality, the Melia Bavi Mountain Retreat phase 1 outdoor swimming pool and restaurant area is the finest witness to principle 1, which preserves the past, harmonizes with the nature, and efficiently fulfills current travel demands (Fig. 4).

Principle 2: Conservation
Impacts of less than 5% on ruins or landscapes may be authorized to maintain the quality of the building in Ba Vi National Park in order to conserve architectural ruins and historic landscapes. Architectural ruins and landscapes were chosen to conserve the existing state as historical legacy, a landscape component from another era. Those are building ruins that have spent time in organic symbiosis with trees, producing a unique, appealing legacy that is certainly an effective tourist, leisure, and exploration destination. Simultaneously, the growth of digital technology might help to completely recall the recollections of certain regions and places. The stone church on the 800 s, erected by Father Paul Seitz between 1932 and 1940, is an example of

Fig. 4 Melia Bavi Mountain Retreat outdoor swimming pool phase 1 [3]

preservation work. The church blended with the plants and the surrounding landscape into a single body over time, and the Church's supremacy and mystery seemed to grow. Despite the fact that no intervention has been made, the church has become a well-known tourist destination (Fig. 5).

Principle 3: New Construction

This is the allowed construction of new structures alongside architectural and landscape ruins. This is a contentious notion in theory, but it is necessary and acceptable in reality in order to effectively promote cultural values, not to leave heritage behind but to provide appropriate circumstances for heritage to fulfill the requirements of current society. New architectural works, on the other hand, should not detract from the value existing architectural and landscape ruins, but rather enrich them. Many sites on the 600, 700, and 800 m heights are eligible for application based on this principle, such as: building new villas next to the preserved ruins; the pine hill area is on the 600 m level (French kindergarten), with enough area and space to build a new high-class resort next to the remaining ruins (Fig. 6).

Principle 4: Restoration

Reconstruction (or re-engineering) of structures from architectural and landscape ruins allows for a 20 to 80% reduction in their impact. Rehabilitation is a strategy for attracting tourists while also creating a feeling of restorative, historical, and educational importance. The restoration is based on the following principles: Overall oriented planning to conserve and promote the value of Ba Vi National Park's architectural ruins and landscapes; on-site evaluation and reliance on credible sources. The colonel's ruins are one of the works that need to be recovered in Ba Vi National

Fig. 5 The stone church on the 800 m height. (*Source* Authors)

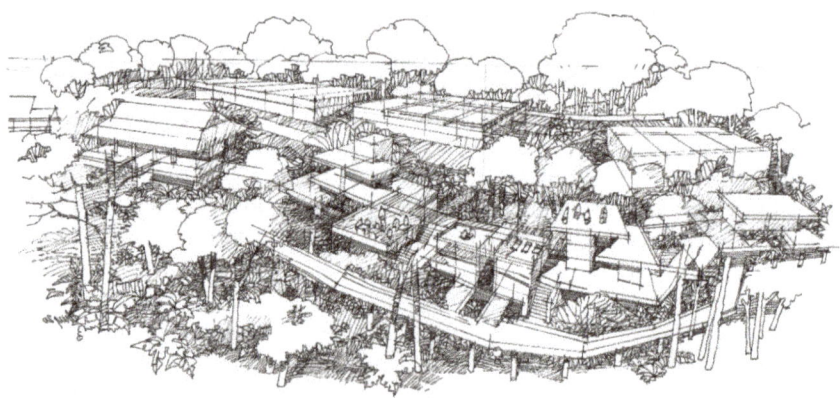

Fig. 6 Construction of a high-class resort in the pine hill area (French kindergarten) at 600 m level [7]

Park, according to French records given by the National Archives Center and the results of the current status study in 2020.

Principle 5: Landscape Conservation and Embellishment
Ba Vi Resort is full of benefits that are difficult to obtain in other resorts surrounding Hanoi. It is located in the conservation area of Ba Vi National Park, which has a diversity of terrain and environment with animals, plants, and a moderate temperature. As a result, in order to build a resort, the national park's current natural landscape must

first be subjected to stringent conservation in order to conserve biodiversity and to educate the ecological environment via travel, experience, and exploration. This is a large-scale landscape with a diversified tropical plant cover, including several varieties of forest trees, lawns, and so on, on various terrain heights, with landscape vistas stretching to the plain. Furthermore, various types of plants have damaged building ruins in the resort throughout time. The symbiotic architectural ruins integrated with trees have created a whole of the organic, complete, and unique architectural landscape that attracts tourists till now. It might be stated that this is a portion of the limited landscape with added value that successfully serves tourism, therefore it is critical to investigate and utilize in the direction of landscape conservation and enhancement.

When used in a tourist resort, this principle allows for the active utilization of the natural landscape's worth, as well as a key role in environmental education. At the same time, it not only increases the amount of green cover in Ba Vi National Forest, but it also promotes biodiversity and its worth. The use of streams and a new landscape lake for the Melia Bavi Mountain Retreat phase 1 complex is a successful example of nature-based landscape conservation and enhancement.

5 Conclusion

The attitude and investment of Melia Bavi Mountain Retreat Phase I towards green, ecological, and humanity, on the basis of conservation and promotion of values, can be observed in the process of surveying and analyzing the existing condition. Nature, culture, and history in the Ba Vi National Park region are all valid and should be protected. This is a viable approach to promoting both the worth of valuable resources and the balance of natural ecosystems and humanities in the Ba Vi National Park region. As a result, encouraging and facilitating wider deployment is critical.

Sustainable development for cultural and ecological resort tourism is a major driver of economic growth and employment creation, including jobs for locals. Encourage new growth without disrupting the environment's biological balance, while also enhancing the natural ecological worth of Ba Vi National Park and boosting Ba Vi's cultural worth. All actively participate in the study process and knowledge of national culture, as well as in educating citizens about their love of nature, the country, and their feeling of civic responsibility for social activities, particularly security and the protection of the living environment.

References

1. National Archives Center I (1914–1951) Archives related to the planning and architecture of the French Ba Vi region
2. Technology Development Company Limited (2008) Research papers on Ba Vi area; Collected materials: Planning drawings, architecture, characters and history related to the French Ba Vi Resort

3. Trung Nghia Construction Design and Survey Company (2010) Topographic measurement map, scale 1/500 of the Administration and Services Division I
4. Technology Development Company Limited (2014) Map of Melia Ba Vi Mountain Retreat project master plan
5. Baschet E Les Grands Dossiers de L'ilustration L'Indochine. Histoire d'un siècle 1843–1944. Ed. Le livre de Paris
6. Thuy VH The sketches of architect

Urban Infrastructure Issues: A Sustainable Alley Development and Construction in Viet Nam

Ngoc Thao Linh Dang, Ngoc Son Truong, Ngoc Toan Truong, and Thi Tra Do

Abstract The optimal planning and use of urban space aims to help solving problems of livelihood, spatial connectivity, increase living area and reducing air pollution. However, the challenges of rapid urbanization place a significant strain on the sustainable development of cities because the government is not prepared to invest heavily in every aspect or character of the city. Many studies have identified alley space as an important component in optimizing the network of urban spaces. Nonetheless, long-term sustainable solutions to alley infrastructure have been recognized as critical to city sustainable development in various literatures, but have been rarely applied in practice. This study examines the current state of alleys in Da Nang, Vietnam while proposing solutions to improve a typical alley space toward sustainable development, addressing environmental and urbanization issues.

Keywords Urban space · Alley space · Ecological architecture · Spatial connectivity · Sustainable development · Space recycling

1 Introduction

In recent years, fast urbanization has always been accompanied by many challenges in large cities, such as the inability of urban living and working spaces to keep up with the rate of urbanization's growth, as well as the degradation of the urban environment [1]. Large cities are predicted to have about 60% of population by 2030 leading urban areas occupy 2% of total global area while emitting up to 40% of its emissions [2]. Therefore, improving living quality of city citizens is always a top priority for which leading to the necessity of a sustainable, green and beautiful living environment.

N. T. L. Dang (✉) · N. S. Truong · N. T. Truong · T. T. Do
The University of Danang, University of Science and Technology (DUT), 54 Nguyen Luong Bang Street, District Lien Chieu, Da Nang City, Vietnam
e-mail: dntlinh@dut.udn.vn

N. S. Truong
Ho Chi Minh City University of Technology (HCMUT), 268 Ly Thuong Kiet Street, District 10, Ho Chi Minh City, Vietnam

© The Author(s), under exclusive license to Springer Nature Singapore Pte Ltd. 2023
J. N. Reddy et al. (eds.), *ICSCEA 2021*, Lecture Notes in Civil Engineering 268,
https://doi.org/10.1007/978-981-19-3303-5_13

Many solutions have been applied to solve this rapid urbanization impacts, especially in developing countries [3]. Urban space is an important factor which needs to be improved towards sustainable development as it helps reducing pollution and saving limited resources [4]. The phrase "space recycling" is recently mentioned as an important strategy towards developing a sustainable city. In particular, the reclaim of alley space is being focused as a new component in many big cities [5].

An alley is a space between buildings that connects the street frontage to service areas on both sides and the back of the building. Alleys vary in form and design – they can be as narrow as a few feet or as wide as forty feet; some are purposefully created, while others are the result of ill-advised construction projects. However, all alleys provide accessing pathway [6]. Alley spaces now play an important role in urban areas, particularly in developed cities like Viet Nam. Despite the fact that it is not explicitly mentioned in the city planning project, alley space has always played a role in the urban framework. These alleys, which vary in size and character, can be found in residential areas, commercial areas, and city centers. It is worth noting that alleys are identified as dark corners in cities because their dirty with poor security. According to survey data, alley spaces attract criminals and easily becomes a breeding ground for infectious diseases [7]. There is no doubt that reclaiming back alley space has become an urgent topic that has been identified as the primary research object in many infrastructure projects, attracting significant attention from the government and its citizens.

This study emphasizes the significance of alley space in urban development. In particular, it describes the current state of alley space in urban areas around the world as well as in Da Nang. Moreover, its outcome is to propose a proper plan to reclaim alley spaces for sustainable urban development in Da Nang city.

2   Literature Review

2.1   Overview of Research in the World

The visual picture of modern cities is heavily influenced by street space. Furthermore, it is one of the most significant characteristics indicating the development of a green and sustainable city that attracts tourists [8]. Alley space was largely defined in previous research as a new and necessary component of developing urban space for long-term growth [4].

Alley areas are being refurbished to entice not just tourists but also locals to explore cities in the United States, according to survey data. Alleys can also be easily linked to become part of the city's green city network, cycling network, or art display. When compared to the roadway network, the government considers that alley spaces provide a different urban spatial experience [9].

Alley reclamation can be observed clearly in the instance of the "Green Alley" initiative, which was also initiated and has been carried out in several major cities

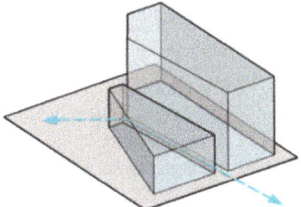

Fig. 1 High density mixed use

across the United States since 2006. "Green Alley" is one of those projects aimed at attaining sustainable urban development, as well as one of the aims discussed in industrialized countries for more than a half-century. Pilot projects began in 2006, and by 2010, more than 100 "green lanes" had been built. It began in Chicago and then spread to San Francisco, Los Angeles, Detroit, Seattle, Minneapolis, and, lastly, Denver (Fig. 1). In short, many studies and projects on "Green Alley" have drawn a lot of attention from the government and its citizens [8].

Alley Oop is a new urban park in Vancouver that welcomes the public to play in a downtown laneway. In a city where the value of land continues to climb dramatically, finding innovative ways to generate and use public space is vital to the success and social well-being of city people. The previously underutilized space has been transformed into a bright, playful space with opportunities to play sports and spend time casually [10].

Many governments of major cities around the world literally concerned about alley space developments in their localities (Table 1). Many successful strategies of enhancing alley space have been applied around the world, including: infrastructural improvement [9], landscape trees [11], operational functions [12]. However, alley spaces are not concerned carefully in developing countries which causes impacts on life quality of society.

2.2 Current Issues in Viet Nam

In general, most of studies have identified the importance of alley space to human life such as preserving the existing traditional culture and connection within communities. It is worth noting that there is lack of research articles focusing on alley space renovation and developement in comparison with its appearance [16]. Alleys now occupy a significant portion of urban areas not only around the world, but also in Viet Nam. These alley spaces are rarely mentioned in city planning projects, but their exist and have become an essential part of cities. In fact, these alleys of various sizes and lengths can be found throughout residential and commercial areas [17].

Firsly, low traffic volume and infrequent occurs in small alleys. The infrastructure of these alleys is degraded easily as it is also not well-invested. The road surface, for example, is concave, and the surrounding neighborhood is unpleasant, unsightly, and

Table 1 Projects and studies on alley space are published

Field	Citation	Title, Author	Description
Government project and manuals	[11]	The Chicago Green Alley Handbook, Chicago Department of Transportation; (2010)	Providing design guidelines and recommendations for green alley located along properties
	[9]	Seattle Integrated Alley Handbook: Activating Alleys for a Lively City; Mary Fialko and Jennifer Hampton; (2011)	This handbook offers varieties of alley development case studies in Seattle, USA
	[13]	What is Green Infrastructure?; U.S. Environmental Protection Agency (2013)	The U.S. Environmental Protection Agency provides a wide range of materials related to green infrastructure, including pavements, alleys development which reduces the impact of drainage during storms
Researches	[14]	Green Alley Programs: Planning for a sustainable urban infrastructure?; J. P. Newell, et al. (2012)	The revival of alley spaces in North American and European cities through government and local policies
	[15]	Resident perceptions of urban alleys and alley greening; Mona, et al. (2010)	Citizen's viewpoints on green alley development. The results highlight the pragmatic relationship between them and their apprehension about these spaces

unsafe. Maintenance is carried out inconsistently, which has a negative impact on the surrounding urban landscape. Secondly, these alleys usually have its drainage system invested unreasonably so that water stasis in the rainy season causing unsanitary and increasing the possibility of spreading infectious diseases. Furthermore, gabbage is not handled carefully, resulting in unsanitary living conditions in the surrounding area. Thirdly, number of alleys are located in hidden area with darkness where people pass through scarily. As a result, it quickly became a criminal hotspot in the city. Moreover, many alleys appear due to shortcomings in the city urban planning as well as topographical factors, thereby causing alley spaces to be skewed, distorted, and unevenly high. Later, these alley spaces seem to encroach and interleave within current urban area which usually obstruces traffic.

2.3 Current State of Alleys in Da Nang

To begin with, there are several models of alleys that needs to be determined in order to make reasonable and appropriate renovation plans. In Da Nang, alleys can be classified into 6 different categories based on its uses and functions:

* High density mixed use You can help us facilitate quick and accurate publication

These alleys are typically found between high-rise buildings such as offices, residential areas, restaurants, retail stores and parking lots. They are used as temporary parking lots as well as pedestrian passageway (Fig. 1).

* Low density mixed use

These alleys are surrounded by low-level residential, restaurant, office, service, retail with less space. This type of alley experiences a fresher atmosphere with more sunlight during the day due to the lower height of the building (Fig. 2).

* Commercial district

The alleys in the commercial district are filled with with numerous retail establishments that can aid in the expansion of the local business community (Fig. 3).

* Multi family residential

This sort of alley is characterized by adjacent multi-unit buildings. Within neighborhoods, alley space has the potential to become pedestrian passageways (Fig. 4).

* Single family residential

Spacious houses are separated by its backyard while it is also a connection point within and between blocks (Fig. 5).

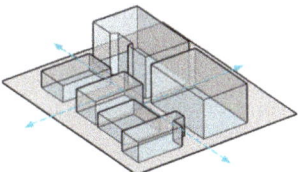

Fig. 2 Low density mixed use

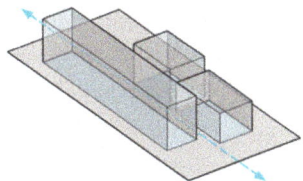

Fig. 3 Commercial district

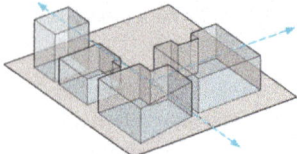

Fig. 4 Multi family residential

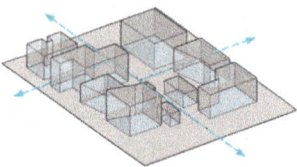

Fig. 5 Single family residential

3 Methodology

3.1 *Research Philosophy*

A case study is a research method that is used to gain an in-depth, multi-faceted understanding of a difficult subject in its real-life setting. It is a well-established research approach that is frequently used in a variety of subjects, particularly the social sciences. A case study can be described in a variety of ways, with the essential principle being the requirement to thoroughly analyze an event or phenomena in its natural setting. As a result, it is frequently referred to as a "naturalistic" design, as opposed to a "experimental" design (such as a randomized controlled trial), in which the investigator attempts control over and manipulation of the variable(s) of interest [18].

According to Yin, case studies can be used to explain, describe, or investigate events or phenomena in the everyday contexts [22]. These can be used to investigate and explain causal relationships and pathways that arise as a result of a new policy initiative or service development, for example. Unlike experimental designs, which aim to test a specific hypothesis by purposefully modifying the environment (like, for example, in a randomised controlled trial giving a new drug to randomly selected individuals and then comparing outcomes with controls).

3.2 *Research Approach*

As a result, the type of research used in this paper can be classified as an applied case study method dedicated to gathering information from published studies in order to select appropriate improvement methods and suggesting conclusions.

3.3 *Case Study*

According to the report, a typical multi-family residential alley at 54 Nguyen Cong Tru Street has been chosen for analysis and remedial measures. In particular, it is situated on the central street between the two Han River bridges and the Dragon Bridge, with a constantly changing road surface. It also connects Nguyen Cong Tru and Nguyen Thong streets with a nearly 300 m (Fig. 6).

There are several passageways that are continuously connected to alleys of varying width. Only bicycles and motorcycles travel at high speeds during the day and at low speeds at night. Furthermore, its infrastructure deteriorates because the road surface is convex and particularly stagnant during the rainy season; waste is not carefully collected; unfriendly, unattractive space; and a lack of light at night (Fig. 6).

4 Recommendation and Discussions

In the present study, "Green Alley" is defined as a succesful camgpain in Chicago [11] as well as Seattle [9] in the USA so that selected solutions that can be studied and applied in renovating the case of 54 Nguyen Cong Tru Street, Da Nang, Vietnam: an examination of pavement, lighting and landscape.

Fig. 6 General view of alley from 54 Nguyen Cong Tru Street, Da Nang

4.1 Sustainable Solutions

* **Pavement:** Permeable pavement is currently an optimal solution in alley improvement and construction since it allows 80 percent of precipitation to flow back into the earth, eliminating waterlogging. Topmix is a particular form of concrete produced by Tarmac. Permeable concrete is a flood-resistant concrete that drains quickly. It's employed as a very absorbent surface coating material that lets leachate soak through rather than collecting on the pavement [19].

In this case, semi-permeable concrete is used with two main layers (Fig. 7). Firstly, it has a permeable surface with a profusion of relatively large pebbles on the surface leading down to a brick base, with looser debris at the bottom. Secondly, drainage systems are attached to the brick and rubble foundation inside the subsoil, allowing excess water to drain into the sewers as well as into the existing ground. To avoid contamination, the underlying soil must be cleaned before its construction (Fig. 7).

* **Light:** Lighting in alley plays a different role compare to street lighting as it can help reducing light pollution as well as encouraging people using public space at night [9]. Therefore, lighting fixtures are chosen to save energy, reduce glare and still ensure to see the stars in the sky at night. These devices are specifically designed to direct light downwards, focusing it where it is needed (Fig. 8). Instead of the yellow light provided by traditional high-pressure sodium bulbs, these lighting fixtures will employ LEDs to produce white light. LED solid-state lighting may provide all lighting characteristics due to its quality, ease of use, reliability, and achievement of the physical limits of electricity-to-light conversion [9]. This will also help people be able to distinguish colors at night [9] (Fig. 9).

* **Landscape:** In the early period of the twenty-first century, the rapid development of the city in both height and width has reduced the expansion of green space within city itself. What is mentioning is that alley spaces often fall into a state of abandonment. If it is possible to protect and maintain the green land system throughout the main roads and alleys, they will play a practical role in forming an attractive landscape environment guide along the city. The existion of natural landscape offers a balanced living environment that enhances air quality while also increasing the appeal of alley areas [20]. Consequently, those small, tall, straight trees with wide-canopy are suitable for the alley area as they reduce the amount of heat shining on

Fig. 7 Topmix semi-permeable concrete structure

Fig. 8 Section drawing of proposed concept with Topmix permeable pavement

Fig. 9 Section drawing of proposed concept with lighting fixtures

Fig. 10 Section drawing of proposed concept with tree system and landscape

pavement surface (Fig. 10). Native plants and trees, which do not require frequent watering and fertilization and are easily suited to the local conditions, should also be examined. Narrow alleys, on the other hand, could benefit from rows of potted plants and windows along the side wall [21].

* **Activity:** Benches are attached along the green area, making the alley an ideal place to sit and read a book or talk with friends. This not only adds to the environmental benefits, but also emphasizes the uniqueness of a specific social space. Furthermore, it is proposed to organize local events at the end of each month in order to attract residents as well as tourists from the surrounding areas, which promoting economic benefits for people in the area. Alley activisim can change the way people engage with it (Fig. 10) [9].

4.2 Renovated Concept Proposal

Authors created a concept proposal for alley development and construction at 54 Nguyen Cong Tru by simulating proposed renovation solutions into perspective drawings. To approach and contribute to sustainable alley development and construction, the government and planners should develop a sustainable and strategic policy

Fig. 11 Perspective drawing of the proposed renovation concept

in construction and management to avoid unnecessary costs, enviromental damage, and other issues (Fig. 11).

5 Conclusion

To summarize, the number of alleys occupies a significant portion of urban areas, particularly in developed cities in Vietnam. These alleys scattered in residential areas and commercial areas with different sizes and characteristics. Despite the fact that alley space has always played a central role in the urban framework, it is not mentioned in city master planning projects.

There is no denial that urban space renovation case studies have been applied succesfully in many large cities in the world, which also addresses problems of urbanization, especially in developing countries. It is an important factor which help reducing pollution, saving natural resources and improving the quality of human life towards sustainable urban development. In particular, renovating alley space is currently a new issue focused by government.

In these recent years, urban infrastructure development in Da Nang is growing rapidly without consideration of public space area within city. In other words, a lack of development planning and mitigation has clearly posed a risk to the degradation of alley space. There are well-known case studies in Chicago that demonstrate the negative consequences of uncontrolled alley development. Therefore, short and long-term solutions are considered to be adapted in Da Nang city's alley infrastructure development as soon as possible.

In the future, authors raise interesting questions for further research in terms of providing quantitative analysis methods and simulations from the software to clarify the efficiency of proposal.

Acknowledgements This work was supported by The University of Danang, University of Science and Technology, code number of Project: **T2020-02-12**

Truong Ngoc Son was funded by Vingroup Joint Stock Company and supported by the Domestic Master/ PhD Scholarship Programme of Vingroup Innovation Foundation (VINIF), Vingroup Big Data Institute (VINBIGDATA), code **[VINIF.2020.TS.07]**

References

1. Chatzimentor A, Apostolopoulou E, Mazaris AD (2020) A review of green infrastructure research in Europe: challenges and opportunities. Landsc Urban Plan 198:103775
2. UN-Habitat (2013) Urban planning for city leaders. United Nations Human Settlements Programme USA
3. Lalbakhsh E (2012) The impact of recycling urban space in sustainable development in developing countries. APCBEE Proc 1:331–334
4. Rehan RM (2013) Sustainable streetscape as an effective tool in sustainable urban design. HBRC J. 9(2):173–186
5. Nur K (2020) Alley activation: genius loci to construct a resilient city. J. Arch. Urba. 44:63–68
6. Hage SA (2008) Alleys: negotiating identity in traditional, urban, and new urban communities. Masters Theses, p 110
7. Nhi LTH (2016) Cải tạo ngõ hẻm trong đô thị theo hướng bền vững. Tạp chí Kiến Trúc 08:17
8. Im J (2019) Green streets to serve urban sustainability: benefits and typology. Sustainability 11(22):6483
9. Fialko M, Hampton J (2011) Seattle integrated alley handbook: activating alleys for a lively city. UW Green Futures Lab, Scan Design Foundation, & Gehl Architects. University of Washington, Seattle, WA
10. Design HAA (2016) More awesome now laneway activations, Vancouver, BC
11. Transportation CDO (2010) The Chicago Green Alley Handbook, Chicago city, USA
12. Transportation BDO (2013) Alley Gating & Greening Program, City of Baltimore
13. Agency USEP (2013) What is Green Infrastructure? USA
14. Newell JP et al (2013) Green alley programs: planning for a sustainable urban infrastructure? Cities 31:144–155
15. Seymour M et al (2010) Resident perceptions of urban alleys and alley greening. Appl Geogr 30(3):380–393
16. Trang PTQ (2009) Văn hóa ngõ phố trong đời sống đô thị tỉnh lị Việt Nam, in Khoa Xã hội học. Trường Đại học Khoa học Xã hội và Nhân văn
17. GIBERT M (2013) Le réseau de ruelles de HCMV au défi de la modernisation : projets locaux, enjeux métropolitains. University Paris Diderot, France
18. Crowe S et al (2011) The case study approach. BMC Med Res Methodol 11(1):100
19. Tarmac L (2020) Topmix Permeable. https://www.designingbuildings.co.uk/wiki/Topmix_Permeable
20. Anh TKHD (2016) Các nguyên tắc hình thành hệ thống không gian xanh đô thị. Tạp chí Quy hoạch Đô thị 24(7)

21. Schutzki RE (2005) A guide for the selection and use of plants in the landscape. Michigan State University
22. Yin RK (2009) Case study research, design and method, vol 4. Sage Publications Ltd., London

Urban Spatial Adaptability to Drought-Flood Coexistence (DFC). A Case Study from Phan Rang-Thap Cham City, Ninh Thuan Province, Vietnam

Nguyen Quoc Vinh, Le Minh Ngoc, Nguyen Trong Khanh, and Le Anh Duc

Abstract Since the Earth was gravely impaired by CC (Climate Change) in the past decades, a remarkable number of researchers have shifted their viewpoints regarding the city. They consider it a social system that gravitates towards a Social-Ecological System (SES). The very new concept of urban resilience focuses on the urban ability to recover from external shocks and attacks and bounce back to a new stable state, which helps narrow down the former broad concept of urban sustainability. Drought and flood are deemed overarching in Vietnam among the natural attacks, especially in Ninh Thuan province. This study, *"Urban spatial adaptability to Drought-Flood Coexistence (DFC), a case study of Phan Rang-Thap Cham City, Ninh Thuan Province"*, aims to assess the city's adaptability to not only grow, exploit and conserve during the attacks of local DFC but also to renew or self-organize its structure and functions after disturbances. During the study period of 1988 to 2020, the urban spatial adaptability to DFC of the examined area is qualitatively and quantitatively measured. Both urban spaces and the DFC of Phan rang-Thap Cham are investigated at interrelated urban scales, from macro to meso scales. The study covers four parts: (i) The concepts of urban spatial resilience and adaptability, (ii) The study methodology, (iii) The context of Phan Rang-Thap Cham City, Ninh Thuan province, and the study results, (iv) Conclusions.

Keywords Resilience · Urban spatial adaptability · Drought- flood coexistence (DFC) · Remote sensing

N. Q. Vinh (✉) · N. T. Khanh
Faculty of Civil Engineering, Ho Chi Minh City University of Technology, VNU-HCM, 740128
Ho Chi Minh City, Vietnam
e-mail: vinh.bmkt@hcmut.edu.vn

L. M. Ngoc
Urban Design and Landscape Architecture, Omgeving, Ho Chi Minh City, Vietnam

L. A. Duc
Faculty of Architecture, Van Lang University, Ho Chi Minh City, Vietnam

© The Author(s), under exclusive license to Springer Nature Singapore Pte Ltd. 2023   177
J. N. Reddy et al. (eds.), *ICSCEA 2021*, Lecture Notes in Civil Engineering 268,
https://doi.org/10.1007/978-981-19-3303-5_14

1 Introduction

With the increasingly severe CC impacts on urban areas, many studies have been taken to analyze the urban adaptive capacity in its ability to revert to a balanced condition and develop sustainably afterwards. The unique concept is the resilience and adaptability of urban space, which are based on ecological sciences. Urban is considered to be an ecosystem named urban ecosystem. The Social-Ecological System (SES) consists of a natural and social ecosystem [1]. This concept emphasizes four stages that an urban has to experience to recover from disturbances of natural disasters before coming back to a steady-state, including (i) growth and exploitation; (ii) conservation; (iii) collapse or release; (iv) renewal and reorganization [2–5].

In the context of Phan Rang-Thap Cham city of Ninh Thuan province, drought and flood are two specific natural phenomena that can strike this area anytime, causing severe damage to society and affecting people's lives. Meanwhile, the urbanization process of Phan Rang-Thap Cham since the 90s has facilitated an expansion in the area of built-up space (including traffic systems and buildings). This type of space is known as the impermeable surface and land surface temperature [6]. Urbanization has also reduced natural space (having hydrological surface and vegetation cover). This matter more or less hinders cities' operations, impedes the ecological networks to provide Ecosystem Services (ES), which are: (i) provisioning (food, genetic resources, biochemical, freshwater, etc.); (ii) regulating (climate, disease, natural hazard protection, water purification, etc.); (iii) supporting (water cycling, provision of habitat, soil formation, and retention, etc.); and (iv) cultural services (knowledge system, education, and inspiration, sense of place, etc.) [7]. Besides, many studies have identified air and land surface temperatures inversely correlated with vegetation cover while positively correlated with impermeable surfaces [6, 8]. They are consequences of drought or an event related to drought and are used as an index to evaluate the drought levels in the research [9, 10]. The above contents are comprehensively researched and analyzed in the context of Phan Rang with four parts: (i) The concepts of urban spatial resilience and adaptability, (ii) The study methodology, (iii) The context of Phan Rang-Thap Cham City, Ninh Thuan province, and the study results, (iv) Conclusions.

2 Resilience and Adaptability of Urban Spaces

2.1 Urban Resilience

Resilience, urban resilience, urban adaptability.

The resilience concept has gone through a long process. Initially, it comes from the Latin root "resalire". Walter William. In his 1882 dictionary, Skeat translated the term into "walking or leaping back".

Ecologist Hollings referred to this concept in 1973 as the ability of a system to cope with external attacks but still retain former structures and functions [11]. In 1996, he developed this concept into engineering and ecological resilience [12]. The first branch concentrates on "stability near an equilibrium steady state, where resistance to disturbance and speed of return to the equilibrium are used to measure the property". The second branch underscores "conditions far from any equilibrium steady state, where instabilities can flip a system into another regime of behaviour - that is, to another stability domain". In this case, resilience can be measured by the magnitude of disturbance that the system absorbs before changing its structure. Many scholars interpret it as an adaptive cycle through four phases: (i) growth and exploitation; (ii) conservation; (iii) collapse or release; (iv) renewal and reorganization [3, 13]. In general, resilience has been used in multiple fields in the last few decades, especially in natural ecology: 'the ability of the system to revive or recover after a shock or external impact' [3, 11]. Some scholars use this word to describe 'the time it takes to reconstruct the system'.

Urban planners began to be aware of this concept in the late 1990s [14]. The consideration of urban - an SES - supported by natural ecosystem services involves: (i) water and food supply, (ii) flooding, drought and disease regulating, (iii) soil and nutrients formation, and (iv) cultural, recreational, spiritual, religious and other intangible benefits [7]. These were combined in urban planning studies. Among the various definitions of urban resilience, the Sendai Framework states: "Resilience is the ability of a system, community or society to resist, absorb, accommodate, adapt to, transform and recover from the effects of a hazard in a timely and efficient manner" [15]. S. Meerow describes: "Urban resilience is the ability of an urban system to maintain or rapidly return to desired functions in the face of a disturbance, to adapt to change and to quickly transform systems that limit current or future adaptive capacity" [16].

2.2 Urban Spatial Adaptability to DFC

Some scholars believe that adaptive urban planning plays an essential role in accelerating the resilience process in urban areas [12]. The literature on resilience in planning has firstly emphasized preparation and mitigation [17]. A decade later, adaptation came into focus while mitigation is insufficient to prevent disturbances from occurring anymore. A study on assessment indexes of urban adaptive capacity from Fleischhauer shows that urban can minimize external dangers with an appropriate spatial structure [18]. Several other studies suggest that open spaces, green spaces, water surfaces, etc., as backup spaces, promote biodiversity. These spaces are a part of an adaptation strategy known for 'nurturing conditions for recovery and renewal after disturbance' [13]. Furthermore, enhancing green space and water surface in urban areas is seen as an approach that helps reduce air and surface temperature [15], mitigate the risk of drought, and increase biodiversity and resilience [19].

Accordingly, the Phan Rang-Thap Cham case research will use the notion of *adaptability*, which is the capacity of urban spaces, including water, vegetation covers, and built-up surfaces, to adapt to DFC during its operation.

3   Research Methodology

A matrix of urban spaces and DFC was built to measure urban spatial adaptability to the DFC of Phan Rang-Thap Cham City. The matrix has rows of urban spatial components, which are basically according to the ecological approach (*water, vegetation, bare land, built-up*), columns of adaptive indicators, and variables. Spatial components are analyzed on urban spaces' adaptive variables/attributes (spatial networks and land covers) and evaluated through adaptive indicators of the study zones. Variables or variable spatial quantities, e.g. biology, physics, or society, can be tracked through structures (including *location, amount, configuration*) [20] and functions to understand, recognize and evaluate urban resilience or adaptability [3, 13]. *Location* is the dominant role; it is therefore considered a *core variable*. With structural components, the first two quantities are qualitatively evaluated depending on the land cover—the distribution of natural functions, the size and the form of that space [20]. The following quantity is quantitatively evaluated from the accessibility and density of intersections of that spatial network [21], and the measurement of the study area. Concerning adaptation, two spatial factors of ecological structures and functions need to consider DFC services. Adaptive indicators to DFC are theoretically examined from the former studies on resilient assessment framework, practically surveyed in Phan Rang-Thap Cham City and systematized for the study context later on, including: compatibility [22], diversity [23–26], connectivity [24, 26–29], robustness [24, 27, 30], redundancy [23–26], and NDVI, LST [10].

The urban spatial adaptability to DFC in Phan Rang-Thap Cham City is investigated from 1988 to 2020, of which the year 1988: desertification started, 2005: urbanization's impacts began, and 2020: studying time. They are examined on multi-scales which relies on the mutual changing speed and affects the scale of the studied objects, as follows: (i) *Catchments*: compatibility of urban systems with ecosystem services (the flood-induced factors such as topography, slope, water resources, and drought elements (meteorology, maximum temperature zone of the entire river basin) were examined); (ii) *Urban:* the level of rationality, flexibility and harmony with the natural condition in the allocation of natural land use functions; (iii) *Urban unit:* the degree of diversity, connection, flexibility and stability in the organization of the natural space (water, vegetation, and barren) and artificial spaces (transportation, buildings) are analyzed and measured through adaptive indicators.

LULC, LST and NDVI in Phan Rang-Thap Cham City. LULC maps of the city gat three different points of time (1988, 2005 and 2020) are examined through R.S. images to explore the land cover transformations. LST and NDVI maps are then extracted, from which we obtained key figures of heat and green spaces, respectively, transformed through the years.

LULC: The multispectral Landsat satellite images of the following table were used (Table 1). The Landsat images were registered to Universal Transverse Mercator projection, zone 49 N, WGS84 Datum, the spatial grid of 30-m resolution. The dates

Table 1 Detailed characteristics of the Landsat Satellite images used in the study

Path/row	Acq. date	Dataset	Producer	Attribute	Type
123,052	26/03/1988	TM	USGS	Ortho, GLS2000	GeoTIFF
123,052	29/03/2005	TM	USGS	Ortho, GLS2000	GeoTIFF
123,052	11/04/2020	OLI-TIRS	USGS	Ortho, GLS2000	GeoTIFF

(*Source* Authors)

of the Landsat images were chosen close to the same vegetation season. The methodology used involves the following stages: (i) image pre-processing, the design of classification scheme, (ii) image classification, (iii) accuracy assessment and analysis of the LULC changes.

- Image pre-processing: including cloud masking, layer stacking, sub-setting by administrative boundary;
- Classification scheme: utilizing four classes (Water, Vegetation, Barren and Urban - Built-Up);
- Image classification: training data was used for Maximum Likelihood Classification (MLC) and accuracy assessment. In this research, colour composites; ratio images (MNDWI, NDBI, NDVI); Google Earth and topographical map were used to generate training data;
- Accuracy assessment: This assessment was performed for the classified maps of all three steps. Error matrices were adopted to assess classification accuracy using two measures: overall and Kappa statistics.

NDVI. NDVI is used to quantify vegetation by measuring the difference between near-infrared (vegetation reflects) and red light (vegetation absorbs) [31]. The NDVI algorithm subtracts the red reflectance values from the near-infrared and divides them by the sum of near-infrared and red bands.

$$NDVI = \frac{(NIR - Red)}{(NIR + Red)} \tag{1}$$

NDVI values are represented as a ratio ranging from -1 to $+1$, negative values represent water, around zero is bare soil, built-up areas, from 0.2 to 0.5 is for shrubs, grassland, and over 0.6 is dense green vegetation.

LST. The final LST (the temperature at the interface of the Earth's surface up to 2–3 m high) can be measured in the following steps: (i) converting the Satellite Digital Number (D.N.) to Radiance value, (ii) conversion of Radiance to At Sensor Temperature, (iii) conversion of Satellite Brightness Temperature into LST.

4 The Research Context and Results

4.1 The Context of Ninh Thuan Province and Phan Rang-Thap Cham City

Ninh Thuan Province. Is the coastal province in the South - Central part of Vietnam, located at the coordinates: 11°18′14″ to 12°09′15″ Northern latitude, 108°09′08″ to 109°14′25″ Eastern longitude.

Terrain: three-quarters of the provincial extent surrounded by mountains (Fig. 1a)

Meteorology: dry (75–77%), hot (26–27 °C), strong wind (2.3–5 m/s, the strongest is of 25 m/s) and high rate of evaporation. Three climate sub-regions include: (1) the coastal area (III) - worst droughts with an average rainfall of around 500–700 mm/year, where Phan rang-Thap Cham city is located, (2) the plain area (II) - drought with rainfall from 750 to 1,200 mm/year, (3) the mountainous area (I) - rainfall of 1,000–1,700 mm/year (Fig. 1b).

Hydrology: The total area of the river basin is 3,092 km², including 46 rivers, mainstreams and four principal aquifers, of which the Dinh river basin covers the majority of the area with 3,000 km², and others 92 km² (Fig. 1c).

Extreme Weather Events: (i) Storms: four times a year, from October to December, (ii) Flash flood: often accompanied by upstream storms, concentrating in the high mountains and riversides, (iii) Drought: due to the hot and dry weather, forest fires often occur which seriously degrade the ecosystem and increase the drought intensity.

Phan Rang-Thap Cham City. It is located downstream of the Dinh River and is the province's political, administrative, economic, cultural, and scientific centre. Natural land: 10,111 ha [10]. Area of construction land: 2,519 ha.

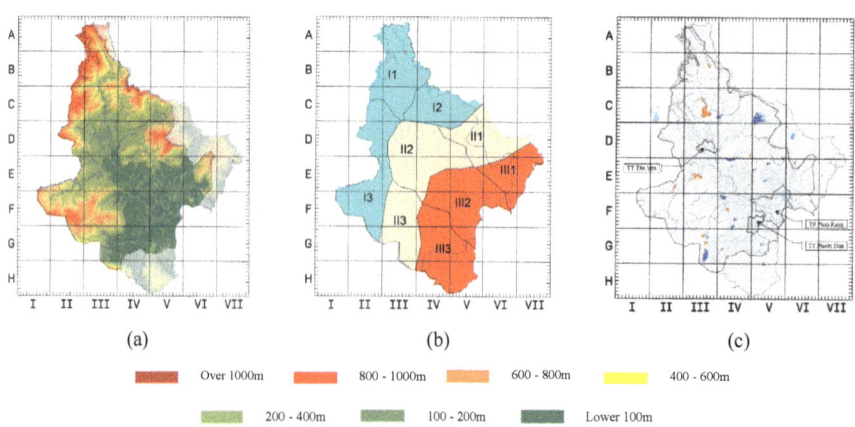

Fig. 1 Map of natural features of river basins in Ninh Thuan province. **a** Topography, **b** Temperature Zoning, **c** Hydrological and reservoirs system. (*Source* Authors, background map from USGS)

The terrain is relatively flat, and the slope is less than 5%. The main water surfaces are the Dinh river, Nai lagoon, some irrigation canals and drainage canals.

Urban Space Transformations of Phan Rang-Thap Cham city in the Period 1988-2020

Socio-economic development has significantly impacted the process of urbanization since the opening policy in 1986, thereby creating a massive and pervasive development wave. It has also led to a conspicuous change in the urban structure regarding connecting frameworks and land covers (Fig. 2a–b-c) during the study period (1988–2020). The hydrological system of urban space has not been changed much, but the green cover, bare land and construction have been transformed (Table 2). The bare land decreased gradually and was replaced by the green cover and built-up area.

Table 2 Changes in the urban area (ha) and Remote Sensing indexes (NDVI, LST) of Phan Rang-Thap Cham (1988–2020)

Urban space		Research period (1988–2020)					
		1988	%	2005	%	2020	%
Natural	Water	1,703	17	2,198	22	1,536	15
	Vegetation	3,641	36	2,895	29	4,049	40
	Barren	3,532	35	2,831	28	1,891	19
Built-up	Impermeable	1,226	12	2,188	22	2,635	26
Total area		10,102	100	10,112	100	10,111	100
R.S. indicators	NDVI	0.25		0.12		0.30	
	LST	23.66		28.06		27.3	

(*Source* Authors, figures extracted from remote sensing images, USGS)

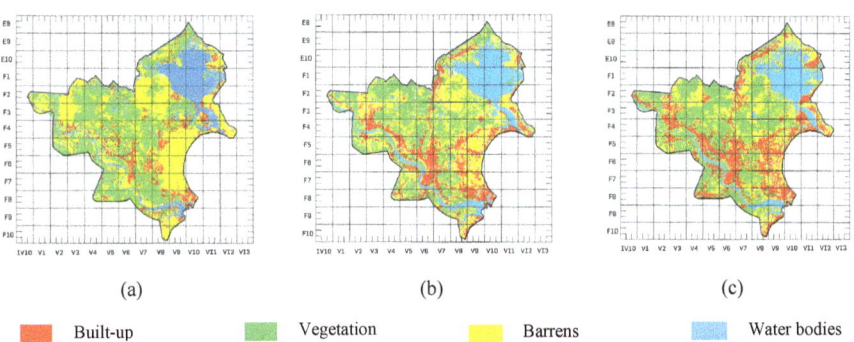

(a)   (b)   (c)

■ Built-up   ■ Vegetation   ■ Barrens   ■ Water bodies

Fig. 2 The transformation process of the urban area of Phan Rang-Thap Cham (1988–2020). **a** 1988, **b** 2005, **c** 2020 (*Source* Authors, background map from USGS)

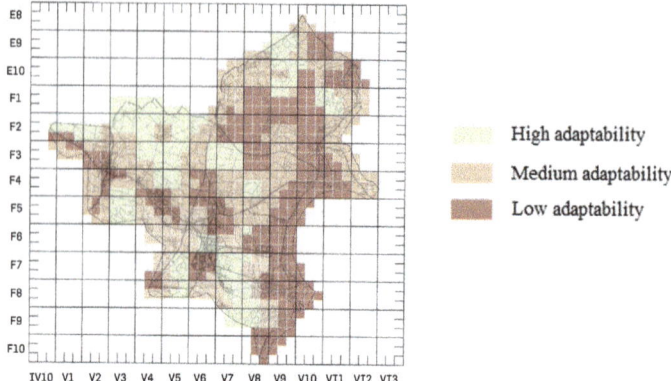

Fig. 3 Urban spatial adaptability to DFC of Phan Rang-Thap Cham City, Ninh Thuan province (*Source* Author)

4.2 Urban Spatial Adaptability of Phan Rang-Thap Cham City to DFC

The adaptability of urban space is evaluated using the same methodology on both scales: urban and urban units. In particular, urban areas are evaluated based on the cities' administrative boundaries of Phan Rang City, while urban units are assessed on pre-selected size plots according to the principles below.

Principles of Choosing a Surveyed Urban Unit
The survey location was selected by principles of the 5-region division of the Burgess model [32] and 05 segments of Dunay in rural–urban transect [33]. It is known in the form of the central core (L), development centre (T.T.), suburban (NO), rural (N.T.), and natural (T.N.). The plots' scales and sizes are based on the theoretical basis of the dimensions of urban cells [34] and are referred to as the size 400 × 400 m (the average size of a residential unit). According to the principle, the examined plots were assumed with two high adaptabilities, two average and one low (Table 3).

Table 3 Principles of choosing a surveyed urban unit

		Spatial range					Adaptability	
		Natural	Rural	Suburban	Center	Core	Level	Legend
Plot ▼	Size ▶	12,000m	12,000m	1,200m	400m	400m		
Range ▶								
	1	TN1	NT1	NO1	TT1	L1	High	●
	2	TN2	NT1	NO2	TT2	L2	High	●
	3	TN3	NT1	NO3	TT3	L3	Average	◒
	4	TN4	NT1	NO4	TT4	L4	Low	○
	5	TN5	NT1	NO5	TT5	L5	Low	○

(*Source* Authors)

Table 4 Assessment framework of urban spatial adaptability to DFC

Urban spaces			Adaptive Indicators of Urban Spaces to DFC								Urban Spatial Adaptability to DFC
Scales	Spatial components	Spatial attributes/adaptive variables	1. Compatibility*	2. Efficiency*	3. Diversity	4. Connectivity*	5. Redundancy	6. Robustness	7. NDVI	8. LST	Adaptability of the studied urban spaces
Basin/Urban/Urban fragments	Water	- **Location***	x						x	x	x
		- *Amount*		x							
		- *Configuration*			x	x	x	x			
	Vegetation	- **Location***	x								
		- *Amount*		x							
		- *Configuration*			x	x	x	x			
	Barren	- *Amount*		x							
	Built-up	- **Location***	x								
		- *Amount*		x							
		- *Configuration*			x	x	x	x			

* *the core adaptive indicators and variables of urban land covers to DFC*
x: the relevance of urban spatial attributes and adaptive indicators

(*Source* Authors)

Adaptive Indicators/Indexes, Variables and Assessment Levels

From the built-in system of indicators and variables mentioned in the study methodology, rating scales are then established from low, average to high levels.

From the above theoretical bases, the core adaptive indicators are selected for the study as follows: *compatibility* (approx. 30%), *connectivity* (approx. 30%), *efficiency* (approx. 20%) and 10% for the rest. The core adaptive variables are chosen as location, amount and configuration. In addition, according to ecosystem services [7], water and vegetation are the core natural spatial components to provide ecosystem services of providing water and regulating drought and flood; meanwhile, built up is the main component to consume the services.

Therefore, the rating levels of *compatibility* reach from low to high when the built-up spaces are located from the more to the less DFC vulnerable areas (the riverine, close to lagoon, etc.). The levels of *connectivity* reach from low to high when urban spaces, particularly water and vegetation, are connected to ecological networks across scales (river, lagoon, etc.). The levels of *efficiency* reach from low to high when water and vegetation are decreased through times, barren and built-up

Table 5 Urban spatial adaptability of Phan Rang-Thap Cham City to DFC in 2020

	Urban Spaces	1. Compatibility*	2. Efficiency*	3. Diversity	4. Connectivity*	5. Redundancy	6. Robustness	7. NDVI	8. LST	Urban Spatial Adaptability to DFC
Phan Rang City – 2020 / **Water**	1. Natural and artificial water networks transformed at the appropriate DFC service locations.	○								
	2. The water area changed through times (↑, ↓, →)		○							
	3. Functions: diversified with dynamic: river, stream/ static: lake, lagoon, etc.				◑					
	4. Configurations: diversified, evenly distributed, compatible with landscape across scales				◑			●	○	◑
	5. Natural surfaces water connected with ecological networks				◑					
	6. Natural surfaces water connected with external areas				◑					
	7. Natural surfaces water connected with internal land covers				◑					
	8. Spare spaces for regulating floods located at appropriate locations.				◑					

(continued)

Table 5 (continued)

Category	Item					
	9. Enough capacity at appropriate to store, cope with floods.					◒
Vegetation	10. Natural and artificial vegetation transformed at the appropriate DFC service locations.	●				
	11. The vegetation area changed through times (↑, ↓, →)		●			
	12. Functions: diversified (natural forest, agriculture, grasslands, etc.) and artificial (parks, urban trees, etc.)			◒		
	13. Configurations: diversified, compatible with landscape across scales.			◒		
	14. Natural and social functions connected with ecological networks				◒	
	15. Natural and social functions connected with external areas				○	
	16. Natural and social functions connected with internal land covers				◒	
	17. Enough capacity at appropriate locations to store, cope with droughts.					◒
Bare	18. The bare land area changed through time (↑, ↓, →).			●		
	19. Built-up transformed at the appropriate DFC services locations (Typopology high, slope from 5-10%).	○				
	20. The built-up area changed through time (↑, ↓, →).		○			
	21. Configurations: diversified (urban areas, urban blocks, small blocks in priority) appropriate with DFC service locations			●		
	22. Built-up (mainly transportation) connected with ecological networks				●	
	23. Built-up (particularly transportation) connected with external areas				●	
	24. Built-up (mainly transportation) connected with internal land covers				●	
	25. Spare spaces at the appropriate DFC service locations					◒
	26. Roads, urban areas, blocks and buildings are strong enough to resist floods.					●

(*Source* Authors)

area are increased through times, when NDVI index is decreased and vice versa for LST. (Table 4).

The Adaptability to DFC of Phan Rang-Thap Cham City

Based on the evaluation method described above, the adaptability results to the DFC of the Phan Rang urban area is summarized as shown in Table 5 and Fig. 3.

The Adaptability of Urban Units to DFC

Based on the assessment results of the adaptability of urban spaces to the DFC of Phan Rang-Thap Cham City, its urban units are also accordingly assessed with the same method and summarized in Table 6.

Table 6 Adaptability to DFC of typical plots in Phan Rang-Thap Cham City. **a** PRTN 1–3, **b** PRNT 1–3, **c** PRNO 1–5, **d** PRNT 2–4, **d** PRTT 2–4, **e** PRL 1–5

(*Source* Authors, background map from Google Earth and USGS)

5   Conclusions

That urban development becomes highly adaptive to natural ecosystems and approaches resilience and sustainability, in general, is on the trend. The drought-flood adaptive urban development, in particular, is a concern not only for the specific province of Ninh Thuan or the like in Central Vietnam but also for many similar regions all over the world. With the relevant concepts analyzed, synthesized and systematized, a framework for assessing the adaptability to DFC of the urban area was established to study a specific case of Phan Rang city.

In conclusion, remarks for urban spatial adaptability to DFC are as follow: (i) The relevant concepts of urban spatial adaptability are the base to study Phan Rang-Thap Cham in the context of interrelated with DFC, (ii) Using R.S. and GIS utilities could help much in separating urban spatial components to examine their adaptability and in studying interrelation between land covers and LST, and (iii) The assessment framework and the adaptability to DFC of the urban area of Phan Rang-Thap Cham City can be the premise for many further studies, that can be applied to planning studies in areas with similar climatic characteristics, iv) The main results of urban spatial adaptability to DFC in Phan Rang-Thap Cham City and urban units are the concrete examples of applying the theories on planning practice to DFC.

The adaptability of Phan Rang-Thap Cham to DFC would help facilitate more future research and complete the principles of organizing urban space adaptive to the natural context and move towards sustainable development.

Acknowledgements The authors wish to acknowledge the assistance and encouragement from the colleagues at the Department of Architecture, Department of Geomatics Engineering, HCMC University of Technology, from Omgeving Company, Vietnam, and from The Faculty of Architecture, Van Lang University.

References

1. Pickett ST, Cadenasso ML, McGrath B (2013) Ecology of the city as a bridge to urban design. In: Resilience in Ecology and Urban Deisgn. Springer, Heidelberg, pp 7–28
2. Folke C (2006) Resilience: the emergence of a perspective for social-ecological systems analyses. Glob Environ Chang 16(3):253–267. https://doi.org/10.1016/j.gloenvcha.2006.04.002
3. Holling CS, Gunderson LH (2002) Panarchy: understanding transformations in human and natural systems, pp 25–62. https://doi.org/10.1016/j.ecolecon.2004.01.010
4. Holling CS, Gunderson LH (2002). Resilience and adaptive cycles. In: Panarchy: understanding transformations in human and natural systems, pp 25–62
5. Walker B, Salt D (2006) Resilience thinking: sustaining ecosystems and people in a changing world
6. Van TT (2011) Study of urban temperature changes under the urbanization impacts by remote sensing methods and GIS: a case study of Hochiminh City (2011)
7. Millenium ecosystem assessment (2005) Millenium ecosystem assessment: ecosystems and human well-being
8. Vinh NQ, Khanh NT, Anh PT (2019) The inter-relationships between LST, NDVI, NDBI. In: Remote sensing to achieve drought resilience in Ninh Thuan, Vietnam
9. Szalińska Wiwiana, Otop Irena, Tokarczyk Tamara (2018) Urban drought. E3S Web Conf 45:00095. https://doi.org/10.1051/e3sconf/20184500095
10. WMO and GWP (2016) Handbook of drought indicators and indices, no 1173
11. Holling CS (1973) Resilience and stability of ecological systems. Ann Rev Ecol Syst 4(1):1–23
12. Holling CS (1996) Engineering resilience versus ecological resilience. Eng Within Ecol Constraints 1996:31–44. https://doi.org/10.17226/4919
13. Holling CS (1985) Resilience of ecosystems: local surprise and global change. In: Global changing proceedings of ICSU symposium, Ottawa, 1984, pp 228–269
14. Mileti D, Henry J (1999) Disasters by design: a reassessment of natural hazards in the United States. Nat Hazards, pp 1–372. ISBN 0-309-51849-0
15. Framework S, Reduction DR (2015) Sendai framework for disaster risk reduction 2015–2030, no. March, pp 1–25
16. Meerow Sara, Newell Joshua P, Stults Melissa (2016) Defining urban resilience: a review. Lands Urban Plan 147:38–49. https://doi.org/10.1016/j.landurbplan.2015.11.011
17. Godschalk DR (2003) Urban hazard mitigation: creating resilient cities. Nat Hazards Rev. https://doi.org/10.1061/(ASCE)1527-6988(2003)4:3(136)
18. Fleischhauer Mark (2008) The role of spatial planning in strengthening urban resilience. In: Pasman Hans J, Kirillov Igor A (eds) Resilience of Cities to Terrorist and other Threats. Springer Netherlands, Dordrecht, pp 273–298. https://doi.org/10.1007/978-1-4020-8489-8_14
19. Crossman ND (2018) Drought resilience, adaptation and management policy (DRAMP) framework. In: Supporting Technoogy Guidelines, no July, p 17

20. Cadenasso ML, Pickett STA, McGrath B, Marshall V (2013) Ecological heterogeneity in urban ecosystems: reconceptualized land cover models as a bridge to urban design. In: Resilience in ecology and urban design, pp 107–129
21. Janssen M, Anderies JM, Elmqvist T, Mcallister RR (2006) Toward a network perspective of the study of resilience in social-ecological systems. Ecol Soc. 11. http://www.ecologyandso ciety.org/vol11/iss1/art15/.
22. McHarg I (1992) Design with nature. John Wiley & Sons Inc., New York
23. Hassler U, Kohler N (2014) Resilience in the built environment. Build Res Inf 42(2):119–129. https://doi.org/10.1080/09613218.2014.873593
24. Sharifi A, Yamagata Y (2017) Urban resilience assessment: multiple dimensions, criteria, and indicators. Adv Sci Technol Secur Appl 2016:259–276
25. Suárez Marta, Gómez-Baggethun Erik, Benayas Javier, Tilbury Daniella (2016) Towards an urban resilience index: a case study in 50 Spanish cities. Sustainability 8(8):774. https://doi. org/10.3390/su8080774
26. Ahern J (2013) Urban landscape sustainability and resilience: the promise and challenges of integrating ecology with urban planning and design. Landsc Ecol 28(6):1203–1212. https:// doi.org/10.1007/s10980-012-9799-z
27. Eraydin A et al (2013) Resilience thinking in urban planning. Resil Think Urban Plan 106:39– 51. https://doi.org/10.1007/978-94-007-5476-8
28. Eraydin A, Taşan-Kok T (2013) Introduction: resilience thinking in urban planning
29. Allan P, Bryant M (2011) Resilience as a framework for urbanism and recovery. J Landsc Archit. https://doi.org/10.1080/18626033.2011.9723453
30. Bruneau M, et al (2003) A framework to quantitatively assess and enhance the seismic resilience of communities. Earthq Spectra. https://doi.org/10.1193/1.1623497
31. Rouse W, Haas RH, Deering DW (1974) Monitoring vegetation systems in the Great Plains with ERTS, NASA SP-351
32. Burgess (2016) The Burgess Urban Land Use Model. Hofstra University
33. Duany A, Talen E (2002) Transect planning. J Am Plan Assoc (2002). https://doi.org/10.1080/ 01944360208976271
34. Bindzárová A (2016) Type and size of urban cell as tools for sustainable urban (Re)development. Procedia Eng. 161:1482–1489. https://doi.org/10.1016/j.proeng.2016.08.614

Construction Management Session

A Research on Quality Management of Feasibility Study Report: A Case Study of Mekong River Delta, Vietnam

Ha-The-Cuong Truong, Han-Hisang Wang, and Son-Ha Nguyen

Abstract Projects in Can Tho, one out of five biggest cities in Vietnam and becomes the most developed city in Mekong River Delta, that reach better outcomes usually derive from private sector whilst most of public projects are in the same patterns like lack of money, escalation in total cost investment, or even design changes during construction period. These issues are the typical wastefulness in public investment of Vietnamese construction industry. Many researches on the issues have been conducted but their research orientation has commonly focused on project implementation phase or closing phase rather than pre-construction phase. It is interesting to note that the three primary phases in a Vietnamese project life cycle are highly likely fit in the three phases in Juran Trilogy Diagram, namely quality planning, quality control and quality improvement. This also concludes that a lower quality of feasibility study report in preparation phase will result in a lot of wasted money in successional phases and vice versa. The overall aim of this research, therefore, is to improve quality in performance of the feasibility study reports of new construction projects for which public budgets are being allocated efficiently. The two main research objectives are to find out factors or attributes affecting quality management of feasibility study reports in public investment affairs of Group-B or higher projects and to integrate appropriately the so-called "best practice" feasibility study model from western countries into Vietnamese construction industry. Thanks to the most important factors, the research shall also develop a quality checklist to determine qualitative indicators of progressing activities as well as suggested to review or make appropriate decisions.

H.-T.-C. Truong (✉)
Master of Construction Engineering and Management, Minh Ha Construction Consultant Co., Ltd., Cau Kho Ward, District 1, Ho Chi Minh City, Vietnam, National Central University, Taoyuan City, Taiwan
e-mail: 106325603@cc.ncu.edu.tw

H.-H. Wang
Institute of Construction Engineering and Management, College of Engineering, National Central University, 300 Zhongda Road, Zhongli District, Taoyuan City, Taiwan

S.-H. Nguyen
Minh Ha Construction Consultant Co., Ltd., Cau Kho Ward, District 1, Ho Chi Minh City, Vietnam, Master of Business Administration of Shu-Te University, Kaohsiung City, Taiwan

© The Author(s), under exclusive license to Springer Nature Singapore Pte Ltd. 2023   193
J. N. Reddy et al. (eds.), *ICSCEA 2021*, Lecture Notes in Civil Engineering 268,
https://doi.org/10.1007/978-981-19-3303-5_15

Keywords Feasibility study · Mekong river delta · Critical success factor ·
Pre-construction phase · Vietnamese construction industry

1   Introduction

Can Tho, one out of five biggest cities in Vietnam, is the most developed city in
Mekong River Delta due to governmental policy in investment prioritization of this
area. The investments in public projects such as Can Tho oncology hospital (500
beds), Can Tho City development and enhancement in urban resilience, and Can
Tho solid waste treatment plant will help to promote socioeconomic development
and increase living standard of local citizens. While projects in Can Tho that reach
higher outcomes usually derive from private sector, most of public projects are in the
same pattern like lack of money (Western capital cultural central project), escalation
in total cost investment (Can Tho riverside embankment project), or even design
changes during construction period (Renovation and reform of Bun Xang Lake).
These are the typical wastefulness in public investment of Vietnamese construction
industry. There also have other types of bigger wastefulness such as project approval,
budget allocation and project implementation, initial design and build in policies or
programs. Although the government of Vietnam has published a lot of legal documents to guide and deal with such problems, the preparation phase in Mekong River
Delta in practice still seems to be overlooked, particularly contents of the feasibility
study is often built in a formalistic manner. Additionally, although many researches
from domestic and international scholars on the above-mentioned issues have been
conducted, their research orientation has commonly focused on project construction phase rather than the pre-construction phase. Therefore, a demand of research
on quality of feasibility study in public investment projects in Mekong River Delta
should be reacted.

2   Literature Review

2.1   Definitions of Feasibility Study

The term of feasibility study has been formulated and broadly developed by some
scholars over the time. Matson [1] and Brockhouse [2] stated that the analyses would
be conducted via this means, the feasibility study, during the process of project planning to evaluate how the operation of a business was with some initial assignments
in technology (like the conveniences, apparatuses, production line, and so on) and
finance (the needed budget, quantities, pricing, salaries, and so on). Abou-Zeid,
Bushraa [3] defined the word "feasibility study" as an economical and accounting
norms that showed up in the beginning of 1960s for the simple definition of accurate
assessments and goings-over so that the possibility of various investment choices

shall be found out by cost-and-benefit calculations to pull out measurements for each choice. Heralova [4] argued that the feasibility study (FS) is an opening study employed quite soon, when thoughts of a construction project has just sprouted. Its objective is to consider if a project is a valuable suggestion and to assist in gauging other available solutions as well. Other extra project documents such as business case, project execution plan, and the strategic brief shall also be developed and benefits by the FS. The effective FS will have the influence on the project completion. Kruger, Jensen [5] contend that the FS acts as a tool in making decisions—a means to conduct analysis of the advantages and disadvantages of project implementation. It can vary from form to form depending on the prospective projects in term of scope. The broad structure of FS can be found out and it will be used to render diverse possibilities in a locale in general or simply do a comparison of a single selection to others in a project in specific. In overall, there are not many differences of opinion in understanding the phrase "feasibility study" in academic environment. It is just an essential tool to make decisions to any construction project proposals. Under Vietnamese construction law, it is defined as a document presenting the contents of study on the necessity for, feasibility and effectiveness of, construction investment in accordance with the selected basic design plan, which serves as a basis for consideration of, and decision on, construction investment [6].

2.2 Contents of FS

A literature review conducted by Jónsson [7] concluded that a theoretically best practice format for a typical FS for public projects should comprise the six major following sections, namely project overview, alternatives, benefits and cost, net present value (NPV), sensitivity analysis, and recommendations. However, the structure of a typical FS under the Article 54 of the Construction Law No. 50/2014/QH13 dated 18/6/2014 (be amended and supplemented by the No. 62/2020/QH14 dated 17/6/2020) includes merely two major sections, namely basic design and other contents. As for basic design, it consists of explanatory documents and basic design drawings. They shall elaborate initial solutions for any project, becoming the footing for setting out design in the upcoming phases.

Regarding other contents of FS, it shall often be supplementary documents and it is used for making the explanation or doing analysis why authorities should invest the construction project. It is also structured into five sections and each section will display the contents as the followings: The first section used to describe briefly background information of the project, including the necessary for investment, investment policy, construction investment objectives, and construction locations and to be-used land area, capacity and form of construction investment. The second section used to describe profoundly the availability of project locale as well as the appropriateness of project proposal. The third section is set up to do necessary analyses and to evaluate possible impacts of external and internal factors when the project proposal shows up. The fourth section will make clear of financial issues of the project. The last section will supplement other relevant contents that suit with project proposal. They may be the distinct requirements from project sponsors to commitments of project owner on tax repayment or extra analyses of demands for products in domestic and international marketplace.

As the above breakdown structure, it is easy to figure out that there are a few similarities and much more disparities in building FS contents between the best practices of western countries and Vietnamese construction laws. While the similarities are rendered in the ways to describe project overview, these disparities show in steps of analyzing benefits and cost, evaluating NPV, and conducting sensitivity analysis under the model of Jónsson [7]. In overall, the way to build up a project FS in European countries is clearer and become more systematically than that of in Vietnam, particularly they focus more on the efficiency of each proposed solutions rather than how many documents you have to submit.

2.3   The Roles of FS in a Project Life Cycle

Matson [1] and Brockhouse [2] believed that the FS appearing at the beginning of the process of project development shall act as a machine which is used to test whether such analyses under technical and economic aspects shall work properly to create feasible products. The study shall additionally expose the responses of business to alteration to those fundamental expectations. Thank to FS, planning developers can draft their opinions on paper, avoiding inaccuracies in project design before their implementation, which can lead to negative effects on project performance. Besides, a FS that can provide essential lessons to reduce dramatically costs of the project [1]. Abou-Zeid, Bushraa [3] contended that thanks to the measurements proposed in the FS, persons who are in charge of making decision should do the comparisons on varied solutions to decide their investment properly. The incorporation of research, data collection, and analysis in a FS is put into consideration on the effectiveness of investments in new technology or projects. The FS, therefore, shall respond to key questions relating to the viability of project's technical and financial aspects, involving project structure and organization and the costs, benefits, and risks as well [5]. A decent FS indicates that the proponents should depict the whole scenario and some various details in how a success project will be made. In a process of official decision-making, the FS also appears to do the risk identification and mitigation, and to detect any potential of law breaching as well. The study is useful in provision of data and analysis documents for project developers to work with sponsors or other expected business in term of strong support [5].

2.4 Definition of Quality Management

PMBOK® Guide—Sixth Edition (2017) indicated that the processes and activities of the performing organization to specify quality policies, objectives, and responsibilities so that the undertaken project will meet the needs are called project quality management. Policy and procedures will be used in deployment of the quality management system with uninterrupted process improvement activities performed throughout, as appropriate. The three main contents of quality management system include quality management planning, quality management, and quality control. Under Point 1 Article 2 of Decree No. 06/2021/ND-CP dated 26 January 2021 of the Vietnamese government, the quality management of construction works means the managerial activities of entities participating in construction activities as regulations in this Decree and other relevant legislations during the process of preparation and investment implementation for construction works, exploration and usage of the works so as to ensure quality and safety of the works.[1] Hence, quality management of the FS shall involve the processes and activities of the performing organization to point out quality policies, objectives, and responsibilities so that the undertaken FS will meet its requirements under current Vietnamese construction regulations in term of legality, quality, and safety satisfaction.

2.5 Critical Successful Factors (CSFs) of a FS

Its definition is in term of "… the limited number of areas, in which results, if they are satisfactory, will ensure successful competitive performance for the organization … … the few key areas where things must go right_ for the business to flourish … areas of activity that should receive constant and careful attention from management … the areas in which good performance is necessary to ensure attainment of [organizational] goals" [8]. There are a lot of researches on CSFs such as exploring CSFs for stakeholder management in construction projects [9], identifying programs CSFs in construction industry [10], factors contributing to successful public private partnerships in road infrastructure investment in Vietnam: stakeholder's perspective [11], and so on. Yang, Shen [9] pointed out three top factors in stakeholder management in Hong Kong construction industry, including: (1) Managing stakeholders with social responsibilities (economic, legal, environmental and ethnical) (2) Exploring the stakeholders' needs and constraints to the project, (3) Communicating with and engaging stakeholders properly and frequently. Kiani, Yousefi [10] found out seven most important CSFs in Iranian construction industry as follows: (1) Support from senior management; (2) Clear and realistic goals, deliverables, benefits of program; (3) Clear and realistic goals of each project aligning strategic objectives; (4) Right cost estimates of program; (5) Effective program cost management; (6)

---

[1] http://vanban.chinhphu.vn/portal/page/portal/chinhphu/hethongvanban?class_id=1&_page=1& mode=detail&document_id=202585.

Aligning and directing program components benefits toward strategic benefits; and (7) Right schedule estimates of program. Nguyen, Smith [11] listed out CSF groups in road infrastructure project in Vietnam, comprising (1) Prevailing environment; (2) Project participants; (3) Project implement ability; (4) Effective procurement; (5) Sound financial package; and (6) Government support. Chan, Scott [12] set up a hypothesis that that "Project success is a function of project-related factors, project procedures, project management actions, human-related factors and external environment and they are interrelated and intra-related." Duy Nguyen, Ogunlana [13] also discovered 20 CSFs in Vietnamese construction industry and ended up with the five main factors, namely (1) Competent project manager; (2) Adequate funding until project completion; (3) Multidiscipline / competent project team; (4) Commitment to project; and (5) Availability of resources. Obviously, there is not much research of CSFs on particular consultancy activity like FS. CSFs of a FS under this research may be in term of factors/attributes which will affect directly the success of conducting a comprehensive FS. These factors possibly appear in performance of project overview, alternatives, benefits and cost, NPV, sensitivity analysis, and recommendations under the best practice model suggested by Jónsson [7]; or in elaboration of the two major set of documents comprise of basic design (explanatory documents, drawings) and other contents under Vietnamese construction laws. This research, therefore, will try to identify all possible factors, which affect severely to quality of the FS under the base of Vietnamese construction law. The research also unveils the relationships among them to help project stakeholders gain a better understanding before making any decision of their investment. The final purpose of compiling any FS in Mekong Delta is to let respective authorities appraise and approve it. Therefore, the success of a FS is to make sure its contents is in accordance with their requirements under the Article 54 of the Construction Law No. 50/2014/QH13 dated 18/6/2014.

3 Research Methodology

The methodology in this research shall include two main parts. The first 10 questions appeared after the completion of a literature review on quality of FS. The first question "Is there any relationship between the quality of FS and construction performance?" is to find the correlation between the quality of the two phases, pre-construction phase and construction phase, and the second question "Is the higher the quality of FS the better managerial skills of the employer and project team have and vice versa?" is to look for the effects of project stakeholders on the quality of FS. The third question "Are there any sustainable development goals involved in the FS contents?" is to seek for due consideration on environmental issues from the FS doers while the forth question "Is there any correlation between quality of FS contents and its contract price?" is to gather the effects of consulting contract price on the quality of FS. The fifth question "Can the quality of FS contents affect procurement phase?" aims to find any interconnected linkage after the completion of FS phase. The next question "Is there any association between project loan agreements and the FS contents?" is to deal with international constraints from third parties to the contents of FS. In order to discover the capability of local consultant in producing the FSs, in other words, human behavior factors, the seventh question "How about the capability of local consultants in preparation of FS contents?" is posed, and the eighth question "Which primary criteria authority bodies usually use to conclude contents of FS are possible or impossible?" is composed to understand critical factors wherein activities of appraisal and approval of FS usually occur from viewpoints of authority bodies. While the ninth question "What are difficulties the employers/ consultants usually encounter as they conduct the reports? If yes, how can they resolve the problems?" is used to diagnose common troubles that FS makers usually confront and their solutions, the last question "May the FS makers use Building Information Modelling or other advanced technology in preparation of the FS contents?" to hunt for ideal solutions in coping with current conditions of making the feasibility studies in Vietnam. The usage of those 10 questions is to interview three experts who are working in construction domains in Mekong River Delta of Vietnam. One expert is working for a planning and investment department in Can Tho, the biggest city in Mekong River Delta. He has much more working experience in public investment and plays a key role of project appraisal. The next expert is a staff of Can Tho Project Management Unit for Construction Investment and he is responsible for performing activities of pre-construction phase of any project in his office, the largest and renowned unit in Mekong Delta until now. The last specialist is a scholar who is doing his researches on construction matters and lecturing in Can Tho University, the third largest and famous university in over Vietnam. An additional interview with the expert who is working for an international consulting construction company in Can Tho also happened. His company plays a key role of project management unit and supervisor for infrastructure projects in Can Tho, Soc Trang and Tra Vinh. The interviewees can offer several success factors of FS in Mekong Delta that shall become valuable information.

The next part is try to develop a questionnaire survey. It took a lot of time to review carefully the literature on CSFs for FS contents and the knowledge of success factors from large projects in the current Vietnamese construction practice undergone by stakeholders. As a result, a questionnaire that comprised 25 questions had appeared. The questionnaire merely deals with the issues of FS contents in Group-B projects or over. The project priority under this research places on civil engineering (hospitals, schools …), infrastructure (residential zones, wastewater treatment zones …) or transportation (national high ways, high speed rails …) with the total amount of investment as from \$45 billion VND to \$2,300 VND or higher (equivalent to from \$1.9 million USD to \$98.7 million USD) to civil engineering projects, from \$120 billion VND to \$2,300 billion VND or higher (equivalent to from \$5.15 million USD to \$98.7 million USD) for transportation projects, and from \$60 billion VND to \$1,500 billion VND or higher (equivalent to from \$2.58 million USD to \$64.4 million USD) to infrastructure projects. Additionally, the primary source of project funding usually derives from state budgets or tied aid credits of international multilateral or bilateral sponsors.

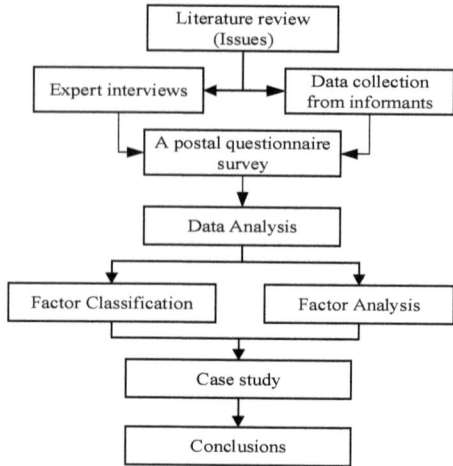

Fig. 1 Overall research methodology

After finalizing the questionnaire, a pilot test to be run by two experts in order to check the wording of questions, identifying ambiguous questions, and testing time of completion. The final questionnaire became neater and more comprehensive with the total number of question of 30. It is ready for delivering to potential respondents to collect needed data. The purpose of the questionnaire is to find out some typical factors affecting to quality of FS reports in Mekong River Delta. Construction companies, authority bodies and other construction-relevant actors in Mekong River Delta like Can Tho, Soc Trang, and Tra Vinh received the questionnaire by emails and other means of communication like Facebook messenger or Zalo application (a common social networking in Vietnam). The analysis of collected data will occur after three weeks of observation, waiting for and receiving survey responses. The software SPSS will prevail to find out CSFs during the time of data analysis beside the activities of factor classification. The final results are validated by case studies of Can Tho general hospital project, and the project of developing Can Tho and enhancing urban resilient capability. The whole process to implement this research study as shown in the Fig. 1.

4 Research Results

4.1 Factors Affecting to Quality Management of FS

There are total 50 factors which are determined to affect to quality management of a FS showed up and they contributed to seven groups of technical, economic, managerial, human behavior, sustainable, legal, and scheduling. Base on the ranking of each factor by counting the highest frequency of selected answer from respondents and the analysis of results from program SPSS 22, they are divided into two groups. The factors, that the percent of cases equal to or greater than 50% will form Group 1, and the others will be Group 2. By doing this, the factor group 1 includes factors like (1st) Experience and capability of producing FS report of consultant, (2nd) Project effects on social security and locale economy, (3rd) Analysis of project effects on surrounding environment and natural resources, (4th) Procedures in preparation of FS report and component reports (resettlement, EIA ...), (5th) Input data, survey data of existing conditions is accuracy and deservedly reliable, (6th) Project economic efficiency, (7th) Effects of initially designated development planning or programs on project demand, (8th) Professional consultant service providers, (9th) Project compliance with initially designated development planning or programs, (10th) The effects of authorities (Governing body, project owner ...), (11th) Participation of top specialists in criticism of all project aspects, (12th) The sum total of project investment, (13th) Managerial skills of project team, (14th) Consultant's capability of report writing, (15th) Outline and duty of producing FS report are made correctly and appropriately, (16th) Investment and research of consultant, (17th) Project objectives are specific, clear, and measurable, (18th) The comprehensiveness of basic design, (19th) Specialist involvement during project appraisal. The results of factor analysis are shown in the Fig. 2 below.

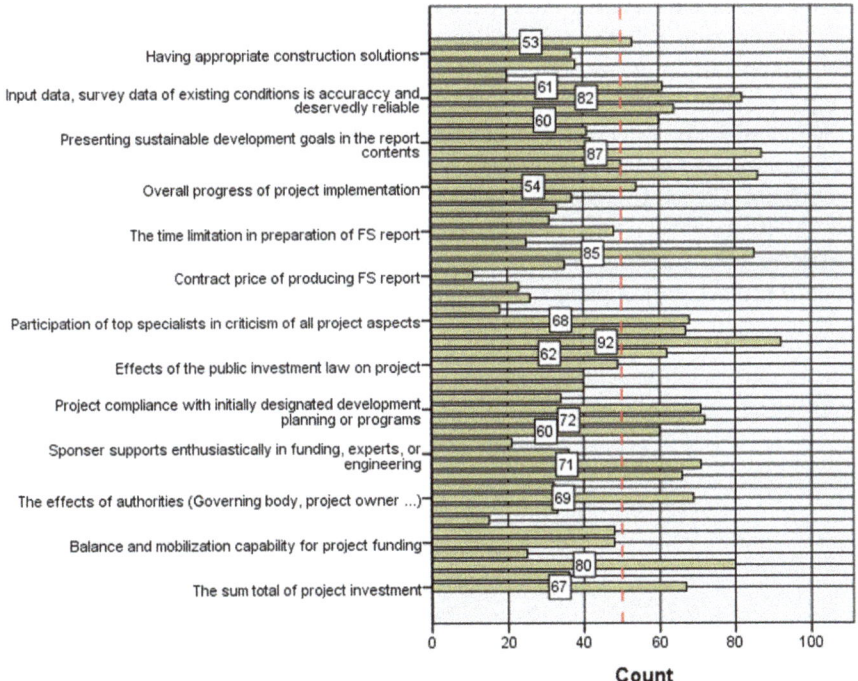

Fig. 2 Horizontal bar chart of all factors

4.2 Development of a Checklist to Qualify FS

Under Vietnamese construction law, the process of performance of FS reports can split into three steps. It starts with the step of recruiting a consulting service provider, and then followed by the step of FS production. The final step in this process is to submit FS products to authorities for their appraisal and approval. Therefore, the 19 most important factors that have been found previously will be filled respectively into those three steps. Among the above-mentioned 19 factors, the factor of "the effects of authorities (Governing body, project owner …)" is duplicated in the first step and the last step primarily because as the rule of thumb the authorities in practice in Mekong River Delta usually show their power in selecting their intended contractors and manipulating desired results of project preparation. Hence, the total number of the factors that are suggested to be used in the quality checklist should be 20, including 18 separate factors and one duplicated factor.

The checklist for quality evaluation as indicated in the Table 1 below will also exploit the Likert scale from the lowest value 1 to the highest value 5 to rate for each factor. The highest score for each factor is 5, so the maximum score for the entire checklist shall be 100. Base on the total score of the checklist, evaluator shall classify evaluated objects by three pillars. As for the first pillar, the total score should be from 80 to 100, it means the quality of FS performance meets sufficiently its requirement. For the second pillar, the total score should be from 60 to less than 80, it means the quality of FS performance meets the requirement after some necessary adjustments. The third pillar, the total score should be less than 60, it means the quality of FS performance does not meet the requirement and it should be rejected. The checklist then is validated by two experts who have participated in the initial phase of the research and by case studies of the two projects: Medical equipment project of Can Tho general hospital (total investment cost: 22,052,169 Euro; Donor: French government, Group-B project, project period: 2013–2014), and Can Tho River embankment—support program to respond to climate change (Total investment cost: 32,935,066 Euro; Donor: Agence Française de Développement, Group-B project, project period: 2015–2018).

Table 1 Checklist to qualify a FS

	A FORM FOR EVALUATION OF FEASIBILITY STUDY REPORT					
	PROJECT NAME .... date ... month ... year ...					
		RATING SCALE				
NO.	EVALUATED CONTENTS	1- Dissatisfied	2-Fairly dissatisfied	3- Acceptable	4-Fairly satisfied	5-Satisfied
I	RECRUITING A CONSULTANT SERVICE PROVIDER (MAXIMUM SCORE IS 20)					
1	Outline and duty of producing FS report are made correctly and appropriately	☐	☐	☐	☐	☐
2	Experience and capability of producing FS report of consultant	☐	☐	☐	☐	☐
3	Professional consultant service providers	☐	☐	☐	☐	☐
4	The effects of authorities (Governing body, project owner ...)	☐	☐	☐	☐	☐
II	FEASIBILITY STUDY PRODUCTION (MAXIMUM SCORE IS 30)					
5	Managerial skills of project team	☐	☐	☐	☐	☐
6	Investment and research of consultant	☐	☐	☐	☐	☐
7	Input data, survey data of existing conditions is accuracy and deservedly reliable	☐	☐	☐	☐	☐
8	Procedures in preparation of FS report and component reports (resettlement, EIA ...)	☐	☐	☐	☐	☐
9	The sum total of project investment	☐	☐	☐	☐	☐
10	Consultant's capability of report writing	☐	☐	☐	☐	☐
III	APPRAISAL AND APPROVAL OF FEASIBILITY STUDY (MAXIMUM SCORE IS 50)					
11	Project objectives are specific, clear, and measurable	☐	☐	☐	☐	☐
12	Project economic efficiency	☐	☐	☐	☐	☐
13	The comprehensiveness of basic design	☐	☐	☐	☐	☐
14	Project effects on social security and locale economy	☐	☐	☐	☐	☐
15	Analysis of project effects on surrounding environment and natural resources	☐	☐	☐	☐	☐
16	Effects of initially designated development planning or programs on project demand	☐	☐	☐	☐	☐
17	Project compliance with initially designated development planning or programs	☐	☐	☐	☐	☐
18	Participation of top specialists in criticism of all project aspects	☐	☐	☐	☐	☐
19	Specialist involvement during project appraisal	☐	☐	☐	☐	☐
20	The effects of authorities (Governing body, project owner ...)	☐	☐	☐	☐	☐
	TOTAL (MAXIMUM SCORE IS 100)	... / 100				
	CONCLUSION: - Meet its requirement: 80 – 100 - Meet the requirement after some necessary adjustments: 60 to <80 - Not meet the requirement: <60	☐ Meet requirements		☐ Meet requirements after adjustments		☐ Not meet requirements
	EVALUATOR (Signature and full name)	CHECKER (Signature and full name)		APPROVER (Signature and full name)		

5 Conclusions and Recommendations

This research has discovered initially 50 factors by secondary data and interview methods, then ended up with 19 important factors by the survey questionnaire. Overall, the identified factors which affects to quality management of FS reports derived primarily from human actors (consultants, project owners, governing bodies, and third parties) as Duy Nguyen et al. [6] concluded in their paper. The procedures of project implementation and the reliability of input data is the main difference in here.

For academia, this research will help to systemize researches relating to pre-construction or pre-investment phase and to build up a checklist for quality measurement of FS in Mekong River Delta. For construction domains, the research found out 19 critical factors and their constraints in the checklist in carrying out completely a typical FS report. They comprise both internal and external effectiveness factors. Scrutinizing on those factors when conducting any FS in Mekong River Delta will help construction domains such as government bodies, project owners, consultants, or third parties to work more efficiently and to make investment decisions more appropriately.

In case there is no limitation in time and budget, this study will be more valuable if the range of data collection is extended to the South of Vietnam or whole country or even the objectives are expanded to private sectors to conduct some interesting comparisons. It is strongly recommended to carry out such researches in the near future to gain better understanding on construction investment procedures in Vietnam as a footing to call for foreign direct investment in Mekong River Delta.

References

1. Abou-Zeid A, Bushraa A, Ezzat M (2007) Overview of feasibility study procedures for public construction projects in Arab countries. Eng Sci 18(1):377–386
2. Brockhouse JW (2010) Vital steps: A cooperative feasibility study guide
3. Chan APC, Scott D, Chan APL (2004) Factors affecting the success of a construction project. J Constr Eng Manag 130(1):153–155
4. Dahie AM, Osman AA, Omar AA (2017) The role of project management in achieving project success: Empirical study from local NGOs in Mogadishu-Somalia. Int J Eng Sci 14844
5. Duncan WR (1996) A guide to the project management body of knowledge
6. Duy Nguyen L, Ogunlana SO, Thi Xuan Lan D (2004) A study on project success factors in large construction projects in Vietnam. Eng Constr Archit Manag 11(6):404–413
7. Gemmill B, Bamidele-Izu A (2002) The role of NGOs and civil society in global environmental governance Global environmental governance: Options and opportunities 77–100
8. Heralova RS (2017) Life cycle costing as an Important contribution to feasibility study in construction projects. Procedia Engineering 196:565–570
9. Jónsson HR (2012) Feasibility analysis procedures for public projects in Iceland. Reykjavík University, Iceland
10. Jónsson HR (2012) Feasibility analysis procedures for public projects in Iceland. Reykjavik University, Master

11. Kiani S, Yousefi V, Yakhchali S, Mellatdust A (2014) Identifying program critical success factors in construction industry. Manag Sci Lett 4(6):1325–1334
12. Kruger CE, Jensen J, Frear C, Yorgey G (2018) Completing a successful feasibility study for an anaerobic digestion project. Washington State University Extension
13. Matson, J. (2000). Cooperative feasibility study guide
14. Naoum S (2012) Dissertation research and writing for construction students. Routledge
15. Nguyen DT, Smith C, Manning M, Nguyen DT (2014) Factors Contributing To Successful Public Private Partnerships (Ppps) In Road Infrastructure Investment In Vietnam: Stakeholder's Perspective. 2014 Asia Conference on Economics & Business Research (ACEB 2014) CONFERENCE PROCEEDINGS.
16. Nguyễn H (2017) Nhận diện 3 lãng phí trong đầu tư công. Báo Đầu tư điện tử
17. Rockart JF (1979) Chief executives define their own data needs. Harv Bus Rev 57(2):81–93
18. Shen L-Y, Tam VW, Tam L, Ji Y-B (2010) Project feasibility study: the key to successful implementation of sustainable and socially responsible construction management practice. J Clean Prod 18(3):254–259
19. XIII, VNA (2014) The construction law. 50/2014/QH13. V. N. A. XIII.
20. Yang J, Shen GQ, Ho M, Drew DS, Chan AP (2009) Exploring critical success factors for stakeholder management in construction projects. J Civ Eng Manag 15(4):337–348
21. Zhang S, Pan F, Wang C, Sun Y, Wang H (2017) BIM-based collaboration platform for the management of EPC projects in hydropower engineering. J Constr Eng Manag 143(12):04017087

Assessing the Effect of Design Risks on the Performance of Design-Build Projects in Vietnam

Thu Anh Nguyen, Vo Thi Dinh Khanh, Sy Tien Do, Phuoc Quy Dao, and Truong-An Pham

Abstract The design risk can have a significant impact on the success of design-build projects. Based on previous research presented by the authors on the identification of 21 design risks in the design-build projects in Vietnam, the design risk factors in design-build projects were classified into five groups: risk of inaccuracy or inadequate design information, inappropriate design risk, inappropriate designer's capacity risk, inappropriate contractor design capacity risk, risk of ambiguity project scale and scope of work. This study analyses the impact of design risks on project performance through three criteria: schedule, cost, and quality. Structural Equation Modelling (SEM) revealed that risk of improper design, risk of improper designer's capacity, risk of ambiguous project scale, and scope of work have a significant and negative impact on the project's performance. The result improves the understanding of design-build contractors to achieve better results thanks to better design risk management.

Keywords Design-build · Design risk · Project performance · Structural Equation Modeling (SEM)

1 Introduction

Design and build (D&B) procurement has been widely used around the world. After several international integration years, Vietnamese contractors have acquired and applied new technology and management methods. However, this procurement brings about several risks for the contractor, requiring the contractor to combine design and build (D&B) in a contract. D&B contractors are responsible for design risks such as design changes, inappropriate design standards, design schedule delays,

T. A. Nguyen · V. T. D. Khanh · S. T. Do · P. Q. Dao · T.-A. Pham (✉)
Faculty of Civil Engineering, Ho Chi Minh City University of Technology (HCMUT), 268 Ly Thuong Kiet Street, District 10, Ho Chi Minh City, Vietnam
e-mail: phamtruongansos@gmail.com

Vietnam National University Ho Chi Minh City, Linh Trung Ward, Thu Duc District, Ho Chi Minh City, Vietnam

contractors' lack of design experts, etc. Those obstacles have a negative influence on the project efficiency, contributed to intricateness for the contractor. This study presents the impaction of the design risk on the performance of the D&B projects, from that proposes an effective managing method for the contractor to improve project effectiveness.

2 Literature Review

According to the author's previous research, the authors have identified 21 design risks of design risk factors in projects implemented D&B procurement method, classified into five risk groups of inaccurate or improper design; improper designer's capacity; improper design capability of the contractor, ambiguity project scale and objectives [1]. Albert PC Chan, Danny CK Ho, and CM Tam [2] have developed a multivariate regression analysis to determine the importance of the factors to the success of the design and build (D&B) on the project. The article has identified that time efficiency, cost and design quality, contractor's capabilities are the main factors contributing to the overall success of D&B projects. Li-Chung Chao [7] used a multivariate regression model to predict project efficiency through variables: cost performance, time, quality of design, and workforce. The previous references only show the general factors influencing project performance but do not explicitly address design risks. This article identifies, analyzes, and assesses the impact of design risks on D&B project performance.

2.1 Risks in Design

Design risks are technical changes in the design that directly affect the project's performance. The implicit threaten is that the design does not satisfy the project requirements. The threaten includes fundamentally flawed designs, not feasible, ineffective, unstable, or below the owner's standards. A poor design can interfere with project progress [1]. The author's previous research has identified 21 design risks in the design-build project in Vietnam. Exploratory factor analysis that these factors could be categorized into five groups consist of the risk of unsatisfactory design (DR1), risk of unsuitable design (DR2), risk of improper designer's capacity (DR3), risk of unsuitable contractor capacity (DR4), and risk of ambiguous project scale and scope of works (DR5) [1], as shown below:

Table 1 Grouping of design risks

Group of design risks	Sign	Design risks
Group 1: Risk of unsatisfactory design (DR1)	R20	Misunderstand in the information exchange between owner and designer
	R22	Delay of design
	R21	Delay in design problem solving
	H4	The design proposal and drawings do not match
	H7	Wasting time for re-checking the design based on the owner's requirements
	H5	Design structural solutions are not feasible
	R8	The contradiction between key designers
	R9	The subjects are not synchronized when combined
Group 2: Risk of unsuitable design (DR2)	H2	Design over budget
	H1	Design late in submitting the owner
	R4	The contractor's design team works under multiple constraints on the project
	R17	The owner disagrees on the source of the leading equipment or materials used
Group 3: Risk of improper designer's capacity (DR3)	R11	Designers lack experience, knowledge to handle problems
	R14	The designer lacks a sense of responsibility
	R10	The designer spirit is not comfortable, not concentrating
	R15	Delay of design approval, the unqualified approval process
Group 4: Risk of unsuitable contractor capacity (DR4)	R1	The contractor does not have a thorough preparation of the design process
	R2	Lack of high-quality designers
Group 5: Risk of ambiguous project scale and scope of works (DR5)	R5	Project scale is continually changing
	R19	Owner's frequent change orders

2.2 Project Efficiency

A practical project must ensure that it meets the investment goals, investor's needs, expected project schedule on budget, and guarantees the quality of the project. Chan [4] proposed criteria to evaluate the design and build (D&B) projects' performance, including objective and subjective categories. Objective issues include time, cost, health and safety, and profitability, while subjective categories include quality, technical performance, functionality, productivity, satisfaction, and environmental sustainability. The project schedule is apprised by exceeding construction progress, construction time, and construction speed. The three indicators to evaluate the cost is construction cost, cost overrun, and unit cost.

3 Research Method

The research method includes five steps: Firstly, the authors summarize statistical criteria to evaluate the D&B project's effectiveness from reference resources. A multi-choice answer sheet is created, then sent to experts (who have worked more than five years) to revise the questionnaire. After being revised to meet the supplement and appropriate Vietnamese situation, the answer sheets are used for the offline interview. Finally, the authors do statistics on the indicators to assess the project's performance based on the survey results for effective assessment models.

In this study, the authors use the 05 levels Likert scale for the questionnaire design. After studying the previous documents and consulting with experts, the study has adjusted and supplemented the criteria to analyze D&B project performance in Vietnam based on design risks. The contents of the survey include two main parts:

- Part A: asking the interviewer's general information: work experience, position, type of their participated project, etc.
- Part B: Evaluating the project's performance indicators with scale: (1). Very low, (2). Low, (3) Medium, (4) High, (5) Very High [5].

In an allocated condition, the complete degree of the general conditions is described as time [4]. On-schedule is one of the success criteria for a D&B project, evaluated by 03 criteria: time overrun, construction speed, construction time. Cost is stated as the extent to which general conditions motivate completing a project within the estimated budget [4]. It is evaluated through the following criteria: The cost difference between the contract value and the payment value, the cost per m^2 of the construction floor, and the cost incurred. The quality is expressed in terms of specification, function, and form. It is defined as the sum of the features required by a product or service to fulfill a specific need [4]. It is assessed through the following criteria: the compatibility between the desired quality of the investor and the actual completion, the satisfaction of the construction standards, the conformity of the work quality after completion.

Fig. 1 The research process

Table 2 Criteria to measure project effectiveness

Criteria	Measurement criteria	Scale
P1. Schedule	P1.1.Schedule variance	(Total time to complete—Total time to complete the plan) / Total time to complete the plan. [3]
	P1.2. Construction speed	Floor area / (finishing time—starting time) [3]
	P1.3. Schedule delay since the delay in material supply	Number of delays due to material supply [4]
P2. Cost	P2.1. The cost difference between the contract value and the payment value	(Total cost of completion—Total cost estimate) / Total cost estimate. [3]
	P2.2. Cost per m^2 of gross floor	Cost of building divided by gross floor [4]
	P2.3 Cost incurred	Accounting for 5%, the highest 10% of the total cost [expert interview]
P3. Quality	P3.1. Compatibility between the quality desired of investor and actual	(1) Very low—(5) Very high
	P3.2 Compliance in construction standards	(1) Very low—(5) Very high
	P3.2 Building Quality Compliance after Completion	(1) Very low—(5) Very high

4   Research Results

There are 180 survey sheets sent to specialists working in the civil engineering field with 169 responses. One hundred fifty-three answer sheets are appropriated for the research. The results are shown in Table 6 Characteristics of surveyed subjects, Appendix.

4.1   *Indicators Affect Project Performance*

The statistical results of the surveying are summarized in Table 7. Table of statistics projects efficiency measurement criteria, Appendix.

4.2   *Correlation Between Groups of Design Risks and Project Performance*

Based on and survey results from previous research, the research builds a model to measure the relationship between the 05 risk concepts. Research using the confirmatory factor method to test the appropriateness of the measurement model.

The significant coefficients to evaluate the appropriateness of the model: chi-square/df coefficient is 1.622 less than 2, model conformity index is 0.862 greater than 0.8, CFI coefficient (0.905) is more significant than 0.9, RMSEA coefficient is 0.067, less than 0.08 [6], all values meet the evaluation threshold, so the model is suitable for CFA analysis. The standardized load coefficients of each observed variable are more significant than 0.5 [6]. The Critical Ratios (CR) of the variables DR1, DR2, DR3, DR4 is greater than 0.7 (Hair et al., 2010); only variable DR5 has CR = 0.547 < 0.7. The Average Variance Extracted (AVE) of the three variables DR1, DR2, DR3, DR5 was 41.6%, 42.3%, 41.8%, and 38.1%, respectively, less than 0.5 [6]. (Table 5).

For the convergent validity assessment of the model, the above criteria must be met; however, this situation is rarely appearing. Therefore, although variables DR1, DR2, DR3, and DR5 have relatively small extraction variances, these are essential characteristics that affect the design process of the project and the effectiveness of the D&B project, so they are still included in the hypothetical model.

The most significant correlation coefficient between the component concepts in the DR2 and DR5 is 0.804 < 0.9 (Hair et al., 2010), so the concepts have discriminant validity. From the results table, we see that all relationships between conceptual

Table 3 Table of results of the normalized load factor, internal reliability, average extraction variance (AVE) of the model

Relationship			Loading	CR	AVE (%)
R20	←	DR1	0.698	0.85	41.6
R22	←	DR1	0.675		
R21	←	DR1	0.698		
H4	←	DR1	0.675		
H7	←	DR1	0.58		
H5	←	DR1	0.635		
R8	←	DR1	0.562		
R9	←	DR1	0.624		
H2	←	DR2	0.677	0.741	42.3
H1	←	DR2	0.802		
R4	←	DR2	0.507		
R17	←	DR2	0.579		
R11	←	DR3	0.771	0.737	41.8
R14	←	DR3	0.679		
R10	←	DR3	0.574		
R15	←	DR3	0.534		
R1	←	DR4	0.713	0.786	65.0
R2	←	DR4	0.890		
R5	←	DR5	0.523	0.547	38.1

Table 4 The correlation coefficient between design risk group

Relationship			Loading	P
DR1	⇔	DR2	0.744	***
DR1	⇔	DR3	0.756	***
DR1	⇔	DR4	0.519	***
DR1	⇔	DR5	0.563	***
DR2	⇔	DR3	0.692	***
DR2	⇔	DR4	0.575	***
DR2	⇔	DR5	0.823	***
DR3	⇔	DR4	0.710	***
DR3	⇔	DR5	0.412	0.004
DR4	⇔	DR5	0.472	0.002

Table 5 The correlation coefficient between design risk group and project performance

Relationship			Loading	S.E	C.R	P
P1	←	DR2	0.394	0.111	3.568	***
P2	←	DR2	0.481	0.140	3.446	***
P3	←	DR2	0.217	0.060	3.636	***
P1	←	DR3	−0.359	0.093	−3.850	***
P2	←	DR3	−0.385	0.106	−3.646	***
P3	←	DR3	−0.283	0.063	−4.481	***
P1	←	DR5	−0.219	0.078	−2.807	0.005
P2	←	DR5	−0.302	0.108	−2.791	0.005

variables have statistical significance < 0.05, concluding that conceptual variables are correlated.

After confirming that structural measures are reliable and valid, the study evaluates the structural model results to test the predictive ability of the model and the relationships between the research variables. From the CFA study results and consideration of the relationships of the conceptual variables, a model was developed to express the relationship between the five risk groups and the project performance. The significant coefficients to evaluate the suitability of the model: chi-square/df coefficient is 1.687 less than 2, model conformity index is 0.796 less than 0.8, CFI coefficient is 0.854 greater than 0.9, RMSEA coefficient is 0.067, less than 0.08 [6], all values meet the evaluation threshold, so the model is well evaluated.

The results of regression coefficients for hypotheses from the model are significant < 0.05. Hence, the importance of the measurement attributes to the model is enhanced.

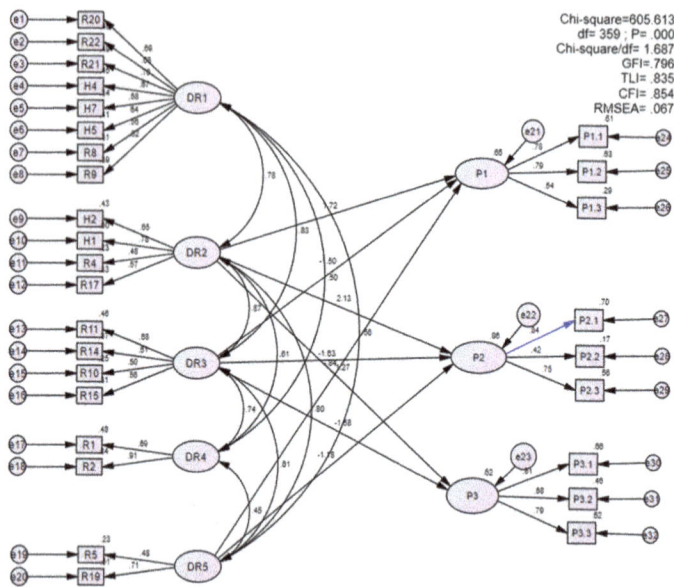

Fig. 2 The theoretical model of the effect of design risks on project performance

Effect of design risks on the effectiveness of the project progress

According to the results of the research model, "project progress" is affected by 03 design risk factors: "Risk of unsatisfactory design", " Risk of improper designer's capacity", "Risk of ambiguous project scale and scope of works". The risk of unsuitable design has the most significant impact (impact factor is 1.72) and directly impacts and has a direct ratio with the project progress. When fewer details of a project are provided to the designer, the more assumptions the designer will make, often lead to design problems that exceed the owner's budget. Investors who have their perception of ideas, equipment, and materials will often be dissatisfied with the proposals of the designs. If the opinions between the two parties are too different, the investor will request the redesign; then, the designer will not be able to re-apply the old design to future problems. Redesign often results in changes in equipment, materials, and implementation details, thus lengthen the project's overall progress. Besides, inappropriate designer's capacity directly affects and has a reciprocal ratio on the project progress (impact factor 1.498). The designers' capacity directly affects the ability to develop the project; the neophyte designer unable to have adequate quality control, leading to document error, inconsistent.

Consequently, the design profile must be updated several times, disrupts the project construction process. The risk of vague project size and scope also impacts project progress (impact factor –0.844). In the D&B contract, the investor and the contractor should be clear on essential issues such as the investor's requirements, project scale, and scope of work. Vaguely defining the work scope is the most common reason for later financial, progress disputes in a D&B project.

The progress of the project is considered as "Schedule variance" (0.78), "construction speed" (0.79), "Schedule delay since the delay in material supply." (0.54). Thereby, to enhance the efficiency of construction progress, it is necessary to reduce the risk of inappropriate design, improve designer capacity, clearly define the project scale and scope of work, investor's requirement.

Effect of design risk on project cost-effectiveness

Based on the results of the research model, it is found that "project cost" is affected by "Risk of unsuitable design", "Risk of improper designer's capacity", "Risk of ambiguous project scale and scope of works". In which, the risk of unsuitable design has the most considerable direct influence on the project's cost-effectiveness (impact factor 2.13). If an inappropriate design is approved for execution, it will increase errors and costs incurred from rework for the contractor. The risk of improper designer's capacity has an inverse, direct effect on the project cost (impact factor 1.63). Designers rely on the available experience and knowledge to make the design and materials suitable for the budget, at the same time design according to standards, control design quality well, minimize errors, and avoid errors. The risk of design rework causes incurred costs for the contractor. The risk of ambiguous project scale and scope of work directly affects the project cost (impact factor 1.18). Increasing the project scale increases the project-incurred cost. The owner requests to change and rework the design to increase the cost specified in the contract.

The project cost performance is evaluated by "The cost difference between the contract value and the payment value" (0.84), "Cost per m^2 of the gross floor" (0.42), and "cost incurred" (0.75). Hence reducing the risk of design mismatches, enhancing personal capacity, clarifying project scale and scope of work are the three main criteria to reduce construction costs for contractors, limit the incurred costs, and increase project cost efficiency.

Effect of design risk on project quality performance

Based on the results of the research model, the effectiveness of "project quality" is affected by "Risk of unsuitable design", "Risk of improper designer's capacity". In which "Risk of improper designer's capacity" has the most considerable negative and direct impact on "project quality" (impact factor –1.58). The lower risk of unsuitable design, the higher quality of the project. Experienced and knowledgeable designers define the nature of the work for the project, propose precise project information requirements for the investor to confirm, able to handle adequate design arising from the investor, design under the standards to ensure the quality of the project Unsuitable design risk directly affects the project quality (impact factor 1.27).

"Project quality" is evaluated by three criteria: "Compatibility between the quality desired of investor and actual" (0.81), "Compliance in construction standards" (0.68), "Building Quality Compliance after" (0.79).

5 Conclusion

Based on the author's previous research, there have been identified 21 design risks in the project design process implemented in the form of D&B and arranged these risks into five groups: Risk of unsatisfactory design; Risk of unsuitable design; Risk of improper designer's capacity; Risk of an unsuitable contractor; Risk of ambiguous project scale and scope of works. The study also analyzed the project performance indicators based on three criteria: progress, cost, and quality.

A theoretical model was established for five design risks and three project performance criteria to assess design risks' effect on project performance. The model results identify three design risks that have a significant direct impact on project performance with a statistically significant level: Risk of unsuitable design that directly affect the project schedule, cost, and quality. The risk of unsuitable designer capacity directly affects and has a reciprocal ratio on the effectiveness of project progress, cost, and quality. The risk of ambiguous project scale and scope of works directly impacts the reciprocal ratio with progress and project cost-effectiveness.

From the results of the survey, several management regulatory implications can be inferred. Firstly, risk factors are often related to each other; a diverse network of interactive risk factors can better reflect the complex nature of their impact on the project objectives. Second, based on the interaction between the factors, contractors, investors can apply reference and propose responses to these risk factors systematically and comprehensively. Third, a properly planned design is critical to a D&B project's success, requiring careful check from designers to check the inputs and outputs of the design carefully. Finally, D&B contractors should build up an experienced and highly spiritual designer team besides enhancing communication and coordination systems among project participants.

Research and explore the impact lines of the design risk factors on the D&B project performance. From a theoretical perspective, the research contributes to the D&B procurement system and enriches risk analysis and management research. Meanwhile, in practical terms, the research results can provide the contractor with a checklist, analysis of risks in the project design process. The study's findings can also help design-build contractors understand how to achieve project success through design risk management from a comprehensive and systematic perspective.

However, the research process also has limitations. First, the study was performed base on respondents' subjective perceptions of the impact of design risk factors on project performance. Second, the study uses only progress, cost, and quality as project effectiveness data and ignores the other aspects. Although those three factors are the project's main issues, other factors related to project performance can be considered.

Research expands the risk management understanding of projects in the design-build project and provides the opportunity and basis for future research. Future research can be performed to explore integrated risk management in the connection between design, supply, and construction of a design-build project.

Acknowledgements This research is fund by Vietnam National University, Ho Chi Minh City (VNU- HCM) under grant number To-KTXD-2020-19. We acknowledge the support of time and facilities from Ho Chi Minh City University of Technology (HCMUT), VNU-HCM for this study.

Appendix

Table 6 Characteristics of surveyed subjects

Characteristics	Classify	Count	Percentage (%)
Experience in the construction industry	Less than 3 years	52	30.8
	3–5 years	58	34.3
	5–10 years	38	22.5
	Over 10 years	21	12.4
The subject's current expertise	Architect	17	10.1
	Structural Design Engineer	60	35.5
	M & E Engineer	10	5.9
	Field Engineer	31	18.3
	Project management	51	30.2
Role when participating in construction design projects	Investor, PMU	24	15.7
	Construction contractor	75	49.0
	Project management consulting unit	4	2.6
	Design consulting unit	47	30.7
	State units	3	2.0
Type of project	Civil and industrial works	148	93.1
	Bridge and road works	7	4.4
	Waterworks	4	2.5
Capital source for project implementation	State capital	38	18.0
	Foreign investment capital	45	21.3
	Private capital	108	51.2
	State and private capital	20	9.5

Table 7 Table of statistics projects efficiency measurement criteria

Criteria	Factor	Results
P1. Schedule	P1.1.Schedule variance	35.3% interviewers assumed actual construction progress delay 4%–6% compared to design plan
	P1.2. Construction speed	45.8% of interviewers assumed the construction volume fluctuates within 10% of the expected volume
	P1.3. Schedule delay since the delay in material supply	35.3% of interviewers assumed it happened 2 times
P2. Cost	P2.1. The cost difference between the contract value and the payment value	38.56% of interviewers assumed the differences between the contract and settlement value: 2%–4%
	P2.2. Cost per m^2 of gross floor	52% of interviewers assumed the average cost is from 5 million VND / m^2—7 million VND / m^2
	P2.3 Cost incurred	41% of interviewers assumed the cost incurred was at an average rate of 1.5%–2.5% of the total costs
P3. Quality	P3.1. Compatibility between the quality desired of investor and actual	44% of interviewers assumed it is 90%–99%
	P3.2 Compliance in construction standards	79% of interviewers assumed it meets the standards in the construction process
	P3.2 Building Quality Compliance after Completion	58% of interviewers assumed the quality of works after completion is appropriate

References

1. Khanh VTĐ, Thu NA (2020) Assessing design risks in the design-build projects in Vietnam. Vietnam J Constr, pp 221–225
2. Chan AP, Ho DC, Tam CM (2001) Design and build project success factors: a multivariate analysis. J Constr Eng Manag 127(2):93–100
3. Ling FYY, Chan SL, Chong E, Ee LP (2004) Predicting the performance of design-build and design-bid-build projects. J Constr Eng Manag 130(1):75–83
4. Chan FN, Lam EW et al (2002) The framework of success criteria for design/build projects. J Manag Eng 18(3):120–128
5. Trọng H, Ngọc CNM (2008) Analyze research data with SPSS. University of Economics Ho Chi Minh City, Hong Duc
6. Hair JF (2013) Multivariate data analysis. Pearson, USA
7. Chao L-C, Hsiao C-S (2012) Fuzzy model for predicting project performance based on procurement experiences. Autom Constr 28:71–81

Developing a Detailed Process to Quantify Risks on Variations of Project Time

Duc-Anh Le, Long Le-Hoai, and Chau Ngoc Dang

Abstract The quantification of project time variations is important to project execution. However, so far, there has been still a lack of studies that focus on specific project schedules. Using structured and semi-structured interviews with contractors, this study develops a process in the planning phase of risk responses, which could help to assess the level of project time variations based on the impact of event chains. The application of the process is also illustrated using a case study, which is a part of the pre-validated actual project. The developed process could help to evaluate the precise impact of risk events with simple handling. The findings of this study could provide practitioners (e.g., project management teams) with a guideline to quantify the impact of risks on the dynamic levels of project time based on the analysis of the sensitivity chart of risk events. Hence, they could not only avoid the phenomena of excessive optimism and pessimism, but also allocate resources more appropriately during project implementation processes to reduce the unexpected impacts of risks.

Keywords Time variations · Project time · Event chains · Process · Monte Carlo

1 Introduction

Risk is a common concern of construction management because of specific distinctive characteristics [4, 14, 18]. Commonly, in a construction project, risks depend on its complexity and the strategic nature of its products. Each project is unique with certain characteristics such as various stakeholders, open production system, long

D.-A. Le · L. Le-Hoai
Faculty of Civil Engineering, Ho Chi Minh City University of Technology (HCMUT), 268 Ly Thuong Kiet Street, District 10, Ho Chi Minh City, Vietnam

Vietnam National University Ho Chi Minh City (VNU-HCM), Linh Trung Ward, Thu Duc District, Ho Chi Minh City, Vietnam

C. N. Dang (✉)
Faculty of Civil Engineering, School of Engineering and Technology, Van Lang University, Ho Chi Minh City, Vietnam
e-mail: chau.dn@vlu.edu.vn

© The Author(s), under exclusive license to Springer Nature Singapore Pte Ltd. 2023   219
J. N. Reddy et al. (eds.), *ICSCEA 2021*, Lecture Notes in Civil Engineering 268,
https://doi.org/10.1007/978-981-19-3303-5_17

time performance, and vulnerable environments by internal and external factors [16]. Risks and uncertainties always appear in any stage from preparation to operation regardless of size, characteristic, complexity, and place of construction projects [14]. Construction projects will receive detrimental consequences such as increasing costs and time delay, even project failure, if there exist failures to respond to risks and uncertainties [14, 18]. Therefore, previous studies have focused on risk management (e.g., risk identification, risk classification, qualitative risk analysis, and quantitative risk analysis), of which risk quantification is considered as one of the most complex issues.

The quantification of project time variations plays an important role in project execution. Although risks and uncertainties have significant influences on decision-making, they are vague, subjective, and uncertain [6]. Therefore, many authors attempted to quantify risks such as assessing the effect on risks [8], predicting the effect on risks [13], analyzing documents to evaluate risks [1], applying fuzzy theory to find probability density functions of risk [12], and applying Monte Carlo on schedule [10]. However, so far, there has been still a lack of studies that focus on specific project schedules. As such, there is a need to develop a process in the planning phase of risk responses, which could help to assess the level of project time variations based on the impact of event chains.

Accordingly, this study aims to help project managers to determine project completion time and milestones with corresponding probabilities and eliminate the probability and level of impact of each risk on a specific project. Based on a proposed process, construction organizations can assess the schedule viability of a project at distinct stages: biding, pre-construction, construction. This study offers a certain practical way to evaluate project schedules. Project managers will comprehend worst-case scenarios, most likely scenarios, and best-case scenarios, especially sensitive to risks. After that, construction firms use the report from evaluation to respond risks and uncertainties. In addition, the actual results of a specific project are compared with the simulation results to prove the reliability of the proposed process. This study explores the applicability of event chain methodology as a technique that can simplify the process of modeling risk and uncertainties while minimizing the effects of cognitive, confirmation, and psychological bias.

2   Literature Review

Many authors have expressed their concerns about quantifying risks on variations of project time. Most of them identified common risks and investigated their correlation. Furthermore, some authors have allocated these risks to stakeholders. Little research has been attempted to predict variations of project time based on quantitation risks.

Many previous studies have focused on risk factors, as well as risk relationships. When identifying and evaluating significant risks to allocate risks for stakeholders, El-sayegh [4] concluded that contractors were more responsible for risks than owners in the UAE construction industry. Likewise, with the aim of classifying risks according to organization and project levels, Subramanyan et al. [15] indicated that risk factors could relate to two groups of participants, who were working on construction sites and at offices, respectively. Such groups' possibilities of appearance were also compared to predict their occurrence in a project. By allocating each construction risk based on contracting terms, Hanna et al. [7] concluded that most severe consequences were usually caused by legal terms. After identifying and classifying risk factors, some authors also developed models to explore their new relationships. Qazi et al. [11] used utility theory and Bayesian belief networks to find the relationship between project complexity and risks to classify risk and mitigate risk in the initial project phase. Reviewing methods, tools, and techniques relating to risk identification, Siraj and Fayek [14] indicated that most studies rely on their risk nature or available risks listed not only to identify but also to classify risks. More recently, Erol et al. [5] explored the relationship between complexity and risk in detail based on quantitative findings.

In addition, several authors have attempted to develop general risk management processes and quantify the risk impact. Del Cano and de la Cruz [2] introduced a general process to manage construction risks from the views of clients and consultants. This process was appropriate for the requirements of other participants because it was easy to be adjusted based on a particular situation. Nasir et al. [10] employed Monte Carlo simulation as well as program evaluation and review technique for schedule risk analysis based on the activity duration value of a range. As a result, time performance was quantified by developing risk evaluation in construction-schedule model. In a research of Jannadi and Almishari [8], a risk assessor model was designed by using a computer program to assess risks for crucial construction activities relating to the probability, severity, and exposure of all hazards. Schatteman et al. [13] employed uncertainty to investigate the effects of schedules by integrating them. After predicting project time performance, they also developed a computer program to manage risks (e.g., identifying, analyzing, and quantifying risk elements). In another study, Choi and Mahadevan [1] employed project-specific documents to analyze risks by using relatively traditional ways like surveys. Sachs and Tiong [12] used fuzzy theory deriving customized probability density functions to quantify risks. Zeynalian et al. [19] developed a risk analysis and management model to consider potential risks based on the identification and probabilities of risks using the Delphi method.

The literature review indicates that the research area on variations of project time is context-specific. The study could contribute some valuable findings to not only Vietnam but also global knowledge. Moreover, in Vietnam as well as other similar developing countries, the quantification of the effects of risks on time variations with regard to different issues of project time has still received little attention. Thus, this study will attempt to fill the gap of research by directly quantifying the impact of risk events. By providing a process of directly quantifying risks on a specific project schedule, the results of this study could be applied to quantify the effects of risk events on project schedule in practice.

3   Research Method

The research framework (Fig. 1) includes three main phases: (1) building a risk list, (2) proposing a quantitative process, and (3) evaluating the process.

In Phase 1, a risk library was first created by reviewing risks encountered after bid selection and before project implementation. After synthesizing all major risks, this study conducted various face-to-face interviews to identify a list of potentially appropriate risk events. In such interviews, 15 experts (including 11 contractors, two consultants, one owner, and one governmental management agency), who had high experience and managerial positions, were invited to participate.

Then, in Phase 2, a quantitative process of risk events that could affect the time of construction works. Specifically, patterns and processes that quantify the risks in previous scholar documents or in construction companies were reviewed. For quantifying risks, a set of criteria were proposed. As a result, a process, which can quantity the effects of risk events on project time, was developed based on experts' perspective.

Finally, in Phase 3, a construction package of bored piles was used as a case study to demonstrate the reliability and appropriateness of the developed process.

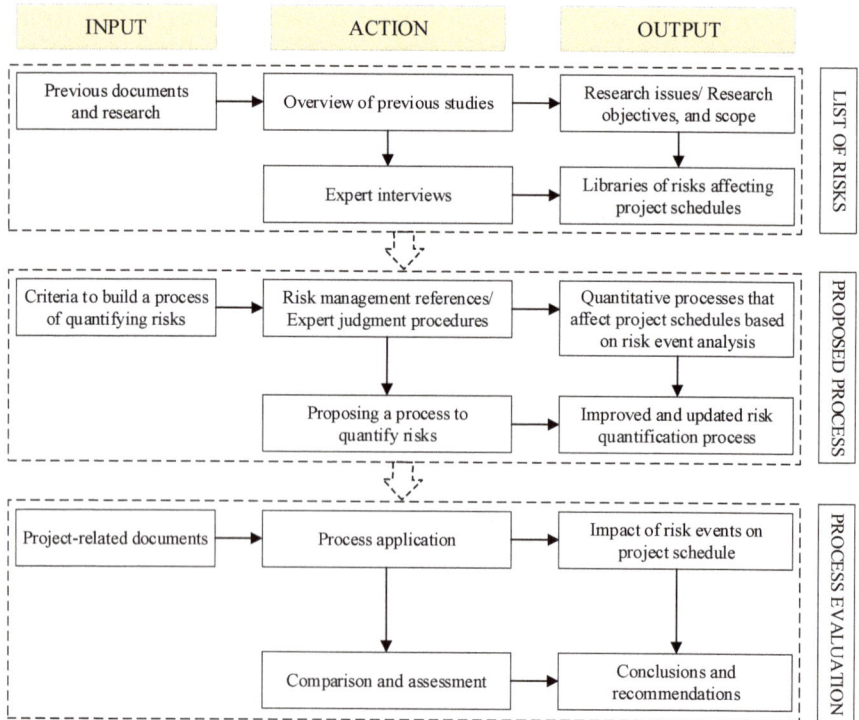

Fig. 1 Research framework

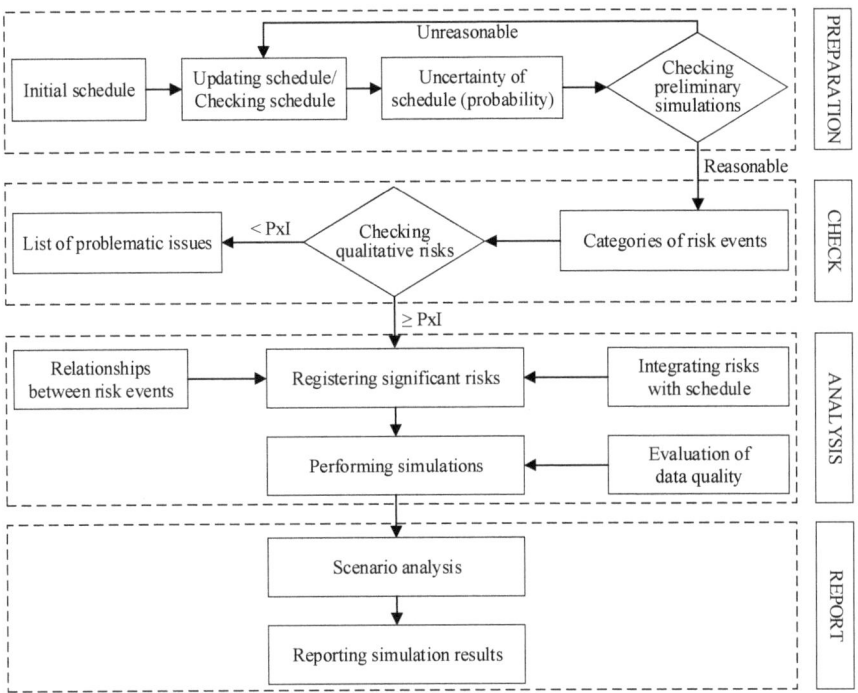

Fig. 2 Proposed process

The simulation results achieved after updated more risk events into this process were compared with the actual values of the package's project schedule.

4 Proposing Detailed Process to Quantify Risks

Some important steps in the proposed process are detailed progress, qualitative analysis of risk events, quantitative risk analysis, and evaluation of results. These steps all appear in other risk management processes such as PMBOK and ISO31000:2018. This study extends such processes by building a detailed process (Fig. 2) with clear decentralized responsibility for each individual in assessment teams of risk events.

4.1 Preparation of Process Development

In this step, the evaluation team should have a detailed schedule with a specific hierarchy of work breakdown structure level 3 (PMBOK). The proposed process adds more parameters with regard to the probability of completing construction tasks to the project schedule created by the MS Project software. In terms of PERT analysis, a triangular distribution function is selected to show the probability of completing construction tasks. If there is adequate information about the project, the Crystal Ball software can be used to identify appropriate time distribution functions for construction tasks. Besides this, further considerations should focus on the reserved time, relationships, constraints, logic, and consistency of construction tasks according to technical specifications. In addition, project schedules need to be established in advance to determine project completion time periods.

Figure 3 shows the schedule of bored piles (i.e. case study), which is established by an anonymous contractor's bidding department based on similar projects' bill of quantities. The construction time for completing 148 piles (from beginning to handovering of foundation items) is 100 days. The project schedule is developed using a planned calendar with seven working days per week, including holidays. Further information about the case study is presented in Table 1.

In Table 2, various probability information of case study's construction tasks is presented with three cases: optimistic (10% probability), most likely (50% probability), and pessimistic (90% probability). It can be seen that most simulated durations of construction tasks are lower than the planned durations (Fig. 3). For example, the pessimistic time (longest duration) of the preparation task is just 10.37 days, which is shorter than its planned time (12 days) in Fig. 3.

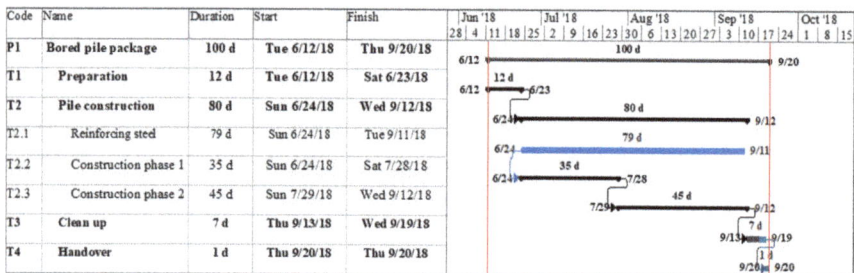

Fig. 3 Schedule of bored piles

Table 1 Characteristics of case study

Characteristic	Detail
Location	Quy Nhon City, Binh Dinh Province
Type	Condotel hotels and apartments
Size	23 floors and a basement
package	Bored pile (diameters of D800 and D1000)
Duration	100 days (June 12, 2018 to September 20, 2018)

Table 2 Results of Monte Carlo simulation

Code	Task	Duration (days)			Start	Finish
		Optimistic (10%)	Most likely (50%)	Pessimistic (90%)		
P1	Bored pile package	88.84	92.26	95.88	12-Jun-18	12-Sep-18
T1	Preparation	8.39	9.36	10.37	12-Jun-18	21-Jun-18
T2	Pile construction	74.53	76.93	79.38	21-Jun-18	6-Sep-18
T2.1	Reinforcing steel	67.46	69.88	72.44	21-Jun-18	30-Aug-18
T2.2	Construction phase1	33.31	34.14	35.06	21-Jun-18	25-Jul-18
T2.3	Construction phase2	41.18	42.76	44.32	25-Jul-18	6-Sep-18
T3	Clean up	4.96	5.01	5.13	6-Sep-18	11-Sep-18
T4	Handover	1	1	1	11-Sep-18	12-Sep-18

4.2 Check of Risk Events

This step aims to classify risk events into two groups based on their P (probability) and I (impact) values (measured on a 5-points Likert scale). Specifically, Group 1 (i.e. list of problematic issues) includes risk events with the calculated PxI values less than the standard PxI value. In Group 2, risk events will have their PxI values larger than the standard PxI value. It should also be noted that the PxI value depends on project to project. For example, if the budget contingency of a project is large, the PxI value is small, thereby more risk events should be considered and analyzed. If project managers just wish to focus on risk events with severe impacts, the PxI value will be larger.

Regarding the above-mentioned case study, the durations of construction tasks are provided by its site/project managers and senior engineers based on their experience with previous similar projects. These durations are given based on optimistic, most likely, and pessimistic conditions. After that, the site management team will assess risk events based on their probability and impact. Risk events which have considerable impacts could be grouped into four categories as follows:

- Event 1 (E1): Design changes (e.g., pile number and length) by design unit's requests,
- Event 2 (E2): Impact of weather and other force majeure issues,
- Event 3 (E3): Subcontractors' capability to supply materials and equipment, and
- Event 4 (E4): Dealing with problems that may arise during underground construction processes.

4.3 Quantitative Analysis

In terms of quantitative analysis, risk events in Group 2 will be considered. Specifically, they are analyzed quantitatively with regard to their impacts on project schedule performance. Risks are integrated with project schedule by assigning risk events' durations (start and finish time) and impacts on construction tasks that are directly affected. Besides this, the relationships between risk events are also considered in terms of possibilities of forming risk event chains, which in turn affect a certain task in the project schedule. After risk event assignment, simulations are performed. Finally, experts will preliminarily assess the simulation results.

Table 3 presents the results of integrating risk events on the case study's project schedule. It can be seen that risk event E1 only affects task T2.1, T2.2, and T2.3. Commonly, risk event E1 happens at the beginning of construction works (0–60%), which may have a day delay due to changing old to new designs in terms of pile number and length. Risk event E2 (e.g., weather) affects all other construction tasks during the whole package (06/2018–09/2018) with a probability of 15% based on previous historical data. Risk event E3 only affects construction tasks necessarily using materials and equipment. Generally, risk event E3 appears at the beginning of construction works (0–80%). Due to previous relationships with subcontractors, the site management team proposes a 30% probability of occurring and a two-days delay for providing more materials and equipment. As risk event E4 cannot be known in advance, its duration is selected as equal to task duration. According to similar

Table 3 Risk integration with project schedule

Code	Task	Risk events				
			Event 1	Event 2	Event 3	Event 4
P1	Bored pile package					
T1	Preparation	Probability Delay Time occurs		15% 1 day 30%-60%		
T2	Pile construction					
T2.1	Reinforcing steel	Probability Delay Time occurs	50% 1 day 0%-60%	15% 2 days 0%-100%	30% 2 days 0%-80%	
T2.2	Construction phase1	Probability Delay Time occurs	50% 1 day 0%-60%	15% 2 days 0%-100%	30% 2 days 0%-80%	30% 2 days 0%-100%
T2.3	Construction phase2	Probability Delay Time occurs	50% 1 day 0%-60%	15% 2 days 0%-100%	30% 2 days 0%-80%	30% 2 days 0%-100%
T3	Clean up	Probability Delay Time occurs		15% 1 day 0%-100%		
T4	Handover					

projects, a 30% probability of happening is proposed for risk event E4 and a two-days duration for troubleshooting.

4.4 Report of Simulation Results

Based on the results of risk integration and simulations, the site management team will focus on analyzing risk events that significantly affect the project schedule (e.g., risks with high sensitivity). Besides this, the site management team also assesses the project implementation time according to optimistic, most likely, and pessimistic conditions. Based on such analyses, reports will be made for the site management team to respond to risks (e.g., avoiding, transferring, mitigating, and ignoring). Such analyses can also be re-performed at other project stages such as periodic reviews, new technical/technological assessments, or allocation of responsibilities among parties.

5 Illustrative Example

The results of simulation (Fig. 4) indicate a 77% chance for the package completed on time (September 20, 2018) and a 50% chance completed on September 16, 2018 (most likely). In pessimistic situations, the package has a 90% chance to be completed on December 22, 2018. On the other hand, in optimistic conditions, the chance for completing the package on September 9, 2018 is 10%. As compared with the initial schedule, risk events extend project schedule performance for two days. Accordingly, the site management team re-assesses the results and add restrict construction at night

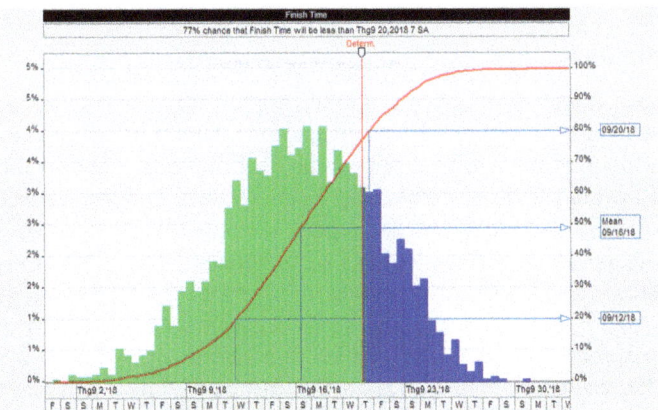

Fig. 4 Planned simulation results

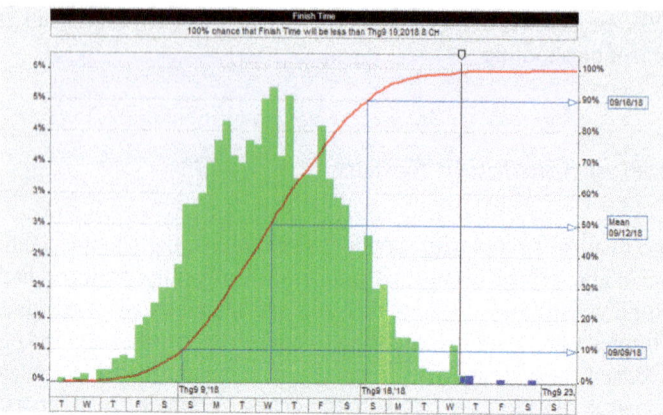

Fig. 5 Upgraded simulation results

(risk event E5) as a further condition, which was not considered for the previous check of risk events.

As shown in Fig. 5, the results of simulation have some changes in completion time milestones: September 9, 2018 (optimistic), September 12, 2018 (most likely), and September 16, 2018 (pessimistic). Despite the addition of risk event E5, the project implementation durations are reduced. This is not so surprising because the simulations use the probability values of risk events without depending on the number of such events.

A comparison with previous similar projects allows the site management team to assess the results of simulation being reasonable. The possibility of completing the package on time is relatively high. Therefore, the site management team considers to accept risk events and keeps following the initial as-planned schedule.

6 Conclusions

Although many previous studies have attempted to quantify risks, there is still a lack of research focusing on assessing risks visually. This study proposes a process of quantifying the impact of risk events on project schedule. The process includes four steps: preparation, check, analysis, and report. The project schedule is considered as the input data, on which some risk events are integrated. The result of the process is a report which includes various important information: project times in optimistic, most likely, and pessimistic conditions and risk events with sensitivity-based rankings. The proposed process is also applied to a case study, which is a bored pile package. The results of simulation show that this case study could have 90% and 50% chances to be completed four and eight days earlier than the initial project schedule, respectively. These results are assessed to be reasonable and, accordingly, the site management team suggest accepting risk events as their risk management strategy.

Despite the above contributions, the limitations of this study should also be noted. Quantifying risk events independently may be a limitation of this study. The chains and interrelationships of risk events are not considered. On the other hand, this study has not attempted to quantify the impact of risk events on project cost due to the difficulties of collecting full data of real project costs. Future studies should focus on quantifying risk events at other project stages and/or the effects of risk events on both project time and cost.

Acknowledgements The authors fully acknowledge the support of time and resources from the affiliated organizations for this study. This research is funded by Vietnam National University Ho Chi Minh City (VNU-HCM) under grant number B2021-20-08. We would like to thank Ho Chi Minh City University of Technology (HCMUT), VNU-HCM for the support of time and facilities for this study.

References

1. Choi HH, Mahadevan S (2008) Construction project risk assessment using existing database and project-specific information. J Constr Eng Manag 134(11):894–903
2. Del Cano A, de la Cruz MP (2002) Integrated methodology for project risk management. J Constr Eng Manag 128(6):473–485
3. Diab MF, Varma A, Panthi K (2017) Modeling the construction risk ratings to estimate the contingency in highway projects. J Constr Eng Manag 143(8):1–9
4. El-Sayegh SM (2008) Risk assessment and allocation in the UAE construction industry. Int J Proj Manag 26(4):431–438
5. Erol H, Dikmen I, Atasoy G, Birgonul MT (2020) Exploring the relationship between complexity and risk in megaconstruction projects. J Constr Eng Manag 146(12):1–14
6. Gonzalez P, González V, Molenaar K, Orozco F (2014) Analysis of causes of delay and time performance in construction projects. J Constr Eng Manag 140(1):1–9
7. Hanna AS, Thomas G, Swanson JR (2013) Construction risk identification and allocation: Cooperative approach. J Constr Eng Manag 139(9):1098–1107
8. Jannadi OA, Almishari S (2003) Risk assessment in construction. J Constr Eng Manag 129(5):492–500
9. Kim HJ, Reinschmidt KF (2011) Effects of contractors' risk attitude on competition in construction. J Constr Eng Manag 137(4):275–283
10. Nasir D, McCabe B, Hartono L (2003) Evaluating risk in construction-schedule model (ERIC–S): construction schedule risk model. J Constr Eng Manag 129(5):518–527
11. Qazi A, Quigley J, Dickson A, Kirytopoulos K (2016) Project complexity and risk management (ProCRiM): Towards modelling project complexity driven risk paths in construction projects. Int J Proj Manag 34(7):1183–1198
12. Sachs T, Tiong RL (2009) Quantifying qualitative information on risks: Development of the QQIR method. J Constr Eng Manag 135(1):56–71
13. Schatteman D, Herroelen W, Van de Vonder S, Boone A (2008) Methodology for integrated risk management and proactive scheduling of construction projects. J Constr Eng Manag 134(11):885–893
14. Siraj NB, Fayek AR (2019) Risk identification and common risks in construction: Literature review and content analysis. J Constr Eng Manag 145(9):1–13

15. Subramanyan H, Sawant PH, Bhatt V (2012) Construction project risk assessment: development of model based on investigation of opinion of construction project experts from India. J Constr Eng Manag 138(3):409–421
16. Taroun A (2014) Towards a better modelling and assessment of construction risk: Insights from a literature review. Int J Proj Manag 32(1):101–115
17. Zayed T, Amer M, Pan J (2008) Assessing risk and uncertainty inherent in Chinese highway projects using AHP. Int J Proj Manag 26(4):408–419
18. Zeng J, An M, Smith NJ (2007) Application of a fuzzy based decision-making methodology to construction project risk assessment. Int J Proj Manag 25(6):589–600
19. Zeynalian M, Trigunarsyah B, Ronagh HR (2013) Modification of advanced programmatic risk analysis and management model for the whole project life cycle's risks. J Constr Eng Manag 139(1):51–59

Development of Novel Hybrid Artificial Intelligent Model for Optimizing Material Supply Chain in Construction Projects

Vu Hong Son Pham and Huynh Chi Duy Nguyen

Abstract On the construction site, the time is rushed so the managers are only concerned about human resources and construction methods to meet the schedule, while paying little attention to construction materials. Because of the lack of concern about material costs, the current cost management of managers is still not really effective. Therefore, this study proposes to use a novel Hybrid Artificial Intelligent Model (HAIM) to optimize the material supply chain save cost for construction contractors. The construction logistics planning method is utilized to find suitable material requirements as well as control the cost of construction materials. The validation analysis shows the advanced searchability of proposed model to traditional and model metaheuristic algorithms. Using HAIM, Managers can optimize construction costs through material demand planning; thereby increasing profits for the project and the company will increase its competitiveness in the construction market.

Keywords Construction logistics planning model · Artificial intelligent · Inventory control · Optimize delivery schedules · Construction materials · Fuzzy logic

1 Introduction

Today's construction materials are all specified by the investor and the value of materials accounts for 50–60% of the total project value: Contractors have to compete fiercely with each other and each other is mainly management costs. Therefore, it is absolutely necessary and important to manage & control the right amount of materials to help limit unnecessary costs properly. Currently, managers tend to buy in bulk to get the best discount from suppliers and this often leads to overloading of the warehouse at the construction site; Material quality is not guaranteed because

V. H. S. Pham · H. C. D. Nguyen (✉)
Faculty of Civil Engineering, Ho Chi Minh City University of Technology (HCMUT), 268 Ly Thuong Kiet Street, District 10, Ho Chi Minh City, Vietnam
e-mail: nhcduy007@gmail.com

V. H. S. Pham
Vietnam National University Ho Chi Minh City, Linh Trung Ward, Thu Duc District, Ho Chi Minh City, Vietnam

© The Author(s), under exclusive license to Springer Nature Singapore Pte Ltd. 2023   231
J. N. Reddy et al. (eds.), *ICSCEA 2021*, Lecture Notes in Civil Engineering 268,
https://doi.org/10.1007/978-981-19-3303-5_18

the storage time is too long. From there, we can see the difficult problem of how managers can optimize material costs; meet progress and limit warehouse overload.

Based on the schedule of construction [1] and [7] use material requirements planning MRP and last planner systems to optimize the material cost, which help manage projects better and increase profits for the company.

Moreover, to manage the risk of material unit price, the article will consider the unit price of the material as fuzzy values with triangular membership function. This study focuses on building and developing a model to find the optimal solution for material costs and have the warehouse size suitable to the actual situation of the construction site.

The remainder of this study is organized as follows. Section 2 presents a literature review. The detailed problem description, notation, and mathematical model are outlined in Sect. 3. An application and numerical analysis are conducted to evaluate the performance of the proposed policies in Sect. 4. Section 5 concludes this study.

2 Literature Study and Review

Roach [14] introduced and developed the Economic Order Quantity model of material costs. [7] Propose a short-term supply chain and proactive material supply model using the method: Last Planner Systems. Mao and Cheng [4], Using an Economic Order Quantity (EOQ) model that takes into account uncertainty about unit prices to optimize material costs. Hsu et al. [3]; Nolz [11] develops construction logistics planning model to optimize construction costs with warehouse constraints and material demand. The construction logistics planning model outperforms the EOQ model because it comprehensively considers the costs related to materials starting from purchase to completion of work on site. Furthermore, in this study we will further examine the flexibility of time between purchase orders for material. Accordingly, this study took into account the risk of material price fluctuations and solved it with triangular membership functions.

Each construction phase has different demands of material, different warehouse sizes and construction site layouts, so this study will combine flexible changes between FOP and warehouse size to bring about more efficiency cost of materials. From that, we see that the problem has many variables and the search space is large, so we need to use a suitable algorithm to effectively solve the problem. This study uses the hybrid algorithm to solve the problem: combination of the dragonfly algorithm [9] and the PSO algorithm [8]. Finally, this study hopes to provide managers with an overview of material costs throughout the construction process. This study will help managers make the most appropriate choice at each stage of construction considering all actual site conditions and company finances.

3   Research Methodology

The research method of the article consists of two main parts: Building a model using the **Construction Logistic Planning Model** (CLP) and using a hybrid algorithm to optimize the objective function. Moreover, the effectiveness of this model will be proved using specific examples.

Objective function: The cost of material is objective function for this research. According to the construction logistics planning model, the cost of material is divided into 4 component costs: Ordering Cost; Financing Cost; Stock-out Cost; Layout Cost.

When optimizing the cost of building materials, supply plan is the most important in the model. So to have an optimal material supply plan, we need to determine the time between purchases and the volume of materials. Fixed Ordering Period (FOP) is the period between purchases.

Today FOP has 2 main trends: FOP = 1 and FOP > 1:

FOP = 1 is also known as buying materials every day. With FOP = 1, materials are purchased every day, there will be no inventory and the cost of burying the material will be zero. With **FOP > 1** then we do not purchase materials every day. The time between purchases will be more than 1 day so there will be a material inventory and an inventory cost will appear. [2, 15].

- Ordering Cost (OC): The cost of every purchase. This cost includes 2 parts of the purchase cost (included the supplier's discount cost) and the cost of transporting the material from the supplier to the construction site.
- Financing Cost (FC) [10]: When materials are in stock, financing costs are incurred. The financing cost can be understood as the cost of the contractor's trade-off when burying capital in inventory. Contractors often tend to buy materials in large quantities to get a good discount from the supplier, but in return, there will be inventory.
- Stock-out Cost (SC) [13]: This is the cost of delay.
- Layout Cost (LC): Above in the ordering cost is the cost of moving the material from the supplier to the warehouse on site. Layout Cost is the cost of transporting materials from the warehouse to the construction floor. Moreover, this cost includes the cost of rearranging the construction site when there are changes in the construction stages.

The construction logistics planning model fully reflects the cost of materials throughout the construction process from the supplier to the site and during the construction phase. From there, choosing a model to optimize the cost of construction materials is a reasonable choice. Using the CLP model, there is a large search space and many complex variables, so we choose the hybrid algorithm DA-PSO to optimize the objective function.

Fuzzy Number: The unit price of non-fixed construction materials always fluctuates over time and depends on external factors a lot and it is difficult to foresee. Therefore, when optimizing the cost of building materials, we must consider the variation in

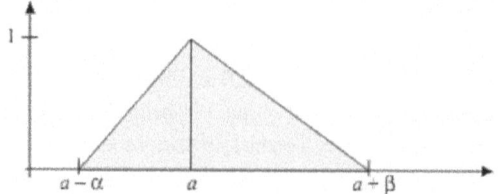

$$A(t) = \begin{cases} 1 - \dfrac{a-t}{\alpha}; & a - \alpha \le t \le a \\[2mm] 1 - \dfrac{a-t}{\beta}; & a \le t \le a + \beta \\[2mm] 0; & t < a \vee t > b \end{cases}$$

Fig. 1 Triangular fuzzy numbers

unit price to match the actual construction cost. In the study proposing the unit price of the material is an uncertain number (consistent with the actual site), we will use the fuzzy number to have a description of the unit price of the material. [4] (Fig. 1).

The process of model: The process of model will show how to build a model to optimize the cost of building materials. This section will guide and demonstrate how to determine the input parameters and how to calculate the construction logistic planning model. In addition, when building a construction logistic planning model, we will proceed to optimize the material cost through the algorithm and detail the chart below.

First of all; we need determine the minimum daily need for materials. The Last Planner System is used to determine the minimum daily need of materials. We take the steps according to the following flowchart to determine the material needs to be used each day [6]:

Each task and each material is carried out as shown in the proecess shown in Fig. 2. In this sequence the manager can determine with certainty the minimum need to use each day to meet the construction schedule.[12].

Step 1: Input parameters.
The objective function of the CLP model has four components:

- OC costs: Ordering Cost includes 2 parts of the cost: material unit price and transportation cost;
- FC cost: Finacing Cost – The profit lost to contractors because of inventory materials instead of investing elsewhere (Material interest);
- SC costs: Stock-out Cost – The costs the contractor lost because of the delay (Late schedule penalties);
- LC costs: Layout Cost represents resource travel costs between facilities and storage areas and site re-arrangement costs (Cost of transport equipment; change in site layout).

Constraints: FOP; storage capacity of material warehouse.
Step 2–3: Initialize the first population. Based on the random population initialization property of the algorithm. We initialize the initial population using the DA algorithm. The individuals in the population will move towards a more potential new location and appear a new optimal point of the population. Each loop in

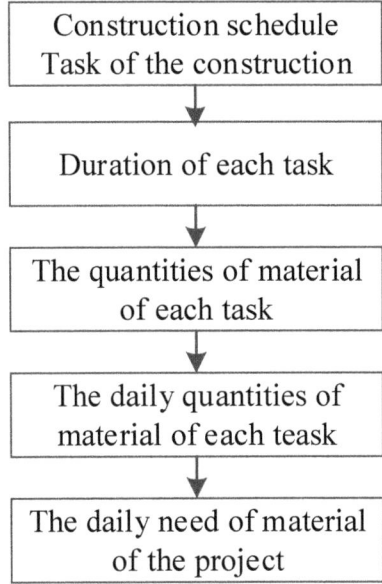

Fig. 2 Flowchart of the daily need for material

algorithm will check conditions of warehouse size and material inventory. Each population is calculated to find the best population in each loop.

$$CLC = \sum OC + FC + SC + LC \tag{1}$$

Step 4–5: Improve the population. In this step, we combine and integrate DA and PSO algorithms together to find more optimal results.[5].

DA Algorithm:

$$\Delta X_{t+1} = (s\,S_i + \alpha\,A_i + c\,C_i + f\,F_i + e\,E_i) + w\Delta X_t \tag{2}$$

$$\Delta X_{t+1} = X_t + \Delta X_{t+1} \tag{3}$$

s;a;c;f;e;w are weights of the algorithm;
PSO:

$$V_i(t) = w(t) \times V_i(t-1) + C_1 r_1 \times [X_i^L - X_i(t-1)] + C_2 r_2 \times [X_i^G - X_i(t-1)]$$
$$X_i(t) = V_i(t) + X_i(t-1)$$
$$\tag{4}$$

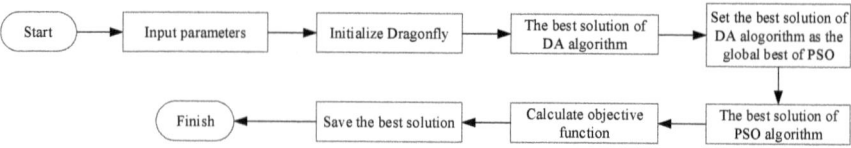

Fig. 3 Flowchart of the optimal algorithm (every iteration)

$+X_i^L = \{ X_{i1}^L, X_{i2}^L, ..., X_{iM}^L \}$ represents the best position the individual achieves after (t-1) loop (called pbest)

$+ X_i^G = \{ X_{i1}^G, X_{i2}^G, ..., X_{iM}^G \}$ represents the best position of the whole population after (t-1) loop (called gbest)

Step 6–7: After finding the optimal result in the population. We will proceed to calculate the objective function and save that optimal plan. And that is the process of a loop. Algorithm ends when the number of iterations reaches the maximum number of iterations (Input parameter) (Fig. 3).

4   Model Applications

Brick is the material of choice for the research. We have selected the small-layout construction to test for this model. Construction materials, after being transported to the construction site, will be properly stored to avoid the effects of weather. The building is located in the city, so a limited warehouse area is the problem. The study will carry out the corresponding different cases between FOP and warehouse size at the construction site. From there, the manager will have the most suitable choice for the actual situation at the project.

According to the contract: Each day is fined 61.5 million VND for late progress.
Daily interest rate: 0.1%/day.
Cage hoist unit price: 50 million VND/month.
Cage hoist unit productivity: 20,000 birck/hour.
The project progress from 24/04/2019 to 09/06/2019 (Table 1):
Input parameters (Table 2):

Table 1   The Schedule of project (brick work)

TAKS/Floor	Duration (days)	Start	Finish
Brick work			
Floor 4	15	24/04/2019	08/05/2019
Floor 5	15	11/05/2019	25/05/2019
Floor 6	15	11/05/2019	25/05/2019
Floor 7	15	26/05/2019	09/06/2019
Floor 8	15	26/05/2019	09/06/2019

Table 2 Input parameters of the model

Unit price of materials purchased (Fuzzy Number)	Brick (thousand brick)	Unit price (million VND / thousand brick)
	0–20	(1; 1.2; 1.4)
	20–50	(1; 1.1; 1.2)
	50–70	(0.7; 0.9; 1.1)

Table 3 Warehouse size and material cost

FOP	1	2	3	5	7	9
Storage (thousand brick)	17	60	70	90	130	140
CLC (million VNĐ)	2570.5	2095.3	1775.5	1551.7	1447.5	1447

Optimal results by DA-PSO algorithm:

Looking at the results we see that the material demand plan increases the time between demand times (Increase FOP), the storage size will have to increase to accommodate the inventory of materials. As the FOP increases, the volume of material demanded will increase so the manager will get a high discount from the supplier. Red line shows the relationship between material cost (left vertical axis) and fixed ordering period (FOP); the blue column shows the warehouse size corresponding to the fixed ordering period (FOP) (Table 3).

The highest proporation of material cost is OC (60–80%); the second-highest pororation of material cost is SC, and the rest is shared between FC and LC. Ordering Cost is the cost that accounts for the highest percentage of material costs. Since then, managers have forgotten about other costs such as FC; SC; LC. Up to the present; The first thing managers think about in terms of material cost optimization is to buy in bulk to get the best discount from the supplier. When the volume of the purchase is too large, what will happen That is inventory, thereby increasing warehousing costs.

Brick is the finishing material selected in this study. In the Finishing phase, the construction site is narrower:

- Unlike in the structure phase, there is only one material type: Steel;
- In the finishing phase a lot of different materials.
- Moreover, the storage capacity in the layout site is still limited because of preparing for the construction of infrastructure.

Therefore, the storage space for bricks is very limited. From there, we can see that buying bricks in large quantities mistakenly getting the supplier's discount is not feasible.

In fact, at the project applied to this research, the contractor who chose the option every day buys bricks to the site and uses them to suit the narrow construction ground of the project. And looking at Fig. 4 we can also see that this is the most expensive choice for the contractor. In this selection of the contractor, the cash flow of the

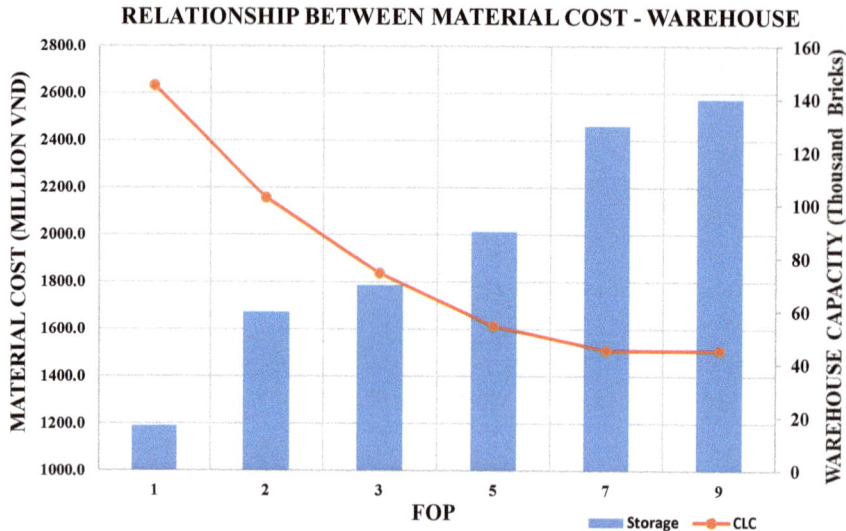

Fig. 4 Relationship between material cost and warehouse capacity

contractor will not be buried capital, but this is the most expensive option to buy the materials and takes time and effort in demand and transport every day.

The optimal result of that problem is the intersection between the relationship curve between FOP - material cost and the column chart showing the warehouse size. Optimizing the result of that problem is FOP = 5 (to the right of the intersection of 2 curves). Why is this the optimal result of the problem?

- First, about the storage issue: Actually the layout site and the ground floor of the building will hold about 70–80 (Thousand Brick). However, we choose FOP = 5 corresponding to the size of warehouse 90 (Thousand Brick) because we will take advantage of the construction floors of wall construction to make temporary storage of bricks. (Make use of 2 - 3 floors as temporary storage; Each floor will hold maximum at 10 (thousand Brick)) When the size of the warehouse increases, all problems are almost solved.
- Second, FOP = 5 means that the quantity of material per demand is greater so the contractor will get a better discount from the supplier than in the case of FOP = 1 that the contractor initially chose.
- Third, FOP = 5, the time between material purchases is also short, thus minimizing spending of a large part of capital on inventories. Therefore, the project's cash flow plan will be better controlled.

Finally, a short FOP helps to reduce the quantity of materials in the inventory, with new materials purchased and constantly updated when needed, so that the product quality is better than if the materials were bought ahead of time and stored long-term.

5 Conclusion

Cost management is a project manager's top priority, but managers have not properly considered the management of construction materials costs. It is hoped that this study will help project contractors and managers to accurately view the cost of construction materials in order to properly manage project costs in a way that will yield the highest profit. The research has proposed flexibility in decision-making, even as contractors choose which to prioritize among warehouse size, material cost and fixed ordering period (FOP). Project managers should also consider actual constraint conditions and many other factors, including objective and subjective factors, while deciding on the optimal cost of materials that match the available warehouse-size in each phase of the construction.

References

1. Behera P, Mohanty RP, Prakash A (2015) Understanding Construction supply chain management. Prod Plan Control 26:1332–1350. https://doi.org/10.1080/09537287.2015.1045953
2. Hong VLALP (2011) Solutions to material delivery management, vol. 9. Ministry of construction.
3. Hsu P-Y, Angeloudis P, Aurisicchio M (2018) Optimal logistics planning for modular construction using two-stage stochastic programming. Autom Constr 94:47–61. https://doi.org/10.1016/j.autcon.2018.05.029
4. Jaśkowski P, Sobotka A, Czarnigowska A (2018) Decision model for planning material supply channels in construction. Autom Constr 90:235–242. https://doi.org/10.1016/j.autcon.2018.02.026
5. Jia Q, Guo Y (2016) Hybridization of ABC and PSO algorithms for improved solutions of RCPSP. J Chin Inst Eng 39:727–734. https://doi.org/10.1080/02533839.2016.1176866
6. Khanh HD, Kim SY (2016) A survey on production planning system in construction projects based on last planner system. KSCE J Civ Eng 20:1–11. https://doi.org/10.1007/s12205-015-1412-y
7. Mao H, Cheng P (2010) Design of material delivery system based on lean construction. ICLEM 2010:1793–1799
8. Marini F, Walczak B (2015) Particle swarm optimization (PSO). A tutorial. Chemom Intell Lab Syst 149:153–165. https://doi.org/10.1016/j.chemolab.2015.08.020
9. Mirjalili S (2016) Dragonfly algorithm: a new meta-heuristic optimization technique for solving single-objective, discrete, and multi-objective problems. Neural Comput Appl 27:1053–1073. https://doi.org/10.1007/s00521-015-1920-1
10. Mitsel AA, Kritski OL, Stavchuk LG (2017) An inventory model with random demand. J Phys: Conf Ser 803:012099. https://doi.org/10.1088/1742-6596/803/1/012099
11. Nolz PC (2020) Optimizing construction schedules and material deliveries in city logistics: a case study from the building industry. Flex Serv Manuf J. https://doi.org/10.1007/s10696-020-09391-7
12. Olivieri H, Seppänen O, Alves TdCL, Scala NM, Schiavone V, Liu M, Granja AD (2019) Survey comparing critical path method, last planner system, and location-based techniques. J Constr Eng Manag 145:04019077. doi:10.1061/(ASCE)CO.1943-7862.0001644
13. Panova Y, Hilletofth P (2018) Managing supply chain risks and delays in construction project. Ind Manag Data Syst 118:1413–1431. https://doi.org/10.1108/IMDS-09-2017-0422

14. Roach B (2005) Origin of the economic order quantity formula; transcription or transformation? Manag Decis 43:1262–1268. https://doi.org/10.1108/00251740510626317
15. Yelin E, Katz P, Banks C (2020) A policy to do better next time: lessons learned from the covid-19 pandemic. ACR Open Rheumatol 2:253–254. https://doi.org/10.1002/acr2.11145

Enhancing Building with Adaptive Design Objects Using Building Information Modelling Towards Consuming Energy Efficacy in Building

Ngoc Son Truong, Duc Long Luong, Ngoc Tri Ngo, Quang-Trung Nguyen, and Ngoc Thao Linh Dang

Abstract Energy consumption in buildings is an important factor in the design and operation of a project as if energy reduction used in a building helps to lower operating costs and its environmental impact. Optimizing architectural design options, building surfaces, and HVAC systems from the beginning of a project using Building Information Modeling (BIM) to reduce energy consumption in the building is a new trend nowadays. BIM is said to be an interactive environment that efficiently manages and analyzes elements related to architecture, structure, electricity, and water design as well as the construction process. Furthermore, there are a variety of building energy modeling (BEM) tools that can import these BIM files and perform energy simulation. The simulation and calculation of the energy model via BIM, in particular, is regarded as a new solution for the construction industry in the future. In this research, energy simulation in buildings using a BIM system is proposed to change those design elements within construction process in order to reduce the Energy Use Intensity (EUI) and energy cost in the building. The result of this research are expected to help architects and building managers improving and enhancing the efficiency of energy used in buildings.

Keywords Building information modeling (BIM) · Building energy modeling (BEM) · Energy use intensity · The energy cost

N. S. Truong (✉) · D. L. Luong
Faculty of Civil Engineering, Ho Chi Minh City University of Technology (HCMUT), 268 Ly Thuong Kiet Street, District 10, Ho Chi Minh, Vietnam
e-mail: tnson.sdh19@hcmut.edu.vn

D. L. Luong · N. T. Ngo · Q.-T. Nguyen
Vietnam National University Ho Chi Minh City, Linh Trung Ward, Thu Duc District, Ho Chi Minh, Vietnam

N. S. Truong
Faculty of Project Management, The University of Danang, University of Science and Technology (DUT), 54 Nguyen Luong Bang Street, District Lien Chieu, Da Nang, Vietnam

N. T. L. Dang
Faculty of Architecture, The University of Danang, University of Science and Technology (DUT), 54 Nguyen Luong Bang Street, District Lien Chieu, Da Nang, Vietnam

© The Author(s), under exclusive license to Springer Nature Singapore Pte Ltd. 2023
J. N. Reddy et al. (eds.), *ICSCEA 2021*, Lecture Notes in Civil Engineering 268,
https://doi.org/10.1007/978-981-19-3303-5_19

241

1 Introduction

Energy consumption in building is an important part in the design and operation process of a project. If the energy used was reduced, it can improve the energy-efficient building use as well as protecting the environment and decresing the operation cost. Optimizing architectural design concept, building facade and HVAC system by using Building Information Modeling (BIM) is now a trendy researches from the very begining of each project [1].

BIM technology has advanced rapidly in the field of engineering construction in recent years, yielding remarkable results in design and construction [2]. Building Information Modelling (BIM) is a trend and interactive environment that combines various tools and software to manage and analyse building construction process such as electricity, water, and building elements. In particular, the simulation and calculation of the energy model through BIM is now considered a new solution for the future of the construction industry. [3].

This research is expected to help improving and enhancing energy efficiency in buildings. Although there have been some studies on BIM model design modifications to increase energy efficiency, there have not been many studies on buildings with complex structures and detailed HVAC systems. As a result, it is obvious that an energy simulation model must be developed in order to change the façade design at the design stage as proposed above.

2 Literature Review

2.1 Building Information Modeling (BIM) and Building Energy Modeling (BEM)

In these recent years, BIM technology has grown rapidly in the field of engineering construction and achieved valuable results [4–8]. Building information model (BIM) is seen as digital environment that allows stakeholders to share product lifecycle updated through 3D building modelling data [9, 10]. However, it has now been developed towards energy simulation, which is more complicated and useful [11]. Buildings consume about 40% of global energy and generate 30% of CO_2 emissions [12, 13]. These numbers are increasing because of a rapid urbanization [14, 15], as a result of which the greenhouse effect and global warming have increased [16]. Therefore, using energy efficiency in buildings is essential to reduce energy costs, environmental impacts and to increase the competitive value of buildings [17–19].

Number of Building Energy Modelling (BEM) tools are capable of importing data files from BIM to perform quality simulation jobs towards efficient building [20], hhowever, they can only be used in significant situations. Besides, many studies have used BIM as an effective power analysis tool from the beginning of the design process [21]. In details, following to the sustainable evaluation, available standard constructed components have been selected to install the energy consumption reduction in building [22]. It has been reported that building sustainable energy consumption can be affected by its direction, however, the integrated optimization model has not been used to evaluate its facade elements effectively [23].

Energy used in building is a dependent of building facade selection so that energy simulation tools have been studied in many researches in regards to predict its efficacy [24]. Nevertheless, most existing simulation tools entail extended time and large computing power to reach optimal designs, and often involve manual interference to conduct the desired iterations. Numerous research have discussed the limitations, barriers and solutions of integrating simulation into the design process [25]. As a result, an integrated energy simulation tools which can result both energy consumption reduction following to its facade selection is now a popular research target.

2.2 Energy Simulation Software Systems and Green Building Studio

Energy modelling is a type of geometry that is used for analysis on building simulation equipment or software, such as Revit. It is a generic model of virtual buildings represented in three-dimensional (3D) space for analysis and calculation by online or offline tools. The design team can use energy modelling to optimize specific elements of the building and prioritize the analysis of the core acoustics. In detail, this returns the most accurate values after analyzing its energy works [26].

Autodesk leverages its cloud to enhance professionals' ability to perform computationally intensive design and work. Designers can use its enormous processing power to perform simulation, which can aid in optimizing their design and making decisions on building weight, costs, safety and energy used [27]. Many software vendors/manufacturers have developed energy simulation systems that can be used to virtually assess a building's efficiency. Autodesk Green Building Studio is cloud-based energy analysis software that can assist users in performing energy analysis while optimizing energy efficiency during the design processs [28].

2.3 Energy Use Intensity (EUI)

Buildings consume a significant amount of energy during their operational phase and it is mostly a significant source of carbon discharge throughout their service life, which has a direct impact on global warming. As a result, it is critical to optimize energy use intensity (EUI) as well as energy cost through the use of building information modelling technology (BIM) to perform energy analysis. [27]. Most studies of EUI (kWh/m2/year) have been carried out and all potential home buyers or residents are required as an energy rating certificate. [29].

$$EUI = \frac{\sum SiteEnergyUse\left(\frac{kWh}{a}\right) - \sum Site\ Renewable\ Energy\ Generation\left(\frac{kWh}{a}\right)}{ModelledFloorarea\left(m^2\right)}$$

(1)

The research objective is to investigate modelling software for energy simulation while evaluating various measures for reducing energy consumption by changing design criteria in order to achieve the best energy-efficient building in terms of EUI and energy cost. In this study, an attempt was made to assess the EUI and energy cost of a school building in Hanoi, Vietnam.

3 Methodology

The target of this research proposes the analysis and simulation of energy used in building based on BIM system with its integrated tools. Besides, it also compares those simulation's results which can help proposing changes within construction process towards efficient energy used.

This research is followed up by 3 different tools including such as BIM (Revit) and other 2 energy simulation tools (Green Building Studio-GBS and Autodesk Insight). And so, author et al. will create an energy model (Energy model) in BIM based on those collected data with the initial set of information: works schedule, HVAC system; types of wall, roof and glass' materials and the ratio of the window opening in different building's facades. From the beginning, Revit software can generate an energy model (first simulation model) for the GBS and gbXML file meanwhile collected energy data is transferred to Data Cloud of Autodesk. The model was then calibrated and simulated adapted to different environmental scenarios of Green Building Studio-GBS and Autodesk Insight software. And so, each of simulation case corresponds to the change of relevant design elements.

Author et al. uses Autodesk Insight software to run those simulations in order to export energy intensity graphs (EUI-KWh / m2 / year) with various alternatives plan. These options are sorted in descending order of EUI value while comparing with each other so as to reach to the most appropriate one. As it can be seen from Fig. 1, this is an energy simulation method that is considered to be effective and highly reliable in assessing energy consumption in those current buildings [11].

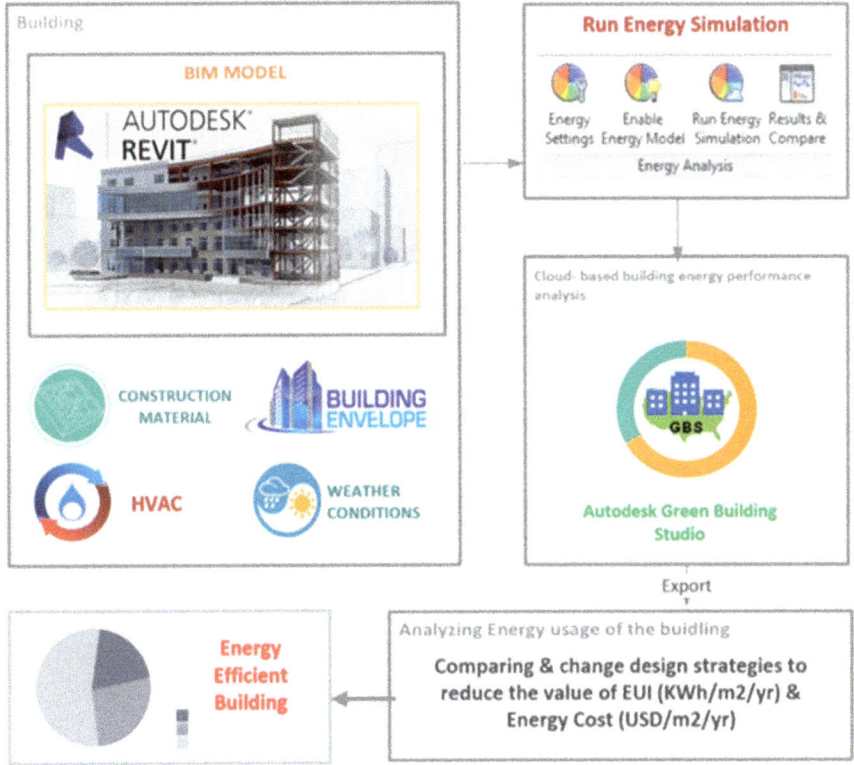

Fig. 1 Flowchart of research methodology

3.1 Case Study and Model Development

Author et al. have selected the actual and simulated project in Me Tri town, Hanoi capital city, Vietnam (Fig. 2). This selected building is currently used for teaching purposes with its attributes shown in Table 1. This selected project is then simulated in 3D on Revit software with the declaration of important parameters such as geographical position, building direction, HVAC system and materials of building's structure.

Weather's data was taken from the weather station 8.8 km away from the building, which could be used suitably during the simulation. According to the result, the highest mean maximum temperature in June is about 35–37.5°C while the lowest is 7.5–10°C in December.

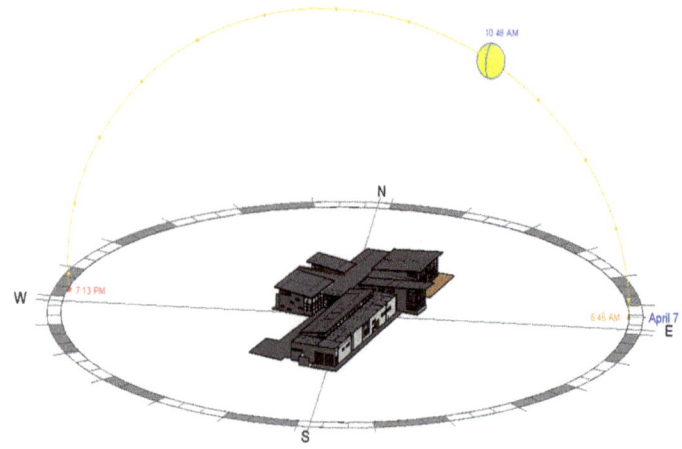

Fig. 2 Case study building

Table 1 Building program

Attributes	Values
Project type	Classroom
Location	Hanoi City, Vietnam
Floor area	1490 m^2
Zones program	Primary school classroom
Working hour	(8AM-8PM) 12 h/6d/w
Num. of people per area	0.0565 ppl/m^2
Window-wall-ratio (Western)	85%
Window-wall-ratio (Eastern)	60%
HVAC	Central VAV, Chiller 5.96
Num. of people per area	0.0565 ppl/m^2

3.2 Energy Simulation in Green Building Studio and Autodesk Insight

According to a designed structural model, an energy model is then created in BIM software with those initial settings such as: working schedule, an appropriated HVAC system intended for office used; using materials and structures in BIM software to serve the energy simulation process (Table 2).

When the simulation ends, engnergy model will be transfered into gbXML file (Fig. 3) and updated to the Cloud of Autodesk. This output helps creating a proper input for Green Building Studio (GBS) and Autodesk Insight to contiure the energy process. Afterthat, GBS will automatically run 247 cases so as to release number of energy intensities (EUI-KWh/m2/year) with different design concept following

Table 2 Defined material

Attributes	Materials	U-value (W/m2 K)
Exterior wall	Standard wall construction - C	0.3967
Interior wall	Block, interior/exterior finish, R-15 plus	0.3861
Window glass type	Single-glazed windows	3.29
Roof Construction	4 in lightweight concrete	1.275

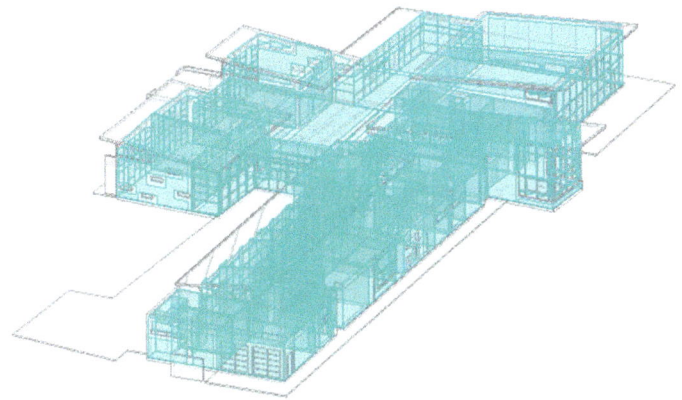

Fig. 3 Conversion model to gbXML

by number of changing variables. According to the result of GBS, the author has clarified 4 important variables that affected to the EUI index of a project such as: *Building orientation, Window-Wall-Ratio (Western), Window Glass Type and Roof Construction.* At last, these outputs from GBS can help choosing those 4 inputs (Table 3) for Autodesk Insight so that it can releases an ideal case with proper information about energy used and its costs.

Table 3 Design parameters

Variable	Attributes	No.of values
Building orientation	0, 45, 90, 135, 180, 225, 270, 315 (o)	8
Window-wall-ratio (Western)	15, 30, 40, 50, 65, 80, 95 (%)	7
Window glass type	Trp LoE, DbI Clr, DbI LoE, SgI Clr	4
Roof construction	R19, R38, R60, R10, R15, 10.25-inch SIP	6

4 Results, Analysis and Discussions

4.1 Scenario Analysis

The case study results, which is clearly seen from Fig. 4, can now be presented, analysed and discussed to understand the relationship between *Building orientation, Window-Wall-Ratio (Western), Window Glass Type and Roof Construction* and energy use intensity (EUI) of the building.

According to the data shown above, the author changes each value of the variables sequently to find out EUI and the energy cost (Tables 4 and 5). And so, corresponding to each variable, the author chooses an option with the lowest EUI index and energy cost. In addition, this result will then combine with other options of those remaining variables to release the final proposal which helps designers getting an ideal model with the lowest EUI and energy cost.

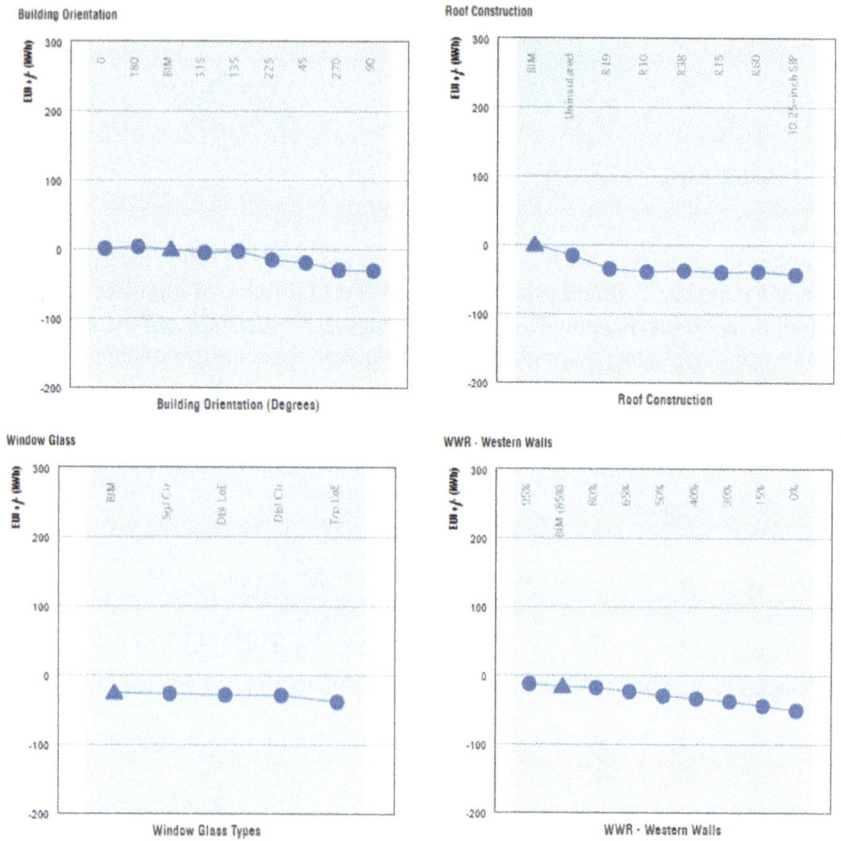

Fig. 4 Impact of vvariable on the EUI index

Table 4 Impacts of building orientation and window glass types on design objectives

	Base case	Building orientation							Window glass type			
		45°	90°	135°	180°	225°	270°	315°	Trp LoE	DbI Clr	DbI LoE	SgI Clr
EUI (KWh/m^2/year)	342	332	**311**	339	346	327	312	337	**300**	328	330	336
Energy cost (USD/m^2/year)	28	27	**26**	28	28	27	26	28	**25**	27	27	28

Table 5 Impacts of window-wall-ratio (western) and roof construction on design objectives

	Base case	Window-wall-ratio (western)							Roof construction					10.25-inch SIP
		15%	30%	40%	50%	65%	80%	95%	R19	R38	R60	R10	R15	
EUI (KWh/m^2/year)	342	**299**	309	315	321	330	339	345	309	307	306	305	304	**301**
Energy cost (USD/m^2/year)	28	**24.9**	25.7	26.2	26.7	27.4	28	28.5	23.9	23.6	23.6	23.7	23.6	**23.4**

In the base case, the front of the building is Western following with the EUI and energy cost is 342 KWh/m^2/year and 28 USD/m^2/year in total. Following to result from Tables 4 and 5, a final plan with ideal variables come up with: building orientation of 90°, Window Glass Type of Trp LoE, Window-Wall-Ratio (Western) of 15% and Roof Construction of 10.25-inch SIP.

5 Validation of Computation Results

Based on the simulation results from Tables 4 and 5, the author finds out the significant impact of design elements on building's EUI and energy cost. After that, author et al. initializes the energy model with 4 ideal variables found in this proposed project (Case study proposal) and compares it with the original case. As a result, it can be seen from Table 6 and Fig. 5 that the EUI and the energy cost are 207 Kwahu/m^2/year and 16.6 USD/m^2/year in total in the case study proposal.

There are two basic parameters the total annual average electricity cost is the electricity unit price of USD 0.072/KWh (at current prices in Vietnam) and the total work area of 1,490 m^2. It is clearly calculated of 41,720 USD, equivalent to 509,580

Table 6 Defined material comparison

Variable	Base case	Case study proposal
Building orientation	0	90
Window-wall-ratio (Western)	85	15
Window glass type	Single-glazed windows	Trp LoE
Roof construction	4 in lightweight concrete	10.25-inch SIP

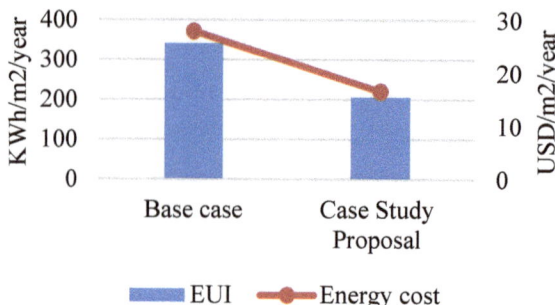

Fig. 5 Design objects' comparison

KWh in the base care, however, it is only 24,734 USD, equivalent to 308,430 KWh in this case study proposal. Thus, comparing to the original case (Base case), the case study proposal saves 16,986 USD/year, corresponding to 201,150 KWh. From the results of calculation and its simulation, this selected proposed plan is an ideal option to use in the construction and repair process towards energy savings and its cost.

6 Conclusion

This research successfully models an existing building into an energy model in order to calculate the EUI energy intensity index as well as the annual energy cost for various scenarios. As a result, this result allows designers to select the most optimal and suitable option for each design option., this result benefits for designers to choose the most optimal and suitable option for each design option.

However, this research fails to select popular design parameters and its alternative materials. In the future, authors propose an energy simulation model combined with a machine learning optimization algorithm which can select variables automatically while building a separate computational model.

The outcome of this proposed plan can also be combined with the LOTUS evaluation criteria, which assists designers in quickly determining the score of Vietnam green building standards (if any) in order to meet investor requirements. Furthermore, the result of this research will be very useful for not only engineers but also architects in designing, consulting and assessing the energy consumption of a building following with its environmental impact.

Acknowledgements **Truong Ngoc Son** was funded by Vingroup Joint Stock Company and supported by the Domestic Master/ PhD Scholarship Programme of Vingroup Innovation Foundation (VINIF), Vingroup Big Data Institute (VINBIGDATA), code **[VINIF.2021.TS.039]**

References

1. Abanda FH, Byers L (2016) An investigation of the impact of building orientation on energy consumption in a domestic building using emerging BIM (building information modelling). Energy 97:517–527
2. Ngu PT et al (2018) Construction Project Quality Management using Building Information Modeling 360 Field. Int J Adv Comput Sci Appl 9:228–233
3. Choi J et al (2016) Development of openBIM-based energy analysis software to improve the interoperability of energy performance assessment. Autom Constr 72:52–64
4. Chan DWM, Olawumi TO, Ho AML (2019) Perceived benefits of and barriers to building information modelling (BIM) implementation in construction: The case of Hong Kong. J Build Eng 25:100764
5. Charef R et al (2019) Building information modelling adoption in the european union: an overview. J Build Eng 25:100777
6. Yuan H, Yang Y, Xue X (2019) Promoting owners' BIM adoption behaviors to achieve sustainable project management. Sustainability 11(14):3905
7. Jalaei F, Jalaei F, Mohammadi S (2020) An integrated BIM-LEED application to automate sustainable design assessment framework at the conceptual stage of building projects. Sustain Cities Soc 53:101979
8. Luque A et al (2020) ADAPTS: an intelligent sustainable conceptual framework for engineering projects. Sensors (Basel) 20(6):1553
9. Nguyen TA, Nguyen PT, Do ST (2020) Application of BIM and 3D laser scanning for quantity management in construction projects. Adv Civ Eng 2020:8839923
10. Nguyen T et al. (2020) Application of BIM in managing the volume of high-rise building walls. In: Materials today: Proceedings
11. Kamel E, Memari AM (2019) Review of BIM's application in energy simulation: tools, issues, and solutions. Autom Constr 97:164–180
12. Allouhi A et al (2015) Energy consumption and efficiency in buildings: current status and future trends. J Clean Prod 109:118–130
13. Chou J-S, Ngo N-T (2016) Smart grid data analytics framework for increasing energy savings in residential buildings. Autom Constr 72:247–257
14. Zhou S et al (2014) Real-time energy control approach for smart home energy management system. Electr Power Compon Syst 42(3–4):315–326
15. Zhao H-X, Magoulès F (2012) A review on the prediction of building energy consumption. Renew Sustain Energy Rev 16(6):3586–3592
16. Ngo N-T (2019) Early predicting cooling loads for energy-efficient design in office buildings by machine learning. Energy Build 182:264–273
17. Tsai M-S, Lin Y-H (2012) Modern development of an adaptive non-intrusive appliance load monitoring system in electricity energy conservation. Appl Energy 96:55–73
18. Chen J et al (2015) Energy demand forecasting of the greenhouses using nonlinear models based on model optimized prediction method. Neurocomputing 174:1087–1100
19. Yaïci W, Entchev E (2016) Adaptive neuro-fuzzy inference system modelling for performance prediction of solar thermal energy system. Renew Energy 86:302–315
20. Shi X et al. (2016) A review on building energy efficient design optimization rom the perspective of architects. Renew Sustain Energy Rev 65(Suppl C):872–884
21. Shoubi MV et al (2015) Reducing the operational energy demand in buildings using building information modeling tools and sustainability approaches. Ain Shams Eng J 6(1):41–55
22. Hay R, Ostertag CP (2018) Life cycle assessment (LCA) of double-skin façade (DSF) system with fiber-reinforced concrete for sustainable and energy-efficient buildings in the tropics. Build Environ 142:327–341
23. Cuce E et al (2014) Optimizing insulation thickness and analysing environmental impacts of aerogel-based thermal superinsulation in buildings. Energy Build 77:28–39
24. Pham A-D et al (2020) The development of a decision support model for eco-friendly material selection in Vietnam. Sustainability 12(7):2769

25. Tzempelikos A, Athienitis AK, Karava P (2007) Simulation of façade and envelope design options for a new institutional building. Sol Energy 81(9):1088–1103
26. Luong DL et al (2020) Building a decision-making support framework for installing solar panels on vertical glazing façades of the building based on the life cycle assessment and environmental benefit analysis. Energies 13(9):2376
27. Mahiwal SG, Bhoi MK, Bhatt N (2021) Evaluation of energy use intensity (EUI) and energy cost of commercial building in India using BIM technology. Asian J Civ Eng 22:877–894
28. Najjar M et al (2019) Integrated optimization with building information modeling and life cycle assessment for generating energy efficient buildings. Appl Energy 250:1366–1382
29. Toronto Green Standard (2018) Energy efficiency report submission & modelling guidelines. E.a.E.D.T.C.P Division, Toronto city

Error Assessment of Point Cloud and BIM Models to Actual Works

Sy Tien Do, Hiep Hoang, and Dat Ho Quang Che

Abstract 3D Laser Scanning technology is being widely applied in the construction industry 4.0 for promoting the digitalizing and BIM modeling procedure, also called Scan-to-BIM. Scan-to-BIM technology helps to capture the continuously existing condition of the project and construct a BIM model in just a short period of time, also, update and store the data of the project throughout construction stages. Scan-to-BIM technology could, however, entail many risks if the users cannot control the errors of the point cloud model and BIM model. Although point cloud collected by 3D laser scanners has high accuracy, there are mistakes during Scan-to-BIM process, which cause errors. Therefore, the 3D BIM model based on the point cloud contains errors of dimension, coordinate, position, and these errors could cause difficulties for construction project management engineers and make the BIM model meaningless. This paper presents the causes of the errors, also analyzes and evaluates the errors of Scan-to-BIM process by the methodology of comparing point cloud and BIM models with physical objects to control the errors and support the next step of using the VR, AR, and MR. The data used in the paper is obtained from a real project which applies Scan-to-BIM technology to verify the theory.

Keywords 3D laser scanning · BIM · Scan-to-BIM · Point cloud · Error assessment

S. T. Do (✉)
Faculty of Civil Engineering, Ho Chi Minh City University of Technology (HCMUT),
268 Ly Thuong Kiet Street, District 10, Ho Chi Minh City, Vietnam
e-mail: sy.dotien@hcmut.edu.vn

Vietnam National University Ho Chi Minh City, Linh Trung Ward, Thu Duc District,
Ho Chi Minh City, Vietnam

H. Hoang · D. H. Q. Che
PortCoast Consultant Corporation, 92 Nam Ky Khoi Nghia, District 1,
Ho Chi Minh City, Vietnam

© The Author(s), under exclusive license to Springer Nature Singapore Pte Ltd. 2023   253
J. N. Reddy et al. (eds.), *ICSCEA 2021*, Lecture Notes in Civil Engineering 268,
https://doi.org/10.1007/978-981-19-3303-5_20

1 Introduction

Nowadays, 3D Laser Scanning technology obtained by laser scanners is becoming a common practice due to the demand for high accuracy in construction work, especially in project management. Documenting as-is conditions of sites and generating BIM model in just a short time conducted by Scan-to-BIM technology is an improvement in project management. Many sites all over the world are using this technology in quality control and quantity control. In Vietnam, Scan-to-BIM is also a potential technology, being adopted at several projects in recent years.

Nevertheless, there are inadvertent mistakes during Scan-to-BIM process, which cause errors. Hence, the 3D BIM model formed by the point cloud contains errors of dimension, coordinate, position, causing obstacles for construction project management engineers and make the BIM model unusable. One of the most important reasons for the deviation is the lack of a particular error assessment method to verify the completeness and accuracy of the data for each phase.

The following qualities should be spotted in a good quality-assessment (QA) method. To begin with, on the basis of project requirements, the method itself should enable the assessment of the accuracy and completeness of the statistics as well as the model. The compulsory level to which the model's details and accuracy should be might be determined by the end uses of the BIM. Moreover, receiving instant outcomes while the project is in progress can be advantageous. Early identification of potential mistakes can also allow such errors to be fixed. Figures, for instance, can be amassed again with different scanning devices, should there be a detection of a calibration error. Furthermore, it is necessary to trace the sources from which those errors appear to address the issue urgently. Finally, a QA procedure is deemed good when it is user-friendly and offers efficiency in time and resource requirements.

In this article, an error assessment process will be offered using both comparison algorithm in a computer program for Phase 1: Quality control for Point Cloud Data, Phase 2: Quality control for 3D BIM Model compare to Point Cloud Data, and physical measurement comparison for Phase 3: Quality control for 3D BIM Model compare to physical object. This paper also points out the causes of the errors, finds solutions, and gives some instructions in order to diminish the deviation as optimal as possible. This article will support further studies on the Error Assessment of Scan-to-BIM process in the future.

2 Literature Review

In recent years, there have been many suggestions for evaluating the quality of the Scan-to-BIM process. The physical measurement method is one of the simplest approaches to control the quality of 3D Laser Scanning process explored by Cheok and Franazsek for the QA of as-is 2D/3D building plans. Although this approach is effective in identifying modeling issues, assessing comprehensively the model's

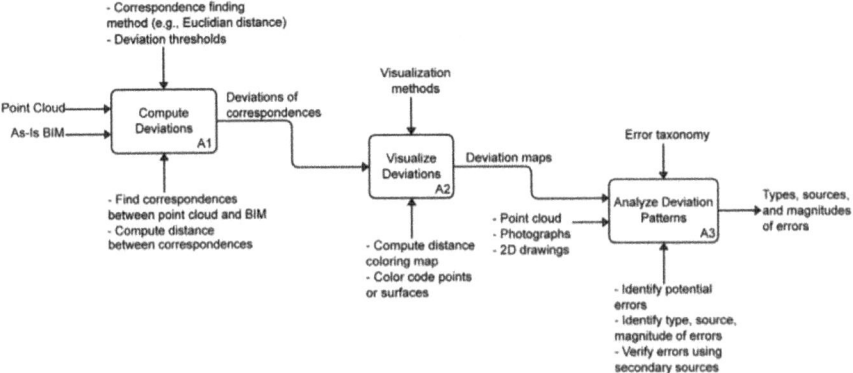

Fig. 1 The deviation analysis method of Engin Burak Anil et al. (2013)

quality is challenging and consuming an enormous of time because it will establish a direct comparison between the model and the physical measurements [1, 2]. Afterward, Tang et al. [3], and Anil et al. [4] introduced the deviation analysis method using correspondence finding method and visualization methods as a way of assessing the quality of as-is BIMs generated from the laser-scanned point cloud data (Fig. 1). The deviation analysis method is capable of recognizing almost six times more errors with more than 40% in time-saving compared to the physical measurement method in the analysis results. However, the current implementation of the deviation analysis method is limited in differentiating errors happening in the plane of the component surfaces [3, 4].

On the aspect of scanning method and quality control, the study of Thu et al. [6] had good potential to exploit scanning technology to civil engineering however they cannot clarify and solve the difference, deviation, and errors to actual work caused by the process [5, 6]. Quality control was also studied by Quang et al. [7] brought many advantages of laser scanning in construction supervising, but the accuracy of the process was not reliable. Based on the workflow for the Scan-to-BIM process suggested by Quang et al. [7], there are three phases needed to be focused on: Digital Terrain Modeling, Creating Model, and Creating BIM Model [7]. In our research, both the physical measurement method and the deviation analysis method will be applied to control the errors for each phase.

3   The Error Assessment Method

The given flowchart (Fig. 2) depicts information about assessing errors throughout three phases namely in Point Cloud Data, in 3D Model vs. Point Cloud Data, and 3D Model vs physical object.

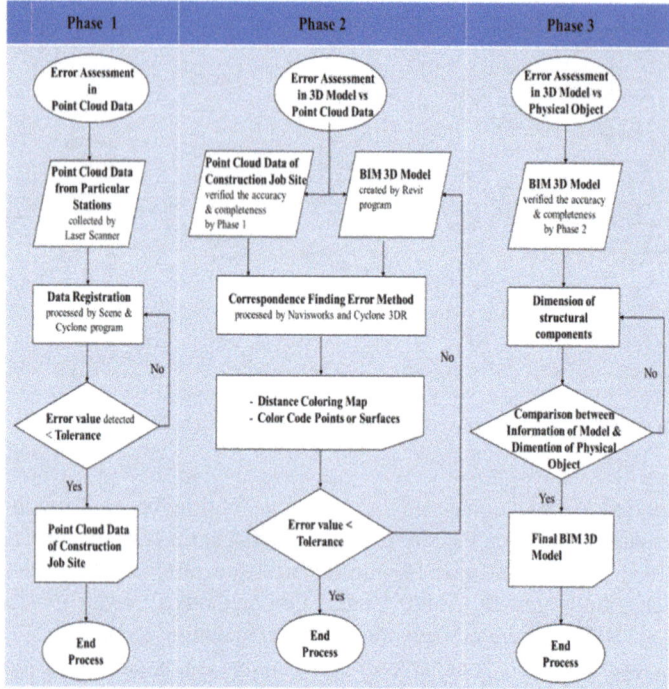

Fig. 2 Error Assessment suggested for Scan-to-BIM process

First and foremost, phase 1 presents the error assessment in Point Cloud data. To be more specific, Point Cloud Data from Particular Stations will conduct the evaluation with the aid of Laser Scanner in gathering the errors. Then, the Data Registration will be processed by the Scene and Cyclone program. At this point, if the Error value detected is not accepted, it will be processed again in the last step. However, if the error value is tolerated, it will be transferred to the Point Cloud Data of Construction Site which is also the end of this process.

In phase 2, the errors accumulated from phase 1 will be evaluated in either Point Cloud Data of Construction Site, which verified the accuracy and completion by phase 1, or BIM 3 M Model created by Revit. The data from both of the previously mentioned will experience a Correspondence Finding Error Method processed by Leica 3DR (Fig. 3), followed by an assessment of Distance Coloring Map and Color Code Points or Surfaces. Similar to phase 1, if the error value is acceptable, it will then be carried on in phase 3. Or else, those errors will go back to the first state of the ongoing phase.

Phase 3 illustrates how the output from phase 2 is evaluated in 3D Model vs. physical object. Once their accuracy and completeness have been verified, those errors will undergo the dimension of structural components. During the comparison between Information of Model and Dimension of physical object, unless those errors are qualified, they will be transferred back to the last period. Those successfully

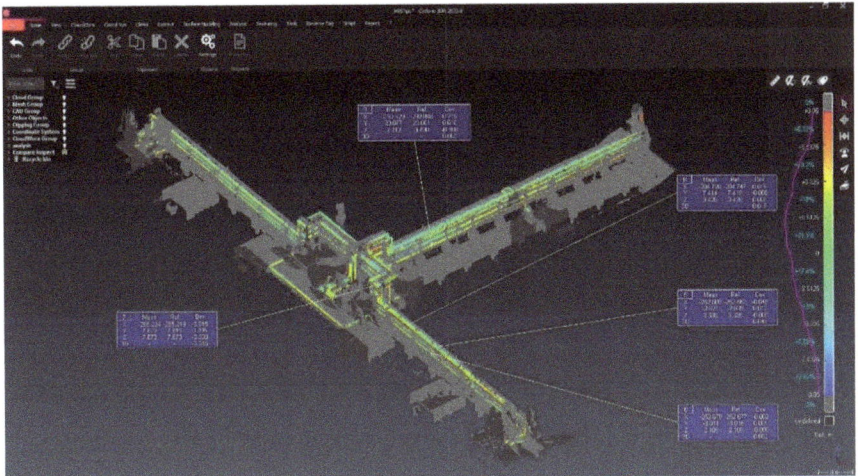

Fig. 3 Error assessment in Phase 2 conducted by 3DR program

passing processes will become the Final BIM 3D Model which is also the conclusive result (Fig. 4).

4 Error Category

Categories and types of error are as shown in Table 1

Error of scanner (1). The first factor which affects the accuracy of the 3D scanning process is the specification of a scanner. The quality and density of the point cloud depend on the type of scanner, and user setting.

Out of field of view (2). When an object is out of a field of view of projector and camera, it is difficult to gather the point of scanning object's digital file [8]. So, it is clear that proper selection of camera and projector is essential for the field of view and scanning with accuracy.

Environment interference (3). The high temperature could affect the directional ray of the laser scan. Furthermore, during the scanning process, multiple moving or stand-still obstacles could intercept the scanning ray and cause an error in the process. The error of Scan to BIM process. From the point cloud, the modeling engineer creates models using Autodesk Revit. By drawing the model, there are errors, inevitable, existing during the process because of the level, skill of the engineer.

Software problem (4). When the scanning process is completed, the points are synchronized among different stations to create a finished point cloud. To optimize, the process is analyzed and linked automatically using algorithms and

Inspect Cloud vs CAD

Theoretical: Default 2
Measure: Convert from Hyosung Ben CRE 1.lgs 1

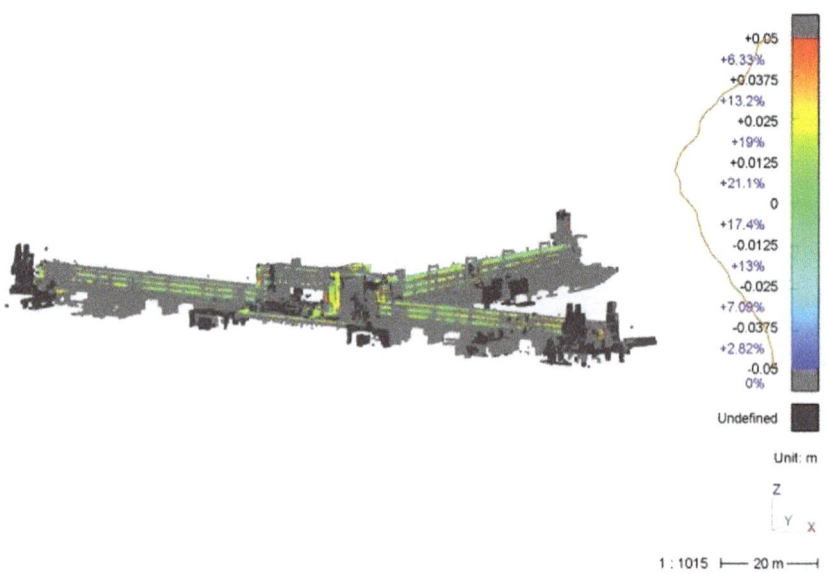

id	name	Meas X (m)	Meas Y (m)	Meas Z (m)	Ref X (m)	Ref Y (m)	Ref Z (m)	Dev X (m)	Dev Y (m)	Dev Z (m)	Dev 3D (m)	Tol- (m
5	Label #5	-252.60924	-2.6267	3.33776	-252.56162	-2.63913	3.33802	-0.04762	0.01243	-0.00026	0.04922	-0.
6	Label #6	-252.67904	-3.01444	2.1093	-252.67707	-3.01504	2.10945	-0.00196	0.00059	-0.00014	0.00206	-0.
7	Label #7	-285.23441	7.62025	7.67273	-285.21898	7.61552	7.67274	-0.01544	0.00473	-0.00001	-0.01615	-0.
8	Label #8	-284.7281	7.41362	3.42582	-284.74667	7.41945	3.42579	0.01857	-0.00583	0.00003	0.01946	-0.
9	Label #9	-292.62863	23.6772	3.31155	-292.84807	23.06127	3.41989	0.21945	0.61593	-0.10834	0.66277	-0.

Fig. 4 Error Assessment in Phase 3 when comparing BIM Model to physical object conducted by 3DR program

applications. The error of the finished point cloud completely depends on the calculation of them. Therefore, developing algorithms and apps will determine the accuracy of the model [8].

Table 1 Categories and types of error

Phase	Category	Type	Error
Collecting point cloud	1. Error of scanner	Type of scanner	Point density and accuracy
		User setting	
	2. Out of field of view	Range of scanner	Point density and accuracy
	3. Environment interference	Heat	Inconsistent ray
		Fog and dust	Disturbed
		Black and shining object	Missing point
		Crystal object	Scattering
		Experience and skill	Missing point
Building model	4. Software problem	Software	Point cloud model accuracy
		Experience and skill	Model accuracy

5 Solutions and Discussions

5.1 Solution

Based on the problems and results gathered from the workflow interviewing some expert individuals, and on-field experiments, there are many solutions to control the error of the model. As the product specifications cannot be modified, the best way to downgrade the error is to focus on the improvement of the scanning procedure and limit the influence of the environment.

Error of scanner (1). The scanner has its specification: distance measurement, laser class, field on view, range, speed, resolution, and most importantly for this article is the accuracy. Therefore, choosing the appropriate scanner decide the quality and accuracy of the point cloud. Furthermore, the 3D scanner also required proper calibration to produce accurate results. By calibration, the process scanner provides more precise data using a standard calibration plate as the specification. **Out of field of view (2).** Laser scanners have scan mode options that adapt to a variety of situations. Understanding the actual condition of the site and selecting the appropriate scan mode improves the quality of the point cloud model. For example, the Leica RTC360 scanner has 3 selectable settings (3/6/12 mm@10 m) with a range of up to 130 m and 2,000,000 pts/sec [10], depending on the type of construction.

Environment interference (3). Choosing the proper time to scan is important because the impact of environmental temperature could affect the error. The operating temperature should be from 18 °C to 30 °C [9] to avoid the scanning ray not being direct.

Software problem (4). For each 3D scanner, it is highly recommended using software and application developed by its own company because it is optimal and data synchronization. The software has its algorithm, and this could decide the precision of the point cloud and BIM model, by choosing the proper software could improve the accuracy of the model.

Besides, it also depends on the level and experience of the modeler. This leads to the required high-quality modeling engineer training.

5.2 Discussion

To the 3D Laser scanning and Scan to BIM process, it is highly required that the product, which is the digital 3D model, as the same as possible to the actual work, and with the development of 3D scanning technology, that requirement became possible.

With the development of technology and in a specific situation like the COVID-19 pandemic, using 3D scanning, scan to BIM the construction, work is helpful and easier in management, information storage.

Scanning for Point Cloud is a time-consuming operation and requires a high budget to have a proper, high-quality scanner. In milestone, this technology brings many benefits, effectiveness, and time-saving. Future research will improve technology more accurately, less time, and decrease cost. With the development of AI, the future scanner could optimize the process, disturbance handler, and saving time.

Creating as-built BIM from Point Cloud includes a serious manual process, which is subject to a personal flaw, thus raising the economical of processing. It also suffered a high cost because it takes a long period. Currently, commercially available BIM tools are not able to publish from the point cloud to the BIM model, this could lead to errors cause by human. Future research or development should focus on automatically create a BIM model.

No standard process is established to determine the proper process with a particular setting needed to meet the required level of detail for different environments and objects to be modeled. The current practice in the industry in this regard is driven by the client's requirements, the higher requirement the higher cost it takes. A worldwide set of standards or specification need to be researched to define the level of detail and required accuracy for the laser scanning operation. This is another crucial area that future research should focus on.

The finished product of this process is the BIM model. This product is applied for further technology like virtual reality (VR), augmented reality (AR), or mixed reality (MR) to create an easier way to the management and quality control. Therefore, error assessment among 3D scanning to Point cloud to BIM model and actual work is necessary. If the point cloud or BIM model contains too much error, the process is meaningless and these errors could cause many problems in practical application. Furthermore, this technology could help maintain, preserve ancient places or heritages with the highest accuracy.

6 Conclusion

Combining two methods and triple checks for the error assessment help to increase the reliability of the final BIM 3D Model. In addition, this suggested process will be conducted in a shorter time and give more comprehensive assessments for the deviation occurring in the Scan-to-BIM process. In other words, the combined method is more effective in the domain requirements of timely, detailed, and thorough quality assessment of the point cloud data and BIM model.

However, this process suggested in this paper has some limitations. This study clarifies some of the problems that 3D laser scanning and Scan to BIM usually suffer. We have studied, researched, and operated 3 different types of scanners, scanning an actual work, analyzing the preliminary point cloud, and build a model using applications. Although we have carried out many experiments, been trained, and studied from other research, we cannot carry out the solutions. Besides, we have given proof that it can improve the process and how much it is better by calculating, comparing it with the actual work. Despite the software and algorithms supporting the process, it still has many errors because of subjective causes during manual modification in some parts of the process. Furthermore, due to limited time, some of the information and documents are restricted this study was only conducted at the level of general processes.

The Error Assessment Method suggested in this paper can be enhanced in future research. We plan to conduct intensive research to prove the efficiency of this QA process comparing to other papers. Furthermore, these related aspects are also necessary to consider: recognizing more QA problems and evaluating the performance of the combined process; carrying out a more detailed assessment of the effectiveness of the technique, and building pattern recognition approaches for automated deviation pattern analysis.

Acknowledgements This project would have been impossible without the financial, equipment, software and data support of the Portcoast Consultant Corporation. Moreover, we would like to thank Mr. Phuoc Le Nguyen Thanh, Mr. Truong Vo Van (Laser Scanning team), Mr. Nhan Doan Viet, Mr. Hung Le Xuan (Modeling team), Mr. Nhan Truong Nguyen Trong (VR, AR, MR team), Mr. Minh Vu Hoang and Mr. Loc Pham Vo (Meshing team).

References

1. Cheok GS, Filliben JJ, Lytle AM (2009) Guidelines for accepting 2D building plans. NIST interagency/internal report (NISTIR), 7638
2. Cheok GS, Franazsek M (2009) Phase III: evaluation of an acceptance sampling method for 2D/3D building plans. NIST interagency/internal report (NISTIR), 7659
3. Tang P, Anil EB, Akinci B, Huber D (2011) Efficient and effective quality assessment of as-is building information models and 3D laser-scanned data. In: International workshop on computing in civil engineering
4. Anil EB, Tang P, Akinci B, Huber D (2013) Deviation analysis method for the assessment of the quality of the as-is building information models generated from point cloud data. Autom Constr 35:507–516
5. Do ST, Nguyen TA, Hoang H, Vu LT, Nguyen VNT, Vu TV, Le PNT, Pham ATT, Dang QM (2019) Integrating point cloud from 3D laser scanning and unmanned aerial vehicle (UAV) equipment in order to collect construction project information modelling. Vietnam J Constr 12:39–42
6. Nguyen TA, Do ST, Hoang H, Nguyen KDT, Huynh HP, Che DHQ, Nguyen DH (2020) Application of augmented reality for simulating 3D model from point cloud and photogrammetry – A study case of construction site inspection. Vietnam J Constr 04:4–8
7. Hoang H, Nguyen TA, Do ST, Vu LT, Nguyen VNT, Vu TV, Le PNT, Pham ATT, Dang QM (2020) Application of 3D Laser scanning technology in creating topographical surveying. Vietnam J Constr 01:3–7
8. Dang QM, Nguyen DH, Le PNT, Vu TV, Che DHQ, Nguyen KDT, Vu MH, Pham TLM, Nguyen NP, Hoang H, Nguyen TA, Do ST (2019) 3D laser scanning-The applications of point cloud model in quality assurance, quality control, and construction site inspecting/laser scan 3D. In: The 5th science and technology symposium for oisp students, Ho Chi Minh City University of Technology
9. Javaid M, Haleem A, Kumar L (2019) Dimensional errors during scanning of product using 3D scanner. In: Advances in engineering design, pp. 727–736
10. Leica Geosystems AG (2018) Leica RTC360 3D Reality Capture Solution

Factors Affecting Employee Retention in Construction: Empirical Study in the Mekong Delta Region

Duc-Anh Le, Long Le-Hoai, Van H. Le, and Chau Ngoc Dang

Abstract Together with the whole country, the construction industry in the Mekong Delta region (in the South of Vietnam) is developing very fast. However, this region faces more difficult challenges in attracting qualified labor forces than other areas. To meet the region's needs, there should be coordination between management units, training units, investors, and construction organizations to provide strategies to attract more construction employees. Using a questionnaire which includes 36 potential factors regarding the retention of employees in construction, 160 responses were collected from different organizations in the Mekong Delta region. The results indicated several important factors which could affect employee retention such as base salary, salary payment term, labor contract, labor market mobility, and opportunities to participate in large construction projects. Using factor analysis, this study also identified eight main influential constructs of employee retention in the Mekong Delta area's construction fields, namely organizational conditions and policies (C1), motivation (C2), working environment (C3), demanding (C4), income (C5), job opportunities (C6), welfare (C7), and job nature (C8). The findings of this study could provide useful information about how to develop and attract qualified human resources in the Mekong Delta region, as well as other similar areas in other developing countries.

Keywords Factors · Employee retention · Human resources · Construction industry · Mekong Delta

D.-A. Le · L. Le-Hoai
Faculty of Civil Engineering, Ho Chi Minh City University of Technology (HCMUT), 268 Ly Thuong Kiet Street, District 10, Ho Chi Minh City, Vietnam

Vietnam National University Ho Chi Minh City (VNU-HCM), Linh Trung Ward, Thu Duc District, Ho Chi Minh City, Vietnam

V. H. Le
Construction Management Graduate Program, Ho Chi Minh City University of Technology (HCMUT), 268 Ly Thuong Kiet Street, District 10, Ho Chi Minh City, Vietnam

C. N. Dang (✉)
Faculty of Civil Engineering, School of Engineering and Technology, Van Lang University, Ho Chi Minh City, Vietnam
e-mail: chau.dn@vlu.edu.vn

J. N. Reddy et al. (eds.), *ICSCEA 2021*, Lecture Notes in Civil Engineering 268,
https://doi.org/10.1007/978-981-19-3303-5_21

1 Introduction

Vietnam's economy is a multi-sector economy that has developed enormously in recent years. From 2002 to 2018, Vietnam has achieved many outstanding economic achievements such as an increase of 2.7 times GPD reaching 2,700 USD per person in 2019, reduction of the poverty rate (the income of poor people is lower than 3.2 USD per day) from 70 to 6% [10]. In particular, the construction industry plays a vital role in the Vietnamese economy [26]. Infrastructure development and urbanization are helping to propel the Vietnamese economy forward [19]. According to the General Statistics Office [14], in 2020, the growth rate of the construction industry increased by 6.76%, higher than those rates in 2011, 2012, and 2013. According to the Ministry of Construction [23], there were thousands of construction projects under construction in the first quarter of 2021 including commercial housings (1,380 projects with 306,053 apartments), social housings (72 projects with 105,971 apartments), and commercial tourism projects (5,180 tourist apartments, 534 tourist villas, and 46 offices combined with accommodation).

Human resources play an important role in many fields including the construction industry [5], which employ a large amount of labor. Human resources may be a significant issue of construction companies due to several reasons. Specifically, construction employees have to work in a hazardous environment for a long time [6, 24]. In addition, they suffer more stress when working with several team members and unethical supervisors [24]. In 2019, there were 21.4% employees in the US construction industry changing their works, which could be the highest change in all industries [35]. In Vietnam, during the 2000–2007 period, employment elasticities with respect to the construction industry output are 0.41% [22]. Most researchers usually focus on employee recruitment and retention, which could be two important aspects of human resources. Chih et al. [8] reviewed several benefits of employee retention such as improving productivity and reducing on-site accidents in construction projects. As a result, not only project objectives (e.g., cost, time, and quality) will be improved, but turnover costs (including recruitment and retaining) will also be reduced. Last but not least, it helps companies to accumulate experience, expertise, and skills. Therefore, if construction companies retain their employees for long terms, they will have more competitive advantages than other organizations [8, 24, 30, 31].

The Mekong Delta region has great advantages in terms of large population, but its economy development is less than other surrounding areas. The population in this region increased an average of approximately 300,000 people per year, reaching 21 million people in 2010 [9]. In 2011, the General Statistics Office [11] announced that there were over 10 million workers in Mekong Delta, which accounted for 20.1% total workers in Vietnam. Undoubtedly, Mekong Delta has become a significant area in Vietnam because of its area and population. However, this region has faced difficult challenges regarding human resources such as labor shortages and low skilled employees. Based on the Ministry of Planning and Investment report in 2012, Ngan et al. [25] noted that the number of workers having technical trainings in Mekong Delta accounted 3.4% in 51.4 million workers. Moreover, according to the results of the 2019 Population and Housing Census in Vietnam, 700,000 workers in Mekong Delta immigrated to the Southeastern region, accounting for more than 50% of all immigrants of the Southeastern area [13]. Nevertheless, the region is facing many human-related problems, of which retaining the workforce is among the most difficult challenges. In 2020, the General Statistics Office [14] announced that the proportion of the trained workforce with degrees and certificates was very low (13.6%). Thus, the Mekong Delta region has to solve the problem of labor shortages as an important issue for its economic development.

This study aims to identify the factors that affect the retention of labour resources in the Mekong Delta region. It also explores the underlying factors behind the workers' decision to stay in this area. It is expected that the results of this study could provide project and firm managers with an overview of how to retain labour forces in the Mekong Delta area's construction fields successfully.

2 Literature Review

Various research endeavors have explored the role of human resources in the construction industry, of which several models were proposed to optimize the resource usages or evaluate the effects of labor forces on global economy. For example, Tabish and Jha [33] found that human factors and management actions play a key role in achieving project success. Liu et al. [20] proposed a practical tool to effectively manage human resources in construction by modeling the demand of construction workforce and assessing the effects of labor market under global economy. Wang et al. [34] developed an artificial neural network model to explore the relationship between motivating factors and human values, and indicated that construction project managers' performance could be improved based on promoting motivators in terms of human values. Recently, Sarihi et al. [28] has proposed a simulation framework for assigning multiskilled, mainly managerial, workforces across multiple projects of a firm, which optimizes its entire allocation process toward minimizing resource usage and fluctuations, maximizing social sustainability, and minimizing costs.

In addition, many other attempts have identified and classified the influencing factors of employee retention in construction. For instance, Avila [3] found that learning and development, work environment, payment, and benefits are four main components of employee retention in construction. Smither [32] developed a management framework for enhancing employee retention and raising profitability and, thence, concluded that employers' care could enhance employees' work quality and career commitments. Huang et al. [16] found that marriage, honored employee status, gender, speed of promotion, relative pay (e.g., inter-firm and intra-firm wages), and economic cycles significantly affect employee retention time. Firm-specific human capital, signaling effects, and wages had a vital role in job retention. Compared with individual-based items, firm-based factors had significantly more pronounced effects on employees' ultimate decisions. Malone and Issa [21] developed predictive models for work-life balance and organizational commitment of women, which could be used to measure employees' satisfaction and short-term and long-term commitments in the US construction industry. Chih et al. [7] found that psychological contract breach results in emotional exhaustion, which in turn predicts employees' turnover intentions and actual turnover. Different from old labors, younger employees tend to leave their organizations and become exhausted more easily. Gupta et al. [15] identified eight factors related to site amenities and labor welfare attributes: health and medical provisions, site services, labor camp facilities, hygiene and sanitation, leave and benefits, social welfare and employment policies, remuneration, and accommodation facilities. They also found that improper letters of intent and delays in land acquisition for setting up proper labor camps could lead to poor quality of site amenities and labor welfare in Indian construction projects. Ayodele et al. [4] identified 26 factors affecting construction workforce turnover, which were grouped into four categories: job nature, external industry environment, construction firm itself, and individual workforce. Karakhan et al. [18] listed eight foundational attributes that could enhance construction workforce sustainability: nurturing, diversity, equity, health and well-being, connectivity, value, community, and maturity. They also suggested many strategies to assess and improve every workforce sustainability attribute.

It can be seen that employee retention is one of the most significant issues in the construction industry worldwide. Undoubtedly, conducting a study about how to enhance employee retention in the current Vietnamese context could also derive some valuable findings contributing to the global knowledge, given the high turnover rate in developing countries' construction industry [7]. Thus, this study is to identify critical influential factors of employee retention in the Mekong Delta area, which is one of the important economic areas in Vietnam. The findings of this study could provide local organizations and investors with a better understanding of how to improve the area's human resources, thereby contributing to the development of its construction industry and local economy.

3 Research Method

3.1 Questionnaire Design

A questionnaire was designed based on more than 30 potential factors of employee retention which were extracted from the literature. Respondents were requested to rate these factors on a scale from 1 = "not significant" to 5 = "very significant". Next, a pilot study was conducted with the participation of 11 experts, who were working in the Mekong Delta area. All experts had at least five years of construction experience, of which three of them had over 10 years of industry experience. With some minor modifications, the questionnaire, including 36 influential factors of employee retention, was finalized and used for data collection.

3.2 Data Collection

Using the hand-delivery method, 200 questionnaires were distributed to respondents working for both construction-related and educational organizations in the Mekong Delta area. A total of 160 valid returned questionnaires accounted for a response rate of about 80% were collected for this study's analyses using the SPSS statistical software. The test yielded a Cronbach's alpha coefficient of internal consistency value of 0.748 (>0.70), which is considered to be reliable.

Out of 160 responses, 51.3% were from contractors, 20.0% were from consultants, 19.4% were from owners, and 9.4 were from educational organizations. About professional fields of construction, 36, 21, 16, 17, and 10% respondents were mainly involved in civil and industrial construction, transportation construction, technical infrastructure construction, irrigation construction, and materials trading, respectively. In terms of industry experience, 20.6% had less than 3 years of experience, 33.1% had 3–5 years of experience, 26.9% had 5–7 years of experience, and 19.4% had more than 7 years of experience.

3.3 Data Analysis

First, mean score method is used to achieve the mean values and rankings of 36 influential factors of employee retention. As there seemingly exist different opinions between contractor and other groups about the importance of employee retention factors, Spearman's rank correlation test is performed to check whether such groups' ranking of these factors is related. Furthermore, t-test is used to investigate whether each item's mean values rated by such groups are different.

Then, factor analysis with varimax rotation method is used to explore the main influential factors of employee retention. Several tests are performed to check whether factor analysis is applicable for 36 influential factors of employee retention. The Kaiser–Meyer–Olkin (KMO) measure of sampling adequacy should be larger than 0.5 [29]. The Bartlett's test of sphericity, which indicates whether the correlation matrix is not an identity matrix, must be significant at the 0.05 level [19]. Furthermore, to signify the model's reliability, all factors' communalities included in the factor model must be above 0.5. As a rule of thumb, factor loadings less than 0.5 are also suppressed.

4 Ranking Influential Factors of Employee Retention

Table 1 present the mean values and rankings of 36 influential factors. In general, all 36 influential factors' mean values are larger than 3, of which up to 20 factors have their mean values over 4. This indicates that they are all important factors which

Table 1 Ranking of factors affecting employee retention

Code	Influential factors	Mean	Standard deviation	Rank	Levene's test		t-test	
					F-value	p-value	t-value	p-value
F1	Base salary	4.56	0.64	1	1.558	0.214	-0.890	0.375
F2	Overtime allowances	4.16	0.76	13	1.115	0.293	-1.428	0.155
F3	Collaborator's salary	3.64	1.04	32	0.009	0.924	-1.104	0.271
F4	Time for salary increase consideration	4.19	0.79	6	0.722	0.397	-0.740	0.941
F5	Salary payment term	4.45	0.73	2	0.831	0.363	-0.413	0.680
F6	Job's danger level	4.16	0.99	11	0.784	0.377	0.669	0.504
F7	Job's heaviness	3.79	1.05	27	0.317	0.574	-0.835	0.405
F8	Job's stability	4.18	0.88	7	1.809	0.181	-0.242	0.809
F9	Overtime	3.74	0.89	28	0.269	0.605	0.271	0.787
F10	Work location	3.79	0.98	26	0.014	0.907	-0.495	0.621
F11	Qualifications and methods of managers	4.17	0.92	8	0.492	0.484	0.371	0.711
F12	Job position	3.83	0.95	24	5.200	0.024	-1.278	0.203
F13	Relationship with colleagues and superiors	4.02	0.95	18	2.321	0.130	-1.091	0.277
F14	Living conditions	3.71	0.96	29	0.008	0.930	0.671	0.503
F15	Job selection and remuneration	4.16	0.80	10	0.098	0.755	0.330	0.742
F16	Environment using foreign languages	3.26	1.18	36	0.354	0.553	-1.417	0.159
F17	Desire of work place close to home	3.67	1.10	30	0.271	0.603	1.172	0.243
F18	Opportunities of postgraduate learning	3.41	1.11	35	0.001	0.978	-0.755	0.452
F19	Rewarding	4.16	0.84	12	2.206	0.139	0.602	0.548
F20	Recognition of efforts and dedication	4.17	0.87	9	0.218	0.641	0.211	0.833
F21	Holding of power	3.66	1.03	31	0.370	0.544	0.103	0.918
F22	Promotion opportunities	4.13	0.87	15	0.003	0.957	0.683	0.496

(continued)

Table 1 (continued)

Code	Influential factors	Mean	Standard deviation	Rank	Levene's test		t-test	
					F-value	p-value	t-value	p-value
F23	Entertainment and traveling	3.54	1.00	33	0.011	0.916	-0.727	0.468
F24	Labor contract	4.39	0.82	3	0.102	0.750	0.234	0.815
F25	Insurance benefits	4.16	0.93	14	0.540	0.464	-0.138	0.890
F26	Occupational safety training	4.04	1.01	17	0.432	0.512	-0.012	0.991
F27	Training to improve professional skills	4.01	0.89	19	0.002	0.968	0.350	0.727
F28	Organization of workplace regulations	3.86	0.97	22	0.006	0.938	0.943	0.347
F29	Development opportunities and vision	4.04	0.89	16	0.219	0.641	0.426	0.670
F30	Labor market mobility	4.21	0.88	4	0.136	0.713	0.103	0.918
F31	Opportunities to participate in large construction projects	4.20	0.85	5	0.044	0.834	-0.259	0.796
F32	Characteristics of technologies and engineering	3.83	0.89	23	0.100	0.752	-0.206	0.837
F33	Living costs	3.88	0.98	21	0.523	0.471	1.176	0.241
F34	Educational quality	4.00	0.88	20	0.378	0.539	-0.895	0.372
F35	Safety and technical training courses	3.82	0.99	25	3.101	0.080	-0.500	0.618
F36	Centers of engineering and construction informatics	3.50	1.04	34	0.230	0.632	-0.152	0.880

could significantly contribute to employee retention in construction. The computed Spearman rank correlation coefficient between contractor and other groups is 0.765 (p-value $= 0.000$), indicating that there is a strong correlation between these groups in ranking the influential factors. In addition, the results of t-test (Table 1) show that the differences of opinion about the mean ratings between these groups are insignificant. In general, the test results together indicate that there is a consensus of these groups on the ratings of 36 influential factors.

Table 2 Top five influential factors

Rank	Overall	Contractor	Other
1	Base salary	Base salary	Base salary
2	Salary payment term	Salary payment term	Salary payment term
3	Labor contract	Labor contract	Labor contract
4	Labor market mobility	Labor market mobility	Overtime allowances
5	Opportunities to participate in large construction projects	Job's danger level	Opportunities to participate in large construction projects

In Table 2, the top five rankings are presented according to three different groups: overall, contractor, and other. All groups share the same perspective about the top three rankings: "base salary" (rank 1), "salary payment term" (rank 2), and "labor contract" (rank 3). It can be seen that these factors directly relate to contract fundamentals. This implies that construction employees in the Mekong Delta area always wish to have stable salary levels with clear and comprehensive contracts with their employers (e.g., construction organizations). Commonly, employers are trying to attract and retain qualified human resources by offering big year-end bonuses and salary increases. This is because money is the best way to gain short-term satisfaction for employees [27]. According to the General Statistics Office [12], employees' income in the Mekong Delta area is very low (just equal one thirds of those in other Southeast regions). Thus, business organizations and authorities should pay more attention to salary and contract issues to enhance employee satisfaction and retention in this area's construction fields.

Table 2 also show some differences between contractor and other groups about the remaining positions. Contractors consider "labor market mobility" and "job's danger level" as the fourth and fifth significant factors while other organizations concern more about "overtime allowances" and "opportunities to participate in large construction projects". These results imply that each organization should have its own strategy to retain employees. Specifically, as one of the main players in construction projects, contractors always need many employees to work for them. However, construction activities always pose high risks and employee turnover has long been, and will continue to be, a significant concern in the construction industry [7]. Accordingly, contractors should create favorable environments, where employees can participate with long-term and safe working conditions. On the other hand, to better retain employees, other organizations should provide them with good overtime allowances and opportunities to participate in large construction projects, which could directly affect employees' satisfaction.

Table 3 Factor analysis results

Code	Influential constructs/ Measuring items	Factor loading	Eigenvalue	% of variance	Cumulative %
C1	*Organizational conditions and policies*		5.688	17.236	17.236
F28	Organization of workplace regulations	0.766			
F29	Development opportunities and vision	0.730			
F32	Characteristics of technologies and engineering	0.725			
F36	Centers of engineering and construction informatics	0.719			
F26	Occupational safety training	0.712			
F35	Safety and technical training courses	0.710			
F27	Training to improve professional skills	0.706			
F34	Educational quality	0.682			
F23	Entertainment and traveling	0.534			
C2	*Motivation*		3.416	10.353	27.589
F19	Rewarding	0.754			
F14	Living conditions	0.660			
F20	Recognition of efforts and dedication	0.637			
F15	Job selection and remuneration	0.616			
F22	Promotion opportunities	0.583			
F21	Holding of power	0.574			
C3	*Working environment*		3.087	9.355	36.944
F9	Overtime	0.820			
F10	Work location	0.783			
F11	Qualifications and methods of managers	0.713			
F13	Relationship with colleagues and superiors	0.540			
F12	Job position	0.529			
C4	*Demanding*		2.597	7.869	44.813
F17	Desire of work place close to home	0.915			
F33	Living costs	0.915			

(continued)

Table 3 (continued)

Code	Influential constructs/ Measuring items	Factor loading	Eigenvalue	% of variance	Cumulative %
F18	Opportunities of postgraduate learning	0.600			
C5	*Income*		2.464	7.467	52.281
F1	Base salary	0.830			
F5	Salary payment term	0.753			
F4	Time for salary increase consideration	0.723			
F2	Overtime allowances	0.603			
C6	*Job opportunities*		1.937	5.870	58.151
F31	Opportunities to participate in large construction projects	0.852			
F30	Labor market mobility	0.842			
C7	*Welfare*		1.864	5.649	63.800
F25	Insurance benefits	0.819			
F24	Labor contract	0.765			
C8	*Job nature*		1.704	5.164	68.964
F6	Job's danger level	0.852			
F7	Job's heaviness	0.745			

5 Main Influential Constructs of Employee Retention

In terms of factor analysis, the initial check showed that three influential factors (i.e., collaborator's salary (F3), job's stability (F8), and environment using foreign languages (F16)) had their communality values less than 0.5 and were excluded. Therefore, 33 remaining factors were considered to be appropriate for factor analysis (with principal component analysis and varimax rotation method). The KMO measure of sampling adequacy was satisfactory with a value of 0.853. The chi-square value of Bartlett's test of sphericity was 3,105 with the significance level at 0.000 indicated that that the correlation matrix was not an identity matrix. The results of factor analysis extracted eight components with the eigenvalues higher than 1 and the total amount of variance explained was 68.96%. These results imply that the extracted eight components could be considered as the main influential constructs of employee retention in the Mekong Delta area's construction fields, namely organizational conditions and policies (C1), motivation (C2), working environment (C3), demanding (C4), income (C5), job opportunities (C6), welfare (C7), and job nature (C8) (Table 3).

Organizational conditions and policies (C1) explained a variance of 17.23%. It comprised of nine influential factors: organization of workplace regulations (F28), development opportunities and vision (F29), characteristics of technologies and engineering (F32), centers of engineering and construction informatics (F36), occupational safety training (F26), safety and technical training courses (F35), training to improve professional skills (F27), educational quality (F34), and entertainment and traveling (F23). Workplaces' regulations, as well as development opportunities and vision, could directly and significantly affect employees' long-term commitment to any organization. In addition, various technologies, engineering, and safety conditions could play a vital role in maintaining and improving organizational competitive advantages. On the other hand, training courses and educational programs can enable employees to be more competent. With different enjoying activities, construction organizations will also improve employees' happiness. These altogether increase employees' satisfaction, thereby enhancing employee retention in the Mekong Delta region's construction fields.

Motivation (C2), with 10.36% of variance explained, included six influential attributes: rewarding (F19), living conditions (F14), recognition of efforts and dedication (F20), job selection and remuneration (F15), promotion opportunities (F22), and holding of power (F21). Rewarding and recognition are commonly applied to promote employees' working spirit in construction organizations. Together with job selection and remuneration, living conditions could be considered as the main reason why employees feel satisfied with any company [2]. Besides these issues, individual demands (e.g., promotion and power) could also play a vital role in enhancing employee retention in many construction environments, including a region like Mekong Delta.

Working environment (C3) appears as the third construct, by which 9.36% of variance was explained. This construct consisted of five influential factors: overtime (F9), work location (F10), qualifications and methods of managers (F11), relationship with colleagues and superiors (F13), and job position (F12). Overtime-related issues are usually a big concern of any employee. As construction is a heavy and high-risk industry, managers' qualifications and methods always play a decisive role in supporting and guiding construction and project teams effectively, thereby enhancing employees' trust. Given that construction activities commonly involve multiple people, employees should have good relationships with colleagues and superiors. Together with work location, job position could significantly affect employees' job satisfaction and long-term commitments to any construction organization in the Mekong Delta region.

Demanding (C4), which accounted for 8.97% of variance, is also an important influential construct in terms of employee retention. Three items contributing to this construct include desire of work place close to home (F17), living costs (F33), and opportunities of postgraduate learning (F18). With most people, working near to home could bring them various benefits (e.g., time savings). In addition, satisfying living costs is important to any employee. In today's fast changing and highly competitive construction environment, learning and development are also important to employees [3]. It can be seen that these factors directly relate to employees' life needs and future aspirations, which could significantly contribute to employee retention in the Mekong Delta region in general and in the construction industry in particular.

Income (C5) explained 7.47% of variance and involves four influential factors: base salary (F1), salary payment term (F5), time for salary increase consideration (F4), and overtime allowances (F2). When signing labor contracts with any employer, most employees concern much about salary and payment issues. In addition, construction organizations should also provide their employees with acceptable time for salary increase consideration. Overtime allowances should also be good enough, as construction activities usually involve hard works. In previous studies (e.g., [3, 16]), some pay-related factors (e.g., salary and variable pay) were found to significantly affect employees' decision of commitment on employers.

Job opportunities (C6) consisted of opportunities to participate in large construction projects (F31) and labor market mobility (F30). With 5.87% of variance, this construct appears to be the sixth important issue affecting employee retention in the Mekong Delta region's construction fields. As compared with other delta areas, this region's infrastructures are relatively underdeveloped, so participating in large-scale construction projects usually a big dream of many employees, especially young engineers. According to Avila [3], employees concern work environments because they want to gain more opportunities to acquire business knowledge. Like other manufacturing industries, the construction industry has also been facing high labor market mobilities. These jointly result in both favourable opportunities and difficult challenges for employees in joining large-scale construction projects, which usually require much experience and appropriate expertise.

Welfare (C7), which includes insurance benefits (F25) and labor contract (F24), accounted for 5.65% of variance explained. In the Mekong Delta region, where per capita income is much lower than many other regions [12], having a stable job with good insurance benefits becomes essential to most employees. It may be because employees, including internal migrants, in this region usually face unstable jobs and even unemployment. This is also similar to other Asian areas, where many internal migrants often face job insecurity due to the lack of labor contracts [17].

Job nature (C8), with 5.16% of variance explained, comprises two influential factors: job's danger level (F6) and job's heaviness (F7). Most people wish to work in good safety conditions, so job-related danger levels usually have a significant impact on employees' job selection and long-term commitments with construction organizations. In fact, construction activities usually pose safety risks and, therefore, maintaining safety is necessary for developing human resource strategies [1]. Besides this, construction activities also involve hard works. Accordingly, jobs' heaviness becomes a critical barrier for construction firms to retain organizational human resources, which always play a pivotal role in determining project and firm successes.

6 Conclusions

This study identified 36 influential factors of employee retention in the Mekong Delta region's construction fields, of which several important factors were highlighted: base salary, salary payment term, labor contract, labor market mobility, and opportunities to participate in large construction projects. Besides this finding, job's danger level and overtime allowances were also emphasized by contractors and other organizations, respectively. This study further found eight main influential constructs of employee retention in the Mekong Delta area, namely organizational conditions and policies (C1), motivation (C2), working environment (C3), demanding (C4), income (C5), job opportunities (C6), welfare (C7), and job nature (C8). These constructs covered a wide range of critical issues for construction organizations to retain their employees successfully.

Despite the above contribution, this study still has some limitations. First, the research results just focus on identifying the influential factors and constructs of employee retention in the Mekong Delta region, whose socioeconomic conditions are not so developed as compared with other delta areas. Accordingly, the findings cannot directly apply to construction organizations in other socioeconomic settings. Second, this study has not investigated the effects of influential factors and constructs of employee retention on organizational growth and performance of construction organizations. Thus, further studies should examine these relationships to support construction organizations in terms of both employee retention and firm development.

Acknowledgements The authors fully acknowledge the support of time and resources from the affiliated organizations for this study. This research is funded by Vietnam National University Ho Chi Minh City (VNU-HCM) under grant number B2021-20-08. We would like to thank Ho Chi Minh City University of Technology (HCMUT), VNU-HCM for the support of time and facilities for this study.

References

1. Ahmadian Fard Fini A, Akbarnezhad A, Rashidi TH, Waller ST (2017) Job assignment based on brain demands and human resource strategies. J Constr Eng Manag 143(5):1–16
2. Allen L, Sewards J (1992) Issues in human resources: managing talent in the 21st century. J Manag Eng 8(4):340–345
3. Avila EA (2001) Competitive forces that drive civil engineer recruitment and retention. Leadersh Manag Eng 1(3):17–22
4. Ayodele OA, Chang-Richards A, González V (2020) Factors affecting workforce turnover in the construction sector: a systematic review. J Constr Eng Manag 146(2):1–24
5. Castañeda JA, Tucker RL, Haas CT (2005) Workers' skills and receptiveness to operate under the Tier II construction management strategy. J Constr Eng Manag 131(7):799–807
6. Cheng MY, Hoang ND (2015) Evaluating contractor financial status using a hybrid fuzzy instance based classifier: Case study in the construction industry. IEEE Trans Eng Manag 62(2):184–192

7. Chih YY, Kiazad K, Zhou L, Capezio A, Li M, Restubog SLD (2016) Investigating employee turnover in the construction industry: a psychological contract perspective. J Constr Eng Manag 142(6):1–9
8. Chih YY, Kiazad K, Cheng D, Emamirad E, Restubog SLD (2018) Interactive effects of supportive leadership and top management team's charismatic vision in predicting worker retention in the Philippines. J Constr Eng Manag 144(10):1–10
9. Collins N, Jones S, Nguyen TH, Stanton P (2017) The contribution of human capital to a holistic response to climate change: learning from and for the Mekong Delta. Vietnam. Asia Pac Bus Rev 23(2):230–242
10. Crowe (2021): Why invest in Vietnam? https://www.crowe.com/vn/vi-vn/insights/doing-business-in-vietnam/doing-business-in-vietnam-2020/why-invest-in-vietnam
11. General Statistics Office (2011): Labor and employment survey report in the first 6 months of 2011. https://www.gso.gov.vn/wp-content/uploads/2019/03/Sach_LDVL_6thang 2011_Vie_In.pdf
12. General Statistics Office (2019): Labor and employment survey report 2018. https://www.gso.gov.vn/wp-content/uploads/2020/02/Lao-dong-viec-lam-2018.pdf
13. General Statistics Office (2020): Results of the 2019 population and housing census. https://www.gso.gov.vn/du-lieu-va-so-lieu-thong-ke/2020/11/ket-qua-toan-bo-tong-dieu-tra-dan-so-va-nha-o-nam-2019/
14. General Statistics Office (2020): Socio-economic situation in the fourth quarter of 2020. https://www.gso.gov.vn/en/data-and-statistics/2021/01/socio-economic-situation-in-the-fourth-quarter-and-the-whole-year-2020/#
15. Gupta M, Hasan A, Jain AK, Jha KN (2018) Site amenities and workers' welfare factors affecting workforce productivity in Indian construction projects. J Constr Eng Manag 144(11):1–11
16. Huang C, Lin HC, Chuang CH (2006) Constructing factors related to worker retention. Int J Manpow 27(5):491–508
17. International Organization for Migration (2005) Migration, development and poverty reduction in Asia. Academic Foundation, Geneva
18. Karakhan AA, Gambatese J, Simmons DR, Nnaji C (2020) How to improve workforce development and sustainability in construction. In: Construction Research Congress 2020: Safety, Workforce, and Education. Reston, pp. 21–30
19. Le-Hoai L, Lee YD, Son JJ (2010) Partnering in construction: investigation of problematic issues for implementation in Vietnam. KSCE J Civ Eng 14(5):731–741
20. Liu J, Love PE, Sing MC, Carey B, Matthews J (2015) Modeling Australia's construction workforce demand: Empirical study with a global economic perspective. J Constr Eng Manag 141(4):1–7
21. Malone EK, Issa RR (2014) Predictive models for work-life balance and organizational commitment of women in the US construction industry. J Constr Eng Manag 140(3):1–10
22. Manning C (2010) Globalization and labour markets in boom and crisis: the case of Vietnam. ASEAN Econ Bull 27(1):136–157
23. Ministry of Construction (2021): The Ministry of Construction announced information on housing and real estate market in the first quarter of 2021. https://moc.gov.vn/vn/tin-tuc/1285/67250/bo-xay-dung-cong-bo-thong-tin-ve-nha-o-va-thi-truong-bat-dong-san-quy-i2021.aspx
24. Nawaz Khan A, Khan NA, Soomro MA (2020) Influence of ethical leadership in managing human resources in construction companies. J Constr Eng Manag 146(11):1–12
25. Ngan C, Sandra J, Pauline S, Nguyen HT (2017) Environmental and human resource development issues in Vietnam: the case study of the Mekong Delta. Can Tho Univ J Sci 07(2017):109–117
26. Nguyen LD, Ogunlana SO, Lan DTX (2004) A study on project success factors in large construction projects in Vietnam. Eng Constr Archit Manag 11(6):404–413
27. Rothbard AT (1998) How to keep employees. J Manag Eng 14(4):21–22
28. Sarihi M, Shahhosseini V, Banki MT (2020) Multiskilled project management workforce assignment across multiple projects regarding competency. J Constr Eng Manag 146(12):1–12

29. Sharma S (1996) Applied Multivariate Techniques. Wiley, New York, NY
30. Sing CP, Love PED, Tam CM (2012) Stock-flow model for forecasting labor supply. J Constr Eng Manag 138(6):707–715
31. Sing CP, Love PED, Tam CM (2012) Multiplier model for forecasting manpower demand. J Constr Eng Manag 138(10):1161–1168
32. Smither L (2003) Managing employee life cycles to improve labor retention. Leadersh Manag Eng 3(1):19–23
33. Tabish SZS, Jha KN (2012) Success traits for a construction project. J Constr Eng Manag 138(10):1131–1138
34. Wang D, Arditi D, Damci A (2017) Construction project managers' motivators and human values. J Constr Eng Manag 143(4):1–10
35. Yildirmaz A, Ryan C, Nezaj J (2019) 2019 State of the workforce report: pay, promotions and retention. ADP Research Institute. https://www.adpri.org/wp-content/uploads/2020/07/19220244/The-State-of-the-Workforce-Full-Research-Report-2019.pdf

Identifying Stakeholder's Behavioral Intentions of Applying BIM to Construction Projects in Vietnam

Thi-Thao-Nguyen Nguyen, Thu Anh Nguyen, and Sy Tien Do

Abstract Nowadays, industrial revolution 4.0 is taking place firmly and affecting many industries and fields. In construction, some new technologies have been developed and applied to the design, construction and management. Building Information Modeling (BIM) is a digital environment that allows stakeholders to quickly share and update project lifecycle information across 3D models containing data. BIM is increasingly being applied in construction projects. However, the results of applying BIM for construction projects in Vietnam are still limited. This study combines the Theory of Planned Behavior (TPB) and the Technology Acceptance Model (TAM) to exploit and explain the stakeholder's behavioral intention when applying BIM to construction projects in Vietnam. Descriptive statistics, Cronbach's Alpha reliability analysis and Correlation analysis with SPSS were used in this study. The results provide proper solutions to raise awareness about the importance of applying BIM among stakeholders in the construction industry in Vietnam through seminars, training, or workshops.

Keywords TPB · TAM · BIM · Behavioral intention · Vietnam

T.-T.-N. Nguyen (✉) · T. A. Nguyen · S. T. Do
Department of Construction Engineering and Management, Faculty of Civil Engineering, Ho Chi Minh City University of Technology (HCMUT), 268 Ly Thuong Kiet, District 10, Ho Chi Minh City, Vietnam
e-mail: nttnguyen.sdh19@hcmut.edu.vn

Vietnam National University Ho Chi Minh City, Linh Trung Ward, Thu Duc City, Ho Chi Minh City, Vietnam

T.-T.-N. Nguyen
Faculty of Project Management, The University of Danang, University of Science and Technology (DUT), 54 Nguyen Luong Bang Street, Lien Chieu District, Da Nang City, Vietnam

© The Author(s), under exclusive license to Springer Nature Singapore Pte Ltd. 2023
J. N. Reddy et al. (eds.), *ICSCEA 2021*, Lecture Notes in Civil Engineering 268,
https://doi.org/10.1007/978-981-19-3303-5_22

1 Introduction

BIM is a system that enables management information integration through compatibility and information sharing at all stages of the project. This technology establishes a collaborative system between different disciplines and enables smooth communication between those fields [3]. Awareness of the advantages of BIM and its implementation can significantly increase project productivity and efficiency [4]. Despite agreements on the applicability and potential benefits of BIM in construction, many people are still unclear about using BIM and the advantages of implementing BIM. Therefore, the acceptance of BIM is still a concern of research around the world [3]. In particular, as the construction industry is becoming more complex, the need for BIM will become more apparent [5].

Research on BIM acceptance models that reflect the stakeholders' opinions is scarce in Vietnam. The objective of this research is to explore BIM adoption mechanisms based on research models of technology acceptance, behavioral intention, including TAM (Technology Acceptance Model) and TPB (Theory of Planned Behavior) [1, 2, 6, 7]. TAM has been widely applied in explaining individuals' acceptance of new technologies and their associated behaviors [8, 9]. There are many modifications for TAM, but the original TAM is easy to apply in different research environments. It has a more solid theoretical base and has full experimental support [6]. TPB (Ajzen, 1991) is developing and improving the Theory of Reasonable Action (TRA). The three main factors that influence intentions are attitudes, subjective norms and perception of behavior control [10].

This study explains the relationship between applying BIM technology and the combination model of TPB and TAM. This research aims to examine stakeholder perceptions when applying BIM technology in Vietnam and test the adoption of BIM using this combined model. Therefore, the research purpose is to propose a fully integrated model of the concepts of TAM and TPB to understand stakeholder behavioral intentions better using BIM.

2 Literature Review

As mentioned above, this study is based on two theories for behavioral intention (TPB and TAM) to consider the decisive factors influencing the intention to use BIM in construction projects.

2.1 Technology Acceptance Model—TAM

The advent of BIM is considered a new technological model. One of the tools helpful in explaining the intention to adopt a new system is TAM—Technology Acceptance

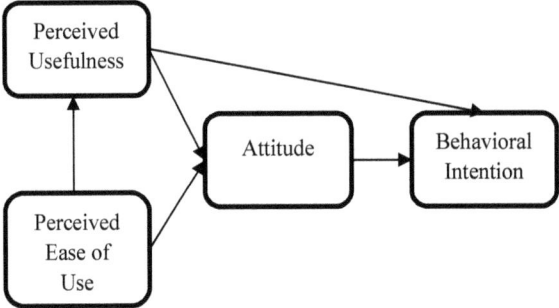

Fig. 1 Technology acceptance model [2]

Model. According to Davis (1989), TAM explains the factors determining the acceptance of new technology. He explained that the success of an information system could be determined by user acceptance, measured by factors including perceived usefulness (PU), perceived ease of use (PEOU), attitudes toward use (ATU) and behavioral intention to use (BIU) [2]. TAM has successfully predicted about 40% of the use of a new system [11]. TAM is modeled and presented in Fig. 1.

PU—Perceived Usefulness: The level at which using BIM technology will improve their productivity (Davis 1989). Several studies have found that PU is strongly correlated with BIU [2, 3, 11–13].

PEOU—Perceived Ease of Use: The level at which a person believes using new technology will be free from effort (Davis 1989). Previous research shows that PEOU is a decisive factor for successful new technological adoption [3, 14].

ATU—Attitude towards use: is measured by beliefs and assessment of the behavior's outcome (Ajzen 1991). Attitude is the level at which a person tends to be positive or negative towards using BIM [13]

2.2　Theory of Planned Behavior—TPB

TPB (Ajzen 1991) is most widely used in explaining and analyzing human behavior. According to TPB, behavioral intention is generally determined by attitudes towards behavior, subjective norms and perceived behavioral control. TPB is modeled and presented in Fig. 2.

SN—Subjective norm: The perception of individuals who should or should not perform the behavior (Ajzen 1991). In other words, SN refers to normative beliefs about expectations from others [10, 15].

PBC—Perceived behavioral control reflect how easy or difficult it is to perform the behavior of interest [1].

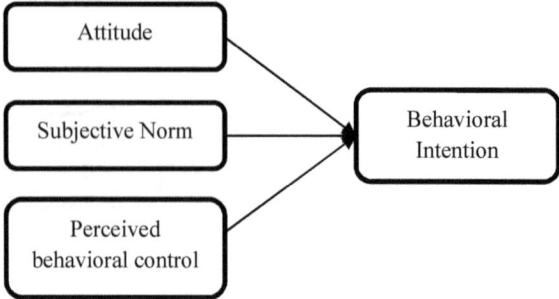

Fig. 2 The theory of planned behavior [1]

2.3 *Behavioral Intention*

Behavioral intention is a measure of the strength of a person who intends to perform a specified behavior [1]. Many researchers expect behavioral intention to have a significant positive effect on new technology adoption [3]. In this study, the behavioral intention using BIM is considered as a dependent variable. Furthermore, its measurements for an individual's intent to accept BIM are the willingness to use BIM tools and information to handle their work or recommend BIM to colleagues.

3 Research Model

In accordance with the research objectives and relevant documents, the proposed model, as shown in Fig. 3 below, is presented mainly based on the combination of Theory of Planned Behavior TPB presented by Ajzen 1991 and Technology Acceptance Model (TAM) presented by Davis 1989. Because BIM is a new technology applied in the construction field, this study model is suitable to explain the factors influencing the behavioral intention to use BIM.

The proposed model—Fig. 3 covers the core structures of both TPB and TAM and is mainly similar to Taylor and Todd 1995 in a study that assessed the critical factors of using and adopting information technology.

The author proposes the following hypotheses to clarify the factors affecting the behavioral intention of stakeholders. Our research hypotheses are presented in Table 1.

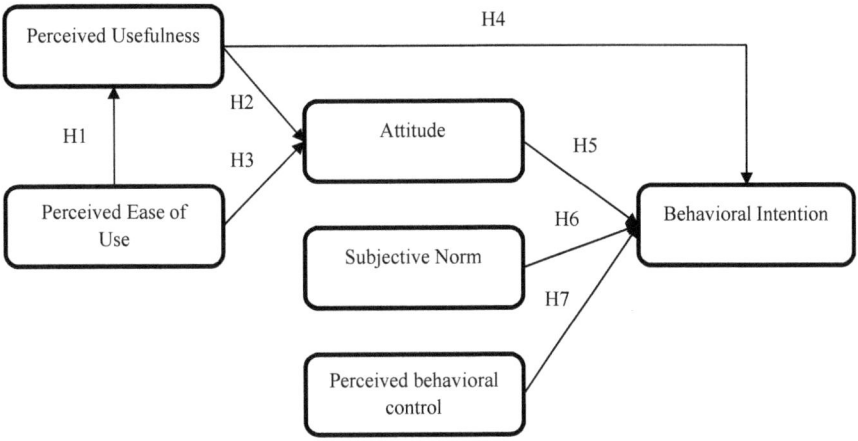

Fig. 3 Proposed model (Base on Taylor and Todd, 1995)

Table 1 Research hypotheses

Hypothesis	Meaning	Sources
Hypothesis H1	Perceived Ease of Use (PEOU) has a direct positive effect on Perceived Usefulness (PU)	[2, 12, 13, 16]
Hypothesis H2	Perceived Usefulness (PU) has a direct positive effect on Attitude towards behavior (ATU)	[2, 12, 13]
Hypothesis H3	Perceived Ease of Use (PEOU) has a direct positive effect on Attitude towards behavior (ATU)	[2, 12, 13]
Hypothesis H4	Perceived Usefulness (PU) has a direct positive effect on behavioral intention (BIU)	[2, 3, 11–13]
Hypothesis H5	Attitude towards behavior (ATU) has a direct positive effect on behavioral intention (BIU)	[1, 2, 12, 13]
Hypothesis H6	Subjective Norm (SN) has a direct positive effect on behavioral intention (BIU)	[12, 17]
Hypothesis H7	Perceived behavioral control (PBC) has a direct positive effect on behavioral intention (BIU)	[2, 3, 11–13]

4 Methods of Data Collection

Secondary data is collected through theoretical analysis and synthesis methods. The theoretical basis is synthesized from many different academic sources with high reliability, while primary data is collected through the survey method. Besides, questionnaires were used to collect data on 200 samples. After recovering and putting into data processing and cleaning, the number of valid samples remaining was 154, accounting for 77% of the distributed sample cards. These valid samples will be further analyzed in the following steps.

Following the questionnaire, respondents are primarily engineers, architects, and managers working in Da Nang City, Viet Nam. As a result, almost all respondents had less than 10 years working experience with only 79.22%. More than 76% of respondents are engineers and architects, while 23.38% of respondents are managers and directors. In particular, respondents are mainly individuals belonging to contractors and design consultants, accounting for 81.17%.

The result of the questionnaire has been collected from 23 items in relation to the various factors of the proposed model. There are five questions related to the advantage of using BIM, such as improving productivity and performance at work, completing tasks more quickly and responding quickly to any changes using BIM. This measurement is adjusted from previous studies [3, 10, 13, 18]. Furthermore, these five questions are also adapted from the Perceived Ease of Use use questions established by previous studies [2, 3, 13, 18]. Furthermore, they also related to ease of understanding and using BIM, ease of exchanging and communicating between stakeholders. In particular, three questions are concerned about users' positive feelings and opinions when using BIM [13, 17]. Three subjective norm questions were considered, including others of meaning [1]. Colleagues, organizations, and governments influence your choices [10, 17]. Next, perceived behavioral control is measured by four questions. These questions related to a personal belief about how difficult or easy it is to give a behavioral intention. For example, people need time to look for information and adequate equipment to use the new technology [10, 17]. The intention to use BIM is evaluated by the 5-point Linkert scale, which includes three questions covering behavioral intention proposed by Ajzen 1991, indicating the possibility that the respondents intend to use and advise colleagues to use BIM [3].

5   Data Analysis and Discussion

5.1   Reliability of Measurement

Based on the measurement scale of variables from the research model, the reliability of the measurement scale of the independent variables and dependent variables is shown in Table 2.

Table 2 indicates that the respondents showed a positive attitude towards BIM on all the six variables: perceived usefulness of BIM (M = 4.074, SD = 0.534)—factor is assessed to have the highest influence on behavioral intent to use BIM, perceived ease of use of BIM (M = 3.674, SD = 0.922), attitude towards using BIM (M = 3.794, SD = 0.858), subjective norm to use BIM (M = 3.565, SD = 0.747), Perceived behavioral control to use BIM (M = 3.361, SD = 0.730), and behavioural intention to use BIM (M = 3.751, SD = 0.474). According to this result, these remaining factors also have a rather great influence on behavioral intention to use BIM with the fluctuation level of 3.361–3.794.

Table 2 Reliability of measurement

Factors	No of OV*	Cronbach's Alpha	Mean	Std
PU	5	0.656	4.074	0.534
PEOU	5	0.817	3.674	0.922
ATU	3	0.913	3.794	0.858
SN	3	0.772	3.565	0.747
PBC	4	0.867	3.361	0.730
BIU	3	0.853	3.751	0.474

*: Observed variables

The reliability of the measurement scale is satisfactory because Cronbach's Alpha coefficient is more significant than 0.6, and the total variable correlation coefficient of the observed variables is more significant than 0.3. Therefore, the data set of 23 observed variables is sufficiently reliable for the next step - the correlation analysis.

5.2 Correlation Analysis

Pearson correlation analysis is one of the essential steps in quantitative analysis. It is known as the best method for measuring relationships among variables. Results are shown in Table 3.

Based on Table 3, the results of the analysis show that many variables are correlated with each other. Correlation analysis of all seven hypotheses is presented, indicating

Table 3 Correlations matrix (N = 154)

		BIU	PU	PEOU	ATU	SN	PBC
BIU	Pearson correlation	1					
	p-value						
PU	Pearson correlation	0.676**	1				
	p-value	0.000					
PEOU	Pearson correlation	0.389**	0.414**	1			
	p-value	0.000	0.000				
ATU	Pearson correlation	0.700**	0.508**	0.460**	1		
	p-value	0.000	0.000	0.000			
SN	Pearson correlation	0.234**	−0.063	−0.108	−0.136	1	
	p-value	0.004	0.439	0.183	0.093		
PBC	Pearson correlation	0.498**	0.424**	0.328**	0.544**	−0.131	1
	p-value	0.000	0.000	0.000	0.000	0.105	

**. Correlation is significant at the 0.01 level (2-tailed)

the p-value and the Pearson correlation coefficient. Table 3 indicates that H1, H2, H3, H4, H5, H6, H7 all have p-values that are less than 0.05 and the Pearson correlation coefficient is 0.414, 0.508, 0.460, 0.676, 0.700, 0.234 and 0.498 are all greater than 0. Specifically, with $r = 0.700$, ATU and BIU have the strongest positive correlations. Meanwhile, the correlation between SN and BIU is relatively weak, with $r = 0.234$. Other correlations such as PU and BIU, PU and ATU, PBC and BIU, PEOU and ATU, PEOU and PU all have a strong positive correlation.

Therefore, these hypotheses are supported by the correlation test. Specifically, based on the analysis results, respondents expressed a positive attitude towards using BIM; they found BIM useful and thought they could afford the time and resources for using BIM. Furthermore, their growing social awareness of BIM greatly influences their behavioral intention. As their Perceived Ease of Use BIM increased, their Perceived usefulness of BIM also increased. As BIM becomes more useful and easier to use, they will take a positive attitude towards its use.

6 Conclusion

In conclusion, the analysis above has clarified the research objective with investigations has helped us better understand the behavioral intentions of stakeholders towards using BIM through factors from the combination of TPB and TAM. This study explores the positive relationship between Perceived usefulness, Perceived Ease of Use, Attitude towards behavior, Subjective norms, Perceived behavioral control of the stakeholder's behavioral intention when applying BIM to construction projects in Vietnam. These factors obviously will lead to the widespread adoption and application of BIM technology in the construction industry in Vietnam.

Raising awareness about applying BIM technology among stakeholders in the Vietnamese construction industry through seminars is essential. For example, short-term courses and training courses on BIM technology at companies or universities should be more enjoyable. In particular, it is possible to add a BIM course to the students' curriculum. Although carefully planned, this research is still limited. This study has not investigated the effects of any external factors. Therefore, future investigations could be built to test individuals' behavioral intentions when applying BIM technology from a larger perspective by extending the model and adding technical skill, environment, and economics. Further studies may be conducted to investigate the user's behavior and experiences with BIM.

Acknowledgements Nguyen Thi Thao Nguyen was funded by Vingroup Joint Stock Company and supported by the Domestic Master/PhD Scholarship Programme of Vingroup Innovation Foundation (VINIF), Vingroup Big Data Institute (VINBIGDATA), code **VINIF.2020.TS.32**. We acknowledge Ho Chi Minh City University of Technology (HCMUT), VNU-HCM for supporting this study.

References

1. Ajzen I (1991) The theory of planned behavior. Organ Behav Hum Decis Process 50(2):179–211. https://doi.org/10.1016/0749-5978(91)90020-T
2. Davis FD (1989) Perceived usefulness, perceived ease of use, and user acceptance of information technology. MIS Q 13(3):319–340. https://doi.org/10.2307/249008
3. Lee S, Yu J, Jeong D (2015) BIM acceptance model in construction organizations. J Manag Eng 31(3). https://doi.org/10.1061/(asce)me.1943-5479.0000252
4. Al-Ashmori YY et al (2020) BIM benefits and its influence on the BIM implementation in Malaysia. Ain Shams Eng J 11(4):1013–1019. https://doi.org/10.1016/j.asej.2020.02.002
5. Abubakar M, Ibrahim YM, Kado D, Bala K (2014) Contractors' perception of the factors affecting building information modelling (BIM) adoption in the nigerian construction industry, pp 167–178
6. Mugo DG, Njagi K, Chemwei B, Motanya JO (2017) The technology acceptance model (TAM) and its application to the utilization of mobile learning technologies. Br J Math Comput Sci 20:1–8. https://doi.org/10.9734/BJMCS/2017/29015
7. Hong SH, Lee SK, Kim IH, Yu JH (2019) Acceptance model for mobile building information modeling (BIM). Appl Sci 9(18). https://doi.org/10.3390/app9183668
8. Mahalingam A (2014) A study on significance of system dynamics approach in understanding adoption of information technology in building construction projects. presented at the Proceedings of the 31st International Symposium on Automation and Robotics in Construction and Mining (ISARC), 08 July 2014
9. Batarseh S, Kamardeen I (2017) The impact of individual beliefs and expectations on BIM adoption in the AEC industry
10. Liao C, Chen J-L, Yen DC (2007) Theory of planning behavior (TPB) and customer satisfaction in the continued use of e-service: an integrated model. Comput Hum Behav 23(6):2804–2822. https://doi.org/10.1016/j.chb.2006.05.006
11. Bradley J (2009) The technology acceptance model and other user acceptance theories. In: Handbook of Research on Contemporary Theoretical Models in Information Systems, pp 277–294. https://doi.org/10.4018/978-1-60566-659-4.ch015
12. Taylor S, Todd P (1995) Assessing IT usage: the role of prior experience. MIS Q 19(4):561–570. https://doi.org/10.2307/249633
13. Acquah R, Eyiah A, Oteng D (2018) Acceptance of building information modelling: a survey of professionals in the construction industry in Ghana. Electron J Inf Technol Constr 23:75–91
14. El-Wajeeh M, Galal-Edeen G, Mokhtar H (2014) Technology acceptance model for mobile health systems
15. Han H, Hsu L-T, Sheu C (2010) Application of the theory of planned behavior to green hotel choice: testing the effect of environmental friendly activities. Tour Manag 31(3):325–334. https://doi.org/10.1016/j.tourman.2009.03.013
16. Chen MC, Chen SS, Yeh HM, Tsaur WG (2016) The key factors influencing internet finances services satisfaction: an empirical study in Taiwan. Am J Ind Bus Manag 06:748–762. https://doi.org/10.4236/ajibm.2016.66069
17. Wu Z, Jiang M, Li H, Luo X, Li X (2021) Investigating the critical factors of professionals' BIM adoption behavior based on the theory of planned behavior. Int J Environ Res Public Health 18(6). https://doi.org/10.3390/ijerph18063022

Implementation of BIM and High Technology in Project Life Cycle

Thu Anh Nguyen and Truong-An Pham

Abstract Building Information Modelling (BIM) has made a considerable contribution over the past few decades regarding information technology applied in the construction industry. Recently, the fourth industrial revolution brings about the tendency toward digital twins when physical objects are digitalized beside the BIM model integrated with Hi-Tech for the digital twin future. The boom in benefits that successfully apply future applications to different phases of the project's life cycle is spread through the application of superior technologies such as BIM 4D construction simulation; BIM 5D controls costs; H-BIM Historic Building Information Modelling, 3D Laser Scanning; Virtual and Augmented Reality applications from design to construction. This paper summarizes the integration of construction 4.0's critical components over the project life-cycle concentrated on three main phases of a standard project.

Keywords BIM · Hi-Tech · Laser scan · Digital twin · Construction 4.0

1 Introduction

According to the BCG report from the year 2016 on the revolutionizing construction, the critical factor influencing construction 4.0 is digitalization. Within ten years of emerging, digitalization has an average annual budget-saving 17 percentage in the design phase, 13.5 percent in the operation phase [1]. When physical objects data are interacted by humans, under sensing layer, networked existing the communication and interaction between intelligent objects [2], this is referred to as a Cyber-physical

T. A. Nguyen · T.-A. Pham (✉)
Faculty of Civil Engineering, Ho Chi Minh City University of Technology (HCMUT), 268 Ly Thuong Kiet Street, District 10, Ho Chi Minh, Vietnam
e-mail: phamtruongansos@gmail.com

T. A. Nguyen
e-mail: tran.pham@hcmut.edu.vn

Vietnam National University Ho Chi Minh City, Linh Trung Ward, Thu Duc District, Ho Chi Minh, Vietnam

environment, which is described as a system that includes both physical and digital or cyber components when technology combines the virtual and physical worlds [3]. CPS has robots and cobots (collaborative robots) for routine and hazardous tasks and UAV and mobile mapping for surveying and actuators, carrying, shifting, and placing [4]. Building Information Modeling (BIM) has been promoted in the building industry in conjunction to shape the construction sector into a practical, environmentally sustainable, and globally competitive competitor [5]. Construction 4.0 reflects the convergence of digital and physical patterns and innovations that will reshape the way environments are engineered and constructed [3].

A variety of emerging technologies can be used, from planning to architecture and engineering to installation and service. The deployment is distributed in four main agents: (1) user interfaces and implementations (big data and analytics, visualization and virtual reality, mobile interfaces, and augmented reality); (2) software platform and monitoring (BIM in the cloud, ubiquitous networking, and tracking); (3) digital/physical integration layer (additive processing, 3D laser scanning); (4) and sensors and facilities (AI, robotic, Unmanned aerial vehicles, embedded sensors) [1]. Building life, BIM, a robust digital technology, is progressively transforming how data is gathered and analyzed from paper-based documentation to object-based modeling. This leapfrog, particularly during pre-construction phases [5], contributed to the new construction context. The adoption of digital technologies in the civil sector also industrial manufacturing (3D printing, off-site manufacture, prefabrication, and assembly), and Cyber-physical systems have resulted in a modern design and construction model.

This study provides an overview implementation of technology evolution in Construction projects over three phases: pre-construction, construction, and post-construction phases [6]. The critical tasks amid three stages will be reckoned to stick with the implementation of Construction 4.0's core components and the civil field's potential as a key enabling technology of the fourth construction. An overview network of the cyber-physical system and digital Ecosystem in the building also be summarized in this research.

2 Research Method

To summarize the implementation of BIM and hi-tech during the project life-cycle, the author carried out three principle steps. Firstly, the authors conduct a comprehensive review of the fourth industry evolution in construction to provide a concise overview of the core component of construction 4.0 mentioned in this paper. The summarization of tasks in a standard project is also reviewed in the second step. Online processing is set up; stick on a keyword set as pre-construction, construction, and post-construction phase before literature. A list of hi-technologies—project task pairs were generated based on the initial literature, software analysis, emerging trends, and construction 4.0 core components. The authors intensively applied all real projects and master thesis experiences with research and deployment in this step.

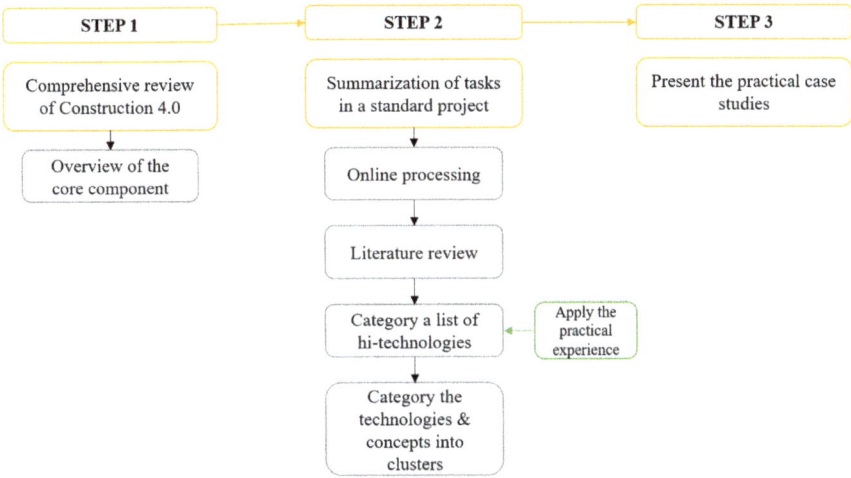

Fig. 1 The flowchart of the research method

After the categorization step, this paper categorizes the technologies and concepts in Construction 4.0 into the clusters they relate to. Finally, the practical case studies are described in the last part of this research. (Fig. 1).

3   Research and Analysis

In "Construction 4.0, an Innovation Platform for the Built Environment," Anil Sawhney defines the combination of the Cyber-Physical System and the Ecosystem called Construction 4.0 [3] (Fig. 2).

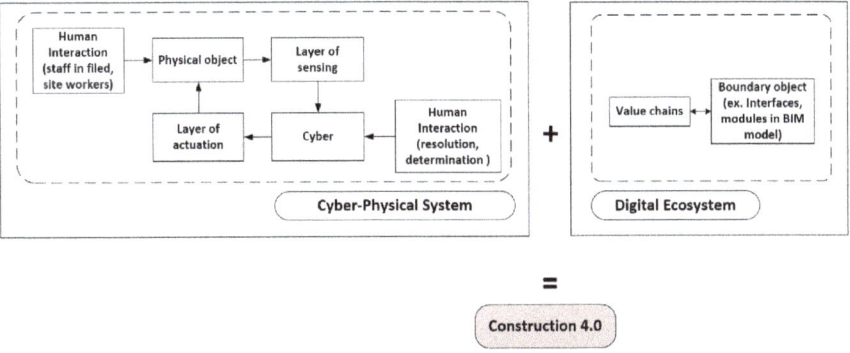

Fig. 2 Core components of the construction 4.0 [3]

With the integration of construction 4.0's core components, there are 03 visible, noticeable transformations in the building environment. First of all, is the upgrade of the construction and industrial production. When the appearance of 3D printing technology, utilities in prefabrication, off-site manufacturing, and automatic machine has gradually merged in the construction industry, it has significantly handled previous difficulties on-site compared to the conventional method. In this transformation, BIM and CDE link each other to deliver the data between physical and digital objects. The second transformation is cyber–biological systems; construction 4.0 uses robots and UAVs for processing, transport, and assembly, actuators to convert digital signals into physical behavior, and sensors and IoT to detect sensitive details about physical objects (including people) from the physical layer.

Digital applications and services: The Digital Ecosystem, built in the Construction 4.0 framework's digital layer, is at the heart of the digital transition. BIM and CDE provide the basis for developing advanced multimedia resources. The Construction 4.0 platform supports the distribution and enterprise method through virtual and laser scanning technologies, cloud computing and artificial intelligence (AI) and data processing, big data and reality capturing, blockchain, emulation, and virtual reality. Though Digital Ecosystems offer the necessary creativity, data specifications and interoperability are critical components of this overall transition. The benefits of the fourth evolution industry in the construction field are un-countable such as generating a prosperous future for creativity; Enhancing suitability; Enhancing the value and reputation of industry; Time & cost saving; Increasing site safety and health; Improving consistency time–cost ability [3, 7–10].

Oesterreich et al. (2016) [11] classified the fourth construction concept and technology into 03 clusters while Sawhey (2020) [3] divided them into 04 clusters modeling and simulating; Smart Construction On-site; Smart Construction Off-site and Digitalization & Virtualization. In this paper, the authors use the entrepreneurial intention concept and categorized them into two criteria: based on the developing stages from BIM to Digital Twin and Big Data Toward Smart Manufacturing of the fourth industry evolution and 4 clusters mentioned below (Fig. 3). For promoting smart manufacturing, both emerging digital twins and big data are essential. With the aid of a digital twin, manufacturers can manage mappings in real-time between physical objects and digital images, paving the way for cyber-physical convergence. The digital twin-driven smart manufacturing can be made rendered sensitive and predictive. In conjunction with the precise measure and forecast capacities of big data, it can benefit from more practical and accurate production control in several respects. They work well together to promote the growth of intelligent manufacturing.

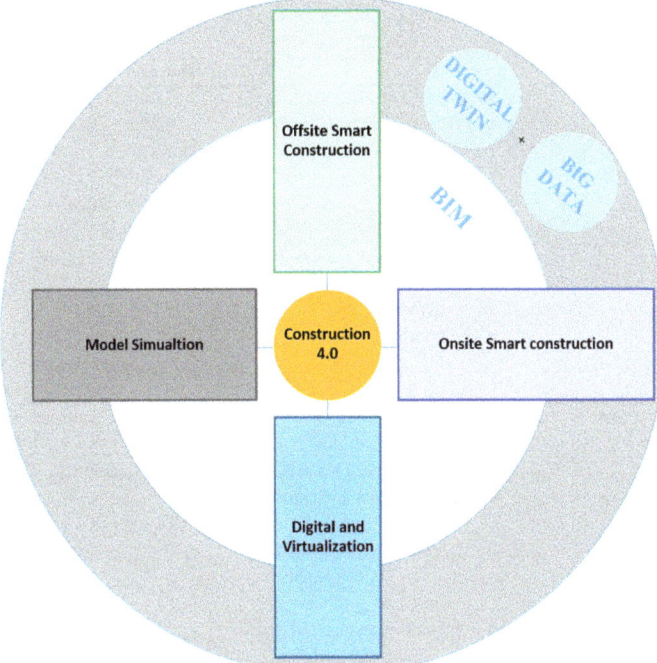

Fig. 3 Four Emerging technology clusters in construction project

It can be briefly said that 16 technological developments strongly impact the future of the construction industry, summarized in Table 2.

Table 1 Construction framework's components in 04 clusters–followed two development stages of the fourth industrial revolution

			1. Model Simulation	2. Digital And Virtualization
Construction 4.0	Digital Twin + Big Data	BIM	Virtual / Augmented Reality (VR/AR) Energy / Construction Simulations	Cloud-Based Project Management Product Life Cycle Management Laser Scanner Mobile Interfaces Adaptive Building Systems Cloud Computing (Web-Based Tools)
			Data-Driven Generative Design Laser Scanner Big Data Analytics (Deep Learning) High-Performance Computing	Facility Management Cyber-Security Actuators BlockChain Smart Home (AI Assistants) Big Data Analytics
			3. On-Site Smart Construction	4. Off-site Smart Construction
Construction 4.0	Digital Twin + Big Data	BIM	Embedded Sensors/RFID Nanotechnology / Advanced Materials Worker with Wearable Sensors Additive Manufacturing Unmanned Aerial Systems Laser Scanner	CDE Modularization / Prefabrication
			Equipment with Sensors Human-Computer Interaction (HCI) Autonomous Robots Machine Learning Artificial Intelligence	Cyber-Physical System Internet of Things Robotic And Automation

The topics illustrated in the Table 3 giving the floating application of BIM and Hi-tech in construction projects over the building life cycle.

Table 2 Core Components and the archaeologists' usage technology in construction 4.0

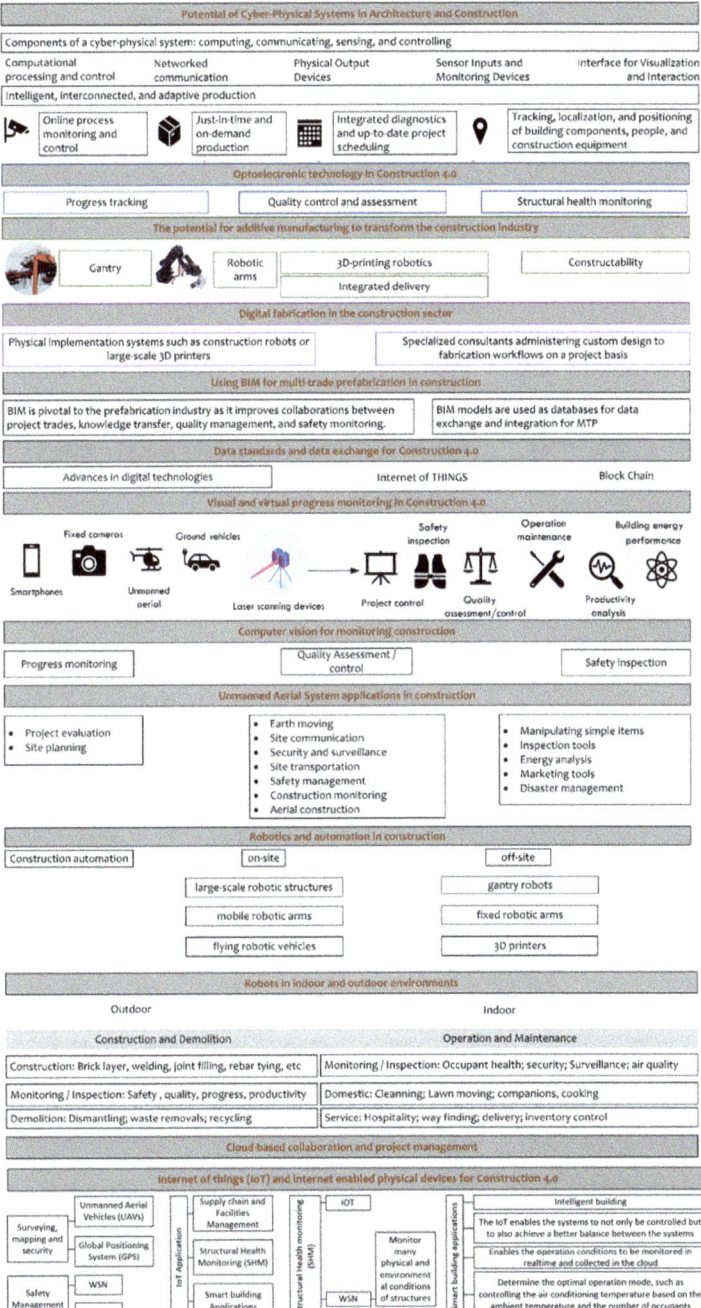

Table 3 The floating application of BIM and Hi-tech in the life cycle construction project

Phase		Results	Limitation
Pre-construction phase			
1	Applying BIM and CDE in construction schedule management [12]	- Identifying whether the use of BIM and CDE strengthens contractor progress monitoring performance - Evaluating the impact of the information exchange on the efficiency of Vietnam's construction progress management	- Due to time and equipment limitations, the research could only be accomplished at the level of general processes and applications for some basic construction tasks;
2	Application of Virtual Reality in safety training in construction at Ho Chi Minh City, Vietnam [13]	- Creating a mobile application by using VR and BIM in safety training in Vietnam - The virtual training room provides the safety training area besides occupational accident situations	In the occupational accident simulation area, 12 experienced labor accidents and safety procurement methods
3	Applied BIM in energy calculations for construction building [14]	- Applying BIM in a case study for analyzing energy usage and proposing application workflow - Comparing 02 solutions to propose one optimum solution for the case study - Walls, partitions, doors, windows, ceilings, and roofs are used to create an Energy Analytical Model (EAM) energy analysis model. Using Green Building Studio tool for energy analysis	- The research results provide construction staff access to BIM concepts, BIM implementation tools, the combination of BIM concepts, and sustainable development in construction based on friendly materials to the environment - More research options are needed to optimize energy efficiency and cost over the life of the building - Apply this process from the concept design stage until the project goes into operation - Provide specific and scientific evidences to motivate the development of environmentally friendly materials
4	Application of Quick Response code in document management	- Embedding QR code into facility assets for facility management, drawings management, safety training in Autodesk Environment	- The process workflow is quite general, and the research basing on document method, this research not applied for any case study
Construction Phase			

(continued)

Table 3 (continued)

Phase		Results	Limitation
1	Research on Application of BIM 360 Docs in managing changes arising in the design and build project [15]	- The application of BIM 360 Docs in the management process - Clarification of the status of conventional change arising in specific D&B project management	- BIM 360 Field is applied for the first time in Vietnam, projects that are implementing BIM 360 Field to support construction quality management in Vietnam are very limited. Therefore, the number of samples collected in the study is also limited - The world has applied BIM 360 Field to quality management for all project participants such as Investor, design consultant, main contractor and subcontractor (Fernandes, 2013). The research study on the application to manage and control internal quality between main contractors and subcontractors, so the study only evaluates the benefits and effectiveness of BIM 360 Field application. in internal quality management between main contractors and subcontractors. The effectiveness of coordination with investors and design consultants has not been evaluated

(continued)

Table 3 (continued)

Phase		Results	Limitation
2	Application of building information modeling (BIM) in managing the volume of high-rise building walls [16]	- Increase the building efficiency - Minimize errors in acceptance work and volume control - Impetus empowering the work's execution	The study only relates the survey to a limited number of employees who are implementing BIM for volume management, so it is not enough to cover the entire survey sample - The research results cannot completely tell the current status of BIM application. In future studies, it is necessary to apply other data collection methods to achieve full sample coverage - The research identifies applying BIM for construction volume management. Therefore, the application of the project in research is still at a simple level. The needs of QS have not been fully evaluated in the actual project. For actual implementation, it takes time to prepare the coordination process between departments as well as integrate the necessary information in the work
Post-construction Phase			
1	BIM and Laser Scanning in Quality management [17]	- Observe the status of conventional methods in QA, QC - Demonstrate an overview of tasks for the integrated application - Identify the benefits and shortcomings of the integration - Define the workflow - Application and evaluation of proposed workflow for real projects - Evaluation of research direction	- The research provides general workflows in applying Laser Scanning in QC QM and a comparison between the conventional method and BIM-Laser Scanning method in processing time, accuracy, cost. The advantages of the emerging technologies out weighted it's disadvantagetages

(continued)

Table 3 (continued)

Phase		Results	Limitation
2	Application of BIM information model in the management and operation of high-rise buildings [18]	- The principal purpose of this research is to evaluate the status of technology implementation in the FM processes and propose the frameworks to build up BIM to support the FM operations - By conducting surveys to FM experts, the research explored the current hand-over processes and the needs for implementing the information platform (IT) to support for FM tasks	- The study only conducted the survey with a limited number of building management units - These processes need to be developed with specific application levels in mind more detailed and divided for many areas of building management - The application model in real works has just started to build Web and Mobile applications at a basic level to test the applicability of the process. However, these applications have not been used in actual works, as well as need to improve more basic features to evaluate the application level more effectively

4 Conclusions

The benefits of BIM in the construction field are undeniable. The application of I4.0 at the sector level can have many effects and improvements. In digital layer: The digital tools such as Unmanned Aerial System; Cloud-based project management; AR. VR; AI directly impacts the Building Information Model; on the other side, Cybersecurity; Big Data and Analytics; Blockchain, and Laser Scan have now support cloud-based Common Data Environment performance. Consideration of the Physical Layer-Construction Site, the integration of robotics and automation; sensors; IoT; and actuator straightly affect Building Information Model.

In comparison, the implementation of additive Manufacturing, equipment with IoT, Offsite Construction all affect both assets under construction and integrated into cloud-based common data environment [10, 19]. The most significant benefit is sustainable growth when I4.0 encompasses a set of advancements, including convergence, real-time resource management, and performance productivity optimizing. On top of that, this integration brings about improved polyvalence and performance I4.0 provides customer-centered production simplicity through automation and robotics. The continuous monitoring and production controls help increase the consistency of goods and services. The results of those benefit are better product life cycle management, physical and digital integration boost the fabrication speed. Improving teamwork as data access increases and the digital and physical layers enhance intercorporate collaboration. Improved and better workplace environment. The workforce would have better and safer working conditions with improved efficiency, real-time event control, practical workstations, and increased job structuration. Creative ways of extracting demand are, for instance, given by I4.0 and modern modes of work. Finally, optimum energy consumption, error mitigation, and other improvements to the ecosystems encourage sustainable operations.

Acknowledgements This research is funded by Japan International Cooperation Agency Project for ASEAN University Network/Southeast Asia Engineering Education Development Network (JICA Project for AUN/SEED-Net) in the framework of Collaborative Education Program (CEP) under Program Contract No. HCMUT CEP 2101.

References

1. Philipp Gerbert SC, Rothballer C (2016) How Technology Is Revolutionizing Construction. Boston Consulting Group. World Economic Forum. https://www.bcg.com/publications/2016/how-technology-is-revolutionizing-construction. Accessed 23 Mar 2021
2. Forum WE (2016) Shaping the future of construction: a breakthrough in mindset and technology. In: World Economic Forum, 2016
3. Sawhney A, Riley M, Irizarry J (2020) Construction 4.0: An Innovation Platform for the Built Environment. Taylor & Francis Group, Routledge
4. Lee EA (2010) CPS foundations. Design Automation Conference; 2010: IEEE
5. Lau SEN, Zakaria R, Aminudin E, Saar CC, Yusof A, Wahid CMFHC (2018) A review of application building information modeling (BIM) during pre-construction stage: retrospective and future directions. In: IOP Conference Series: Earth and Environmental Science, 2018. IOP Publishing
6. Latiffi AA, Mohd S, Kasim N, Fathi MS (2013) Building information modeling (BIM) application in Malaysian construction industry. J Constr Eng Manag 2(4A):1–6
7. Alcácer V, Cruz-Machado V (2019) Scanning the industry 4.0: a literature review on technologies for manufacturing systems. Int J Eng Sci Technol 22(3):899–919
8. Liao Y, Deschamps F, Loures EDFR, Ramos LFP (2017) Past, present and future of Industry 4.0-a systematic literature review and research agenda proposal. Int J Prod Res 55(12):3609–3629
9. Pereira AC, Romero F (2017) A review of the meanings and the implications of the industry 4.0 concept. Procedia Manuf 13:1206–1214
10. Oesterreich TD, Teuteberg F (2016) Understanding the implications of digitisation and automation in the context of industry 4.0: a triangulation approach and elements of a research agenda for the construction industry. Comput Ind 83:121–139
11. Duy LT (2018) Applying building information modelling (BIM) and common data environment (CDE) in construction schedule management [Master Thesis]. HoChiMinh City University of Technology
12. Tan LT (2015) Application of Virtual Reality in safety training in construction at Hochiminh city Viet Nam [Master Thesis]. HoChiMinh City University of Technology
13. Duc TH (2016) Applied BIM in Energy calculations for construction building [Master Thesis]. HoChiMinh City University of Technology
14. Tin CM (2017) Research on Application of BIM 360 Docs in managing changes arising in the design and build project [Master Thesis]. HoChiMinh City University of Technology
15. Phong VD (2018) Application of building information modelling (BIM) in managing the volume of high-rise building walls
16. Thu NA, Sy DT, Truong-An P, Cuong NM (2020) Application of BIM and 3D Laser Scanning for Quantity Surveying and Quanlity Management in Construction Projects. Construction Digitalisation for Sustainable Development, 2020

17. Ninh THH (2016) Application of BIM information model in the management and operation of high-rise buildings–Truong Huu Ha Ninh
18. Kagermann H, Wahlster W, Helbig J (2013) Recommendations for implementing the strategic initiative Industrie 4.0: Final report of the Industrie 4.0 Working Group. Forschungsunion, Berlin, Germany
19. Justin Rose VL, Milon T, Cappuzzo A: Sprinting to Value in Industry 4.0. Boston Consulting Group; [accessed]. https://www.bcg.com/publications/2016/lean-manufacturing-technology-digital-sprinting-to-value-industry-40

Integration of H-BIM, Virtual Reality, and Augmented Reality in Digital Twin Era - A Case Study in Cultural Heritage

Thu Anh Nguyen, Sy Tien Do, Truong-An Pham, Dai Huu Nguyen, and Hiroshi Tamura

Abstract This research has digitized the 100-year-old Hung King Temple in Ho Chi Minh City, Vietnam, for heritage conservation and tourism promotion. H-BIM and 3D laser digitization is regarded as a steppingstone toward heritage preservation. While laser scanning allows surveyors to identify complex existing geometries with great precision in a short period, BIM reserves the information model for future operations and maintenance. Scan-to-HBIM has emerged as a game-changing technology for transferring data to a digital module and processing it to create spatial objects. Tourism is becoming more interested in Virtual Reality (VR) and Augmented Reality (AR). Travel involvement in VR technology is an unavoidable direction to catch up with the development of tourism. These integrations are assisting in ushering in the future Digital Twin city era. This study focuses on clarifying the integration of Hi-Tech, such as VR, AR into H-BIM point cloud model for the virtual public show, in which the Temple of King Hung is chosen for the case study. A new approach involving VR and AR vision algorithms is also presented to promote the city's value by promoting digital tourism, making valuable cultural heritage and tourism accessible through technology.

Keywords Temple of Hung King · Scan-to-HBIM · AR · VR · Cultural heritage

1 Introduction

According to the World Tourism Organization data, following strong growth in recent years, tourism growth has exceeded the global average since 2014 [1]. In Vietnam, the

T. A. Nguyen · S. T. Do (✉) · T.-A. Pham · D. H. Nguyen
Faculty of Civil Engineering, Ho Chi Minh City University of Technology (HCMUT), 268 Ly Thuong Kiet Street, District 10, Ho Chi Minh City, Vietnam
e-mail: sy.dotien@hcmut.edu.vn

Vietnam National University Ho Chi Minh City, Linh Trung Ward, Thu Duc District, Ho Chi Minh City, Vietnam

H. Tamura
Department of Civil Engineering, Yokohama National University, Yokohama, Kanagawa, Japan

© The Author(s), under exclusive license to Springer Nature Singapore Pte Ltd. 2023
J. N. Reddy et al. (eds.), *ICSCEA 2021*, Lecture Notes in Civil Engineering 268,
https://doi.org/10.1007/978-981-19-3303-5_24

total contribution of Travel & Tourism to GDP is 7.0% of the total economy in 2019 [2]. In 2017 Hue Monuments received over three million tourists, generating over VND 320 billion to ticket sales and increasing economic growth substantially [3]. The mentioned statistic indicates that if be adequately preserved and used. Legacy may be a valuable source of income for tourism, economy, and society in general, particularly long-standing and sustainable. Therefore, it can be seen that the preservation of local cultural heritage is necessary.

Remote sensing data and images had transformed the understanding of ancient sites and heritages. Including in intensely researched landscapes, LiDAR/airborne laser scanning has been utilized to obtain the often-faint topography remains of previously unknown locations. Several studies applied this innovation to the management of archeologic heritage, emphasizing how Laser Scanning may help nominate historical artifacts and surroundings to the World Heritage lists [4, 5]. Besides, BIM for heritage assets or H-BIM is a relatively new academic field of research, and the adoption by professionals does not seem very popular [6].

The Temple of Hung King, located in District 1, Ho Chi Minh City, Vietnam is classified as a monument in 2015. The nearly 100-year old relic has just been ranked by the People's Committee for conservation [7]. Many people are unaware of these historic relics because of the visitor opening day limit and the Temple's hidden location [8]. Therefore, the Temple of Hung King is chosen for the case study in this research. A Laser Scanner is employed to get the Temple's spatial info to digitalize temple information. This study aims to highlight the integration of Hi-Tech for virtual public displays such as VR and AR in H-BIM point cloud model. After Scan-to-HBIM processing, the BIM model is integrated with AR and VR for the promoting and developing of virtual tours application. The application purposes are promoting local public consciousness concerning heritage places. The importance of heritage monuments is increasing as a result of new experiences brought forth by technology. This can be additionally considered a pilot project investigation of the practicability of innovative technology to come up with virtual tourism.

2 Literature Review

Since the mid-1990s, 3D Laser Scanning has been a robust data-collecting technology with substantial growth [9]. The shift toward standardization of BIM forms plays an essential role in consideration of laser-scanning implementation. By integrating high-quality digital survey data sets, BIM can represent existing historical structures and conduct in-depth research and analysis of proposed interventions [8]. In the Architecture, Engineering, and Construction (ACE) industry, UASs have been employed successfully in a variety of applications, including cultural heritage conservation [10, 11], traffic surveillance and landslide monitoring [12], quality management [13], quantity management [14] and city planning [15].

Besides the upward trend of BIM and 3D Laser Scanning technology, Virtual Reality (VR) and Augmented Reality (AR) is becoming a useful management tool for complicated Cultural Heritage. The direct fruition of the virtual model through a completely immersive VR tour might pave the way for creative development in architectural upkeep by increasing visitor comprehension of the tangible cultural heritage [16]. The potentiality of VR applications to promote tourism knowledge Cultural Heritage is proven by developing a significant number of projects in this research area [16]. It has been demonstrated that the integration of AR is of great help to improve the development of tourism [17]. Various aspects of augmented tourism experiences had been academically investigated the AR conceptualization, AR characteristics, and framework for augmented virtual tourism [18].

These new paradigms affect all aspects of our lives and shape how we communicate, learn, and approach the world around us. Whether this is a mobile application, an online catalog, or a social media exchange. One of the technologies offering new ways to interact is enhanced reality (AR). This technology has enormous potential for cultural heritage promotion and preservation [19].

Promoting local cultural heritage generates prospects for tourism development [20], allowing localities around the country to benefit from tourism development and contribute to poverty alleviation. Virtual Reality developers can ultimately build a program to simulate future situations and prepare for future cases [21]. In Vietnam, the projects Complex of Hue Monuments - Tomb of Tu Duc (2020) [22], and Vietnam Complex of Monuments - An Dinh Palace, Vietnam (2021) [23]; had been public for sharing open Heritage 3D spatial source and virtual 3D point cloud tour. Map3D visitHcmc.vn had published a 360 image tour for 16 attractive tourist places at Ho Chi Minh city, and Hung King Temple is included [24].

So that, this research uses a 3D Laser Scanner to obtain heritage information to develop virtual tours, promoting local public consciousness about heritage places. It is also regarded as a pilot project investigating the feasibility of using technology to generate virtual tourism in Ho Chi Minh City.

3   Research Methodology

3.1   Research Process

Figure 1 presents the research method for applying digitalization trends in cultural heritage in integrating emerging technology, preserving cultural inheritance, and

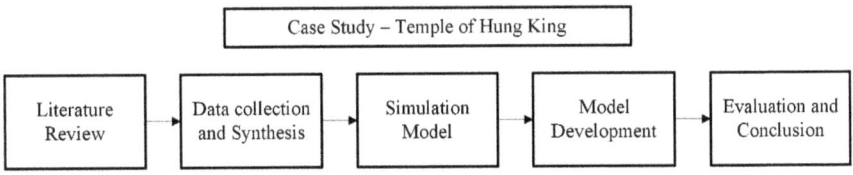

Fig. 1 Research process for applying digitalization trends in cultural heritage

promoting the city's real worth in the future. Overall, this research's five main steps are comprehensive, beginning with a documentary survey for literature review. Research materials relevant to heritage preservation and tourism promotion also the integration of BIM and Laser Scanning in heritage conservation. Then, a small online study on the internet to find the appropriate case study is compressed. At this step, the priority in choosing the pilot project for the case study is the heritage located in Hochiminh city, which is instead recognized, instantly forgotten, easily approach for data collection, small-scale, and therefore suitable for this research. After determining the appropriate cultural heritage—Temple of Hung King—the authors sifted on Google Maps and Google Earth to plan the on-site data collection. A draft on-site station position, an instrument checklist, and an enforce responsibility list are carried out. Right after the point cloud data of the Temple of Hung King is collected, it is processed to create a completed registration point cloud model. The point cloud model is then utilized for model simulation and development, merging with Hi-Tech integration for the public showcase, enhancing people's awareness about the Temple. The overall workflow from scanning to modeling and upgrading for the public showcase is provided within the software list in use.

3.2 Project Information

The Temple of Hung Kings is one of the oldest places to worship King Hung Vuong in Ho Chi Minh City; currently located at 2, Nguyen Binh Khiem Street, Ben Nghe Ward, District 1, Ho Chi Minh City, Vietnam. Hung King Temple is a holy tabernacle where Hung King, Viet Nam's first king, is worshipped by other valiant generals in history. In 2012, UNESCO designated the practice of worshipping the Hung Kings in Phu Tho province as part of the world's intangible cultural heritage. The Temple has the same architecture as the temples in Hue, with a square layout, a roof stacked with matches, adding a front porch, forming three curved roofs. Decorative motifs have dragons and phoenixes according to the royal form. This reasearch is the cooperation of BIMLab-HCMUT in the Spring of 2021, comes as part of AUN's project on conservation monument buildings; it also being essential documentation of buildings and digital assets to promote the destination through the development of AR, VR technologies.

3.3 Instrument and Software

Because of the limit on the number of open days at Hung King temple, besides the regulations are not allowed to stick targets on the heritage, the point cloud acquisition time requires restricted time-collection. It means that any of the obtained scan data must be registered on the same day. In other words, after the point cloud data is acquired, it must be checked and registered on-site to make sure no data is damaged, unusable; in this case, it should be detected for additional data collection and replacement in the following day. As the area of Hung King Temple is around 400–600 m^2 (Google Earth) while the scanning range of Trimble X7 is 0.6–80 m [25], the Trimble X7 scanner is suitable for point cloud acquisition in this case study. This scanner provided a full in-field registration, allowing users to check if the quality of the point cloud meets the registration requirement and make sure the auto-orientation is completed. This research used the software Trimble Perspective for the controller support the 3D data visualization and processing.

4 Results

4.1 Overall Workflow

The workflow consists of three main steps, as shown in Fig. 2. Heritage model after simulated through Scan-to-HBIM process in integrated with AR, VR for model development, the process of manipulating point cloud data is divided into on-site point cloud acquisition, point cloud interpretation, and demand-based application.

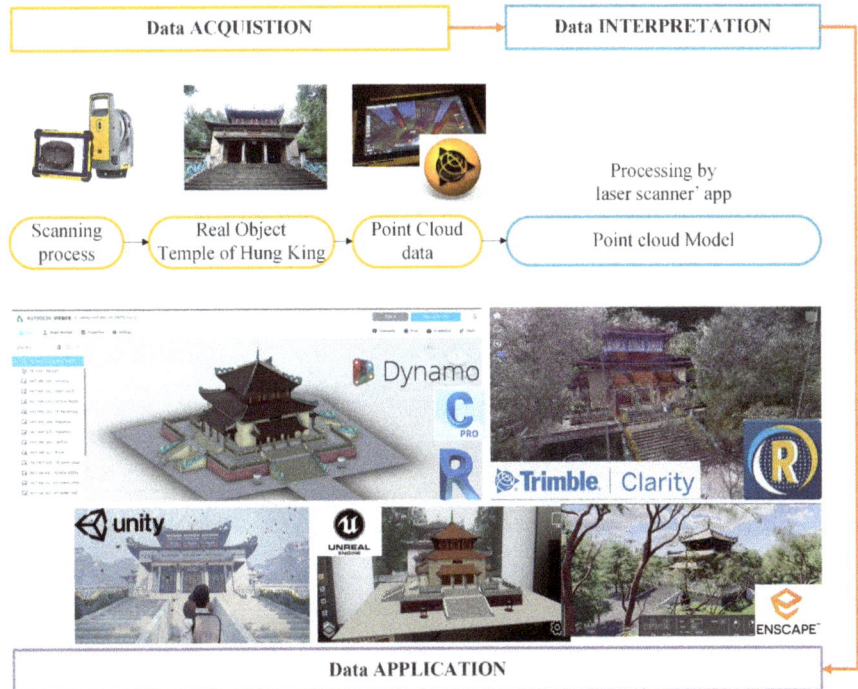

Fig. 2 General workflow from scan to HBIM and Hi-Tech integration

To clarify the investigation's objectives, scanning technology users must determine the project scope based on the job requirements. They must also understand the standards and model requirements in order to carry out the work.

In step 1—Data acquisition, a Trimble X7 3D laser scanner was used to survey the building and the construction element object. The objectives are acquired and saved in a raw point cloud format, which contains errors and noise. Then point cloud data be performed quality control by processing the raw point cloud data using the Trimble application for filtering to retain the necessary additional information to remove the noise (unnecessary point cloud). After processing and registration, the point cloud data is called the Point cloud model; this step is named Data Interpretation step. Based on the HBIM model and the integration of Hi-Tech, the application is developed and assessed its adoption. The design of the 3D objects that served as the mobile application and PC game content was a preliminary phase of the project.

4.2 Integration of Hi-Tech in Heritage Tourism

The product of integrating information into the BIM model procedure (see Fig. 3), the last stage of the Scan-to-HBIM process, is the usage of finished products. The products can be developed, processed, enhanced for apps representing mobile device models, VR glasses with Virtual Reality, and technologies with AR. A complete model may even be used to place individuals in a play set to explore, visit, and entertain. Technology users can access information fully using current data, manage it via the same data environment and restore damaged item details with 3D printing

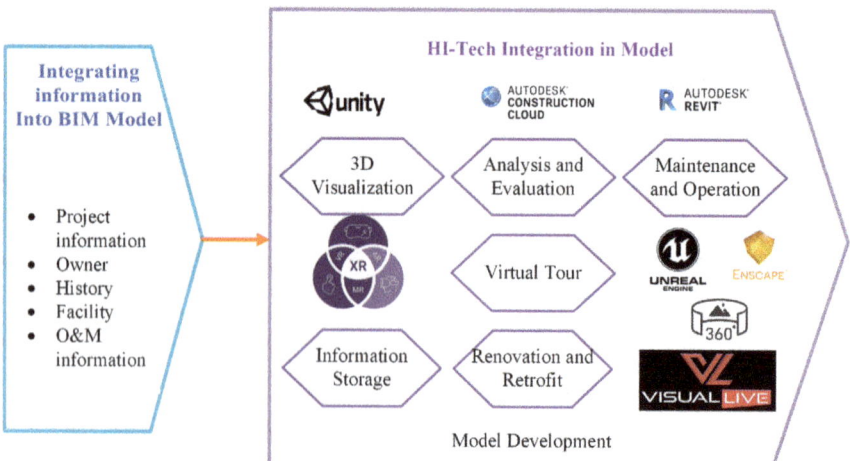

Fig. 3 Integration of Hi-Tech in HBIM model

technology. The collected 3D point cloud data is used for heritage operation and maintenance by applying BIM and Scan-to-HBIM processes in the subsequent research. Furthermore, the emerging Hi-tech such as AR, VR support enhancing the Temple's popularities, bringing about an authentic feeling experience for visitors. In this study, there were three different applications with different scenarios and be served for various purposes.

The BIM-ready Revit model offers a geometrically correct model for facilities management, project coordination, and future design. The built Revit 3D model is the final deliverable product because the project is subject to a BIM workflow from the start. When the BIM model is conducted, any following applications are based on it, so it is essential to spend time and effort in the BIM model. For an ancient building, historical, architectural, and structural information is critical; the management of the above information must be ensured from the beginning. The model is then converted into two-dimensional (2D) information in drawings, planning, sections, elevations, and so on. The model's size and details are a source of constant contention. To ensure that the 2D outputs are suitable for the operation and maintenance (O&M) phase, they should be supplemented with high-level detail. With the colored point cloud model obtained from the scanning step, even historical objects can be easily reconstructed with Revit, at least in terms of size. However, the best method to reconstruct architectural details and objects displaying ancient traces is to construct from mesh using photogrammetry. The reason is with Revit tools, it is hard to establish the textures of details and color of time. The main task of the BIM model here is to manage information for O&M. The most important aspects of any BIM project are information and access to it. It is envisaged that the BIM would be effectively used as a single source of reliable survey information. Over time, new information might be added as galleries become unoccupied, exposing architectural elements and formerly hidden functions.

With the good management of information in the BIM-ready model, the development of any following applications is simple. A virtual model of a historical monument can have various values such as scientific, education, and history depending on its information. It can also be used to assess the evolution of the environment by comparing 3D reconstructions made at various times during the object's life. It is thus simple to compare the present with the past and assess and quantify the changes brought about by time [16].

All BIM information is listed and categorized by how it is imported into Autodesk Viewer (Fig. 4), allowing visitors to access and collaborate on the same model on any device or operating system, supporting travelers to access the BIM data specified in the project quickly. Image targets represent images that Vuforia Engine (Fig. 5) can detect and track by comparing natural features extracted from camera images with a database of known target resources. With this technology, visitors can experience the virtual reality model of the Hung King Temple building only through smartphones and photo tracking applications.

In terms of visualization, with Enscape, BIM modeling and visualization could be united. Enscape has experimented with Virtual Reality (VR) (Fig. 6). Thanks to Virtual Reality, visitors now have a full 360-degree view, which allows them to

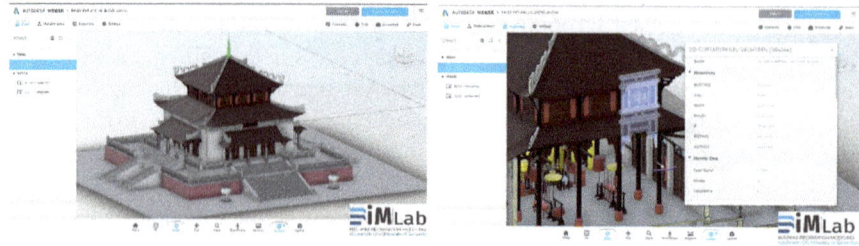

Fig. 4 HBIM model in autodesk viewer display

Fig. 5 HBIM-AR model in vuforia image target apps

Fig. 6 HBIM-VR model in escape display

get a sense of space and design and visualize a project's actual and spatial scale, particularly monuments and historical heritage projects.

In a futuristic scenario, the technical staff will simulate the start and arrangement of the restoration monument using a combination of VR and BIM models. These virtual operations, which are based on reality-based models and managed through a BIM process, will aid in cost and duration prediction and simulation of future scenarios, and so on. VR and AR models will never replace the actual live tour. Still, they will certainly increase the visitor's understanding of the Cultural Heritage and support studies and scientific analyses.

5   Conclusion

To promote local public awareness of historical sights using the intelligence model obtained to produce virtual tours. As a result of new technological experiences, the value of heritage monuments is growing. This study can also be regarded as a pilot project study of the feasibility of virtual tourism with innovative technologies. To achieve this process, BIM is considered as an integrated information transit, and the digital control of existing buildings is one of the primary O&M solutions. Since the Scan-to-BIM approach is perfectly adapted to buildings beyond the age of 20 years, because many existing structures are created with 2D planning and lack adequate BIMs, it should be highlighted that these applications have demonstrated an increase in tourists' access to intangible heritage. The application accomplishes the goals set forth. They are providing appealing material while meeting usability standards, resulting in a favorable engagement experience. AR will become more available in the future years, and people will be increasingly willing to use it. In the sphere of cultural heritage, organizations and businesses began to employ this technology to give more realistic experiences for tourists and museum visitors.

The mapping of data coordinates is vital for constructing information data models and communication with geographic information systems (GIS) and general cultural heritage maps; however, artificial targets should be used to improve the accuracy of control point selection. We recommend checking the response of the laser to the target before use because some laser scanners may exhibit unexpected behavior when dealing with reflective objects, which can cause considerable measurement inaccuracies. Due to the impact of the pandemic, we have not been able to use mesh and texture for the display products and some architectural details in the project.

This study has not yet examined user adoption perspectives. Future research needs to survey whether technology experiencers' perceptions and awareness have changed after using the product. Also, investigate if the popularity of the Hung King Temple has increased and the historical, architectural values of this Temple have increased. Application user satisfaction with the application's user interface, music, or historical information.

Acknowledgements This research is funded by the Japan International Cooperation Agency Project for ASEAN University Network/Southeast Asia Engineering Education Development Network (JICA Project for AUN/SEED-Net) in the framework of Collaborative Education Program (CEP) under Program Contract No. HCMUT CEP 2101.

References

1. (UNWTO) WTO (2015): UNWTO Annual Report 2015. Spain
2. Council WTT (March 2021): Economic Impact Reports; VIET NAM 2021 Annual Report
3. Linh D (2018) Sustainable tourism development is associated with conservation and promotion of heritage values. Nhan Dan. https://nhandan.vn/du-lich/phat-trien-du-lich-ben-vung-gan-voi-bao-ton-phat-huy-gia-tri-di-san-322835

4. Megarry WP, Davenport BA, Comer DC (2016) Emerging applications of lidar/airborne laser scanning in the management of world heritage sites. Conserv Manag Archaeol Sites. 18(4):393–410
5. López FJ, Lerones PM, Llamas J, Gómez-García-Bermejo J, Zalama E (2018) A review of heritage building information modeling (H-BIM). Multimodal Technol Interact 21
6. Historic E (2018) 3D Laser Scanning for Heritage: Advice and Guidance on the Use of Laser Scanning in Archaeology and Architecture: Historic England Swindon
7. Son T, Vuong H (2015) Temple in Zoo and Botanical Garden is classified as a relic. VN EXPRESS; [accessed]. https://vnexpress.net/den-hung-vuong-o-thao-cam-vien-duoc-xep-hang-di-tich-3241090.html
8. Dien L (2016). Hung Temple in Saigon Zoo and Botanical Garden is "besieged". Tuoitre.vn; [accessed]. https://tuoitre.vn/den-hung-trong-thao-cam-vien-sai-gon-bi-bao-vay-1170957.htm
9. Pfeifer N, Briese C (2007) Laser scanning–principles and applications. In: GeoSiberia 2007–International Exhibition and Scientific Congress; 2007: European Association of Geoscientists & Engineers
10. Koutsoudis A, Vidmar B, Ioannakis G, Arnaoutoglou F, Pavlidis G, Chamzas C (2014) Multi-image 3D reconstruction data evaluation. J Cult Herit 15(1):73–79
11. Uysal M, Toprak A, Polat N (2013) Photo realistic 3D modeling with UAV: Gedik ahmet pasha mosque in afyonkarahisar. Int Arch Photogramm Remote Sens Spat Inf Sci ISPRS Arch 5:W2
12. Barmpounakis E, Geroliminis N (2020) On the new era of urban traffic monitoring with massive drone data: the pNEUMA large-scale field experiment. Transp Res Part C: Emerg Technol 111:50–71
13. Nguyen TA, Nguyen PT, Do ST (2020) Application of BIM and 3D laser scanning for quantity management in construction projects. Adv Civ Eng 2020
14. Nguyen TA, Do ST, Pham T-A, Nguyen MC (2020) Application of BIM and 3D laser scanning for quantity surveying and quality management in construction projects. CDSD Tranforming Through Innovation; 2020; Ha Noi, Vietnam
15. Schroeder JJ (1985) Restaurant critics respond: we're doing our job. Cornell Hotel Restaur Adm Q 25(4):56–63
16. Fassi F, Mandelli A, Teruggi S, Rechichi F, Fiorillo F, Achille C (2016) VR for cultural heritage. In: De Paolis L, Mongelli A (eds) Augmented Reality, Virtual Reality, and Computer Graphics. AVR 2016. LNCS, vol 9769. Springer, Cham. https://doi.org/10.1007/978-3-319-40651-0_12
17. Kerstetter DL, Confer JJ, Graefe AR (2001) An exploration of the specialization concept within the context of heritage tourism. J Travel Res 39(3):267–274
18. Yovcheva Z, Buhalis D, Gatzidis C (2013) Engineering augmented tourism experiences. In: Cantoni L, Xiang Z (eds) Information and Communication Technologies in Tourism 2013, pp 24–35. Springer, Berlin, Heidelberg. https://doi.org/10.1007/978-3-642-36309-2_3
19. Höllerer T, Feiner S (2004) Mobile augmented reality telegeoinformatics: location-based computing and services
20. Nayyar A, Mahapatra B, Le D, Suseendran G (2018) Virtual reality (VR) & augmented reality (AR) technologies for tourism and hospitality industry. Int J Eng Sci Technol 7(2.21):156–160
21. Yung R, Khoo-Lattimore C (2019) New realities: a systematic literature review on virtual reality and augmented reality in tourism research. Curr Issues 22(17):2056–2081
22. CyArk (2020): Complex of Hué Monuments–Tomb of Tu Duc, Vietnam. Pointcloud.hcmgis.vn; [accessed]. https://openheritage3d.org/project.php?id=n06n-qa49
23. CyArk (2020): Complex of Hué Monuments–An Dinh Palace, Vietnam. Pointcloud.hcmgis.vn; [accessed]. https://openheritage3d.org/project.php?id=4z5b-vz23
24. Ho Chi Minh City 3d/360 Tourism. Map3d.visithcmc; [accessed]. https://map3d.visithcmc.vn/
25. Geospatial Trimble X7 3D LASER SCANNING SYSTEM. geospatial.trimble.com; [accessed]. https://geospatial.trimble.com/sites/geospatial.trimble.com/files/2020-10/Datasheet%20-%20Trimble%20X7%20-%20English%20%28US%29%20-%20Screen.pdf

Multi-objective Optimization of Time—Cost—Environmental Impacts in Roadway Construction Projects: A Case Study in Vietnam

Truc Thi Minh Huynh, Trung-Viet Tran, and Anh-Duc Pham

Abstract Greenhouse gas emissions (GHGs) have become and are becoming an important environmental issue in all industries worldwide. The construction industry has also begun to consider evaluating greenhouse gas emissions to assess the viability and prioritization of projects in a sustainable manner. The creation of emissions during the construction stage of roadway projects is mainly due to the use of fossil fuels for the operation of construction machinery. Therefore, the need for research on the efficiency of using equipment in the transport construction towards reducing the environmental impact is increasingly urgent. The main objective of this study is to analyze and propose solutions for the time - cost - GHGs emission optimization problem. The research process will go through four main parts as follows: The first is the literature review; Next is the development of analytical model; Then is solving the problem by a multi-objective optimization algorithm; The final is discussing on results and delivering the conclusions. Data from the National Highway No.20 through Lam Dong Province will be used as a case study.

Keywords Multi-objective optimization · Time · Cost · Environmental impact · Roadway project

1 Introduction

Green House Gases (GHGs) emissions have been an important environmental issue since the 1997 Kyoto Protocol [14]. In recent years, the international community has been making progress in reducing the activities that cause global climate change. Inhence, many strict limits have been issued to reduce the amount of released carbon

T. T. M. Huynh (✉) · A.-D. Pham
Faculty of Project Management, The University of Danang-University of Science and Technolohy, Danang, Vietnam
e-mail: htmtruc@dut.udn.vn

T.-V. Tran
Faculty of Road and Bridge Engineering, The University of Danang-University of Science and Technolohy, Danang, Vietnam

© The Author(s), under exclusive license to Springer Nature Singapore Pte Ltd. 2023   313
J. N. Reddy et al. (eds.), *ICSCEA 2021*, Lecture Notes in Civil Engineering 268,
https://doi.org/10.1007/978-981-19-3303-5_25

on the Earth [7]. The construction industry has also begun to consider GHG assessments to appraise the feasibility and priorities of projects in a sustainable manner [6]. In all fields of the construction industry, however, most of the national research and effort has focused on assessing GHG emissions from buildings.

As reported by the University of Washington in sustainability indicator system for transportation named "Greenroads", the amount of consumed energy per mile in the U.S. is equal to the consumption of 50 households in the same year in 2011. In 2008, the U.S. had about 8.5 million miles of asphalt lanes and about 600,000 bridges, spending more than 25% of the total energy consumed in the U.S. for the whole year 2005 [10]. Indeed, the construction industry is largely responsible for global GHG emissions, especially the transportation field, due to its reliance on fossil fuel consumption to convert it into the energy driving heavy equipments. Burning fossil fuels produces CO_2 in the air - the most harmful ingredient in GHGs. The generation of GHGs from construction equipment is much greater in terms of average emissions than vehicles because of the differences in fuel type, engine engineering and engine power [15].

For transportation infrastructure construction projects, the on-site operation of heavy equipments and plants releases the majority of the total project's GHGs. The category, life-span and power of the equipment as well as the type of fuel used to operate the machine greatly affect the rate of emissions. Excavators, bulldozers, graders, soil trucks, spreaders and rollers can produce more emissions than other construction equipments per hour [1]. Because of the complexity of the specifications, the selection of machinery and equipment should be considered more carefully in order to minimize the emissions from the projects. However, construction contractors have to consider completion project schedule, project costs that are still the most important project goals. Therefore, the optimization problem of time—cost—GHG emissions from the construction projects becomes one of the most important and necessary issues for contractors in the context of sustainable development today. This raises the need for optimization studies based on the factors that affect these objectives.

This research applies Non-dominated Sorting Genetic Algorithm II (NSGA-II) to solve the problem. The NSGA-II algorithm is built on the combination of two algorithms, including the genetic algorithm (GA) to develop optimal points and the distance crowding algorithm to produce reasonable distribution of optimal solutions on the Pareto curve [4]. This study focuses on analyzing the affect of the selection of productivity options and the number of machines for each construction activity to the optimized level of time—cost—emissions. Notically, the scope of this study focuses on the period of operating machinery and equipment for construction until project completion. Accordingly, this study has three specific research objectives. The first is the quantitative analysis of the influences of productivity and the machines' number to the project time, project budget, and the emissions generated from construction activities. The second is to develop an application framework of NSGA-II algorithm to solve this optimal problem. The last is to propose solutions for selecting machine options to help contractors achieve the optimizied time—cost—emission goals.

The research process includes the following important steps. The first step is an overview study of the environmental impact of traffic construction in general as well as the environmental impact from the use of machinery in particular. The second step is to develop a NSGA-II application framework. The third step is to analyze the model. The last is to analyze results and recommendations to help the construction contractor choose the optimal machinery solutions.

2 Literature Review

The problem of analyzing the factors affecting the emissions of construction machinery—equipment is also studied by many authors. [9] developed a method of linking construction with emissions and recording the effect of engine type, equipment classification, equipment type and productivity on emissions caused by use. device using model NONROAD (EPA). This study also claims that excavation of large amounts of soil causes the largest percentage of emissions compared to remaining construction works [9]. [8] compared the costs, energy consumption and emissions of excavations by different soil types. In this study, the author uses the machine productivity and cost data available in the RS Means software for comparison. Research suggests that excavating clay causes more environmental impacts than other soils such as sand, gravel and sandy clay. This study also concludes that increasing engine power also causes more fuel consumption and more CO_2 emissions. Accordingly, the highest amount of fuel and emissions is recorded when using an engine with a capacity of 350 horsepower for each type of land. In addition, the study emphasizes the importance of choosing the right engine size excavator while optimizing costs, energy consumption and emissions [8].

Various optimization models have been proposed to help in equipment selection for specific construction tasks. For example, [11] studied a mathematical model to choose a loader that maximizes equipment productivity at a time. [12] studies the application of AHP method to capture human preferences in the selection of construction equipment. A similar approach based on the AHP method is also studied by [3, 5, 13]. However, the number of researches on the choice of construction machine options considering the exhaust gas factor from the use of the machine is limited. [2] this study proposes a decision-making support model for construction enterprises to help evaluate construction machine options while still considering emissions from equipment use as well as supporting policy makers on penalties for the release of GHGs to the environment [2]. However, this study has not mentioned and solved the problem of selecting construction machines to optimize both project progress objectives—construction costs with GHGs emissions; especially for road works. Therefore, through the process of general analysis, the issue of supporting methods for selecting machine options considering progress—cost—environmental impact objectives should be promoted with research attention.

3 Model Development

3.1 Decision Variables

In construction projects with a high degree of mechanization such as construction projects, the duration of the work depends on the type of machine, the quantity and the effective operation time of construction machinery and equipment [11]. This study considers the type and the number of vehicles as the decision variables for the optimization problem. Due to the calculation's complexity and data collection, this study has not analyzed the effect of the effective operatation level to the project time. (1) Machinery group: includes different types of machinery and equipment (productivity, unit price, different fuel consumption level) with the ratio of quantity between machines meeting technical requirements; for example, the ratio of the number of dumped trucks in proportion to one excavator to ensure there is no waiting time and in accordance with the machinery specifications. (2) The number of vehicles: depends on the number of available machinery and equipment of the contractor.

3.2 Objective Functions

Objective 1: Minimizing the project time $Tp \to min$

$$Tp = max(F_i) = max(S_i + D_i) \text{ with } i = \overline{1, Q} \tag{1}$$

whereas, Q is number of construction activities; S_i is starting time of activity i, F_i is finishing time of activity i; D_i (days) is duration of activity i,

$$D_i = d_i.w_i \tag{2}$$

w_i is quantity of activity i, d_i (days/ activity quantity) is unit time of activity i. If the project time depends on machinery operation k, $d_i = d_i^k$ with d_i^k is computed by the Eq. (3):

$$d_i^k = \frac{P_i^k}{N_i^k} \tag{3}$$

whereas P_i^k, N_i^k: is productivity and number of the machine k to complete the activity i. If the project time depends on labour working, time to complete the activity i depends on workers' working performance. In other words, the relationship between d_i and the two decision variables is discrete.
Objective 2: minimizing the project cost $C_p \to min$

$$C_p = C_D + C_I = \sum_{i=1}^{Q} (C_i^M + C_i^L + C_i^E) + C_o + b.T_p \tag{4}$$

whereas, C_p is the project completing cost, C_D is direct cost, C_I is indirect cost.

$$C_D = \sum_{i=1}^{Q} (C_i^M + C_i^L + C_i^E) \tag{5}$$

With C_i^M is material cost, C_i^L labor cost, C_i^E is equipment cost to complete the activity i; C_i^M, C_i^L, C_i^E are calculated using Eqs. (6), (7), (8):

$$C_i^M = w_i.M_i \tag{6}$$

whereas, M_i is unit material cost of the activity i;

$$C_i^L = w_i.L_i \tag{7}$$

whereas L_i is unit labor cost of the activity i.

$$C_i^E = w_i.d_i^k.R_i^k = w_i.\frac{P_i^k}{N_i^k}.R_i^k \tag{8}$$

whereas R_i^k is cost per day of the machinery group k to complete the activity i.

$$C_I = C_o + b.T_p \tag{9}$$

whereas C_o is preliminary cost of the project; b is average indirect cost per day.

Objective 3: minimizing the project's emissions $GHG_p \rightarrow min$

$$GHG_p = \sum_{i=1}^{Q} w_i.P_i^k.GHG_i^k \tag{10}$$

whereas GHG_p is the total emission (kg-CO2e), GHG_i^k is the average emission per day of the machinery group k to complete the activity i and is computed by the Eq. (11):

$$GHG_i^k = \left[\sum_u F_u.EF_u\right]_i^k \tag{11}$$

With u is the number of fuel type used by the machinery group k to complete the activity i; F_u is consumption per day of fuel type u used by the machinery group k to finish the activity i. Data of F_u is refered in the technical machinery and equipment catalogue. EF_u is the emission conversion factor of the energy source type u (Table 1).

Table 1 The emission conversion factor of energy sources (EPA, 2011)

Fuel type	Unit	Kg-CO$_2$e/litre, kWh
Diesel	Lit	2.663
Electricity	kWh	0.7

The Eq. (10) becomes the Eq. (12):

$$GHG_p = \sum_{i=1}^{Q} w_i . P_i^k . \left[\sum_u F_u . EF_u \right]_i^k \tag{12}$$

3.3 Constraints

The optimization problem has specific constraints:

(1)   The constraint about productivity of machinery group P_i^k:

$$\left(P_i^k\right)^{min} \leq P_i^k \leq \left(P_i^k\right)^{max} \tag{13}$$

(2)   The constraint about the number of machinery group N_i^k:

$$1 \leq N_i^k \leq \left(N_i^k\right)^{max} \tag{14}$$

(3)   The constraint about the starting time of the first activity:
       This study supposes the starting time will on day number 0.

$$S_1 = 0 \tag{15}$$

(4)   The constraint about the relationships of activities:
       For the simplicity, this study supposes the finish-to-start relationship.

$$S_t + d_t + lag_{t,i} \leq S_i; \forall t \in \{P\} \tag{16}$$

Wheras $\{P\}$ is the proceed activities of the activity i; $lag_{t,i}$ is the late starting time of the activity i and the activity t.

(5)   The constraint

$$P_i^k \in R | P_i^k > 0 \tag{17}$$

$$N_i^k \in Z | N_i^k > 0 \tag{18}$$

3.4 Apply the NSGA-II Algorithm to Solve the Problem

NSGA-II is multi-objective optimization algorithm based on non-dominated sorting. At first offspring population is created by using the parent population. The two populations are combined together to form population of size 2 N. Then a non–dominated sorting is used to classify the entire population. After that the new population is filled by solutions of different fronts, one at a time. The filling starts with the best non-dominated front and continues with solutions form other fronts until the population size of N is reached. This study utilizes the SolveXL tool to solve the multi-objective optimization problem. The SolveXL tool is an add-in which can solve multi-objective optimization problems using the NSGA-II algorithm. The application process consits of 3 following stages: The first is to generate the number of machinery alternatives for each activity. The next is to establish the properties (variables, constraints, objective functions) of optimization problem in Ms. Excell. The project related parameters including: number of activities; quantity of each activity; the relationships of activities; late-start time; number of different machinery groups for each activity; economic - technical parameters of vehicles such as productivity, fuel consumption level, operating costs; economic parameters of the project such as material unit cost of each activity, labor unit cost of each activity; the average emission per day of the machinery groups for every activity. Then algorithm related parameters are required to select, including population size, crossover type, crossover rate mutator type, mutation rate, number of runs, number of generations.

4 Case Study

The study uses data of the new construction package of National Highway number 20 running through Lam Dong province (km 189 + 00 to km 191 + 00) as an illustration for the research model. The project data (quantity, activities, unit material cost, unit labor cost, working relationship, number of vehicle units are assumed by the author based on equipment capacity). Such data is collected from the Bidding Documents of the Contractor, the Construction Estimation Norm that is issued by the Vietnam Ministry Of Construction (VMOC), machinery and equipment catalogue. The data on fuel consumption level for machine options in the case study is referenced in Appendix Circular 11/2019/TT-BXD on Investment Cost Management Construction issued by the VMOC.

The combination of these two decision variables creates machinery alternatives. Example: The activity "Excavation of roadbeds": based on the contractor's machinery and equipment capabilities, it is possible to design 3 machinery alternativess with different capacities, but generally include two main types of equipment: excavators and bulldozers. In order to determine the cost and amount of fuel used to operate the machine, the unit price and the fuel consumption level of each alternative should also be provided (Table 2). For this example, assuming the maximum number of

Table 2 Machinery specifications for the activity "Excavation of the road foundation"

Machinery group	Machinery type	Capacity		Productivity (working shift/ 100 m³)	Unit price (VND/working shift) (8 h/working shift)	Fuel consumption level (litre/working shift)
1	Front shovel	1.25	m³	0.307	2,691,116	83
	Crawler dozer 5 Rear dump truck	110 7	CV T	0.068 1.45	1,765,118 1,215,806	46 46
2	Front shovel	1.6	m³	0.268	3,812,327	113
	Crawler dozer 5 Rear dump truck	110 10	CV T	0.068 1.11	1,765,118 1,893,175	46 57
3	Front shovel	2.3	m³	0.245	4,842,769	138
	Crawler dozer 5 Rear dump truck	110 12	CV T	0.068 1.01	1,765,118 2,185,819	46 65

machinery groups that can be mobilized is 3, then this example has 9 machine alternatives for the activity "Excavation of the ground".

There are 12 activities in this case study, consisting of: (1) Excavation of roadbeds; (2) Roadbed soil filling, tightness K95; (3) Soil filling, tightness K98; (4) Compressing, tightness K95; (5) Shovel, transport aggregates from storage yard to spread lower layer, 2 km; (6) Lower 2-layer aggregate foundation, 25 cm thick; (7) Shovel, transport aggregates from storage yard to spread upper layer, 2 km; (8) Upper 1-layer aggregate foundation, 18 cm thick; (9) Take coat 1,0 kg/m²; (10) Asphalt concrete pavement spreading with medium-grain asphalt, the thickness of the pavement has been pressed for 7 cm; (11) Take coat 0,5 kg/m²; (12) Asphalt concrete pavement spreading with fine grain asphalt, the thickness of the pavement has been pressed for 5 cm. The author assumes 12 activities in this project have a "finish to start" relationship and do not allow a late start-up between jobs (Table 3). If the contractor does not apply the construction segmentation, the shortening of project progress depends on shortening the construction time of the works on the ring. In addition, because all 12 activities in the study case have a high degree of mechanization, the execution time of the tasks is determined by the operational efficiency of construction equipment.

Table 3 Project properties of the studied case

Activity	Quantity	Material unit cost (VND)	Labor unit cost (VND)	GHG$_i$ (kg-Co2e)	Precedessors	Lag$_i$ (days)
1	4630 m^3	–	1,106,336	43.53	–	–
2	2622 m^3	–	2,285,956	231.68	1FS	0
3	3120 m^3	–	289,108	223.69	2FS	0
4	1821 m^3	–	289,108	90.54	3FS	0
5	2238 m^3	–	–	860.15	4SS	1
6	1576 m^3	47,438,934	823,847	625.81	4FS, 5FS	0
7	1611 m^3	–	–	860.15	6SS	2
8	1135 m^3	51,030,512	902,308	479.34	6FS, 7FS	0
9	6304 m^2	1,646,837	64,008	258.31	8FS	0
10	6322 m^2	14,859,924	561,000	359.51	9FS	0
11	6304 m^2	709,965	64,008	258.31	10FS	0
12	6304 m^2	11,545,386	407,000	359.51	11FS	0

The algorithm parameters applied for this case study include: the population size is 100; the crossover type is single one point; the crossover rate is 0.8; the mutation type is simple by gene; 10 runs; the number of generation in every run is 100; the random seed value is 371,618,932. Table 4 presents the global optimized solution. Basing on the optimized solution, detail of the optimized machinery alternatives for each activity is convinced. For example, the best machinery solution is the alternative 3 with the number of machienry group needed to be mobilized is 3 to complete the activity 1. It means that the project should operate 3 front shovels (capacity of 1.6 m^3), 3 crawler dozers (capacity of 110CV), and 15 rear dump trucks (capacity of 10 T) to finish the activity "Excavation of roadbeds". Overall, such machinery and equipment mobilization for the total project could help the project achieve the optimized objectives of time, cost, and emissions. In detail, the optimized time of completing the project is 52.4 days with the best project cost of 4,362,122,592 VND, releasing 78,515 kg-CO2e of GHG emissions. Figure 1 describes the distribution of solutions.

Table 4 The optimized machienery solution of the studied case

Activity	Alternative	Number	Machinery description
1	3	3	3 Front shovels (1.6 m^3); 3 Crawler dozers (110CV); 15 Rear dump trucks (10 T)
2	3	3	6 Tamper machine; 3 Double-drum steel wheel vibratory roller (25 T); 3 Crawler dozer (110cv); 15 Rear dump truck (10 T)
3	2	3	3 Double-drum steel wheel vibratory roller (25 T); 3 Crawler dozer (110cv); 15 Rear dump truck (10 T)
4	2	2	2 Pneumatic tire roller (16 T)
5	1	3	3 Front shovels (1.6 m^3); 3 Crawler dozers (110CV); 15 Rear dump trucks (10 T)
6	1	3	3 Crawler dozer (110cv); 3 Grader (110cv); 3 Double-drum steel wheel vibratory roller (25 T); 3 Pneumatic tire roller (16 T); 3 Smooth drum steel wheel roller (10 T); 3 Water truck (5 m^3)
7	1	3	3 Front shovels (1.6 m^3); 3 Crawler dozers (110CV); 15 Rear dump trucks (10 T)
8	1	3	3 Crawler dozer (110cv); 3 Grader (110cv); 3 Double-drum steel wheel vibratory roller (25 T); 3 Pneumatic tire roller (16 T); 3 Smooth drum steel wheel roller (10 T); 3 Water truck (5 m^3)
9	2	3	3 Bitumen spreaders (190cv); 3 compressors (600 m^3/h)
10	1	3	3 Asphalt mixing plants (25 T/h); 3 Front shovels (0,6 m^3); 3 Crawler dozers (110CV); 24 Rear dump trucks (7 T); 3 Asphalt spreaders (130-140cv); 3 Smooth drum steel wheel roller (10 T); 3 Pneumatic tire roller (16 T)
11	2	3	3 Bitumen spreaders (190cv); 3 compressors (600 m^3/h)
12	1	3	3 Asphalt mixing plants (25 T/h); 3 Front shovels (0,6 m^3); 3 Crawler dozers (110CV); 24 Rear dump trucks (7 T); 3 Asphalt spreaders (130-140cv); 3 Smooth drum steel wheel roller (10 T); 3 Pneumatic tire roller (16 T)

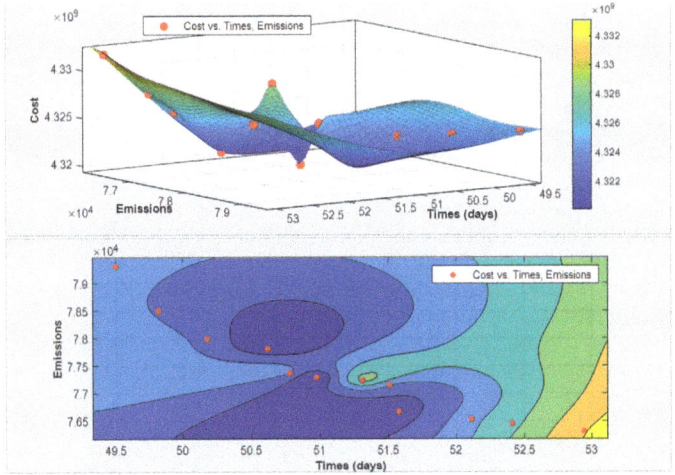

Fig. 1 The distribution of solutions

5 Conclusions and Recommendations

This study proposed the time-cost-emission optimization model in selecting the optimized machienery and equipment solutions for transportation construction projects. The mathematic problem including decision variables, constraints, and objective functions is developed. The NSGA-II algorithm is applied to solve the problem. Overall, this model consists of three main stages. The first stage is to set up the specific input parameters. Next, the objective functions of the project duration, project cost and emissions from the project are calculated. Finally, the NSGA-II algorithm is used to find the most optimal solution. The case study is conducted for the construction package of 02 km of National Highway 20 through the district of Lam Dong province. The model proposed in this study could support transportation construction enterprises to select the best machinery and equipment solutions optimizing time and cost, while still considering GHG emissions released into environment. In hence, this study has concrete contribution to the sustainable development of construction industry.

References

1. Miller-Hooks E, Melanta S, Avetisyan H (2010) Tools to support GHG emissions reduction: a regional effort. Part 1–carbon footprint estimation and decision support. Maryland. State Highway Administration
2. Avetisyan HG, Miller-Hooks E, Melanta S (2012) Decision models to support greenhouse gas emissions reduction from transportation construction projects. J Constr Eng Manag 138(5):631–641

3. Cheung S-O, Lam T-I, Wan Y-W, Lam K-C (2001) Improving objectivity in procurement selection. J Manag Eng 17(3):132–139
4. Deb K, Pratap A, Agarwal S, Meyarivan T (2002) A fast and elitist multiobjective genetic algorithm: NSGA-II. IEEE Trans Evol Comput 6(2):182–197
5. Hastak M, Halpin DW (2000) Assessment of life-cycle benefit-cost of composites in construction. J Compos Constr 4(3):103–111
6. Holton I, Glass J, Price ADF (2010) Managing for sustainability: findings from four company case studies in the UK precast concrete industry. J Clean Prod 18(2):152–160
7. IPCC (2007): IPCC Fourth assessment report: Climate change 2007. Intergovernmental Panel on Climate Change. In I. P. o. C. Change (Ed.)
8. Lewis P, Hajji A (2012) Estimating the economic, energy, and environmental impact of earthwork activities construction research congress 2012
9. Marshall SK, Rasdorf W, Lewis P, Frey HC (2012) Methodology for estimating emissions inventories for commercial building projects. J Archit Eng 18(3):251–260
10. Muench ST et al (2011) Greenroads Manual v1.5. University of Washington
11. Peurifoy RL, Schexnayder CJ (2002). Equipment and Methods Construction Planning (6th ed.). McGraw-Hill, New York
12. Shapira A, Goldenberg M (2005) Development of systematic process and practical model for equipment selection in construction projects: National Building Research Institute, Technion, Haifa, Israel
13. Skibniewski MJ, Chao LC (1992) Evaluation of advanced construction technology with AHP method. J Constr Eng Manag 118(3):577–593
14. UNFCC (1997): United Nations: Framework Convention to Climate Change. Kyoto Protocol
15. UW (1997) Rep. MS-12: Heavy-Duty Engine Emissions in the Northeast Northeast States for Coordinated Air Use: Univeversity of Washington

Novel Tendering Perspective for Encouraging Bidder Effort in PPP Projects

Vu Hong Son Pham and Kim Anh Phan

Abstract For PPP projects, the lowest price is not the top criteria for bidding winning, instead, the quality issue is a leading concern by the employer or government. Thus, contractors need to spend a lot of effort and expense in the bidding stage to balance profitability and quality, in order to reduce incurring costs in the construction phase. Encouraging high effort of the tenderer in the procurement stage is a critical problem based on the owner's perspective. Whereas, bid compensation is one of the suggestions which should be analyzed the feasibility in many scenarios. In this paper, the objective is to assess how to use effectively bid compensation in a particular case that has involving the owner, strong contractor, and normal contractor. Game-theoretic is applied to set up the modeling concept of bid compensation, as well as, explain strategic interaction between players. To solve the model is to solve the Nash Equilibrium solution. For purposes, this research should compare among models with the increasing number of bidders to provide the effectiveness of the bid compensation decision by the owner. Moreover, it will also be measured the weight of bid compensation amount that impacts to depth investing of bidder's choice.

Keywords PPP projects · Contract · Bid compensation · Bidding · Game theory

1 Introduction

The building sector is one of the most vital parts of the economy in each country, in which, the production of public projects is essential in socio-economic development. Nevertheless, a budgetary challenge faced by the government for investing in community projects [25], from that point, forcing to find effective ways in sharing risks and profitability during execution. Public-Private Partnerships (PPPs) have become the

V. H. S. Pham · K. A. Phan (✉)
Faculty of Civil Engineering, Ho Chi Minh City University of Technology (HCMUT), 268 Ly Thuong Kiet Street, District 10, Ho Chi Minh City, Vietnam
e-mail: anh.phan.imp19@hcmut.edu.vn

Vietnam National University Ho Chi Minh City, Linh Trung Ward, Thu Duc, Ho Chi Minh City, Vietnam

best choice to take a place of traditional contracts. Nowadays, PPPs are being interesting by multi-stakeholder as financial investors, contractors, consultants, suppliers, sub-contractors, and others; as well as, there are being facilitated by the authority.

In the book of Public-Private Partnerships: The Worldwide Revolution in Infrastructure Provision and Project Finance [5], the definition of public-private partnerships are as *"Public-private partnerships are arrangements whereby private parties participate in or provide support for, the provision of infrastructure, and a PPP project results in a contract for a private entity to deliver public infrastructure-based services."* In addition, that is the blending of various functions such as design, build, procurement, maintenance, and (or) operation into a unique contract [4, 5]. Occasionally, for reaching sustainability targets, Public-private partnerships are considerate a potential vehicle [5, 6, 27].

Despite applying PPP form in many fields and worldwide, PPP's successes have been limited with a number of failed cases. Comparing with conventional forms, PPP projects are high complexity and risk in the long-term, thus, preparation needs to be required deeply [3]. Following to pre-qualification assessment process of interested investors by the government, the tender invitations will be sent to qualify concessionaires for the next step. Their proposal has to prepare and submit to the authority which involves searching costs for consulting, design, and research of market in the pre-tender stage. As KPMG reports in 2010, that costs account for 1.5–2% of the total cost on average. Thus far, these initial costs are a barrier for drawing consortia due to uncertainty in awarding the contract [12]. Other important obstacles for using PPPs are extended time, high costs for transaction, a lack of competition and transparency, which causes to inefficiencies and ineffectiveness in tendering processes [24, 24]. Soomro and Zhang [9] focused on the number of initiating failure mechanisms at each project stage, it may be concluded that the project feasibility and procurement stages are the most critical for public sector personnel to trigger a number of failure mechanisms.

The environment among contractors has been considered as high competition. One of the most their competitive aspects is the contractor selection process. Moreover, competitive bidding is considered a legal requirement for projects that used national budget or loan capital, in an attempt to against the squandering of public funds and prevent abuses, for example, fraud, waste, and favoritism, that belongs state organization's charge [17]. While price competition is used to priority in a construction bidding environment, the lowest bid price is the common method to select the winning bidder in many countries [15]. However, for PPP projects, the requirements for associated project risks, project planning, schematic design, regulations during the project construction, operation, and transfer are requisite at the early stage to implement and perform the project. Thus, encouraging high effort of the tenderers in the procurement stage is a critical problem based on the public sector's perspective. Whereas, bid compensation is one of the suggestions which should be analyzed the feasibility in many scenarios. In this paper, the objective is to assess how to use effectively bid compensation in a particular case that has involving the owner, strong contractor, and normal contractor. Game-theoretic is applied to set up the modelling concept of bid compensation, as well as, explain strategic interaction between players.

2 Overview and Research Methodology

Accordance to the book, namely Game Theory-Analysis of Conflict, written by Myerson (1991), the definition of game theory can be "the study of mathematical models of conflict and cooperation between intelligent rational decision-makers". From the 1950s to the present, game theory has been experienced in whole fields of social science, logic, systems science, and computer science as well. In construction management, game theory has been common to used to explain and predict outcomes such as problems in the bidding process [11, 16, 23] subcontractor cooperation [1], risk management [13], renegotiation [20]. And competitive bidding in the electricity market [18].

So far, PPPs have been attracted researchers' attention in the academic publication of the construction field. As bid compensation and financial renegotiation issues related to project procurement and contract administration in PPP policies studied based on the game theory [21]; even, a case study of Taiwan High-Speed Rail was applied to illustrate the theoretical renegotiation. Ahmed et al. [11] identified the possibility of the winner's curse in single-stage bidding and multistage bidding and compared their degree through the non-cooperative game theory concept. Li et al. [26] used an improved alternating offer model in the construction stage to determine the risk allocation between two players involved in public-private partnership (PPP) projects by using a model of a bargaining game. Jin et al. [7] understood and provided the framework for the renegotiation process in PPP projects users paid, the aim of developing the bargaining game model is to adjust optimally values between the concession period and government guarantee for the beginning of a renegotiation stage, whereas, concession price model of renegotiation is represented by coalitional game. Bayat et al. [10] coordinated the theory of fuzzy and the theory of bilateral bargaining game to optimize value between the length of concession duration and capital frame in build–operate–transfer (BOT) agreements. Shang and Abdel [8] supported the owner to propose the modelling of payment mechanisms for PPPs related transportation by developing the Stackelberg game model, the model contributes to maximized aim for the overall performance of the project in order to get social benefits while that maximize the profit for the interest of private investment as well. Assaad et al. [14] modelling the bidding-decision making process based on 982 US public construction projects collected and then comparing the performance of three learning algorithmic game theory approaches with two bidding strategies as winning more projects and reducing the winner's curse situation.

For the bid compensation issue, it is quite limited in competitive construction bidding studies. According to Ho [19], modelling among homogeneous contractors in costly bid preparation are analyzed, the author concluded that is less effective in very complexity or very simple project. Ho and Hsu [22] examined the concept of bid compensation with heterogeneous contractors, the paper illustrated that bid compensation can be effective depends on particular scenarios of contractor selection. Following to De Clerck and Demeulemeester [3], besides searching about particular characteristics of PPP competitive tender and impacting of competition-advanced policies of government to bidding outcome, the author measured the influence of compensation factor to the unsuccessful bidders for their research efforts and concluded that the compensation seems to be inefficiency in low-risk projects, easily result in opportunistic deed in bidding.

Nevertheless, up to now, the strategic interaction between bidders and the owner in the bid compensation model has not been analyzed. Thus, game theory has been considered the best mechanism to analyze players' behavior and support to making-decision. Nash equilibrium (NE) solution needs to be used to explain the best response for all parties. Thus, this study aimed to establish a game model involving an owner and bidders to develop appropriate strategies that may help an owner analyze the effectiveness of this approach and quantify whether bid compensation has an impact on the effort that a contractor puts into the bid submission process.

3   Modelling Development and Analysis

3.1   Model Assumption: Heterogeneous Bidders with Complete Information

In the construction market, varying strengths of construction firms are common as scale, the capability of finance and technical, fame. Depending on these characteristics, the companies can be selected by the owner through the pre-qualification process. In there, the contractors have experience or knowledge in similar projects that is a remarkable advantage over other competitors. Overall, the companies own a huge amount of important resources (material and human), show stronger capability for technology and finance, or get greater renown than competitive others, that is regarded superior, defined as strong contractors in this paper. The owner tends to prefer these companies, thus, their chance of bid winning is higher also. Another group is normal contractors, thereby, they are less powerful in competition and low the ability of bid winning.

With the current great stride in information technology, the information of projects and bidders are easily acquired. Furthermore, for large-scale projects, the number of contractors in bidding is small because of high initial requirements, hence, the owner and bidders know clearly the attendee about the power of competition, imperfection, the expected profit, and project cost. Consequently, it is reasonable for assuming complete information modelling. Under completeness of information, each player identifies other competitors who are strong or regular, as well as, they also calculate their project payoffs and others.

3.2 Bidding Strategies of Bidders and Owner

For bidders, the assumed types of effort are high and average levels, denoted by H and A, respectively. Level A of effort is established as the level that meets the typical requirements from the tenderer and does not incur any suggestions for improving quality purpose. Conversely, the level H of effort is set as the level that will provide an alternative proposal and incur extra cost, which is indicated as E, to heighten the quality of a proposal, where the enhancement is recognized by an effective proposal evaluation system. Technically, the evaluation criteria and their respective weights specified in the requirement have been converted from the standard of quality.

For the owner, it has two strategies including compensation and no bid compensation, denoted by S and No S, respectively. The S strategy is compensation offered by the owner to a second-rank bidder for their extra effort in the proposal. On the contrary, the No S strategy is not applied compensation for the unsuccessful bidder.

3.3 Payoffs

The payoffs of the model are set up based on the following components: (1) P—The highest profit of the contractor shall be achieved; (2) S—Bid compensation is assumed for the second rank bidder; (3) E—For the sake of simplicity, it is assumed that E will be separated with the tender price thereby the contractor need to consider the price-quality competition to win the contract for conducting high effort; (4) α—it is assumed that if the Owner chooses strategy "no S" so bidder play extra effort with ratio α (αE).

3.4 Modelling of 3 Players: 1 Owner—1 Strong Bidder—1 Regular Bidder

It is considered that there are 8 possible Nash equilibrium for an owner, strong and regular bidders as (S, H, H), (S, H, A), (S, A, H), (S, A, A), (No S, H, H), (No S, H, A), (No S, A, H) and (No S, A, A). Due to the strong bidder takes advantage to compare with the regular bidder and always wins the bid (by 100% of probability) if they choose the same level of effort (H or A) or strong bidder play with level H. Regular bidder has a winning chance with 50% of probability at (A, H) case. Thus, the possible payoffs of players are shown in Table 1.

Impact of Bid Compensation
According to Fig. 1, the strong bidders always tend to play full effort by 100% probability whether bid compensation and its amount or not. The reason is that the strong bidder would like to hold the winning chance steady in the game of bidding. In

Table 1 Payoffs for three-players with owner, one strong bidder, one regular bidder

Bidder 2-regular		H		A	
Bidder 1-strong		H	A	H	A
Owner	S	$E-S$	$E/2-S$	$E-S$	$-S$
		$P-E$	$P/2+S/2$	$P-E$	P
		$S-E$	$P/2+S/2-E$	S	S
	No S	αE	$\alpha E/2$	αE	0
		$P-\alpha E$	$P/2$	$P-\alpha E$	P
		$-\alpha E$	$P/2-\alpha E$	0	0

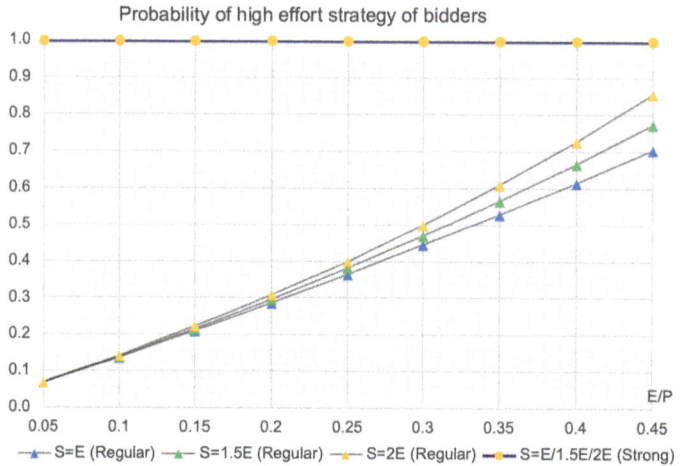

Fig. 1 The chance of playing high effort of bidders for the circumstance of one strong bidder involving

contrast, the amount of bid compensation impacts regular bidders in inputting extra effort decisions, and probability can increase by 15% from $S = E$ to $S = 2E$ (Fig. 1).

3.5 Modelling of Four-Players: 1 Owner—2 Strong Bidders—1 Regular Bidder

The modelling of four-players are 16 possible Nash Equilibrium included (S, 2H, H), (S, H, A, H), (S, A, H, H), (S, 2A, H), (S, 2H, A), (S, H, A, A), (S, A, H, A), (S, 2A, A), (No S, 2H, H), (No S, H, A, H), (No S, A, H, H), (No S, 2A, H), (No S,

Table 2 Payoffs for four-players with owner, two strong bidder, one regular bidder

Bidder 3-*regular*		H				A			
Bidder 2-*strong*		H		A		H		A	
Bidder 1-*strong*		H	A	H	A	H	A	H	A
Owner	S	E − S	E − S	E − S	E/3 − S	E − S	E − S	E − S	−S
		P/2 + S/2 − E	S/2	P − E	P/3 + S/3	P/2 + S/2 − E	S	P − E	P/2 + S/2
		P/2 + S/2 − E	P − E	S/2	P/3 + S/3	P/2 + S/2 − E	P − E	S	P/2 + S/2
		−E	S/2 − E	S/2 − E	P/3 + S/3 − E	0	0	0	0
	No S	αE	αE	αE	αE/3	αE	αE	αE	0
		P/2 − αE	0	P − αE	P/3	P/2 − αE	0	P − αE	P/2
		P/2 − αE	P − αE	0	P/3	P/2 − αE	P − αE	0	P/2
		−αE	−αE	−αE	P/3 − αE	0	0	0	0

2H, A), (No S, H, A, A), (No S, A, H, A), (No S, 2A, A). The expected payoffs are shown in Table 2.

Impact of Bid Compensation

Following Fig. 2, in the model with two strong bidders, the probability of regular bidder is always zero for S = E to S = 2E. It indicates that the regular bidder is unwilling to input more effort in the proposal, even, when the owner offers the amount double the extra cost (S = 2E) but it still can not impact to regular bidder's decision due to the lower competition level than strong bidders. For strong bidders in Fig. 2, the probability of playing high effort is over 90% for S = E to S = 2E at the case of E/P = 0.05, and it tends to drastically decrease to 12% and 50% for S = E and

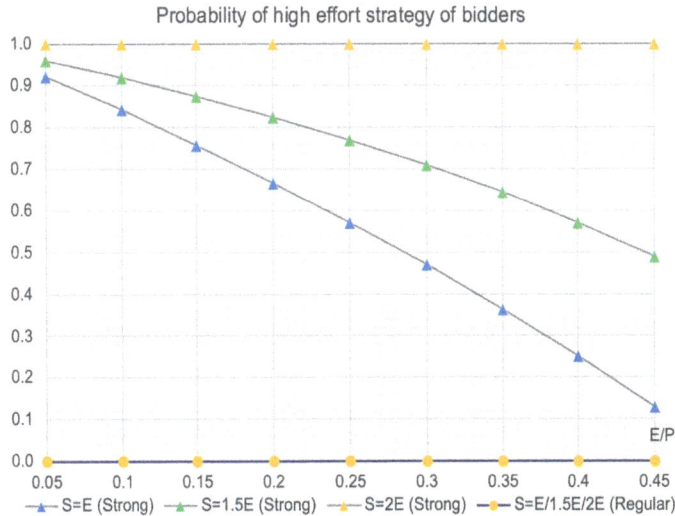

Fig. 2 The chance of playing high effort of bidders for the circumstance of two strong bidders involving

S = 1.5E, respectively. However, with S = 2E, the 100% probability is unchanged from E/P = 0.05 to 0.45. From that point, it is found that the bid compensation and its amount are significant impacts to strong bidder's strategy in the game having two strong bidders. Based on the desired probability in Fig. 2, the owner can define how much is affordable bid compensation to correspond with the level of project complexity.

According to Fig. 3, with the case E/P = 0.35, the probability of bid compensation strategy (v*) of α = 0.1 are 95%, 80% and 70% for S = E, S = 1.5E and S = 2E, respectively. Similarity, with the case E/P = 0.45, which is shown in Fig. 4, the v*

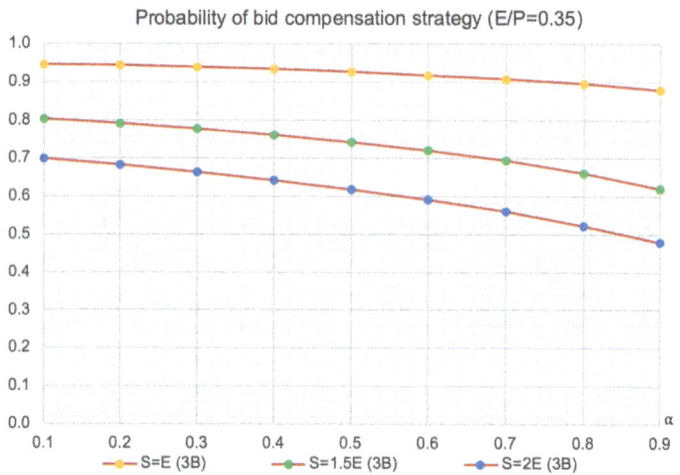

Fig. 3 The probability of bid compensation strategy of the owner, E/P = 0.35

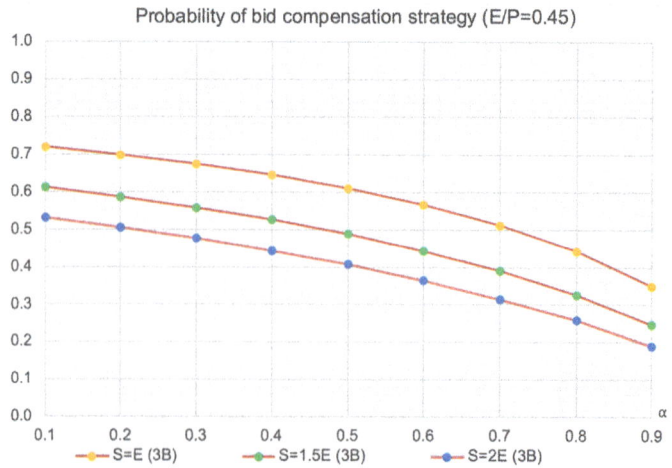

Fig. 4 The probability of bid compensation strategy of the owner, E/P = 0.45

of $\alpha = 0.1$ are 73%, 62% and 53% for S = E, S = 1.5E and S = 2E, respectively. Besides, the v* tends to decrease when a increases to 0.9 and sharply fall at E/P = 0.45 comparing to E/P = 0.35, which means the probability of bid compensation strategy has drop trend if the difference of putting more effort in bidder's proposal between S and No S strategy is trivial. The chance of using bid compensation by the owner tends to decrease in the condition of the increasing amount of S and the increasing of E/P ratio. It is concluded that the bid compensation strategy is influenced by the complex level of the project and ratio factor with the extra cost (E).

3.6 Comparison the Effectiveness of Bid Compensation of Three-Players Game and Four-Players Game

The comparative result among two models with one strong bidder and two strong bidders is shown in Figs. 5 and 6. In three-players game with one strong bidder, it has a possibility of choosing strategy H of the regular bidder, and bid compensation influence the probability because the bidder has a chance of bid winning or get a stipend for the effort. Regardless, in four-players game with two strong bidders, it is rare a chance to win the bidding, thus, that bidder does not prefer to play high effort strategy even bid compensation double extra cost, which is shown in Fig. 5. Considering the strong bidder strategy, the contractors always choose to play high effort to achieve the best opportunity for winning in three-players case. But, in four-players game with two strong bidders, this probability is lower and only increases if the compensation tends to increase higher than the extra cost. As a result, it is

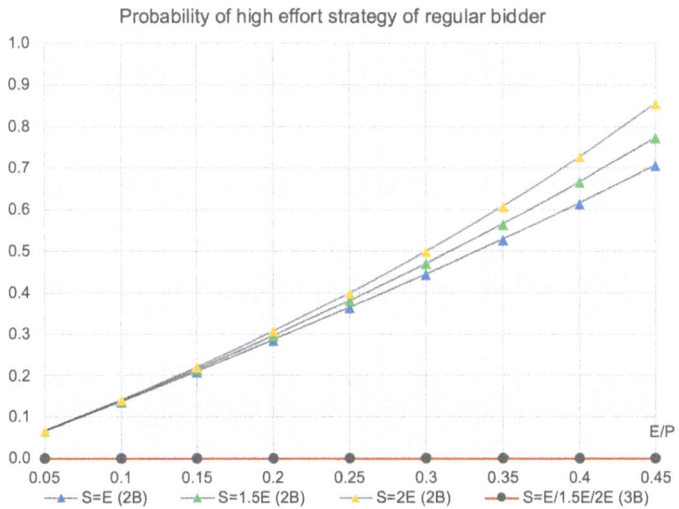

Fig. 5 Comparison regular bidder behavior in three-players game and four-players game

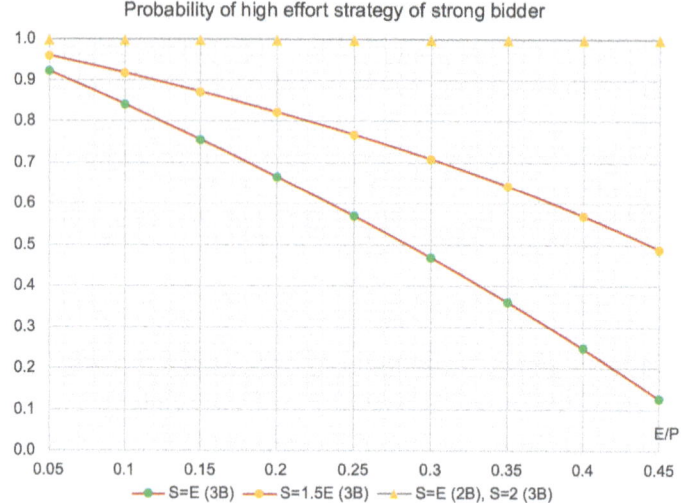

Fig. 6 Comparison strong bidder behavior in three-players game and four-players game

suggested that the owner can offer compensation in three-players game with a strong bidder if the owner wishes to encourage the proposal of the regular bidder in high-quality and the reasonable stipend is equal to extra cost. In four-players game with two strong bidders, it is considered to offer a compensation amount corresponding with the project complexity to promote the effort of strong bidders (Fig. 6).

The comparative results among two models with one strong bidder and two strong bidders are shown in Figs. 5 and 6. After analyzing the impact of bid compensation for each model and combining results, which is shown in Figs.1, 2, 3, 4, 5 and 6, primary outcomes can be listed as follows:

First outcome: In the three-players games with one strong bidder, the paying of bid compensation only impact the probability of putting more effort of the regular bidder. Thus, the bid compensation can be ineffective as the owner's desire in the bidding model with two bidders only, because of easily causing to waste.

Second outcome: In the four-players games with two strong bidders, the effectiveness of bid compensation becomes better because its magnitude can drive the higher probability of choosing strategy H of strong bidders. Besides, to achieve the desired level for the probability of playing high effort, it should consider the complex degree of the project to offer an affordable amount of compensation.

Third outcome: According to Figs. 3 and 4, the slope of the probability curve for bid compensation strategy by owner is larger when the project becomes extremely complex. Moreover, the possibility of offering compensation tends to drop due to increasing of α. This finding provides the basis to support the bidder in indicating the owner's strategy.

4   Conclusion

The bid compensation is suggested as the novel tendering for encouraging bidder's effort in PPPs is important. In this paper, the objective is to measure the effectiveness of bid compensation strategies among the owner, strong bidder, and regular bidder under multiple circumstances. For the owner, the impact of misjudging the effectiveness of the bid compensation cause them to lose the opportunity of taking other measures to promote higher efforts in project planning, and tendering preparation. Otherwise, if contractors fail to make use of the bid compensation, they may lose the opportunity of receiving a stipend to make up a part of the expense for their effort. Under a more general setting in the condition of heterogeneous bidders, the use of bid compensation can be effective as the owner's desire.

In this paper, the bid compensation model is built on the basis of game-theoretic analysis. The effectiveness of bid compensation assessment depends on many factors in different scenarios of project procurement. Note that since this model is based on a distinctive set of presumptions mentioned at the threshold of the model, thus, the result provides the trend and analyzes the relationship between parameters only. The use of the model should consider the applicability of the assumptions to the problems. Also note that the validity model is for both owner and contractor.

From the owner's point of view, the bid compensation should not pay to the unsuccessful bidder for the game having two-bidders. Nevertheless, for the game having three-bidders, the strong bidder's probability of playing high efforts is influenced by the bid compensation and its amount comparing with extra effort cost. It is recommended that the owner should use a bid compensation strategy when there are more than three bidders participated and define bid compensation magnitude to achieve the desired effect. Whereas, from the contractor's viewpoint, the complexity of the project, the magnitude of bid compensation, and the ratio of playing extra effort are three major factors that impact to owner's probability of using a bid compensation strategy.

The bid compensation model is built on necessary simplification of assumptions, such as the advantages of strong bidders and information completeness, to study economic behavior, as well as, draw valuable implications from complex circumstances. In future work, the research should be applied artificial intelligence tools to optimize parameters in payoffs, as well as, developed the general modelling for all possible scenarios. Hence, the model should be considered the suitability of utilization in uncertain cases. In addition, the empirical survey should be updated to verify the effectiveness of the incentive mechanism in motivating higher quality of the proposal.

Aknowledgement   We would like to thank Ho Chi Minh City University of Technology (HCMUT), VNU-HCM for the support of time and facilities for this study

References

1. Javanmardi A, Abbasian-Hosseini SA, Liu M, Hsiang SM (2018) Benefit of cooperation among subcontractors in performing high-reliable planning. J Manag Eng 34(2):04017062. https://doi.org/10.1061/(asce)me.1943-5479.0000578
2. Chan APC, Lam PTI, Chan DWM, Cheung E, Ke Y (2010) Potential obstacles to successful implementation of public-private partnerships in beijing and the hong kong special administrative region. J. Manag. Eng. 26(1):30–40. https://doi.org/10.1061/(asce)0742-597x(2010)26:1(30)
3. De Clerck D, Demeulemeester E (2016) Creating a more competitive PPP procurement market: game theoretical analysis. J Manag Eng 32(6). https://doi.org/10.1061/(ASCE)ME.1943-5479.0000440
4. EPEC (2011) The guide to guidance: how to prepare, procure and deliver PPP projects. European PPP expertise centre, Luxembourg. www.eib.org/epec/g2g/index.htm
5. Grimsey D, Lewis M (2007) Public private partnerships: The worldwide revolution in infrastructure provision and project finance. Edward Elgar, Cheltenham, UK
6. Hodge GA, Greve C, Boardman AE (eds) (2010) International handbook on public-private partnerships. Edward Elgar, Cheltenham, UK
7. Jin H, Liu S, Li J, Liu C (2020) A game-theoretic approach to developing a concession renegotiation framework for user-pays PPPs. Int J Constr Manag 20(6):642–652. https://doi.org/10.1080/15623599.2020.1738003
8. Shang L, Abdel Aziz AM, Stackelberg, (2020) Game theory-based optimization model for design of payment mechanism in performance-based PPPs. J Constr Eng Manag 146(4):1–13. https://doi.org/10.1061/(ASCE)CO.1943-7862.0001806
9. Soomro MA, Zhang X (2016) Evaluation of the functions of public sector partners in transportation public-private partnerships failures. J. Manag. Eng. 32(1):04015027. https://doi.org/10.1061/(asce)me.1943-5479.0000387
10. Bayat M, Khanzadi M, Nasirzadeh F, Chavoshian A (2020) Financial conflict resolution model in BOT contracts using bargaining game theory. Constr Innov 20(1):18–42. https://doi.org/10.1108/CI-12-2018-0099
11. Ahmed MO, El-Adaway IH, Coatney KT, Eid MS (2016) Construction bidding and the winner's curse: game theory approach. J Constr Eng Manag. 142(2):1–9. https://doi.org/10.1061/(ASCE)CO.1943-7862.0001058
12. Carrillo P, Robinson H, Foale P, Anumba C, Bouchlaghem D (2008) Participation, barriers, and opportunities in PFI: The united kingdom experience. J Manag Eng 24(3):138–145. https://doi.org/10.1061/(asce)0742-597x(2008)24:3(138)
13. Xiang P, Zhou J, Zhou X, Ye K (2012) Construction project risk management based on the view of asymmetric information. J Constr Eng Manag 138(11):1303–1311. https://doi.org/10.1061/(asce)co.1943-7862.0000548
14. Assaad R, Ahmed MO, El-adaway IH, Elsayegh A, Siddhardh Nadendla VS (2021) Comparing the impact of learning in bidding decision-making processes using algorithmic game theory. J Manag Eng 37(1):04020099. https://doi.org/10.1061/(asce)me.1943-5479.0000867
15. Awwad R, Ammoury M (2019) Owner's perspective on evolution of bid prices under various price-driven bid selection methods. J Comput Civ Eng 33(2):04018061. https://doi.org/10.1061/(asce)cp.1943-5487.0000803
16. Liu RY, Hu Y (2007) Analysis of bidding collusion problem based on game theory. 2007 Int Conf Wirel Commun Netw Mob Comput WiCOM, pp. 5300–5302. https://doi.org/10.1109/WICOM.2007.1299
17. Rowles D, Cahalan S. (2020) Local government procurement laws—Who the heck is a 'responsible bidder'? https://www.sgrlaw.com/local-government-procurement-laws-who-the-heck-is-a-responsible-bidder/. Accessed 22 Apr. 2020
18. Abapour S, Mohammadi-Ivatloo B, Tarafdar Hagh M (2020) Robust bidding strategy for demand response aggregators in electricity market based on game theory. J Clean Prod 243. https://doi.org/10.1016/j.jclepro.2019.118393

19. Ho SP (2005) Bid compensation decision model for projects with costly bid preparation. J Constr Eng Manag 131(2):151–159. https://doi.org/10.1061/(ASCE)0733-9364(2005)131:2(151)
20. Ping Ho S (2006) Model for financial renegotiations in public-private partnership projects and its policy implications: a game theory approach. J Constr Eng Manag 132(7):678–689
21. Ho SP (2009) Government policy on PPP financial issues: bid compensation and financial renegotiation. Policy, finance: management public-private partnerships, pp. 267–300. https://doi.org/10.1002/9781444301427.ch15.
22. Ho SP, Hsu Y (2014) Bid compensation theory and strategies for projects with heterogeneous bidders: a game theoretic analysis. J Manag Eng 30(5). https://doi.org/10.1061/(ASCE)ME.1943-5479.0000212.
23. Chen TC, Lin YC, Wang LC (2012) The analysis of BOT strategies based on game theory - Case study on Taiwan's high speed railway project. J Civ Eng Manag. 18(5):662–674. https://doi.org/10.3846/13923730.2012.723329
24. Dixon T, Pottinger G, Jordan A (2005) Lessons from the private finance initiative in the UK: benefits, problems and critical success factors. J Prop Invest Financ 23(5):412–423. https://doi.org/10.1108/14635780510616016
25. Liu T, Wang Y, Wilkinson S (2016) Identifying critical factors affecting the effectiveness and efficiency of tendering processes in Public-Private Partnerships (PPPs): a comparative analysis of Australia and China. Int J Proj Manag 34(4):701–716. https://doi.org/10.1016/j.ijproman.2016.01.004
26. Li Y, Wang X, Wang Y (2017) Using bargaining game theory for risk allocation of public-private partnership projects: insights from different alternating offer sequences of participants. J Constr Eng Manag 143(3):04016102. https://doi.org/10.1061/(asce)co.1943-7862.0001249
27. Yescombe ER (2007) Public-private partnerships: principles of policy and finance. Butterworth-Heinemann, London

Proposing the Performance Assessment Model for Coastal Projects in Vietnam Toward Sustainable Development

Ngan-Hanh Pham-Nguyen, Truc Thi Minh Huynh, and Minh-Huy Nguyen

Abstract Today, coastal construction project is one of the types of projects attracting a lot of domestic and foreign investment capital in the coastal regions of Vietnam. However, apart from the qualified projects, there are also many coastal projects having poor quality, failing to meet the requirements, encountering risks, facing problems affecting environmental pollution, causing coastal erosion, changing marine geological strata, or more seriously, there may be problems related to occupational safety, human life. It is necessary to propose a model to evaluate the performances of coastal construction projects, from which this study will come up with measures to solve and improve the efficiency of these projects. Important criteria in evaluating the performances of coastal projects are investigated using Exploratory Factor Analysis (EFA) method. The Importance Performance Analysis (IPA) method is utilized to identify those attributes of coastal projects that are most in need of improvement or that are candidates for possible cost-saving conditions without significant detriment to overall quality.

Keywords Coastal project · Performance management model · Importance performance analysis · Sustainable development

1 Introduction

Today, coastal construction project is one of the types of projects attracting a lot of domestic and foreign investment capital in the coastal regions of Vietnam. Coastal projects are mainly built for tourism development in resions, as well as serving the government's policy of marine economic development, driving the

N.-H. Pham-Nguyen · T. T. M. Huynh
Faculty of Project Management, The University of Danang–University of Science and Technolohy, Danang, Vietnam

M.-H. Nguyen (✉)
Danang Investment and Construction Project Management Board of Transportation Works, Danang People's Committee, Danang, Vietnam
e-mail: huyitb@gmail.com

© The Author(s), under exclusive license to Springer Nature Singapore Pte Ltd. 2023   339
J. N. Reddy et al. (eds.), *ICSCEA 2021*, Lecture Notes in Civil Engineering 268,
https://doi.org/10.1007/978-981-19-3303-5_27

dramatic economy transformation of coastal provinces, contributing to the renovation and socio-economic development, promoting the image and beauty with bold sea nuances. However, apart from the qualified projects, there are also many coastal projects having poor quality, failing to meet the requirements, encountering risks, facing problems affecting environmental pollution, causing coastal erosion, changing marine geological strata, or more seriously, there may be problems related to occupational safety, human life … Therefore, it is necessary to propose a model including the consumption indicators to evaluate the effectiveness of the coastal construction projects, from which this study will come up with measures to solve and improve the efficiency of these projects.

Vietnam has 3260 km of coastline stretching from North to South. In particular, the Central Coast region is considered to have the harshest climate of the coastal region of Vietnam, however, this area also has many favorable characteristics for the development of coastal constructions. Especially, Da Nang city is a dynamic and developing area with many construction projects with bold marine nuances. For that reason, in this study, the author has chosen Da Nang as the representative of the coastal area of Vietnam to conduct surveys for the research process.

In this study, important indicators in evaluating the effectiveness of coastal projects will be concerned, to identify and analyse which factors have influence to the coastal construction projects using Exploratory Factor Analysis (EFA) method. After that, the Importance Performance Analysis (IPA) method is applied to develop the evaluation model with explored critical criteria. The model is helpful to identify those attributes of coastal projects that are most in need of improvement or that are candidates for possible cost-saving conditions without significant detriment to overall quality, ensuring a stable and long-term development of coastal construction.

2 Overview of the Project Performance Evaluation Criteria

This study reviews previous research documents, combining with opinions of experts in the relevant field, to select the survey subjects properly, collect and analyse the data of respondents [1–5]. The materials used in this research are mainly from foreign research. Currently, there are not many domestic studies on evaluating the performance of coastal projects, which are very developed in central coastal areas of Vietnam, especially in Da Nang city. Therefore, the author is interested in specific factors, thereby building a questionnaire to survey experts and stakeholders in the coastal projects, to propose an appropriate performance assessment model for those projects. Table 1 presents evaluation criteria investigated in this research.

Table 1 Criteria to evaluate project performances

No	Criteria	Description
1	Schedule	Schedule of implementation and completion of coastal project
2	Cost	The cost for implementing the entire coastal project includes costs incurred due to other factors
3	Quality	Participants ensure the quality of coastal projects
4	Construction team's satisfaction	Ensure satisfaction of stakeholders directly involved in the project
5	End-user's satisfaction	Meet the requirements of the construction end-users, suitable for the coastal environment
6	Legal (Labor Safety)	The level of compliance with legal regulations related to the project as well as the degree of resolving other legal issues related to the contract, labor safety, disputes (if any) in the projects …
7	Quality workmanship	The qualifications of the parties involved in the construction of coastal projects
8	Personal growth	Skills development in supervision, management, administration and implementation of the parties involved in coastal projects
9	Specification	Meet the requirements for specifications of coastal projects
10	Functional parameter	Meet the functional parameter requirements of a coastal project
11	Safety and Health	During the project implementation, there was no accident or major injury to the participants and residents around the project. Having a reasonable construction plan in the rainy season …
12	Technology	Using modern technology and innovative technology for the implementation of coastal projects
13	Environmental performance	Impacts on the environment, such as impacts on sand, beaches, air, etc
14	Operate	Activities and operations of the project of the management unit, the parties directly involved in the operation of the project
15	Commercial profits	Profits when the project is completed

3 Research Methodology

This research applies the Importance Performance Analysis—IPA method. The IPA model was proposed by Martilla and Jame in 1977. Importance–Performance Analysis (IPA) is a simple and useful technique for identifying those attributes of a product or service that are most in need of improvement or that are candidates for possible cost-saving conditions without significant detriment to overall quality. To this end, a two dimensional IPA grid displayed the results of the evaluation about importance and performance of each relevant attribute [6]. In order to implement, complete and evaluate the effectiveness of construction and use of coastal projects, it is necessary to base on many evaluation indicators such as construction schedule, cost, environment, profitability, ... The importance of the indicators influences the assessment of whether the project is effective and meets the set requirements.

By conducting the literature review, the author has proposed a questionnaire survey consisting of 15 evaluation criteria listed in Table 1. The 5-level Likert scale is used to evaluate opinions about level of importance of indicators with the following values: 1—strongly disagree, 2—disagree, 3—Neither agree or disagree, 4—Agree, 5—strongly agree. The questionnaire has three essential parts, including the introduction to author, the research method and purpose; the information of the surveyed person; an introduction to the 5-level scale and answering questions. An example question to get opinions on the level of agreement of survey participants about the importance of evaluation indicator is as follows: Do you think the *Schedule/Cost/Quality/....* indicator is important for the performance assessment of coastal construction projects? This study also applies several methods as a whole to achieve the research goal including: Cronbach's Alpha coefficients to assess the reliability of scales [7]; mean value analysis (MVA) to rank the relative importance of each criterion [8]; exploratory factor analysis (EFA) to shorten variables and use the average value to reduce the indicators into more important ones [8].

During the analysis, the authors refer to the study of Hoang Trong và Chu Nguyen Mong Ngoc on the expected sample size, the minimum sample size is equal to 4–5 times the total number of observed variables [9]. Because the survey has 15 questions so that the minimum sample size will be 60–75 samples. The method of random sampling is applied, stratified in combination with the criteria including: working position, types of coastal constructions, number of years of experience. In order to assess indicators affecting to the efficiency of coastal construction projects, the study has conducted surveys of people working in construction industry in Da Nang city representing for coastal provinces in Vietnam due to its typical characteristics. The method of data collection is implemented by quantitative method through the questionnaire distribution with a sample size of 80 responses from construction companies in Da Nang city. The total of 80 responses were collected, of which 71 responses were accepted for analysis, and 09 responses were rejected due to invalid. General information of the repondents is shown in the Table 2. Survey results show that all respondents are stakeholders of coastal construction projects. The respondents have a high level of expertise with 69% having over 3 years of working experience. Due to this characteristic, we can see that the participants are qualified to answer the questions of the proposed research.

Table 2 Respondents' characteristics

Role of survey objects	Frequency	Percent
Investor	11	15.49%
State management agency on investment and construction	8	11.27%
Contractors	23	32.39%
Project management consultancy unit	9	12.68%
Project management board	13	18.31%
Design consultancy unit	7	9.86%
TOTAL	*71*	*100.00%*
Professional experience of survey subjects	**Frequency**	**Percent**
< 3 years	22	30.99%
3–5 years	10	14.08%
5–7 years	6	8.45%
> 7 years	33	46.48%
TOTAL	*71*	*100.00%*
Types of coastal construction	**Frequency**	**Percent**
Hotel, Resort	58	81.69%
House	4	5.63%
Public works (parks, …)	3	4.23%
Port works, works against coastal erosion	4	5.63%
Road construction works	2	2.82%
TOTAL	*71*	*100.00%*

4 Results and Discussion

4.1 Cronbach's Alpha Test Results

To show the reliability of the collected data, as well as the reliability of the scale, SPSS software is used to test Cronbach's Alpha coefficients. Specifically, in this study, using Cronbach's Alpha coefficients to check the consistency that the items on the scale are correlated with each other (correlation between questionnaires and correlation between the total score of the whole set of questionnaires and scores for each question), the results are shown in Table 3. Because the result of Cronbach's Alpha is 0.874 that is greater than 0.6, so that it can be understand that the scale system is built to ensure good quality with 15 variables.

Table 3 Summary of indicators by importance level

No	Criteria	Ranking	Mean	Std. deviation
3	Quality	1	4.59	0.729
1	Schedule	2	4.45	0.858
2	Cost	3	4.39	0.783
15	Commercial profits	4	4.38	0.781
6	Legal (Labor safety)	5	4.24	0.746
11	Safety and health	6	4.18	0.639
13	Environmental performance	7	4.15	0.804
5	End-user's satisfaction	8	4.07	0.662
14	Operate	9	4.03	0.632
4	Construction team's satisfaction	10	4.01	0.665
9	Specification	11	3.99	0.521
10	Functional parameter	12	3.97	0.506
7	Quality workmanship	13	3.83	0.697
12	Technology	14	3.82	0.661
8	Personal growth	15	3.75	0.603

4.2 Analyzing the Average Value of Evaluation Indicators in Coastal Construction Projects

In order to rank the indicators, the study used the mean value to align the indicators according to the importance level, which affects the performance of coastal construction projects at Da Nang. The results are shown in Table 3.

From the summary table, it can be seen that the amplitude of evaluating indicators from 1 to 5 helps the respondents to clearly express their assessment on the surveyed indicators. The average of the indicators evaluated ranged from 3.75 to 4.59, the dispersion of the data expressed in the standard deviation at a low level (highest is 0.858), showing that in general, survey participants converged with their views and agreed with the proposed indicators.

4.3 Exploratory Factor Analysis (EFA)

After the first running EFA, variable number 14 is needed to be removed from the next analysis because this variable has factor loading (FL) value less than 0.5. Result of the second running EFA with 14 remaining variables is shown in Table 4.

The result of KMO (0.741) satisfies condition $0.5 < KMO < 1$, so EFA is suitable for actual data. The result of Bartlett's test (568,627) with significance level Sig. <

Table 4 Final analysis of the exploratory factors

No	Criteria	Factor loading
1	Schedule	0.812
2	Cost	0.859
3	Quality	0.675
4	Construction team's satisfaction	0.678
5	End-user's satisfaction	0.847
6	Legal (Labor safety)	0.750
7	Quality workmanship	0.755
8	Personal growth	0.866
9	Specification	0.716
10	Functional parameter	0.765
11	Safety and health	0.785
12	Technology	0.599
13	Environmental performance	0.728
15	Commercial profits	0.775

0.05, proves that the data used for analysis is completely appropriate. Thus observed variables are linearly correlated with representative factors.

4.4 Two-Dimensional Importance Analysis

Two-dimensional importance analysis is a technique derived from importance performance analysis, which is typically used to measure service operations, the importance indices of customer satisfaction, and performance ratings [8]. In this study, two-dimensional importance analysis was used to measure the importance of the indicators for evaluating the performance of coastal construction projects. Two-dimensional importance analysis was performed by combining factor loadings and mean values. The factor loadings were displayed graphically on the x-axis, and mean values were displayed on the y-axis. Indicators with higher-than-average mean values and loading coefficients were considered more crucial and explanatory indicators (Quadrant I). Indicators with lower-than-average mean values and lower loading coefficients were considered less important and explanatory indicators (Quadrant III). Indicators that exhibited high factor loadings but low mean values, or that exhibited high mean values but small factor loadings, were considered explanatory or important indicators, and were included in Quadrants IV and II, respectively [8]. The study will focus on analyzing 4 evaluation indicators in Fig. 1, which are in the first ¼ angle - have higher means and factor loading values than average, highly explanatory and play very important roles in the performance evaluation model of coastal projects.

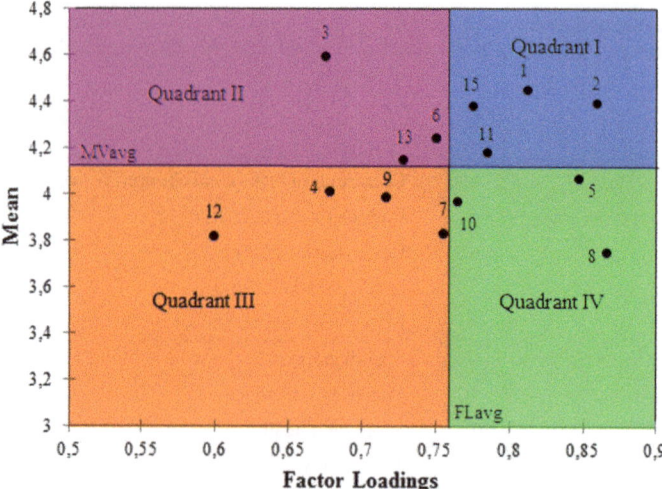

Fig. 1 Performance evaluation model for coastal construction projects

Indicator 1-Schedule (mean = 4.45, FL = 0.812), indicator 2-Cost (mean = 4.39, FL = 0.859), indicator 15-Commercial profits (mean = 4.38, FL = 0.785), and indicator 11-Safety and Health (mean = 4.11, FL = 0.775) are 4 criteria considered as the top critical indicators to assess performances of coastal projects which are located in Quadrant I. The result of indicator "Schedule" (mean = 4.45, FL = 0.812) shows that schedule is the most important aspect of project implementation and completion schedule, when coastal projects are heavily influenced by natural conditions. If schedule is not guaranteed, it may affect the quality of construction materials, which in turn will affect the effectiveness of the coastal construction project. Costs during the project implementation, including costs incurred due to factors of the coastal environment, can be considered as one of the decisive indicators to the effectiveness of a coastal project. When the cost is guaranteed, it will be a premise for other indicators such as the schedule or the quality of the project to be guaranteed. The result of Commercial profits represents that the stakeholders focus on the profits they receive after completing a coastal construction project. A coastal project that suffers a loss of capital and fails to reach the expected profitability also reduces the effectiveness of constructión process. The indicator of Safety and Health is recognized as the fourth important criterion in evaluating the coastal project performance. Coastal projecst should not only concerned with occupational safety and health of stakeholders directly involved in the project, but also interested in impacts caused by the projects to residents around them to prevent accidents or injuries during the construction of the project.

5 Conclusion

The study has surveyed and identified the importance of the indicators proposed in the model used to evaluate the performance of coastal projects in Danang City representing for coastal provinces in Vietnam. To facilitate the analysis of the results and suggest recommendations, the study has drawn two-dimension charts based on factor loadings and mean values. Based on the research results, the author proposes a number of recommendations to improve efficiency in the construction process of coastal projects. Firstly, due to natural characteristics and the participation of many stakeholders, different materials and equipments and specific characteristics of the coastal projects, so it is necessary to set up quality assurance system and construction schedule at the beginning of the project implementation. The quality assurance system must be developed and complied with by all parties involved. Secondly, there should be a careful plan to check, review, and select investors, project management consultants, supervision consultants that are qualified and capable. Human factors need to be paid attention such as qualifications, management skills, supervision, experience in operating coastal construction projects. Thirdly, it's necessary to check and review carefully the specifications and functional parameters of the project during the construction process. This could help the project can withstand the impacts of climate and coastal environment and ensure that there are no excessive errors affecting labor safety in the project construction. Fourthly, protective measures and safety equipment should be arranged and equipped appropriately for the safety of workers and people around the project (especially during the rainy and stormy season). Fifthly, it should to closely supervise the operation of coastal projects and constructions to meet the needs of customers and users. The construction of coastal projects must be highly practical, both serving people well, and contributing to the creation of marine values of coastal areas without endangering natural conditions. Sixthly, there should be more regulations on handling construction behaviors affecting the marine environment, the natural structure of soil, sand, and other buildings around the project. Finally, it should propose more regulations on the application of appropriate construction and management methods, optimizing the capabilities of human resources, equipments, terrain, topography, natural conditions and characteristics of construction industry.

In terms of managerial implications, the performance assessment model coastal projects in Vietnam plays an important role in the construction of coastal works. Thereby, it helps the project stakeholders have a basis to choose the appropriate strategy, consider which criteria are the foundation for the coastal project to achieve the highest performance, meet the requirements and goals of coastal project. The above recommendations partly help stakeholders to develop basic plans, from which they can focus on developing important indicators that affect the performance of coastal projects, while also can build a long-term strategy towards sustainable development.

Future studies need to analyze more about the correlation and mutual influence between the indicators, investigate more subjects in other coastal localities across the country, helping to complete the performance assessment model of coastal projects in Vietnam close to reality and high applicability.

References

1. Chan APC, Chan APL (2004) Key performance indicators for measuring construction success. Benchmark Int J 11(2):203–221
2. Dejaco M, Re Cecconi F, Maltese S (2016) Key performance indicators for building condition assessment. J Build Eng 9:17–28
3. Oor S-U-R, Ogunlana S (2010) Beyond the 'iron triangle': Stakeholder perception of key performance indicators (KPIs) for large-scale public sector development projects. Int J Proj Manag 28:228–236
4. Griffith AF, Gibson GE, Hamilton MR, Tortora AL, Wilson CT (1999) Project success index for capital facility construction projects. J Perform Constr Facil 13(1):39–45
5. Ashley DB, Lurie CS, Jaselskis EJ (1987) Determinants of construction project success: a process view. Proj Manag J 18(2):69–79
6. Martilla JA, James JC (1977) Importance-performance analysis. J Mark 41:77–79
7. SPSS software (2021) Statistical package, version 20. SPSS Inc., Chicago. http://www.spss.com
8. Chou J-S, Pramudawardhani D (2015) Cross-country comparisons of key drivers, critical success factors and risk allocation for public-private partnership projects. Int J Project Manage 33(5):1136–1150
9. Trong H, Ngoc CNM (2008) Research data analysis with SPSS. Hong Duc Publishing House, Ho Chi Minh City

Simulation of Vessel Collision Scenario Using Photogrammetry and 3D Laser Scanning-A Case Study at the Container Terminal

Sy Tien Do, Hiep Hoang, Nhan Nguyen Trong Truong, and Dat Ho Quang Che

Abstract Currently, transportation on the waterway in Vietnam is being increased considerably to meet the booming economy's demand. There are many river ports and seaports constructed within the two initial decades of the twenty-first century. The comprehensive studies of the ports system planning, hydrodynamic model for the channels were studied. However, specific research, such as inspection work after vessel collision is not yet fully estimated. This paper presents a study on the vessel collision of a container terminal located in Southern Vietnam. The inspection work was carried out by using innovative survey technologies, namely, 3D laser scanning, and photogrammetry. The simple-to-use process for simulation of vessel collision was then proposed. The evaluation and testing of structural components and operating equipment will then be performed meticulously.

Keywords 3D laser scanning · Vessel collision · Photogrammetry · Container terminal · Simulation

1 Introduction

Mekong delta is the biggest delta in Vietnam, which contains a dense network of about 2,360 rivers. In which, the waterway transportation in both cases of inland waterway and international waterway is mainly distributed in the four river systems, namely, Dinh An, Soai Rap, Sai Gon-Vung Tau, and Cai Mep-Thi Vai. There is approximately 41% of the country's total inland waterway freight has been transported from the Mekong Delta region and Ho Chi Minh City through the Soai Rap, and Cai Mep-Thi

S. T. Do (✉)
Faculty of Civil Engineering, Ho Chi Minh City University of Technology (HCMUT), 268 Ly Thuong Kiet Street, District 10, Ho Chi Minh City, Vietnam
e-mail: sy.dotien@hcmut.edu.vn

Vietnam National University Ho Chi Minh City, Linh Trung Ward, Thu Duc District, Ho Chi Minh City, Vietnam

H. Hoang · N. N. T. Truong · D. H. Q. Che
PortCoast Consultant Corporation, 92 Nam Ky Khoi Nghia, District 1, Ho Chi Minh City, Vietnam

© The Author(s), under exclusive license to Springer Nature Singapore Pte Ltd. 2023
J. N. Reddy et al. (eds.), *ICSCEA 2021*, Lecture Notes in Civil Engineering 268,
https://doi.org/10.1007/978-981-19-3303-5_28

Vai rivers [1]. Due to the booming economy of Viet Nam, vessel transportation and the number of ports has been rapidly increased, which leads to the risk of vessel collisions with the ports that may occur frequently. In reality, during the period from 2015 to 2019, three collisions between vessels and wharf occurred in the Cai Mep-Thi Vai river, namely, the Phu My Port, Cai Mep International Terminal (CMIT), and SP-SSA International Container Terminal (SSIT). For surveying the collision scene, the traditional method using the camera, Vernier calipers, geodetics equipment was applied, e.g. Phu My Port, Cai Mep International Terminal (CMIT) [2, 3].

Recently, 3D laser scanning technology has been widely applied to many different professions. Based on the laser technology platform, it enables us to accurately collect an enormous number of measurable data points of visible objects within a short period [4]. Along with the common application of Building Information modeling (BIM) in the construction industry, the data of 3D laser scanning has also been transferred into BIM models, herein, called Scan-to-BIM [5]. For construction BIM, Faro showed that 3D laser scanning can be used for the survey, supervising construction, as-built documentation, and facility management. One of its widespread applications is the reconstruction of Heritage buildings by combining with the Photogrammetry method [6, 7]. This combination technique also appears in the monitoring of terrain surveys or geotechnical issues [8]. In addition, it has been served regularly in road safety and accident reconstruction [9, 10]. For the Maritime industry, 3D laser scanning and Photogrammetry technique have been commonly used for modeling the ship or vessel hull [11, 12]. Although 3D laser scanning has many such applications, its applications are not yet popular in Vietnam, especially in the Coastal and Marine works.

This study aims to investigate the damaged mechanism of the wharf using 3D laser scanning and photogrammetry techniques at a site of the Cai Mep-Thi Vai river. A simple-to-use process for application of scan-to-BIM into investigation work of damaged wharf caused by Vessel collision is then proposed. This process is expected to contribute a guideline for Port and Coastal engineers to collect the precise field data for inspection work of Vessel collision.

2 Literature Review

2.1 The 3D Laser Scanning Technique

The 3D scanning technique was firstly developed in the 1960s to accurately capture visible objects [13]. However, it has just only got popular in engineering during the last two decades of the twentieth century. From 1993 to 1994, the first 3D scanners were commercially produced by REPLICA and Cyra Technologies. Currently, there are several companies such as Faro, Leica, and Trimble that have been entering this market for distribution strategy besides pioneering their products.

The 3D laser scanning technology uses the laser to capture objects' surfaces for creating the model containing countless points called "Point cloud" [13]. A point cloud is a set of data points in a 3D coordinate system (x, y, z), commonly contained color parameters. Depending on the aptitude of the scanners and measurement method, the measurement accuracy is commonly less than 10 mm. Based on the website of Leica Geosystem and the datasheet of Leica P50 and Leica RTC360, the accuracy of the Leica-P50 scanner and Leica-RTC360 scanner is 3 mm and 1.9 mm (Leica RTC360 3D Reality Capture Solution), respectively. Also mentioned by FARO and Stormbee corporations on their website and datasheet, the accuracy of the Faro S350A on the tripod provided by FARO and combined with UAV Stormbee of Stormbee corporation is 1.0 mm (Faro - Focus Laser Scanner) and less than 20 mm, respectively.

2.2 Photogrammetry Technique

Photogrammetry is the science or art of obtaining reliable measurements by means of photographs. The term photogrammetry was developed from the Greek words phos or phot, which refers to light, gramma, which means something is drawn or written, and metrein, the noun of measure. In detail, the visible objects are firstly captured in lots of images. Based on the similarity and overlapping density between the captured images, the 2D or 3D digital model of objects is then created. Herein, the digital data points within the images' overlapping zone can be assigned 3D coordinates by using the collinearity equation defining the relationship between object and image coordinates [14]. The first generation called Plane Table Photogrammetry was developed in the nineteenth century by several authors. Therein the measurements were made on a map on a table. Since the twentieth century, the second and third generation has been developed using the Analog and Analytical technique, respectively. Analog Photogrammetry uses mechanical, optical, and electrical components for re-creating the measurable 3D model in 3D space. Whereas, in Analytical Photogrammetry, the 3D modeling is mathematical (not re-create) and the measurements are made in 2D images assisted with digital processing. The latest generation is Digital photogrammetry, is being used popularly since the first decade of the twenty-first century. The principle of this method bases on Analytical Photogrammetry and digital images are used to process it. The way photogrammetric procedures are implemented from the traditional plane table photogrammetry into Analog photogrammetry, through analytical photogrammetry into Digital photogrammetry.

Nowadays, photogrammetry is considered the best technique for the processing of image data, being able to deliver at any scale of application accurate, metric, and detailed 3D information with estimates of precision and reliability of the unknown parameters from the measured image correspondences (tie points).

3 Research Methodology

3.1 Process Description

The inspection of the damaged wharf includes the estimation of working conditions of structural units and equipment operating on the wharf. Herein, the Ship to Shore Container Crane is commonly operated on the wharf of the container terminal. For the first issue, the existing condition of damaged structures needs to be investigated; and then, vessel collision should also be re-simulated to provide the scenarios for structural analysis. To solve the second issue, the displacement of the rails shall be estimated. Therefore, the two research problems, namely, simulation of vessel collision and checking displacement of rails are given in this study.

The workflow for simulation of vessel collision was developed. In which, the field data collection is firstly conducted to provide the input data for the simulation process. At the incident scene, the damaged objects such as structural components of the wharf and equipment operating on the wharf shall be recorded by cameras. Afterward, the characteristics and navigation history tracks of an induced collision vessel, the as-built report of the damaged wharf, technical specification, and warranty information of equipment shall also be collected.

After completing the field data collection, the wharf and vessel shall be re-built as the 3D digital models. Herein, the scanned data from 3D laser scanners and 3D BIM model-based on as-built drawings shall be combined to create the 3D digital model of the wharf, whereas the 3D digital model of the damaged vessel is re-built by using the Photogrammetry technique. It is worth noting that the point cloud data and 3D BIM models must use the same landmark coordinates system.

3.2 Simulation of the Vessel Collision with the Wharf

The simulation of the vessel collision process is implemented based on the 3D digital models of the vessel and wharf. According to navigation history tracks, the speed of the vessel and its collision direction is firstly estimated. The vessel is then slowly moved how its laceration fully penetrates the wharf-damaged area. It means that the surface of the laceration touched that of the damaged area modeled by point cloud data. Also, the structural components and the operating equipment are estimated by combined BIM and Photogrammetry technique.

3.3 Checking Displacement of Rails

The 3D-BIM model of rails is reconstructed using as-built drawings first. Along with that, the existing rails are also scanned by 3D laser scanning technology. Based on the scanning data and the as-built 3D BIM model, the displacement of the rails shall be analyzed by specialized Scan-to-BIM software called Cyclone 3DR of Leica.

4 A Case Study in a Container Terminal

4.1 Site Description

SP—SSA International Container Terminal (SSIT) located in the downstream area of Nga Tu ditch, in Phuoc Hoa commune, Tan Thanh district, Ba Ria—Vung Tau province, Southern Viet Nam. It was designed for capable of accommodating vessels of up to 160,000 DWT (for maximum vessels of 80,000 DWT at present). Since 2014, it has been put into operation and the total cargo throughput reached 200,000 TEU/year in 2018.

In the early afternoon of June 05, 2019, the bulk carrier with partially loaded 57,000 DWT moved on Vung Tau—Thi Vai Channel (Cai Mep area) and struck to SSIT Wharf. This vessel firstly collided with the barge mooring at the berth and then struck to the berth. The main characteristics of the vessel are given in Table 1. Therein, the vessel is a bulk carrier with it weighs approximately 50,000 DWT in case of the full-load capacity.

After the collision, the damaged components of the wharf are presented in Fig. 1a and 1b. On the top side and front view of the wharf, the broken concrete zones on the desk topping, beams, fender block, and three damaged fenders were observed. Whereas, one fallen pile, one cracked pile, three damaged fenders were also detected on the bottom of the wharf. Additionally, their locations on the wharf are schematized in Fig. 2. Herein, a fallen pile located at the axis of A46, which is denoted as Pile (A46), and the other cracked pile located at the axis of B46-1 also denoted as Pile (B46-1) is shown in Fig. 3a and 3b.

Table 1 The characteristics of vessel involved in the accident

Symbol	Unit	Vessel involved in the accident
L_{OA} (length overall of ship)	m	189.99
L_{BP} (Length between perpendiculars)	m	185.00
B_M (width)	m	32.26
D_V (depth of vessel)	m	18.00
C_C (cargo capacity)	DWT	58,414
D_L (Loaded draff)	m	13.067

(a) The scene of vessel collision   (b) The damaged wharf

Fig. 1 Photos of collision

Fig. 2 Front view of the collision area

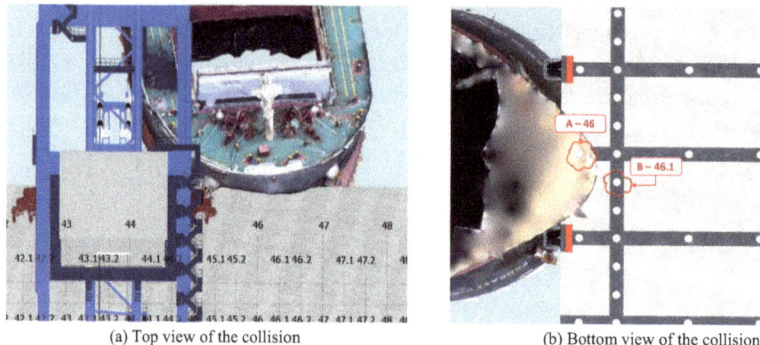

(a) Top view of the collision   (b) Bottom view of the collision

Fig. 3 Simulation of vessel using photogrammetry method

4.2 Reconstruction of the Accident Using 3D Laser Scanning and 3D Photogrammetry

The investigation of wharf-damaged structures was firstly conducted using a 3D laser scanner. This device is named Focus Laser Scanner version S350A developing by

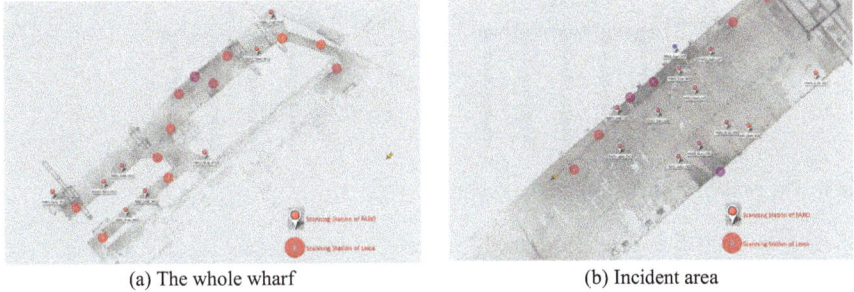

(a) The whole wharf   (b) Incident area

Fig. 4 Laser scanning layout

FARO company. By using the laser technique, the S350A Focus enables capturing complex objects fast, straightforward, and highly accurate with a measurement range up to 350 m. At the site collision, 32 scanning stations were conducted within the incident area and 57 scanning stations were then implemented for the whole wharf. After scanning, the existing objects were digitized as billion points, which are called point clouds. Herein, each point contains the coordinate, elevation, and color. The scanning layout and scanned results are given in Fig. 4a and 4b. For the rails of the STS crane, the Leica-RTC360 was selected. Herein, the Leica-RTC 360 can quickly make 3D reality capture with a measuring rate of up to 2 million points per second and complete the colored 3D in under 2 min. The range accuracy of Leica RTC 360 is 1.0 mm + 10 ppm, moreover, it also contains the automated target less field registration (based on VIS technology) and the double scanning regime which denoise automatically the effect of moving objects on scanned data. Hence, this device is suitable for scanning in detail the components required strictly for their displacement. The scanners were located on both sides of the rails to scan the whole surface of the rails. In fact, there is a total of 24 scanning stations of Leica-RTC 360 were conducted.

A great number of digital photos of the vessel involved in the collision were captured by the Canon 6D Mark II camera. The camera was located at fixed positions on the wharf, barges moving around the damaged vessel, and on the damaged vessel. Based on the digital pictures, the 3D digital model of the vessel was then created by using Reality Capture software. Along with that, the Autodesk software, which is named Revit version 2020, was used to create the wharf's 3D-BIM model. Additionally, the AutoCAD 3D version 2020 and Leica 3DR version 2020 were applied to estimate the rails' displacement.

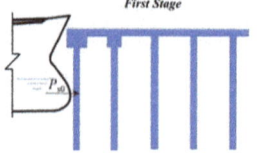 

Fig. 5 The predictive scenario of vessel cooling

5 Results and Discussion

5.1 Vessel Collision Scenarios

The 3D digital model of the damaged vessel and the 3D-BIM as-built model of the wharf combined with the point cloud model. Based on the navigation history tracks, the speed, and incense moving of the vessel were determined. Then, using these calculated parameters, the re-modeled vessels were slowly moved into the wharf. This process was manually implemented using Autodesk Revit 2020 and Autodesk Navisworks 2020 software. In which, the vessel was firstly touched on the pile A46, thereafter, continue hit to the wharf. The collision stages are then predicted in Fig. 5. Herein, the vessel impacts concentrated force on Pile and Superstructure and their magnitude shall be estimated in accordance with AASHTO (2017).

5.2 Checking of the Deflection of the Rails of STS Crane

The dimensions of the rails of the STS crane, including the elevation on top of the rails, were selected from the as-built drawings of the rail's installation reports. Subsequently, the as-built model was carried out using these dimensions. For tracking of defection, point clouds obtained by the Leica-RTC360 scanner were firstly processed by Leica 3DR software for making the level contour lines of rails. Thereafter, these results were combined with the as-built model of rails. Consequently, the gap between those is the deflection of rails. It should be noted that the as-built model and point cloud data were assigned with the same coordinate system. Therein, the coordinate system was measured from the permanent control points, which are the base points to accurately refer to any existing locations. The establishing coordinate system from permanent control points determines the exact of the point cloud model.

The result of the comparison between the scanned data and the as-built model at a typical cross-section of rails is as shown in Fig. 6. In which, the baseline (grey line) of the rail is the measured line obtained from points cloud data. And then, the gap between the baseline and the as-build model is shown through the red and navy-blue regions. Herein, the red and navy-blue color represents the inward and outward displacement of the measured line compared with the as-built one, respectively.

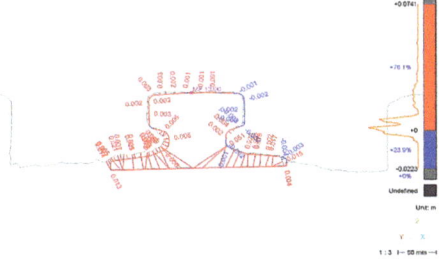

Fig. 6 Deviation of as-built model and mesh model

6   Conclusions

By using a 3D laser scanner from FARO, the actual size of components was captured with an accuracy of up to 2 mm. The deformation of STS crane rails after a collision could be checked by using point cloud data obtained from the 3D laser scanners and the as-built drawings. Accordingly, these results would be used for the estimation of STS crane working conditions. The collision scenario was reliably simulated by combining 3D laser scanning, photogrammetry, and as-built drawings. Accordingly, these results would probably be provided the rational calculation scheme for the wharf structures inspection.

In conclusion, the BIM process should be applied from the initial stage of the Port construction process. In which, the 3D laser scanning technology is recommended for digitizing as-built buildings, structures, or industrial facilities through the whole project stages.

Acknowledgements   The data supporting this study was provided by Portcoast Consultant Corporation. We are grateful to the engineers of Portcoast and SP-SSA International Terminal, for the valuable investigated data.

References

1. PortCoast (2013) Logistics development scheme in Ba Ria-Vung Tau province
2. PortCoast (2015) Detailed design for berth structure repair of Cai Mep international terminal
3. PortCoast (2017, 2018) Assessment of damage level of the segment 103 m; Survey, design dismantling, and repair design of segment 103 m of Phu My Port
4. Jaime G, Zalama E (2012) Automated registration of 3D scans using geometric features and normalized color data. Comput Civ Infrastruct Eng 00:1–14
5. Bosché F, Ahmed M, Turkan Y, Haas CT, Haas R (2015) The value of integrating Scan-to-BIM and Scan-vs-BIM techniques for construction monitoring using laser scanning and BIM: the case of cylindrical MEP components. Autom Constr 49:201–213
6. Wehr A, Lohr U (1999) Airborne laser scanning - an introduction and overview. ISPRS J Photogramm Remote Sens 54:68–82

7. Barber D, Mills J, Bryan P (2002) Laser scanning and photogrammetry: 21st century metrology. In: The international archives of the photogrammetry, remote sensing and spatial information sciences, pp. 1–8

8. Bitelli G, Dubbini M, Zanutta A (2004) Terrestrial laser scanning and digital photogrammetry techniques to monitor landslide bodies. In: The international archives of the photogrammetry, remote sensing and spatial information sciences

9. Fraser CS, Hanley HB (2004) Developments in close-range photogrammetry for 3D modelling: the iwitness example. In: Processing and visualization using high-resolution imagery, pp. 1–4

10. Parry I, Miht H, Cripps N (2003) Investigating the use of simulation model non-linear (SIMON) for the "virtual testing" of road humps

11. Menna F, Troisi S (2014) Photogrammetric 3D modelling of a boat's hull. In: Proceedings of the optical 3-d measurement techniques VIII, pp. 1–8

12. Kawasaki J, Miyoshi J (2016) Measuring method hull of small fishing boats by using laser telemeter. In: 2016 techno-ocean, IEEE, Kobe, pp. 260–263

13. Edl M, Mizerák M, Trojan J (2018) 3D Laser scanners: history and applications. Int Scient J Simul 4(4):1–5

14. Aber JS, Marzolff I, Ries JB, Aber SEW (2019) Small-format aerial photography and UAS imagery - Principles, techniques, and geoscience applications, 2nd edn. Elsevier, the Netherlands

Construction Material Session

Activation of Nanoparticle and Alkaline Environment on Fly Ash Geopolymer Mortar

Tan No Nguyen, Huu Quoc Phong Le, and Anh Tuan Le

Abstract The goal of this study is to evaluate the influences of nanoparticles on the mechanical property of fly ash geopolymer mortars. The nano-silica (NS) and nano-alumina (NA) were added to geopolymer mortar with a various amount of nanoparticle-to-fly ash in range from 5 to 25% by weight, respectively. In this research, the different NS/NA ratios in range from 0.5 to 2, by weight, were investigated. For the geopolymerization process, the alkaline solutions (AS) including sodium hydroxide (SH) and sodium silicate solution (SS) were mixed with the SH/SS ratio of 1, by weight. The different ratios of alkaline liquid to fly ash in range from 0.4 to 0.6, by weight, were investigated. The results were focused on the geopolymerization between nanoparticles and alumino-silicate resource by strength in ambient temperature condition. The compressive strength of geopolymer mortar can be up to 22.62 MPa. The suitable mix proportion can be obtained by amount of 5% nanoparticle by fly ash weight, a ratio of NS/NA as 0.5, and ratio of AS/FA as 0.5, by weight, respectively.

Keywords Geopolymer mortar · Nanoparticle · Alumino-silicate · Alkaline liquid · Ambient temperature

T. N. Nguyen (✉)
Faculty of Civil Engineering, Industrial University of Ho Chi Minh City (IUH), 12 Nguyen Van Bao, Ward 4, Go Vap District, Ho Chi Minh City, Vietnam
e-mail: nguyentanno@iuh.edu.vn

H. Q. P. Le
Falculty of Civil Engineering, Can Tho University of Technology, Can Tho, Vietnam

A. T. Le
Faculty of Civil Engineering, Ho Chi Minh City University of Technology (HCMUT), 268 Ly Thuong Kiet Street, District 10, Ho Chi Minh City, Vietnam

Vietnam National University Ho Chi Minh City, Linh Trung Ward, Thu Duc District, Ho Chi Minh City, Vietnam

© The Author(s), under exclusive license to Springer Nature Singapore Pte Ltd. 2023
J. N. Reddy et al. (eds.), *ICSCEA 2021*, Lecture Notes in Civil Engineering 268,
https://doi.org/10.1007/978-981-19-3303-5_29

1 Introduction

Ordinary Portland cement (OPC) has been widely used in the construction industry field. The production of OPC contributes the high percentage of greenhouse gas emissions, which is approximately 5–7% global CO_2 emissions [10, 17]. Geopolymer was coined by Davidovits [5] as a potential alternative material to OPC due to an environmentally friendly material and excellent mechanical properties. For example, the production of 1 tonne of kaolin based geopolymeric-cement generates 0.180 tonnes of CO_2 compared with 1 tonne of CO_2 for Portland cement, i.e. six times less [6]. Besides, Geopolymer obtained early age strength [7, 13] or good resistance to acid and fire [2, 3].

Geopolymer are generally synthesized from polymerisation of solid aluminosilicates with alkaline environment. Aluminosilicate materials have been known as supplementary cementitious materials in making geopolymer such as fly ash [11, 15], metakaolin [8], rice husk ash [12], or blast furnace slag [3, 9]. These materials are high contents of SiO_2 and Al_2O_3 to replace Portland cement in construction fields. When solid Al-Si source materials come into contact with alkaline solution like NaOH, KOH or $Ca(OH)_2$, leaching of both alumina and silica species starts leading to hardened geopolymer [4]. There were several studies using a small amount of nanoparticle added to geopolymer mortars between 0.5 and 10 percent to improve structural properties of geopolymer [1, 10]. To have a better understanding of compressive strength of fly ash based geopolymer mortars with the addition of nano materials, this study investigated (1) the effect of alkaline environment on geopolymerization, (2) the effect of various percentage of each nano material in partial replacement of fly ash in geopolymer pastes and (3) the effect of mixing of nano-SiO_2 and nano-Al_2O_3 on the compressive strength of mortars.

2 Materials and Methods

2.1 Materials

The raw materials such as fly ash (FA), sand (S), Zeolite (Zeo), silicafume (SF), nano-silica (NS), nano-alumina (NA) and alkaline liquid (AL) are used in mix proportion, as seen in Fig. 1.

a) Fly ash   b) Sand   c) Nano-silica   d) Nano-alumina

e) Sodium hydroxide   f) Sodium silicate

Fig. 1 Raw materials

Table 1 Chemical compositions of fly ash, nano-silica and nano-alumina

Materials	SiO_2	Al_2O_3	Fe_2O_3	K_2O	Na_2O	CaO	MgO	TiO_2	LOI*
FA (wt. %)	51.70	31.90	3.48	0.68	0.34	1.21	0.81	–	9.63
SF (wt. %)	95.3	–	–	1.20	0.35	–	–	–	3.15
NS (wt. %)	99.60	0.20	0.02	0.01	–	0.02	–	0.02	0.15
NA (wt. %)	–	99.8	–	–	–	–	–	–	0.2

* Loss on ignition

Fly ash (FA) is low-calcium (class F) in according to ASTM C618. This FA has specific gravity is 2.5 and total alumino-silica content is about 83.6%, by weight. The specific surface area is 344 m^2/kg by Blain method.

The specific gravity and bulk density of sand are 2.62 and 1.32 kN/m^3, respectively. The fineness modulus is 2.1.

On the other hand, specific gravity and bulk density of silica fume are 2.2 and 1.32 kN/m^3, respectively. The chemical composition with 95.3% SiO_2, an average particle size of 8 μm are used.

The nano particles are composition with 99.6% and 99.8% in pure NS and NA powder, respectively. The average particle size range is 10−50 nm for NS and NA, respectively.

The alkaline liquid combination of sodium silicate and sodium hydroxide in solution is used. The sodium silicate solution included Na_2O and SiO_2 about 36–38% is mixed to sodium hydroxide 10 Molar (Tables 1 and 2).

Table 2 Mix proportion

Mix	S/ FA	FA %	SF/NS %	NA %	(Si + Al) %	Si/Al
G0	3	25	0	–	20.85	1.61
GS1		23.75	1.25	–	21.06	1.78
GS2		22.5	2.50	–	21.27	1.96
GS3		21.25	3.75	–	21.47	2.17
GS4		20	5.00	–	21.68	2.40
GS5		18.75	6.25	–	21.89	2.66
GA1	3	23.75	–	1.25	21.06	1.39
GA2		22.5	–	2.50	21.27	1.20
GA3		21.25	–	3.75	21.47	1.04
GA4		20	–	5.00	21.68	0.91
GA5		18.75	–	6.25	21.89	0.79
GS1A1	3	23.75	0.63	0.63	21.06	1.57
GS2A2		22.5	1.25	1.25	21.27	1.52
GS3A3		21.25	1.88	1.88	21.47	1.48
GS4A4		20	2.50	2.50	21.68	1.44
GS5A5		18.75	3.13	3.13	21.89	1.40
G1S1A2	3	23.75	0.42	0.83	21.06	1.50
G2S1A2		22.5	0.83	1.67	21.27	1.40
G3S1A2		21.25	1.25	2.50	21.47	1.31
G4S1A2		20	1.67	3.33	21.68	1.23
G5S1A2		18.75	2.08	4.17	21.89	1.16
G1S2A1	3	23.75	0.83	0.42	21.06	1.63
G2S2A1		22.5	1.67	0.83	21.27	1.65
G3S2A1		21.25	2.50	1.25	21.47	1.67
G4S2A1		20	3.33	1.67	21.68	1.69
G5S2A1		18.75	4.17	2.08	21.89	1.71

2.2 Experimental Testing

The test sets up for the investigation of fly ash based geopolymer mortar as seen in
Fig. 2. In the mix proportion of mortar, the ratio of FA and RS is 1–3 by weight.
Alkaline liquid is mixed in range from 0.4 to 0.6 by FA weight to react in geopolymer
processing. The replacement of nano particles in range from 5 to 25%, by weight in
fly ash are investigated. The ratios of NS and NA in range from 0.5 to 2, by weight
are used.

The raw materials are mixed to produce dry mixing. After that, activated liquid
is added to produce fresh mortar. The specimens are cured in room condition. The
compressive strength at 7 days is tested in accordance with ASTM C39.

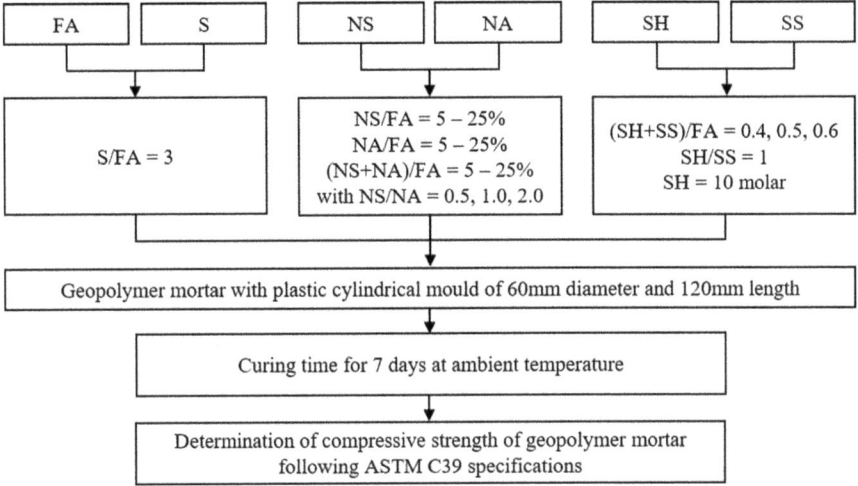

Fig. 2 Preparation for geopolymer mortar

3 Results and Discussion

3.1 Effects of Alumino—Silicate Content and Alkaline Solution on Geopolymerization

Figure 3 shows the trend of compressive strength of fly ash geopolymer mortar depending on the alkaline liquid-fly ash (AL/FA) ratios of 0.4, 0.5, and 0.6, by weight, and the content of $Na_2O/(SiO_2$ and Al_2O_3) ratio ($N/(Si + Al)$). With the same amount of nanomaterial, the increase in AL/FA ratios caused the increasing $N/(Si + Al)$ content. In all cases of mixtures, the compressive strength of geopolymer mortars at 0.5 AL/FA ratio (about 11 MPa) was higher than that of geopolymer mortars at 0.4 and 0.6 AL/FA ratios (about 1 MPa and 8 MPa, respectively). The results were indicated that the 0.5 AL/FA ratio caused higher dissolution of fly ash leading to high reactivity with fly ash resulting in gel formation of aluminosilicate geopolymer for higher strength.

3.2 Effects of Replacement of Fine Particles on Reaction of Geopolymer

To investigate the influence of each nanomaterial on compressive strength of geopolymer mortars, fly ash (FA) was replaced partially by silicafume (SF), nano-silica (NS) or nano-alumina (NA), and the optimum value of 0.5 AS/FA ratio was used in all mixtures. It is clearly seen that the strength of SF and NA specimens based

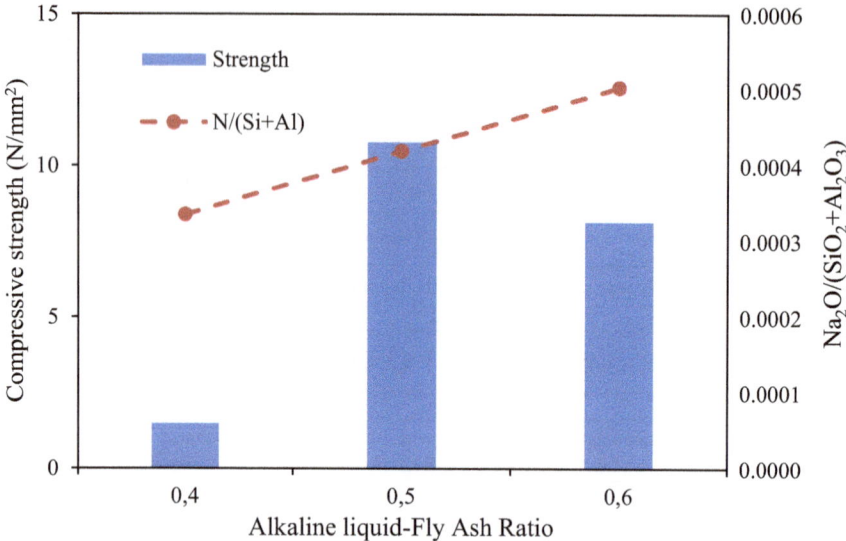

Fig. 3 Compressive strength of geopolymer mortar at different alkaline solution-fly ash ratios

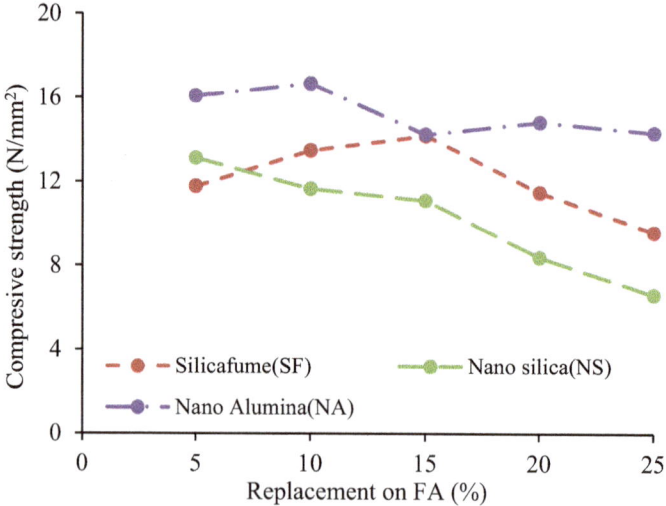

Fig. 4 Compressive strength with various fine particles replacement on fly ash

geopolymer was higher than that of NS specimens given in Fig. 4. The strength could be up to 16 N/mm² by replacement of 10% NA in FA binder. With higher in NA replacement, the strength was slightly reduced. The mixture with silicafume was shown that the strength could be up to 14 N/mm² by replacement of NS about 15%

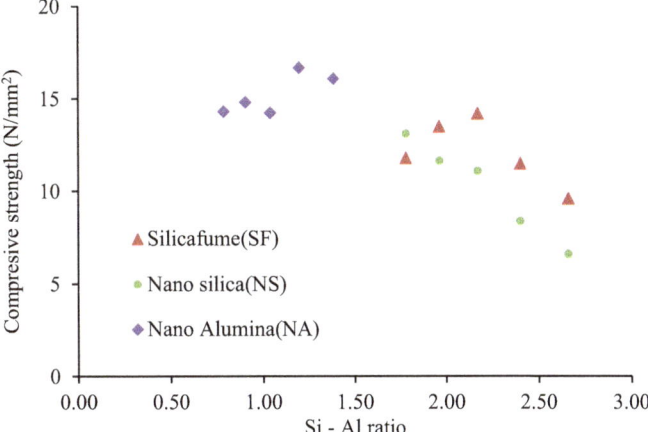

Fig. 5 Behaviour between Si-Al ratio and product of poly-geopolymer

by weight in FA. However, the strength was tent to reduce with higher in NS replacement. The mixture with NS was shown the suitable replacement in FA was 5% by weight. The higher NA replacement was reduced in strength of geopolymer mortar.

To have better understanding of poly-geopolymer product, Fig. 5 shows the relationship between the compressive strength of nano-material mixtures and Si-Al ratios. It could be seen that the optimum ratio of Si-Al ratio of equal was observed with NA mixture. In addition, with the same ratio of Si-Al, the mixture with SF was obtained strength to be almost higher to compare to its with NS.

Figure 6 was shown that the effect of SF mixing with NA on strength of mortar was significantly higher to compare to NS mixing with NA. The highest strengths were about 22 N/mm^2 and 17 N/mm^2 by replacement of 10% SF-NA and 15% NS-NA in FA binder, respectively. It was indicated that SF and NS could be reacted to NA in FA binder to geopolymerization. Hence, the addition of nanoparticles served as fillers in the empty spaces led to reduce porosity or improve density of geopolymer samples [1, 16]. In other words, the appropriate SiO_2 and Al_2O_3 contents enhanced geopolymeric reaction to form CSH or CASH, and NASH gels causing the higher compressive strength of geopolymer mortars [1, 14].

3.3 Effects of Nano-Silica and Nano-Alumina on Reaction of Geopolymer

Compressive strength of fly ash geopolymer mixes containing various ratios of nanoparticles was shown in Fig. 7. The results were illustrated that the highest compressive strength of geopolymer pastes containing nano-silica and nano-alumina

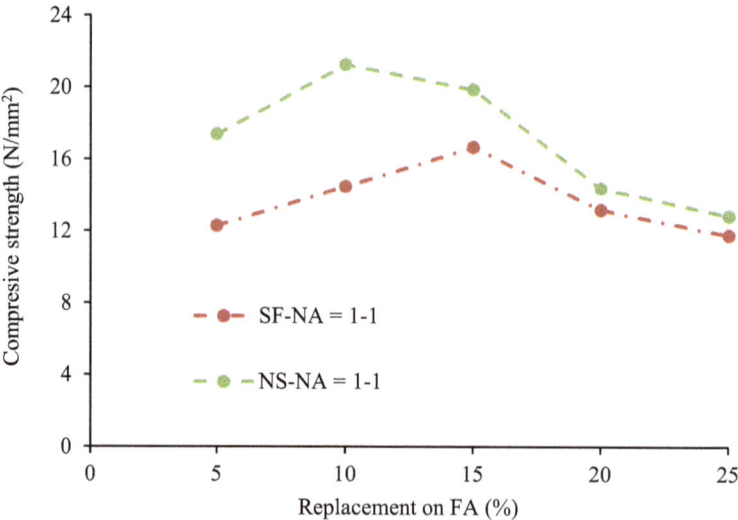

Fig. 6 Effect of various silica particle and NA on strength of mortar

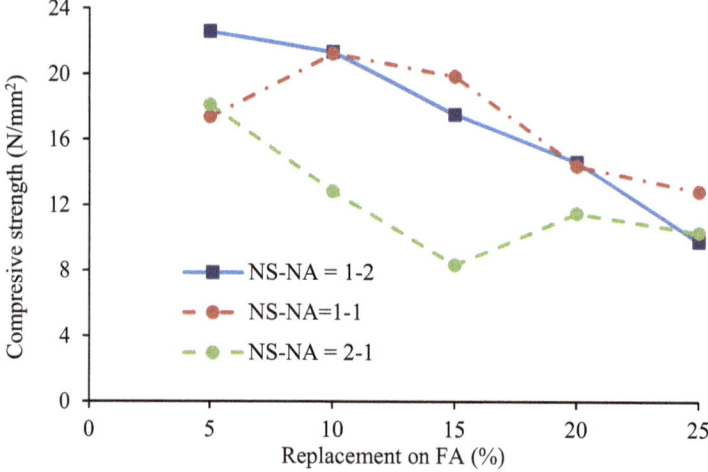

Fig. 7 Effect of various ratio of NS and NA on strength of mortar

(NS/NA) ratios of 0.5, 1.0, 2.0, were about 23 MPa, 22 N/mm^2, and 18 MPa, respectively, when 5% nanoparticle to FA ratio was mixed. With nanoparticle content increased from 5 to 25% by weight of fly ash, the strength of geopolymer samples relatively decreased. It is noteworthy that the strength of geopolymer mortars was enhanced by adding nano particles in range from 5 to 15%. Moreover, the higher compressive strength of geopolymer mortars was obtained by higher in nano particles. It was explained that adding nanoparticles in alumino—silicate binder could be high homogeneity of geopolymer matrix and the high increase in the geopolymeriza-tion [1, 14]. The growth of geopolymer gel was caused to improve the compressive strength of mortars containing higher NA. The same trend has been found by [14] for adding nanoparticles to fly ash based geopolymer.

4 Conclusions

This research was presented the influence of alkaline activator, silicafume, nano-silica and nano-alumina on the compressive strength of fly ash based geopolymer mortars. Based on the results of this study, the following conclusions could be indicated:

Fly ash with alumino—silicate content could be obtained by alkaline liquid—fly ash ratio in range from 0.4 to 0.6, by weight. Strength of mortar was about 10 N/mm^2 cured at room temperatures. With the high silica content, silicafume and nano-silica were added to fly ash to improve strength of geopolymerization.

The reaction of alumino—silicate was improved by using of nanoparticle in alkaline environment. The addition of nano-silica to fly ash ratio of 5%, by weight, and nano-alumina to fly ash ratio of 5–10%, by weight, based geopolymer resulted in the strength of about 12 N/mm^2 and 16 N/mm^2, respectively.

The highest strength of mortars could be obtained up to 23 N/mm^2 by 5% nanoparticles replacement in fly ash binder and 1–2 ratio of nano silica—nano alumina, respectively. The strength of fly ash geopolymer mortars could be improved by mixing suitable nano-silica and nano-alumina ratio.

Acknowledgements We would like to thank Ho Chi Minh City University of Technology (HCMUT) for the support of time and facilities for this study.

References

1. Assaedi H, Shaikh FUA, Low IM (2015) Characteristics of nanosilica-geopolymer nanocom-posites and mixing effect. Int J Chem Mol Eng 9(12):1471–1478
2. Bakharev T (2005) Resistance of geopolymer materials to acid attack. Cem Concr Res 35(4):658–670
3. Cheng TW, Chiu JP (2003) Fire-resistant geopolymer produced by granulated blast furnace slag. Miner Eng 16(3):205–210
4. Comrie DC, Kriven WM (2012) Composite cold ceramic geopolymer in a refractory application. Ceram Trans 153:211–225
5. Davidovits J (2005) Geopolymer chemistry and sustainable development. The Poly(sialate) terminology: A very useful and simple model for the promotion and understanding of green-chemistry. Geopolymer Green Chemistry and Sustainable Development Solutions, Geopolymer 1:9–17
6. Khale D, Chaudhary R (2007) Mechanism of geopolymerization and factors influencing its development: A review. J Mater Sci 42:729–746
7. Kim B, Lee S (2020) Review on characteristics of metakaolin-based geopolymer and fast setting. J Korean Ceram Soc 57(4):368–377

8. Kwon H-M, Nguyen TN, Le TA (2009) Improvement of the strength of acrylic emulsion polymer-modified mortar in high temperature and high humidity by blast furnace slag. KSCE J Civ Eng 13(1):23–30
9. Naskar S, Chakraborty AK (2016) Effect of nano materials in geopolymer concrete. Perspect Sci 8:273–275
10. Nguyen KT, Nguyen QD, Le TA, Shin J, Lee K (2020) Analyzing the compressive strength of green fly ash based geopolymer concrete using experiment and machine learning approaches. Constr Build Mater 247:118581
11. Nguyen T-N, Le A-T, Nguyen M-T (2017) Factors influencing strength development in soft soil clay mixed rice husk ash based geopolymer. Adv Exp Mech 2:153–158
12. Pham KVA, Nguyen TK, Le TA, Han SW, Lee G, Lee K (2019) Assessment of performance of fiber reinforced geopolymer composites by experiment and simulation analysis. Appl Sci 9(16):3424
13. Phoo-ngernkham T, Chindaprasirt P, Sata V, Hanjitsuwan S, Hatanaka S (2014) The effect of adding nano-SiO_2 and nano-Al_2O_3 on properties of high calcium fly ash geopolymer cured at ambient temperature. Mater Des 55:58–65
14. Rattanasak U, Chindaprasirt P (2009) Influence of NaOH solution on the synthesis of fly ash geopolymer. Miner Eng 22(12):1073–1078
15. Stefanidou M, Papayianni I (2012) Influence of nano-SiO_2 on the Portland cement pastes. Compos B Eng 43(6):2706–2710
16. Turner LK, Collins FG (2013) Carbon dioxide equivalent (CO_2-e) emissions: A comparison between geopolymer and OPC cement concrete. Constr Build Mater 43:125–130
17. Zidi Z, Ltifi M, Ben Ayadi Z, El Mir L (2019) Synthesis of nano-alumina and their effect on structure, mechanical and thermal properties of geopolymer. J Asian Ceram Soc 7(4):524–535

An Evaluation of Chloride Ion Penetration of Reinforced Concrete with Surface Coating Materials Exposed for 25 Years Under Coastal Environments

Sachie Sato, Masaru Kakegawa, and Yoshihiro Masuda

Abstract In this study, reinforced concrete specimens with three surface coating materials were exposed to a coastal environment for 25 years to investigate their long-term durability and salt penetration control effect. The specimens were fabricated simultaneously and analyzed at 4.8 years and eight years. The exposure tests are conducted in the coastal area of Hokkaido, in northern Japan, at 40 m from the coastline, and the annual chloride ion flux is 7.1 mg/dm^2/day on the seaward side of specimens. Chloride ions that penetrated the concrete are analyzed by potentiometric titration to determine the number of chloride ions and the penetration depth. We compared the amount of chloride ion penetrating to each material used for repairing and coating. The change in the shielding effect of the surface coating materials is also determined based on the change over time from the past experimental results. As a result, the coating material S (multi-layer textured coat) has the most practical effect on preventing chloride ion penetration.

Keywords Chloride ion penetration · Reinforced concrete · Surface coating · Cross-section repair

1 Introduction

In recent years, various initiatives have been carried out globally to transition to a recycling-oriented society. In the architectural sector, importance is placed on repair, restoration, and maintenance techniques to achieve a longer lifespan and ensure long-term soundness in structures.

S. Sato (✉)
Faculty of Architecture and Urban Design, Tokyo City University, 1-28-1 Tamazutsumi, Setagaya, Tokyo, Japan
e-mail: s-sato@tcu.ac.jp

M. Kakegawa
Taiheiyo Materials Corporation, Tabata ASUKA Tower, 6-1-1, Tabata, Kita-ku, Tokyo, Japan

Y. Masuda
Utsunomiya University, 7-1-2, Yoto, Utsunomiya, Tochigi, Japan

This study fabricated specimens on which a surface covering was applied. The cross-sectional repair was performed, replicating reinforced concrete structures in a coastal environment subject to salt damage caused by external chlorides. Exposure testing was performed to evaluate the durability of different construction methods.

The exposure testing was conducted in a coastal area of Hokkaido, with analysis performed on specimens subjected to exposure for about 25 years. Results in this study were compared with past survey results1 recorded at 4.8 and 8 years of exposure [1]. Our previous study considers the effects of deterioration according to the number of years of exposure and the effects of surface coating methods and cross-sectional repair methods on durability and corrosion in rebar [2]. This study focused on the concentration of chloride ions through concrete and cross-sectional repair materials.

In this paper, among the items studied in testing, this paper considers the effects of chloride ion content at each depth level from the surface and the impact of repairs and coating material in suppressing chloride ion penetration.

2 Experimental Overview

Table 1 shows the types of materials used in the repairs. Table 2 details the content and properties of the concrete; Table 3 shows the mix proportion of the concrete. The 28-day strength of concrete is 30.8 N/mm². Figure 1 is a structural schematic of the RC specimens. Two reinforcement bars were embedded in the concrete, with portions that have been partially repaired using cross-sectional repair materials.

These repaired specimens were left at the exposure test site for predetermined lengths of time (43.025 N, 140.53 E, 40 m from coastline, and the annual chloride ion flux is 7.1 mg/dm²/day[3]: Fig. 3). At 4.8, 8, and 25 years, several specimens were broken apart for analysis of chloride ion content.

Figure 2 explains the notation of the specimen names used in the Results section. Two-symbol IDs denote RC specimens.

Samples for measurement of chloride ion content were collected at the positions shown in the cross-sectional diagram of Fig. 1. The samples were crushed with a

Table 1 Repair materials

Process	Symbol	Type
Cross-sectional repair	N	None
	CM	Cement mortar
	PS	SBR* polymer cement mortar
	PI	SBR* polymer cement mortar with anti-rust additive
	LE	Lightweight epoxy mortar
Surface coating	N	None
	L	Thin textured coat
	S	Multi-layer textured coat

*SBR: styrene-butadiene rubber

Table 2 Concrete content and properties

Materials	Properties
Cement	Ordinary Portland cement Density: 3.16 g/cm^3
Fine aggregate	River sand Density 2.62 g/cm^3, F.M.2.64
Coarse aggregate	Crushed hard sandstone Density 2.64 g/cm^3, F.M.6.71
Air entraining agent	Natural resinate
Rebar	Round steel bars with mill scale and acetone defatting: φ13 mm, SR235 (SR24) (JIS G 3112)

Table 3 Concrete mix composition

Target slump (mm)	Target air content (%)	W/C (%)	S/a (%)	Water	Cement	Fine aggregate	Coarse aggregate	AE (*cement weight (%))
				(kg/m^3)				
180	4.0	65.0	48.0	185	285	862	940	0.020

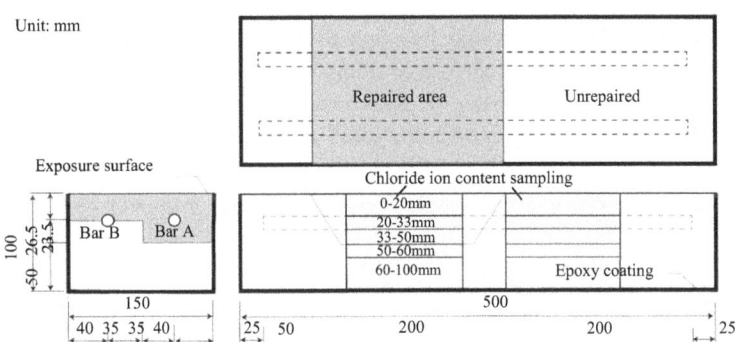

Fig. 1 Schematic of RC specimen

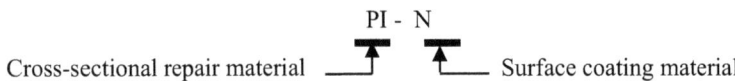

Fig. 2 Specimen notation

jaw crusher, dried at 105 °C, and pulverized with a vibration disk mill. All chloride ions were extracted from the pulverized samples using JISA1154 (Methods of Test for Chloride Ion Content In Hardened Concrete), and the chloride ion content was analyzed using potentiometric titration.

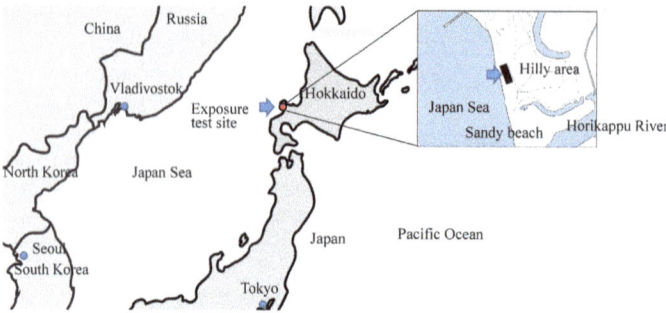

Fig. 3 Exposure test site

3 Results and Discussion

Figure 4a, b shows the relationship between the degree of penetration of chloride ion content and distance from the specimen surface. Variations were significant at 4.8 and 8 years, with higher salt content in some cases than in the 25-year samples. However, multiple specimens were prepared for each standard and were extracted for analysis at each of the ages of materials, and natural environmental conditions were not identical for all specimens due to individual differences among specimens and their installation positions; this may be a factor behind the variations. The results for non-repaired sections without surface covering (N) can be considered equivalent to N-N among the samples. Variation tended to increase near the position of the rebar (26.5 mm from the surface). Moreover, cases in which a significant difference in chloride ion content occurred between repaired portions and non-repaired portions in PI and LE were noteworthy, indicating a possibility of a decline in adhesion to concrete.

The measurement results for 25 years show chloride ion content generally decreasing with distance from the surface. In some samples, the position at which maximum chloride content is reached was observed to move inward from the outermost layer, which seems like a skin effect.

It was found that when a surface coating material is applied, the amount of chloride ion penetration into the concrete is lower, with S, in particular, showing a high salt shielding effect. The cross-sectional repair material found that the amount of chloride in the surface layer was high for CM (cement mortar), but permeation and diffusion in the depth direction tended to be small. In LE, chloride ions permeated relatively profoundly in surface coatings L and N. It is likely that the corrosion area ratio is high for LE-N and LE-L.

Permeation of chloride ions was compared for different surface coating materials and was calculated for finishing materials as the salt shielding effect ratio Cp using the following formula.

$$Cp = \frac{\Sigma C_{NN} - \Sigma C_{Fini}}{\Sigma C_{NN}} \times 100(\%) \tag{1}$$

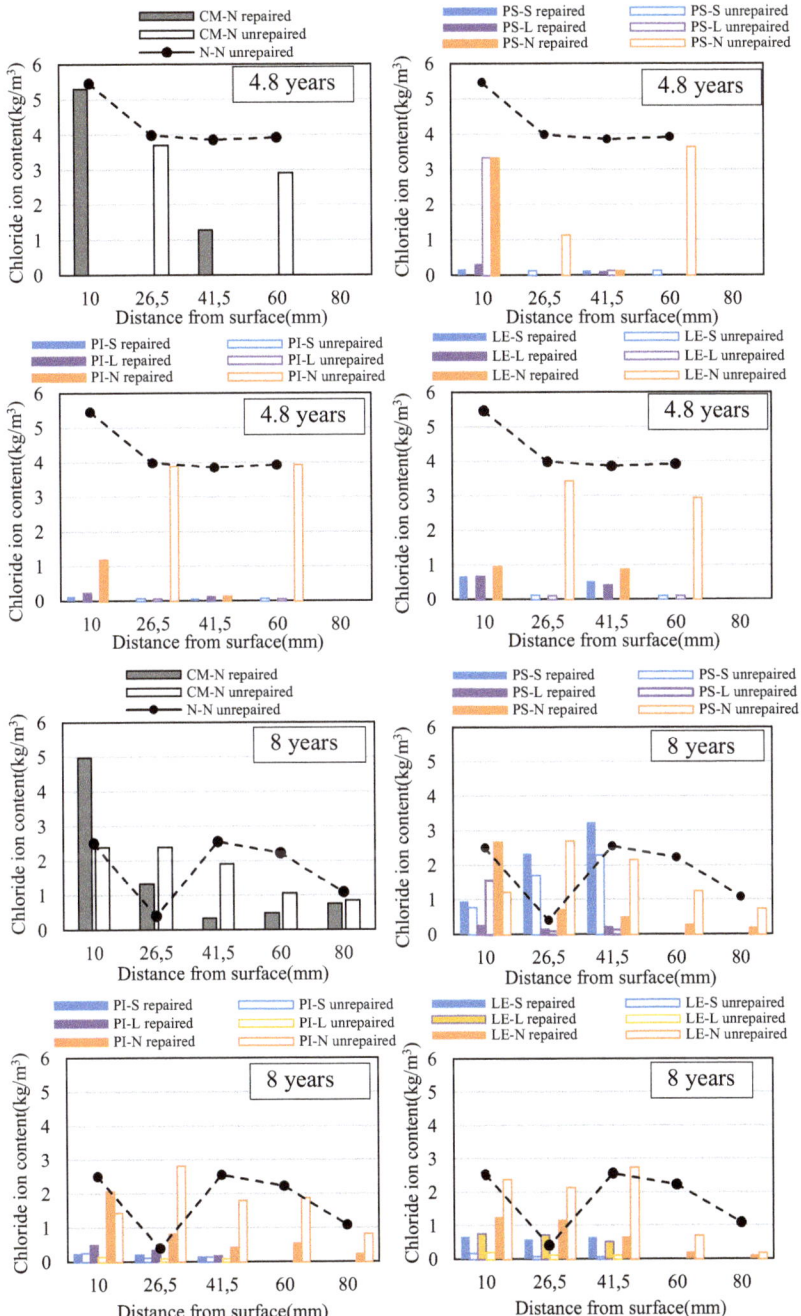

Fig. 4a Chloride ion content (4.8 and 8 years)

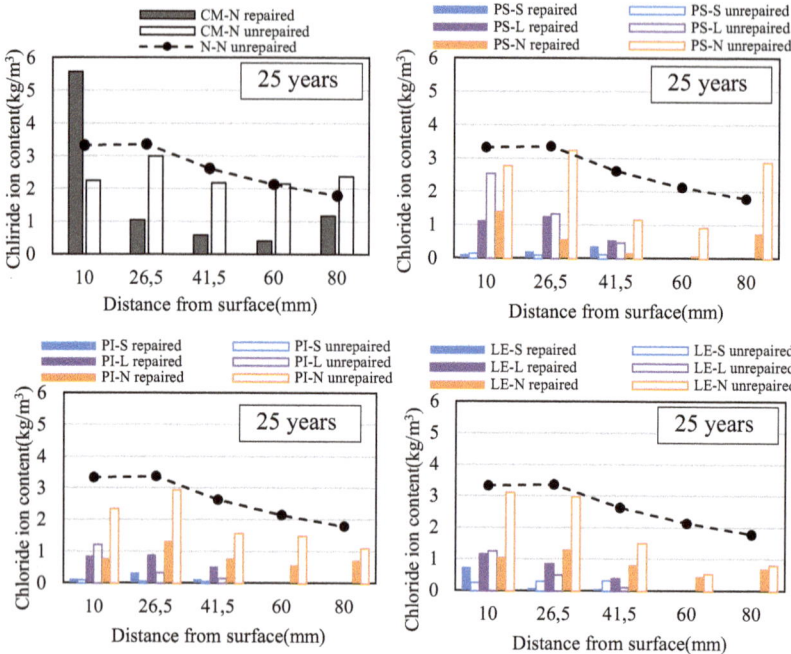

Fig. 4b Chloride ion content (25 years)

where, ΣC_{NN} is total chloride ion content (kg/m^3) at each depth in the case of no finishing material/no repair (N), and ΣC_{Fini} is total chloride ion content (kg/m^3) in the non-repaired portion of the specimens that have surface coating material.

Figure 5 shows the results of the calculation of the chloride ion shielding ratio Cp in non-repaired portions. From the figure, it can be seen that S retains high salt shielding capability even after 25 years. Compared to S, L shows a significant decrease over time, particularly between 8 and 25 years; maintenance after about 10 years should be considered necessary.

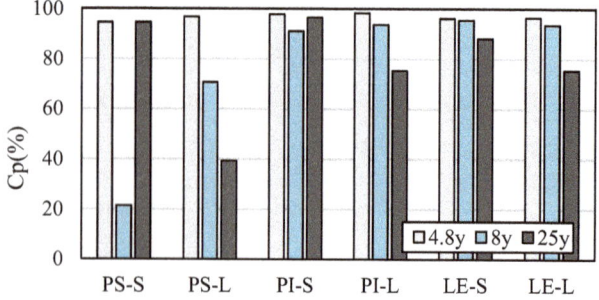

Fig. 5 Chloride ion shielding effect ratios

4 Conclusion

This study analyzed the degree of chloride ion permeation in rebar in RC specimens exposed to a coastal area for 25 years and evaluated different repair methods. It was found that multi-layer finishing coating materials have the most significant effect. The chloride ion permeation inhibitory effect due to types of surface coating material was outstanding, but the inhibitory effect declined as surface coating material deteriorated over time, and it is thought that repairs are required after roughly ten years. In addition, after 25 years of chloride penetration, the chloride ions contents in the area near the surface were observed to be a "skin effect" in the area near the surface.

Acknowledgements We received cooperation in this study from HACHIYO Consultant Co., Ltd. and from Yusuke Usui, a Tokyo City University student writing a graduation thesis at the time. We would like to thank everyone involved. This research was carried out with the support of a JSPS Grant-in-Aid for Scientific Research (16K0659, 19K04695).

References

1. Sato K, Masuda Y et al (2002) Outdoor exposure test on evaluation of repair reinforced concrete, Summaries of technical papers of annual meeting. Arch Inst Japan A-1:625–626 (In Japanese)
2. Sato,S, Masuda Y, Kakegawa M (2020) Long-term performance of repairs to reinforced concrete exposed to coastal conditions, XV International Conference on Durability of Building Materials and Components (DBMC 2020, Barcelona)
3. Sakihara K et al (2015) Study on evaluation of air born chloride ions resistivity of PVC siding. <Results of the fifth year of exposure> Proceeding of the architectural research meetings, Kyushu Chapter, Architectural Institute of Japan 54, pp. 73–76(In Japanese)
4. Andrade C, Diez JM, Alonso C (1997) Mathematical modeling of a concrete surface 'skin effect' on diffusion in chloride contaminated media. Adv Cem Based Mater 6(2):39–44

Applicability of Electrical Resistance Method to Moisture Transfer Measurement in Paste Using Mineral Admixture

Takahiro Aoki, Yuko Ogawa, and Kenji Kawai

Abstract Concrete is a porous material which has the ability to hold moisture stably inside. Since moisture is closely related to the deterioration of concrete structures, it is important to understand the moisture transfer in concrete in order to maintain the concrete structure appropriately. The electrical resistance method is one of non-destructive methods to investigate the moisture in concrete and the moisture transfer can be assessed with time by using one specimen. However, the electrical resistance may be affected by mineral admixtures in concrete. Although previous studies on the effects of mineral admixtures on the electrical resistance have been conducted, there are still many unclear points. Therefore, the purpose of this study is to assess the applicability of the electrical resistance method to moisture transfer measurement in paste containing mineral admixtures by focusing on the moisture transfer in the drying process. The moisture transfer in the hardened cement paste using fly ash and blast furnace slag with water-to-binder ratios of 0.35 and 0.55 was investigated by the electrical resistance method. Stainless steel rods of 0.9 mm in diameter were placed at an interval of 4 mm in the specimen. The calibration test was also conducted to obtain the relationship between the electrical resistivity and the internal relative humidity of the specimen. The result showed that the specific resistance of the specimen containing mineral admixtures was higher than that of the specimen without mineral admixtures. It was observed that at the drying period of 70 days, the internal relative humidity at a depth of 4 mm from the exposed surface was almost the same as the surrounding relative humidity. It is possible to know the change in the internal relative humidity distribution in the cement paste using mineral admixtures during the drying process by performing the calibration test appropriately.

Keywords Moisture transfer · Electrical resistance · Mineral admixture · Stainless steel rod · Relative humidity

T. Aoki · Y. Ogawa · K. Kawai (✉)
Civil and Environmental Engineering Program, Graduate School of Advanced Science and Engineering, Hiroshima University, 1-4-1 Kagamiyama, Higashihiroshima, Japan
e-mail: kkawai@hiroshima-u.ac.jp

© The Author(s), under exclusive license to Springer Nature Singapore Pte Ltd. 2023   379
J. N. Reddy et al. (eds.), *ICSCEA 2021*, Lecture Notes in Civil Engineering 268,
https://doi.org/10.1007/978-981-19-3303-5_31

1   Introduction

Concrete is a porous material which has the ability to hold moisture stably inside. Since moisture is closely related to the deterioration of concrete structures, it is important to understand the moisture transfer in concrete in order to maintain the concrete structure appropriately [1]. There are various methods to assess the moisture transfer in concrete. The electrical resistance method is one of non-destructive methods to investigate the moisture in concrete, and the moisture transfer can be assessed with time by using one specimen. It is a practical method that can assess rapid change of moisture and its change over time. Using this method, moisture transfer in cement paste during moisture and water absorption [5] and moisture transfer in mortal after exposure to high temperature [6] have been examined. However, the electrical resistance may be affected by mineral admixtures in concrete. It has been shown that the use of an admixture densifies the pore structure and changes the electrical resistance [4]. Also, it was shown that the difference in the resistivity of the material itself also affects the measured value [7]. Although the previous studies on the electrical resistance have been conducted, there are many unclear points about the effect of using mineral admixtures on the measured values. Therefore, the purpose of this study is to assess the applicability of the electrical resistance method to moisture transfer measurement in paste containing mineral admixtures by focusing on the moisture transfer in the drying process.

2   Experimental Outlines

2.1   Materials and Specimens

Ordinary Portland cement (OPC) conforming to JIS R 5201 (density: 3.16 g/cm^3, specific surface area: 3360 cm^2/g) was used as a cement in this study, while fly ash (FA) classified as Type II in JIS A 6201 (density: 2.23 g/cm^3, specific surface area: 3530 cm^2/g) and ground granulated blast furnace slag (BFS) classified as 4000 in JIS A 6206 (density: 2.91 g/cm^3, specific surface area: 4170 cm^2/g) were used as mineral admixtures. Cement paste specimens with water to binder ratios (W/Bs) of 0.35 and 0.55 were prepared. The replacement ratios of FA and BFS were 20% and 50%, respectively. The specimen for the moisture transfer test had a dimension of 40 × 35 × 80 mm, and stainless electrodes with a diameter of 0.9 mm were embedded inside the specimen at an interval of 4 mm, as shown in Fig. 1. For the calibration test specimen to obtain the relationship between the internal relative humidity of the specimen and the specific resistance in the drying process, stainless electrodes were embedded in the center of the specimen having a size of 40 × 35 × 8 mm at an interval of 4 mm, as shown in Fig. 2. The specimen for the pore size distribution test was prepared with the same dimensions as the specimen for the moisture transfer test, 40 × 35 × 80 mm. All the specimens were prepared at 20 °C, demolded 24 h

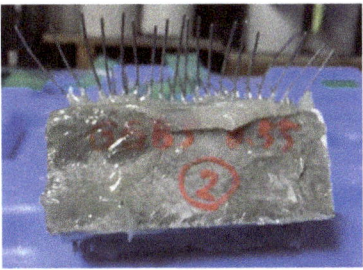

Fig. 1 Specimen for moisture transfer test

Fig. 2 Specimen for calibration test

after casting, and cured in water at 40 °C for 55 days to ensure sufficient hydration of the binders. Cement paste specimens containing no mineral admixtures were also prepared, which were demolded 24 h after casting and cured in water at 20 °C for 27 days. Hereafter, specimens cured for 28 days using no mineral admixtures are denoted by OPC28, and specimens cured for 56 days using no mineral admixtures, FA and BFS are denoted by OPC56, FA56, and BFS56, respectively.

2.2 Experimental Methods

In the calibration test, the specimen after curing were stored in a container at a temperature of 20 °C and relative humidity of 98, 85, 75, 70, or 60% which was prepared with the saturated salt method. After the mass of the specimen became constant, the specific resistance was measured.

In the moisture transfer test, all surfaces of the specimens except for an exposed surface (40 × 35 mm) were coated with epoxy resin after curing, and then the specimens were placed at a temperature of 20 °C and a relative humidity of 60%.

The specific resistance was measured after given periods. The specific resistance in both tests was measured at AC 1 kHz and 1 V using an LCR meter [2]. The measured specific resistance was converted into the specific resistance using Eq. 1 [3].

$$\rho = \frac{R}{Sf} = R / \left\{ \log\left(\frac{d}{a}\right) / (\pi \times l) \right\} \tag{1}$$

where R: electrical resistance (kΩ), a: radius of stainless steel rods (mm), d: interval of stainless-steel rods (mm), l: length of current-carrying part (cm), ρ: specific resistance (k$\Omega \cdot$ m), Sf: geometrical factor (In this study, it was calculated as 0.0863.).

The specific resistance of OPC, FA, and BFS themselves was also measured. The sample was packed in a glass container, and the electrodes were arranged so that the interval was 4 mm and the current-carrying part was 35 mm, similarly to the conditions in the moisture transfer test.

In the pore size distribution test, the specimen was crushed into 2.5–5 mm at the moisture transfer test periods of 0 days and 56 days. They were soaked in acetone for 24 h to stop further hydration and dried in a vacuum desiccator for another 24 h before the measurement. The sample at the start of the moisture transfer test was collected from the center of the specimen, while the samples at the moisture transfer test period of 56 days were obtained from the depths of 0–6, 6–14, and 14-22 mm from the exposed surface.

3   Experimental Results and Discussion

Figure 3 shows the results of the pore size distribution test. In the case of specimens containing FA, the total pore volume decreased as the W/B decreased, and the pore

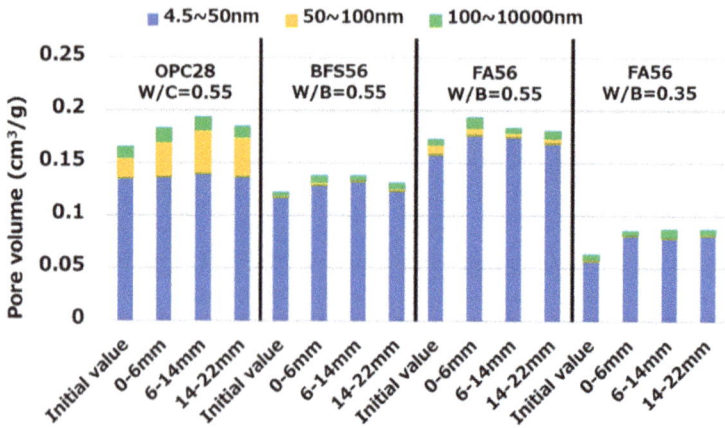

Fig. 3 Pore volume in each specimen

distribution shifted to a smaller diameter. This tendency was also observed for the other cases, and it was confirmed that the paste specimen with the lower W/B had denser microstructure. The total pore volume of FA56 with W/B = 0.55 was almost the same as that of OPC28, but the volume of the smaller pore (4.5–50 nm) in FA56 with W/B = 0.55 was larger than that in OPC28. This indicates that the FA56 with W/B = 0.55 had a denser microstructure at the later age, which can be due to the pozzolan reaction of FA. Also, the total pore volume of BFS56 was significantly smaller than that of OPC28, and it can be due to the latent hydraulicity of BFS.

Figure 4 shows the specific resistance of OPC, BFS, and FA themselves. The specific resistance of FA was the highest, followed by BFS and OPC. As in previous studies, it was shown that the difference in the resistivity of the material itself may affect the measured value [7].

Figures 5, 6, 7 and 8 show the changes over time in the specific resistance in the moisture transfer test. The number of days in the figure represents the duration of the moisture transfer test. As the drying period elapses, the specific resistance near the exposed surface increases, which indicates that the drying process can be assessed by the electrical resistance method for the paste including FA or BFS.

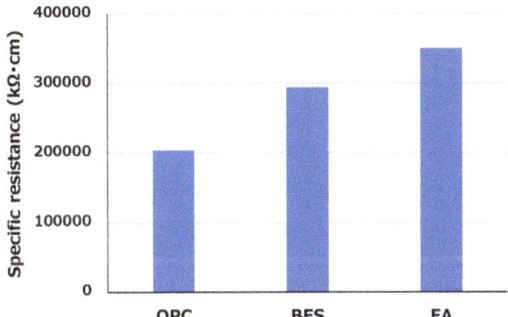

Fig. 4 Specific resistance of each material

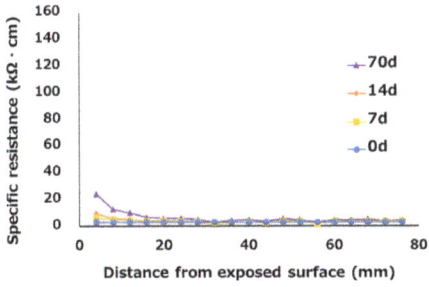

Fig. 5 Change in specific resistance of specimen OPC28, W/B = 0.55

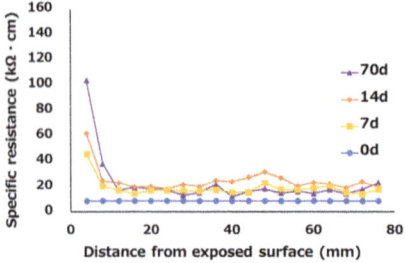

Fig. 6 Change in specific resistance of specimen BFS56, W/B = 0.55

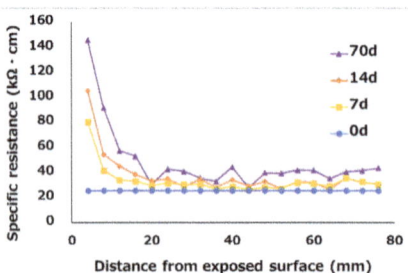

Fig. 7 Change in specific resistance of specimen FA56, W/B = 0.55

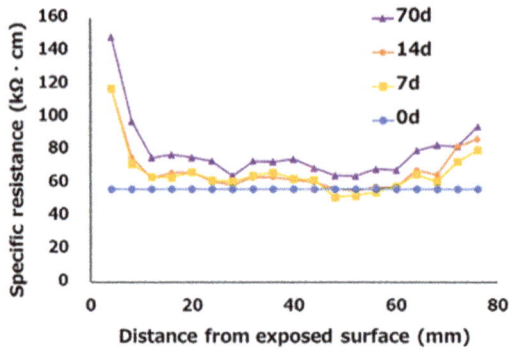

Fig. 8 Change in specific resistance of specimen FA56, W/B = 0.35

The specific resistances of FA56 and BFS56 with W/B = 0.55 were high at the start of the moisture transfer test (0 d) when compared to OPC28. This may be because the specific resistances of FA and BFS themselves are high as shown in Fig. 4.

Comparing the differences depending on the W/B for FA56 as shown in Figs. 7 and 8, the specific resistance of FA56 with W/B = 0.35 was higher than that with W/B = 0.55 at the start of the moisture transfer test. However, as the drying time elapsed, the increase of the specific resistance near the exposed surface of FA56 with W/B =

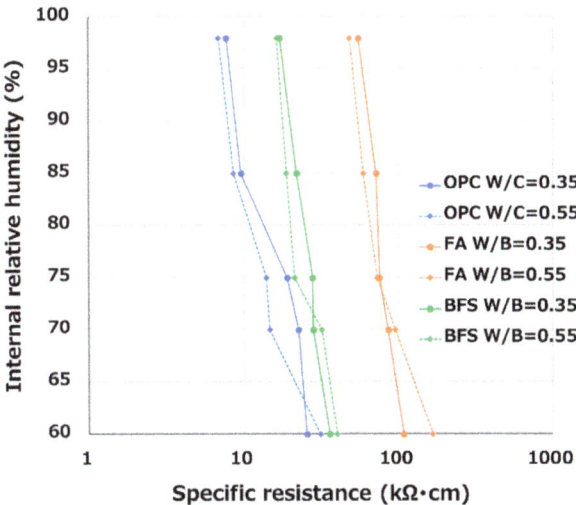

Fig. 9 Relationship between internal humidity and specific resistance

0.55 was higher than that with W/B = 0.35. It is indicated that the paste specimen with W/B = 0.35, a low water-to-binder ratio, had a high specific resistance at the start of drying due to the dense pore structure, whereas the increase in the specific resistance by drying was low due to the less susceptibility to drying. This tendency was also observed for OPC28 and BFS56.

Figure 9 shows the relationship between the internal relative humidity of the paste specimen and specific resistance obtained by the calibration test for 70 days. The FA56 for both W/Bs had the highest resistivity at the same internal relative humidity, followed by BFS56 and OPC28. Since this shows the same tendency as the result of the specific resistance of the material in Fig. 4, it is suggested that the difference in the specific resistance of the material has an effect on this result. Besides, although the differences in the W/B were not significant, the specific resistance for W/C = 0.35 tended to be higher than W/C = 0.55 at high internal relative humidity, while the specific resistance for W/C = 0.55 tended to be higher than W/C = 0.35 at low relative humidity.

Figure 10 shows the change in the internal relative humidity (FA56, W/B = 0.55) with time which is calculated from the specific resistance shown in Fig. 7 and the calibration test result. It can be observed that the internal relative humidity at the depth of 4 mm from the exposed surface at the test period of 70 days was almost the same as the humidity of the surrounding environment (60% RH).

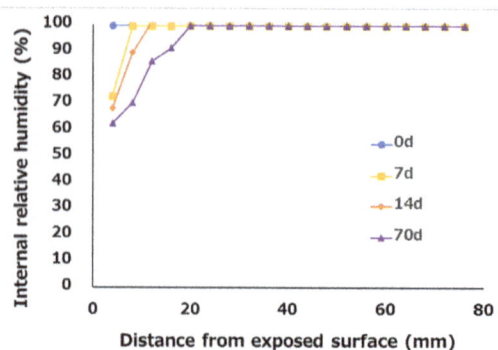

Fig. 10 Change of internal relative humidity FA56, W/B = 0.55

4 Conclusions

(1) It was found that the moisture transfer in the drying process can be assessed by the electrical resistance method for the paste using the admixture such as FA and BFS, similarly to the cement paste using only OPC.

(2) The specific resistance of admixtures was higher than that of OPC, and the specific resistance of FA was higher than that of BFS. This difference also affected the specific resistance of the paste specimen containing admixtures.

(3) At the start of the moisture transfer test, the specific resistance for the specimen with W/B = 0.35 was higher than that with W/B = 0.55. However, as the drying proceeded, the increase in the specific resistance near the exposed surface of the specimen with W/B = 0.55 was higher because the specimen with W/B = 0.55 can be susceptible to drying due to its porous microstructure.

(4) When the moisture transfer test was conducted for 70 days under an environment at a relative humidity of 60%, the internal relative humidity of the specimen using FA with W/B = 0.55 at the depth of 4 mm from the exposed surface reached the humidity of the surrounding environment.

References

1. Ishida T, Chaube PR, Kishi T, Maekawa K (1997) Modeling of pore water content in concrete under generic drying wetting conditions. J Jpn Soc Civ Eng 564(35):199–209
2. Kashima T, Kawano H, Watanabe H, Tanaka Y (1999) Fundamental study on electrical resistance for assessing durability of concrete. Proc Jpn Concr Inst 21(2):895–900
3. Katsura O, Yoshino T, Tahata M, Kawata E (1994) Estimation of water content in concrete by resistivity and temperature. Proc Jpn Concr Inst 16(1):735–740
4. Kurumisawa K, Nawa T (2016) Effect of fly ash fly ash on conductivity of hardened cement paste. Cem Sci Concr Technol 70:230–235

5. Mizoguchi M, Kitagawa T, Daungwilailuk T, Ogawa Y, Kawai K (2018) An investigation on moisture and water absorption in cement paste with electrical resistance method. In: International Conference on Durability of Concrete Structures, pp 581–586
6. Mizoguchi M, Kitagawa T, Daungwilailuk T, Ogawa Y, Kawai K (2020). Investigation on moisture transfer in mortar after exposure to high temperature. In: Reddy J, Wang C, Luong V, Le A (eds) ICSCEA 2019. LNCE, vol 80, pp 563–572. Springer, Singapore. https://doi.org/10.1007/978-981-15-5144-4_53
7. Seki H, Miyata K, Kitamine H, Kaneko Y (1992) Experimental study on permeability of concrete based on electrical resistance. J Jpn Soc Civ Eng 451(17):49–57

Application of Recycled Fine Aggregate from Construction and Demolition Materials for Mortar

Thuy Ninh Nguyen, Viet Hai Vo, Tuyet Giang Vo, and Tuan Anh Le

Abstract Construction and demolition materials are collected during the process of constructing and demolishing buildings. These materials are a major solid waste stream in Vietnam, with a relatively large amount released every year. In this study, the waste materials such as concrete, mortar, bricks, and ceramic tile are investigated to use as recycled fine aggregates (RFA) in mortar. Several tests were conducted to determine fineness modulus, porosity, flow, setting time, and strength for RFA, fresh and hardened mortar mixes, and evaluate the effects of RFA on the properties of mortar. The results indicated that the particle size distributions of concrete and mortar wastes were suitable for coarse grading, whereas those of bricks and ceramic tile wastes were for fine grading. The mortar mixes containing concrete waste showed dominant properties compared to other mixes. The properties of mixes with mortar and tile wastes were moderate, except tile waste had the initial setting time of fresh mixes extended. Brick waste was the most defective and suggested not to be used with high content.

Keywords Recycled fine aggregate · Demolition materials · Mortar · Setting time · Flexural strength

1 Introduction

Waste management is one of the most challenging problems of developing countries and Vietnam. The Vietnamese government adopted several legislative requirements for waste treatment. Among the main types of wastes, construction, and demolition waste (CDW) has important concern. CDW is one of the heaviest and most voluminous waste streams generated in large cities such as Ha Noi, Hai Phong, and Ho

V. H. Vo · T. G. Vo · T. A. Le (✉)
Faculty of Civil Engineering, Ho Chi Minh City University of Technology, VNU-HCM, Ho Chi Minh, Vietnam
e-mail: latuan@hcmut.edu.vn

T. N. Nguyen · V. H. Vo · T. G. Vo · T. A. Le
Faculty of Civil Engineering, VNU-HCM, Ho Chi Minh, Vietnam

Chi Minh City. Therefore, the recycling of wastes from the processes of construction, renovation, relocation, extension, maintenance, and demolition of buildings and structures is urgently required [1].

The construction field is amongst the most environmental impacting in the world, consuming high amounts of raw materials and energy. The challenges and opportunities of CDW recycling are economically and environmentally viable constituting a solution not only for the reduction in the volume of waste deposited into landfills but also the reduction in the extraction of raw materials. In building construction, sand is the most commonly used fine aggregate in cementitious materials [2, 3]. Vietnam faces a shortage of sand after years of over-exploitation for the construction sector. The sand dredged from the Mekong Delta is mostly used as materials for construction in Ho Chi Minh City area. At the current rate of natural sand exploitation, domestic supplies could be depleted by the near future. To deal with the problem, some provinces in the South are looking at a solution of producing artificial sand from sedimentary rock. Besides, there is another sustainable solution of using recycled aggregate (RA) from CDW. The recycling processes include initial inspection, crushing, magnetic separation, and vibrating screen is shown in Fig. 1 [4].

From an environmental perspective, the use of fine RA as sand in cementitious materials can bring several advantages such as reducing sand mining and preventing

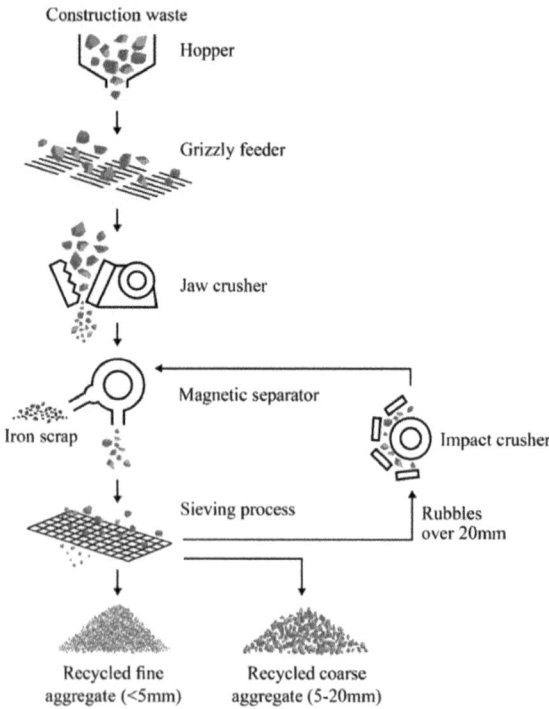

Fig. 1 Process of recycling aggregate from demolished construction [4]

illegal deposits and landfill of the fine fractions of CDW. The composition is mainly dependent on its sources such as concrete, masonry, bricks, pavements, etc., which can vary by construction. In some places, clayed bricks are the main building material, while in others, masonry, tile, and cement board (cemboard) are the main materials. Building and construction technology has a strong effect on the chemical composition of CDW [5–8].

This research aims to investigate the characteristics of recycled fine aggregate (RFA) from CDW and their effects on mortar. Sieve analysis and testing for volumetric characteristics were performed for RFA. Then, the properties of mortar were determined by setting time, compression, and flexure tests to evaluate and compare the effects of RFA types on fresh and hardened mortar.

2 Materials

Portland cement, type I, was used throughout this research. Based on ASTM C33 [9], The porosity and fineness modulus of river sand were determined as 48.4% and 2.3, respectively. In this research, the recycled fine aggregates (RFA) were obtained from demolished concrete (C), masonry (M), brick (B), and ceramic tile (T). The gradation of RFA and mixtures is illustrated in Fig. 2 and detailed in Table 1.

The gradation of RFA was determined according to ASTM C33 [9]. In the mix proportion of mortar, the ratio of cement and sand is 1:3 by mass. On the other hand, sand is replaced by RFA and their mixtures in mortar. The fineness modulus and porosity of RFA were tested according to ASTM C136. Also, the flow, setting time, and strength of mortar mixes were tested in accordance with ASTM C807, ASTM C109, and ASTM C348, respectively.

3 Results and Discussion

3.1 Influence of RFA and Mixtures on the Characteristics

In this section, the effect of RFA and mixtures on the fineness modulus and porosity of fine aggregate was presented. In Fig. 3, it can be considered RFA and their mixtures as fine sand with fineness moduli range from 1.7 to 2.5. As compared to the river sand, most of the RFA has finer gains due to lower porosity and fineness modulus values. Concrete waste and brick waste showed the highest and lowest fineness modulus of 2.5 and 1.7, respectively, which is correlated to the hardness. As processing in the crushers, feeble material tends to be broken and crushed finer, hence, has lower fineness modulus. The fineness modulus varies when two or more types of RFA are blended and depends on the content of each which participated. The porosity, as seen in Fig. 4, shows a relation to the gradation (Fig. 1) and fineness modulus (Fig. 3).

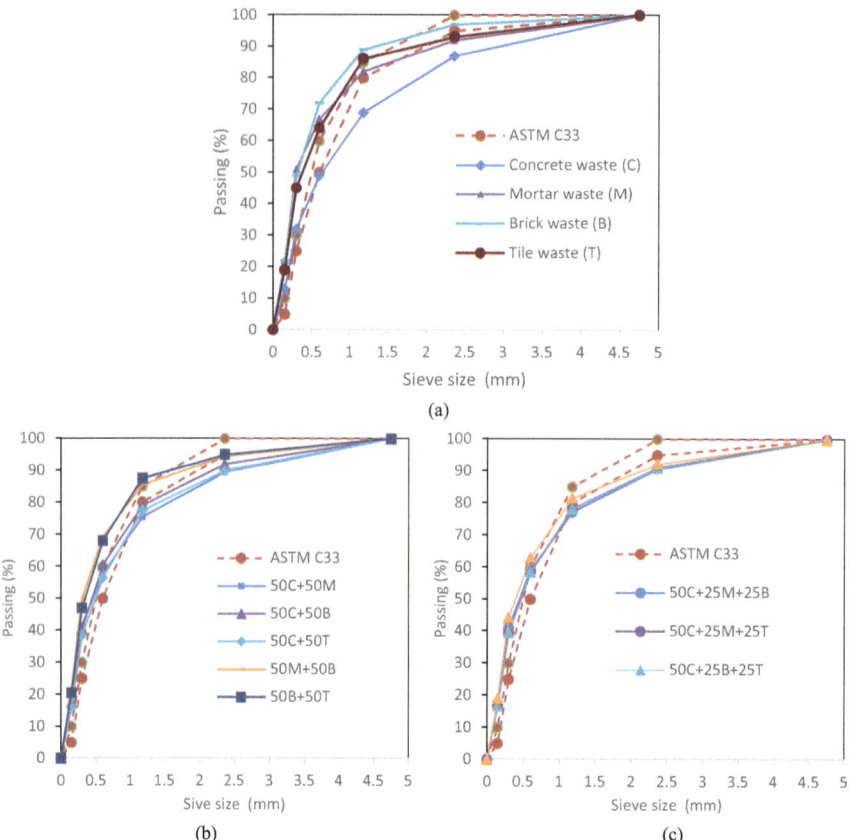

Fig. 2 Grain size distribution of **a** RFA, **b** Mix 1 – Mix 6, and **c** Mix 7 – Mix 10

Concrete, mortar wastes, and their mixtures are less porous than the rest since brick and tile wastes are also porous in the grain.

3.2 Influence of RFA on Fresh Mortar

The properties of fresh mortar are shown in Figs. 5, 6, and 7, such as flow, initial setting time, and setting time. The flow has an almost similar trend with the porosity. The greater porosity of RFA, which has higher water absorption, leads to the lower flow of the mortar mixture. It indicates that mixing water was absorbed and cause a decrease in workability of fresh mix, especially the brick and tile wastes.

The time of setting test was performed by modified Vicat needle. Initial setting time is the time when the mortar loses its plasticity and Setting time is the duration from the initial to final setting time that the mortar stiffens to a defined consistency

Table 1 Grain size distribution of RFA

Mixture	Mix 1	Mix 2	Mix 3	Mix 4	Mix 5	Mix 6	Mix 7	Mix 8	Mix 9	Mix 10
Concrete (%)	50	50	50	–	–	–	50	50	50	25
Mortar (%)	50	–	–	50	50	–	25	25	–	25
Brick (%)	–	50	–	50	–	50	25	–	25	25
Ceramic Tile (%)	–	–	50	–	50	50	–	25	25	25
Sieve (mm)	Passing (%)									
4.75	100	100	100	100	100	100	100	100	100	100
2.36	89.5	92.0	90.0	94.5	92.5	95.0	90.8	91.0	91.0	92.3
1.18	75.5	79.0	77.5	85.5	84.0	87.5	77.3	78.3	78.3	81.5
0.60	58.0	60.5	56.5	69.5	65.5	68.0	59.3	58.5	58.5	63.0
0.30	41.5	40.5	38.5	50.0	48.0	47.0	41.0	39.5	39.5	44.3
0.15	17.5	17.5	16.0	22.0	20.5	20.5	17.5	16.8	16.8	19.0

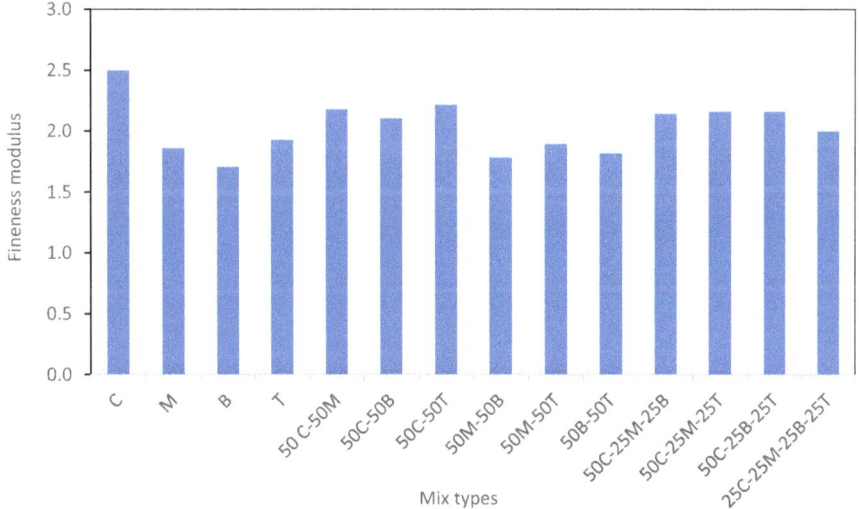

Fig. 3 The fineness modulus values of RFA and mixtures

[10]. In terms of initial setting time (see Fig. 6), the mixtures containing concrete, mortar, brick, and tile wastes take 185, 205, 220, 230 minutes to begin setting, respectively. Different from other mixes, the mixtures with tile waste took longer to initiate setting while its setting time was short as seen in Fig. 7. As compared to brick, tile has a denser structure to resist water permeability. From the authors' view,

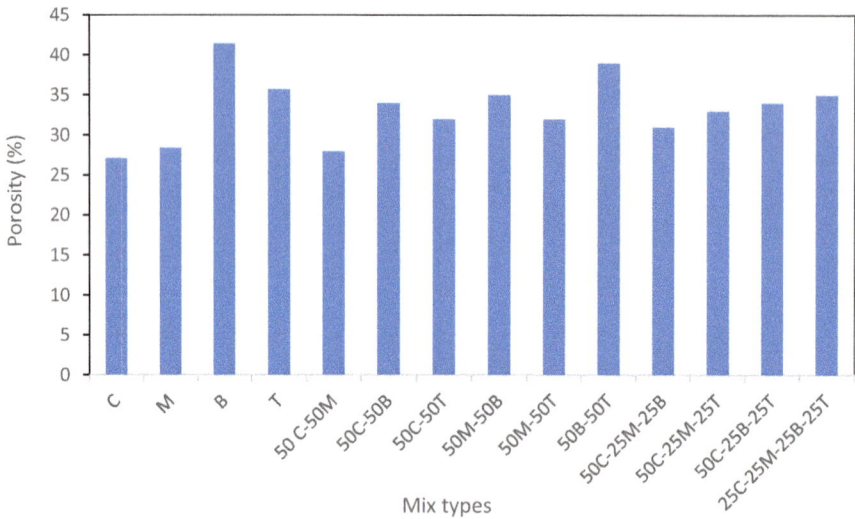

Fig. 4 The porosity values of RFA and mixtures

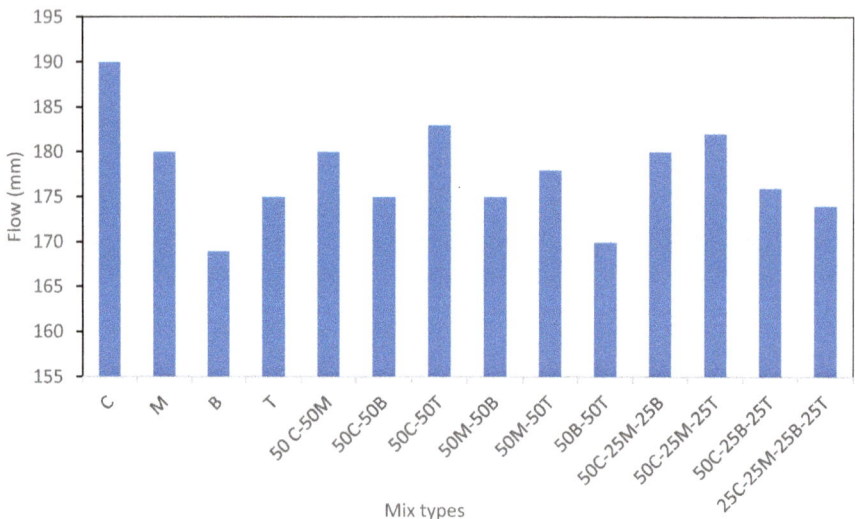

Fig. 5 Effect of RFA and mixtures on the flow of fresh mortar

it may be deduced that tile waste adsorbed and retained water on the surface so that its mixture could extend the time for mixing but still maintain good workability and the rate of hydration.

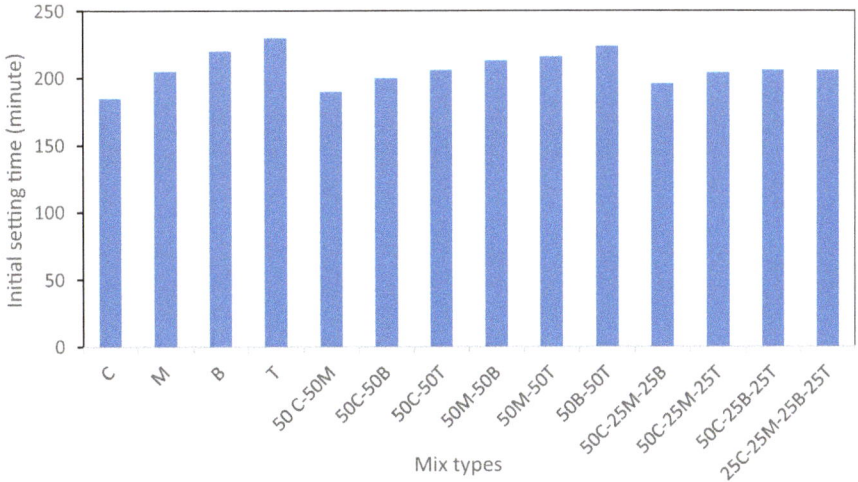

Fig. 6 Effect of RFA and mixtures on the initial setting time of mortar

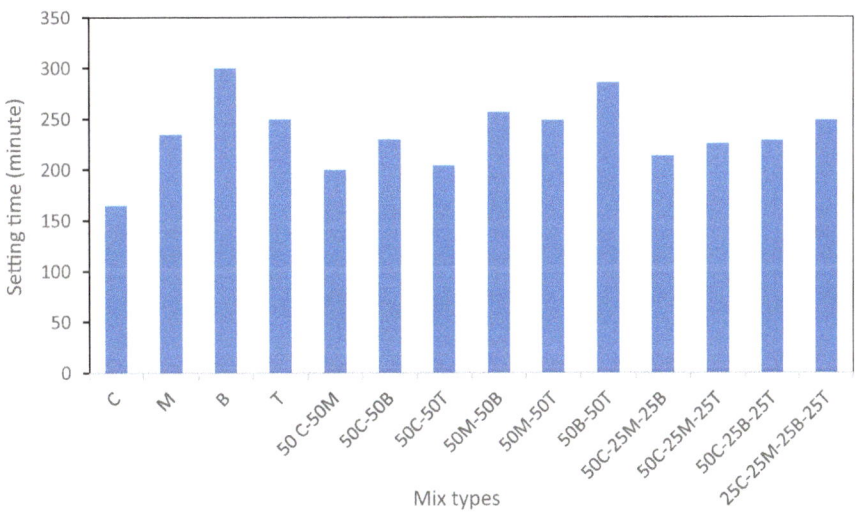

Fig. 7 Effect of RFA and mixtures on the setting time of mortar

3.3　Influence of RFA on Strength Mortar

Figures 8 and 9 show the compressive and flexural strength of hardened mortar with different types of RFA. It is considered that the longer the setting time of the mortar mixture, the lower the strength of the hardened one as comparing with Fig. 7. The longer setting time indicates that there is a retarding in the hydration process, that might cause by the water absorption and retaining mechanism of RFA.

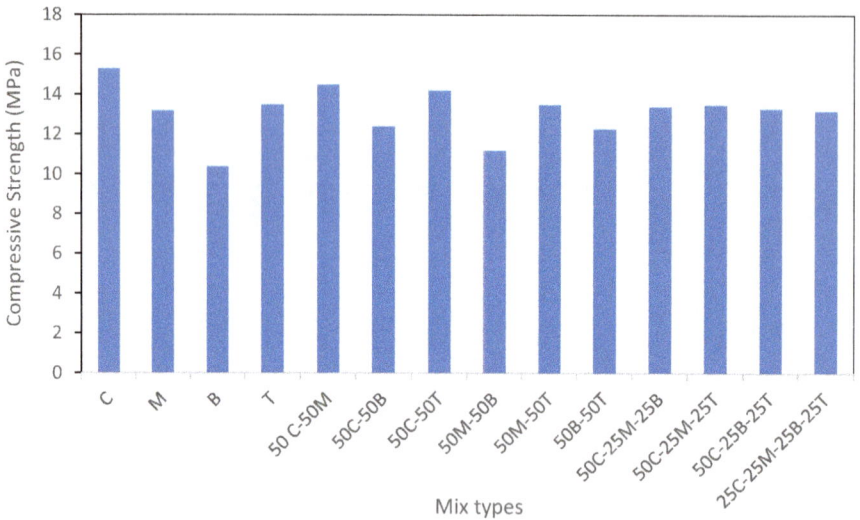

Fig. 8 Effect of RFA and mixtures on the compressive strength of mortar

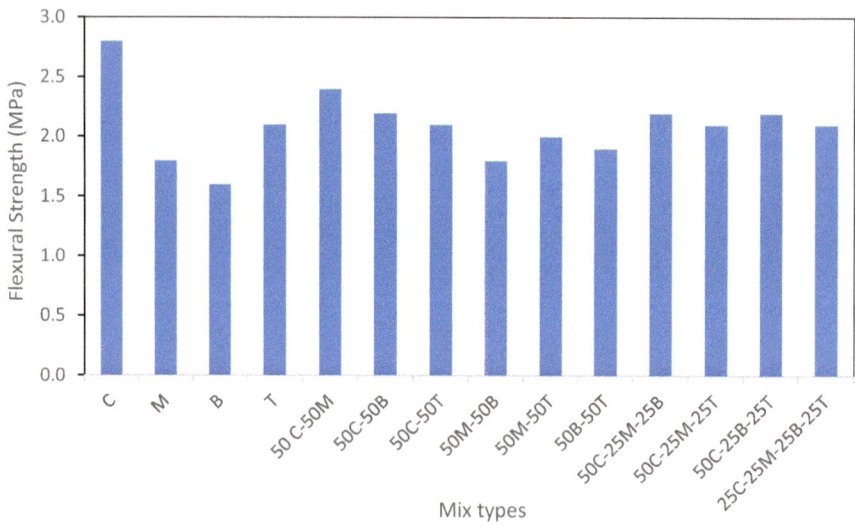

Fig. 9 Effect of RFA and mixtures on the flexural strength of mortar

In summary, mixtures with concrete waste or contain concrete waste show prominent characteristics and properties in a mortar among all. Besides, mixtures with brick waste or 50% of brick waste mostly have properties in fresh and hardened mortar below the average.

4 Conclusion

This paper is to perform research on the effect of RFA and mixtures on the properties of fresh and hardened mortar. Various mixtures with different types of RFA were manufactured for laboratory experiments to figure out their fresh and hardened properties. Based on the testing results and evaluations, some conclusions are drawn. It is suggested that concrete waste is the most qualified demolition material using as a fine aggregate in mortar due to the high fineness modulus, low porosity, which lead to good workability and high strength of mortar. Brick waste showed the unqualified attributes and recommended blending less than 50 percent in the mixture of RFA. Mortar and tile wastes are at the average and able to mix with concrete waste to diversify recycled material inputs. In addition, tile waste could be combined with other RFA to prolong the initial setting time for fresh mixtures.

The characteristics of RFA, the properties of fresh and hardened mortar with RFA and mixtures showed rational relations. It indicates the confidence of these findings and the applicability to further research for fine aggregate in concrete.

Acknowledgments This research was funded by Vietnam National University Ho Chi Minh City (VNU-HCM) under grant number C2021-76-02.

References

1. Ngo KT, Tran HS, Le VP, Nguyen XH, Nguyen TK, Vu VH, Tran VC (2018) Study on current situation of construction and demolition waste management in Vietnam. STCE J - NUCE 12(7):107–116
2. Hoornweg D, Bhada-Tata P (2012) What a waste: a global review of solid waste management. World Bank, Washington, DC
3. Malia M, Brito J, Pinheiro MD, Bravo M (2013) Construction and demolition waste indicators. Waste Manag Res 31(3):241–255
4. Eguchi K, Teranishi K, Nakagome A, Kishimoto H, Shinozaki K, Narikawa M (2007) Application of recycled coarse aggregate by mixture to concrete construction. Constr Build Mater 21:1542–1551
5. Park SS, Kim SJ, Chen K, Lee YJ, Lee SB (2018) Crushing characteristics of a recycled aggregate from waste concrete. Constr Build Mater 160:100–105
6. Velay-Lizancos M, Martinez-Lage I, Azenha M, Granja J, Vazquez-Burgo P (2018) Concrete with fine and coarse recycled aggregates: e-modulus evolution, compressive strength and non-destructive testing at early ages. Construct Build Mater 193:323–331
7. Behera M, Minocha A, Bhattacharyya S (2019) Flow behavior, microstructure, strength and shrinkage properties of self-compacting concrete incorporating recycled fine aggregate. Construct Build Mater 228:116819
8. Evangelista L, Guedes M (2019) Microstructural studies on recycled aggregate concrete. In: New trends in eco-efficient and recycled concrete, Woodhead publishing series in civil and structural engineering, pp. 425–451
9. ASTM C33-18 (2018) Standard specification for concrete aggregates. ASTM International, West Conshohocken, PA
10. ASTM C807–08 (2008) Standard test method for time of setting of hydraulic cement mortar by modified vicat needle. ASTM, West Conshohocken, PA

Effect of Duration and Temperature of Initial Curing on Compressive Strength of Cement Paste Containing Fly Ash Activated by Sodium Sulfate

My Nhan Le, Thanh Hang Dang, Phuong Quyen Nguyen Dinh, and Phuong Trinh Bui

Abstract Fly ash (FA) is considered as a supplementary cementitious material that can help enhance long-term mechanical properties and durability of concrete. However, the compressive strength at early ages of FA concrete is low owing to slow pozzolanic reaction of FA. To overcome this problem, the present paper focuses on studying duration and temperature of initial curing for cement paste containing FA activated by sodium sulfate (Na_2SO_4) in order to promote the pozzolanic reaction. Three replacements of FA were 0, 20, and 40% by mass of binder while the Na_2SO_4 contents were 0 and 4% by mass of binder. After casting for 24 h, the cement pastes were cured at various temperatures (i.e., 27 ± 2, 40, and 80 °C) for 3 h and 6 h in a drying oven. The results showed the use of Na_2SO_4 as an activator increased compressive strength of the pastes regardless of the FA replacements and curing ages. Initial temperature curing reduced the compressive strength at the age of 28 days of the pastes with FA activated by Na_2SO_4. Long duration of high temperature curing negatively affected the compressive strength at the age of 28 days of the pastes. Consequently, long duration and high temperature of initial curing reduced the compressive strength at later ages of the cement pastes with FA activated by Na_2SO_4.

Keywords Activation · Compressive strength · Fly-ash cement paste · Initial curing · Sodium sulfate

1 Introduction

The use of fly ash (FA) released from coal-fired power plants for concrete production is gradually increasing in the modern construction industry. The FA is often used

M. N. Le · T. H. Dang · P. Q. N. Dinh · P. T. Bui (✉)
Faculty of Civil Engineering, Ho Chi Minh City University of Technology (HCMUT), 268 Ly Thuong Kiet Street, District 10, Ho Chi Minh, Vietnam
e-mail: buiphuongtrinh@hcmut.edu.vn

Vietnam National University Ho Chi Minh City, Linh Trung Ward, Thu Duc District, Ho Chi Minh, Vietnam

as a supplementary cementitious material to replace cement and to improve some properties of the concrete. However, using FA to replace a part of cement makes the compressive strength at the early ages of the concrete lower when compared with the traditional concrete [9]. It was reported that the compressive strength at the age of 7 days of cement pastes with FA was lower than that of cement pastes without FA [9]. When low-calcium FA replacements were 50 and 60% by mass, the compressive strength of the cement pastes at the ages of 7, 28, and 90 days was lower than that of the cement pastes without FA [7]. It results from the slow pozzolanic reaction of FA in cement concrete [1, 3].

To overcome the above drawback, Lee et al. [8] revealed that the chemical activators play important roles in promoting the pozzolanic reaction of FA in cement concrete. The authors conducted the experiments to enhance the compressive strength of FA cement pastes with the addition of chemical activators such as sodium sulfate (Na_2SO_4). With the addition of Na_2SO_4 at 0 and 4% by mass, the compressive strength of cement pastes with 40% FA replacement increased and was generally equal to that of cement pastes without FA at the age of 28 days [8]. A comparison of the compressive strength between the FA cement pastes (cement pastes with 40% FA replacements, denoted by Fa40) and the plain cement paste without FA (denoted by Fa0) was carried by Bui et al. [2] when adding Na_2SO_4 to the pastes. The addition of 4% Na_2SO_4 resulted in a higher increase in the compressive strength of the cement pastes with 20 and 40% FA replacements (denoted by Fa20 and Fa40, respectively) when compared with that of the plain cement paste (Fa0) at early ages [2].

On the other hand, a study of the influence of various curing temperatures (10, 20, 50, and 80 °C) on compressive strength at the ages of 1, 3, 7, and 28 days of the high-performance concrete was carried by Flatr et al. [6]. The experimental results showed that the compressive strength of the concrete increased when the curing temperature increased. Nevertheless, the compressive strength of concrete cured at 80 °C at the ages of 7, 14, and 28 days was lower than that of the others cured at 20 and 50 °C [6].

In brief, most of existing studies have focused on the effect of Na_2SO_4 activator on properties of the cement pastes cured at ambient temperature, but not at high temperature. Therefore, the aim of this study was to investigate the effect of duration and temperature of initial curing on the compressive strength of cement paste with FA activated by Na_2SO_4.

2 Experimental

2.1 Materials

Type I Portland cement supplied by Ha Tien cement factory and Class F FA were used as cementitious materials. The physical properties of Type I Portland cement

Table 1 Physical properties of Type I Portland cement and FA

Physical properties		Testing methods	Cement	FA
The compressive strength (MPa)	3 days	ASTM C109/C109M-16a	20.3	–
	7 days		27.0	–
Setting time (min)	Initial time	TCVN 6017:1995	180	–
	Final time		210	–
Percentage retained on 0.090-mm sieve (%)		TCVN 4030:2003	0.1	–
Percentage retained on 0.045-mm sieve (%)			–	3.8
Density (g/cm^3)		TCVN 4030:2003	3.1	2.2

– : not measured

Table 2 Chemical compositions (% by mass) of Type I Portland cement and Class F FA

Compositions	SiO_2	Fe_2O_3	Al_2O_3	CaO	K_2O	Na_2O	SO_3	MgO	LOI
Cement	21.98	2.53	4.3	62.57	0.01	0.37	1.89	1.56	2.89
FA	51.6	9.3	20.9	0.4	–	–	–	1.4	5.8

LOI: loss on ignition; –: not measured

and FA are presented in Table 1. The chemical compositions of the cementitious materials are listed in Table 2.

In addition, tap water conforming to TCVN 4506:2012 was used to produce cement paste. Sodium sulfate (Na_2SO_4) was used in the study as an activator. Density of Na_2SO_4 was 2.66 g/cm^3.

2.2 Mixture Proportions

A low water-to-binder ratio of 0.3 was selected to improve the initial compressive strength of FA cement paste. The replacement ratios of Portland cement by FA were 0 (Fa0), 20% (Fa20), and 40% (Fa40) by mass. The dosages of Na_2SO_4 were 0 and 4% by mass of cementitious materials, based on the study of Bui et al. [2]. Table 3 shows the mass proportions of all pastes in this study.

2.3 Mixing Procedure

A mixing procedure was carried out to mix the compounds. First, a chemical solution was made from Na_2SO_4 activator and water. Second, a dry mixture of cement and FA was made at an extremely slow speed in a lab mixer. Final, water or the chemical solution was added to the dry mixture for mixing to obtain the paste.

Table 3 The mass proportions of all pastes

Mixture proportion	Water	Cement	FA	Na_2SO_4
Fa00Na0	0.30	1.00	0.00	0.00
Fa00Na4	0.30	1.00	0.00	0.04
Fa20Na0	0.30	0.80	0.20	0.00
Fa20Na4	0.30	0.80	0.20	0.04
Fa40Na0	0.30	0.60	0.40	0.00
Fa40Na4	0.30	0.60	0.40	0.04

2.4 Specimen Preparation

After mixing, the pastes were immediately cast in 50-mm cube-shaped molds in laboratory conditions. After 24 h of casting, the pastes (including Fa0 and Fa40) were cured at: (1) 40 °C and (2) 80 °C for 3 or 6 h in a laboratory oven. After initial temperature curing for 3 or 6 h, the pastes were cured at 27 ± 2 °C until testing the designated compressive strength. Additionally, the reference pastes (including Fa0, Fa20, and Fa40) were cured at 27 ± 2 °C to evaluate the effects of high temperature curing on the compressive strength of FA cement pastes with and without Na_2SO_4.

2.5 Compressive Strength Test

A hydraulic compression machine was used for testing the compressive strength of the cement pastes at testing ages (i.e., at 1, 3, 7, and 28 days) as per TCVN 6016:2011. The value of compressive strength of each mixture proportion at each age was average value of three corresponding specimens.

3 Results and Discussion

3.1 Effect of Na₂SO₄ on Compressive Strength Development

Effects of the addition of Na_2SO_4 activator on the compressive strength of both plain cement paste and FA cement paste are shown in Fig. 1. Overall, the compressive strength of all cement pastes increased with time, regardless of FA replacement and Na_2SO_4 addition. It confirms that the cement hydration and the reactions in the paste with and without FA replacement continuously proceeded. Additionally, the compressive strength of Fa40Na0 specimen was lower than that of Fa20Na0, and that of Fa20Na0 was also lower than that of Fa00Na0 at 3 and 7 days. It indicates that the increases in FA replacement reduced the compressive strength at the early

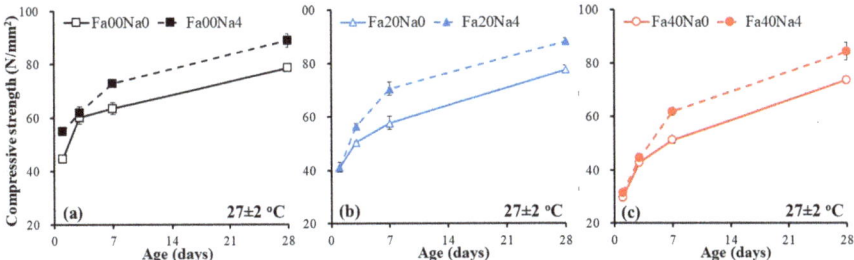

Fig. 1 Effect of Na$_2$SO$_4$ on compressive strength development of cement pastes with 0 **a** , 20% **b**, and 40% **c** FA replacements and cured at 27 ± 2 °C

ages (i.e., at 1 and 3 days). This tendency was also found in a previous study by Baert et al. [3]. The slow pozzolanic reaction of FA at early ages could be a reason for the reduction of the compressive strength [4] and a low cement content in the FA cement paste. Nevertheless, the compressive strength of Fa20Na0 specimen was generally equal to that of Fa00Na0 and even slightly higher than that of Fa40Na0 at 28 days. Duran-Herrera et al. [5] also found that the difference between plain cement pastes and FA cement pastes was potentially due to the pozzolanic reaction of FA that happened in such cement pastes at 28 days.

An existing study reported that the addition of Na$_2$SO$_4$ from 1 to 4% by mass remarkably increased the compressive strength of FA cement paste with a water-to-binder ratio of 0.485 at early ages (i.e., at 1 and 3 days) [8]. In this study, when compared with the cement paste with 0% Na$_2$SO$_4$, the compressive strength of Fa00Na4 specimen was slightly higher at 1, 3, and 7 days and notably higher at 28 days (see Fig. 1 (a)). There was also slightly higher compressive strength of Fa20Na4 and Fa40Na4 specimens than that of Fa20Na0 and Fa40Na0 specimens at the early ages (i.e., at 1 and 3 days) and notably higher at 7 and 28 days (see Figs. 1 (b) and (c)), respectively. The increases the compressive strength of Fa20Na4 and Fa40Na4 specimens were more significant than that of Fa0Na4 specimen from the age of 1 to 7 days. It implies that the addition of 4% Na$_2$SO$_4$ to Fa20 and Fa40 specimens led to the higher increase in compressive strength compared with Fa0 specimen at early ages. It is explained that the addition of Na$_2$SO$_4$ to the cement pastes contributes to early formation of ettringite which filled up the voids and thereby increased the compressive strength of the cement pastes, as well as promoted the pozzolanic reaction of FA [2].

3.2 Effect of Initial Temperature Curing on Compressive Strength

Figures 2 (a) and (b) show the effects of various initial temperatures curing (i.e., at 27 ± 2, 40, and 80 °C) on compressive strength of cement pastes with 0 and 4%

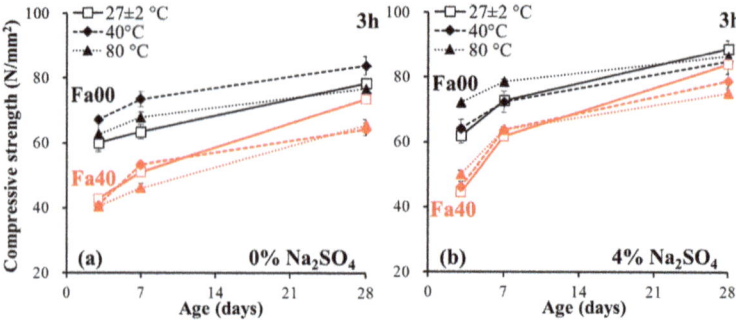

Fig. 2 Effect of initial temperature curing on compressive strength of cement pastes with 0 **a** and 4% **b** Na₂SO₄ and cured at various temperatures for 3 h

Na₂SO₄ for 3 h, respectively. From the Fig. 2 (a), the compressive strength of plain cement pastes with 0% Na₂SO₄ (Fa00Na0) cured at 40 or 80 °C was higher than that of corresponding pastes cured at 27 ± 2 °C at the ages of 3, 7, and 28 days, except the lower compressive strength of the paste cured at 80 °C at 28 days. Meanwhile, the compressive strength of Fa40Na0 specimens cured at 40 and 80 °C were nearly the same at 3 days when compared with the corresponding pastes cured at 27 ± 2 °C. The compressive strength of such specimens cured at 40 and 80 °C was significantly lower when compared with that cured at 27 ± 2 °C at the age of 28 days. Additionally, the other cement pastes with 4% Na₂SO₄ (i.e., Fa00Na4 and Fa40Na4) cured at 40 or 80 °C had the slightly higher compressive strength than that of corresponding pastes cured at 27 ± 2 °C at 3 and 7 days but slightly lower at 28 days, as seen in Fig. 2 (b). These tendencies were also observed in case of the pastes cured at high temperatures for 6 h, as seen in Figs. 3 (a) and (b). It is explained that the high temperature not only promoted the cement hydration but also caused crack formations in the matrix [10]. Additionally, another reason is that high temperature curing increased the rate

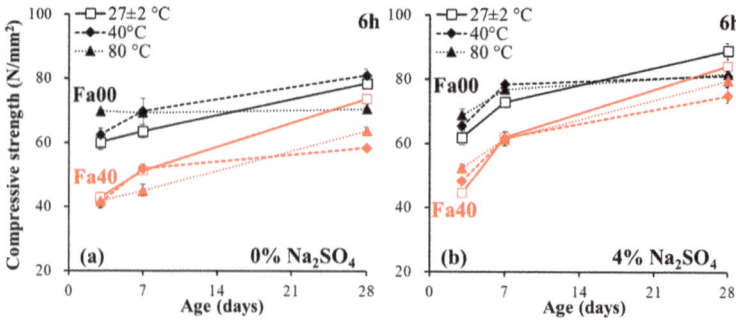

Fig. 3 Effect of initial temperature curing on compressive strength of cement pastes with 0 **a** and 4% **b** Na₂SO₄ and cured at various temperatures for 6 h

of cement hydration process, leading to formation of a thick layer on the surface of the cement pastes and thereby preventing the hydration process [10].

Regarding Fa40Na0 specimens, the compressive strength of the specimens cured at 27 ± 2, 40, and 80 °C for 3 and 6 h was nearly the same at 3 days, as seen in Figs. 2 (a) and 3 (a). It indicates that the negative effects of crack formation in the pastes cured at high temperature could be balance to the acceleration of cement hydration caused by high temperature.

3.3 Effect of Duration of Initial Curing on Compressive Strength

Effects of duration of initial curing on the compressive strength of cement pastes at the ages of 3, 7, and 28 days are shown in Figs. 4 and 5. The compressive strengths of plain cement paste and cement paste with 40% FA replacement cured at 40 °C for 3 h were slightly lower than those of corresponding pastes cured at 40 °C for 6 h at 3 days, but it showed a significantly higher compressive strength from 7 to 28 days, except Fa00Na0 specimen. Regarding Fa00Na0 specimens, the compressive strength of the pastes cured at 40 °C for 3 h was higher than that of the pastes cured at 40 °C for 6 h not only at the early ages (i.e., at 3 and 7 days) but also at 28 days

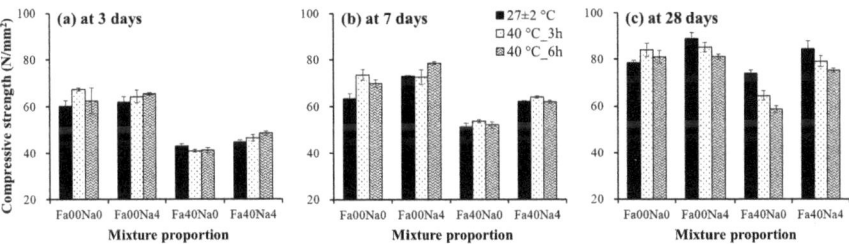

Fig. 4 Effect of duration of initial curing on compressive strength of cement pastes at 3 **a**, 7 **b**, and 28 **c** days and cured at 27 ± 2 and 40 °C

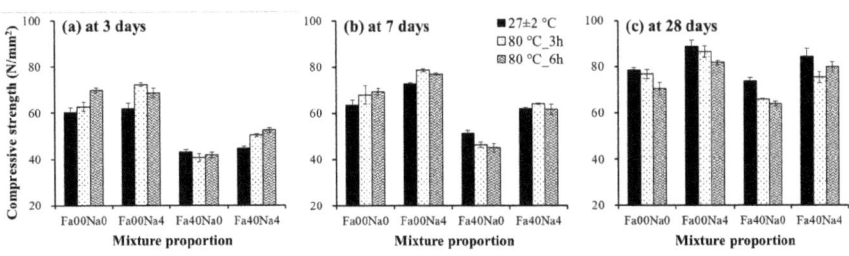

Fig. 5 Effect of duration of initial curing on compressive strength of cement pastes at 3 **a**, 7 **b**, and 28 **c** days and cured at 27 ± 2 and 80 °C

(see Fig. 4). Similarly, for curing at 80 °C, the compressive strength of cement pastes (i.e., Fa00Na0, Fa40Na0, and Fa40Na4) cured for 3 h were slightly lower than that of corresponding cement pastes cured for 6 h at 3 days but that increased remarkably from 7 to 28 days. It indicates that long duration of high temperature curing negatively affected the compressive strength at the age of 28 days of the pastes regardless of Na_2SO_4 addition.

4 Conclusions

High temperature curing contributed to an increase in compressive strength of the cement pastes with and without FA at the early ages but reduced the compressive strength of such pastes at the later ages (i.e., at 28 days).

Long duration of high temperature curing negatively affected the compressive strength at the age of 28 days of the pastes with and without Na_2SO_4.

Consequently, long duration and high temperature of initial curing led to the reduction of the compressive strength at later ages of the cement pastes with FA activated by Na_2SO_4. A further microstructure analysis of the cement paste with FA activated by Na_2SO_4 should be done to clarify the interactive reactions in such pastes cured at high temperature.

Acknowledgements The authors acknowledge the support of time and facilities from Ho Chi Minh City University of Technology (HCMUT), VNU-HCM for this study.

References

1. Bui PT, Ogawa Y, Nakarai K, Kawai K (2016) Effect of internal alkali activation on pozzolanic reaction of low-calcium fly ash cement paste. Mater Struct 49(8):3039–3053
2. Bui PT, Ogawa Y, Kawai K (2020) Effect of sodium sulfate activator on compressive strength and hydration of fly-ash cement pastes. J Mater Civ Eng 32(6):04020117
3. Baert G, Poppe AM, De Belie N (2008) Strength and durability of high-volume fly ash concrete. Struct Concr 9(2):101–108
4. Berry EE, Hemmings RT, Cornelius BJ (1990) Mechanisms of hydration reactions in high volume fly ash pastes and mortars. Cem Concr Compos 12(4):253–261
5. Durán-Herrera A, Juárez CA, Valdez P, Bentz DP (2011) Evaluation of sustainable high-volume fly ash concretes. Cem Concr Compos 33(1):39–45
6. Flatr J, Broukalova I (2019) Influence of curing temperature on the mechanical properties of high- performance concrete. IOP Conf Ser Mater Sci Eng 583:012011
7. Lam L, Wong YL, Poon CS (2000) Degree of hydration and gel/space ratio of high-volume fly ash/cement systems. Cem Concr Res 30(5):747–756
8. Lee CY, Lee HK, Lee KM (2003) Strength and microstructural characteristics of chemically activated fly ash-cement systems. Cem Concr Res 33(3):425–431

9. Teja KK, Rao BK (2018) Strength and durability of high volume fly ash concrete. Inter J Civ Eng Tech 9(6):109–116
10. Wang M, Hu Y, Wang Q, Tian H, Liu D (2019) A study on strength characteristics of concrete under variable temperature curing conditions in ultra-high geothermal tunnels. Constr Build Mater 229:116989

Effect of Molarity of Potassium Hydroxide Solution on Heat Resistance of Fly Ash-Slag Based Geopolymer

Apriany Saludung, Takumu Azeyanagi, Yuko Ogawa, and Kenji Kawai

Abstract Designing a building material with high fire resistance is crucial for minimizing the impact of catastrophic fire. Geopolymer, a new type of binder synthesized by the reaction of aluminosilicate sources with an alkaline solution, has been extensively studied as it offers comparable or even high durability when compared to ordinary Portland cement (OPC). The concentration of the alkaline solution is one of the important factors that affect the durability; however, there is very little data on the heat resistance of geopolymer activated with potassium hydroxide. Thus, this study aims to investigate the effect of molarity of potassium hydroxide on the heat resistance of fly ash-slag blended geopolymer paste. The potassium hydroxide (10, 12, and 14 M) and sodium silicate solutions were prepared as activator solutions. The initial heat curing at 70 °C for 24 h was applied. In addition, an OPC specimen was also prepared for comparison. After curing for 28 days, some specimens were exposed to high temperatures (500, 750, and 950 °C). The compressive strength of unexposed and exposed geopolymer and OPC paste specimens was measured. The results showed that the increase of molarity of potassium hydroxide solution increased the heat resistance of geopolymer in terms of retained compressive strength. Moreover, the geopolymer specimens could maintain higher compressive strength than that of OPC paste when exposed to high temperatures up to 950 °C. This is attributed to the mineral composition of geopolymer, which is different from OPC paste, as studied by X-ray diffraction analysis.

Keywords Geopolymer · Potassium hydroxide · Molarity · Compressive strength · Heat resistance

A. Saludung · T. Azeyanagi
Department of Civil and Environmental Engineering, Graduate School of Engineering, Hiroshima University, 1-4-1 Kagamiyama, Higashihiroshima City, Japan

Y. Ogawa · K. Kawai (✉)
Civil and Environmental Engineering Program, Graduate School of Advanced Science and Engineering, Hiroshima University, 1-4-1 Kagamiyama, Higashihiroshima City, Japan
e-mail: kkawai@hiroshima-u.ac.jp

1 Introduction

Geopolymer material produced by the reaction of aluminosilicate sources with alkaline solutions has emerged as an alternative binder to ordinary Portland cement (OPC), essentially because of low energy consumption and low carbon dioxide emission when using industrial by-products in its production. The by-products used in the synthesis are mainly fly ash and ground granulated blast furnace slag (GGBS). Geopolymer produced from only fly ash is associated with a very long setting time and low compressive strength; therefore, the addition of slag is crucial to improve these properties [15]. The reaction mechanism of geopolymerization has been explained by some authors. The geopolymerization process starts with the dissolution of the solid aluminosilicate source by alkaline hydrolysis that releases $[SiO_4]^-$ and $[AlO_4]^-$ tetrahedral units. These tetrahedral units are linked with each other by sharing oxygen atoms to form polymeric bonds. The last process involves the polycondensation, where the geopolymer gel sets and forms a three-dimensional aluminosilicate network [4, 6, 18]. In fly ash/slag blends, the C-S-H phase and aluminosilicate gel co-exist in the system, which contributes to strength improvement [15].

On the other hand, designing buildings with good fire resistance is required in many sectors due to the experiences of fire disasters. Even though many buildings have been constructed with non-combustible materials such as brick, concrete, and steel, they cannot completely prevent the outbreak of fire and the building from collapsing. Some buildings are equipped or covered with heat-insulating materials, which can be very expensive. Geopolymer materials possess the potential as a fire-retardant due to their ceramic-like properties [3]. As shown by several previous studies, geopolymer material can maintain high compressive strength after exposure to high temperatures compared to OPC [8, 17, 20]. For instance, Kong and Sanjayan [8] compared the compressive strength of OPC paste with fly ash geopolymer paste after exposure to various temperatures. They found that OPC paste almost totally lost its compressive strength after being heated at 400 °C, while geopolymer paste exhibited high compressive strength after exposure to a temperature of 800 °C, which was 5.4% higher than its initial strength at room temperature. Another study by Zhang et al. [20] also compared the compressive strength of OPC and geopolymer pastes. They found that OPC paste completely lost its compressive strength after exposure to 800 °C, while geopolymer still maintained its compressive strength after exposure to this temperature. The reason behind this is because of the re-crystallization of geopolymer at a very high temperature which leads to the formation of the analogies to natural minerals having extremely high fire resistance [5]. Moreover, some studies also claimed that geopolymer could maintain its strength at high temperatures due to the sintering effect [7, 10].

The fire resistance of geopolymer is affected by several factors, such as the type of raw materials [9, 14, 20], curing conditions [1], and alkali cation types [1, 11]. The combination of sodium silicate and sodium hydroxide is the most common activator solution used in geopolymer production. Although many studies have been conducted on the heat resistance of geopolymers, there is very little data on the heat resistance of geopolymer activated with potassium hydroxide (KOH). In addition, the molarity of alkali hydroxide solution can significantly affect the properties of geopolymer [2, 15]. Thus, this study aims to investigate the effect of the molarity of KOH solution on the heat resistance of fly ash-slag blended geopolymer paste. As a comparison, a heat resistance test on OPC paste was also conducted.

2 Materials and Method

2.1 Materials and Specimen Preparation

This study utilized fly ash (Blaine fineness $= 3{,}550$ cm^2/g, density $= 2.24$ g/cm^3) which conforms to JIS A 6201, GGBS (Blaine fineness $= 4{,}170$ cm^2/g, density $= 2.91$ g/cm^3) which conforms to JIS A 6206, and OPC (Blaine fineness $= 3{,}340$ cm^2/g, density $= 3.16$ g/cm^3). Their chemical compositions are shown in Table 1. Sodium silicate solution consisting of 13.4 mass% Na_2O, 27.3 mass% SiO_2, and 59.3 mass% H_2O, and potassium hydroxide solution were utilized as alkaline solutions. The potassium hydroxide solutions were prepared in 10, 12, and 14 M by dissolving the potassium hydroxide pellets in water to form one liter solution. These solutions were prepared one day before mixing. The ratio of alkaline liquid to binder (l/b) and the ratio of sodium silicate to potassium hydroxide (SS/KH) were maintained constant at 0.52 and 1.35 by mass, respectively. The detail of the mix proportion is shown in Table 2.

The geopolymer paste was synthesized as follows: fly ash and GGBS were first dry-mixed together in a pan mixer for 3 min. Alkaline solutions were added to the dry materials and mixed rapidly and continuously for another 3 min. Subsequently, the fresh geopolymer paste was cast into cylindrical plastic molds ($\Phi50 \times 100$ mm).

Table 1 Chemical composition of raw materials. LOI is loss of ignition at 1000 °C

Component (mass%)	SiO_2	Al_2O_3	Fe_2O_3	CaO	K_2O	TiO_2	MgO	Na_2O	SO_3	LOI
Fly ash	64.44	20.65	4.18	2.25	1.53	1.19	0.58	N/A	N/A	2.9
GGBS	35.45	14.06	0.27	43.78	0.23	0.56	5.84	0.24	0.62	0.05
OPC	19.89	5.19	3.07	64.79	0.36	N/A	1.26	0.31	1.95	2.65

Table 2 Mix proportion

Sample ID	Fly ash (g)	GGBS (g)	OPC (g)	Alkali solution/binder ratio	Na_2SiO_3/KOH	Water/cement ratio	Na_2SiO_3 solution (g)	KOH solution (g)	Water (g)
Geo-10 M	385	315	0	0.52	1.35	–	210	156	–
Geo-12 M	385	315	0	0.52	1.35	–	210	156	–
Geo-14 M	385	315	0	0.52	1.35	–	210	156	–
OPC paste	0	0	700	–	–	0.45	0	0	315

On the other hand, OPC paste was prepared with a water to cement ratio of 0.45 in accordance with the above mixing procedure. Just after casting, both OPC and geopolymer specimens were cured in an oven at 70 °C for 24 h under sealed condition. After oven curing, the specimens were stored at ambient temperature (approximately 20°C), demolded after aging for three days, and stored in a sealed plastic bag to prevent alkalinity loss or carbonation before the tests.

2.2 Heat Resistance Test

After curing for 28 days, specimens were exposed to high temperatures at 500, 750, and 950 °C by placing them into an electric furnace with a heating rate of 20 °C/min. When the targeted temperature was reached, it was maintained for 1 h. Afterward, the furnace was turned off and let it cool down to room temperature. Some specimens were left unexposed for comparison.

2.3 Characterization and Measurement

The compressive strength of unexposed and exposed geopolymer paste and OPC paste specimens at the age of 28 days was evaluated following JIS R 5201. It was conducted using a universal testing machine with a compression capacity of 250 kN and a loading rate of 0.2 mm/min. The test was repeated for three specimens for each mixture, and an average was taken.

The phase of geopolymer and OPC product was characterized by X-ray diffraction (XRD). It was conducted using an X-ray diffractometer in the 2-theta range of 10–65° with a Cu Kα X-ray source.

The microstructure of unexposed and exposed geopolymer specimens was examined using scanning electron microscopy (SEM). It was conducted at 15 kV of accelerated voltage. The samples were coated with platinum.

3 Results and Discussion

3.1 Compressive Strength

The compressive strength of the specimens unexposed and exposed to high temperatures was measured and compared. Figure 1(a) shows that the increase of KOH molarity leads to the increase of initial compressive strength. The improvement is attributed to the fact that the increase in molarity increases the dissolution rate of the aluminosilicate precursors due to the high pH of high concentration of alkali solution

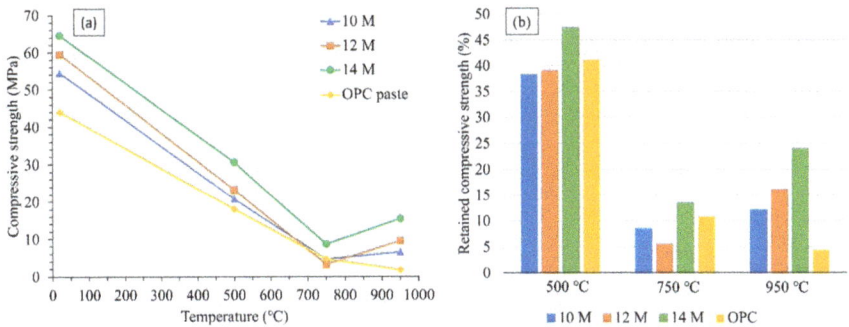

Fig. 1 Compressive strength of geopolymer paste as a function of temperature **a** and retained compressive strength after heat exposure **b**

[4, 12], resulting in more gel formation and denser microstructure. In addition, it can be seen that the compressive strength of all the geopolymer specimens is higher than that of OPC paste.

After exposure to high temperatures, all the specimens exhibited a significant loss of compressive strength due to the decomposition of C-(A)-S-H gel. Similar to C-S-H in OPC paste, C-(A)-S-H in geopolymer is the main source of compressive strength development, and therefore, the decomposition of C-(A)-S-H in geopolymer and C-S-H in OPC paste will lead to a significant loss of compressive strength. Further heating at a higher temperature, all the specimens continuously lost their strength; however, when geopolymer specimens were heated up to 950 °C, the compressive strength slightly increased and became higher than the retained compressive strength of specimens after exposure to 750 °C.

Increasing the molarity improves the heat resistance of geopolymer. Figure 1(b) shows the percentage of retained compressive strength of geopolymer and OPC paste. Although the three series of geopolymer present a similar tendency of strength loss, the geopolymer with 14 M KOH shows the highest retained compressive strength corresponding to approximately 24% of its initial strength after heating at 950 °C. In contrast, the OPC specimen retains only 4.5% of its initial compressive strength after heating at 950 °C.

3.2 Phase Changes

Figure 2 represents the XRD patterns of unexposed and exposed OPC paste and geopolymer paste. Only the XRD pattern of geopolymer activated with 14 M KOH was shown since all the geopolymer specimens present similar phase changes after exposure to high temperatures. It can be seen in Fig. 2(a) that OPC paste mainly consists of portlandite, with some low-intensity peaks of C_2S (belite), C_3S (alite), ettringite, and calcite. At 500 °C, the portlandite peak was still observable, and it

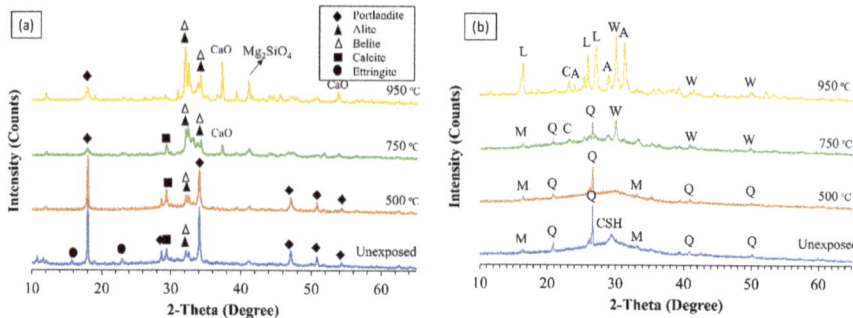

Fig. 2 XRD patterns of unexposed and exposed OPC pastes **a**, and geopolymer paste activated with 14 M KOH **b**. (Q = quartz, M = mullite, C = cristobalite, W = wollastonite, A = akermanite, L = leucite)

started to decrease considerably at 750 °C. At 750 °C, a CaO peak was formed, and considerable amounts of C_2S and C_3S were observed. Calcite peak disappeared at 950 °C, which indicates a complete decomposition of calcite at a temperature between 750 and 950 °C. The peak intensity of CaO and C_3S increased considerably with the increase of exposure temperature at 950 °C. This is attributed to the further decomposition of portlandite, calcite, and C-S-H.

Figure 2(b) represents the XRD pattern of geopolymer paste (14 M). In the unexposed specimen, the mullite ($Al_6Si_2O_{13}$) and quartz (SiO_2) peaks are originally from fly ash [16]. After exposure to 500 °C, these peaks remain, while the C-(A)-S-H peak almost vanished due to dehydroxylation. Unlike the OPC specimen, further heating at 750 °C, new crystalline peaks such as wollastonite ($CaSiO_3$) and cristobalite (SiO_2) were observed in geopolymer, proving the re-crystallization of this material into secondary mineral phases. Moreover, at 950 °C, other new crystalline phases such as leucite ($KAlSi_2O_6$) and akermanite ($Ca_2Mg[Si_2O_7]$) were observed. The presence of these two major phases is believed to contribute to the slight increase in the compressive strength after exposure to 950 °C. The effect of molarity on phase change was insignificant because all the geopolymer specimens present similar phase changes at high temperatures; however, increasing the molarity increases the potassium content in the paste, and therefore, the amount of leucite formed after heating at 950 °C probably increased.

3.3 Microstructure

The geopolymer specimen showing the highest retained compressive strength after exposure to high temperatures was also characterized by SEM, and the result is shown in Fig. 3. In the unexposed specimen, it was evident that geopolymeric gel was found to co-exist with the calcium aluminate silicate gel (C-(A)-S-H), resulting in a compact microstructure.

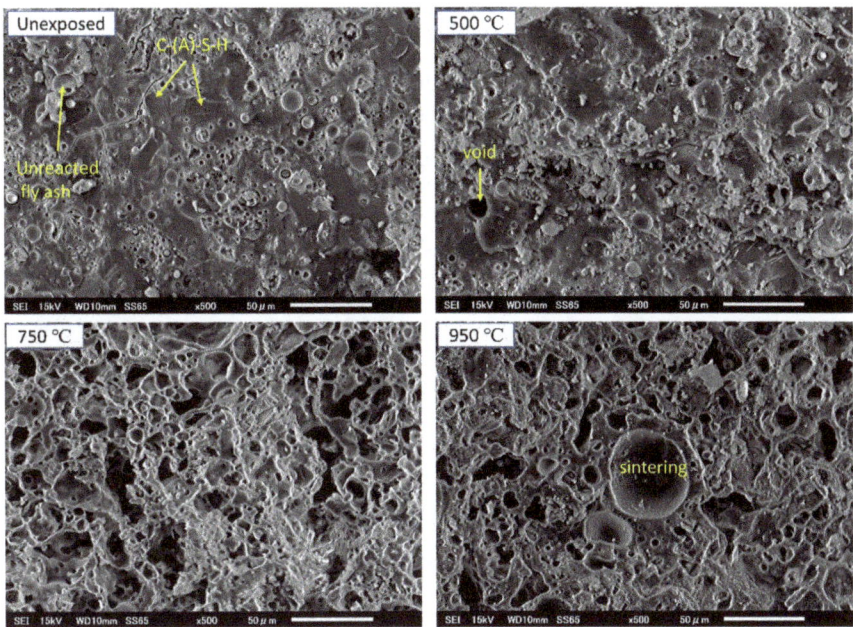

Fig. 3 SEM images of unexposed and exposed geopolymer paste activated with 14 M KOH

After heating at 500 °C, the deterioration occurs due to the impact of dehydration and decomposition of the gels, resulting in a coarser microstructure compared to the unexposed sample. According to previous studies, the dehydration of the C-S-H in hardened OPC commences at approximately 110 °C and becomes completely decomposed at a temperature range of 600–800 °C [13, 19]. The dehydration of C-(A)-S-H in geopolymer is estimated to occur at a temperature similar to C-S-H in OPC. Further heating at 750 °C, the microstructure becomes highly porous, resulting in very low compressive strength. However, after heating at 950 °C, there was evidence of sintering. This could be the reason for slightly increasing the compressive strength from 750 to 950 °C of exposure temperature.

4 Conclusion

In this study, the effect of KOH molarity on the heat resistance of geopolymer exposed to 500, 750, and 950 °C was investigated. The following conclusions can be drawn from the experimental results.

1. The increase of molarity of KOH solution increases the initial compressive strength of geopolymer pastes.

After exposure to high temperature, 14 M KOH solution performed the highest retained compressive strength, followed by 12 and 10 M. The compressive strength of all the geopolymer samples exposed to 950 °C was higher than that exposed to 750 °C, which is attributed to the sintering effect and formation of new crystalline phases.

2. In comparison with the OPC paste, geopolymer pastes showed better heat resistance in terms of compressive strength loss. Unlike geopolymer which recrystallized at high temperature, OPC paste presents an increase in the amount of C_2S and C_3S and the formation of CaO as the result of the thermal decomposition of C-S-H and calcium hydroxide.

Geopolymer made from fly ash and GGBS, activated with sodium silicate and potassium hydroxide solutions has been proven to have better performance than OPC paste when exposed to high temperatures,thus promising the use of geopolymer as an alternative material to OPC.

References

1. Bakharev T (2006) Thermal behaviour of geopolymers prepared using class F fly ash and elevated temperature curing. Cem Concr Res 36:1134–1147
2. Budh CD, Warhade NR (2014) Effect of molarity on compressive strength of geopolymer mortar. Int J Civ Eng Res 5:2278–3652
3. Davidovits J (2017) Geopolymers: ceramic-like inorganic polymers. J Ceram Sci Technol 8:335–350
4. Duxson P, Fernández-Jiménez A, Provis JL, Lukey GC, Palomo A, Van Deventer JSJ (2007) Geopolymer technology: the current state of art. J Mater Sci 42:2917–2933
5. Duxson P, Lukey GC, Van Deventer JSJ (2006) Evolution of gel structure during thermal processing of Na-geopolymer gels. Langmuir 22:8750–8757
6. Hajimohammadi A, Provis JL, Van Deventer JSJ (2010) Effect of alumina release rate on the mechanism of geopolymer gel formation. Chem Mater 22:5199–5208
7. Jaya NA, Al Bakri Abdullah MM, Ghazali CMR, Hussain M, Hussin K, Ahmad R (2016) Kaolin Geopolymer as precursor to ceramic formation. In: MATEC Web Conference
8. Kong DLY, Sanjayan JG (2010) Effect of elevated temperatures on geopolymer paste, mortar and concrete. Cem Concr Res 40:334–339
9. Kong DLY, Sanjayan JG, Sagoe-Crentsil K (2007) Comparative performance of geopolymers made with metakaolin and fly ash after exposure to elevated temperatures. Cem Concr Res 37:1583–1589
10. Kuenzel C, Grover LM, Vandeperre L, Boccaccini AR, Cheeseman CR (2013) Production of nepheline/quartz ceramics from geopolymer mortars. J Eur Ceram Soc 33:251–258
11. Lahoti M, Wong KK, Tan KH, Yang EH (2018) Effect of alkali cation type on strength endurance of fly ash geopolymers subject to high temperature exposure. Mater Des 154:8–19
12. Nath SK, Kumar S (2019) Role of alkali concentration on reaction kinetics of fly ash geopolymerization. J Non Cryst Solids 505:241–251
13. Peng GF, Huang ZS (2008) Change in microstructure of hardened cement paste subjected to elevated temperatures. Constr Build Mater 22:593–599
14. Rickard WDA, Temuujin J, Van Riessen A (2012) Thermal analysis of geopolymer pastes synthesised from five fly ashes of variable composition. J Non Cryst Solids 358:1830–1839
15. Saha S, Rajasekaran C (2017) Enhancement of the properties of fly ash based geopolymer paste by incorporating ground granulated blast furnace slag. Constr Build Mater 146:615–620

16. Saludung A, Ogawa Y, Kawai K (2018) Microstructure and mechanical properties of FA/GGBS-based geopolymer. In: MATEC Web Conference. https://doi.org/10.1051/matecconf/201819 501013
17. Saxena SK, Kumar M, Singh NB (2017) Fire resistant properties of alumino silicate geopolymer cement mortars. Mater Today Proc 4:5605–5612
18. Singh B, Ishwarya G, Gupta M, Bhattacharyya SK (2015) Geopolymer concrete: a review of some recent developments. Constr Build Mater 85:78–90
19. Tantawy MA (2017) Effect of high temperatures on the microstructure of cement paste. J Mater Sci Chem Eng 05:33–48
20. Zhang HY, Kodur V, Qi SL, Cao L, Wu B (2014) Development of metakaolin-fly ash based geopolymers for fire resistance applications. Constr Build Mater 55:38–45

Effect of Partial Replacement of Cement by Waste Sludge from Water Supply Plant on Compressive Strength and Water Absorption of Hardened Concrete

Ngoc Duy Vo, Ba Tien Mai, Ngoc Phi Long Duong, Phuong Trinh Bui, Xuan Loc Luu, and Duc Thang Vu

Abstract In this study, effect of partial replacement of cement by waste sludge from water supply plant on compressive strength and water absorption of hardened concrete was investigated to evaluate potential use of such waste for concrete production. After drying at 110 °C in a laboratory oven, the sludge was grounded and sieved to have a particle size of less than 0.14 mm to partially replace cement at levels of 0, 10, 20, and 30% by mass. Mixture proportion of reference concrete with 0% sludge replacement in which a desired slump and compressive strength at 28 days were 6 ± 2 cm and 45 MPa, respectively was designed according to Bolomey-Skramtaev method. All concretes with a water-to-binder ratio of 0.53 were prepared. The results showed that the waste sludge decreased slump of fresh concrete when the mixing water amount of all concrete mixture proportions was kept constant. The higher the waste sludge replacement, the lower the slump of fresh concrete. The partial replacement of cement by waste sludge in a range of 10–30% decreased the compressive strength of hardened concrete by 15.10–47.17% when compared with the reference concrete at the age of 28 days. The sludge replacements at 10 and 20% by mass increased the water absorption by 3.82 and 93.51%, respectively. Consequently, replacing 10% of cement with the sludge for concrete production was beneficial in terms of not only ensuring designed slump, compressive strength, and water absorption at the age of 28 days but also decreasing carbon dioxide emission from cement production and utilizing the waste sludge from the water supply plants towards sustainable development.

N. D. Vo · B. T. Mai · N. P. L. Duong · P. T. Bui (✉) · X. L. Luu
Faculty of Civil Engineering, Ho Chi Minh City University of Technology (HCMUT), 268 Ly Thuong Kiet Street, District 10, Ho Chi Minh City, Vietnam
e-mail: buiphuongtrinh@hcmut.edu.vn

Vietnam National University Ho Chi Minh City, Linh Trung Ward, Thu Duc District, Ho Chi Minh City, Vietnam

D. T. Vu
Saigon Water Corporation, Ho Chi Minh City, Vietnam

© The Author(s), under exclusive license to Springer Nature Singapore Pte Ltd. 2023   419
J. N. Reddy et al. (eds.), *ICSCEA 2021*, Lecture Notes in Civil Engineering 268,
https://doi.org/10.1007/978-981-19-3303-5_35

Keywords Cement replacement · Compressive strength · Hardened concrete ·
Waste sludge · Water absorption

1 Introduction

Concrete is one of the artificial construction materials that have the most important influence on the development of buildings and infrastructure of all countries in
the world [5]. However, traditional concrete production is not an environmentally
friendly process. According to Mahasenan et al., the production of cement used to
make concrete released 2.2 billion tons of carbon dioxide (CO_2) into the environment
in 2016 [5]. On the other hand, the demand for fresh water plays a crucial role and is
gradually increasing to meet human demands. This results in continuous operation
of water supply and treatment plants. During the operation of these plants, a huge
amount of waste sludge seriously affecting the productivity and quality of water is
formed from the water treatment process. An increase in waste sludge has become a
growing concern of the water plants because such waste is often discharged directly
into the environment [3]. Therefore, it is necessary to utilize the waste sludge to
eliminate its negative impacts to society and environment.

Some countries such as China, America, and France have focused on using waste
sludge for construction [2, 4, 6, 15]. Tang et al. made lightweight aggregate by using
waste sludge from the Shihmen Reservoir for concrete and masonry productions and
found that the concrete using lightweight aggregates from such sludge had lower
thermal and electrical conductivity coefficients than the conventional concrete using
traditional lightweight aggregate [6]. In addition to lightweight aggregate production,
the waste sludge was utilized to make Portland cement clinker. The use of sludge
in the production of cement clinker was completely beneficial in terms of strength
development of cement-based mortar and helped to reduce production costs [2].
Zhao et al. also replaced cement by uncontaminated marine sediments for mortar and
concrete productions. The experimental results indicated that slump and compressive
strength of the concrete samples decreased with an increase in sludge replacement
[15].

Briefly, the use of waste sludge for partially replacing cement in concrete industry
is expected to not only reduce CO_2 emissions from cement production, but also lower
the construction cost and save landfill spaces for waste sludge treatment. Nevertheless, a few studies of replacement of cement by waste sludge from water supply
plants have been done. Therefore, the aim of this study was to investigate the effect
of partial replacement of cement by waste sludge from water supply plants on the
compressive strength and water absorption of hardened concrete to evaluate potential
use of such waste for concrete production.

2 Experiments

2.1 Materials

Cementitious materials used in this study included cement and waste sludge. The cement used was type I Portland cement supplied by Nghi Son Company conforming to TCVN 2682:2009 [7]. The waste sludge (see Fig. 1 (a)) was taken from Saigon Water Corporation (SAWACO) in Ho Chi Minh City, Vietnam. After drying at 110 °C in a laboratory oven, the sludge was grounded and sieved to have a particle size of less than 0.14 mm (see Fig. 1 (b)). Their physical properties and chemical compositions are listed in Tables 1 and 2. Density, residual percentage on 0.09-mm sieve, and compressive strength of the cement were tested as per TCVN 4030:2003 [10] and TCVN 6016:2011 [12], respectively. Strength activity index with Portland cement

Fig. 1 **a** Waste sludge at water supply plant and **b** ground waste sludge

Table 1 Physical properties of cement and sludge

Materials	Density (g/cm³)	Residual percentage on 0.09-mm sieve (%)	Compressive strength (MPa) at		Activity index (%) at 7 days	Setting time (min)		Standard consistency (%)
			3 days	28 days		Initial	Final	
Cement	3.10	0.3	37.10	59.20	–	150	180	33
Sludge	2.50	12.8	–	–	79	–	–	55

– : not measured

Table 2 Chemical compositions of cement and sludge (% by mass)

Compositions (%)	SiO₂	Fe₂O₃	Al₂O₃	CaO	MgO	Na₂O	K₂O	SO₃	LOI
Cement	14.00	2.00	20.50	63.00	1.50	0.09	0.65	2.40	2.72
Sludge	41.90	30.20	17.50	5.26	1.18	0.00	1.30	0.33	0.42

LOI: Loss on ignition

Table 3 Physical properties of fine and coarse aggregates

Materials	Density (g/cm³)	Bulk density (kg/m³)
Natural river sand	2.63	1640
Crushed stone	2.80	1670

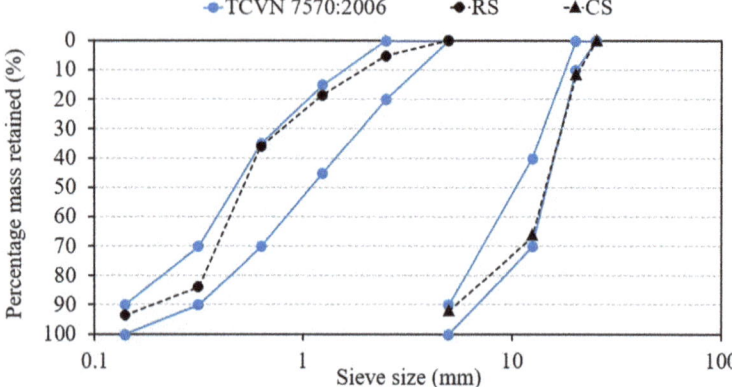

Fig. 2 Grading curve of aggregates compared with TCVN 7570:2006

was tested for the sludge as per TCVN 6882:2016 [14]. Setting time and standard consistency of materials were also measured as per TCVN 6017:2015 [13].

Natural river sand (RS) with fineness modulus of 2.1 was used as fine aggregate whereas crushed stone (CS) with a maximum particle size of 20 mm was used as coarse aggregate. The physical properties and particle size distribution of fine and coarse aggregates are shown in Table 3 and Fig. 2, respectively. In addition, tap water was used to mix concrete conforming to TCVN 4506:2012 [11].

2.2 Concrete Mixture Proportion

Mixture proportions of reference concrete with 0% sludge replacement of cement was designed according to the Bolomey-Skramtaev method in which a desired slump and compressive strength at 28 days were 6 ± 2 cm and 45 MPa, respectively. All concretes with a water-to-binder ratio of 0.53 and crushed stone amount of 1382 kg were prepared and shown in Table 4. Three levels of replacement of Portland cement by waste sludge were 10, 20, and 30% by mass. The replacement by mass changed the total volume of cementitious materials, leading to the change of sand volume. As a result, the replacement of waste sludge led to the change of sand amount in each mixture proportion as seen in Table 4.

Table 4 Mixture proportion of all concretes

Mixture proportions	Waste sludge replacement (%)	Unit: kg					Water-to-binder
		Cement	Sludge	Sand	Crushed stone	Water	
WS0	0	390	0	460	1382	205	0.53
WS10	10	351	39	452	1382	205	0.53
WS20	20	312	78	444	1382	205	0.53
WS30	30	273	117	436	1382	205	0.53

2.3 Test Procedure

After mixing the concrete components, slump of fresh concrete was immediately tested according to TCVN 3106:1993 [8] to evaluate effect of partial replacement of cement by waste sludge on consistency of fresh concrete.

After slump test, the concrete mixtures were cast into cubic specimens with a size of 100 mm. All specimens were demoulded after 24 h for casting and cured in a water bath until testing age. The compressive strength of hardened concrete was performed at the ages of 3, 7, and 28 days as per TCVN 3118:1993 [9].

In addition to compressive strength, water absorption and void volume of hardened concrete with 0, 10, and 20% waste sludge replacements were measured at the age of 28 days as per ASTM C642-13 [1].

3 Experimental Results and Discussion

3.1 Effect of Partial Replacement of Cement by Waste Sludge from Water Supply Plant on Slump of Fresh Concrete

Figure 3 (a) shows effect of partial replacement of cement by waste sludge from water supply plant on slump of fresh concrete. It is clear that the waste sludge replacement reduced the slump of fresh concrete. The higher the waste sludge replacement, the lower the slump of fresh concrete. It can be explained that the waste sludge had higher water absorption than cement through the higher standard consistency (see Table 1). The reduction of slump of fresh concrete containing uncontaminated marine sediments was also reported in the study of Zhao et al. [15]. However, replacing cement with waste sludge at levels of 10 and 20% still ensured designed slump of fresh concrete of 6 ± 2 cm.

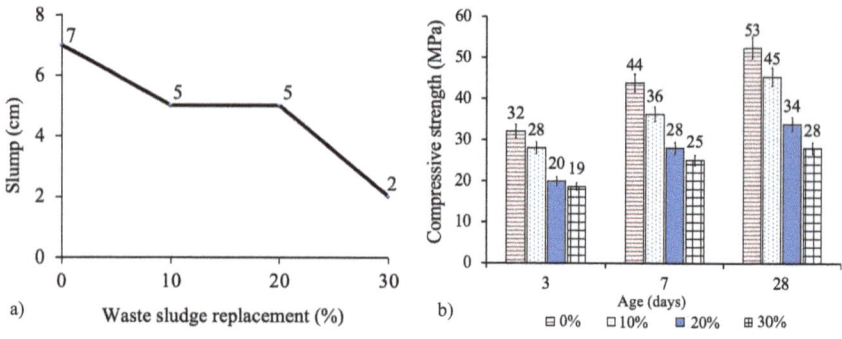

Fig. 3 Effect of waste sludge replacement on **a** slump of fresh concrete and **b** compressive strength of hardened concrete at the ages of 3, 7, and 28 days

3.2 Effect of Partial Replacement of Cement by Waste Sludge from Water Supply Plant on Compressive Strength of Hardened Concrete

The compressive strength of hardened concretes with various waste sludge replacements at the ages of 3, 7, and 28 days is presented in Fig. 3 (b). Generally, the compressive strength of all concretes increased with time. It indicates that the hydration of cement continuously proceeded. In addition, the compressive strength of hardened concrete decreased with an increase of waste sludge replacement. The partial replacements of cement with the waste sludge in the range of 10–30% decreased the compressive strength of hardened concrete by 15.10–47.17% when compared with the reference concrete at the age of 28 days. Liu et al. found that the compressive strength of concrete decreased gradually with the increase in drinking waste sludge ash as a cement replacement in concrete [4]. It is explained that the waste sludge has a lower reactivity than cement (see Table 1), resulting in slower pozzolanic reaction of the sludge when compared with the cement hydration [4]. However, the compressive strength of hardened concrete containing 10% waste sludge in the present study was 45 MPa at the age of 28 days, ensuring the desired compressive strength.

3.3 Effect of Partial Replacement of Cement by Waste Sludge from Water Supply Plant on Water Absorption and Void Volume of Hardened Concrete

The water absorption and void volume of hardened concretes with various waste sludge replacements at the age of 28 days is presented in Fig. 4. It can be seen that the replacements of cement by waste sludge at levels of 10 and 20% by mass increased the water absorption at the age of 28 days by 3.82 and 93.51%, respectively.

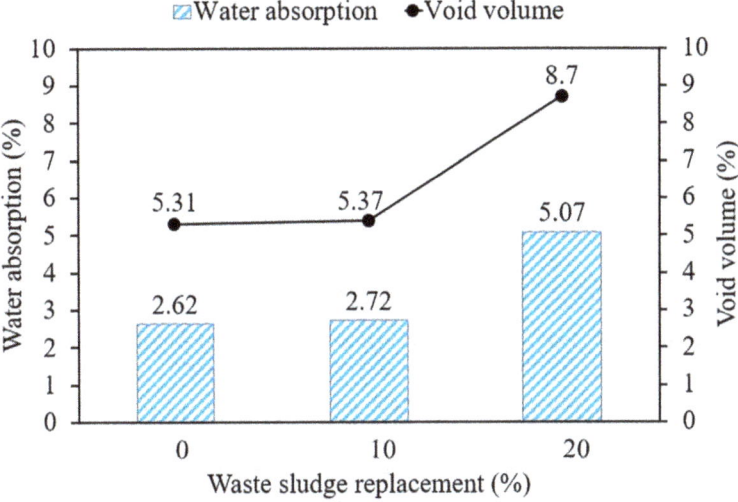

Fig. 4 Water absorption and void volume of hardened concretes at the age of 28 days

The sludge replacement at levels of 10 and 20% by mass also increased void volume of the hardened concrete at the age of 28 days by 1.13 and 63.84%, respectively. Generally, the higher the sludge replacement, the higher the water absorption and void volume of hardened concrete.

A relationship between compressive strength and water absorption, and that between compressive strength and void volume of hardened concrete at the age of 28 days are also shown in Fig. 5. It is obvious that the compressive strength of hardened concrete has a negative relationship with the water absorption as well as the void volume of the concrete. The higher the compressive strength of hardened concrete, the lower the water absorption and void volume.

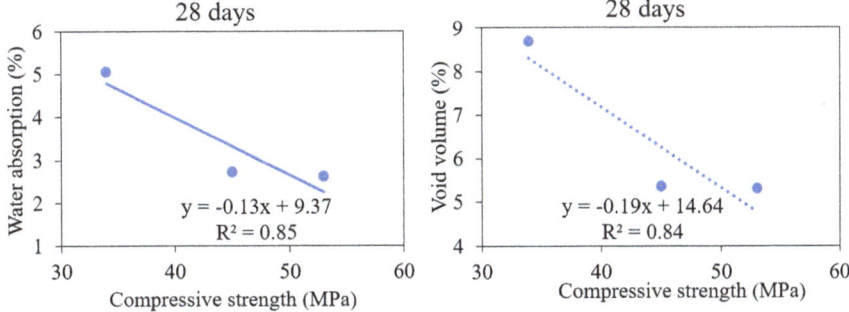

Fig. 5 a Relationship between compressive strength and water absorption and **b** that between compressive strength and void volume of hardened concrete at the age of 28 days

4 Conclusions

Based on experimental results, following conclusions can be drawn:

- The partial replacements of cement by waste sludge reduced slump of fresh concrete due to the high standard consistency of such waste. The higher the waste sludge replacement, the lower the slump of fresh concrete.
- The partial replacements of cement by waste sludge in the range of 10–30% by mass reduced the compressive strength up to 28 days and increased the water absorption and void volume at 28 days of hardened concrete.

Consequently, replacing 10% of cement with the sludge for concrete production was beneficial in terms of not only ensuring designed slump, compressive strength, and water absorption of hardened concrete at the age of 28 days but also decreasing CO_2 emission from cement production and utilizing the waste sludge from the water supply plants towards sustainable development. Further studies of the pozzolanic reaction of sludge activated by chemical activators in order to replace Portland cement by a high amount of waste sludge for concrete production, and long-term properties and microstructure analysis of cement-based concrete containing waste sludge should be done.

Acknowledgements The authors acknowledge the support of time and facilities from Ho Chi Minh City University of Technology (HCMUT), VNU-HCM for this study.

References

1. ASTM (2013) ASTM C642:13 Standard Test Method for Density, Absorption, and Voids in Hardened concrete. American Society for Testing and Materials, Pennsylvania
2. Aouad G, Laboudigue A, Gineys N, Abriak NE (2012) Dredged sediments used as novel supply of raw material to produce Portland cement clinker. Cem Concr Compos 34(6):788–793
3. Khadka RB, Khanal AB (2008) Environmental management plan (EMP) for Melamchi water supply project, Nepal. Environ Monit Assess 146:225–234
4. Liu Y, Zhuge Y, Christopher WKC, Phuong NP, Li D, Oh JA, Siddique R (2021) The potential use of drinking water sludge ash as supplementary cementitious material in the manufacture of concrete blocks. Resour Conserv Recycl 168:105291
5. Mahasenan N, Smith S, Humphreys K (2003) The cement industry and global climate change: current and potential future cement industry CO2 emissions. In: Proceedings of the 6th International Conference on Greenhouse Gas Control Technologies, vol 2, pp 995–1000, Kyoto, Japan
6. Tang CW, Chen HJ, Wang SY, Spaulding J (2011) Production of synthetic lightweight aggregate using reservoir sediments for concrete and masonry. Cem Concr Compos 33(2):292–300
7. TCVN (2009) TCVN 2682:2009 Portland cements–Specifications. Ministry of Science and Technology, Vietnam
8. TCVN (1993) TCVN 3106:1993 Fresh heavyweight concrete–Method for slump test. Ministry of Science and Technology, Vietnam
9. TCVN (1993) TCVN 3118:1993 Heavyweight concrete–Method for determination of compressive strength. Ministry of Science and Technology, Vietnam

10. TCVN (2003) TCVN 4030:2003 Cement–Test method for determination of fineness. Ministry of Science and Technology, Vietnam
11. TCVN (2012) TCVN 4506:2012 Water for concrete and mortar–Technical specification. Ministry of Science and Technology, Vietnam
12. TCVN (2011) TCVN 6016:2011 Cements–Test methods–Determination of strength. Ministry of Science and Technology, Vietnam
13. TCVN (2015) TCVN 6017:2015 Cements–Test methods–Determination of setting time and soundness. Ministry of Science and Technology, Vietnam
14. TCVN (2016) TCVN 6882:2016 Mineral additive for cement. Ministry of Science and Technology, Vietnam
15. Zhao Z, Benzerzour M, Abriak NE, Damidot D, Courard L, Wang D (2018) Use of uncontaminated marine sediments in mortar and concrete by partial substitution of cement. Cem Concr Compos 93:155–162

Effect of Polypropylene Fiber Addition to Bottom Layer on the Mechanical Properties of Functionally Graded Concrete Containing Fly Ash

Minh Luan Vo, Phuong Trinh Bui, and Quoc Viet Dang

Abstract The present study focuses on effect of polypropylene (PP) fiber addition to bottom layer on the mechanical properties of functionally graded concrete (FGC) containing fly ash (Fa). The FGC specimens were prepared with two concrete layers having the same thickness. The top layer of FGC specimens was composed of normal concrete or Fa concrete without PP fiber, whereas the bottom layer was made from normal concrete, Fa concrete, or Fa concrete with 0.3% PP fiber. A constant water-to-cementitious materials ratio of 0.36 was applied for all mixtures. Fly ash was used to partially replace ordinary Portland cement by the proportion of 20% (by mass). The results showed that the flexural strength and impact resistance were significantly increased at the age of 28 days, whereas the compressive strength was slightly enhanced due to the PP fiber addition to the bottom concrete layer of FGC specimens. Functionally graded concrete using normal concrete in a top layer and Fa concrete reinforced with PP fiber in a bottom layer was an optimal material considering not only effectively economic aspect but also beneficial mechanical properties in terms of compressive strength, flexural strength, and impact resistance up to the age of 28 days.

Keywords Functionally graded concrete · Compressive strength · Flexural strength · Impact resistance · Polypropylene fiber

M. L. Vo · P. T. Bui (✉)
Faculty of Civil Engineering, Ho Chi Minh City University of Technology (HCMUT), 268 Ly Thuong Kiet Street, District 10, Ho Chi Minh City, Vietnam
e-mail: buiphuongtrinh@hcmut.edu.vn

Vietnam National University Ho Chi Minh City, Linh Trung Ward, Thu Duc District, Ho Chi Minh City, Vietnam

Q. V. Dang
Faculty of Bridge and Road Construction, Mientrung University of Civil Engineering (MUCE), 24 Nguyen Du Street, Ward 7, Tuy Hoa City, Phu Yen Province, Vietnam

© The Author(s), under exclusive license to Springer Nature Singapore Pte Ltd. 2023
J. N. Reddy et al. (eds.), *ICSCEA 2021*, Lecture Notes in Civil Engineering 268,
https://doi.org/10.1007/978-981-19-3303-5_36

1 Introduction

More than 25 billion tons of concrete are yearly consumed. Hence, it is assumed that concrete is the most widely used construction material due to its advantages such as low cost, wide application, high strength, and durability [2, 6]. Nowadays, concrete is used as a material in almost residential and infrastructure facilities. However, concrete has some ecological disadvantages through its production, transportation, or utilization. Ordinary Portland cement (OPC) is the main material in the composition of concrete. A large amount of carbon dioxide was released into the atmosphere during OPC production [10].

To eliminate carbon dioxide emissions and make concrete eco-friendly sustainable, supplementary cementitious materials such as fly ash (Fa) and ground granulated blast furnace slag have been used to replace OPC for concrete production. Fly ash is a by-product released from the operation of coal-fired power plants and it often contains large amounts of active SiO_2 and Al_2O_3 which react with $Ca(OH)_2$ in cement matrix to produce the further C–S–H gels [11]. The reaction between active SiO_2, Al_2O_3 and $Ca(OH)_2$ is called the pozzolanic reaction of Fa. It is reported that the mechanical properties and durability of concrete with 10–40% Fa replacement could be enhanced at later ages, but not at early ages [11–13]. The reactivity of Fa at the early ages is slower than the hydration of cement. Hence, concretes containing Fa show lower in mechanical properties than those without Fa at the early ages [11, 13].

On the other hand, monolayer concrete is inherently brittle and has low tensile strength, crack resistance, and deformation properties [11]. Generally, the normal concrete is prone to crack due to external forces [10]. Therefore, recent studies have focused on the improvement of concrete performance by innovative solutions such as the incorporation of steel fiber, carbon fiber, rubber fiber, and polymer fiber [4, 9]. The fiber often plays a role as a reinforcement, bridging the cracks, and arresting crack propagation. Fiber amount is one of the main factors affecting the properties of concrete. The high amount of fiber should not be applied due to the poor dispersion as well as the balling effect of fiber, resulting in the concrete performance reduction [7, 9].

Besides, functionally graded concrete (FGC) has been recently explored as a new type of concrete because it is produced from various layers of concrete such as normal concrete reinforced by steel fiber, recycled aggregate concrete, and so on. Functionally graded concrete could perform higher compressive and flexural strengths as well as impact resistance in comparison to monolayer concrete [3, 9]. Almost studies of FGC using Fa or polypropylene (PP) fiber individually were conducted. However, a few studies of FGC incorporating Fa and PP fiber have been done. Therefore, this present study suggested the addition of Fa and PP fiber to a bottom layer of FGC with the expectation of not only reduction in the cement content but also improvement of mechanical performance of concrete. The effectiveness of this technique was evaluated on the compressive and flexural strengths as well as impact resistance in this study.

2 Experiments

2.1 Materials

The cementitious materials used in the present study included Type-I OPC and Type-F Fa, which were supplied from Nghi Son cement company and Duyen Hai thermal power plant, respectively. Both OPC and Fa conformed to TCVN 2682:2009 and TCVN 10302:2014, respectively. The density and residual percentage on 0.09-mm sieve of OPC were 3.1 g/cm^3 and 0.3%, respectively, whereas those of FA were 2.2 g/cm^3 and 7.0%, respectively. The chemical compositions of OPC and Fa are shown in Table 1.

River sand (RS) taken from the Dong Nai river and crushed stone (CS) were used as fine and coarse aggregates, respectively. River sand had a fineness modulus of 2.55, density of 2.65 g/cm^3, and water absorption of 0.82%. Meanwhile, maximum size, density, and water absorption of CS were 20 mm, 2.69 g/cm^3, and 0.35%, respectively. The particle size distributions of RS and CS compared with TCVN 7570:2006 via sieve analysis are shown in Table 2.

Polypropylene fiber used in the present study had a density of 0.91 g/cm^3, diameter of 0.02 mm, length of 6–12 mm, and high mechanical strength (tear and tensile strength of 500 N/mm^2), conforming to TCVN 12393:2018. In addition, tap water in accordance with TCVN 4506:2012 was used to cast all concrete mixtures. Lignosulfonate-based Sikament R4 superplasticizer (SP) originated from the Sika group was also used for all mixtures to enhance the workability.

Table 1 Chemical compositions of OPC and Fa

Cementitious materials	Chemical compositions (% by mass)								
	SiO_2	Al_2O_3	Fe_2O_3	CaO	K_2O	Na_2O	SO_3	MgO	LOI
OPC	24.0	4.0	2.0	63.0	0.6	0.1	2.4	1.5	2.7
Fa	45.5	20.5	17.4	5.4	5.7	0.3	0.5	0.5	4.8

LOI: loss on ignition

Table 2 Sieve analysis of fine and coarse aggregates

Fine aggregate			Coarse aggregate		
Sieve size (mm)	Cumulative mass retained (%)	TCVN 7570:2006 (%)	Sieve size (mm)	Cumulative mass retained (%)	TCVN 7570:2006 (%)
5.0	0	0	20	9.3	0–10
2.5	12.7	0–20	12.5	48.2	–
1.25	23.5	15–45	10	68.7	40–70
0.63	39.1	35–70	5	90.8	90–100
0.315	86.9	70–90	–	–	–
0.14	91.7	90–100	–	–	–
			–		

2.2 Concrete Mixture Proportion

Figure 1 presents five types of FGC specimens composed of various concrete mixtures in the present study. The types of FGC concrete included FGC specimens using normal concrete without PP fiber (A), FGC specimens using Fa concrete without PP fiber (B), FGC specimens using normal concrete and reinforced with PP fiber in a bottom layer (C), and FGC specimens using Fa concrete and reinforced with PP fiber in a bottom layer (D), and FGC specimens using normal concrete in a top layer and Fa concrete reinforced with PP fiber in a bottom layer (E). It is noted that FGC specimens in the present study consisted of two concrete layers with the same thickness.

For concrete mixture proportions, the normal concrete (NC) with a water-to-cementitious materials ratio (W/C) of 0.36 was designed according to ACI 308R-01. The target slump of 8 ± 2 cm and compressive strength at 28 days of 70 MPa were chosen according to the previous studies [3, 9, 14]. Based on the recommendation of previous studies [13, 14], the replacement of OPC with Fa at 20% by mass of cementitious materials was done for Fa concrete (FAC) and 0.3% PP fiber was added to the concrete in a bottom layer of FGC specimens. Table 3 shows the concrete mixture proportions of all FGC specimens.

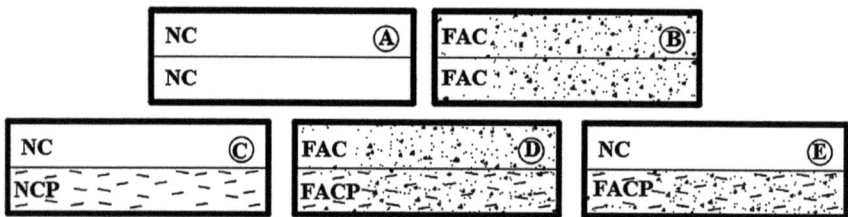

Fig. 1 FGC specimens considered in the present study

Table 3 Mixture proportions of concrete

Mixture proportions	W/C	Unit (kg)					PP (% $V_{concrete}$)	Slump - top layer (cm)	Slump - bottom layer (cm)
		Water	Cement	Fa	RS	CS			
NC	0.36	183	508	0	772	969	0	8.0	8.5
FAC	0.36	183	406	102	742	969	0	14.0	16.0
NCP	0.36	183	508	0	772	969	0.3	–	7.0
FACP	0.36	183	406	102	742	969	0.3	–	11.0

% $V_{concrete}$: % by volume of concrete; Fa: Fly ash; RS: River sand; CS: Crushed stone; -: not measured

2.3   Slump Test and Specimen Preparation

The slump was conducted for all mixtures after mixing as per TCVN 3106:1993. After the slump test, fresh concrete of each mixture was immediately cast into steel molds in two layers according to TCVN 3105:1993. It should be noted that the bottom concrete layer was first cast. Then, the top concrete layer was continuously cast after approximately 45 min to eliminate the casting time effect on the bonding between two concrete layers of FGC specimens. The placing surface of specimens was then covered with a plastic sheet to avoid the evaporation of water and carbonation. Nine cubic specimens with dimensions of $150 \times 150 \times 150$ mm and three beam specimens with dimensions of $100 \times 100 \times 400$ mm for each mixture proportion were cast for compressive and flexural strength measurements in the present study, respectively. Meanwhile, three cylindrical specimens with 150 mm in diameter and 63 mm in height of each mixture proportion were cast for impact resistance test. After 24 h of casting, all specimens were demolded and then stored in water until testing ages.

2.4   Compressive Strength, Flexural Strength, and Impact Resistance Tests

The compressive strength of each FGC specimen was tested at 3, 7, and 28 days using a hydraulic compression machine in accordance with TCVN 3118:1993. The flexural strength of each FGC specimen at the age of 28 days was tested by Instron machine in accordance with TCVN 3119:1993, ASTM C1609 and the limit flexural toughness was calculated following a previous study of Dong et al. [5].

   Impact resistance (strength) test was carried out by using drop weight method (the ball falling mass testing) recommended by ACI 544.2R–89. This value could be used to evaluate the effectiveness of PP fiber inside concrete specimens. The value could be also used to design the structures that are affected by dynamic loads such as pavement, breakwater, and so on. The impact resistance of each FGC specimen was tested after polishing the surface of concrete specimens at the age of 28 days.

3   Results and Discussion

3.1   Slump

Slump of fresh concrete of each mixture proportion is shown in Table 3. It can be seen that the Fa replacement led to the increase in the slump of fresh concrete even though the mixing water amount was not changed. The increase in the slump of fresh concrete due to Fa replacement was also observed in a previous study by Naresh et al. [13]. Nevertheless, the slump of fresh concrete was remarkably decreased since PP fiber was added. This is also observed in an existing study [7]. The incorporation of PP fiber affected the flexibility of aggregates in the concrete mixture. Consequently, the interlocking and friction between PP fiber and aggregates were increased, resulting in a decrease in the slump of fresh concrete [7].

3.2 Compressive Strength

Figure 2(a) shows the compressive strength of FGC specimens at 3, 7, and 28 days. It can be seen that the Fa replacement resulted in the reduction of compressive strength of concrete at the early ages. As a result, FGC containing Fa (FAC/FAC) was lower in compressive strength than that of reference FGC (NC/NC) at the ages of 3 and 7 days. It is due to the dilution effect and slow reactivity of Fa [2]. However, the compressive strength of FAC/FAC specimens was significantly increased and almost the same as that of reference FGC at 28 days. The reason is probably due to the additional generation of C–S–H and C–A–S–H gels from the pozzolanic reaction of Fa in corresponding FAC/FAC specimens [11]. Otherwise, the incorporation of PP fiber was found to slightly increase the compressive strength of the FGC specimens. For example, the compressive strength of FGC specimens with PP fiber addition (NC/NCP, FAC/FACP, and NC/FACP) at 28 days was approximately 0.7–3.6% higher than that of the FGC specimens without PP fiber (NC/NC and FAC/FAC). The difference in the compressive strength of FGC specimens with PP fiber incorporation (NC/NCP, FAC/FACP, and NC/FACP) was negligible irrespective of Fa replacement. This is also observed in the previous study in which there was not a significant difference in compressive strength of FGC specimens using OPC with and without steel fiber reinforcement in the bottom layer [9].

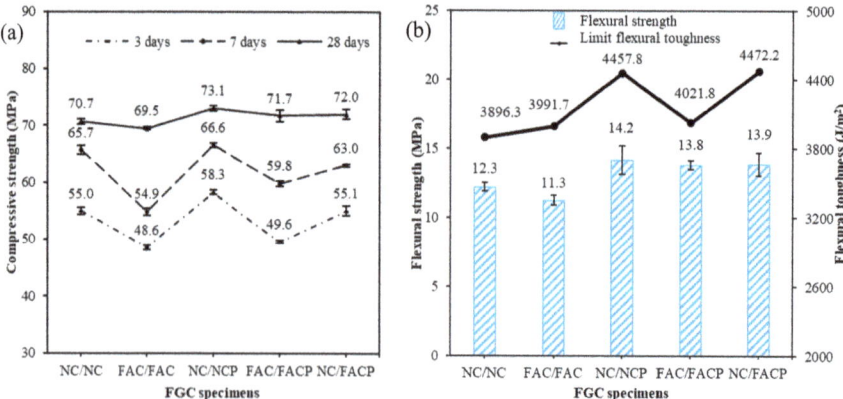

Fig. 2 Compressive strength at 3, 7, and 28 days **a** and flexural strength and impact resistance at 28 days **b** of FGC specimens

3.3 Flexural Strength and Flexural Toughness

Figure 2(b) shows the flexural strength and limit flexural toughness of all FGC specimens at 28 days. It is clear that the flexural strength of FGC specimens reinforced with PP fiber in the bottom layer was higher than that of FGC specimens without PP fiber. In particular, NC/NCP, FAC/FACP, and NC/FACP specimens were approximately 12.3–15.4% higher in flexural strength than NC/NC and FAC/FAC specimens. The distribution of PP fiber in FGC specimens created a connection network and increased the flexural strength of the FGC specimens. This is in good agreement with the previous studies [4, 8, 9]. The replacement of Fa led to the decrease in the flexural strength of concrete, which was 8.1% lower than that of the reference concrete (NC/NC). This reduction could be compensated by the PP fiber incorporation. Similar to compressive strength results, the difference in the flexural strength of FGC specimens with PP fiber addition (NC/NCP, FAC/FACP, and NC/FACP) was negligible regardless of Fa replacement. According to Fig. 2(b), the limit flexural toughness of FGC specimens was improved since the bottom layer of FGC specimens was reinforced with PP fiber. In particular, NC/NCP, FAC/FACP, and NC/FACP specimens were approximately 2.4–14.4% higher in limit flexural toughness than NC/NC and FAC/FAC specimens. This could be explained due to the enhancement of bond between aggregates and cement pastes as well as the prevention of crack propagation for FGC specimens containing PP fiber [5]. In addition, the maximum limit flexural toughness of FGC was observed for NC/FACP specimen with a value of 4472.2 J/m^2.

3.4 Impact Resistance

Figure 3 shows the impact resistance of FGC specimens at 28 days. It is clear that the FGC specimens with PP fiber addition (NC/NCP, FAC/FACP, and NC/FACP) were

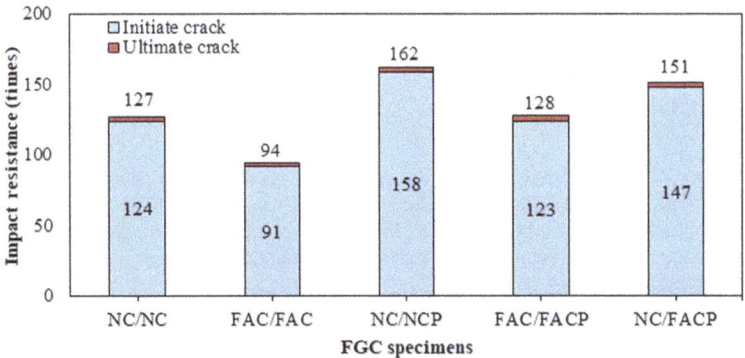

Fig. 3 Impact resistance of FGC specimens at 28 days

superior in the impact resistance compared to the corresponding FGC specimens without PP fiber (NC/NC and FAC/FAC). For example, the impact resistance of NC/NCP specimen was 27.5% higher than that of NC/NC one while FAC/FACP specimen had higher impact resistance than FAC/FAC one by 36.2%. The FGC specimen using NC in a top layer and FAC reinforced with PP fiber in a bottom layer (NC/FACP) had a higher impact resistance than the NC/NC, FAC/FAC, and FAC/FACP specimens by 18.9, 60.0, and 17.9%, respectively. It was due to PP fiber incorporation in FGC specimens distributing and creating a connection network, leading to the increase in the impact resistance [1].

4 Conclusion

Based on the experimental results, the following conclusions can be drawn:

– There was no significant difference in the compressive strength results between FGC without PP fiber and FGC with PP fiber addition to a bottom layer.
– After 28 days, compared with FGC without PP fiber, FGC with PP fiber addition to a bottom layer performed better in flexural strength, limit flexural toughness, and impact resistance, regardless of the Fa replacement.
– Functionally graded concrete using NC in a top layer and FAC reinforced with PP fiber in a bottom layer (NC/FACP) is recommended as the optimal material considering the mechanical properties of concrete up to the age of 28 days. Properties of FGC after 28 days should be further investigated to evaluate the effect of Fa on long-term mechanical properties and durability of FGC reinforced with PP fiber.

Acknowledgements We acknowledge the support of time and facilities from Ho Chi Minh City University of Technology (HCMUT), VNU-HCM for this study. We also acknowledge the help in the experimental work of students of the Department of Construction Materials, Faculty of Civil Engineering, HCMUT, VNU-HCM.

References

1. Bhogayata AC, Arora NK (2018) Impact strength, permeability and chemical resistance of concrete reinforced with metalized plastic waste fibers. Constr Build Mater 161:254–266
2. Celik K, Meral C, Mancio M, Mehta PK, Monteiro PJM (2014) A comparative study of self-consolidating concretes incorporating high-volume natural pozzolan or high-volume fly ash. Constr Build Mater 67:14–19
3. Chan R, Liu X, Galobardes I (2020) Parametric study of functionally graded concretes incorporating steel fibres and recycled aggregates. Constr Build Mater 242:118186
4. Do TMD, Lam TQK (2021) Nonlinear analysis of multi-layer steel fiber reinforced concrete beams. J Consruct 4:58–63

5. Dong S, Zhou D, Ashour A, Han B, Ou J (2019) Flexural toughness and calculation model of super-fine stainless wire reinforced reactive powder concrete. Cem Concr Compos 104:103367
6. Gagg CR (2014) Cement and concrete as an engineering material: an historic appraisal and case study analysis. Eng Fail Anal 40:114–140
7. Hasan AH, Maroof NR, Ibrahim YA (2019) Effects of polypropylene fiber content on strength and workability properties of concrete. Polytech J 9(1):7–12
8. Karahan O, Durak U, Llkentapar S, Atabey II, Atiş CD (2019) Resistance of polypropylene fibered mortar to elevated temperature under different cooling regimes. Constr Mag 18(2):386–397
9. Liu X, Yan M, Galobardes I, Sikora K (2018) Assessing the potential of functionally graded concrete using fibre reinforced and recycled aggregate concrete. Constr Build Mater 171:793–801
10. Makul N (2020) Advanced smart concrete–a review of current progress, benefits and challenges. J Clean Product 274:122899
11. Mehta A, Siddique R (2017) Properties of low-calcium fly ash based geopolymer concrete incorporating OPC as partial replacement of fly ash. Constr Build Mater 150:792–807
12. Menéndez E, Sanjuán MÁ, García-Roves R, Argiz C, Recino H (2020) Sustainable and durable performance of pozzolanic additions to prevent alkali-silica reaction (ASR) promoted by aggregates with different reaction rates. Appl Sci 10(24):9042
13. Naresh J, Lavanya B, Kumar KS (2021) A study on bond strength of normal concrete to high volume fly ash concrete. IOP Conf Ser Mater Sci Eng 1057:012082
14. Ramjan S, Tangchirapat W, Jaturapitakkul C, Ban CC, Jitsangiam P, Suwan T (2021) Influence of cement replacement with fly ash and ground sand with different fineness on alkali-silica reaction of mortar. Mater 14(6):1528

Effect of Pre-crack Length on the Moisture Distribution of Asphalt Concrete Pavement in a Short-Term Rain Event

Tan Hung Nguyen, Truong Phu Nguyen, Minh Triet Pham, and Trong-Phuoc Huynh

Abstract This study was conducted to assess the moisture distribution in a pre-cracked asphalt concrete pavement in a short-term rain event by finite element method via the diffusion process. The asphalt concrete pavement with three different pre-crack lengths was used to investigate. The results show that the moisture distribution zone increases according to the time of the rain event and the pre-crack length of the asphalt concrete pavement. It is highlighted that the proposed model can effectively assess the moisture distribution in a pre-cracked asphalt concrete pavement. Moreover, this study also found that the pre-crack length affected the moisture distribution in the pre-cracked asphalt concrete pavement significantly. The pavement with higher pre-crack length had a higher increasing rate of moisture saturation with time. In the future, further research should be conducted to investigate the moisture distribution in a pre-crack asphalt concrete pavement in terms of different conditions such as material, environment, and traffic load.

Keywords Moisture distribution · Asphalt concrete pavement · Finite element method

1 Introduction

During the service life of asphalt concrete pavements (ACP), they are subjected to traffic loads and environmental deteriorations such as moisture, freeze/thaw cycle, chemical compounds, etc. [1–3]. These factors cause damage to the ACP and form

T. H. Nguyen · T. P. Nguyen
Department of Civil Engineering, Can Tho University of Technology, 256 Nguyen Van Cu Street, Ninh Kieu District, Can Tho City 94000, Vietnam

M. T. Pham
Department of Aerospace Engineering, Pusan National University, 2 Busandaehak-ro 63beon-gil, Geumjeong-gu, Busan 46241, Republic of Korea

T.-P. Huynh (✉)
Department of Civil Engineering, College of Engineering Technology, Can Tho University, Campus II, 3/2 Street, Ninh Kieu District, Can Tho City 94000, Vietnam
e-mail: htphuoc@ctu.edu.vn

the different types of pavement distresses such as cracking, bleeding, corrugation, and shoving [4]. Among these types of distresses, cracking is a major problem mainly related to moisture damage, which is considered one of the major distresses of ACP. Firstly, this damage can be explained by the flow of water through the interconnected pore of the ACP, known as striping. ACP is the loss of the adhesive bonding force between the asphalt binder and the aggregate [5]. In this case, the moisture damage of ACP depends largely on the components of the asphalt materials such as a binder, aggregate, and air void [6]. Secondly, the moisture in humid environments can penetrate into the ACP as a vapor [7]. It is reported that for the ACP with low porosity, the moisture damage happened by moisture diffusion [8], which is also considered as one of the most important reasons for moisture damage [9].

To date, there have been a few studies investigating moisture diffusion and its effects on moisture distribution and damage in ACP. Kassem et al. [7] measured the moisture diffusion of asphalt mixtures to evaluate the moisture damage of the asphalt mixtures. In their study, three kinds of asphalt mixture with aggregate designed by sandstone, limestone, and gravel, were evaluated. The moisture diffusion was evaluated by moisture diffusion coefficient, which was estimated by Mitchell's equation. The results showed that asphalt mixture with a higher moisture diffusion coefficient having lower resistance to moisture damage. The mixture with gravel obtained lower resistance to moisture damage than the one with limestone did. The authors suggested that the moisture diffusion coefficient could be used to evaluate the resistance to moisture damage of the asphalt mixture. In another study, Apeagyei et al. [10] investigated the moisture damage of ACP having different aggregates. Based on the results, it can be seen that the asphalt mixture with better aggregate performance exhibited higher resistance to moisture damage. The conclusion was made that the moisture damage strongly depended on the aggregate performance. Additionally, this study has also been recommended the appropriate aggregate used for asphalt mixture to provide high resistance to moisture damage. Furthermore, the validation of experimental measurement and numerical simulation of moisture distribution by diffusion in ACP was carried out by Hénon et al. [11]. In their study, the diffusion coefficient was determined by experiment based on manual in ASTM E96 method. The results indicated a good agreement between the experimental and numerical simulation for the diffusion coefficient. Furthermore, the study pointed out that the moisture diffusion coefficient significantly affects the moisture-related damage process of ACP. Recently, Cammarata et al. [12] observed the moisture distribution in a pre-cracked ACP through the finite element method. The prediction of moisture distribution in the pre-crack ACP during rain events was carried out. The results showed that the pre-crack length in an ACP developed rapidly due to the moisture damage after a short-time rain event. They recommended that the pre-crack in the ACP should be sealed as soon as possible to avoid moisture damage. Based on the literature, it can be seen that there have been a few studies that investigated moisture distribution in ACP, especially for the ACP with a pre-crack. There is an identified gap for the study of moisture distribution in a pre-cracked ACP.

To fill this gap, this present study was carried out to analyze the moisture distribution in a pre-cracked ACP with different pre-crack lengths for a short-term rain event. To observe the effect of pre-crack length on the moisture distribution in a pre-cracked ACP, three different pre-crack lengths (18, 20, and 22 cm) in the ACP were selected to analyze. The finite element method (FEM) was used to model the pavement and assess the moisture distribution. In this study, the crack in the ACP was assumed to be a single mode and straight.

2 Methodology

2.1 Governing Equation of Moisture Distribution

In this study, it is assumed that the ACP has a pre-crack, and the moisture caused by rain events is absorbed in the pre-crack ACP through the diffusion process. To assess the flux of moisture according to the depth of the pre-crack, the traditional one-dimensional diffusion equation corresponding to Fick's first law [13] is derived as follows:

$$q_i = -D\frac{\partial m}{\partial x_i} \tag{1}$$

where x_i is the ith coordinate direction, q_i is the flux of moisture in the coordinate direction, D is the diffusion coefficient, and m is the mass of moisture.

According to the studies of Lehner [14] and Apeagyei et al. [15], the moisture absorbed in the ACP is not followed the equation Eq. (1). To accurately model the moisture distribution in a pre-crack, Fick [13] substituted his first law's equation into the mass conservation's equation and formulate a new equation:

$$\nabla \cdot (D\nabla m) = \frac{\partial m}{\partial t} \tag{2}$$

This equation is referred to Fick's second law, which shows the relationship of moisture distribution according to time t. Equation (2) is a partial differential equation (PDE) that can be used to calculate the moisture distribution in a pre-cracked ACP. For the analysis, it is noticed that the diffusion coefficient D is determined via a series of experiments.

2.2 Finite Element Analysis

In the present study, a model of the pre-cracked ACP followed the configurations of Cammarata et al. [12] was used. Figure 1 shows the computational domain and boundary conditions.

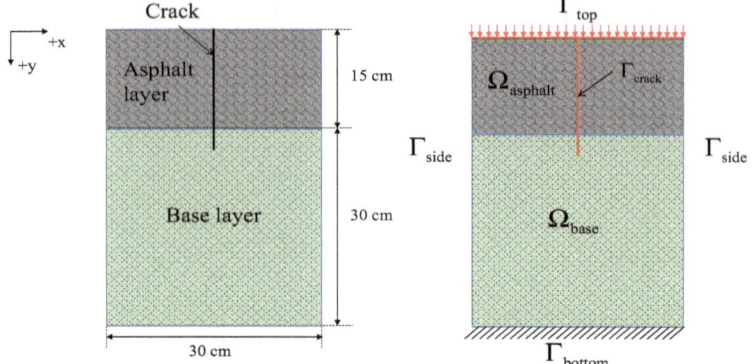

Fig. 1 Computational domain and boundary conditions of a pre-cracked ACP, after Cammarata et al. [12]

The boundary conditions were mathematically formulated as follows:

$$\begin{cases} m(t) = 0 & \text{on } \Gamma_{\text{bottom}} \\ D\frac{\partial m}{\partial t} \cdot \overrightarrow{n} = 0 \text{ on } \Gamma_{\text{side}} \\ D\frac{\partial m}{\partial t} \cdot \overrightarrow{n} = 1 \text{ on } \Gamma_{\text{top}} \cup \Gamma_{\text{crack}} \end{cases} \tag{3}$$

To solve the PDE presented in Eq. (2), both sides of the equation with a trial function v were calculated and perform integration by parts. This technique is well-known as Galerkin's projection in FEM. Assuming the diffusion coefficient D to be a constant, the resulting weak form of Eq. (2) can be expressed as follows:

$$\int_{\Omega} \frac{\partial m}{\partial t} v \, d\Omega + D \int_{\Omega} \nabla m \cdot \nabla v \, d\Omega = \int_{\Gamma_{\text{bottom}}} v \, d\Gamma + \int_{\Gamma_{\text{crack}}} v \, d\Gamma \quad \forall v \in \mathbb{V} \tag{4}$$

with the function space V is defined as $\mathbb{V} = \left\{ v \in H^1(\Omega) : v|_{\Gamma_{\text{top}}} = 0 \right\}$, and

$$D = \begin{cases} D_{\text{asphalt}} & \text{in } \Omega_{\text{asphalt}} \\ D_{\text{base}} & \text{in } \Omega_{\text{base}} \end{cases} \tag{5}$$

In this study, to observe the effects of pre-cracked length on the moisture distribution in a pre-cracked ACP, diffusion coefficients D_{asphalt} and D_{base} obtained from the studies of Cammarata et al. [12] were used to analyze. More specifically, the diffusion coefficients of the asphalt layer and the base layer are $D_{\text{asphalt}} = 5.97e^{-5}$ cm^2/hr and $D_{\text{base}} = 6.84$ cm^2/hr, respectively. To model the pre-cracked ACP and analyze the moisture distribution in the ACP system, FEM was used to solve Eq. (4). The model was developed on an open-source python library, FEniCS (version 1.5). Details of the open-source code can be referred to in the literature [16]. The numerical model was discretized into 10,800 triangle elements with 5551 nodes. Instead of doing mesh independence studies, the second-order meshes were used to ensure the

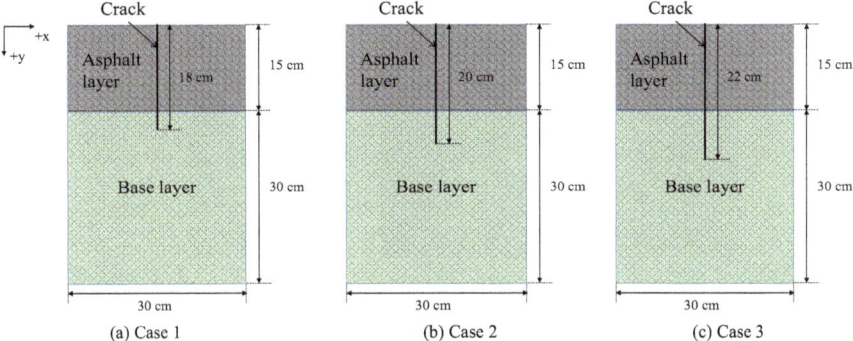

Fig. 2 The geometry of the pre-cracked ACP, after Cammarata et al. [12]

accuracy of the FEM models. Figure 2 shows models of the ACP with three different pre-crack lengths used for observation in this study.

3 Results and Discussion

The results of moisture distribution in the pre-cracked ACP in each case according to the time of rain event are displayed in Figs. 3, 4 and 5.

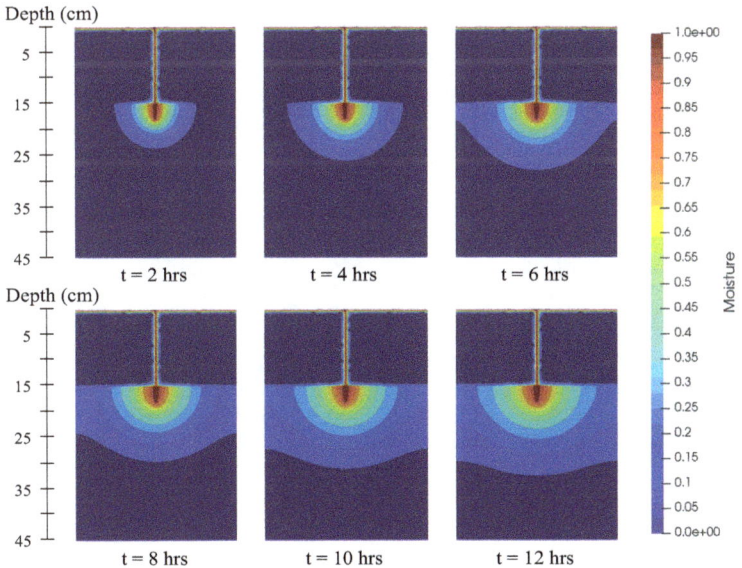

Fig. 3 Moisture distribution for pre-crack length of 18 cm (Case 1)

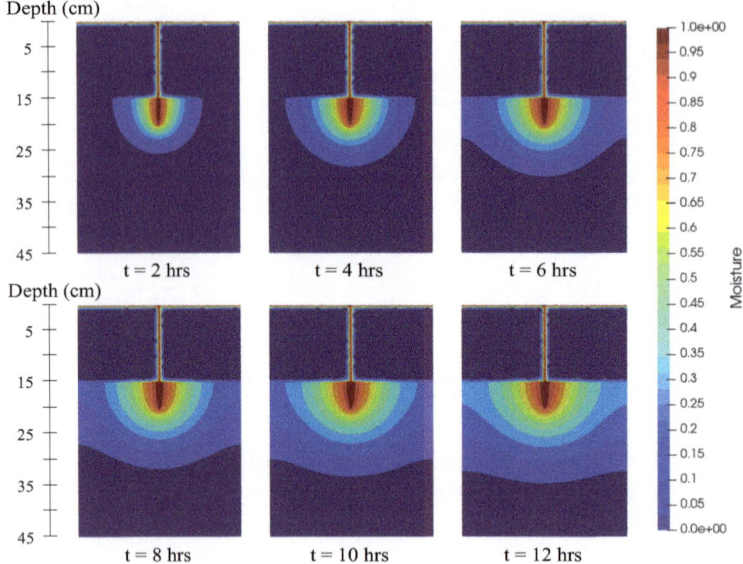

Fig. 4 Moisture distribution for pre-crack length of 20 cm (Case 2)

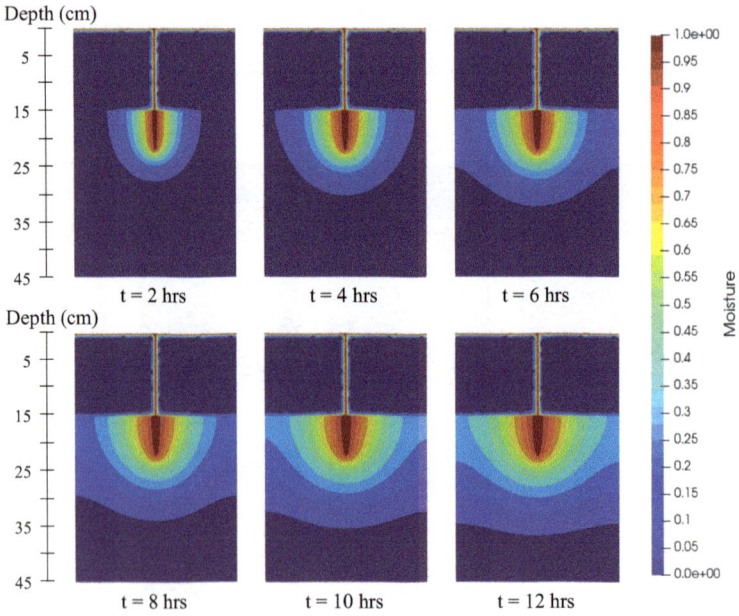

Fig. 5 Moisture distribution for pre-crack length of 22 cm (Case 3)

The results show that, for each case, as the time of rain event increases, the zone of moisture distribution becomes larger. It can be explained that, as the rain happens, it provides water on the surface of the pavement system, which can go down to the pre-crack spatial. At the pre-crack spatial, water could transport around by the diffusion process. Therefore, as the time of rain event increases, the zone of moisture distribution becomes larger. To verify the accuracy of the present FEM results, the comparison of the moisture distribution results between this study and the literature was carried out. This study created a model of the pre-cracked ACP based on the configurations in the study of Cammarata et al. [12]. It is highlighted that the results of moisture distribution according to time in this study are similar to that in the study of Cammarata et al. [12]. This finding inferred that the developed model can effectively assess the moisture distribution in a pre-cracked ACP. From the figures above, it can be seen as the pre-crack length increases, the zone of moisture distribution becomes larger. For instance, by moisture diffusion, at 12 h of rain event, a pre-crack length of 18 cm could transport water to the depth about 33 cm from the surface, while the one of 20 and 22 cm could do the same thing to the depth of about 35 and 37 cm, respectively.

On the other hand, to observe the saturation at a point, the result of saturation at the redpoint, which is marked in Fig. 6, is plotted. This point was selected to show the saturation because it was the highest position that moisture could be spread in the base layer. The results of the saturation respecting to time of rain event for the observation point are shown in Fig. 7 below. This figure indicates that when the pre-crack length increases, the saturation of the observation point increases. For example, at 12 h, as the pre-crack length goes up 2 cm from 18 to 20 cm, the saturation rises about 0.06, from 0.27 to 0.21. Considering the increasing rate of saturation according to time, the current study found that the pavement with higher pre-crack length showed a higher increasing rate. A possible explanation for this might be that this study assumed the diffusion coefficient was constant according to time. It is noted that the saturation at the red point behaves linearly according to time. The reason can be explained that

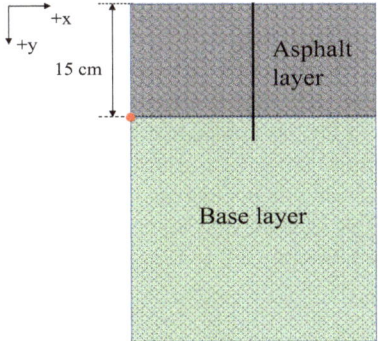

Fig. 6 Observation point for saturation

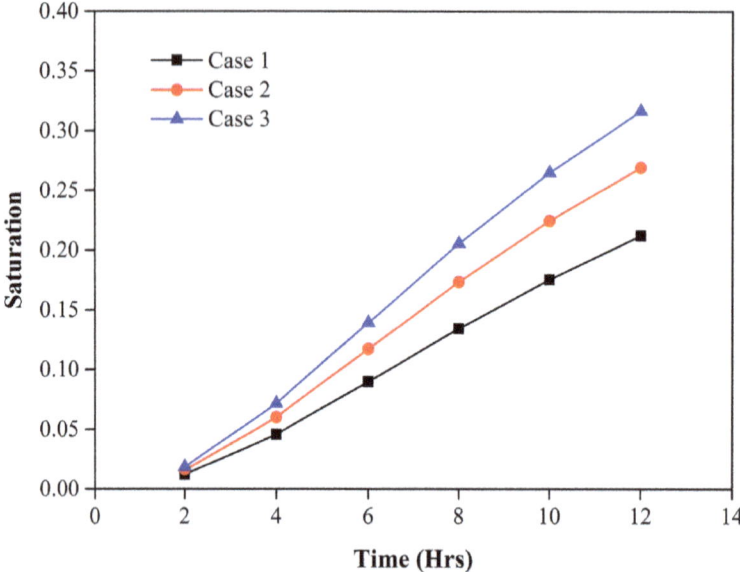

Fig. 7 Saturation of the observation point respecting to time of rain event

the moisture distribution's equation was assumed to have the same form as the heat transfer's equation, which was a linear differential equation.

4   Conclusions

This study was conducted to assess the moisture distribution in a pre-cracked ACP during a short-term rain event via the diffusion process. The ACP with three different pre-crack lengths was evaluated. To model and predict the moisture distribution in a pre-cracked ACP, FEM was used. Based on the results, the conclusions were drawn as the time of rain event and the pre-crack length increase, the zone of moisture distribution becomes larger. Also, the pre-crack length has a significant effect on the moisture distribution in the pre-cracked ACP. When considering the increasing rate of saturation according to time, the current study found that the pavement with higher pre-crack length showed a higher increasing rate. In the future, further research should be conducted to investigate the development of moisture distribution in terms of different conditions such as material, environment, and traffic load.

References

1. Breakah TM, Bausano JP, Williams RC (2009) Integration of moisture sensitivity testing with gyratory mix design and mechanistic-empirical pavement design. J Transp Eng 135(11):852–857
2. Wang W, Wang L, Xiong H, Luo R (2019) A review and perspective for research on moisture damage in asphalt pavement induced by dynamic pore water pressure. Constr Build Mater 204:631–642
3. Nguyen TH, Ahn J, Lee J, Kim JH (2019) Dynamic modulus of porous asphalt and the effect of moisture conditioning. Materials 12(8):1230
4. McDonald T, McDonald P (2010) Guide to Pavement Maintenance. iUniverse
5. Hicks RG (1991) Moisture damage in asphalt concrete. Transportation Research Board
6. Cooley JL, Brown ER, Maghsoodloo S (2001) Developing critical field permeability and pavement density values for coarse-graded Superpave pavements. Transp Res Rec 1761(1):41–49
7. Kassem E, Masad E, Bulut R, Lytton R (2006) Measurements of moisture suction and diffusion coefficient in hot-mix asphalt and their relationships to moisture damage. Transp Res Rec 1970(1):45–54
8. Chen C, Williams RC (2014) Water flow simulation and analysis in HMA microstructure. J Traffic Transp Eng 1(5):362–370
9. Kringos N, Scarpas A (2005) Raveling of asphaltic mixes due to water damage: computational identification of controlling parameters. Transp Res Rec 1929(1):79 87
10. Apeagyei AK, Grenfell JR, Airey GD (2015) Influence of aggregate absorption and diffusion properties on moisture damage in asphalt mixtures. Road Mater Pavement Des 16:404–422
11. Hénon FE, Carbonell RG, Desimone JM (2002) Effect of polymer coatings from CO2, on water-vapor transport in porous media. AIChE J 48(5):941–952
12. Cammarata J, Hariharan N, Allen D, Little D (2020) A study of moisture-induced cracking during a short-term rain event in a pre-cracked asphalt concrete pavement with an expansive base layer. Int J Pavement Eng 21(10):1180–1190
13. Fick A (1995) On liquid diffusion. J Membr Sci 100(1):33–38
14. Lehner F (1979) On the validity of Fick's law for transient diffusion through a porous medium. Chem Eng Sci 34(6):821–825
15. Apeagyei AK, Grenfell JR, Airey GD (2015) Application of Fickian and non-Fickian diffusion models to study moisture diffusion in asphalt mastics. Mater Struct 48(5):1461–1474
16. Alnæs M et al (2015) The FEniCS project version 1.5. Arch Numer Softw 3(100):9–23

Effect of Recycled Coarse Aggregate Treatment Using Cement – Sodium Bicarbonate Slurry on Compressive Strength of Hardened Concrete

Quoc Hai Nguyen Phan, Hoang Tu Vo, Phuong Trinh Bui, and Ngoc Thanh Nguyen

Abstract Recycled concrete aggregate (RCA) from demolished constructions is a promising replacement to natural aggregate (NA) for concrete production. Nevertheless, the adhered old mortar on RCA particles is porous and weak, resulting in reducing mechanical properties of concretes. This study focused on using cement – sodium bicarbonate ($NaHCO_3$) slurry to treat RCA in order to enhance compressive strength of concrete. RCA was disposed by soaking in the slurries consisting of 4% cement and 2, 4, 6, or 8% $NaHCO_3$ for 48 h. Four RCA replacements were 0, 25, 50, and 75% by volume. All concrete specimens were cured in water before compression test at designated days (i.e., 3, 7, and 28 days). Untreated RCA concrete was prepared for comparative purposes. Outcomes indicate that the treatment by cement – $NaHCO_3$ slurry significantly reduced water absorption and crushing value of treated RCA when compared with the untreated one. The compressive strength of concrete with untreated RCA was lower than that of concrete with NA. Compressive strength of concrete with treated RCA had intelligibly growth at all ages when compared with that of concrete with the untreated one. In general, this solution is an efficient method in improving RCA properties through the reductions of its water absorption and crushing value, resulting in an increase in compressive strength of concrete.

Keywords Cement – sodium bicarbonate slurry · Chemical treatment · Crushing value · Recycled concrete aggregate · Water absorption

1 Introduction

Recently, numerous of construction and demolition (C&D) waste rapidly steps up with the growth of urbanization. It is evaluated that approximately 900 million tons of

Q. H. N. Phan · H. T. Vo · P. T. Bui (✉) · N. T. Nguyen
Faculty of Civil Engineering, Ho Chi Minh City University of Technology (HCMUT), 268 Ly Thuong Kiet Street, District 10, Ho Chi Minh City, Vietnam
e-mail: buiphuongtrinh@hcmut.edu.vn

Vietnam National University Ho Chi Minh City, Linh Trung Ward, Thu Duc District, Ho Chi Minh City, Vietnam

C&D waste are discharged in Europe, USA, and Japan [9]. Meanwhile, a consumption of natural stone as common aggregate for concrete production has been gained to 40 billion tons annually [12]. As reported by Tong et al. [21], the sum of C&D waste in Vietnam reached to 12.802 million tons in 2008 and was estimated to be 22.352 million tons in 2015. Therefore, recycled concrete aggregate (RCA) derived from C&D waste is one of the promising materials to solve the lack of natural materials for applications in numerous construction fields such as asphalt concrete, Dallas Houston Metro Area project [22], towards sustainable development. However, the disadvantage of RCA use is that RCA contains adhered old mortar which is approximately 70% by mass [25]. Water absorption of RCA is in range from 4 to 5.2% while that of natural aggregate (NA) is from 0.5 to 1% [5]. Moreover, the interfacial transition zone of RCA is weak and contains micro cracks [4], contributing to reducing the properties of concrete with RCA.

Several chemical methods have been applied to enhance the properties of RCA. A method of pre-soaking RCA with acid compound in removal of mortar was conducted by Tam et al. [13]. However, this method was quite inefficient because the pH of concrete was low, leading to the electrochemical corrosion of steel in the concrete structures [3]. Pacheco and Ding [7] reported that the strong sulphuric and hydrochloric acid might lead to the negative effects on durability of concrete. Grinding solution is executed to remove the adhered mortar on the surface of RCA. Unfortunately, this solution could create micro-fissures in RCA because of the collision between particles when grinding [10]. An accelerated carbonation by curing RCA in elevated gaseous CO_2 environment was also done. However, this approach has many drawbacks such as the long curing period and high cost of additional equipment [23]. Nguyen et al. [6] focused on using silica fume for treating RCA and found an improvement of the performances of concrete with treated RCA. Nevertheless, this method depends on the pozzolanic reaction of silica fume which occurs slowly at early ages [23].

Therefore, the purpose of this study was to investigate the RCA treatment by using cement – $NaHCO_3$ slurry and evaluate the effect of the treatment as the new method on enhancing compressive strength of concretes to utilize RCA for concrete production in Vietnam towards construction sustainability.

2 Experiments

2.1 Materials

Ordinary Portland cement (C) with grade of 40 MPa supplied from Ha Tien company conformed to TCVN 2682:2009 [14] and its chemical compositions are listed in Table 1. Both NA and RCA were used as coarse aggregates while river sand (RS) was used as fine aggregate. The untreated RCA was collected from concrete plants and thereby crushed into particles with the size range from 5–20 mm to replace the

Table 1 Chemical compositions of cement (% by mass)

Compositions (%)	SiO$_2$	Fe$_2$O$_3$	Al$_2$O$_3$	CaO	MgO	Na$_2$O	K$_2$O	SO$_3$	LOI
Content	21.52	2.84	4.54	63.47	1.42	0.39	0.03	1.81	2.72

LOI: Loss on ignition.

NA. The densities of C, NA, RCA, and RS were 3.10, 2.70, 2.54, and 2.73 g/cm^3, respectively. The water absorptions of the NA and untreated RCA were 0.21 and 4.05%, respectively. The crushing values of NA and untreated RCA were 4.7 and 15.7%, respectively. The particle size distributions of RS, NA, and untreated RCA (RCA-UT) are shown in Fig. 1(a). It indicated that the untreated RCA (see Fig. 1(b)) was weaker, and had higher water absorption and finer particles than the NA.

After crushing, RCA was separated in two type, including the untreated and treated one, in order to examine the effect of treatment by using cement – NaHCO$_3$ slurries on its water absorption and crushing value. The treated RCA was immersed in the solution including 4% cement and the various amounts of NaHCO$_3$ as shown in Table 2. Afterwards, treated RCA was dried in air to reach saturated - surface dry (SSD) condition before casting. The particle size distribution (see Fig. 1(a)), water absorption, and crushing value of the treated RCA (RCA-T) were tested as per TCVN 7572–2:2006 [15], TCVN 7572–4:2006 [16], and TCVN 7572–11:2006

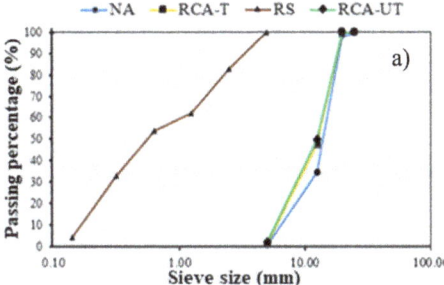

Fig. 1 Particle size distribution of aggregates **a** and untreated RCA after crushing **b**

Table 2 Concentration of cement – NaHCO$_3$ slurry

Symbol	Concentration of slurry (% by mass)		Components (g)			
	C	NaHCO$_3$	C	NaHCO$_3$	W	RCA
N2	4%	2%	40	20	1000	1000
N4	4%	4%	40	40	1000	1000
N6	4%	6%	40	60	1000	1000
N8	4%	8%	40	80	1000	1000

C: cement; W: water; NaHCO$_3$: sodium bicarbonate; RCA: recycled concrete aggregate.

Table 3 Mixture design

Proportion	W/C	RCA replacement (% by volume)	Components (kg)					% by mass of cement
			C	W	RS	NA	RCA	WRA
R0	0.35	0	480	168	848	989	0	0.9
R25-UT	0.35	25	480	168	848	742	233	0.9
R50-UT	0.35	50	480	168	848	495	465	0.9
R75-UT	0.35	75	480	168	848	247	698	0.9
R25-T	0.35	25	480	168	848	742	233	0.9
R50-T	0.35	50	480	168	848	495	465	0.9
R75-T	0.35	75	480	168	848	247	698	0.9

W: water; C: cement; RS: river sand; NA: natural aggregate; RCA: recycled concrete aggregate; WRA: water-reducing agent; W/C: water-to-cement ratio.

[17], respectively. In addition, tap water (W) and water-reducing agent (WRA) were employed in this study.

2.2 Mixture Design

A control mixture with 0% RCA was designed as per TCVN 10306:2014 [18] and ACI 318–11 [1]. Three replacements of NA by untreated and treated RCA were 25, 50, and 75% by volume. All mixture proportions with a constant water-to-cement ratio (W/C) of 0.35 were prepared, as shown in Table 3. The number after each R shows the percentage of replacement of NA by RCA. UT shows the untreated RCA and T shows the treated RCA. The dosage of WRA of 0.9% by mass of cement was employed to remain the designed slump of fresh control concrete at 80 ± 20 mm and to evaluate the effect of RCA treatment on slump of fresh concrete.

2.3 Slump and Compression Tests

The slump of fresh concrete was immediately tested after mixing all components together, as per TCVN 3106:1993 [19]. After the slump test, the fresh concrete was casted into steel cubic moulds with dimensions of $100 \times 100 \times 100$ mm. Then, all the specimens were covered with polyethylene sheets and cured at 30 ± 2 °C. After 24 h of casting, all specimens were demoulded and thereby cured into the water bath.

Three specimens of each mixture proportion at ages of 3, 7, and 28 days were tested to determine the compressive strength of concrete as per TCVN 3118:1993 [20].

3 Experimental Results and Discussion

3.1 Effect of Treatment on Water Absorption and Crushing Value of RCA

Figures 2(a) and (b) show the water absorption and crushing value of RCA after the treatment by using various concentrations of slurry for 48 h. It is clear that the treatment decreased water absorption of RCA when compared with the untreated RCA (UT). According to Fig. 2(a), the water absorption of RCA treated by the slurry with 4% $NaHCO_3$ (N4) was 3.3%. The water absorption of RCA treated by the slurry with 4 and 6% $NaHCO_3$ (N4 and N6, respectively) was nearly the same. Meanwhile, the water absorption of RCA treated by the slurry with 8% $NaHCO_3$ (N8) was higher than that of RCA treated by the slurry with 4 or 6% $NaHCO_3$. This could be explained that a high amount of $NaHCO_3$ could lead to calcium bicarbonate $(Ca(HCO_3)_2)$ formation, which was easy to dissolve in water [2, 24].

According to Fig. 2(b), the crushing value after the treatment by using various concentrations of slurry for 48 h was lower than that of the untreated RCA (UT). This result was similar to the report of Radević et al. [8]. The crushing value of RCA treated by the slurry with 4 and 6% $NaHCO_3$ (N4 and N6, respectively) was nearly the same (i.e., 10.0 and 9.9%, respectively). The improvements of water absorption and crushing value of the treated RCA were due to the $CaCO_3$ formation which filled pores in the adhered mortar in RCA surface [8]. From the experiment results, it can be found that the treatment by using the slurry with 4% cement and 4% $NaHCO_3$ for 48 h was the optimum method for improving the water absorption and crushing value of the RCA. Therefore, the treated RCA by using the slurry with 4% $NaHCO_3$ was used for the concrete production in the present study.

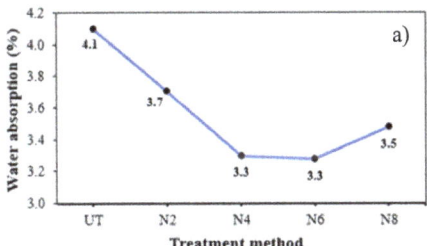 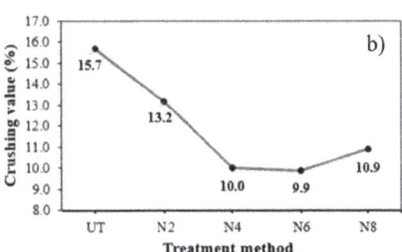

Fig. 2 Water absorption **a** and crushing value **b** of untreated and treated RCA by using various concentrations of slurry for 48 h

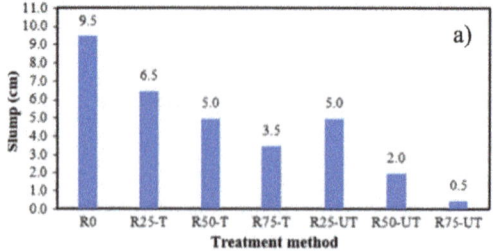

Fig. 3 Effects of untreated RCA and treated RCA on slump of fresh concrete **a** and the formation of $CaCO_3$ in treated RCA **b**

3.2 Effect of RCA Treatment on Slump of Fresh Concrete

Figure 3(a) shows the effect of untreated and treated RCA on slump of fresh concrete. Perceptibly, the slump of fresh concrete with untreated RCA (R25-UT, R50-UT, R75-UT) was lower than that of the control concrete with 100% NA (R0). This tendency was reported in the previous study of Safiuddin et al. [11]. It is mainly explained that the morphology of untreated RCA was very rougher and more irregular than that of NA [11]. In addition, the adhered old mortar in the untreated RCA contained a high number of pores, leading to a high water absorption capacity. Moreover, the untreated RCA had the finer particles than the NA, as seen in Fig. 1(a), resulting in the lower slump of fresh concrete with untreated RCA.

The slump of fresh concrete with treated RCA (R25-T, R50-T, R75-T) was slightly higher than that of fresh concrete with untreated one (R25-UT, R50-UT, R-75-UT) but was still lower than that of the control concrete (R0), regardless of the treatment method. It could be explained that the treated RCA had lower water absorption (see Fig. 2(a)), and more plain and smoother particles than the untreated RCA. Furthermore, the formation of $CaCO_3$ which dispersed cement particles (see Fig. 3(b)), improving rheological of paste and increasing workability of fresh concrete [26].

3.3 Effect of RCA Treatment on Compressive Strength of Concrete

Figure 4 shows the effect of untreated and treated RCA on compressive strength of concrete at the ages of 3, 7, and 28 days. Data demonstrated the concrete with RCA had the lower compressive strength when compared with the control concrete (R0) at all ages, regardless of the RCA treatment. The lower compressive strength was due to the higher crushing value of untreated RCA and treated RCA (15.7 and 10.0%, respectively) than that of NA (4.7%).

Moreover, the concretes with untreated RCA (R25-UT, R50-UT, R75-UT) had the lower compressive strength at all ages when compared with those with treated

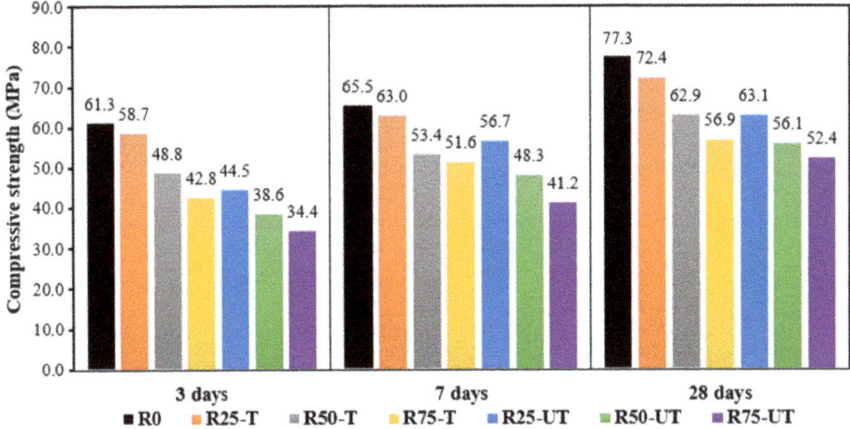

Fig. 4 Effects of untreated and treated RCA on compressive strength of hardened concrete at 3, 7, and 28 days

RCA. This could be due to $CaCO_3$ formation in ITZ between RCA and adhered mortar (as seen in Fig. 3(b)) which might be the factor of this improvement [24].

4 Conclusions

The RCA replacement reduced slump of fresh concrete because the RCA had finer particles and higher water absorption than the NA. However, the use of RCA treated by the slurry of 4% cement and various concentrations of $NaHCO_3$ led to the higher slump of fresh concrete than that of fresh concrete with the untreated RCA.

The compressive strength of concrete containing RCA at the ages of 3, 7, and 28 days was lower when compared with that of concrete with 100% NA, regardless of the RCA treatment. The higher the RCA replacement, the lower the compressive strength of concrete at all ages.

The treatment by using slurry with 4% cement and 4% $NaHCO_3$ was effective in improving water absorption and crushing value of RCA, leading to an enhancement of compressive strength of concrete.

Acknowledgements The authors acknowledge the support of time and facilities from Ho Chi Minh City University of Technology (HCMUT), VNU-HCM for this research. The authors appreciate Mr. Lam Ven Phan and Mr. Le Dai Thanh, Ha Tien 1 Cement Joint Stock Company and LePhan Construction Co., Ltd. for material support.

References

1. American Concrete Institute (2011) ACI318-11 Building Code Requirement for Structural Concrete and Commentary. ACI Committee, Farmington Hills, MI
2. Fernández BM, Simon SJR, Hills CD, Carey PJ (2004) A review of accelerated carbonation technology in the treatment of cement–based materials and sequestration of CO2. J Hazard Mater 112:193–205
3. Hou R, Li S, Ding Y (2018) Experimental study on physicochemical and mechanical properties of mortar subjected to acid corrosion. Adv Mater Sci Eng 2018:1–11
4. McNeil K, Kang THK (2013) Recycled concrete aggregate: a review. Int J Conc Struct Mater 7:61–69
5. Ngo SCN (2004) High–strength structural concrete with recycled aggregates. Diss Uni South Queens 2004:1–112
6. Nguyen HAT, Dinh NNH, Bui PT (2020) Effect of surface treatment of recycled concrete aggregate by cement–silica fume slurry on compressive strength of concrete. J Mater Eng Struct 7:591–596
7. Pacheco TF, Ding Y (2013) Handbook of recycled concrete and demolition waste. Civil. Struc. Eng. 1st edn. Woodhead Publishing. E-book. Accessed 28 Mar 2021
8. Radević A, Despotović I, Zakić D, Orešković M, Jevtić D (2018) Influence of acid treatment and carbonation on the properties of recycled concrete aggregate. Chem Ind Chem Eng Q 24:23–30
9. Shi C, Wu Z, Cao Z, Ling TC, Zheng J (2018) Performance of mortar prepared with recycled concrete aggregate enhanced by CO2 and pozzolan slurry. Cem Concr Comp 86:130–138
10. Shi C, Li Y, Zhang J, Li W, Chong L, Xie Z (2016) Performance enhancement of recycled concrete aggregate–a review. J Clean Prod 112:466–472
11. Safiuddin M, Alengaram UJ, Salam MA, Jumaat MZ, Jaafar FF (2011) Properties of high–workability concrete with recycled concrete aggregate. Mater Resour 14:248–255
12. Tam VWY, Soomro M, Evangelista ACJ (2018) A review of recycled aggregate in concrete applications (2000–2017). Constr Build Mater 172:272–292
13. Tam VWC, Tam CM, Le KN (2007) Removal of cement mortar remains from recycled aggregate using pre-soaking approaches. Resour Conser Recycl 50:82–101
14. TCVN (2009) TCVN 2682:2009 Portland cements–Specifications. Institute of Building Materials, Ministry of Science and Technology, Vietnam
15. TCVN (2006) TCVN 7572-2:2006 Aggregate for concrete and mortar–Test methods–Part 2: Determination of particle size distributions. Ministry of Science and Technology, Vietnam
16. TCVN (2006) TCVN 7572-4:2006 Aggregates for concrete and mortar–Test methods–Part 4: Determination of apparent specific gravity, bulk specific gravity and water absorption. Ministry of Science and Technology, Vietnam
17. TCVN (2006) TCVN 7572-11:2006 Aggregate for concrete and mortar–Test methods–Part 11: Determination of crushing value (ACV) and softening coefficient coarse aggregate. Ministry of Science and Technology, Vietnam
18. TCVN (2014) TCVN 10306:2014 High strength concrete–Proportional design with cylinder sample. Ministry of Science and Technology, Vietnam
19. TCVN (1993) TCVN 3106:1993 Fresh heavyweight concrete–Method for slump test. Ministry of Science and Technology, Vietnam
20. TCVN (1993) TCVN 3118:1993 Heavyweight concrete–Method for determination of compressive strength. Ministry of Science and Technology, Vietnam
21. Tong TK, Nguyen VT, Dang TTH, Tran TVN (2018) Current status of construction and demolition waste management in Vietnam: challenges and opportunities. Int J Geo 15:23–29
22. Yoon CH, Taeyoung Y, Intai K (2011) The application of recycled concrete aggregate for hot mix asphalt base layer aggregate. J Civ Eng 15:473–478
23. Yang J, Shaban WM, Su H, Kim HM, Li L, Xie J (2019) Quality improvement techniques for recycled concrete aggregate: a review. J Adv Concr Tech 17:151–167

24. Zhan BJ, Xuan DX, Zeng W, Poon CS (2019) Carbonation treatment of recycled concrete aggregate: effect on transport properties and steel corrosion of recycled aggregate concrete. Cem Concr Comp 104:103–306
25. Zhang J, Shi C, Li Y, Pan X, Poon CS, Xie Z (2015) Performance enhancement of recycled concrete aggregates through carbonation. J Mater Civ Eng 27:1–7
26. Zli M, Abdullah MS, Saad SA (2015) Effect of calcium carbonate replacement on workability and mechanical strength of Portland cement concrete. Adv Mater Res 1115:137–141

Estimation of Strength Development of Concrete Using Dielectric Measurements

Ryota Isobe, Sachie Sato, Norihiko Kurihara, and Seiichi Sudo

Abstract This study found the possibility of estimating the strength development of concrete during the initial curing period using dielectric measurements. In concrete structures, the strength development due to hydration progress during the initial curing period is a critical process and determines the subsequent performance. The compressive strength of concrete is typically measured by compressing small cylindrical specimens that has been poured. However, since the strength development of concrete is greatly affected by such as surroundings and heat of hydration, the actual construction sites require a direct and non-destructive measurement method for the concrete structures. In this study, we conducted basic experiments to estimate the initial strength development by evaluating the hydration progress of mortar specimens using dielectric measurements. The results showed the dielectric measurements could evaluate the free and bound water state because there was the correlation between the time variation of dielectric relaxation time and the material age.

Keywords Rate of hydration · Dielectric measurement · Curing time · Strength development

1 Introduction

In concrete structures, initial curing is a critical process that determines the subsequent performance. For a concrete member to perform adequately, it is necessary to perform appropriate curing after pouring the concrete at the construction site. However, since the initial strength development is affected by temperature and humidity, fluctuations are likely to occur at actual construction sites. Thus, the impact of external environmental factors on the initial curing period differs depending on the

R. Isobe (✉) · S. Sato · N. Kurihara
Faculty of Architecture and Urban Design, Tokyo City University, 1-28-1 Tamazutsumi, Setagaya, Tokyo, Japan
e-mail: g2181604@tcu.ac.jp

S. Sudo
Faculty of Science and Engineering, Tokyo City University, 1-28-1 Tamazutsumi, Setagaya, Tokyo, Japan

© The Author(s), under exclusive license to Springer Nature Singapore Pte Ltd. 2023   459
J. N. Reddy et al. (eds.), *ICSCEA 2021*, Lecture Notes in Civil Engineering 268,
https://doi.org/10.1007/978-981-19-3303-5_39

type of admixture and cement. Also, it has to confirm whether the initial curing has been adequate. A sufficiently long wet curing period should be used, or the compressive strength should be verified by cylindrical specimens cast at the same time as the concrete. However, a more rational method is required at the construction site.

The strength of concrete develops as the hydration reaction of cement progress and can be evaluated by increasing or decreasing the amount of free water and bound water in the concrete if the concrete can be regarded as sealed before the formwork is removed. I.e., if the amount of moisture inside the concrete can be quickly evaluated, it will be possible to evaluate the performance of concrete during the initial curing process.

Although previous studies have attempted to evaluate the relationship between free water and electrical resistance or capacitance, no attempt has been made to assess the change in bound water by focusing on the movement of water molecules [1–3].

In the present study, we conducted basic experiments to estimate the hydration progress of concrete by evaluating the state of free and bound water in mortar specimens using dielectric measurements.

2 Outline of the Experiment

Two series of experiments were conducted to clarify the effects of the water-cement ratio, materials used, and concrete temperature conditions on the dielectric measurements. In Series 1, the effect of the water-cement ratio on the dielectric spectra was investigated, and in Series 2, the effect of different aggregate types and curing temperatures on the dielectric spectra were investigated.

In Series 1, mortars with the same materials and different water-cement ratios were prepared and examined at 20 °C. Table 1 shows the parameters investigated their values for Series 1.

In Series 2, mortars were prepared using blast furnace cement type B with three different types of fine aggregates. Table 2 shows the parameters for Series 2. In Series 2, two curing temperatures were considered: an ambient temperature of 20 °C and the temperature using an incubator and cooler at 5 °C.

Table 1 Series 1 test parameters

Parameter	Value
Types of cement	Ordinary portland cement
Water to cement ratio	0.45, 0.55, 0.65
Sand to cement ratio	2.0
Test age (day)	1, 2, 3, 5, 7, 14, 91
Curing temperature (°C)	20 ± 3

Table 2 Test parameters for Series 2

Parameter	Value
Water to cement ratio	0.55
Sand to cement ratio	3.0
Curing Temperature(°C)	20, 5
Air content (%)	4.5 ± 0.5
Test age (day)	2, 4, 7, 14, 28, 91

Table 3 Materials used in the experiment

	Material	Symbol	Details
Series 1	**Cement**	**OPC**	Ordinary Portland cement: density of 3.16 g/cm^3
	Fine aggregate	**S**	Natural mountain sand: density (saturated surface-dry) of 2.62 g/cm^3, F.M.2.51
Series 2	**Cement**	**BB**	Blast furnace slag cement type B: density of 3.04 g/cm^3
	Fine aggregate	**S1**	Crushed sandstone: density (saturated surface dry) of 2.61 g/cm^3, F.M.2.98
		S2	Crushed sandstone: density (saturated surface dry) of 2.64 g/cm^3, F.M.3.05
		S3	Natural mountain sand: density (saturated surface-dry) of 2.63 g/cm^3, F.M.3.11
	Admixture	**Ad**	AE water reducing agent: lignin sulfonic acid compound with poly-carboxylic acid ether
		A	AE agent: modified rosin acid compound-based anionic surfactant

The materials used in the experiment are listed in Table 3. In Series 2, the air content was adjusted with an AE water reducer and an AE agent to be $4.5 \pm 0.5\%$.

In all series, the specimens were cured in a sealed condition and demolded at the specified curing temperature and age before being tested. The compressive strength was measured according to JIS R 5201 (physical test method for cement).

The technique of dielectric measurements evaluates the dielectric polarization of atoms or molecules in a material by observing the frequency dependence of the complex permittivity. The complex permittivity ϵ^* is obtained by formula: $\epsilon^* = \epsilon' - j\epsilon''$ where ϵ' is a relative permittivity, ϵ'' is a dielectric loss, j is an imaginary unit. The amplitude of the dielectric spectrum is proportional to the number of dipoles in a unit volume, and the peak frequency of the dielectric loss spectrum reflects the time scale of the rotational motion of the dipoles. Dielectric measurements are one of the most reliable methods for investigating the state of water molecules because the water molecules have a large electric dipole moment. The dielectric spectra of the mortar were measured in the frequency range of 40 to 10 GHz. Dielectric measurements were carried out for frequencies ranging from 100 to 10 GHz using a time domain reflectometry (TDR) and for frequencies ranging from 40 to 110 MHz using an impedance analyzer (4294 A, Agilent). Flat end coaxial electrodes were used for both dielectric measurement sensors. The coaxial electrodes were inserted immediately after casting into the surface of the specimens to a depth of 1 cm. In addition to the specimens' age, the maturity used to evaluate the samples at different curing temperatures.

3　Experimental Results and Discussion

Figure 1 shows the relationship between the experimental results of the compressive strength of the prepared mortar specimens and the maturity. The maturity M(°C day) is obtained by Eq. (1)

$$M = \Sigma(T \times time(day)), \tag{1}$$

where T is the temperature. The datum temperature is defined as 0 °C.

The compressive strength F (N/mm^2) is obtained by Eq. (2)

$$F = a \ln(M) + b, \tag{2}$$

where M is the maturity, a and b are experimental constants. Figure 1 shows the regression curve given by Eq. (2).

Table 4 lists the values of the coefficients a and b in Eq. (2).

Figure 2 shows the dielectric spectra of the mortar with W/C0.55 for Series 1 in the MHz frequency range. The dielectric loss peak reflects the rotational motion of the bound water as observed in the measurement frequency range. This loss peak frequency shifts to a lower frequency with increasing curing time. This peak frequency shift after 74 h is small compared to the early stage of curing, where 74 h is approximately three days and M is 61.7 °C•D at 20 °C curing. It should be noted that the amplitude of the loss peak is constant.

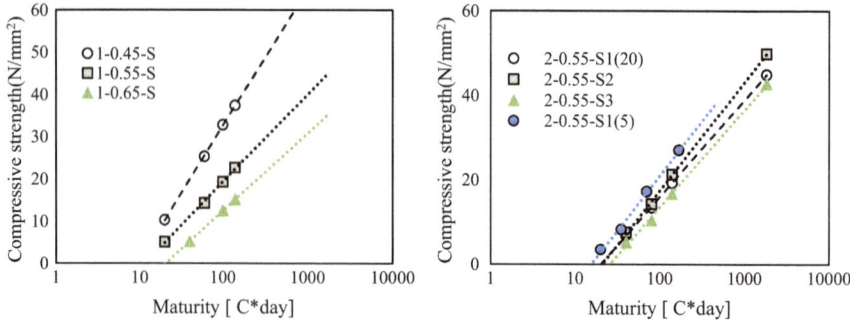

Fig. 1 The relationship between strength development and maturity

Table 4 Values of the experimental constants

	a	B
1-0.45-S	13.95	− 31.56
1-0.55-S	8.96	− 21.88
1-0.65-S	7.93	− 23.92
2-0.55-S1(20)	9.98	− 29.96
2-0.55-S1(5)	11.26	− 30.74
2-0.55-S2	11.15	− 33.91
2-0.45-S3	9.98	− 32.32

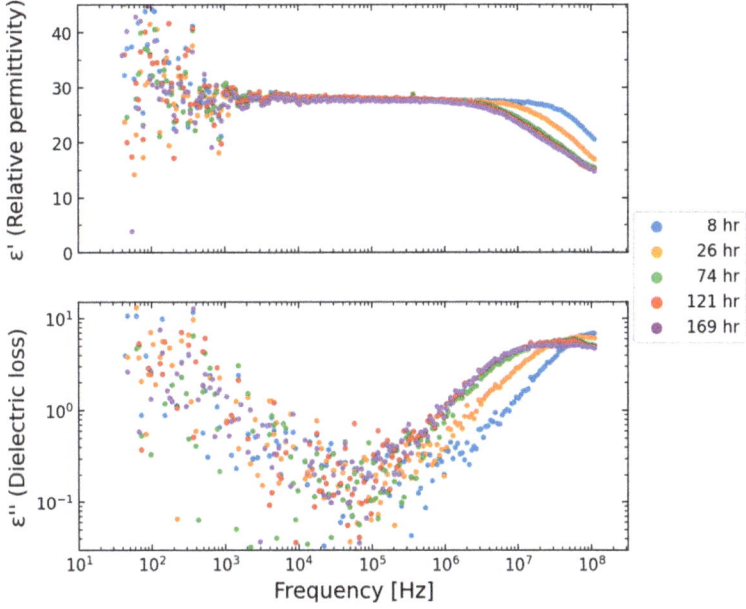

Fig. 2 Dielectric spectra of 1-0.55-S with varying curing

Figure 3 shows the dielectric spectra of the mortar with fine aggregate s1 for Series 2 in the GHz frequency range. The dielectric loss peak reflecting the rotational motion of free water is observed in the measured frequency range. The amplitude of the loss peak frequency decreases with increasing curing time. However, the shift of the peak frequency is minimal.

Figure 4 shows the relationship between the maturity and relaxation time τ(s) of the bound water for the case of fine aggregate s1 in Series 2. The relaxation time τ is obtained by formula: $\tau = 1/2\pi f_m$ where f_m is dielectric loss peak of bound water. The relaxation time increases with the progress of the hydration reaction and the strength development. When the curing temperature is low, the relaxation time

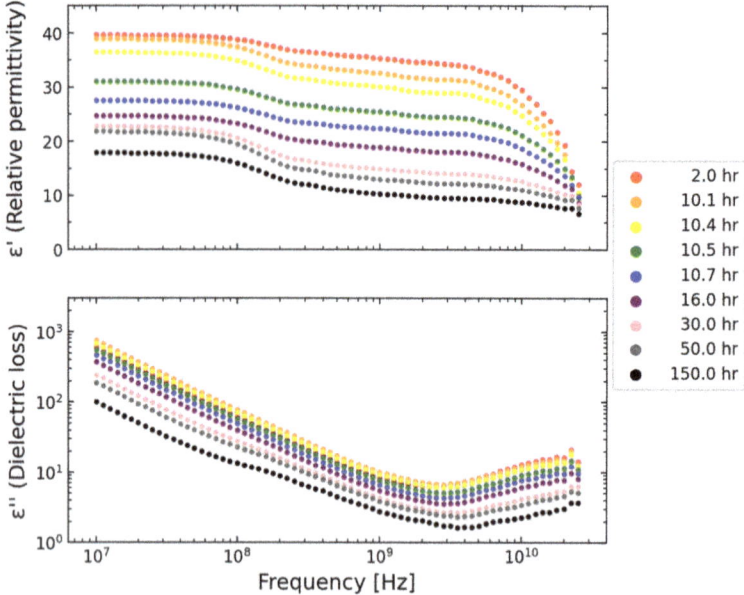

Fig. 3 Dielectric spectra for 2-0.55-S1(20) with varying curing time

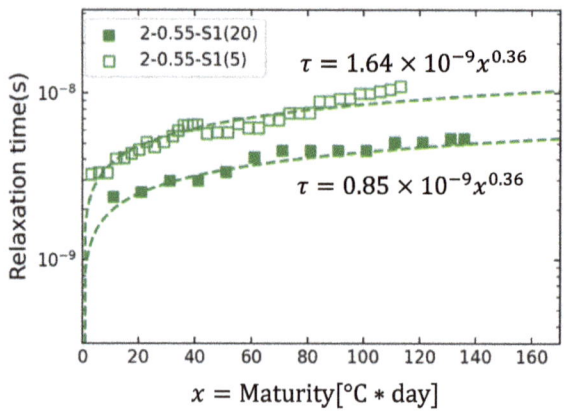

Fig. 4 The relationship between relaxation time and maturity for the two curing temperatures temperatures (2-0.55-S1(20), 2-0.55-S1(5))

increases as the relaxation time is correlated with the apparent activation energy. In the regression equation shown in Fig. 4, the rate of increase is proportional to the maturity to the power of 0.36 regardless of the curing temperature, which is consistent with the results. This suggests that the same formulation with the maturity can be used even when cured under different curing temperatures. Since the increment of the maturity was the same, this relationship was expressed as the time variation

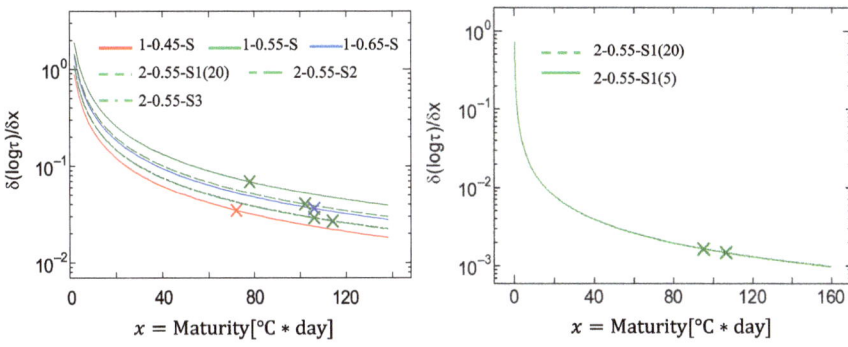

Fig. 5 The relationship between the time variation of relaxation time and maturity

Table 5 Maturity and age (days) at 50% of the 28-day strength

	Maturity [°C*day]	Age (days)
1-0.45-S	73.3	3.7
1-0.55-S	80.2	4.0
1-0.65-S	107.0	5.3
2-0.55-S1(20)	106.2	5.3
2-0.55-S1(5)	92.7	18.5
2-0.55-S2	108.2	5.4
2-0.45-S3	119.4	6.0

of relaxation time, and the relationship with the maturity is shown in Fig. 5. In order to examine the possible evaluation of the initial curing using Eq. (2) and the coefficients given in Table 4, the maturity and age were calculated at which 50% of the strength of the 28-day-old material is developed, with the results shown in Table 5. Based on the results of Table 5, an X mark was plotted at the point where each curve in Fig. 5 intersects the corresponding maturity. As a result, similar results were confirmed for the time variation of relaxation time corresponding to the maturity at which 50% of the 28-day strength was developed. The same results were obtained for different curing temperatures by evaluating the maturity. These results indicate the possible monitoring of the progress of the hydration reaction during the curing period, although it is necessary to investigate the variation among specimens and the reproducibility under different temperature conditions.

4 Conclusion

This study conducted basic experiments with mortar to estimate the initial strength of concrete using dielectric measurements. The results showed that the dielectric measurements can evaluate the state of free and bound water. In addition, it was shown that the time variation of relaxation time to the target initial strength was unaffected by water to cement ratio, fine aggregate, or curing temperature.

Acknowledgements We received cooperation in this study from Rikuya Sugawara, a Tokyo City University student writing a graduation thesis at the time. This research was supported by an industry-university collaboration project with Tokyu Construction. Part of this research was carried out with the support of a JSPS Grant-in-Aid for Scientific Research (16K0659, 19K04695). We would like to thank everyone involved.

References

1. Kutsukake F, Yuasa N, Birumachi M, Yuki H (2013) Expermental examination concerning meaning of indicated value of holding high frequency capacity type moisture meter to concrete. Summaries of Technical Papers of Annual Meeting, Aug, pp 1233–1234
2. Shiraishi M, Nonaka H, Satou D, Kidokoro K, Kawase M, Onozuka Y (2019) Preliminary experiments of high frequency capacity type moisture meter to concrete. Summaries of Technical Papers of Annual Meeting Sep, pp 669–670
3. Wada T (2018) Mesurement result of high frequency capacity type moisture meter to concrete. Summaries of Technical Papers of Annual Meeting Sep, pp 1041–1042

Service Life Prediction of Reinforced Concrete Coastal Structures in Vietnam Subjected to Chloride Ingress

Quynh Chau Truong, Chinh Van Nguyen, Charbel-Pierre El Soueidy, and Emilio Bastidas-Arteaga

Abstract Chloride-induced corrosion on reinforcing steel is one of the main causes that may shorten the service life of concrete structures. Recent studies have found that the deterioration process of these structures is highly influenced by material properties, environmental factors (exposure conditions) and climatic conditions (such as temperature, humidity, seasonal variations, etc.). For infrastructure in coastal areas of Vietnam, it has been observed that corrosion occurs within 10 to 30 years of service. The purpose of this paper is to carry out the service life prediction of reinforced concrete (RC) structures in coastal areas in Vietnam. A model simulation of chloride penetration into concrete before and after repair and under various climatic conditions is developed in FreeFEM++ software to assess the effects on structure service life. A case study is conducted in three coastal cities in Vietnam where the annual temperature and humidity vary. The results show that chloride ingress into concrete is highly influenced by weather conditions. In addition, some maintenance strategies, that are appropriate with the practical conditions in each area in Vietnam, are recommended to prolong the service life of corroding RC structures.

Keywords Chloride penetration · Climatic conditions · Durability · Vietnam · Modelling

Q. C. Truong (✉) · C.-P. E. Soueidy
Institution for Research in Civil and Mechanical Engineering, CNRS UMR 6183, University of Nantes, Nantes, France
e-mail: quynh-chau.truong@etu.univ-nantes.fr; tqchau@dut.udn.vn

Q. C. Truong · C. Van Nguyen
The University of Danang-University of Science and Technology, 54 Nguyen Luong Bang, Danang, Vietnam

E. Bastidas-Arteaga
Laboratoire Des Sciences de l'Ingénieur Pour l'Environnement (LaSIE)-UMR CNRS 7356, Université de La Rochelle, La Rochelle, France

© The Author(s), under exclusive license to Springer Nature Singapore Pte Ltd. 2023
J. N. Reddy et al. (eds.), *ICSCEA 2021*, Lecture Notes in Civil Engineering 268,
https://doi.org/10.1007/978-981-19-3303-5_40

1 Introduction

Steel corrosion in concrete is the primary cause of deterioration of RC structures and eventual reduction of service life. The high alkalinity of the cement paste surrounding the reinforcing steel induces a reactively passive state, but chloride ingress breaks down the passive film on steel reinforcing bars and initiates corrosion. Previous studies indicate that the structural deterioration process subject to chloride ingress is highly affected by material properties, concrete cover thickness, exposure environment and climatic conditions including humidity and temperature [1].

Many models have been developed to simulate the chloride ingress in RC for service life prediction. A simple and general model based on Fick's second law has been commonly adopted, in which the diffusion coefficient and surface chloride content are constant in a saturated environment [18]. However, experimental work has proved that those parameters are time dependent [9, 11] and so a time dependent chloride diffusion model has been proposed [8]. Moreover, the chloride ingress is even more complex in an unsaturated environment, that involves the interaction of diffusion, convection, and chemical phenomena. In response more comprehensive models of chloride ingress have been established [1, 10] that consider coupled heat transfer and moisture transport relationships and chloride binding capacity among those parameters mentioned above.

Vietnam is a tropical country where the climate is hot and humid. The RC structures located along the more than 3,200 km of coastline are vulnerable to corrosion attack due to a high content of chloride ions. The corrosive environments of RC structures can be classified in 3 zones as described in TCVN 12,041: 2017 [17], including the atmospheric zone (XS_1), submerged zone (XS_2), tidal zone and splash zone (XS_3). The deterioration of RC structures in the coastal areas of Vietnam is recognised as a serious problem for building and infrastructure maintenance. It has been reported that steel corrosion in RC structures begins after 10 to 30 years of service, depending on concrete strength, concrete permeability, chloride resistance, cement type, cement content, construction method and maintenance history. Over their service life, the repair cost of these structures is estimated to be 70% of the initial construction cost.

The aim of this study is to investigate the effect of varying environmental conditions in Vietnam on chloride ingress into concrete which leads to service life reduction for RC structures. This paper uses the chloride ingress model developed by Emilio Bastidas et al. [1] to predict the corrosion initiation time and service life of repaired RC structures in coastal areas in Vietnam with the support of FreeFEM++ software. Three coastal cities in the Northern area (Hai Phong), Central area (Danang) and Southern area (Ho Chi Minh) of Vietnam are selected. Based on the results, a proper maintenance strategy for RC structures in each coastal area is recommended.

2 Model of Chloride Ingress and Model of Maintenance

The modelling approach proposed by Bastidas-Arteaga et al. [1] was applied to simulate chloride penetration in concrete. In this method chloride ion transfer is treated using a convection–diffusion coupling. However, this study considers constant surface chloride concentration, temperature, and humidity. When humidity is constant, only the diffusion phenomenon occurs. In this case, the one-dimensional chloride ingress process, considering temperature, humidity and time, is presented in Eq. (1):

$$\frac{\partial C_{fc}}{\partial t} = D_c^* \left(\frac{\partial^2 C_{fc}}{\partial x^2} \right) \tag{1}$$

where D_c^* represents the apparent chloride diffusion coefficients:

$$D_c^* = \frac{D_c}{1 + \left(\frac{1}{w_e} \right) \left(\frac{\partial C_{bc}}{\partial C_{fc}} \right)} \tag{2}$$

where w_e is the amount of free water and $\frac{\partial C_{bc}}{\partial C_{fc}}$ is the chloride binding capacity of cement [12]. The isotherm used to estimate the binding capacity in this study is Langmuir isotherm [7]:

$$C_{bc} = \frac{\alpha \ C_{fc}}{1 + \beta \ C_{fc}} \tag{3}$$

where C_{bc} is the bound chloride concentration, C_{fc} is the free chloride concentration, $\alpha = 0.1185$ and $\beta = 0,09$ are material coefficients for a C_3A content of 8% [7].

D_c in Eq. 2 is the effective chloride diffusion coefficient depending on the temperature, concrete age and humidity [10, 16].

$$D_c = D_{c,ref} . k_1(T) . k_2(t) . k_3(h) \tag{4}$$

$D_{c,ref}$ is a reference diffusion coefficient which has been measured in standard conditions.

$k_1(T)$ is the factor accounting for temperature effect:

$$k_1 = \exp \left[\frac{U_c}{R} \left(\frac{1}{T_{ref}} - \frac{1}{T} \right) \right] \tag{5}$$

$k_2(t)$ is the time effect factor:

$$k_2(t) = \left(\frac{t_{ref}}{t} \right)^{m_c} \tag{6}$$

$k_3(h)$ is the factor of relative humidity effect:

$$k_3(h) = \left[1 + \frac{(1-h)^4}{(1-hc)^4}\right]^{-1} \tag{7}$$

For repairing corrosive concrete, "cover rebuilding" repair is chosen, and this method is implemented by reconstructing the chloride-contaminated concrete cover by several techniques (formed concrete, dry shotcrete, wet shotcrete, and manual repair). In this case, chloride penetration should be modelled before and after repair to evaluate the effectiveness of repair strategy on the durability of RC structures. The interaction of the physical and chemical mechanisms of chloride penetration before repair can be modelled with Eq. 1. However, after repair chloride penetration requires consideration of the interaction of two sources of chloride ions, i) the chloride content on the concrete surface exposed to the environment that penetrates the whole concrete and ii) the distribution of residual chloride ions in the original, aged concrete. These residual chlorides will diffuse into the new concrete layer as well as being redistributed in the remaining concrete layer. In the end, the resulting diffusion of chloride ions in the original and repaired concrete is a combination of chloride ingress from the exposed concrete surface and the chloride redistribution inside the concrete [4, 14].

3   Case Study

3.1   Problem Description and Basic Assumptions

The software FreeFEM++ is used to calculate the simplified one-dimensional model of chloride ingress under varying weather conditions as described above. The maintenance model is then developed to observe the response of RC structures to a chosen repair strategy. Other analysis assumptions are also considered in this study:

- The chloride content in the concrete is zero at the start.
- The structure is exposed to a marine atmospheric zone and located 0.1 km from the shoreline. Chlorides are carried by salt spray from seawater and mixed in the humid air.
- The surface chloride content is assumed to be constant during the lifetime of structures, $C_s = 2.95$ kg/m^3 concrete [15].
- The concrete is modelled as 8% C$_3$A Ordinary Portland Cement (OPC), the cement content of 400 kg/m^3 and water/cement ratio of 0.5 meet Vietnamese standards for RC structures in a marine atmospheric zone [17].
- Concrete used in structure repair is the same quality as the original concrete.

Table 1 The model parameters of the chloride penetration and maintenance models

Parameter	Symbol	Units	Value
Activation energy of the chloride diffusion process	U_c	kJ/mol	41.8
Gas constant	R	J/ (mol K)	8.314
Reference temperature in K	T_{ref}	K	296
Aging factor	m_c		0.15
Humidity at which D_c drops halfway	h_c		0.75
Reference time of exposure	t_{ref}	day	28
Reference chloride diffusion coefficient	$D_{c,ref}$	m²/s	3×10^{-11}

Table 2 Annual temperature and humidity in three cities in Vietnam

Locations	Annual Temperature (°C)	Annual humidity (%)
Haiphong [5]	23.3	85
Danang [2]	25.8	81.6
Ho Chi Minh City [6]	27.4	78

- Service life is 50 years, and the corrosion initiation starts when the chloride content at the position of rebar exceeds a critical value $C_{rit} = 0.5\%$ by weight of cement (about 2.0 kg/m³ concrete) [3].
- Concrete cover is 50 mm ($X_c = 50$ mm).

The model parameters and their values used for chloride ingress model and maintenance model are presented in Table 1.

Climatic conditions are based on three coastal cities in Vietnam with varying temperature and humidity data: Haiphong (Northern Vietnam where the climate is in the subtropical zone, high temperature and humidity), Danang (Central Vietnam, located in the tropical monsoon climate, high temperature with less fluctuation), Ho Chi Minh City (Southern Vietnam, sub-equatorial climate, high and quite stable temperature during the year). The data of annual temperature and humidity in three mentioned-above cities is presented in Table 2.

3.2   Results and Discussion

Service Life Prediction of Unrepaired Concrete Structures

Chloride concentration at rebar depth ($X_c = 50$ mm) in unrepaired concrete structures over a 50-year service life was modelled and is shown in Fig. 1 for the three coastal cities. The surface chloride concentration for all study locations was assumed to be the same and constant. It is observed that the relative humidity and temperature have a significant effect on the progression of free chloride concentration over 50 years. Considering a critical chloride content of 2 kg/m3, the estimated time to corrosion

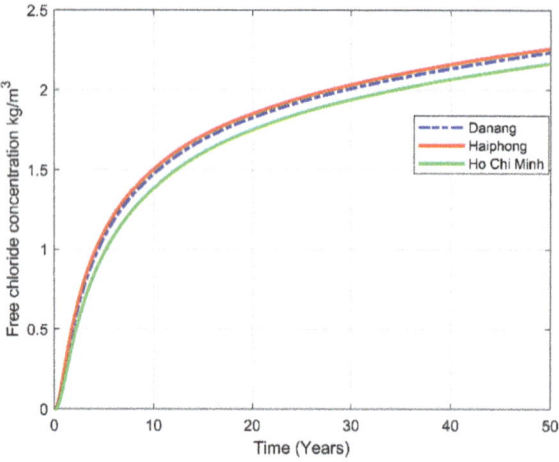

Fig. 1 Chloride profiles at the cover depth of 50 mm in three coastal cities

initiation in Haiphong and Danang are 27.6 years and 29.2 years respectively, while it is 34 years for RC structures in Ho Chi Minh City. Although the average annual temperature in Ho Chi Minh City is recorded as the highest, the time to initiate corrosion on RC infrastructure in this area is the longest. This may be attributed to the lower annual humidity in Ho Chi Minh City compared to the two remaining cities.

Service Life Prediction of Repaired Concrete Structures

Chloride ingress is now modelled considering concrete repair occurs during structure service life. The parameters in Table 1 were used and chloride concentration is again assessed at a depth of 50 mm. A maintenance plan may consider several repair factors including the number of repair cycles, the replacement concrete depth and repair interval (time between two repairs). The first repair strategy considered is the removal and replacement of 50 mm of surface concrete at 15 and 18-year intervals.

Figures 2a, b present 50-year chloride concentration with repair intervals of 15 years and 18 years and a 50-mm repair depth. By taking a threshold chloride content of 2 kg/m^3 at a cover thickness of 50 mm, it is observed that the chloride profiles of Ho Chi Minh satisfy the requirement of reducing the risk of corrosion initiation for multiple repairs. Hence the 50-mm replacement depth and the 18-year maintenance interval are well-adapted for marine structures in Ho Chi Minh City. Meanwhile, in Haiphong, those above repair strategies are not suitable in preventing the risk of corrosion initiation, the repair interval for Haiphong should be shorter, approximately 10 to 13 years. For Danang, it is necessary to address the maintenance issue in the last years of service life when the corrosion rate is faster, and the repair should be conducted at an earlier age than planned.

According to previous studies [13], replacement depth is one of the main factors influencing the effectiveness of maintenance and repairing with a deeper depth is

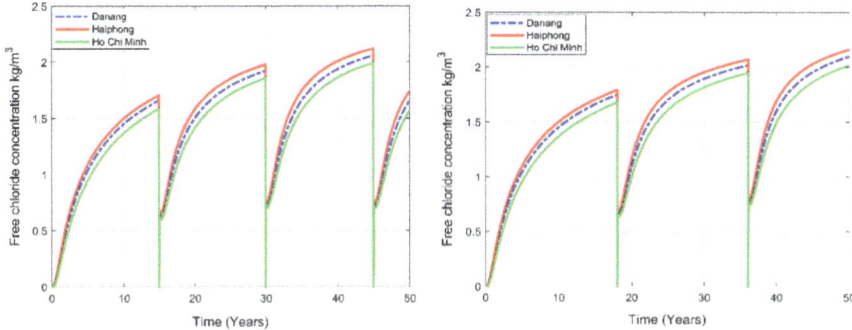

Fig. 2 Chloride profiles at $X_c = 50$ mm with repair interval of **a** 15 years, **b** 18 years ($X_p = 50$ mm)

proved by previous studies to be more effective. Hence, a 60-mm deeper repair depth is considered in this section, with repair intervals of 15- and 18-years. Figures 3a, b show that when considering the threshold chloride level of 2 kg/m³ at the rebar depth, both repair strategies are appropriate for RC structures in Danang and Ho Chi Minh City. In which case the less frequent repair interval of 18 years may be preferred from both economic and environmental aspects as it has less expected number of repairs. For Haiphong, the 18-year interval appears to be a better solution compared to the 50-mm replacement depth. However, a comprehensive maintenance plan should evaluate additional repair intervals and replacement depths to achieve the optimization of protecting structures from corrosion and the maintenance cost.

In summary, temperature and humidity have a significant effect on the service life of RC structures exposed to chloride-heavy environments. The conditions of high temperature and high humidity in most of cities in Vietnam would accelerate the chloride penetration into concrete and shorten the service life of structures before and after repair. Particularly in coastal areas of Northern and Central Vietnam, where

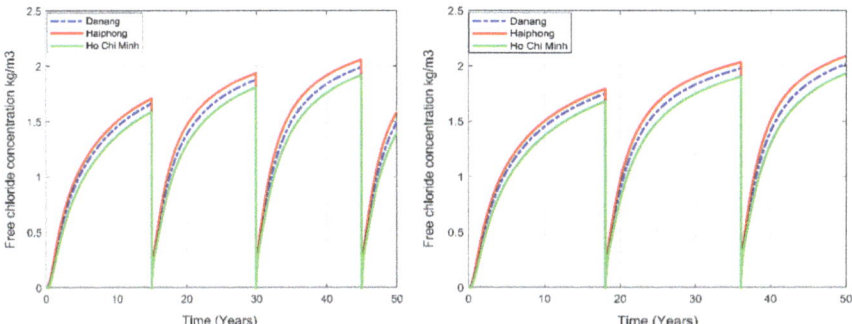

Fig. 3 Chloride profiles at $X_c = 50$ mm with repair interval of **a** 15 years, **b** 18 years ($X_p = 60$ mm)

the average annual humidity and temperature are quite high compared to other locations, maintenance policy and repair strategy should be well planned to ensure the serviceability and durability of chloride contaminated RC structures. Besides cover rebuilding method, it is recommended to include additional corrosion protection methods such as water concrete coating, anti-corrosion coating for reinforcement, or application of supplementary cementitious materials to optimize the effectiveness of maintenance and enhance the durability of structures. Moreover, the current standards for concrete quality and concrete cover depth to prolong the operational life of RC structures subjected to chloride ingress must be complied with.

4   Conclusions

This paper presents the influence of different weather conditions in the three coastal cities of Haiphong, Danang and Ho Chi Minh City on the durability of RC structures exposed to a chloride environment. It is shown that chloride penetration is highly affected by climate conditions through temperature and humidity, in which chloride ingress is more sensitive with humidity rather than temperature. Accordingly, Haiphong with highest humidity have highest risk of corrosion initiation after 27.6 years, followed by Danang with 29.2 years and Ho Chi Minh City with 34 years. Climate conditions also affect the response behaviour of RC structures in each location when applying one repair strategy for structures exposed to chloride attack. Particularly, for 50 mm concrete replacement depth of corroding concrete structures, the 18-year maintenance interval are well-adapted for marine structures in Ho Chi Minh City. Meanwhile, in Haiphong, those above repair strategies are not suitable in preventing the risk of corrosion initiation. For Danang, it is necessary to consider the maintenance issue in the last years of service life and the repair should be conducted at an earlier age than planned. Therefore, it is necessary to consider the local environmental conditions in developing a maintenance and repair strategy for RC structures.

Further improvements are in progress and to be considered:

- Implementing sensitivity analysis and stochastic modelling of chloride penetration to estimate the probability of corrosion initiation.
- Modelling of chloride ingress under real weather conditions in Vietnam with hourly or daily temperature and humidity as input
- Surveying the evolution in time of chloride penetration into RC structures located in submerged zone or tidal zone in coastal areas in Vietnam.

References

1. Bastidas-Arteaga E, Chateauneuf A, Sánchez-Silva M, Bressolette PH, Schoefs F (2011) A comprehensive probabilistic model of chloride ingress in unsaturated concrete. Eng Struct 33(3):720–730
2. Da Nang. (n.d.). Wikipedia. https://en.wikipedia.org/wiki/Da_Nang. Accessed 30 Mar 2021
3. DuraCrete (2000) Probabilistic performance-based durability design of concrete structure. In: The European Union-Brite EuRam III
4. Eldho CA, Jones S, Nanayakkara O, Xia J (2016) Performance of concrete patch repairs: from a durability point of view. In: Proceedings of the 5th International Conference on the Durability of Concrete Structures. International Conference on the Durability of Concrete Structures
5. Hai Phong. (n.d.). Wikipedia. https://en.wikipedia.org/wiki/Haiphong. Accessed 30 Mar 2021
6. Ho Chi Minh City. (n.d.). https://en.wikipedia.org/wiki/Ho_Chi_Minh_City. Accessed 30 Mar 2021
7. Luping T, Nilsson LO (1993) Chloride binding capacity and binding isotherms of OPC pastes and mortars. Cem Concr Res 23(2):247–253
8. Maage JCM, Helland S (1995) Practical non-steady state chloride transport as a part of a model for predicting the initiation period. In: Proceedings of the 1st International RILEM Workshop, pp 398–406
9. Mangat PS, Molloy BT (1992) Factors influencing chloride-induced corrosion of reinforcement in concrete. Mater Struct 25(7):404–411
10. Martín-Pérez B, Pantazopoulou SJ, Thomas MDA (2001) Numerical solution of mass transport equations in concrete structures. Comput Struct 79(13)1251–1264
11. Mustafa MA, Yusof KM (1994) Atmospheric chloride penetration into concrete in semitropical marine environment. Cem Concr Res 24(4):661–670
12. Nilsson LO, Massat M, Tang L (1994) Effect of non-linear chloride binding on the prediction of chloride penetration into concrete structures. In: ACI Symposium Publication, p 145
13. Petcherdchoo A (2018) Probability-based sensitivity of service life of chloride-attacked concrete structures with multiple cover concrete repairs. Adv Civ Eng 2018:1–17
14. Rahimi A, Reschke T, Westendarp A, Gehlen C (2014) Chloride transport in concrete structural elements after repair. In: Fib Symposium 2015 Copenhagen, S, pp 1–14
15. Rodney WM (2001) On the service life modelling of Tasmanian concrete bridges
16. Saetta AV, Scotta RV, Vitaliani RV (1993) Analysis of chloride diffusion into partially saturated concrete. ACI Mater J 90(5):441–451
17. TCVN 12041:2017 (2017) Concrete and reinforced concrete structures—General requirements for design durability and service life in corrosive environments
18. Tuutti K (1982) Corrosion of steel in concrete. Swedish Cement and Concrete Institute

Evaluation on Engineering Properties and Microstructure of Green Mortar Incorporating Steel Slag as Fine Aggregate

Trong-Phuoc Huynh, Hoang-Phong Huynh, Tri-Khang Lam, and Minh-Thien Do

Abstract The use of industrial by-products as alternative materials to naturally sourced materials has been attracted researchers in the world. Following this trend, an experimental evaluation on the effect of using steel slag aggregate (SLA) to replace river sand (RS) on the engineering properties and microstructure of the mortar was performed. The SLA was used to replace RS at levels of 0–100% (an interval of 20%, by volume). The engineering properties of the mortars including compressive strength, water absorption, drying shrinkage, and ultrasonic pulse velocity (UPV) were tested up to the age of 56 days. The microstructure of all mortar samples was also observed. Test results showed that the replacement of <60% RS by SLA negatively affected the properties of the mortars. Increasing the replacement levels to 60–80% enhanced the performance of the mortars. However, the mortar sample with 100% SLA did not show any improvement in its properties. As expected, all of the mortar samples prepared for this study achieved the compressive strength values of all above the target strength of 10 MPa. In which, the 80% SLA sample obtained the highest compressive strength value of 17.4 MPa and 18.9 MPa at 28 and 56 days, respectively. This result was furthermore confirmed through the microstructure observation of the mortars. Consequently, SLA was found as a potential alternative to RS in the production of construction mortars.

Keywords Green mortar · Steel slag aggregate · Compressive strength · Ultrasonic pulse velocity · Microstructure analysis

T.-P. Huynh (✉) · H.-P. Huynh · M.-T. Do
Department of Civil Engineering, College of Engineering Technology, Can Tho University, Campus II, 3/2 Street, Ninh Kieu District, Can Tho City 94000 , Vietnam
e-mail: htphuoc@ctu.edu.vn

T.-K. Lam
School of Graduate, Can Tho University, Campus II, 3/2 Street, Ninh Kieu District, Can Tho City 94000 , Vietnam

© The Author(s), under exclusive license to Springer Nature Singapore Pte Ltd. 2023   477
J. N. Reddy et al. (eds.), *ICSCEA 2021*, Lecture Notes in Civil Engineering 268,
https://doi.org/10.1007/978-981-19-3303-5_41

1 Introduction

Minimization of industrial waste disposal is one of the environmental considerations that encourage the reuse of such waste material in many countries of the world. Million tons of steel slag (SS) are annually generated by the steelmaking industry [1, 2]. Particularly in Vietnam, total SS quantity generated in the 2005–2010 period was under 2 million tons/year and then jumped to 5 million tons in 2018 [3]. Estimation statistic shows that the amount of SS will increase to about 10 and 15 million tons in 2025 and 2030, respectively [3]. Averagely, producing one ton of steel releases about 130–200 kg of SS, which is generated during the separation of the molten steel from impurities in a steelmaking furnace [4]. A large portion of SS has been dumped or landfilled without appropriate treatment, occupying a large number of lands and causing serious environmental impacts [1]. Therefore, the use of SS in civil engineering applications can alleviate the need for its disposal and limit the use of natural resources toward the sustainable development goal [2]. Several studies have proved that SS can be reused in various applications in the construction industry, such as using coarse SS as fine and coarse aggregates in concrete [5, 6], using fine SS as a cementitious material for cement replacement [7], using SS as an aggregate in the production of construction bricks [8] and mortar [9], etc.

Although previous studies have demonstrated a high potential application of SS in manufacturing different types of construction materials [5–9], so far, the reuse of locally sourced SS in the production of construction mortar has been very limited in Vietnam. In recent years, the demand for natural aggregate, especially river sand (RS) for the manufacturing of mortar and concrete has been increasing rapidly in Vietnam. However, the demand for RS was higher than its reserve and the overexploitation of RS caused many drawbacks to the environment and human life [10]. Therefore, the local government has limited the exploitation of RS sources and encouraged the use of alternative materials (e.g. locally available industrial by-products) in construction activities [10]. So that, to overcome the abovementioned issues and provide more information to the literature resource, this experimental study was conducted. The use of locally sourced SS as fine aggregate to either partial or full replacement of natural aggregate as well as the effect of SS aggregate amount on the engineering properties and microstructure of the mortars were evaluated. It is noted that the target strength of the mortar designed for this study was limited to ≥ 10 MPa at 28 days (grade M10 as stipulated by TCVN 4314:2003), where the mortar is suitable for use as building/plastering materials in the finishing stage of the construction.

2 Experimental Details

2.1 Materials and Mixture Proportions

In this study, binder materials used for the preparation of mortar samples included type-PCB40 cement, class-F fly ash (FA), and ground granulated blast-furnace slag (GGBFS) in compliance with TCVN 6260:2009, TCVN 10302:2014, and TCVN 11586:2016, respectively. The specific gravities and primary chemical compositions of these materials are shown in Table 1. It could be seen that cement comprises mainly SiO_2 and CaO, whereas FA primarily consists of SiO_2 and Al_2O_3 and the major compositions of GGBFS are SiO_2, Al_2O_3, and CaO.

Besides, RS and SLA in saturated surface dry condition were used as fine aggregates in the mortar mixtures. Their particle size distribution (PSD) and physical properties are presented in Fig. 1. The scanning electron microscope (SEM) image of the SLA particle is presented in Fig. 2. The sand-size SLA particles had subrounded to angular shapes with rough surface textures. Distinct asperities and edges were visible in angular, bulky particles and a heterogeneous porous structure was also

Table 1 Specific gravities and chemical compositions of cement, FA, and GGBFS

Materials	Specific gravity	Compositions (% by mass)					
		SiO_2	Fe_2O_3	Al_2O_3	CaO	MgO	Others
Cement	2.84	23.5	3.7	6.0	59.9	2.0	5.0
FA	2.14	59.2	6.1	26.7	1.1	0.9	6.1
GGBFS	2.85	35.9	0.3	13.0	38.1	8.0	4.7

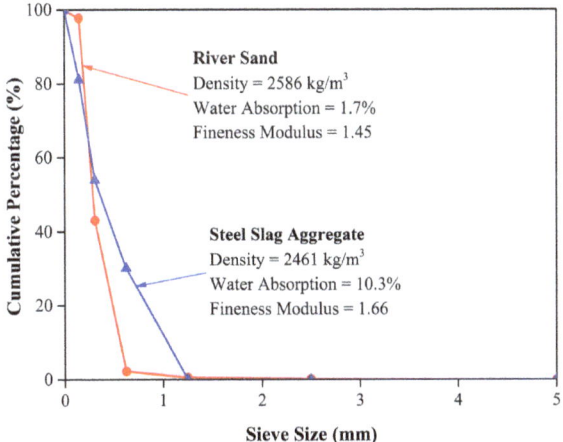

Fig. 1 PSD and physical properties of RS and SLA

Fig. 2 SEM morphology of the SLA particle

observed on the surface of some particles. This supported the significantly higher water absorption rate of SLA in comparison to that of the RS.

This study applied the densified mixture design algorithm (DMDA) as previously described by Chen et al. [11] for designing the mixture proportions of the green mortars. DMDA uses pozzolanic materials (e.g. FA and GGBFS) in combination with cement as binder materials to ensure the long-term strength and durability of the mortars. Thus, the amount of FA and GGBFS were determined through DMDA calculations. A superplasticizer (SP) was used to reduce water and ensure the worka-bility of the mortar mixtures. To investigate the effect of replacing RS by SLA on the performance of the green mortars, six mortar mixtures were designed with the RS replacement levels of 0% (SLA00), 20% (SLA20), 40% (SLA40), 60% (SLA60), 80% (SLA80), and 100% (SLA100) (by volume). Based on the laboratory pre-trials, a water-to-binder (w/b) ratio was kept constant at 0.8 for all mortar mixtures to obtain the target strength of $\geq$ 10 MPa at 28 days as above mentioned. The ingredients for each mixture were presented in Table 2.

Table 2 Mixture Proportions of Green Mortars

Mixtures	w/b	Material Proportions (kg/m³)						
		Cement	FA	GGBFS	RS	SLA	Water	SP
SLA00	0.8	166	129	71	1440	0	293	7.8
SLA20		166	129	71	1152	274	293	7.8
SLA40		166	129	71	864	548	293	7.8
SLA60		166	129	71	576	822	293	7.8
SLA80		166	129	71	288	1096	293	7.8
SLA100		166	129	71	0	1370	293	7.8

2.2 Sample Preparation and Test Methods

All of the ingredients as shown in Table 2 were mixed homogeneously using a laboratory mixer and then different sizes of samples were prepared for various tests. The compressive strength test was conducted at 1, 7, 14, 28, and 56 days on the samples of 40 × 40 × 160 mm following TCVN 3121–11:2003. Drying shrinkage of the mortars was monitored for up to 56 days using the prism samples of 25 × 25 × 285 mm as per TCVN 8824:2011. The water absorption and UPV of the mortars at 28 and 56 days were measured on the cubic samples of 50 × 50 × 50 mm and cylindrical samples of Ø100 × 200 mm according to TCVN 3121–18:2003 and TCVN 9357:2012, respectively. Furthermore, the microstructure of the mortar samples at 56 days was analyzed through their SEM images.

3   Results and Discussion

3.1 Compressive Strength

The effect of the replacement of RS by SLA on the compressive strength development of the mortar samples is presented in Fig. 3. It could be observed that the compressive strength of all mortar samples developed with curing time, demonstrating the continuous development of cement hydration with curing time. At the age of 28 days, the compressive strength values of mortar samples containing 20, 40, and 100% SLA were about 10.4, 10.7, and 13.5 MPa. These values were about 73.4, 75.8, and 95.6%

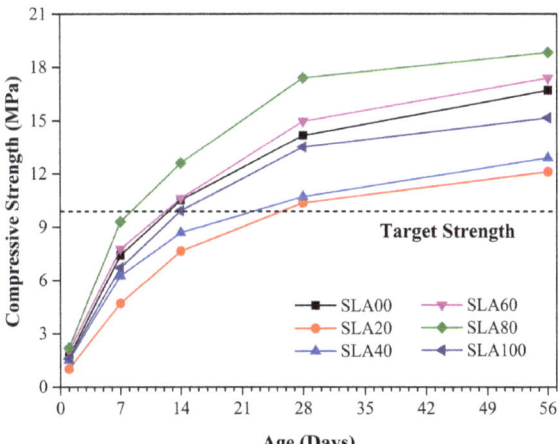

Fig. 3 Compressive strength of mortar samples

as compared to the compressive strength value of the non-SLA sample. The reduction in compressive strength of the mortars could be attributed to the increase in void volume as increasing the amount of SLA in the mixtures [12]. Besides, the SLA particle was larger than the RS particle (Fig. 1). Therefore, the excessive amount of SLA at a fully RS replacement level might result in a non-homogeneous mixture and poor filling effect, leading to the decline in mortar strength [13]. In contrast, the mortar samples incorporating 60 and 80% SLA exhibited an increased compressive strength value in comparison with that of the SLA-free mortar. A similar strength pattern was observed for all mortar samples at 56 days, indicating that the appropriate levels of replacing RS by SLA were 60–80%. The possible reasons for this phenomenon were the contribution of high angularity and rough surface of SLA particles (Fig. 2), which improved the adhesion between the aggregate and hydrated paste, consequently led to increasing the compressive strength [14, 15]. Also, the internal curing effect of SLA promoted the cement hydration as well as the pozzolanic reaction of fine SLA particles, increasing the compressive strength of the SLA60 and SLA80 samples [13, 15]. Moreover, Zeghichi [13] suggested that the filling effect took advantage of an appropriate SLA content, creating a denser matrix and thus increasing the compressive strength of the mortars.

3.2 Water Absorption

The effect of replacing RS by SLA on the water absorption of the mortar samples is presented in Fig. 4. It is found that replacing RS with SLA affected the water absorption of the mortars insignificantly regardless of the replacement level. As a result, the water absorption values of all mortar samples at 28 days were in the range

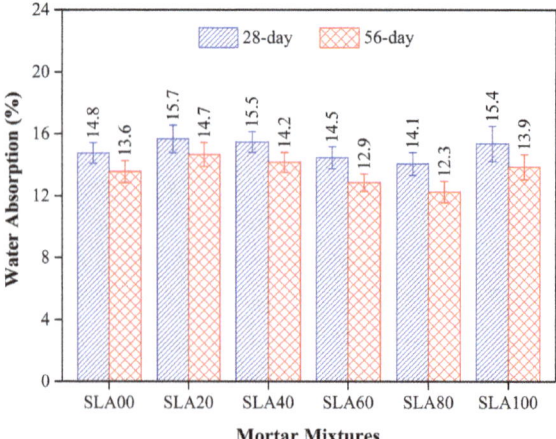

Fig. 4 Water absorption of mortar samples

of $14 \pm 2\%$ and the water absorption values of the mortars at 56 days were about 6.8–14.6% lower than that of the mortars at 28 days. It was due to the continuous growth of the cement hydration with curing time. Following the trend of compressive strength development, water absorption values of the mortar samples with 20, 40, and 100% SLA substitution were slightly increased. Whereas, water absorption values of the samples containing 60 and 80% SLA slightly reduced as compared to that of the no SLA mortars. The rise in water absorption of the SLA20, SLA40, and SLA100 samples was attributable to the significantly higher water absorption capacity of SLA in comparison with that of RS. Thus, as above-mentioned, the incorporation of more SLA was associated with the higher void volume [12] and non-homogeneous mixtures [13], which consequently increased the water absorption of the mortars. Conversely, the slight reduction in water absorption of the SLA60 and SLA80 samples may be attributed to the denser internal structure of the mortars that contributed by the filler effect, internal curing effect, and pozzolanic reaction of the SLA when it was available at an appropriate amount in the mixture [13–15].

3.3 Drying Shrinkage

Figure 5 describes the development of drying shrinkage of the mortars from 0 to 56 days, indicating the change in length of the mortar samples. It could be observed that the content of SLA caused a strong impact on the change in length of the samples.

The mortar samples tended to shrunk at a higher degree as increasing the SLA content in the mixtures and the change in length was more significant at the mortars containing 60–100% SLA. The process of drying shrinkage is caused by the evaporation of water by hydrostatic tension from the small capillary pores of the paste

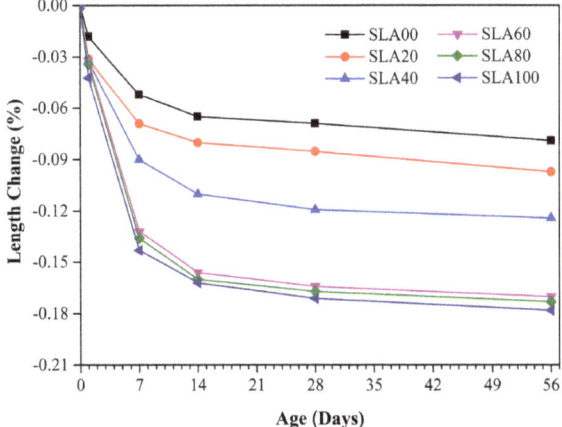

Fig. 5 Drying shrinkage of mortar samples

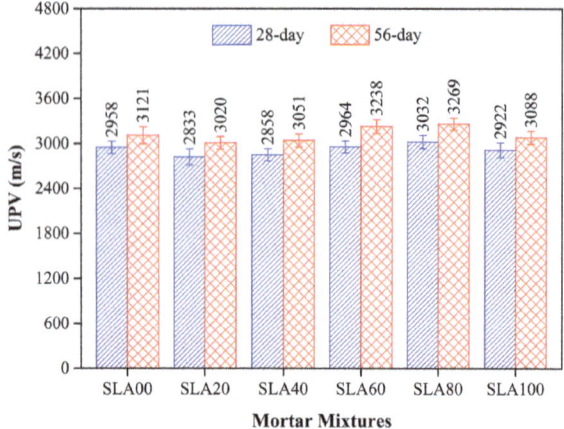

Fig. 6 UPV values of mortar samples

[16]. Thus, the higher shrinkage of the mortars with the higher SLA content may be attributed to the increase in total porosity due to the increase in the SLA amount [12]. Also, since the drying shrinkage strongly depends on the absorption of the aggregate [17], the incorporation of more SLA with high absorption capacity (as shown in Fig. 1) increased the drying shrinkage of the mortars.

3.4 Ultrasonic Pulse Velocity

UPV is a non-destructive test that has been widely used to assess the durability performance of mortar [18]. In this study, the UPV values of all mortar samples at 28 and 56 days are given in Fig. 6. As a result, the UPV values of the mortars at 28 days and 56 days ranged from 2833 to 3032 m/s and from 3020 to 3238 m/s, respectively. Higher UPV values at the age of 56 days indicated a continuous growth of cement hydration with time. This study found that the incorporation of SLA as a substitution for RS affected the UPV values of the mortar samples insignificantly. The UPV values of the mortars containing 20, 40, and 100% SLA were slightly smaller than that of the SLA00 sample, while an increase in UPV values was found that the SLA60 and SLA80 samples. The UPV test results were in good agreement with the compressive strength results as the higher strength was associated with the higher UPV value (see Figs. 3 and 6).

3.5 Microstructure Analysis

The microstructures of the mortar samples at 56 days are displayed in Fig. 7. It could be observed that more pores and less-dense microstructures were observed

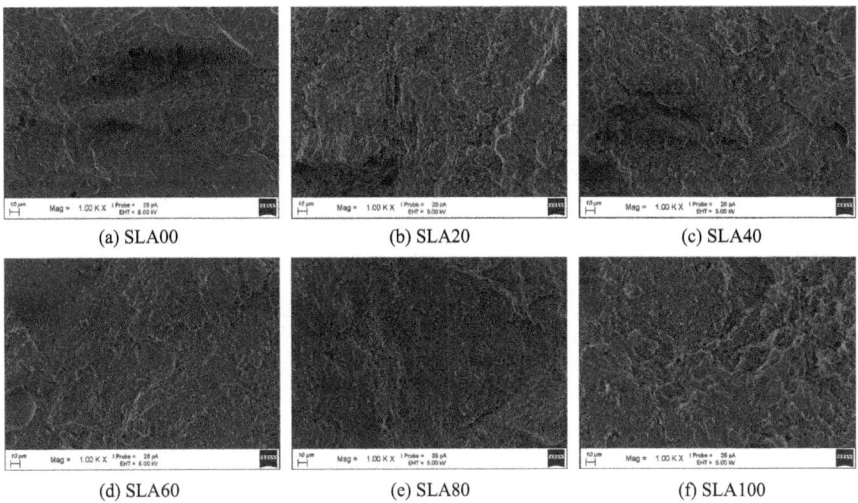

| (a) SLA00 | (b) SLA20 | (c) SLA40 |
| (d) SLA60 | (e) SLA80 | (f) SLA100 |

Fig. 7 SEM micrographs of mortar samples at 56 days

in the SEM micrographs of the mortars with 20, 40, and 100% SLA contents, and the fracture patterns were also propagated in these samples. On the other hand, the mortar samples incorporating 60 and 80% SLA contents exhibited relatively denser microstructures with fewer voids/pores, resulting in lower water absorption and consequently higher strength and UPV values, as discussed previously in this study (see Figs. 3, 4, and 6).

4 Conclusions

Based on the experimental outcomes, it could be concluded that replacing RS with SLA at levels below 60% caused the reduction in compressive strength and UPV while the drying shrinkage and water absorption of the mortars were increased. These properties were positively improved by increasing the replacement levels to 60–80%. However, the 100% SLA sample did not show any improvement in its properties as compared to the non-SLA sample.

As expected, all of the mortar samples had compressive strength values above the target strength of 10 MPa. In which, the 80% SLA sample exhibited the highest compressive strength value of 17.4 MPa and 18.9 MPa at 28 and 56 days, respectively. This result was furtherly confirmed through the SEM morphologies of the mortars. Consequently, SLA was found as a potential alternative to RS in the production of green mortars.

References

1. Yüksel İ (2016) A review of steel slag usage in construction industry for sustainable development. Environ Dev Sustain 19(2):369–384
2. Yildirim IZ, Prezzi M (2011) Chemical, mineralogical, and morphological properties of steel slag. Adv Civ Eng 2011:463638
3. VLXD.org (2020) Solutions for managing, processing, and using iron and steel slag in Vietnam (in Vietnamese). https://vatlieuxaydung.org.vn. Accessed 9 March 2021
4. Furlani E, Maschio S (2016) Long term compression strength of mortars produced using coarse steel slag as aggregate. Adv Civ Eng 2016:3431249
5. Yang JW, Wang Q, Yan PY, Zhang B (2013) Influence of steel slag on the workability of concrete. Key Eng Mater 539:235–238
6. Olonade KA, Kadiri MB, Aderemi PO (2015) Performance of steel slag as fine aggregate in structural performance of steel slag as fine aggregate in structural concrete. Niger J Technol 34:452–458
7. Wang Q, Yang J, Yan P (2013) Cementitious properties of super-fine steel slag. Powder Technol 245:35–39
8. Shih PH, Wu ZZ, Chiang HL (2004) Characteristics of bricks made from waste steel slag. Waste Manage 24(10):1043–1047
9. Faraone N, Tonello G, Furlani E, Maschio S (2009) Steelmaking slag as aggregate for mortars: Effects of particle dimension on compression strength. Chemosphere 77(8):1152–1156
10. Phung T (2017) Warning to 2020 there is no more sand (in Vietnamese). https://tuoitre.vn/canh-bao-den-nam-2020-khong-con-cat-de-xay-dung-1362706.htm Accessed 9 March 2021
11. Chen YY, Bui LAT, Hwang CL (2013) Effect of paste amount on the properties of self-consolidating concrete containing fly ash and slag. Constr Build Mater 47:340–346
12. Washington State DOT (2015) WSDOT strategies regarding use of steel slag aggregate in pavements. A report to the State Legislature in response to 2ESHB 1299, Construction Division Pavements Office, Washington, D.C.
13. Zeghichi L (2006) The effect of replacement of naturals aggregates by slag products on the strength of concrete. Asian J Civ Eng 7(1):27–35
14. Yu X, Tao Z, Song TY, Pan Z (2016) Performance of concrete made with steel slag and waste glass. Constr Build Mater 114:737–746
15. Guo Y, Xie J, Zheng W, Li J (2018) Effects of steel slag as fine aggregate on static and impact behaviours of concrete. Constr Build Mater 192:194–201
16. Wongkeo W, Thongsanitgarn P, Chaipanich A (2012) Compressive strength and drying shrinkage of fly ash-bottom ash-silica fume multi-blended cement mortars. Mater Des 1980–2015(36):655–662
17. Choi SY, Kim IS, Yang EI (2020) Comparison of drying shrinkage of concrete specimens recycled heavyweight waste glass and steel slag as aggregate. Materials 13:5084
18. Lafhaj Z, Goueygou M, Djerbi A, Kaczmarek M (2006) Correlation between porosity, permeability and ultrasonic parameters of mortar with variable water/cement ratio and water content. Cem Concr Res 36:625–633

Frost Damage Resistance of Portland Blast-Furnace Slag Cement Concrete Cured Internally by Using Roof-Tile Waste Aggregate

Taishi Kirimoto, Yuko Ogawa, and Kenji Kawai

Abstract The purpose of this study was to clarify the relationship between the internal curing effect of roof-tile waste aggregate and frost damage resistance, and to investigate the effects of the replacement ratio of roof-tile waste aggregate as well as curing conditions on the frost damage resistance of concrete. Five types of concrete using Portland blast-furnace slag cement were prepared with a water-to-cement ratio of 0.50 and different replacement ratios of roof-tile waste coarse aggregate (20% and 50% by volume) and roof-tile waste fine aggregate (23% and 58% by volume). Concrete specimens were cured at 20 °C under sealed condition for 3, 7, or 28 days, and they were subsequently exposed to air at 20 °C and 60%R.H. until the age of 28 days. Then, the tests for resistance of concrete to freezing and thawing were carried out for 300 cycles in water, and the relative dynamic modulus of elasticity as well as the mass loss was assessed at the designated periods. As a result, it was found that sufficient frost damage resistance could be secured when the replacement ratio of roof-tile waste aggregate was approximately 20% by volume, and the improvement in frost damage resistance was observed especially when fine aggregate was replaced.

Keywords Roof-tile waste aggregate · Internal curing · Curing condition · Frost damage resistance

1 Introduction

Generally, curing is conducted for concrete immediately after casting in a humid environment in order to promote the cement hydration. However, external curing water sometimes cannot reach the inside of concrete sufficiently. In addition, blast-furnace slag cement concrete needs extended curing period of 2 days compared with ordinary Portland cement concrete for keeping enough durability [1]. Internal curing, which is expected to solve this problem, is a curing method in which part of the aggregate is replaced with saturated internal curing material to supply internal

T. Kirimoto · Y. Ogawa · K. Kawai (✉)
Civil and Environmental Engineering Program, Graduate School of Advanced Science and Engineering, Hiroshima University, Higashihiroshima City, Japan
e-mail: kkawai@hiroshima-u.ac.jp

water into the cement paste continuously, resulting in cement hydration promotion. As an internal curing material, lightweight aggregate (LWA) and super absorbent polymer (SAP) are well known, and it has been reported that these internal curing materials can make the pore structure of concrete denser and improve the strength and durability of the concrete [3]. However, some studies indicated that the compressive strength of concrete containing LWA or SAP was the same as or lower than that of the reference concrete [2]. The lower strength of concrete with LWA can be caused by the high crushing value of LWA.

Currently, roof-tile waste aggregate has been proposed as an internal curing material. Roof-tile waste aggregate is made by crushing off-specification roof tiles. In the Chugoku region of Japan, a kind of roof-tile, "Sekisyu kawara" is produced, and approximately 13,000–16,000 tons of off-specification products are generated annually, and recycling of the off-specification products is desired. Roof-tile waste aggregate has a higher water absorption than conventional aggregates and a lower crushing value than conventional internal curing materials. It has been reported to be effective in improving the performance of Portland blast-furnace slag cement concrete [6].

On the other hand, the use of internal curing materials with high water absorption increases the amount of freezable water in concrete, which is generally concerned about lowering the frost damage resistance of concrete [7]. However, it has been reported that the densification of the microstructure due to the internal curing effect may improve the frost damage resistance of concrete [8]. In addition, the difference in internal curing effect between coarse aggregate and fine aggregate of waste roof-tiles is still not clear. Therefore, the purpose of this study was to clarify the relationship between the internal curing effect of roof-tile waste aggregate and frost damage resistance, and to investigate the effects of the replacement ratio of roof-tile waste aggregate as well as curing conditions on the frost damage resistance of concrete.

2 Experimental Procedures

2.1 Materials

Portland blast-furnace slag cement type B (BB), which conforms to JIS R 5211 [5] and contains 30%-60% ground granulated blast furnace slag, was used as cement. Table 1 shows the chemical compositions of BB. The density of the cement was 3.04 g/cm^3 while the Blaine fineness was 3860 cm^2/g. Crushed quartz-porphyry was used as the conventional fine and coarse aggregates in the concrete. Roof-tile waste coarse aggregate (RWCA) and roof-tile waste fine aggregate (RWFA) were used

Table 1 Chemical compositions (%) of BB

SiO$_2$	Fe$_2$O$_3$	Al$_2$O$_3$	CaO	MgO	K$_2$O	Na$_2$O	SO$_3$	LOI
25.95	1.92	8.14	53.73	4.30	0.41	0.24	2.05	1.71

Table 2 Physical properties of aggregates

	Crushed quartz-porphyry		Roof-tile waste	
	fine aggregate	coarse aggregate	fine aggregate	coarse aggregate
Notation	S	G	RWFA	RWCA
Density in saturated surface-dry condition (g/cm^3)	2.61	2.62	2.28	2.26
Water absorption (%)	1.05	0.67	9.75	9.08
Aggregate size (mm)	0–5	5–20	1–5	5–13

as internal curing materials in a saturated surface-dry condition after 7-day water immersion. For these aggregates, the physical properties are listed in Table 2 and the particle size distributions are shown in Fig. 1. Images of roof-tile waste aggregates are also shown in Fig. 2.

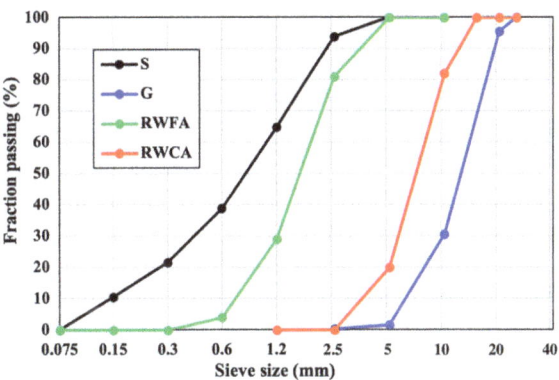

Fig. 1 Particle size distributions of aggregates

      (a) RWFA   (b) RWCA

Fig. 2 Images of roof-tile waste aggregates

Table 3 Mixture proportions and properties of fresh concrete

	Unit content (kg/m³)						AE (C × %)	WR (C × %)	Properties of fresh concrete		
	W	C (BB)	S	RWFA	G	RWCA			Slump (cm)	Air content (%)	Temperature (°C)
BBC	170	340	783	0	977	0	0.0050	0.0025	12.0	5.2	27.8
G20	170	340	783	0	781	170	0.0030	0.0035	11.0	4.1	24.0
G50	170	340	783	0	490	423	0.0030	0.0035	11.0	4.0	24.0
S23	170	340	603	157	977	0	0.0060	0.0020	12.0	4.8	21.2
S58	170	340	329	397	977	0	0.0060	0.0020	10.5	4.7	21.7

2.2 Mixture Proportions

Five mixtures were prepared in this study, as presented in Table 3. The water-to-cement ratio (W/C) and the unit water content of all the mixtures were fixed at 0.50 and 170 kg/m³, respectively. A mixture named BBC was the reference concrete to compare with other concretes using roof-tile waste aggregate. Mixtures G20 and G50 denote concretes in which the replacement ratios of RWCA were 20% and 50% by volume, respectively, while mixtures S23 and S58 denote concretes in which the replacement ratios of RWFA were 23% and 58% by volume, respectively. The amount of internal curing water in G20 and S23 was 14.10 kg/m³, while that in G50 and S58 was 35.15 kg/m³. Besides, chemical admixtures including a water-reducing agent (WR) and an air-entraining agent (AE) were used to achieve the slump and air content of 10.0 ± 2.0 cm and 4.5%±0.5% as the targets for all the fresh concretes in the material design.

2.3 Curing Conditions

Concrete specimens were covered with aluminum adhesive tape and wet fabrics immediately after casting to avoid water evaporation. The concrete specimens were cured in their molds in a controlled room at 20 °C and 60%R.H. for 3 (3D), 7 (7D), or 28 days (Sealed). After the curing, the specimens for 3D and 7D were subsequently exposed to air at 20 °C and 60%R.H. until the age of 28 days.

2.4 Frost Damage Resistance

For each mixture and curing condition, two specimens of 100 × 100 × 400 mm were prepared. Test for frost damage resistance of concrete was started at the age of 28 days according to the method A (Freezing and thawing in water) of JIS A

1148 [4] for 300 freeze-thaw cycles. The freeze-thaw cycles consisted of alternately lowering the temperature of the center of specimen from 5 °C to −18 °C and raising it from −18 °C to 5 °C. The duration of each cycle was 3 h. The primary resonance frequency of the deflection vibration and the mass of the specimens were measured after every 30 freeze-thaw cycles, and the relative dynamic modulus of elasticity, durability factor (DF) and mass change ratio were calculated. The DF calculated by using Eq. 1 is an index to evaluate the frost damage resistance of concrete, and generally, if this value is 60 or higher, it is judged to have sufficient frost damage resistance. The mass change ratio was determined using Eq. 2.

$$DF = \frac{P \times N}{M} \tag{1}$$

where DF is the durability factor, P is the relative dynamic modulus of elasticity after N cycles (%), N is the number of freeze-thaw cycles where the relative dynamic modulus of elasticity is 60% or 300 freeze-thaw cycles (whichever is smaller), and M is 300 freeze-thaw cycles.

$$W_n = \frac{w_0 - w_n}{w_0} \times 100 \tag{2}$$

where W_n is the mass change ratio after n cycles (%), w_n is the mass of specimen after n cycles (g), and w_0 is the mass of specimen in 0 cycles (g).

2.5 Mass Change During Exposure to Air

Mass change during exposure to air was measured using two specimens of $100 \times 100 \times 400$ mm prior to the test for frost damage resistance in order to assess the water content in concrete. On the assumption that all the mass loss during exposure to air was due to water evaporation, the amount of water in the concrete before the frost damage resistance test was determined using the theoretical water content in each mixture and the amount of mass loss during exposure to air. The theoretical water content in each mixture is listed in Table 4. The remaining water content in the concrete was defined as the sum of the amount of water absorbed up to the maximum value of the mass gain at the initial freeze-thaw cycles and the amount of water before the frost damage resistance test as expressed by Eq. 3. That is, the remaining water content in the concrete includes the free water that can freeze during the frost damage resistance test and the water in hydrates.

$$Remaining\ water\ content\ in\ the\ concrete = m_t - m_l + m_a\ (\text{kg/m}^3) \tag{3}$$

Table 4 Theoretical water content in each mixture

| | Unit content (kg/m^3) | | | | |
	W	Water in S	Water in G	Internal curing water	Total
BBC	170	8.20	6.50	0	184.70
G20	170	8.20	5.22	14.10	197.52
G50	170	8.20	3.31	35.15	216.66
S23	170	6.32	6.50	14.10	196.92
S58	170	3.45	6.50	35.15	215.10

where m_t is the theoretical water content (kg/m^3), m_l is the amount of mass loss during exposure to air (kg/m^3), and m_a is the amount of water absorbed up to the maximum value of the mass gain at the initial freeze-thaw cycles (kg/m^3).

Water in S, water in G and internal curing water are calculated from the water absorption and the unit content of each material.

3 Results and Discussion

3.1 Frost Damage Resistance

The results of DF calculated from the relative dynamic modulus of elasticity are shown in Fig. 3. All the specimens except for G50-Sealed and S58 had higher DF than 60, and had sufficient frost damage resistance. Especially, the DF of S23 for all the curing conditions was higher than that of BBC. However, when the replacement ratio of RWFA was approximately 50%, the DF was lower than 50, and S58 concretes were not resistant to frost damage regardless of curing conditions. On the other hand,

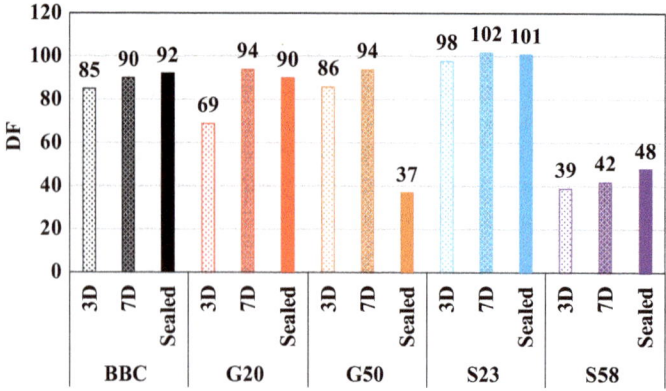

Fig. 3 DF calculated from the relative dynamic modulus of elasticity

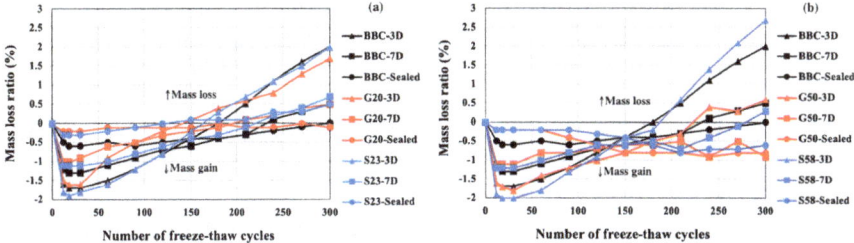

Fig. 4 Results of the mass change ratios: **a** the replacement ratio of roof-tile waste aggregate was approximately 20%, **b** the replacement ratio of roof-tile waste aggregate was approximately 50%

G50 had high frost damage resistance when they were exposed to air after sealed curing for a certain period of time.

Figures 4a and b show the mass change ratios for the cases where the replacement ratio of roof-tile waste aggregate was approximately 20% and 50%, respectively. The mass gain at the initial freeze-thaw cycles and the mass loss ratio afterwards were larger for the specimens with a shorter sealed curing period, and this tendency was more significant for the specimens using RWFA than RWCA. When the fine aggregate was partially replaced by RWFA, the roof-tile waste aggregates existing closer to the surface of the concrete could be degraded by the freeze-thaw action, resulting in inferior scaling resistance. On the other hand, the scaling resistance was not degraded by using roof-tile waste aggregate when the concrete was sufficiently cured under sealed condition.

3.2 Remaining Water Content in Concrete

Figure 5 shows the relationship between the DF and the remaining water content in concrete, and the remaining water content is shown in Table 5. As illustrated in Fig. 5, when the replacement ratio of roof-tile waste aggregate was approximately 50%, the DF was less than 60 and the amount of remaining water in concrete tends to be large. On the other hand, in the cases of G50-3D and G50-7D, the evaporated water content in concrete during exposure to air (m_l) was larger than the amount of water absorbed at the initial freeze-thaw cycles (m_a) as shown in Table 5. It might imply that the reduction in the remaining water content in the concrete resulted in high resistance to frost damage of the concrete using RWCA even if the replacement ratio of roof-tile waste aggregate was approximately 50%. Previous study using LWA showed that the best way to produce concrete with high frost damage resistance is to use LWA with a low water content of less than 5% [7]. On the other hand, the roof-tile waste aggregate used in this study has a water content of 9%–10%, but it was observed that the frost damage resistance can be secured or improved by adjusting the replacement ratio or reducing the remaining water content in the concrete by exposure to air after appropriate curing.

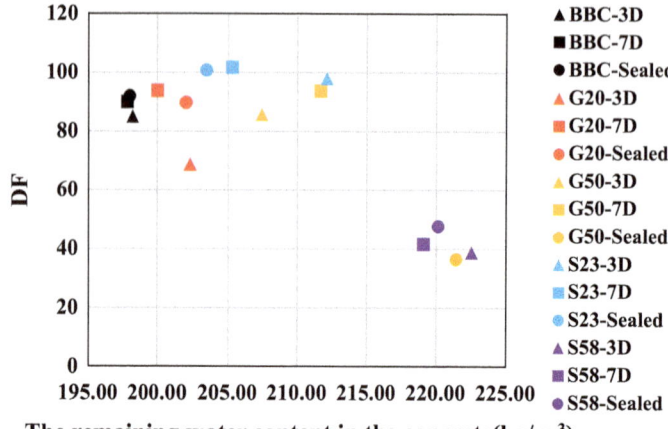

The remaining water content in the concrete(kg/m³)

Fig. 5 Relationship between DF and remaining water content in the concrete

Table 5 Remaining water content in the concrete

Specimens		m_t (kg/m³)	m_l (kg/m³)	m_a (kg/m³)	Remaining water content in the concrete $(m_t\text{-}m_l + m_a)$ (kg/m³)
BBC	3D	184.70	25.77	39.25	198.18
	7D	184.70	15.91	29.00	197.79
	Sealed	184.70	0	13.25	197.95
G20	3D	197.52	31.72	36.50	202.30
	7D	197.52	20.29	22.75	199.98
	Sealed	197.52	0	4.50	202.02
G50	3D	216.66	48.22	39.00	207.44
	7D	216.66	29.01	24.00	211.65
	Sealed	216.66	0	4.75	221.41
S26	3D	196.92	25.32	40.50	212.10
	7D	196.92	16.91	25.25	205.26
	Sealed	196.92	0	6.50	203.42
S58	3D	215.10	36.32	43.75	222.53
	7D	215.10	22.05	26.00	219.05
	Sealed	215.10	0	5.00	220.10

4 Conclusions

1) The durability factor was higher than 60 in both cases of using RWCA and RWFA, when the replacement ratio of roof-tile waste aggregate was approximately 20% by volume. Moreover, the improvement in frost damage resistance was observed when using RWFA.

2) When the replacement ratio of roof-tile waste aggregate was more than 50%, the DFs in almost all the cases were lower than 60. However, when the concrete cured under sealed condition for a certain period followed by exposure to air, the DF of concrete even using RWCA at the replacement ratio of 50% was higher than 60.

3) The scaling resistance was not degraded by using roof-tile waste aggregate when the concrete was sufficiently cured under sealed condition.

Acknowledgements The authors would like to express their gratitude to Tokuyama Corporation for their cooperation in the tests for frost damage resistance in this study.

References

1. Dan Y, Iyoda T, Ohtsuka Y, Sagawa Y, Hamada H (2009) The relationship between curing condition and durability on concrete using blast-furnace slag cement. J Jpn Soc of Civ Eng E 65(4):431–441
2. de Sensale GR, Goncalves AF (2014) Effects of fine LWA and SAP as internal water curing agents. Int J Concr Struct Mater 8(3):229–238
3. Hasholt MT, Jespersen MH, Jensen OM (2010) Mechanical properties of concrete with SAP. Part I: Development of compressive strength. International RILEM Conference on Use of Superabsorbent Polymers and Other New Additives in Concrete pp 117–126
4. JIS A 1148 (2019) Freeze-thaw test method for concrete. Japanese Standards Association, Tokyo
5. JIS R 5211 (2019) Portland blast-furnace slag cement. Japanese Standards Association, Tokyo
6. Sato R, Shigematsu A, Nukushina T, Kimura M (2011) Improvement of properties of Portland blast furnace cement type B concrete by internal curing using ceramic roof material waste. J Mater Civ Eng 23(6):777–782
7. Tachibana D, Imai M (1994) Improving techniques for frost damage resistance of high-strength lightweight concrete. J Jpn Soc Civ Eng 496:51–60
8. Yildirim ST, Meyer C, Herfellner S (2015) Effect of internal curing on the strength, drying shrinkage and freeze-thaw resistance of concrete containing recycled concrete aggregates. Constr and Build Mater 91:288–296

Influence of Superplasticizer on Workability and Early Strength of Fly Ash/Blast Furnace Slag Based Geopolymer Mortar

Kemala Indriani, Apriany Saludung, Yuko Ogawa, and Kenji Kawai

Abstract Recent studies have shown the presence of ground granulated blast-furnace slag (GGBS) in fly ash based geopolymer induced denser microstructures and better compressive strength. However, the inclusion of GGBS gives rapid setting time despite its benefit in terms of mechanical properties and could cause difficulty in handling the geopolymer in its application such as troubles with delivery time of fresh mix geopolymer concrete. This study aims to investigate the effect of polycarboxylate based superplasticizer, which is commonly used in Portland cement concrete to improve the workability and achieve appropriate compressive strength, on the fresh properties and early strength of geopolymer mortar. Geopolymer mortar was composed of fly ash and GGBS acting as binder with sodium silicate and 12 M of sodium hydroxide solution as alkaline activator. Alkaline to binder and fly ash/GGBS ratios were kept constant at 0.5 and 1.5, respectively. Superplasticizer was added into alkaline solution before casting at various dosages of 0%, 1%, 2%, 4%, and 8% by the binder mass and all the specimens were cured in two regimes of heat curing and 20°C curing. The results have shown that the inclusion of superplasticizer maintained the workability over the time and may delay the setting of geopolymer mortar. Moreover, the mixture with 8% superplasticizer had dense microstructure and comparable compressive strength to that without superplasticizer at the age of 7 days.

Keywords Superplasticizer · Geopolymer · Workability · Compressive strength

1 Introduction

Portland cement is one of the most used construction materials in the world, and consequently the Portland cement industry contributes around 8% of worldwide

K. Indriani · Y. Ogawa · K. Kawai (✉)
Civil and Environmental Engineering Program, Graduate School of Advanced Science and Engineering, Hiroshima University, 1-4-1 Kagamiyama, Higashihiroshima, Japan
e-mail: kkawai@hiroshima-u.ac.jp

A. Saludung
Department of Civil and Environmental Engineering, Graduate School of Engineering, Hiroshima University, 1-4-1 Kagamiyama, Higashihiroshima, Japan

© The Author(s), under exclusive license to Springer Nature Singapore Pte Ltd. 2023   497
J. N. Reddy et al. (eds.), *ICSCEA 2021*, Lecture Notes in Civil Engineering 268,
https://doi.org/10.1007/978-981-19-3303-5_43

emission of CO_2 [2]. This substantial amount of greenhouse gases harms the environment despite its excellent properties for durable concrete structures, for instance around 0.7 tons of CO_2 is released to produce 1 ton of cement [10]. Therefore, reduction of Portland cement use is required in near future to decrease its negative impact to the environment. Geopolymer is one of the alternatives to ordinary Portland cement (hereinafter called OPC), firstly proposed by Davidovits, who derived geopolymer cement from aluminosilicate sources such as fly ash and metakaolin, mixed together with alkaline activator and yielded polymeric Si-O-Al bonds through chemical reaction [3, 4]. The geopolymer has been investigated over decades where fly ash and ground granulated blast furnace slag (hereinafter called GGBS) are commonly used as aluminosilicate sources. Geopolymer concrete aims to be an eco-friendly concrete that does not cause harmful emissions. Thus, sustainable development goals through an environmentally conscious construction could be achieved in the future by geopolymer.

As mentioned previously, GGBS is one of aluminosilicate sources generally used for geopolymer, and the presence of GGBS in geopolymer induced dense and refined pore structure which gives compactness and improved compressive strength of geopolymer compared to fly ash based geopolymer [8, 15]. However, it inversely affects the setting time of geopolymer. It was reported that the inclusion of GGBS in geopolymer gives rapid setting time despite its benefit in terms of mechanical properties [5, 16]. The initial and final setting times of geopolymer significantly decreased with the increase of GGBS content in geopolymer [12], and low workability was indicated as the decrease in slump and flow value with the increase in the GGBS content [12, 16]. Furthermore, this behavior could cause difficulty in handling geopolymer in its application such as troubles with delivery time of fresh mix geopolymer concrete.

Besides, studies on utilization of superplasticizer in geopolymer were conducted to tackle the rapid setting time with GGBS inclusion. Generally, superplasticizer used in OPC concrete aims at high workability without excessive addition of water, and thus appropriate compressive strength can be obtained. Polycarboxylate based superplasticizer was found to retard the setting time of $Ca(OH)_2/Na_2SO_4$ activated geopolymer paste [1]. In addition, the study concluded the improvement in compressive strength was indicated regardless of the type of superplasticizer [1]. Moreover, another study showed improvement in compressive strength with the increase in the dosage of superplasticizer [6]. Conversely, it was found that the compressive strength at early age was decreased by the superplasticizer addition regardless of its type [13]. Jithendra and Elavenil [9] also indicated increasing superplasticizer dosage in geopolymer led to the decrease in compressive strength, whereas the slump value was improved. With respect to the results of previous studies, further investigations are necessary to clarify the influence of superplasticizer on the mechanical and fresh properties of geopolymer.

Thus, the present study aims to investigate the effect of polycarboxylate based superplasticizer, which is commonly used in OPC concrete to improve the workability and achieve desired compressive strength, on the fresh properties and early strength of fly ash/GGBS based geopolymer.

2 Experimental Details

2.1 Materials and Mix Proportion

Low calcium fly ash and GGBS were used as precursor to synthesize geopolymer mortar, and the chemical compositions of both precursors are shown in Table 1. The density of fly ash and GGBS were 2.24 and 2.91 g/cm^3, respectively. Sodium silicate and sodium hydroxide acted as alkaline activators which obtained locally in Japan. Sodium silicate is colorless liquid with high viscosity which has chemical compositions presented in Table 2 and it was diluted with water to make sodium silicate solution. Meanwhile, sodium hydroxide in pellets form with 98% purity was used in this study. Sodium hydroxide pellets with density of 2.13 g/cm^3 was diluted with water to make 12 M of sodium hydroxide solution. The ratio of sodium silicate to sodium hydroxide was set to 2.0 by mass and both solutions were prepared separately one day prior to mixing. Quartz porphyry crushed sand (density $= 2.61$ g/cm^3, fineness modulus (FM) $= 2.72$) was selected for fine aggregate in the present study and used in saturated surface dry (SSD) condition. Furthermore, polycarboxylate based superplasticizer with density of 1.06 g/cm^3 was added at various dosages from 0% to 8%.

In this study, five mixtures of geopolymer mortar were prepared with different dosages of superplasticizer which were 0% (control mix), 1%, 2%, 4%, and 8% by binder mass as shown in Table 3. Alkaline to binder and fly ash/GGBS ratios were selected through experimental evaluation and kept constant at 0.5 and 1.5, respectively. Mix design in the present study was conducted with the absolute volume method considering the density and specific gravity of materials. The aggregate

Table 1 Chemical compositions of aluminosilicate sources

Composition (%)	SiO_2	Al_2O_3	Fe_2O_3	CaO	K_2O	TiO_2	MgO	Na_2O	SO_3	LOI
Fly Ash	64.44	20.65	4.18	2.25	1.53	1.19	0.58	–	–	2.90
GGBS	35.45	14.06	0.27	43.78	0.23	0.56	5.84	0.24	0.62	0.05

LOI: Loss on ignition

Table 2 Chemical compositions of sodium silicate

Properties	Value
SiO_2 (%)	35.60
Na_2O (%)	17.50
H_2O (%)	46.90
Specific gravity	1.69

Table 3 Mix proportion of geopolymer mortar (kg/m^3)

Mix	Binder		Alkaline Activator		Water		SP*	Fine aggregate	l/s
	Fly Ash	GGBS	SS gel*	NaOH pellets	For SS solution	For NaOH solution			
BS40-0%	360	240	55	32.43	145	67.57	–	1,253	0.36
BS40-1%	360	240	55	32.43	139	67.57	6	1,253	0.36
BS40-2%	360	240	55	32.43	133	67.57	12	1,253	0.36
BS40-4%	360	240	55	32.43	121	67.57	24	1,253	0.36
BS40-8%	360	240	55	32.43	97	67.57	48	1,253	0.36

*SS gel: SiO_2 + Na_2O in sodium silicate; SP: Superplasticizer

content was also determined by volume, then moisture content was measured to calculate surface water of aggregate which is further used to modify the water content in geopolymer mixture. For the addition of admixture in this study, the liquid to solid (l/s) ratio was set to 0.36 for all the mixtures to obtain sufficient workability and same proportion of geopolymer mixture. Liquid comprises the sum of water from alkaline solutions, superplasticizer, and surface water of fine aggregate. Meanwhile, solid comprises the sum of solid contents in alkaline activators and binders. Liquid to solid ratio was kept constant by modifying the water content for dilution of sodium silicate based on the measured moisture content of sand at the time of mixing.

2.2 Experimental Methods

The sequences of mixing geopolymer mortar are as follows. The dry materials including fly ash, GGBS and sand firstly were mixed together for 3 min. After that, sodium hydroxide solution was added gradually to the mixture, followed sequentially by sodium silicate and superplasticizer in 2 min. Lastly, all the materials were mixed for another 2 min. The specimens were cast in cylindrical molds of 50 mm diameter and 100 mm height. All the specimens were sealed with plastic adhesive tape and cured in two regimes, which were heat curing and 20 °C curing. For the heat curing, specimens were firstly cured at 70 °C for 24 h, and then stored at 20 °C. Meanwhile, for the 20 °C curing, specimens were placed at 20 °C right after casting. All the specimens were demolded right before the compressive strength test.

The workability of geopolymer was assessed with the flow test which was repeatedly conducted after casting by the time when the geopolymer was set. Flow diameter was measured immediately after the mortar has spread over 15 falling motions on the average value. The compressive strength test was performed to evaluate the early strength of geopolymer with the addition of superplasticizer at the ages of 3, 7, and 28 days with a capacity of 250 kN and 0.2 mm/min of loading rate. Both flow test and compressive strength test were conducted according to JIS R5201. Furthermore, microstructure observation was carried out by scanning electron microscopy (SEM) to study the morphology of geopolymer. The SEM analysis was conducted on hardened mortar subtracted from specimens after the compressive strength test at the age of 7 days. It was coated with platinum and imaged with an accelerating voltage of 15 kV.

3 Results and Discussion

3.1 Workability

Figure 1 indicates the influence of superplasticizer addition on the workability of fresh fly ash/GGBS based geopolymer mortar. The flow diameter of control mix (BS40-0%) decreased rapidly over time, although the initial flow diameter at the time of 5 min after casting was the highest. Meanwhile, the flow diameter of the specimens with superplasticizer regardless of the dosages decreased gradually and tended to maintain the workability of fresh geopolymer mortar. For example, both BS40-4% and BS40-8% specimens exhibited flow diameter over 110 mm even 15 minutes after casting, while for the case of BS40-0% was 101 mm. The results indicate that superplasticizer may delay the setting time of geopolymer mortar, and it is consistent with previous studies [1, 7] which stated that polycarboxylate based superplasticizer can retard the setting of fresh geopolymer regardless of the type of alkaline activator used.

Other studies reported that the workability which was evaluated using the slump value of geopolymer was increased linearly with the addition of superplasticizer [1, 9, 13]. According to those studies, the addition of superplasticizer was calculated by binder mass without further modification of the water content of geopolymer mixture and other proportions were kept constant. Hence, liquid content of geopolymer

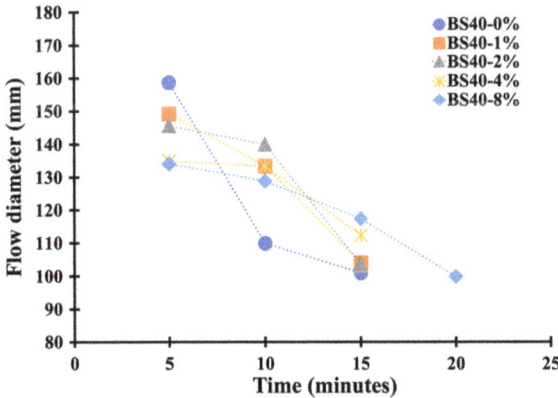

Fig. 1 Effect of superplasticizer on workability of geopolymer with function of time

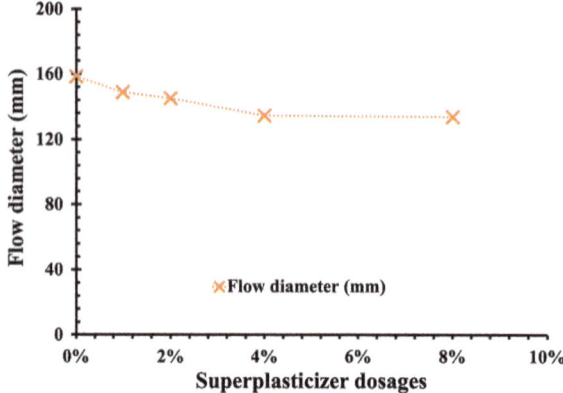

Fig. 2 Effect of superplasticizer dosages on workability of geopolymer

mixture increased with the increase of superplasticizer dosages. Therefore, the improvement in the workability of geopolymer was indicated in the previous studies. On the other hand, the workability of fresh geopolymer mortar in this study showed opposite results to the previous studies. As shown in Fig. 2, flow diameter decreased with the increase in the superplasticizer dosage. Superplasticizer is a liquid form admixture of which the addition requires the modification of the water content to keep the mixture volume in OPC mixture. Similarly to this, the addition of super-plasticizer in geopolymer mixture should modify the water content in geopolymer mixture in order to maintain the volume of mixture proportion.

3.2 Compressive Strength

The compressive strength of fly ash/GGBS based geopolymer mortar in various dosages of superplasticizer are presented in Fig. 3. For the heat curing condition as can be observed in Fig. 3a, the compressive srength at the ages of 3 and 7 days decreased with the increase in the dosage of superplasticizer up to 4%, while the compressive strength of BS40-8% subjected to heat curing was slightly lower than the control mix, but it was higher than that of the other mixes containing superplasticizer. Meanwhile, for the 20 °C curing condition as shown in Fig. 3b, the compressive strengths of the specimens containing superplasticizer were lower than that of the control mix especially at the age of 28 days, and the decrease in the compressive strength with the increase in the dosage of superplasticizer was not observed. Focusing on the case of BS40-8%, it retards the setting of geopolymer more, while it shows comparable compressive strength to the control mix under both curing conditions at early ages of both 3 and 7 days. This result may support a previous study [14] which stated that a higher dosage of superplasticizer was needed due to the lower adsorption of the superplasticizer on alkali activated slag paste compared to OPC paste. However,

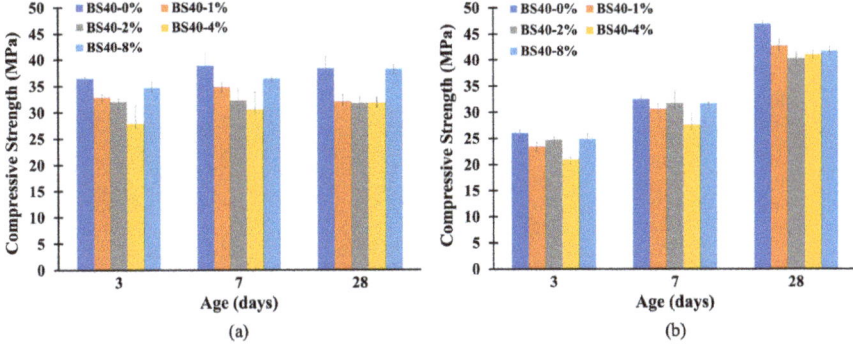

Fig. 3 Compressive strength of fly ash/GGBS based geopolymer mortar with various superplasticizer dosages under **a** heat curing; **b** 20 °C curing at the ages of 3, 7, and 28 days

the variation on these results may be attributed to the instability of superplasticizer in multi-compound alkaline activator and high alkaline solution [13]. Therefore, further investigation on the effect of superplasticizer on the compressive strength of geopolymer is required.

3.3 Microstructure Observation

Figure 4 shows the SEM images of fly ash/GGBS based geopolymer mortar containing superplasticizer at the age of 7 days. Micropores and unreacted fly ash particles were observed in the specimens containing 2% and 4% of superplasticizer as shown in Fig. 4b and Fig. 4c, respectively. Meanwhile, the specimen at a higher superplasticizer dosage of 8% demonstrated in Fig. 4d had more homogeneous and denser microstructure which was similarly to the specimen without superplasticizer (Fig. 4a). This result may indicate that the increase of superplasticizer dosage enhanced the microstructure, hence resulting in the high compressive strength compared to the other specimens at low superplasticizer dosages. It was also reported in previous study that the presence of superplasticizers reduces agglomerated particles in geopolymer [11]. According to these results, denser microstructure observed in BS40-8% may not only have enhanced the mechanical properties of geopolymer mortar but also have retarded the setting of geopolymer.

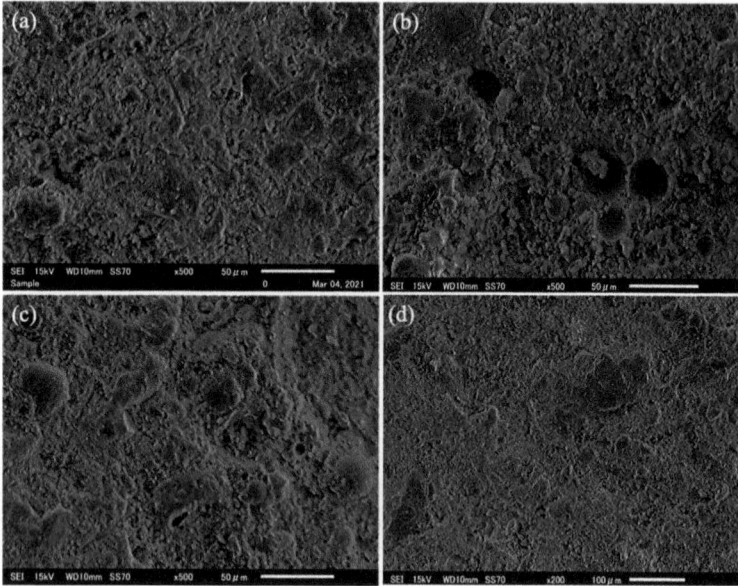

Fig. 4 SEM images of fly ash/GGBS based geopolymer mortar containing **a** 0%, **b** 2%, **c** 4% and **d** 8% of superplasticizer at age of 7 days

4 Conclusions

The effect of polycarboxylate based superplasticizer on the workability and early strength of fly ash/GGBS based geopolymer mortar was investigated. Based on the experimental results, the following conclusions of the present study can be drawn.

1. The increase in superplasticizer dosages on fly ash/GGBS based geopolymer mortar decreased the flow diameter of fresh fly ash/GGBS based geopolymer mortar. However, the increase in superplasticizer may retard the setting of geopolymer mortar.
2. The inclusion of superplasticizer shows variation on the compressive strength of fly ash/GGBS geopolymer mortar which may be attributed to the instability of superplasticizer in alkaline solution. The decrease in the compressive strength of heat curing specimen was observed with the increase in the dosage of super-plasticizer up to 4% at the ages of 3 and 7 days. However, 8% inclusion of superplasticizer showed comparable compressive strength at early age for both curing conditions and retarded the setting of geopolymer more.
3. Microstructure observation through SEM showed denser microstructure in the specimen with 8% inclusion of superplasticizer, resulting in the higher compressive strength than the other specimens containing superplasticizer.

References

1. Alrefaei Y, Wang YS, Dai JG, Xu QF (2020) Effect of superplasticizers on properties of one-part $Ca(OH)_2/Na_2SO_4$ activated geopolymer pastes. Constr Build Mater 241:117990
2. Andrew RM (2018) Global CO2 emissions from cement production 1928–2017. Earth Syst Sci Data 10:2213–2239
3. Davidovits J (2013) Geopolymer Cement a review. Geopolymer Sci. Tech, pp. 1–11
4. Davidovits J (1994) Properties of geopolymer cements. Geopolymer Institute, Saint-Qeuntin p 1–19
5. Elyamany HE, Abd Elmoaty AEM, Elshaboury AM (2018) Setting time and 7-day strength of geopolymer mortar with various binders. Constr Build Mater 187:974–983
6. Gupta N, Gupta A, Saxena KK, Shukla A, Goyal SK (2020) Mechanical and durability properties of geopolymer concrete composite at varying superplasticizer dosage. Mater Today Proc 44:12–16
7. Jang JG, Lee NK, Lee HK (2014) Fresh and hardened properties of alkali-activated fly ash/slag pastes with superplasticizers. Constr Build Mater 50:169–176
8. Jindal BB (2018) Feasibility study of ambient cured geopolymer concrete—A review. Adv Concr Constr 6:387–405
9. Jithendra C, Elavenil S (2019) Role of superplasticizer on GGBS based geopolymer concrete under ambient curing. Mater Today Proc 18:148–154
10. Kawai K, Sugiyama T, Kobayashi K, Sano S (2005) Inventory data and case studies for environmental performance evaluation of concrete structure construction. J Adv Concrete Tech 3:435–456
11. Li S, Zhang J, Li Z, Gao Y, Liu C (2021) Feasibility study of red mud-blast furnace slag based geopolymeric grouting material: Effect of superplasticizers. Constr Build Mater 267:120910
12. Nath P, Sarker PK (2014) Effect of GGBS geopolymer on setting, workability and early strength properties of fly ash geopolymer concrete cured in ambient condition. Constr Build Mater 66:163–171
13. Nematollahi B, Sanjayan J (2014) Effect of different superplasticizers and activator combinations on workability and strength of fly ash based geopolymer. Mater Des 57:667–672
14. Palacios M, Houst YF, Bowen P, Puertas F (2009) Adsorption of superplasticizer admixtures on alkali-activated slag pastes. Cem Concr Res 39:670–677
15. Saludung A, Ogawa Y, Kawai K (2019) Effect of GGBS Addition on properties of fly ash-based geopolymer at high temperature. Proc Jpn Concr Inst 41:1991–1996
16. Xie J, Wang J, Rao R, Wang C, Fang C (2019) Effects of combined usage of GGBS and fly ash on workability and mechanical properties of alkali activated geopolymer concrete with recycled aggregate. Compos Part B Eng 164:179–190

Mechanical Properties and Length Change of High Strength Concrete Containing Coal Bottom Ash as an Internal Curing Agent

Phuong Trinh Bui, Trong Huu Nguyen, Van Chi Vo, and Xuan Loc Luu

Abstract Construction demand for high rise buildings is gradually increasing at several developing countries, leading to an over-exploitation of river sand which causes environmental impacts such as riverbank erosion. Coal bottom ash, which is a by-product of coal fired power plants, is considered as one of alternative materials to natural river sand for concrete production to diminish environmental impacts and conserve natural resources. Therefore, the aim of the present study was to investigate mechanical properties and length change of high strength concrete internally cured by coal bottom ash to evaluate the effectiveness of replacement of river sand by this material. In experiment, five replacements of natural river sand by coal bottom ash under saturated-surface dry condition were 0, 25, 50, 75, and 100% by volume. A water-to-cement ratio of 0.34 was employed for all concrete mixtures. Mechanical properties (including compressive strength and flexural strength), and length change (i.e., shrinkage) of high strength concretes cured under sealed condition were tested up to age of 56 days. Results showed that the replacement of natural river sand by coal bottom ash reduced compressive strength of the concretes at the age of 1 day. The 50% replacement of coal bottom ash increased compressive and flexural strengths of the concretes after aging for 7 days. The replacement reduced shrinkage of the high strength concretes up to the age of 56 days. Consequently, the coal bottom ash under saturated-surface dry condition is found to be a potential alternative to natural river sand to produce internally cured high strength concrete.

Keywords Coal bottom ash · Compressive strength · Flexural strength · Internal curing · Shrinkage

P. T. Bui (✉) · T. H. Nguyen · V. C. Vo · X. L. Luu
Faculty of Civil Engineering, Ho Chi Minh City University of Technology (HCMUT), 268 Ly Thuong Kiet Street, District 10, Ho Chi Minh City, Vietnam
e-mail: buiphuongtrinh@hcmut.edu.vn

Vietnam National University Ho Chi Minh City, Linh Trung Ward, Thu Duc District, Ho Chi Minh City, Vietnam

1 Introduction

A demand of construction for buildings and infrastructure is gradually increasing at developing countries such as Vietnam, leading to an increase in use of concrete. As a result, an exploitation of natural resources (i.e., river sand and crushed stone as fine and coarse aggregates, respectively) for concrete production is elevated. The exploitation of natural river sand (RS) for the concrete is one of main factors causing environmental impacts such as riverbank erosion. Therefore, it is necessary to seek other materials as alternatives to RS in construction industry.

Coal bottom ash (CBA), a by-product of coal-fired power plants, has been considered as one of alternative materials to RS for concrete production [7] because of its particle size distribution which is nearly the same as that of RS [1]. An effective utilization of CBA as RS replacement can diminish the environmental problems and save the natural resources [7]. In addition to having angular and irregular shape, the CBA is lighter, more porous, and brittle than the RS [7]. The CBA has high water absorption capacity which resulted in a reduction of workability and bleeding of concretes [7]. Bai et al. investigated the effect of replacement of RS by oven dried CBA at levels of 0, 30, 50, 70, and 100% by mass on strength and drying shrinkage of concretes with fixed water-to-cement ratios and fixed workability [1]. For the fixed water-to-cement ratios, the compressive strength and the drying shrinkage reduced with an increase in the CBA amount. For the fixed workability, the compressive strength was comparable with that of the control concrete without CBA while the drying shrinkage increased with the increase in the CBA amount up to 30% replacement level [1]. Singh and Siddique reported that the compressive and splitting tensile strengths of concretes containing oven dried CBA were nearly the same or slightly lower at early ages, and almost comparable to those of control concrete containing no CBA after aging for 90 days [7]. Briefly, there was no significant effect of replacement of RS by oven dried CBA on the compressive strength of the concretes [1, 7].

On the other hand, curing process is a necessary technology to ensure mechanical properties of concrete. A suitable curing helps cement to continuously hydrate and develop its strength. External curing is one of traditional curing methods which supply external water to cure the concrete. Nevertheless, external water penetrates the surface of concretes with a low water-to-cement ratio only several millimeters [4]. This results in an ineffective curing for such concretes. Internal curing with prewetted lightweight and porous materials was proposed to overcome the disadvantage of external curing [2]. Internal curing is a curing process for concrete from the inside out via supplying additional internal water from lightweight and porous materials, instead of part of mixing water [2, 4]. The lightweight and porous materials have high water absorption capacity and can release water to the matrix of the concrete during its self-desiccation. Internal water released from the lightweight and porous materials contributes to an increase in degree of cement hydration and thereby an improvement of the strength and a reduction of autogenous shrinkage of the concrete [2]. In general, internal curing agents commonly used are prewetted lightweight aggregate, porous ceramic waste aggregate, and so on [2–4]. Meanwhile, there are a few studies in terms of properties of high strength concrete internally cured by CBA.

The aim of the present study was therefore to investigate replacements of RS by CBA under saturated-surface dry condition to evaluate the effectiveness of internal curing with the CBA on mechanical properties (including compressive strength and flexural strength), and shrinkage of high strength concrete with a low water-to-cement ratio, to utilize the CBA from the coal-fired power plants, and diminish environmental impacts caused by the over-exploitation of RS in Vietnam.

2 Experiments

2.1 Materials

A cement supplied by Nghi Son company was type I Portland cement in accordance with ASTM C150. The cement had a density of 3.15 g/cm^3, Blaine fineness of 3630 cm^2/g and compressive strength at 28 days of 46.2 N/mm^2. Table 1 shows chemical compositions of the cement.

Fine and coarse aggregates prepared in the present study were RS, CBA, and crushed stone, respectively. The CBA collected from the Duyen Hai coal-fired power plant in Tra Vinh province was utilized to replace the RS as shown in Fig. 1a. Table 2 shows physical properties of RS, CBA, and crushed stone in accordance with TCVN

Table 1 Chemical compositions of cement

Compositions (% by mass)	SiO$_2$	Fe$_2$O$_3$	Al$_2$O$_3$	CaO	MgO	SO$_3$	K$_2$O	Na$_2$O	LOI
Content	19.70	3.15	4.57	61.80	1.84	2.34	0.713	0.0348	2.38

LOI: loss on ignition

Fig. 1 CBA and RS used in the present study **a** and particle size distributions of aggregates **b**

Table 2 Physical properties of aggregates

Physical properties	Size (mm)	Fineness modulus	Density (g/cm^3)	Water absorption (%)
RS	0.14–5	2.40	2.6	5.61
CBA	0.14–5	3.28	1.73	20.86
Crushed stone	5–20	–	2.69	0.36

– : not measured

7570:2006. Figure 1b shows their particle size distributions. According to test results of fineness modulus and particle size distribution as shown in Table 2 and Fig. 1b, respectively, it is clearly observed that the particles of CBA were coarser, and CBA had extremely higher water absorption capacity than RS.

In addition, tap water in accordance with TCVN 4506:2012 was prepared to mix concrete mixtures. Water-reducing agent in accordance with TCVN 8826:2011 was employed to ensure desired slump of fresh control concrete containing 100% RS of 8 ± 2 cm.

2.2 Concrete Mixture Proportion

Mixture proportion of the control concrete containing 100% RS was designed as per TCVN 10306:2014. The replacements of RS by CBA under saturated-surface dry condition were 0, 25, 50, 75, and 100% by volume. All the concrete had a water-to-cement ratio of 0.34 and a water amount of 178 kg to produce high strength concrete. The constant dosage of water-reducing agent was 1.2% by mass of the cement to investigate effect of CBA on slump of fresh concrete. Table 3 shows all mixture proportions of concretes in the present study.

Table 3 Mixture proportions of concrete

Mixture proportions	CBA replacement (% by volume)	Unit (kg)					Water-reducing agent (% by mass of cement)	Slump (cm)
		Water	Cement	RS	CBA	Crushed stone		
BA0	0	178	524	682	0	1060	1.2	7.0
BA25	25	178	524	511	114	1060	1.2	8.0
BA50	50	178	524	341	227	1060	1.2	11.5
BA75	75	178	524	170	341	1060	1.2	14.0
BA100	100	178	524	0	454	1060	1.2	16.5

2.3 Slump Test and Specimen Preparation

After mixing all components of each mixture proportion as shown in Table 3 via a mechanical mixer, its slump was measured as per TCVN 3106:1993. After slump test, fresh concrete of each mixture proportion was immediately cast into steel molds in two layers according to TCVN 3105:1993. The placing surface of specimens was then covered with plastic sheet to prevent water evaporation and carbonation.

2.4 Compressive Strength and Flexural Strength Tests

After casting for 24 h, twelve cubic specimens with dimensions of 150 × 150 × 150 mm and nine prism specimens with dimensions of 100 × 100 × 400 mm of each mixture proportion were demolded and then cured under sealed condition until testing ages. The compressive strength of each mixture proportion at the ages of 1, 7, 28, and 56 days was tested via a hydraulic compression machine in accordance with TCVN 3118:1993. Meanwhile, the flexural strength of each mixture proportion at the ages of 7, 28, and 56 days was tested in accordance with TCVN 3119:1993.

2.5 Length Change Test

After casting for 24 h, two prism specimens with dimensions of 100 × 100 × 400 mm of each mixture proportion were demolded. All surfaces of the specimens were then covered with epoxy resin to prevent not only moisture evaporation from hardened concretes to environment but also moisture penetration from environment to them. The procedure of length measurement via the vertical displacement of concrete specimens was carried out as per 22 TCN 60–84. Measurement time begun after coving the epoxy resin for 1 day (i.e., after aging for 2 days) and the measurement of the specimens was done up to the age of 56 days. During the measurement, all the specimens were placed at temperature in the range of 27–29 °C. The length change of each mixture proportion at each age was the average of two concrete specimens.

3 Results and Discussion

3.1 Slump

Slump of fresh concrete of each mixture proportion is shown in Table 3. The replacement of RS by CBA under saturated-surface dry condition increased slump of fresh concrete when the mixing water amount was kept constant. The higher the CBA replacement, the higher the slump of fresh concrete. The increase in slump of fresh concrete was also observed in a previous study of Zhang and Poon [9]. It indicates that the CBA under saturated-surface dry condition did not absorb mixing water in fresh concrete in the present study while the CBA under oven dry condition absorbs a part of mixing water, leading to reduce slump of fresh concrete as reported in the previous study [7]. The increase in slump was due to the CBA having the coarser particles than the RS (see Fig. 1b). It indicates that the moisture state and particle size of the CBA significantly affected the slump of fresh concrete.

3.2 Compressive Strength

Effect of replacement of RS by CBA under saturated-surface dry condition on compressive strength at the ages of 1, 7, 28, and 56 days of hardened concrete cured under sealed condition is shown in Fig. 2a. The compressive strength of all concretes increased with time. It demonstrates that the cement hydration in all the concretes continuously proceeded with time. In general, the compressive strength of the hardened concretes with the CBA (BA25, BA50, BA75, and BA100) was lower that of the control concrete without CBA (BA0) at an early age (i.e., at 1 day). The reason could be due to the use of porous CBA which increased the pore volume in the concrete. In addition, the coarser particles of the CBA than those of RS (see Fig. 1b) increased the excess water in these mixture proportions with the fixed mixing water amount. As a result, the slump of fresh concrete containing the CBA increased (see Table 3) and the corresponding compressive strength at early age (i.e., at 1 day) decreased. Singh and Siddique reported that the compressive strength of the concretes with CBA as RS replacement slightly reduced at 7 days and even at 28 days [7]. It was explained that the pozzolanic activity of CBA was slow at early ages and started after 28 days of curing age [7]. The reduction of compressive strength was also observed for the BA25, BA75, and BA100 specimens at later ages (i.e., at 7, 28 days) and even at 56 days. At the age of 56 days, the effect of internal curing and pozzolanic reaction of CBA on increasing the strength did not overcome the negative effect of excess water

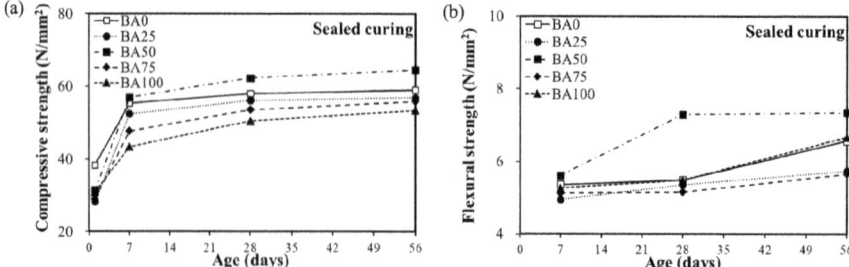

Fig. 2 Effect of CBA replacement on compressive strength **a** and flexural strength **b** of hardened concrete under sealed curing

in these mixture proportions and porous property of CBA. Meanwhile, the compressive strength of the BA50 specimen was higher slightly at 7 days and significantly at 28 and 56 days when compared with that of BA0 specimen. It indicates that the internal curing with CBA promoted the cement hydration and its pozzolanic reaction, thereby leading to a significant increase in compressive strength of the BA50 specimen. Kim and Lee also found that the enhancement of hydration by CBA was due to (1) the internal water released from CBA into the cement paste and (2) the high pozzolanic reactivity of CBA [5].

3.3   Flexural Strength

Effect of replacement of RS by CBA under saturated-surface dry condition on flexural strength at the ages of 7, 28, and 56 days of hardened concrete cured under sealed condition is shown in Fig. 2b. The flexural strength of all concretes increased with time. It confirms again that the cement hydration in all the concretes continuously proceeded with time. The flexural strength of the hardened concretes with the 50% CBA replacement was higher that of the control concrete without CBA (BA0) at all ages. It indicates 50% replacement with CBA as an internal curing agent was the optimum for increasing the flexural strength of the concrete. Meanwhile, the others with CBA replacements (BA25 and BA75) had the lower flexural strength than that without CBA (BA0). The reason could be due to the excess water in these mixture proportions and porous-brittle characteristics of CBA which negatively reduced the flexural strength of concrete. The flexural strength of BA100 specimen had nearly the same as that of the BA0. This could be due to the effectiveness of internal curing with CBA which exceeded the negative effect of excess water and porous-brittle characteristics of CBA in BA100 specimen.

3.4   Length Change

Effect of replacement of RS by CBA under saturated-surface dry condition on length change of hardened concretes with and without CBA is shown in Fig. 3.

When compared with the BA0 specimen, the BA25 specimen had the lower shrinkage up to the age of 7 days. Although the shrinkage of the BA50 and BA75 specimens was higher at early ages (i.e., at 3 days), their shrinkage was lower at later ages (i.e., after 3 days) than that of the BA0 specimen. The reduction of shrinkage by the CBA was also found in some previous studies [5]. Meanwhile, the shrinkage of the BA100 and BA0 specimens was nearly the same up to 7 days. The higher the CBA replacement, the higher the shrinkage of the concretes up to the age of 7 days. The higher shrinkage of the concretes with the higher content of CBA may be attributed to the increase in porosity due to the increase in the CBA amount [6]. The increase in shrinkage of concrete with CBA could be due to the refinement of

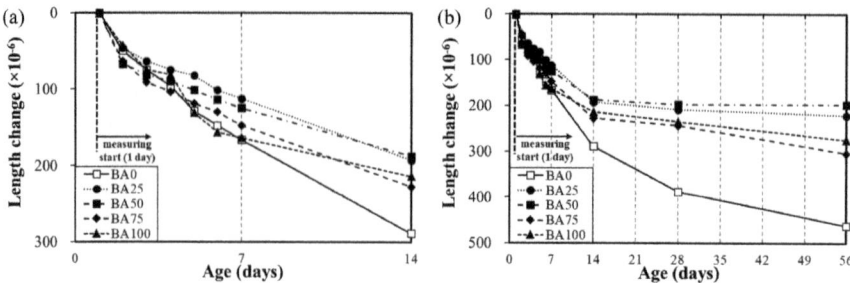

Fig. 3 Effect of CBA replacement on length change up to 14 **a** and 56 **b** days of hardened concrete

pore size distribution resulting from the internal curing effect. The refinement leaded to a further increase in capillary tension and more concentration of the cement paste in concrete [8]. Pore structure of such concretes should be investigated in future works to strengthen an explanation on their shrinkage. According to Fig. 3b, the shrinkage of the concretes with CBA was significantly lower than that of the control concrete after the age of 14 days. The shrinkage of the BA specimens was the lowest. This is because of the effectiveness of internal curing with CBA.

4 Conclusion

Based on the experimental results, the following conclusions can be drawn:

– With the same mixing water amount, the coal bottom ash (CBA) under saturated-surface dry condition increased slump of fresh concrete due to its coarser particles than those of natural river sand (RS). The higher the CBA replacement, the higher the slump of fresh concrete.
– The replacement of RS by CBA under saturated-surface dry condition slightly decreased compressive strength of hardened concretes cured under sealed condition at early age (i.e., at 1 day). However, the 50% replacement of CBA increased compressive and flexural strengths of the concretes after aging for 7 days and reduced shrinkage of the concretes up to the age of 56 days.

Consequently, the CBA under saturated surface-dry condition as an internal curing agent could replace 50% by volume of RS as fine aggregate to play a positive role in increasing compressive strength and reducing shrinkage of the concrete. Future works in terms of a chemical approach and analysis of microstructure of the concretes should be done to evaluate the effect of pozzolanic reaction of the CBA on their mechanical properties and durability.

Acknowledgements We acknowledge the support of time and facilities from Ho Chi Minh City University of Technology (HCMUT), VNU-HCM for this study.

References

1. Bai Y, Darcy F, Basheer PAM (2005) Strength and drying shrinkage properties of concrete containing furnace bottom ash as fine aggregate. Constr Build Mater 19(9):691–697
2. Bentz DP, Weiss WJ (2011) Internal curing: a 2010 state-of-the-art review. Report no 7765, US Department of Commerce, National institute of standards and technology internal report (NISTIR), Gaithersburg, Maryland.
3. Bui PT, Ogawa Y, Nakarai K, Kawai K (2017) Internal curing of Class-F fly-ash concrete using high-volume roof-tile waste aggregate. Mater Struc 50:203
4. Castro J, De Le Varga I, Golias M, Weiss WJ (2010) Extending internal curing concepts (using fine LWA) to mixtures containing high volumes of fly ash. In: Proceedings of 2010 concrete bridge conference: achieving safe, smart & sustainable bridges, Phoenix, pp. 24–26
5. Kim HK, Lee HK (2018) Hydration kinetics of high-strength concrete with untreated coal bottom ash for internal curing. Cem Concr Compos 91:67–75
6. Ratchayut K, Somnuk T (2008) Properties of self-compacting concrete in corporating bottom ash as a partial replacement of fine aggregate. ScienceAsia 34:87–95
7. Singh M, Siddique R (2016) Effect of coal bottom ash as partial replacement of sand on workability and strength properties of concrete. Clean Product 112:620–630
8. Sellevold EJ (1987) The function of silica fume in high strength concrete. In: Proceedings of international conference on utilization of high strength concrete, Norway, pp. 39–50
9. Zhang B, Poon CS (2017) Internal curing effect of high volume furnace bottom ash (FBA) incorporation on lightweight aggregate concrete. J Sustain Cement-Based Mater (SCBM) 6:366–383

Microscopic Investigation on Interfacial Transition Zone in Fly Ash Cement Hydrates Using Roof-Tile Waste Aggregate

Kazuma Okamoto, Phat Tan Huynh, Yuko Ogawa, and Kenji Kawai

Abstract Previous studies reported that the strength reduction in fly ash concrete (FAC) can be improved by internal curing with roof-tile waste aggregate (RWA). However, how much the internal curing by RWA improved the compressive strength of FAC was different, and there was no clear explanation on it. The purpose of this study is to clarify the effect of the internal curing in the FAC by examining the range of the internal curing effect with RWA in the interfacial transition zone (ITZ) around coarse aggregate. Paste and concrete specimens in which part of cement was replaced with fly ash were prepared. For paste specimens, one aggregate particle was placed in the center of the specimen. Vickers hardness around the aggregate in the paste specimen and the compressive strength of the concrete specimen were investigated. As a result, the range of Vickers hardness improvement around RWA changed depending on the replacement ratio of fly ash. This indicates that the internal curing effect differs depending on the mixture proportion of paste matrix such as the replacement ratio of fly ash. Furthermore, it was observed that the compressive strength of concrete increased as the thickness of ITZ around RWA decreased.

Keywords Interfacial transition zone · Roof-tile waste aggregate · Internal curing · Vickers hardness · Fly ash concrete

K. Okamoto · Y. Ogawa · K. Kawai (✉)
Civil and Environmental Engineering Program, Graduate School of Advanced Science and Engineering, Hiroshima University, 1-4-1 Kaganiyama, Higashihiroshima, Japan
e-mail: kkawai@hiroshima-u.ac.jp

P. T. Huynh
Depertment of Civil and Environmental Engineering, Graduate School of Engineering, Hiroshima University, 1-4-1 Kaganiyama, Higashihiroshima, Japan

Faculty of Civil Engineering, Ho Chi Minh City University of Technology (HCMUT), 268 Ly Thuong Kiet Street, District 10, Ho Chi Minh, Vietnam

Vietnam National University Ho Chi Minh City, Linh Trung Ward, Thu Duc District, Ho Chi Minh, Vietnam

© The Author(s), under exclusive license to Springer Nature Singapore Pte Ltd. 2023   517
J. N. Reddy et al. (eds.), *ICSCEA 2021*, Lecture Notes in Civil Engineering 268,
https://doi.org/10.1007/978-981-19-3303-5_45

1 Introduction

Fly ash, which is a by-product of coal-fired power generation, is desired to be used as an admixture for concrete from the viewpoint of resource recycling and reduction of environmental impact. However, the utilization of fly ash concrete (FAC) has not been promoted well due to problems such as a decrease in the initial strength. Internal curing can be one of the solutions against this issue, and some studies investigated the internal curing effect on the concrete strength. Although it has been reported in some previous studies that the strength of FAC can be improved by the internal curing with roof-tile waste aggregate (RWA) [1] [4], the internal curing effect on concrete properties was different depending on conditions such as the replacement ratio of fly ash. To explain the difference in the internal curing effect, it is necessary to clarify the effect of internal curing on FAC for the mixture proportion design such as an appropriate replacement ratio of RWA.

Concrete has a very porous structure called interfacial transition zone (ITZ), and it is closely related to the durability of concrete [6]. That is, improving the brittle structure of ITZ leads to more durable concrete. Also, fly ash has pozzolanic reactivity, and the reaction could improve ITZ for a long term when the FAC is cured sufficiently. For example, Golewski reported that mature fly ash concrete forms ITZ with a denser structure than the concrete without fly ash, resulting in low permeability and high durability [2]. Considering that the internal curing with RWA can promote the reaction of fly ash, the FAC may have high ITZ improvement, resulting in improving the FAC strength. Therefore, it is important to investigate the internal curing effect on the structure of ITZ in fly ash concrete.

Based on these backgrounds, the purpose of this study is to examine the range of the internal curing effect with RWA, focusing on the ITZ around coarse aggregate embedded in the fly ash cement paste.

2 Experimental Outlines

2.1 Materials and Specimen Preparation

High-early-strength Portland cement meeting the standard values of the Japanese Industrial Standards JIS R 5210 (Portland cement) was used in this study. Low-calcium fly ash meeting the standard values of Type II as per JIS A 6201 (Fly ash for use in concrete) was also used as cementitious material. Crushed quartz-porphyry was used as the conventional fine and coarse aggregates in the concrete. These fine and coarse aggregates met the standard values as per JIS A 5005 (Crushed stone and manufactured sand for concrete). Roof-tile waste aggregate derived from crushed roof-tile waste was also used as an internal curing agent, partially replacing the coarse aggregate. It was used in a saturated surface-dry condition after 7-day water immersion. Table 1 lists the physical properties of coarse aggregates.

Table 1 Physical characteristics of the materials

Notation	Density in saturated surface-dry condition (g/cm^3)	Water absorption (%)	Aggregate size (mm)
Crushed stone	2.62	0.67	5–20
Roof-tile waste aggregate	2.26	9.08	5–13

Paste and concrete specimens with a water-to-binder ratio (W/B) of 0.40 were prepared with the fly ash replacement ratios of 0% (FA0) and 20% (FA20). Paste specimen with the fly ash replacement ratio of 40% (FA40) was also prepared. For every paste specimen, one coarse aggregate of crushed stone or RWA was embedded. Concrete specimens were prepared by replacing 0% and 10% of the coarse aggregate volume by the RWA. Tables 2 and 3 list the mix proportions of concrete and paste specimens, respectively. Paste was cast into an aluminum mold having an inner size of $21 \times 21 \times 20$ mm, and one RWA or crushed stone was embedded in the center of the specimen. Concrete specimen was a cylindrical specimen of $\Phi 100 \times 200$ mm. All the specimens were prepared at 20°C, and then cured at 20°C under sealed condition until the test ages.

Table 2 Mixture proportion of concrete

Notation	W/B	Replacement ratio		Unit content (kg/m^3)						
		FA* (mass%)	RWA (vol.%)	W	C	FA*	S	G2010	G1505	RWA
FA0G0	0.40	0	0	165	413	0	805	503	412	0
FA0G10			10						313	86
FA20G0		20	0		330	83	787	492	402	0
FA20G10			10						324	86

*FA: fly ash

Table 3 Mixture proportion of paste

Notation	W/B	FA* replacement ratio (%)	Aggregate
FA0-N	0.40	0	Crushed stone
FA0-T			RWA
FA20-N		20	Crushed stone
FA20-T			RWA
FA40-N		40	Crushed stone
FA40-T			RWA

*FA: fly ash

2.2 Test Methods

Degree of reaction of fly ash. The degree of fly ash reaction in paste specimens was measured at 7, 28 and 91 days by using selective dissolution method (SDM) proposed by Ohsawa et. al. [5]. The sample used for measuring the degree of fly ash reaction was obtained around a coarse aggregate in the paste specimen. The procedure of SDM is as follows:

Firstly, a 1-g powder sample was put in a pre-weighed centrifuge tube and weighed precisely, and 30 mL of 2 mol/L HCl solution was added in the tube. Then, the tube was placed in a water bath at 60°C for 15 min to promote the reaction. It is noted that during this step, the solution was stirred occasionally with a glass rod. The tube was removed from the bath and centrifuged at 4000 rpm for 30 s to separate solid and liquid phases. The liquid phase was decanted as much as possible without disturbing the solid residue in the bottom. After that, the tube was filled with hot water and centrifuged again before the liquid phase was decanted. This operation was repeated 3 times.

Secondly, the tube was then filled with 30 mL of 5% Na_2CO_3 solution, placed in a water bath at 80°C for 20 min and stirred continuously. The tube was then centrifuged at 4000 rpm for 1 min. The liquid phase was decanted and the residue was washed with hot water and centrifuged. This washing was repeated 3 times. The tube with the residue was dried at 105°C for 24 h and then weighed.

The degree of fly ash reaction is calculated using the following Eq. 1:

$$\alpha = 1 - \left\{ X \frac{(1 - I_g')}{k_2} / (1 - I_g)k_1 \right\} \tag{1}$$

α = degree of fly ash reaction (%), k_1 = original fraction of fly ash in ignited base (g), k_2 = residue extracted of a 1-g fly ash (g), X = residue extracted of the hydrated fly ash in a 1-g hydrated sample (g), I_g = loss on ignition of the hydrated sample (g), $I_{g'}$ = loss on ignition of the residue extracted (g).

Vickers hardness. The Vickers hardness around a coarse aggregate in the paste specimen including ITZ was measured at the ages of 7, 28 and 91 days. The paste specimen was cut into plate samples with a thickness of approximately 5 mm, and the samples were soaked in acetone to stop hydration and dried in a vacuum desiccator for 24 h. After that, a 21 × 21 mm surface to be tested was polished with No. 400 and No. 800 abrasive compounds, ultrasonically cleaned with 2-propanol, and then vacuum-dried for 1 day. The Vickers hardness was measured by using a microhardness tester. The hardness from the interface between the aggregate and paste to 5 mm was measured in 3 to 5 directions. The Vickers hardness is calculated using the following Eq. 2:

$$HV = \frac{F}{S} = \frac{2F \sin(\theta/2)}{d_1 d_2} = 1.8544 \frac{F}{d_1 d_2} \tag{2}$$

HV = the Vickers hardness (kgf/mm^2), F = test force applied on the specimen (kgf), θ = angle of diamond pyramid ($^\circ$), d_1 and d_2 = diagonal length of the indentation (mm), S = indentation surface area (mm^2).

Compressive strength. Compressive strength tests were conducted at the ages of 7, 28 and 91 days in accordance with JIS A 1108 (Method of test for compressive strength of concrete). Three cylindrical specimens having 100-mm diameter and 200-mm height were used for each condition.

3 Results and Discussions

3.1 Degree of Reaction of Fly Ash

Figure 1 shows the degree of reaction of fly ash around a coarse aggregate embedded in the paste specimen for FA20 and FA40. The degree of fly ash reaction in the paste around RWA showed the higher value than that around crushed stone. It implies that the internal curing by RWA promoted the pozzolanic reaction around the RWA. Considering that ITZ in concrete is generally well known to be a porous structure mainly formed by crystals of Ca(OH)$_2$ [1], it is expected that ITZ structure can be improved by replacing Ca(OH)$_2$ with C-S–H due to the pozzolanic reaction promoted by the internal curing with RWA. On the other hand, at 7 days, FA40-N showed a higher degree of reaction of fly ash than FA40-T. In FA40, the water-cement ratio was high, and the amount of Ca(OH)$_2$ precipitated around the crushed stone was large, which may have promoted the fly ash reaction. More detailed examination is required in the future.

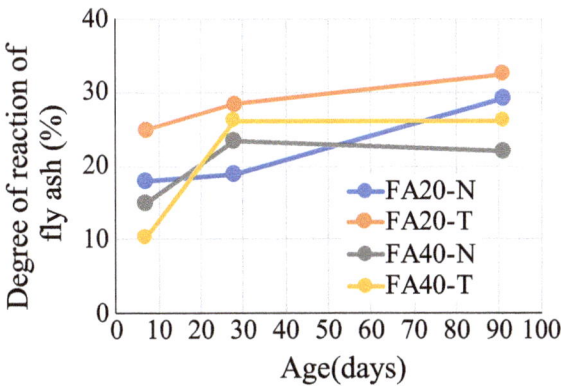

Fig. 1 Degree of reaction of fly ash

3.2   Extent and Range of Internal Curing Effect with RWA

Figure 2a–c show the distribution of the Vickers hardness around coarse aggregate in paste specimen. It can be seen that the hardness in the range up to approximately 100 μm from the coarse aggregate is lower than that in the farther area, which means that the thickness of ITZ is approximately 100 μm. In addition, it can be confirmed that the thickness of ITZ around RWA is thinner than that around crushed stone. This tendency was also observed for other ages and fly ash replacement ratio. This suggests that ITZ is improved by the internal curing with RWA.

Figure 3a–c show the Vickers hardness of paste around RWA normalized by that around crushed stone for each fly ash replacement ratio. The normalized hardness is higher as it is closer to the aggregate, and it is closer to 1 as the distance from the aggregate is longer. The tendency was more remarkable as the fly ash replacement ratio was higher. This indicates that the internal curing effect of RWA is remarkable in the vicinity of the aggregate, and this effect becomes smaller in the farther area from the aggregate. The Vickers hardness around RWA was more than 1.05 times as high as that around crushed stone in the range of 1 mm for FA0, 300 μm for FA20 and 100 μm for FA40 at the age of 7 days. That is, the extent of the internal curing became narrower as the fly ash replacement ratio increased. Internal curing is a curing method in which water is supplied by the humidity gradient inside the specimen, and it is

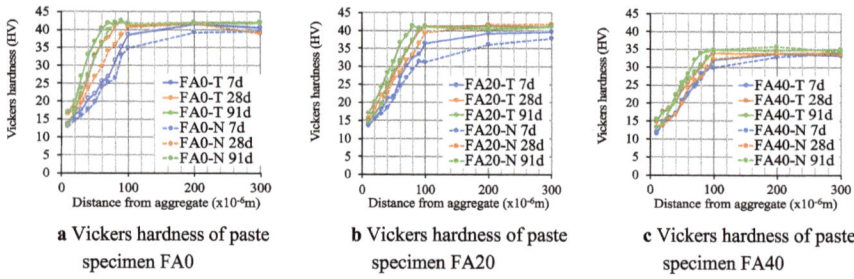

a Vickers hardness of paste specimen FA0   **b** Vickers hardness of paste specimen FA20   **c** Vickers hardness of paste specimen FA40

Fig. 2 **a** Vickers hardness of paste specimen FA0 **b** Vickers hardness of paste specimen FA20 **c** Vickers hardness of paste specimen FA40

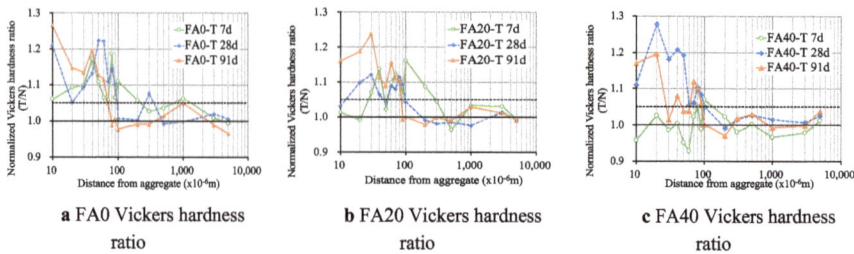

a FA0 Vickers hardness ratio   **b** FA20 Vickers hardness ratio   **c** FA40 Vickers hardness ratio

Fig. 3 **a** FA0 Vickers hardness ratio **b** FA20 Vickers hardness ratio **c** FA40 Vickers hardness ratio

considered that the extent of the internal curing is narrowed because the replacement of fly ash increases the internal relative humidity [3].

The increase in normalized hardness due to RWA from aggregate to 100 μm was very large in all the mixture proportions. When cement is replaced with fly ash, the increase in normalized hardness due to RWA from aggregate to 100 μm became more pronounced after 28 days. This might be because the internal curing effect of RWA promoted the fly ash reaction for a long time. In other words, when cement is replaced with fly ash, the range of the area affected by internal curing was narrowed, but the improvement of hardness near the aggregate including ITZ was large.

3.3 Relationship between Compressive Strength and ITZ

Figure 4 shows the compressive strength of concrete. The strength was improved by using RWA regardless of the fly ash replacement ratio, and especially for FA20, and the strength increased remarkably with the age. This result is consistent with the remarkable improvement in the hardness of ITZ for a long time.

Figure 5 shows the relationship between the strength of the concrete specimen including RWA (FA0G10 and FA20G10) and the thickness of ITZ around RWA embedded in the paste specimen (FA0-T and FA20-T) as well as the relationship between the strength of the concrete specimen without RWA (FA0G0 and FA20G0) and the thickness of ITZ around conventional aggregate embedded in the paste specimen (FA0-N and FA20-T). As the thickness of ITZ around RWA decreases, the compressive strength of the concrete including RWA increases significantly when compared to the specimen using only conventional aggregate. In particular, the tendency was significant on FA20. This suggests that, especially when fly ash is used, the range of the internal curing effect of RWA is narrow, but the ITZ thickness is significantly reduced, resulting in the increase in the compressive strength.

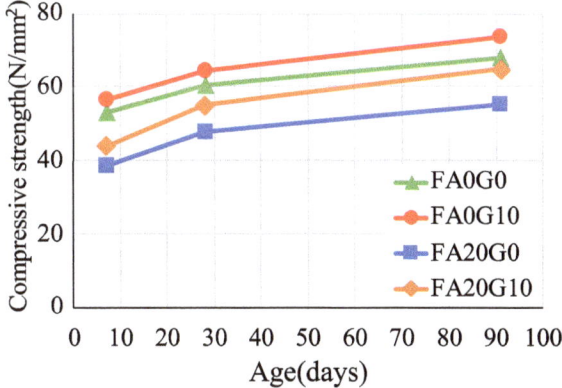

Fig. 4 Compressive strength of concrete

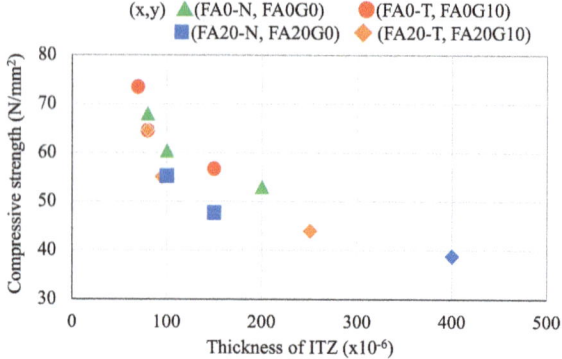

Fig. 5 Relationship between compressive strength and thickness of ITZ

4 Conclusions

- The Vickers hardness around RWA was more than 1.05 times as high as that around crushed stone in the range of 1 mm for FA0, 300 μm for FA20 and 100 μm for FA40 at the age of 7 days. The improvement in hardness near the aggregate in the specimen containing fly ash was more significant than that in the specimen without fly ash, especially at the later age.
- The thickness of ITZ around RWA was thinner than that around crushed stone regardless of the fly ash replacement ratio. This improvement of ITZ around RWA could lead to the increase in the compressive strength of concrete.
- In the case of FA20, the increase in compressive strength of concrete using RWA and the decrease in ITZ thickness around RWA were remarkable when compared to FA0.
- The degree of reaction of fly ash around RWA was higher than that around conventional coarse aggregate.

References

1. Bui PT, Ogawa Y, Nakarai K, Kawai K, Sato R (2017) Internal curing of Class-F fly-ash concrete using high-volume roof-tile waste aggregate. Mater Struct 50(203):1–12
2. Golewski LG (2018) Evaluation of morphology and size of cracks of the Interfacial Transition Zone (ITZ) in concrete containing fly ash (FA). J Hazard Mater 357:298–304
3. Izumo K, Nawa T (2003) Shrinkage strain of mortar using fly ash under two different temperature. Doboku Gakkai Ronbunshu 60(739):221–236
4. Ogawa Y, Bui PT, Kawai K, Sato R (2020) Effects of porous ceramic roof tile waste aggregate on strength development and carbonation resistance of steam-cured fly ash concrete. Constr Build Mater 236:117462

5. Ohsawa S, Sakai E, Daimon M (1999) Reaction ratio of fly ash in the hydration of fly ash-cement system. Cem Sci Concr Technol 53:96–101
6. Uchikawa H (1995) Influence of interfacial structure between cement paste and aggregate on the quality of hardened concrete. Concr J 33(9):5–17

Physical and Mechanical Properties of Lightweight Fly Ash/GGBS Based Geopolymer Mortar Foamed with Hydrogen Peroxide

Tuan Anh Le, Tan Khoa Nguyen, and Thuy Ninh Nguyen

Abstract Vietnam is one of the countries in Southeast Asia that possesses many fly ash and ground granulated blast furnace slag. They are known as by-products from the thermal power plant and steel industries with rich aluminosilicate content to activate with alkaline liquid in geopolymer. In this research, hydrogen peroxide (H_2O_2) is used as a foaming agent in lightweight geopolymer mortar. The experimental work with mixed proportions of fly ash, ground granulated blast furnace slag, alkaline solution, fine aggregate, and foam agent is conducted to investigate lightweight geopolymer mortar properties. As a result, the flow diameter and flow time depends on the sodium solution to geopolymer solid ratio and fly ash to ground granulated blast furnace slag ratio. Besides, the higher percentage of foam agent and higher content of GGBS result in a more significant expansion degree of lightweight geopolymer mortar. On the other hand, the compressive strength of specimens decreases with the increase of foam agent and content of GGBS. In conclusion, the lightweight geopolymer mortar foamed with hydrogen peroxide shows a high potential to produce lightweight panels in construction.

Keywords Lightweight · Geopolymer mortar · Fly ash · Ground granulated blast furnace slag · Hydrogen peroxide · Flowability · Strength

T. A. Le (✉)
Faculty of Civil Engineering, Ho Chi Minh City University of Technology, 268 Ly Thuong Kiet Street, District 10, HCMC, Vietnam
e-mail: latuan@hcmut.edu.vn

T. A. Le · T. N. Nguyen
Vietnam National University Ho Chi Minh City, Linh Trung Ward, Thu Duc District, HCMC, Vietnam

T. K. Nguyen
Institute of Research and Development, Duy Tan University, Da Nang, Vietnam

Faculty of Civil Engineering, Duy Tan University, Da Nang, Vietnam

J. N. Reddy et al. (eds.), *ICSCEA 2021*, Lecture Notes in Civil Engineering 268,
https://doi.org/10.1007/978-981-19-3303-5_46

1 Introduction

The emission of greenhouse gases has emerged as a severe problem that dramatically affects global climate change. Greenhouse gases emission was caused by various sectors, including transport, energy supply, residential, businesses, industrial processes, agriculture, etc. Calcining limestone in the production of Portland cement concrete is responsible for a large amount of CO_2 emission with up to 10% of CO_2 emission globally [1]. The manufacture of geopolymer material was reported to cut carbon footprint by 90% compared to PCC production [1].

Geopolymer products are formed through alkaline liquid reacted with silica and alumina contained in aluminosilicate precursors. One of the first geopolymer cement, PYRAMENT, was able to withstand the weight of a commercial plane only after 6 h of curing [2]. Depending upon local resources and availability, solid aluminosilicate precursors can be in natural forms such as fly ash (FA), metakaolin, ground granulated blast furnace slag (GGBS), red mud, waste glass, etc. [5]. Geopolymers are porous materials with many and different shapes of pores [3]. The attraction of geopolymer material comes from their high strength, good thermal and chemical resistance and excellent durability, which are useful for practical applications. Geopolymers, including dense and lightweight types, are used in the civil engineering, chemical and nuclear industries. Lightweight geopolymer can be applied to various applications such as filters, catalyst supports [7], thermal insulators and heat exchangers [4].

There is a lot of studies about lightweight geopolymer materials. However, it lacks information about the physical and mechanical properties of FA/GGBS based geopolymer mortar foamed with hydrogen peroxide. In this study, the effects of aluminosilicate sources, alkaline solution (sodium solutions) and foam agent on flowability, expansion degree and compressive strength are investigated. It is noted that hydrogen peroxide is used as a foaming agent in the mixture to produce lightweight geopolymer mortar.

2 Experimental

2.1 Materials

In this study, the experimental work of lightweight geopolymer mortar was carried out using fly ash (FA) and ground granulated blast furnace slag (GGBS) as source material, sodium solution as an activated liquid and fine aggregate. The specific gravity and specific surface of FA are 2.45 and 320 m^2/kg, respectively. For GGBS, the specific gravity and specific surface are 2.81 and 340 m^2/kg, respectively. The

Table 1 Chemical compositions of FA and GGBS

Oxide (%)	SiO_2	Al_2O_3	CaO	MgO	Fe_2O_3	SO_3	$K_2O + Na_2O$	LOI
GGBS	38.4	17.3	34.5	4.32	1.21	0.42	1.31	2.54
FA	51.3	31.7	1.11	0.84	3.45	0.24	1.05	10.31

LOI: Loss of Ignition

amount of aluminosilicate for FA and GGBS is 83 and 55.7%, respectively. Chemical compositions of FA and GGBS are shown in Table 1.

The sodium solution (SS) is a combination of sodium silicate and sodium hydroxide in the solution. The sodium silicate solution included Na_2O and SiO_2 about 36–38% is mixed to sodium hydroxide solution 10 Molar (M). The ratio of SiO_2 to Na_2O by mass was approximately 2. Besides, river sand is used as fine aggregate. The specific gravity and bulk density of sand are 2.45 and 1290 kg/m^3, respectively. The fineness modulus is 1.72. The hydrogen peroxide (H_2O_2) is used as a foaming agent in this research. The hydrogen peroxide is mixed as an aqueous solution of about 20 to 40% concentration by weight (dosage of H2O2).

2.2 Test method

In the mix proportion of geopolymer mortar, the ratio of source materials (FA + GGBS) and sand is 1:3 by weight. The ratio of sodium solutions and source materials is mixed in the range from 0.8 to 1.1 by weight. Moreover, the source material combines FA and GGBS in three ratios 100:0, 50:50 and 0:100% by weight. Dosage of hydrogen peroxide agent is investigated from 1 to 3% by volume and at 20% concentration. The mixed proportion of geopolymer mortar is shown in Table 2. In this Table, the FA/S is the ratio between …., and SS/GS is the ratio between sodium solution and geopolymer solid, which is the total of FA and GGBS, by mass.

Geopolymer paste was prepared by mixing FA and GGBS with a sodium solution. Then, fine aggregate is added to this paste to create geopolymer mortar. After mixing for 5 min, hydrogen peroxide is added and mixed thoroughly for another 1 min. After the addition of a foaming agent, the mixture is poured into a plastic mold. The poured mixture is allowed to expand for 5 min due to the release of gases. The specimen is sealed and cured at ambient conditions. The specimens are tested compressive strength on 28-day.

Table 2 Mix proportion of geopolymer mortar

Mix	FA/S	FA	GGBS	SS/GS
F1S01	1:02	100	0	0.8
F1S11		50	50	
F0S11		0	100	
F1S02	1:02	100	0	0.9
F1S12		50	50	
F0S12		0	100	
F1S03	1:02	100	0	1
F1S13		50	50	
F0S13		0	100	
F1S04	1:02	100	0	1.1
F1S14		50	50	
F0S14		0	100	

To evaluate the properties of lightweight geopolymer mortar, the standards ASTM C1437, and C39 were employed to determine flowability, and compressive strength of lightweight geopolymer mortar. The expansion rate was determined by the change of height in measuring cylinder as shown in Fig. 1.

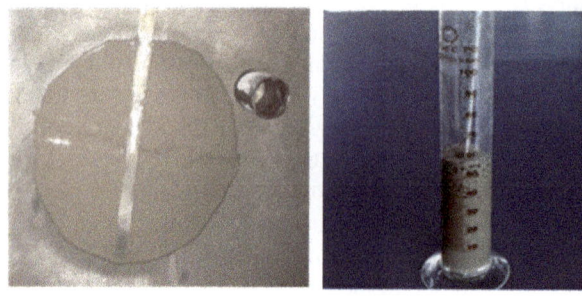

a. Flow test   b. Expansion test   c. Compression test

Fig. 1 Testing programs of this study

3 Results and discussion

3.1 Effect of Geopolymer Binder on the Flow of Lightweight Geopolymer Mortar

In this research, geopolymer binder is mixed by replacement of GGBS in fly ash. The results are shown that the flow diameter of lightweight geopolymer mortar is increased by GGBS adding. According to Fig. 2, at the same SS/GB ratio value, the flow diameter increases from 12–20% when the GGBS content is added from 0–100%. This result is consistent with the previous research [6] that is rheological of slag-based geopolymer mortar reduces with the addition of fly ash. It means the flow diameter of geopolymer mortar using only GGBS is highest. The second level belongs to geopolymer mortar using blended FA and GGBS, and the mortar using only FA has the lowest value of flow diameter. Besides, at the same ratio of FA/GGBS, the flow diameter of specimens increases with the SS/GB ratio increase.

On the other hand, based on Fig. 3, the flow time increases about 15% by GGBS adding in binder. The higher alkaline liquid results in a higher flow time. It can be noted that the flowability of geopolymer mortar not only increased by alkaline activator—binder ratio but also depended on the type of aluminosilicate resource material.

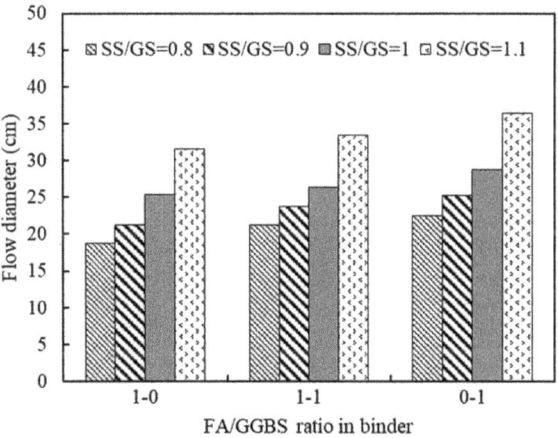

Fig. 2 Relationship between flow diameter and various geopolymer binder

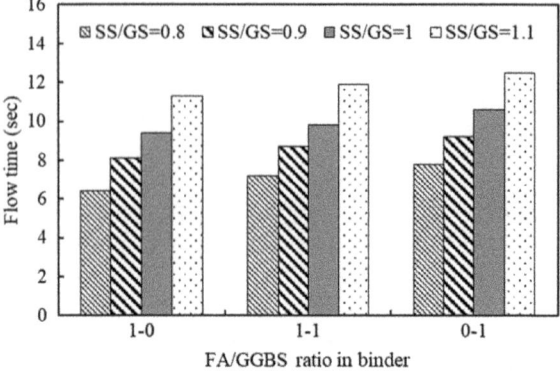

Fig. 3 Relationship between flow time with various geopolymer binder

3.2 Effect of Foam Agent and Source of Aluminosilicate on Expansion Degree of Lightweight Mortar

The geopolymer binder mixed fly ash and slag can be reacted with a foaming agent to obtain a pore structure in mortar. As seen in Fig. 4, the expansion degree of lightweight geopolymer mortar increases with the increase of GGBS content and volume of foam agent in the range from 1 to 3%. For example, at the range of 1% of foam, the expansion degree of specimen is 24%. When GGBS is added from 50 to 100%, expansion increases from 27–32%. This trend is also the same for those mixtures using 2 and 3% foaming agents. From these facts, the foaming agent with hydrogen peroxide solution can work with geopolymer mortar to produce air bubble formation.

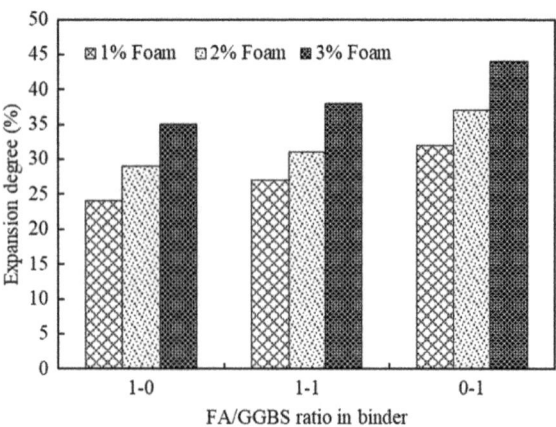

Fig. 4 Relationship between FA/GGBS ratio and expansion degree of lightweight mortar

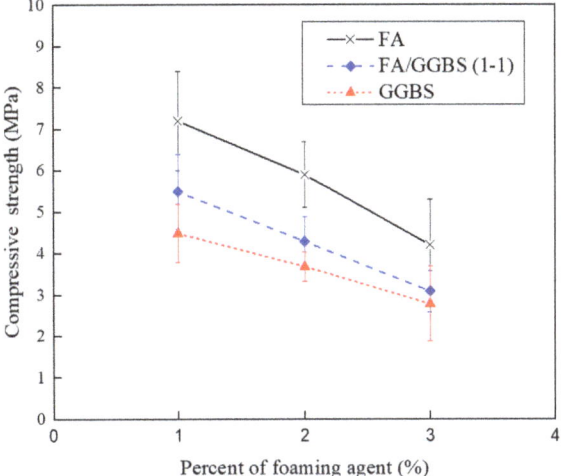

Fig. 5 Relationship between foam agent, source materials and strength of lightweight mortar

With an increasing pores volume in mixture, the expansion degree of specimen increases about 1.3 times for all mixtures. It can be concluded that the expansion degree of lightweight geopolymer mortar belongs to a type of aluminosilicate and volume of foam agent.

3.3   Effect of Foam Agent and Source of Aluminosilicate on the Strength of Lightweight Geopolymer Mortar

In this section, the effect of foam agent and source of aluminosilicate on compressive strength of lightweight geopolymer mortar are evaluated. As shown in Fig. 5, the compressive strength of specimens decreases with the increase of foam agents. For all three categories of mixtures, the compressive strength decreases about 1.6 to 1.7 times when the usage of foam agents increases from 1 to 3%. Furthermore, the specimens using only FA show the highest compressive strength, while the specimens using only GGBS have the lowest strength. The mixtures using FA as source material has higher compressive strength, about 50–60%, than the mixtures using GGBS. The reason might come from the content of aluminosilicate from the source material. Based on Table 1, the content of aluminosilicate of FA is higher than GGBS (83.6 compared with 55.7%). With a higher content of aluminosilicate, the geopolymerization in the mixtures using FA produces more products than the mixture using GGBS.

Besides, the relationship between expansion degree and compressive strength of lightweight geopolymer mortar is illustrated in Fig. 6. According to Fig. 6, the

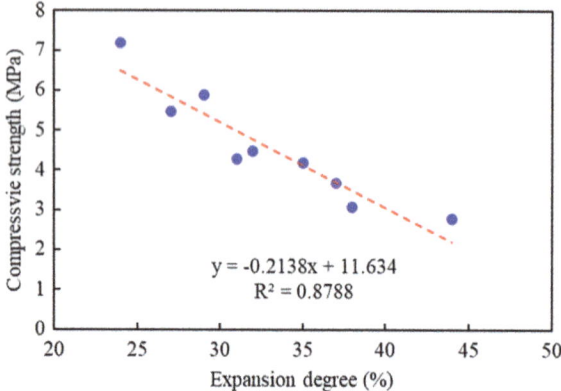

Fig. 6 Relationship between expansion degree and strength of lightweight mortar

compressive strength of lightweight mortar decreases with the increase of expansion degree. When the expansion degree changes from 24 to 44%, the compressive strength reduces 2.57 times from 7.2 MPa to 2.8 MPa. It can be explained that the compressive strength of mortar depends on the pores of specimens' structure. With a more considerable expansion degree, more pores are created. As a result, the strength of lightweight mortar decreases.

4 Conclusion

The research on the properties of FA/GGBS based geopolymer mortar foamed with hydrogen peroxide has some results as following:

- The flow diameter and flow time of geopolymer mortar firmly belong on SS/GS and FA/GGBS ratio.
- The expansion degree of lightweight geopolymer mortar depends on the type of aluminosilicate and volume of foam agent. A higher percentage of foam agent and higher content of GGBS result in a more significant expansion degree of lightweight geopolymer mortar.
- The compressive strength of specimens decreases with the increase of foam agent and content of GGBS. Also, the strength of specimens reduces when the expansion degree increases. The lightweight geopolymer mortar foamed with hydrogen peroxide can produce lightweight panels in construction from these above results.

Acknowledgements This research is funded by Vietnam National University Ho Chi Minh City (VNU-HCM) under grant number B2020-20-01.

References

1. Davidovits J (1994) Geopolymers: Man-made rock geosynthesis and the resulting development of very early high strength cement. J Mater Educ 16:91–137
2. Davidovits PJ (2002) 30 years of successes and failures in geopolymer applications. Market trends and potential breakthroughs . In: Geopolymer 2002 conf, pp. 1–16
3. Duxson P, Provis JL, Lukey GC et al (2005) Understanding the relationship between geopolymer composition, microstructure and mechanical properties. Colloids Surf A Physicochem Eng Asp 269:47–58. https://doi.org/10.1016/j.colsurfa.2005.06.060
4. Kamseu E, Nait-Ali B, Bignozzi MC et al (2012) Bulk composition and microstructure dependence of effective thermal conductivity of porous inorganic polymer cements. J Eur Ceram Soc 32:1593–1603. https://doi.org/10.1016/j.jeurceramsoc.2011.12.030
5. Luukkonen T, Abdollahnejad Z, Yliniemi J et al (2018) One-part alkali-activated materials: A review. Cem Concr Res 103:21–34
6. Roy B, Laskar AI (2021) Rheological behavior of geopolymer mortar with fly ash, slag and their blending. In: Adhikari S, Dutta A, Choudhury S (eds) Advances in structural technologies. Springer Singapore, Singapore, pp 99–110
7. Sazama P, Bortnovsky O, Dĕdeček J et al (2011) Geopolymer based catalysts-new group of catalytic materials. Catal Today 164:92–99. https://doi.org/10.1016/j.cattod.2010.09.008

Study of the Properties of Concrete Using Mixed Recycled Coarse Aggregates of Different Qualities

Naoki Kawata, Sachie Sato, and Hiroyuki Tanano

Abstract In Japan, research on recycled aggregates has been conducted since the 1970s. The JIS standard compiles the findings from 2005 to 2007 and revised from 2018 to 2019. In these standards, the quality of recycled aggregate is divided into three levels: Classes H, M, and L. The applications of aggregates other than high-quality H are limited. However, Class-H aggregate has not been widely used because of the increased cost and energy required to produce better-quality recycled aggregate. One solution to this problem is to use a recycled aggregate of medium quality formed by mixing ordinary aggregate and recycled aggregate (mostly low-quality recycled aggregate) rather than using recycled aggregate alone in concrete. In this study, we investigate the strength and durability of concrete made from mixtures of recycled aggregates of different quality and normal aggregate. We confirm that the quality of concrete is higher when low-quality recycled aggregate mix ordinary aggregate than when alone use moderate-quality recycled aggregate.

Keywords Concrete · Recycled aggregate · Compressive strength · Drying shrinkage · Freeze-thaw tests

1 Introduction

In recent years, the lack of availability of aggregates for concrete has become a problem, and securing aggregate resources has become critical. Furthermore, the recycling of dismantled concrete blocks has been identified as an environmental problem. The use of recycled aggregate produced by processing concrete blocks has attracted attention and, in Japan, JIS and various other studies are being conducted to expand the use of recycled aggregate.

N. Kawata (✉) · S. Sato · H. Tanano
Department of Architecture, Tokyo City University, 1-28-1 Tamazutsumi, Setagaya, Tokyo, Japan
e-mail: naoki.kawata1998@gmail.com

H. Tanano
Building Research Institute, 1 Tachihara, Tsukuba, Ibaraki, Japan

© The Author(s), under exclusive license to Springer Nature Singapore Pte Ltd. 2023
J. N. Reddy et al. (eds.), *ICSCEA 2021*, Lecture Notes in Civil Engineering 268,
https://doi.org/10.1007/978-981-19-3303-5_47

JIS classifies recycled aggregates as high-quality recycled aggregate, H; medium-quality recycled aggregate, M; and low-quality recycled aggregate, L. Recycled aggregate H is classified as JIS A 5306. JIS A 5308 (Ready-mixed concrete) of Annex A (Aggregate for ready-mixed concrete) includes Recycled aggregate H and use for ready-mixed concrete like other aggregates. However, the greater cost of producing high-quality recycled aggregate is an issue. On the other hand, medium- and low-quality recycled aggregates (M and L) can reduce energy and cost during production. However, JIS A 5308 Annex does not incorporate it, and under the Building Standard Law (Article 37) limits their application in buildings. The standards for recycled aggregate M specified in JIS A 5022 Annex A have density, water absorption, impurity content, etc. [1] It is known that these qualities have a significant influence on the durability of recycled aggregate concrete. In addition, recycled aggregate concrete M permit to mix of recycled aggregate M or recycled aggregate L with ordinary aggregate as specify in JIS A 5308 Annex A, and the combination of these materials also know to have a significant effect on durability. However, it has not been quantitatively clarified to what extent the quality of recycled aggregate and the combination of aggregates affect the durability of recycled aggregate concrete M.

In this study, experimental investigations were conducted to quantitatively evaluate the effects of aggregate quality and aggregate combination on mechanical properties and durability, which are essential for using recycled aggregate concrete M in building structures.

2 Summary of Experiments

2.1 Tests on Recycled Aggregate

Table 1 shows the items of the aggregate test and the standard outline of the test method. Table 2 shows the aggregate types. The freezing and thawing resistance test is a test method conducted on recycled aggregates. Recycled aggregates immersed in water in a container were subjected to freezing and thawing cycles and every ten

Table 1 Test methods for recycled aggregate

Test	Method
Impurities	JIS A 5021 Annex B
Oven-dry density, Water absorption	JIS A 1110
Fine-grain amount	JIS A 1103
Grade of aggregate	JIS A 1102
Shape of aggregate (solid volume percentage for shape determination)	JIS A 5005
Freeze–thaw resistance	JIS A 5022 Annex D

Table 2 Types of aggregate and ratio of usage

Types of Coarse Aggregate	Symbol	Aggregate Ratio (%)
Ordinary	O	100
Recycled M	M	100
Recycled L	L	100
Ordinary: Recycled M	MH	50:50
Ordinary: Recycled L	ML	50:50

cycles conduct sieve tests. This test was conducted up to thirty cycles to calculate the FM freezing resistance index defined by the following equation:

$$FM = FMa - FMb \tag{1}$$

where FM: Frost damage index of coarse aggregate (FM freezing damage index);
   FMa: Fineness modulus of coarse aggregate before the freeze–thaw test; and
   FMb: Fineness modulus of coarse aggregate after the freeze–thaw test.

2.2 Examination of Concrete

Table 3 shows the types and combinations of aggregates. Table 4 shows the basic mixing conditions of concrete. Table 5 shows a summary of the tests of mechan-

Table 3 Types of concrete and aggregate combinations

Coarse aggregate	Mixture name
O	OR
M	MM
MH	MH
ML	ML

Table 4 Mixture condition

Factors	Levels
Types of cement	Blast furnace
Water to cement ratio	59%
Water content	175 kg/m^3
Sand–total aggregate ratio	49%
Design standard strength	27 N/mm^2

Table 5 Test methods for the concrete

Test	Method	Test age
Compressive strength	JIS A 1108	4 w, 13 w, (1 y)
Drying shrinkage	JIS A 1129–2 (contact gauge method)	7, 14, 21, 28, 35, 42, 49, 56, 63, 70, 77, 84, 91 days
Freeze–thaw resistance [2, 3]	JIS A 1148	30, 60, 90, 120, 150, 180, 210, 240, 270, 300 days

ical properties and durability of concrete. Recycled aggregate provided by Seiki Corporation and Higuchi Sangyo Co.

3 Results and Discussion

3.1 Aggregate Test Results

Figure 1 shows the results of the impurity content test for each aggregate. It can seem that it is M, not L, that contains the highest amount of impurities. Since the results show that it expects that the quality of concrete using M would be the lowest.

Figures 2 and 3 show the results of the absolute dry density and water absorption tests for each aggregate. The figures show the results of experiments conducted by the authors and the standard values given in the manufacturer's reports and JIS. The standard values for dry density are the lower limit, and the standard values for water absorption are the upper limit. The water absorption of MH and ML is smaller than that of M. Therefore, it is considered that MH and ML have greater freeze–thaw resistance than the M used in this study.

Figure 4 shows the particle size distribution curve obtained from the sieving test. Table 6 shows the test results for the fineness modulus of each aggregate. There was no significant difference in the mass fraction passing through each sieve for any of the aggregates and the coarse grain fraction not observed significant change, suggesting that the effect of grain size on strength and other properties is small.

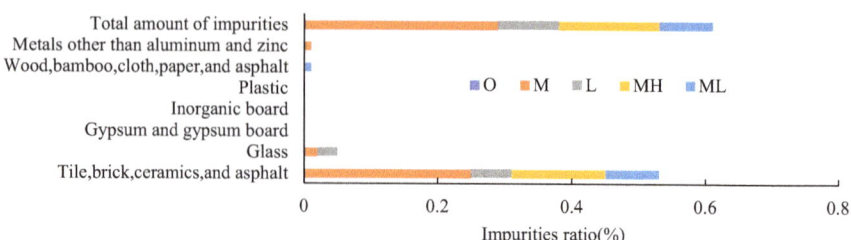

Fig. 1 Ratios of impurities in coarse aggregate

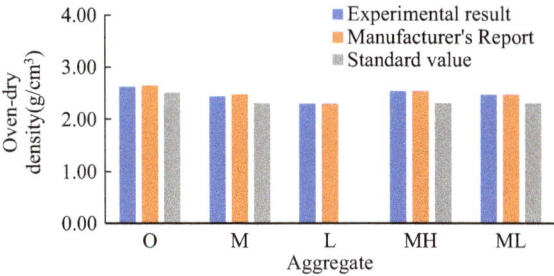

Fig. 2 Oven-dry densities

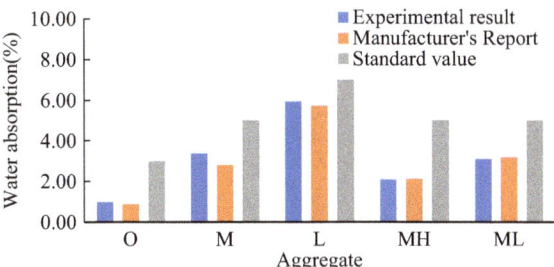

Fig. 3 Water absorption test results

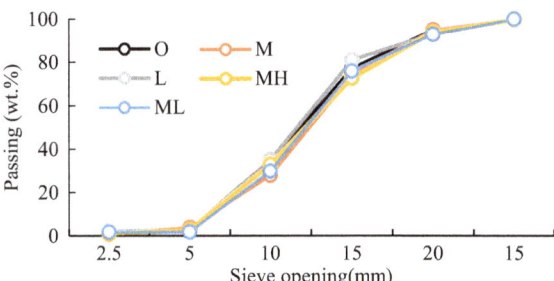

Fig. 4 Particle size distribution

Table 6 Results of F.M

Aggregate	Experimental result	Manufacturer's report
O	6.87	6.57
M	6.95	6.78
L	6.79	6.56
MH	6.96	6.6
ML	6.93	6.66

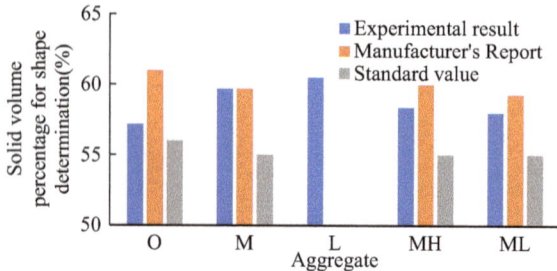

Fig. 5 Results for solid volume percentage for shape determination

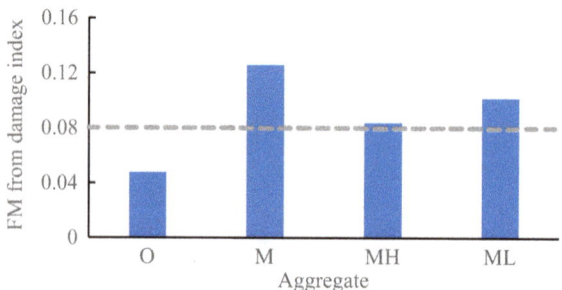

Fig. 6 Frost damage index for coarse aggregate

Figure 5 shows the test results for the solid volume percentage for shape determination of each aggregate. O, which has the best quality, had a lower actual percentage than M and L. It was also lower than those of MH and ML, mixed with each aggregate. However, according to the report from the aggregate manufacturer, the solid volume percentage of the ordinary coarse aggregate was 61%. It is therefore essential to increase the number of tests appropriately for recycled aggregate because of the large variation in each production lot.

As shown in Fig. 6, for M to use as a frost-damage-resistant product, the FM frost damage index must be less than 0.08. In this study, the FM frost damage index of recycled coarse aggregate M was much higher than 0.08. The FM frost damage index of recycled coarse aggregates MH and ML could be lower than that of each coarse aggregate alone, but it could not be lower than 0.08. From the above, it is considered necessary to study the mixing ratio of the mixed coarse aggregates.

3.2 Compressive Strength

Compressive strength and Young's modulus were obtained from specimens that were 28 days old.

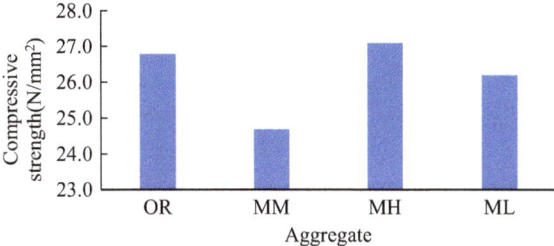

Fig. 7 Compressive strength of concrete

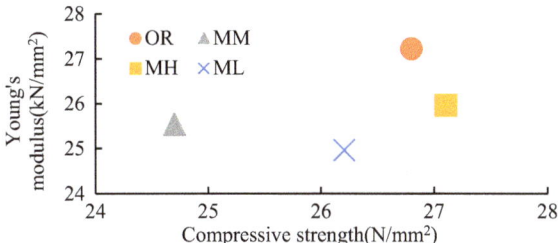

Fig. 8 Relationship of compressive strength and Young's modulus

From Fig. 7, the compressive strength of recycled coarse aggregate concrete M was the lowest. This may be due to a large number of impurities in the recycled coarse aggregate M and the presence of low-strength areas in some parts of the concrete, and this heterogeneity may have reduced the strength of the entire concrete. From Fig. 8, Young's modulus was smaller for recycled coarse aggregate concrete ML than for recycled coarse aggregate concrete M, which had the lowest compressive strength. Also, the compressive strength of ordinary coarse aggregate concrete was smaller than that of recycled coarse aggregate concrete MH, but Young's modulus was larger. This may be because young's modulus increased in proportion to the unit volume mass.

3.3 Drying Shrinkage

Drying shrinkage tests conduct based on the contact gauge method specified in JIS A 1129–2.

Figure 9 shows the relationship between age after drying and length change rate and Fig. 10 shows the relationship between age after drying and mass change rate. From Fig. 9, comparing the recycled coarse aggregate concrete M (20-MM) with the concrete using ordinary coarse aggregate (20-OR), it was confirmed that the length change rate (dimensional change rate due to drying shrinkage) was smaller for the

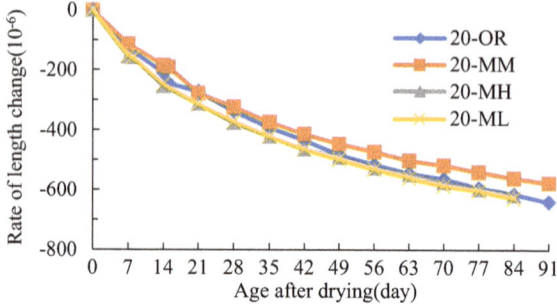

Fig. 9 Rate of length change

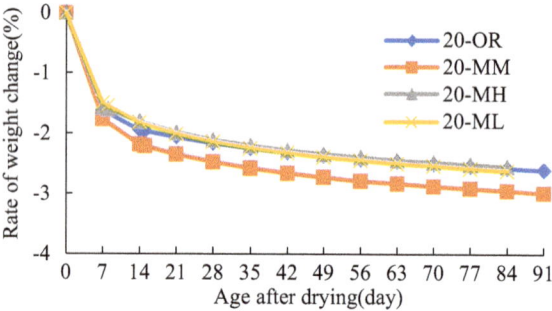

Fig. 10 Rate of weight change

recycled coarse aggregate concrete M. This may have been due to the higher water absorption of the recycled coarse aggregate M by the remaining mortar. On the other hand, the rate of change in the length of recycled coarse aggregate concrete MH (20-MH) and recycled coarse aggregate concrete ML (20-ML) was almost the same, indicating that the quality of the coarse aggregate used had almost no effect. From Fig. 10, the rate of mass change of recycled coarse aggregate concrete MM was the largest. In general, the rate of change in length correlates with the rate of change in mass, but in this study observe since the opposite trend, further verification is necessary.

3.4 Results of Freeze–Thaw Tests

The simple freezing and thawing test for aggregates is a test method that newly establish when JIS A 5022 was revised in 2018. Repeatedly freezing and thawing samples conduct the test in a standard household freezer. [2, 3].

Figure 11 shows the relationship between the number of cycles and the relative dynamic elastic modulus and Fig. 12 shows the relationship between the number of

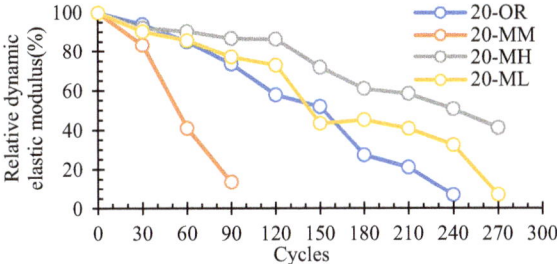

Fig. 11　Relative dynamic elastic modulus

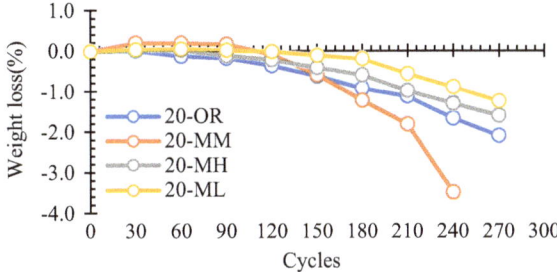

Fig. 12　Weight loss (%)

cycles and the mass loss ratio. From Fig. 11, the relative dynamic elastic modulus was below 60% at 60 cycles for recycled coarse aggregate concrete M, which is the fastest decrease. This result shows that applications where resistance to frost damage require to consider difficult to use in recycled coarse aggregate M. On the other hand, the relative dynamic modulus of elasticity of recycled coarse aggregate MH remained at 85% or higher until 120 cycles, confirming that it has the same or better freeze–thaw resistance as normal concrete. In addition, it finds that recycled coarse aggregate L could improve freeze–thaw resistance by mixing with normal coarse aggregate. The mechanism of this effect needs to investigate in the future.

Figure 12 shows that the mass of concrete with recycled coarse aggregate increased up to approximately 100 cycles after the start of the freeze–thaw test. In general, the mass decreased along with the relative dynamic modulus of elasticity, but the concrete with recycled coarse aggregate observe the opposite trend.

Figure 13 shows the condition of the surface of normal concrete and recycled coarse aggregate concrete M specimens when the relative dynamic modulus of elasticity was less than 20%. Although the relative dynamic modulus of elasticity is almost the same, the condition of the specimen surface is significantly different. It can infer that the inner structure, as well as the surface layer of normal concrete, is severely damaged about to with concerning the relative dynamic modulus of elasticity. On the other hand, in the case of recycled coarse aggregate concrete M, the surface layer was only slightly damaged but the internal structure was severely damaged, indicating

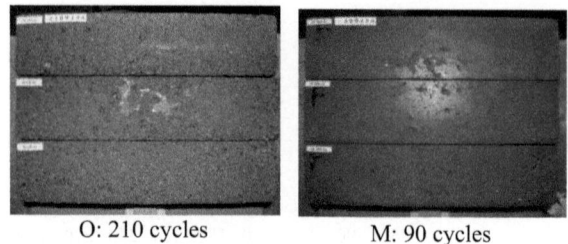

O: 210 cycles   M: 90 cycles

Fig. 13 Surfaces of the specimen when under 20% of relative dynamic elastic modulus

that there was a difference in damage between the surface layer and the internal structure of the specimen. These results suggest that the cracks and voids caused by the internal structure damage are larger in the recycled coarse aggregate concrete than in the normal concrete, including the initial defects, and the mass of water that penetrated these defects exceeds the mass of the missing concrete structure due to the damage.

4 Conclusion

This study investigates the properties of concrete using various mixing recycled coarse aggregate obtained by blending recycled coarse aggregate M and L with normal aggregate. The results showed that the aggregate properties, mechanical properties of concrete, and freezing and thawing tests were superior to those obtained when 100% recycled coarse aggregate M was used. The mechanism of this effect needs to investigate in detail in the future. As for the drying shrinkage, the mass-loss rate did not correlate with the amount of drying shrinkage, and further study needs.

References

1. Yoda Kazuhisa, Onodera Toshiyuki, Shintani Akira, Kawanishi Taichiro (2006) Influences of the quality of recycled coarse aggregate on concrete properties. Ann Proceed Concr Eng 28(1):1457–1462
2. Katahira Hiroshi, Watanabe Hiroshi (2005) Development of a simple test method to assess freeze-thaw resistance of recycled aggregate. Ann Proceed Concr Eng 27:1351–1356
3. Katahira Hiroshi, Watanabe Hiroshi (2006) Study on rational judgment method of a simple test method to assess freeze-thaw resistance of recycled aggregate. Ann Proceed Concr Eng 28:1433–1438

The Enhancement of Recycled Aggregate Concrete by Steel Fiber Reinforcement for Rigid Pavement

Thuy Ninh Nguyen, Tuan Anh Le, and Hai Viet Vo

Abstract The use of demolition waste as recycled aggregate (RA) in concrete is a high demand for sustainable developments; however, there are drawbacks to replacing RA with natural aggregate. The purpose of this study is to evaluate the application of RA to rigid pavements with steel fibers (SF) as reinforcement due to its advantages in improving tensile strength and reducing cracks. Compression tests were conducted on concrete samples cast with several contents of SF and RA to determine their desirable contents. Flexible strength was determined using the four-point bending test to show the effects of SF on the drawbacks of recycled aggregate concrete (RAC). As a result, the concrete mixture was less workable when adding SF and/ or RA. Although the compressive strength and flexure strength of RAC were subsequently decreased when a part of natural aggregate was replaced by RA, some of them were recovered after increasing a small amount of SF. In summary, it would be a possibility of applying SF to RAC pavement for lessening damages and prolong its life span.

Keywords Recycled aggregate · Steel fibers · Rigid pavement · Compressive strength · Flexible strength

1 Introduction

With an increasing demand for construction while the current resource of natural aggregate is limited, leading to high prices, and not guaranteed quality of aggregate. To reduce the exploitation of natural resources, it is urgent to find an alternative material, which would be environmentally friendly and economical [1]. One of the proposed methods to reduce construction solid wastes is to utilize them to

T. N. Nguyen
Faculty of Civil Engineering, VNU-HCM, Ho Chi Minh City, Vietnam

T. A. Le · H. V. Vo (✉)
Faculty of Civil Engineering, Ho Chi Minh University of Technology, VNU-HCM, Ho Chi Minh City, Vietnam
e-mail: haivo@hcmut.edu.vn

© The Author(s), under exclusive license to Springer Nature Singapore Pte Ltd. 2023
J. N. Reddy et al. (eds.), *ICSCEA 2021*, Lecture Notes in Civil Engineering 268,
https://doi.org/10.1007/978-981-19-3303-5_48

produce concrete for construction works or others. For this reason, some studies have been carried out and shown that the compressive strength of concrete decreases as including recycled aggregate (RA) in a concrete mix, specifically, when replacing over 50% of course aggregate with RA, the strength of recycled aggregate concrete (RAC) decreased by 16.6% and decreased by 26.4% when using 100% RA for RAC, compared to concrete with only natural aggregates [2]. It is due to the physical properties of RA affect the properties of concrete. The basic characteristics such as shape, surface texture, pore-volume, and water absorption of RA are often significantly different from those of natural coarse aggregate because of the presence of a concrete layer, which is loose and contains impurities in concrete. Safiuddin et al. also highlighted the influence of RA on the properties of fresh and hardened concrete after hardening as well as its durability. RA can also affect the workability of fresh concrete since it has great angularity, surface roughness, absorbency, and porosity compared to natural aggregates [3]. The effect of RA on the properties of hardened concrete depends on its source, classification, and physical properties. According to a few published studies, the mechanical properties of hardened concrete using RA decreased, however, with the RA replacement of less than 30% (by mass) of course aggregate the decrease was significantly [4, 5]. The above issues show that the reuse of construction solid waste and proposing research on using recycled aggregate for concrete are potentially feasible and practical. Meanwhile, fibers such as steel fiber, glass fiber, PP plastic fiber, etc. have been studied and shown to have advantages such as minimizing the cracking, increasing strength and durability for concrete, especially concrete road surface with a reasonable amount for each type of steel fiber with different lengths and shapes [6, 7].

The main purpose of this study is to evaluate the ability of steel fibers of strengthening RAC as its natural aggregate replaced by recycled one. Slump and unconfined compression tests were conducted to determine the workability of concrete mixture and the effect of RA and SF on the strength of hardened concrete, respectively. For the tensile strength, a 4-point bending test was performed to assess pre- and post-cracking behaviors of concrete with steel fiber reinforcement.

2   Materials and Methodology

Cement PC40 (Type I) was used in this study. RA, coarse and fine aggregates had their properties tested according to TCVN7572:2006 [8] and they are shown in Table 1. Concrete waste was broken off from specimens obtained concrete work testing and demolition waste to get RA. Then, RA was washed and sieve to the size of 5 to 20 mm. Table 2 shows the gradation of course aggregate and RA.

The hooked-end steel fibers (SF) Dramix 3D 80/60BG with 60 mm in length, 0.75 mm in diameter, and strength of 1225 MPa, were used in this study. When mixed into concrete, SF will be randomly and relatively dispersed in the concrete mixture. The addition of steel fibers to the concrete mix is primarily intended to control cracks, due to the friction, adhesion, and hooked-ends of SF and concrete [7]. However, the

Table 1 Properties of aggregates

Properties	Fine agg.	Natural aggregate	RA
Specific gravity (g/cm³)	2.72	2.70	2.3
Bulk density (kg/m³)	1392	1526.5	1206
Water absorption (%)	0.95	0.21	4.2
Fineness modulus	2.167	–	–

Table 2 Gradation of course aggregate and RA

Sieve size (mm)	Cumulative percent retained (%)	
	RA	Course agg.
20	7.50	1.93
12.5	55.33	65.84
10	79.56	86.98
5	100	100

dosage of fibers depends on the mixing technique and the workability of a concrete mixture. Hence, the optimal content of SF was determined and its effect on RAC was investigated. SF was dosed as kilogram per cubic meter of concrete.

The aggregates were prepared by washing and air-drying to remove dust and deleterious materials, then stored in room condition prior to mixing. The mix proportioning was based on ACI mix design [9], and shown in Table 3, with the specified compressive strength of 65 MPa and target slump of 8–9 cm. All components were quantified as the proportion and mixed in a forced-action concrete mixer. For a good

Table 3 The proportion of concrete mixtures

Sample	RA (%)	Cement (kg)	Natural aggregate (kg)	SF (kg/m³)	RA (kg)	Fine aggregate (kg)	Water (L)	Additive (L)
Control	0	480	990	0	0	835	168	4.32
SF35	0		990	35	0			
SF50	0		990	50	0			
SF65	0		990	65	0			
RA25	25		743	0	235			
RA50	50		495	0	469			
RA75	75		248	0	704			
RA25SF50	25		743	50	235			
RA50SF50	50		495	50	469			
RA75SF50	75		248	50	704			

dispersion, the aggregates and SF were first mixed in the dampen mixer, cement, and water with water-reducing admixture were then added. The whole mixing process was less than 2 min followed by the slump test and casting. To prevent shrinkage by water evaporating, plastic sheets were used to cover the surface of the mixes imme-diately after casting. There were 3 replicates for the compressive strength test and 2 for the bending test per each mix. The specimens were cured by submergence until the testing time of 7 and 28 days.

The compression test was conducted according to TCVN 3118:1993 [10] by an auto compression machine with cubic specimens, 15 × 15× 15 cm. The two flat opposite surfaces of the specimen were placed in contact with the upper and lower boards as seen in Fig. 1a. The compression rate was set at 8 daN/cm² per second and maximum load at failure was recorded. Specimens with the size of 10 × 10 × 40 cm were prepared for the four-point bending test in accordance with TCVN 3119:1993 [11] (Fig. 1b). The load application was parallel with the exposed surface of the specimen as casting and with the rate of about 0.8 daN/cm² per second. The load and displacement were recorded until the specimen broke and split apart. The flexural strength value, R_u, was calculated by Eq. 1.

$$R_u = \frac{Pl}{bh^2} \tag{1}$$

where: P = maximum load (kN),
  l = distance between 2 lower bearing pads (cm),
  b, h = width and height of the cross section (cm).

(a)   (b)

Fig. 1 **a** compression test, **b** bending test

3 Results and Discussion

3.1 Compression Test

According to the slump values shown in Fig. 2, both RA and SF tend to lower the workability of the concrete mixture. In Fig. 3a, the slump seems to slide down significantly as there was more than 50 kg of SF in the mix. This might lead to the decrease of compressive strength of mix F65 compared to mix F50 as seen in Fig. 3. With the current mixing technique, SF with a relatively excessive content in the mixture can cause tangle and form bundles that harden the casting work and reduce the homogeneity. Besides, the inadequate amount of SF would not affect and even slightly decrease the compressive strength due to its interference in mixing. On the other hand, the compressive strength of concrete decreases when the content of RA replacement increases. It correlates to the low workability since RA has an angular

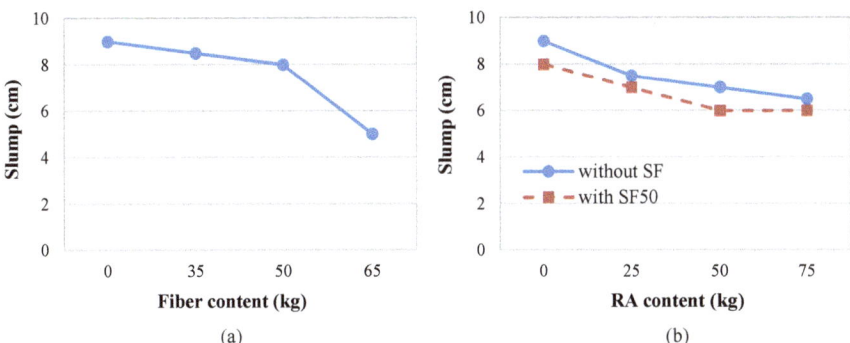

Fig. 2 The effect of **a** SF and **b** RA on the workability of concrete mixtures

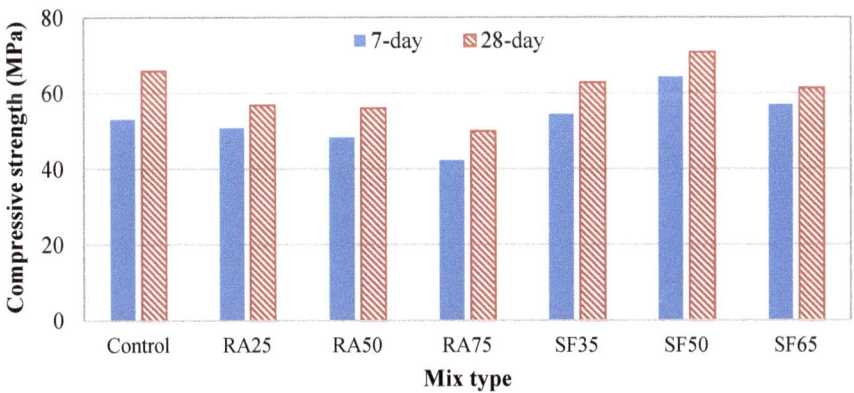

Fig. 3 The compressive strength values of concrete containing RA and SF

shape and high water absorption as shown in Table 1. It fortified the point that SF with an appropriate amount, about 50 kg, in this case, can improve to some extent the adverse effect of RA. It can also be seen in Fig. 3 that there seems to be a certain influence on the strength development of concrete, by both RA and SF, as compared to the control mix.

3.2 Flexure Strength Evaluation

Figure 4 shows the flexure strength of concrete at 28-day curing with different contents of SF. As seen in Fig. 4a, SF constantly increased the flexure strength of concrete by its linking and bonding mechanism with concrete. The increment continued as over 50 kg was added, which is contrary to that of the compressive strength. As a result, the SF content of 50 kg was suggested to be appropriate so far. In addition, Fig. 4b indicates that the concrete with SF can maintain residue strength post-cracking, and accordingly, the failure occurred gradually as compared to the control mix.

The effect of the SF content of 50 kg on the flexure strength and compressive strength of RAC at different content of RA is shown in Fig. 5. According to TCVN 3119:1993 [11], tensile strength is equivalent to 0.58 flexure strength. Rigid pavements have a regular design with a flexure strength of 4.5 MPa at least [12]. In the case of this study, the flexure strength value decreased as each RA content increases and was below the design requirement (Fig. 5a). It was due to the weak bonding between mixed-in RA and the mortar component in concrete. Whereas, with the same RA content, the flexure strength was improved significantly when SF was supplemented in the concrete mixes. In the case of the compressive strength, as seen in Fig. 5b, its improvement by SF was insignificant compared to the declination by RA. By modifying RAC with SF, the strength of RAC was compensated, especially the flexure

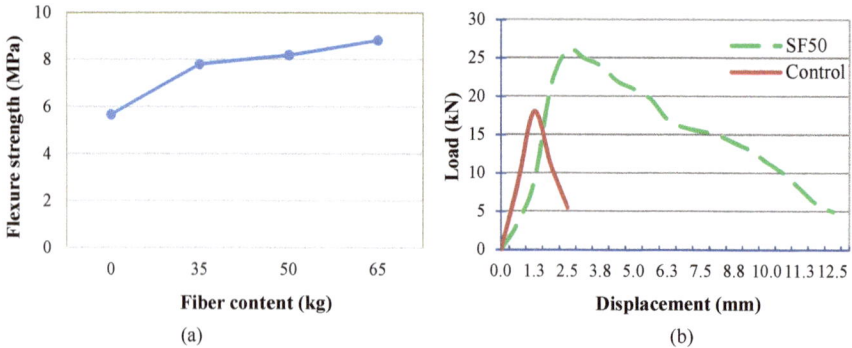

(a)   (b)

Fig. 4 The effect of SF on **a** the flexure strength and **b** load–displacement curve of hardened concrete

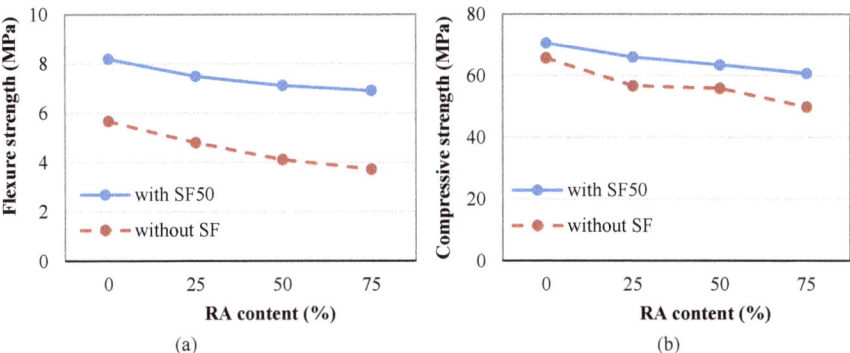

Fig. 5 The effect of SF on the **a** flexure strength and **b** compressive strength of RAC

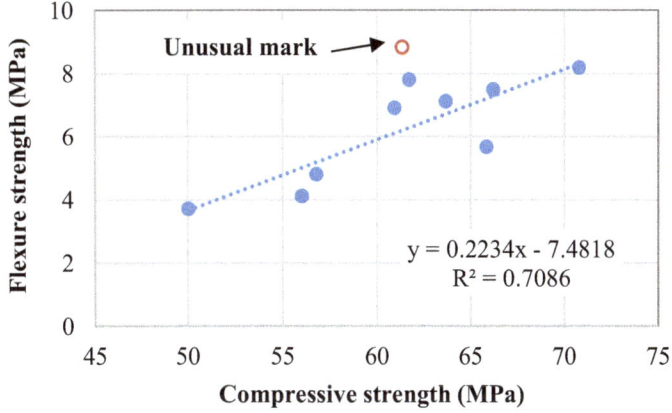

Fig. 6 Relationship between the compressive strength and flexure strength

strength, which plays an important role in the bearing capacity and durability of rigid pavements.

In Fig. 6, there is an unusual mark, mix SF65, which exceeded the appropriate content of SF. Therefore, as the unusual mark is ignored, the linear relationship between the compressive strength and flexure strength has a fair regression coefficient (R^2) of nearly 0.71, which can be applied for estimations of flexure strength for rigid pavement testing.

4 Conclusions

The objective of this research is to suggest a combination of RA and SF for the design of concrete mixture. Various concrete mixtures with different RA and SF contents were manufactured for laboratory experiments to determine the degree of effectiveness of RA and SF dosages. The following conclusions are drawn based on the results of compression and bending tests for concrete mixes.

With the addition of SF or RA, the workability of fresh concrete reduces, and it is significant when the content of SF exceeds a certain amount in concrete mixtures. In terms of this study, the SF content of 50 kg per cubic meter of concrete is suggested to be appropriate for strength improvement and acceptable workability. On the other hand, RAC has strength decreased significantly as the proportion of RA substitution increases. Consequently, with the reinforcement of SF, the bearing capacity of RAC is compensated for long life-span rigid pavements, especially the flexure strength and post-cracking endurance. It is recommended combinations of different types of fiber and RA as fine aggregate for further investigations.

Acknowledgements This research was funded by Vietnam National University Ho Chi Minh City (VNU-HCM) under grant number C2021-76-02.

References

1. Ngo TK, Lam ND, Nguyen TT, Pham KQ, Ngo KV (2020) Utilizing recycled aggregate and autoclaved aerated concrete to develop paving block concrete. STCE J–NUCE 14(4V):106–117
2. McGinnis MJ, Davis D, Rosa A, Weldon BD, Kurama YC (2017) Strength and stiffness of concrete with recycled concrete aggregates. Constr Build Mater 154:258–269
3. Safiuddin M, Alengaram UJ, Rahman MM, Salam MA, Jumaat MZ (2013) Use of recycled concrete aggregate in concrete: a review. J Civ Eng Manag 19(6):796–810
4. Anderson KW, Uhlmeyer JS, Russell M (2009) Use of recycled concrete aggregate in pccp: Literature search, Special report WA-RD 726.1, Washington State Department of Transportation, Washington D.C
5. Yang K, Chung H, Ashour A (2008) Influence of type and replacement level of recycled aggregates on concrete properties. ACI Mater J 105:289–296
6. Chan R, Santana RA, Oda AM, Paniguel RC, Vieira LB, Figueiredo AD, Galobardes I (2019) Analysis of potential use of fibre reinforced recycled aggregate concrete for sustainable pavements. J Clean Prod 218:183–191
7. Marcalikova Z, Racek M, Mateckova P, Cajka R (2020) Comparison of tensile strength fiber reinforced concrete with different types of fibers. Procedia Struct Integr 28:950–956
8. TCVN (2006) Aggregates for concrete and mortar - Test methods, TCVN 7572-1-20:2006, Ministry of Science and Technology, Hanoi, Vietnam
9. ACI (2009) Standard practice for selecting proportions for normal, heavyweight and mass concrete. ACI 211.1-91 reapproved, American Concrete Institute, Michigan
10. TCVN (1993) Heavyweight concrete - Method for determination of compressive strength. TCVN 3118:1993, Ministry of Science and Technology, Hanoi, Vietnam
11. TCVN (1993) Heavyweight concrete - Method for determination of flexural tensile strength. TCVN 3119:1993, Ministry of Science and Technology, Hanoi, Vietnam
12. Hansen W, Jensen EA, Mohr P (2001) The effects of higher strength and associated concrete properties on pavement performance. FHWA-RD-00-161, Federal Highway Administration, Washington D.C

Utilization of Waste Incineration Bottom Ash as Fine Aggregate in the Production of Terrazzo Tiles for Pavement

Trong-Phuoc Huynh, Huy-Phuong Phan, Van-Hien Pham, Van-Anh Ngo, and Hoang-Tung Luu

Abstract Turning solid waste materials into useful construction materials has been attracted many researchers in the world. This study investigated the possibility of using waste incineration bottom ash (IBA) from local incineration plants in the production of terrazzo tiles for pavement. In which, the IBA was used as a crushed sand substitution at levels of 0–100 vol.% (interval of 25%). Test results show that all of the terrazzo tiles produced in this investigation exhibited consistent dimensions and nice shapes without visible defects. The incorporation of IBA reduced the flexural, whereas increased the surface water absorption and surface abrasion values of the tiles. However, all of the terrazzo tiles regardless of the IBA content met the requirements of TCVN 7744:2013 and were classified as Type-I, which is the highest quality of terrazzo tiles for external use. Moreover, scanning electron micrographs of the tiles well supported their engineering properties. Research results further demonstrated great potential in manufacturing terrazzo tiles using IBA as a fine aggregate.

Keywords Terrazzo tile · Incineration bottom ash · Flexural strength · Surface abrasion · Microstructure

1 Introduction

Along with modernization, the Earth is facing many problems. One of which is domestic solid waste (garbage). For many decades, domestic solid waste-related

T.-P. Huynh (✉) · H.-P. Phan · V.-H. Pham
Department of Civil Engineering, College of Engineering Technology, Can Tho University, Campus II, 3/2 Street, Ninh Kieu District, Can Tho City 94000 , Vietnam
e-mail: htphuoc@ctu.edu.vn

V.-A. Ngo
Department of Transportation Engineering, College of Engineering Technology, Can Tho University, Campus II, 3/2 Street, Ninh Kieu District, Can Tho City 94000 , Vietnam

H.-T. Luu
Department of Chemical Engineering, College of Engineering Technology, Can Tho University, Campus II, 3/2 Street, Ninh Kieu District, Can Tho City 94000 , Vietnam

© The Author(s), under exclusive license to Springer Nature Singapore Pte Ltd. 2023   555
J. N. Reddy et al. (eds.), *ICSCEA 2021*, Lecture Notes in Civil Engineering 268,
https://doi.org/10.1007/978-981-19-3303-5_49

problems have been faced worldwide. The generation of a huge quantity of solid waste without proper treatment methods causes serious environmental pollution and harms human being health. To limit these problems, waste treatment plants have been built in many areas to handle and treat solid waste. A waste incineration plant was also completed built and located in Can Tho City, Southern Vietnam. As reported by the local Government, the average amount of domestic solid waste received daily is more than 453 tons (accounting for about 70% of daily household waste generated in Can Tho City) and the total amount of domestic solid waste treated from 8/2018 to 8/2020 was about 317,000 tons [1]. In addition, the amount of domestic solid waste in big cities of Vietnam such as Ha Noi, Ho Chi Minh, and Da Nang cities accounts for a large proportion. Statistics in 2017 showed that the total domestic solid waste generated in Ha Noi and Ho Chi Minh cities was 7500 and 8700 tons/day, respectively [2]. The amount of domestic solid waste tended to increase shortly with an estimated rate of about 800 tons/year in 2030 [2]. Therefore, a huge amount of incineration bottom ash (IBA) will be released after burning the abovementioned quantity of domestic solid waste. If a proper treatment method is offered, this IBA can be considered as a potential source of raw materials for construction.

Previous studies have been conducted in the world, showing that IBA was not a hazardous waste after a proper treatment [3] and it can be utilized in pavement structure [4], as aggregate in asphalt concrete [5], as aggregate in cement concrete [6], used in cement mortar [7], etc. The reuse of IBA brings positive effects on the environment, limits both the exploitation and excessive use of natural resources, and reduces the negative impact on human life. So far, domestic and foreign studies related to the utilization of IBA as fine aggregate in the development of terrazzo tiles for pavement have been limited. Therefore, this study was performed to evaluate the possibility of producing terrazzo tiles using locally sourced IBA. Also, the effect of replacing crushed sand with IBA on the engineering properties and microstructure of the terrazzo tiles was evaluated based on Vietnamese standards.

2 Materials and Experimental Methods

2.1 Materials

The terrazzo tiles were prepared using type-PCB40 blended Portland cement, class-F fly ash (FA) from the thermal power plant, ground granulated blast furnace slag (hereafter called slag), IBA sourced from a local incineration plant, crushed sand, crushed stone, and water. Characteristics of cement, slag, and FA are shown in Table 1. The concentration of heavy metal in the IBA was checked before using with the results as shown in Table 2. Based on the regulations of QCVN 07:2009/BTNMT, IBA was a non-hazardous material that can be re-used as a sourced material for making terrazzo tiles. Properties of the aggregates used in this study were given in Fig. 1 and Table 3. It is noted that IBA particles exhibited a porous structure (Fig. 2),

Table 1 Specific Gravities and Chemical Compositions of Cement, Slag, and FA

Materials	Specific gravities	Chemical compositions (% by mass)					
		SiO_2	Al_2O_3	Fe_2O_3	MgO	CaO	Others
Cement	2.84	23.5	6.0	3.7	2.0	59.9	4.9
Slag	2.85	35.9	13.0	0.3	7.9	38.1	4.8
FA	2.14	59.2	26.7	6.1	0.9	1.1	5.9

Table 2 Concentration of Heavy Metal in the IBA

Concentration (mg/L)	Cu	Cr (VI)	Cd	Pb	Ni	Zn
IBA	<0.01	<0.003	0.0003	<0.0007	<0.001	<0.015
QCVN 07:2009/BTNMT	–	≤ 5	≤ 0.5	≤ 15	≤ 70	≤ 250

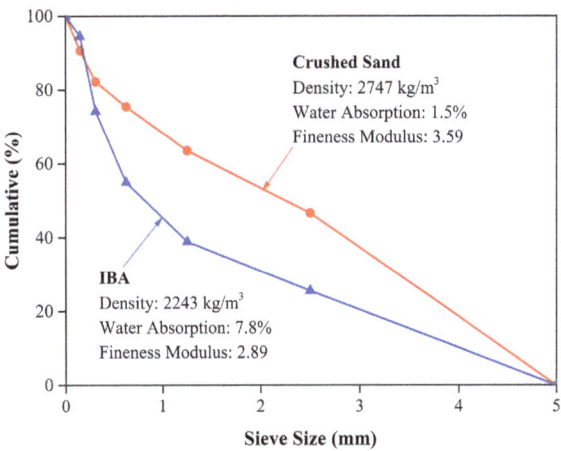

Fig. 1 Grain size distribution and physical properties of fine aggregates

Table 3 Properties of Coarse Aggregate

Materials	Density (kg/m^3)	Water absorption (%)	Maximum diameter (mm)
Crushed Stone	2764	0.48	9.5

leading to a significantly higher water absorption rate of IBA in comparison with crushed sand (see Fig. 1).

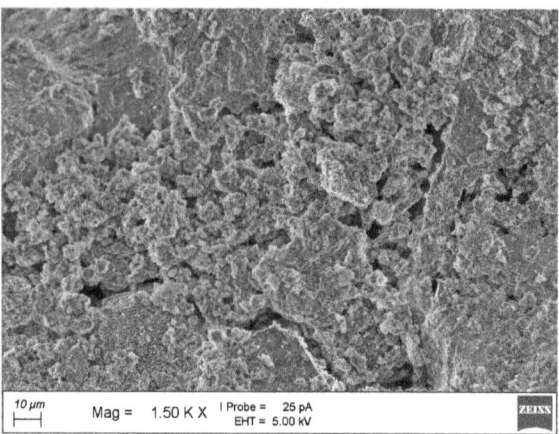

Fig. 2 SEM morphology of IBA particles

Table 4 Ingredient Proportions for the Preparation of Terrazzo Tile Samples (kg/m³)

Mix designation	Cement	Slag	FA	IBA	Crushed sand	Crushed stone	Water
B00	456	195	90	0	689	664	237
B25	456	195	90	141	517	664	237
B50	456	195	90	281	345	664	237
B75	456	195	90	422	172	664	237
B100	456	195	90	563	0	664	237

2.2 Mixture Proportions

In this study, the controlled terrazzo tile mixture namely B00 was designed using densified mixture design algorithm (DMDA). The major contribution of DMDA as well as the procedures for mix design by DMDA were previously described by Chen et al. [8] and Hwang and Hung [9]. Based on the B00 mixture, the crushed sand was then replaced by IBA at 25, 50, 75, and 100% (by volume). It is noted that the aggregates used were in saturated surface dry condition. A water/binder ratio of 0.32 was applied for all terrazzo tile mixtures. The ingredient proportions of all terrazzo tile mixtures were shown in Table 4.

2.3 Sample Preparation and Test Methods

The terrazzo tile samples (Fig. 3) were prepared by the following procedures as briefly described in Fig. 4. The terrazzo tile specimens were subjected to various test programs including dimensions and visible defects (at 28 days, TCVN 7744:2013),

Fig. 3 Terrazzo tile samples produced in this study

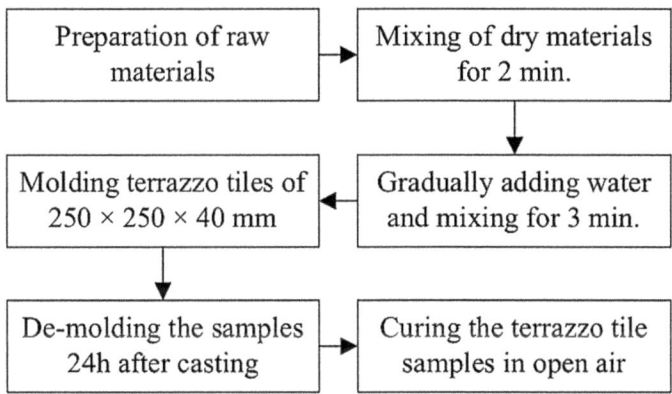

Fig. 4 Procedures for the preparation of terrazzo tiles

SWA (at 28 and 56 days, TCVN 7744:2013), SA (at 28 and 56 days, TCVN 6065:1995), FS (at 7, 28, and 56 days, TCVN 6355-3:2009), and microstructure observation using a scanning electron microscope (SEM). An average value of five terrazzo tile specimens was taken and reported as the final result.

Table 5 Dimensions of Terrazzo Tiles

Mix designation	Length (mm)	Width (mm)	Height (mm)
B00	249.0	248.2	41.3
B25	248.6	248.1	40.4
B50	248.6	248.6	41.4
B75	248.9	248.2	41.1
B100	248.7	248.6	40.5
Allowable Tolerance (TCVN 7744:2013), mm	250 ± 2	250 ± 2	40 ± 3

3   Results and Discussion

3.1   Dimensions and Visible Defects

Dimensions of the terrazzo tiles were measured to confirm compliance with TCVN 7744:2013, with the results shown in Table 5. In addition, all terrazzo tiles showed a nice shape with no visible defects (see Fig. 3). These results demonstrate that the tiles conform to TCVN in terms of dimensions and visible defects. It is noted that the deformities of the plastic molds under the repeated use caused a slight variation in the dimensions of the tiles.

3.2   Flexural Strength

FS is commonly used to assess the technical quality of the terrazzo tiles during their service life. Figure 5 presents the growth in FS of the terrazzo tiles over time. It is found that the FS of all terrazzo tiles increased with curing ages and the FS values reduced with increasing IBA content. The same FS development patterns were found at all curing times (7, 28, and 56 days) that the highest FS value was obtained from the controlled terrazzo tile (B00 mix). The FS value was then reduced when using IBA to replace crushed sand in the terrazzo tile mixtures and the higher the replacement level the lower the FS value of the tiles (see Fig. 6). These lower FS values were caused primarily by (1) the presence of porous IBA particles (see Fig. 2), which reduced the structural density of the tiles; (2) the slow pozzolanic reaction of IBA [10] was also considered as one of the reasons for reducing FS of the tiles, especially when the terrazzo tiles comprised of high IBA volume; and (3) the lower strength of the IBA particles in comparison with crushed sand particles was another possible reason for the lower FS of the IBA tiles [11]. On the other hand, it is interesting to see that the FS of the terrazzo tiles was kept increasing after 28 days, which was attributable to both the internal curing effect of the highly porous IBA particles and the pozzolanic reaction of very IBA particles [12]. Overall, the results clearly illustrated that the terrazzo tile samples exhibited excellent FS and that all of the tiles regardless of

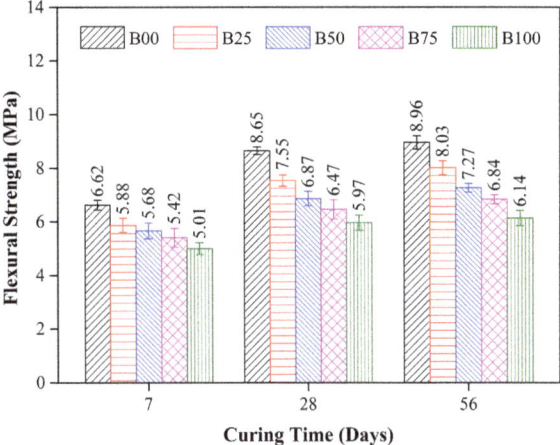

Fig. 5 FS of terrazzo tiles

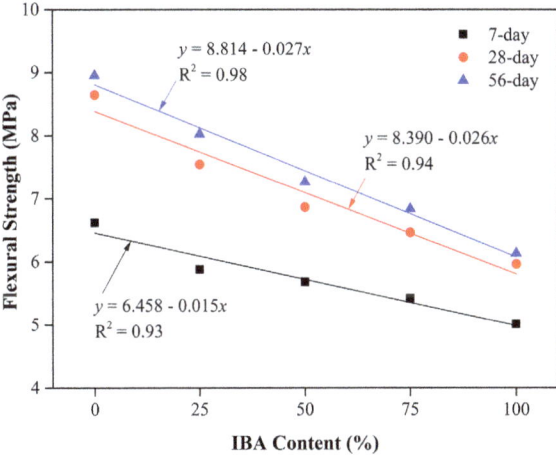

Fig. 6 Correlation between IBA content and FS of the terrazzo tiles

the IBA content were classified as Type-I (with the FS value of at least 5 MPa at 28 days) terrazzo tiles for external use following TCVN 7744:2013. Furthermore, these results confirm the correlation between IBA content and FS of the terrazzo tiles (see Fig. 6).

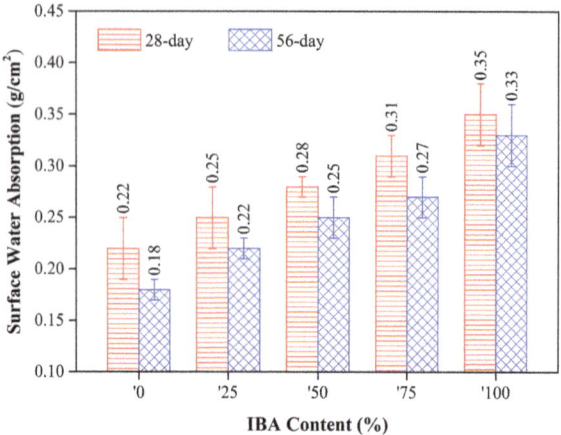

Fig. 7 SWA of terrazzo tiles

3.3 Surface Water Absorption

SWA is one of the factors affecting the durability of terrazzo tiles. SWA is the action of surface tension on capillaries that allows the transport of liquid in porous solids [13]. Thus, the lower the water infiltration the greater the durability of the tiles. Figure 7 shows the result of the SWA test at 28 and 56 days for the terrazzo tiles incorporating different IBA contents. As a result, the SWA values of the terrazzo tiles at 28 and 56 days were in the ranges of 0.22–0.35 g/cm^2 and 0.18–0.33 g/cm^2, respectively. Thus, all of the terrazzo tiles had SWA values of below 0.4 g/cm^2, which is the level required under TCVN 7744:2013 for Type-I terrazzo tiles. In other words, the terrazzo tiles produced in this study satisfied the national standard requirement in terms of SWA. The reduced absorption capacity of the terrazzo tiles at 56 days in comparison to that at 28 days indicated the reduced void volume of the internal structure of the tiles, which is attributable to the pozzolanic reaction at later ages, generating more hydration products to fill the voids [14]. Moreover, the SWA values were found to be increased proportionally with the IBA content in the terrazzo tile mixtures. As aforementioned, this phenomenon could be explained by the porous nature of IBA particles (see Fig. 2).

3.4 Surface Abrasion

SA is also an indicator of the strength of the terrazzo tiles. The SA of the tile is characterized by the mass loss of the tile's surface during the grinding process, corresponding to a distance of 600 m (TCVN 6065:1995). Therefore, the lower the abrasion level, the stronger the terrazzo is. The results of the SA test for the terrazzo

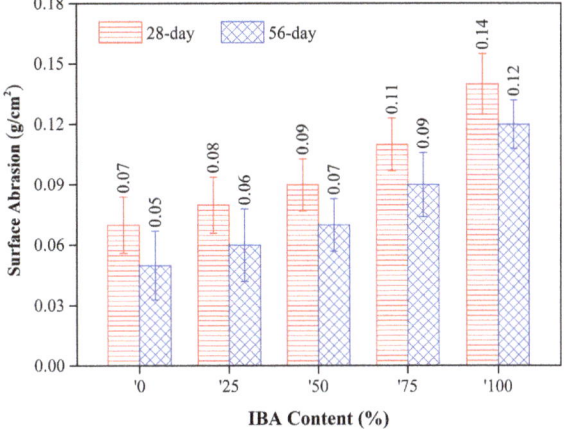

Fig. 8 SA of terrazzo tiles

tiles at 28 and 56 days are shown in Fig. 8. It can be observed that the SA values of the terrazzo tiles at 28 days ranged from 0.07 g/cm² to 0.14 g/cm². These values were then correspondingly reduced to 0.05–0.12 g/cm² at 56 days. All of the terrazzo tiles registered SA values of far below 0.4 g/cm², which is the level required under TCVN 7744:2013 for Type-I terrazzo tiles. A similar trend for the abovementioned properties of the terrazzo tiles that the incorporation of IBA resulted in increasing SA value of the tiles. This finding could be explained through the reduction in strength of the tiles as above discussion (Sect. 3.2). The results of SA were in good agreement with the FS measurements of the tiles that the decline in strength led to the loss in abrasion resistance of the terrazzo tiles [12].

3.5 Microstructure Analysis

Figure 9 displays the SEM micrographs of terrazzo tiles with different IBA contents. It could be observed that the incorporation of IBA as crushed sand substitution introduced more voids within the terrazzo tile structure, more significant at high replacement levels, resulting in the less compactness of the internal structure. As a result, the growth in SWA and the loss in FS and SA resistance were observed.

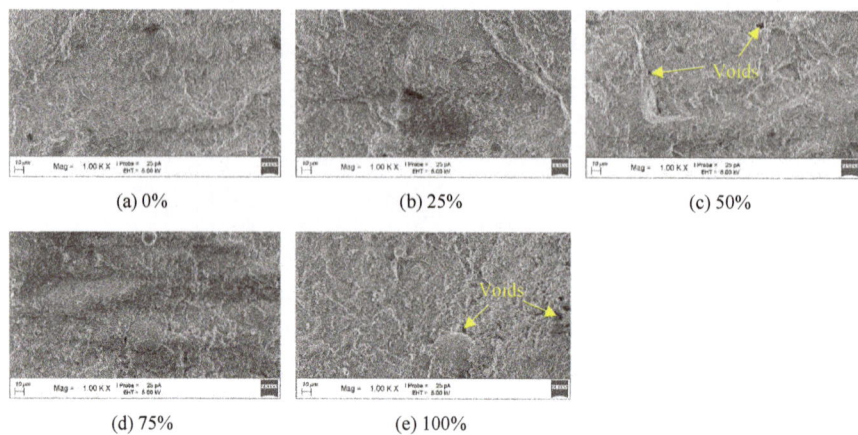

(a) 0%   (b) 25%   (c) 50%

(d) 75%   (e) 100%

Fig. 9 SEM micrographs of terrazzo tiles with various IBA contents

4 Conclusions

Based on the experimental outcomes, the following conclusions can be drawn: (1) IBA was a non-hazardous waste, which can be reused as a construction material; (2) The terrazzo tiles produced in this study showed consistent dimensions and nice shapes with no visible defects; (3) The use of IBA to replace crushed sand in the terrazzo tiles caused a reduction in FS and an increase in both SWA and SA. Also, SEM micrographs of the terrazzo tiles well supported the obtained engineering properties; (4) At 28 days, the terrazzo tiles obtained the values of FS, SWA, and SA in the ranges of 5.97–8.65 MPa, 0.22–0.35 g/cm^2, and 0.07–0.14 g/cm^2, respectively. These values well conformed to the TCVN 7744:2013 requirements for Type-I terrazzo tiles and further demonstrated great potential in producing terrazzo tiles using IBA as a fine aggregate. Hence, the IBA can be reused to either partially or fully replace crushed sand in the tiles depending on the specific requirements for each application purpose.

References

1. Ministry of Industrial and Trade of the Socialist Republic of Vietnam (2020) Can Tho waste incineration plant has been confirmed to complete environmental protection works (in Vietnamese). https://moit.gov.vn. Accessed 14 Mar 2021
2. National Environment Report (2017) Solid waste. Ministry of Natural Resources and Environment, Hanoi, Vietnam (in Vietnamese)
3. Nguyen HM, Huynh TP, Le TP, Ngo VA, Chau MK, Le N (2021) Recycling of waste incineration bottom ash in the production of interlocking concrete bricks. J Sci Technol Civil Eng NUCE 15(2):101–112

4. Forteza R, Far M, Segui C, Cerda V (2004) Characterization of bottom ash in municipal solid waste incinerators for its use in road base. Waste Manage 24(9):899–909

5. Eymael MMT, Wijs WD, Mahadew D (1994) The use of MSWI bottom ash in asphalt concrete. Stud Environ Sci 60:854–862

6. Müller U, Rübner K (2006) The microstructure of concrete made with municipal waste incinerator bottom ash as an aggregate component. Cem Concr Res 36(8):1434–1443

7. Saikia N, Mertens G, van Balen K, Elsen J, van Gerven T, Vandecasteele C (2015) Pre-treatment of municipal solid waste incineration (MSWI) bottom ash for utilisation in cement mortar. Constr Build Mater 96:76–85

8. Chen YY, Bui LAT, Hwang CL (2013) Effect of paste amount on the properties of self-consolidating concrete containing fly ash and slag. Constr Build Mater 47:340–346

9. Hwang CL, Hung MF (2005) Durability design and performance of self-consolidating lightweight concrete. Constr Build Mater 19:619–626

10. Li XG, Lv Y, Ma BQ, Chen QB, Yin XB, Jian SW (2012) Utilization of municipal solid waste incineration bottom ash in blended cement. J Clean Prod 32:96–100

11. Cheng A (2012) Effect of incinerator bottom ash properties on mechanical and pore size of blended cement mortars. Mater Des 36:859–864

12. Shen P, Zheng H, Xuan D, Lu JX, Poon CS (2020) Feasible use of municipal solid waste bottom ash in ultra-high performance concrete. Cement Concr Compos 114:103814

13. Uzoegbo HC (2020) Dry-stack and compressed stabilized earth-block construction. In: Nonconventional and Vernacular Construction Materials, Woodhead Publishing Series in Civil and Structural Engineering, 2nd edn. Elsevier, pp 305–350

14. Liu J, Xing F, Dong B, Ma H, Pan D (2014) Study on water sorptivity of the surface layer of concrete. Mater Struct 47:1941–1951

Geomatics Session

Create 3D Models from Photos Captured by Sony Alpha 7 Mark 2 Digital Camera

Anh Thu Thi Phan and Xuan Phuc Mai

Abstract Creating a 3D model of the object is necessary for assessing the current status, exploitation, management, and information storage of the construction. Using photographs to create 3D models is a common practice. The photos of Solar BK construction are captured for generating the 3D point cloud using COLMAP open source application using structure from motion algorithm. However, the created cloud is not the right size and needs to be adjusted to the original size. In this study, the authors propose using the average scale to re-establish the construction's size. The average scale is calculated from the ten measured edges chosen horizontally and vertically around the building. The result obtains a dense point cloud model with nearly 2.7 million points. After correcting the point cloud, the lengths of the checked edges have an absolute error of no more than 3 cm from the actual size and a relative error of less than 1:80. The resulted point cloud is sent into AutoCAD software to reconstruct the 3D model of the construction manually. The results show the ability to create 3D models of objects from photos taken by digital cameras. This result is suitable for applications that do not require too high accuracy, such as building status management or building data visualization.

Keywords 3D models · Digital camera · Point cloud · Photogrammetry

1 Introduction

Many structures, especially ancient architecture works, were built in ancient times for several hundred years. Over that time, the constructions were degraded, damaged, and intact due to many objective and subjective factors. Along with the development of science and technology, digitizing construction works is essential for managing

A. T. T. Phan (✉) · X. P. Mai
Faculty of Civil Engineering, Ho Chi Minh City University of Technology (HCMUT), 268 Ly Thuong Kiet Street, District 10, Ho Chi Minh City, Vietnam
e-mail: ptathu@hcmut.edu.vn

Vietnam National University Ho Chi Minh City, Linh Trung Ward, Thu Duc District, Ho Chi Minh City, Vietnam

© The Author(s), under exclusive license to Springer Nature Singapore Pte Ltd. 2023
J. N. Reddy et al. (eds.), *ICSCEA 2021*, Lecture Notes in Civil Engineering 268,
https://doi.org/10.1007/978-981-19-3303-5_50

and storing building information. Therefore, the management of project information for the restoration and repair work, retaining the characteristics of each project is essential and meaningful. For this works, a 3D model of the building is required to serve the restoration of the works, 3D visualization, surveying the current status, the performance of the construction, or adding details from the model to ensure harmony with the existing details [1–3].

To determine the shape and size of the construction, direct measurement methods using total station give high measurement accuracy. However, the discrete measuring points or the measuring edges do not give an overview of the object and the spatial correlation between the detailed measurement points. In addition, the number of measurement points is too small; therefore, it is not easy to visualize the construction's status in detail.

The 3D models of buildings are essential input data for generating BIM models. Therefore, there are many methods of reconstructing the 3D model of the existing object. For new constructions, 3D models can be reconstructed from design drawings using specific software. However, for many existing constructions, it is not possible to obtain design drawings for this process. Currently, 3D laser scanning is a leading technique that can collect millions of data points in minutes; thereby, 3D models of buildings or objects can be created easily and quickly [4–6]. Laser scanning is a trendy technical technique in the world and is applied in many industries, not only in construction. However, in Vietnam, 3D laser scanning is not widespread yet. Some units in our country have pioneered in accessing and mastering this technique but still have faced particular difficulties. This barrier comes from the fact that the cost of equipment and accompanying software is still relatively high compared to the general income level in Vietnam and the requirement of technicians with high professional qualifications. Therefore, the use of laser scanners for small projects is uneconomical. In addition, not all objects can be favorable for setting up the scanner. For the aerial scanning system, it is not possible to see the details on the lateral side of the construction.

Besides the 3D laser scanning method, the photogrammetric technique's structure from motion method (SfM) also results in point cloud models. Photos can be captured with conventional digital cameras, and photography can be done without a complicated design process. The requirement is to ensure the overlap and roughly the same image scale. This method is a flexible indirect data collection method. Therefore, photogrammetry is considered a valuable solution for the 3D reconstruction of small construction instead of using laser scanning devices. Point clouds are generated from photographs and using paid software or open-source software. This method reduces the cost, and data processing is also easier for general users who need to use point cloud models.

Moreover, the photogrammetric technique still has its advantages compared to 3D laser scanning, such as: producing point clouds with color information or using photos taken at different times, collecting data flexibly. In this study, a 3D model of the building is generated from photos captured by the Sony Alpha 7 Mark 2 digital camera using the SfM technique. The point cloud is adjusted to the actual scale based on the average scale determined from several measured edges. The target building locates in Ho Chi Minh City. The generated model's accuracy is evaluated by comparing its edges with the corresponding measured edge on the target construction.

2 Data Acquisition

2.1 Photos Acquisition

For this study, the Solar BK construction located at the campus of Ho Chi Minh City University Technology -VNU is chosen. This construction is a small construction with four main slides. It covers an approximate area of 150 m². The construction has two floors with a slanted roof, and the decorative details around the building are simple.

The set of input photos for the structure from motion (SfM) technique must ensure that the photographs cover the entire building [7]. Furthermore, to ensure that requirement, it is better to plan to capture photos, such as to estimate the number of photos to be taken. It is recommended to take more photos than necessary to avoid missing corners or parts of the building. When taking photos of construction surfaces, it is advisable to avoid staying in one position and only changing the direction of view to take pictures. The camera should be moved parallel to the surface of the building and let the aiming direction be perpendicular to the surface of the building. This method also applies when the surface is vertical. It is recommended to use a lens with a fixed focal length, do not change the focus or enlarge the frame during the shooting process. The aperture of the lens should be fixed during the shooting process. A large aperture must be opened (usually from f/5.6 to f/8) to ensure a clear photo. For the photos that are next to each other to connect, the photos are superimposed on each other. Moreover, the overlap in both directions must be more significant than 80%.

The Sony alpha 7-mark digital camera is used to conduct data collection. It has featured 24 Megapixels (6000 × 4000 pixels) Sigma 35 f2.4 (Fig. 1). The camera parameters are adjusted for precise and bright images: aperture f/8, speed 1/50 s, and ISO 100. The data set was taken on July 18, 2020, from 8:00 to 8:30 am with

(a)   (b)   (c)

Fig. 1 Camera and captured photos **a** Sony alpha 7-mark digital camera, **b** photo of Solar BK building and Jeanne d'Arc church in Ho Chi Minh City

Table 1 Measured edge's length

ID	Measured length (m)	ID	Measured length (m)	ID	Measured length (m)
1	2.680	7	0.352	13	3.038
2	2.142	8	2.136	14	2.420
3	1.154	9	0.591	15	2.903
4	3.776	10	1.300	16	2.146
5	2.187	11	2.143	17	1.170
6	2.183	12	2.685	18	0.662

relatively sunny weather to collect photos of the Solar BK construction. However, after trying to process the data set unsuccessfully due to missing some corners of the building, extra photos were capture on July 25, 2020, with the same parameters. As a result, 398 photos were captured to cover the surrounding of this building.

2.2 Directly Measured Data

Immediately after taking the photos, the direct measurement was carried out. Many edges were randomly selected around the building. Selected edges were sharp edges and were easily visible on the photos to be checked on the model. The selected edges were of different sizes and distributed in both horizontal and vertical directions. The dimensions of the edges were measured directly with tape. As a result, 18 edges were measured around the Solar BK building (Table 1).

3 Methodology

3.1 Structure from Motion

Structure from motion (SfM) is a photogrammetric technique for estimating three-dimensional structures from two-dimensional image sequences. This technique can reproduce the 3D image of an object, a person, or a sizeable spatial landscape. SfM is used to estimate 3D models from sequences of overlapping 2D images by solving the linear triangulation. SfM is the ability to process out-of-order and heterogeneous image sets without prior knowledge of camera parameters. Although data of camera position or ground control point can be added and used for achieving more accurate results. At the first step, scale-invariant feature transform (SIFT) and its variants algorithms are applied for identifying and fitting key points in SfM SIFT generates numerical descriptors for each point in each image. These descriptors are constant in any scale or orientation, suitable for identifying points or objects in images taken from different perspectives and under different conditions. Then the coherence-based feature matching is checked using a rough structure of 2D photos and the relative position of the critical point until a sufficient number of photos and points are matched. SfM is often used to define the entire reconstruction process, from the image to the thick point cloud. In the second step, SfM is applied to provide camera parameters and estimate sparse point clouds. Then, multi-view stereo reconstruction algorithms (MVS) are used in the next step to increase the number of points of the point cloud.

3.2   Point Cloud Generation

Due to the limitation of the data processing capacity of the personal computer, the photo sets were divided into subprojects to facilitate the data processing. Input data was processed by COLMAP open-source application using SfM algorithm through feature detection and extraction steps; join features, verify geometrical elements, and model the point cloud to give the final result a dense point cloud. Then the Cloud Compare open-source application was used to remove noisy points. The point clouds were aligned together to create the complete point cloud of the construction by applying Affine transformation. Specifically, the dataset consisting of 398 photos was divided into two parts, each part consisting of 199 photos. Part (I) included photos of three sides of the building without obstacles, and part (II) was the remaining face hidden by the tree. As a result, the total number of detected features of the part (I) in all 199 images was 2,114,962 points. The whole point cloud modeling process was completed in 83.5 min. The model after removing points is only 1,813,593 points. For part (II), only 184/199 input images were used in the sparse point cloud modeling process (Fig. 2). The point cloud model had 854,261 points after removing noise

Fig. 2 Point cloud generating process **a** keypoint matching **b** Camera Pose Estimation with SfM **c** point cloud of the part (I), **d** part (II), and **e** completed point cloud of Solar Bk construction BK

points. After obtaining the point clouds, the Align tool of Cloud Compare software (based on the Affine transform for three-dimensional space) was used to combine the point clouds based on the coordinates of 4 pairs of tie points. The process was started by taking the point cloud of part I like the root to pair it with the part II point cloud. Finally, the point cloud of the whole construction was generated (Fig. 2).

3.3 Point Cloud Correlation

After obtaining the point cloud of the construction, the directly measured edges were measured again on the corresponding point clouds. The average scale of the first ten edges (ID 1 to 10), evenly distributed in both directions and across the faces of the building, was taken. The average ratio is calculated using the formula (1)

$$m_{tb} = \sum_{i=1}^{n} m_i/n \tag{1}$$

where m_i is the ratio between the length measured on the model and measured directly at the construction of each edge; n is the number of selected edges. Then the point cloud was corrected by applying that scale. The eight remaining edges were used to check the absolute errors.

3.4 3D Reconstruction

After adjusting the point cloud with the average ratio to scale up the point cloud to its actual size, the cloud data was converted to ASCII (.txt) format. Then, these point clouds were added to Autodesk Recap and saved in Autodesk's point cloud format (*.rcs or *.rcp) for easy opening in AutoCAD software to reconstruct 3D models. Reconstructing 3D models process was manually done by using Auto CAD software. Dimensions of the building details were explicitly noted.

4   Result and Discussion

In this study, the captured images are guaranteed to have 80% overlap to ensure successful image matching. In addition, the photo scale must also be similar. The process of taking photos around the subject is done with an almost stable observed distance. In this way, the scales among the photos do not vary too much. The process of processing image data created point cloud is done by open-source software COLMAP. From the captured data, the data processing is done for two small projects

Table 2 The number of input photos and the number of points of the point cloud of each part	Part	Number of the input image	Number of generated points
	I	199	1,813,593
	II	199	854,261

because of the limitations of personal computers before linking them together. In part II, the construction is hidden by the tree. Some photos can not be extracted with the correct key points for the photo-matching process; therefore, only 184 per 199 photos are used for the sparse point generation process. The dense point cloud of each part is successfully generated. Then, connecting the independent point clouds into a complete model for the whole building is done by Cloud Compare open-source application. The number of points for each cloud is significant, and the color of points is retained to best support the visualization of objects (Table 2). The roof surface is unclear because there are no captured photos above. However, the size of the roof can be determined correctly on the cloud (Figs. 2 and 3). The process of point cloud alignment is done through an affine transformation. The transformation parameters are calculated from a set of 4 pairs of points that overlap in two adjacent point clouds. The result is the complete point cloud of the building (Fig. 2).

For photographic techniques, a cloud model is created with a different scale to the object's size. Therefore, these point clouds need to be brought back to their actual

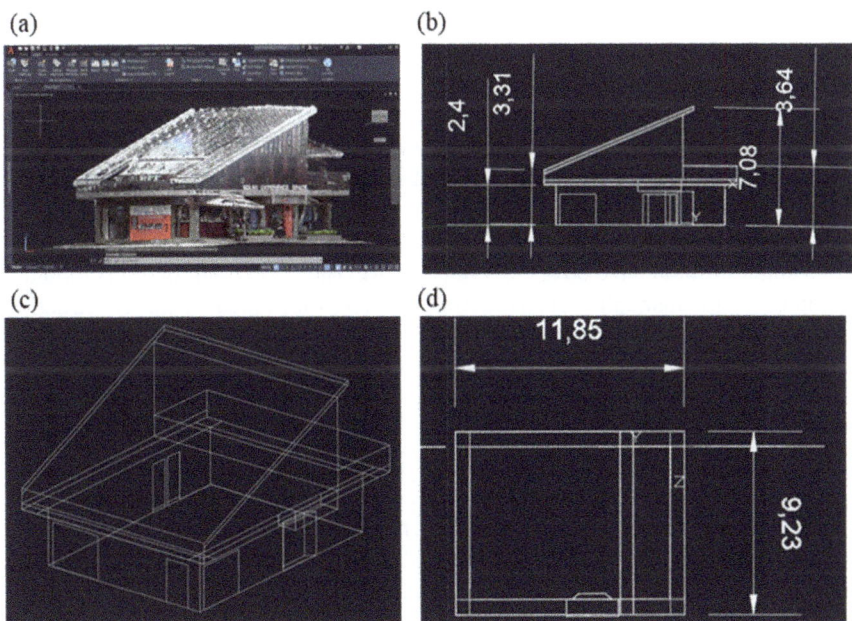

Fig. 3 Reconstructing 3D model of Solar BK construction using Auto CAD **a** actual point cloud with colors, **b** isometric view, **c** top view, and **d** side view

Table 3 The absolute error of checked edges

ID	Measured length (m)	The distance measured in point cloud (m)	Absolute errors (m)	Relative errors
11	2.143	2.123	−0.02	1:107
12	2.685	2.657	−0.028	1:95
13	3.038	3.069	0.031	1:98
14	2.42	2.45	0.03	1:80
15	2.903	2.917	0.014	1:207
16	2.146	2.137	−0.009	1:238
17	1.17	1.159	−0.011	1:106
18	0.662	0.655	−0.007	1:94

size. In some cases, it is impossible to make precise measurements of the feature points to correct the cloud to its actual size. Therefore, the authors propose to use the average scale with the assumption that the scale of three dimensions is uniform. Calculating the average scale from 10 measured edges in different directions on the building is used to correct the actual size of the Solar BK building cloud. The other eight edges are used for checking absolute and relative errors (Table 3). Based on the results, the maximum absolute error of the eight edges is 3 cm and belongs to the edge measured close to the ground surface. The cause of the error is that the number of edges used to calculate the ratio is not enough to cover all directions, and the error in the measurement process can be considered one of the reasons. The image matching process may create the wrong position of the 3D point because of image distortion.

In general, the quality of the generated cloud has a dense density of points (approximate 2.7 million points), showing details of objects. Although there are some noise points due to the automatic image stitching process, the ability to create a point cloud model from image data taken by conventional digital cameras has been found with a small absolute size error (less than 3 cm). Then, the point cloud is inputted to AutoCAD for manually reconstructing the 3D point cloud of the construction (Fig. 3).

5   Conclusion

In this study, the point cloud of constructions has been generated from photos capture by Sony alpha 7-mark digital camera. The point cloud scales up to actual size using an average scale computed from the edge's length. The absolute error of the point cloud model after being corrected is approximately several centimeters. The results satisfy purposes that do not require high accuracy. Moreover, the use of an average scale to calibrate the point cloud model can be used as an alternative to control points. For small projects, this method is feasible with low error and economic efficiency.

Errors can come from data collection, such as taking photos, measuring the field, and processing data when setting the software's parameters.

For photographing subjects, the time to complete depends on the weather conditions at the time of the shooting and the equipment's capabilities. The completion time depends on the computer's hardware configuration for point cloud modeling from images taken through COLMAP software. It is better to reduce the data processing time by applying other methods or processes in the future. As a result of this study, the 3D model of the object is reconstructed using Auto CAD. This process takes much time because unneeded points make the low down of computer speed. It is better to reduce the number of unneeded points before performing the reconstructing process.

Acknowledgements This study was supported by Ho chi Minh City University of Technology-VNU-HCM under grant number SVKSTN-2021-KTXD-14.

References

1. Fangi G, Malinverni ES, Tassetti AN (2013) The metric documentation of cham towers in Vietnam by spherical photogrammetry. https://doi.org/10.5194/isprsannals-II-5-W1-121-2013
2. To T, Nguyen D, Tran G (2015) Automated 3D architecture reconstruction from photogrammetric structure and motion: A case study of the 'one pilla' Pagoda, Hanoi, Vietnam. Int Arch Photogramm Remote Sens Spatial Inf Sci 40(7W3):1425–1429. https://doi.org/10.5194/isprsarchives-XL-7-W3-1425-2015
3. Stathopoulou EK, Georgopoulos A, Panagiotopoulos G, Kaliampakos D (2015) Crowdsourcing lost cultural heritage. ISPRS Ann Photogramm Remote Sens Spatial Inf Sci 2(5W3):295–300. https://doi.org/10.5194/isprsannals-II-5-W3-295-2015
4. Singh SP, Jain K, Ravibabu Mandla V (2014) A newapproach towards image based virtual 3D city modeling by using close range photogrammetry. ISPRS Ann Photogramm Remote Sens Spatial Inf Sci 2(5):329–337. https://doi.org/10.5194/isprsannals-II-5-329-2014
5. Barrile V, Bilotta G, Nunnari A (2017) UAV and computer vision, detection of infrastructure losses and 3D modeling. ISPRS Ann Photogramm Remote Sens Spat Inf Sci 4(4W4):135–139. https://doi.org/10.5194/isprs-annals-IV-4-W4-135-2017
6. Yastikli N, Özerdem OZ (2017) Architectural heritage documentation by using low cost UAV with fisheye lens: Otag-i Humayun in Istanbul as a case study. ISPRS Ann Photogramm Remote Sens Spatial Inf Sci 4(4W4):415–418. https://doi.org/10.5194/isprs-annals-IV-4-W4-415-2017
7. Iglhaut J, Cabo C, Puliti S, Piermattei L, O'Connor J, Rosette J (2019) Structure from motion photogrammetry in forestry: a review. Curr For Rep 5(3):155–168. https://doi.org/10.1007/s40725-019-00094-3

Estimating $PM_{2.5}$ Mass Concentration from MODIS AOD Products in Ho Chi Minh City, Vietnam

Phan Hong Danh Pham, Dang Khoa Le, Thi Minh Trang Nguyen, and Vu Hien Phan

Abstract Air pollution, especially $PM_{2.5}$ mass concentration is one of big problems in Ho Chi Minh City in recent years. This study focuses on deriving a linear regression model based on a relationship between ground-level $PM_{2.5}$ mass concentration measurements and satellite aerosol optical depth (AOD) values. The $PM_{2.5}$ measurements were collected from 25 ground stations in the inner city while atmospheric AOD values were extracted from Moderate Resolution Imaging Spectroradiometer (MODIS) green and blue band images. The observed period was from January 1, 2020 to May 31, 2020. As a result, the multivariable linear regression model was built from the sub-dataset of observations from the 20 ground stations. Correlation between $PM_{2.5}$ mass concentration measurements and MODIS blue band and green band AOD values is relative high, that of 0.85, and RMSE of 6.439 ($\mu g/m^3$). The remain subdataset of observations from the 5 ground stations were used to validate the model, and it indicated a correlation coefficient of 0.88 and a RMSE of 5.567 ($\mu g/m^3$). The result model is expected to be applied for deploying air quality monitoring systems derived from satellite observations and understanding geospatial distribution of $PM_{2.5}$ mass concentration in Ho Chi Minh City, Vietnam.

Keywords $PM_{2.5}$ · MODIS · AOD · Linear regression · Air polution

1 Introduction

Since the 1990s, air pollution, high blood pressure, smoking and high blood sugar have been considered four major global health risk factors [1]. They caused about 4.9 million global deaths in 2017. Therein, air polution significantly reduces millions of people's quality of life, leaving them with many disabilities and other health

P. H. D. Pham · D. K. Le · T. M. T. Nguyen · V. H. Phan (✉)
Department of Physics, International University, Linh Trung, Thu Duc, Ho Chi Minh City, Vietnam
e-mail: phvu@hcmiu.edu.vn

Vietnam National University Ho Chi Minh City, Linh Trung, Thu Duc, Ho Chi Minh City, Vietnam

© The Author(s), under exclusive license to Springer Nature Singapore Pte Ltd. 2023
J. N. Reddy et al. (eds.), *ICSCEA 2021*, Lecture Notes in Civil Engineering 268,
https://doi.org/10.1007/978-981-19-3303-5_51

burdens. According to IQAir, Vietnam has experienced some of the worst air pollution recorded in 2017 with $PM_{2.5}$ mass concentrations higher than 10 $\mu g/m^3$ when the mortality and disease burden rates were 50,232 and 1.38, respectively [2]. $PM_{2.5}$ is one of air pollutants, including particulate matter (PM_{10}, $PM_{2.5}$), nitrogen oxides (NO_x), ground-level ozone (O_3), hydrocarbons (HC) and volatile organic compounds (VOC), sulphur dioxide (SO_2), and carbon monoxide (CO) [3]. There are several methods for air quality monitoring from ground measurements such as modelling the distribution of emission sources from a comprehensive emission inventory [4], or calculating daily particulate pollutant levels by the gravimetric method based on PM_{10} and $PM_{2.5}$ measurements from handheld devices along main roads in the inner city [5]. In 1975, Griggs discovered that atmospheric aerosol optical depth (AOD) values can be determined from satellite data based on scattering and absorption of shortwave radiation in the atmosphere [6]. Therefore, optical satellite imagery has become potential data sources for estimation on mass concentration of air pollutants.

Based on the relationship between ground-level $PM_{2.5}$ measurements and satellite image-derived AOD values, popular methods of $PM_{2.5}$ estimation that have been used are: two-variable regression, multi-variable regression, artificial intelligence, and surface aerosol concentration modelling. At present, the regression model is a highly effective method. Experience models on PM_{10} or $PM_{2.5}$ estimation were proposed using local ground-level measurements and AOD values derived from optical satellite images, specially Moderate Resolution Imaging Spectroradiometer (MODIS) data products [7–15] and Landsat image sources [16–20]. In 2009, Green et al. made a comparison of GOES and MODIS AOD to Aerosol Robotic Network (AERONET) AOD and $PM_{2.5}$ mass on land surface of Interagency Monitoring of Protected Visual Environments at Bondville, Illinois, and indicated that the correlation between AOD and $PM_{2.5}$ mass concentration was highest in autumn and lowest in winter [21]. The results indicated that local experience models have achieved good correlation coefficients.

For the last 30 years, Ho Chi Minh City (HCMC) has become one of the fastest growing cities in the world. At present, however, it has suffered serious air pollution from that. In this study, we exploited the MODIS blue and green band AOD products in combination with the ground-level $PM_{2.5}$ measurements of the PAMAir air quality network to reveal an experience model of $PM_{2.5}$ estimation in HCMC and surrounding. The result indicates a good correlation fittings and is expected to be applied for deploying air quality monitoring systems in this region.

2   Materials and Methods

2.1   Study Area

HCMC is one populated and modern city in Vietnam with about 9 million citizens in 2019 [22]. It occupies a total area of 2061 km^2, extending from $106°22'E$ to

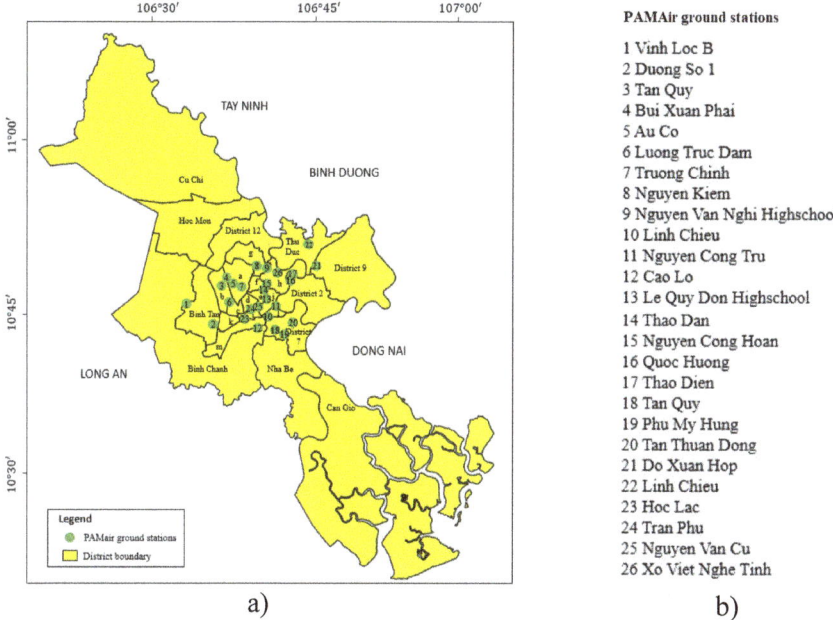

PAMAir ground stations

1 Vinh Loc B
2 Duong So 1
3 Tan Quy
4 Bui Xuan Phai
5 Au Co
6 Luong Truc Dam
7 Truong Chinh
8 Nguyen Kiem
9 Nguyen Van Nghi Highschool
10 Linh Chieu
11 Nguyen Cong Tru
12 Cao Lo
13 Le Quy Don Highschool
14 Thao Dan
15 Nguyen Cong Hoan
16 Quoc Huong
17 Thao Dien
18 Tan Quy
19 Phu My Hung
20 Tan Thuan Dong
21 Do Xuan Hop
22 Linh Chieu
23 Hoc Lac
24 Tran Phu
25 Nguyen Van Cu
26 Xo Viet Nghe Tinh

a)   b)

Fig. 1 **a** The Ho Chi Minh City superimposed with the PAM air ground station layer and **b** The list of PAM Air ground stations' name in HCMC

106°54′E and 10°10′N to 10°38′N. It is the center of the southern key economic region contributing about 45% to the economy country, including HCMC, Binh Phuoc, Tay Ninh, Binh Duong, Dong Nai, Ba Ria-Vung Tau, Long An and Tien Giang. HCMC is a paramount transportation hub connecting surrounding provinces as well as serving as an international gateway, as shown in Fig. 1. The city is tropical and monsoonal with two seasons: a rainy season from May to November and a dry season from December to April.

2.2 Input Data

MODIS MCD19A2 Product The MODIS MCD19A2 data product is MAIAC algorithm-based level-2 gridded aerosol optical thickness over land surface. The MCD19A2 product is availabe on the USGS Earth Data website [23]. Additionally, it can be retrieved by Google Earth Engine as well. The MCD19A2 provides daily MODIS blue band (0.47 µm) and green band (0.55 µm) AOD images at 1 km resolution, as presented in Fig. 2. Currently, AOD is retrieved at high altitudes of below 4.2 km. The observed period is in the dry season, from January 1 to May 31, 2020. To cover HCMC, we need one scene at the location of 'h28v07'. The images are referenced to the sinusoidal datum and stored in the hdf format. In this study, the

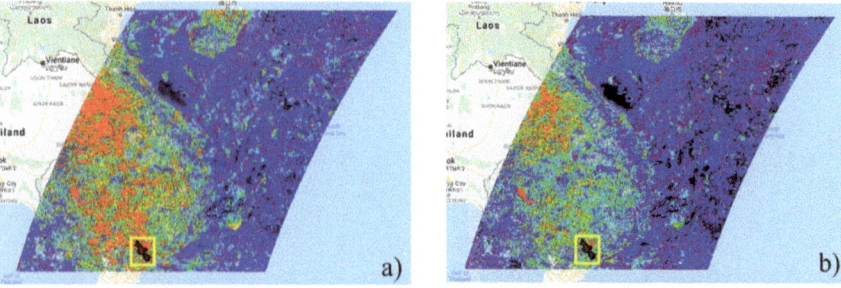

Fig. 2 The MODIS MCD19A2 product of **a** blue band and **b** green band AOD on January 5, 2020. The yellow box shows the study area HCMC

AOD values were derived from MODIS Terra satellite data, corresponding to the acquisition time at 10:30 AM in HCMC.

PAM Air Ground-Level PM$_{2.5}$ Measurements PAM Air company provided hourly ground-level PM$_{2.5}$ data of the 25 stations from January 1 to December 31, 2020 in HCMC, except of the Vinh Loc B station. The gound stations' location and name are shown in Fig. 1. The PM$_{2.5}$ mass concentration was measured by the PAS-OA320 sensor, monitoring outdoor air quality [24]. The device provides near real-time measurements of air temperature, humidity, PM$_{10}$ and PM$_{2.5}$.

2.3 The Relationship Between PM$_{2.5}$ Mass Concentration and AOD

Radiance from a pixel on the Earth's surface received by the satellite sensor is a combination of surface reflection and scattering by gases and particles in the atmosphere. By making assumptions notably including a derived surface reflectivity and aerosol characteristics such as single scattering albedo and phase function, relative to angular distribution of scattered light and mainly a function of particle size, AOD can be estimated [21]. The MODIS instrument was designed for aerosol retrievals and makes use of data at 7 wavelengths from 0.47 to 2.13 μm, which provides some information about aerosol particle size in addition to AOD. The MODIS algorithm accounts for surface reflectivity and assumed aerosol and atmospheric characteristics. Seasonally and spatially varying aerosol characteristics derived from Aerosol Robotic Network measurements over various sites are used to assign an aerosol type in the MODIS land algorithm. AOD products are retrieved from the observed mean radiance over a 10 km^2 area that is aggregated from 400 subpixels with 500 m resolution by neglecting the brightest 50% and darkest 20% of pixels.

AOD derived from satellite measurements is being investigated for predicting ground-level PM$_{2.5}$ concentrations. Green et al. reported that the MODIS AOD products have potential for predicting PM$_{2.5}$ mass in summer and autumn, but perform less well in winter and spring [19]. Therein, for the three GOES, MODIS and AERONET AOD datasets, the authors used the simple linear regression to determine correlation between ground-level PM$_{2.5}$ measurements and AOD values. In this study, we analyzed the relationship between ground-level $PM_{2.5}$ mass concentration and MODIS blue band AOD_{blue} and green band AOD_{green} values using the multivariable linear regression, following to Eq. (1).

$$PM_{2.5} = \alpha + \beta_1 \times AOD_{blue} + \beta_2 \times AOD_{green} \tag{1}$$

3 Data Processing

3.1 Data Preparation

In fact, the MODIS AOD product is not available on all days during the observed period and each available image may not cover the full study area, containing no-data pixels due to clouds or bad weather conditions. Additionally, a consecutive day MODIS AOD image occasionally has some duplicated value pixels which need to be removed. Moreover, the PAM Air PM$_{2.5}$ measurements are also missed on some days. For each ground station, the average PM$_{2.5}$ mass concentration from 10:00 to 11:00 AM per day was computed while two MODIS blue band and green band AOD values were extracted, as shown in Table 1. As a result, the total number of 1016 observations was collected as input data for the model, as described in Table 2.

3.2 Data Filtering

The input dataset was divided to two groups, including a sub-dataset of 873 observations from 20 ground stations for building the model and another one of 143 observations from 5 ground stations for validating the model. Firstly, the multivariable linear regression from the raw sub-dataset shows that correlation between PM$_{2.5}$ mass concentrations and blue band and green band AOD values is low, $r = 0.40$, due to the presence of many anomalies. Thus, an analysis on the dataset using SPSS was performed and a lower and upper boundary filter with a confidence level of 95% was applied, as illustrated in Fig. 3.

Table 1 The average PM$_{2.5}$ mass concentration from 10:00 to 11:00 AM corresponding to MODIS blue band and green band AOD values on January 5, 2020 in HCMC

Available stations	PM$_{2.5}$ (μm/m^3)	AOD$_{blue}$	AOD$_{green}$
Au Co	40.745	0.562	0.401
Bui Xuan Phai	45.040	0.617	0.441
Cao Lo	46.125	0.851	0.618
Thao Dan	39.895	0.665	0.475
Do Xuan Hop	38.225	0.597	0.426
Duong So 1	47.170	0.813	0.591
Linh Chieu	39.205	0.462	0.325
Nguyen Cong Tru	34.845	0.603	0.430
Nguyen Cong Hoan	28.615	0.630	0.450
Tan Quy Dong	32.495	0.543	0.383
Nguyen Khoai	27.015	0.619	0.442
Nguyen Kiem	39.350	0.550	0.393
Phu My Hung	38.255	0.590	0.421
Quoc Huong	36.045	0.669	0.478
Tan Quy Dong	35.780	0.676	0.482
Tan Quy	40.640	0.563	0.402
Tan Thuan Dong	32.485	0.484	0.342
Thao Dien	33.565	0.654	0.467
Tran Phu	41.670	0.661	0.472
Le Quy Don Highschool	38.405	0.628	0.449
NVN Secondaryschool	36.815	0.527	0.371
Nguyen Van Cu	32.995	0.661	0.472
Xo Viet Nghe Tinh	31.295	0.646	0.461

Table 2 The total number N of observations from January 1 to May 31, 2020

Variable	N	Max	Min	Mean	StdEv
PM$_{2.5}$ (μg/m^3)	1016	109.480	5.915	33.857	14.844
AOD$_{blue}$	1016	2.080	0.019	0.429	0.305
AOD$_{green}$	1016	1.621	0.013	0.306	0.227

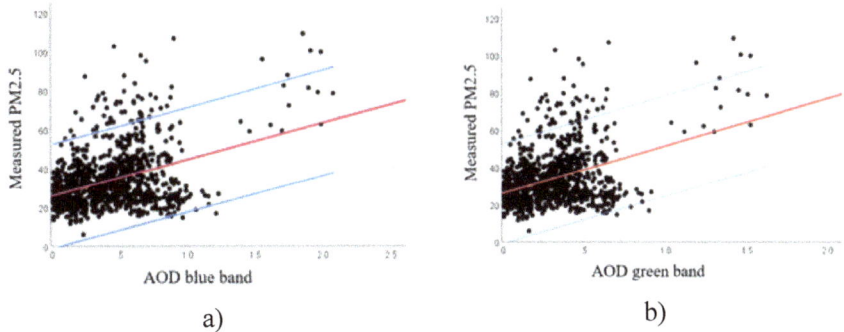

Fig. 3 Using the lower and upper boundary filter with a confidence level of 95% based on the distribution of PM$_{2.5}$ measurements and **a** blue band and **b** green band MODIS AOD values

4 Result and Discussion

4.1 Estimating Ground-Level PM$_{2.5}$ Mass Concentration

Based on the filtered observations, a correlation between ground-level PM$_{2.5}$ measurements and MODIS AOD$_{blue}$ and AOD$_{green}$ values was high with a correlation coefficient of 0.85 and a RMSE of 6.439 ($\mu g/m^3$), as presented in Table 3. The multivariable linear regression was presented by Eq. (2). For validating the model, the observations from the 5 ground stations was substituted into Eq. (2), and the result indicated a correlation coefficient of 0.88 and a RMSE of 5.567 ($\mu g/m^3$), as described in Fig. 4a.

$$PM_{2.5} = 17.378 + 37.450 \times AOD_{blue} + 5.637 \times AOD_{green} \qquad (2)$$

The relationship between ground-level PM$_{2.5}$ measurements and MODIS blue band and green band AOD values was shown in Eq. (2). Accordingly, the two MODIS blue band and green band AOD images on January 5, 2020 were substituted into the model. As a result, the PM$_{2.5}$ mass concentration in the whole HCMC was estimated, as illustrated in Fig. 4b. The observations on this day were used to verify the model, and the result indicated correlation of 0.63 and a RMSE of 5.149 ($\mu g/m^3$). Subsequently, estimated PM$_{2.5}$ mass concentration was approximate from 20 to 50 ($\mu g/m^3$) in the whole HCMC, but it presented a range of (30–35) ($\mu g/m^3$) in most of the study area in green color. Several areas in blue tone were corresponding to

Table 3 The multivariable linear regression on observations of the 20 ground stations

Sub-dataset	N	α	β$_1$	β$_2$	r	RMSE ($\mu g/m^3$)
Raw	873	27.967	−95.667	153.760	0.40	13.570
Filtered	438	17.378	37.450	5.637	0.85	6.439

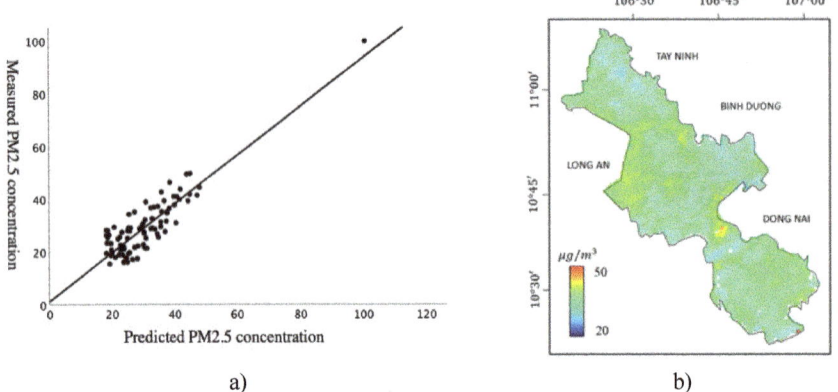

Fig. 4 **a** The scattergram of the five ground stations' observations and **b** The spatial distribution of the estimated PM$_{2.5}$ mass concentration on January 5, 2020 in HCMC

lower PM$_{2.5}$ mass concentration of (20–30) (μg/m^3) while few areas in red tone were corresponding to higher concentration of (40–50) (μg/m^3). Therefore, the air quality of PM$_{2.5}$ mass concentration is at the moderate and unhealthy for sensitive group levels on January 5, 2020 in HCMC.

4.2 Discussion

The model presented that the correlation coefficient between ground-level PM$_{2.5}$ measurements and MODIS AOD values of 0.85 is relative high and similar to Xie's correlation coefficient of 0.9 [14] and Hua's correlation coefficient of 0.88 [15]. In addition, these linear models were recommended to be strongly sensitive to ground-level PM$_{2.5}$ measurements which were affected significantly by weather conditions and environment at stations and their surroundings. Moreover, accuracy of PM$_{2.5}$ sensors are considered because they need to be cleaned with a time interval of about three months. Therefore, ground-level PM$_{2.5}$ datasets were potential to contain a large number of anomalies. In this study, to gain the model with high correlation, approximate 40% of the observations were removed after applying the lower and upper boundary filter and then matching ground-level measurements to satellite AOD values.

5 Conclusion

The paper presented the relationship between ground-level PM$_{2.5}$ mass concentration and MODIS blue band and green band AOD values in HCMC from January 1 to

May 31, 2020. The multivariable linear regression model was retrieved from the observations filtered by the lower and upper boundary with a confidence level of 95%. Although the correlation coefficient is high, the model should be improved by experiencing ground-level PM$_{2.5}$ measurements in a long-term period. The model shows a useful method to understanding the distribution of PM$_{2.5}$ mass on HCMC. Thus, the result is expected to to be applied for deploying air quality monitoring systems derived from satellite observations.

Acknowledgements We thank the International University—Vietnam National University, Ho Chi Minh City for providing the 2020 Student Project Fund for this work (the number of SV2020-SE-01) and the PAM Air company for sharing the hourly PM$_{2.5}$ mass concentration measurements in Ho Chi Minh City.

References

1. IHME, Global Burden of Disease (GBD). http://www.healthdata.org/gbd/2019. Accessed 31 Jul 2021
2. IQAir, World Air Quality. https://www.iqair.com/. Accessed 31 Jul 2021
3. Air Quality in Europe. https://www.airqualitynow.eu/pollution_home.php. Accessed 31 Jul 2021
4. Ho QB, Vu HNK, Nguyen TT et al (2019) A combination of bottom-up and top-down approaches for calculating of air emission for developing countries: a case of Ho Chi Minh City, Vietnam. Air Qual Atmos Health 12:1059–1072. https://doi.org/10.1007/s11869-019-00722-8
5. Hien TT, Chi NDT, Nguyen NT, Vinh LX, Takenaka N, Huy DH (2019) Current status of fine particulate matter (PM$_{2.5}$) in Vietnam's most populous city, Ho Chi Minh City. Aerosol Air Qual Res 19:2239–2251. https://doi.org/10.4209/aaqr.2018.12.0471
6. Griggs M (1975) Measurements of atmospheric aerosol optical thickness over water using ERTS-1 data, 1975. J Air Pollut Control Assoc 25(6):622–626. https://doi.org/10.1080/000 22470.1975.10470118
7. Engel-Cox JA, Holloman CH, Coutant BW, Hoff RM (2004) Qualitative and quantitative evaluation of MODIS satellite sensor data for regional and urban scale air quality. Atmos Environ 38:2495–2509
8. Mamun MI, Islam M, Mondol PK (2014) The seasonal variability of aerosol optical depth over Bangladesh based on satellite data and HYSPLIT model. Am J Remote Sens 2(4):20
9. Xin J, Zhang Q, Wang L, Gong C, Wang Y, Liu Z, Gao W (2014) The empirical relationship between the PM$_{2.5}$ concentration and aerosol optical depth over the background of North China from 2009 to 2011. Atmos Res 138:179–188. https://doi.org/10.1016/j.atmosres.2013.11.001
10. Nguyen TTN et al (2015) Particulate matter concentration mapping from MODIS satellite data: a Vietnamese case study. Environ Res Lett 10:095016
11. You W, Zang Z, Zhang L, Li Y, Pan X, Wang W (2016) National-scale estimates of ground-level PM$_{2.5}$ concentration in China using geographically weighted regression based on 3 km resolution MODIS AOD. Remote Sens 8(3):184. https://doi.org/10.3390/rs8030184
12. Butt MJ, Assiri ME, Ali MA (2017) Assessment of AOD variability over Saudi Arabia using MODIS deep blue products. Environ Pollut 231:143–153. https://doi.org/10.1016/j.envpol. 2017.07.104
13. Xu X, Zhang C (2020) Estimation of ground-level PM$_{2.5}$ concentration using MODIS AOD and corrected regression model over Beijing, China. PLoS ONE 15(10):e0240430. https://doi. org/10.1371/journal.pone.0240430

14. Xie Y, Wang Y, Zhang K, Dong W, Lu B, Bai Y (2019) Daily estimation of ground-level $PM_{2.5}$ concentrations over Beijing using 3km resolution MODIS AOD. Environ Sci Technol 49:12280–12288. https://doi.org/10.1021/acs.est.5b01413

15. Hua Z, Sun W, Yang G, Du Q (2019) A full-coveragae daily average $PM_{2.5}$ retrieval method with two-stage IVW fused MODIS C6 AOD and two-stage GAM model. Remote Sens 11:1558. https://doi.org/10.3390/rs11131558

16. Sifakis N, Paronis D (1998) Quantitative mapping of air pollution density using Earth observations: a new processing method and application on an urban area. Int J Remote Sens 19:3289–3300

17. Hadjimitsis DG, Clayton CRI (2009) Determination of aerosol optical thickness through the derivation of an atmospheric correction for short-wavelength Landsat TM and ASTER image data: an application to areas located in the vicinity of airports at UK and Cyprus. Appl Geomat J 1:31–40

18. Nadzri O, Mohd ZMJ, Lim HS (2010) Estimating particulate matter concentration over arid region using satellite remote sensing: a case study in Makkah, Saudi Arabia. Mod Appl Sci 4:131–142

19. Thi Van T, Hang Hai N, Quoc Bao V, Duong Xuan Bao H (2018) Remote sensing-based aerosol optical thickness for monitoring particular matter over the city. In: Proceedings of the 2nd International Electronic Conference on Remote Sensing, MDPI, Basel, Switzerland, 22 March–5 April 2018. https://doi.org/10.3390/ecrs-2-05175

20. Nguyen NH, Tran VA, Pham QV, Nguyen TB, Vu VH (2018) Determining PM10 model in Hanoi using landsat 8 oli and ground-measured dust data. VNU J Sci Earth Environ Sci 34(1)

21. Green M, Kondragunta S, Ciren P, Xu C (2009) Comparison of GOES and MODIS aerosol optical depth (AOD) to aerosol robotic network (AERONET) AOD and IMPROVE $PM_{2.5}$ mass at Bondville, Illinois. J Air Waste Manage Assoc 59(9):1082–1091. https://doi.org/10.3155/1047-3289.59.9.1082

22. Ho Chi Minh City Government. https://eng.hochiminhcity.gov.vn/. Accessed 31 Jul 2021

23. USGS Earthdata. MODIS MCD19A2v006 . https://lpdaac.usgs.gov/products/mcd19a2v006/. Accessed 31 Jul 2021

24. PAM Air, sensor. https://pamair.org/sensor/. Accessed 31 Jul 2021

Extracting Ground Points and Generating Digital Elevation Model (DEM) from Point Clouds from Point Clouds

Anh Thu Thi Phan, Quoc Thai Phan, and Anh Khoa Viet Nguyen

Abstract Digital elevation models (DEM) is essential information for leveling, surveying in building construction. There are many methods for generating DEM. Because of consuming much time and labor, directed methods are replaced by many indirect methods for generating DEM. In this study, DEM is proposed to be generated from point clouds rapidly collected by laser scanning or photogrammetry. Two datasets are used for checking the proposed data processing process. As a result, two ground point cloud areas were extracted. Many ground points in the steepest areas cannot be extracted. For a flat area with less than 10° of slope angle, the results of ground point extraction are good. The surface is divided into many equal grid cells for generating DEM, and the grid cell's elevation is computed from the point's elevation inside it. Finally, DEMs are displayed as TINs by creating Delaunay triangle networks.

Keywords DEM · KDE · Ground point cloud · TIN

1 Introduction

Ground data collection is critical in topographic surveying works, requiring accuracy depending on the purposes. As a result, the ground surface is described as digital elevation models (DEM), digital terrain models (DTM). These models serve many tasks such as leveling, surveying in building construction. From the past, direct measurement methods using a total station or other surveying methods will give the high accuracy of the collected points. However, these methods need much time, effort, and labor for ground data collection, and the direct measurement method cannot be applied in an unreachable location.

A. T. T. Phan (✉) · Q. T. Phan · A. K. V. Nguyen
Faculty of Civil Engineering, Ho Chi Minh City University of Technology (HCMUT), 268 Ly Thuong Kiet Street, District 10, Ho Chi Minh City, Vietnam
e-mail: ptathu@hcmut.edu.vn

Vietnam National University Ho Chi Minh City, Linh Trung Ward, Thu Duc District, Ho Chi Minh City, Vietnam

© The Author(s), under exclusive license to Springer Nature Singapore Pte Ltd. 2023   589
J. N. Reddy et al. (eds.), *ICSCEA 2021*, Lecture Notes in Civil Engineering 268,
https://doi.org/10.1007/978-981-19-3303-5_52

With the rapid development of science and technology, the collection of ground data by 3D laser scanning is becoming popular for many different purposes [1–3]. Laser scanning is a data collection technique that uses a laser sensor to scan an object's surface to record the distance and intensity of the reflected points. This technique can quickly collect millions of data points and generate an accurate scale of the object's point cloud. Significantly, all data has been established in complete 3D space, whereby subsequent processing is entirely on the 3D model. The data obtained from the laser scan is a point cloud. The points' coordinates (X, Y, Z) can be determined from an obtained distance using principles such as time of flight and phased shift.

In addition, photogrammetry is another method for generating a 3D point cloud [4, 5]. The point cloud's accuracy depends on the photo resolution and the applied method for generating the point clouds. UAV-based point cloud generation is getting familiar for normal users thanks to cheap equipment and open sources applications for data processing.

The research on point cloud data has been applied quite early. There have been many studies in processing point cloud data by many algorithms, serving many different purposes [6–13]. Because of dense points, point cloud processing is complicated. Ground points extraction is the first step for point cloud processing. Depending on the study purpose, the ground points can be used to generate DEM or DTM, and non-ground points can be used for further analysis. In general, the ground points can be extracted from point clouds generated by laser scanning and photogrammetry methods. Because of the importance of DEM and the advantages of dense point clouds, this study aims to conduct automatic ground data separation from the obtained point clouds for generating DEM. In detail, two-point clouds generated by laser scanning and photogrammetry technique are used for data processing.

2   Data Acquisition

In this study, two sets of point clouds acquired by two different methods are used. Each dataset is acquired in specific areas with specific topographical features (Fig. 1). Specifically, dataset one (HCM-VNU) was collected directly at the Ho Chi Minh City University of Technology- VNU campus in Ho Chi Minh City with Leica BLK 360 laser scanner. In this area, the topographical surface of the survey area is relatively flat, with a low slope of fewer than 10°. The Leica BLK 360 laser scanner uses a laser source with a wavelength of 830 nm, a scanning field of 360°, with a speed of about 360,000 points/second. Dataset two (RGB_North_railArea1) is downloaded at an accessible point cloud website. The point cloud is generated by applying the photogrammetric technique, the dataset describes the information along a rail track, with the topographic surface of the survey area having significant variation, and two-point clouds are stored in LAS type format. The characteristics of the point cloud are shown in Table 1.

(a)   (b)

Fig. 1 The point cloud used in this study. **a** point cloud collected by the laser scanner and **b** point cloud generated by photogrammetric technique

Table 1 The characteristics of the point cloud

Data	Size (Mb)	Number of points	Density (points/m^2)	Slope
HCM-VNU	906	36.533.580	2235	Less than 10°
RGB_North_railArea1	730	22.536.697	32	Less than 22°

3 Methodology

3.1 Ground Point Extraction

In this project, ground points are filtered from the point clouds based on the characteristics of the cloud dataset. With the ground points obtained after classification, sample data points according to grid cells are computed. Then, the irregular triangle network (TIN) is generated to represent the DEM model. For this, the point cloud is reorganized in a quadtree structure. In detail, the data set is divided into many cells. Then, the Kernel Density Estimation (KDE) is applied to extract the elevation histogram of each cell. Ground points have a lower elevation than objects on the surface. Therefore, the ground points candidates are separated with the condition:

$$Z_{point} \in lowpeak(KDE) \pm \frac{binsize}{2} \tag{1}$$

where low peak (KDE) is the lowest peak of the histogram generated by KDE and binsize is the histogram bin size depending on the characteristics of each voxel's points.

After applying KDE to extract the ground point candidates, the resulting point clouds still contained many non-ground noise points. Therefore, the region growing method is applied to extract ground points from this new dataset. The filtering process starts from a multiplier for each voxel. The group of ground points is grown from the seed pixel by adding in neighboring points that are similar to increase the number of ground points. Finding the neighbors point is done by checking all points of eight

Table 2 The characteristics of the ground point cloud and chosen grid size

Data	Data processing time for extracting ground point	Number of extracted ground point	Density (point/m^2)	Grid size (m)
HCM-VNU	3 h 30	18.861.113	413	0.5
RGB_North_railArea1	4 h 30	5.691.996	42	5

neighbor voxels and the target voxel. The set of neighboring points is considered according to the shortest distance to the seed point. The surface is created according to the normal vector, the third eigenvector of the set of neighboring points just considered. Points are defined as ground points when the distance to the created plane does not exceed a threshold. This threshold depends on the topographic surface of the dataset. The points that satisfy the given conditions are sent to the ground point class. The process will iterate for the entire dataset. This process takes several hours for point cloud extraction (Table 2). As a result, several million ground points have been extracted for each data set.

3.2 DEM Generation

After extracting ground points, the number of points is enormous with high spatial resolution. For generating digital elevation modeling (DEM), the terrain surface is divided into small cells of equal size (Table 2). At each cell, there will be an elevation point representing the entire cell. So to build a digital elevation model, the ground point data set is divided into a grid. The grid cell's elevation is the average elevation of all points in that grid cell. Then, the DEM is displayed with calculated elevation by generating the Delaunay triangle network (TIN model).

$$grid cell' selevation = \frac{1}{n} \sum_{i=1}^{n} Z_{P_i} \tag{2}$$

where n is the number point inside the target grid cell and Z_{P_i} is the third component of point P_i's coordinates.

In addition, the dataset will be imported to Cloud Compare open-source application for extracting ground points using the CSF algorithm [13]. Then the TIN model is also built on Cloud Compare. The results of ground point data separation of the two methods are compared directly to each other to serve as a basis for evaluating the effectiveness of the proposed method used in the study.

4 Result and Discussion

In this study, the data is organized in a quadtree structure, and the cloud data is divided into voxels. For each cell, a histogram of the point's elevation is plotted using the KDE filter. The size of the bin (bin size) is determined based on the characteristics of the data set. From there, the ground point candidates are extracted. After applying the KDE algorithm for extracting ground point candidates, the results show that most non-ground points (trees, houses, …) have been removed (Fig. 2a and b). However, some minor details have not been removed yet. Therefore, the growing region algorithm is applied to extract the ground points from its candidates. As a result, the ground point cloud has been extracted (Fig. 2c and d). In detail, more than eighteen million ground points have been extracted in dataset one (Table 2). Besides, the dataset is also processed by using Cloud Compare open-source application with Cloth Simulation Filter (CFS) algorithm to obtain similar results.

The ground point clouds obtained by the two methods are overlaid for direct comparison. From the results, the proposed data process gives the results closed to the results of the CFS algorithm. Each algorithm has different advantages and disadvantages, specifically with both algorithms; with dataset one, the terrain variation is insignificant (slope <10°), both algorithms give similar results. Cloudcompare software's Cloth Simulation Filter (CSF) algorithm runs very fast with good results. Nevertheless, with large terrain difference datasets, such as dataset two with a slope greater than 20°, the data of survey areas is not seamless; the proposed data process does not give results as good as applying the CFS algorithm. In detail, many ground

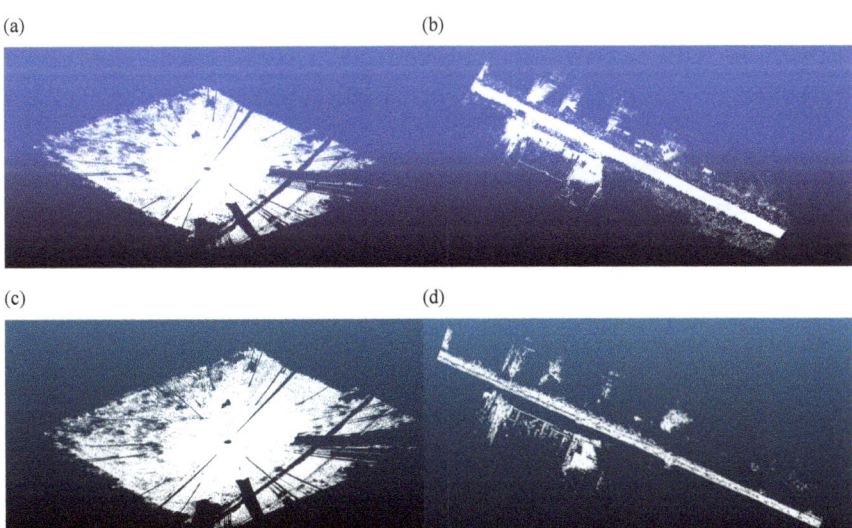

(a) (b)

(c) (d)

Fig. 2 The results of extracting ground points. The columns show dataset one (a,c) and (b,d) dataset two. The rows show ground point candidates extracted using KDE (a,b) and ground point (c,d)

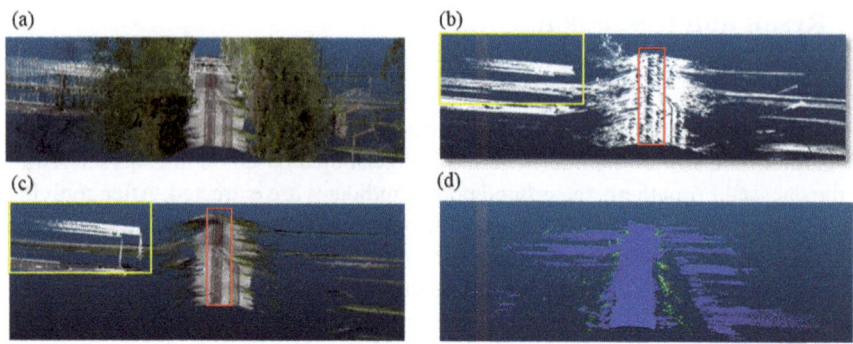

Fig. 3 The results of extracting ground points in dataset two using the proposed method and CFS'algorithm added in cloud compare open-source applications. **a** the original point cloud, **b** ground points extracted by the proposed method, **c** ground points extracted by CSF's algorithm, and **d** the overlaying of two ground point clouds with the overlap displayed in blue color

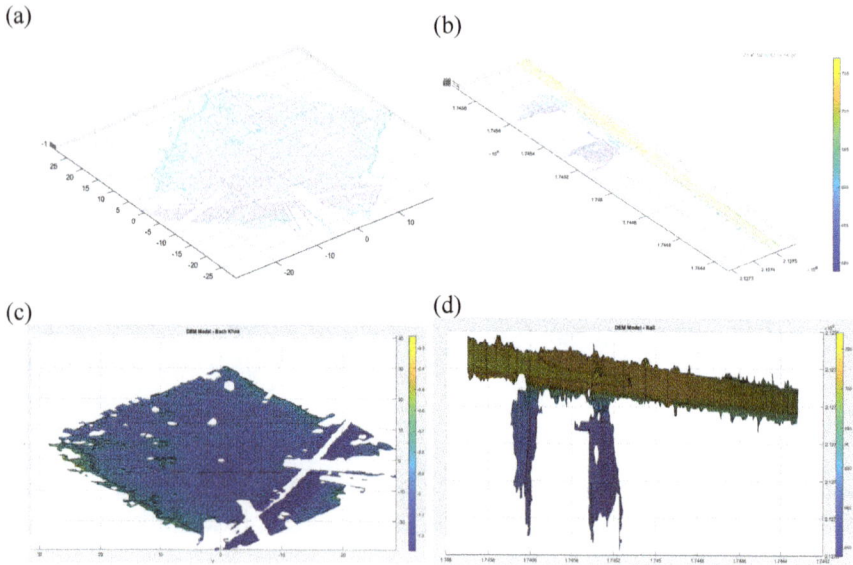

Fig. 4 The results of generating DEM. The columns show dataset one (a,c) and (b,d) dataset two. The rows show DEM (a,b), and TIN (c,d)

points cannot be extracted inside the yellow rectangular by the proposed method (Fig. 3), whereas, CFS algorithm cannot remove the non-ground point inside the red rectangular. As a result of an overlaying two-point cloud, the proposed method cannot extract the ground point at two slopes covered by many trees.

For generating DEM, the terrain surface is divided into an equal grid cell. Each grid has a presented elevation. The grid size corresponds to the accuracy of the DEM model that needs to be built. Since the resulted point cloud containing ground points has a very dense density; therefore, each grid cell will contain many elevation points. It is necessary to compute the represented value for each grid cell. In this study, the area inside the grid is considered a flat area. The grid cell elevation is computed as the average elevation of all points inside it (Fig. 4a and b). Then, the digital elevation model is represented through the TIN model from the elevation point data set representing the grid cells (Fig. 4).

5 Conclusion

This study focuses on generating DEM from point clouds. The data process consists of two main steps: extracting ground points and generating a digital elevation model DEM. With the assumption that the input datasets ensure the necessary accuracy. The implementation process includes organizing data in quadtree structures, classifying ground data through kernel density function estimation (KDE), and using a region-growing algorithm for the automatic extraction of ground points. Finally, DEM and TIN are generated. The process was performed on two different data sets with different topographical characteristics. As a result, ground points are extracted. The proposed method has minor limitations with data with significant terrain differences. For example, in dataset two, with the most significant slope up to 22°, ground information may be missing in places with significant slopes. Then, the sure is divided into a grid. The elevation of the grid is computed as the average elevation of all points inside it. Finally, the digital elevation model (DEM) is represented by generating irregular triangles networks (TIN).

Ground point clouds obtained by the proposed method and ground separation algorithm using the CFS tool in Cloudcompare software are overlaid for visual evaluation [8]. The obtained results show that the algorithms of both methods can automatically separate ground data from point cloud datasets. In the case of steep slopes area, the CFS algorithm solves the better ground point extraction results. From the results of the visual comparison, the proposed method is effective for extracting ground points in flat areas. DEM is generating by assuming the flat area of each grid cell. This assumption is not suitable for slope areas. It is better to simulate a surface for each grid cell in the near future to get the best fit of the DEM to the actual ground surface.

Acknowledgements This study was supported by Ho chi Minh City University of Technology-VNU-HCM under grant number SVKSTN-2021-KTXD-15.

References

1. Schäfer T, Weber T, Kyrinovič P, Zámečniková M. Deformation measurement using terrestrial laser scanning at the hydropower station of Gabcikovo. In: Proceedings of INGEO 2004 and FIG Regional Central and Eastern European conference on engineering surveying, Bratislava, Slovakia, 11–13 November 2004
2. Lindenbergh R, Pfeifer N, Rabbani T (2005) Accuracy analysis of the Leica HDS3000 and feasibility of tunnel deformation monitoring. In: Proceedings of ISPRS WG III/3, III/4, V/3 Workshop "Laser scanning 2005", Enschede, the Netherlands, 12–14 September 2005
3. Lindenbergh R, Pfeifer N (2005) A statistical deformation analysis of two epochs of terrestrial laser data of a lock. In: Proceedings of optical 3D measurement techniques, vol II, pp 61–70. Vienna, Austria
4. Schaich M (2013) Combined 3D scanning and photogrammetry surveys with 3D database support for archaeology & cultural heritage. A practice report on ArcTron´s Information System aSPECT3D. Photogrammetric, Week 13, pp 233–246
5. Bolognesi M, Furini A, Russo V, Pellegrinelli A, Russo P (2014) Accuracy of cultural heritage 3D models by RPAS and terrestrial photogrammetry. ISPRS Int Arch Photogramm Remote Sens Spatial Inf Sci XL–5:113–119
6. Zhang K, Bi W, Zhang X, Fu X, Zhou K, Zhu L (2015) A new Kmeans clustering algorithm for point cloud, Int J Hybrid Inf Technol. https://www.researchgate.net/publication/283784835_A_New_Kmeans_Clustering_Algorithm_For_Point_Cloud
7. Chang Y-C, Ayman H, Lee DC, Yom J-H (2008) Automatic classification of LiDAR data into ground and non-ground points. Int Arch Photogramm Remote Sens Spatial Inf Sci XXXVII:B4. https://www.isprs.org/proceedings/XXXVII/congress/4_pdf/81.pdf
8. Rodríguez-Cuenca B, Garcia-Cortes S, Ordóñez C, Alonso MC (2015) Automatic detection and classification of pole-like objects in urban point cloud data using an anomaly detection algorithm. Remote Sens. https://www.researchgate.net/publication/282313644_Automatic_Detection_and_Classification_of_Pole-Like_Objects_in_Urban_Point_Cloud_Data_Using_an_Anomaly_Detection_Algorithm
9. Hui Z, Cheng P, Ziggah YY, Nie Y (2018) A threshold-free filtering algorithm for airborne LiDAR point clouds based on expectation Maximization. Int Arch Photogramm Remote Sens Spatial Inf Sci XLII-3
10. Zhu Q, Li Y, Hu H, Wu B (2017) Robust point cloud classification based on multi-level semantic relationships for urban scenes. ISPRS J Photogramm Remote Sens 129:86–102
11. Lodha SK, Fitzpatrick DM, Helmbold DP (2003) Aerial LiDAR data classification using AdaBoost
12. Fröhlich C, Mettenleiter M (2004) Terrestrial laser scanning—new perspectives in 3D surveying. Int Arch Photogramm Remote Sens Spatial Inf Sci XXXVI(8/W2)
13. Zhang W et al (2016) An easy-to-use airborne LiDAR data filtering method based on cloth simulation. Remote Sens 8(6):501. https://www.researchgate.net/profile/Wuming_Zhang2

Mangrove Classification Using an Integration of Radar and Optical Images of Sentinel 1 and 2: A Case Study of Can Gio, Ho Chi Minh City

Vu Hien Phan, Tan Nhat Le, and Ngan Truong Nguyen

Abstract Mangrove forests are of great importance to coastal communities, providing not only a source of food and resources but also protecting coastlines, preventing erosion and regulating our climate. However, mangroves have been changed significantly due to deforestation and restoration during the last decades. At present, remote sensing has been widely proven to be essential in monitoring and mapping mangrove forest. In this study, we exploited an integration of radar and optical images of Sentinel 1 and 2 to classify mangroves in the Can Gio district, Ho Chi Minh City. Sentinel-1 images were collected in February 2021 while Sentinel-2 images were in January 2021. Based on an analysis of training samples, a decision tree diagram was designed to classify the Can Gio mangroves with four major plants, consisting of nypa palm, rhizophoraceae, avicennia and ceriops tagal. The result presented a spatial distribution of the mangrove types inside the Can Gio mangrove forest and their sparse appearance in combination with other vegetation in the Can Gio district. The classifier for the Can Gio mangroves obtain an overall accuracy of 80% and Cohen's Kappa coefficient of 0.75. The decision tree model on the integration of radar and optical images of Sentinel 1 and 2 was expected to contribute a large inventor of classification algorithms and to be effectively applied for mangrove classification.

Keywords Mangroves · Decision tree classifier · Radar and optical image integration · Sentinel 1 · Sentinel 2

V. H. Phan (✉)
Department of Physics, International University, Linh Trung, Thu Duc, Ho Chi Minh City, Vietnam
e-mail: phvu@hcmiu.edu.vn

T. N. Le · N. T. Nguyen
Faculty of Civil Engineering, University of Technology, Ward 14, District 10, Ho Chi Minh City, Vietnam

V. H. Phan · T. N. Le · N. T. Nguyen
Vietnam National University Ho Chi Minh City, Linh Trung, Thu Duc, Ho Chi Minh City, Vietnam

© The Author(s), under exclusive license to Springer Nature Singapore Pte Ltd. 2023   597
J. N. Reddy et al. (eds.), *ICSCEA 2021*, Lecture Notes in Civil Engineering 268,
https://doi.org/10.1007/978-981-19-3303-5_53

1   Introduction

Mangroves are forest ecosystems that occur in saline coastal environments where a tropical or subtropical climate is present [1]. With their salt tolerant roots mangrove ecosystems are able to survive in intertidal areas where a combination of salt sea water and fresh river water is present. Mangroves are valuable ecological and economic resource and they serve as a very important factor in coastal protection. However, mangroves are under big threat and many have already been lost in the last decades [2].

Remote sensing has been proven to be a valuable tool in analyzing and monitoring mangroves [3]. Many studies use NDVI values for monitoring and mapping mangrove changes [4]. Also false color composites are used to distinguish between very high NIR and lower NIR response [5]. Smooth textures indicate mangrove from other land cover if they are densely distributed. Dense mangrove forests give higher NIR response than sparsely distributed and mixed mangrove areas [6]. Also visual interpretation is still a highly used method for mangrove mapping. However, confusion between mangroves and other vegetation is the most commonly reported source of classification error. Though, these traditional methods already give classification accuracies of mangrove classes ranging from 75 to 95% for producer's and user's accuracies [7]. More specific information on spectral properties of mangrove are based on the differences within the different species.

At present, exploiting optical images for monitoring and detecting forest change has been popular. For example, in Google Earth Engine the Mangrove Forest of the World (2000) is a globally available raster dataset, visualized. This database is made using Landsat satellite data from the year 2000. More than 1000 Landsat scenes were classified using hybrid supervised and unsupervised digital image classification techniques [8]. Recently exploiting radar and optical remote sensing data has been experienced to create forest inventory [9]. Although many different methods have been used, this study will focus in using the radar Sentinel-1 and optical Sentinel-2 images for a better knowledge in mangrove mapping. Mangroves will be classified following to a decision tree diagram designed from samples of the Can Gio mangrove forest, Ho Chi Minh City (HCMC).

2   Materials and Methods

2.1   Study Area

Can Gio is a coastal district, located to the southeast of HCMC, from (106°46′12″E, 10°22′14″N) to (107°00′50″E, 10°40′00″N), as shown in Fig. 1a. The region has a monsoon tropical climate, with two distinct seasons including the rainy season from May to October and the sunny season from November to consecutive April. It occupies a total natural area of 71,316 ha, of which more than 70% is the area of

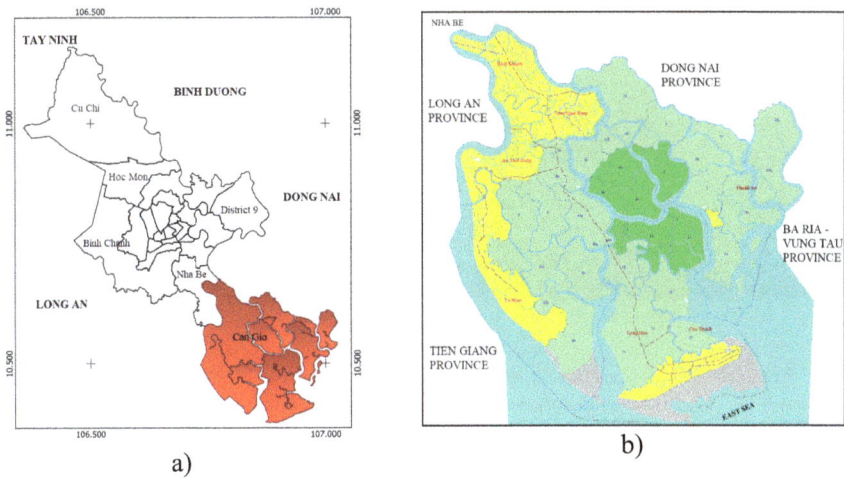

Fig. 1 **a** The location of the Can Gio, HCMC and **b** The zonation of the Can Gio biosphere reserve [11]

mangroves and rivers [10]. The Can Gio mangrove forest plays an important role in national defense. Coastal protection is one of main functions of the mangrove forest.

The Can Gio mangrove forest was recognized by UNESCO as a biosphere reserve on January 21, 2000. The biosphere reserve includes the entire area of the Can Gio protection forest and the remaining administrative area of the Can Gio district, divided into three zones [11]. Figure 1b describes the zonation of the biosphere reserve, including i) the core zone, green colored, mainly preserving the mangrove ecosystem of both planted and natural forests, ii) the buffer zone, light green colored, essentially restoring ecosystems based on the dominant species and protecting the core zone, and iii) The transition zone, yellow colored, encouraging economic development models. The Can Gio mangrove forest has some typical mangrove species such as Sonneratia caseolaris, Rhizophora apiculata Blume, Rhizophora mucronata, Avicennia marina Vierh, Bruguiera gymnorrhiza Savigny, Lumnitzera littorea, etc. but there are four major species, including Nypa Palm, Rhizophoraceae, Avicennia, and Ceriops Tagal.

2.2 Data Preparation

Sentinel-1 Radar Images

Sentinel-1 is the first mission of the Copernicus Programme satellite constellation by the European Space Agency [12]. The mission is able to provide data quickly in the event of disaster monitoring, composed of a constellation of two satellites sharing the same orbit plane. They carry a C-band synthetic-aperture radar (SAR) instrument which provides a collection of data in all-weather, day or night. This instrument has a spatial resolution of down to 5 m, a swath of up to 400 km and a temporal resolution

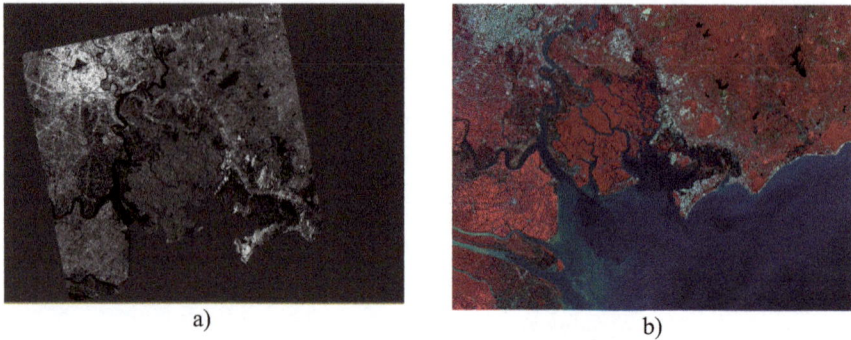

a)   b)

Fig. 2 After the data pre-processing, **a** the Sentinel-1 sigma0 VV band image on February 27, 2021 and **b** the Sentinel-2 color-infrared composite image on January 7, 2021, covering the Can Gio, HCMC

of 12 days. Sentinel-1 data products are free and open at the Copernicus Open Access Hub [13]. In this study, the data product at SAR Level-1 with Ground Range Detected (GRD) on February 7, 2021 was exploited. The Sentinel-1A image scene was named 'S1A_IW_GRDH_1SDV_20210227T11', covering the whole Can Gio mangrove forest. The product was in dual polarization (VH and VV) at a 10 m spatial resolution and in the Sentinel Standard Archive Format for Europe (SAFE) format. A data pre-processing of four steps was applied using the SNAP software, including i) raster subset for the area study, ii) radiometric calibration to convert digital number values to backscattering coefficients of sigma0 (σ^0), gamma0 (γ^0) and beta0 (β^0), iii) speckel filtering, and iv) geometric correction. Figure 2a presented the sigma0 band (σ^0_{VV}) of the VV polarizaton data after the data pre-processing.

Sentinel-2 Optical Images

The Copernicus Sentinel-2 mission aims at monitoring variability in land surface conditions and Earth's surface changes [14]. The sentinel-2 satellites contribute to ongoing multispectral observations with a revisited time of 5 days supporting a broad range of applications such as agricultural monitoring, emergencies management, land cover classification or water quality. The ESA provide the Sentinel-2 data products free of charge at the Copernicus Open Access Hub [13]. These products were at three spatial resolutions of 10, 30 and 60 m and were referenced to the UTM/WGS84 coordinate system. To cover the whole Can Gio, two Sentinel-2A image scenes located at 'T48PXS' and 'T48PYS', captured on January 7, 2021 were used. These products were at the Level-2A data processing, presenting reflectance values at the bottom of the atmosphere. In this study, 10 m spatial resolution bands consisting of B2 (490 nm), B3 (560 nm), B4 (665 nm) and B8 (842 nm) were exploited. A data processing was performed on these bands to obtain spectral reflectance values in this study area, including i) band stacking for each scene, ii) mosaicking two scenes, and iii) converting digital number (DN) value to reflectance. Figure 2b presented an infrared color composite of the reflectance images. Additionally, a vegetation index image, NDVI, was transformed from B3 and B4 reflectance images, to extract vegetation coverage.

Subsequently, all the images of backscattering coefficients, reflectance values and vegetation index were clipped with the Can Gio boundary. They were used as input data for making a decision tree model for mangrove classification.

Data Sampling

A sample dataset of 1498 pixels was collected from a field trip with 105 locations and from a Sentinel-2 true color composite images with 44 locations spreadly over the whole study area. The land cover of the region was classified into seven types, consisting of Nypa Palm, Rhizophoraceae, Avicennia, Ceriops Tagal, other vegetation, no vegetation, and water surface. The field trip photos of the four major species and their canopy structures on the Sentinel-2 true-color composites were presented in Fig. 3. The sample dataset was divided to two groups for decision tree design and verification, as shown in Table 1.

Fig. 3 The field trip photos (a, b, c, d) and the canopy structures on the Sentinel-2 true-color composite image (e, f, g, h) of Nypa Palm, Rhizophoraceae, Avicennia, and Ceriops Tagal, respectively, in the Can Gio mangrove forest

Table 1 The sample dataset of 1498 pixels divided into the two sub-datasets A and B, for designing and verifying a decision tree for the Can Gio mangrove classification, respectively

Land cover types	Nypa palm	Rhizo-phoraceae	Avicen-nia	Ceriops tagal	Other vegetation	No vegetation	Water surface
Sub-dataset A	51	313	112	57	45	79	301
Sub-dataset B	38	64	46	29	44	125	194
Total	89	377	158	86	89	204	495

2.3 Designing a Decision Tree Diagram

The sub-dataset A was used as a training sample for the decision tree. Accordingly, based on the locations of the training sample, ranges of pixel values of the land cover types on all the input images in the Can Gio were determined, as shown in Table 2. The table indicated that water surface has no returned-signal, corresponding to all backscattering coefficients derived from the Sentinel-1 data equaling zero.

Firstly, the NDVI image was used to classify the land cover into three basic groups, including water surface, no vegetation and vegetation classes. The no vegetation class consists of various features, such as roads, buildings, bare soil, or mudflats. Then, major mangrove species and other vegetation were determined special ranges of the backscattering coefficients of γ_{VH}^0 and β_{VV}^0, and the reflectance values of B3, see red numbers in Table 2, based on an analysis of their spectral scattering distribution. Subsequently, the decision tree diagram was designed to classify the land cover into seven classes including water surface, no vegetation, Ceriops Tagal, Rhizophoraceae, Avicennia, Nypa Palm, and other vegetation, orderly, as described in Fig. 4. In this study, the maximum pixel value of each land cover type was used as a threshold to separate one class and others.

3 Results and Discussion

According to the decision tree diagram, an algorithm was built to process the input images of backscattering, reflectance, and vegetation index derived from the radar Sentinel-1 and optical Sentinel-2 images. Figure 5 presented the classified image of the land cover of the Can Gio while Table 3 described the areas of the land cover types. The result indicated that the Avicennia specie, magenta colored, spreads regularly over the whole region. The Rhizophoraceae specie, orange colored, occupies a largest area of vegetation cover, but mostly has a convergent appearance inside the buffer and core zone of the mangrove forest biosphere reserve. The Ceriops Tagal specie, red colored, occupies a small area and mostly distribute along big river branches. The Nypa Palm specie, blue colored, also occupies a small area and appears as small shrubs sparsely over the whole Can Gio.

For verifying the classified image, the sub-dataset B of samples was applied by confusion matrix and accuracy statistics. The result indicated that the classified image obtained an overall accuracy of 80% and Cohen's Kappa coefficient of 0.75. The thresholds based on the maximum values of the land cover ranges could create errors in the classified image. Additionally, the sample size also affected on the resulting image. Therefore, to obtain Cohen's Kappa higher coefficient, the training sample should be larger.

Table 2 Ranges of pixel values of the land cover types on all the input images in the Can Gio District

Land cover Type	Water surface		No vegetation		Ceriops tagal		Rhizopho-raceae		Avicennia		Nypa palm		Other vegetation	
	Min	Max	Min	Max	Min	Max	Min	Max	Min	Max	Min	Max	Min	Max
σ^0_{VH}	0.000	0.000	0.005	0.230	0.008	0.033	0.014	0.085	0.025	0.087	0.014	0.055	0.014	0.088
γ^0_{VH}	0.000	0.000	0.006	0.284	0.009	0.040	0.017	0.104	0.031	0.107	0.017	0.067	0.018	0.109
β^0_{IVH}	0.000	0.000	0.008	0.392	0.013	0.056	0.024	0.148	0.044	0.152	0.024	0.096	0.025	0.151
σ^0_{VV}	0.000	0.000	0.012	0.354	0.023	0.063	0.061	0.257	0.061	0.299	0.078	0.199	0.073	0.328
γ^0_{VV}	0.000	0.000	0.015	0.436	0.028	0.077	0.076	0.316	0.075	0.366	0.096	0.242	0.090	0.402
β^0_{VV}	0.000	0.000	0.020	0.617	0.039	0.108	0.105	0.446	0.106	0.517	0.136	0.346	0.126	0.569
B2	0.035	0.053	0.088	0.267	0.010	0.020	0.005	0.017	0.011	0.029	0.012	0.020	0.015	0.031
B3	0.050	0.076	0.108	0.284	0.034	0.052	0.020	0.035	0.038	0.067	0.032	0.045	0.042	0.077
B4	0.020	0.069	0.101	0.298	0.010	0.021	0.007	0.018	0.016	0.030	0.015	0.022	0.020	0.043
B8	0.000	0.010	0.112	0.315	0.220	0.332	0.188	0.319	0.238	0.365	0.274	0.363	0.239	0.470
NVDI	−0.995	−0.534	−0.193	0.428	0.830	0.934	0.831	0.953	0.815	0.895	0.857	0.912	0.756	0.905

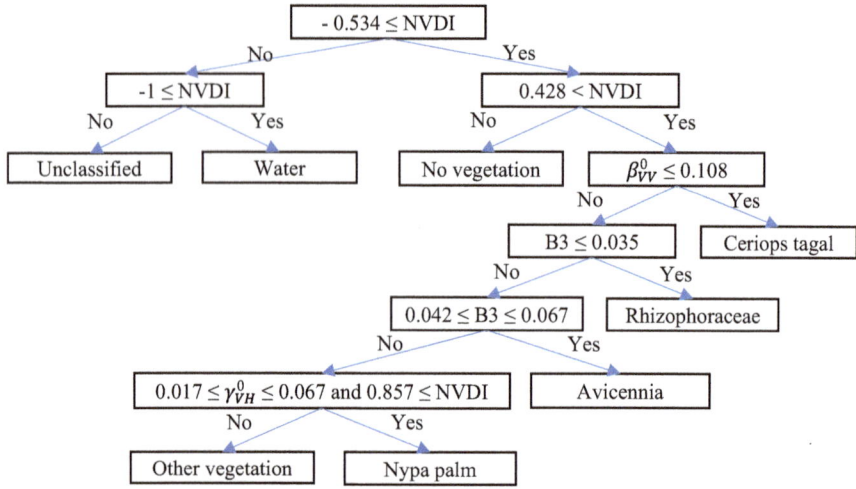

Fig. 4 The decision tree diagram for the Can Gio mangrove classification using the integration of radar Sentinel-1 and optical Sentinel-2 images

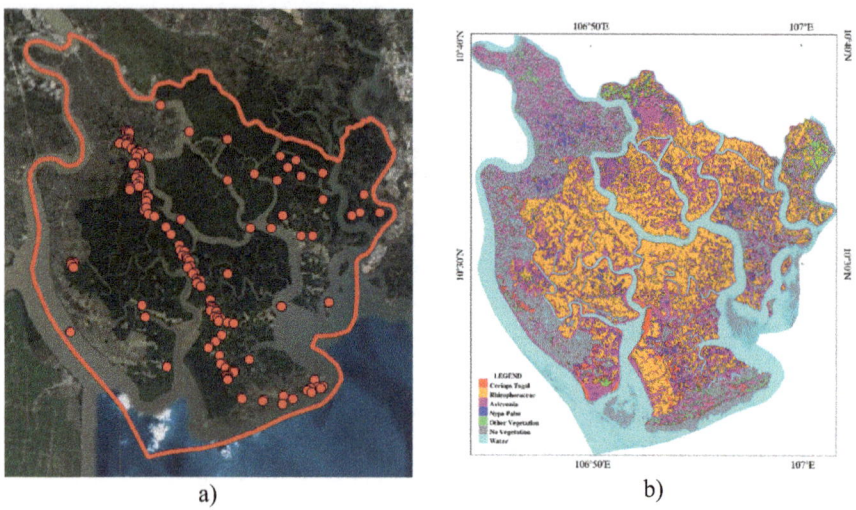

a)   b)

Fig. 5 **a** The true-color composite from Sentinel-2 multispectral images covering the Can Gio, captured on January 7, 2021, and **b** The land cover of the Can Gio classified by the decision tree from the integration of radar Sentinel-1 and optical Sentinel-2 images on February 27, 2021 and January 7, 2021, respectively

Table 3 The areas of the land cover types of the Can Gio District, computed from the classified image

Land cover type	Nypa palm	Rhizo-phoraceae	Avicen-nia	Ceriops tagal	Other vegetation	No vegetation	Water surface	Total
Area (ha)	5176.95	19,813.57	9422.02	2942.55	3621.36	12,690.58	18,224.34	71,891.33

4 Conclusion

The Sentinel-1 radar and Sentinel-2 optical satellite missions were chosen as remote sensing data to forest mapping due to both good spatial and temporal resolutions. The decision tree classifier was created and applied to classify the mangroves in the Can Gio, HCMC. The result showed the separability of different mangrove densities and also between the mangrove species Nypa Palm, Rhizophoraceae, Avicennia, and Ceriops Tagal with different reflectance properties. A fusion of optical with radar data is found to be the good input for discriminating mangrove and its properties in the area of the Can Gio. Classification results of 80% overall accuracy are obtained with this method compared to validation ground-truth data. The result provided understanding for mangrove mapping. The decision tree classification on the integration of radar and optical images was expected to contribute a large inventor of classification algorithms and to be effectively applied for mangrove classification.

References

1. The Geographer Online. https://www.thegeographeronline.net/3-managing-coastal-margins.html. Accessed 31 Jul 2021
2. Alongi DM (2002) Present state and future of the world's mangrove forests. Environ Conserv 29(03):331–349
3. Kuenzer C, Bluemel A, Gebhardt S, Quoc TV, Dech S (2011) Remote sensing of mangrove ecosystems: a review. Remote Sens 3:878–928. https://doi.org/10.3390/rs3050878
4. Khairuddin B, Yulianda F, Kusmana C, Yonvitne (2016) Degradation mangrove by using Landsat 5 TM and Landsat 8 OLI image in Mempawah Regence, West Kalimantan Province year 1989–2014. Procedia Environ Sci 33:460–464. https://doi.org/10.1016/j.proenv.2016.03.097
5. Pujiono E, Kwak DA, Lee WK, Sulistyanto KSR, Lee JY, Lee SH, Park T, Kim MI (2013) RGB-NDVI color composites for monitoring the change in mangrove area at the Maubesi Nature Reserve, Indonesia. For Sci Technol 9(4):171–179. https://doi.org/10.1080/21580103.2013.842327
6. Reddy S, Agrawal M, Prasar RC (2016) Automatic extraction of mangrove vegetation from optical satellite data. ISPRS Arch XLI-B8
7. Heumann BW (2011) Satellite remote sensing of mangrove forests: recent advances and future opportunities. Prog Phys Geogr 35(1):87–108
8. Giri C, Ochieng E, Tieszen LL, Zhu Z, Singh A, Loveland T, Masek J, Duke N (2011) Status and distribution of mangrove forests of the world using earth observation satellite data. Glob Ecol Biogeogr 20:154–159

9. Fernandez-Ordonez Y, Soria-Ruiz J, Leblon B (2009) Forest inventory using optical and radar remote sensing. In: Jedlovec G (ed) Advances in geoscience and remote sensing, chapter 26. InTech
10. Can Gio District Government. https://cangio.hochiminhcity.gov.vn/. Accessed 31 Jul 2021
11. Can Gio mangrove biosphere reserve. https://www.rungngapmancangio.org/. Accessed 31 Jul 2021
12. Sentinel-1. https://sentinels.copernicus.eu/web/sentinel/missions/sentinel-1. Accessed 31 Jul 2021
13. Copernicus Open Access Hub. https://scihub.copernicus.eu/. Accessed 31 Jul 2021
14. Sentinel-2. https://sentinel.esa.int/web/sentinel/missions/sentinel-2. Accessed 31 Jul 2021

Geotechnical Engineering Session

Assessing Landslide Susceptibility in Korea Using a Deep Neural Network

Ba-Quang-Vinh Nguyen, Thanh-Hai Do, and Yun-Tae Kim

Abstract Rainfall is a key triggering factor for landslides. Most of landslides in Korea were triggered by heavy rainfall. In this study, we used a deep neural network (DNN) to assess landslide spatial probability at Mt. Hwangnyeong, Busan, Korea. The results was validated based on 26 landslides using a receiver operating characteristic (ROC) curves. The areas under the curve (AUC) of the success-rate curve and predicted-rate curve showed that the proposed model was successful in predicting the spatial probability of landslide at Mt. Hwangnyeong. In addition, the DNN model was compared to the infinite slope model and showed better performance than the infinite slope model. The performance of the DNN model at three different activation functions were also compared to select the optimum function. This result showed that the DNN model with ReLu function has the best accuracy. A classified landslide susceptibility (CLS) map was established from the landslide spatial probability map by the geometrical interval method. A statistical test was performed and indicated that the classified landslide susceptibility map had statistical significance.

Keywords Deep neural network · Infinite slope model · Landslide susceptibility · Rainfall

B.-Q.-V. Nguyen (✉)
Department of Civil Engineering, International University, Quarter 6, Linh Trung Ward, Thu Duc City, Ho Chi Minh City, Vietnam
e-mail: nbqvinh@hcmiu.edu.vn

T.-H. Do
Faculty of Civil Engineering, Ho Chi Minh City University of Technology (HCMUT), 268 Ly Thuong Kiet Street, District 10, Ho Chi Minh City, Vietnam

Y.-T. Kim
Ocean Engineering Department, Pukyong National University, Busan, Korea

B.-Q.-V. Nguyen · T.-H. Do
Vietnam National University Ho Chi Minh City, Linh Trung Ward, Thu Duc District, Ho Chi Minh City, Vietnam

© The Author(s), under exclusive license to Springer Nature Singapore Pte Ltd. 2023   609
J. N. Reddy et al. (eds.), *ICSCEA 2021*, Lecture Notes in Civil Engineering 268,
https://doi.org/10.1007/978-981-19-3303-5_54

1 Introduction

Rainfall-induced landslide is one of a major natural hazards in Korea that significantly affected people lives and facilities [5, 15, 20–23, 30]. Landslides can be effected by many conditioning factors (CFs) [8]. Therefore, an assessing landslide spatial probability using real landslide events and the CFs is an urgent task in decreasing and preventing damages from landslides.

There are various methods for assessing landslide spatial probability, including heuristic method [1, 27], physical-based model [16, 20, 21], and probabilistic method [4, 19]. In the probability-based approach, several different machine learning (ML) techniques are used for predicting landslide spatial probability. In which logistic regression [2, 4, 11, 26], support vector machine [18, 24, 25, 28, 29], decision trees [3, 7, 25], artificial neural network [8, 9, 12, 19] were most popular used. In addition, deep neural network (DNN), which is one of the novel branches in the ML techniques, has recently received more attention. Therefore, this study used a deep neural network architecture for assessing landslide spatial probability at Mt. Hwangnyeong, Busan, Korea.

The landslide spatial probability map was validated using a receiver operating characteristic (ROC) curve then was compared to the infinite slope model. A classified landslide susceptibility (CLS) map including five susceptible classes: very low, low, moderate, high, and very high were generated by a geometrical interval classifier. Besides that, to evaluate statistical significance of the CLS map a statistical test was carried out. The results indicated that the used DNN architecture is a promising method in the landslide spatial probability assessment.

2 Study Area

Mt. Hwangnyeong where has a dense population is located at the center of Busan. There were several large and small landslides in the past. On September 10, 1999, a large-scale landslide occurred at the road incision at the entrance of Mt. Hwangnyeong Tunnel. The size of the active slope is about 130 m in length, about 50 m in width, and about 20–30 m in thickness, moving up to 17 m in horizontal distance and flowing down to the end of the four-lane road [10]. This landslide caused 1 death, 3 serious injuries, and heavy traffic congestion in Busan. In addition, Mt. Hwangnyeong has major facilities and densely populated areas along the periphery of the mountain, has many hiking trails and forest roads, and contains numerous hermitage and sports facilities. Damage-sensitive elements exist throughout the mountain, and landslides can recur at any time.

26 landslide events at Mt. Hwangnyeong were collected based on satellite images and Google Earth (Fig. 1). These landslide events were considered as a landslide inventory for predicting landslide spatial probability in this study area. A random

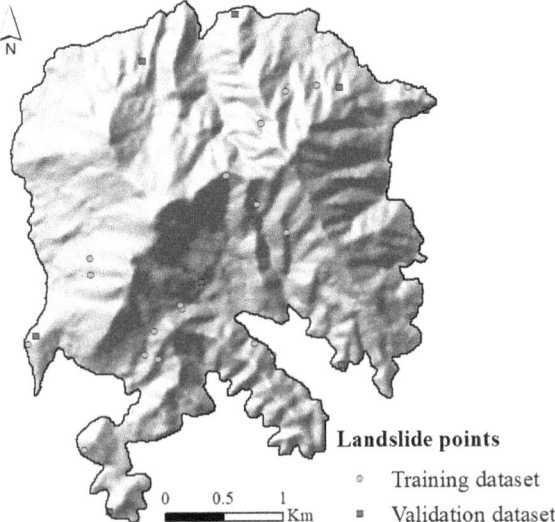

Fig. 1 Study areas and landslide inventory

separation of the 26 landslides was carried out and showed 70% of the landslides were used for training dataset and the remainder for validation dataset (Fig. 1).

3 Methodology

A DNN architecture was used to assess landslide spatial probability at Mt. Hwangnyeong. DNN model is one of the machine learning techniques which has been popularly used in regression and classification [14].

Structure of the used DNN model includes an input layer, three hidden layers, and an output layer. The number of neurons in the input layers is the same as the number of selected landslide conditioning factors (CFs) whereas the output layers included two neurons that infer landslide or non-landslide classes. The number of neurons in hidden layers and the iteration number of training are dependent on specific training data. In the present study, these values have been obtained from trial-and-error process to obtain optimum results and avoid over-fitting problem. The procedure used for landslide spatial probability mapping in this study is shown in Fig. 2a including three main steps.

Step 1: multicollinearity test was conducted for detecting the multicollinearity among the [6]. The tolerances (TOL) and variance inflation factor (VIF) are two factors for assessing the multicollinearity among the CFs. When VIF > 10 or tolerance < 0.1, it shows a potential multicollinearity problem in the dataset

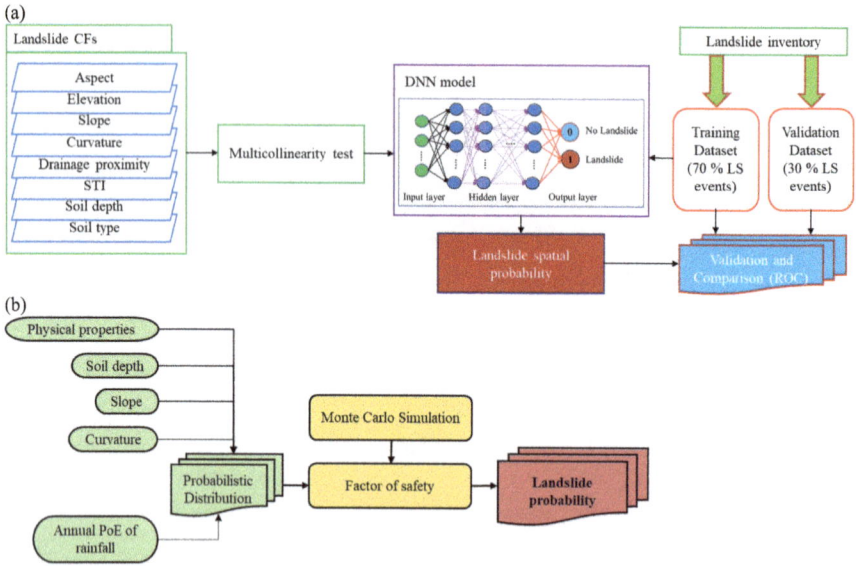

Fig. 2 **a** The flow chart of the DNN model, **b** The flow chart of the physical-based model

[13]. Therefore, the CFs having the VIF above 10 or the TOL lower 0.1 are multicollinearity and should be removed.

Step 2: landslide spatial probability map in the study area was established by DNN model using training dataset of landslide inventory and selection CFs in step 1.

Step 3: the result from the used DNN model was validated using a receiver operating characteristic (ROC) curve.

In addition, landslide spatial probability was also established using a physical-based model which is based on the infinite slope model and Monte Carlo simulation (MCs). The main process of the physical-based model for predicting landslide spatial probability was presented in Fig. 2b. This procedure can be summarized in two steps.

$$FS = \frac{c + \left[(\gamma H - \gamma_w h) \cos^2 \alpha\right] \tan \phi}{\gamma H \sin \alpha \cos \alpha} \tag{1}$$

where c is a cohesion (kN/m^2), ϕ is a friction angle ($^\circ$), γ is an unit weight of soil (kN/m^3), H is a soil depth (m), γ_w is an unit weight of water (kN/m^3), h is a saturated depth (m), α is a slope angle, g is a gravity acceleration (m/s^2).

4 Data Preparation

Figure 3 shows 8 CFs used for the landslide spatial probability mapping at Mt. Hwangnyeong. A 10 × 10 m digital elevation model (DEM) obtained from the National Geographic Information Institute (NGII) was used to derive the topographic and hydrologic factors. The topographic factors including aspect, elevation, slope, and curvature were shown in Fig. 3a–d, respectively. The aspect was divided into 9 classes: (1) flat, (2) north, (3) northeast, (4) east, (5) southeast, (6) south, (7) southwest, (8) west, and (9) northwest. The elevation ranges from 65.00 to 424.65 m a.m.s.l. Slopes belong the range of 0.02° to 47.11°. Curvature ranges from -12.35 to 10.99.

The hydrological factors presented in Fig. 3e–f include the drainage proximity (Fig. 3e) calculated using the Euclidean function in a GIS environment and the stream

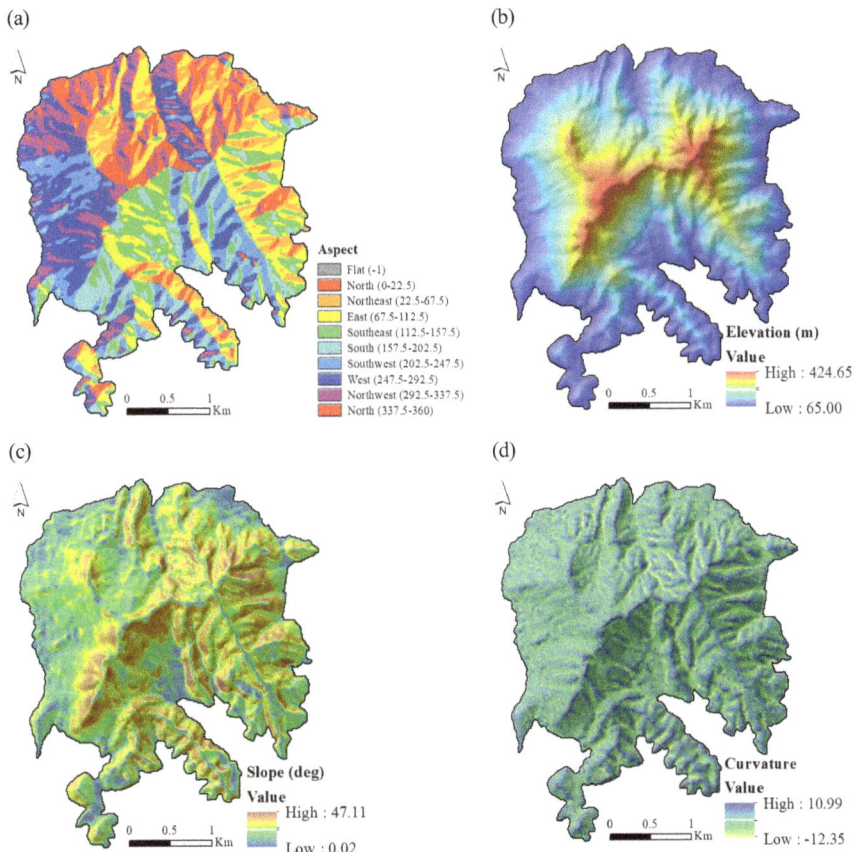

Fig. 3 Conditioning factors: **a** aspect, **b** elevation, **c** slope, **d** curvature, **e** drainage proximity, **f** STI, **g** soil depth, **h** soil type

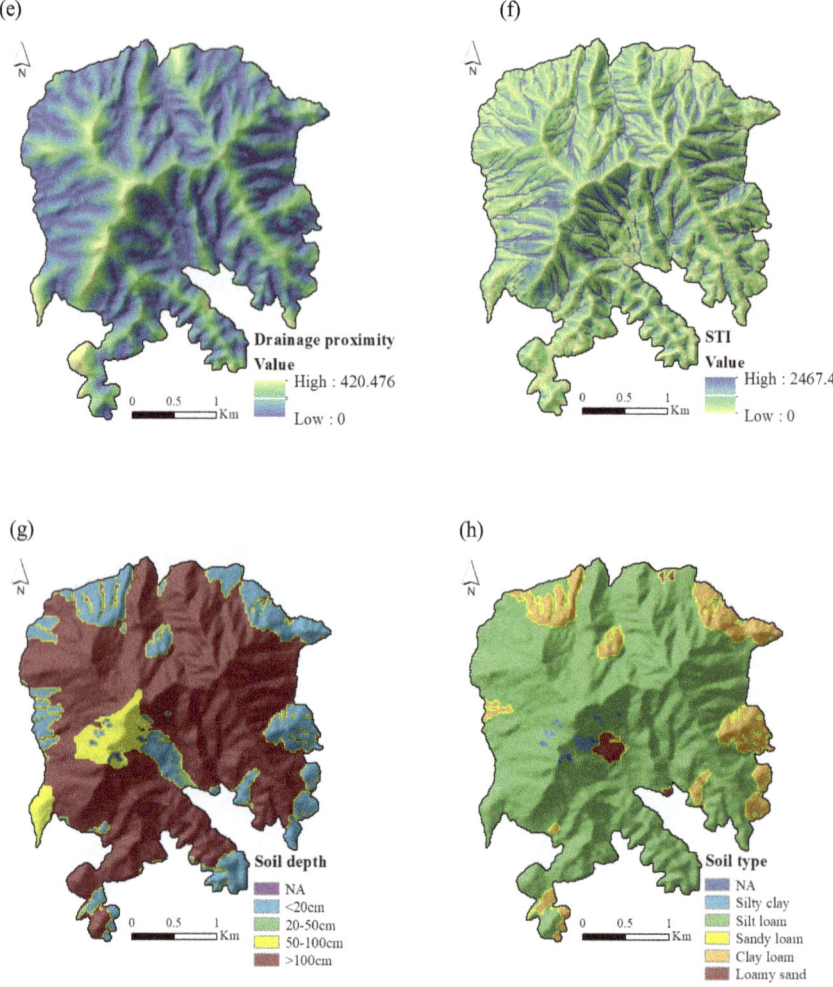

Fig. 3 (continued)

transport index (STI) (Fig. 3f) measured the sediment transport capacity of overland flow.

This study area soil depth was obtained from the National Institute of Agriculture Sciences (NIAS) and comprised five classes: (1) NA, (2) < 20 cm, (3) 20–50 cm, (4) 50–100 cm, and (5) > 100 cm (Fig. 3g).

The soil type at the study area was provided by the National Institute of Agriculture Sciences (NIAS) and included six classes: (1) NA, (2) Silty clay, (3) Silt loam, (4) Sandy loam, (5) Clay loam, (6) Loamy sand (Fig. 3h).

Table 1 Physical properties of soil [20]

Properties	COV	COV Reference	Mean value	Standard deviation
Unit weight (γ, kN/m^3)	0.08	Hammitt (1966), Harr (1987), Wolff (1994)	18.1	1.448
Cohesion (c, kN/m^2)	0.275	Bakker (2004)	10.2	2.805
tan (friction angle) (tan (ϕ))	0.15	Bakker (2004)	0.412	0.062
Hydraulic conductivity (k_s, mm/h)	0.9	Harr (1987), Wolff (1994)	28.8	25.92
Porosity (n)	-	-	0.45	–

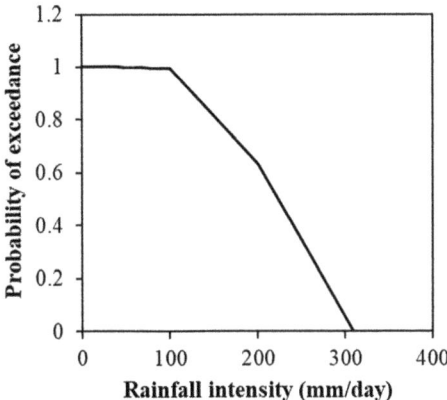

Fig. 4 Rainfall intensity and annual probability of exceedance [20]

There have not had any researches evaluating landslide spatial probability at Mt. Hwangnyeong before. Therefore, there is no information about soil physical properties and rainfall at the time when landslides occurred in this area. To overcome this difficulty the soil physical properties and rainfall data in this study were obtained from the study of [20] that assessed landslide probability at Mt. Umyeon, Korea. Table 1 presents the soil physical properties while Fig. 4 shows rainfall annual probability of exceedance used in this research.

5 Results and Discussions

Table 2 shows the results of the multicollinearity test among 8 CFs. The correlation among the CFs was assessed using VIF and TOL values. The VIF values range from 1.026 to 1.523, TOL values from 0.656 to 0.975. These results are satisfied critical values (VIF > 10 or TOL < 0.1) and prove that there is no multicollinearity among the 8 landslide CFs.

Table 2 Multicollinearity test result

Conditioning factors	VIF	TOL
Aspect	1.026	0.975
Elevation	1.523	0.656
Slope	1.382	0.724
Curvature	1.269	0.788
Drain proximity	1.431	0.699
STI	1.308	0.765
Soil depth	1.456	0.687
Soil type	1.075	0.930

Figure 5a presents the landslide spatial probability obtained from the DNN model. The landslide probability ranges from 0 to 1 presented the likelihood of landslide occurrence from low to high. The landslide probability from the physical model was presented in Fig. 5b. The distribution of landslide probability was totally different

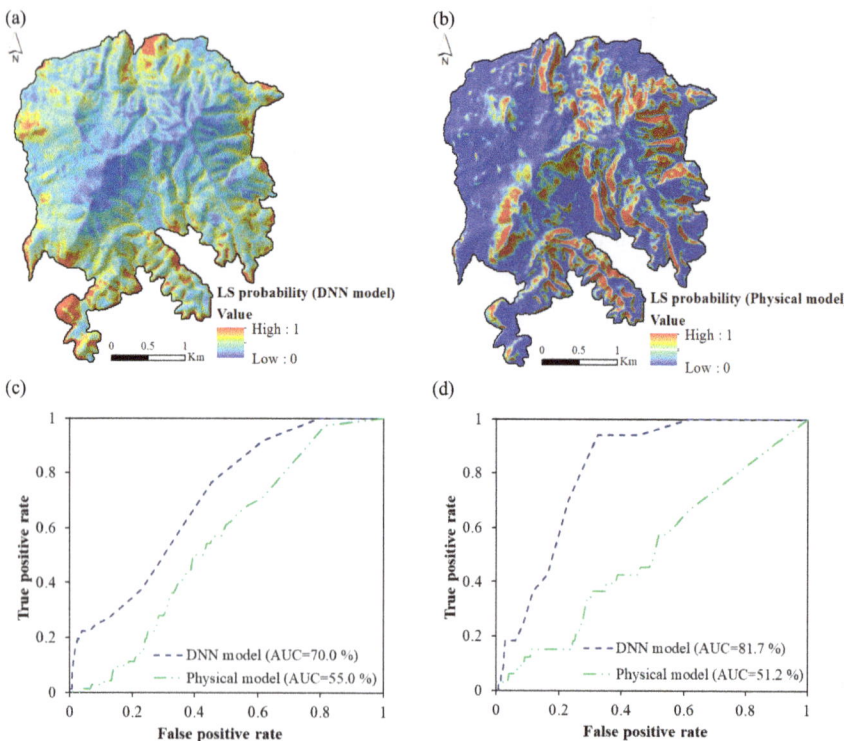

Fig. 5 **a** Landslide spatial probability from DNN model (LS: landslide), **b** Landslide spatial probability from the physical model (LS: landslide), **c** Success-rate curve, (d) Predicted-rate curve

between the DNN model and the physical model. In the physical model, landslide occurrence mostly depends on the slope, due to the high landslide probability distributed on the areas with the steep slope. For the DNN model, the landslide probability was established based on the landslide inventory and various CFs. Therefore, the distribution of landslide probability was not dependent in one any factor.

Figure 5c and d show the performance analysis results of the DNN and the physical models with training data and validation data. The AUC values of the success-rate curves (Fig. 5c) were 70.0% for the DNN model and 55.0% for the physical model. Figure 5d presents the predicted-rate curves with the AUC values of 81.7% and 51.2% for the DNN model and the physical model, respectively. These AUC values indicated that the DNN model was successful in landslide spatial probability mapping. In addition, the DNN model showed better accuracy than the physical model. These observations proved the efficiency of using the DNN model for mapping landslide spatial probability using real landslide events and the CFs.

A Classified landslide susceptibility (CLS) map was generated from the landslide spatial probability map by geometrical interval technique. Five susceptible levels of landslide occurrence including very low, low, moderate, high, and very high were defined and presented in Fig. 6a. The distribution of each susceptibility class was shown in Fig. 6b. It can be observed that very low class has the largest area, following by low, moderate, high, and very high, respectively.

To assess the statistical significance of the CLS map, a Chi-square test was conducted. The Chi-square test assumed that the area of landslide susceptibility classes was completely random at the significant level of 5%. The results show that Chi-square value of 26.943 exceed 3.841 and all p values are lower than 0.05. Thus, the null hypothesis is rejected and the CLS map could be considered as statistically significant.

Activation function is the core of the neural network's architecture, therefore the selection of the suitable activation function in a neural networks is an important task

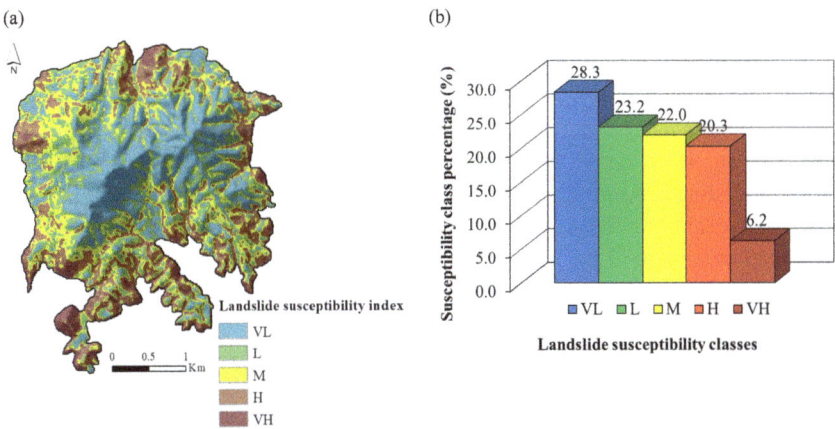

Fig. 6 **a** Classified landslide susceptibility map, **b** percent of area of susceptibility class

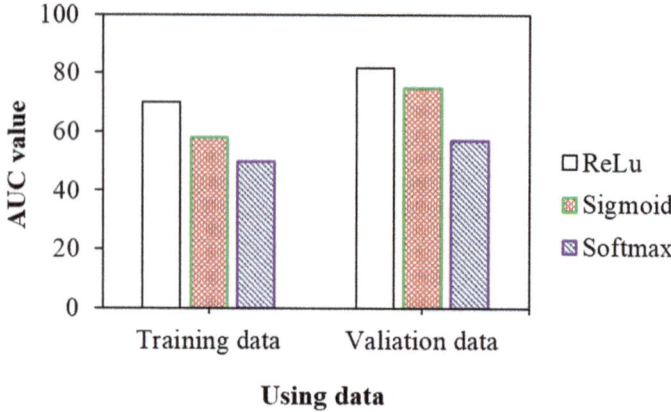

Fig. 7 AUC values of the DNN model at different activation functions

which can affect the accuracy of the neural networks [22, 23]. This study carried out a comparison of the performance of the DNN model at three different activation functions such as ReLu, sigmoid and softmax. Figure 7 presents the AUC values of the DNN model using ReLu, sigmoid and softmax activation functions. The results in Fig. 7 showed that for both training data and validation data, the AUC value of DNN model with ReLu activation function is greater than the DNN architectures using sigmoid and softmax. This was also recognized in the studies of [22, 23] that proved that the performance of the DNN architectures using ReLu was more efficient than those using sigmoid and softmax.

6   Conclusions

This study used a Deep Neural Network to assess the spatial probability of landslide at Mt. Hwangnyeong, Busan, Korea. The landslide spatial probability map was established and then validated using ROC curve. The AUC values of ROC curves were 70.0% for training dataset and 81.7% for validation dataset. These results proved the efficiency of the DNN model in landslide spatial probability mapping.

A comparison in the performance between the DNN model and the physical model was carried out. This comparison indicated the major improvement of the DNN model compared to the physical model.

A classified landslide susceptibility map including five susceptibility classes (very low, low, moderate, high, and very high) was generated from the spatial probability map using the geometrical interval method. In addition, the Chi-square test was conducted and showed that the CLS map had statistical significance.

A comparison in the performance of the DNN model using different activation functions was carried out. This result indicated a using ReLu function gives a better performance than other activation functions.

References

1. Abella EAC, Van Westen CJ (2007) Generation of a landslide risk index map for cuba using spatial multi-criteria evaluation. Landslides 4:311–325
2. Akgun A (2012) A comparison of landslide susceptibility maps produced by logistic regression, multi-criteria decision, and likelihood ratio methods: a case study at İzmir Turkey. Landslides 9(1):93–106
3. Alkhasawneh MS, Ngah UK, Tay LT, Isa M, Ashidi N, Al-Batah MS (2014) Modeling and testing landslide hazard using decision tree. J Appl Math 2014: 929768
4. Bai S-B, Wang J, Lü G-N, Zhou P-G, Hou S-S, Xu S-N (2010) GIS-based logistic regression for landslide susceptibility mapping of the Zhongxian segment in the three gorges area China. Geomorphology 115(1–2):23–31
5. Bai S, Wang J, Thiebes B, Cheng C, Yang Y (2014) Analysis of the relationship of landslide occurrence with rainfall: a case study of Wudu County China. Arab J Geosci 7(4):1277–1285. https://doi.org/10.1007/s12517-013-0939-9
6. Bui DT, Lofman O, Revhaug I, Dick O (2011) Landslide susceptibility analysis in the hoa binh province of vietnam using statistical index and logistic regression. Nat Hazards 59(3):1413
7. Bui DT, Ho T-C, Pradhan B, Pham B-T, Nhu V-H, Revhaug I (2016) GIS-based modeling of rainfall-induced landslides using data mining-based functional trees classifier with adaboost, bagging, and multiboost ensemble frameworks. Environ Earth Sci 75(14):1101
8. Bui DT, Tuan TA, Klempe H, Pradhan B, Revhaug I (2016) Spatial prediction models for shallow landslide hazards: a comparative assessment of the efficacy of support vector machines, artificial neural networks, kernel logistic regression, and logistic model tree. Landslides 13(2):361–378
9. Chen W, Pourghasemi HR, Zhao Z (2017) A GIS-based comparative study of dempster-shafer, logistic regression and artificial neural network models for landslide susceptibility mapping. Geocarto Int 32(4):367–385
10. Choi JC, Paik IS (2002) Study on analysis for factors inducing the whangryeong mountain landslide. J Eng Geol 12(2):137–150
11. Das I, Stein A, Kerle N, Dadhwal VK (2012) Landslide susceptibility mapping along road corridors in the Indian Himalayas using bayesian logistic regression models. Geomorphology 179:116–125
12. Ermini L, Catani F, Casagli N (2005) Artificial neural networks applied to landslide susceptibility assessment. Geomorphology 66(1–4):327–343. https://doi.org/10.1016/j.geomorph.2004.09.025
13. Hair JF, Black WC, Babin BJ, Anderson RE (2009) Multivariate data analysis. Upper Saddle River, NJ [etc.], vol. 24. Pearson Prentice Hall, New York, p. 899.
14. Haykin SS (2009) Neural networks and learning machines/Simon Haykin. Pearson, Upper Saddle River, NJ
15. Igwe O, Mode W, Nnebedum O, Okonkwo I, Oha I (2014) The analysis of rainfall-induced slope failures at Iva Valley area of Enugu State Nigeria. Environ Earth Sci 71(5):2465–2480. https://doi.org/10.1007/s12665-013-2647-x
16. Jibson RW (2011) Methods for assessing the stability of slopes during earthquakes—A retrospective. Eng Geol 122(1–2):43–50
17. LeCun Y, Bengio Y, Hinton G (2015) Deep learning. Nature 521(7553):436
18. Lee S, Hong S-M, Jung H-S (2017) A support vector machine for landslide susceptibility mapping in Gangwon Province Korea. Sustainability 9(1):48

19. Lee S, Hong S-M, Jung H-S (2018) GIS-based groundwater potential mapping using artificial neural network and support vector machine models: the case of Boryeong city in Korea. Geocarto Int 33(8):847–861
20. Nguyen VBQ, Kim YT (2020) Rainfall-earthquake-induced landslide hazard prediction by monte carlo simulation: a case study of MT. Umyeon in Korea. KSCE J Civ Eng 24(1):73–86
21. Nguyen B, Lee S, Kim Y (2020) Spatial probability assessment of landslide considering increases in pore-water pressure during rainfall and earthquakes: Case studies at Atsuma and Mt Umyeon. CATENA 187:104317
22. BQV Nguyen YT Kim (2021) Regional-scale landslide risk assessment on Mt. Umyeon using risk index estimation Landslides https://doi.org/10.1007/s10346-021-01622-8
23. Nguyen BQV, Kim YT (2021) Landslide spatial probability prediction: a comparative assessment of naïve Bayes, ensemble learning, and deep learning approaches. Bull Eng Geol Environ. https://doi.org/10.1007/s10064-021-02194-6
24. Peng L, Niu R, Huang B, Wu X, Zhao Y, Ye R (2014) Landslide susceptibility mapping based on rough set theory and support vector machines: A case of the three gorges area, China. Geomorphology 204:287–301
25. Pradhan B (2013) A comparative study on the predictive ability of the decision tree, support vector machine and neuro-fuzzy models in landslide susceptibility mapping using GIS. Comput Geosci 51:350–365
26. Tsangaratos P, Ilia I (2016) Comparison of a logistic regression and Naïve Bayes classifier in landslide susceptibility assessments: The influence of models complexity and training dataset size. CATENA 145:164–179
27. Van Westen CJ (2000) The modeling of landslide hazards using GIS. Surv Geophys 21(2–3):241–255
28. Xu C, Xu X, Dai F, Saraf AK (2012) Comparison of different models for susceptibility mapping of earthquake triggered landslides related with the 2008 Wenchuan earthquake in China. Comput Geosci 46:317–329
29. Yilmaz I (2010) Comparison of landslide susceptibility mapping methodologies for Koyulhisar, Turkey: conditional probability, logistic regression, artificial neural networks, and support vector machine. Environ Earth Sci 61(4):821–836
30. Zhou J-W, Cui P, Fang H (2013) Dynamic process analysis for the formation of Yangjiagou landslide-dammed lake triggered by the Wenchuan earthquake, China. Landslides 10(3):331–342. https://doi.org/10.1007/s10346-013-0387-3

Investigation of Permeability of Dredging Sand Mixing Cement and Bentonite

Bich Thi Luong, Phong Duy Nguyen, Hoang-Hung Tran-Nguyen, and Khanh Duy Tuan Nguyen

Abstract The hydraulic conductivity of soilcrete specimens created from dredging sand mixing with cement and bentonite expects to be relatively low to apply as impermeable cores for embankments, dams, containment walls, and so on. Hydraulic conductivity of dredging sand samples taken in Dong Thap province mixed with a cement content of $300\,kg/m^3$, various bentonite contents of 25, 50, 75, and $100\,kg/m^3$, respectively, was conducted in the laboratory. The permeability tests followed the ASTM D5856 and D5084 standards for more than 3 months. The results indicate that: (1) The hydraulic conductivity of the dredging sand mixed with cement was lower 1000 times than that of the unmixed sand; (2) the hydraulic conductivity of the sand mixed with cement and bentonite was lower than that of the sand mixed with cement and slightly increases with increasing in bentonite contents; (3) the hydraulic conductivity of soilcrete decreases with increasing in curing times; (4) the hydraulic conductivity of soilcrete was identical with hydraulic gradients; (5) the hydraulic conductivity of soilcrete varied from 4.86×10^{-9} m/s to 1×10^{-10} m/s.

Keywords Hydraulic conductivity · Permeability · Soilcrete · Dredging sand · Bentonite

1 Introduction

The natural clay with low hydraulic conductivity (e.g., $< 10^{-9}$ m/s) is commonly used as impermeable cores for embankments, dams, and landfills. Clay cores are

B. T. Luong (✉) · P. D. Nguyen · K. D. T. Nguyen
Graduate Students, Faculty of Civil Engineering, Ho Chi Minh City University of Technology (HCMUT), 268 Ly Thuong Kiet Street, District 10, Ho Chi Minh City, Vietnam
e-mail: ltbich.sdh19@hcmut.edu.vn

H.-H. Tran-Nguyen
Faculty of Civil Engineering, Ho Chi Minh City University of Technology (HCMUT), Ho Chi Minh City, Vietnam

B. T. Luong · P. D. Nguyen · H.-H. Tran-Nguyen · K. D. T. Nguyen
Vietnam National University Ho Chi Minh City, Linh Trung Ward, Thu Duc, Ho Chi Minh City, Vietnam

© The Author(s), under exclusive license to Springer Nature Singapore Pte Ltd. 2023
J. N. Reddy et al. (eds.), *ICSCEA 2021*, Lecture Notes in Civil Engineering 268,
https://doi.org/10.1007/978-981-19-3303-5_55

economical because of a natural material and simple construction. However, suitable clays for seepage cutoff are increasingly rare [1, 12].

Sand-bentonite mixture can be potentially replaced to natural clays as an impermeable material for earth levees [1, 2, 11, 15, 18]. Bentonite or montmorillonite is a swelling material. Bentonite mixing sand decreases significantly the hydraulic conductivity of unmixed sand [1, 2, 18]. The hydraulic conductivity of sand-bentonite mixture (k_s) depends on moisture, bentonite contents, and uniform distribution of bentonite in the mixture [11]. Xu et al. [18] showed that k_s decreased sharply from 1×10^{-6} m/s to 1×10^{-10} m/s as sandy soil mixed with a bentonite content of 5%. However, as bentonite content exceeding 5%, k_s of the mixture decreased insignificantly. Cowland and Leung [6] concluded that the permeability of the sand-bentonite mixture reached 1×10^{-9} m/s at a bentonite content of 7%. Sällfors and Öberg-Högsta [15] proposed that the bentonite content should be from 4 to 13% to achieve impermeability for economic construction. However, the sand-bentonite mixture can induce surface cracks at high bentonite contents and low water content increased the permeability of the mixture (Esener 2005 cited from [1]).

Cement is also utilized with a small content of 5–15% as an additive in sand-bentonite mixtures (soilcrete) to increase strength, to decrease permeability, and to decrease surface cracks [1, 9]. The hydraulic conductivity of sand mixed with 10% bentonite plus 5% cement mixture was lower 10 times than that of sand mixed with 10% bentonite [1]. Iravanian [9] founded that a soilcrete specimen created from 80% sand, 15% bentonite, and 5% cement achieved a hydraulic conductivity of 10^{-9} m /s at 28 days of age.

Dredged sand materials from the bottom of rivers are local materials with low cost applied mainly for rising elevations of highway embankments in the Southern Vietnam. Earth levees have often failed during annual flood seasons. Seepage flows washes sand particles via linked voids and cracks due to lack of compaction. The void spaces inside an earth levee have developed, enlarged, and eventually caused the collapses of an earth levee which is a typical failure in the Mekong delta. Soilcrete walls are highly potential solutions to protect earth levees in the annual floods. The mechanical properties of soilcrete have been well studied but soilcrete permeability has still limited database in Vietnam. The literature review has revealed that dredging sand mixed with cement and bentonite reduced permeability but the reduction of soilcrete permeability was inconsistent with cement contents and bentonite contents. Soilcrete permeability of 10^{-9} m/s or lower is considered an impermeability material to create seepage cutoff walls for earth levees. For the dredging sand taken in the Mekong delta, its fine grain sizes and organic content may cause unexpected soilcrete permeability. Therefore, this study aims at a better understanding of permeability behaviors of specimens molded from the dredging sand sample mixing cement with bentonite added. Relevant cement contents plus appropriate bentonite additives to mix with the dredging sand samples to create suitable low permeability of soilcrete for seepage cutoff walls need to be determined for earth levee reinforcement in the Mekong delta. Several soilcrete specimens were made in the laboratory to conduct permeability tests. The permeability of all soilcrete specimens was investigated for permeability variation with time, cement contents, and bentonite contents. This paper attempts to propose proper bentonite contents to achieve relatively low permeability of soilcrete made from the dredging sand taken in the Mekong delta. Further, this study makes an effort to build a permeability database gradually for soilcrete created from the dredging sand taken in the Mekong delta.

2 Materials and Testing Methods

2.1 Materials

Dredging sand was taken in Dong Thap Province, which is one of the thirteen provinces in the Mekong delta, Vietnam. The dredging sand samples were carefully stored and tested for key properties. The optimum water content (w_{op}) and maximum dry unit weight of sand samples were determined by the standard proctor compaction test (ASTM D698). The key properties of the dredging sand samples are given in Table 1.

Bentonite was used for this study meeting the API SPEC 13A standard. The typical properties of bentonite are printed in Table 2.

Blended Portland cement PCB40 was utilized to mix with dredging sand following the TCVN 6260:2009 (Vietnam standard). The PCB40 properties are showed in Table 3.

Table 1 The key parameters of the dredging sand samples

Optimum water content w_{op} (%)	Maximum dry unit weight γ_{dmax} (kN/m3)	Organic content OC (%)	pH
15.15	15.55	6.76	6.7

Table 2 The bentonite properties

Density ρ (g/cm³)	Percent of fine mass passing the sieve 0.075 mm (%)	Liquid limit LL (%)	Water content w (%)
0.9	80	440	10

Table 3 The Portland PCB40 cement properties

Compressive strength, (MPa)		Setting time, (minutes)		Fine fraction * (%)	Stability of volume (mm)	SO₃ (%)
3 days	28 days	Starting	Ending			
≥ 18	≥ 40	≥ 45	≤ 420	≤ 10	≤ 10	≤ 3.5

*Percent fine remaining on the sieve ≥ 0.09 mm

Table 4 The maximum allowable contents (mg/L) in water used to mix with dry cement

Total dissolved salts	Sulfate $(SO_4)^{2-}$	Chloride $(Cl)^-$	Non-dissolvable solids
10.000	2.700	3.500	300

Tap water was used for creating soilcrete specimens and testing permeability. The key properties of tap water are suitable the TCVN 4506:2012 (Vietnam standard) and presented in Table 4.

2.2 Specimens

Unmixed sand specimen. The natural dredging sand samples brought to the laboratory were dried before mixing with tap water to obtain a sand sample at the optimum water content of 15.15%. Sand at the w_{op} was compacted with the equivalent energy of 600 kN-m/m^3 in-cylinder plastic mold with dimensions of $D \times H = (62 \times 140)$ mm [3]. At the end of a compacted sand specimen process, the 2 porous stones and geotextiles were covered to prevent fine particles washing out during a permeability test. The specimen was saturated in water under a vacuum pressure of −80 to − 90 kPa within 24 h.

Soilcrete specimens. Soilcrete specimens created from dredging sand mixing cement at a cement content of 300 kg/m^3 and various bentonite contents of 25, 50, 75, and 100 kg/m^3, respectively, in cylindrical plastic molds with 65 mm in height and 62 mm in diameter. The water to binder ratio ($w{:}BC$) of 0.7:1 was applied to created soilcrete specimens in the laboratory. The binders can be cement only or cement - bentonite mixture. The parameters of the soilcrete specimens are presented in Table 5. The procedure to create the soilcrete specimens was conducted following (1) dry sand was mixed with water to reach the optimum water content of 15.15%;

Table 5 Mix proportions of specimens used for testing

Specimen ID	Specimen quantity	Cement content (kg/m^3)	Bentonite content (kg/m^3)	Mass of soil sample at moisture 15.15% (g)	Water to binder ratio $w{:}BC$	Mass of cement (g)	Mass of bentonite (g)	Mass of water (g)
B0	2	300	0	350	0.7	58.9	0	41.23
B25	2	300	25	350	0.7	58.9	4.9	44.7
B25b	1	300	25	350	0.7	58.9	4.9	44.7
B50	2	300	50	350	0.7	58.9	9.8	48.1
B75	2	300	75	350	0.7	58.9	14.7	51.5
B100	2	300	100	350	0.7	58. 9	19.6	55

(2) dry cement and bentonite with suitable mass were mixed with the moist sand sample within 5 min, amount of water calculated following a *w:BC* ratio of 0.7:1 was added to the above mixture, then uniformly mixed for 5 min; (3) the mixture was placed into a plastic mold in 3 layers and compacted by a vibrating compactor to eliminate air bubbles as much as possible in a total maximum time of 30 min; (4) all specimens were covered by plastic wrap and immersed in water for 2 days or more; (5) soilcrete specimens were removed from the molds and saturated under a vacuum pressure of −80 to −90 kPa for at least 24 h before carrying out the permeability tests.

2.3   Permeameter and Permeability Tests

A rigid wall permeameter was used to determine hydraulic conductivity (k_{soil}) of the compacted sand specimens with the falling head-constant tailwater method (ASTM D5856) [4]. Low hydraulic gradients of 2 to 5 were applied to avoid leakage along the rigid wall. For soilcrete specimens, the hydraulic conductivity (k_s) was tested by the falling head-constant tailwater method under hydraulic gradients of 30 to 45 or the falling head-rising tailwater method under hydraulic gradients of 100–150 controlled by air pressure on the flexible wall permeameter. This permeameter was designed basing on the ASTM D5084 [5]. Schematics of the permeameter was shown in Fig. 1. The process of installing specimens into the permeameter was done underwater for fully saturated specimens. Cell water pressure was higher than water head-in pressure

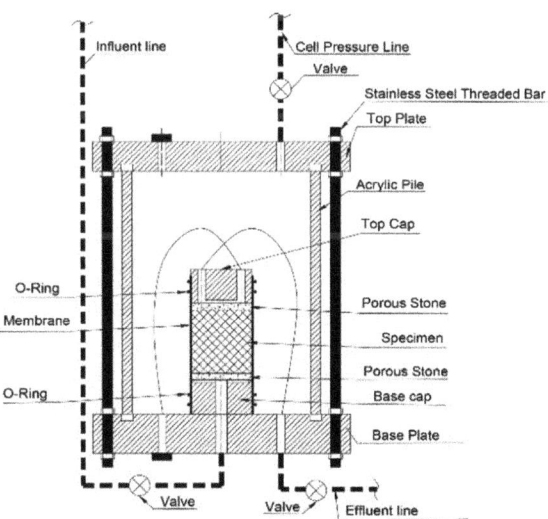

Fig. 1 Flexible wall permeameter

about 10 to 20 kPa. Data was recorded every day to investigate the change of soilcrete hydraulic conductivity with time.

Equations 1 and 2 were used to calculate hydraulic conductivity at the laboratory room temperature for the falling head-constant tailwater and the falling head-rising tailwater methods, respectively. Hydraulic conductivity at the standard temperature of 20°C, k_{20}, converted by Eq. (3).

$$k = 2.303 \frac{aL}{At} . \log \frac{h_1}{h_2} \tag{1}$$

$$k = 2.303 \frac{aL}{2A.t} . \log \frac{h_1}{h_2} \tag{2}$$

$$k_{20} = R_T \times k \tag{3}$$

where k—hydraulic conductivity at a room temperature (m/s), L—length of specimen (m), A—cross sectional area of specimen (m^2), a—cross sectional area of head-in pipe (m^2), ($a_{in} = a_{out} = a$), $t = t_1 - t_2$—reading duration (seconds) at head-in of h_1 and h_2, h_1—head loss in at reading time t_1 (m), h_2—head loss in at reading time t_2 (m), R_T—converting ratio (ASTM D5084).

3 Experimental Results and Discussions

Soilcrete specimens made from dredging sand mixing cement with or without bentonite were carried out the permeability tests. The hydraulic conductivity data were carefully obtained and analyzed. Relations between measured hydraulic conductivity varying with time, bentonite content, hydraulic gradient were presented and discussed in the following.

3.1 Hydraulic Conductivity of Unmixed Sand and Soilcretes

The hydraulic conductivity of the untreated sand specimen, bentonite soilcrete specimens made from the dredging sand at a cement content of 300 kg/m^3, and bentonite contents of 0, 25, 50, 75, and 100 kg/m^3, respectively, at an age of 28 days are displayed in Fig. 2. The results indicate that k_s of sand mixing with a cement content of 300 kg/m^3 reduced significantly from 10^3 to 10^4 times comparing to the unmixed sand, and to decrease more by adding bentonite. Alkaya and Esener [1] and Iravania [9] reported the similar conclusions. For the sand mixing cement, the cement hydration reactions produced calcium-silicate-hydrate (CSH) gel products filling the space between sand particles, reducing porosity, leading to reduction of k_s. The sand mixing

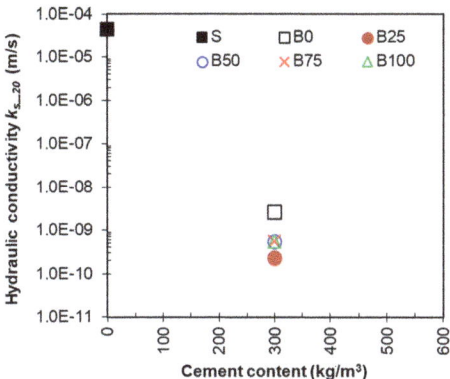

Fig. 2 k_{soil} of the compacted sand and k_s of soilcrete specimens

cement and bentonite has the two key chemical reactions which are (1) The hydration reactions producing CSH and Ca(OH)$_2$ and (2) the pozzolanic reactions between Ca^{2+} and pozzolan (SiO$_2$ and Al$_2$O$_3$). The pozzolanic reactions generate calcium-aluminate-hydrate (CAH), calcium-silicate-hydrate (CSH), and calcium-aluminum–silicate-hydrate (CASH). These products were in the form of gels and quickly occupy void spaces in a specimen. Therefore, k_s of the bentonite soilcrete specimen was lower than that of soilcrete specimen [9].

3.2 k_s *Versus the Curing Time*

Hydraulic conductivity of all soilcrete specimens created from dredging sand, cement, and bentonite reduced sharply up to 60% for the first two weeks and then declined gradually (Fig. 3). This result is similar to other researchers [8, 17]. Hydration of cement and ion exchanges between Ca^{2+} and pozzolan in bentonite took much time to complete [10], cementitious products from these reactions formed continually to fill the soilcrete pores [9]. Consequently, the permeability of bentonite soilcrete decreased with increasing in curing times.

3.3 k_s *Versus the Hydraulic Gradient*

Figure 4 presents the relationship between the hydraulic conductivity of two bentonite soilcrete specimens (B25, B25b) and hydraulic gradient values. Both specimens were created from dredging sand, a cement content of 300 kg/m^3 and a bentonite content of 25 kg/m^3. Hydraulic gradients of 40 and 132 were applied to the B25 and B25b during permeability tests, respectively. Figure 4 showed k_s of soilcrete apparently

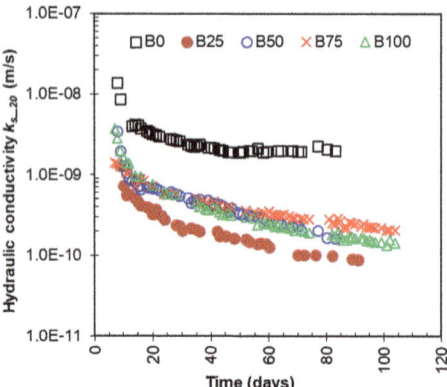

Fig. 3 k_s versus curing times

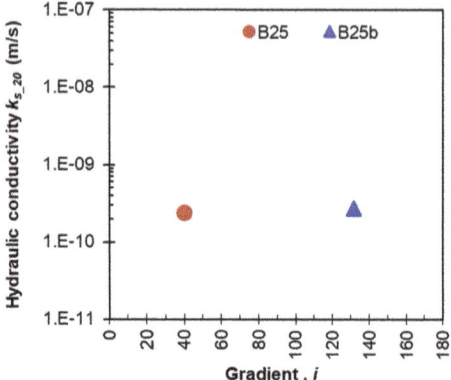

Fig. 4 k_s versus hydraulic gradients

independent of the hydraulic gradient. The similar findings were reported by other authors [7, 14].

3.4 k_s *Versus Bentonite Contents*

Effect of bentonite contents on hydraulic conductivity was evaluated by comparing k_s of all soilcrete specimens made from sand mixing cement content of 300 kg/m³ and bentonite contents of 25, 50, 75, and 100 kg/m³, respectively. Figue 5 shows that k_s of bentonite soilcrete specimens were lower than k_s of soilcrete specimen at the same cement content, k_s at a bentonite content of 25 kg/m³ was lower than those at higher bentonite contents of 50, 75, and 100 kg/m³, respectively. The result agrees well with other researchers [1, 9]. Bentonite has high specific surface area which

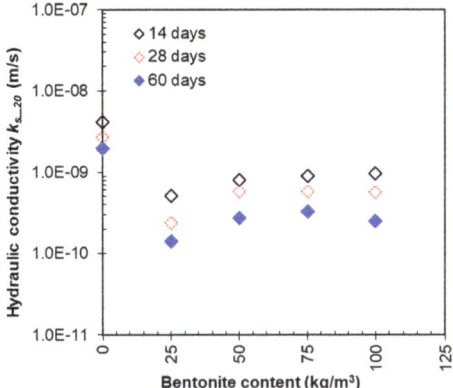

Fig. 5 k_s versus bentonite contents

allows to absorb more water and minimizes movement of water flow [1]. Besides, the reactions among pozzolan in bentonite and Ca^{2+} ion from the hydration reactions created gel products to lead less void spaces in soilcrete specimen to cause reduction of hydraulic conductivity [13].

Further, k_s of specimens at higher bentonite contents of 50, 75, and 100 kg/m³, respectively, increased fairly. The amount of bentonite was excessive the optimum content. Free bentonite particles cause swelling and form gels around sand grains leading to increase in the size of aggregate and the volume of voids in a specimen [16].

4 Conclusions

Several soilcrete and bentonite soilcrete specimens were molded in the laboratory from the dredging sand taken in Dong Thap province mixing with a cement content of 300 kg/m³ and various bentonite contents of 25, 50, 75, and 100 kg/m³, respectively. The hydraulic conductivity of the all specimens was investigated by the falling head-constant tailwater and the falling head-rising tailwater methods on flexible wall permeameters. The rigid wall permeameter was used to determine the hydraulic conductivity of the compacted sand specimen. The results suggest the following findings.

1. The hydraulic conductivity of soilcrete was lower than that of untreated soil from 10^3 to 10^4 times.
2. The hydraulic conductivity of bentonite soilcrete was lower than that of soilcrete. However, the hydraulic conductivity of bentonite soilcretes increasing with higher bentonite content.
3. The hydraulic conductivity of all soilcrete decreasing with curing time.

4. The hydraulic conductivity of bentonite soilcrete was independent of the hydraulic gradient (40–132).
5. The hydraulic conductivity of bentonite soilcrete from 4.86×10^{-9} m/s to 1×10^{-10} m/s.

Acknowledgements This research is funded by Vietnam National University in Ho Chi Minh City (VNU-HCM) with the grand No. B2018-20-04 and managed by Ho Chi Minh City University of Technology (HCMUT). The authors acknowledge the significant supports.

References

1. Alkaya D, Esener AB (2011) Usability of sand-bentonite-cement mixture in the construction of unpermeable layer. Sci Res Essays 6(21):4492–4503
2. Ameta NK, Wayal AS (2008) Effect of Bentonite on Permeability of Dune Sand. Electron J Geotech Eng 13(A):8
3. American Society for Testing and Materials (1998) Standard test method for laboratory compaction characteristics of soil using standard effort (12,400 ft-lbf/ft^3 (600 kN-m/m^3). ASTM D 698, 8 p
4. American Society for Testing and Materials (1995) Standard test method for measurement of hydraulic conductivity of porous material using a rigid wall, compaction mold permeameter. ASTM D5856, 8 p
5. American Society for Testing and Materials (2010) Standard test method for measurement of hydraulic conductivity of saturated porous material using a flexible wall permeameter. ASTM D5084, 24 p
6. Cowland JW, Leung BN (1991) A field trial of a bentonite landfill liner. Waste Manage Res 9(1):277–291
7. Gueddouda MK, Lamara M, Abou-bekr N, Taibi S (2010) Hydraulic behaviour of dune sand-bentonite mixtures under confining stress. Geomech Eng 2(3):213–227
8. Helson O, Eslami J, Beaucour A, Noumowe A, Gotteland P (2018) Hydro-mechanical behaviour of soilcrete through a parametric laboratory study. Constr Build Mater 166:657–667
9. Iravania A. (2015) Hydro-mechanical properties of compacted sand-bentonite mixtures enhanced with cement. Ph.D. thesis, Eastern Mediterranean University
10. Kamruzzaman AHM (2002) Physico-chemical and engineering of cement treated Singapore marine clay. M.E. Thesis, National University of Singapore
11. Kenney T, Van-Veen WA, Swallow MA, Sungaila MA (1992) Hydraulic conductivity of compacted bentonite-sand mixtures. Can Geotech J 29(3):364–374
12. Martirosyan V, Yamukyan M (2018) Comparative study of behaviour of soil and soil-bentonite mixtures for the construction of impermeable barriers. Int J Sci Eng Res 2(3):12–21
13. Nontananandh S, Yoobanpot T, Boonyong S (2005) Scanning electron microscopic investigation of cement stabilized soil. In: Proceedings of 10th national conference on civil engineering, Chonburi-Thailand, pp. 23–26
14. Picandet V, Rangeard D, Perrot A, Lecompte T (2011) Permeability measurement of fresh cement paste. Cem Concr Res 41:330–338
15. Sällfors G, Öberg-Högsta AL (2002) Determination of hydraulic conductivity of sand-bentonite mixtures for engineering purposes. Geotech Geol Eng 20(1):65–80
16. Taha OME, Taha MR (2007) Volume change and hydraulic conductivity of soil-bentonite mixture. Jordan J Civ Eng 9(1):43–58

17. Tran-Nguyen HH, Nguyen KTD, Nguyen TT (2020) Permeability of soilcrete specimens made from the mekong delta's soft clay mixed with cement slurry. Geo-congress 2020:751–758
18. Xu S, Wang Z, Zhang Y (2011) Study on the hydraulic conductivity of sand-bentonite mixtures used as liner system of waste landfill. Adv Mater Res 194–196:909–912

Performance Analysis of Pile Under Negative Skin Friction by Load-Transfer Method

Van Qui Lai, Suraparb Keawsawasvong, Thanh Hai Do, Quoc Thien Huynh, Quoc Viet Tran, and Huu Thoi Tra

Abstract This paper presented a performance analysis of pile under negative skin friction by load-transfer method. In the analysis, the nonlinear t-z curves were applied to described the relationship between mobilized load and displacement at pile shaft and pile tip. A new iterative algorithm was proposed to determine the behavior of the pile. The analysis results were compared with the unified method. The results showed the proposed load-transfer method can be correctly performed the behavior of the piles more than the other.

Keywords Load-transfer · Negative skin friction · Axially loaded pile

1 Introduction

Piles is the option for the foundation plan of high-rise building, special in areas having thickness of soft soil layer. In research as well as in application, issues of pile and pile group are pile capacity, pile settlement, working of pile group, etc. [1, 2]. In that, negative skin friction is the big issues in predicting pile performance. There has many researches about negative skin friction [3, 4]. However, the previous works rarely focus on effect of applied load on negative skin friction. In Viet Nam, the practical engineering usually uses the comments from TCVN 10304–2014 and Unify method [5] to analysis pile considering negative skin friction.

V. Q. Lai (✉) · T. H. Do · Q. V. Tran
Faculty of Civil Engineering, Ho Chi Minh City University of Technology (HCMUT), 268 Ly Thuong Kiet Street, District 10, Ho Chi Minh, Vietnam
e-mail: lvqui@hcmut.edu.vn

Vietnam National University Ho Chi Minh City, Linh Trung Ward, Thu Duc, Ho Chi Minh, Vietnam

S. Keawsawasvong
Department of Civil Engineering, Thammasat School of Engineering, Thammasat University, Pathumthani 12120, Thailand

Q. T. Huynh · H. T. Tra
Institute of Research and Development, Duy Tan University, Danang 550000, Vietnam

© The Author(s), under exclusive license to Springer Nature Singapore Pte Ltd. 2023   633
J. N. Reddy et al. (eds.), *ICSCEA 2021*, Lecture Notes in Civil Engineering 268,
https://doi.org/10.1007/978-981-19-3303-5_56

Load-transfer methods were so successful in predicting the pile performance [6, 7]. For instance, load-distribution curve at pile head, or load distribution along pile length. Besides, load-transfer could be applied to analysis pile under negative skin friction. However, there rarely focused on negative skin friction issues. Thus, aim of this paper presented a load-transfer method in analysing pile under negative skin friction. Investigation of the effect of applied load on pile under negative skin friction was made. The results were compared to Unified method [5]

2 Theory and methodology

2.1 Negative Skin Friction in Pile Analysis

In general, pile is moved down to the ground under axially load and the pile usually settles larger than settlement of surrounding soil. Because of this movement, the friction between the pile shaft and the soil surrounding is occurred in the upward direction. However, in some cases, the soil settles more than settlement of pile and there has friction that occurred on pile shaft in the downward direction. This frictional drag is known as the negative skin friction, as shown in Fig. 1. The negative skin friction adds more load on the pile and reduces the pile bearing capacity. There are many reasons that lead to the occasion of the negative skin friction. The usual reason is that the piles installed in soft ground. However, the soft ground is greatly consolidated by high back-fill layer or reduce underground water.

According to the experiment results in a research [8], Fig. 2 shows that the axial force on the pile would be raise along with the depth and start decreasing from a

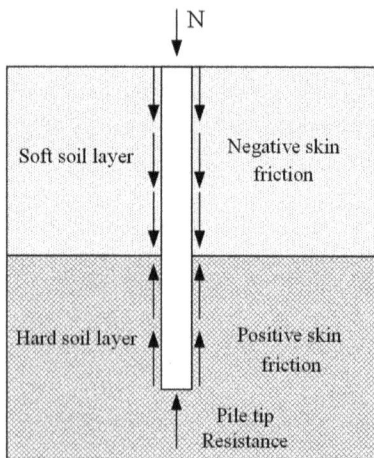

Fig. 1 The occasion of negative skin friction surrounding the pile

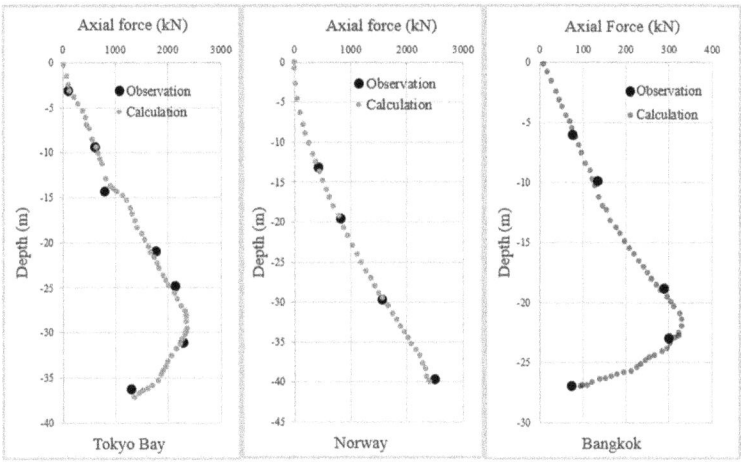

Fig. 2 Distribution of negative skin friction with depth [8]

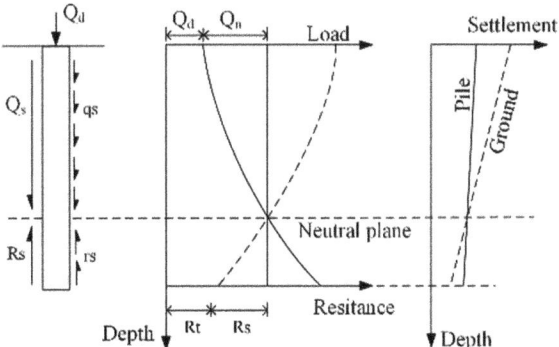

Fig. 3 Determination the neutral plan based on the principle of force balance

special position to the pile's base. The plane that crosses that special position is known as a "neutral plane". This plane divides the ground into the positive frictional zone and negative frictional zone. The neutral plane is also the position that the settlement of the pile and the ground are equalled. In another way, the neutral plane is also the equilibrium position between upper load and lower load (Fig. 3).

2.2 Unified Method

Fellenius [5] proposed the method, unified method, to determine the negative skin friction pile by using the principle of equilibrium, as shown in Fig. 3. Whereas, R_t and R_s is the ultimate bearing capacity of the base pile and the shaft pile sequentially.

Q_{ult} is the total ultimate bearing capacity and determined by Eq. (1):

$$Q_{ult} = Q_u = R_s + R_t \tag{1}$$

When the ultimate pile base resistance is mobilized, the pile axial force Q_z is determined by Eq. (2) and showed by the dashed line in Fig. 3

$$Q_z = Q_{ult} - \int A_s r_s dz \tag{2}$$

where, A_s is the pile's shaft area, r_s is the unit positive friction.

The Fig. 3 presented that when the negative skin friction takes place, the axial force distributed from the pile top to the neutral plane could be calculated by the Eq. (3) and the axial force distributed from the neutral plane to the pile base could be calculated by the Eq. (4). Based on the principle of continuity of force, the value of the axial forces at the neutral plane were equivalent. In other words, the sum of the axial forces above and below the neutral plane were balanced.

$$Q_z = Q_d + \int A_s q_n dz \tag{3}$$

$$Q_z = R_t + \int A_s r_s dz \tag{4}$$

where, q_n is the unit negative friction.

2.3   Load-Transfer Method

The load-transfer method firstly suggested by Seed and Reese [9]. The pile was divided into many element piles. The interaction between the pile and the soil was simulated by the interaction of the small element piles and the ground. This thing was based on the simple relationship between the mobilizing reaction and the displacement of the ground surrounding the pile (this relationship is known as the t-z curve). Principles of load-transfer method was shown in Fig. 4

There are many suggestions for the equation of the t-z curve [10]. According to some recent study of Tirawat and Lai [6], the Eq. (5) and (6) were used to analyse negative skin friction by load-transfer method.

$$\tau_{,i} = a_{,i}\left(1 - e^{-b_{,i}.w_{,i}}\right) \tag{5}$$

$$P^b_{,n} = a^b\left(1 - e^{-b^b.w_{,n}}\right) \tag{6}$$

where: $a_{,i}$; $b_{,i}$; a^b; b^b - the load-transfer curve's parameters in the i element pile; $w_{,i}$ -the displacement of the pile or the ground surrounding the i element pile;

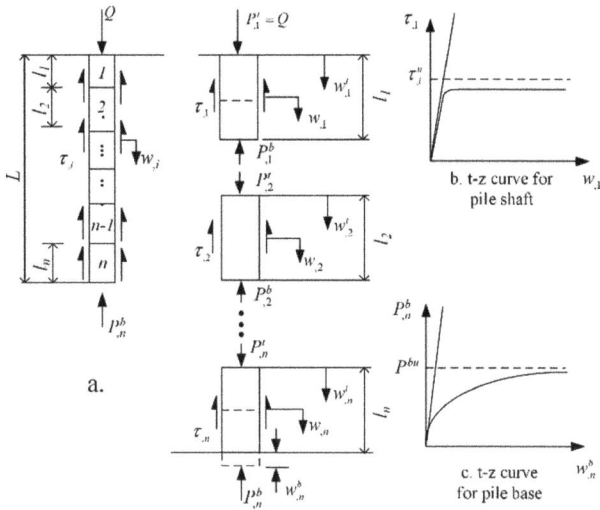

Fig. 4 Concept of load—transfer method; **a** pile segments; **b** t-z for pile shaft; **c** t-z for pile base

The meaning of $a_{,i}$, $b_{,i}$ are shown in Fig. 4b. Where $a_{,i}$ is the maximum unit friction in the i element pile, $a_{,i}.b_{,i}$ are the initial slope of the load-transfer curve. Similarly, the relationship between the mobilized load and the displacement at pile base, including two parameter a^b, b^b, was showed in Fig. 4c.

In the first studies, the $a_{,i}$;$b_{,i}$; a^b; b^b, parameters of the load-transfer curves (t-z curves) were specified from the experiment. However, based on the proposed method of Randolph and Worth [11], these parameters can be determined:

$$a_{,i} = \frac{\tau_{,i}^f}{R} \tag{7}$$

$$b_{,i} = \frac{G_{s,i}}{a_{,i} r_0 ln\left(\frac{r_m}{r_0}\right)} \tag{8}$$

$$a^b = \frac{P^{bf}}{R} \tag{9}$$

$$b^b = \frac{4G_{sb}r_0}{a^b(1 - \vartheta_{sb})} \tag{10}$$

Where: r_0—the radius of the circular pile; r_m—the distance from the pile to the zone that was not affected by the pile's settlement; $G_{s,i}$, G_{sb}—shear modulus of the soil surrounding the pile and below the pile base.; $\tau_{,i}^f$—the ultimate unit friction of the i element pile; P^{bf}—the mobilized force at the pile base; R—the coefficient,

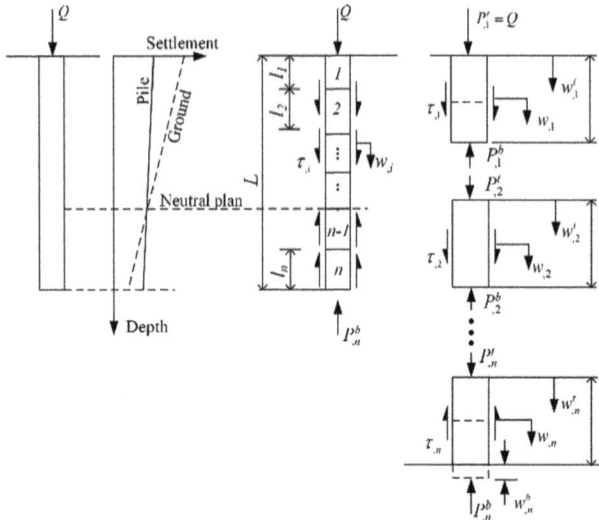

Fig. 5 The concept of load—transfer method in analysing negative skin friction

which value 0.85–0.9. The value of $\tau_{,i}^f, P^{bf}$ can be calculated by some formulas proposed by Tirawat and Lai [6]

Based on the negative skin friction theory, when the settlement of the ground is greater than the pile, the top part of the pile creates an opposite direction of friction and that force's value is adequate to the positive friction, Fig. 5. So, the behaviour at the pile shaft determined by a load-transfer method is completely suitable to describe the development of the negative skin friction at the pile shaft. The Fig. 6 presents an algorithm of the proposed load-transfer method into the analysis of the negative skin friction of a vertical load applied pile.

3 Verification

A hypothesis cases referred to a real project in District 6, Ho Chi Minh City, Viet Nam, was used for analysis. The geology of the studied case, including three soil layers, was shown in Fig. 7. In that, Layer 1—Soft clay had the thickness of 10.5 m; Layer 2—Hard clay had the thickness of 6 m; Layer 3—Stiff clay had the thickness of 9.8 m. The concrete pile with diameter $D = 0.4$ m, pile length 20 m (including 1 section 11 and 1 section 9 m). Elastic module of concrete is 2.65×10^6 ton/m². The elevation of the top of the pile is $+ 0.000$ m, the groundwater level is -0.700 m. X is the thickness of the embankment layer.

For verification, the paper focused on analysing pile loaded by difference value and surrounding soil consolidated by the thickness of back-fill layer. The cases of analysis were summarized in Table 1. It is noted that, to applied the proposed load-

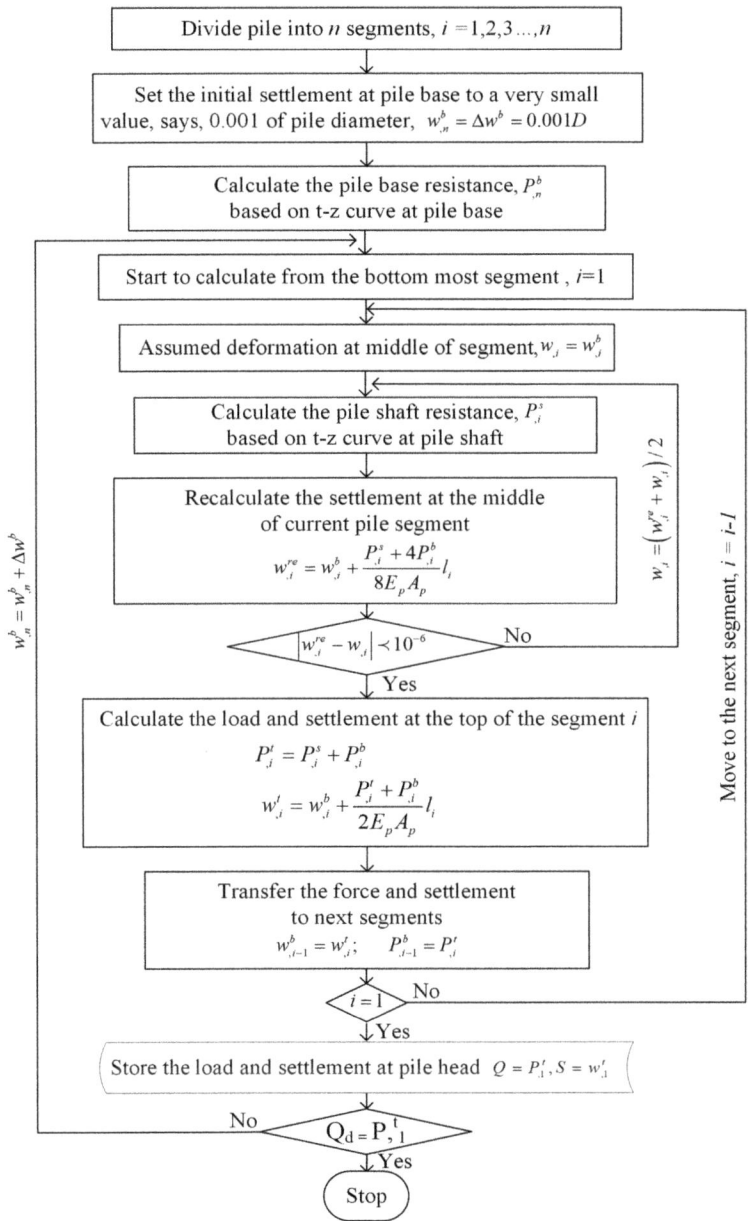

Fig. 6 Algorithm to analysis the negative friction using load—transfer method

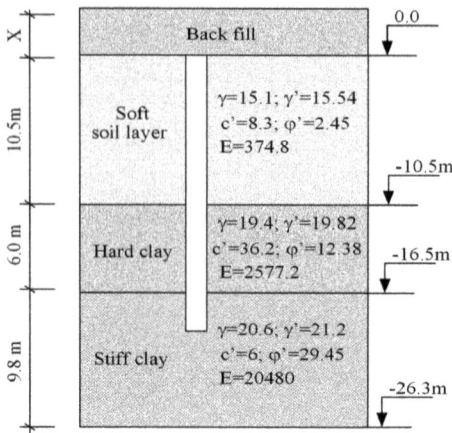

Fig. 7 Geology of studied cases

Table 1 The studied cases

Cases	Case 1	Case 2	Case 3
Applied load on pile top	0 (kN)	50 (ton)	100 (ton)
Thickness of back-fill (X)	1 m		

transfer method, the consolidation settlement of ground was firstly determined. In this study, the consolidation settlement of ground can be determined by below equation

$$U_z = \beta \sum \frac{p_i h_i}{E_i} \qquad (11)$$

Where: U_z—the consolidation settlement, β consolidation coefficient ($\beta = 0.8$), h_i—the thickness of layer i, E_i—the young modulus of layer.

The results were showed in Figs. 8 and 9. The Fig. 8 showed the settlement on the pile head for studied cases. It can be seen that the predicted settlements of proposed load-transfer method was little larger than those from unified method. In Fig. 9, the load distribution with depth was investigated. In case a, b there had a huge difference between the proposed method and unified method. In case c, the results between two methods were the same. In these cases, the proposed method can be performed more correctly than unified method. Because of with small applied load on pile head, the resistance in the bottom pile still not be mobilized. And the unified method used the full mobilized resistance at pile base was not suitable. This thing also explained for the small difference between results showed in Fig. 8. Besides that, from the Fig. 9, it can be seen that, the neutral plan was moved upper when applied load on pile head was increased. This thing can be explained that when applied load was increased, the settlement of pile was increased. So that, the neutral plane where the pile settlement and the ground settlement equal, was moved up to the ground surface.

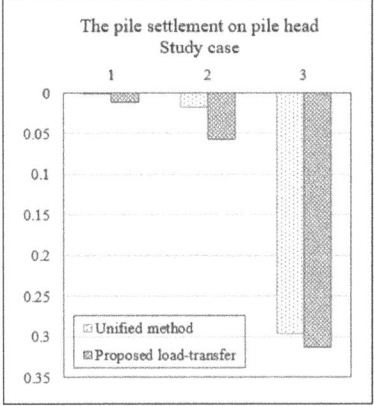

Fig. 8 Settlements of piles in studied cases

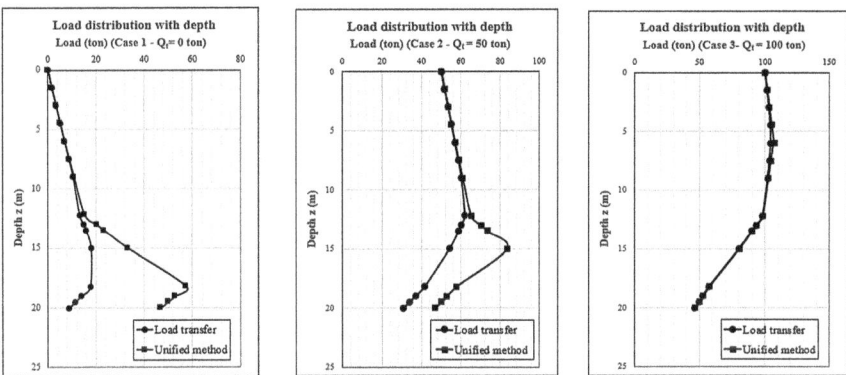

Fig. 9 Load distribution with depth under applied load and negative skin friction

4 Conclusion

The paper presented a proposed load-transfer method in analysing the pile under negative skin friction. By considering the effect of applied load on pile head, the analysis results from proposed method were compared with the results from the unified method. The comparison showed that, the proposed method can be successful performed the behaviour of pile under negative skin friction. Furthermore, the proposed method was better than unified approach in performing the pile under small applied load on pile head and negative skin friction.

Acknowledgements We would like to thank Ho Chi Minh City University of Technology (HCMUT), VNU-HCM for the support of time and facilities for this study.

References

1. Memar M, Seyed MAZ, Amir HV (2020) Effect of pile cross-section shape on pile group behaviour under lateral loading in sand. Int J Phys Model Geotech 20(5):308–319
2. Nguyen TD, Lai VQ, Phung DL, Duong TP (2019) Shaft resistance of shaft-grouted bored piles and barrettes recently constructed in Ho Chi Minh city. Geotech Eng J SEAGS & AGSSEA 50(3):155–162
3. Terzaghi K (1967) Remarks made during the opening session at the institution of civil engineers, from theory to pratice in soil mechanics. Wiley, New York
4. Peck DF (1976) Operant conditioning and physical rehabilitation. Eur J Behav Anal Modif 3(1):158–164
5. Fellenius BH (1989) Unified design of piles and pile groups. Transp Res Rec 1169:75–82
6. Boonyatee T, Lai VQ (2017) A non-linear load transfer method for determining the settlement of piles under vertical loading. Int J Geotech Eng 14(1):1–12
7. Lai VQ, Huynh QT, Do TH, Nguyen TG (2020) Performance analysis of axially loaded piles by load transfer method: a case study in Ho Chi Minh city. Lect Notes Civ Eng 80:757–766
8. Indraratna B, Balasubramaniam AS, Phamvan P, Wong YK (1992) Development of negative skin friction on driven piles in soft Bangkok clay. Can Geotech J 29(3):393–404
9. Seed H, Reese L (1957) The action of clay along friction piles. J Geoteh Engng 504:92
10. Fellenius BH (2013) Discussion of "A simplified nonlinear approach for single pile settlement analysis." Can Geotech J 50(6):685–687
11 Randolph MF Wroth CP (1978) Analysis of deformation of vertically loaded Piles. J Geotech Eng Div ASCE 104(12)

Solutions to Reduce Negative Effect of Saline Soil Cement Column in Ben Tre province

Thanh Hai Do and Van Qui Lai

Abstract It is normally known that a negative effect on the strength characteristics of soil cement sample. This paper focuses on the effect of salt and bottom ash on the strength development of saline soil cement mixing in Ben Tre province. The saline soil cement samples (SSC) in this study consisted of saline soil, cement and bottom ash for recycling dredged soil in this area. Several pairs of samples with various bottom ash contents (i.e., 0, 50, and 100%) to have specimens of non-salt and salt-rich dredged soil with symbol as SSC-N and SSC-S, respectively. In this study, several series of unconfined compression test were carried out with curing time increase. It is found that bottom ash content decrease the negative effect in soil cement samples. In case of without bottom ash, at 7 and 28 days of curing time, the strength of SSC-N has approximately higher than that of SSC-S. It is found that the percent of strength reduction due to salt concentration is approximately 23% at 7 days and 11% at 28 days of curing time. It is worth to note that inclusion of bottom ash into soil admixture gives benefits of increasing shear strength due to pozzolanic reaction. This application is not environmentally sound, and societal demand to recycle waste material in sustainable development in Ben Tre province.

Keywords Saline soil · Strength characteristics · Bottom ash · Ben Tre province

1 Introduction

Nowadays, soil cement is popular using for improving the soft soil in Viet Nam. In practice, the dredged soil has been dumped in waste disposal sites in the sea. This practice, however, is not environmentally friendly, and therefore there has been increasing social demands to reuse the dredged soil in construction projects. However,

T. H. Do (✉) · V. Q. Lai
Faculty of Civil Engineering, Ho Chi Minh City University of Technology (HCMUT), 268 Ly Thuong Kiet Street, District 10, Ho Chi Minh City, Vietnam
e-mail: dthai@hcmut.edu.vn

Vietnam National University Ho Chi Minh City, Linh Trung Ward, Thu Duc District, Ho Chi Minh City, Vietnam

© The Author(s), under exclusive license to Springer Nature Singapore Pte Ltd. 2023
J. N. Reddy et al. (eds.), *ICSCEA 2021*, Lecture Notes in Civil Engineering 268,
https://doi.org/10.1007/978-981-19-3303-5_57

the mixture always have negative effect on strength development because of high salt concentration [7].

Numerous studies have been carried out on the mechanical properties of soil–cement mixture, the effects of additives, the behavior of single soil–cement columns, and so forth. Lin [8] reported that water/cement ratio had a great effect on the bearing capacity of the composite foundation through the unconfined compression tests of soil–cement columns in laboratory. Feng et al. [3] studied the characteristics of soft mud treated by cement and found that the treated soft mud with small cement content can effectively reduce settlements. By X-ray diffraction analysis, Mohamed [9] explained main causes for deterioration of marly soils and discovered more palygorskite in marly soils when stabilized with lime or cement. Wang et al. [13] performed a quantitative analysis of microstructures in soil–cement through image processing techniques. Gu et al. [4] investigated structural characteristics of cemented soils in a quantitative way. Ning et al. [10] carried out laboratory tests to analyze the effects of environmental factors, such as water, chemical solution, and PH value, on mechanical properties of soil–cement. However, most researchers have focused on behavior of common soft soils improved by cement while a few research has been performed on marine soft soils. Alqasimi [1] studied the proportion of soil cement mixing for marine sandy soil. Pei and Wu [11] proposed methods to select additive agents for marine soft soils treated by cement through laboratory tests. Chew et al. [2] evaluated the properties of marine soils improved by cement through electronic microscopy scanning and X-ray diffraction tests. Xu et al. [12] conducted field tests to investigate the behavior of soil–cement columns in marine soft soils. Stabilized dredged soil mixed with cement, bottom ash had a good advantage not only for recycling bottom ash but also for improving its strength characteristics. In these compositions cement is always added to enhance the strength, but the effect of salt concentration on the strength development is not investigated.

This paper focuses on the effect of salt and bottom ash on the strength development of saline soil cement mixing. The saline soil cement samples (SSC) in this study consisted of saline soil, cement and bottom ash. dredged soil is leached firstly by washing method, and then mixed with various bottom ash contents (i.e., 0, 50, and 100%) to have several pairs of non-salt and salt-rich dredged soil with symbol as SSC-N and SSC-S, respectively. Unconfined compression strengths of these specimens were determined under different curing times.. In this study, in order to investigate the saline effects of dredged soil on the strength and stiffness, several series of unconfined compression test were carried out on the specimens with various mix ratios and curing times.

2 Materials

The geotechnical properties of the dredged soil, taken from the construction site of Ben Tre province, are shown in Table 1. The natural water content of the dredged soil is 78.5% and its plasticity index about 36.7. The dredged soft clay is mostly

Table 1 Properties of dredged soil

Water content (%)	Liquid limit (%)	Plastic limit (%)	Specific gravity	USCS	Salinity (g/l)
78.5	62.0	25.3	2.70	CL	15

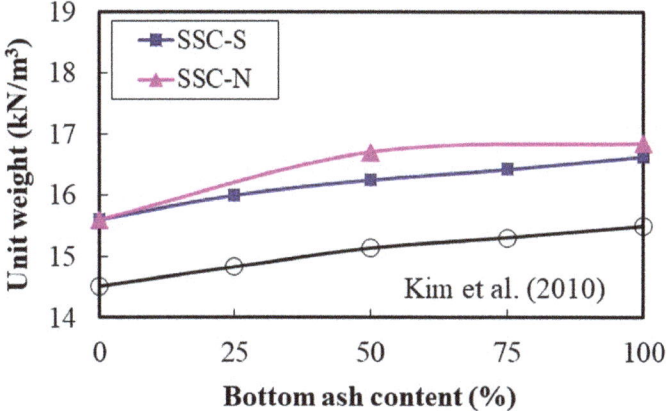

Fig. 1 Bulk unit weight with various bottom ash contents and curing times

classified as CL according to the Unified Soil Classification System. Initial salinity of dredged soil is 22.5 (g/l). Dredged soil is washed several times until the salinity decrease to almost 0.0 (g/l,) so called non-salt dredged soil.

As an additive agent, ordinary Portland cement was used in this study and its specific gravity is 3.15. Bottom ash from Pha Lai power plant in Hai Duong province was selected for use in this study. The specific gravity obtained for bottom ash is 2.1. Bottom ash used for this study is composed of particle sizes ranging from 0.022 to 4.75 mm (passing US standard #4 sieve) as in Fig. 1. It appears to be a poorly graded sand-like material. Bottom ashes were analyzed for chemical composition via X-ray fluorescence spectrometer. The bottom ash used contains nearly 59.2% SiO_2, 21.5% Al_2O_3 and 4.73% CaO, respectively. With the removal of particles larger than 4.75 mm, a large surface area is available to react with cement additive.

3 Experimental Program

To evaluate the effect of bottom ash on the strength development of mixture, test specimens were prepared at different percentages of bottom ash content (i.e., 0 to 100%) while water content and cement content are fixed as shown in Table 2.

The mixing machine is used need to calibate carefully before conducting. The mixing procedure is followed Vietnamese Standard TCVN 9403:2012. This fuction

Table 2 Mixing and testing conditions

Mixtures	Parameters	Content[a] (%)
	Cement	12
	Water	120
SSC-S (salt-rich dredged soil)	BA#4	0, 25,50,75, 100
SSC-N (non-salt dredged soil)	BA#4	0, 50, 100
	Curing time (days)	7, 28

[a] Content is presented by weight in relation to the dry weight of the dredged soil

can be chosen in this machine with manual procedure in order to have workability mixture.

With this water content, the mixture has a suitable workability with the flow value in range of 20 ± 5 cm [5], each specimen can be poured into PVC mold with the 50 mm diameter and 100 mm height without compaction effort. The mold together with specimen was wrapped to prevent moisture loss, and then, placed for curing inside the room having a temperature of $20 \pm 5°C$ for 7 and 28 days of curing time as shown in Table 2. After curing, each specimen was removed from its mold, and then subjected to unconfined compression test with speed of 1.0 mm/min.

4   Experimental Results

4.1   Physical Properties

Figure 1 shows the plot of unit weight versus bottom ash content and curing time. The bottom ash content of 0, 25, 50, 75 and 100% and curing times of 7, 28, and 60 days were the main conditions set out during the specimen preparation. It is demonstrated that the unit weight increase with the increasing bottom ash content and slightly increase with curing time. For a certain curing time, the reasonable explanation could be given due to the addition of bottom ash with a lot of particle size, which could be attributed to the increasing amount of cementing products, being formed.

Figure 2 shows the plot of the post-curing water content, which is the ratio of the weight of water to the weight of solid after 28 days of curing, versus bottom ash content. The water content of SSC-S has higher value than those of SSC-N. The reason that the water content decreases with increasing bottom ash content is the increased water consumption due to the increasing amount of cementing products resulting from the pozzolanic reaction (eventually increased bulk unit weight as shown in Fig. 1). Thus, the non-salt dredged soil has higher hydration and pozzolanic reaction than salt-rich dredged soil in mixtures. These results show the same trends with the results of Kim et al. [5] but SSC mixtures have higher value of water content at a given bottom ash content due to without air foam addition.

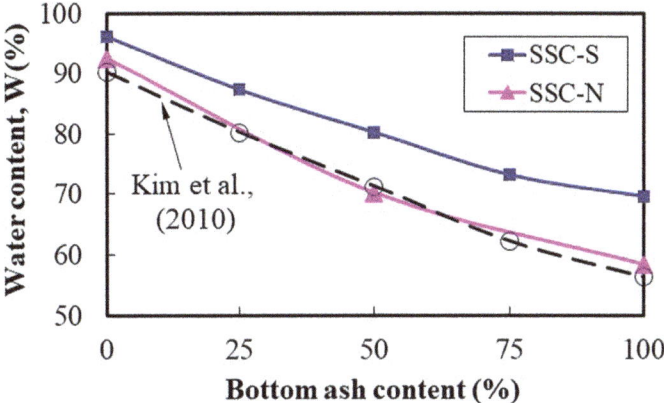

Fig. 2 Water content changes of mixtures of SSC-S and SSC-N

The initial void ratio of mixture after curing can be expressed as follows:

$$e_{0m} = \frac{G_{s,m}(1 + W_m)\gamma_w}{\gamma_m} - 1 \tag{1}$$

Where Gs,m is the specific gravity of mixture, Wm is the water content of the mixture after curing, γ_m is the unit weight of the mixture after curing.

Post-curing initial void ratio significantly decreases with the increasing bottom ash content at 28 days in Fig. 3. For the all results, SSC-N always have smaller value of post-curing initial void ratio because the more ettringite is formed in these mixture.

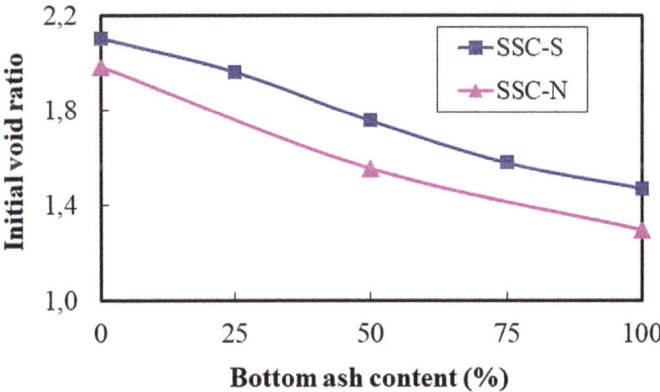

Fig. 3 Void ratio of mixtures of SSC-S and SSC-N

4.2   Unconfined Compression Strength

The stress-strain behaviors of samples cured at 7 and 28 days are investigated after testing. In these results, it was shown that the same shape correspond to specimens with the same bottom ash content, and also the same trend to all specimens. As obtained in these results, the maximum compressive strength of sample increases with increase in bottom ash content. These results are similar to results of Kim et al. [5]. The increase in shear strength due to addition of bottom ash can be explained mainly by two mechanisms. Firstly, the development of friction between granular materials in the soil mixture can mobilize better resistance against shear. Secondly, the pozzolanic reaction in the mixture can cause more bond strength.

Unconfined compressive strengths of samples with various bottom ash contents at 7 and 28 days of curing time are presented in Fig. 4. This result indicates that the addition of bottom ash can reduce the effect of salt on cement-soil mixtures because the high content of $SiO_2 + Al_2O_3$ ($59.2 + 21.5 = 80.7\%$) increases the pozzolanic reaction, reduce the effect of salt on mixing with cement. It is shown that the effect of salt and bottom ash on the strength of composite material without bottom ash content of 100%. It is found that the percent of strength reduction due to salt concentration is approximately 23% at 7 days and 11% at 28 days of curing time.

It is found that modulus of elasticity was developed similarly to strength development for the addition of bottom ash in Fig. 5. The results of this study present that $E_{50} = (57-120)q_u$, regardless of bottom ash contents and curing time. The E_{50}/q_u is larger than reported value in soil mixing with air foam, which ranged from 40 to 80 [5]. This larger range indicates that the secant modulus of soil mixing can be predicted using E_{50}/q_u ratio obtained from the test have some effect of salinity.

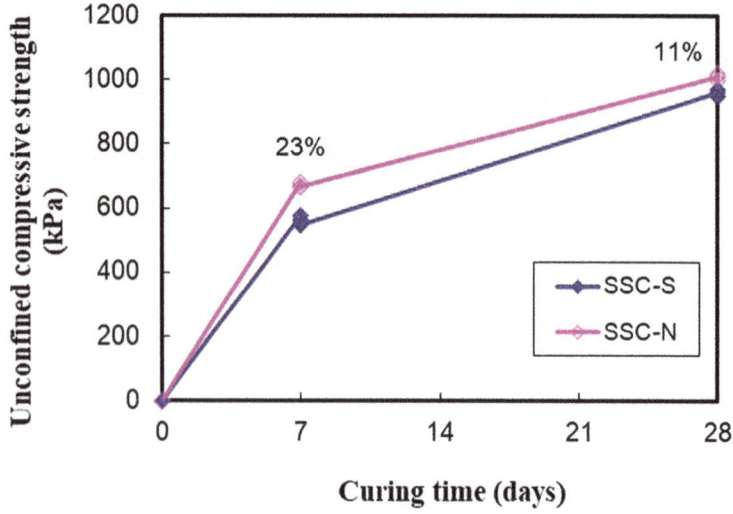

Fig. 4 Strength development of SSC-S and SSC-N with bottom ash content of 100%

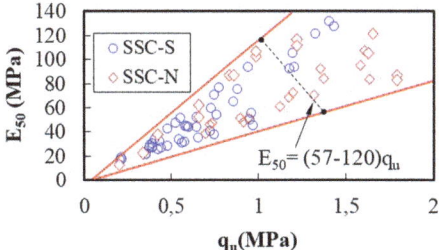

Fig. 5 Relationship between the secant modulus and unconfined compressive strength

However, the unconfined compressive strength (q_u) of SSC-S samples is smaller than SSC-N with same cement content. Thus, the same effect to modulus behaviour.

5 Conclusions

This paper investigated the effects of bottom ash on the strength development of dredged soils mixing with cement and bottom ash. Several kinds of SSC specimens were prepared with various bottom ash contents with non-salt and salt-rich dredged soil. Some conclusion can be drawn as:

- The unit weight of SSC specimens increases as bottom ash content increases in salt-rich stabilized dredged soil. However, water content, and initial void ratio decrease as bottom ash content increases.
- It is found that salt concentration of clayey soil contributes not only to change the physical properties of cement-stabilized soil but also to reduce significantly the strength of mixture. The improvement of bond strength is effective only on secant modulus. And the ratios of secant modulus to unconfined compressive strength of three mixtures are almost same, and in range of (46–100), regardless of mixing condition and curing time.
- It is found that the percent of strength reduction due to salt concentration is approximately 23% at 7 days and 11% at 28 days of curing time. It is very valuable to used waste bottom ash in reducing the negative effect on strength development of salt-rich cement soil. This application is not environmentally sound, and societal demand to recycle waste material in sustainable development.

Acknowledgments This research is funded by Vietnam National University Ho Chi Minh City (VNU-HCM) under grant number **C2021-20-40**

References

1. Alqasimi AM (1993) Improving engineering properties of Sabkha soils using Portland cement. PhD Thesis, University of South Carolina.
2. Chew SH, Kamruzzaman AHM, Lee FH (2004) Physicochemical and engineering behavior of cement treated clays. J Geotech Geoenviron Eng 130(7):696–706
3. Feng TW, Lee JY, Lee YJ (2001) Consolidation of soft mud treated with small cement content. Eng Geol 59:327–335
4. Gu M, Liu S, Hong Z, Yu X (2005) Quantifying research on structural characteristics of cemented soils. Rock Soil Mech 26(11):1862–1866
5. Kim YT, Ahn J, Han WJ, Gabr MA (2010) Experimental evaluation of strength characteristics of stabilized dredged soil. J Mat Civ Eng 22(5):539–544
6. Kumar S, Stewart J (2003) Evaluation of Illinois pulverized coal combustion dry bottom ash for use in geotechnical engineering applications. J Energy Eng 129(2):42–55
7. Phuong LH, Hai DT, Phan V (2013) Effect of soil salinity on strength of soil cement. Proceed Southern Inst Water Res Resear Vietnam 15(1):254–260
8. Lin Q (1989) Research on the tests of cement–soil pile compound foundation. Master Thesis, Zhejiang University, Hangzhou, China
9. Mohamed AMO (2000) The role of clay minerals in marly soils on its stability. Eng Geol 57(3–4):193–203
10. Ning B, Si L, Liu B (2005) Fracturing behaviours of cemented soil under environmental erosion. Chin J Rock Mech Eng 24(10):1778–1782
11. Pei X, Wu J (2000) Selection of cement additive in treating marine soft soil with DCM. Chin J Geotech Eng 22(3):319–322
12. Xu C, Dong T, Ye G (2006) Application of cement deep mixing method in Lianyungang marine soft soil foundation. Rock Soil Mech 27(3):495–498
13. Wang Q, Chen H, Cai K (2003) Quantitative evaluation of microstructure features of soil contained some cement. Rock Soil Mech 24(1):12–16

Stability Factors of Cantilever Sheet Pile Walls in Clays by Using Finite Element Method

Suraparb Keawsawasvong, Van Qui Lai, Quoc Thien Huynh, and Chung Nguyen Van

Abstract This paper presents a parametric study of the stability of cantilever sheet pile walls in clays by using the plane strain finite element analysis. In the finite element analysis, the cantilever sheet pile wall is modeled by using plate elements while the clay is modeled by using volume elements and obeys the Tresca failure criterion. The cantilever sheet pile wall has the excavated height and the embedded length. The clay has the unit weight and the undrained shear strength. By using the strength reduction method, the safety factor of the cantilever sheet pile wall at the limit state can be obtained. The results presented in the paper are illustrated in the form of design charts of dimensionless parameters, which are the relationships between the stability factor and the wall embedded length ratio. The proposed design charts of the stability factors can be employed to preliminarily estimate the embedment of the cantilever sheet pile walls that are widely constructed in practice.

Keywords Sheet pile · Excavation · Stability · Plane strain · Finite element method

S. Keawsawasvong
Department of Civil Engineering, Thammasat University, Thammasat School of Engineering, Pathumthani 12120, Thailand

V. Q. Lai (✉)
Faculty of Civil Engineering, Ho Chi Minh City University of Technology (HCMUT), 268 Ly Thuong Kiet Street, District 10, Ho Chi Minh City, Vietnam
e-mail: lvqui@hcmut.edu.vn

Vietnam National University Ho Chi Minh City, Linh Trung Ward, Thu Duc District, Ho Chi Minh City, Vietnam

Q. T. Huynh
Institute of Research and Development, Duy Tan University, Danang 550000, Vietnam

C. N. Van
Faculty of Civil Engineering, HCMC University of Technology and Education, Ho Chi Minh City, Vietnam

© The Author(s), under exclusive license to Springer Nature Singapore Pte Ltd. 2023   651
J. N. Reddy et al. (eds.), *ICSCEA 2021*, Lecture Notes in Civil Engineering 268,
https://doi.org/10.1007/978-981-19-3303-5_58

1 First Section

Retaining walls are the structures that are commonly used to retain lateral earth pressure induced by soil movements. The cantilever sheet pile wall is one of the retaining wall systems that can be employed during underground constructions to maintain the stability of soils. Designs of sheet pile walls are very important in order to ensure an adequate safety factor. In general, the analyses of sheet pile walls in clays have been often based on the classical limit equilibrium method (LEM) by assuming that there are the equilibrium forces between active and passive lateral earth pressures on both sides of a cantilever sheet pile wall. This technique has been generally known as Rankine's method [1]. However, in Rankine's method, the effect of soil heave is not properly taken into account. The finite element method (FEM) is the method that better than the LEM since the FEM considers both effects from the soil heave and the equilibrium of the lateral earth pressures. In this study, the plane strain FEM is employed to determine the stability of cantilever sheet pile walls in clays. It is noted that, the deep excavation problems were general studied by FEM [2–8].

The sheet pile wall has the excavated height (H) and the embedded length (D) as shown in Fig. 1. In general, the stability solutions of retaining walls in geotechnical engineering are presented in the form of the stability factor denoted by $\gamma H/s_u$, where γ is soil unit weight, and s_u is the undrained shear strength of clay. Some classical solutions of the stability factors for excavation works with supporting structures were proposed by Terzaghi [9], Bjerrum, and Eide [10], and O'Rourke [11]. For an unsupported vertical open cut without supporting structures, Martin [12] employed the lower bound and upper bound limit analyses to determine a very accurate bound

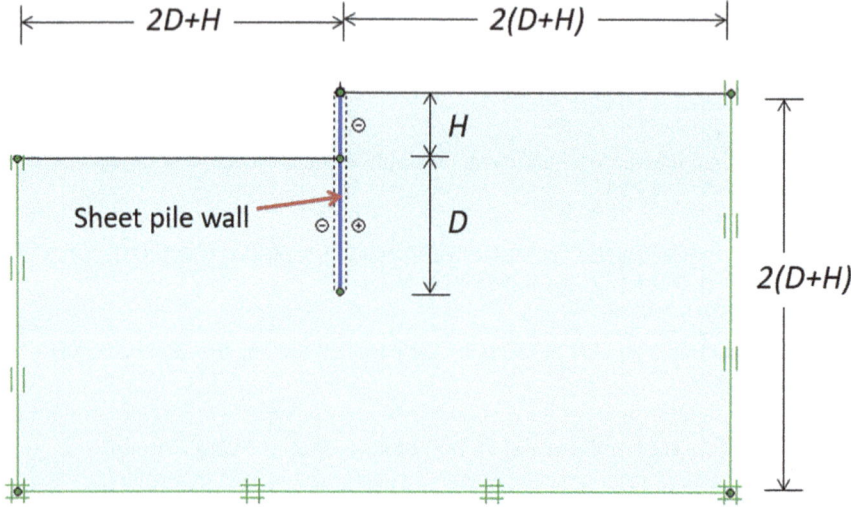

Fig. 1 The geometry of a cantilever sheet pile wall by using Plaxis2D

solution of this problem as N = 3.77 in which this value can be considered as the exact solution of an unsupported vertical open cut in clay. For more complex problems of cantilever retaining walls in clays, Teeravong [13] also carried out some FEM results of cantilever retaining walls. However, the existing solutions by Teeravong [13] do not cover all possible ranges of the ratios between the embedded length and the excavated height (or D/H). The paper presents the stability factors of cantilever sheet pile walls in clays by considering the complete range of D/H varying from 0 to 10.

2　Problem Formulation and Use of FEM

In this paper, the two-dimensional plane strain finite element software, namely Plaxis2D [14], is employed to analyze the undrained stability of cantilever sheet pile walls in clays in which the range of *D/H* is in between 0 to 10. Figure 1 shows the model of a cantilever sheet pile wall in Plaxis 2D. A cantilever sheet pile wall has a non-dimensional thickness and is modeled by plate elements. Note that the cantilever wall has no bracing system. The material model of clay is the Tresca material with the associated flow rule. The clay has the undrained strength (s_u) and the unit weight (γ). In the FEM, the undrained Young's modulus E_u is defined as $E_u/s_u = 300$ and the undrained Poisson's ratio is 0.495. The material properties of sheet pile walls correspond to the sheet pile types IV. It should be noted that the thickness of the wall does not have any influence on the stability of cantilever walls since Young's modulus of the wall is set to be very high corresponding to the properties of steel [14]. As a result, the properties of the wall can be considered to be very rigid comparing to the properties of soil. The interface elements are employed to model the interaction planes between the sheet pile wall and the clay (at both sides around the wall). The roughness of the interface elements for all simulations is set to be 0.667, which is a typical value for soil-structure interface [14]. The boundary displacement conditions on both left and right sides of the domains are zero movements in the horizontal direction $(u_x = 0)$ while the vertical movements are set to be free $(u_y \neq 0)$. At the bottom plane of the domain, both horizontal and vertical displacements are enforced to zero $(u_x = u_y = 0)$ corresponding to have no movement. At the top ground surface and the plane at the excavated side, the boundary displacement conditions are set to be free, which can be moved in both vertical and horizontal directions $(u_x \neq 0, u_y \neq 0)$. The sizes of the domains are chosen to be sufficiently large, where the plastic yielding zones are not allowed to intersect the right, left, and bottom boundaries of the domains. The selected sizes of the domains are explicitly shown in Fig. 1, which are applied to all simulations of cantilever sheet pile walls in this study.

　　In Plaxis2D, the clay is discretized by 15 nodded triangular elements while the plate element and soil-structure interface element consist of 10 nodes. In order to ensure that the number of meshes used in this study is sufficient, several trials and errors of mesh generations of the problem of the cantilever sheet pile wall with *D/H* = 2 are carried out. By plotting the stability factors of this problem with various numbers of meshes automatically generated by the software, it can be found from Fig. 2a that

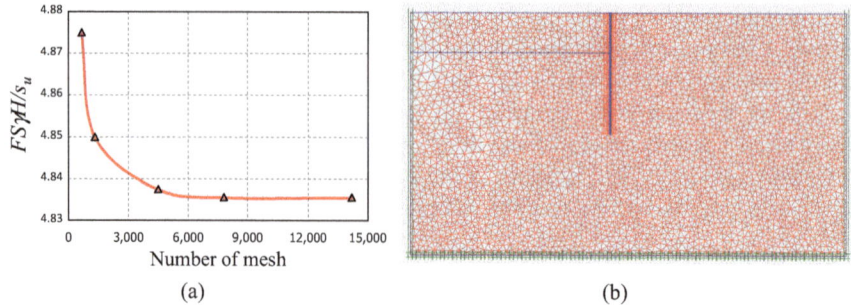

Fig. 2 **a** Stability factors with various numbers of mesh; **b** an example of mesh generation (about 7,800 elements)

the sufficient number of meshes can be successfully obtained when the number of meshes converges to approximately 7,800 elements. Thus, the meshes of all FEM models in this study are strictly controlled to be not less than 7,800 elements. An example of mesh generation (about 7,800 elements) is shown in Fig. 2b. The failure analyses of cantilever sheet pile walls are performed by using the strength reduction method in Plaxis2D. By using this reduction method, the value of safety factor (FS) can be achieved when the FS value converges to a constant value which is generally considered to be the final result of the strength reduction method. Note that the input dimensional parameters in this study are H, D, γ, s_u, and the output parameter is FS. By using the technique of dimensionless analysis [15], the dimensionless parameters in this study involve one dimensionless input parameter, which is the embedment ratio D/H, and one output dimensionless parameter, which is the stability factor $FS\gamma H/s_u$.

3 Results and Discussions

Figure 3 show the relationship between $FS\gamma H/s_u$ and D/H. The stability factors of the cantilever sheet pile wall obtained from FEM are in the range of $FS\gamma H/s_u = 3.8$ to 5.0 for $D/H = 0$ to 10. In the case of $D/H = 0$ (or an unsupported vertical open cut), the FEM solution, which is $FS\gamma H/s_u = 3.77$, is exactly equal to the exact solution reported by Martin [12]. Note that if the stability factor $FS\gamma H/s_u$ from any excavation projects is less than 3.77, any retaining wall system may not be requested since the unsupported vertical open cut can be sufficiently used during construction sequences. In addition, the results presented in this study is in good agreement with the existing solutions reported by Teeravong [13] for the range of $D/H = 0.5$ to 2. This indicates that the finite element models in this study are correctly performed. From Fig. 3, it is found that when D/H is more than 4, the stability factor seems to approximately converge to $FS\gamma H/s_u = 5.0$. It can be implied that an increase of $D/H > 4$ has a very small effect on an increase of $FS\gamma H/s_u$. Any increment of the

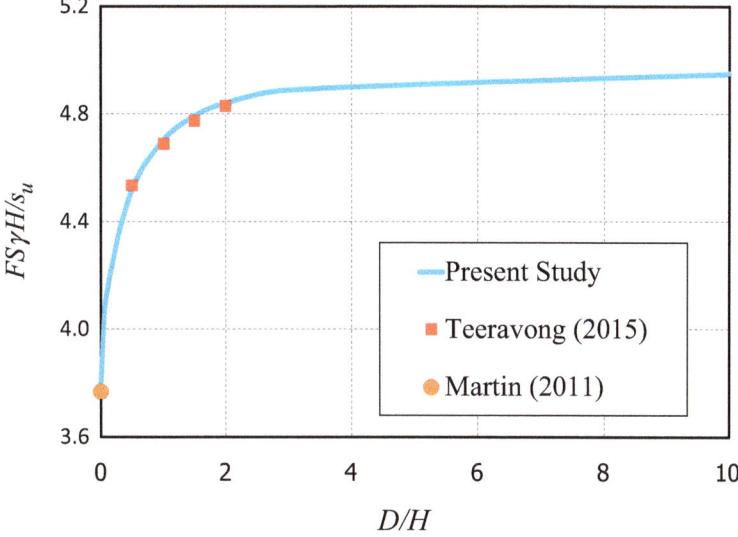

Fig. 3 The relationship between $FS\gamma H/s_u$ and D/H

embedded depth of the wall may not further strengthen the stability of the sheet pile wall. Other types of walls (e.g., piled wall or diaphragm wall) or the bracing systems are necessarily required in the excavation works that need more values of $FS\gamma H/s_u$ > 5.0. The proposed design charts in Fig. 3 can be used to preliminarily predict the stability factor or the excavated depth of the cantilever sheet pile walls in practice.

The predicted failure mechanisms including the deformed meshes and the incremental shear strains are shown in Figs. 4 and 5, respectively. For the deformed meshes, two cases including $D/H = 2$ and 10 are selected to demonstrate the collapse modes of the cantilever sheet pile walls. It can be clearly seen from Fig. 4 that, in the case of small D/H, the cantilever sheet pile wall is plunged and translated horizontally into the excavated side, where the rotational mechanism of the wall can be clearly observed. For the case of very high D/H, the wall is only translated horizontally without the rotational mechanism since the wall system is mainly failed from the effect of soil heave. From this incident, the limitation of using a cantilever sheet pile

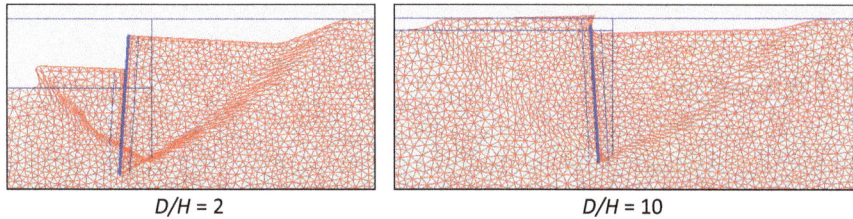

Fig. 4 Comparisons of deformed meshes between two cases of D/H

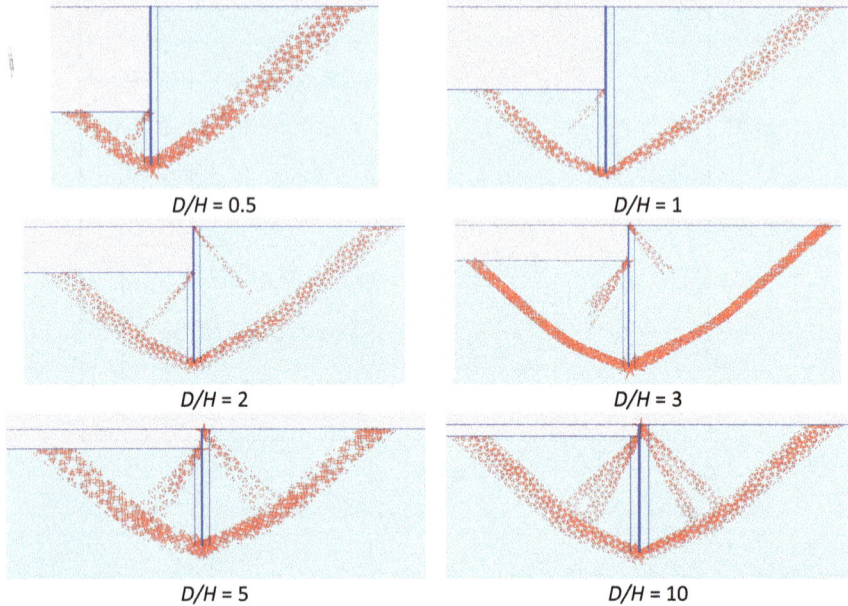

Fig. 5 Comparisons of incremental shear strains for different ratios of D/H

wall is $FS\gamma H/s_u = 5$ since there is no benefit on increasing the embedded depth D/H more than 4. The other systems such as the diaphragm wall or piled wall with bracing systems may be employed instead of using the cantilever sheet pile wall when dealing with the cases of $D/H > 4$.

The incremental shear strains can reveal the failure slip lines as shown in Fig. 5. The comparison of incremental shear strains is demonstrated in Fig. 5 for different ratios of D/H. Generally, the shapes of the failure slip lines are quite similar to the classical circular-arc failures encountered in slope stability problems. It can be seen from all cases shown in Fig. 5 that the failure slip lines begin from the tip of the wall and extend to the excavated side and the ground surface. An increase of D/H significantly results in an increase in the sizes of the failure slip lines. The predicted failure mechanism presented in this study can be used to predict the scales of the failures when the sheet pile wall system collapses.

4 Conclusions

The stability factors of cantilever sheet pile walls in clays are numerically derived by using the plane strain finite element analysis. In this study, the strength reduction method in Plaxis2D is employed to obtain the safety factor of the cantilever sheet pile wall at the limit state. The results are presented in the form of dimensionless parameters, which is the relationship between the stability factor $FS\gamma H/s_u$ and the embedment ratio D/H. The effects of the embedment ratio on failure mechanisms of cantilever sheet pile walls in clays are also demonstrated in the paper.

Acknowledgements This research was supported by the Faculty of Engineering Research Fund, Thammasat University. We would like to thank Ho Chi Minh City University of Technology (HCMUT), VNU-HCM for the support of time and facilities for this study.

References

1. Das BM (2014) Principle of foundation engineering, 8th edn. Cengage Learning, Boston
2. Huynh QT, Lai VQ, Boonyatee T, Keawsawasvong S (2021) Behavior of a deep excavation and damages on adjacent buildings: a case study in Vietnam. Transp Infrastruct Geotechnol 8:1–29
3. Huynh QT, Lai VQ, Tran VT, Nguyen MT (2019) Analyzing the settlement of adjacent buildings with shallow foundation based on the horizontal displacement of retaining wall. In: Long PD, Dung NT (ed) Geotechnics for sustainable infrastructure development, Springer, Heidelberg, pp 313–320
4. Huynh QT, Lai VQ, Tran VT, Nguyen MT (2019) Back analysis on deep excavation in the thick sand layer by hardening soil small model. In: ICSCEA. Springer, Heidelberg, pp 659–668
5. Lai VQ, Le MN, Huynh QT, Do TH (2019) Performance analysis of a combination between D-wall and secant pile wall in upgrading the depth of basement by Plaxis 2D: a case study in Ho Chi Minh city. In: ICSCEA. Springer, Heidelberg, pp 745–755
6. Huynh QT, Lai VQ, Boonyatee T, Keawsawasvong S (2022) Verification of soil parameters of hardening soil model with small-strain stiffness for deep excavations in medium dense sand in Ho Chi Minh City. Vietnam Innov Infrastruct Solut 7:15
7. Huynh QT, Lai VQ, Shiau J, Keawsawasvong S, Mase LZ, Tra HT (2022) On the use of both diaphragm and secant pile walls for a basement upgrade project in Vietnam. Innov Infrastruct Solut 7:17
8. Lai VQ, Huynh QT, Vo NH, Van C N (2020). Study on the influence of diaphragm wall on the behavior of pile raft foundation. In: 2020 5th international conference on green technology and sustainable development (GTSD), pp. 387–391
9. Terzaghi K (1943) Theoretical soil mechanics. Wiley, New York
10. Bjerrum L, Eide O (1956) Stability of strutted excavations in clay. Geotechnique 6:115–128
11. O'Rourke TD (1993) Base stability and ground movement prediction for excavations in soft clay, retaining structures. Thomas Telford, London.
12. Martin CM (2011) The use of adaptive finite-element limit analysis to reveal slip-line fields. Geotech Lett 1:23–29
13. Teeravong K (2008) Undrained stability of piled retaining walls in clay. Master Thesis, Chulalongkorn University, Bangkok, Thailand
14. Brinkgreve RBJ (2012) Plaxis2D. Plaxis, the Netherlands
15. Butterfield R (1999) Dimensional analysis for geotechnical engineering. Géotechnique 49:357–366

Studying the Wave Propagation in the Soft Soil Medium Subjected to Low-Frequency Dynamic Effects

Tham Hong Duong and Thoi Huu Tra

Abstract Dynamic effects on structures are complicated topics, both in theoretical and experimental aspects. This article aims to study more about the characteristics of the waves that originate from a low-frequency source of vibration, propagate through soil medium, and come to attack a receiver building in the vicinity. The virtual experimental method uses a finite element Plaxis 2D model in which the soil medium is compressible, subjected to a low-frequency vibration (i.e., pile driving and an impact loading in horizontal direction, both excitations have the same frequency f = 1.5 Hz). The mechanism of reflecting the waves from the source to the surrounding medium and the ground surface is studied by examining the direction of displacement vectors in different angles of propagation direction. The target focuses on determining the direction of propagation, reflects from the pile tip to the ground surface, and studying the additional settlement of the receiver footing due to the dynamic effects in the far-field. The former is to understand more about the decay of the body waves, and the mechanism of detaching the body waves to surface waves. Some preliminary results on free-domain displacement indicate that the longitudinal P-wave travels from the pile tip to along a line of which the angle is between 72 to 76° to the horizontal direction, and the transversal S-waves are detected at about 18 to 40° to the horizontal direction, and the surface R-wave may travel at a longer distance up to 50 m. The propagation velocity of R-wave in soft soil is calculated by dividing the distance by the time between two arbitrary points in the time-domain displacement, equals 35.5 m/s.

Keywords Dynamic effects · Far-field · Low-frequency response · Additional settlement

T. H. Duong (✉)
Faculty of Civil Engineering, Ho Chi Minh City Open University, 35–37 Ho Hao Hon Street, District 1, Ho Chi Minh, Vietnam
e-mail: tham.dh@ou.edu.vn

T. H. Tra
Master of Engineering, Ho Chi Minh City University of Technology (HCMUT), Ho Chi Minh, Vietnam

1 Introduction

Dynamic effects are vibrational action originating from a source and propagating through the medium to attack a receiver objective. A typical kind of dynamic effect is pile driving which is popular in construction activities for buildings. In some specific cases, pile driving is cheap and efficient for increasing the bearing capacity for the foundation, so this procedure is still applicable. Piles are cast in factories, installed into soil by dropping the hammer with a driving frequency of a few cycles per second. As such the frequency is rather low as compared to that of machinery vibration which operates at thousands of cycles per second. By applying an impact load exerted on the pile head, this driving force causes an axial wave propagating along the structure shaft; if the weight of the falling part is heavier than that of the pile, it makes the pile move downward, together with the displaced soil around the pile. Vibration running thru the shaft can move along the longitudinal axis of the structure and bounding from the pile tip to the head, causing tensile stress in localities around the pile head; this stress is calculated to the square root of the falling height of the hammer [1, 2]. The maximum stress at the top of a pile is tensive stress to some extent causes cracks in reinforced concrete pile subjected to impact driving force. The energy provided from the hammer is available at some percent at hammer/pile ratio and its loss is due to partly delivering to cushion, friction effects…etc. Besides, waves are radiating from the pile tip to the surrounding soil medium, running spherically, and propagating up to the ground surface, coming to stir some receiver foundations of the buildings in the vicinity. This is a typical kind of dynamic effect which in some specific cases, intense vibration can cause serious damages to these facilities [2]. A lot of studies were conducted to determine the responsive vibration of the soil medium during the process of a wave traveling and the change in physical–mechanical properties under vibration [1, 3], etc. Many prior research works studied intensively the dynamic effects of pile driving on the properties of soil medium. According to Smith and Chow [4], from the dynamic analysis of pile driving, the drivability of a solid prismatic pile is assessed by a numerical model in which the inertia effects of the soil around a pile, and the viscous nature of the soil are taken into account; a series of curves expressing the relationship between the penetrating rate (also called blow count) and the assumed static resistance of the pile [2]. In soil mass subjected to vibration, the excess pore water pressure also increases to several times of the cohesion especially at the surrounding surface of the pile as a result of the decrease in effective stress, and the coefficient of horizontal consolidation and the properties of soil compressibility are changed simultaneously [1, 7]. For mitigating the negative impact of the vibration due to pile driving, several measures of screening the vibration are suggested [4].

However, the mechanism of wave propagation thru the soil medium is still questionable. This article will study the wave propagation through soil medium originating from a source of low-frequency vibration, due to pile driving and a lateral impact load, coming to stir a receiver footing.

2 Theoretical background

Pile driving causes ground waves generated by the shaft and toe of the structure. The ground moves retrogressively in three dimensions, and the wavefront is a spherical one. From a specific point in the soil medium (e.g. pile tip) the two-dimension (planar) representation of the wavefront is shown in Fig. 1 below:

P-wave or compression/tension wave causes the soil be compressed or displaced along the direction in which soil particle motion coincides with the direction of P-wave transmission. Shear waves, including SV and SH in the vertical direction (symbol V) and horizontal direction (symbol H) respectively, cause the soil particle to move perpendicular to the direction of wave transmission. P-wave and S-wave are the first variety of wave types: body wave for implying that these waves propagate through the body of the soil medium.

The second variety of wave types is the surface wave or Rayleigh wave (R-wave). The motion at a distant point on the ground surface apart from the vibration source, associated with the damping effect of the soil material leads to a very complicated retrogressive motion to be analyzed (Fig. 1b). R- wave which is traveling on the ground surface produces motion both in the vertical and horizontal direction and grows maximum in amplitude at a point near the ground surface as shown in Fig. 1c.

Body waves, including P-wave and S-wave, diminish at a distance much shorter than the surface R-wave. Partly because the latter travels only in presence of a boundary, meanwhile the formers are suppressed by the damping of the soil medium. The most important characteristic is the R-wave which has the smallest velocity of propagation, and the most of energy [1]. In the halfspace of soil medium, Dowding [1] suggests that the R-wave travels on a surface and radiates the energy over a cylindrical surface instead of a spherical one as the body waves, i.e. P-wave and S-wave. And the distance from the source of the low-frequency vibration to the origin of reflective R-wave can be illustrated by the following Fig. 2 [1]:

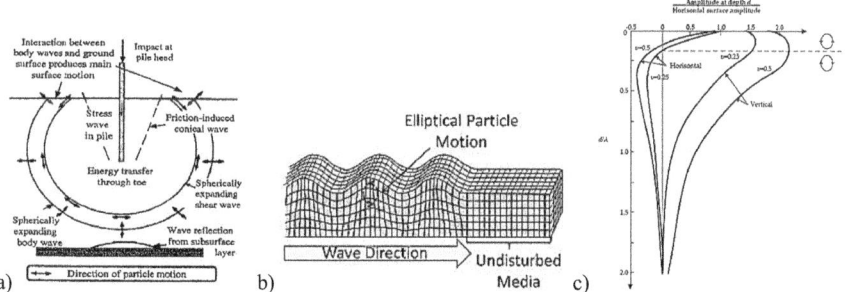

Fig. 1 **a** Wave front with displacement direction (arrows); **b** Deformation of the soil surface due to R-wave; **c** amplitude of R-wave as a function of depth, Poisson's ratio and wave length [1]

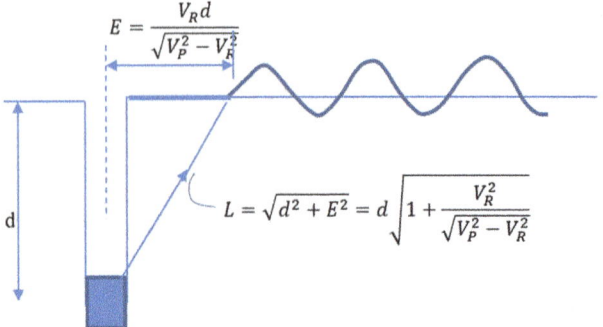

Fig. 2 The location of the origin of Rayleigh wave [1]

3 Model for this study

In finite element models, properties of the soil material are prescribed appropriately; the input data are based upon relevant dynamic tests in the laboratory, especially for determining the damping and the velocity of wave propagation [1, 4, 5, 8]. The Plaxis model for this study is modified from the Plaxis tutorial instruction manual. The soil of the study site is prescribed in Table 1 [6]. Two models will be studied: with a vertical driving load, and a horizontal one at the same load amplitude and frequency. For pile subjected to a lateral impact, 60 m far-field with absorbent boundaries is sufficient for R-wave propagation. The pile is modeled by the plate element.

An impact load with amplitude $P_o = 50$ kN. In the case of vertical impact load, the wave emitted from the pile tip travel rapidly in some direction to the ground surface. By investigating several trial lines from the pile tip, and if the displacement vectors indicate a compression/traction motion along these lines, the direction of P-wave propagation is determined. The hardening model used in the model for soil at the pile tip partially implies that the pile driving attains the densest state. On the other hand, if the displacement vectors are perpendicular to the line of wave propagation, proving a transversal motion along the line, then the S-wave is recognized.

For studying the effect of vibration on the construction works in the vicinity, a receiver footing subjected to a uniformly distributed charge $p = 20$ kPa, located at the

Table 1 Soil properties of the Hardening Soil model in this study [6]

Layer	Thickness (m)	γ_{sat}	Cohesion c (kPa)	$\varphi(^o)$	E (kPa)	E_{50}^{ref} (kPa)	E_{oed}^{ref} (kPa)	E_{ur}^{ref} (kPa)	m
Clay (MC)	11	18.0	12	3	1.25e4	-	-	-	-
Sand (HS)	14	20.0	1	31	5e4	5e4	5e4	1.5e5	0.5

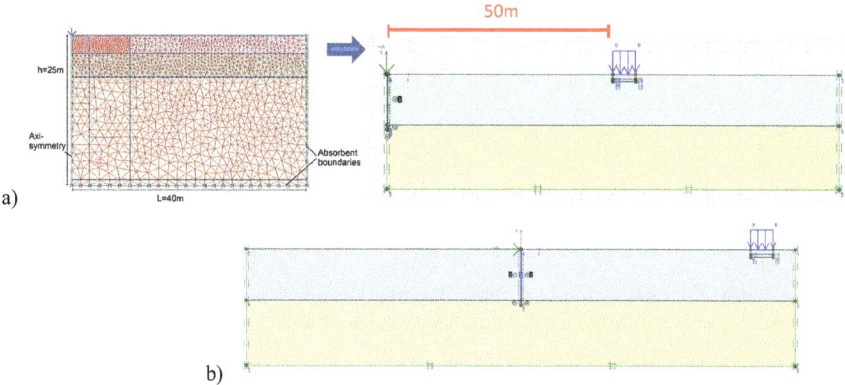

a)

b)

Fig. 3 The Plaxis model to be modified for studying the vibration due to **a** pile driving; **b** pile subjected to a lateral impact load (Ramadan, 2019 [5])

distance of 50 m apart from the pile centerline is modeled by a 0.6-m-thick circular plate, locating at 1 m below the ground surface (Fig. 3).

4 Results

4.1 Vibrational response of a particle in the free domain

Responses of points in the ground surface (i.e. particle in the free domain) are plotted for calculating the damping and wave propagation velocity. Figure 4 is for pile driving in the vertical direction and Fig. 5 is for the pile subjected to a horizontal impact load.

The response of the soil at a depth 1 m below the ground surface, where the displacements are expected to attain the maximum in the far-field of R-wave, is shown in Fig. 6 below:

Results shown in Fig. 6 implies that under a very low-frequency excitation, the impact load exerted from left to right, the displacements U_x and U_y at a depth 1 m

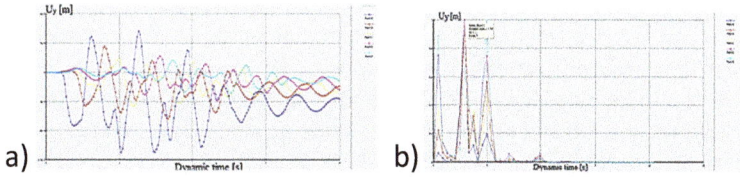

a)

b)

Fig. 4 Time-domain response at different particles on the ground surface; **a** displacement U_y in meters; **b** Frequency-domain analysis for displacement at various points

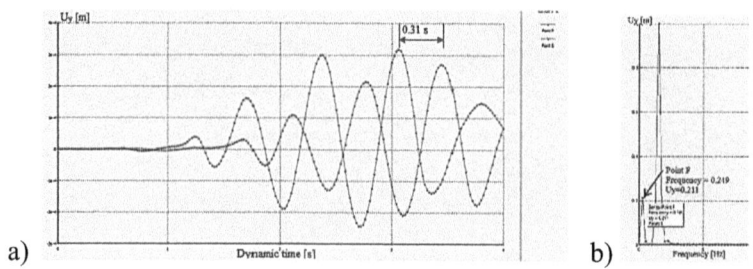

Fig. 5 Responses at points laying more than 30 m from the pile subjected to a horizontal impact load. **a** Time-domain recordings; **b** low-frequency response of displacement f = 0.25 Hz

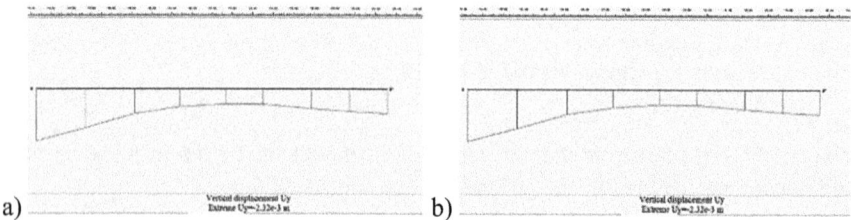

Fig. 6 The displacement in **a** vertical U_y, and **b** horizontal U_x direction of points on the surface 1 m below the ground surface, 30 to 40 m apart from the source

below the ground surface tend to enlarge downward and direct from right to left, with the extreme displacements are of $-2,44.10^{-3}$ mm and $-2,32.10^{-3}$ mm, respectively. This trend of displacement of the ground surface results in an additional settlement of the receiver footing.

4.2 Direction of wave propagation

Based on the displacement vectors laying on a line, the direction of wave propagation could be determined. The P-wave travels in a direction in which all the displacement vectors lay on a line; the S-wave on the contrary travels in a direction perpendicular to the axis of the line.

P-wave. . This wave travels along a line making an angle of 72° to the horizontal direction in the case of pile driving (Fig. 7a, upper), and causes a movement downward for the pile tip. In the case of a pile subjected to a horizontal impact load, the P-wave reflects from the pile tip an angle of 72–76° to the ground surface and causes heave (Fig. 7b, upper figure).

S-wave. The shear wave travels in the soil medium to the angle of 18° to the horizontal direction (Fig. 7a, lower). A horizontal impact load exerted on the pile

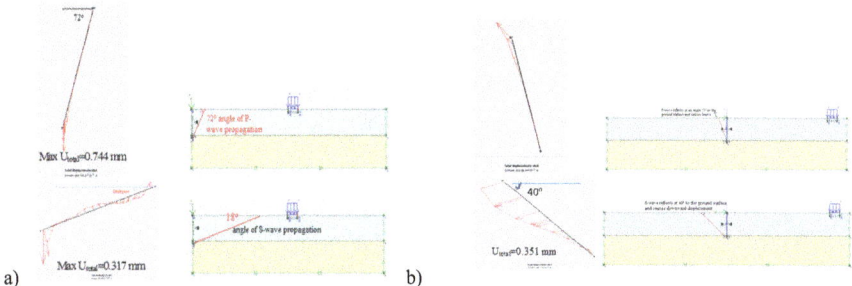

Fig. 7 Direction of the P- wave and S-wave propagation **a** case of the vertical driving force **b** case of the horizontal impact load

head causes the pile tip to move to the opposite direction and the vibration through the soil body at an angle of nearly 40° to the ground surface. But the displacement field in case of exerting laterally at the pile head seems to be mostly one-sided displacement (Fig. 7b, lower) as compared to that of pile driving vertically (Fig. 7a, lower). No P-wave travels right on the ground surface but R-wave at the distance exists. The built-in tool of the Plaxis yields $V_S = 54.27$ m/s for S-wave, and $V_S = 101.5$ m/s for P-wave.

R-wave. For calculating the velocity of the R-wave, the distance between different points in far-field (i.e., point F, 29.5 m, and G, 40.5 m apart from the pile) and the time recording in Fig. 5a, in which the distance between two peaks of responsive curves is calculated. The velocity of R-wave approximately equals 35.5 m/s.

5 Discussion

- This study is conducted by using the given data in [6]. In the first case of pile driving with low frequency $f = 1.5$ Hz, there are many strange modes of vibrational response (Fig. 4b). In the problem of a pile subjected to horizontal impact load with the same driving frequency, a single low-frequency of the response, $f = 0.25$ Hz is recognized. This can be explained that the plate model created the planar front of the wave traveling to the surrounding medium, instead of a spherical front.
- Results of the angle of wave propagation in soil medium in Plaxis numerical model agreed well with the previous research works [1, 5] (Fig. 3). The angle of longitudinal P-wave reflection from the pile tip, in both cases of pile driving vertically and impact loading horizontally is between 72 to 76°. The velocity of the P-wave is given by the built-in tool of Plaxis software, or $V_P = 101.5$ m/s, $V_S = 54,27$ m/s, and $V_R = 35.5$ m/s measured in the response (Fig. 5), the angle is 70° to the horizontal direction. The distance where the P-wave attenuates is $E = 4.4$ m ($d = 12$ m).

- The velocity of the R-wave is better calculated by using the results of the problem of pile driving. According to Massarch et al. [4], the R-wave would dominate at the distance of d_{crit} which is determine by the equation below:

$$d_{crit} = L_{pile} \tan(\theta_{crit}) = L_{pile} \tan\left(\sin^{-1}\left(\frac{V_S}{V_P} \right) \right) \qquad (1)$$

As such, if V_S=54.27 m/s and V_P=101.5 m/s, then the distance of R-wave will be E=7 m, and the angle of R-wave reflection equals approximately 58°. This is a good point for the numerical model in this study.
- With a smaller value of soil modulus E, for instance, E = 3000 kPa, the result yields a steeper angle of P-wave propagation, up to 80° to the horizontal direction. As such, the wavelength of the R-wave would be smaller (e.g., $V_R \approx 28$ m/s). These findings agreed with the prior research works [9, 10].

6   Conclusion

A model is developed for studying the wave propagation in soil medium subjected to a low-frequency excitation. Two working conditions are investigated, one is of pile driving in the vertical direction, and the second is of the pile subjected to an impact load in the horizontal direction. The angle of P-wave reflection is around 70–76° to the horizontal direction, 18° for S-waves. This agrees well with the theory in which the longitudinal wave attenuates first, then the transversal shear wave, and finally the R-wave. Frequency-domain analysis indicates a very low frequency of response $f = 0.25$ Hz, together with the displacement both in vertical and horizontal directions due to pile driving result the additional settlement of the receiver footing. The velocity of ground surface R-wave can be estimated by examining the response curves at two points in the far-field. This R-wave has the lowest velocity as compared to that of the P-wave and S-waves and travels at a distance of up to 50 m from the source of vibration. This study could be further to give a better understanding of the effects of construction activities on the buildings in the vicinity.

References

1. Dowding Charles H. (1996) Construction vibration, Prentice Hall international series in civil engineering and engineering mechanics. ISBN 0-13-299108-X
2. Fleming WGK, Weltman AJ, Randolph MF, Elson WK (1994) Piling Engineering, the, 2nd edn. Wiley, New York and Toronto
3. Degrande G, De Roeck G, Van den Broeck P, Smeulders D (1998) Wave propagation in layered dry, saturated and unsaturated poroelastic media. Int J Solids Struct 35(34–35):4753–4778
4. Massarsch KR (2000) Settlements and damage caused by construction-induced vibrations. In: Proceedings of the international workshop in wave, Bochum, Germany, pp 299–315

5. Mohammed Ramadan M (2019) Influence of construction induced vibrations on soil and adjacent structures, Master Thesis, Tanta University. doi: https://doi.org/10.13140/RG.2.2.19085.59360
6. Bentley Communities Homepage (2019) Plaxis 2D tutorial manual build 10097, pp 160–168. https://www.bentley.com. Accessed 05 Mar 2020
7. Randolph MF, Carter JP, Wroth CP (1979) Driven piles in clay – the effects of installation and subsequent consolidation. Géotechnique 29(4):361–393
8. Sebastiano Foti (2000) Multi-station methods for geotechnical characterization using surface waves. PhD dissertation, Politechnico di Torino. Corpus ID: 107967655, https://doi.org/10.6092/Polito/Porto/2497212
9. Duong TH, Tra TH (2020) Studying the wave propagating characteristics of a defected semi-rigid structure (near-field and far-field). In: Proceedings of the 1st international conference in structural health monitoring and structure engineering (SHM & SE), Ho Chi Minh City Open University, December, vol. 148, Springer, Singapore, pp. 161–172
10. Whyley PJ, Sarsby RW (1992) Ground borne vibration from piling. Ground Eng 25(4):32–37

The Correlation Between SPT and CPT Result in Sandy Soils in Ho Chi Minh City

Nhat Truyen Phu, Thanh Long Vo, and Ba Vinh Le

Abstract This study establishes the correlation of test results in site from SPT and CPT in Ho Chi Minh City. Based on the soil investigation at the Phu Tho Hyppodrome, District 11, Ho Chi Minh City from the ground to 40 m depth, every 2 m/test SPT of 12 boreholes and 15 holes CPTu in the area of 10,000m2, the author collected, classified and evaluated geotechnical conditions and statistic to establish the correlation of test results in site from SPT and CPT. When the correlation function $q_c = 3748.1 \times \exp[(-0.126z^2 + 7.006z - 47.711) \times N_{30} \times 10^{-3}]$ is applied, the cost and time of geotechnical investigation will be greatly reduced, especially in CPT. This provides geotechnical engineers with the estimation of SPT based on the CPT results and vice versa, creates favorable conditions for the economic design in the state of equipment, economic limit or construction plan unfavorable.

Keywords SPT · CPT · Correlation · Sandy soil · HoChiMinh city

1 Introduction

Since the early twentieth century, when the Standard Penetration Test (SPT) is discovered and used by C.R. Gow in North America and then the emergence of Cone Penetration Test (CPT) in the Netherlands, the application of both experiments have become widely popular, especially SPT and they were considered an integral part of the process. geological investigation and construction. But over time with the

N. T. Phu (✉) · T. L. Vo
Faculty of Geology And Petroleum Engineering, Ho Chi Minh City University of Technology (HCMUT), 268 Ly Thuong Kiet Street, District 10, Ho Chi Minh City, Vietnam
e-mail: pntruyen@hcmut.edu.vn

B. V. Le
Faculty of Civil Engineering, Ho Chi Minh City University of Technology (HCMUT), 268 Ly Thuong Kiet Street, District 10, Ho Chi Minh City, Vietnam

N. T. Phu · T. L. Vo · B. V. Le
Vietnam National University Ho Chi Minh City, Linh Trung Ward, Thu Duc District, Ho Chi Minh City, Vietnam

© The Author(s), under exclusive license to Springer Nature Singapore Pte Ltd. 2023
J. N. Reddy et al. (eds.), *ICSCEA 2021*, Lecture Notes in Civil Engineering 268,
https://doi.org/10.1007/978-981-19-3303-5_60

strong development of scientific and technical advances, the experiments have gradually improved and upgraded. Especially for the CPT experiment, there is a drastic improvement in technology, so the cost of this experiment also increases, meanwhile, the implementation of SPT gradually becomes a tradition with little improvement. so the price is also much lower.[1–4]

Until 1961, people only started to learn about the correlation between the results of the two above experiments, this was intended to lower costs in performing CPT experiments or using CPT results to test proof of SPT results and facilitate calculations when facing limited equipment and economic conditions …

Today, to find the correlation between CPT and SPT, one can also find by establishing regression equations with specialized software for each soil type in different locations. Although this is not general, but thanks to the details, the results are more accurate and closer to the reality of the research area.

There have been various research evaluating the correlation between N30 number and ground soil parameters. For example, Akca (2003) evaluated the correlation between the SPT results and the CPT results for UAE regional geology [5]. In this article, the author compares the difference between soils with carbonate content in the UAE region with soils with silicate content of other authors. Akca's results on the coefficient $n = q_c / N$ are somewhat higher than the previous studies [1–3]. The reason is analyzed by the cementation of soil particles, or the soil with coarse shells and shells.

Kara, O. (2010) established the correlation between N30 and qc for hererogeneous soil in Turkey by the function $qc = aN^x$ in which a and x are constants defined by soil types [6]. The results showed that the correlation coefficients were quite low, ranged from 0.29 to 0.51 even though the data were filtered.

Tarawneh (2014) determined the correlation between the SPT experimental results and CPT by the multi-linear regression method [7]. The results obtained are 6 correlation functions between N from SPT experiment and variables qc, fs from CPT experiment and effective stress with correlation coefficient R^2 ranges from 0.829 to 0.839.

Using the value q_c or f_s alone to find the correlation with N_{30} is not really reasonable. Because in the CPT experiment, q_c is the static resistance of the ground, fs is the static resistance to friction in units of the shoot against the soil, and q_c and f_s are determined by the CPT penetration. land. As for the SPT experiment, the value of the N_{30} closed hammer was determined that when the SPT piercing head is closed to the ground, the soil will act on the SPT head, including 2 corresponding forces, one is penetration resistance, the other is frictional force. both provide the same energy against the cut-through of the SPT head and are converted to the N_{30} hammer number.

Therefore, in this study, the authors bases on the above principle to begin to reformulate the correlation between SPT field test results and CPT. In addition, the authors also assessed correlation according to traditional methods to compare with previous studies.

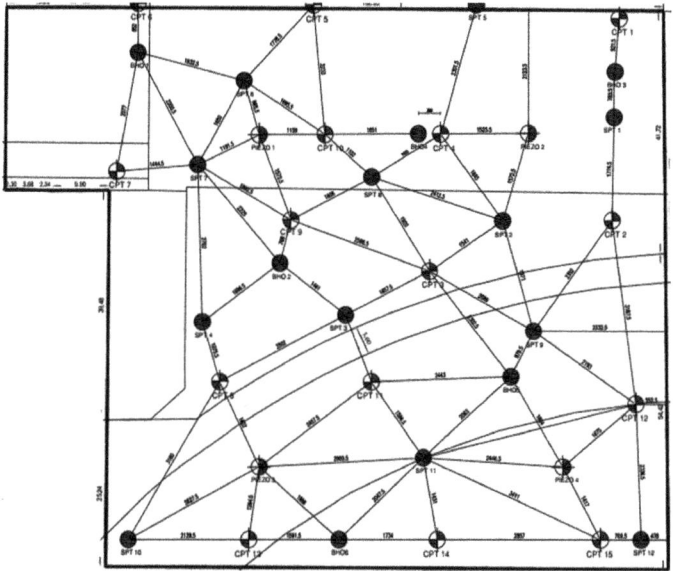

Fig. 1 Layout diagram of SPT and CPT borehole

2 Overview of the Study Area

2.1 Location

This article uses geological data of the Phu Tho Racecourse Project, District 11, Ho Chi Minh City in an area of 10,000m² [8] with total of 15 CPTu test positions and 12 SPT test positions (Fig. 1). SPT tests are performed every 2 m within a depth range of 50 m, while CPTu testing is conducted with electronic equipment that measures pore water pressure and parameters such as qc, fs, u are recorded every 1 cm.

2.2 The Stratigraphy

Geology of the study area consists of 3 main layers. The top layer is sandy clay with laterite gravel, consisting of sub-layers of A, 1A, 1B, and 1C with depth ranges from 4.4 m to 12.6 m. The second layer is a clayed sand yellowish gray with white spots, with a relatively large thickness from 23.2 m to 31.2 m. The third layer is the clay layer distributed below the second layer to the bottom of the survey borehole (50 m).

3 Methods of Analysis

In this paper, the SPT–CPT correlation is built exclusively in layer 2 since sandy soil is distributed relatively uniformly within the area and the thickness is relatively great. We ignored the remained layers due to the small thickness as well as the risk producing inaccurate results. To find the best correlation, we only selected data from 10 closest pairs of SPT and CPT boreholes to perform statistics and correlation calculation. The selected CPT data (qc, fs) are the average values within a depth range of 30 cm where the equivalent N–values were measured in layer 2. To establish the SPT–CPT relationships, we employed the following 4 common regression functions:

- Linear multiple regression equation (y = ax + b),
- One-variable exponential regression (y = aebx),
- One-variable power regression equation (y = axb).
- Polynomial regression equation (y = ax2 + bx + c).

and use the correlation coefficient R^2 to evaluate the relationship between the factors x and y ($0 \leq R^2 \leq 1$). The larger R^2, the higher the relationship.

Although was chosen from the closest boreholes, those pairs of N value–qc and N value–fs data still have anomalies where the SPT–CPT data shows unreasonable difference. Therefore, to get fairer results, we carried out filter process to eliminate the anomalies of these geological data using rather simple rules:

- We only receive regression functions with index R2 ≥ 0.5.
- Filter by a one-variable power regression relationship between N30 and qc;
- In a data sheet, maximum 2 data can only be filtered out;
- The eliminated data must be obvious anomalies;
- After filtering, the anomaly does not significantly change the graph shape;
- The post–filter R2 must be increased by more than 0.1;
- Number of data filtered should not exceed 5% of the total number.

4 Results and Discussions

4.1 Results with Raw Data

The N_{30}–q_c (kPa) and N_{30}–f_s (kPa) correlation of raw data in layer 2 is illustrated in Fig. 2a and 2b, respectively. As can be seen in Fig. 2a, among 4 regression equation, power regression equation produces optimal value of R^2 up to 0.4665. Also, we observes a trend of very concentrated, non-discrete and isotropic data. Thus, we believe that the qc and N_{30} values are very likely to have a strong correlation to each other after being filtered which will be described in the following section. Unlike N_{30}–q_c, the correlation between N_{30} and f_s is quite weak (Fig. 2b) with trivial R^2 values.

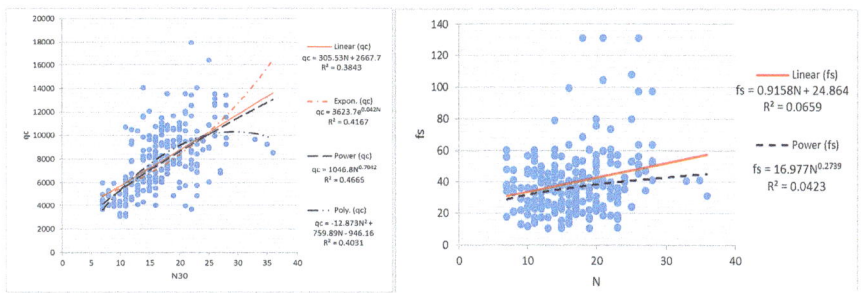

Fig. 2 Correlation between qc vs N30 (a - left) and fs vs N30 (b- right) in soil layer 2

4.2   Results with Filtered Data

From the statistic data, we can draw graph in Fig. 3. From the obtained graph, we notice that the filtered data results in R^2 increased significantly. Typically, the one-variable exponential regression equation gives $R^2 = 0.7255$, which is very high compared with the data that was not filtered. Thus, we can see clearly that the tip penetration resistance qc of the soil in the CPT experiment has a strong correlation with the number of N_{30} hammers in the SPT experiment.

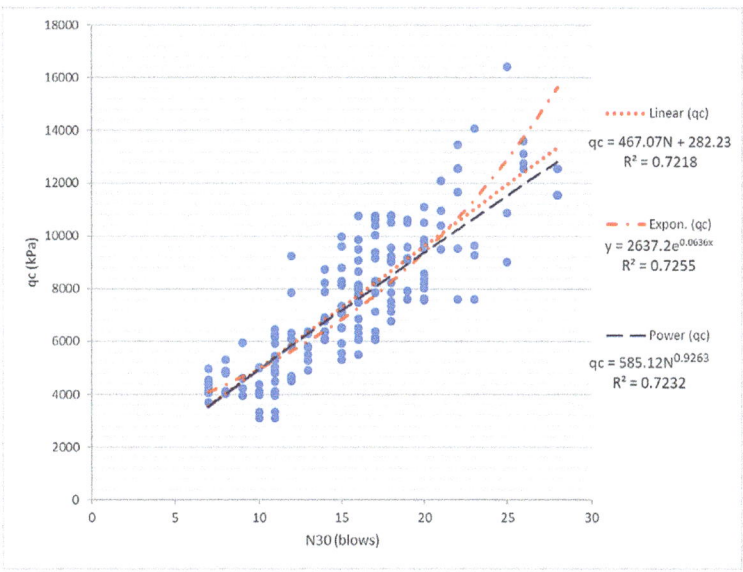

Fig. 3 Correlation between qc with N30 in soil layer 2 with filtered data

5 Results of new method

In this section, the author bases on the above principle to begin to re-formulate the correlation between SPT field test results and CPT. With the initial assumption, we have the formula (Eq. 1):

$$N_{30} = f(q_c) + h(f_s) \tag{1}$$

Rearrange Eq. (1):

$$\left(1 - \frac{h(f_s)}{f(q_c) + h(f_s)}\right) \times N_{30} = f(q_c) \tag{2}$$

$$\frac{h(f_s)}{f(q_c) + h(f_s)} = \frac{\overline{S}_s \times f_s}{\overline{S}_s \times f_s + S_c \times q_c} = \frac{S_s \times N_{30}^x \times f_s}{S_s \times N_{30}^x \times f_s + S_c \times q_c}$$

$$= \frac{f_s}{f_s + \frac{S_c}{S_s \times N_{30}^x} \times q_c} = \frac{f_s}{f_s + y \times N_{30}^{-x} \times q_c} \tag{3}$$

$$\frac{f_s}{f_s + y \times N_{30}^{-x} \times q_c} \approx \frac{f_s}{f_s + q_c} \times N_{30}^a \times b \tag{4}$$

Set equation

$$\delta(x) \approx \frac{f_s}{q_c + f_s} \times N_{30}^a \tag{5}$$

From Fig. 4 we have the correlation between z and $\delta(z)$:

$$\delta(z) = 0.0123z^2 - 0.6828z + 0.9134 \tag{6}$$

Let $\delta(z)$ into the Eq. (2) we get the following equation:

$$(1 - \delta(z) \times b) \times N_{30} \approx f(q_c) \tag{7}$$

Set equation

$$\zeta = (1 - \delta(z) \times b) \times N_{30} \tag{8}$$

From Fig. 5 we have have the correlation between qc (kPa) and ζ

$$q_c = 3748.1 \times e^{0.054\zeta} \tag{9}$$

where ζ is equation with N30 and z are variations:

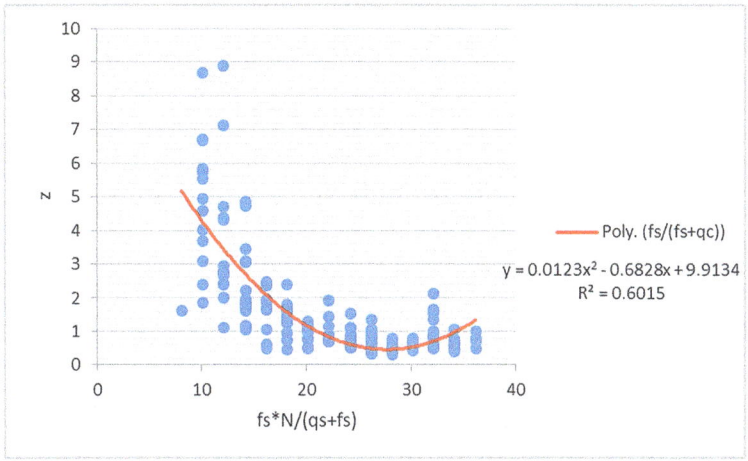

Fig. 4 Correlation between z and $\frac{f_s}{q_c+f_s} \times N_{30}^{-1.4} \times 10^4$

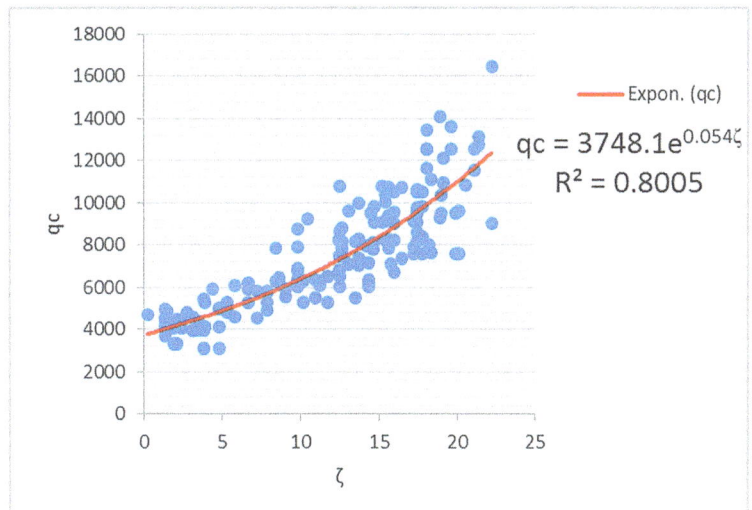

Fig. 5 Correlation between q_c (kPa) and ζ

$$\zeta = (1 - (0.0123z^2 - 0.6828z + 0.9134) \times 0.19) \times N_{30} \qquad (10)$$

$$\zeta = (-2.337z^2 + 129.732z - 883.546) \times 10^{-3}) \times N_{30} \qquad (11)$$

At that time, Eq. (12) was rewritten as follows:

$$q_c = 3748.1 \times e^{[(-0.126z^2+7.006z-57.711) \times N^{30} \times 10^{-3}]} \qquad (12)$$

Table 1 Geotechnical properties of soils in study area [8]

Layer No	Moisture content	Unit weight		Specific Gravity	Saturation	Porosity	Void ratio	Atterberg limit				Compression Modulus	Friction Angle	Cohesion	Clay	Silt	Sand	Gravel
		Wet	Dry					Liquid limit	Plastic limit	Plasticity Index	Liquidity Index				<0.005 (mm)	0.05–0.005 (mm)	2.0–0.05 (mm)	10.0–2.0 (mm)
	W	γ	γ_d	Gs	Sr	n	eo	WL	WP	IP	IL	E(100–200)	φ	C				
	%	kN/m3	kN/m3	kN/m3	%	%		%	%	%		kPa	Degree	kPa	%	%	%	%
1A	25.09	18.95	15.15	27.15	86.05	44.20	0.79	33.43	18.97	14.46	0.42	5649	12°52'	23.8	25.8	24.6	48.3	1.5
1B	20.32	20.09	16.71	27.22	88.00	38.60	0.63	31.13	15.45	15.67	0.31	6766	14°5'	26.1	28	23.7	37.1	11.3
1C	21.02	19.97	16.50	27.22	88.07	39.37	0.65	33.00	16.07	16.92	0.29	7150	14°2'	25.8	26.2	23.4	49.8	0.7
2	19.49	20.17	16.89	26.74	89.29	36.82	0.58	23.23	16.83	6.40	0.41	12,768	24°45'	10.7	9	8.7	77.1	5.3
3A	20.08	20.76	17.29	27.37	93.86	36.83	0.59	43.57	21.90	21.67	0.01	8670	17°18'	59.6	50.1	32.2	17.6	0

where: z (m) is test depth.

6 CONCLUSIONS

In general, the correlation between SPT and CPT data in traditional ways is only average for the surveyed area. With Eq. (12) established in this research, the correlation results are very approximate to reality with the correlation coefficient $R^2 = 0.8005$. To be able to give correlation results for higher R^2 coefficients, it is necessary to adopt modern equipment and more data sets. Thus, geotechnical engineer can use CPT results to verify the SPT results and vice versa, facilitating the design calculation when facing limited equipment and economic conditions. or the construction site is not favourable.

Acknowledgements This research is funded by Ho Chi Minh City University of Technology – VNU-HCM under grant **number T-ĐCDK-2020-69**

References

1. Lunne T, Robertson PK, Powell JJM (1997) Cone penetration testing in geotechnical practice. Blackie Academic, EF Spon/Taylor & Francis Publication, New York, 1997, 312 pp
2. Schmertmann JH (1978) Guidelines for cone penetration test, performance and design. Federal Highway Administration Report FHWA-TS-78–209, Washington, DC
3. Robertson PK (1990) Soil classification using the cone penetration test. Can Geotech J 27(1):151–158
4. Robertson PK (2009) CPT interpretation – a unified approach. Can Geotech J 46:1–19
5. Akca N (2003) Correlation of SPT–CPT data from the United Arab Emirates. Eng Geol 67(3–4):219–231
6. Kara O, Gunduz Z (2010) Correlation between CPT and SPT in Adapazari, Turkey. In: Proceedings of the 2nd international symposium on cone penetration testing, Huntington Beach, California, pp 2–18
7. Tarawneh B (2014) Correlation of standard and cone penetration tests for sandy and silty sand to sandy silt soil. Electron J Geotech Eng 19:6717–6727
8. Vo Thanh Long, Engineering geological survey results of Phu Tho Racecourse Project, District 11, Ho Chi Minh City, 2015

Using Inverse Technique to Evaluate the Effect of Capillary Barrier on Hydraulic Properties of Unsaturated Multi-Layered Sand from 1-D Desaturation Test

Nam Viet To, Ky Viet Nguyen, Hai Hong Dao, and Truyen Nhat Phu

Abstract A numerical study was used to estimate the applicability of the inverse analysis approach to assess the hydraulic properties of unsaturated soils under capillary barrier effect in a one – dimensional outflow experiment. A sample structure comprising a 400 mm thick layer of fine sand overlying a 400 mm thick layer of medium sand was set up in a Plexiglas cylindrical tube. The drying experiment was carried out to determine cumulative flux, soil resistivity and soil suction during experiment. Simultaneously, the unsaturated hydraulic properties were also calculated by a finite element program and an inversion program using measured value of soil suction and the outflow rate from desaturation test as input data. To evaluate the applicability of the inverse parameter estimation method, the comparison between computed and measured values of the unsaturated hydraulic properties of the layered sands was conducted. The results indicate that the inverse analysis based on the 1-D desaturation experiment can be used to evaluate the capillary barrier effect of unsaturated layered soils with reasonable accuracy…

Keywords Capillary barriers · Unsaturated soil · Inverse analysis · 1-D column test · FEM

1 Introduction

Capillary barrier is the phenomenon that appears when soil water moves from the smaller to larger pore zones; it is formed at the interface between these zones and has long been known. Capillary barrier relates to various problems of soils such as seepage, water contamination, water storage, slope stability, water resource management and waste management in different fields of hydrogeological engineering,

N. V. To (✉) · K. V. Nguyen · H. H. Dao · T. N. Phu
Faculty of Geology And Petroleum Engineering, Ho Chi Minh City University of Technology (HCMUT), 268 Ly Thuong Kiet Street, District 10, Ho Chi Minh City, Vietnam
e-mail: tovietnam@hcmut.edu.vn

Vietnam National University Ho Chi Minh City, Linh Trung Ward, Thu Duc District, Ho Chi Minh City, Vietnam

© The Author(s), under exclusive license to Springer Nature Singapore Pte Ltd. 2023   679
J. N. Reddy et al. (eds.), *ICSCEA 2021*, Lecture Notes in Civil Engineering 268,
https://doi.org/10.1007/978-981-19-3303-5_61

environmental engineering, geotechnical engineering, agricultural engineering, soil science, etc. [9].

Many researches have been conducted to show the effective use of capillary barrier to reduce inward oxygen transport and the migration of pore water into the underlying waste [3], prevent water infiltration in arid and semi-arid conditions [10, 8], resist the capillary rise and stop salinization from underlying sources, thus permitting salt-sensitive plants to develop well [12], protect steep slope from rainfall-induced slope instability [11], etc.

In unsaturated soils, the capillary barrier can be comprehended by observing the unsaturated hydraulic properties of two different soils [2]. The hydraulic properties of unsaturated soil are represented by two relationships: volumetric water content - soil suction (θ–ψ) and hydraulic conductivity - soil suction (k-ψ). The θ–ψ relationship is called the soil–water characteristic curve (SWCC), which plays the most important role in the description and design of a capillary barrier [6]. However, proposed methods for achieving reliable unsaturated hydraulic properties of soils are still difficult. Therefore, it is necessary and essential to find simple and acceptable approaches to accurately estimate the hydraulic properties of unsaturated layered soils [9].

During recent decades, numerical methods and mathematical analysis have become potential tools to estimate the unsaturated hydraulic properties of soils effectively and simply. The objective of this study was to estimate the reliability of a numerical method for determining the hydraulic properties of unsaturated layered sands under the capillary barrier effects [4]. To serve the objective of this study, a numerical simulation was carried out combining finite element code and inversion code to estimate unsaturated hydraulic parameters for the capillary barrier using observed data of soil suction and the outflow rate from one-step desaturation test as input data [5]. Additionally, a performance of comparison between modeled and measured hydraulic properties was also conducted to estimate the satisfactory technique for inverse parameter appraisement of unsaturated hydraulic properties [9].

2 Materials and Methods

2.1 Materials and Experimental Methods

In this study, two sands used to perform the capillary barrier of vertical heterogeneous soil are medium sand and fine sand with summarized properties in Table 1.

Table 1 The material properties

Material	Gs	Cu	Cc	D50 (mm)	emax	emin
Medium sand	2.66	1.65	0.94	0.55	0.99	0.60
Fine sand	2.7	1.73	0.98	0.18	1.22	0.73

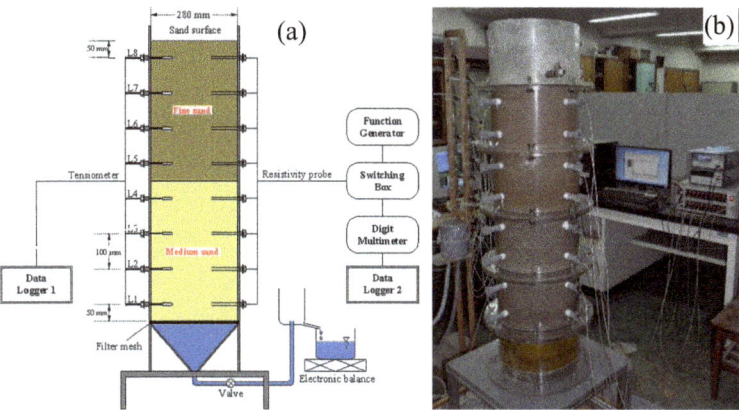

Fig. 1 a The experimental model, **b** the column test in laboratory

A sample structure comprising a 40 cm thick layer of fine sand overlying a 40 cm thick layer of medium sand was put up in a Plexiglas column (Height: 800 mm, Diameter: 280 mm) [9]. At every 100 mm, the tensiometers and electrical resistivity probes (ERP) were installed along the soil column to measure the soil suction and soil saturation respectively. An electronic balance was set up to get the cumulative outflow value of the soil column by weighing and automatic recording. Output signals of all the implements were recorded automatically using a data logger system, which can record at every given time-interval (Fig. 1).

Using calibration curves created by using Archie's law [1] for two sands, the soil saturation at eight locations in the soil column can be calculated at any time during the desaturation test.

To avoid a change in water–air surface tension, the environment in the laboratory was always maintained at 20 °C [7]. In this research, only one-step outflow test was carried out, therefore, the water table was kept at the same bottom level of the soil column during the experiment.

3 Numerical Method

3.1 Forward Simulation for Unsaturated Flow

Based on Darcy's law, in 1931, Lorenzo A. Richards published the equation for unsaturated flow in porous media with condition of one-dimension transient flow and isotropic flow, the Darcy equation is written as:

$$\frac{\partial \theta}{\partial t} - \frac{\partial}{\partial z}\left[k(\theta)\left(\frac{\partial h}{\partial z} + 1\right)\right] = 0 \tag{1}$$

where θ is the volumetric water content, t is time, z is the depth which is zero at bottom of the sample and negative direction is downward, k is the hydraulic conductivity, h is the pressure head.

The hydraulic properties of unsaturated soils are described by using the Van Genuchten model (1980)[14]:

$$\Theta = \frac{\theta - \theta_r}{\theta_s - \theta_r} = \left(1 + \left(\frac{P_g - P_l}{P_0}\right)^{\frac{1}{1-m}}\right)^{-m} \tag{2}$$

$$k_r = \sqrt{\Theta}\left(1 - \left(1 - \Theta^{1/m}\right)^m\right)^2 \tag{3}$$

where Θ is the normalized water content, θ is the volumetric water content, θ_r is the residual water content; θ_s is the saturated water content, $P_g - P_l$ is suction head, P_0 and m is the curve fitting parameter, k_r is relative hydraulic conductivity.

3.2 Nonlinear Optimization

A non-linear optimization method used to minimize the objective function was the L2 error norm, which expresses the difference between observed and predicted data by using the least–square solution for the parameters in Eqs. (2) and (3) [13].

In the inverse analysis, when many measurements are available, the system of equation can be expressed:

$$\underline{y} = \underline{\underline{h}}.\underline{x} \Leftrightarrow \underline{y}_0 + \Delta\underline{y} = \underline{\underline{h}}(\underline{x}_0 + \Delta\underline{x}) \tag{4}$$

where $\underline{x}$ are unknown soil properties, $\underline{\underline{h}}$ is the transformation matrix which is $h_{ij} = \partial y_i/\partial x_j$.

The error for the measurements is established between measured value $y^{<m>}$ (saturation, pressure head, outflow, displacements, stresses, etc.) and predicted value

$$e = \underline{\underline{W}}(\underline{y}^{<m>} - \underline{y}) = \underline{\underline{W}}[\underline{y}^{<m>} - (\underline{y}_0 + \Delta\underline{y})] \tag{5}$$

Therefore, the objective function (Γ) to be minimized becomes:

$$\Gamma = \underline{e}^T.\underline{e} = [\underline{\underline{W}}.\underline{y}^{<m>} - \underline{\underline{W}}.\underline{\underline{h}}.\underline{x}_0 - \underline{\underline{W}}.\underline{\underline{h}}.\Delta\underline{x}]^T.[\underline{\underline{W}}.\underline{y}^{<m>} - \underline{\underline{W}}.\underline{\underline{h}}.\underline{x}_0 - \underline{\underline{W}}.\underline{\underline{h}}.\Delta\underline{x}] \tag{6}$$

where $\underline{x}_0$ is initial guess of unknown variables, $y^{<m>}$ is the measurement data and $\underline{\underline{W}}$ is the weight factor (where W_{ij} = weight value, $W_{ij} = 0$ for i $\neq$ j).

In the optimization process, the minimization method for the objective function and the sensitivity analysis are embedded into a multi¬level algorithm. Minimization starts with the initial guess following van Genuchten parameters P_0 and m. The values of initial guess parameters are repeatedly updated until a certain stopping criterion is satisfied, or the convergence criterion is achieved. At each step, the initial parameter values for the minimization are set by the result of the previous step to the new parameterization.

4 Results & Discussions

For convenience in data processing, the data used as input data for analysis and simulation were selected at 2 locations in each soil layer (L2, L4 in medium sand layer; L5 and L8 in fine sand layer), especially at location nearby the layer interface, so that the data set must fully represent the information required for the compilation of the SWCC and input value for simulation.

4.1 Experimental Results

Under the high positive pore water pressure, the water in the soil column immediately flows out with high velocity right after opening the valve at the bottom of the column. When water in the soil drains, the pore water pressure will decrease at four chosen locations (Fig. 2). The pore water pressure at all locations in the soil column tends to get closer to the hydrostatic line (final condition) over time.

In desaturation experiment for sand column, the surface settlement is very small during the test, so the void ratio of soil is assumed to be constant throughout the experiment. This is consistent with the assumption of the Richards equation with no movement of soil particles [9].

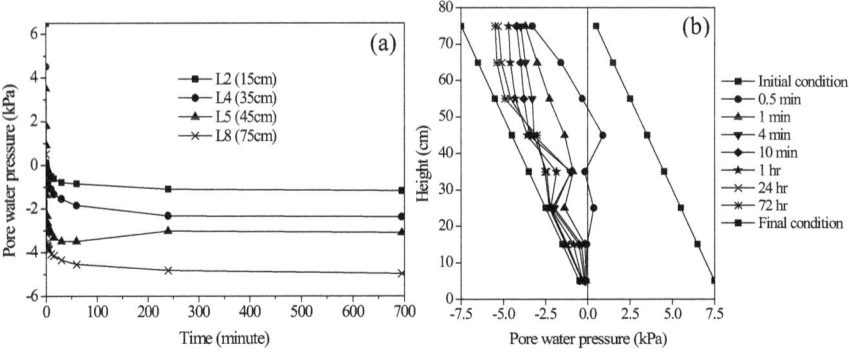

Fig. 2 **a** Pore-water pressure vs. time at four different locations, **b** Pore-water pressure profile

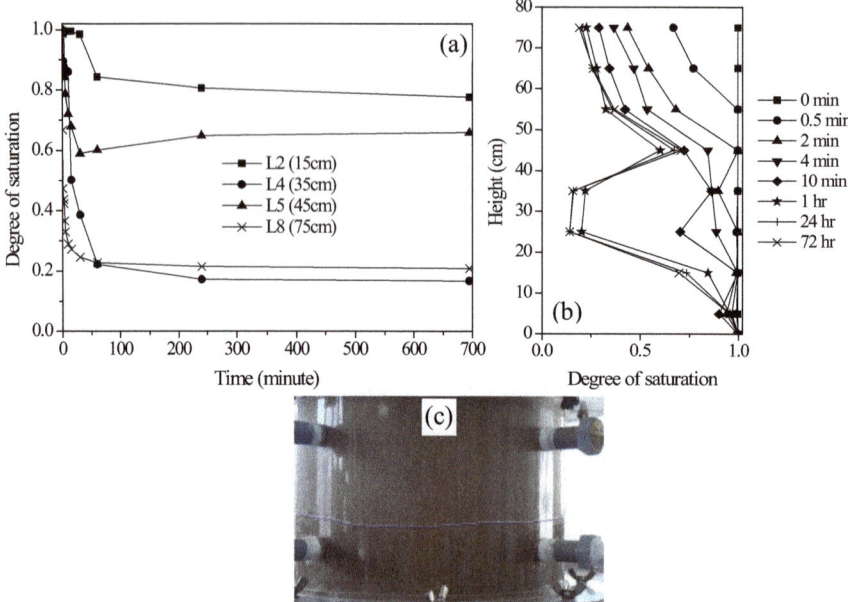

Fig. 3 a, b Water saturation profile at L2, L4, L5 and L8 during the test, **c** Water remaining at interface between two sands in laboratory test

The degree of saturation of soil for four chosen locations at any time during the experiment was calculated using a calibration curve for relation between soil resistivity and soil saturation [1]. Figure 3 shows that at the beginning of the experiment, the degree of saturation in the soil column at locations was uniformly equal to 99%. The soil saturation at different locations started to decrease only after the pore water pressure at those locations approaches to air-entry values or when the capillary fringe passes through the profile.

When the unsaturated porous sand with a fine grained sand overlays on a medium-grained sand, the capillary barrier phenomenon is developed during experiment [15]. The gradual decrease in degree of saturation across the interface from 65% (L5) in the fine sand layer to 15.8% (L4) in the medium sand layer (Fig. 3b) was obtained, that value shows that around the junction zone between the two soil pore systems, there is always an intermediate retention property. The nearly saturated liquid flow across the interface explains the lack of marked discontinuity in water saturation that would be expected during unsaturated flow in layered soils. In addition, the permeability barrier effect was limited to a small increase in the fine sand. The capillary barrier was also the main cause of high hysteresis at location around the interface of two layers (Figs. 2 and 3).

4.2 Numerical Results

In this study, the van Genuchten parameters P_0 and m ($P_0 = 1.4$ and $m = 0.8$ for medium sand; $P_0 = 3.0$ and $m = 0.8$ for fine sand) are referenced from previous independent studies [9] as the initial guess to reduce the number of iterations in the calculation of unsaturated hydraulic properties. The value of P_0 and m were calculated by solving iteratively the general flow equation and result values of the parameters after each iteration was used as initial guess for the next loop. This process is repeated until the solution agrees with measured data. After 9 iterations, inversion results converged to $m_m = 0.9$, $P_{0m} = 1.54$ for medium sand and $m_f = 0.92$, $P_{0f} = 3.83$ for fine sand.

The comparison between calculated data and observed data of pore water pressure shows the well agreement (Fig. 4). This figure indicates that at location of L2, L4, L5 and L8, the calculated data of pore water pressure dropped more quickly than observed data at the beginning of testing and then progressively decreased with time during testing. The final predicted values of pore water pressure at four locations were all lower than measured values at the end of the experiment.

The Fig. 5a also shows a good result of calculated data of soil saturation in comparison with measured data. However, contrast to pore water pressure, the predicted saturation dropped more slowly than measured values right after beginning of testing and gradually decrease to get value lower than experimental values at the end of experiment. Collecting data of pore water pressure and degree of saturation at the same location during the experiment can plot the SWCC (Fig. 5b). The difference of the

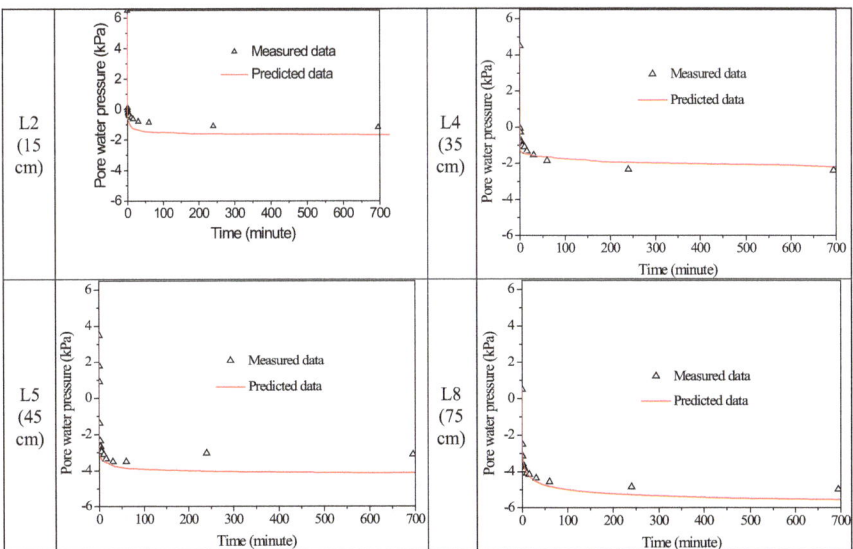

Fig. 4 Comparison between calculated and observed data of pore water pressure at four chosen locations.

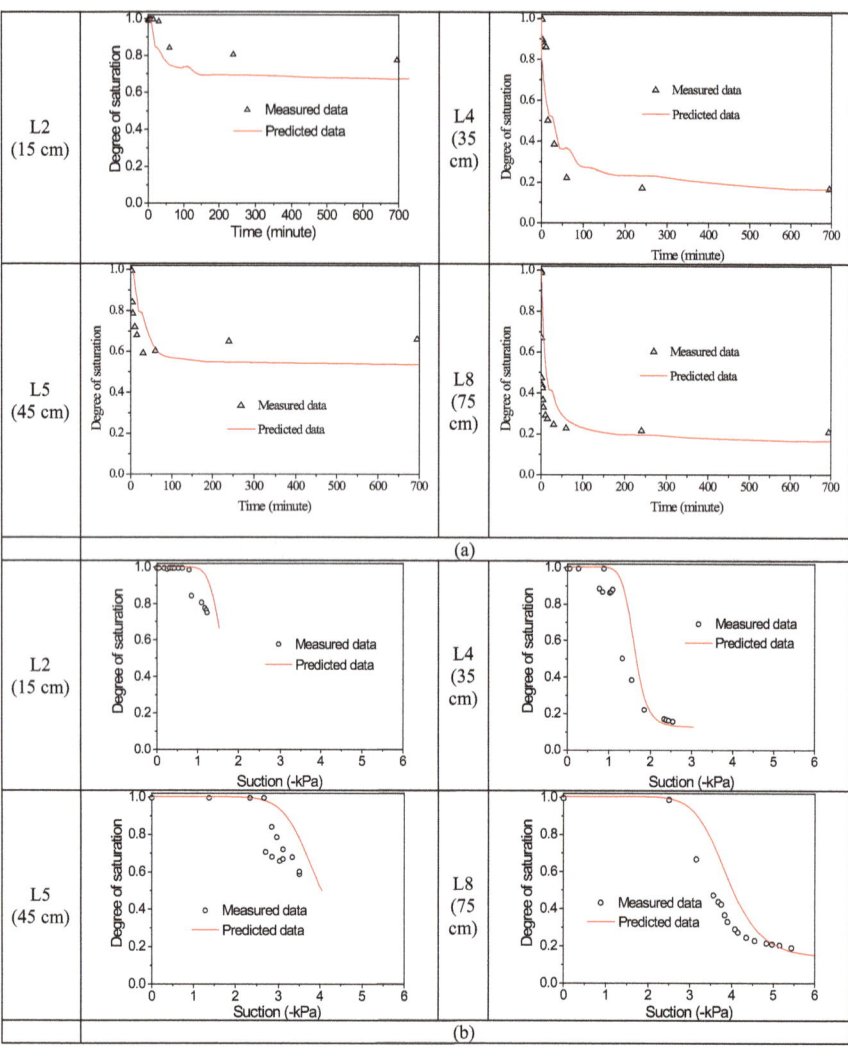

Fig. 5 Comparison between simulated and experimental data of hydraulic properties at four chosen locations:(a) Degree of saturation vs time, (b) SWCC.

pore size between fine sand layer and medium sand layer in soil column and considerable changes of pore water pressure at bottom of soil column causes the discrepancy between predicted and measured data (Figs. 4 and 5). In other words, when the fine sand layer overlying medium sand layer in 1-D desaturation experiment, the capillary barrier effect is the main cause of the difference between modeled values and observed values. Therefore, the capillary barrier effect can be controlled by the contrast in the hydraulic properties of the layer and the soil formation [15]. The bottom medium-sand deposits start to drain before the fine-sand layer because the

Table 2 RMSE from calculated and observed values

	Pore-water pressure (kPa)				Saturation			
Position	L2	L4	L5	L8	L2	L4	L5	L8
RMSE	0.57	0.62	1.02	0.55	0.065	0.052	0.11	0.045
Average	0.6		0.785		0.059		0.077	

medium sand has lower air-entry value than the fine sand. This results in a capillary barrier effect with slow water drainage across the interface, water retention in the fine sand layer and prevent the pore water pressure and water saturation in area affected by capillary barrier effect from reaching static conditions even after long periods of time.

In order to evaluate the difference between the calculated value and observed values of unsaturated hydraulic properties, the root mean square error (RMSE) was conducted for simulated and measured values of suction and water saturation of soils. Table 2 shows the accuracy of comparison between the measured values and modeled values with good results at locations of L2, L4 and L8. The hysteresis phenomenon strongly affects the calculation results, therefore, at locations influenced by the strongest hysteresis, the difference between predicted and measured data will be highest, which was shown with highest values of RMSE at L5 (Table 2, Fig. 7b [9]).

5 Conclusions

The issue of hydraulic properties for unsaturated layered sand showed the acceptable performance of inverse analysis in predicting the unsaturated hydraulic properties of sands under capillary barrier effect. The comparisons between predicted and measured hydraulic properties in layered soil column are less agreed than the comparisons in the homogeneous soil column. When the layered soil profiles consist of single homogeneous soil layers, the straightforward calculation can be conducted using known hydraulic parameters of single homogeneous soil to compute the unsaturated hydraulic properties of sand. The straightforward calculation shows better results and requires much shorter time to estimate the unsaturated hydraulic properties than the inverse analysis.

In the experiment, at location slightly above the interface, besides the capillary barrier effect, the hysteresis effect also mainly governed to final results and iterative steps for the convergence in inversion analysis. Therefore, in this case, the multi-step outflow condition is more appropriate than one-step outflow condition.

This study showed that inversion method could be a potential indirect approach to estimate hydraulic properties of unsaturated soils under the effect of the capillary barrier.

Since the laboratory experiment with layered sand and with more initial boundary conditions was carried out, this study should be extended to further research for different soils with various sedimentation and much more initial boundary conditions.

Acknowledgements This research is funded by Vietnam National University HoChiMinh City (VNU-HCM) under grant number C2021-20-38.

References

1. Archie GE (1942) The electrical resistivity log as an aid in determining some reservoir characteristics. Petroleum Transactions of AIME 146:54–62
2. Bahar A, Alexandre RC (2016) Influence of capillary barrier effect on biogas distribution at the base of passive methane oxidation biosystems: Parametric study. Waste Manage 63:172–187
3. Bussiere B, Aubertin A, Chapuis RP (2003) The behavior of inclined covers used as oxygen barriers. Can Geotech J 40:512–535
4. Dagmar T, Jiri M (2010) The influence of artificial sealing on the capillary barrier's function. Waste Manage 30:125–131
5. Durner W, Schultze EB, Zurmuhl T (1997) State-of-the-art in inverse modeling of inflow/outflow experiments. In: Characterization and measurement of the hydraulic properties of unsaturated porous media, Proceedings of International Workshop, Riverside, CA. October 22–24, edited by MT van Genuchten, FJ Leij, L Wu, pp. 661–681
6. Fredlund DG, Rahardjo H (1993) Soil mechanics for unsaturated soils. Wiley, NewYork
7. Kechavarzi C, Soga K (2002) Determination of Water Saturation Using Miniature Resistivity Probes During Intermediate Scale and Centrifuge Multiphase Flow Laboratory Experiments. Geotechnical testing journal, GTJODJ 25(1):95–103
8. Khire MV, Benson CH, Bosscher PJ (1999) Field data from a capillary barrier and model predictions with UNSAT-H. J Geotech Geoenviron Eng 7(5):18–27
9. Nam TV, Min TK, Shin HS (2013) Using inverse analysis to estimate hydraulic properties of unsaturated sand from one-dimension outflow experiments. Eng Geol 164:163–171
10. Nyhan J, Schofield T, Starmer R (1997) A water balance study of four landfill cover designs varying in slope for semi-arid regions. J. Envir. Quality 26:1385–1392
11. Rahardjo H, Kim Y, Gofar N, Leong EC, Wang CL, Wong JLH (2018) Field instrumentations and monitoring of GeoBarrier System for steep slope protection. Transportation Geotechnics 16:29–42
12. Rooney DJ, Brown KW, Thomas JC (1998) The effectiveness of capillary barriers to hydraulically isolate salt contaminated soils. Water Air Soil Pollut 104:403–411
13. Santamarina JC, Fratta D (2005) Discrete Signals and Inverse Problems. Wiley, New York
14. van Genuchten MT (1980) A Closed-Form Equation for Predicting the Hydraulic Conductivity of Unsaturated Soils. Soil Science. Society. Am. J 44:892–898
15. Zornberg JG, Bouazza A, McCartney JS (October 2010) Geosynthetic capillary barriers: State-of-the-knowledge. Geosynth Int 17(5):273–300

Port and Coast Sessions

Assessing Wave Attenuation by Mangrove Forest in Bac Lieu Province Using XBeach

Nguyen Kiet and Nguyen Danh Thao

Abstract Mangroves naturally protect the coast against high waves, storm surges, sea level rise and erosion by dissipating the energy of the incoming waves. As the importance of mangrove forest is earning attention recently, it is necessary to assess the ability of mangroves in reducing waves. This paper analyzes the reduction of wave energy as well as wave height due to mangrove forests at the coast of Bac Lieu province based on the XBeach 1D numerical model. The factors that have most effect on the wave reduction efficiency of mangroves are forest density, forest band width, incident wave height and water depth. These results show that the minimum band width of mangrove forest should be equal to 500 m, 450 m, 250 m and 150 m with respect to the case of no mangrove, sparse, medium and high density in order to reduce approximately 95% of the energy of the incoming wave.

Keywords Mangroves · XBeach 1D · Wave attenuation · Bac Lieu province · Numerical model

1 Introduction

Mangroves play an important role in protecting the coastline from erosion and storm surge by dissipating wave energy. In particular, the interlaced root systems and trunks of mangroves have the ability to reduce the incident wave height and tidal effects [1]. Therefore, the assessment of the role of mangroves in coastal protection is getting more attention when climate change is becoming more imperative which causes natural disasters and negative impacts on coastal areas.

Regarding this problem, there are three main research directions, which are physical model, numerical model and field survey. Typical for the physical model are

N. Kiet · N. D. Thao (✉)
Faculty of Civil Engineering, Ho Chi Minh City University of Technology (HCMUT), 268 Ly Thuong Kiet Street, District 10, Ho Chi Minh City, Vietnam
e-mail: ndthao@hcmut.edu.vn

Vietnam National University Ho Chi Minh City, Linh Trung Ward, Thu Duc District, Ho Chi Minh City, Vietnam

the results showing that the arrangement of mangroves does not significantly affect wave reduction (less than 10%) [2]. Ismail et al. [3] demonstrated that the roots of mangroves dissipate more waves than the stems and foliage.

The second research direction is to conduct field trips to measure mangrove parameters, wave data and topography in the selected survey area. Horstman et al. [4] determined that the wave height will decrease by nearly 90% in the first 100 m when waves propagate into the mangroves in the coastal area of Thailand. Mazda conducted measurements on the Vinh Quang coast, in the north of Vietnam, and concluded that the wave reduction of mangroves is highly dependent on the type of mangrove [5, 6]. However, the disadvantage of this research direction is the limitation of measurement data and the wave attenuation is actually a combination of both mangrove trees and effect of forest floor topography [7].

The third research methodology is to use numerical models to simulate the impact of the incoming waves propagating into mangrove areas. The numerical model can analyze the initial conditions separately and assess the impact of these above factors, addressing limitations from the field surveys or physical measurements. Mangroves in Vietnam have many diverse species, concentrated mainly in the coastal areas of the North and the South which have attracted many studies using numerical model. Burger [8] used the SWAN model to simulate the propagation of waves into mangroves in Thai Binh province. Phuoc and Massel [9] developed a numerical model for mixed mangroves with variable water depth, obtained some results about the very effective role of mangrove roots in sediment accumulation, contributing to the prevention and control of mangroves erosion in the coastal area of Can Gio.

This paper focuses on the coastal area of Bac Lieu province which receives a large source of sediment from the Mekong River. This has all the favorable factors for the clonal multiplication and development of mangrove system. The main species of mangroves in Bac Lieu are Rhizophora, Avicennia and Bruguiera [10]. Particularly, the Rhizophora accounts for the largest proportion, so it can be used as a representative tree for the numerical model. The results from field data show that the parameters of mangrove forests can be classified according to the forest density [11] as in Table 1.

XBeach model (developed by TUDelft University to simulate wave propagation and interaction between hydrodynamic factors) is used and taken into account for 4

Table 1 Mangroves structure of the studied area

Density	Root			Trunk			Canopy		
	Nv (root/m2)	bv (cm)	ah (m)	Nv (trunk/m2)	bv (cm)	ah (m)	Nv (canopy/m2)	bv (cm)	ah (m)
Sparse	15	1	0	0.3	20	2	20	1	10
Medium	45	2	1	0.6	45	5	110	1	12
Dense	70	3	1	0.9	75	8	450	1	12

N_v: number of roots, trunks and canopies per square meter; b_v: diameter of roots, trunks, canopies; a_h: height of roots, stems and canopies.

cases of density: sparse vegetation (**C1**), medium vegetation (**C2**), dense vegetation (**C3**) and no vegetation (**C4**). Accordingly, the factors that have the most significant impact on the wave reduction efficiency of mangroves in Bac Lieu province are analyzed.

2 Methodology

2.1 Input data

The input parameters are taken from the Delft Dashboard model, divided into 3 main groups: bottom topography, mangroves parameters and hydrodynamics (waves & tides). According to Mendez and Losada [12], the interaction of waves and coastal vegetation depends mainly on wave height, wave period and wave direction. In this study, the model only considers the main wave direction that is perpendicular to the shore for the XBeach 1D case study.

The wave height, wave peak period and water level are taken according to the data of Typhoon Linda which occurred in October & November 1997. This storm is considered the "worst" in the Mekong Delta region [13].

According to Fig. 1b, the significant wave height $H_{s,max}$ was recorded up to 3 m as well as the largest wave peak period $T_{p,max}$ was 10 s around the end of October and the beginning of November 1997 with the tidal range is around 2.5 m. In the model, 5 cases of wave height, 4 cases of wave peak period and 9 cases of water depth due to tides were recorded with respect to each case of mangrove tree density (sparse, medium, dense vegetation and no vegetation), then generate totally 180 input parameter combinations.

2.2 Governing Equations

XBeach model is built based on the Wave Action Balance [14]:

$$\frac{\partial A}{\partial t} + \frac{\partial C_{gx} A}{\partial x} = -\frac{D_w + D_f + D_v}{\sigma} \tag{1}$$

where: $A = S_w(x,t)/\sigma(x,t)$; t is the time; S_w – wave energy density; σ – relative wave frequency; C_{gx} – velocity of wave group; D_w – wave energy reduced due to wave breaking; D_f – wave energy decreases due to bottom friction; D_v – wave energy decreases due to the presence of mangroves. In Eq. (1), the right-hand side is the values D_f, D_w, D_v representing the wave energy dissipation process. In addition, generating sources such as wind will be ignored. Calculated results from the Eq. (1) are the values of wave radiation stress - which are also the input parameters for the system of Eqs. (2) and (3). Output values such as: water level, wave velocity and wave height are calculated according to the two-dimensional system of Shallow Water Equations is written as:

$$\frac{\partial \eta}{\partial t} + \frac{\partial U^L h}{\partial x} = 0 \tag{2}$$

$$\frac{\partial U^L}{\partial t} + U^L \frac{\partial U^L}{\partial x} = g \frac{\partial \eta}{\partial x} + \frac{-\tau_{bx} + F_w + F_v}{\rho h} \tag{3}$$

where: η - water level over time t; h – water depth; U^L – Lagrange wave velocity; τ_{bx} - bottom shear stress; F_v – resistance force due to mangroves; F_w – wave force due to wave radiation stress. In order to evaluate the wave reduction efficiency of mangroves, wave reduction ratio r (m^{-1}) according to the formula Mazda [5] is used:

$$r = \frac{\Delta H}{H \Delta x} = \frac{H_1 - H_2}{H_1} \times \frac{1}{\Delta x} \tag{4}$$

where: H_1 – wave height at the starting point of mangrove; H_2 – wave height after moving a distance $\Delta x = 100$ m.

2.3   Model Setup

In the mathematical model of XBeach, the bottom topography needs to satisfy the requirements of offshore boundary conditions, as follows: ratio of significant wave height H_s and water depth h, ratio of wave group velocity C_g and wave velocity C. In order to satisfy these two conditions, the model's calculation domain has a length of 12 km from the offshore boundary to the starting point of the mangrove forest. The mangrove tree model in XBeach is shown in 3 parts representing the roots, trunk and canopy as illustrated in Fig. 2.

The computational grid resolution increases gradually from offshore boundary to shore in order to shorten the computational time as well as the results have to be enough accuracy. To optimizing the computational grid size, there are 5 cases with the smallest grid cells at the shore are considered: 1 m, 2 m, 5 m, 8 m and 10 m. The deviation results of short wave height H_s, water level η, wave force F_x and long wave height of 5 cases are shown in Fig. 3. Deviation values of the 5-m-grid-cell case are less than 10% and computational time is roughly 4 min per case, which are acceptable accuracy and feasible running time. Therefore, the 5-m-grid-cell should be used in XBeach model.

In order to validating the model, the value of significant wave height from 31st Oct, 1997 to 04th Nov, 1997 is considered because the maximum significant wave height reached over 2.5 m in this period. The value of H_s need to be reduced to simulate significantly the measured data. Therefore, the formula f_w of Uchiyama [15] is used

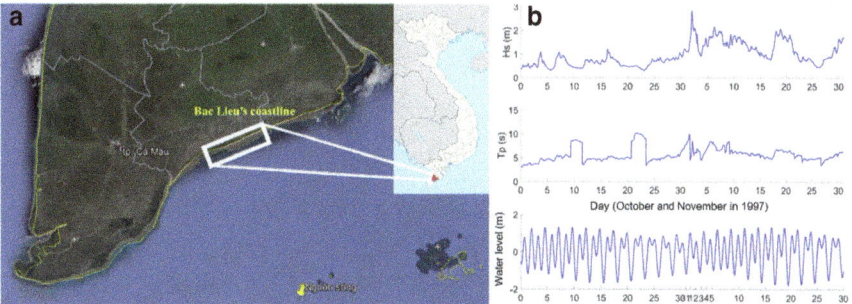

Fig. 1a Studied area in the coastal area of Bac Lieu province. (Fig. 1**b** Source: Google Earth Pro) and Wave and tidal data were recorded from Typhoon Linda at the coast of Con Dao in October and November 1997 at coordinates (8°30'N, 106°E) (Source: ERA – Interim and Delft Dashboard)

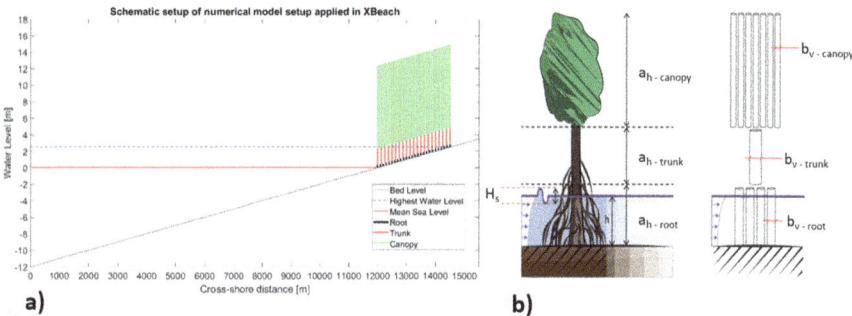

Fig. 2a The mathematical model of XBeach, mangrove forest is the cylinder with diameter, height and density shown in Table 1; 2**b** The parameters of Rhizophora is illustrated in [11]

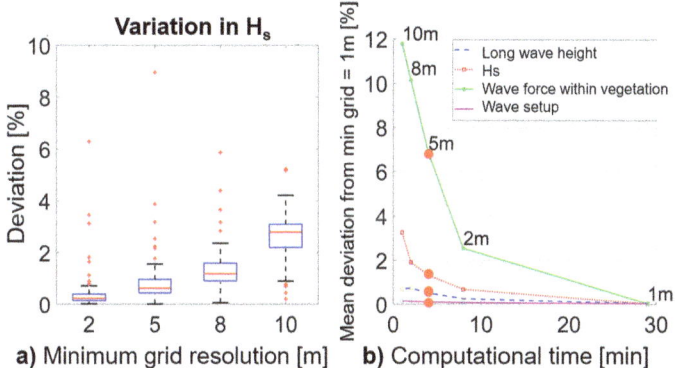

Fig. 3 Deviation values of long wave height, H_s, wave force due to vegetation and water setup

to determine ($\mathbf{f_w = 0.011}$) instead of default value of $f_w = 0$ and is used to calibrate the model in this area within the mentioned period.

3 Results and Discussions

Factors affecting wave reduction of mangroves are forest density, incident wave height, wave peak period, forest band width, bottom friction and water depth which are specifically analyzed through the results of the model.

Wave energy reduction due to mangroves: According to Eq. (1), the wave energy dissipated when propagating ashore depends on three factors: dissipation due to wave breaking, bottom friction and mangroves. Calculated results in the typical case (wave height $H_s = 3$ m; wave peak period $T_p = 10$ s; water level $= 2.5$ m; forest band width $= 2500$ m) are $D_v = 427$ (W/m^2), $D_f = 0.57$ (W/m^2) and $D_w = 2.7$ (W/m^2). The energy dissipated by bottom friction D_f and breaking wave D_w accounts for a very small proportion of only about 0.5% of the total energy dissipated due to all three factors ($D_w + D_f + D_v$).

From formula (5) [12, 16], it can be seen that the wave energy reduction due to mangroves is a function of wave height and water depth: $D_v = f(H_s, h \dots)$

$$D_v = \left(\frac{gk}{2\sigma}\right)^3 \frac{\rho C_D b_v N_v sinh^3 kh + 3sinhkh}{6\sqrt{\pi}kcosh^3 kh} \left(\frac{H_s}{\sqrt{2}}\right)^3 \tag{5}$$

where: b_v – section diameter; k – number of waves.

Wave height & water depth: Fig. 4a, dissipation wave energy D_v is proportional to wave height and inversely proportional to water depth, which is also consistent with Mazda et al. [6]. Figure 4b shows the relationship of D_v and water depth h. In considering the cases with the same wave height H_s from 0.6 m to 0.7 m and H_s from 0.8 m to 0.9 m, the water depth increases with respect to the wave energy decreased since D_v tends to decrease.

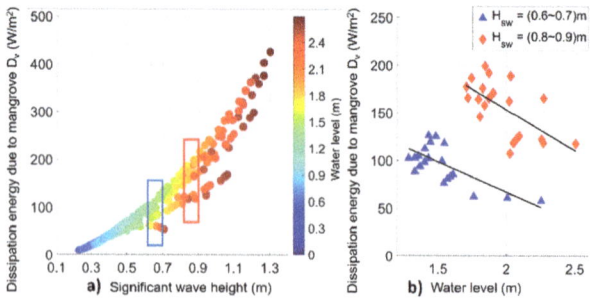

Fig. 4 Dissipation energy due to mangrove, significant wave height and water depth

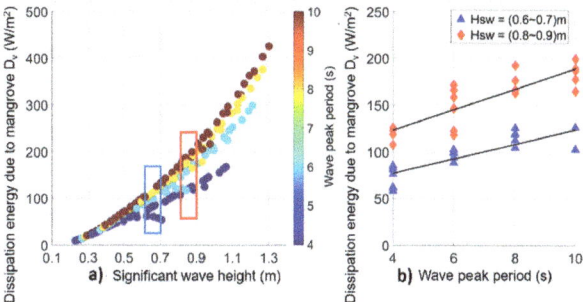

Fig. 5 Dissipation energy due to mangrove and wave peak period

Wave peak period: Considering the equivalent wave height value in Fig. 5, the case with a larger wave peak period give a higher value of D_v. Consequently, it can be concluded that the wave peak period is proportional to the dissipation wave energy D_v.

Density: The r wave reduction ratio is calculated according to formula (4) for 3 cases of different density and 1 case of no vegetation. As shown in Fig. 6a, the wave reduction ratio increases gradually with the density of sparse, medium and dense vegetation. Results from XBeach model are compared with previous studies in Fig. 6b. Regarding the case of dense vegetation (**C3**), wave reduction ratio $r_{XBeach} = [7.1/8.9] \times 10^{-3} (m^{-1})$ is equivalent to the result of $r_{Bao} = [4/12] \times 10^{-3} (m^{-1})$ [17]. Similarly for the case of sparse vegetation (**C1**), $r_{XBeach} = [0.7/3.3] \times 10^{-3} (m^{-1})$ value is not significantly different from field measured value $r_{Mazda} = [0.1/2.2] \times 10^{-3} (m^{-1})$ [5]. In terms of no mangroves (**C4**), wave reduction ratio $r_{XBeach} = [0.1/1.7] \times 10^{-3} (m^{-1})$ and studies by Mazda et al. [6], Quartel et al. [18] has the value $r_{Mazda-Quartel} = [0.5/2] \times 10^{-3} (m^{-1})$. Thus, it can be seen that the value of r

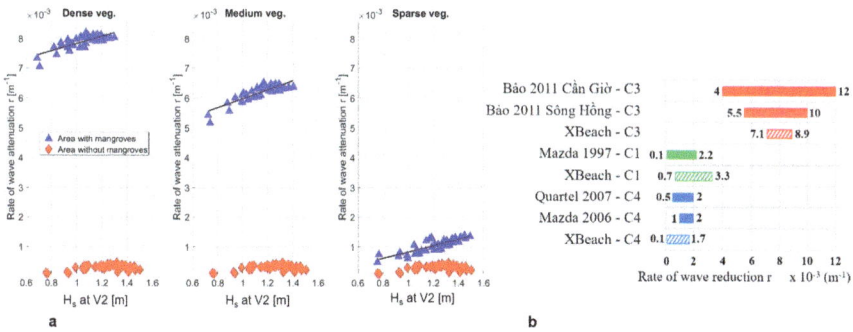

Fig. 6a Rate of wave reduction in mangrove and non-mangrove areas for three different densities, where V2 is the starting point of mangrove forest. Figure 6b The comparison of rate of wave reduction in previous studies and XBeach model

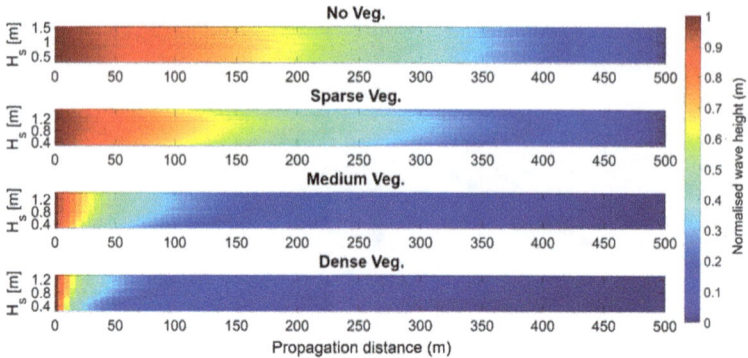

Fig. 7 Normalized wave heights with respect to 4 cases of density

wave reduction ratio obtained from the XBeach model is consistent with the results obtained in some previous studies.

In order to clarify the relationship between density and mangrove band width, wave height is normalized to the value [0 1], corresponding to the minimum value of H_s at x = 500 m and the maximum value of H_s at x = 0 m:

$$H_i^{normalised} = \frac{H_i - min(H)}{max(H) - min(H)} \tag{6}$$

where: H – data set of wave height; i – index of each element in the set; min(H) – the smallest element in the set; max(H) – the largest element in the set.

Figure 7 shows that the wave reduction of the mangrove depends very much on the density factor. In the absence of vegetation, wave propagates approximately 300 m to reduce the incident wave height by 50%. Regarding sparse vegetation, the distance of wave propagation should be equal to 200 m. In terms of medium and dense vegetation, the required mangrove band widths are 70 m and 30 m respectively. In order to reduce wave height for sparse vegetation, the width of mangrove should be larger than the dense vegetation. Higher density of mangrove only needs smaller band width to mitigate the same percentage of wave height reduction.

4   Conclusions

The mangroves are very effective in reducing the incident wave height. The wave energy reduced by mangroves is significantly dependent on initial wave height, water depth before mangrove, density and forest band width. Specifically, the density, the forest band width, the incident wave height and the wave peak period are proportional to the wave reduction level of mangrove trees. In contrast, the water depth also has a great impact, which is inversely proportional to wave reduction.

The wave reduction ratio obtained from the mathematical model is relatively accurate compared with some previous studies, especially in the case of forests with dense vegetation, sparse vegetation and no mangroves. The wave reduction rate is proportional to the forest density, the higher the forest density, the greater the wave reduction rate. Regarding the case of without mangrove, the wave reduction rate is very low in the first 100 m, mainly due to phenomena such as bottom friction or breaking waves when propagating ashore.

Density and band width are closely related to wave reduction in Bac Lieu province. The higher density forests need smaller band widths to reduce the same ratio of wave height reduction. In order to entirely dissipate wave energy (over 95%), the minimum width of the mangrove strip is roughly 500 m, 450 m, 250 m and 150 m respectively for the cases of no vegetation, sparse vegetation, medium vegetation and dense vegetation.

Acknowledgements This research is funded by Vietnam National University HoChiMinh City (VNU-HCM) under grant number **562-2020-20-03**.

References

1. Kathiresan K, Rajendran N (2005) Coastal mangrove forests mitigated tsunami. Estuar Coast Shelf Sci 65(3):601–606
2. Hashim AM, Catherine SMP (2013) A laboratory study on wave reduction by mangrove forests. APCBEE Proc 5:27–32
3. Ismail H, Abd Wahab AK, Alias NE (2012) Determination of mangrove forest performance in reducing tsunami run-up using physical models. Nat Hazards 63:939–963
4. Horstman EM, Dohmen-Janssen CM, Narra PMF, van den Berg NJF, Siemerink M, Hulscher SJMH (2014) Wave attenuation in mangroves: A quantitative approach to field observations. Coast Eng 94:47–62
5. Mazda Y, Magi M, Kogo M, Hong PN (1997) Mangroves as a coastal protection from waves in the Tong King delta. Vietnam. Mangroves and Salt marshes 1(2):127–135
6. Mazda Y, Magi M, Ikeda Y (2006) Wave reduction in a mangrove forest dominated by Sonneratia sp. Wetlands Ecol Manage 14:365–378
7. Vu Duy Vinh (2015) Applying numerical model to assess role of mangrove forest in wave attenuation in Hai Phong coastal area. J Mar Sci Technol 15(1):67–76
8. Burger B (2005) Wave attenuation in mangrove forests. Thesis, Delft University of Technology, M. Sc
9. Phuoc VLH, Massel SR (2006) Experiments on wave motion and suspended sediment concentration at Nang Hai, Can Gio mangrove forest. Southern Vietnam. Oceanologia 48(1):23–40
10. Hanh NH, Manh DQ, Sang TV, Ket CB (2019) Restoration and development of mangroves in the coastal area of the Mekong Delta. Institute of Ecology for Building Protection, Electronic Journal of Forest and Environmental Protection (in Vietnamese)
11. Janssen MPJM (2016) Flood hazard reduction by mangroves. Master Thesis, Tudelft Repository, 103 pages
12. Mendez FJ, Losada IJ (2004) An empirical model to estimate the propagation of random breaking and nonbreaking waves over vegetation fields. Coast Eng 51(2):103–118

13. Takagi H, Anh LT, Thao ND (2017) 1997 Typhoon Linda storm surge and people's awareness 20 years later: Uninvestigated worst storm event in the Mekong Delta. Natural Hazards Earth System Science Discussion, European Geosciences Union
14. Holthuijsen LH, Booij N, Herbers THC (1989) A prediction model for stationary, short-crested waves in shallow water with ambient currents. Coast Eng 13(1):23–54
15. Yusuke Uchiyama JC (2010) Wave–current interaction in an oceanic circulation model with a vortex-force formalism: Application to the surf zone. Ocean Model 34(1–2):16–35
16. Dalrymple RA, Kirby JT, Hwang PA (1984) Wave diffraction due to areas of energy dissipation. J Waterw Port Coast Ocean Eng 110:67–79
17. Bao TQ (2011) Effect of mangrove forest structures on wave attenuation in coastal Vietnam. Oceanologia 53:807–818
18. Quartel S, Kroon A, Augustinus PGEF, Van Santen P, Tri NH (2007) Wave attenuation in coastal mangroves in the red river delta. Vietnam. J. Asian Earth Sci. 29:576–584

Mapping Tidal Harmonic Constant Map from Vung Tau – Bac Lieu, Viet Nam by Using a Numerical Model in Curvilinear Coordinate

Tran Thi Kim, Nguyen Thi Thu Hong, Nguyen Khac Thanh Long, Nguyen Ky Phung, and Nguyen Thi Bay

Abstract Harmonic constituents are used to predict tide levels, each constituent represents a distinct factor contributing to the tidal regime. In this paper, a hydraulic model in the curvilinear coordinates is applied to simulate eight tidal constituents (M_2, S2, O_1, K_1, N_2, K_2, P_1 and Q_1 from Vung Tau to Bac Lieu, Vietnam. The hydraulic model with the two-dimensional orthogonal curvilinear grid has the advantage of increasing the accuracy in the results at the domain boundary. The numerical method of this model derives from the solution of the Reynolds system of equations averaged over the depth in the curvilinear coordinate systems. The model verification is implemented based on the equilibrium of the tidal currents of energy. The results of this model are used to mapping tidal constituents from Vung Tau to Bac Lieu, Vietnam. From Vung Tau to Bac Lieu, the amplitude and phase values of the tidal constituents K_1 are higher than those from Vung Tau to Tra Vinh and approximately as high as those in Bac Lieu with the amplitude of 0.5 m and a phase range from 345^0 to 350^0. Meanwhile, the amplitude values of the constituent O_1 fluctuate within the value of 0.3 m–0.31 m, the phase values are in the range $270^0 - 285^0$. The values of the constituent M_2 is the highest (0.8 m–0.9 m) which is double that of the constituent S_2.

T. T. Kim · N. T. T. Hong
Faculty of Marine Resource Management, Ho Chi Minh City University of Natural Resources and Environment, 236B Le Van Sy Street, Ward 1, Tan Binh District, Ho Chi Minh City, Vietnam

T. T. Kim
Institute for Environment and Resources, Vietnam National University, 142 To Hien Thanh Street, Ward 14, District 10, Ho Chi Minh City, Vietnam

N. K. T. Long · N. T. Bay (✉)
Faculty of Civil Engineering, Ho Chi Minh City University of Technology (HCMUT), 268 Ly Thuong Kiet Street, District 10, Ho Chi Minh City, Vietnam
e-mail: ntbay@hcmut.edu.vn

Vietnam National University Ho Chi Minh City, Linh Trung Ward, Thu Duc District, Ho Chi Minh City, Vietnam

N. K. Phung
Institute for Computational Science and Technology, SBI Building, Quang Trung Software City, Tan Chanh Hiep Ward, District 12, Ho Chi Minh City, Vietnam

© The Author(s), under exclusive license to Springer Nature Singapore Pte Ltd. 2023   701
J. N. Reddy et al. (eds.), *ICSCEA 2021*, Lecture Notes in Civil Engineering 268,
https://doi.org/10.1007/978-981-19-3303-5_63

Keywords Harmonic constant · Tidal constituent map · Curvilinear coordinate

1 Introduction

Each harmonic constituent has characteristic amplitude and phase, describing how much that one constituent dominates the tidal signal for the site as a whole. Many models using predicting tidal were constructed such as U-Tide [1], T- Tide [2], LSEIM [3], normal time–frequency transform (NTFT) [4] was applied by Su et al. [5], the China coastal (FVCOM) [6–9], the Princeton Ocean Model (POM) [10–12]. Guohong Fang et al. (1999) [13] simulated the principal tidal constituents M_2, S_2, K_1 and O_1 in the South China Sea, Gulf of Tonkin and Gulf of Thailand by using the numerical model. Minh N.N. et al. (2013) calculated characterize tidal dynamics in the Gulf of Tonkin using a high-resolution model and combination of all available data [14]. Furthermore, for Mekong delta, Hung P.M. et al. (2019) simulated the mechanisms of the overall tidal wave propagation in the South China Sea with specific attention for the South China Sea [15]. For coastal areas, tidal currents are affected by changes in the depths and morphology of these areas which have not been clarified from previous studies. For the analyses mentioned above, we identified the harmonic constants map of 8 main tidal constituents from Vung Tau to Bac Lieu, Vietnam by using a numerical model, in which, the bottom friction component changed according to the bottom level. The hydraulic model with the two-dimensional orthogonal curvilinear grid has the advantage of increasing the accuracy in the results at the domain boundary.

2 Material and Method

2.1 Material

The study area is the coastal area from Vung Tau to Bac Lieu (Fig. 1a), geographically located from 106.009° to 106.84° East longitude and 8.91° to 9,63° North latitude.

The domain consists of four boundaries: three liquid boundaries are the East Sea boundaries; the solid boundary is the coastline from Vung Tau to Bac Lieu. A computational mesh is a curved-perpendicular mesh with a mesh size of 130 * 155 cells with dx, dy in the range of 300 m to 500 m.

Topography: The topographic data is collected in 2010 from the Southern Institute of Water Resources Research, Vietnam (SIWRR). The parameters in the model are set up in Table 1, Time step (dt): 72 s, Coriolis: 2.4096 × 10–5, the density of seawater: 1026 kg/m3. The friction bed coefficients for four tidal constituents were set up changing the water depth with the range of 0.015 to 0.058 for the coastal area (< 20 m) which as the shore, the greater friction and the constant with 0.006 for the offshore area (20 m–88 m).

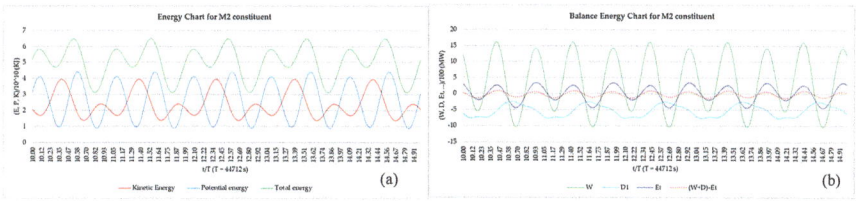

Fig. 1 The simulation results of the energy of the constituent M_2 (a); and the simulation results of the tidal constituents energy balance: M_2, S_2, K_1 and O_1 (The orange line represents the results of variation in energy in the domain (Et), the green line is total energy across the liquid boundary (W), the blue line represents loss of energy caused by bottom friction (D_1) and the red color is energy balance ((W + D)-Et) (b)

Table 1 The Parameters in the model

Tidal Constituents	Sigma	Period (s)	Friction bed coefficient for coastal area
K1	7.26×10^{-5}	86,148	0.024 to 0.058
O1	6.76×10^{-5}	92,952	0.021 to 0.055
P1	7.25×10^{-5}	86,652	0.021 to 0.053
Q1	6.50×10^{-5}	96,732	0.015 to 0.048
M2	14.05×10^{-5}	44,712	0.014 to 0.048
S2	14.54×10^{-5}	43,200	0.011 to 0.045
N2	1.38×10^{-5}	45,576	0.018 to 0.043
K2	1.46×10^{-5}	43,092	0.012 to 0.049

Boundary conditions: The harmonic constant data for the four tidal constituents were extracted from the DTU10 Global Tide Model for the three open boundaries at sea.

2.2 The Governing Equations in the Curvilinear Coordinate System

The governing equations in the curvilinear co-ordinates (ξ, η) are constructed based on the Reynolds equation in Cartesian coordinates [16]:

$$
\begin{cases}
p_\tau + gHJ^{-1}\left(g_{22}\varsigma_\xi - g_{12}\varsigma_\eta\right) = \Psi_1 \\
q_\tau + gHJ^{-1}\left(g_{11}\varsigma_\eta - g_{12}\varsigma_\xi\right) = \Psi_2 \\
JH_\tau + p_\xi + q_\eta = 0
\end{cases}
\tag{1}
$$

In which: $\tau = t$; $p = JUH$; $q = JVH$; $H = h + \varsigma$
U, V- the "contravariant" base vectors of the curvilinear coordinate system

$$U = J^{-1}\left(uy_\eta - vx_\eta\right); \quad V = J^{-1}\left(-vy_\xi - vx_\xi\right) \tag{2}$$

$$\Psi_1 = \Psi_{a1} + \Psi_{t1} + \Psi_{k1}; \quad \Psi_2 = \Psi_{a2} + \Psi_{t2} + \Psi_{k2} \tag{3}$$

$\Psi a1$, $\Psi a2$—the nonlinear component in curvilinear coordinate system ξ, η
Ψ_{t1}, Ψ_{t2}—Friction bed component; Ψ_{k1}, Ψ_{k2}—the Coriolis component.

$$\Psi_{a1} = -\left[(pU)_\xi + (pV)_\eta + JH\left(U^2\Gamma_{11}^1 + 2UV\Gamma_{12}^1 + V^2\Gamma_{22}^1\right)\right]; \tag{4}$$

$$\Psi_{a2} = -\left[(qU)_\xi + (qV)_\eta + JH\left(U^2\Gamma_{11}^2 + 2UV\Gamma_{12}^2 + V^2\Gamma_{22}^2\right)\right]; \tag{5}$$

$$\Psi_{t1} = -\frac{K}{H}|u|p; \quad \Psi_{t2} = -\frac{K}{H}|v|q; \tag{6}$$

$\Gamma_{i,j}^k$ – Christoffel symbol type $-$ II is defined as follows : $\Gamma_{i,j}^k = \dfrac{\partial e_i}{\partial \xi^j} e^k$. (7)

e_i; e_j—the base vectors of the curvilinear coordinate system ξ, η; e_i,e_j–the "covariant" base vectors $\vec{e}$ [17–19].
The mesh is generated based on the Poisson Eqs. (8) [18, 20].

$$\Delta^2\xi = P(\xi, \eta); \quad \Delta^2\eta = Q(\xi, \eta) \tag{8}$$

where: P and Q are the control functions. The solution of this equation system in the domain (ξ, η) is:

$$L(r) = g_{22}r_{\xi\xi} - 2g_{12}r_{\xi\eta} - g_{11}r_{\eta\eta} = -J^2\left(\Pr_\xi + Qr_\eta\right) \tag{9}$$

J: the "Jacobian" of the transformation. $J = x_\xi x_\eta + y_\xi y_\eta; \quad 0 \neq J < \infty$

$$r = xi + yj; \quad g_{22} = x_\eta^2 + y_\eta^2 = |r_\eta|^2;$$
$$g_{12} = x_\xi x_\eta + y_\xi y_\eta = r_\xi r_\eta; \quad g_{11} = x_\xi^2 + y_\xi^2 = |r_\xi|^2 \tag{10}$$

Boundary conditions: The boundary conditions at the liquid boundaries give the form of oscillation ζ of each tidal constituent: $\zeta = A\cos(\sigma t - g)$, in which: A, σ, g°: Amplitude, frequency and phase of the tidal oscillation. The boundary conditions at the solid boundaries [18]: p = 0 on ξ = const and q = 0 on η = const.

Energy balance equation: In define the energy in the study area, the total energy is a combination between kinetic energy and potential enery. These components are calculated in the below Eq. 11- Eq. 13 while two energy dissipation due to bottom friction and mesh smoothing is identify by using Eq. 13 [16].

$$K = \frac{1}{2}\rho \iint_{\Omega^*} J^{-1}H^{-1}p^2 d\Omega^*; \quad P = \frac{1}{2}\rho g \iint_{\Omega^*} J\varsigma^2 d\Omega^* \tag{11}$$

$$W = \rho \int_{\eta_0} q\left(\frac{J^{-2}H^{-2}p^2}{2} + g\varsigma\right) d\varsigma; \tag{12}$$

$$D_1 = -K\rho \iint_{\Omega^*} J^{-2}H^{-3}p^3 d\Omega^*; \quad D_2 = \frac{2}{dt} \iint_{\Omega^*} \gamma\left(P^*\Delta^2 p + Q^*\Delta^2 q\right); \tag{13}$$

In which: K- Kinetic Energy; P - Potential energy; W - Energy flows across the liquid boundary;

D_1 - velocity of energy dissipation due to bottom friction; D_2 - velocity of energy dissipation when smoothing.

P*,Q*: "convariant" component of velocity vector. $P^* = g_{11}U + g_{12}V$; $Q^* = g_{21}U + g_{22}V$ (14).

Equation 4 are integrated with alternating direction implicit method on C-Arakov grid. The mesh node is at the boundary, on which the velocity component is perpendicular to the boundary. In the algorithm, using the semi-implicit diagram: "gradient" the water level is calculated according to the implicit diagram, while the nonlinear component is solved by an explicit diagram [19, 21].

3 Results

3.1 Tidal Energy Balance

The results showed that the energy variation of the semi-diurnal constituent M_2 are the most important, followed by the tidal constituents K_1, O_1 and S_2. As the coast of area is influenced by the mixed regime of semi-diurnal tides, the greatest tidal energy is M_2, followed by the tidal constituents K_1, O_1 and S_2. The highest total energy of constituent M_2 values about 6.6×10^{10} kJ (in which, the value of the potential energy is 4.3×10^{10} kJ while that of kinetic energy is lower, at about 4×10^{10} kJ). For the tidal constituents S_2, the highest values of total energy E, the potential energy P and the kinetic energy are 0.82×10^{10} kJ, 0.58×10^{10} kJ and 0.42×10^{10} kJ, respectively. The energy values of the tidal constituents N_2, K_2, P_1 and Q_1 are recorded lower than those of the tidal constituents M_2, K_1, O_1 and S_2. The model verification is implemented based on the equilibrium of the tidal currents of energy. In general, the oscillations of the isolines are sequences of sinusoidal. The oscillation of the total energy (W) and loss of energy caused by bottom friction (D1) has an opposite phase while the total energy oscillates in phase with variation in energy in the domain (Et). The energy balance oscillation in the variable region around axis '0' represents the energy balance in the region. The homogeneity of the oscillation of the energy inflow in the region and the energy variation is substantial for all four constituents, and the energy loss because friction is virtually negligible.

3.2 The Harmonic Constant Maps

These harmonic constants map was constructed from the results of the simulation of the amplitude and phase of the four main tidal constituents: K_1, O_1, M_2, S_2, N_2, K_2, P_1 and Q_1. For the enhanced, the results had been recorded after the 4th period, when the simulation of harmonic constituent oscillation was stable. Therefore, the data extracted from amplitude and phase values at the 5th period ensure the accuracy of the displayed results.

Next, these results above were compared to the harmonic constants of 2 tidal stations Vung Tau (X: 726,308.9, Y: 1,142,999.7) and Con Dao (X: 679,710.8, Y: 958,386.1) which are taken from the archives of the Institute of Oceanology, Chinese Academy of Sciences, from the British Admiralty Tide and Tidal Stream Tables, it shows that there is a small difference between observated and simulated data (Table 2).

The amplitude fluctuation of the constituent M_2 is from 0.1 m and 0.9 m, which reaches the highest amplitude value of the eight constituents. However, the phase is virtually unchanged for the whole region (30^0). In the whole region, the amplitude value increases when approaching the shoreline with the highest amplitude value at Bac Lieu (0.9 m) (Fig. 2a).

Similarly, the amplitude and phase isolines of the constituent S_2 are oriented similar to the M_2 wave, the amplitude value is only half of the amplitude value of the constituent M_2, the amplitude value increases from offshore to shore. In the Bac Lieu and Soc Trang coastal, amplitude values reach maximum, compared to the whole region (0.45 m). The phase values of the constituent S_2 fluctuate from 70^0 in Vung Tau to 140^0 in Bac Lieu and the isolines are almost perpendicular to the coastline (Fig. 2b).

For the diurnal constituent K_1, the amplitude values are in the range 0.35 m - 0.4 m and the value increases when approaching the shoreline. The phase values

Table 2 Comparison between simulation and observation of the harmonic constants for four constituents

Place	Vung Tau				Con Dao			
	Sim		Obs		Sim		Obs	
	H (m)	g(0)	H (m)	g(0)	H (m)	g(0)	H (m)	g(0)
M2	0.83	45	0.79	63	0.7	50	0.79	81
S2	0.32	80	0.3	111	0.32	110	0.27	142
O1	0.35	257	0.46	277	0.27	275	0.46	288
K1	0.4	234	0.61	268	0.42	350	0.64	288

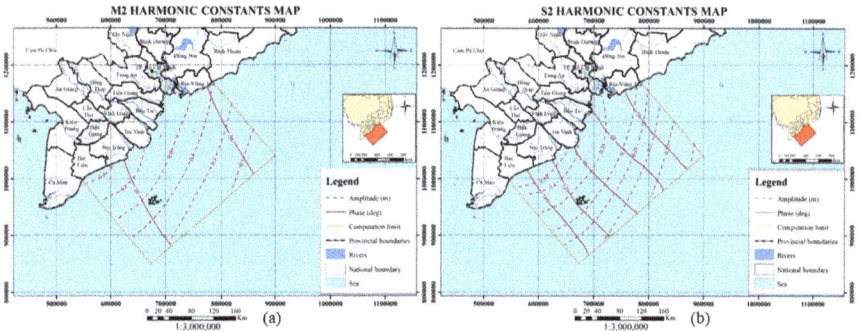

Fig. 2 The harmonic constants map of the constituents M_2 **a** and S_2 **b**

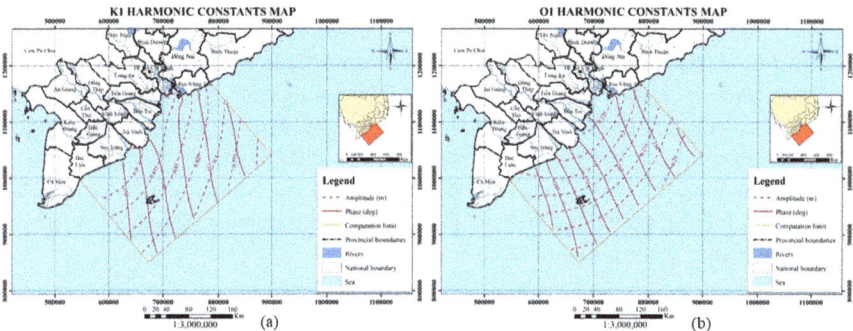

Fig. 3 The harmonic constants map of the constituents K_1 **a** and O_1 **b**

of the constituent K_1 increase from 324^0 (Vung Tau) to 350^0 (coastal of Bac Lieu). From Vung Tau—Ganh Hao, the phase values are higher than those for eight tidal constituents and the amplitude value is only lower than M_2 constituent (Fig. 3a). Meanwhile, the amplitude and phase isolines of the constituent O_1 are oriented similar to constituent K_1, the amplitude and phase value of wave O_1 ((amplitude 0.25 m– 0.325 m) and phase (245^0–285^0). This is the constituent with the lowest amplitude value in four main tidal constituents (K_1, O_1, M_2 and S_2) with little fluctuation (0.075 m) (Fig. 3b).

Furthermore, the amplitudes and phases of four constituents K_2, N_2, P_1 and Q_1 are small. While the amplitude values of K_2 range from 0.03 m to 0.07 m which has the lowest amplitude value in eight tidal constituents, the constituent K_2 value phases increase sharply from 20^0 (Vung Tau) to 120^0 (in Bac Lieu coastal). The values of

Table 3 The Resume of harmonic constants values for eight constituent

Name		The harmonic constants values for eight constituent							
		K1	O1	M2	S2	K2	N2	P1	Q1
Amplitude (m)	Min	0.35	0.25	0.1	0.1	0.03	0.08	0.13	0.055
	Max	0.5	0.325	0.9	0.45	0.07	0.16	0.19	0.07
Phase (0)	Min	325	245	30	70	20	120	230	230
	Max	350	285	30	140	120	200	260	260

amplitudes of the constituent N_2 are in the range 0.08 m–0.16 m and the phase values in the range 120^0–200^0. From Vung Tau—Ganh Hao, the amplitude value is largest in Bac Lieu coastal (0.16 m) and the largest phase value is in Bac Lieu and Soc Trang coastal with 200^0. Regarding the constituent P_1, the amplitude values range from 0.13 m to 0.19 m, which is one of the highest amplitude in the four extra tidal constituents. However, the phase values are the opposite; from offshore to longshore the phase value increases from 230^0 to 260^0. Meanwhile, the amplitude and phase isolines of constituent Q_1 are oriented similar to constituent P_1, the amplitude fluctuates very insignificantly with values from 0.055 m–0.07 m and gradually increases when approaching the shoreline. The phase values of the constituent Q_1 are in the range 230^0–260^0 for the whole region. The results of harmonic constants values for eight constituents are resume in Table 3.

4   Conclusion

The results showed that the energy variation of the semi-diurnal constituent M_2 is the most important, followed by the tidal constituents K_1, O_1 and S_2 while the amplitudes and phases of four constituents K_2, N_2, P_1 and Q_1 are negligible. The model verification is implemented based on the equilibrium of the tidal currents of energy. As the coast of the simulation area is influenced by the mixed regime of semi-diurnal tides, the greatest tidal energy is M_2, followed by the tidal constituents K_1, O_1 and S_2. From Vung Tau to Bac Lieu, the amplitude values of the tidal constituents M_2 is the highest (0.9 m), which is double that of the constituent S_2, but the phase is virtually unchanged for the whole region (30^0). Meanwhile, the amplitude values of the constituent S_2 fluctuate within the value of 0.1 m–0.45 m, the phase values are in the range 70^0–140^0. Regarding diurnal constituent, the amplitude values of the constituent K_1 are higher than the constituent O_1 (0.35 m–0.5 m), the phase values are from 325^0 to 350^0.

Acknowledgements Tran Thi Kim was funded by Vingroup Joint Stock Company and supported by the Domestic Master/ PhD Scholarship Programme of Vingroup Innovation Foundation (VINIF), Vingroup Big Data Institute (VINBIGDATA).

References

1. Codiga DL (2011) Unified tidal analysis and prediction using the UTide Matlab functions
2. Pawlowicz R, Beardsley B, Lentz S (2002) Classical tidal harmonic analysis including error estimates in MATLAB using T_TIDE. J Compu Geosci 28(8):929–937
3. Li S et al (2019) Tidal harmonic analysis and prediction with least-squares estimation and inaction method. J Estuarine, Coastal Shelf Sci 220:196–208
4. Xue Z, Liu JP, Ge Q (2011) Changes in hydrology and sediment delivery of the Mekong River in the last 50 years: connection to damming, monsoon, and ENSO. J Earth Surf Process Landforms 36(3):296–308
5. Su X et al (2014) Long-term polar motion prediction using normal time–frequency transform. J Geodesy 88(2):145–155
6. Zhang Z et al. (2020)Correction to: A FVCOM study of the potential coastal flooding in apponagansett bay and clarks coveDartmouth Town 103(3):2811–2812
7. Chen C, Liu H (2003) Beardsley RC (2003) An unstructured grid, finite-volume, three-dimensional, primitive fuations ocean model: application to coastal ocean and estuaries. Journal of atmospheric oceanic technology 20(1):159–186
8. Chen C et al. (2012) Current separation and upwelling over the southeast shelf of Vietnam in the South China Sea. J Geophys Res: Oceans 117(C3):C03033
9. Chen C et al. (2007) A finite volume numerical approach for coastal ocean circulation studies: Comparisons with finite difference models. J Geophys Res: Oceans 112(C3):C03018
10. Xu S et al. (2015) POM. gpu-v1. 0: a GPU-based Princeton ocean model. Geosci Model Develop 8(9):2815–2827
11. Jordi A, Wang D-P (2012) sbPOM: A parallel implementation of Princenton Ocean Model. Environmental Modelling Software 38:59–61
12. Ezer T, Mellor GL (2000) Sensitivity studies with the North Atlantic sigma coordinate Princeton ocean model. Dynamics of Atmospheres Oceans 32(3–4):185–208
13. Fang G et al (1999) Numerical simulation of principal tidal constituents in the South China Sea, Gulf of Tonkin and Gulf of Thailand. J Continental Shelf Research 19(7):845–869
14. Minh NN et al (2014) Tidal characteristics of the gulf of Tonkin. J Continental Shelf Research 91:37–56
15. Phan HM et al (2019) Tidal wave propagation along The Mekong deltaic coast. J Estuarine, Coastal Shelf Science 220:73–98
16. Bay NT, Phung NK (1998) Some study results for the tide in Tonkin gulf. In Proceedings of the 4th National Conference of Marine Science and Technology, Hanoi, Vietnam
17. Massel SR (1989) Hydrodynamics of coastal zones. Elsevier
18. Thompson JF, Warsi ZU, Mastin CW (1985) Numerical grid generation: foundations and applications, vol. 45. North-holland, Amsterdam
19. Androsov A et al (1995) Open boundary conditions for horizontal 2-D curvilinear-grid long-wave dynamics of a strait. J Advances in water resources 18(5):267–276
20. Fletcher C (1991) Computational Techniques for Fluid Dynamics [Russian translation], vol 2. Mir, Moscow
21. Androsov A et al (2002) Numerical modelling of barotropic tidal dynamics in the strait of Messina. J Advances in water resources 25(4):401–415

Structure Session

A General Framework of Higher-Order Shear Deformation Theory for Free Vibration Analysis of Functionally Graded Microplates

Van-Thien Tran and Trung-Kien Nguyen

Abstract This paper proposes a unified higher-order shear deformation microplate model for vibration analysis of functionally graded materials. The theory is developed from fundamental equations of the elasticity theory and modified couple stress theory, from which many different frameworks of size-dependent plate models are recovered. In order to capture the size effects, the modified couple stress theory with one independent length-scale parameter is used. The solution field is approximated by bi-directional series in which hybrid shape functions are proposed, then the stiffness and mass matrix are explicitly derived. Numerical results are presented for different configurations of material distribution, side-to-thickness ratio, size-scale-thickness ratio and boundary conditions on the natural frequencies of functionally graded microplates.

Keywords Series-type solutions · Vibration · Functionally graded microplates · Modified couple stress theory

1 Introduction

The recent development of functionally graded (FG) microplates with continuous variations of materials in a required direction led to a large potential application in engineering field. However, the behaviours of these structures at small scales require advanced computational models to capture size effects. The earlier experimental works revealed that the classical elasticity theory could not accurately predict responses of microstructures at small scale, advanced computations theories with length scale parameters have been therefore developed with different approaches. A number of researches has been performed to accurately predict static and vibration behaviours of FG microplates in which the modified coupled stress theory (MCT) is mostly used. The MCT initiated by Yang et al. [1] is known as the simplest theory

---

V.-T. Tran · T.-K. Nguyen (✉)
Faculty of Civil Engineering, Ho Chi Minh City University of Technology and Education, 1 Vo Van Ngan Street, Thu Duc City, Ho Chi Minh City, Viet Nam
e-mail: kiennt@hcmute.edu.vn

© The Author(s), under exclusive license to Springer Nature Singapore Pte Ltd. 2023   713
J. N. Reddy et al. (eds.), *ICSCEA 2021*, Lecture Notes in Civil Engineering 268,
https://doi.org/10.1007/978-981-19-3303-5_64

accounted for the size effects in which only one material length scale parameter associated with rotation gradient is accounted in the constitutive equations. Based on the MCT, many researches have been developed for static and dynamic analysis of FG microplates, only some representative references are herein cited. Tsiatas [2] investigated static analysis of isotropic micro-plates using classical plate theory (CPT). Ma et al. [3] analysed bending and free vibration responses of FG microplates using first-order shear deformation theory (FSDT). Reddy and Kim [4] presented geometrically nonlinear analysis of FG microplates based on a higher-order shear deformation theory (HSDT). Thai et al. [5, 6] studied size-dependent effects on static and vibration responses of FG microplate by using the HSDT. He et al. [7] used a refined HSDT for analysis of FG microplates. A brief literature review on the behaviours of FG microplates showed that though there is a number of studies have been performed in predicting static and dynamic behaviours of the FG microplates using the MCT with different plate theories, this complicated problem needs to be studied more further.

The objective of this paper is to propose a unified size-dependent plate model based on a general HSDT framework and MCT for analysis of FG materials. It is developed from fundamental equations of the elasticity theory and modified couple stress one. The governing equations of motions are derived from Lagrange's equations and then bi-directional series-type solutions with hybrid shape functions are proposed. Numerical results are presented for different configurations of material distribution, side-to-thickness ratio and boundary conditions on natural frequencies of the FG microplates. New results presented in this study can be of interests to the scientific and engineering community in the future.

2   Theoretical formulation

Consider a FG rectangle microplate in the coordinate system (x_1, x_2, x_3) with sides $a \times b$ and thickness h as shown in Fig. 1. The plate is composed of a mixture of ceramic and metal materials whose properties vary continuously in the thickness direction. The effective material properties of FG microplates can be approximated

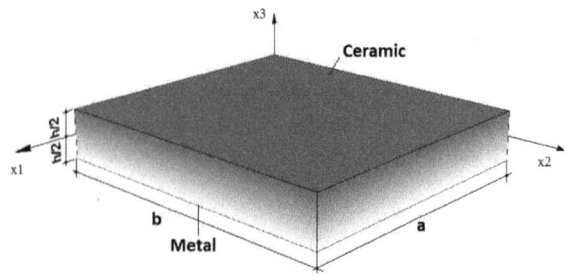

Fig. 1 Geometry of a FG microplate

by the following expressions [6, 7]:

$$P(x_3) = (P_c - P_m)\left(\frac{2x_3 + h}{2h}\right)^p + P_m \tag{1}$$

where P_c and P_m are the properties of ceramic and metal materials, respectively, such as Young's moduli E, mass density ρ, Poisson's ratio ν; p is the power-law index which is positive and $x_3 \in [-h/2, h/2]$.

2.1 Modified couple stress theory

The total potential energy of the FG microplate is obtained by:

$$\Pi = \Pi_U - \Pi_K \tag{2}$$

where Π_U, Π_K are the strain energy and kinetic energy of the FG microplate, respectively. Based on the MCT, the strain energy of the system Π_U is given by [1, 6]:

$$\Pi_U = \int_V (\sigma\varepsilon + m\chi)dV \tag{3}$$

where ε, χ are strains and symmetric rotation gradients, respectively; σ is Cauchy stress; m are higher-order stress corresponding with strain gradients χ. The components of strain ε_{ij} and strain gradients χ_{ij} are defined as follows:

$$\varepsilon_{ij} = \frac{1}{2}\left(u_{i,j} + u_{j,i}\right), \chi_{ij} = \frac{1}{4}\left(u_{n,mj}e_{imn} + u_{n,mi}e_{jmn}\right) \tag{4}$$

where e_{imn} are Knonecker delta and permutation symbol, respectively. The components of stress are calculated from constitutive equations as follows:

$$\sigma_{ij} = \lambda\varepsilon_{kk}\delta_{ij} + 2\mu\varepsilon_{ij}, m_{ij} = 2\mu l^2\chi_{ij} \tag{5}$$

where λ, μ are Lamé constants; l is length scale parameter. The kinetic energy of the FG microplate Π_K is expressed by:

$$\Pi_K = \frac{1}{2}\int_V \rho(x_3)\left(\dot{u}_1^2 + \dot{u}_2^2 + \dot{u}_3^2\right)dV \tag{6}$$

where $\rho(x_3)$ is mass density of the FG microplate; $\dot{u}_1 = u_{1,t}$, $\dot{u}_2 = u_{2,t}$, $\dot{u}_3 = u_{3,t}$ are velocities in x_1-, x_2- and x_3- directions, respectively.

2.2   Unified kinematics of FG microplates

For simplicity purpose, the effects of transverse normal strain are neglected in the sequel, i.e. $u_3(x_1, x_2, x_3) = u_3^0(x_1, x_2)$ where $u_3^0(x_1, x_2)$ is transverse displacement at the mid-surface of the FG microplate. Moreover, it is supposed that the transverse shear stresses are expressed in terms of the transverse shear forces as follows:

$$\sigma_{13} = f_{,3}(x_3)Q_1(x_1, x_2), \sigma_{23} = f_{,3}(x_3)Q_2(x_1, x_2) \tag{7}$$

where $f(x_3)$ is a higher-order term whose first derivative satisfies the free-stress boundary condition at the top and bottom surfaces of the plates, i.e. $f_{,3}\left(x_3 = \pm\frac{h}{2}\right) = 0$; the comma in subscript is used to indicate the derivative of variable that follows. Additionally, transverse shear strains are linearly related to the membrane displacements $u_1(x_1, x_2, x_3)$, $u_2(x_1, x_2, x_3)$ and transverse one $u_3^0(x_1, x_2)$ by:

$$\gamma_{13} = u_{1,3} + u_{3,1}^0 = \frac{\sigma_{13}}{\mu} = \frac{f_{,3}Q_1}{\mu}, \gamma_{23} = u_{2,3} + u_{3,2}^0 = \frac{\sigma_{23}}{\mu} = \frac{f_{,3}Q_2}{\mu}, \tag{8}$$

where $\mu(x_3) = \frac{E(x_3)}{2(1+\nu)}$ is the shear modulus. Furthermore, integrating Eq. (8) in x_3- direction leads to a general displacement field of the FG microplate as follows:

$$u_1(x_1, x_2, x_3) = u_1^0(x_1, x_2) - x_3 u_{3,1}^0 + \Psi(x_3)Q_1(x_1, x_2) \tag{9a}$$

$$u_2(x_1, x_2, x_3) = u_2^0(x_1, x_2) - x_3 u_{3,2}^0 + \Psi(x_3)Q_2(x_1, x_2) \tag{9b}$$

$$u_3(x_1, x_2, x_3) = u_3^0(x_1, x_2) \tag{9c}$$

where $\Psi(x_3) = \int\limits_0^{x_3} \frac{f_{,3}}{\mu(x_3)} dx_3$. Moreover, it is known that the shear forces can be expressed in terms of the rotation (φ_1, φ_2) and gradients of the transverse displacement as follows:

$$Q_1(x_1, x_2) = H^s\left(\varphi_1 + u_{3,1}^0\right), Q_2(x_1, x_2) = H^s\left(\varphi_2 + u_{3,2}^0\right) \tag{10}$$

where $H^s = k^s \int_{-h/2}^{h/2} \mu(x_3)dx_3$ is the transverse shear stiffness of the FG microplates; $k^s = 5/6$ is shear coefficient factor. Substituting Eq. (10) into Eq. (9) leads to a general higher-order shear deformation FG microplate theory as follows:

$$u_1(x_1, x_2, x_3) = u_1^0(x_1, x_2) + \Phi_1(x_3)u_{3,1}^0 + \Phi_2(x_3)\varphi_1(x_1, x_2) \tag{11a}$$

$$u_2(x_1, x_2, x_3) = u_2^0(x_1, x_2) + \Phi_1(x_3)u_{3,2}^0 + \Phi_2(x_3)\varphi_2(x_1, x_2) \tag{11b}$$

$$u_3(x_1, x_2, x_3) = u_3^0(x_1, x_2) \tag{11c}$$

where $\Phi_1(x_3) = H^s \Psi(x_3) - x_3$, $\Phi_2(x_3) = H^s \Psi(x_3)$. Subsituting Eq. (11) into the strains and strain gradients in Eq. (4), the strains $\boldsymbol{\varepsilon}^T = \left[\boldsymbol{\varepsilon}^{(i)} \; \boldsymbol{\varepsilon}^{(s)} \right]$ are obtained as follows:

$$\boldsymbol{\varepsilon}^{(i)} = \boldsymbol{\varepsilon}^{(0)} + \Phi_1(x_3)\boldsymbol{\varepsilon}^{(1)} + \Phi_2(x_3)\boldsymbol{\varepsilon}^{(2)}, \; \boldsymbol{\varepsilon}^{(s)} = \Phi_3(x_3)\boldsymbol{\varepsilon}^{(3)} \tag{12}$$

where $\Phi_3(x_3) = H^s \Psi_{,3}$ with $\Psi_{,3}(x_3) = \frac{f_{,3}(x_3)}{\mu(x_3)}$ and,

$$\boldsymbol{\varepsilon}^{(0)} = \left\{ \begin{array}{c} \varepsilon_{11}^{(0)} \\ \varepsilon_{22}^{(0)} \\ \gamma_{12}^{(0)} \end{array} \right\} = \left\{ \begin{array}{c} u_{1,1}^0 \\ u_{2,2}^0 \\ u_{1,2}^0 + u_{2,1}^0 \end{array} \right\}, \; \boldsymbol{\varepsilon}^{(1)} = \left\{ \begin{array}{c} \varepsilon_{11}^{(1)} \\ \varepsilon_{22}^{(1)} \\ \gamma_{12}^{(1)} \end{array} \right\} = \left\{ \begin{array}{c} u_{3,11}^0 \\ u_{3,22}^0 \\ 2u_{3,12}^0 \end{array} \right\},$$

$$\boldsymbol{\varepsilon}^{(2)} = \left\{ \begin{array}{c} \varepsilon_{11}^{(2)} \\ \varepsilon_{22}^{(2)} \\ \gamma_{12}^{(2)} \end{array} \right\} = \left\{ \begin{array}{c} \varphi_{1,1} \\ \varphi_{2,2} \\ \varphi_{1,2} + \varphi_{2,1} \end{array} \right\} \tag{13a}$$

$$\boldsymbol{\varepsilon}^{(3)} = \left\{ \begin{array}{c} \gamma_{13}^{(0)} \\ \gamma_{23}^{(0)} \end{array} \right\} = \left\{ \begin{array}{c} \varphi_1 + u_{3,1}^0 \\ \varphi_2 + u_{3,2}^0 \end{array} \right\} \tag{13b}$$

Moreover, the symmetric rotation gradients are given by:

$$\chi_{ij} = \frac{1}{2}\left(\overline{\theta}_{i,j} + \overline{\theta}_{j,i}\right) \tag{14}$$

where $\overline{\theta}_i$ is determined in terms of the displacements u_i as follows:

$$\overline{\theta}_1 = \frac{1}{2}\left(u_{3,2} - u_{2,3}\right) = \frac{1}{2}\left(u_3, 2^0 - \Phi_{1,3}u_3, 2^0 - \Phi_{2,3}\varphi_2\right) \tag{15a}$$

$$\overline{\theta}_2 = \frac{1}{2}\left(u_{1,3} - u_{3}, 1\right) = \frac{1}{2}\left(-u_3, 1^0 + \Phi_{1,3}u_3, 1^0 + \Phi_{2,3}\varphi_1\right) \tag{15b}$$

$$\overline{\theta}_3 = \frac{1}{2}\left(u_{2,1} - u_{1}, 2\right) = \frac{1}{2}\left[u_2, 1^0 - u_1, 2^0 + \Phi_2\left(\varphi_{2,1} - \varphi_{1,2}\right)\right] \tag{15c}$$

Substituting Eq. (15) into Eq. (13), the rotation gradients are expressed as follows:

$$\boldsymbol{\chi} = \boldsymbol{\chi}^{(0)} + \Phi_{1,3}\boldsymbol{\chi}^{(1)} + \Phi_{2,3}\boldsymbol{\chi}^{(2)} + \Phi_{1,33}\boldsymbol{\chi}^{(3)} + \Phi_{2,33}\boldsymbol{\chi}^{(4)} + \Phi_2\boldsymbol{\chi}^{(5)} \tag{16}$$

where $\boldsymbol{\chi}^T = \left[\chi_{11} \; \chi_{22} \; 2\chi_{12} \; \chi_{33} \; 2\chi_{13} \; 2\chi_{23} \right]$ and,

$$\boldsymbol{\chi}^{(0)} = \frac{1}{2} \begin{Bmatrix} u_3, 12^0 \\ -u_3, 12^0 \\ u^0_{3,22} - u^0_{3,11} \\ 0 \\ u^0_{2,11} - u^0_{1,12} \\ u^0_{2,12} - u^0_{1,22} \end{Bmatrix}, \boldsymbol{\chi}^{(1)} = \frac{1}{2} \begin{Bmatrix} -u_3, 12^0 \\ u_3, 12^0 \\ u^0_{3,11} - u^0_{3,22} \\ 0 \\ 0 \\ 0 \end{Bmatrix}, \boldsymbol{\chi}^{(2)} = \frac{1}{2} \begin{Bmatrix} -\varphi_{2,1} \\ \varphi_{1,2} \\ \varphi_{1,1} - \varphi_{2,2} \\ \varphi_{2,1} - \varphi_{1,2} \\ 0 \\ 0 \end{Bmatrix}$$

(17a)

$$\boldsymbol{\chi}^{(3)} = \frac{1}{2} \begin{Bmatrix} 0 \\ 0 \\ 0 \\ 0 \\ -u^0_{3,2} \\ u^0_{3,1} \end{Bmatrix}, \boldsymbol{\chi}^{(4)} = \frac{1}{2} \begin{Bmatrix} 0 \\ 0 \\ 0 \\ 0 \\ -\varphi_2 \\ \varphi_1 \end{Bmatrix}, \boldsymbol{\chi}^{(5)} = \frac{1}{2} \begin{Bmatrix} 0 \\ 0 \\ 0 \\ 0 \\ \varphi_{2,11} - \varphi_{1,12} \\ \varphi_{2,12} - \varphi_{1,22} \end{Bmatrix}$$

(17b)

Furthermore, the stresses and strains of FG microplates are related by constitutive equations as follows:

$$\boldsymbol{\sigma}^{(i)} = \begin{Bmatrix} \sigma_{11} \\ \sigma_{22} \\ \sigma_{12} \end{Bmatrix} = \begin{bmatrix} Q_{11} & Q_{12} & 0 \\ Q_{12} & Q_{22} & 0 \\ 0 & 0 & Q_{66} \end{bmatrix} \begin{Bmatrix} \varepsilon_{11} \\ \varepsilon_{22} \\ \gamma_{12} \end{Bmatrix} = \boldsymbol{Q}^{(i)}_\varepsilon \boldsymbol{\varepsilon}^{(i)},$$

$$\boldsymbol{\sigma}^{(o)} = \begin{Bmatrix} \sigma_{13} \\ \sigma_{23} \end{Bmatrix} = \begin{bmatrix} Q_{55} & 0 \\ 0 & Q_{44} \end{bmatrix} \begin{Bmatrix} \gamma_{13} \\ \gamma_{23} \end{Bmatrix} = \boldsymbol{Q}^{(o)}_\varepsilon \boldsymbol{\varepsilon}^{(s)}$$

(18a)

$$\boldsymbol{m} = \begin{Bmatrix} m_{11} \\ m_{22} \\ m_{12} \\ m_{33} \\ m_{23} \\ m_{13} \end{Bmatrix} = 2\mu l_1^2 \begin{bmatrix} 1 & 0 & 0 & 0 & 0 & 0 \\ 0 & 1 & 0 & 0 & 0 & 0 \\ 0 & 0 & 1 & 0 & 0 & 0 \\ 0 & 0 & 0 & 1 & 0 & 0 \\ 0 & 0 & 0 & 0 & 1 & 0 \\ 0 & 0 & 0 & 0 & 0 & 1 \end{bmatrix} \begin{Bmatrix} \chi_{11} \\ \chi_{22} \\ \chi_{12} \\ \chi_{33} \\ \chi_{23} \\ \chi_{13} \end{Bmatrix} = \alpha_\chi \boldsymbol{I}_{6 \times 6} \boldsymbol{\chi}$$

(18b)

where $\alpha_\chi = 2\mu l^2$, $Q_{11} = \frac{E(x_3)}{1-v^2}$, $Q_{22} = \frac{E(x_3)}{1-v^2}$, $Q_{12} = \frac{vE(x_3)}{1-v^2}$, $Q_{44} = Q_{55} = Q_{66} = \mu = \frac{E(x_3)}{2(1+v)}$.

2.3 Energy Principle

In order to derive the equation of motion, Hamilton's principle is used:

$$\int_{t_1}^{t_2} (\delta \Pi_U - \delta \Pi_K) dt = 0$$

(19)

where $\delta\Pi_U$, $\delta\Pi_K$ are the variations of strain energy and kinetic energy, respectively. The variation of the strain energy of FG microplates derived from Eq. (3) as follows:

$$\delta\Pi_U = \int_A (\sigma\delta\varepsilon + \mathbf{m}\delta\chi)dA = \int_A \left[\mathbf{M}_\varepsilon^{(0)}\delta\varepsilon^{(0)} + \mathbf{M}_\varepsilon^{(1)}\delta\varepsilon^{(1)} + \mathbf{M}_\varepsilon^{(2)}\delta\varepsilon^{(2)} + \mathbf{M}_\varepsilon^{(3)}\delta\varepsilon^{(3)}\right.$$
$$\left. + \mathbf{M}_\chi^{(0)}\delta\chi^{(0)} + \mathbf{M}_\chi^{(1)}\delta\chi^{(1)} + \mathbf{M}_\chi^{(2)}\delta\chi^{(2)} + \mathbf{M}_\chi^{(3)}\delta\chi^{(3)} + \mathbf{M}_\chi^{(4)}\delta\chi^{(4)} + \mathbf{M}_\chi^{(5)}\delta\chi^{(5)}\right]dA$$
$$(20)$$

where the stress resultants are given by:

$$\left(\mathbf{M}_\varepsilon^{(0)}, \mathbf{M}_\varepsilon^{(1)}, \mathbf{M}_\varepsilon^{(2)}\right) = \int_{-h/2}^{h/2} (1, \Phi_1, \Phi_2)\sigma^{(i)}dx_3, \quad \mathbf{M}_\varepsilon^{(3)} = \int_{-h/2}^{h/2} \Phi_3\sigma^{(o)}dx_3 \quad (21a)$$

$$\left(\mathbf{M}_\chi^{(0)}, \mathbf{M}_\chi^{(1)}, \mathbf{M}_\chi^{(2)}, \mathbf{M}_\chi^{(3)}, \mathbf{M}_\chi^{(4)}, \mathbf{M}_\chi^{(5)}\right) = \int_{-h/2}^{h/2} (1, \Phi_{1,3}, \Phi_{2,3}, \Phi_{1,33}, \Phi_{2,33}, \Phi_2)mdx_3 \quad (21b)$$

These stress resultants can be expressed in terms of the strains and its gradients as follows:

$$\left\{\begin{array}{c} \mathbf{M}_\varepsilon^{(0)} \\ \mathbf{M}_\varepsilon^{(1)} \\ \mathbf{M}_\varepsilon^{(2)} \\ \mathbf{M}_\varepsilon^{(3)} \end{array}\right\} = \left[\begin{array}{cccc} \mathbf{A}^\varepsilon & \mathbf{B}^\varepsilon & \mathbf{B}_s^\varepsilon & 0 \\ \mathbf{B}^\varepsilon & \mathbf{D}^\varepsilon & \mathbf{D}_s^\varepsilon & 0 \\ \mathbf{B}_s^\varepsilon & \mathbf{D}_s^\varepsilon & \mathbf{H}_s^\varepsilon & 0 \\ 0 & 0 & 0 & \mathbf{A}_s^\varepsilon \end{array}\right] \left\{\begin{array}{c} \varepsilon^{(0)} \\ \varepsilon^{(1)} \\ \varepsilon^{(2)} \\ \varepsilon^{(3)} \end{array}\right\} \quad (22a)$$

$$\left\{\begin{array}{c} \mathbf{M}_\chi^{(0)} \\ \mathbf{M}_\chi^{(1)} \\ \mathbf{M}_\chi^{(2)} \\ \mathbf{M}_\chi^{(3)} \\ \mathbf{M}_\chi^{(4)} \\ \mathbf{M}_\chi^{(5)} \end{array}\right\} = \left[\begin{array}{cccccc} \mathbf{A}^\chi & \overline{\mathbf{B}}^\chi & \overline{\mathbf{B}}_s^\chi & \overline{\overline{\mathbf{B}}}^\chi & \overline{\overline{\mathbf{B}}}_s^\chi & \mathbf{B}_s^\chi \\ \overline{\mathbf{B}}^\chi & \mathbf{B}^\chi & \mathbf{D}_s^\chi & \mathbf{E}^\chi & \mathbf{E}_s^\chi & \mathbf{F}^\chi \\ \overline{\mathbf{B}}_s^\chi & \mathbf{D}_s^\chi & \mathbf{H}_s^\chi & \mathbf{G}_s^\chi & \overline{\mathbf{I}}^\chi & \overline{\mathbf{J}}^\chi \\ \overline{\overline{\mathbf{B}}}^\chi & \mathbf{E}^\chi & \mathbf{G}_s^\chi & \overline{\overline{\mathbf{D}}}^\chi & \overline{\overline{\mathbf{D}}}_s^\chi & \mathbf{K}_s^\chi \\ \overline{\overline{\mathbf{B}}}_s^\chi & \mathbf{E}_s^\chi & \overline{\mathbf{I}}^\chi & \overline{\overline{\mathbf{D}}}_s^\chi & \overline{\overline{\mathbf{H}}}_s^\chi & \mathbf{L} \\ \mathbf{B}_s^\chi & \mathbf{F}^\chi & \overline{\mathbf{J}}^\chi & \mathbf{K}_s^\chi & \mathbf{L} & \mathbf{H}_s^\chi \end{array}\right] \left\{\begin{array}{c} \chi^{(0)} \\ \chi^{(1)} \\ \chi^{(2)} \\ \chi^{(3)} \\ \chi^{(4)} \\ \chi^{(5)} \end{array}\right\} \quad (22b)$$

where the stiffness components of the FG microplates are defined as follows:

$$\left(\mathbf{A}^\varepsilon, \mathbf{B}^\varepsilon, \mathbf{D}^\varepsilon, \mathbf{H}_s^\varepsilon, \mathbf{B}_s^\varepsilon, \mathbf{D}_s^\varepsilon\right) = \int_{-h/2}^{h/2} \left(1, \Phi_1, \Phi_1^2, \Phi_2^2, \Phi_2, \Phi_1\Phi_2\right)Q_\varepsilon^{(i)}dx_3, \quad A_s^\varepsilon = \int_{-h/2}^{h/2} \Phi_3^2 Q_\varepsilon^{(o)}dx_3$$
$$(23a)$$

$$\mathbf{A}^\chi, \overline{\mathbf{B}}^\chi, \overline{\mathbf{B}}_s^\chi, \overline{\overline{\mathbf{B}}}^\chi, \overline{\overline{\mathbf{B}}}_s^\chi, \mathbf{B}_s^\chi = \int_{-h/2}^{h/2} \left(1, \Phi_{1,3}, \Phi_{2,3}, \Phi_{1,33}, \Phi_{2,33}, \Phi_2\right)\alpha_\chi \mathbf{I}_{6\times6}dx_3 \quad (23b)$$

$$\left(\overline{\boldsymbol{D}}^{\chi}, \overline{\boldsymbol{D}}_s^{\chi}, \overline{\boldsymbol{E}}^{\chi}, \overline{\boldsymbol{E}}_s^{\chi}, \overline{\boldsymbol{F}}_s^{\chi}\right) = \int_{-h/2}^{h/2} \Phi_{1,3}\left(\Phi_{1,3}, \Phi_{2,3}, \Phi_{1,33}, \Phi_{2,33}, \Phi_2\right)\alpha_{\chi} \boldsymbol{I}_{6\times6} dx_3$$

(23c)

$$\left(\overline{\boldsymbol{H}}_s^{\chi}, \overline{\boldsymbol{G}}_s^{\chi}, \overline{\boldsymbol{I}}^{\chi}, \overline{\boldsymbol{J}}^{\chi}\right) = \int_{-h/2}^{h/2} \Phi_{2,3}\left(\Phi_{2,3}, \Phi_{1,33}, \Phi_{2,33}, \Phi_2\right)\alpha_{\chi} \boldsymbol{I}_{6\times6} dx_3 \qquad (23d)$$

$$\left(\overline{\overline{\boldsymbol{D}}}^{\chi}, \overline{\overline{\boldsymbol{D}}}_s^{\chi}, \overline{\boldsymbol{K}}_s^{\chi}, \overline{\overline{\boldsymbol{H}}}_s^{\chi}, \overline{\boldsymbol{L}}^{\chi}, \boldsymbol{H}_s^{\chi}\right) = \int_{-h/2}^{h/2} \left(\Phi_{1,33}^2, \Phi_{1,33}\Phi_{2,33}, \Phi_{1,33}\Phi_2, \Phi_{2,33}^2, \Phi_{2,33}\Phi_2, \Phi_2^2\right)$$

$$\alpha_{\chi} \boldsymbol{I}_{6\times6} dx_3$$

(23e)

The variation of kinetic energy $\delta \prod_K$ derived from Eq. (6) is calculated by:

$$
\begin{aligned}
\delta \prod_K &= \frac{1}{2} \int_V \rho(\dot{u}_1 \delta \dot{u}_1 + \dot{u}_2 \delta \dot{u}_2 + \dot{u}_3 \delta \dot{u}_3) dV \\
&= \frac{1}{2} \int_A \left[I_0\left(\dot{u}_1^0 \delta \dot{u}_1^0 + \dot{u}_2^0 \delta \dot{u}_2^0 + \dot{u}_3^0 \delta \dot{u}_3^0\right) + I_1\left(\dot{u}_1^0 \delta \dot{u}_{3,1}^0 + \dot{u}_{3,1}^0 \delta \dot{u}_1^0 + \dot{u}_2^0 \delta \dot{u}_{3,2}^0 + \dot{u}_{3,2}^0 \delta \dot{u}_2^0\right) \right. \\
&\quad + J_2\left(\dot{u}_{3,1}^0 \delta \dot{\varphi}_1 + \dot{\varphi}_1 \delta \dot{u}_{3,1}^0 + \dot{u}_{3,2}^0 \delta \dot{\varphi}_2 + \dot{\varphi}_2 \delta \dot{u}_{3,2}^0\right) + K_2(\dot{\varphi}_1 \delta \dot{\varphi}_1 + \dot{\varphi}_2 \delta \dot{\varphi}_2) \\
&\quad \left. + J_1\left(\dot{u}_1^0 \delta \dot{\varphi}_1 + \dot{\varphi}_1 \delta \dot{u}_1^0 + \dot{u}_2^0 \delta \dot{\varphi}_2 + \dot{\varphi}_2 \delta \dot{u}_2^0\right) + I_2\left(\dot{u}_{3,1}^0 \delta \dot{u}_{3,1}^0 + \dot{u}_{3,2}^0 \delta \dot{u}_{3,2}^0\right) \right] dA
\end{aligned}
$$

(24)

where $I_0, I_1, I_2, J_1, J_2, K_2$ are mass components of the FG microplates which are defined as follows:

$$(I_0, I_1, I_2, J_1, J_2, K_2) = \int_{-h/2}^{h/2} \left(1, \Phi_1, \Phi_1^2, \Phi_2, \Phi_1\Phi_2, \Phi_2^2\right)\rho dx_3 \qquad (25)$$

3 Series-Type Solutions of FG Microplates

Based on the Ritz method, the membrane and transverse displacements, rotations $\left(u_1^0, u_2^0, u_3^0, \varphi_1, \varphi_2\right)$ of the FG microplate can be expressed in terms of the series of approximation functions and associated values of series as follows:

$$\left\{u_1^0(x_1, x_2, t), \varphi_1(x_1, x_2, t)\right\} = \sum_{i=1}^{n_1} \sum_{j=1}^{n_2} \left\{u_{1ij}(t), x_{ij}(t)\right\} R_{i,1}(x_1) P_j(x_2) \qquad (26a)$$

$$\left\{u_2^0(x_1, x_2, t), \varphi_2(x_1, x_2, t)\right\} = \sum_{i=1}^{n_1} \sum_{j=1}^{n_2} \left\{u_{2ij}(t), y_{ij}(t)\right\} R_i(x_1) P_{j,2}(x_2) \qquad (26b)$$

$$u_3^0(x_1, x_2, t) = \sum_{i=1}^{n_1} \sum_{j=1}^{n_2} u_{3ij}(t) R_i(x_1) P_j(x_2) \tag{26c}$$

where $u_{1ij}, u_{2ij}, u_{3ij}, x_{ij}, y_{ij}$ are variables to be determined; $R_i(x_1), P_j(x_2)$ are the shape functions in x_1-, x_2- direction, respectively. As a result, five unknowns of the plate only depend on two shape functions. It should be noted that the accuracy, convergence rates and numerical instabilities of the Ritz solution depends on the construction of the shape functions, which was discussed in details in [8–11]. The functions $R_i(x_1)$ and $P_j(x_2)$ are constructed to satisfy the boundary conditions (BCs) in which the simply-supported and clamped–clamped BCs are followed:

- Simply supported (S): $u_2^0 = u_3^0 = \varphi_2 = 0$ at $x_1 = 0, a$ and $u_1^0 = u_3^0 = \varphi_1 = 0$ at $x_2 = 0, b$
- Clamped (C): $u_1^0 = u_2^0 = u_3^0 = \varphi_1 = \varphi_2 = 0$ at $x_1 = 0, a$ and $x_2 = 0, b$

The combination of simply-supported, clamped boundary conditions on the edges of the plate leads to the different BCs as follows: SSSS, CSCS, CCSS, CCCC which will be considered in the numerical examples. Substituting Eq. (26) into Eqs.(20), (24) and then the subsequent results into Eq. (15) lead to the characteristic equations of motion of the FG microplates as follows: $\mathbf{Kd} + \mathbf{M\ddot{d}} = \mathbf{0}$ where $d = \begin{bmatrix} u_1 & u_2 & u_3 & x & y \end{bmatrix}^T$ is the displacement vector to be determined; $\mathbf{K} = \mathbf{K}^\varepsilon + \mathbf{K}^\chi$ is the stiffness matrix which is composed of those of the strains $\mathbf{K}^\varepsilon$, symmetric rotation gradients $\mathbf{K}^\chi$; M is the mass matrix. These components are given more details as follows:

$$\mathbf{K}^\zeta = \begin{bmatrix} \mathbf{K}^{\zeta 11} & \mathbf{K}^{\zeta 12} & \mathbf{K}^{\zeta 13} & \mathbf{K}^{\zeta 14} & \mathbf{K}^{\zeta 15} \\ {}^T\mathbf{K}^{\zeta 12} & \mathbf{K}^{\zeta 22} & \mathbf{K}^{\zeta 23} & \mathbf{K}^{\zeta 24} & \mathbf{K}^{\zeta 25} \\ {}^T\mathbf{K}^{\zeta 13} & {}^T\mathbf{K}^{\zeta 23} & \mathbf{K}^{\zeta 33} & \mathbf{K}^{\zeta 34} & \mathbf{K}^{\zeta 35} \\ {}^T\mathbf{K}^{\zeta 14} & {}^T\mathbf{K}^{\zeta 24} & {}^T\mathbf{K}^{\zeta 34} & \mathbf{K}^{\zeta 44} & \mathbf{K}^{\zeta 45} \\ {}^T\mathbf{K}^{\zeta 15} & {}^T\mathbf{K}^{\zeta 25} & {}^T\mathbf{K}^{\zeta 35} & {}^T\mathbf{K}^{\zeta 45} & \mathbf{K}^{\zeta 55} \end{bmatrix} \text{ with } \zeta = \{\varepsilon, \chi\} \tag{27}$$

where the components of stiffness matrix $\mathbf{K}^\varepsilon$ are defined as follows:

$$K_{ijkl}^{\varepsilon 11} = A_{11}^\varepsilon T_{ik}^{22} S_{jl}^{00} + A_{66}^\varepsilon T_{ik}^{11} S_{jl}^{11}, \quad K_{ijkl}^{\varepsilon 12} = A_{12}^\varepsilon T_{ik}^{02} S_{jl}^{20} + A_{66}^\varepsilon T_{ik}^{11} S_{jl}^{11}$$

$$K_{ijkl}^{\varepsilon 13} = B_{11}^\varepsilon T_{ik}^{22} S_{jl}^{00} + B_{12}^\varepsilon T_{ik}^{02} S_{jl}^{20} + 2B_{66}^\varepsilon T_{ik}^{11} S_{jl}^{11}$$

$$K_{ijkl}^{\varepsilon 14} = B_{s11}^\varepsilon T_{ik}^{22} S_{jl}^{00} + B_{s66}^\varepsilon T_{ik}^{11} S_{jl}^{11}, \quad K_{ijkl}^{\varepsilon 15} = B_{s12}^\varepsilon T_{ik}^{02} S_{jl}^{20} + B_{s66}^\varepsilon T_{ik}^{11} S_{jl}^{11}$$

$$K_{ijkl}^{\varepsilon 22} = A_{22}^\varepsilon T_{ik}^{00} S_{jl}^{22} + A_{66}^\varepsilon T_{ik}^{11} S_{jl}^{11}, \quad K_{ijkl}^{\varepsilon 23} = B_{12}^\varepsilon T_{ik}^{20} S_{jl}^{02} + B_{22}^\varepsilon T_{ik}^{00} S_{jl}^{22} + 2B_{66}^\varepsilon T_{ik}^{11} S_{jl}^{11}$$

$$K_{ijkl}^{\varepsilon 24} = B_{s12}^\varepsilon T_{ik}^{20} S_{jl}^{02} + B_{s66}^\varepsilon T_{ik}^{11} S_{jl}^{11}, \quad K_{ijkl}^{\varepsilon 25} = B_{s22}^\varepsilon T_{ik}^{00} S_{jl}^{22} + B_{s66}^\varepsilon T_{ik}^{11} S_{jl}^{11}$$

$$K_{ijkl}^{\varepsilon 33} = D_{11}^{\varepsilon} T_{ik}^{22} S_{jl}^{00} + D_{12}^{\varepsilon}\left(T_{ik}^{02} S_{jl}^{20} + T_{ik}^{20} S_{jl}^{02}\right) + D_{22}^{\varepsilon} T_{ik}^{00} S_{jl}^{22} + 4D_{66}^{\varepsilon} T_{ik}^{11} S_{jl}^{11}$$
$$+ A_{s44}^{\varepsilon} T_{ik}^{00} S_{jl}^{11} + A_{s55}^{\varepsilon} T_{ik}^{11} S_{jl}^{00}$$

$$K_{ijkl}^{\varepsilon 34} = D_{s11}^{\varepsilon} T_{ik}^{22} S_{jl}^{00} + D_{s12}^{\varepsilon} T_{ik}^{20} S_{jl}^{02} + 2D_{s66}^{\varepsilon} T_{ik}^{11} S_{jl}^{11} + A_{s55}^{\varepsilon} T_{ik}^{11} S_{jl}^{00}$$

$$K_{ijkl}^{\varepsilon 35} = D_{s12}^{\varepsilon} T_{ik}^{02} S_{jl}^{20} + D_{s22}^{\varepsilon} T_{ik}^{00} S_{jl}^{22} + 2D_{s66}^{\varepsilon} T_{ik}^{11} S_{jl}^{11} + A_{s44}^{\varepsilon} T_{ik}^{00} S_{jl}^{11}$$

$$K_{ijkl}^{\varepsilon 44} = H_{s11}^{\varepsilon} T_{ik}^{22} S_{jl}^{00} + H_{s66}^{\varepsilon} T_{ik}^{11} S_{jl}^{11} + A_{s55}^{\varepsilon} T_{ik}^{11} S_{jl}^{00}$$

$$K_{ijkl}^{\varepsilon 45} = H_{s12}^{\varepsilon} T_{ik}^{02} S_{jl}^{20} + H_{s66}^{\varepsilon} T_{ik}^{11} S_{jl}^{11}, \quad K_{ijkl}^{\varepsilon 55} = H_{s22}^{\varepsilon} T_{ik}^{00} S_{jl}^{22} + H_{s66}^{\varepsilon} T_{ik}^{11} S_{jl}^{11} + A_{s44}^{\varepsilon} T_{ik}^{00} S_{jl}^{11} \quad (28)$$

with noticing that:

$$T_{ik}^{rs} = \int_0^a \frac{\partial^r R_i}{\partial x_1^r} \frac{\partial^s R_k}{\partial x_1^s} dx_1, \quad S_{jl}^{rs} = \int_0^a \frac{\partial^r S_j}{\partial x_2^r} \frac{\partial^s S_l}{\partial x_2^s} dx_2 \quad (29)$$

The components of stiffness matrix $\mathbf{K}^{\chi}$ are defined as follows:

$$K_{ijkl}^{\chi 11} = \frac{A^{\chi}}{4}\left(T_{ik}^{22} S_{jl}^{11} + T_{ik}^{11} S_{jl}^{22}\right), \quad K_{ijkl}^{\chi 12} = -\frac{A^{\chi}}{4}\left(T_{ik}^{22} S_{jl}^{11} + T_{ik}^{11} S_{jl}^{22}\right)$$

$$K_{ijkl}^{\chi 13} = \frac{\overline{B}^{\chi}}{4}\left(T_{ik}^{02} S_{jl}^{11} - T_{ik}^{11} S_{jl}^{02}\right), \quad K_{ijkl}^{\chi 14} = \frac{1}{4}\left(B_s^{\chi} T_{ik}^{22} S_{jl}^{11} + B_s^{\chi} T_{ik}^{11} S_{jl}^{22} - \overline{B}_s^{\chi} T_{ik}^{11} S_{jl}^{02}\right)$$

$$K_{ijkl}^{\chi 15} = \frac{1}{4}\left(\overline{B}_s^{\chi} T_{ik}^{02} S_{jl}^{11} - B_s^{\chi} T_{ik}^{22} S_{jl}^{11} - B_s^{\chi} T_{ik}^{11} S_{jl}^{22}\right), \quad K_{ijkl}^{\chi 22} = \frac{A^{\chi}}{4}\left(T_{ik}^{22} S_{jl}^{11} + T_{ik}^{11} S_{jl}^{22}\right)$$

$$K_{ijkl}^{\chi 23} = \frac{\overline{B}^{\chi}}{4}\left(T_{ik}^{11} S_{jl}^{02} - T_{ik}^{02} S_{jl}^{11}\right), \quad K_{ijkl}^{\chi 24} = \frac{1}{4}\left(\overline{B}_s^{\chi} T_{ik}^{11} S_{jl}^{02} - B_s^{\chi} T_{ik}^{22} S_{jl}^{11} - B_s^{\chi} T_{ik}^{11} S_{jl}^{22}\right)$$

$$K_{ijkl}^{\chi 25} = \frac{1}{4}\left(B_s^{\chi} T_{ik}^{22} S_{jl}^{11} + B_s^{\chi} T_{ik}^{11} S_{jl}^{22} - \overline{B}_s^{\chi} T_{ik}^{02} S_{jl}^{11}\right)$$

$$K_{ijkl}^{\chi 33} = \frac{1}{4}\left(A^{\chi} - 2\overline{B}^{\chi} + \overline{D}^{\chi}\right)\left(T_{ik}^{00} S_{jl}^{22} - T_{ik}^{20} S_{jl}^{02} - T_{ik}^{02} S_{jl}^{20} + T_{ik}^{22} S_{jl}^{00} + 2T_{ik}^{11} S_{jl}^{11}\right)$$
$$+ \frac{1}{4}\overline{\overline{D}}^{\chi}\left(T_{ik}^{00} S_{jl}^{11} + T_{ik}^{11} S_{jl}^{00}\right)$$

$$K_{ijkl}^{\chi 34} = \frac{1}{4}\left[\left(\overline{B}_s^{\chi} - \overline{D}_s^{\chi}\right)\left(T_{ik}^{20} S_{jl}^{02} - T_{ik}^{22} S_{jl}^{00} - T_{ik}^{11} S_{jl}^{11}\right) + \overline{K}_s^{\chi}\left(T_{ik}^{20} S_{jl}^{11} - T_{ik}^{11} S_{jl}^{20}\right) + \overline{\overline{D}}_s^{\chi} T_{ik}^{11} S_{jl}^{00}\right]$$

$$K_{ijkl}^{\chi 35} = \frac{1}{4}\left[\left(\overline{B}_s^{\chi} - \overline{D}_s^{\chi}\right)\left(-T_{ik}^{00} S_{jl}^{22} + T_{ik}^{02} S_{jl}^{20} - T_{ik}^{11} S_{jl}^{11}\right) + \overline{\overline{D}}_s^{\chi} T_{ik}^{00} S_{jl}^{11} + \overline{K}_s^{\chi}\left(T_{ik}^{11} S_{jl}^{20} - T_{ik}^{20} S_{jl}^{11}\right)\right]$$

$$K_{ijkl}^{\chi 44} = \frac{1}{4}\left[\overline{H}_s^{\chi}\left(T_{ik}^{22} S_{jl}^{00} + 2T_{ik}^{11} S_{jl}^{11}\right) + \overline{\overline{H}}_s^{\chi} T_{ik}^{11} S_{jl}^{00}\right.$$
$$-\overline{L}^{\chi}\left(T_{ik}^{11} S_{jl}^{20} + T_{ik}^{11} S_{jl}^{02}\right) + H_s^{\chi}\left(T_{ik}^{22} S_{jl}^{11} + T_{ik}^{11} S_{jl}^{22}\right)\right]$$

$$K_{ijkl}^{\chi 45} = \frac{1}{4}\left[\overline{L}^{\chi}\left(T_{ik}^{11}S_{jl}^{20} + T_{ik}^{02}S_{jl}^{11}\right) - \overline{H}_{s}^{\chi}\left(T_{ik}^{02}S_{jl}^{20} + T_{ik}^{11}S_{jl}^{11}\right) - H_{s}^{\chi}\left(T_{ik}^{11}S_{jl}^{22} + T_{ik}^{22}S_{jl}^{11}\right)\right]$$

$$K_{ijkl}^{\chi 55} = \frac{1}{4}\left[\overline{H}_{s}^{\chi}\left(2T_{ik}^{11}S_{jl}^{11} + T_{ik}^{00}S_{jl}^{22}\right) + \overline{\overline{H}}_{s}^{\chi}T_{ik}^{00}S_{jl}^{11}\right.$$

$$\left.-\overline{L}^{\chi}\left(T_{ik}^{20}S_{jl}^{11} + T_{ik}^{02}S_{jl}^{11}\right) + H_{s}^{\chi}\left(T_{ik}^{11}S_{jl}^{22} + T_{ik}^{22}S_{jl}^{11}\right)\right] \tag{30}$$

The components of mass matrix $\mathbf{M}$ are given by:

$$M_{ijkl}^{11} = I_0 T_{ik}^{11}S_{jl}^{00}, \ M_{ijkl}^{13} = I_1 T_{ik}^{11}S_{jl}^{00}, \ M_{ijkl}^{14} = J_1 T_{ik}^{11}S_{jl}^{00},$$

$$M_{ijkl}^{22} = I_0 T_{ik}^{00}S_{jl}^{11}, \ M_{ijkl}^{23} = I_1 T_{ik}^{00}S_{jl}^{11}, \ M_{ijkl}^{25} = J_1 T_{ik}^{00}S_{jl}^{11}$$

$$M_{ijkl}^{33} = I_0 T_{ik}^{00}S_{jl}^{00} + I_2\left(T_{ik}^{11}S_{jl}^{00} + T_{ik}^{00}S_{jl}^{11}\right), \ M_{ijkl}^{34} = J_2 T_{ik}^{11}S_{jl}^{00}, \tag{31}$$

$$M_{ijkl}^{35} = J_2 T_{ik}^{00}S_{jl}^{11}, \ M_{ijkl}^{44} = K_2 T_{ik}^{11}S_{jl}^{00}, \ M_{ijkl}^{55} = K_2 T_{ik}^{00}S_{jl}^{11}$$

It is worth noticing that for free vibration analysis, by denoting $\mathbf{d}(t) = \mathbf{d}e^{i\omega t}$ where ω is the natural frequency of the FG microplates and $i^2 = -1$ is imaginary unit, the natural frequency can be hence derived from the following the equation $(\mathbf{K} - \omega^2 \mathbf{M})\mathbf{d} = \mathbf{0}$.

4 Numerical Examples

In this section, numerical examples are carried out to investigate free vibration behaviours of FG microplates with different BCs in which the shear function $f(x_3) = x_3 - 4x_3^3/3h^2$ is selected. The FG microplates are supposed to be made of ceramic material Al_2O_3 and metal one Al whose properties are given as follows: Al_2O_3 ($E_c = 380$ GPa, $\rho_c = 3800$ kg/m^3, $\nu_c = 0.3$), Al ($E_m = 70$ GPa, $\rho_m = 2702$ kg/m^3, $\nu_m = 0.3$). Preliminary convergence study of the series solution showed that the number of series $n_1 = n_2 = n = 10$ can be considered as the convergence point of the solution field, therefore this value will be used in the sequel computations. Moreover, for convenience, the following normalized parameter are used in the numerical examples:

$$\overline{\omega} = \frac{\omega a^2}{h}\sqrt{\frac{\rho_c}{E_c}} \tag{32}$$

In order to verify the accuracy of the present theory, Table 1 introduces normalized fundamental frequencies of Al/Al$_2$O$_3$ FG microplates in which the solutions are calculated with different values of the power index $p = 1, 2, 5, 10$, side-to-thickness ratio $a/h = 10, 20$, ratio of thickness-to-material scale $h/l = \infty, 5, 2.5, 5/3, 1.25, 1$, and four boundary conditions (SSSS, CSCS, CCSS, CCCC). The results obtained from the present theory with SSSS boundary condition has been compared with thosed of He et al. [7] based on the MCT and IGA-HSDT theory, Thai and Kim [6] using the IGA method, TSDT and MCT. It can be seen that there are good agreements

Table 1 Normalized fundamental frequencies of Al/Al$_2$O$_3$ FG microplates

BCs	a/h	p	Theory	h/l					
				∞	5	2.5	5/3	1.25	1
SSSS	20	1	Present	4.5141	4.9608	6.0715	7.5784	9.3065	11.1115
			TSDT-MCT [6]	4.5228	4.9568	6.0756	7.5817	9.2887	11.1042
			HSDT-MCT [7]	4.5228	4.9568	6.0756	7.5817	9.2887	11.1042
		2	Present	4.1328	4.5055	5.5080	6.8791	8.4182	10.0537
			TSDT-MCT [6]	4.1100	4.5006	5.5082	6.8661	8.4062	10.0450
		5	Present	3.8949	4.1938	5.0048	6.1394	7.4333	8.8334
			TSDT-MCT [6]	3.8884	4.2005	5.0199	6.1457	7.4397	8.8286
		10	Present	3.7641	4.0336	4.7303	5.7438	6.8892	8.1474
			TSDT-MCT [6]	3.7622	4.0323	4.7488	5.7453	6.9013	8.1494
	10	1	Present	4.4186	4.8537	5.9779	7.4863	9.1982	11.0127
			TSDT-MCT [6]	4.4192	4.8526	5.9664	7.4619	9.1537	10.9511
		2	Present	4.0071	4.4029	5.4160	6.7853	8.3177	9.9588
			TSDT-MCT [6]	4.0090	4.4006	5.4071	6.7580	8.2863	9.9101
		5	Present	3.7704	4.0825	4.9086	6.0476	7.3324	8.7213
			TSDT-MCT [6]	3.7682	4.0876	4.9169	6.0447	7.3338	8.7135
		10	Present	3.6376	3.9110	4.6355	5.6433	6.8060	8.0434
			TSDT-MCT [6]	3.6368	3.9162	4.6464	5.6487	6.8030	8.0448
CSCS	10	1	Present	5.9217	6.7270	8.6699	11.1709	13.9332	15.4952
		2	Present	5.3641	6.0886	7.8423	10.0972	12.5893	15.2033
		5	Present	5.0089	5.5972	7.0417	8.9330	11.0458	13.2752
		10	Present	4.8179	5.3319	6.6071	8.2965	10.1959	12.2104
CCSS	10	1	Present	6.2571	7.1981	9.4486	12.3090	15.4451	18.7177
		2	Present	5.6642	6.5115	8.5418	11.1216	13.9519	16.9038
		5	Present	5.2629	5.9547	7.6375	9.8094	12.2142	14.7385
		10	Present	5.0546	5.6600	7.1471	9.0888	11.2542	13.5382
CCCC	10	1	Present	7.5972	8.9737	11.9334	15.7186	19.8229	24.0786
		2	Present	6.9314	8.0884	10.8045	14.2073	18.7789	21.7420
		5	Present	6.3776	7.3691	9.6230	12.4991	15.6417	18.9349
		10	Present	6.1521	6.9859	8.9859	11.5569	14.4000	17.3824

bewteen the models. Moreover, it is observed from Table 1 that the fundamental frequencies decrease with an increase of the power index p. This can be explained by the fact that with an increase of p led to an decrease of the volume fraction of ceramic and stiffness of FG microplates. The effect of the power index p on the natural frequencies of Al/Al$_2$O$_3$ FG microplates is also plotted in Fig. 2 for different values of ratio of thickness-to-material scale $h/l = 5$, 2.5, 5/3, 1.25, 1, and SSSS boundary condition. It can be seen from this figure that the highest and lowest curves correspond to $h/l = 1$ and 5, respectively, and there are large deviations of these curves. The variations of normalized fundamental frequencies with respect to the ratio of thickness-to-material scale h/l are observed in Table 1 and Fig. 3 for Al/Al$_2$O$_3$

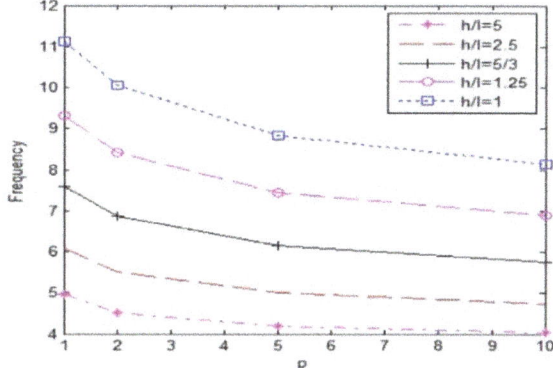

Fig. 2 Variation of normalized fundamental frequencies with respect the power index p ($a/h = 20$, SSSS)

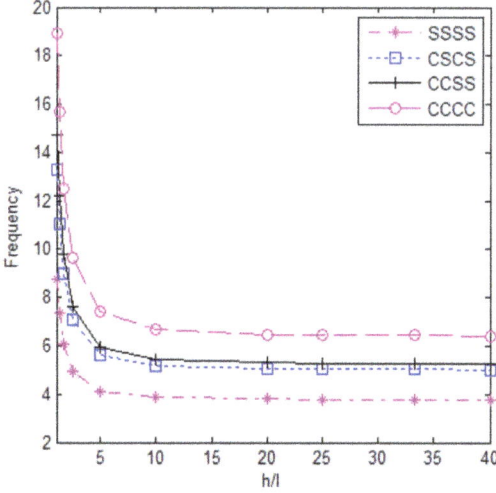

Fig. 3 Variation of normalized fundamental frequencies with respect to h/l ($p = 5$, $a/h = 10$)

FG microplates with $p = 5$ and $a/h = 10$. The graphs are displayed with four boundary conditions (SSSS, CSCS, CCSS, CCCC) within which it is observed that the natural frequencies decrease with the increase of h/l up to the value of $h/l = 10$ from which the curves become flatter and size effects can be hence neglected.

5   Conclusions

A unified higher-order shear deformation microplate model for vibration analysis of functionally graded materials has been proposed in this paper. The present theory is developed from fundamental equations of the elasticity theory and modified couple stress theory in which many different frameworks of size-dependent plate models could be recovered. In order to capture the size effects, the modified couple stress theory with one independent length-scale parameter is used. The solution field is approximated by bi-directional series in which hybrid shape functions are proposed, then the stiffness and mass matrix are explicitly derived. Numerical examples are investigated for different configurations of material distribution, side-to-thickness ratio, size-scale-thickness ratio and boundary conditions on the natural frequencies of FG microplates. The obtained results shows that the size effect is important and needs to be accounted for the computations. The size effect leads to an increase of the stiffness and natural frequency of the FG plates. The proposed unified size dependent plate model presents the accuracy and efficiency in predicting vibration behaviours of FG microplates.

Acknowledgements This research is funded by Vietnam National Foundation for Science and Technology Development (NAFOSTED) Under Grant No. 107.02-2018.312.

References

1. Yang F, Chong ACM, Lam DCC, Tong P (05/01 2002) Couple stress based strain gradient theory for elasticity. Int J Solids Struct 39:2731–2743
2. Tsiatas GC (2009/06/15/ 2009) A new Kirchhoff plate model based on a modified couple stress theory. Int J Solids Struct 46(13):2757–2764
3. Ma HM, Gao XL, Reddy JN (2011/08/01 2011) A non-classical Mindlin plate model based on a modified couple stress theory. Acta Mech 220(1):217–235
4. Reddy JN, Kim J (2012/02/01/ 2012) A nonlinear modified couple stress-based third-order theory of functionally graded plates. Compos Struct 94(3):1128–1143
5. Thai H-T, Vo TP (2013/02/01/2013) A size-dependent functionally graded sinusoidal plate model based on a modified couple stress theory. Compos Struct 96:376–383
6. Thai H-T, Kim S-E (2013/02/01/ 2013) A size-dependent functionally graded Reddy plate model based on a modified couple stress theory. Compos Part B: Eng 45(1):1636–1645
7. He L, Lou J, Zhang E, Wang Y, Bai Y (2015/10/15/ 2015) A size-dependent four variable refined plate model for functionally graded microplates based on modified couple stress theory. Compos Struct 130:107–115

8. Mantari JL, Canales FG (2016/09/15/ 2016) Free vibration and buckling of laminated beams via hybrid Ritz solution for various penalized boundary conditions. Compos Struct 152:306–315
9. Moreno-García P, Dos Santos JV, Lopes H (03/24 2017) A review and study on Ritz method admissible functions with emphasis on buckling and free vibration of isotropic and anisotropic beams and plates. Arch Comput Methods Eng 25:785–815
10. Nguyen T-K, Truong-Phong Nguyen T, Vo TP, Thai H-T (2015/07/01/ 2015) Vibration and buckling analysis of functionally graded sandwich beams by a new higher-order shear deformation theory. Compos Part B: Eng 76:273–285
11. Nguyen T-K, Vo TP, Nguyen B-D, Lee J (2016/11/15/ 2016) An analytical solution for buckling and vibration analysis of functionally graded sandwich beams using a quasi-3D shear deformation theory. Compos Struct 156:238–252

An Experimental Study on Factors Affecting the Coefficients of Shrinkage, Creep, and Crack Occurrence During Basement Construction in High-Rise Buildings

Phuc Binh An Nguyen, Quang Duy Tran, Quang Thai Bui, Quoc Viet Nguyen, Anh Tu Ta, Minh Quoc Le, Minh Duc Le, and Van Hai Luong

Abstract This paper is an experimental study on factors affecting the coefficients of shrinkage, creep, and crack occurrence during basement construction in high-rise buildings. Recently, the construction phase is one of the most important steps in the development of a building. Therefore, this paper will analyze the causes, conditions, and effects of concrete cracks during the construction of basements in high-rise buildings by using testing in laboratories and on-site. Finally, some recommendations and suggestions for the control of basement cracks during construction will also be presented.

Keywords Shrinkage · Creep · Basement cracks

1 Introduction

During the construction of basements, especially adjacent buildings, project contractors have checked and applied some solutions to minimize the risks of basement cracks. However, the cracks still appear, and the number of cracks varies in each project. There are many factors relating to reinforced concrete cracks, including

P. B. A. Nguyen · Q. D. Tran · Q. T. Bui · Q. V. Nguyen · A. T. Ta · M. Q. Le
RICONS Construction Investment JSC, 53-55 Ba Huyen Thanh Quan Street, Vo Thi Sau Ward, District 3, Ho Chi Minh City, Vietnam

M. D. Le
Golden Base Construction JSC, 29 Noi khu My Toan 2 Street, Tan Phong Ward, District 7, Ho Chi Minh City, Vietnam

V. H. Luong (✉)
Faculty of Civil Engineering, Ho Chi Minh City University of Technology (HCMUT), 268 Ly Thuong Kiet Street, District 10, Ho Chi Minh City, Vietnam
e-mail: lvhai@hcmut.edu.vn

Vietnam National University Ho Chi Minh City (VNU-HCM), Linh Trung Ward, Thu Duc District, Ho Chi Minh City, Vietnam

subjective and objective factors. When concrete cracks appear, external components will break into the reinforcement as well as the internal structure, leading to damage in the long term.

In concrete, it is important that the shrinkage coefficients can affect the durability of concrete structures. Generally, the shrinkage is classified as drying shrinkage, plastic shrinkage, autogenous shrinkage, chemical shrinkage, carbonation shrinkage, and temperature shrinkage [1].

Jing Chen [1] analyzed the different curing conditions affecting the shrinkage of concrete. There are significant influences of different curing conditions on the shrinkage performance of concrete. The variation rule for 28-day shrinkage of concrete is consistent with that of 180-day shrinkage. The concrete cured under standard conditions has less shrinkage than that cured under natural indoor conditions. However, it has more shrinkage than that cured under totally enclosed conditions [2]. Other researches focus on the mass concrete cracks of the basement foundation. Feng [2] studied the application of the control measures for mass concrete cracks in the basement foundation of a high-rise building and analyzed their construction effects.

Other researches, Juwen [3] used FEA (Finite Element Analysis) conducted on the temperature fields of the thick raft foundation and applied methods for cracking prevention. The field test results proved the applicability of a suite of temperature feedback regulation measures. Lam [4] analyzed a three-dimensional thermal model in the program Midas Civil to determine the early-age temperature evolution of a massive foundation with 2 steps of pouring concrete (after 72 h and 144 h). The temperature differences at the core of each concrete pour with respect to the concrete outer portion, which induce a risk of through cracking in the structure body or surface, were determined.

Guo [5] presented the method for cracking prevention of larger volume concrete foundations of high-rise buildings, including active and passive control methods. And the method was applied in a field case study of one high-rise building in Shanghai. Reducing the amount of cement in the concrete as possible, selecting low hydration cement, incorporating water-reducing, retarding agents, and fly ash can effectively control the dehiscence of concrete. Optimizing the design of the foundation and improving edge restrained conditions can slow down shrinkage stress and secondary stress in the concrete to some degree.

A study of Zichun [6], on a practical project, showed that the primary task is to control the selection of raw materials and determine raw materials. In a case study – Dalian World Finance Center project, Zhang [7] summarized the integrated construction technology of super high-rise mass concrete slab, including key issues such as mix design technology, pouring time selection, construction elements, and temperature testing technology. After 60 days of curing completion, no temperature cracks were found on the concrete surface, thus the mass concrete construction quality has been guaranteed.

So far, there has not been much research about basement cracks in Vietnam. But the study [8] reports the situation of basement seepage in 115 projects, which are located in Ho Chi Minh City. The basement crack is one of many reasons for the basement seepage, other than the waterproofing solution.

This paper's research factors related to the construction sequences include the computation, laboratory testing, and site testing as below:

- Affecting different displacements between foundations, considering the simultaneous working of the underground water pressure. This issue is considered the specificity of the project, including the geology, location, and issues related to the overall construction solution by the contractor.
- Affecting the concrete shrinkage, the basement boundary restraints.
- Affecting the difference in temperature of the mass foundation.

This paper also presented the experiments to determine the factors affecting the coefficient of shrinkage, creep, and the occurrence of cracks during the construction of basements in high-rise buildings in Ho Chi Minh City. Finally, some recommendations in terms of materials, construction methods, and construction sequence at the construction site are made. Measures to minimize the influence of cracks appearing in the basements of high-rise buildings are also included.

2 Experience and Testing Results

2.1 Project Information

The study project is located in district 1, HCM City, as indicated in Fig. 1. It is a complex apartment building with more than 10,000m^2 of area. It has four basements constructed by the semi-top-down method. Basement construction sequences are

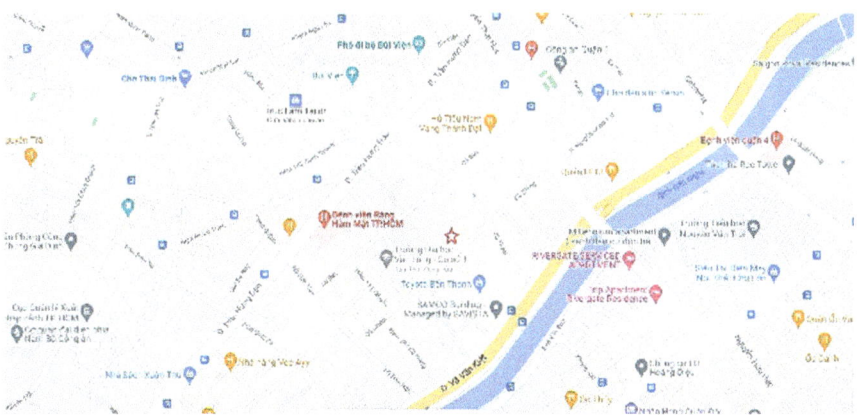

Fig. 1 Location of project

from the ground floor to basement 1, then basement 3, before constructing basement 2; and finally basement 4, which is near -15.6 m deep from the ground level.

The basement 4 structural details are a 0.6-m thickness of the slab and a 2.8-m thickness of mass foundation. The concrete grade is B35 grade, following the Vietnamese standard.

The research has the following steps as below:

- Calculating the basement structure with the FEM software. Using the design document for modeling the structure which considers the construction factor (included the construction load, construction progress) and the geology of the project (water level and soil parameter).
- Considering factors related to concrete suppliers, casting method, and concrete grade in the project. Clarify and control some issues that cause concrete cracks, such as the concrete aggregates, the ratio of water and cement, concrete curing, environmental temperature, etc.
- On-site testing to compare and analyze the data between on-site and structural modeling.

2.2 Shrinkage Test Result

Table 1 shows that the shrinkage factor increased at a rapid rate during the first 7 days, then gradually and slowly increased in the next 14 days. Plastic shrinkage (condensation) is volume shrinkage caused by the hydration reaction before the concrete solidifies and forms molecular chains. Plastic shrinkage occurs within about 3 to 12 h after the concrete is cast. Since the concrete remains in a plastic state, this condensation is known as plastic shrinkage. The condensate size is about 1% of the absolute cement mass, increasing with the water-cement ratio. Figure 2 demonstrates the onsite test result of concrete shrinkage at 3, 7, and 28 days, respectively. The results showed the shrinkage factor on the entire slab at basement 4.

Table 2 shows the comparison of concrete shrinkage results between the laboratory and onsite tests. The actual onsite results show that the shrinkage is quite different from the lab results.

Table 1 Test results of shrinkage factor

Shrinkage	Days							
	1	3	7	10	14	21	23	28
‰	0.000	0.016	-0.032	-0.042	-0.074	-0.110	-0.124	-0.148

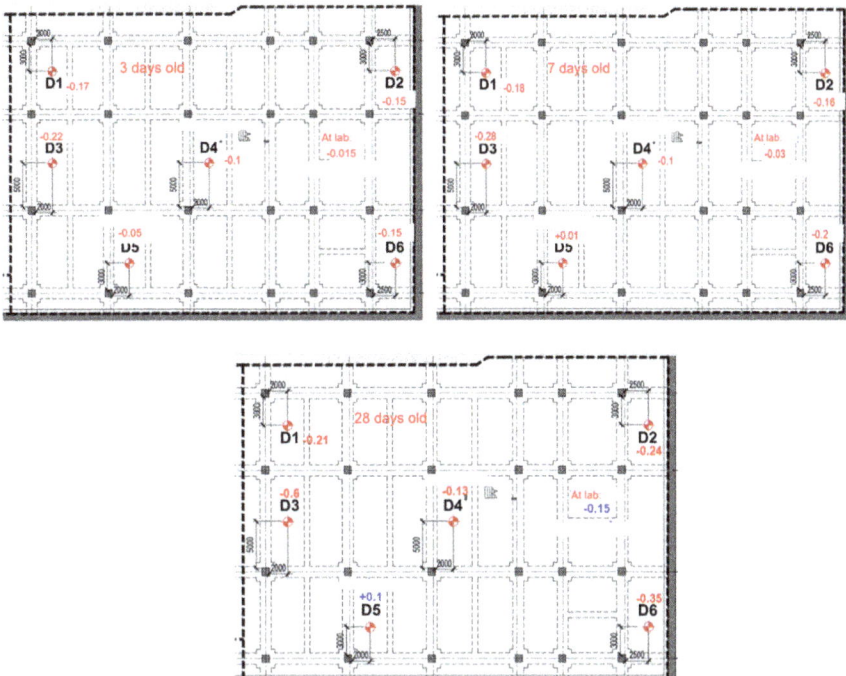

Fig. 2 Onsite test result of concrete shrinkage at 3, 7, and 28 days

Table 2 Concrete shrinkage result comparison

Days old	Lab result (‰)	Onsite result (‰)
3	0.015	0.05 – 0.22
7	0.030	0.10 – 0.28
28	0.145	0.13 – 0.60

3   Numerical Results

3.1   Computational Model

The analysis crack width ranges from 0.20 to 0.45 mm.

The actual crack width ranges from 0.36 to 0.65 mm.

The largest position onsite has a crack of 0.65 mm adjacent to the raft foundation (Fig. 3).

Cracks often appear at the following locations:

- Along the junction between the foundations, beams, and the slab
- Slab area close to diaphragm wall

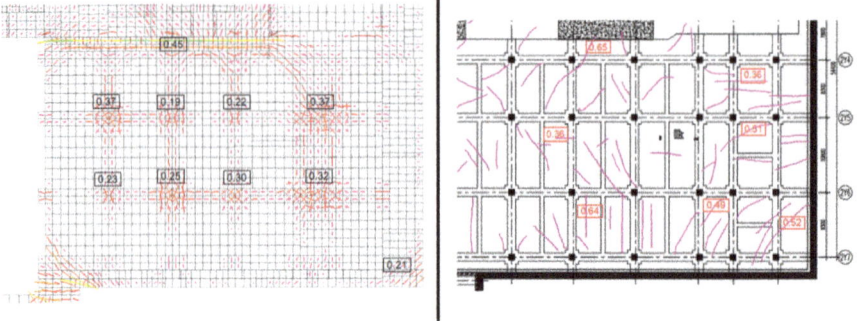

Fig. 3 Crack result comparison between model and on-site (mm)

3.2 Assessment

Through the analysis results, the actual crack locations compared with the analytical model are quite similar, mainly at the joint between the foundation, beam, and slab. However, the calculated crack width is quite small compared to that on-site. The reason is that the model analyzes under ideal conditions, e.g., all positions in the basement have the same shrinkage factor. Furthermore, the mechanism and characteristics of the cement hydration process are not accurately described. The tensile stress in concrete is increased due to water evaporation, and the temperature difference between the concrete and the external environment has not been clearly expressed either. Each location will have a shrinkage difference.

After casting the concrete slab, in around 1 to 5 days, the shrinkage and temperature of the concrete block will vary greatly and be very sensitive due to the cement hydration process.

Many factors may affect the amplitude of shrinkage, for example:

- Curing conditions: humidity, ambient temperature, watering maintenance, etc.
- Materials, aggregate: quality of materials, fineness of cement, water/cement ratio, chemical additives, etc.
- Construction impacts: installing, dismantling scaffolding, moving MEP equipment with heavy loads, etc.
- Reinforcing ratio: to increase the safety factor, design consultants often arrange an excessive amount of reinforcement, leading to the small spacing between steel bars. These tiny spacings will cause concrete segregation and decrease the concrete quality.

The ideal environment for concrete is moisture: the concrete must be kept moist for as long as possible after pouring. After the concrete is poured, even though it has hardened, the internal hydration process continues to achieve maximum concrete strength. In a very dry environment, the water in the concrete evaporates quickly, there is not enough amount left for the hydration process, and the concrete strength may stop developing and cause cracking.

Usually, the temperature increases during curing. Cement hydration and early strength development become rapid. If curing occurs at a low temperature, final strength will not be affected, although strength development is slow. But if the temperature is below zero degrees Celsius, the cement hydration will stop, and the strength will not only stop growing but the cement structure will be destroyed due to water condensation as well (freezing and thawing).

The size of cement particles directly affects the hydration, curing process, strength, and heat of hydration. The finer the cement particles are, the larger the area in contact with water is. As a result, hydration will be rapid, setting and hardening will be accelerated, and initial strength will be high. However, if the cement particles are too small, it is easy to react with water and calcium dioxide in the air to damage the cement preservation. If the cement is too fine, its shrinkage is very large during curing. Typically, the particle size of cement is in the range of 0.007–0.2 mm.

Inappropriate storage will cause the cement to be damp. The cement grain surfaces combine with moisture to cause caking because of hydration, severely reducing strength. Slow hydration and carbonation will occur due to the influence of water and CO_2 in the air, despite good storage. Strength decreases from 10 to 20% after 3 months, from 15 to 30% after 6 months, and then by 25 to 40% after 1 year, so the effective preservation time of cement is 3 months and should not be left for a long time.

Due to the combination of the above factors, the actual cracks will be larger and the cracking rules will also be more complicated. In fact, the heat of cement hydration only fluctuates greatly in the early period of time. In the long run, the cracks will tend to stabilize and close up when the tensile stress in the concrete decreases over time.

4 Conclusions and Recommendations

In summary, the main causes of floor cracking are the heat of cement hydration, self-shrinkage, and dry shrinkage of concrete. The occurrence of cracks belongs to the inevitable nature of concrete and therefore it is difficult to limit absolutely.

In order to partially limit the occurrence of cracks, it is necessary to implement concrete curing measures in accordance with the conditions of temperature, humidity, and impact loads in basement construction. In addition, it is also necessary to limit the amount of cement in the concrete mix, arrange expansion joints, or place additional steel to avoid the impact of shrinkage on concrete.

Acknowledgements We acknowledge the support of time and facilities from Ho Chi Minh University of Technology (HCMUT), VNU-HCM for this study. This study was also supported by Sungshin, Viet Han & Viet Duc concrete supplier companies and Golden Base Construction company.

References

1. Chen J, Li XH, Wang FL, Xu FL (2013) Influence of different curing conditions on shrinkage of concrete. Adv Mater Res 838–841:36–41. https://doi.org/10.4028/www.scientific.net/AMR.838-841.36
2. Feng HH, Wang X (2010) Causes and control measures of mass concrete crack of high-rise building basement foundation slab. Adv Mater Res 163–167:1609–1613. https://doi.org/10.4028/www.scientific.net/AMR.163-167.1609
3. Ju Y, Lei H (2019) Actual temperature evolution of thick raft concrete foundations and cracking risk analysis. Adv Mater Sci Eng 2019:1–11. https://doi.org/10.1155/2019/7029671
4. Van Tang L, Trong Nguyen C, Bulgakov B, Ngoc Pham A (2018) Composition and early-age temperature regime in massive concrete foundation. MATEC Web Conf 196:04017. https://doi.org/10.1051/matecconf/201819604017.
5. Guo JJ, Wang W (2010) Methods for crack prevention of large volume concrete foundation of high-rise building. Appl Mech Mater 29–32:305–309. https://doi.org/10.4028/www.scientific.net/AMM.29-32.305
6. Mao ZC, Wang D, Liu YB (2015) Measures to control the temperature of the concrete cracks in foundation mass. Appl Mech Mater 723:309–312. https://doi.org/10.4028/www.scientific.net/AMM.723.309
7. Zhang GJ, Li HN, Shi CQ (2011) Study on mass concrete slab construction technology of Dalian World Finance Center. Appl Mech Mater 137:175–180. https://doi.org/10.4028/www.scientific.net/AMM.137.175
8. Hung ND, Long LH, Tam NM (2016) Assessing the current status of waterproof in civil construction. J Const :5

An Improved Approach for Damage Identification in Plate-Like Structures Based on Modal Assurance Criterion and Modal Strain Energy Method

Thanh-Cao Le, Van-Sy Bach, Chi-Thien Nguyen, Manh-Hung Tran, and Duc-Duy Ho

Abstract The modal strain energy (MSE) method is proven in many research pieces as an effective tool for damage detection in plate-like structures. However, in some complicated damage scenarios, the damage detection results are not accurate. This study presents an improved two-stage approach based on the Modal Assurance Criterion (MAC) and MSE method to boost the efficiency of MSE. Firstly, a new damage indicator named Modal Strain Energy Damage Index (MSEDI), based on MAC, is developed to locate potential damage elements more accurately in the first stage. Then, the damaged zones in plate structures are determined in both locations and extents in the second stage by minimizing an objective function using a genetic algorithm (GA) algorithm. Two different objective functions, global MSE, and local MSE are considered to examine their efficiencies on the performance of the GA algorithm. Finally, the feasibility of the proposed approach is investigated by numerical examples on a plate comprising multiple damage scenarios and different boundary conditions. The obtained results indicate that even in highly complex damage scenarios, the proposed improvement can identify the actual damage sites and estimate the extent of damage with high precision. The numerical results also show that the computational cost of the optimization process using the objective function based on local MSE change is much lower than that using the objective function based on global MSE change.

Keywords Damage identification · Genetic algorithm · Modal Assurance Criterion · Modal Strain Energy · Plate-like structures

T.-C. Le · V.-S. Bach · C.-T. Nguyen · M.-H. Tran · D.-D. Ho (✉)
Faculty of Civil Engineering, Ho Chi Minh City University of Technology (HCMUT), Ho Chi Minh, Vietnam
e-mail: hoducduy@hcmut.edu.vn

Vietnam National University Ho Chi Minh City, Ho Chi Minh, Vietnam

T.-C. Le · V.-S. Bach
Faculty of Civil Engineering, Nha Trang University, Khanh Hoa Province, Vietnam

© The Author(s), under exclusive license to Springer Nature Singapore Pte Ltd. 2023   737
J. N. Reddy et al. (eds.), *ICSCEA 2021*, Lecture Notes in Civil Engineering 268,
https://doi.org/10.1007/978-981-19-3303-5_66

1 Introduction

During the last three decades, the development of vibration-based damage detection methods has attracted interest from many researchers. Besides, vibration-based damage identification technologies have been widely used in mechanical, aerospace, and civil applications. In particular, the modal strain energy (MSE) method has been proven to be highly effective for damage detection in structures. Stubbs et al. (1995) [9] firstly applied the MSE method to identify the damage in beam structures. Then, Cornwell et al. (1999) [2] have expanded MSE for the plate structures. Hu and Wu (2008) [5] verified the MSE method using experimental vibration results to detect surface cracks in a thin, isotropic aluminum plate with the free boundary condition. Fu et al. (2016) [4] established a two-step procedure using a combination of MSE and response sensitivity analysis to identify damage in the plate structures using isotropic homogeneous material. Le and Ho (2020) [7] proposed a two-step process for damage identification in plates, using only the vertical displacement component of the mode shape. Besides, optimization algorithms have been applied for model updating to solve the problem of structural optimization and damage identification. Among optimization tools, the genetic algorithm (GA) is used very popularly in optimization problems [1, 3, 8].

As an effort to fill in the research gaps mentioned above, this study proposes an improved two-stage approach based on the Modal Assurance Criterion (MAC) and MSE method. Firstly, MAC values are calculated for all of the mode shapes that modal data can be collected. Then, some mode shapes with the most significant MAC value are selected as the input for calculating a new damage indicator based on MAC (MSEDI), named Modal Strain Energy Damage Index. The advantage of this development is to increase identification accuracy and reduce false alarms appearing in its prediction when considering the complex damage scenarios. Secondly, GA is applied to determine the extent of detected damaged elements and help eliminate the false alarms (if any) obtained in the first stage. The importance of selecting a suitable mode shape is examined by considering various combinations of mode shapes with different MAC values. Moreover, the numerical example is performed for a concrete plate with multiple damages.

2 Damage Identification Approach Based on Modal Assurance Criterion and Modal Strain Energy Method

See Fig. 1.

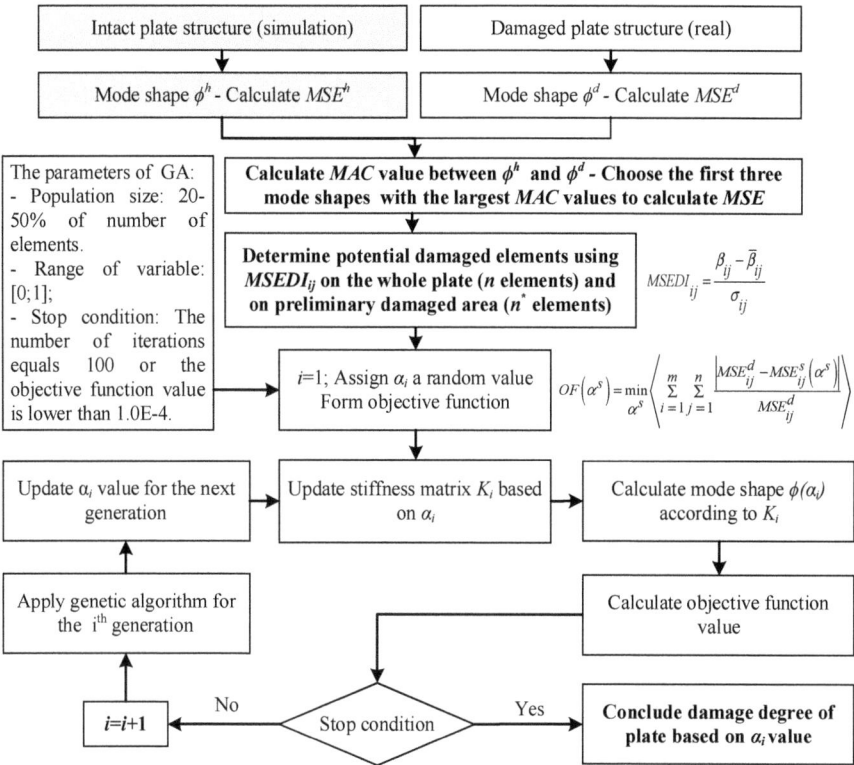

Fig. 1 Flow chart of the proposed damage identification approach

2.1 Modal Assurance Criterion (MAC)

The MAC is used to determine the similarity of two mode shapes. In order to identify the correlation between two any i^{th} and j^{th} mode shapes of two sets of vector $\{\phi^A\}$ and $\{\phi^B\}$, MAC is calculated as follow:

$$MAC(i,j) = \frac{\left(\{\phi^A\}_i^T \{\phi^B\}_j\right)^2}{\left(\{\phi^A\}_i^T \{\phi^A\}_j\right)\left(\{\phi^B\}_i^T \{\phi^B\}_j\right)} \quad (1)$$

If the mode shapes are identical, the MAC will have a value of one or 100%. If the mode shapes are very different, the MAC value will be close to zero. The MAC criterion is mainly used for detecting the occurrence of damages in the structure. However, the MAC could not determine the location and extent of the damage.

2.2 Damage Identification Approach

An improved damage detection approach is proposed in this study based on the theory of the MSE method, GA [7], and MAC, an. First, MAC values are used to choose the appropriate mode shapes as the input for the damage identification algorithm. Second, the MSE method is used to locate the damaged element. Finally, GA is deployed to identify the extent of the damage. The flow chart of the proposed approach is shown in Fig. 1.

3 Numerical Verification

3.1 Properties of Plate

A concrete plate is $2000 \times 2000 \times 150$ mm in size, four edges are fixed, as depicted in Fig. 2. Figure 3 shows the first six mode shapes. The material properties and damage scenario are given in Table 1.

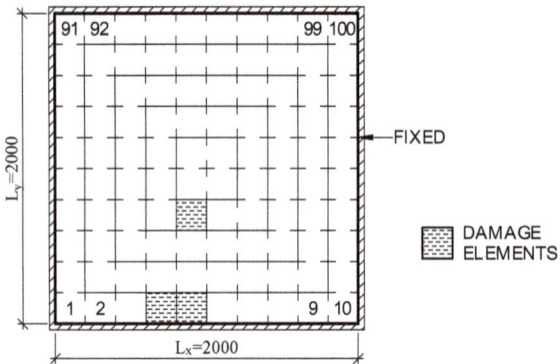

Fig. 2 Mesh of 10×10 elements for the plate and location of damaged elements

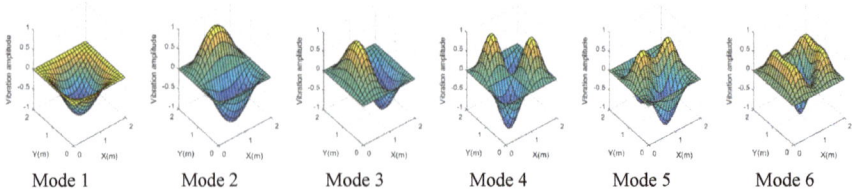

Mode 1   Mode 2   Mode 3   Mode 4   Mode 5   Mode 6

Fig. 3 The first six mode shapes for intact state

Table 1 Material properties and damage scenario

Material properties	
Young's modulus	20GPa
Poisson's ratio	0.2
Mass density	2400 kg/m^3
Damage scenario	
Damage element	Damage extent
4th element	40%
5th element	45%
35th element	20%

Table 2 The natural frequencies, MAC values

Mode	Mode 1	Mode 2	Mode 3	Mode 4	Mode 5	Mode 6
f_{intact} (Hz)	172.61	338.46	340.35	484.70	579.81	585.82
f_{damage} (Hz)	170.48	332.99	339.57	481.88	572.11	584.01
Δf (%)	1.2	1.6	0.2	0.6	1.3	0.3
MAC	1.000	0.950	0.951	0.998	0.984	0.988

3.2 Modal Analysis

A finite element model of the plate was established by using MATLAB software. Nine-node isometric quadrilateral element was used to model the plate obeying the Mindlin plate theory [7]. The plate's free vibration was analyzed for both intact and damaged states. Figure 2 shows the first six mode shapes. The natural frequencies, MAC values are summarized in Table 2. The differences from the natural frequencies of both states are minimal, about 0.3–1.6%. The MAC values of all modes are almost equal to 1; the smallest value is 0.95.

3.3 Damage Location Identification Results

In the first step, the preliminary damaged zone with the size 1000 × 1000 mm is localized by the application of MSEDI on the whole plate. Then MSEDI is deployed again on this local damaged zone to eliminate more wrong alarms. The damage threshold has been chosen to be 30% to eliminate noise effects [6]. Compared to the previous studies, the critical advantage in this study is that the MAC value is used as the indicator for choosing the most effective mode shapes as the input of the MSE method. The effectiveness of this improvement was investigated in two cases. For case 1, a single-mode and a combination of two or more modes of all the first six modes were used. For case 2, a combination of only two or three modes with the highest values of MAC (i.e., modes 1, 4, and 6) was used. The damage location identification results of both two cases on the preliminary damaged zone are shown in Tables 3 and 4.

The damage location identification results show that: modes 1, 4, and 6 give the best location identification results. These modes can successfully locate the damaged element with minor false alarms; meanwhile, modes 2, 3, and 5 cannot locate the damaged elements. Especially, mode 1 has the most significant value of MAC (i.e., 1.000), next is mode 4 (i.e., 0.998), and then is mode 6 (i.e., 0.988). While MAC values of three modes 2, 3, and 5 are lower a little bit, with 0.950, 0.951, and 0.984,

Table 3 The damage location identification results for case 1

Mode	Damage index chart	Damage elements	Right (Wrong)	Mode	Damage index chart	Damage elements	Right (Wrong)
1		3-**4**-**5**-24-25-**35**	3 (3)	1 + 2		3-**4**-**5**-15-24-33-**35**	3 (4)
2		3-14-15-16-44	0 (5)	1 + 2 + 3		3-**4**-**5**-14-15-**35**	3 (3)
3		6-12-26-32-36-42-46	0 (7)	1 + 2 + 3 + 4		3-**4**-**5**-14-**35**	3 (2)
4		**5**-13-14-**35**	2 (2)	1 + 2 + 3 + 4 + 5		3-**4**-**5**-14-15-**35**	3 (3)
5		34-42	0 (2)	1 + 2 + 3 + 4 + 5 + 6		**4**-**5**-14-15-**35**	3 (2)
6		**4**-**5**-6-**35**	3 (1)				

Table 4 The damage location identification results for case 2

Mode	Damage index chart	Damage elements	Right (Wrong)	Mode	Damage index chart	Damage elements	Right (Wrong)
1 + 4		3-**4**-**5**-14-25-**35**	3 (3)	1 + 6		**4**-**5**-25-**35**	3 (1)
1 + 4 + 6		**4**-**5**-**35**	3 (0)				

Table 5 Effect of damage threshold on damage location identification results

Threshold	Damage index chart	Damage elements	Right (Wrong)	Threshold	Damage index chart	Damage elements	Right (Wrong)
10%		3–4-5–14-25–35	3 (3)	40%		4–5-35	3 (0)
20%		4–5-5–35	3 (1)	50%		5–35	2 (-1)
30%		4–5-35	3 (0)	60%		5–35	2 (-1)

respectively. On the other hand, the combination of two or more first mode shapes shows the ability of damage localization, but some wrong elements were identified. In the case of using three modes with the high values of MAC, the identification results are boosted significantly. Specifically, using the combination of two or three first modes in both cases also locate real damaged elements. However, in the case of using a combination of two first modes with high MAC values, there are only three wrong elements compared to four wrong elements in the first case. Moreover, the identification result is entirely accurate in the case of using a combination of three modes with the most significant MAC values (i.e., modes of $1 + 4 + 6$).

In this study, the effect of the damage threshold on the damage identification results was also investigated. The threshold varied from 10 to 60% of the maximum value of MSEDI. The combination of the three modes (i.e., modes of $1 + 4 + 6$) was used as the input to calculate MSEDI. The results in Table 5 show that: with the damage threshold lower than 30%, the proposed method can locate the damaged elements (i.e., elements of 4, 5, 35) with some false alarms; with the damage threshold of 30–40%, the proposed method can successfully locate the damaged elements with minor false alarms; with the damage threshold larger than 40%, there is an element (i.e., element of 4) misidentified from the damage index chart. Therefore, a threshold of 35% was recommended for the best damage localization.

3.4 Damage Extent Identification Results

In the second step, the GA was used to determine the extent of damage [7]. The effectiveness of the damage extent identification algorithm is evaluated in two cases: for case 1, using the combination of three modes no matter what MAC values are; for case 2, using the combination of three first modes that have the high values of MAC. When the stop condition is met, the following two values are suggested to evaluate: (1) the number of iterations performed when the algorithm stopped; (2) the error in the damage extent identification when the stop condition is satisfied. The damage

Table 6 The damage extent identification results

Case 1: Estimated damage extent (Error (%))			Case 2: Estimated damage extent (Error (%))	
Number of iterations	1	100	1	38
3th element	0.267 (26.69)	0.219 (21.92)		
4th element	0.031 (92.31)	0.147 (63.19)	0.341 (14.73)	0.4 (0.00)
5th element	0.741 (64.71)	0.43 (4.37)	0.456 (1.42)	0.45 (0.00)
14th element	0.158 (15.84)	0.089 (8.90)		
15th element	0.24 (23.96)	0.082 (8.22)		
35th element	0.299 (49.70)	0.197 (1.60)	0.171 (14.59)	0.2 (0.01)
Objective function value	3.16E-01	7.66E-05	3.16E-01	7.66E-05

extent identification results of both cases are listed in Table 6. The results in Table 6 show that: with the combination of the same number of mode shapes, the damage extent identification results are significantly different in the two cases. In case 1, the GA cannot correctly identify the damage extent even after 100 iterations. Meanwhile, in case 2, when using the three modes with a high value of MAC (i.e., modes of 1 + 4 + 6), the proposed method gives the accurate damage extent identification results after 38 iterations.

4 Conclusions

In this study, a damage identification method using MAC values and the MSE method was improved successfully to boost the accuracy of the origin MSE method. From the analysis results, the essential conclusions were drawn as follows:

(1) The proposed method can accurately detect damages in like-plate structures with complex damage scenarios when using the appropriate mode shapes and damage threshold. The first step locates the preliminary damaged sites. The second step identifies the exact extent of the elements.
(2) The damage identification result depends entirely on the used mode shapes. Each mode is sensitive to some locations of damage. Therefore, the combination of some mode shapes is better to identify damages than using individual mode. MAC has been proved to be an effective indicator to choose the modes as the input for the MSE method.
(3) Using the three modes with high MAC values with a damage threshold of 35% is good enough to accurately detect the location and the extent of the damaged elements in like-plate structures.

Acknowledgements We would like to thank Ho Chi Minh City University of Technology (HCMUT), VNU-HCM for the support of time and facilities for this study.

References

1. Chou JH, Ghaboussi J (2001) Genetic algorithm in structural damage detection. Comput Struct 79(14):1335–1353
2. Cornwell P, Doebling SW, Farrar CR (1999) Application of the strain energy damage detection method to plate-like structures. J Sound Vib 224:359–374
3. Friswell MI, Penny JET, Garvey SD (1998) A combined genetic and eigensensitivity algorithm for the location of damage in structures. Comput Struct 69(5):547–556
4. Fu YZ, Liu JK, Wei ZT, Lu ZR (2014) A two-step approach for damage identification in plates. J Vib Control 22(13):3018–3031
5. Hu HW, Wu CB (2008) Nondestructive damage detection of two dimensional plate structures using modal strain energy method. J Mech 24:319–332
6. Le TC, Ho DD (2015) Damage detection in plate-like structures using modal strain energy-based approach. Review of ministry of construction 6:100–105
7. Le TC, Ho DD (2020) Structural damage identification of plates using two-stage approach combining modal strain energy method and genetic algorithm. Proc. of the international conference on modern mechanics and applications 2020
8. Marano GC, Quaranta G, Monti G (2011) Modified genetic algorithm for the dynamic identification of structural systems using incomplete measurements. Comput Aided Civ Infrastruct Eng 26(2):92–110
9. Stubbs N, Kim JT, Farrar CR (1995) Field verification of a nondestructive damage localization and severity estimation algorithm. Proc. of 13th International Modal Analysis Conference 1:210–218

Bond Behavior Between Glass Fiber-Reinforced Polymer (GFRP) Bars and Saline Water–Sand Concrete

Hai La-Hong, Binh Thanh Nguyen, Dien Ngoc Vo-Le, Hue Ngoc Pham, and Long Nguyen-Minh

Abstract This paper presents an experimental study to investigate the bonding behaviour between GFRP bars and the concrete made from saline water and sand. The experimental program included twenty-four specimens of the concrete made from saline water and sand as an experimental group and six specimens of common concrete as a control group. The major survey parameters include the compressive strength of concrete (25 MPa and 35 MPa) and the diameter of GFRP reinforcing bar (12 mm and 16 mm). The specimens are examined under the pull-out testing method. The research results show that the average bonding strength of GFRP bars in the concrete made of salt-contaminated water and sand has only slightly been decreased from 5.6 to 7.2% compared to that in fresh water- sand concrete, and the decline degree of this bonding strength in 12 mm-diameter GFRP bars is more obvious than that in the 16 mm- diameter bars. The increase of concrete strength from 29.28 MPa to 35.56 MPa significantly enhances the average bonding strength of the 12 mm GFRP bar (18.7%), but only a slight improvement has been recorded in the 16 mm GFRP (6.3%). Increasing the GFRP bar diameter from 12 to 16 mm boosts the average bonding strength from 11.7 to 35.4%.

Keywords Bond strength · Saline water and sand concrete · GFRP bars · Concrete strength

H. La-Hong · D. N. Vo-Le · L. Nguyen-Minh (✉)
Faculty of Civil Engineering, Ho Chi Minh City University of Technology (HCMUT), 268 Ly Thuong Kiet Street, District 10, Ho Chi Minh City, Vietnam
e-mail: nguyenminhlong@hcmut.edu.vn

B. T. Nguyen
Department of Construction, 03 Cach Mang Thang 8 Street, An Hoi Ward, Ben Tre, Ben Tre, Vietnam

D. N. Vo-Le
Tien Giang University (TGU), 119 Ap Bac, My Tho, Tien Giang, Vietnam

H. N. Pham
People's Committee, Nguyen Van Linh Street, Hamlet 3, Ward 1, Nga Nam Town, Soc Trang Province, Vietnam

1 Introduction

To save fresh water resources and partially solve the scarcity of river sand for producing concrete, using saline water and sea-sand for mixing concrete seems to be a topical issue and viable alternative, especially in saline areas and islands, and this solution can significantly decrease construction investment cost. Although its potential advantages, the use of seawater and sea-sand in the concrete mix is currently prohibited in design standards because the presence of a high amount of chlorides (Cl^-) in seawater is the main cause of the reinforcement corrosion in concrete structures [1, 2]. Indeed, seawater has an average total salinity of 3.5%, of which 78% is sodium chloride (NaCl) [3]. Another concern, sulfate ions (SO_4^{-2}) in seawater interact with the hydration compounds of cement to create an expanding mineral (Ettringite) with the increase in its volume approximately 2.86 times, and this phenomenon can cause internal stress in concrete. However, the existing studies report that salt in seawater has no significant negative impact on the characteristics of the hardened concrete [4, 5]. Therefore, the proposed solution is to combine seawater concrete with non-corrosive reinforcement such as fiber-reinforced polymer (FRP) bars in concrete structures to prevent corrosion, and this solution can be reasonable in both technical and economic perspectives [3].

The adhesive bonding between FRP reinforcement and concrete is the key factor in determining the working length, anchor length and lap length of FRP reinforcement, as well as in determining the bearing capacity and cracking behaviour of the concrete structures using FRP rebar. However, current studies on the bond behavior of GFRP bars in saline water–sand or seawater sea-sand concrete are still very limited. In terms of the study of Dong et al. [6], which presents an experimental study on the bond durability of basalt fiber–reinforced polymer (BFRP) bars, steel–fiber reinforced polymer (FRP) composite bars (SFCB), and steel bars embedded in seawater sea-sand concrete (SWSSC) and based on a simulated ocean environment. The effects of the environment type, exposure period (3, 6, and 9 months), and reinforcement type on bond durability were investigated. The test results showed that after 9 months of exposure, for the BFRP bars, the bond stress retention was 92% in the wet-dry cycling environment and 78% in the immersion environment; for the SFCBs, the retention in the wet-dry cycling environment was 100% and that in the immersion environment was 90%. Regarding Soares et al. [7], the study assessed the bond mechanism between GFRP bars and seawater concrete, the influence of concrete age, rod diameter and anchorage length were investigated. The test results highlighted that the bond mechanism between the GFRP rod and the concrete is not only frictional but also has a mechanical adhesion component conferred by the rod ribs. The larger GFRP rob diameter provided the higher values of maximum pull-out force and shear strength, and the use of seawater had no severe impact on the bond behaviour between the GFRP and concrete. A study on the bond behaviour of FRP bars in MPC (magnesium potassium phosphate cement) seawater concrete was done by Sun et al. [8]. The effects of reinforcing bars, type of concrete and mixing water on the bond behaviour of FRP and steel bars were investigated and discussed. The results showed that the

MPC concrete increases the bond strength of BFRP and GFRP bars by 51.06% and 24.42%, respectively, compared with that in Portland cement (PC) concrete. Using seawater in MPC concrete can enhance the bond strength of GFRP bar by 13.75%.

This experimental research analyses the bonding behaviour between GFRP rebar and the concrete made from saline water and sand. In this study, the water and sand were taken from the saline area, and there was a mix of river sand and saline sand with an appropriate ratio for casting concrete specimens. In addition, the strain of GFRP bars in the concrete was also determined through strain gauges which were attached on the surface of GFRP bars. The experimental program included twenty-four specimens of the concrete made from saline water and sand as an experimental group and six specimens of common concrete as a control group. The major survey parameters include the compressive strength of concrete (25 MPa and 35 MPa) and the diameter of the GFRP reinforcing bar (12 mm and 16 mm).

2 Experimental Program

2.1 Materials

The cement used in this study was a commercial product (PCB40, INSEE Power-s). The saline water was delivered from the coastal area in Ben Tre province, Vietnam, and its chemical composition is shown in Table 1. The saline sand has the fineness modulus of 1.33 which is a combination of river sand (fineness modulus of 2.4) and saline sand (fineness modulus of 0.8) with a corresponding ratio of 1/3 and 2/3. The saline sand with the fineness modulus of 0.8 was also taken from the similar area. The stone with 20 mm maximum size was used as coarse aggregate.

The GFRP bars used in this study were produced by the manufacturer in Vietnam with their technical characteristics presented in Table 2. The diameters of GFRP bars

Table 1 Chemical composition of saline water

Chloride (Cl^-)	Sulfate (SO^{2-}_4)	Total dissolved solids	pH at 28^0C	Salinity
(mg/L)	(mg/L)	(mg/L)		(%)
12,592.62	2023.39	23,063.6	7.2	2.34

Table 2 Properties of GFRP bars provided by manufacturer

Bar type	Diameter, d_b	Surface treatment	Ultimate tensile strength, f_{fu}	Elastic modulus, E_f	Ultimate strain, ε_{fu}
	(mm)		(MPa)	(GPa)	(%)
12-GFRP	12 ± 0.5	Spiral ribs	800	45	$1 \div 3$
16-GFRP	16 ± 0.5		800	45	$1 \div 3$

Table 3 Proportions of saline water–sand concrete

Concrete type	Cement	Coarse aggregate	Saline sand	Saline water	Slump
	(kg/m^3)	(kg/m^3)	(kg/m^3)	(l/m^3)	(cm)
Group A (25 MPa)	360	1226	547	195	7 ± 1
Group B (35 MPa)	402	1218	562	195	7 ± 1

were 12 mm and 16 mm, and their surface was treated with spiral ribs. The tensile strength of the reinforcing bars was tested in accordance with ASTM D-3916—94D (1996), and three specimens were considered for each test point. The test results indicated that the average tensile strength was 804 MPa for the 12 mm-GFRP bar and 719.52 MPa for 16 mm-GFRP bar.

2.2 Mix Proportions of Saline Water and Sand Concrete

Two concrete types were investigated in this study because they were popularly used in the civil construction. The concrete mix proportions were established in accordance with the guidance of the cement manufacturer for the grades of common concrete, and these mix proportions were also selected to cast saline water–sand concrete specimens (Table 3). The compressive strength test was performed on concrete cubes (150 × 150x 150 mm) at the age of 28 days. Three specimens were considered for each test point. The test results showed that the average compressive strength ($f_{c,cube}$) of saline water–sand concrete cubes was 29.28 MPa and 35.56 MPa corresponding to concrete types of group A and B.

2.3 Test Setup and Procedure

The bond behaviour between GFRP bars and concrete was tested using a direct pull-out test according to ACI 440.03R-04 (2004). In this study, the experimental program included twenty-four specimens of the concrete made from saline water and sand as an experimental group and six specimens of common concrete as a control group (Table 4). The major survey parameters include the compressive strength of concrete (25 MPa and 35 MPa) and the diameter of GFRP reinforcing bar (12 and 16 mm).

Pull-out test specimens are shown in Fig. 1. The total length of GFRP bars was 800 mm, and the designed bond length of 60 and 80 mm were cast into concrete cubes. This embedded length was employed to meet the requirement that the bond length (L_d) should be at least five times the diameter (d_b) of the bar ($L_d = 5d_b$). The PVC tubes were pre-fixed at the designated location on the reinforcing bar before casting concrete. To avoid debonding between the bar and the PVC tube, epoxy resin was inserted between the tube and the bar to ensure that there was no gap in-between.

Table 4 Experimental specimens

Specimen		Nominal diameter of GFRP bars, d_b (mm)	Dimension of concrete cubes (cm)	Quantity	Note
Group A	G12-FW-FS-A	12	12 × 12 x 12	01	Fresh water–sand
	G12-SW-SS-A	12	12 × 12 x 12	06	Saline water–sand
	G16-FW-FS-A	16	16 × 16 x 16	01	Fresh water–sand
	G16-SW-SS-A	16	16 × 16 x 16	06	Saline water–sand
Group B	G12-FW-FS-B	12	12 × 12 x 12	02	Fresh water–sand
	G12-SW-SS-B	12	12 × 12 x 12	06	Saline water–sand
	G16-FW-FS-B	16	16 × 16 x 16	02	Fresh water–sand
	G16-SW-SS-B	16	16 × 16 x 16	06	Saline water–sand
Total				**30**	

Note FW: fresh water, FS: fresh sand, SW: saline water, SS: saline sand

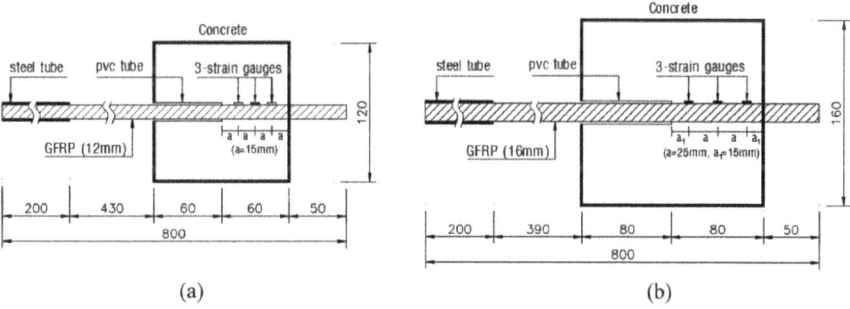

(a)   (b)

Fig. 1 Pull-out test specimens **a** schematic drawing of 12 mm-GFRP specimens, **b** schematic drawing of 16 mm-GFRP specimens

The loading ends of the FRP bar were embedded in a 200 mm long steel tube filled with epoxy resin to directly connect to the anchor of the test machine. The surface of the GFRP bar was installed by 3 strain gauges to determine the deformation of the bar at the measurement position. Figure 2 shows a schematic diagram of the pull-out test setup. The pull-out specimen was positioned in a universal tension–compression test machine that is screw-driven with a maximum load capacity of

Fig. 2 Direct pull-out test setup

1000 kN. The displacement and strain of the GFRP bar were recorded and processed through the Data Logger TDS303 device and computer. The load was applied under a displacement-controlled mode with a displacement rate of 2 mm/min.

3 Experimental Result and Discussion

3.1 Failure Mode

Through a series of direct pull-out tests, the test results were obtained and summarized in Table 5. The test result shows that there were 3 main types of failure modes, including: (a) bar pull-out from concrete, (b) concrete splitting, and (c) concrete rupture (Fig. 3). The failure patterns strongly affect the pulling force of specimens, meaning that the pull-out force of the "bar pull-out from concrete" specimens was greater than that of the specimens damaged by concrete splitting and rupture. Test result also indicates that the samples using common concrete tended to "bar pull-out from concrete" while almost saline water–sand concrete samples were damaged by concrete splitting and rupture. This revealed that there may be a difference in structure between saline water–sand concrete and common concrete, and further studies are absolutely necessary in order to clarify this claim. At the failure stage, the damage of the resin and tendon layer on the surface of the GFRP bars were observed.

3.2 Bond Strength—Slip Length Relationship

The experimental result shows that the bar pull-out failure goes through three distinct stages in the whole pull-out process, namely, the linear, nonlinear, and softening

Table 5 Results of pullout test and failure modes

No	Specimen	$f_{c,cube}$ (MPa)	P_{max} (kN)	τ_{ave} (MPa)	τ_{ave} (group) (MPa)	$S_{LVDT,ave}$ (mm/mm)	$S_{LVDT,ave}$ (group) (mm/mm)	$\varepsilon_{bar,u}$ (‰)	$\varepsilon_{bar,u}$ (group) (‰)	Failure modes
	Group A									
1	G12-FW-FS-A	31.05	20.62	9.12	9.12	0.051	0.051	6.70	6.70	bar pull-out from concrete
2	G12-SW-SS-A (1)	29.28	19.95	8.82	8.61	0.037	0.037	5.80	5.65	concrete rupture
3	G12-SW-SS-A (2)		18.76	8.30		0.035		6.00		concrete rupture
4	G12-SW-SS-A (3)		19.95	8.82		0.038		5.50		concrete rupture
5	G12-SW-SS-A (4)		19.40	8.58		0.038		4.80		concrete rupture
6	G12-SW-SS-A (5)		20.50	9.07		0.039		6.56		concrete rupture
7	G12-SW-SS-A (6)		18.27	8.08		0.037		5.24		concrete splitting
8	G16-FW-FS-A	31.05	47.50	11.82	11.82	0.043	0.043	10.69	10.69	bar pull-out from concrete
9	G16-SW-SS-A (1)	29.28	46.50	11.57	11.66	0.047	0.043	9.11	10.33	concrete splitting ara>
10	G16-SW-SS-A (2)		46.70	11.62		0.043		9.84		concrete splitting
11	G16-SW-SS-A (3)		46.98	11.69		0.040		10.07		concrete splitting
12	G16-SW-SS-A (4)		47.14	11.73		0.044		10.55		concrete splitting
13	G16-SW-SS-A (5)		47.96	11.93		0.048		11.90		bar pull-out from concrete
14	G16-SW-SS-A (6)		45.95	11.43		0.039		10.50		concrete splitting
	Group B									
15	G12-FW-FS-B (1)	37.09	25.77	11.40	11.01	0.059	0.063	7.05	6.95	concrete splitting

(continued)

Table 5 (continued)

No	Specimen	$f_{c.cube}$	P_{max}	τ_{ave}	τ_{ave} (group)	$S_{LVDT, ave}$	$S_{LVDT, eve}$ (group)	$\varepsilon_{bar, u}$	$\varepsilon_{bar, u}$ (group)	Failure modes
16	*G12-FW-FS-B (2)*		*24.02*	*10.62*		*0.066*		*6.85*		*concrete rupture*
17	G12-SW-SS-B (1)	35.56	23.50	10.39	10.22	0.056	0.053	6.03	6.35	concrete rupture
18	G12-SW-SS-B (2)		22.50	9.95		0.052		5.95		concrete rupture
19	G12-SW-SS-B (3)		22.23	9.83		0.048		5.90		concrete splitting
20	G12-SW-SS-B (4)		24.39	10.79		0.056		7.22		bar pull-out from concrete
21	G12-SW-SS-B (5)		23.50	10.39		0.053		6.57		concrete splitting
22	G12-SW-SS-B (6)		22.50	9.95		0.052		6.42		concrete splitting
23	*G16-FW-FS-B (1)*	*37.09*	*50.34*	*12.52*	*12.30*	*0.050*	*0.058*	*14.50*	*14.00*	*bar pull-out from concrete*
24	*G16-FW-FS-B (2)*		*48.50*	*12.07*		*0.066*		*13.50*		*concrete splitting*
25	G16-SW-SS-B (1)	35.56	49.34	12.28	12.40	0.049	0.047	11.11	12.57	concrete splitting
26	G16-SW-SS-B (2)		47.06	11.71		0.044		10.72		concrete splitting
27	G16-SW-SS-B (3)		53.38	13.28		0.052		14.20		bar pull-out from concrete
28	G16-SW-SS-B (4)		51.76	12.88		0.049		13.90		bar pull-out from concrete
29	G16-SW-SS-B (5)		46.50	11.57		0.040		11.20		concrete splitting
30	G16-SW-SS-B (6)		50.89	12.66		0.048		14.30		bar pull-out from concrete

Note $f_{c.cube}$: concrete compressive strength, P_{max}: maximum pullout force, τ_{ave}: average bond strength, $\tau_{ave} = \frac{P_{max}}{\pi d_b L_d}$, d_b: diameter of GFRP bar, L_d: embedded length, $S_{LVDT, ave}$: average slip length, $\varepsilon_{bar, u}$: ultimate strain of GFRP bar

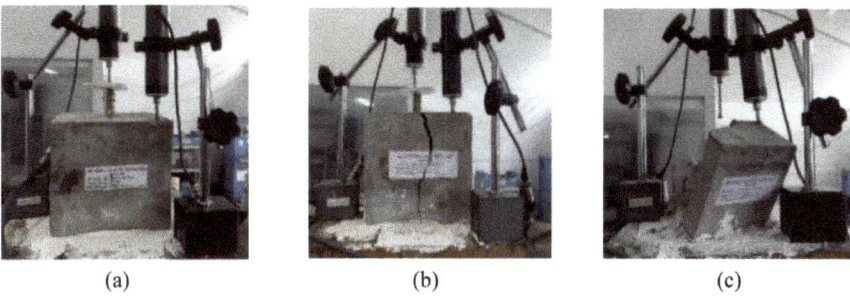

Fig. 3 Failure mode of specimens **a** bar pull-out from concrete, **b** concrete splitting, **c** concrete rupture

stages (Fig. 4). At the initial stage, the bond stress was increased linearly and dramatically with small slip. The bond stress at this linear stage is mainly determined by the chemical adhesion and mechanical interlock between the bar and the concrete [8]. However, a lower stiffness response was observed at the beginning of the curve in the specimens. As the load was increased continuously, the curve shifted from linear to nonlinear and then up to the maximum bond stress. At this stage, the mechanical interlock and friction forces mainly controlled the bond stress, and the micro cracks initiated on the concrete around the bars. After the peak bond stress being reached, the curve started to decline until the bar was pulled out from the concrete. At the failure stage, the maximum average bond strength achieved from 8 to 13 MPa, and the average slip length was from 0.04 to 0.06 mm/mm.

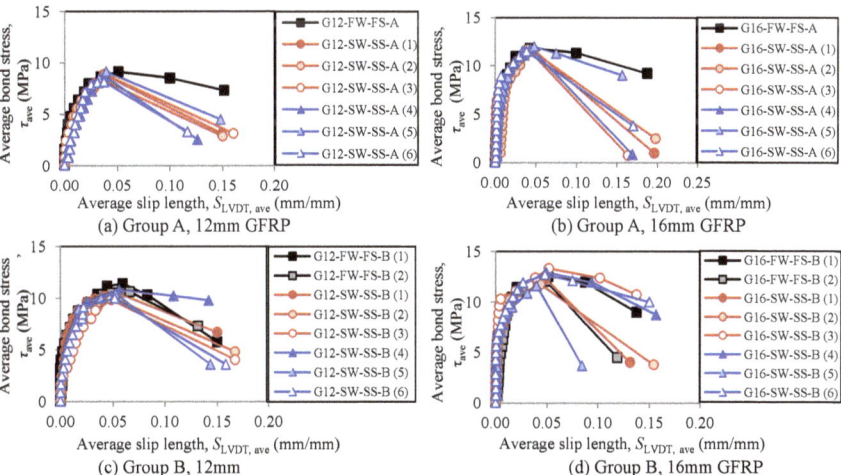

Fig. 4 Average bond strength—average slip length relationship

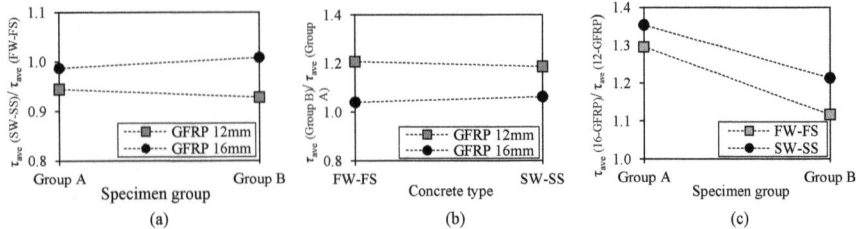

Fig. 5 The ratio of the average bonding strength of (**a**) saline water–sand concrete specimen to control specimen, (**b**) group B specimen to group A specimen, and (**c**) 16 mm-GFRP bar specimen to 12 mm-GFRP bar specimen

3.2.1 Effect of Saline Water and Sand

As shown in Fig. 5a, the average bond strength of 12 mm-GFRP bar in saline water–sand concrete slightly declined by 5.6 and 7.2% corresponding to group A and B compared to common concrete. This can be explained by the fact that fresh water–sand concrete has higher compressive and tensile strength than saline water–sand concrete (5.7 and 3.7% at the age of 28 days, respectively). However, in case of 16 mm-GFRP bar, the effect of saline water–sand on the bond strength was not obvious, the decline degree was only approximately from 0.8 to 1.3%. Similarly, in comparison to control concrete, the average slip length of 12 mm-GFRP bar in the concrete made of salt-contaminated water and sand was smaller from 5.7 to 26.6%, and the influence of saline water–sand on the slip length was inconsiderable for 16 mm-GFRP bar.

3.2.2 Effect of Concrete Compressive Strength

The increase of concrete strength from 29.28 MPa to 35.56 MPa significantly enhanced the average bonding strength for the 12 mm GFRP bar (20.7%), but only a slight improvement has been recorded in the 16 mm GFRP (6.3%), and the increase degree of this bonding strength in 12 mm-diameter GFRP bar is more obvious than that in the 16 mm- diameter bar (Fig. 5b). At the failure stage, the increase of concrete strength increased the average slip length from 23.2 to 40.8% for the 12 mm GFRP bar, and from 8.2 to 15.9% for the 16 mm GFRP bar.

3.2.3 Effect of GFRP Bar Diameter

According to Fig. 5c, in terms of group A specimens, the increase of GFRP bar diameter from 12 to 16 mm plummeted the average bonding strength to 29.6% for common concrete specimens, and to 35.4% for saline water–sand concrete specimens. In case of group B specimens, the average bonding strength increased from

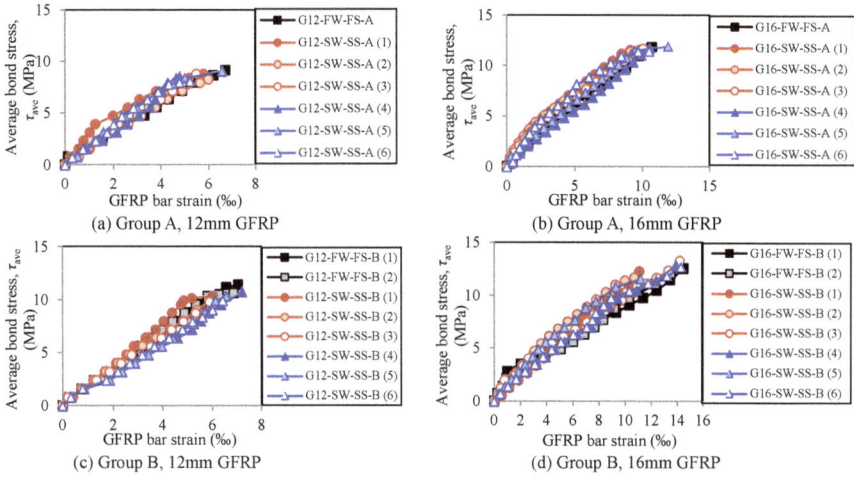

Fig. 6 Average bond strength—GFRP bar strain relationship

11.7 to 21.3% when 16 mm-GFRP bar was used instead of 12 mm-GFRP bar. This result shows that the diameter had more obvious influence on the average bonding strength of GFRP bar in the concrete made from saline sea-sand than that of GFRP bar in common concrete. Besides, the similar growth of GFRP bar diameter decreased the slip length from 10.8 to 20.5%.

3.3 GFRP Bar Strain

As shown in Fig. 6, the effect of saline water-sand and the increase of the concrete compressive strength seemed to be really not obvious to the GFRP bar strain. The maximum strain value of 12 mm-GFRP bar was from 5.65 to 6.95‰ (25.7–31.6% ε_{fu}, of which ε_{fu} is the ultimate strain of GFRP bar). In case of 16 mm-GFRP bar, the strain value achieved from 10.33 to 14‰ (46.9–63.6% ε_{fu}). In contrast, the increase of GFRP bar diameter significantly enhanced the GFRP bar strain for both saline water–sand and control specimens. This is resulted from the significant increase of the bond strength when the GFRP bar diameter was changed from 12 to 16 mm. Specially, the strain value of the GFRP bar boosted from 59.6 to 82.3% and from 98% to 101.4% for group A and B specimens, respectively.

4 Conclusion

Based on the test results, the following main conclusions can be drawn:

- The average bonding strength of GFRP bars in the concrete made of salt-contaminated water and sand has only slightly been decreased from 5.6 to 7.2% compared to that in fresh water- sand concrete, and the decline degree of this bonding strength in 12 mm-diameter GFRP bars is more obvious than that in the 16 mm- diameter bar.
- The increase of concrete strength from 29.28 MPa to 35.56 MPa significantly enhances the average bonding strength for the 12 mm GFRP bar (20.7%), but only a slight improvement has been recorded in the 16 mm GFRP (6.3%). The diameter had more obvious influence on the average bonding strength of GFRP bar in the concrete made from saline sea sand than that of GFRP bar in common concrete. Increasing the GFRP bar diameter from 12 to 16 mm boosts the average bonding strength from 11.7% to 35.4%.
- The maximum strain value of GFRP bar was from 5.65 to 14‰ (25.7–63.6% the ultimate strain). The increase of GFRP bar diameter from 12 to 16 mm significantly enhanced the GFRP bar strain by more than 59.6%.

References

1. Angst U, Elsener B, Larsen CK, Vennesland Ø (2009) Critical chloride content in reinforced concrete-a review. Cem Concr Res 39(12):1122–1138
2. Fernandez I, Herrador MF, Marí AR, Bairán JM (2016) Structural effects of steel reinforcement corrosion on statically indeterminate reinforced concrete members. Mater Struct 49(12):4959–4973
3. Younis A., Ebead UA., Nanni A (2017) A perspective on seawater/FRP reinforcement in concrete structures. In Proceedings of the Ninth International Structural Engineering and Construction Conference, Resilient Structures and Sustainable Construction
4. Younis A, Ebead U, Suraneni P, Nanni A (2018) Fresh and hardened properties of seawater-mixed concrete. Constr Build Mater 190:276–286
5. Nishida T, Otsuki N, Ohara H, Garba-Say ZM, Nagata T (2013) Some considerations for applicability of seawater as mixing water in concrete. J Mater Civ Eng 27(7):B4014004
6. Dong ZQ, Wu G, Zhao XL, Lian JL (2018) Long-term bond durability of fiber-reinforced polymer bars embedded in seawater sea-sand concrete under ocean environments. J Compos Const 22(5):04018042
7. Soares S, Freitas N, Pereira E, Nepomuceno E, Pereira E, Sena-Cruz J (2020) Assessment of GFRP bond behaviour for the design of sustainable reinforced seawater concrete structures. Constr Build Mater 231:117277
8. Sun W, Zheng Y, Zhou L, Song J, Bai Y (2020) A study of the bond behavior of FRP bars in MPC seawater concrete. Adv Struct Eng 1369433220956816:1–14

Collapse Scenario of a Tested Steel Moment Frame on the Shaking Table

Tran Tuan Nam

Abstract This paper provides a summary of column strength deterioration sequence during the collapse test of a steel moment frame on the shaking table. The study addresses the interaction of column axial force and bending moment capacity. Axial force developed in each column varies during the excitation. As a result, column bending moment capacity changes because of interaction with column axial force. On the other hand, local buckling occurring during pre-collapse time also weakens column strength. The deteriorated strength inherited after several pre-collapse load cycles determines the sequence of column failure when the building collapse happens. Global instability state is recognizable when the development of story drift contradicts to the decrease of building overturning moment. The building ends up with collapse after exceeding this state.

Keywords Shaking table est · Steel moment frame · Column strength deterioration

1 Introduction

Shaking table is a useful facility to test the seismic behavior of structures. As such, a shaking experiment on a full-scale steel building was conducted in Hyogo, Japan. This test whose overview was given in [1] was a part of an experimental project focused on steel buildings conducted at E-Defense—the world largest shaking table facility. The building specimen is a full-scale four-story steel moment frame whose plan dimension is 6 × 10 m and the height is 14.3 m. Figure 1 shows the building specimen before and after the test. The building was shaken to collapse with a soft-story mechanism caused by deterioration of first-story columns due to local buckling at column ends.

The test aimed to evaluate the structural and functional performance of the steel building under design-level ground motions. Moreover, the test also examined the

T. T. Nam (✉)
Faculty of Civil Engineering, HUTECH University, Ho Chi Minh City, Vietnam
e-mail: tt.nam@hutech.edu.vn

© The Author(s), under exclusive license to Springer Nature Singapore Pte Ltd. 2023
J. N. Reddy et al. (eds.), *ICSCEA 2021*, Lecture Notes in Civil Engineering 268,
https://doi.org/10.1007/978-981-19-3303-5_68

Fig. 1 The steel building before and after the shaking test

safety margin against collapse under very large ground motions. The ground acceleration histories recorded at the JR Takatori station during the 1995 Kobe earthquake (herein referred to as Takatori motion) were used as the input for the shake table experiments. Various increasing scaled ground motions were applied in the building specimen until collapse.

The experimental responses of the specimen building under the weak, strong and collapse shaking levels were formerly reported in [2–4]. Some other researchers [5–10] numerically simulated and analysed the building responses during the shaking test. Almost studies focused on the physics of structural behavior, or examined local behavior of beams, panel zones and columns. Behavior of non-structural components observed in the test was particularly described in [11]. However, there have not been any studies mentioning the detailed sequence of column failure and summarizes the scenario of the building collapse, as will be discussed in this paper.

2   Deterioration of Column Strength During Pre-collapse Time

Displacement procedure of the building in 100% Takatori case can be divided into two stages: pre-collapse stage (0 ~ 5.71 s, when the building still remains stable) and collapse stage (after 5.71 s, when the building destabilizes and finally collapses). During the pre-collapse time, the building moves mostly in 45° diagonal direction (Fig. 2). Thus, we can assess column strength capacity via bending moment with respect to 45° direction, namely M_{45}. Seismic axial force and bottom bending moment behavior of two typical columns A1 and B3 located at two opposite corners are selected to demonstrate in Fig. 3.

Time-history of seismic axial force due to the movement of the building during the excitation (excluding gravity force) is shown in Fig. 3a. Low-pass filtered curves show the consistent opposite effect of pulling and pushing between two columns.

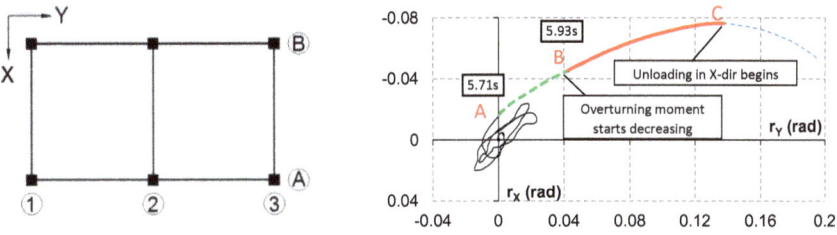

Fig. 2 Story drift orbit (100% Takatori, 1st story)

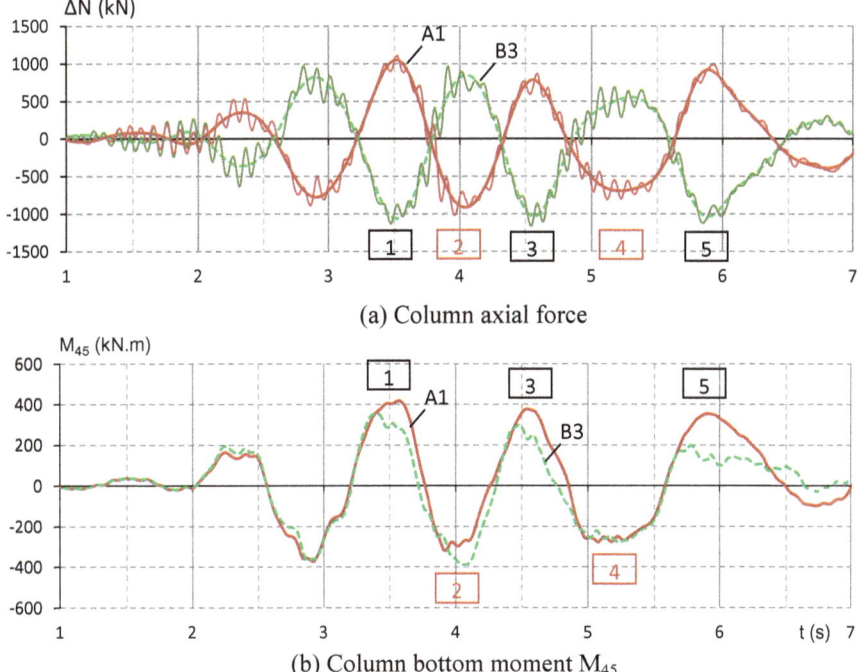

(a) Column axial force

(b) Column bottom moment M_{45}

Fig. 3 Typical column behavior (100% Takatori, 1st story) **a** Column axial force, **b** Column bottom moment M_{45}

As a result, moment capacity of each column is deteriorated in turns whenever that column is subject to compression. Bending moment M_{45} of column A1 primarily deteriorates on negative side while degradation of column B3 mostly performs on positive side (Fig. 3b).

This situation is firmly related to the timing of seismic compression force developed in columns, throughout five stages denoted by 1 to 5 in Fig. 3, in which column A1 deteriorates in stages 2 and 4 when it is subjected to compression; whereas

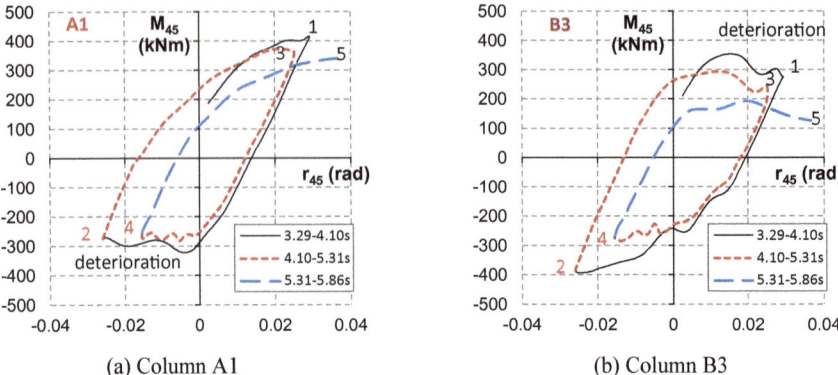

Fig. 4 Hysteretic behavior between M_{45} and r_{45} (100% Takatori, 1st story) **a** Column A1, **b** Column B3

column B3 is weakened in stages 1, 3 and 5 because of the same compression condition. On the other hand, it is visible that after several loading cycles, column moment capacity cannot exceed the deteriorated capacity in previous cycles. This confirms the possibility of local buckling occurrence during pre-collapse time.

Hysteretic relation between M_{45} and related story drift ratio r_{45} along this direction is given in Figs. 4a, b for column A1 and B3, respectively. Three separate hysteretic cycles (with timing durations of 3.29 ~ 4.10 s, 4.10 ~ 5.31 s, 5.31 ~ 5.86 s, respectively) in accordance with three main cycles prior to collapse time are demonstrated in the figure, showing opposite deterioration sides of moment capacity between column A1 and B3.

3 Sequence of Column Failure During Collapse Time

Due to different positions on the building plan, all 6 columns get different effects of seismic axial force; hence get different deterioration states of column moment capacity after pre-collapse load cycles as mentioned above. Consequently, when the building gets instability, each column individually will have varied shear force capacity, resulting in different failure sequence of columns during collapse time.

Regarding column axial force during collapse time, as can be seen in Fig. 5a, the low-pass filtered time-history curve of column seismic axial force (*continuous lines, primary axis*) shows the changing completely consistent to the building displacement. From point A (5.71 s), the time when the building starts moving forward, compression force keeps developing in column B3. On the contrary, tension force is growing in column A1. However, even though story drift displacement is still growing up, column axial forces stop increasing and reverse back at point B (5.93 s). The decrease of building overturning moment (*dashed line, secondary axis*) clarifies this situation.

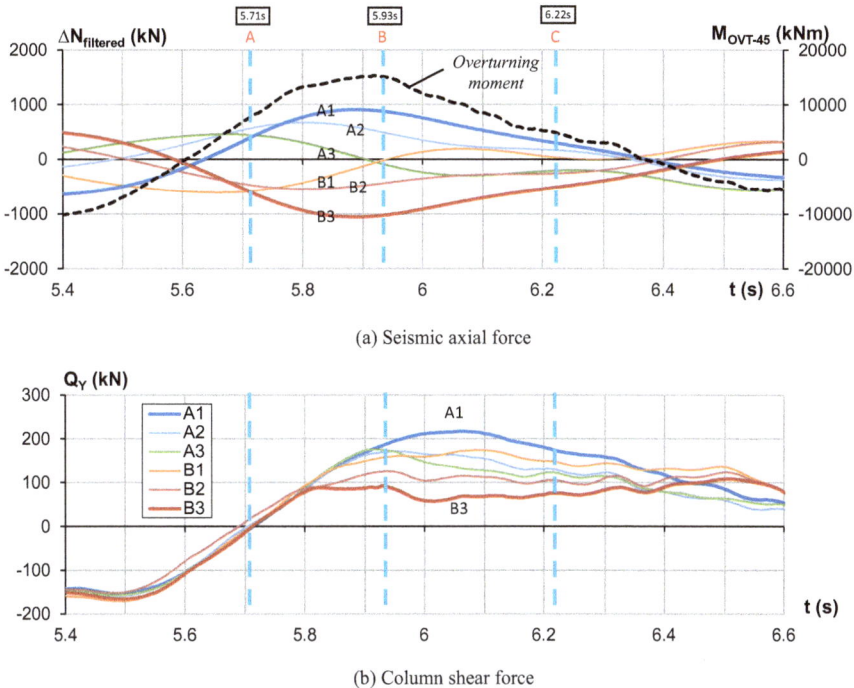

(a) Seismic axial force

(b) Column shear force

Fig. 5 Time-history of all column axial and shear forces (100% Takatori, 1st story) **a** Seismic axial force, **b** Column shear force

Deterioration sequence of column shear is obviously shown in Fig. 5b which plots comparative behavior of column shear Q_Y. Column B3 reaches to its capacity very early because the deterioration inherited from previous cycles is largest. As deformation increases, local buckling occurs more significantly. As a result, it is the hinging due to critical local buckling that causes damage to column B3. The failure sequence then proceeds in turns to columns B2, A3, A2, B1 and finally A1.

Summarily, in order to easily follow behavior trend of column moment capacity in accordance with the story drift orbit during the collapse time, Fig. 6 shows collaborated graphs of column bottom moments $M_{Y\text{-bottom}}$ and $M_{X\text{-bottom}}$ tracing the same extracted segment A-B-C-D (5.71 ~ 6.50 s) of story drift orbit. According to [10], the analytical results agree well with recorded behavior for all 6 columns.

Time sequence of local instability and global instability collaborated with story drift angle orbit is summarized and listed in Table 1 and Fig. 7. Story drift orbit of the upper 2nd story is also plotted in this figure, showing that unloading happens to upper stories after point B (5.93 s) while the frame in the 1st story is still yielding and moving forward because of global instability.

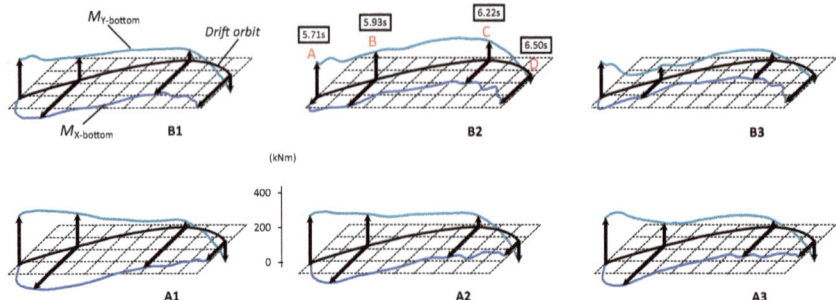

Fig. 6 Column bottom moment (100% Takatori, 5.71–6.50 s)

Table 1 Time sequence of building behavior during collapse

t (sec)	θ_X (rad)	θ_Y (rad)	Point	Member behavior	Global Behavior
5.71	-0.016	0.000	A		Start of awkward building displacement
5.85	-0.033	0.021	#1	Instability of column B3	
5.93	-0.045	0.043	B		Overturning moment reaches to peak and then decreases and reverses sign. Upper stories other than 1st story begin unloading, while 1st story drift still keeps increasing. (Global instability)
6.09	-0.068	0.095			Local impact occurs between steel angle (used as targets for measuring story drift) and supporting table, causing disturbance on floor response acceleration
6.16	-0.074	0.118	#2	Instability of column B2	
6.22	-0.076	0.137	C		Maximum story drift θ_X of 1st story, and then unloading in X-axis begins
6.24	-0.075	0.143	#3	Instability of column A3	
6.33	-0.070	0.166	#4	Instability of column A2	
6.35	-0.068	0.170			Base shear $Q_\alpha = 0$ (collapse)
6.36	-0.067	0.172	#5	Instability of column B1	
6.42	-0.061	0.184	#6	Instability of column A1	

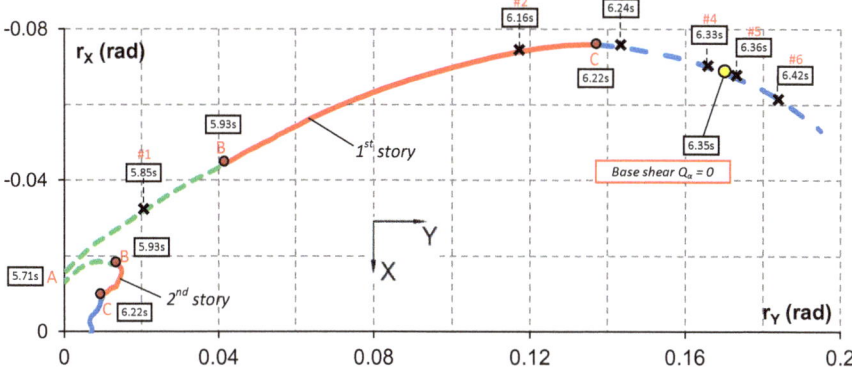

Fig. 7 Story drift orbit of the building during collapse time (100% Takatori)

4 Conclusions

From the results obtained in this study, some conclusions can be drawn.

- In order to understand the experimental collapse behavior, it is necessary to study local responses and deteriorations of members subjected to biaxial bending, and relate them with different stages of the global responses including the progress and mechanism of collapse.
- Experimental results showed totally different deteriorating patterns of biaxial bending moments among all six columns because their axial force magnitudes and variations differed considerably due to the column locations.
- The column deteriorating and damaging sequence is clarified and detailed from these perspectives, in relation with the global response of base shear and story drift displacement during collapse excitation level.
- The cyclic change of displacement direction caused the two-directional accumulated deterioration of columns, reflected by the fact that earlier column damage due to local buckling in the X direction subsequently reduced the capacity in the Y direction.

Acknowledgements This study is a part of *"NEES/E-Defense collaborative research program on steel structures"*. The author acknowledges Prof. Kazuhiko Kasai (Tokyo Institute of Technology, Japan)—team leader for the overall program, and Dr. Bruce F. Maison (California) for their enthusiastic technical support.

References

1. Kasai K, Ooki Y, Motoyui S, Takeuchi T, Sato E (2007) E-Defense tests on full-scale steel buildings, part 1—experiments using dampers and isolators. Proceedings of Structural Congress, ASCE, Long Beach, USA
2. Suita K, Matsuoka Y, Yamada S, Shimada Y, Tada M, Kasai K (2009) Experimental procedure and elastic response characteristics of shaking table test—complete collapse test of full-scale 4-story steel building part 1 (in Japanese). J Struct Constr Eng, Trans AIJ 635:157–166
3. Yamada S, Suita K, Matsuoka Y, Shimada Y (2009) Elastoplastic responses and process leading to a collapse mechanism—complete collapse test of full-scale 4-story steel building part 2 (in Japanese). J Struct Constr Eng, Trans AIJ 644:1851–1859
4. Shimada Y, Suita K, Yamada S, Matsuoka Y, Tada M, Ohsaki M, Kasai K (2010) Collapse behavior on shaking table test—complete collapse test of full-scale 4-story steel building part 3 (in Japanese). J Struct Constr Eng, Trans AIJ 653:1351–1360
5. Tada M, Tamai H, Ohgami K, Kuwahara S, Horimoto A (2008) Analytical simulation utilizing collaborative structural analysis system. Proceedings of the 14th World Conference on Earthquake Engineering, Beijing
6. Pan P, Ohsaki M, Zhang J (2008) Collapse analysis of 4-story steel moment-resisting frames. Proceedings of the 14th World Conference on Earthquake Engineering, Beijing
7. Yu YJ, Tsai KC, Weng YT, Lin BZ, Lin JL (2010) Analytical studies of a full-scale steel building shaken to collapse. Eng Struct 32:3418–3430
8. Isobe D, Han WS, Miyamura T (2013) Verification and validation of a seismic response analysis code for framed structures using the ASI-Gauss technique. Earthq Eng Struct Dyn 42(12):1767–1784
9. Miyamura T, Yamashita T, Akiba H, Ohsaki M (2014) Dynamic FE simulation of four-story steel frame modeled by solid elements and its validation using results of full-scale shake-table test. Earthq Eng Struct Dyn 44(9):1449–1469
10. Kasai K, Nam TT, Maison B (2016) Structural collapse correlative analysis using phenomenological fiber hinge elements to simulate two-directional column deteriorations. Earthq Eng Struct Dyn 45(10):1581–1601
11. Matsuoka Y, Suita K, Yamada S, Shimada Y, Akazawa M, Matsumiya T (2009) Evaluation of seismic performance of exterior cladding in full-scale 4-story building shaking table test (in Japanese). J Struct Constr Eng, Trans AIJ 641:1353–1361

Damage Identification for Steel Frame Structures Using Two-Step Approach Combining Modal Strain Energy Method and Genetic Algorithm

Van-Sy Bach, Thanh-Cao Le, Chi-Thien Nguyen, Manh-Hung Tran, Minh-Nhan Pham, and Duc-Duy Ho

Abstract In this study, the location and extent of damage in steel frame structures are identified using a two-step technique that combines the modal strain energy (MSE) method and a genetic algorithm (GA). In the first step, the damaged elements in the frame are identified using Modal Strain Energy-based Index (MSEBI). This indicator is determined from the change of modal strain energy value of intact and damaged states. To improve the precision of the damage localization results, four first bending mode shapes are combined to calculate MSEBI. In the second step, the damage extents are determined accurately by GA with the objective function based on the MSE change. A numerical verification on a steel plane frame with four different damaged scenarios. The results show that the proposed approach gives a high accuracy in identifying the position and severity of damage for frame structures with different damage scenarios.

Keywords Damage identification · Structural health monitoring · Modal strain energy · Genetic algorithm · Steel frame structures

1 Introduction

Structural damages are inevitable in the operation process of engineering structures due to various environmental and mechanical conditions. These damages then reduce the strength of the structure, and severe damages can even lead to the collapse of the system. Therefore, structural health monitoring (SHM) field plays a vital role in guaranteeing the regular performance and safety of the structural systems [1]. Recently, nondestructive SHM technologies (e.g., impedance-based and vision-based

V.-S. Bach · T.-C. Le · C.-T. Nguyen · M.-H. Tran · M.-N. Pham · D.-D. Ho (✉)
Faculty of Civil Engineering, Ho Chi Minh City University of Technology (HCMUT), Ho Chi Minh City, Vietnam
e-mail: hoducduy@hcmut.edu.vn

Vietnam National University Ho Chi Minh City, Ho Chi Minh City, Vietnam

V.-S. Bach · T.-C. Le
Faculty of Civil Engineering, Nha Trang University, Nha Trang, Khanh Hoa, Vietnam

© The Author(s), under exclusive license to Springer Nature Singapore Pte Ltd. 2023
J. N. Reddy et al. (eds.), *ICSCEA 2021*, Lecture Notes in Civil Engineering 268,
https://doi.org/10.1007/978-981-19-3303-5_69

methods), especially vibration-based methods, have attracted many researchers. Among vibration-based SHM methods, the modal strain energy (MSE)-based method is highly efficient for damage detection in structures because of its high sensitivity to damage [4–6]. Besides, optimization algorithms have been deployed to solve structural optimization problems and damage identification. The genetic algorithm (GA) is an effective tool for solving optimization problems based on a natural selection process that mimics biological evolution [2, 3, 7].

This research aims to develop a two-step procedure for structural damage identification, in which a damage index based on MSE combined with the GA to detect the position and extent of damaged elements in frame structures. The general idea of this method is to deploy damage identification in a two-step procedure. The first step uses a damage index sensitive to the change vibration properties to localize the potentially damaged elements. Then, the GA is applied to these damaged areas in the next step to identify the damaged ratio of the frame. The localization of the potential damaged area in the first step helps to improve diagnostic results. Finally, the proposed two-step procedure is verified by a numerical model of a plane steel frame with some different damaged scenarios.

2 Theoretical Basis

2.1 Modal Strain Energy Method

The MSE theory was presented firstly by Stubbs et al. to identify damages in the Euler–Bernoulli beam structure [9]. Although using only mode shapes and elemental stiffness matrices as inputs, this method can locate the damage with a stiffness reduction of even 10%. The MSE change is sensitive to damage and can be considered as an effective indicator for localizing the damaged areas. The MSE of the eth element of the ith mode of the structure is given by:

$$MSE_i^e = \frac{1}{2}\phi_i^{eT} K^e \phi_i^e \tag{1}$$

where K^e is the stiffness matrix of the eth element, and ϕ_i^e is the displacement vector of the eth element of the ith mode.

2.2 Modal Strain Energy-Based Damage Detection Method

Some damage indexes based on MSE change are used to successfully identify the occurrence and the location of damages in structures, such as MSECR (Modal Strain Energy Change Ratio), MSEBI (Modal Strain Energy-Based Index), MSEEI (Modal

Strain Energy Equivalence Index). Among these indexes, the MSEBI index has proven to be a highly effective indicator for detecting damage [8]. Then, the five-step procedure to identify damage based on MSEBI is proposed as below:

Step 1: Extract mode shape data from the modal analysis of two states: healthy state and damaged state.
Step 2: Calculate the MSE value of each element as Eq. 1 and the total MSE value of the whole frame with n elements:

$$MSE_i = \sum_{e=1}^{n} MSE_i^e \tag{2}$$

Step 3: Determine the fractional MSE value of each element corresponding to the ith mode:

$$FMSE_i^e = \frac{MSE_i^e}{MSE_i} \tag{3}$$

Step 4: Considering m modes, the average fractional MSE value for the eth element is defined as follows:

$$\overline{FMSE^e} = \frac{1}{m} \sum_{i=1}^{m} FMSE_i^e \tag{4}$$

Step 5: Calculate MSEBI index of each element:

$$MSEBI^e = \max\left\{0; \; \frac{\overline{FMSE_{da}^e} - \overline{FMSE_{in}^e}}{\overline{FMSE_{in}^e}}\right\} \tag{5}$$

where $\overline{FMSE_{da}^e}$, $\overline{FMSE_{in}^e}$ are average fractional MSE values of the eth element for damaged and intact state, respectively. If $MSEBI^e > 0$, the eth element is considered as damaged; otherwise, the eth element is undamaged.

2.3 Genetic Algorithm

Structural damages are considered as changes in the structural geometric or material properties. Consequently, these changes affect modal parameters of the whole structure (e.g., mode shape, frequency). For frame structures, the damage is assumed simply by decreasing the stiffness of an element. For example, if an element is damaged 10%, its flexural stiffness is reduced by 10%. The following equation describes the damage ratio or the stiffness reduction ratio of the eth element:

$$EI_{da}^e = \left(1 - \alpha^e\right)EI_{in}^e \tag{6}$$

where EI_{da}^e, EI_{in}^e are the bending stiffness of the eth element of the damaged and healthy state, respectively; α^e is the damaged level of the eth element. Then, the damaged level of the whole frame needed to be identified is given by a vector α:

$$\alpha = \left\{\alpha^1 \ \alpha^2 \ \ldots \ \alpha^k\right\} \tag{7}$$

where k is the number of damaged elements in the frame.

The objective function based on the change of MSE value of the structure is defined as:

$$OF(\alpha^s) = \min_{\alpha^s}\left[\sum_{e=1}^{m}\sum_{i=1}^{n}\left|\frac{\left(\phi_i^{eT}K^e\phi_i^e\right)_{da} - \left(\phi_i^{eT}K^e\phi_i^e\right)_{\alpha^s}}{\left(\phi_i^{eT}K^e\phi_i^e\right)_{da}}\right|\right] \tag{8}$$

where α^s is the variable corresponding to a assumed damaged level of the frame; $\left(\phi_i^{eT}K^e\phi_i^e\right)_{da}$, $\left(\phi_i^{eT}K^e\phi_i^e\right)_{\alpha^s}$ are the MSE value of the eth element corresponding to the damaged state and the assumed damaged level α^s.

The GA is used as an optimization tool to find the variable α^s that gives the minimum value of the objective function in Eq. 8. A vector α^s giving a smaller value of objective function is regarded as more accurate in damage extent estimation. The workflow of the two-step damage identification procedure is presented in Fig. 1.

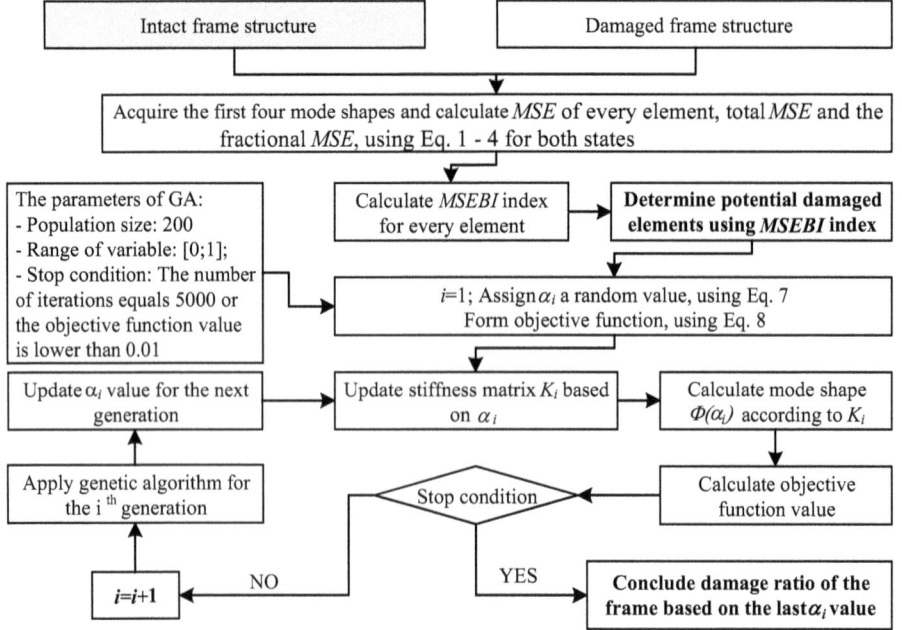

Fig. 1 Workflow of the two-step damage detection procedure

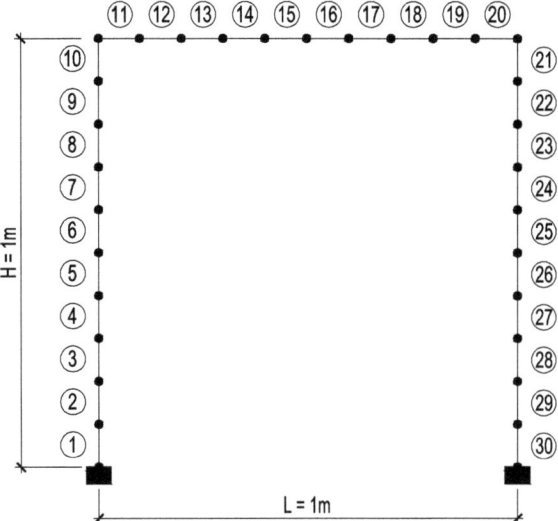

Fig. 2 FE model of steel frame structure

3 Numerical Verification

3.1 Properties of Steel Frame

The finite element analysis (FEA) of a steel 2D frame is used to verify the effectiveness of the proposed procedure. The material properties and four damaged scenarios are presented in Fig. 2 and Table 1. The frame is divided into 30 elements with 31 nodes in the FE model.

3.2 Modal Analysis

The FE model of the frame was established by using MATLAB software. The first four mode shapes are extracted from modal vibration for both intact and damaged states, as shown in Fig. 3. Next, the natural frequencies of the healthy frame are compared to results using SAP2000 to study the convergence of FEA. Table 2 shows that the natural frequency differences are minimal, about 0.01–0.3%.

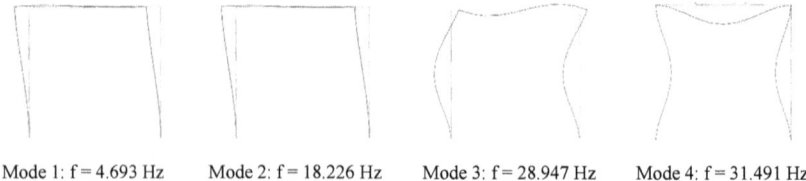

Mode 1: f = 4.693 Hz   Mode 2: f = 18.226 Hz   Mode 3: f = 28.947 Hz   Mode 4: f = 31.491 Hz

Fig. 3 The first four mode shapes of healthy frame

Table 1 Material properties and damage scenario

Material properties		
Modulus of elasticity	20 GPa	
Poisson's ratio	0.3	
Mass density	7,670 kg/m^3	
Cross section area of beam mm^2	40.5×10^6	
Cross section area of column mm^2	50.5×10^6	
Damage scenarios		
Case	Damage element	Damage extent (%)
1	11st	10
2	1st, 11st	20, 10
3	1st, 11st, 15th	20, 10, 10
4	1st, 5th, 11st, 15th	30, 10, 20, 15

Table 2 The natural frequencies of healthy frame

Mode	Mode 1	Mode 2	Mode 3	Mode 4
f (Hz)	4.693	18.226	28.947	31.491
$f_{SAP2000}$ (Hz)	4.689	18.223	28.872	31.481
Δf (%)	0.078	0.012	0.262	0.031

3.3 Damage Location Identification Results

In the first step, the MSEBI is calculated to locate the potential damaged elements. The effectiveness of this index was investigated in four damage scenarios. For case 1 and case 4, the damage chart locates exactly the damaged element with no wrong detected element. However, there are some false alarms with the 4th element for case 2 and the 5th element for case 3. These false elements will be eliminated in the next step, using the GA procedure. Figure 4 shows the damage location identification results of four cases.

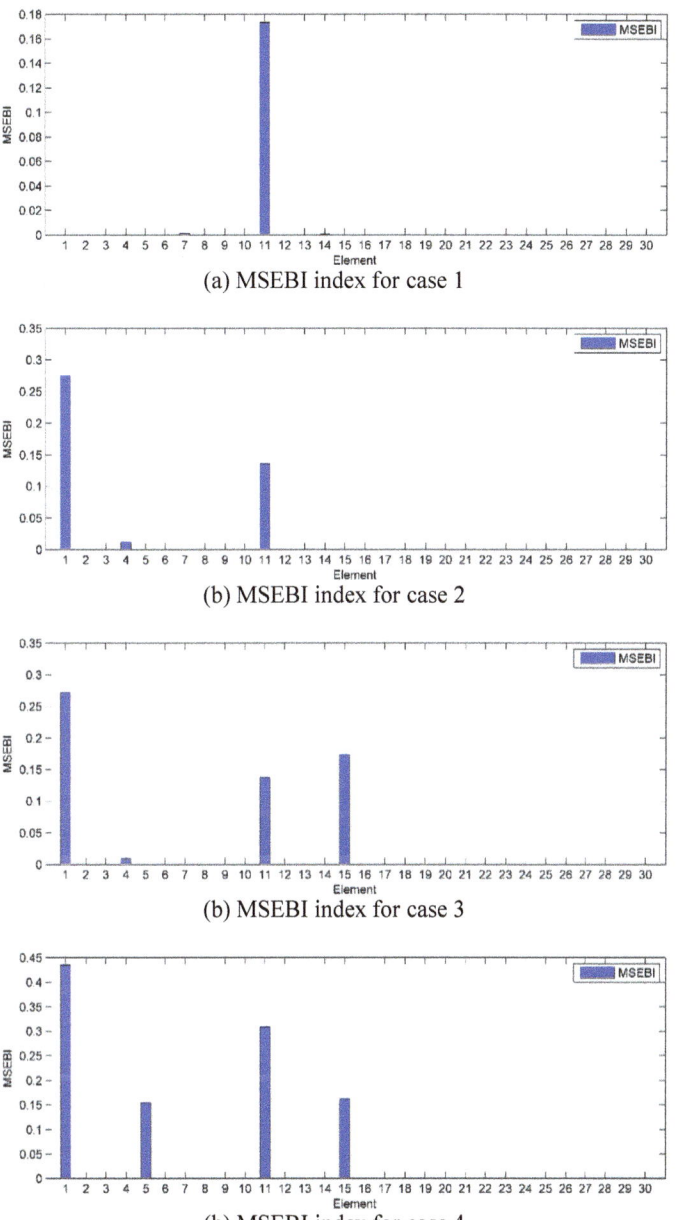

(a) MSEBI index for case 1

(b) MSEBI index for case 2

(b) MSEBI index for case 3

(b) MSEBI index for case 4

Fig. 4 Damage location identification results based on MSEBI index **a** MSEBI index for case 1, **b** MSEBI index for case 2, **c** MSEBI index for case 3, **d** MSEBI index for case 4

Table 3 The damage extent identification results

Case	1	2		3			4			
Index of element	1	1	11	1	11	15	1	5	11	15
Real damage extent (%)	10	20	10	20	10	10	30	10	20	10
Estimated damage extent (%)	10.14	20.05	10.04	19.99	10	9.88	29.98	10.01	19.95	9.88
Error (%)	1.4	0.25	0.4	0.05	0	1.2	0.07	0.1	0.25	1.2

Table 4 The convergent speed of GA procedure

Case	1	2	3	4
Number of generations	10	17	34	39
Number of iterations	550	900	1750	2000
Objective function value	0.0084	0.005	0.0063	0.009
Time of analysis (s)	14.63	22.87	38.14	44.86

3.4 Damage Extent Identification Results

In the second step, the GA was deployed to determine the extent of the potentially damaged elements. The stopping condition used in this study is that a maximum iteration number equals 5000, or the objective function value is smaller than 0.01. When the stopping condition is satisfied, the number of iterations required for convergence to solution and the error in the damage extent identification are used to evaluate the effectiveness of the procedure. The damage extent identification results of four damaged cases are listed in Table 3.

As shown in Table 3, the damage extent estimation results are significantly accurate in all four cases, with the maximum error at only 1.4%. The convergent speed of GA is evaluated by the number of iterations performed when the stop condition is met. This speed depends on the number of variables, as shown in Table 4. The GA takes more time to get the convergent solution when the number of damaged elements increases.

4 Conclusions

In this study, a two-step procedure using MSEBI indicator and GA was developed successfully to identify the position and extent of multiple damages in steel 2D frames. From the results, three remarkable conclusions were drawn as follows:

1. The proposed approach accurately detected damaged elements in steel 2D structures with complex damaged scenarios when combining the first four mode

shapes. The first step locates the potentially damaged elements. Then, the second step determines exactly the damaged extent of these elements.

2. The MSEBI index combined with GA using an objective function based on the MSE change gave the high accuracy of damaged element identification for many different damage scenarios.
3. The convergent speed of the GA depends on the number of damaged elements detected from the first step. Therefore, increasing the accuracy of damage localization results helps the GA get the convergent solution faster.

Acknowledgements We acknowledge the support of time and facilities from Ho Chi Minh City University of Technology (HCMUT), VNU-HCM for this study.

References

1. Avci O, Abdeljaber O, Kiranyaz S, Hussein M, Gabbouj M, Inman DJ (2021) A review of vibration-based damage detection in civil structures: from traditional methods to machine learning and deep learning applications. Mech Syst Signal Process 147:107077
2. Chou J, Ghaboussi J (2001) Genetic algorithm in structural damage detection. Comput Struct 79(14):1335–1353
3. Friswell M, Penny J, Garvey S (1998) A combined genetic and eigensensitivity algorithm for the location of damage in structures. Comput Struct 69(5):547–556
4. Fu Y, Liu J, Wei Z, Lu Z (2014) A two-step approach for damage identification in plates. J Vib Control 22(13):3018–3031
5. Hu H, Wu C (2008) Nondestructive damage detection of two dimensional plate structures using modal strain energy method. J Mech 24(4):319–332
6. Le TC, Ho DD, Huynh TC, Bach VS (2021) Crack detection in plate-like structures using modal strain energy method considering various boundary conditions. Shock Vib 2021:1–17
7. Marano GC, Quaranta G, Monti G (2011) Modified genetic algorithm for the dynamic identification of structural systems using incomplete measurements. Comput-Aided Civ Infrastruct Eng 26(2):92–110
8. Seyedpoor S (2012) A two stage method for structural damage detection using a modal strain energy based index and particle swarm optimization. Int J Non-Linear Mech 47(1):1–8
9. Stubbs N, Kim JT, Farrar CR (1995) Field verification of a nondestructive damage localization and severity estimation algorithm. Proc. of 13th International Modal Analysis Conference 1:210–218.

Designing Pre-engineering Steel Beam Using American and Vietnamese Standards

Van-Cua Bui and Phu-Cuong Nguyen

Abstract In this study, a steel beam of a Pre-Engineering Building (PEB) constructed in Binh Duong is analyzed and designed according to two American (AISC 360–10) and Vietnamese (TCVN 5575–2012) standards. The authors analyze and design beam structures of a PEB divided into two segments. Software SAP2000 is employed for analyzing the internal forces of the structure. Excel spreadsheet is established for designing steel beams using TCVN 5575–2012, while software SAP2000 is employed directly for designing steel beams using AISC 360–10. The weight of beams is estimated from the cross-section design results of two standards, then the cost of PEB beams is calculated based on their weight. The obtained results of beam design according to AISC 360–10 is more economical than TCVN 5575–2012 up to approximately 36% of the cost based on the mass of steel beams.

Keywords Structural analysis · Steel beam · Structural design · Pre-engineering building · SAP2000 · AISC 360–10 · TCVN 5575–2012

1 Introduction

Designing pre-engineering steel beams according to American standards AISC 360–10 is a solution that shows many optimal advantages compared to previous traditional pre-engineering steel beam design. In Vietnam, there are two common standards used to design Pre-Engineering Buildings (PEB): TCVN 5575–2012 [1] and AISC 360–10 [2]. These two standards are completely different in theory as well as the calculation method, which leads to different results. AISC 360–10 is combined between classical mechanical theory and experimental test to find the most suitable formulas for structural design. AISC 360–10 is usually updated and improved to be more relevant to reality. So in terms of science, this is a very reliable practical standard in the design in the world, so it is completely safe to use.

V.-C. Bui · P.-C. Nguyen (✉)
Advanced Structural Engineering Laboratory, Department of Structural Engineering, Faculty of Civil Engineering, Ho Chi Minh City Open University, Ho Chi Minh City, Vietnam
e-mail: cuong.pn@ou.edu.vn; henycuong@gmail.com

© The Author(s), under exclusive license to Springer Nature Singapore Pte Ltd. 2023   777
J. N. Reddy et al. (eds.), *ICSCEA 2021*, Lecture Notes in Civil Engineering 268,
https://doi.org/10.1007/978-981-19-3303-5_70

Currently, the trend of construction of PEBs in Vietnam is very popular because it brings many economic benefits. According to calculations, PEBs are very economical and environmentally friendly compared to conventional reinforced concrete and steel frames. In addition, steel buildings also save foundation structures due to their self-weight. PEBs have a longer life, after the end of design life, most PEBs will be moved to a recycling center where they are melted and used for other purposes instead of being dumped, thereby minimizing construction waste into the environment.

During the research process, it is pointed out that the cost optimization is better when applying AISC 360–10 standard to design pre-engineering steel beams compared to TCVN 5575–2012 standard of Vietnam. Figure 1 represents steel beams designed according to TCVN 5575–2012. Design sections are larger than those of AISC 360–10 as shown in Fig. 2. In this study, the column section is assumed to be I(380–600) × 200 × 10 × 12. We use the FEM commercial software SAP2000 for analyzing the internal forces inside the PEB. After that, the authors would like to do the problem of comparing the steel beam design of PEB using TCVN 5575–2012 and AISC 360–10 standard.

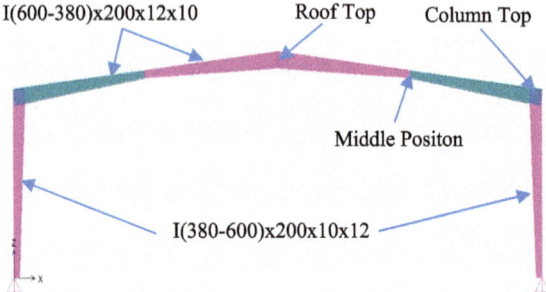

Fig. 1 Geometry of PEB is designed according to TCVN 5575–2012 [1]

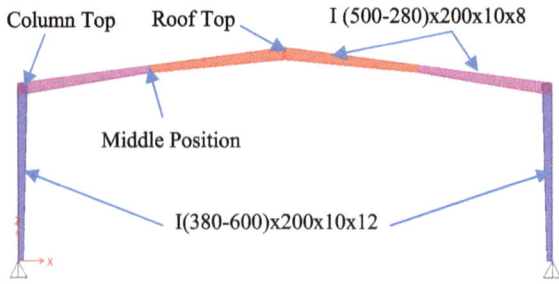

Fig. 2 Geometry of PEB is designed according to AISC 360–10 [2]

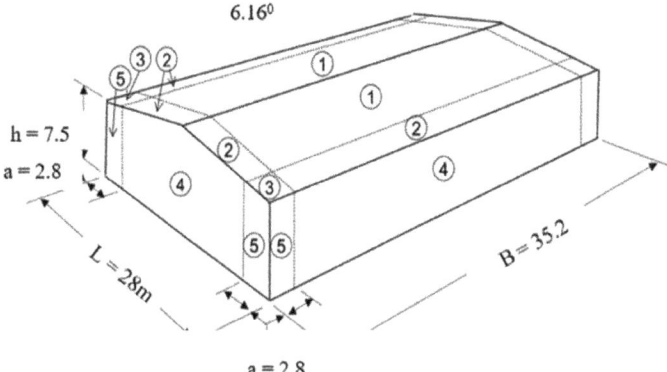

Fig. 3 Geometry of the investigating building

2 Structural Analysis

2.1 Analysis Modeling and Loadings

In this paper, the authors use the real PEB designed by Truong Phu steel structure joint stock company, constructed in 2020 at Bau Bang district, Binh Duong province, Vietnam. The geometry of the building is redrawn by the authors as Fig. 3. Roof live load is factored to be 2.92 kN/m. Looking up the TCVN 2737–1995 [3] and the ASCE 7–10 [4], we have wind zone A and wind zone D, respectively. The roof slope of the building is equal to 10.8% (6.16°). Roof top height (Hr) is 9.0 m. Drop height (He) is 7.5 m. Wind-catching width "a" is defined by MBMA 2012 [5]. The extent of the boundary area is determined through the value a = min (10% of the small side size of the building site and 0.4 h) and must not be less than [4% of the small side size of the construction site and 3 ft (0.924 m)] choose a ≤ 10% of width B = 10% × 28 = 2.80 m.

2.2 Materials and Methods

Steel materials used for the research work are pre-engineering W-sections beams. Steel material is CT38, elastic modulus of steel is E = 2.1e + 08 kN/m^2, Poisson ratio is 0.3, yield stress of steel is f_y = 23 kN/cm^2, ultimate strength of steel is f_u = 38 kN/cm^2. In this study, we use SAP2000 [6] software for structural analysis using both American and Vietnamese standards. Excel spreadsheet is established for designing steel beams using TCVN 5575–2012, while software SAP2000 is employed directly for designing steel beams using AISC 360–10. Load combinations for the design of both standards are listed in Table 1.

Table 1 Statistical table of load combination of TCVN 5575–2012 and AISC 360–10

TCVN 5575–2012	AISC 360–10	
− Load Combination (LC) 1 for Strength	Load Combination ASCE 7–10/ASD for Strength	
+ LC 1: D + LL	+ 1. 1D	+ 11. 1D + 0.45WA2L + 0.75Lr
+ LC 2: D + RL	+ 2. 1D + 1Lr	+ 12. 1D + 0.45WA2R + 0.75Lr
+ LC 3: D + LE	+ 3. 1D + 0.6WA1L	+ 13. 1D + 0.45WB1 + 0.75Lr
+ LC 4: D + RW	+ 4. 1D + 0.6WA1R	+ 14. 1D + 0.45WB2 + 0.75Lr
− Load Combination 2	+ 5. 1D + 0.6WA2L	+ 15. 0.6D + 0.6WA1L
+ LC 5: D + 0.9LL + 0.9RL	+ 6. 1D + 0.6WA2R	+ 16. 0.6D + 0.6WA1R
+ LC 6: D + 0.9LL + 0.9LW	+ 7. 1D + 0.6WB1	+ 17. 0.6D + 0.6WA2L
+ LC 7: D + 0.9LL + 0.9RW	+ 8. 1D + 0.6WB2	+ 18. 0.6D + 0.6WA2R
+ LC 8: D + 0.9RL + 0.9LW	+ 9. 1D + 0.45WA1L + 0.75Lr	+ 19. 0.6D + 0.6WB1
+ LC 9: D + 0.9RL + 0.9RW	+ 10. 1D + 0.45WA1R + 0.75Lr	+ 20. 0.6D + 0.6WB2
− Envelope of (LC1, LC2, …, LC9)	− Envelope of (1, 2, 3, …, 19, 20)	

3 Internal Force Analysis and Design Beam

3.1 Internal Forces of Beams

Results of internal force analysis of pre-engineering steel beams from SAP2000 v22 software are illustrated in Fig. 4 for the moment diagram and in Fig. 5 for the shear diagram. Table 2 summarizes the results of internal force of beams for designing following both TCVN 5575–2012 and AISC 360–10 including bending moments, shear forces, and axial forces. From Table 2, Fig. 4, and Fig. 5, it can be seen that the bending moment at the roof top of TCVN (167.08 kNm) is larger than one of

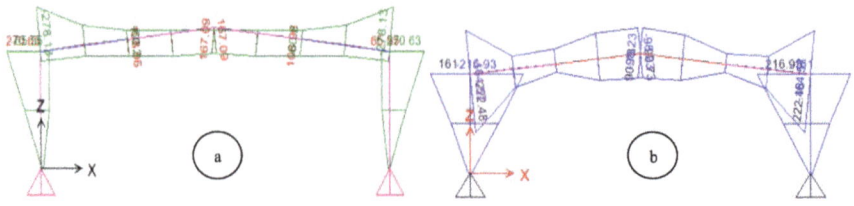

Fig. 4 Envelope moment chart **a** TCVN 5575–2012, **b** AISC 360–10

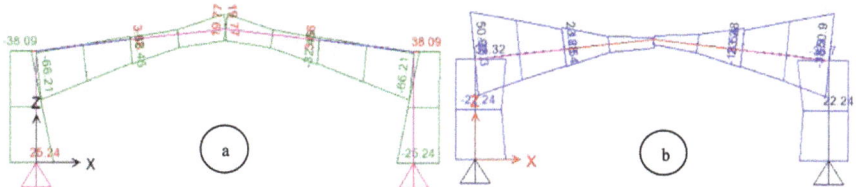

Fig. 5 Envelope shear force chart **a** TCVN 5575–2012, **b** AISC 360–10

Table 2 Results of internal forces for the beam

Position	Moment (kNm)		Axial force (kN)		Shear force (kN)	
	TCVN	AISC	TCVN	AISC	TCVN	AISC
Column top	− 278.13	222.48	− 47.58	41.05	− 66.21	50.89
Middle position	108.96	− 55.79	− 43.27	41.05	− 33.45	22.24
Roof top	167.08	− 98.23	− 38.98	41.05	19.77	23.27

AISC (98.23 kNm), the difference is about 70.1%. The shear force at column top of TCVN (278.13 kNm) is larger than one of AISC (222.48 kNm), the difference is about 30.1%. It can be concluded that the internal forces of the frame in both standards have considerable differences, so the result of the design is different too.

3.2 Designing Beams According to TCVN 5575–2012

After analyzing and calculating beams using SAP2000 v22 software according to TCVN 5575–2012. Each beam segment has the same size as presented in Table 3. The W-section beam of I(600–380) × 200 × 12 × 10 is the optimal design result according to TCVN 5575–2012 satisfy both strength and service conditions.

Table 3 Beam section according to TCVN 5575–2012

	h (mm)	b_f (mm)	t_f (mm)	t_w (mm)	L (m)	I_x (cm^4)	W_x (cm^3)
Column top	600	200	12	10	7.04	57,415	1914
Middle position	380	200	12	10	7.04	20,011	1053
Roof top	600	200	12	10	7.04	57,415	1914

Table 4 Beam section according to AISC 360–10

Content	h (mm)	b_f (mm)	t_f (mm)	t_w (mm)	L (m)	I_x (cm^4)	W_x (cm^3)
Column top	500	200	10	8	7.04	31,383	1255
Middle position	280	200	10	8	7.04	8462	604
Roof top	500	200	10	8	7.04	31,383	1255

3.3 Designing Beams According to AISC 360–10

With design analysis according to American standard AISC 360–10, after many changes of beam cross-section, the authors obtain the optimal design result (I(500–280) × 200 × 10 × 8) for AISC 360–10 as shown in Table 4. The chosen W-section of I(500–280) × 200 × 10 × 8 satisfies conditions of strength and service. It can be seen that the design of beam W-section using AISC 360–10 is obtained the smaller size comparing with the design using TCVN 5575–2012.

3.4 Horizontal Displacement of Column Top

The horizontal displacement at column top for the most dangerous load combination is of 0.0634 m generated by SAP2000 v22 [6] using TCVN 5575–2012 as plotted in Fig. 6a. The horizontal displacement is U1 = 0.0634 m, which is smaller than the allowed horizontal displacement of 0.075 m according to TCVN 5575–2012. Figure 6b shows the maximum horizontal displacement of the frame using AISC 360–10. The horizontal displacement at column top is U1 = 0.0439 m, which is smaller than H/60 (7.5 m/60, 0.125 m) according to MBMA 2012 [5]. It can be seen that the horizontal displacement using TCVN 5575–2012 is 1.44 times larger than the one of AISC 360–10.

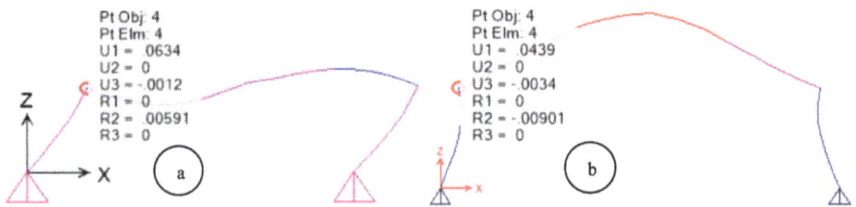

Fig. 6 Horizontal displacement of column top (m) **a** TCVN 5575–2012, **b** AISC 360–10

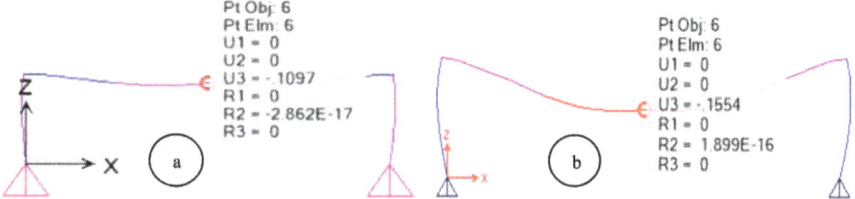

Fig. 7 Deflection of beam (m) **a** TCVN 5575–2012, **b** AISC 360–10

3.5 Deflection of Beam

Figure 7a shows the deflection of a beam of U3 = 0.1097 m using TCVN 5575–2012. This deflection satisfies the allowed deflection of 0.112 m. The beam deflection using AISC 360–10 is U3 = 0.1554 m as shown in Fig. 7b. This deflection is smaller than the allowed deflection of L/180 (28.0 m/180, 0.156 m) by MBMA 2012 [5]. It can be seen that the beam deflection using TCVN 5575–2012 is smaller than the one of AISC 360–10 about 0.71 times since the designed beam W-section using AISC 360–10 is smaller the one of TCVN 5575–2012.

3.6 Discussion

The design result of the steel beam according to TCVN 5575–2012 is I(600–380) × 200 × 12 × 10. The total weight of the steel beam is 2.092 Ton. While the design result of the steel beam according to AISC 360–10 is I(500–280) × 200 × 10 × 8. The total weight of the steel beam is 1.536 Ton. It can be concluded that designing the PEB using the American standard is an optimal solution in terms of cost better than the one of Vietnamese standard, thereby helping investors save considerable cost for PEB projects.

4 Conclusion

The authors finish the work of analysis and design for beams of a pre-engineering steel building using both standards of TCVN 5575–2012 and AISC 360–10. According to the results of analysis and calculation in this study, the design of beam cross-section according to AISC 360–10 is more optimal than TCVN 5575–2012 by nearly 36% of the cost. From the results of the study, it can be concluded that the American standard AISC 360–10 is safe and saving for designing the PEBs in Vietnam. Further studies can apply the nonlinear analysis [8–13] to design PEB structures using both these standards because the nonlinear analysis is an accurate approximation better than the linear analysis learned in traditional undergraduate program.

Acknowledgements The authors gratefully acknowledge the financial support provided by the Scientific Research Fund of Ho Chi Minh City Open University, Vietnam. (No. E2020.08.1).

References

1. TCVN (2012) Steel structures—Design standard, TCVN 5575–2012, Ministry of Construction, Hanoi, Vietnam
2. AISC (2010) Specification for structural steel buildings, AISC 360–10, American Institute of Steel Construction, America
3. TCVN (1995) Loads and Effects Design Standard, TCVN 2737–1995, Ministry of Construction, Hanoi, Vietnam
4. ASCE (2010) Minimum design loads for buildings and other structures, ASCE 7–10. American Society of Civil Engineers, Structural Engineering Institute, America
5. MBMA (2012) Metal Building Systems Manual, MBMA 2012, Metal Building Manufacturers Association, America
6. SAP2000 (2022) Analysis Reference Manual, CSI Berkeley (CA, USA). Computer and Structures, USA
7. Đoàn ĐK (2010) Thiết kế kết cấu thép (Theo quy phạm Hoa Kỳ AISC/ASD), NXB Xây Dựng, Vietnam
8. Nguyen P-C, Kim S-E (2015) Second-order spread-of-plasticity approach for nonlinear time-history analysis of space semi-rigid steel frames. Finite Elem Anal Des 105:1–15
9. Nguyen P-C, Kim S-E (2016) Advanced analysis for planar steel frames with semi-rigid connections using plastic-zone method. Steel Compos Struct 21(5):1121–1144
10. Nguyen P-C, Kim S-E (2017) Investigating effects of various base restraints on the nonlinear inelastic static and seismic responses of steel frames. Int J Non-Linear Mech 89:151–167
11. Nguyen P-C, Kim S-E (2018) A new improved fiber plastic hinge method accounting for lateral-torsional buckling of 3D steel frames. Thin-Walled Struct 127:666–675
12. Nguyen P-C, Tran TT, Nghia-Nguyen T (2021) Nonlinear time-history earthquake analysis for steel frames. Heliyon 7(8):e06832
13. Nguyen P-C, Tran TD (2021) Impacts of residual stress and shear deformation on 2D steel frames using fiber plastic hinge element: nonlinear behavior and strength. SN Appl Sci 3(7):686

Dynamic Response of FGM Plate Under Thermal Load by Using Moving Element Method

Minh Thi Tran, Quang Sy Tran, and Van Hai Luong

Abstract The dynamic response of the functionally graded material (FGM) plate under moving load considering the effect of temperature is investigated. The FGM plate properties are graded in the thickness direction based on the power-law distribution in terms. The Mindlin plate theory is employed to calculate the shear strain of the plate. The properties of the FGM plate such as the elastic modulus and coefficient of expansion due to effect of temperature are modeled. It is assumed that temperature variation is only in the direction of the plate thickness and that temperature field is assumed to be constant in the plane. Temperature distribution function along the plate thickness can be obtained by solving the governing equation of temperature transfer. Both mechanical strain and temperature-induced strain are employed to calculate the plate strain. Comparison study with previous results were performed to verify the effectiveness of the proposed model. Effects of temperature, moving load, plate thickness, and foundation parameters on the FGM plate are examined. It is found that they have a significant effect on the displacement of the plate. As to be expected, the plate displacement depends on the temperature, volume fraction index as well as velocity, and magnitude of moving load, being larger when they are larger.

Keywords FGM · MEM · Temperature · Moving load

1 Introduction

Recently, research and application of new materials with different properties to gradually replace traditional materials is the trend of the times. Functionally graded material (FGM) is a new generation of composite material introduced by a group of Japanese scientists in 1984. FGM has overcome the stress concentration and delamination phenomenon of composite materials. Thanks to its superior properties, FGM is

M. T. Tran · Q. S. Tran · V. H. Luong (✉)
Faculty of Civil Engineering, Ho Chi Minh City University of Technology (HCMUT), Ho Chi Minh City, Vietnam
e-mail: lvhai@hcmut.edu.vn

Vietnam National University Ho Chi Minh City, Linh Trung Ward, Ho Chi Minh City, Vietnam

© The Author(s), under exclusive license to Springer Nature Singapore Pte Ltd. 2023 785
J. N. Reddy et al. (eds.), *ICSCEA 2021*, Lecture Notes in Civil Engineering 268,
https://doi.org/10.1007/978-981-19-3303-5_71

applied in many fields, such as aviation (making aircraft fuselages), medicine (manu-facturing teeth, artificial bones), national defense (bulletproof armor), in the energy industry (insulation panels, turbines, reactors), and in construction. FGM plates are usually made from a mixture of ceramic and metal, which is an isotropic but not homogeneous material. Therefore, the use of this material is becoming increasingly popular around the world.

Time-varying and location-varying load effects on structures are related to building dynamics. This research direction has great attraction and has received the attention of many scientists. The model of a plate structure on a viscoelastic foundation subjected to moving loads has many practical applications. Many studies on plate response under moving loads have been published using various analytical methods.

First of all, the analytical method is usually employed to solve the dynamic prob-lems. Uymaz et al. [1] investigated the vibrations of the square FGM plates with different boundary conditions using the Ritz method and the Chebyshev displacement formula.

Above mentioned method gives an accurate solution but it is difficult and can lead to complex problems such as systems with many degrees of freedom, accelerated motion, or considering nonlinear response. Therefore, the Finite Element Method (FEM) is commonly used to overcome the above weaknesses. Shahidzadeh et al. [4] analyzed the response of square FGM plates with different boundary conditions using Abacus and Fortran software. Sarada Prasad Parida et al. [3] analyzed the dynamic response of FGM beams in a high-temperature environment using the FEM.

However, for structural problems with large lengths, they are assumed to be infinitely long plates. Meanwhile, the computational model by the FEM is finite, so the load will quickly reach the boundary position of the model. To overcome this limitation, the structural model must have a large enough length, which increases the calculation time and requires high computer configuration, so the moving element method (MEM) was born to solve the problem of analyzing the response of beams and slabs subjected to moving loads. Koh et al. [2] proposed the MEM for train-track dynamics. Luong et al. [5] presented the MEM for dynamic analysis of the FGM plates on the Pasternak foundation subjected to harmonic load and moving load. However, response analysis of FGM plates under moving load considering effect of temperature using the MEM has not been studied.

This paper presents the results of a computational study to investigate the dynamic response of the FGM plate under moving load due to effect of temperature using the MEM. The effects of various factors such as temperature, moving load, plate thickness, and foundation parameters on dynamic response of the FGM plate will be examined.

2 Theoretical Formulations

2.1 Mechanical Properties of the FGM Plate

The volume ratio of the FGM component materials varies continuously with the thickness of the plate.

$$P(z) = (P_c - P_m)(z/h + 1/2)^n + P_m. \tag{1}$$

where P_c- properties of ceramic materials; P_m- properties of metallic materials; $P(z)$- property of FGM at any z coordinate along plate thickness; n- volume fraction index; h- plate thickness.

2.2 The FGM Plate on the Pasternak Foundation is Subjected to Moving Loads Under the Effect of Temperature

The temperature field is assumed to be constant in the plane and varies only with the thickness of the plate. The elastic modulus (E) and the coefficient of thermal expansion (α) are assumed to be temperature-dependent.

$$E(z, T) = (E_c(T) - E_m(T))(z/h + 1/2)^n + E_m(T), \tag{2}$$

$$\alpha(z, T) = (\alpha_c(T) - \alpha_m(T))(z/h + 1/2)^n + \alpha_m(T). \tag{3}$$

When working in a temperature environment, the material constants are also functions of the absolute temperature. The material constants are expressed as:

$$P(T) = P_0(P_{-1}T^{-1} + 1 + P_1T + P_2T^2 + P_3T^3). \tag{4}$$

where P_0, P_{-1}, P_1, P_2, P_3- constants depending on the type of material; T- Survey temperature in Kelvin (^{0} K).

The temperature distribution function along the plate thickness can be obtained by solving the equation for the temperature through the thickness as follows:

$$-\frac{d}{dz}\left[\kappa(z)\frac{dT}{dz}\right] = 0. \tag{5}$$

When considering the effect of temperature, the strain field is determined as follows:

$$\{\varepsilon\} = \{\varepsilon\}^m + \{\varepsilon\}^T. \tag{6}$$

where $\{\varepsilon\}^m$ - Strain caused by mechanical load; $\{\varepsilon\}^T$ - strain caused by temperature. The load formed by temperature is presented by:

$$\begin{bmatrix} N_x^T & M_x^T \\ N_y^T & M_y^T \\ N_{xy}^T & M_{xy}^T \end{bmatrix} = \int_{-h/2}^{h/2} \begin{bmatrix} Q_{11} & Q_{12} & 0 \\ Q_{21} & Q_{22} & 0 \\ 0 & 0 & Q_{66} \end{bmatrix} \begin{Bmatrix} \alpha(z,T) \\ \alpha(z,T) \\ 0 \end{Bmatrix} \begin{bmatrix} 1 & z \end{bmatrix} \Delta T dz = \int_{-h/2}^{h/2} \begin{Bmatrix} 1 \\ 1 \\ 0 \end{Bmatrix} \frac{E(z,T)\alpha(z,T)\Delta T}{1-v} \begin{bmatrix} 1 & z \end{bmatrix} dz. \tag{7}$$

The motion equation of the FGM plate element is written in the moving coordinate system (r,s) on the Pasternak foundation considering effect of temperature.

$$\mathbf{M}^{(e)}\ddot{\mathbf{d}}^{(e)} + \mathbf{C}^{(e)}\dot{\mathbf{d}}^{(e)} + \mathbf{K}^{(e)}\mathbf{d}^{(e)} = \mathbf{P}^{(e)}. \tag{8}$$

The mass matrix, damper matrix stiffness matrix, and load vector of the moving FGM plate element are:

$$\mathbf{M}^{(e)} = \mathbf{m} \int_{\Omega^{(e)}} \mathbf{N}^T \mathbf{N} \det \mathbf{J} d\xi d\eta, \tag{9}$$

$$\mathbf{C}^{(e)} = -2\mathbf{m}v \int_{\Omega^{(e)}} \mathbf{N}^T \mathbf{N}_{,r} \det \mathbf{J} d\xi d\eta + c_f \int_{\Omega^{(e)}} \mathbf{N}_w^T \mathbf{N}_w \det \mathbf{J} d\xi d\eta, \tag{10}$$

$$\mathbf{K}^{(e)} = \int_{\Omega^{(e)}} \left\{ (\mathbf{B}_m)^T \ (\mathbf{B}_b)^T \ (\mathbf{B}_s)^T \right\} \begin{bmatrix} \mathbf{D}_m & \mathbf{D}_{mb} & 0 \\ \mathbf{D}_{mb} & \mathbf{D}_b & 0 \\ 0 & 0 & \mathbf{D}_s \end{bmatrix} \begin{Bmatrix} \mathbf{B}_b \\ \mathbf{B}_b \\ \mathbf{B}_s \end{Bmatrix} \det \mathbf{J} d\xi d\eta$$
$$+ \mathbf{m}V^2 \int_{\Omega^{(e)}} \mathbf{N}^T \mathbf{N}_{,rr} \det \mathbf{J} d\xi d\eta + k_{wf} \int_{\Omega^{(e)}} \mathbf{N}_w^T \mathbf{N}_w \det \mathbf{J} d\xi d\eta$$
$$- k_{sf} \int_{\Omega^{(e)}} \left(\mathbf{N}_w^T \mathbf{N}_{w,rr} + \mathbf{N}_w^T \mathbf{N}_{w,ss} \right) \det \mathbf{J} d\xi d\eta - c_f V \int_{\Omega^{(e)}} \mathbf{N}_w^T \mathbf{N}_{w,r} \det \mathbf{J} d\xi d\eta, \tag{11}$$

$$\mathbf{P}^{(e)} = \int_{\Omega^{(e)}} \mathbf{N}^T \mathbf{b}(r,s) \det \mathbf{J} d\xi d\eta - \int_{\Omega^{(e)}} \left(\left[(B_m)^T \right] \begin{Bmatrix} N_x^T \\ N_y^T \\ N_{xy}^T \end{Bmatrix} + \left[(B_b)^T \right] \begin{Bmatrix} M_x^T \\ M_y^T \\ M_{xy}^T \end{Bmatrix} \right) \det \mathbf{J} d\xi d\eta. \tag{12}$$

Table 1 Material parameters of the FGM plate

Material	Young's modulus E (Gpa)	Poisson's ratio v	Density ρ (kg/m³)
Zirconia (ZrO$_2$)	$E_c = 151$	0.3	$\rho_c = 3000$
Aluminum (AL)	$E_m = 70$		$\rho_m = 2702$

Table 2 Dimensionless displacement at the center of the FGM plate according to the element size

Results	The volume fraction index (n)			
	0	**5**	**10**	**∞**
Present	−2.1942	−2.7713	−2.8824	−3.1660
Luong [5]	−2.1941	−2.8172	−2.9048	−3.1661
Difference (%)	0.00	−1.62	−0.77	0.00

where **m**—mass matrix; **N**, **N**$_w$—shape function matrix, vector; **B**$_m$, **B**$_b$, **B**$_s$- membrane, bending, shear strain gradient matrix; **D**$_m$, **D**$_{mb}$, **D**$_b$, **D**$_s$—material matrices; **J**—Jacobi matrix; k_{wf}, k_{sf}, c_f- parameters of Pasternak foundation; V velocity of the moving load; **b**(r, s)- load vector in (r,s) coordinate symtem.

3 Verification of Results

Consider the FGM plate (length $L = 40$ (m), width $B = 10$ (m), thickness $h = 0.3$ (m)) with simple supported two and free supported two long sides. The plate resting on the Pasternak foundation is subjected to a load $P = 10^4$ (N) moving with a velocity $V = 20$ (m/s) along the x-axis of the plate. The material parameters of the plate are shown in Table 1 and the parameters of the foundation are: $k_{wf} = 10^7$ (N/m³), $k_{sf} = 10^5$(N/m), and $c_f = 10^4$ (Ns/m³). The plate is divided into element size of 1 (m) × 1 (m). Dimensionless displacement is defined by the formula: $\overline{w} = w(L/2, B/2) \times E_m \times h^3$.

Table 2 shows the comparison study of proposed present to previous results. It can be seen that there is virtually no difference in the results. This comparison study clearly illustrated that the proposed MEM is accurate and should be used for next study.

4 Numerical Results

Consider the FGM plate (length $L = 20$ (m), width $B = 10$ (m), thickness $h = 0.1$ (m)) with simple supported two short sides and free supported two long sides. The plate resting on the Pasternak foundation is subjected to a load $P = 10^6$ (N), moving with a velocity $V = 20$ (m/s) along the x-axis of the plate. The Poisson's

Table 3 Parameters depend on the type of material (Si₃N₄/SUS304)

Material	Parameter	P_o	P_{-1}	P_1	P_2	P_3	$P(300°K)$
Si₃N₄	E (Pa)	3.49E+11	0	−3.07E-04	2.16E-07	−8.95E-11	3.22E+11
	α (1/K)	5.87E-06	0	9.10E-04	0	0	7.48E-06
	ρ (kg/m³)						2370
	κ (W/mK)						9.19
SUS304	E (Pa)	2.01E+11	0	3.08E-04	−6.53E-07	0	2.08E+11
	α (1/K)	1.23E-05	0	8.09E-04	0	0	1.53E-05
	ρ (kg/m³)						8166
	κ (W/mK)						12.04

ratio, volume fraction index, and temperature respectively are $v = 0.3$, $n = 1$, and $T = 400(°K)$. The material parameters of the plate are shown in Table 3 and the parameters of the foundation are: $k_{wf} = 10^7$ (N/m³), $k_{sf} = 10^5$ (N/m), and $c_f = 10^4$ (Ns/m³). The plate is divided into elements of 1 (m) × 1 (m) size.

4.1 Convergence Study of Time Step

Figure 1 shows the maximun vertical displacement of FGM plate related to the variation of time step such as 0.1(s), 0.025(s), 0.005(s), 0.0025(s), 0.001(s) (a temperature environment of $400°$ K). The results show that time step is decreased, the results is rapidly converged. The result at $\Delta t = 0.0025$(s) is found to be virtually the same as compared to that at $\Delta t = 0.001$(s). Thus, the time step of 0.0025(s) should be employed to next study.

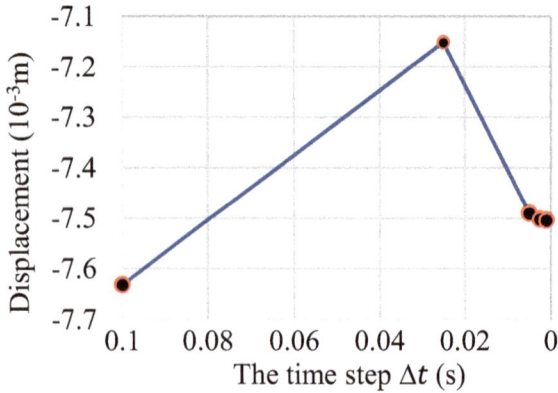

Fig. 1 The convergence of displacement with the time step Δt

4.2 The Effect of Temperature

Figure 2 shows the displacement of the FGM plate for the case of the initial temperature being room temperature ($T_0 = 300^0$ K) and the temperature variations acting equally on both sides of the plate $T_c = T_m = 500^0$ K, 700^0 K, 1000^0 K. It can be seen that the displacement of the plate increases when the absolute temperature increases.

Figure 3 shows maximum displacement w of the FGM plate for the two cases of temperature acting on (a) the top surface $T_c = 350^0$ K, 400^0 K, 450^0 K, 550^0 K and (b) the lower surface of the plate $T_m = 350^0$ K, 400^0 K, 450^0 K, 550^0 K, respectively. Note that the initial temperature is room temperature $T_0 = 300^0$ K. It is observed that when absolute temperature increases, the maximum displacement increases.

This comparison study clearly illustrated that temperature has a significant influence on the dynamic response of the FGM plate. It shows the difference in displacement when the same temperature is applied to each material on the top and bottom of the plate, which is the basic characteristic of FGM materials when each component material is linear but the plate material formed is nonlinear.

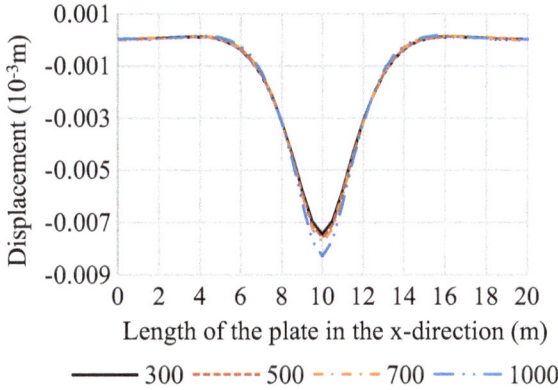

Fig. 2 Displacement of the FGM plate when the temperature is applied to both sides

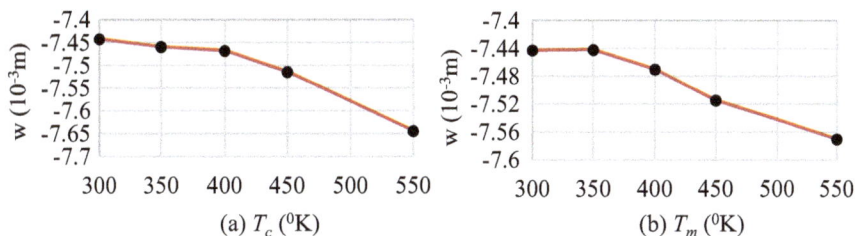

Fig. 3 Maximum displacement of the FGM plate when temperature is applied to the top/bottom surface

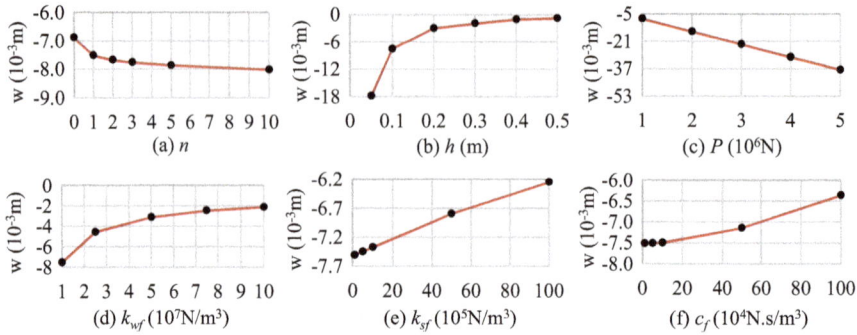

Fig. 4 Maximum displacement of the FGM plate with variations of parametes

4.3 Parameters of Plate, Moving Force, and Foundation

Figures 4 shows maximum displacement w of the FGM plate for the cases of parameters such as (a) volume fraction index n, (b) plate thickness h, (c) magnitude of force P, (d) foundation stiffness k_{wf}, (e) foundation shear strength k_{sf}, and (f) damper coefficient c_f at a temperature of 400^0 K. It is found that they have significant effect on the plate displacement. When volume fraction index, magnitude of force increase and plate thickness, foundation parameters decrease, the maximum displacement increases. This result is completely consistent with the physical properties of the structure. Therefore, parameters should be chosen reasonably to ensure that the displacement of the plate does not exceed the allowable displacement in technical design.

5 Conclusion

The paper is concerned with the moving element analysis of the FGM plate due to effect of temperature under moving load. The plate displacement is determined by Mindlin plate theory and the plate displacement equation is established by the principle of virtual work. The temperature is assumed to be constant in the plane and varies only in the direction of the plate thickness. The temperature of any point on the plate is determined by the temperature distribution function. Temperature not only changes the properties of the material, especially the modulus and coefficient of expansion, but also causes strain of the plate besides mechanical strain. Effect of temperature is employed to consider the cases of acting on both sides, on the top or bottom of the plate. The numerical results show that the temperature has a significant effect on the plate displacement.

The proposed MEM has significant computational efficiency compared to FEM as it overcomes the limitations of the FEM method by attaching the moving elements to a coordinate system that moves at the same speed as the load, thereby reducing the time considerable computation on the computer.

Besides the temperature, the volume fraction index, the plate thickness, the foundation parameters as well as magnitude of moving load were also investigated. It is found that they have significant effect on the plate displacement.

Acknowledgements This research is funded by Vietnam National University Ho Chi Minh City (VNU-HCM) under Grant No. B2022-20-03: "Developing numerical methods for the analysis of dynamic responses of thick plates under dynamic and thermal load".

References

1. Uymaz B, Aydogdu M (2007) Three-dimensional vibration analysis of functionally graded plates under various boundary conditions. J Reinf Plast Compos 26(18):1847–1863
2. Koh CG, Ong JSY, Chua DKH, Feng J (2003) Moving element method for train-track dynamics. Int J Numer Meth Eng 56:1549–1567
3. Parida SP, Jena PC (2019) FGM Beam analysis in dynamical and thermal surroundings using finite element method. Materials Today, part 7, 18:3676–3682
4. Shahidzadeh Tabatabaei SJ, Fattahi AM (2020) A finite element method for modal analysis of FGM plates. An Int J 30:1–2
5. Luong VH, Cao TNT, Lieu QX, Nguyen XV (2020) Moving element method for dynamic analysis of functionally graded plates resting on Pasternak foundation subjected to moving harmonic load. Int J Struct Stab Dyn 20 (1):2050003–1–2050003–25

Dynamic Responses of Composite Sandwich Plate Under Moving Load

Tan Ngoc Than Cao and Van Hai Luong

Abstract This paper presents the dynamic analyses of composite sandwich plate resting on a Winkler viscoelastic foundation under moving load using the Multi-layer Moving Plate Method (MMPM). The governing motion equations of composite sandwich plate are established in a moving coordinate system joined to the moving load. The prominent advantage is that the load is static in this coordinate system, which prevents the updating of locations of load owing to the change of the contact points with the elements. To verify the accuracy of the MMPM, the dynamic responses of composite sandwich plate are examined. Next, the effects of load's velocity on the dynamic responses of composite sandwich plates are studied.

Keywords Composite sandwich plate · Multi-layer moving plate method · Dynamic responses · Moving load

1 Introduction

Laminated composite plate structures have been extensively used in engineering because of their prominent characteristics such as higher strength, lighter weight, and hardness. The dynamic responses of composite plate were presented in the studies of Ghafoori and Asghari [2], Luong et al. [4]. The significant benefit of single-laminated composite plate would be broadened to complex plate systems such as a double-plate system. The free vibrations of double-plate system connected by an elastic layer were presented in the studies of Oniszczuk [5] and Zhang et al. [7]. However, the dynamic

T. N. T. Cao (✉)
College of Technology, Can Tho University (CTU), Can Tho, Vietnam
e-mail: ctnthan@ctu.edu.vn

V. H. Luong
Faculty of Civil Engineering, Ho Chi Minh City University of Technology (HCMUT), 268 Ly Thuong Kiet Street, District 10, Ho Chi Minh City, Vietnam

Vietnam National University Ho Chi Minh City, Linh Trung Ward, Thu Duc, Ho Chi Minh City, Vietnam

J. N. Reddy et al. (eds.), *ICSCEA 2021*, Lecture Notes in Civil Engineering 268,
https://doi.org/10.1007/978-981-19-3303-5_72

responses of laminated composite double-plate system subjected to moving load have not been carried out yet.

In the dynamic analyses of structures subjected to moving load, Finite Element Method (FEM) encounters complication that is the position of moving load has to be renewed at each time step in the calculation process. In the attempt to overcome above drawback, Koh et al. [3] proposed the Moving Element Method (MEM) for investigating the dynamic responses of the train-track system. Recently, Cao et al. [1] presented the Multi-layer moving plate method (MMPM) for studying the dynamic responses of a double-plate system under a moving load.

This paper presents the extension of the MMPM to analyze the dynamic responses of composite sandwich plate resting on a Winkler viscoelastic foundation under moving loads. The mass, damping, and stiffness matrices of multi-layer moving plate element of the composite sandwich plates are established. Next, the dynamic responses of composite sandwich plates are examined to prove the accuracy of the MMPM. Finally, the influences of load' velocity on dynamic responses of the composite sandwich plates are performed.

2　Formulation and Methodology

2.1　Weak Form for Composite Sandwich Plate Resting on a Viscoelastic Foundation

A composite sandwich plate is consisted of two parallel composite laminate plates joined by a connected layer as shown in Fig. 1. Upper plate is subjected to a moving load P which travels along the longitudinal middle line with velocity V The lower plate rests on a viscoelastic Winkler foundation. The connected layer and foundation are modeled by vertical spring stiffness k_{w1}, k_{w2} and damping c_1, c_2, respectively. Both upper and lower plates have the same width B, length L, and thickness $h_i (i = 1, 2)$.

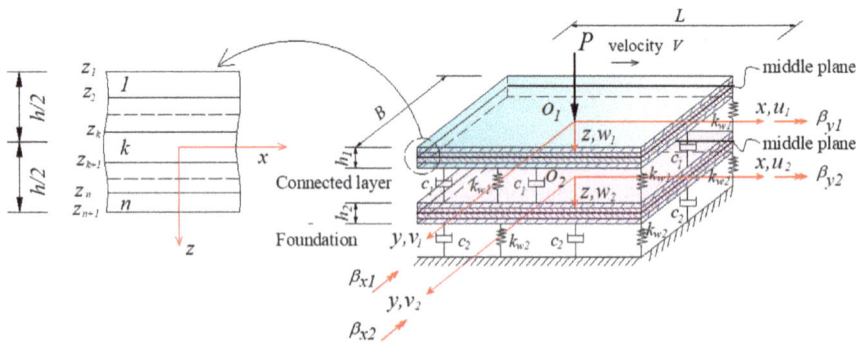

Fig. 1　Model of the composite sandwich plate resting on a viscoelastic winkler foundation

For convenience, $i = 1, 2$ stands for upper and lower plates, respectively. The reference planes are chosen at middle planes of two plates that occupy the domains Ω_1, $\Omega_2 \subset R^2$.

The displacements at any point of the ith($i = 1, 2$) plate using the first-order shear deformation theory are written as

$$u_i(x, y, z) = u_{0i}(x, y) + z\beta_{xi}(x, y); \qquad v_i(x, y, z) = v_{0i}(x, y) + z\beta_{yi}(x, y);$$
$$w_i(x, y, z) = w_{0i}(x, y) \tag{1}$$

where $u_{0i}(x, y)$, $v_{0i}(x, y)$ and $w_{0i}(x, y)$ are the membrane and vertical displacements of middle plane, respectively; β_{xi}, β_{yi} are the middle plane rotations around y-axis and x-axis, respectively.

The linear strain components of plates is given as

$$\begin{Bmatrix} \varepsilon_{xxi} \\ \varepsilon_{yyi} \\ \varepsilon_{xyi} \end{Bmatrix} = \begin{Bmatrix} u_{0i,x} \\ v_{0i,y} \\ u_{0i,y} + v_{0i,x} \end{Bmatrix} + z \begin{Bmatrix} \beta_{xi,x} \\ \beta_{yi,y} \\ \beta_{xi,y} + \beta_{yi,x} \end{Bmatrix} = \varepsilon_{0i} + z\kappa_i;$$

$$\begin{Bmatrix} \gamma_{xzi} \\ \gamma_{yzi} \end{Bmatrix} = \begin{Bmatrix} w_{i,x} + \beta_{xi} \\ w_{i,y} + \beta_{yi} \end{Bmatrix} = \gamma_i \quad (i = 1, 2) \tag{2}$$

where ε_{0i}, κ_i and γ_i $(i = 1, 2)$ are respectively membrane strain, bending strain, and shear strain of upper and lower plate.

The governing motion equations of upper and lower plates are obtained as follow

$$\int_{\Omega_1} \delta\varepsilon_{p1}^T \bar{\mathbf{D}}_1 \varepsilon_{p1} d\Omega_1 + \int_{\Omega_1} \delta\gamma_1^T \bar{\mathbf{D}}_{s1} \gamma_1 d\Omega_1 + \int_{\Omega_1} \delta\mathbf{u}_1^T \mathbf{m}_1 \ddot{\mathbf{u}}_1 d\Omega_1$$

$$+ \int_{\Omega_1} \delta w_1^T k_{w1}(w_1 - w_2) d\Omega_1 + \int_{\Omega_1} \delta w_1^T c_1(\dot{w}_1 - \dot{w}_2) d\Omega_1 \tag{3}$$

$$= \int_{\Omega_1} \delta\mathbf{u}_1^T \mathbf{b}_1(x, y, t) d\Omega_1$$

$$\int_{\Omega_2} \delta\varepsilon_{p2}^T \bar{\mathbf{D}}_2 \varepsilon_{p2} d\Omega_2 + \int_{\Omega_2} \delta\gamma_2^T \bar{\mathbf{D}}_{s2} \gamma_2 d\Omega_2 + \int_{\Omega_2} \delta\mathbf{u}_2^T \mathbf{m}_2 \ddot{\mathbf{u}}_2 d\Omega_2$$

$$- \int_{\Omega_2} \delta w_2^T k_{w1}(w_1 - w_2) d\Omega_2 - \int_{\Omega_2} \delta w_2^T c_1(\dot{w}_1 - \dot{w}_2) d\Omega_2 + \int_{\Omega_2} \delta w_2^T k_{w2} w_2 d\Omega_2 \tag{4}$$

$$+ \int_{\Omega_2} \delta w_2^T c_2 \dot{w}_2 d\Omega_2 = \int_{\Omega_2} \delta\mathbf{u}_2^T \mathbf{b}_2(x, y, t) d\Omega_2$$

where $\mathbf{u}_i = \left[u_i, v_i, w_i, \beta_{xi}, \beta_{yi} \right]^T (i = 1, 2)$ is the displacement vector at any point of the upper and lower composite plates, respectively; single dot and double dots over a symbol specify velocity and acceleration, respectively; $\mathbf{m}_i = \rho_i h_i diag \left[1\ 1\ 1\ h_i^2/12\ h_i^2/12 \right] (i = 1, 2)$ is the mass matrix of composite plate with ρ_i being mass density; $\mathbf{\varepsilon}_{pi} = \left[\mathbf{\varepsilon}_{0i}\ \mathbf{\kappa}_i \right]^T (i = 1, 2)$; flexural matrix $\overline{\mathbf{D}}_i (i = 1, 2)$ and shear matrix $\overline{\mathbf{D}}_{si} (i = 1, 2)$ of composite are defined as in Reddy [6].

The load vectors $\mathbf{b}_1(x, y, t)$ and $\mathbf{b}_2(x, y, t)$ in Eqs. (3) and (4) are respectively defined as

$$\mathbf{b}_1(x, y, t) = \left[0\ 0\ P\delta(x - Vt)\delta(y - 0)\ 0\ 0 \right]^T;$$
$$\mathbf{b}_2(x, y, t) = \left[0\ 0\ 0\ 0\ 0\ 0 \right]^T \tag{5}$$

where δ denotes the Dirac delta function, and t is the travel time of moving load.

2.2 Multi-layer Moving Plate Method (MMPM)

In this method, the composite sandwich plate is discretized into a number of multi-layer moving plate elements as shown in Fig. 2. A coordinates system (r, s) is attached to the moving load, and thus, it moves at the same velocity of the moving load. The correlation between moving coordinates system (r, s) and fixed coordinates systems (x, y) is expressed as

$$r = x - Vt; \quad s = y \tag{6}$$

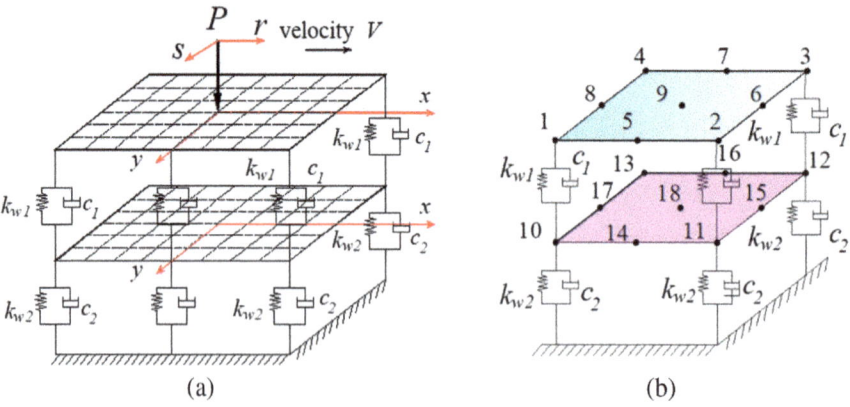

(a)   (b)

Fig. 2 a Discretization of a composite sandwich plate into multi-layer moving plate elements; **b** A multi-layer moving plate element

In the coordinate system (r, s), Eqs. (3) and (4) are rewritten as follow

$$\int_{\Omega_1} \delta\varepsilon_{p1}^T \bar{\mathbf{D}}_{1}\varepsilon_{p1} d\Omega_1 + \int_{\Omega_1} \delta\gamma_1^T \bar{\mathbf{D}}_{s1}\gamma_1 d\Omega_1 + \int_{\Omega_1} \delta\mathbf{u}_1^T \mathbf{m}_1 (V^2 \frac{\partial^2 \mathbf{u}_1}{\partial r^2} - 2V \frac{\partial^2 \mathbf{u}_1}{\partial r \partial t} + \frac{\partial^2 \mathbf{u}_1}{\partial t^2}) d\Omega_1$$

$$+ \int_{\Omega_1} \delta w_1^T k_{w1}(w_1 - w_2) d\Omega_1 - \int_{\Omega_1} \delta w_1^T c_1 (V \frac{\partial w_1}{\partial r} - V \frac{\partial w_2}{\partial r}) d\Omega_1 + \int_{\Omega_1} \delta w_1^T c_1 (\frac{\partial w_1}{\partial t} - \frac{\partial w_2}{\partial t}) d\Omega_1$$

$$= \int_{\Omega_1} \delta\mathbf{u}_1^T \mathbf{b}_1 (r, s, t) d\Omega_1$$

(7)

$$\int_{\Omega_2} \delta\varepsilon_{p2}^T \bar{\mathbf{D}}_{2}\varepsilon_{p2} d\Omega_2 + \int_{\Omega_2} \delta\gamma_2^T \bar{\mathbf{D}}_{s2}\gamma_2 d\Omega_2 + + \int_{\Omega_2} \delta\mathbf{u}_2^T \mathbf{m}_2 (V^2 \frac{\partial^2 \mathbf{u}_2}{\partial r^2} - 2V \frac{\partial^2 \mathbf{u}_2}{\partial r \partial t} + \frac{\partial^2 \mathbf{u}_2}{\partial t^2}) d\Omega_2$$

$$- \int_{\Omega_2} \delta w_2^T k_{w1}(w_1 - w_2) d\Omega_2 + \int_{\Omega_2} \delta w_2^T c_1 (V \frac{\partial w_1}{\partial r} - V \frac{\partial w_2}{\partial r}) d\Omega_2 - \int_{\Omega_2} \delta w_2^T c_1 (\frac{\partial w_1}{\partial t} - \frac{\partial w_2}{\partial t}) d\Omega_2$$

$$+ \int_{\Omega_2} \delta w_2^T k_{w2} w_2 d\Omega_2 + \int_{\Omega_2} \delta w_2^T c_2 (-V \frac{\partial w_2}{\partial r} + \frac{\partial w_2}{\partial t}) d\Omega_2 = \int_{\Omega_2} \delta\mathbf{u}_2^T \mathbf{b}_2 (r, s, t) d\Omega_2$$

(8)

in which, $\mathbf{b}_1(r, s, t) = \begin{bmatrix} 0 & 0 & P\delta(r)\delta(s) & 0 & 0 \end{bmatrix}^T$ is the load vector of upper plate and $\mathbf{b}_2(r, s, t) = \begin{bmatrix} 0 & 0 & 0 & 0 & 0 \end{bmatrix}^T$ is the load vector of lower plate in moving coordinates system (r, s).

Figure 2b shows a multi-layer moving plate element which is consisted of two layers of serendipity quadrilateral nine-node (Q9) elements. Each node has 5 DOFs, and this element has 18 nodes with 90 DOFs. The vector of displacement of element $\mathbf{d}^e$ is defined as

$$\mathbf{d}^e = \left[\underbrace{u_1^1 \; v_1^1 \; w_1^1 \; \beta_{r1}^1 \; \beta_{s1}^1 \; \cdots \; u_1^9 \; v_1^9 \; w_1^9 \; \beta_{r1}^9 \; \beta_{s1}^9}_{\text{upper plate}} \; \underbrace{u_2^{10} \; v_2^{10} \; w_2^{10} \; \beta_{r2}^{10} \; \beta_{s2}^{10} \; \cdots \; u_2^{18} \; v_2^{18} \; w_2^{18} \; \beta_{r2}^{18} \; \beta_{s2}^{18}}_{\text{lower plate}} \right]_{90 \times 1}^T$$

(9)

Using the shape functions, vectors of displacement, membrane, bending and shear strains of composite plates are respectively written as

$$\mathbf{u}_i = \mathbf{N}_i \mathbf{d}^e; \quad w_i = \mathbf{N}_{wi} \mathbf{d}^e; \quad \varepsilon_{pi} = \begin{bmatrix} \varepsilon_{0i} & \kappa_i \end{bmatrix}^T = \mathbf{B}_p \mathbf{d}^e = \begin{bmatrix} \mathbf{B}_{mi} & \mathbf{B}_{bi} \end{bmatrix}^T \mathbf{d}^e;$$
$$\gamma_i = \mathbf{B}_{si} \mathbf{d}^e \; (i = 1, 2)$$

(10)

where

$$
\mathbf{N}_1 = \left[\begin{array}{ccccc|ccccc}
N_j & 0 & 0 & 0 & 0 & 0\,0\,0\,0\,0 \\
0 & N_j & 0 & 0 & 0 & 0\,0\,0\,0\,0 \\
0 & 0 & N_j & 0 & 0 & 0\,0\,0\,0\,0 \\
0 & 0 & 0 & N_j & 0 & 0\,0\,0\,0\,0 \\
0 & 0 & 0 & 0 & N_j & 0\,0\,0\,0\,0
\end{array}\right]_{\substack{5\times9(j=1\div9) \quad 5\times9 \\ 5\times90}}
\;;\quad
\mathbf{N}_2 = \left[\begin{array}{ccccc|ccccc}
0\,0\,0\,0\,0 & N_j & 0 & 0 & 0 & 0 \\
0\,0\,0\,0\,0 & 0 & N_j & 0 & 0 & 0 \\
0\,0\,0\,0\,0 & 0 & 0 & N_j & 0 & 0 \\
0\,0\,0\,0\,0 & 0 & 0 & 0 & N_j & 0 \\
0\,0\,0\,0\,0 & 0 & 0 & 0 & 0 & N_j
\end{array}\right]_{\substack{5\times9 \quad 5\times9(j=1\div9) \\ 5\times90}}
\tag{11}
$$

$$
\mathbf{N}_{w1} = [\,\underbrace{0\ 0\ N_j\ 0\ 0}_{5\times9(j=1\div9)}\ \underbrace{0\ 0\ 0\ 0\ 0}_{5\times9}\,]_{1\times90}\;;\quad
\mathbf{N}_{w2} = [\,\underbrace{0\ 0\ 0\ 0\ 0}_{5\times9}\ \underbrace{0\ 0\ N_j\ 0\ 0}_{5\times9(j=1\div9)}\,]_{1\times90}
\tag{12}
$$

$$
\mathbf{B}_{m1} = \left[\begin{array}{ccccc|ccccc}
N_{j,r} & 0 & 0\,0\,0 & 0\,0\,0\,0\,0 \\
0 & N_{j,s} & 0\,0\,0 & 0\,0\,0\,0\,0 \\
N_{j,s} & N_{j,r} & 0\,0\,0 & 0\,0\,0\,0\,0
\end{array}\right]_{\substack{5\times9(j=1\div9)\quad 5\times9 \\ 3\times90}}
\;;\quad
\mathbf{B}_{m2} = \left[\begin{array}{ccccc|ccccc}
0\,0\,0\,0\,0 & N_{j,r} & 0 & 0\,0\,0 \\
0\,0\,0\,0\,0 & 0 & N_{j,s} & 0\,0\,0 \\
0\,0\,0\,0\,0 & N_{j,s} & N_{j,r} & 0\,0\,0
\end{array}\right]_{\substack{5\times9(j=1\div9)\quad 5\times9 \\ 3\times90}}\;;
$$

$$
\mathbf{B}_{b1} = \left[\begin{array}{ccccc|ccccc}
0\,0\,0 & N_{j,r} & 0 & 0\,0\,0\,0\,0 \\
0\,0\,0 & 0 & N_{j,s} & 0\,0\,0\,0\,0 \\
0\,0\,0 & N_{j,s} & N_{j,r} & 0\,0\,0\,0\,0
\end{array}\right]_{\substack{5\times9(j=1\div9)\quad 5\times9 \\ 3\times90}}
\quad
\mathbf{B}_{b2} = \left[\begin{array}{ccccc|ccccc}
0\,0\,0\,0\,0 & 0\,0\,0 & N_{j,r} & 0 \\
0\,0\,0\,0\,0 & 0\,0\,0 & 0 & N_{j,s} \\
0\,0\,0\,0\,0 & 0\,0\,0 & N_{j,s} & N_{j,r}
\end{array}\right]_{\substack{5\times9(j=1\div9)\quad 5\times9 \\ 3\times90}}\;;
$$

$$
\mathbf{B}_{s1} = \left[\begin{array}{ccccc|ccccc}
0\,0 & N_{j,r} & N_i & 0 & 0\,0\,0\,0\,0 \\
0\,0 & N_{j,s} & 0 & N_i & 0\,0\,0\,0\,0
\end{array}\right]_{\substack{5\times9(j=1\div9)\quad 5\times9 \\ 2\times90}}
\;;\quad
\mathbf{B}_{s2} = \left[\begin{array}{ccccc|ccccc}
0\,0\,0\,0\,0 & 0\,0 & N_{j,r} & N_i & 0 \\
0\,0\,0\,0\,0 & 0\,0 & N_{j,s} & 0 & N_i
\end{array}\right]_{\substack{5\times9(j=1\div9)\quad 5\times9 \\ 2\times90}}
\tag{13}
$$

Substituting Eq. (10) into Eqs. (7) and (8), the motion equations of upper plate element and lower plate element are written in short forms as

$$
\mathbf{M}_1^e \ddot{\mathbf{d}}^e + \mathbf{C}_1^e \dot{\mathbf{d}}^e + \mathbf{C}_1^e \mathbf{d}^e = \mathbf{P}_1^e; \quad \mathbf{M}_2^e \ddot{\mathbf{d}}^e + \mathbf{C}_2^e \dot{\mathbf{d}}^e + \mathbf{C}_2^e \mathbf{d}^e = \mathbf{P}_2^e
\tag{14}
$$

with the element mass $\mathbf{M}_1^e$, $\mathbf{M}_2^e$, element damping $\mathbf{C}_1^e$, $\mathbf{C}_2^e$ element stiffness $\mathbf{K}_1^e$, $\mathbf{K}_2^e$ matrices, and element load vector $\mathbf{P}_1^e$, $\mathbf{P}_2^e$ of upper plate and lower plate are respectively presented as

$$
\mathbf{M}_1^e = \mathbf{m}_1 \int_{\Omega_1^e} \mathbf{N}_1^T \mathbf{N}_1 d\Omega_1^e; \quad \mathbf{M}_2^e = \mathbf{m}_2 \int_{\Omega_2^e} \mathbf{N}_2^T \mathbf{N}_2 d\Omega_2^e
\tag{15}
$$

$$
\mathbf{C}_1^e = -2\mathbf{m}_1 V \int_{\Omega_1^e} \mathbf{N}_1^T \mathbf{N}_{1,r} d\Omega_1^e + c_1 \int_{\Omega_1^e} \mathbf{N}_{w1}^T \mathbf{N}_{w1} d\Omega_1^e - c_1 \int_{\Omega_1^e} \mathbf{N}_{w1}^T \mathbf{N}_{w2} d\Omega_1^e
\tag{16}
$$

$$\mathbf{C}_2^e = -2\mathbf{m}_2 V \int_{\Omega_2^e} \mathbf{N}_2^T \mathbf{N}_{2,r} d\Omega_2^e + c_1 \left(\int_{\Omega_2^e} \mathbf{N}_{w2}^T \mathbf{N}_{w2} d\Omega_2^e \right.$$

$$\left. - \int_{\Omega_2^e} \mathbf{N}_{w2}^T \mathbf{N}_{w1} d\Omega_2^e \right) + c_2 \int_{\Omega_2^e} \mathbf{N}_{w2}^T \mathbf{N}_{w2} d\Omega_2^e \tag{17}$$

$$\mathbf{K}_1^e = \int_{\Omega_1^e} \mathbf{B}_{p1}^T \bar{\mathbf{D}}_1 \mathbf{B}_{b1} d\Omega_1^e + \int_{\Omega_1^e} \mathbf{B}_{s1}^T \bar{\mathbf{D}}_{s1} \mathbf{B}_{s1} d\Omega_1^e + \mathbf{m}_1 V^2 \int_{\Omega_1^e} \mathbf{N}_1^T \mathbf{N}_{1,rr} d\Omega_1^e$$

$$+ k_{w1} \left(\int_{\Omega_1^e} \mathbf{N}_{w1}^T \mathbf{N}_{w1} d\Omega_1^e - \int_{\Omega_1^e} \mathbf{N}_{w1}^T \mathbf{N}_{w2} d\Omega_1^e \right) + c_1 V \left(\int_{\Omega_1^e} \mathbf{N}_{w1}^T \mathbf{N}_{w2,r} d\Omega_1^e \right.$$

$$\left. - \int_{\Omega_1^e} \mathbf{N}_{w1}^T \mathbf{N}_{w1,r} d\Omega_1^e \right) \tag{18}$$

$$\mathbf{K}_2^e = \int_{\Omega_2^e} \mathbf{B}_{p2}^T \bar{\mathbf{D}}_2 \mathbf{B}_{p2} d\Omega_2^e + \int_{\Omega_2^e} \mathbf{B}_{s2}^T \bar{\mathbf{D}}_{s2} \mathbf{B}_{s2} d\Omega_2^e + \mathbf{m}_2 V^2 \int_{\Omega_2^e} \mathbf{N}_2^T \mathbf{N}_{2,rr} d\Omega_2^e$$

$$+ k_{w1} \left(\int_{\Omega_2^e} \mathbf{N}_{w2}^T \mathbf{N}_{w2} d\Omega_2^e - \int_{\Omega_2^e} \mathbf{N}_{w2}^T \mathbf{N}_{w1} d\Omega_2^e \right) + c_1 V \left(\int_{\Omega_2^e} \mathbf{N}_{w2}^T \mathbf{N}_{w1,r} d\Omega_2^e \right.$$

$$\left. - \int_{\Omega_2^e} \mathbf{N}_{w2}^T \mathbf{N}_{w2,r} d\Omega_2^e \right) + k_{w2} \int_{\Omega_2^e} \mathbf{N}_{w2}^T \mathbf{N}_{w2} d\Omega_2^e - c_2 V \int_{\Omega_2^e} \mathbf{N}_{w2}^T \mathbf{N}_{w2,r} d\Omega_2^e \tag{19}$$

$$\mathbf{P}_1^e = \int_{\Omega_1^e} \mathbf{N}_1^T \mathbf{b}_1(r, s, t) d\Omega_1^e; \quad \mathbf{P}_2^e = \int_{\Omega_2^e} \mathbf{N}_2^T \mathbf{b}_2(r, s, t) d\Omega_2^e \tag{20}$$

in which, $(\cdot)_{,r}$ is the first partial derivatives and $(\cdot)_{,rr}$ is second partial derivatives with respect to r.

Finally, the mass, damping, stiffness matrices, and load vector of a multi-layer moving plate element are written as

$$\mathbf{M}_e = \mathbf{M}_1^e + \mathbf{M}_2^e; \quad \mathbf{C}_e = \mathbf{C}_1^e + \mathbf{C}_2^e; \quad \mathbf{K}_e = \mathbf{K}_1^e + \mathbf{K}_2^e; \quad \mathbf{P}_e = \mathbf{P}_1^e + \mathbf{P}_2^e \tag{21}$$

The motion equation of composite sandwich plate is established as

$$M\ddot{d} + C\dot{d} + Kd = P \tag{22}$$

where $\ddot{\mathbf{d}}$, $\dot{\mathbf{d}}$, and $\mathbf{d}$ are global acceleration, global velocity, and global displacement vectors, respectively; $\mathbf{M}$ is global mass matrix; $\mathbf{C}$ is global damping matrix; $\mathbf{K}$ is global stiffness matrix; and $\mathbf{P}$ is the global load vector.

3 Numerical Results

In this section, a simply supported square composite sandwich plate with geometric parameters is defined as $L/B = 1$ and $h_1/L = h_2/L = 0.01$. The upper and lower composite laminated plates have four layers ($0^0/90^0/90^0/0^0$), the material characteristics parameters and moving load are defined as $E_2 = 10.3$GPa $E_1/E_2 = 40$, $G_{12}/E_2 = G_{13}/E_2 = 0.6$, $G_{23}/E_2 = 0.5$, $\upsilon_{12} = 0.25$, $\rho_1 = \rho_2 = 1600$kg/m^3, $P = 10^4$N, $V = 40$m/s. The connection layer and foundation parameters are as follow: $k_{w1} = K_{w1}E_2h_1^3/L^4$, $k_{w2} = K_{w2}E_2h_2^3/L^4$. Figure 3 plots the deflection of longitudinal middle line of the upper plate with two connection layer stiffness coefficients $K_{w1} = 0$ (free connection) and $K_{w1} = 100$. In this example, the foundation parameters are given as $K_{w2} = \infty$ (rigidity foundation), $c_1 = c_2 = 0$. Figure 3 illustrates that the obtained results in this study and published results of Luong et al. [4] agree well. Thus, the accuracy of MMPM for analyzing the response of composite sandwich plate under a moving load is verified.

Next, the effects of load velocity on the responses of composite sandwich plate are investigated and shown in Fig. 4. In this example, three load's velocities are respectively $V_1 = 40$m/s, $V_2 = 60$m/s and $V_3 = 80$m/s. The connection layer and foundation parameters are given as $K_{w1} = K_{w2} = 100$ and $c_1 = c_2 = 10^4$. As can be seen from Fig. 4, the deformed shapes of upper plate and lower plate are asymmetric. In addition, the trends of responses of upper plate and lower plate are dissimilar when the load's speed increases. In case of upper plate, when the load's velocity increases from $V_1 = 40$m/s to $V_3 = 80$m/s, the peak value of deflection decreases from -7.68×10^{-3} mm to -6.56×10^{-3} mm. On the contrary, that of lower plate increases from -3.48×10^{-3} mm to -4.30×10^{-3} mm as the velocity of load increases in this range.

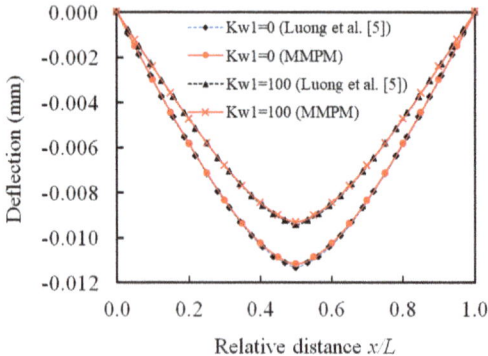

Fig. 3 Deflection of longitudinal middle line of upper plate in the composite sandwich plate with two connection layer stiffness coefficients $K_{w1} = 0$ and $K_{w1} = 100$

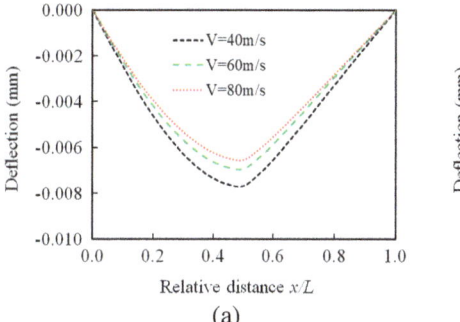 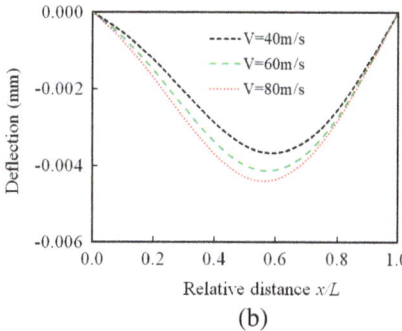

(a)　　　　　　　　　　　　　　(b)

Fig. 4 Deflection of longitudinal middle line of plates with various load's velocities: **a** upper plate, **b** lower plate

4 Conclusion

The extension of Multi-layer Moving Plate Method (MMPM) for analyzing the dynamic responses of composite sandwich plate resting on a viscoelastic Winkler foundation is presented in this study. The formulations of mass matrix, damping matrix, and stiffness matrix of composite sandwich plate element are derived. First, the analysis of dynamic response of composite sandwich plate are considered to examine the accuracy of the MMPM. The comparisons show that obtained results in this study agree excellently with those of published results. Next, the effects of load's velocity on the dynamic responses of the composite sandwich plate structure are investigated.

Acknowledgements This research is funded by Can Tho University (CTU) under Grant Number T2021-02: 'Development of multi-layer moving plate method for dynamic analysis of composite sandwich plate resting on viscoelastic foundation subjected to a moving load'.

References

1. Cao TNT, Reddy JN, Lieu QX, Nguyen XV, Luong VH (2019) A multi-layer moving plate method for dynamic analysis of viscoelastically connected double-plate systems subjected to moving loads. Adv Struct Eng 24(9):1798–1913
2. Ghafoori E, Asghari M (2010) Dynamic analysis of laminated composite plates traversed by a moving mass based on a first-order theory. Compos Struct 92:1865–1876
3. Koh CG, Ong JSY, Chua DKH, Feng J (2003) Moving element method for train-track dynamics. Int J Numer Methods Eng 56:1549–1567
4. Luong VH, Nguyen TT, Liu CR, Phung PP (2014) A cell-based smoothed finite element method using three-node shear locking free mindlin plate element (CS-FEM-MIN3) for dynamic response of laminated composite plates on viscoelastic foundation. Bound Elem Methods Eng 42:8–19

5. Oniszczuk Z (2000) Free transverse vibrations of an elastically connected rectangular simply supported double-plate complex system. J Sound Vib 236(4):595–608
6. Reddy JN (2004) Mechanics of laminated composite plates -Theory and analysis, 2nd edn. CRC Press Inc., New York
7. Zhang Y, Shi D, He D, Shao D (2021) Free vibration analysis of laminated composite double-plate structure system with elastic constraints based on improved fourier series method. Shock Vib 2021:1–25. https://doi.org/10.1155/2021/8811747

The Influence of Ambient Temperature on the Curvature of Displacement in Prestressed Concrete Girders

Tuan Minh Ha, Saiji Fukada, Phuoc Trong Nguyen, and Duc-Duy Ho

Abstract When external factors disrupt the test's measurement signals, it is challenging to determine damage inside a structure. The purpose of this study is to provide preliminary suggestions about measurement noise when displacement curvature is used as an input for damage detection. Specifically, this study examined the effect of ambient temperature on the displacement curvature of a prestressed concrete (PC) girder during six months. The findings demonstrate that external temperature affects the vertical movement of the PC girders. Curvature estimations based on experimental data revealed a proportionate correlation between the curvature and the ambient temperature. The amplitude of the curvature grew as the ambient temperature rose. The calculated curvature changes between -0.0264 mm^{-1} and -0.0379 mm^{-1} when the temperature rises by one degree. Additionally, the regression analysis establishes that changes in ambient temperature may account for around 81–93 percent of the variation in displacement curvature. Thus, the ambient temperature was not the only factor influencing this change.

Keywords PC Girder · Curvature of displacement · Temperature effect

---

T. M. Ha (✉)
HUTECH University, Ho Chi Minh City, Vietnam
e-mail: hm.tuan@hutech.edu.vn

S. Fukada
Faculty of Geosciences and Engineering, Kanazawa University, Kanazawa, Japan

P. T. Nguyen
Ho Chi Minh City Open University, Ho Chi Minh City, Vietnam

D.-D. Ho
Faculty of Civil Engineering, Ho Chi Minh City University of Technology (HCMUT), 268 Ly Thuong Kiet Street, District 10, Ho Chi Minh City, Vietnam

Vietnam National University Ho Chi Minh City, Linh Trung Ward, Thu Duc District, Ho Chi Minh City, Vietnam

© The Author(s), under exclusive license to Springer Nature Singapore Pte Ltd. 2023   805
J. N. Reddy et al. (eds.), *ICSCEA 2021*, Lecture Notes in Civil Engineering 268,
https://doi.org/10.1007/978-981-19-3303-5_73

1   Introduction

For several years, most studies have concentrated on employing structural responses to identify degradation features (occurrence, location, and severity). However, few studies have been undertaken on the effects of environmental variables on structural feature changes (dynamic and static responses). In terms of dynamic reactions, Cornwell et al. [1] discovered that the natural frequencies of the Alamosa Canyon Bridge fluctuated between 4.7 and 5% when the temperature of the deck changed by around 22 °C each day. Peeters and De Roeck [2] concluded two years later that the first four vibration frequencies of the Z24-bridge varied by 14–18% over ten months. Furthermore, with the exception of the second mode, the frequencies of all the modes tested fell as the temperature increased. Additionally, the authors observed that vibration frequencies rose when the temperature was reduced to 0 °C or below. According to research conducted by Yong Xia et al. [3], natural frequencies also decreased as the temperature increased. Ha et al. [4] predicted the decreasing percentage of the initial bending frequency of a full-scale girder to be within 6% prior to the yielding load in 2016.

Additionally, only a few studies have been conducted to examine the influence of environmental circumstances on temperature-induced displacement curvature. Camber [5], also known as curling [6], is a well-documented reaction to temperature variations in various concrete components. Curling in concrete pavements [7], prestressed concrete bridge girders [5], concrete slab track [8], and concrete crosstie [6] have been examined. This study aims to determine the effect of ambient temperature on the displacement curvature of a prestressed concrete (PC) girder. Vertical displacement and air temperatures were recorded on a six-monthly basis. After collecting data, linear regression models were used to examine it. Finally, this study makes recommendations about measurement noise when displacement curvature is used as an input for damage detection.

2   Description of the PC-Girder Investigated

A simply supported PC girder was made and subjected to environmental conditions outside the laboratory. It was 9600 mm long with a 200 mm overhang at either end. The target girder is constructed with sixteen PC steel bars (12φ 7 mm) in the longitudinal bar configuration. Additionally, D-10 shear reinforcement bars are positioned around PC steel bars to serve as stirrups, as seen in Fig. 1. The static elastic modulus of the steel reinforcement D10 bars and PC cables is estimated to be 210,000 N/mm^2 with a Poisson ratio of 0.15. Additionally, PC cables exposed to 1100 N/mm^2 prestress are described with a yield stress of 1580 N/mm^2. Table 1 summarizes the concrete characteristics.

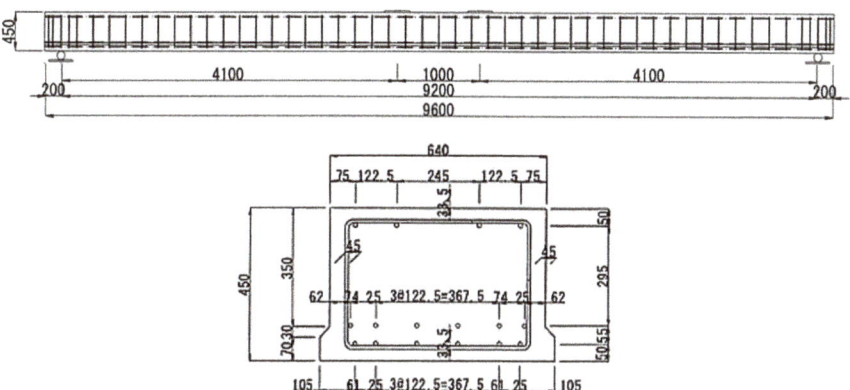

Fig. 1 Diagram of a PC girder

Table 1 Material parameters of concrete

Static elastic modulus	33,000	N/mm^2
Poisson's ratio	0.15	
Tensile strength	3	N/mm^2
Compressive strength	60	N/mm^2

3 Experimental Procedure

A simply supported PC girder was subjected to external weather conditions outside the laboratory. Figure 2 illustrates the test setup used to examine the upward movement. Seven displacement meters were positioned at various positions under the girder with a constant spacing of 1.15 m to yield temperature-induced displacements, as seen in Fig. 3. Additionally, the ambient temperature was monitored using a thermometer during the test. The signal was gathered automatically from the sensors once

Fig. 2 Layout of measured points under the PC girder

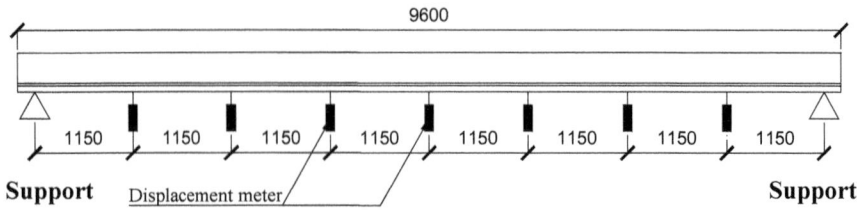

Fig. 3 Layout of the experiment (Unit: mm)

every 30 min, 24 h a day, to monitor camber change caused by diurnal temperature variations. Continuous measurements were taken throughout a six-month period.

4 Procedure for Determining Curvature

Based on the mechanics of materials theory, the correlation of displacement curvature $1/\rho$ (ρ is the radius of curvature) with deflection u is illustrated by,

$$\frac{1}{\rho} = \frac{\frac{\partial^2 u}{dx^2}}{\left[1 + \left(\frac{\partial u}{dx}\right)^2\right]^{3/2}} = \frac{M}{EI} \tag{2}$$

where M denotes the bending moment, E denotes the modulus of elasticity, and I denotes the moment of inertia of the cross-section. The curvature may be approximated as follows, neglecting the second order of the second derivative of the deflection u in (2):

$$\frac{1}{\rho} \approx \frac{\partial^2 u}{dx^2} \tag{3}$$

Additionally, the following relationship exists between curvature, bending moment, and stiffness:

$$\frac{\partial^2 u}{dx^2} = \frac{M}{EI} \tag{4}$$

Equation 4 establishes a relationship between displacement curvature and stiffness. Thus, any modification in the structural stiffness of an observable segment due to degradation results in a change in the curvature at that place. Throughout the monitoring period, the displacements of many indicated sites under the goal girder were continually monitored. The deformation equation (u) may then be determined by fitting a polynomial curve of 4th order using given nodal displacements. By using the second derivative of the deformation equation, the curvature of the girder may be determined at any position. The following technique for conducting the experiment and analyzing the results is proposed:

- Seven sensitive displacement meters are used to determine the temperature-induced displacement of the girder.
- Assuming measurable displacements in the coordinate geometry of u and x, the deformation equation (u) may be approximated by fitting a fourth order polynomial curve to the provided nodal displacements, as shown below.

$$[x, u] = [(x_1, u_1), (x_2, u_2), \cdots, (x_n, u_n)] \tag{5}$$

$$u = \sum_{i=0}^{n} a_{i+1} x^i = a_1 + a_2 x + a_3 x^2 + a_4 x^3 + \cdots + a_{n+1} x^n \tag{6}$$

- The displacement curvature equation may then be determined by examining the deformation equation's second order derivative.

$$k = \frac{1}{\rho} = \frac{\partial^2 u}{dx^2} = \sum_{i=2}^{n} (i-1)(i)a_{i+1} x^{i-2} = 2a_3 + 6a_4 x + \cdots$$
$$+ (n-1)na_{n+1} x^{n-2} \tag{7}$$

- As a result, the displacement curvature may be determined at any position.

5 Changes in Temperature-Related Displacement Curvature

This section analyzes the temperature-induced displacement curvatures of the girder. The relationships between displacement curvatures and temperature are well established using the collected data. When examining measured data as a function of temperature, a link between the predicted curvature and ambient temperature may be noticed (see Fig. 4a–b). These findings demonstrated that there are strong relationships between displacement curvature and measured ambient temperatures. Additionally, a linear regression analysis was used to evaluate the influence of temperature on displacement curvature in this case. The linear equation is expressed in terms of (8).

$$f = \alpha + \beta t + \varepsilon \tag{8}$$

where f denotes the displacement curvature and α and β denote the line's intercept and slope, respectively. ε is the component that contains errors. The temperature of

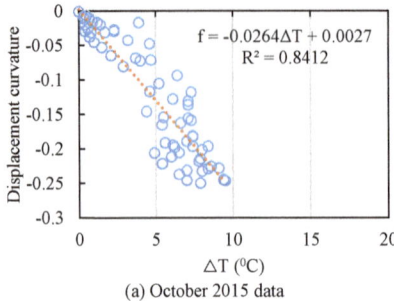
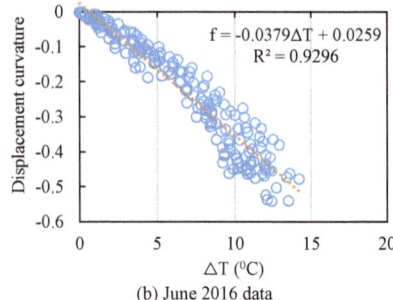

(a) October 2015 data   (b) June 2016 data

Fig. 4 A direct proportionality between mid-span displacement curvature and PC girder temperature

Table 2 Displacement curvature vs. temperature regression coefficients

Time of data collection	slope β	intercept α	Determination coefficient R^2
2015–10	–0.0264	0.0027	0.8412
2016–3	–0.0253	0.0156	0.8273
2016–4	–0.03	0.0024	0.8422
2016–5	–0.0324	0.0341	0.9262
2016–6	–0.0379	0.0259	0.9296

the surrounding environment is used as an independent variable. R is used to do the analysis in this case.

By examining the gradient values of the linear regression lines in Table 2, it is determined that when the ambient temperature is increased by 1 °C, the displacement curvature of the PC girder changes by roughly -0.0264 mm^{-1} and -0.0379 mm^{-1}. Additionally, about 82 percent of the curvature variance may be attributed to changes in the ambient temperature. As a result, the temperature was not the only factor affecting the static structural reaction. As a result, it is possible to argue that additional factors such as ambient humidity, linear temperature gradients in the concrete, equipment noise, and the calculation procedure all affect the displacement curvature, resulting in unanticipated inaccuracies in the measurement field. Furthermore, as seen in Fig. 4, the mid-span displacement curvature fluctuates less in October when the temperature is slightly below 10 °C, than in June when the temperature difference during the day is significant (15 °C). In October, the displacement curvature at the span's center is computed precisely and ranges from 0 to about -0.25 mm^{-1}. This range represents about 46% of the findings from the June monitoring data, with the maximum curvature change value of -0.54 mm^{-1}.

The residual examinations, which include normal distribution verification, homogeneity of variation testing, linearity testing, the Durbin-Watson test for independence, outliers testing, and influential observations, are used to determine the proposed linear regressive model's reliability on the obtained data. Figure 5

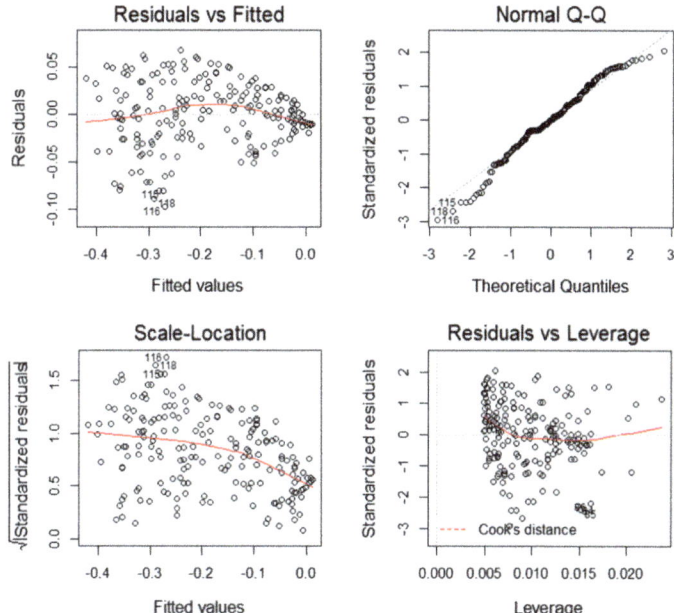

Fig. 5 Results of residual analysis of curvature data in June 2016

illustrates the results of the investigation of linear regression's assumptions. The quantile-quantile (Q-Q) plots indicate a satisfactory fit, with residuals values ranging approximately 0. As such, linear regression models are applicable in this situation.

6 Conclusions

It is essential to identify the variation in structural responses caused by structural degradation from changes caused by environmental variables in order to conduct a reliable evaluation of structural damage characteristics. The study's findings are presented below.

- More than ~82% of the curvature change could be considered by the difference in the ambient temperature.
- When the ambient temperature is increased by 1 °C, the displacement curvature of the objective girder fluctuates between −0.0264 mm^{-1} and −0.0379 mm^{-1}.
- When evaluating the observed data as a function of temperature, there is a linear association between displacement curvature and ambient temperature.

Correlations between displacement curvature and ambient temperature presented in this work aid in the evaluation of a technique for identifying structural deterioration. As a result, the impacts of external variables on practice should be treated as unanticipated mistakes. Additionally, the current results are limited to the temperature range described in this study. As a result, more study should be carried.

Acknowledgements The Regional Innovation Ecosystems Program of the Japanese Ministry of Education, Culture, Sports, Science, and Technology funded this study at Professor Saiji Fukada's laboratory at Kanazawa University in Kanazawa City.

References

[1] Cornwell P, Farrar CR, Doebling SW, Sohn H (1999) Environmental variability of modal properties. Exp Tech 23(6):45–48
[2] Peeters B, De Roeck G (2001) One-year monitoring of the Z24-bridge: Environmental effects versus damage events. Earthq Eng Struct Dyn 30(2):149–171
[3] Xia Y, Hao H, Zanardo G, Deeks A (2006) Long term vibration monitoring of an RC slab: Temperature and humidity effect. Eng Struct 28(3):441–452
[4] Ha TM, Fukada S, Arima N, Moriyama N, Miyashita T (2016) Study on vibration and structural performance of PC Girder removed due to salt damage. Proceedings of 14th East Asia-Pacific Conference on Structural Engineering and Construction, 680–681
[5] Barr PJ, Stanton JF, Eberhard MO (2005) Effects of Temperature Variations on Precast, Prestressed Concrete Bridge Girders. J Bridg Eng 10(2):186–194
[6] Wolf HE, Qian Y, Edwards JR, Dersch MS, Lange DA (2016) Temperature-induced curl behavior of prestressed concrete and its effect on railroad crossties. Constr Build Mater 115:319–326
[7] Armaghani JM, Larsen TJ, Smith LL (1987) Temperature response of concrete pavements. Transp Res Rec 1211:23–33
[8] Ren J, Yang R, Wang P, Yong P, Wen C (2014) Slab upwarping of twin-block slab track on subgrade-bridge transition section: parameter study and repair method. Transp Res Rec J Transp Res Board 2448:115–124

Effect of Steel Fiber on Resistance of Ultra High Performance Fiber Reinforced Concrete Plates Under Impact Load

Thi Hai Vinh Chu, Duc Vinh Bui, and Viet Tue Nguyen

Abstract Ultra High Performance Concretes are significantly improving the bearing capacity of the structures under static as well as dynamic loads. An investigation on the behavior of Ultra High Performance Fiber Reinforced Concrete (UHPFRC) plates subjected to impact load was conducted in this study with three main factors of steel fiber, the strength of the concrete, and the amount of rebar. At the material level, a concrete grade of C120 and a steel fiber content of 1.0, 1.5, and 2.0% by volume respectively were used. All the samples have dimensions L500 mm x W500 mm x T80 mm, they are divided into two groups of with and without reinforcement. The experimental model is set up with a free-falling load acting at the center of the upper surface of the plate. Experimental results show that the steel fiber content directly affects the mechanical properties of UHPFRC such as compressive strength, flexural strength, fracture energy, and toughness of the material. The crack propagation corresponding to the impact loading is influenced by the amount of steel fiber in the concrete.

Keywords Ultra high performance concrete · Steel fiber · Impact loading · Fracture energy

1 Introduction

The addition of ultra-thin, high-strength steel fibers to UHPC results in improved toughness and strain energy transmission when the structure is loaded [4, 9, 10]. Hassan et al. [2] pointed out that the Ultra High Performance Fiber Reinforced

T. H. V. Chu · D. V. Bui (✉)
Faculty of Civil Engineering, Ho Chi Minh City University of Technology (HCMUT), 268 Ly Thuong Kiet Street, District 10, Ho Chi Minh City, Vietnam
e-mail: vinhbd@hcmut.edu.vn

Vietnam National University Ho Chi Minh City, Linh Trung Ward, Thu Duc District, Ho Chi Minh City, Vietnam

V. T. Nguyen
Institute of Structural Concrete, Graz University of Technology, Styria, Austria

© The Author(s), under exclusive license to Springer Nature Singapore Pte Ltd. 2023
J. N. Reddy et al. (eds.), *ICSCEA 2021*, Lecture Notes in Civil Engineering 268,
https://doi.org/10.1007/978-981-19-3303-5_74

Concrete (UHPFRC) tensile strength increased 1.69 times compared with the control sample when there is an addition of 2% steel fiber. The maximum deformation in the axial tensile test of UHPFRC was about 1.5–3.0 ‰, while this value for non-steel fiber UHPC was about 0.15–0.25‰. However, the steel fiber composition does not significantly affect the compressive strength and elastic modulus of UHPFRC. Besides, other studies have shown that steel fiber content, fiber shape, and concrete pouring direction also significantly affect the behavior of the material [5, 11–13].

Jin-Young Lee et al. [6] impact testing for 2-sided panels with concrete strength from 25 to 180 MPa, using two different types of steel fibers. The addition of the steel fiber also reduced deflection and vibrations in the plate. Also, the mixing between 2 different fibers tended to dissipate energy more quickly in the study of R. Yu et al. [14]. M. Orgass et al. [7] commented that the flexural strength of UHPFRC increases with increasing steel fiber content, especially the bearing capacity increases when mixing 2 types of steel fibers into concrete at the rate of 1% short fiber and 1% long fiber. Mixing these two fibers also affected the beam's post-cracking behavior, increasing the toughness of the beam.

In this study, the test samples are made of UHPC grade 120 with a steel fiber content of 1.0, 1.5, and 2.0% by volume. The basis tests to determine the mechanical properties of the material are conducted, the experimental model with a mass falling to the surface of the plates simulated for impact load, the behavior of the plates are investigated. The results of this study are intended to be applied to structures subjected to impact and blast loading.

2   Mechanical Properties of UHPFRC

The basic materials for UHPC include Portland cement type I, silica fume, silica powder, and superplasticizer, which are shown in Table 1. In this study, the steel fibers used are straight, then diameter and length are $\phi 0.2$ mm $\times$ L13 mm respectively, the tensile strength of steel fiber is 2850 MPa by manufacture. To determine the mechanical properties of UHPC, the test program is set up with 4 groups of samples

Table 1   Ingredients of UHPC in $1m^3$

Number	Materials	Unit	Grain size	Quantity / 1m3
1	Cement	kg	25 μm	600
2	Sand	kg	0,3–0,8 mm	477
3	Silica fume	kg	0,15 μm	132
4	Silica powder	kg	11 μm	228
5	Coarse aggregate	kg	1,25–8,0 mm	735
6	Water	lít		161
7	Superplasticizer	kg		17

are signed as follows: N00, N10, N15, and N20 correspond to the steel fiber content of 0; 1.0; 1.5, and 2.0%.

2.1 The Compressive Strength (f_c)

Cylindrical samples have a diameter ϕ150 and height H300 mm are made, two surfaces of the samples are ground flat before compression. Compressive strength is presented in Table 2, as can see that, groups of N20 samples (with 2.0% steel fiber) have the highest compressive strength of 133.17 MPa; the other samples (N15, N10, N00) have lower strength with the results respectively 124.94 MPa; 122.69 MPa; 118.79 MPa (Fig. 1).

Besides, the strength of concrete develops very early, the compressive strength at 3 days is approximate 70–75% compared with 28 days and this value increases to 85% at 14 days. At the collapse stage, the samples with the steel fiber remained almost initial shape as shown in Fig. 2.

Table 2 Compressive strength of sample groups

Symbol	Compressive strength, f_c (MPa)				The destructive form of the samples
	3 days	7 days	14 days	28 days	
N00	86.45	95.08	101.12	118.79	The sample is broken
N10	92.88	98.34	106.32	122.69	The sample is cracked, almost retaining its original shape
N15	87.80	101.21	108.02	124.94	
N20	91.85	104.76	114.16	133.17	

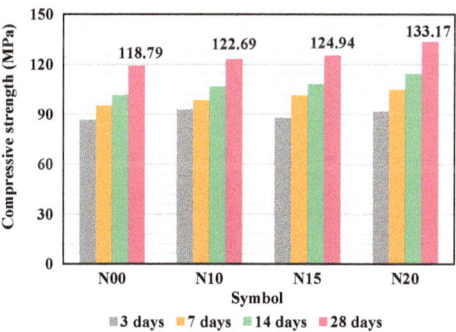

Fig. 1 Compressive strength of sample groups

Fig. 2 Sample shape at the end of the testing

2.2 *Flexural Strength (f_{ct})*

The flexural strength is determined by the bending test (Fig. 3) according to JCI-S-002–2003 [3]. The dimensions of samples are 75 × 100 mm × 450 mm, respectively to wide × height of the cross-section and length. The experimental results of the sample groups are presented in Table 3. Whereby, the flexural strength of the sample group without steel fiber is 9.36 MPa, the sample is broken suddenly at the maximum load (P_{max}), as illustrated in Fig. 4. However, when steel fibers are added to the concrete matrix, the flexural strength of the material increases significantly, are 14.84 MPa; 21.80 MPa; and 26.94 MPa respectively to groups N10, N15, and N20. In all these groups, during the initial period of the test program, the deflection is very small, as shown in Fig. 5. It can be seen that, at the peak load, the deflection of the sample is less than 1 mm. In the post-peak branch, the loading speed is maintained

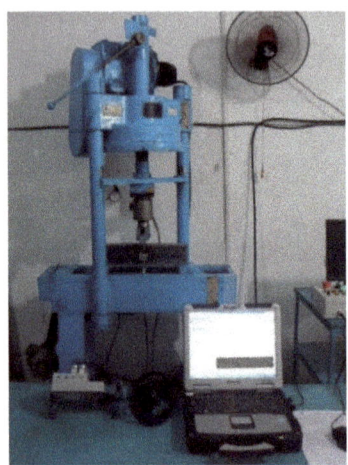

Fig. 3 Bending test

Table 3 Results of the bending test

Symbol	P_{max} (N)	f_{ct} (MPa)
N00	4823	9.36
N10	7648	14.84
N15	11,231	21.80
N20	13,885	26.94

(a) The sample group does not have steel fibers

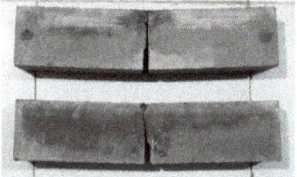

(b) The sample group has steel fibers

Fig. 4 Beam shape after bending

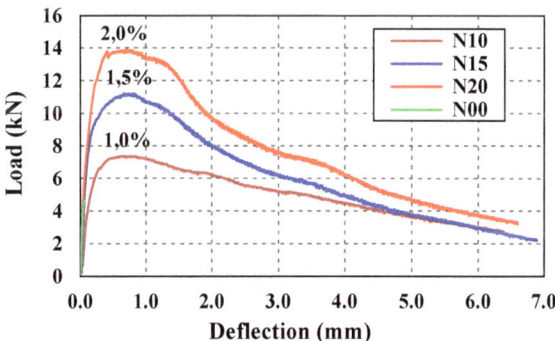

Fig. 5 Force and deflection diagram

at about 0.05–0.1 mm/min, the deflection and crack width gradually increase. At the end of the test, the beams with steel fiber are not fully collapsed and show ductile failure.

Table 4 Fracture energy of notched beam

Symbol	δ_L (mm)	P_L (N)	$D_{BZ,b}$ (Nmm)	$D^f_{BZ,2}$ (Nmm)	$D^f_{BZ,3}$ (Nmm)	G_F (N/mm)
N10	0.05	1678	207.1	3956.1	16,888.5	3.83
N15	0.05	3836	478.5	5916.9	23,866.2	5.38
N20	0.05	5227	651.8	7323.4	29,550.6	6.68

2.3 Fracture Energy (G_F)

The fracture energy of concrete G_F (N/mm) is defined as the energy required to cause cracking per unit area, which represents the toughness of the concrete and the degree of propagation capacity when the structure is bearing. The G_F is calculated to standard RILEM TC 162-TDF [8].

The value $D^b_{BZ,b}$, $D^f_{BZ,2}$, and $D^f_{BZ,3}$, in Table 4 expressed the energy absorption of the concrete and steel fiber sections, these values are calculated from the force–displacement diagram in Fig. 5. Considering the effect of steel fiber content to absorb energy, the N20 group has the largest value ($D^f_{BZ,3}$ = 29,550.6 N/mm), increasing 23.8% compared to the group N15 ($D^f_{BZ,3}$ = 23,866.2 N/mm) and 74.9% group N10 ($D^f_{BZ,3}$ = 16,888.5 N/mm). Furthermore, the energy absorption level of the steel fiber gives a much superior value to the energy absorption level of the concrete part ($D^b_{BZ,b}$). From here, it can be observed that increasing the content of steel fibers leads to increase absorption and energy transmission in UHPFRC, thereby increasing the toughness of the material.

3 Impact Test of Bending Plates

3.1 Test Samples

An experimental program is set up with 9 groups of samples corresponding to three changing criteria: strength of concrete, the amount of steel fiber, and the ratio of rebar. Accordingly, in the symbol of the sample group, UHPFRC is denoted as U and High—strength concrete (fc = 76 MPa), symbolized by H. Fiber content used is 1.0; 1.5; and 2.0%, steel reinforcement used is $\phi 8@100$.

The dimensions of the impact test plate are 500 mm long × 500 mm wide × 80 mm thickness, the test diagram is shown in Fig. 6. Before testing, the surface of the specimen is ground flat and painted with white color so that cracks can be observed. The impact load is generated by the pneumatic cylinder system. Under the action of air, the piston will move up and down vertically, causing the weight to fall and hit the test sample. The weight of a cylindrical mass is 33 kg, the surface in contact with the sample is spherical. The height of drop from the bottom of the

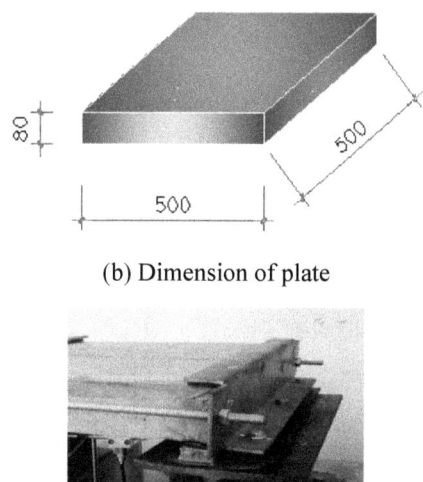

(b) Dimension of plate

(a) Experimental framework and sample   (c) *Support on the opposite side*

Fig. 6 Impact testing frame

mass to the surface plate is 460 mm. The plate is clamped on a steel I-beam on the opposite 2 sides and 2 sides are free as shown in Fig. 6c.

3.2 Experimental Results

a. **Deformation on the sample surface**

To assess the strain on the surface of the UHPFRC plate when subjected to impact loads, strain gauges are used and glued at the bottom of the plate, 40 mm from the center of the plate in the direction perpendicular to the support.

The test diagram for the non-reinforced plates is shown in Fig. 7a, sample N10-U with 1.0% fiber, the strain on the plate gives the maximum value. From the 4th threshing onwards, a sample strain value of 2.0% steel fiber gives the minimum strain value (0.99°/$_{oo}$). The impact load is an active loading in a short time and suddenly so the behavior of the plate is not clear in the first stage, but in almost the loading process the strain of the N20 group is smallest.

For the samples of the reinforced plate in Fig. 7b, the strain of the plate has more clearly, whereby the N20S-U group has the smallest strain (0.37°/$_{oo}$), followed by the N15S-U group with 0.824°/$_{oo}$ and N10S-U group with a maximum strain value is 1.38°/$_{oo}$.

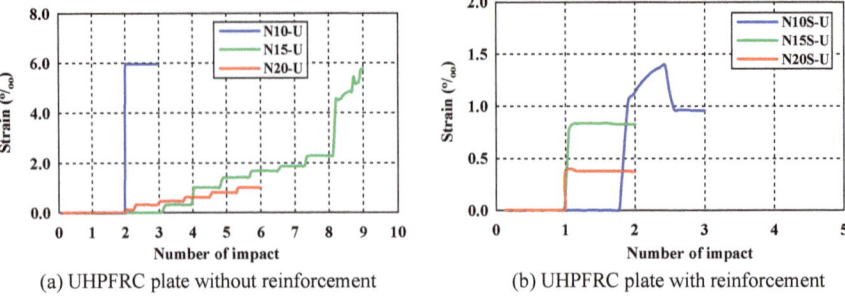

(a) UHPFRC plate without reinforcement   (b) UHPFRC plate with reinforcement

Fig. 7 The strain on the plate surface

Thus, through the 2 figures, it can be seen that when the content of steel fiber and reinforced is increased in the plate, the strain on the surface of the concrete plate is significantly reduced.

b. Crack development the plate

During the experiment process, the crack development under impact load is observed to evaluate the behavior of the plate, the measured values are listed in Table 5.

Thus, comparing sample groups with the same steel fiber content, it is found that if the concrete strength increases, the cracking of the plate will decrease. For example, group N10-H, 1.0% steel fiber, measured crack width is 1.25 mm, while the crack width of N10-U sample is 0.762 mm. Besides, if the fiber content is increased by 2% then the crack width of the plate is the smallest value, 0.190 mm, which is 75% reduction from the plates with a fiber content of 1.0%. Figure 8 shows the cracks

Table 5 Cracking width of sample groups

No	Symbol	Concrete Grade	The volume of steel fiber (%)	Reinforcement	Cracking width (mm)	
					The parallel direction of the support	The perpendicular direction of the support
1	N10-H	C60	1.0	–	1.250	0.305
2	N10-U	C120	1.0	–	0.762	0.152
3	N15-H	C60	1.5	–	0.786	0.125
4	N15-U	C120	1.5	–	0.406	0.102
5	N20-H	C60	2.0	–	0.203	0.051
6	N20-U	C120	2.0	–	0.190	0.076
7	N10S-U	C120	1.0	φ8@100	0.330	0.156
8	N15S-U	C120	1.5	φ8@100	0.279	0.134
9	N20S-U	C120	2.0	φ8@100	0.178	0.063

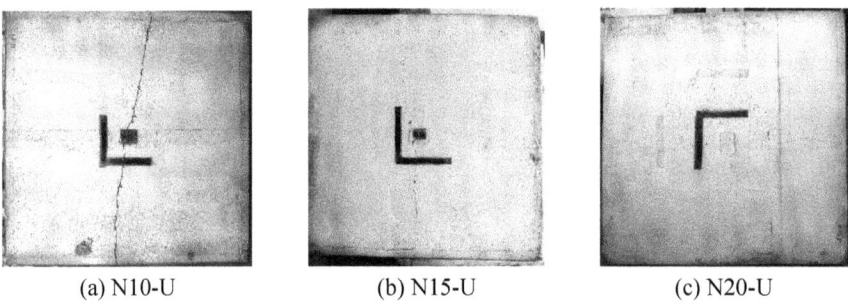

| (a) N10-U | (b) N15-U | (c) N20-U |

Fig. 8 The cracking of the sample surface varied in fiber content

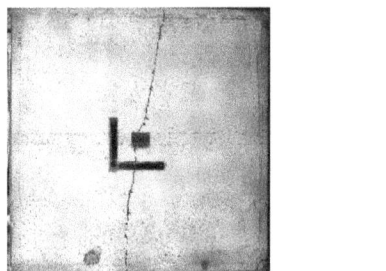 

(a) 1,0% steel fiber, without reinforcement   (b) 1,0% steel fiber, with reinforcement

Fig. 9 Crack widths of the group with and without reinforcement

on the plate surface with different fiber content. It can be explained by the bridging principle of steel fiber, the stress transfer from the concrete to the fiber contributes to reducing the width of a single crack and forming multiple cracks. This occurs until the fiber is pulled from the concrete [1]

Besides, adding reinforcement $\phi 8 @ 100$ into the plate, the crack width also tends to decrease in Fig. 9. For example, in group N10-U, 1.0% fiber, without reinforcement, the crack width is 0.762 mm, while in the group N10S-U, the crack decreases to 0.330 mm, corresponding to a 56% reduction. It can be seen that the reinforcement plays a role in spreading stress in the concrete, avoiding local damage to the sample, thereby reducing the width of cracks in the surface. After the end of the experiment, the plate remains in its original shape, undamaged.

4 Conclusion

The experimental program is set up to evaluate the behavior of UHPFRC plates under the impact loads. The sample groups have 1.0; 1.5, and 2.0% fiber content, with and without reinforcement $\phi 8 @ 100$ (2 layers). From the experimental results, some conclusions can be drawn as follows:

1. The steel fiber content does not affect the compressive strength much, the difference in values is not high (about 6–8%), but it significantly affects the flexural strength of the material.
2. As the steel fiber content in concrete increases, the energy absorption of the material increases. Samples groups with 2% steel fibers found that the greatest absorption of energy.
3. The strain on the surface plate is smaller when increasing the fiber content and the amount of reinforcement used. However, the behavior of the plate when judged by the strain gauge results is not clear in the sample groups.
4. Observation of growth and crack formation evaluates the behavior of the plate. The crack width of the plate decreases gradually as the fiber content increases. Also, adding reinforcement to the plate tends the distribution of stress on the plate to be even, so the crack width is smaller when compared to the plate without reinforcement.

With the results obtained from the experiments, it can be seen that UHPFRC can be used effectively for structures subjected to special loads such as impact loads or explosive loads thanks to the strength and toughness of the material.

References

1. Afroughsabet V, Biolzi L, Ozbakkaloglu T (2016) High-Performance Fiber-Reinforced Concrete: A review. J Mater SCi, Science Business New York 51(14):6517–6551
2. Hassan AMT, Jones SW, Mahmud GH (2012) Experimental test methods to determine the uniaxial tensile and compressive behavior of Ultra High Performance Fiber Reinforced Concrete (UHPFRC). Constr Build Mater 37:874–882
3. JCI-S-002 (2003) Method of test for load-displacement curve of fiber reinforced concrete by use of notched beam. in Japan Concrete Institute Standard
4. Kang ST, Lee Y, Park YD, Kim JK (2010) Tensile fracture properties of an Ultra High Performance Fiber Reinforced Concrete (UHPFRC) with steel fiber. Compos Struct 92:61–71
5. Kang ST, Lee BY, Kim JK, Kim YY (2011) The effect of fiber distribution characteristics on the flexural strength of steel fiber-reinforced ultra high strength concrete. Constr Build Mater 25:2450–2457
6. Lee JY, Yuan TF, Yoo DY, Yoon YS (2017) Benefits of using fiber on impact resistance of FRC slabs. MATEC Web of Conferences (EACEF 2017)
7. Orgass M, Klug Y (2004) Fibre reinforced ultra-high strength concretes. in international symposium on Ultra High Performance Concrete. Kassel, Germany
8. Rilem TC (2002) 162—TDF: Test and design methods for steel fiber reinforced concrete. Bending test. Final Recommendation. Mater Struct 35:579–582
9. Schmidt M, Fehling E (2004) Ultra-high-performance concrete: Research, development, and application in Europe. in International Symposium on UHPC, Kassel
10. Wille K, Naaman AE (2012) Ultra-high performance concrete and fiber reinforced concrete: Achieving strength and ductility without heat curing. Mater Struct 45:309–324
11. Wua Z, Shi C, He W, Wua L (2016) Effects of steel fiber content and shape on mechanical properties of Ultra High Performance Concrete. Constr Build Mater 103:8–14

12. Yoo DY, Lee JH, Yoon YS (2013) Effect of fiber content on mechanical and fracture properties of Ultra high Performance Fiber Reinforced cementitious composites. Compos Struct 106:742–753
13. Yoo DY, Banthia N (2016) Mechanical properties of Ultra-High-Performance Fiber-Reinforced Concrete: A review. Cement Concr Compos 73:267–280
14. Yu R, van Beers L, Spiesz P, Brouwers HJH (2016) Impact resistance of a sustainable Ultra-High Performance Fibre Reinforced Concrete (UHPFRC) under pendulum impact loadings. Constr Build Mater 107:203–215

Finite Element Simulation on Flexural Behavior of RC Slabs Using Coupled Damage-Plasticity Microplane Model

Anh Khac Le Vo, Thai Binh Nguyen, Thi Nguyen Cao, and Van Hai Luong

Abstract This paper focuses on applying and evaluating the available Coupled Damage-Plasticity Microplane (CDPM) model equipped with ANSYS software to represent the non-linear behavior of concrete material. The non-linear behavior of a reinforced concrete slab, subjected to a monotonically increasing load and designed according to the ACI 318-08 standard, is analyzed by the Finite Element Method. The three-dimensional finite element model of the three-point-bending test is established in ANSYS/Workbench. The three-dimensional eight-node hexahedral elements (CPT215) are used for simulating the non-linear behavior of the concrete member part, whereas the three-dimensional embedded elements (REINF264) are employed to model the reinforcement behavior of steel bars. The identification of parameters of the CDPM model for the reinforced concrete control slab is summarized and proposed based on an extensive literature survey. The results of the analysis of the load-deflection at the mid-span curve are presented and validated by the experimental results from the previous study to show the rationality and feasibility of the proposed finite element model.

Keywords Finite element simulation · Coupled damage-plasticity microplane model · Flexural behavior of RC slab · CPT215 element · REINF264 element

A. K. Le Vo · T. B. Nguyen (✉) · V. H. Luong
Faculty of Civil Engineering, Ho Chi Minh City University of Technology (HCMUT), 268 Ly Thuong Kiet Street, District 10, Ho Chi Minh City, Vietnam
e-mail: tbnguyen@hcmut.edu.vn

Vietnam National University Ho Chi Minh City (VNU-HCM), Linh Trung Ward, Thu Duc District, Ho Chi Minh City, Vietnam

T. N. Cao
Faculty of Engineering, Department of Civil Engineering, Tien Giang University (TGU), 119 Ap Bac Street, Ward 5, My Tho City, Tien Giang Province, Vietnam

© The Author(s), under exclusive license to Springer Nature Singapore Pte Ltd. 2023
J. N. Reddy et al. (eds.), *ICSCEA 2021*, Lecture Notes in Civil Engineering 268,
https://doi.org/10.1007/978-981-19-3303-5_75

1 Introduction

Finite element analysis of concrete structures such as beams, columns, and slabs is a challenging task because concrete is a strongly heterogeneous material that exhibits complex non-linear mechanical behavior [4]. Many constitutive models for the non-linear response of concrete have been proposed in many previous investigations [11, 12]. The stress-strain relationship is commonly used to describe the compression and tension behavior of concrete materials. The most common form of numerical modeling of material behavior is continuum modeling. These models formulate a high-order tensorial relationship between the input (i.e., strain) and the response (i.e., stress) of the material. Although these models work perfectly for linear-elastic materials, their non-linear capabilities are limited. In contrast to continuum models, the so-called Microplane model uses a different approach. This means that the constitutive material laws are applied to various oriented planes, forming a unit-sphere instead of a definite cube [7]. The newer 3-D eight-node CPT215 element, which was developed by this CDPM model in ANSYS software, will be a promising element to simulate the non-linear behavior of concrete structures. This element can eliminate mesh sensitivity and numerical instabilities [15]. In this paper, the three-dimensional finite element model of the three-point-bending test is established in ANSYS/Workbench software. The CPT215 elements are used for discretization of the concrete part to simulate the non-linear behavior of the concrete material, whereas the embedded REINF264 elements are employed to model the reinforcement behavior of steel bars. Based on an extensive survey, the identification of CDPM model parameters for the reinforced concrete control slabs is summarized and proposed. The load-displacement at the midspan curve is presented. The simulation results are validated by the experimental data carried by Cao et al. [2] to exhibit the rationality and feasibility of the proposed finite element (FE) model.

2 Three-Point-Bending Test

The reinforced concrete slab was designed under the ACI 318–08 standard [1] and cast in the laboratory by Cao et al. [2]. The slab had the design as described in Fig. 1. The slab had a dimension of $1000 \times 900 \times 150$ mm and was reinforced longitudinally in flexure with 10 mm diameter steel bars. The concrete cover had a depth of 20 mm. The slab was designed with concrete strength of 24 MPa. They were cured for 28 days before conducting the flexural test [2].

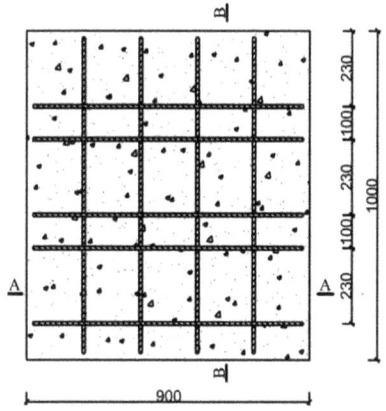
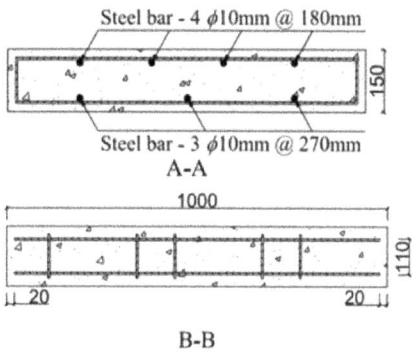

Fig. 1 RC slab design [2]

3 Material Modeling

3.1 Concrete Material

The concrete is modeled via the CDPM model, which is equipped with the ANSYS software. This CDPM model is based on research by Zreid and Kaliske [13–15], to overcome the numerical instability and pathological mesh sensitivity to which strain-softening materials such as the Microplane model are susceptible [9]. A detailed description of the model can be consulted at [7]. However, this model is only accessible through console commands and doesn't show up in ANSYS/Workbench. The CDPM model requires 15 parameters, shown in Table 1 [9]. Based on an extensive literature survey, the parameters of the CDPM model can be determined as follows:

(i) Elasticity parameters: The concrete elastic modulus E and the Poisson's ratio v can be identified from the elastic region of the material stress-strain curve. In the present work, Poisson's ratio is 0.2, and Young's modulus (E_c) for concrete is calculated as the ACI 318-08 code [1].

$$E_c = 4700\sqrt{f'_c} \quad (\text{MPa}) \tag{1}$$

where f'_c is the peak stress; the corresponding strain ε_c can be calculated as Eq. 2, proposed by Feenstra and Borst [3]:

$$\varepsilon_c = \frac{4f'_c}{3E_c} \tag{2}$$

(ii) Plasticity parameters: The parameters f_{uc}, f_{bc}, f_{ut} can be found experimentally. In cases of lacking testing data, the relations [5] can be used if f_{uc} is known [9]:

Table 1 The parameters of the CDPM model

Parameter Subtype	Parameter	Description	Units
Elasticity	E	The modulus of elasticity	MPa
	v	Poisson's ratio	–
Plasticity			
Drucker-Prager yield function	f_{uc}	Uniaxial compressive strength	MPa
	f_{bc}	Biaxial compressive strength	MPa
	f_{ut}	Uniaxial tensile strength	MPa
Compression cap	σ_V^C	The abscissa at the point of intersection between the compression cap and the Drucker-Prager yield function	MPa
	R	The ratio between the major and minor axes of the cap	–
Hardening	D	Hardening material constant	MPa
	R_T	Tension cap hardening constant	–
Damage	γ_{t0}, γ_{c0}	Tension and compression damage thresholds	–
	β_t, β_c	Tension and compression damage evolution constants	–
Nonlocal	c	Nonlocal interaction range parameter	mm^2
	m	Over-nonlocal averaging parameter	–

$$f_{bc} = 1.15 f_{uc} \quad \text{(MPa)} \tag{3}$$

$$f_{ut} = 1.4(f_{uc}/10)^{2/3} \quad \text{(MPa)} \tag{4}$$

In this study, the slab was cast by using ready-mixed concrete with a concrete grade of 24 MPa. The compressive strength obtained from the tests performed on standard cylinders (150×300 mm) was 24.5, carried out by Cao et al. [2]. Therefore, the uniaxial compressive strength f_{uc} for the slab is 24.5 MPa.

The intersection points between the compression cap and the Drucker-Prager function σ_V^C are more challenging to find. If this data is unavailable, it can be estimated as [9]:

$$\sigma_V^C = -\frac{2}{3} f_{bc} \quad \text{(MPa)} \tag{5}$$

The parameter R is the ratio between the major and minor axes of the cap. A detailed description is found at [15]. R is typically equal to two.

The hardening parameter can be calculated as Eq. 6, proposed by Nguyen and Houlsby [6].

$$f_{c0} = \frac{f'_c(E_c + D)}{E_c} - D\varepsilon_c \ (\text{MPa}) \tag{6}$$

where f_{c0} is found to be about 30% of the ultimate compressive stress f'_c. D is the hardening parameter. E_c and ε_c are young's modulus and strain at peak stress are calculated as Eqs. (1) & (2).

(iv) The damage parameters: the parameter R_T, β_t and γ_{t0} are identified from a uniaxial cyclic tensile test. In the absence of the testing data, $R_T = 1$, $\beta_t = 1.5\beta_c$ can be used as starting values. The tension damage threshold γ_{t0} is often set to zero, as softening in tension starts almost immediately after the elastic limit [9]. The values of the damage variables are determined based on the proposition of Zreid and Kaliske [15] and are determined to fit the force-displacement curves experimentally.

(v) The nonlocal parameters: The over-nonlocal averaging parameter m is a numerical parameter, where > 1 regularizes the solution. m is typically equal to 2.5 [9]. The range of nonlocal parameter c is interpreted as the resultant size of the damage zone. Typically, the size should have a reasonable ratio to the size of the entire structure.

The material parameters for concrete material in the simulation are shown in Table 2. The APDL commands for concrete material in ANSYS/Workbench are also presented in Table 3.

Table 2 The CDPM model's input parameters for concrete material

Parameters	Value
E (MPa)	23,200
v	0.2
f_{uc} (MPa)	24.5
f_{bc} (MPa)	28.2
f_{ut} (MPa)	2.5
σ_V^C (MPa)	−18.8
D (MPa)	40,000
R	2
R_T	1
γ_{c0}	2×10^{-5}
γ_{t0}	0
β_c	8,000
β_t	12,000
c (mm^2)	300
m	2.5

Table 3 APDL commands for concrete material in ANSYS software

SC1-24 MPa	Description
MP,EX,SC1,23200	! Define Elasticity Modulus E.
MP,NUXY,SC1,0.2	! Define Poisson's ratio.
TB,MPLA,SC1,,,DPC	! Define Drucker-Prager.
TBDATA,1,24.5,28.2,2.5,1,40000,-18.8	! Define $f_{uc}, f_{bc}, f_{ut}, R_T, D, \sigma_V^C$.
TBDATA,7,2,0,2e-5,12000,8000 TB,MPLA,SC1,,,NLOCAL	! Define $R, \gamma_{t0}, \gamma_{c0}, \beta_t, \beta_c$.
TBDATA,1,300,2.5	! Define nonlocal parameters
	! Define c, m.

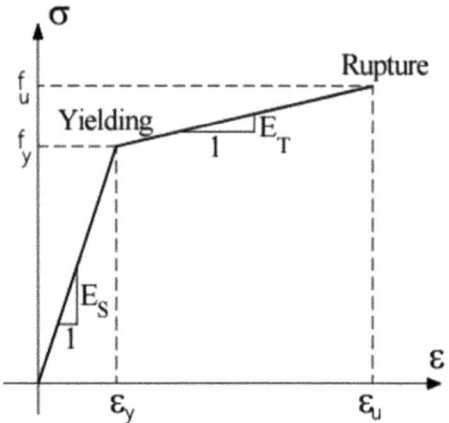

Fig. 2 Bilinear stress-strain relationships for steel reinforcement [8]

3.2 Steel Reinforcement Material

The experimental program [2] used steel reinforcement of grade SD40 with a diameter of 10 mm. The yield stress and tensile strength were 390 and 560 MPa, respectively. In this study, the steel bars in the finite element model are modeled with a bilinear stress-strain relationship as shown in Fig. 2. In the experimental program, the yielding point is defined as 390 MPa, the Young's modulus (E_S) is equal to 200 GPa, and the tangent modulus for the plastic part is taken as $E_T = 0.005E_S = 1000$ MPa [8]. Also, Poisson's ratio is set to 0.3 for all reinforcing bars.

4 Finite Element Modeling

The three-dimensional finite element model of the three-point-bending test of an RC slab is established in the ANSYS/Workbench software. The CPT215 elements

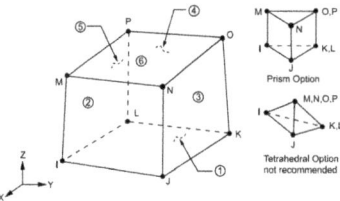

Fig. 3 The Structural Solid Geometry of CPT215 [10]

are used for the concrete part to simulate the non-linear behavior of concrete material by the CDPM model, whereas the REINF264 elements are employed to model the reinforcement behavior of the steel bars. The following sections present in detail the CPT215 element, REINF264, geometry, finite element meshing, boundary conditions, and applied loadings of the simulated RC slab.

4.1 CPT215 Element Description

The CDPM model is only available with coupled pore-temperature elements for the three-dimensional elements CPT214, CPT215, and CPT216 as well as for the two-dimensional elements CPT211, CPT212, and CPT213 [7]. In this study, the CPT215 element is used in the simulation modeling of the RC slab instead of the three-dimensional SOLID185 element by default in the program ANSYS.

The CPT215 element is a three-dimensional eight-node coupled pore-pressure-thermal mechanical solid element (shown in Fig. 3). The CPT215 element has translations in the nodal X, Y, and Z directions in addition to two pore-pressure and temperature degrees of freedom per node. Also, it has the capability of material elasticity, stress stiffening, large deflection, and large strain [10].

4.2 REINF264 Element Description

For structural reinforcing analysis, use REINF264 with a standard 3-D link, beam, shell, and solid element (referred to here as *the base element*) to provide extra reinforcing to those elements. The element is suitable for simulating reinforcing fibers with arbitrary orientations. Each fiber is modeled separately as a spar that has only uniaxial stiffness or conductivity. The REINF264 element can be specified with multiple reinforcing fibers. The nodal locations, degrees of freedom, and connectivity of the REINF264 element are identical to those of the base element. The REINF264 has plasticity, stress stiffening, creep, large deflection, and large strain capabilities [10]. The REINF264 element, which is used together with the 8-node CPT215 solid element, is shown in Fig. 4.

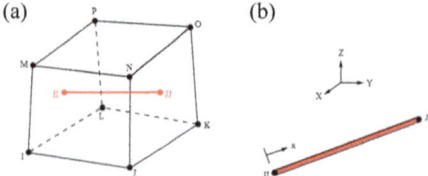

Fig. 4 (a) REINF264 Geometry applied to a 3-D 8-Node Solid element; (b) REINF264 Coordinate System [10]

Table 4 Details of the meshes adopted in the analysis

FE mesh size (Cube element size)	Mesh 1 (40 mm)	Mesh 2 (35 mm)	Mesh 3 (30 mm)	Mesh 4 (25 mm)	Mesh 5 (20 mm)
Concrete elements	1,248	1,950	2,700	4,320	9,200
Reinforcement elements	221	262	298	361	442
Concrete nodes	1,755	2,604	3,552	5,453	11,016
Reinforcement nodes	451	533	605	731	893
Total elements	1,469	2,212	2,998	4,681	9,642
Total nodes	2,206	3,137	4,157	6,184	11,909
Elapsed time (mins)	3	3	6	21	177

4.3   Geometry and Meshing

In order to decrease the elements/nodes in the model to save computer resources, the 3-D finite element model of a half-RC slab (used with the symmetry face) is simulated as shown in Fig. 5. The concrete material part is discretized by several cube elements. To investigate the numerical stabilities and the mesh sensitivity associated with the CPT215 element, the finite element analysis of the flexural half-RC slab is implemented by five different meshes. The details of those meshes which are adopted to discretize the half- RC slab are described in Table 4.

4.4   Boundary Conditions and Applied Loadings

The boundary conditions of the model are applied according to the particularities of the test setup. Two supports were used in the location as shown in Fig. 6, and the equidistant from the edge of the slab was 100 mm. In the present study, an implicit finite element non-linear quasi-static analysis based on displacement control was applied by applying a downward vertical direction (Y-direction) displacement (maximum of 8 mm as per the experiment study) on the middle line of the half-RC slab.

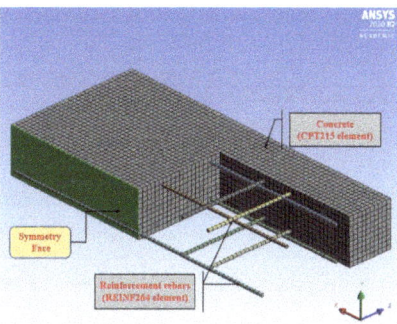

Fig. 5 Finite element model for a half-RC slab

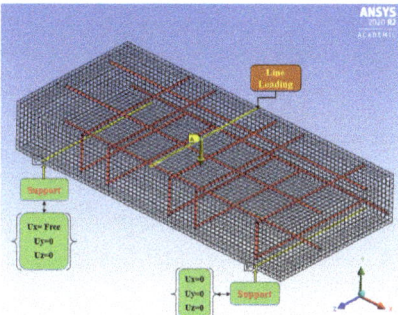

Fig. 6 Boundary conditions and location of applied displacement loading for the FE model

5 Numerical Results

The profile of load-deflection at the mid-span relationships of the five meshes is reported in Fig. 7, along with the experimental data implemented by Cao et al. [2]. It is seen that the shape of the load-deflection at mid-span curves from the computed numerical results are almost indistinguishable among the five meshes. Therefore, it can be observed that the CDPM model equipped with the CPT215 element can eliminate the mesh sensitivity and numerical instabilities.

It is also seen that the current proposed FE model (Mesh 5) yields a load-deflection curve that agrees very well with the experimental results generated by Cao et al. [2]. It reaches the initial cracking load of around 60 kN, while this value in the experiment was 80 kN. The peak load of all five meshes is shown in Table 5 together with the experimental data [2]. The difference in the ultimate loading between the proposed FE model (i.e., 102.27 kN – mesh 5) and the experiment (i.e., 108.52 kN) is about 5.7%. It can be seen that the peak load from simulation is lower than the peak load from the experiment study, implying that the numerical modeling yields safe results.

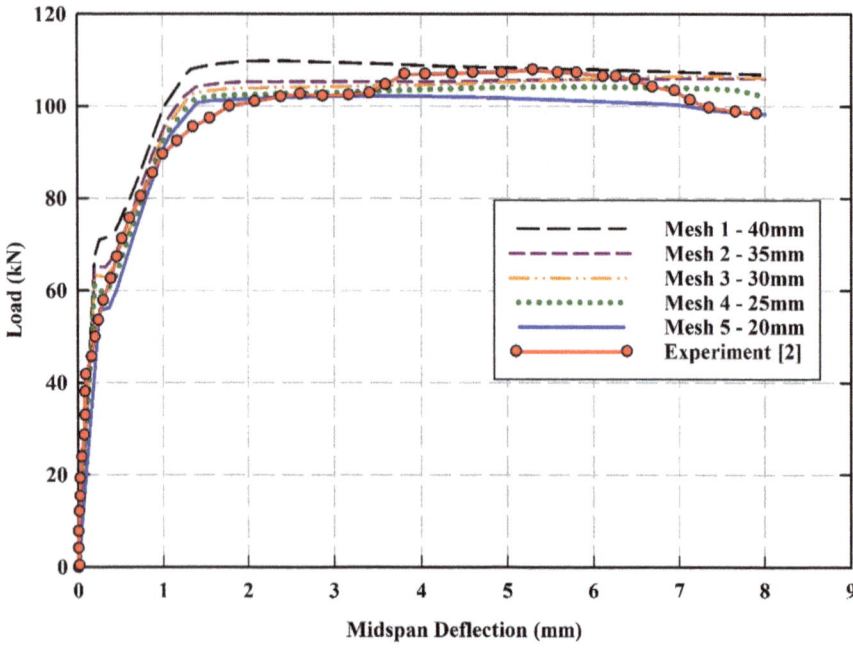

Fig. 7 The comparison of the load and mid-span deflection relationship between the numerical results and the experimental data of the RC slab

Table 5 The peak load comparison between the numerical results and the experimental data of the RC slab

	Experiment [2]	Mesh 1 (40 mm)	Mesh 2 (35 mm)	Mesh 3 (30 mm)	Mesh 4 (25 mm)	Mesh 5 (20 mm)
Peak load (kN)	108.52	109.86	108.11	106.6	104.3	102.3

Figure 8 describes the pattern of concrete failure. A typical flexural failure with small cracks at the mid-span of the RC slab is observed in the experimental test [2]. Although the FE model could not describe the exact crack line, it indicated the flexural failure pattern or the cracking area similar to the observation in the flexural test.

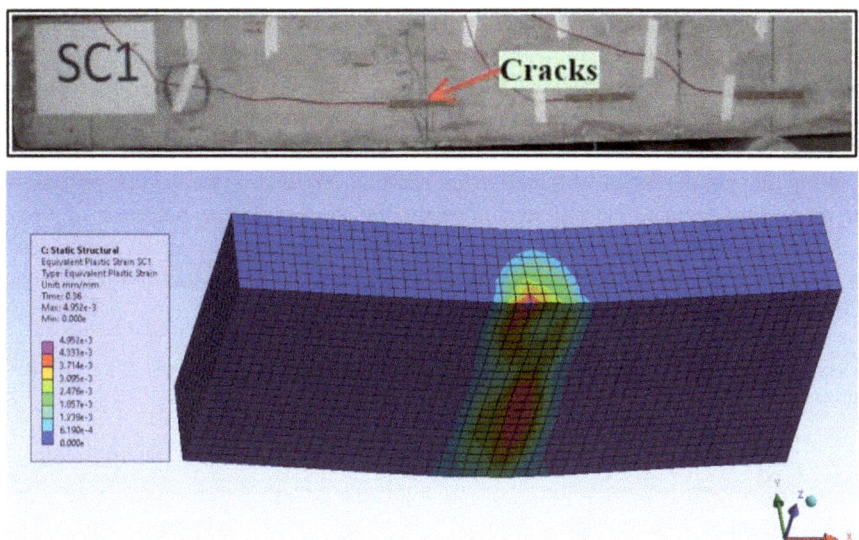

Fig. 8 The comparison of the pattern of concrete failure between the numerical results and the experimental data of the RC slab

6 Conclusions

An efficient finite element model capable of simulating the flexural behavior of an RC slab has been established successfully using ANSYS/Workbench software. In the proposed FE model, the CPT215 elements are used to simulate the non-linear behavior of concrete following the CDPM model, whereas the embedded REINF264 elements are employed to model the reinforcement behavior of the steel bars. The comparison between numerical and experimental results shows quite satisfactory results. The results from the proposed FE model tend to be safer compared to those of the experimental study.

From a convergence study, it has been found that the proposed FE model yields stable numerical solutions and hardly noticeable differences among the results from all five meshes. The success of this proposed FE model must be attributed to the great contribution of the CDPM model, which is equipped with the newer CPT215 element to describe the non-linear behavior of concrete material. It should be noted that the authors have tried to analyze this problem by using the combination of SOLID65 and REINF264 elements. However, it has not been successful. The proposed CDPM model parameters for the considered RC slab are rational with respect to the experimental study of the considered RC slab.

Besides, the embedded REINF264 elements are utilized to simulate the reinforcement behavior of rebar, allowing the user to discrete the model more easily. Normally, the external nodes (corner nodes) of the hex elements (used to model the concrete) must superimpose the nodes of the beam elements (used to model the steel reinforcement, e.g., the LINK188 element). This guarantees that the deformations of steel and concrete are compatible. However, the nodal locations, degrees of freedom, and connectivity of the embedded REINF264 element are identical to those of the base element. Therefore, those compatible conditions do not need to be carried out. As a result, the finite element meshing task is uncomplicated.

With the good results obtained from simulation in this study, the proposed FE model promises to be efficient in analyzing the non-linear behavior of flexural RC slabs in more complicated scenarios. For example, the RC slabs are repaired/reinforced with FRP materials by using externally bonded reinforcement (EBR) or near-surface mounted (NSM) techniques.

Acknowledgements We acknowledge the support of time and facilities from Ho Chi Minh University of Technology (HCMUT), VNU-HCM for this study.

References

1. ACI (2008) ACI 318–08: Building code requirements for structural concrete and commentary. America: American Concrete Institute, p 345
2. Cao NT, Pansuk W, Torres L (2015) Flexural behavior of fire-damaged RC slabs repaired with near-Surface mounted (NSM) carbon fiber reinforced polymer (CFRP) rods. J Adv Concr Technol 13(1):15–29
3. Feenstra PH, Borst RD (1996) A composite plasticity model for concrete. Int J Solids Struct 33(5):707–730
4. Grassl P, Xenos D, Nystrom U, Rempling R, Gylltoft K (2013) CDPM2: A damage-plasticity approach to modeling the failure of concrete. Int J Solids Struct 50(24):3805–3816
5. Jiang H, Zhao J (2015) Calibration of the continuous surface cap model for concrete. Finite Elem Anal Des 97:1–19
6. Nguyen GD, Houlsby GT (2008) A coupled damage-plasticity model for concrete based on thermodynamic principles: Part I: model formulation and parameter identification. Int J Numer Anal Met 32:353–389
7. Pirker M (2020) Non-linear seismic analysis of a concrete gravity dam using a microplane material model. Graz University of Technology, The final project for Master's degree
8. Pham DD, Nguyen PC, Nguyen DL, Le HA (2020) Simulation of concrete-filled steel box column. In Reddy J, Wang C, Luong V, Le A (eds) ICSCEA 2019. Lecture Notes in Civil Engineering, vol 80. ICSCEA 2019. Springer, Singapore
9. Release 2020 (2020) ANSYS mechanical APDL material reference. ANSYS Inc, pp 116–125
10. Release 2020 (2020) ANSYS Mechanical APDL Theory Reference. ANSYS Inc
11. William KJ, Warnke EP (1975) Constitutive model for the triaxial behavior of concrete. In: Proceedings of the International Assoc. for Bridge and Structural Engineering 19:1–30
12. Wang T, Hsu T (2001) Nonlinear finite element analysis of concrete structures using new constitutive models. Comput Struct 79:2781–2791
13. Zreid I, Kaliske M (2014) Regularization of microplane damage models using an implicit gradient enhancement. Int J Solids Struct 51:3480–3489
14. Zreid I, Kaliske M (2016) Implicit gradient formulation for microplane drucker-prager plasticity. Int J Plasticity 83:252–272
15. Zreid I, Kaliske M (2018) Gradient enhanced plasticity-damage microplane model for concrete. Comput Mech 62:1239–1257

Golden Ratio Application in the Optimization of Cold-Formed Steel Sections

Tran-Trung Nguyen, Phu-Cuong Nguyen, and Hoang Thao Phuong Nguyen

Abstract The golden ratio is a ratio used popularly and is a standard to evaluate the beauty in architecture. This ratio appears in the fundamental rule and satisfies sustainable conditions to form and structural geometries in nature. This study will approve a valuable characteristic of the golden ratio when applied for optimizing the cold-formed steel sections' geometries. The optimized problem is established by combining the parametric design software, such as Rhino grasshopper and 3D Karamba, with an applied Octopus. The post-optimized geometric sections archive symmetric forms and high strength when a typical C section combines each other based on the golden ratio. Besides, the proposed sections are evaluated with the experimental results using a finite element analysis to increase the proposed approach's reliability. This study performs reliable and robust parametric design tools, especially with every analysis step in the optimized process. We will create collected data to build a predicted model about geometries and the strength curves of structures using Machine Learning Tools in later research.

Keywords Cold-formed steel · Golden ratio · Finite element method · Optimization · Rhino grasshopper

1 Introduction

One of the keywords that are now most searched for steel materials in the construction industry is Cold-formed Steel (CFS). It is the perfect material for prefabricated components for construction and assembly in construction, is one of the construction solutions that reduce the investment costs as progress is significantly increased. On the other hand, CFS has been entirely integrated with the design software, with

T.-T. Nguyen · H. T. P. Nguyen
Faculty of Architecture, Van Lang University, Ho Chi Minh City, Vietnam

P.-C. Nguyen (✉)
Advanced Structural Engineering Laboratory, Department of Structural Engineering, Faculty of Civil Engineering, Ho Chi Minh City Open University, Ho Chi Minh City, Vietnam
e-mail: cuong.pn@ou.edu.vn; henycuong@gmail.com

design parameters related to the cross-sectional section, and this is one of the significant benefits it brings in the design and construction process for its quickness and efficiency.

The typical cross-section of CFS is commonly found when integrated into C and Z software. The sections on efficiency when joining the load have been verified, and the design instructions are very detailed by Eurocode3 [1] and Eurocode5 [2] standards. With the mentioned cross-sections, which have yet to show the capabilities that CFS materials bring fully, the problem of optimizing cross-sectional sections of CFS C and Z has been promoted today. Schafer and Peköz [3] studied the residual stresses related to characterizing geometric imperfections and were necessary to complete advanced analysis and parametric studies of CFS members successfully. Considering the effect of the CFS geometry imperfections subjected to loads, Haidarali's study [4] has solved this problem through the FEM model of CFS beam under local and distortional buckling. Besides, Jun Ye's series of studies [5, 6] revolve around developing and improving the CFS channel section efficiency by the Particle Swarm Optimization method.

On the other hand, studies to examine the unstable strength of sections C and Z used CFS materials in cross-bracing for stiffness enhancement [7] or studies that incorporate analysis and experiments of CFS frames subjected to seismic effects [8]. About two years ago, the CFS section optimization problem developed strongly with new and more efficient algorithms such as Big Bang- Big Crunch [9], helping CFS beams achieve serviceability and ultimate limit states. To contribute to this new trend, we will also perform the cross-section optimization problem of CFS beams by setting the parametric cross-section that considers a term used in the architecture's beauty; it is the golden ratio. This optimization problem is solved by combining Rhino Grasshopper and Karamba3D with Multi-Objective Evolutionary Optimization, Octopus. CFS beams with optimal cross-sections will be verified and analyzed by using FEM software, ABAQUS. The input data for the problem from optimization to analysis and the results are finally referenced from Mojtabaei's research [9].

2 Golden Ratio Parameters Design in the Optimization Problem

2.1 General about Golden Ratio

It can be said: The Golden Ratio is merely a mathematical ratio that can be found in nature, in famous paintings, and of course, in design, presented in Fig. 1a. Strictly speaking, the golden ratio is the standard ratio between the elements of a design, carefully measured to fit the eyes of those who enjoy the most creative works. People often use the symbol phi (φ) in the Greek alphabet as a golden ratio convention. An example of a golden ratio is the following: The row is divided into two parts: The longer one is called a, the shorter one is called b. The golden ratio here is the same

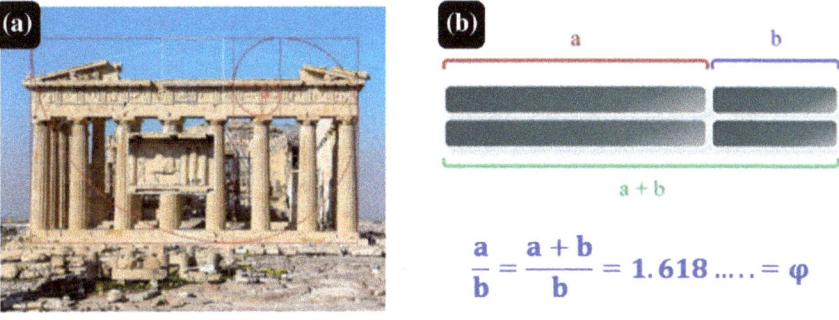

Fig. 1 a Golden proportions are apparent in the remaining structure of the nearly 2500-year-old Parthenon; **b** How to calculate the golden ratio in two straight lines a and b [10].

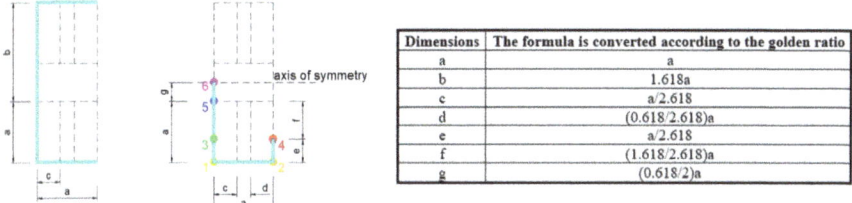

Fig. 2 Establish the typical C section of CFS beams with the golden ratio's control points

as The sum of the two lines (a) and (b) divided by line (a) is equal to the line ratio (a) divided by the line ratio (b). If that ratio is 1.618, then bingo already has the gold standard ratio, as shown in Fig. 1b.

2.2 *Establish Golden Ratio Parameters in the Cross-section*

Refer to Olsen's research ideas on [11] using the golden ratio as a scientific function, along with the straight line scaling as shown in Fig. 1b. The choice section of CFS is a typical C section with dimensions converted to the control points as 1, 2, 3, 4, 5, 6 points assigned to the Golden ratio, shown in Fig. 2.

2.3 *The Process of Optimization*

The controls have been built and shown in Fig. 2, using the Grasshopper tool in Rhino to construct the geometry for the characteristic section C of the CFS beam. The beam structure in this study is referenced from previous research by Mojtabaei [9]. It is produced by cold-rolling or press braking process from coil width of steel plate L

= 453 (mm) and steel thickness t = 1.8 (mm). These parameters are unchanged by default.

The objective functions used in state 3 described in Fig. 3, such as deflection and moment functions, are calculated according to the beam diagram under evenly distributed load with the boundary conditions at the start and end positions are fixed. So deflection and moment values are defined following the proposal equations in the study [9].

$$f(x) = \frac{M_{Ed,ser} L^2}{8EI_{fic}} \tag{1}$$

where $f(x)$ is the deflection of the CFS beam; L is the length of the beam and is kept constant with a value equal to 5000 (mm); $M_{Ed,ser} = 0.7M_{max}$ to ensure that each calculated cross-section reaches the maximum compressive stress in serviceability design based on EC3 [1]; I_{fic} is the effective second moment of area based on the service load, introduced to determine in [9]; E is modulus of CFS material and it's value equal to 210 (GPa).

The M_{max} value is determined base on the highest absolute value of stress (σ) within the beam length. This value is calculated following the equation mentioned in [9]:

$$\sigma = \frac{M_{max}}{W_{eff}} = f_y \tag{2}$$

where W_{eff} is the section modulus for the Ultimate Limit State (ULS). Besides, the stress within the beam length reaches the highest value corresponding to local buckling, and σ equal to the yield stress, f_y.

The results obtained from the above optimization process, described in Fig. 4.

3   Verified by FEM

3.1   Numerical Modeling

To verify the reliability of the proposal cross-section of the CFS beam, a modeling FEM is established by using the ABAQUS software and described in Fig. 5.

3.2   Material Properties and Constitutive Models

The characteristic parameters relate to the CFS material referred to [9], such as elastic modulus (E), yield stress (f_y), and poisson's ratio (μ) equal to 210 (GPa), 450 (MPa),

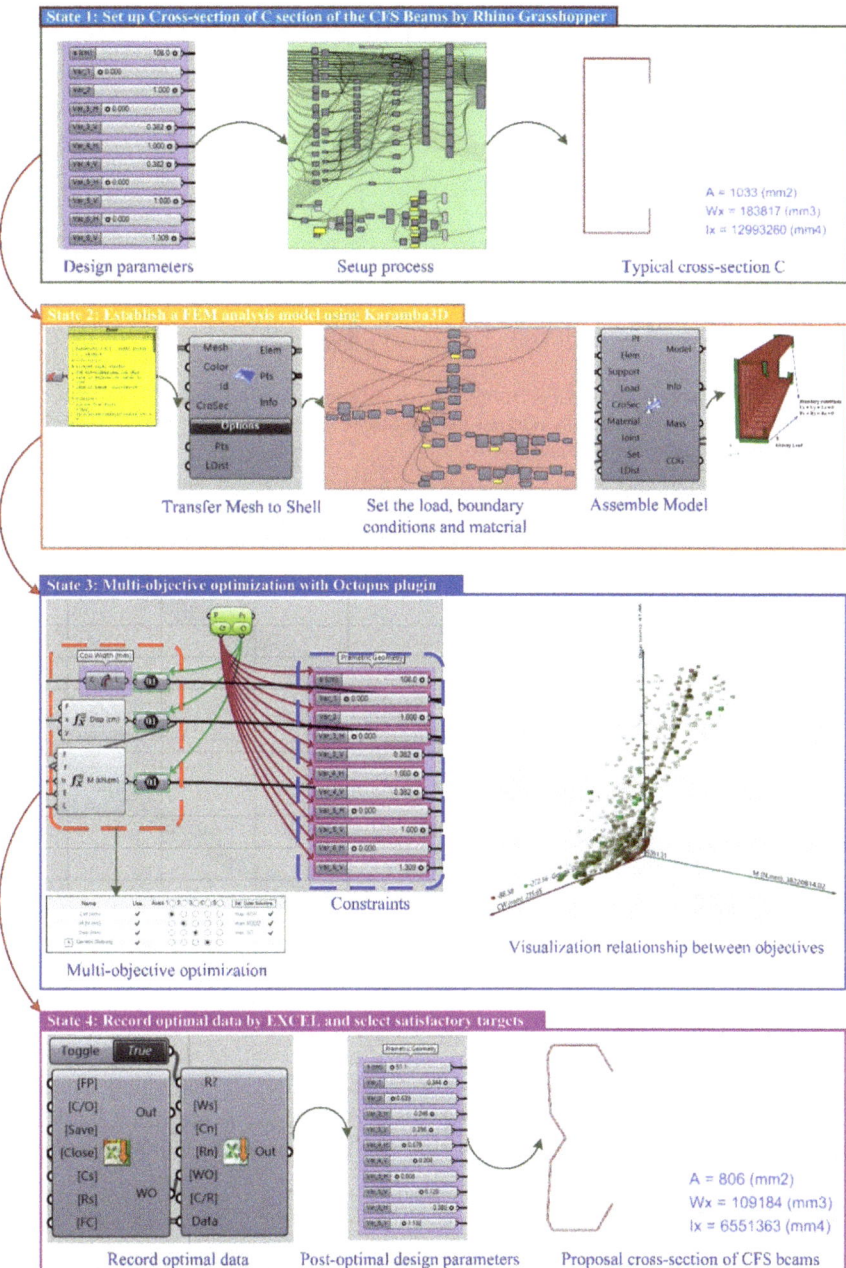

Fig. 3 Optimization process when combining rhino grasshopper and karamba3D for the cross-sectional problem of CFS beams

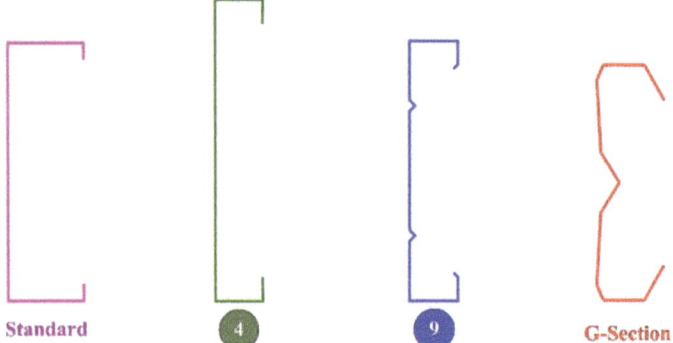

Fig. 4 The cross-section of the CFS Beams use for comparison in this study. The standard, #4, and #9 prototypes have dimensions referred to Mojtabaei's study [9]. The G-section is a proposal cross-section assigned to the Golden ratio

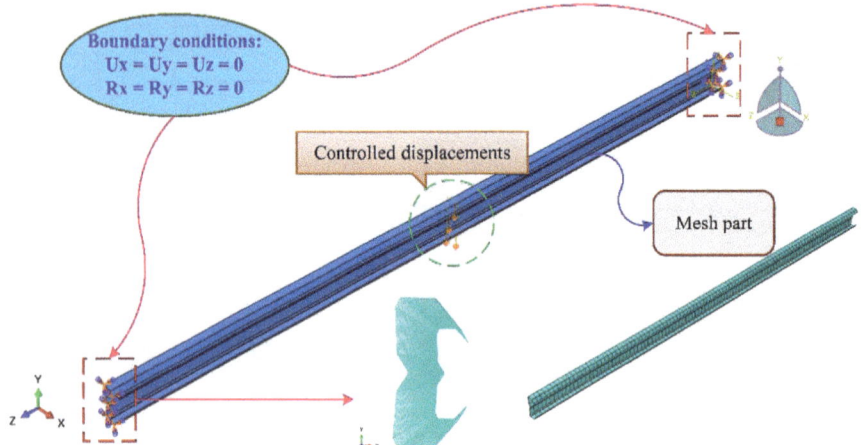

Fig. 5 The FE models subject to pure bending moment with boundary conditions

and 0.3, respectively. The most simple consideration for strain hardening, as shown in Fig. 6, is the elastic, linear hardening model, where E_{sh} is the strain hardening modulus. This model takes strain hardening into account, as provided in Eurocode 3 Annex C [1].

4 Results and Discussions

After the model setting process in ABAQUS with the material model mentioned in subsection 3.2, the selected results are presented as the relationship between the

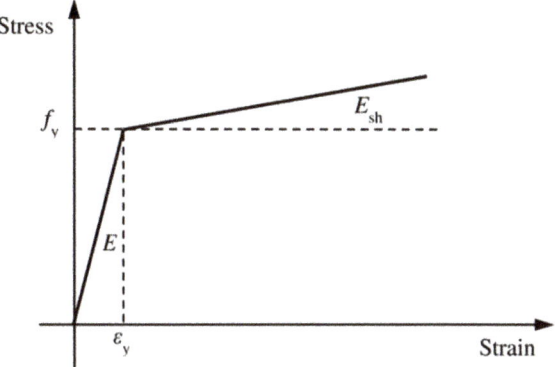

Fig. 6 Elastic linear-hardening model of CFS material [12]

moment and deflection between the span of CFS beams because it is easy to compare with previous studies related to the optimal cross-sectional problem in Mojtabaei' study [9].

Based on the description in Fig. 7, the proposed cross-section considering the golden ratio, G-Section, gives the curvature of the relationship between the moment and deflection at the curve's peak position close to the top curve the optimum proto-type 4 [9] achieved. On the other hand, when considering the developed deflection to about 18 (mm), the G-section's curve is not good because the moment value is too low. This deviation may arise from the material model used in the validation simulation is not suitable, but it is necessary to propose a material curve model compatible with CFS materials. If observing more, when the deflection between the

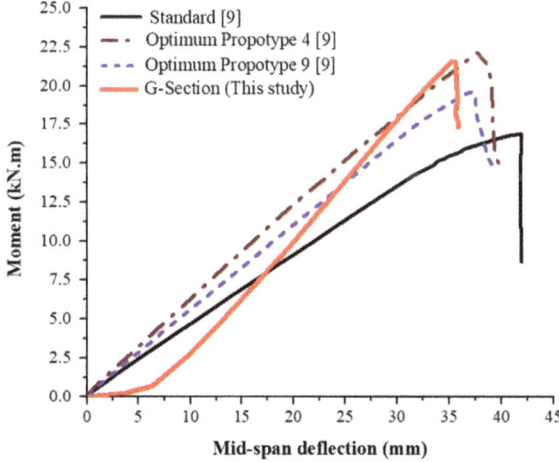

Fig. 7 Moment versus and mid-span deflection curves of the CFS beam's cross-sections

beam reaches over 18 (mm), the G-section's torque value increases significantly, and the maximum deflection of the CFS beam with the section G is the smallest. Points when considering adding the golden ratio in a cross-sectional section of CFS beam structure.

5 Conclusions

The Rhino Grasshopper is a robust tool for establishing complexity forms design in Architecture. It is mainly combined with a parametric design tool, Karamba3D, to create an optimal analysis process effectively.

Assignment of the Golden ratio in cross-section of structural beams carries many exciting things related to structural forms. Especially in construction nowadays, sustainable trends are always needed requirements.

The multi-objects optimization problems used optimal plugins that interactive in the Rhino Grasshopper, Octopus, become more accessible than programming languages for the optimal problems.

Acknowledgements The authors gratefully acknowledge the financial support provided by the Scientific Research Fund of Ho Chi Minh City Open University, Vietnam. (No. E2020.08.1).

References

1. Eurocode CEN (2016) Eurocode 3: Design of steel structures-Part 1–1: General rules and rules for buildings. European Committee for Standardization, Brussels
2. Eurocode CEN (2016) Eurocode 5: Design of steel structures–Part 1–5: plated structural elements. CEN. European Committee for Standardization, Brussels
3. Schafer B, Peköz T (1998) Computational modeling of cold-formed steel: Characterizing geometric imperfections and residual stresses. J Constr Steel Res 47:193–210
4. Haidarali MR, Nethercot DA (2011) Finite element modelling of cold-formed steel beams under local buckling or combined local/distortional buckling. Thin-walled structures 49:1554–1562
5. Ye J, Hajirasouliha I, Becque J, Eslami A (2016) Optimum design of cold-formed steel beams using particle swarm optimisation method. J Constr Steel Res 122:80–93
6. Ye J, Hajirasouliha I, Becque J, Pilakoutas K (2016) Development of more efficient cold-formed steel channel sections in bending. Thin-walled structures 101:1–13
7. Yerudkar DS, Vesmawala GR (2018) Strength and buckling of cold-formed steel laterally unbraced stiffened C and Z sections in Proceedings of the Institution of Civil Engineers-Structures Buildings 171(3):216–225
8. Mojtabaei SM, Kabir MZ, Hajirasouliha I, Kargar M (2018) Analytical and experimental study on the seismic performance of cold-formed steel frames. J Constr Steel Res 143:18–31
9. Mojtabaei SM, Ye J, Hajirasouliha I (2019) Development of optimum cold-formed steel beams for serviceability and ultimate limit states using Big Bang-Big Crunch optimisation. Eng Struct 195:172–181

10. Meisner GB (2018) The Golden ratio: The divine beauty of mathematics. Race Point Publishing.
11. Olsen SA (2017) Golden ratio beauty as scientific function (Lebenswelt. Aesthetics philosophy of experience, no. 11).
12. Yun X, Gardner L (2017) Stress-strain curves for hot-rolled steels. J Constr Steel Res 133:36–46

Integration of Sensors for Improved Damage Identification

Abdurehman M. Bali, Hiroshi Katsuchi, Hitoshi Yamada, and Hiroshi Tamura

Abstract Bridges are valuable assets to a society to overcome barriers of connection. The dependence on the bridge increases the demand for safety and reliability of the structure. However, the operational lifetime of bridges will be affected by the integration of bridge elements and their resistance to deterioration and damage. In this paper, the identification of damage in bridges using the influence line is presented. This approach is based identification and localizing of damage from a Modified Displacement-Based Index (MDBI) value. The effect of transverse position of the vehicle on the measurement of displacement sensor while crossing the bridge is addressed. Additionally, the road roughness effect has been considered to create a noise environment on the measured displacement response. The integration of strain sensor with displacement sensor to improve the damage identification have been addressed. The effect of noise has been greatly reduced by introducing strain based normalizing factor. Two case scenarios have been presented and the comparison of the previous DBI method and the MDBI method have been presented. In both cases, the MDBI successfully located the damage location against the odds in dynamic response effect and the noise.

Keywords Modified DBI · Displacement-based index · Damage identification · Sensor integration

1 Introduction

Bridges are valuable assets to a society to overcome barriers of connection. The social, economic, and political development of a country is majorly dependent on the transportation access, from which bridges play the most important role on land transportation. The operational lifetime of bridges will be affected by the integration of bridge elements and their resistance to deterioration and damage. Therefore, the

A. M. Bali (✉) · H. Katsuchi · H. Yamada · H. Tamura
Yokohama National University, 79-1 Tokiwadai, Hodogaya, Yokohama, Japan
e-mail: bali-muleta-dw@ynu.ac.jp

safety and reliability of the structure shall be maintained to achieve a prolonged service life.

Bridge condition assessment have been one of the key elements in the bridge management system. It uses several approaches to determine the current state of the bridge. Mostly, visual inspection is used due to its simplicity. However, the limitation with this approach creates a search for more advanced methods.

Structural health monitoring is one among a handful of bridge condition assessment techniques. It has gained attention due to its ability to quantitatively evaluate damage to the structures. Nevertheless, its application towards the practicability in bridge management system, where several bridges will be administered, is not exploited.

2 Damage Detection Approach

2.1 Displacement-Based Damage Detection

Several approaches have been proposed to detect damage within a structure. The focus of this paper is towards improving damage detection by using the bridges displacement response at the time of start of operation, i.e., initial status of the bridge, and the current bridge displacement response, i.e., damaged status of the bridge. Ono R. et al. [3] proposed the displacement-based index (DBI) method for damage detection.

The proposed method [3], states that the displacement influence line observed at point C while the load moves from support A to support B is equal to the displacement shape of the beam when a unit load is applied at point C. Hence, applying this concept for a moving load, the following equation can be derived. Readers are advised to refer to the work by Ono R. et al. [3] for full derivation of the method.

$$[x, u_i] = \left[(x_1, u_{i1}), \ldots, (x_j, u_{ij}), \ldots, (x_n, u_{in})\right] \tag{1}$$

$$[x, u_d] = \left[(x_1, u_{d1}), \ldots, (x_j, u_{dj}), \ldots, (x_n, u_{dn})\right] \tag{2}$$

where x is the loading position; j is the loading position index; u_i and u_d are the displacement influence lines at initial state and damaged state, respectively. The DBI at point C can be computed by standardizing the absolute difference Δu between the sound and deteriorated displacements.

$$\Delta u(j) = |u_d(j) - u_i(j)| \tag{3}$$

$$DBI(j) = max\left[0, \frac{\Delta u(j) - \mu}{\sigma}\right] \tag{4}$$

where μ and σ represent the mean and standard deviation of Δu, respectively. DBI indicates the structures state, DBI $= 0$ indicating undamaged state.

2.2 Modified Displacement-Based Damage Detection

The DBI method proposed by Ono R. et al. [3] identifies damage from comparison of the displacement influence line at the time of initial state and damaged state of the bridge. However, the effect of road roughness is not considered. When a vehicle crosses a bridge with a significant speed (more than 10 m/s) the dynamic effect will be higher [1], which will compromise the effect of damage in the bridge. It is indicated that the DBI estimation is affected by sever degradation of the bridge. Furthermore, in case of narrow damaged area the damage indicator locates multiple damage locations which are effects of weak noise resistance of the estimation.

In this research the damage estimation of DBI method is improved by introducing the normalization factors, i.e., change in strain generated from strain sensors and distance factor that take into consideration the position of the vehicle with respect to the position of the displacement sensors to account the effect of noise on the displacement measurement points. Hence the proposed improvement on the DBI method is presented in Eq. 5, 6, 7 and 8 below.

$$MDBI = \sum_{s=1}^{S} \frac{DBI_s * r_{ds}}{r_{es}} \tag{5}$$

where, $MDBI$ is the modified DBI, $S = \{s_1, s_2, \ldots, s_n\}$ is displacement measurement points, DBI_s is the estimated value by displacement sensor s, r_{ds} is the location factor, r_{es} is strain-based factor.

$$r_{ds} = \frac{1}{d_{vs}} \tag{6}$$

where, r_{ds} is the location factor, d_{vs} is the Euclidean distance between the moving vehicle and the displacement measurement points.

$$r_{es} = w_s(\varepsilon^d - \varepsilon^i) \tag{7}$$

$$w_s = \frac{1}{d_{gs}} \tag{8}$$

where, r_{es} is the strain-based factor, w_s is weighing factor for measurement point s, d_{gs} the Euclidean distance between the strain measurement point and the displacement measurement point location, ε^i, ε^d are strain response measurement generated from strain sensors in the longitudinal direction at the time of initial state and damaged state, respectively.

2.3 Transverse Position of the Moving Load

In comparison of the state of the structure before and after the occurrence to a damage, the position of the vehicle matters. The response of individual girder is sensitive to the vehicle's transvers position (TP) while crossing the bridge. To mitigate this problem, an approach of "Nothing-on-road Bridge Weigh-in-Motion" [6], is adopted to identify the transverse position of the moving load. Therefore, the position of the moving load in the transvers direction during recording the response of the bridge in damaged state will be used to correlate with the corresponding record of the intact state for better comparison.

To identify the transverse position (TP) of the vehicle, a series of locations along the bridge's transverse direction covering all possible positions where the vehicle may be present is first assumed and for each assumed TP, the measured response of the weighing station is used to identify the axel weights. The TP that minimizes the error in the least square estimation stated in Eq. 9 will be the true TP. The full derivation of this method can be found from the work by Yang Yu, et al. [6].

$$E_m = \sum_{i=1}^{n-1} \sum_{k=1}^{T} \left(M_{i,k}{}^P - M_{i,k}{}^m \right)^2 \tag{9}$$

where, E_m is the error function, $M_{i,k}^P$, $M_{i,k}^m$ are the predicted and measured bending moments for i^{th} non-weighing station at time instant k respectively.

3 Parametric Studies

3.1 Target Bridge Modeling

The selected target bridge is a 21-m single span concrete-steel composite bridge. The bridge is selected from the research by Amir Gheitasi et al. [2] where numerical simulation and validation at the system-level behavior of the composite steel girder bridges have been studied based on a scale laboratory investigation performed at the University of Nebraska.

The reinforced concrete deck is supported by three I-shaped steel girders. Total width of the bridge is 7.92 m. The thickness of the reinforced concrete slab deck is 190.5 mm while the height of the steel girder is 1400 mm. The bridges girders are simply supported on both ends. Furthermore, girders are connected by end cross

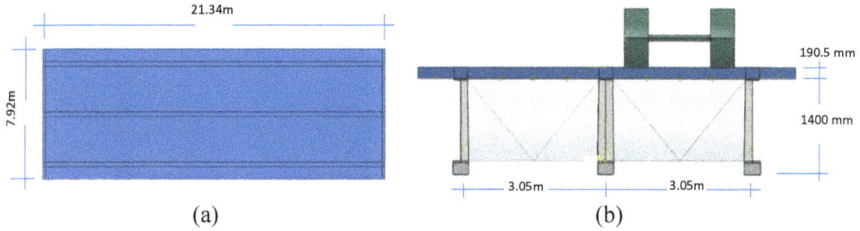

Fig. 1 Target bridge information. **a** Bridge dimension. **b** Bridge cross section

Table 1 Material properties of the bridge

Parameters	Concrete	Steel Girder	Re-Bar
FE Element	Solid	Solid	Beam
Density (Kg/m3)	2400	7850	7850
f' / Fy (Mpa)	39	240	496
fc / Fu (Mpa)	2.3	400	827
E (Mpa)	29,560	200,000	181,000
v	0.18	0.30	0.30

beams, and three intermediate cross beams. Details of the bridge dimensions are expressed in the Fig. 1.

The finite element model is executed by Abaqus software. 3D solid elements are used to model the concrete deck slab and I-shaped steel girder structures. The reinforcement bars are modeled with beam elements while the bracings are wire elements. The material property used to model the bridge is listed in Table 1. The bridge model contains a total of 39,189 elements and 56,412 nodes.

3.2 Bridge-Vehicle-Interaction

Bridge subjected to a moving load will be affected by the dynamic load generated from the vehicle. The variation of the load transferred to the road surface is a function of several factors like: vehicle mass, speed, vehicle suspension and roughness of the road surface.

Road surface is continuously distributed in a random trend, which affects dynamic behavior of both vehicle and bridge. Hence, at the interaction point, the deterioration of the road will have higher impact in altering the dynamics of the vehicle body and the bridge [1, 4, 5]. Therefore, the effect of road roughness is introduced by assuming the condition of the road is good, and the roughness effect in the right wheel and left wheel are identical. The artificial road profile is then generated from a stochastic representation as a function of Power Spectral Density. The following equation is used to generate the artificial road profile from ISO classification:

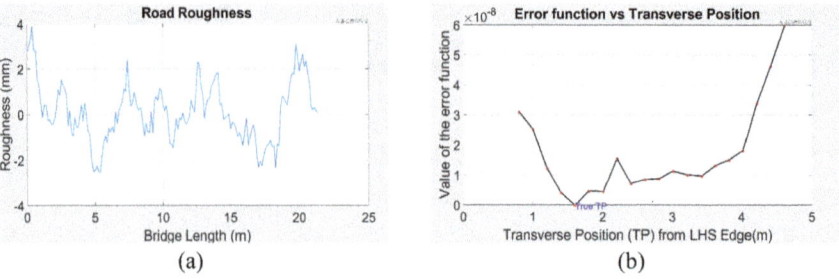

Fig. 2 **a** Road roughness generated **b** True TP identified from least square estimation

$$h(x) = \sum_{i=0}^{N} \sqrt{\Delta n}.2^k.10^{-3}.\left(\frac{n_0}{i.\Delta n}\right).\cos(2\pi.i.\Delta n.x + \varphi_i) \tag{10}$$

where: x is arbitrary position along the bridge length. $\Delta n = (n_{max} - n_{min})/ N$, $n_{max} = 5 Hz$ and $n_{min} = 0.1 Hz$ are the upper and lower cut-off frequencies, respectively. $k = 3$ is the road profile classification; $n_0 = 0.1$ cycle/m; φ_i is the random phase angle uniformly distributed from 0 to 2π. The generated road roughness is presented in the Fig. 2(a) below.

4   Results, Discussion and Conclusion

The results of a parametric study to examine the applicability of the proposed improvement on DBI method have been discussed.

4.1   Transverse Position Identification

Identifying the transverse position of the moving load while recording the displacement measurement data at the time of intact state and damaged state is crucial to find the contribution of each measurement station to damage location estimation and reduce effect of noise. Figure 2(b) shows the identification of true TP of the vehicle at the time to of consideration. The value with minimum estimation error is the true transverse position of the vehicle.

4.2 Damage Identification

In this section, the comparison of damage identification with the DBI method and the MDBI method is presented and compared with two damage scenarios varying in damage location and size as shown in Fig. 3(a) and (b). The arrangement of sensor under the bridge is shown in Fig. 3(c).

The displacement response shown in Fig. 4(a) and (b) shows the effect of the damage on the response continues after the damaged area. Hence, as we can observe from Figs. 5(a) and 6(a) that this phenomenon created a higher noise that compromises the identification of damage by the conventional DBI method in these scenarios. While MDBI method shown on Figs. 5(b) and 6(b) the noise effect is significantly reduced and the damaged locations are clearly identified.

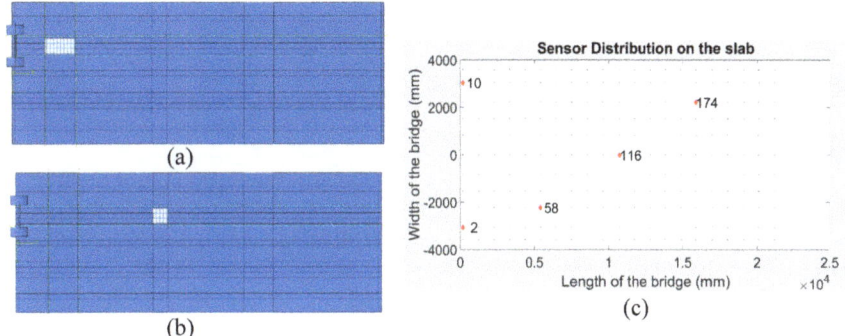

Fig. 3 **a** Location of damage in case 1. **b** Location of damage in case 2. (c) Sensor position under the bridge

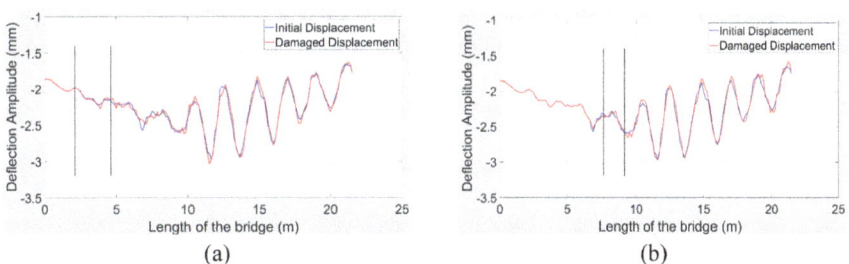

Fig. 4 **a** Displacement response at mid-span for case 1 **b** Displacement response at mid-span for case 2

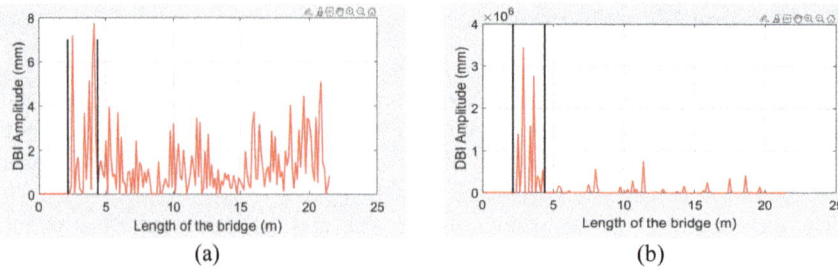

Fig. 5 **a** DBI method damage identification for case 1 **b** MDBI method damage identification for case 1

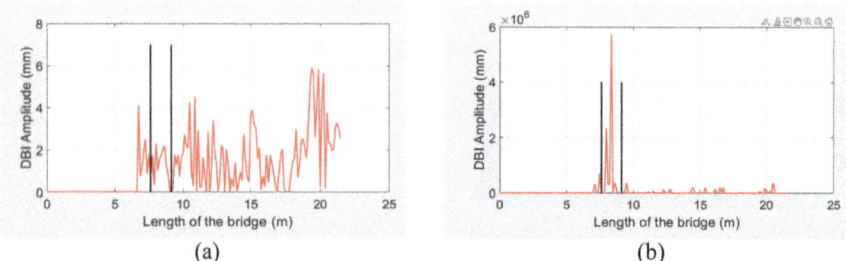

Fig. 6 **a** DBI method damage identification for case 2 **b** MDBI method damage identification for case 2

4.3   Conclusion

The modified DBI method clearly located the damage position even though the displacement response shows sever noise interference after the mid span. Furthermore, when considering the road roughness, the dynamics of the bridge changes after the occurrence of damage. Hence the DBI method experience higher noise interference while the modified DBI clearly locate the location of damage against the noise effect. For narrow area of damage, the DBI method is greatly compromised by the noise while the modified DBI method successfully located the damaged position.

References

1. Agostinacchio M, Ciampa D, Olita S (2015) The vibrations induced by surface irregularities in road pavements – a Matlab approach. Eur Transp Res Rev 6:267–275
2. Gheitasi A, Harris DK (2014) Overload flexural distribution behavior of composite steel girder bridges. J Bridg Eng 20(5):671
3. Ono R, Ha TM, Fukada S (2019) Analytical study on damage detection method using displacement influence lines of road bridge slab. J Civ Struct Heal Monit 9:565–577

4. Sun Z, Nagayama T, Su D, Fujino Y (2016) A damage detection algorithm utilizing dynamic displacement of bridge under moving vehicle. Shock Vib 2016:1–9
5. Sun Z, Zou Z (2016) Towards an efficient method of predicting vehicle-induced response of bridge. Eng Comp Int J CA Eng Soft 33(7):2067–2089
6. Yu Y, Cai CS, Deng L (2018) Nothing-on-road bridge weigh-in-motion considering the transverse position of the vehicle. Struct Infrastruct Eng 14:1108–1122

Investigation on the Performance of Non-uniformly, Discretely and Continuously CFRP Confined Square Reinforced Concrete Columns under Concentric and Eccentric loads

A. D. Mai, M. N. Sheikh, and M. N. S. Hadi

Abstract The behaviour of non-uniformly, discretely and continuously CFRP confined square reinforced concrete (RC) columns under concentric and 25 mm eccentric axial loads was investigated in this study. Eight square RC columns with a height of 800 mm and a side length of 150 mm were cast and tested. Two columns were reference RC specimens, two columns were non-uniformly confined with CFRP rings, two columns were discretely confined with three plies of CFRP rings and two columns were continuously confined with three plies of CFRP. The experimental results showed that non-uniform and continuous CFRP confinement significantly increased the load-carrying capacity and ductility of the square RC columns while discrete CFRP confinement resulted in a considerable enhancement of the load-carrying capacity and a significant improvement in the ductility. The improvement in the load-carrying capacity of the square RC columns due to non-uniform CFRP confinement was less than that due to continuous CFRP confinement but larger than that due to discrete CFRP confinement. However, the improvement in the ductility of the square RC specimens due to non-uniform CFRP confinement was larger than that due to discrete and continuous CFRP confinement under concentric axial load. It was also revealed that the brittle failure of CFRP confined square RC columns could be avoided by the application of non-uniform CFRP confinement.

Keywords Carbon fibre · CFRP · Non-uniform confinement · FRP confinement · Eccentric load

A. D. Mai (✉)
Faculty of Project Management, The University of Da Nang - University of Science and Technology (DUT), 54 Nguyen Luong Bang street, Lien Chieu District, Da Nang, Vietnam
e-mail: maduc@dut.udn.vn

M. N. Sheikh · M. N. S. Hadi
School of Civil, Mining and Environmental Engineering, University of Wollongong, NSW, Australia

© The Author(s), under exclusive license to Springer Nature Singapore Pte Ltd. 2023   857
J. N. Reddy et al. (eds.), *ICSCEA 2021*, Lecture Notes in Civil Engineering 268,
https://doi.org/10.1007/978-981-19-3303-5_78

1 Introduction

Fiber reinforced polymer (FRP) composites have been popularly applied in the construction industry over the last three decades owning to its advantages consisting of high strength-to-weight ratio, free from corrosion, high stiffness and ease of construction [2, 3, 9, 13, 19]. One of popular applications of FRP is to strengthen/retrofit deficient reinforced concrete (RC) columns in which RC columns were externally confined with FRP (FRP jacket) [17]. The lateral confinement exerted by the FRP jacket constrains the dilation of concrete, resulting in a significant improvement in the load-carrying capacity and ductility of RC columns [12, 14, 21]. However, the FRP confinement on square columns is much less effective than the FRP confinement on circular columns [4, 10, 13]. This is attributed to the fact that the distribution of the lateral confining pressure exerted by FRP is non-uniform over the square cross-section while uniform over the circular cross-section [20].

A plenty of investigations have been undertaken to examine the mechanism and performance of FRP confined square concrete columns [8, 20]. However, most of these investigations focused on the response of FRP confined square concrete columns under concentric axial load [20] and a few of these investigations concerned about the response of FRP confined square concrete columns under eccentric axial loads [8]. The research outcomes revealed that the higher axial load eccentricity, the lower FRP confinement effectiveness. This results from the reduction of the effective confinement area of the column cross-section. It should be mentioned that RC columns in practice are likely subjected to the eccentric loads because of accidental load eccentricity, the position of columns in structure and material/ geometry imperfection of the columns. Thus, extensive research studies on eccentrically loaded FRP confined square concrete columns are needed to provide a deep understanding in the influence of the axial load eccentricity on the performance of FRP confined square concrete columns.

FRP confinement can be continuously, discretely and non-uniformly constructed on the square concrete columns, as illustrated in Fig. 1. For continuously FRP confined square columns, the FRP materials are used to wrap along the whole the column height. For discretely FRP confined square columns, the FRP rings/strips are used to discretely wrap along the height of the specimens with a certain clear spacing between two successive rings/strips. Discrete FRP confinement has several advantages over continuous FRP confinement in which discrete FRP confinement could prevent the possible air-void between FRP material and column surface, requires less FRP and adhesive and could construct easily on-site. For non-uniformly FRP confined square columns, two types of FRP rings consisting of primary and secondary FRP rings are used to alternatively wrap along the height of the columns. Non-uniform FRP confinement can be considered as a combination between continuous confinement and discrete confinement. Non-uniform FRP confinement may be used to replace discrete confinement to avoid the strain localization which is usually occurred in the non-confined zones of discretely FRP confined concrete columns.

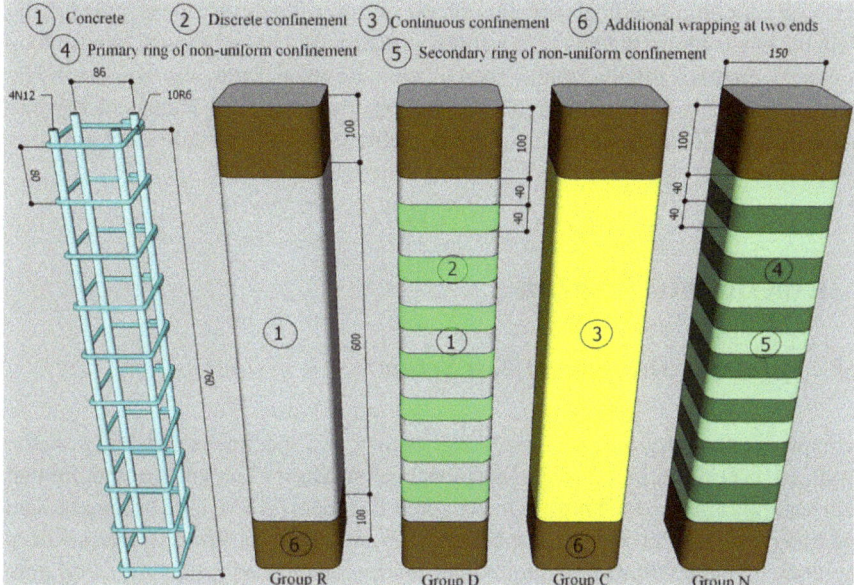

Fig. 1 Details of specimens

The influence of non-uniform, continuous and discrete FRP confinement on the performance of the square RC columns under eccentric axial loads has been examined in some research studies. Saljoughian and Mostofinejad [16] and Mai et al. [11] evaluated the influence of discrete FRP confinement on the square RC columns and revealed that discrete FRP confinement resulted in a slight improvement in the load-carrying capacity but a substantial improvement in the ductility of the square RC columns. Mai et al. [11] also reported that the improvement in the load-carrying capacity and ductility of square RC columns owning to discrete FRP confinement was lower than that of square RC columns owning to continuous FRP confinement. The influence of non-uniform FRP confinement on the performance of the square RC columns under eccentric axial loads was evaluated in Mai et al. [15]. It has been found that the load-carrying capacity improvement of the square RC columns owning to non-uniform confinement was less than that owning to continuous confinement while the ductility improvement of the square RC columns owning to non-uniform confinement was larger than that owning to continuous confinement.

A literature review showed that no study has examined the effectiveness of non-uniform FRP confinement on the square RC columns in comparison to discrete and continuous FRP confinement. Thus, the objective of this study is to experimentally and analytically examine the response of non-uniformly FRP confined square RC columns in comparison to the response of continuously and discretely FRP confined square RC columns. In this study, the behaviour of the square RC columns non-uniformly, continuously and discretely confined with CFRP under eccentric and concentric axial loads was firstly presented. Next, the analytical ultimate axial loads and corresponding bending moments of the columns were also determined using available stress-strain models of constitutive materials and rectangular stress block analytical method. Finally, the analytical ultimate axial loads and corresponding bending moments were compared with the experimental axial loads and bending moments to evaluate the accuracy of the available stress-strain models in estimating the load carrying capacity of FRP confined square RC columns.

2 Experimental Program

2.1 Specimen Design and Test Setup

An experimental program consisted of eight square RC specimens with 150 mm side length and 800 mm height (Fig. 1). Each specimen had four 12 mm diameter deformed steel bars (4N12) and ten 6 mm diameter strain steel bars (10R6) used as longitudinal and transverse reinforcement, respectively. To maintain 20 mm concrete cover at two ends of the column, the longitudinal reinforcement had a height of 760 mm. The transverse reinforcement was placed along the longitudinal reinforcement with 80 mm center-to-center spacing. The specimen had 20 mm concrete cover and 20 mm corner radius.

The specimens were separated into four groups of two identical specimens. The first group's specimens (Group R) were non-confined specimens, which were considered as control specimens. The second group's specimens (Group D) were discretely confined with three plies of CFRP ring. The width of CFRP ring was 40 mm, which was chosen to be equal to the clear spacing between two successive CFRP rings. One CFRP ring was placed in the specimen mid-height, which was in the middle of two successive steel ties. The third group's specimens (Group C) were continuously confined with three plies of CFRP. The last group's specimens (Group N) were non-uniformly confined with CFRP rings. Two types of CFRP rings consisting of the primary and secondary rings were used for non-uniform confinement. The primary rings had four plies of CFRP while the secondary rings had two plies of CFRP. The number of CFRP rings was selected to make sure that the average number of two successive rings was equivalent to the number of CFRP plies of continuous FRP confinement. One of the primary CFRP rings was placed at the specimen mid-height. The primary CFRP rings were placed in the middle of two successive CFRP strips.

Two specimens of each group were subjected to concentric and 25 mm eccentric axial load. The axial loads applied on the columns were generated by a compression testing machine with 5 tons capacity. To generate eccentric axial load, a pair of loading heads was placed at the specimen's two ends. The details of the loading heads could be found in Mai et al. [11]. A laser triangulation and two LVDT were utilized to capture lateral and axial and deformation of tested specimens, respectively.

Each specimen was labeled by two parts which were separated by a dash. The first part represented the confinement scheme (either letter R, D, C or N). The second

Table 1 Configuration of test specimens

Column	$n_{pf}\left(w_{pf}\right)$ (mm)	$n_{sf}\left(w_{sf}\right)$(mm)	$P_{u,e}$(kN)	Δ_u(mm)	δ_u(mm)	$M_{u,e}$(kN.m)	λ	$P_{u,t}$(kN)	$M_{u,t}$(kN.m)
R-0	0(0)	0(0)	993.5	2.8	–	–	1.5	932.4	0
R-25			630.3	2.2	2.5	17.3	1.3	562.2	14.1
D-0	3(40)	0(0)	1204.7	14.9	–	–	6.8	1085.8	0
D-25			742.8	3.1	4.7	22.1	2.6	708.4	17.7
C-0	3(40)	3(40)	1614.5	20.48	–	–	8.2	1413.6	0
C-25			876.6	4.3	5.8	27	4.3	869.2	21.7
N-0	4(40)	2(40)	1530	20.7	–	–	8.9	1309.9	0
N-25			848	4.2	5.7	26	4.3	821.3	20.5

Note $n_{pf}\left(w_{pf}\right)$ indicates the number of CFRP plies and width of the primary CFRP ring; $n_{sf}\left(w_{sf}\right)$ indicates the number of CFRP plies and width of the secondary CFRP ring; Δ_u and δ_u indicate the axial deformation and lateral deformation, respectively, at the experimental ultimate axial load $\left(P_{u,e}\right)$; $P_{u,t}$ indicate the analytical ultimate axial load; $M_{u,e}$ and $M_{u,t}$ indicate the experimental and analytical bending moments, respectively

part of the specimen label represented the axial load eccentricity (either number 0 or 25). The details of the specimen label are presented in Table 1.

2.2 Specimen Preparation and Material Properties

The formwork was made from plywood panels. The specimen's rounded corners were produced by attaching Styrofoam, which has a radius of 20 mm, at the corners of formwork. The specimens were made from one batch of ready-mix concrete and then maintain the moist by watering every working days and wrapping with plastic sheets and wet hessian rugs. The epoxy resin was mixed with hardener with a ratio of 5:1 was utilized as binder for FRP wrapping. To make sure the epoxy to dry, the specimens after confinement were left at laboratory temperature for a minimum of 7 days.

Three standard concrete cylinders were utilized for measuring the compressive strength of concrete at 28-days as recommended in AS 1012.9:2014 [6], which was 36 MPa. The yield tensile strengths of R6 and N12 steel bars were 517 and 568 MPa, respectively, which were measured based on three steel samples of each steel type as recommended in AS 1391-2007 [5]. The FRP had the modulus of elasticity, tensile strength, corresponding tensile strain of 240.43 GPa, 3726 MPa and 1.55%, respectively, which were determined based on ASTM D3039/D3039M-14 [7].

2.3 Experimental results and discussions

As illustrated in Fig. 2(a), all the specimens showed a similar response until the crack of concrete, which usually occurs when concrete attains 70% of its compressive strength. Specimen R-0 experienced an abrupt drop of the axial load after the peak

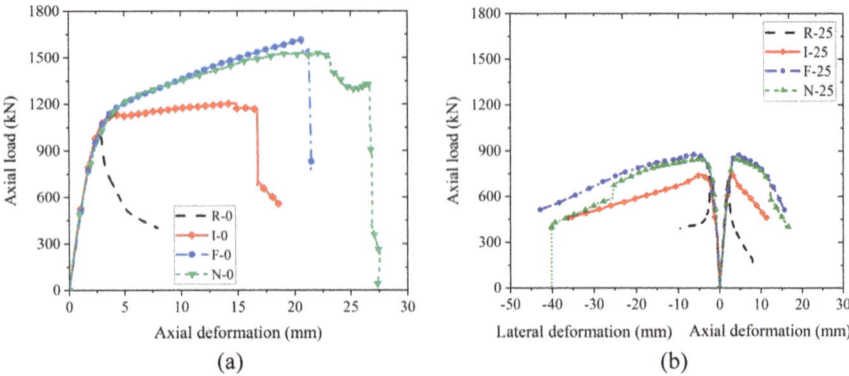

Fig. 2 Behaviour of specimens under: **a** concentric axial load and **b** eccentric axial load

axial load (ultimate axial load). Unlike Specimen R-0, Specimens D-0, C-0 and N-0 experienced a significant rise in the axial load after passing the peak axial load of unconfined specimen (Specimen R-0). This was because of the CFRP confinement effect, which prevented the dilation of concrete after cracking. The gradient of the second portion of the axial load-axial deformation of Specimen C-0 was larger than that of Specimen N-0, indicating that the confinement effect of continuous FRP confinement was larger than that of non-uniform CFRP confinement. The gradient of the second portion of the axial load–axial deformation of Specimen D-0 was much smaller than that of Specimens C-0 and N-0, indicating that the confinement effect of discrete CFRP confinement was much smaller than that of continuous and non-uniform CFRP confinement. It is worth noting that Specimen C-0 and D-0 experienced brittle failures, which were associated with an abrupt drop of the axial loads owning to the rupture of CFRP. In contrast to Specimens C-0 and D-0, Specimen N-0 experienced a ductile failure, which was associated with a slight drop of the axial load after obtaining the peak load. The slight decrease in the axial load of Specimen N-0 was because of the rupture of the secondary CFRP rings. The rise in the axial load of Specimen D-0 after experiencing a decrease in the axial load, which resulted from the confinement effect of the primary CFRP rings, provided a warning sign before the failure of the structure. As illustrated in Table 1, the ductility of Specimen N-0 was highest, followed by the ductility of Specimens C-0 and D-0, respectively. The experimental results also show that square RC specimen attained significant improvement in the ductility under concentric axial load by strengthening with CFRP.

The behaviour of specimens with 25 mm axial load eccentricity is showed in Figure 2(b). As illustrated in Fig. 2(b), Specimen R-25 underwent an abrupt drop of the axial load resulting from the crushing of concrete on the compression side. Unlike Specimen R-25, Specimens D-25, C-25 and N-25 underwent a significant rise in the axial load after passing the peak axial load of Specimen R-25, which revealed the confinement effect of CFRP on the square RC specimen under eccentric axial load. After attaining the peak axial load, Specimens D-25, C-25 and N-25 experienced a slight loss in the axial load. Specimen C-25 obtained the highest load-carrying capacity, followed by Specimens N-25, D-25 and R-25, respectively. Under eccentric axial load, Specimen C-25 attained higher ductility compared to Specimens N-25 because all confined specimens experienced ductile failures under eccentric axial load, which were associated with a slight decline in the axial load after obtaining the peak load.

3 Axial Load and Bending Moment

The experimental bending moments $(M_{u,e})$ of the specimens subjected to eccentric axial load were determined using Eq. 1. The analytical axial loads $(P_{u,t})$ and bending moments $(M_{u,t})$ of the specimens were computed using a combination between the stress-strain models of constitutive materials and rectangular stress block.

$$M_{u,e} = P_{u,e}(\delta_p + e) \tag{1}$$

The stress and strain of concrete without FRP confinement were obtained using the stress-strain model proposed by Yang et al. [18] proposed for unconfined concrete, as showed in Eq. 2.

$$f_c = \left[\frac{(\beta_1 + 1)(\varepsilon_c/\varepsilon_{co})}{(\varepsilon_c/\varepsilon_{co})^{\beta_1+1} + \beta_1} \right] f_{co} \tag{2}$$

The stress and strain of concrete with FRP confinement were determined using the stress-strain model suggested by ACI 440.2R-17 [1] for concrete with FRP confinement, as showed in Eq. 3.

$$\sigma_c = \begin{cases} E_c\varepsilon_c - \frac{(E_c - E_2)^2}{4f_{co}}\varepsilon_c^2, \ for 0 \le \varepsilon_c \le \varepsilon_t \\ f_{co} + E_2\varepsilon_c, \ for \varepsilon_t \le \varepsilon_c \le \varepsilon_{cu} \end{cases} \tag{3}$$

The axial loads and bending moment of test specimens under eccentric and concentric axial loads are summarized in Table 1. As illustrated in Table 1, the $P_{u,t}$ and $M_{u,t}$ agreed well with the $P_{u,e}$ and $M_{u,e}$, respectively. This indicated that the combination between constitutive materials and rectangular stress block resulted in reasonable results of the axial loads and bending moment of the square RC specimens discretely, continuously and non-uniformly confined with FRP.

4 Conclusions

The following concluding remarks can be given based on the experimental results and analytical calculations.

Square RC columns attained a significant improvement in the load-carrying capacity and ductility by strengthening with CFRP using discrete, continuous and non-uniform confinement. The improvement in the load-carrying capacity of square RC columns under eccentric and concentric axial loads due to continuous CFRP confinement was higher than that of square RC columns due to non-uniform confinement. However, the improvement in the ductility of square RC columns under concentric axial load due to non-uniform CFRP confinement was higher than that of square RC columns due to continuous CFRP confinement. Non-uniform CFRP confinement was also effective to provide a warning signal before the failure of structure.

Discrete confinement led to a considerable improvement in the load-carrying capacity and substantial improvement in the ductility of the square RC columns under eccentric and concentric axial loads. However, the load-carrying capacity and ductility enhancement of discrete CFRP confinement was lower than that of non-uniform and continuous CFRP confinement.

The $P_{u,t}$ and $M_{u,t}$ of square RC specimens discretely, continuously and non-uniformly confined with CFRP agreed well with the experimental results, indicating that the combination of the stress-strain models of constitutive materials and rectangular stress block resulted in a reasonable results of axial loads and bending moment.

Acknowledgments The first author thanks the University of Da Nang for funding this study using Funds for Science and Technology Development via the project number of B2021-DN02-02.

References

1. ACI (American Concrete Institute) (2017) Guide for the design and construction of externally bonded FRP systems for strengthening concrete structures. ACI 4402R-17, Farmington Hills
2. Ahmad H, Sheikh MN, Hadi MNS (2021) Behavior of GFRP bar-reinforced hollow-core polypropylene fiber and glass fiber concrete columns under axial compression. J Build Eng. 44:103245
3. Ahmad J, Yu T, Hadi MNS (2021) Behavior of GFRP bar reinforced geopolymer concrete filled GFRP tube columns under different loading conditions. Struct 33:1633–1644
4. Alelaimat RA, Sheikh MN, Hadi MNS (2021) Behaviour of square concrete filled FRP tube columns under concentric, uniaxial eccentric, biaxial eccentric and four-point bending loads. Thin-Walled Struct 168:108252
5. AS (Australian Standard) (2007) Metallic materials-Tensile testing at ambient temperature. AS 1391-2007, Sydney
6. AS (Australian Standard). (2014) Methods of testing concrete: method 9: compressive strength tests-concrete, mortar and grout specimens. AS 1391-2007, Sydney
7. ASTM (American Society for Testing and Materials) (2017) Standard test method for tensile properties of polymer matrix composite materials. ASTM D3039, West Conshohocken

8. Hadi MNS, Widiarsa IBR (2012) Axial and flexural performance of square RC columns wrapped with CFRP under eccentric loading. J Compos Constr 16(6):640–649
9. Lam L, Teng JG (2003) Design-oriented stress-strain model for FRP-confined concrete. Constr Build Mater 17(6–7):471–489
10. Mai AD, Sheikh MN, Hadi MNS (2018) Influence of the location of CFRP strips on the behaviour of partially wrapped square reinforced concrete columns under axial compression. Struct 15:131–137. PubMed PMID: WOS:000442733500010
11. Mai AD, Sheikh MN, Hadi MNS (2018) Investigation on the behaviour of partial wrapping in comparison with full wrapping of square RC columns under different loading conditions. Constr Build Mater 168:153–168 PubMed PMID: WOS:000430756900015
12. Mai AD, Sheikh MN, Hadi MNS (2020) Experimental and analytical investigations on the effectiveness of non-uniform CFRP wrapping on circularised RC columns. Struct Infrastruct Eng 17(8):1125–1140
13. Mai AD, Sheikh MN, Hadi MNS (2020) Failure envelopes of square and circularized RC columns discretely confined with CFRP. Constr Build Mater. 261:119937
14. Mai AD, Sheikh MN, Hadi MNS (2021) Strain model for discretely FRP confined concrete based on energy balance principle. Eng Struct 241: 112489
15. Mai AD, Sheikh MN, Yamakado K, Hadi MNS (2020) Nonuniform CFRP Wrapping to Prevent Sudden Failure of FRP Confined Square RC Columns. J Compos Constr 24(6):04020063
16. Saljoughian A, Mostofinejad D (2016) Corner strip-batten technique for FRP-confinement of square RC columns under eccentric loading. J Compos Constr 20(3):04015077. PubMed PMID: WOS:000379984000019
17. Yang J, Lu S, Wang J, Wang Z (2020) Behavior of CFRP partially wrapped square seawater sea-sand concrete columns under axial compression. Eng Struct 222:111119
18. Yang KH, Mun JH, Cho MS, Kang THK (2014) Stress-strain model for various unconfined concretes in compression. ACI Struct J. 111(4):819–826
19. Yu T, Zhao H, Ren T, Remennikov A (2019) Novel hybrid FRP tubular columns with large deformation capacity: concept and behaviour. Compos Struct 212:500–512
20. Zeng JJ, Guo YC, Gao WY, Li JZ, Xie JH (2017) Behavior of partially and fully FRP-confined circularized square columns under axial compression. Constr Build Mater 152:319–332. PubMed PMID: WOS:000411545600030
21. Zhao H, Ren T, Remennikov A (2021) Behaviour of FRP-confined coal rejects based backfill material under compression. Constr Build Mater 268:121171

K-Fold Cross-Validation Technique for Predicting Ultimate Compressive Strength of Circular CFST Columns

Tran-Trung Nguyen and Phu-Cuong Nguyen

Abstract Concrete-Filled Steel Tube (CFST) columns are a type of composite structure that has been widely used nowadays. It is known for its ability to withstand ultimate compressive strength and its efficient responsiveness in buildings with a modern appearance that transcends large spatial spans. Many design standards of many countries have also proposed calculation formulas, but design equations meet limitations due to experimental tests. If experimental data is large, specifically 663 specimens of CFST columns, determining the critical compressive strength for each sample is reliable and convenient in the case of Machine Learning prediction. In this study, prediction models of Supervised Learning approaches in Machine Learning such as Linear Regression, Logistic Regression, Linear Support Vector Regressor, Random Forest, Decision Tree, and KNN will be performed effectively in predicting the ultimate strength of CFST columns with only experimental raw data. The results obtained by this study show different influences in using the prediction models of the supervised learning approaches. Besides, to increase the stability of the prediction models, a cross-validation technique called K-fold is used with the data set divided into two parts, including 80% for training data and 20% for test data.

Keywords CFST Columns · Decision Tree · KNN · Linear Regression · Machine Learning · Prediction Models · Supervised Learning

1 Introduction

Columns are an essential part of buildings. It receives both horizontal and vertical loads for the establishment, so cross-sectional dimensions and the strength of the

T.-T. Nguyen
VanLang University, Faculty of Architecture, Ho Chi Minh City, Vietnam

T.-T. Nguyen · P.-C. Nguyen (✉)
Advanced Structural Engineering Laboratory, Department of Structural Engineering, Faculty of Civil Engineering, Ho Chi Minh Open University, Ho Chi Minh City, Vietnam
e-mail: cuong.pn@ou.edu.vn; henycuong@gmail.com

materials creating it must be large enough. But according to today's modern architectural style, there are spaces where practical use requires a large aperture and height of each floor, but the column size must not be too large. With such challenging requirements of modern architecture, a concept of composite column type was born: Concrete-Filled Steel Tube (CFST) columns.

The unique feature of this column type is that it can withstand considerable compressive strength with an optimal cross-section compared to traditional reinforced concrete columns. This advantage has used the horizontal compression effect of concrete through the outer steel shell. The design analysis of the ultimate compressive strength of this column, especially with a circular cross-section, has been presented very specifically through empirical formulas in the standards; specifically, two commonly used standards are ACI 318-14 [1] and Eurocode 4 [2]. Besides the mentioned standards, some studies [3, 4] have used advanced analytical methods by using FEM software to determine the critical compressive strength and behavior of CFST columns. The authors of these studies improved the material models of concrete and steel to give more accurate results. But if with a large amount of data, the above methods will not be suitable because the fluctuations related to the geometry of the column, associated with the strength of the material in the data set, are not fully controlled. Therefore, a new idea is formed, which is to use computer science with big data processing, which today is considered the world's trend in the sciences and society in general. Analysis and design of construction structures, particularly Machine Learning (ML), is a branch of Artificial Intelligence (AI).

ML is a field that uses statistical techniques to give computers the ability to "learn" with data in two forms, controlled and uncontrollable learning. The applications of ML are very diverse in many fields such as Retail, Travel, Healthcare, Finance, Media, etc., especially in the construction engineering industry. The application and improvement of optimization algorithms in ML to generate prediction models or related decision problems. In predicting the critical compressive strength of CFST columns by ML, the method has been used increasingly popularly with incredible application speed. For example, Jayalekshmi's research [5] has improved the prediction formula for compressive strength of CFST circular columns based on Artificial Neural Network (ANN) with reliable results compared with other analytical models. Continuing with ANN, Zarringol's study [6] performed concentric and eccentric compressive strength prediction of two columns with circular and square cross-sections, considering the strength reduction factor derived from the Monte Carlo Simulation (MCS) model. The rapid development of algorithms and technical systems in extensive data analysis has resulted in almost perfect accurate prediction models. The problem of the difference between the prediction models in the supervised learning approach of ML in the specific issue here is determining the maximum compressive strength value of the CFST circular column has not been considered.

This study will adapt prediction models such as Linear Regression, Logistic Regression, Linear Support Vector Regressor, Random Forest, Decision Tree, and KNN. The ultimate compressive strength of short circular CFST columns will be predicted from ML models established from a raw data set of 663 specimens. K-fold cross-validation technique is used to increase the stability of the prediction models,

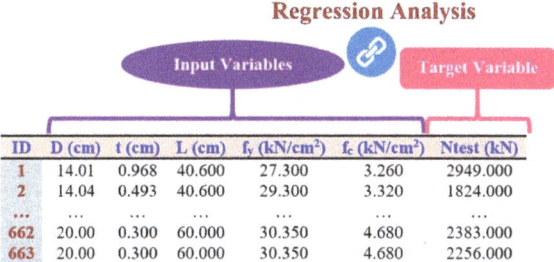

Fig. 1 Input data for regression analysis

i.e., avoid the cases of too much noise or the prediction model with low bias and variance.

2 Supervised Learning Approaches

The problem mentioned in this study is to predict the ultimate compressive strength of short circular CFST columns, which is the problem with the target variable. Therefore, the supervised learning approach is suitable for the target to be solved. An essential tool for modeling and analysis is Regression analysis. It helps investigate the relationship between a dependent variable (target) and independent variables (predictor), as shown in Fig. 1. Figure 1 shows input data of 663 specimens, including five independent variables and one dependent variable, Ntest. Ntest is the ultimate compression strength of CFST columns collected from experimental tests in the literature.

Choosing a suitable prediction model for the problem is mentioned in the Introduction and presented a more general view between the prediction models. From that, an available picture helps to choose the model that will be appropriate. More appropriate and accurate, this study proposes using algorithms for prediction modeling with advantages and limitations of each algorithm, as summarized in Fig. 2.

.

3 Setting up a Prediction Model Using Machine Learning

How to implement a project in general and an individual problem, mainly using ML, the main steps to be performed are as follows: i) How is the problem solved? The issue needs to predict the value of the critical compressive strength of short circular CFST columns. ii) The actual test data set to determine the number of specimens, and in this study, 663 short CFST column specimens are used that were referenced from the research [4]. iii) Exploring the data to see if there is anything unusual (missing data,

Predictive Models	Advantages	Limits
Linear Regression	• Easy to understand an algorithm • High popularity • Fast algorithm speed • It gives optimal results when the relationship between the independent variable and the dependent variable is linear	• Very sensitive to interference • Cannot represent complex models • Not suitable for modeling nonlinear relationships
Logistic Regression	• Do not make assumptions about the distribution of classes in the feature space • Easily scalable to multiple classes (regressive polynomials) • Quick training • High accuracy for many simple data sets • Responsiveness to overfitting • Can interpret model coefficients as indicators of feature importance	• Only works with data types where two classes are close to linearly separable, but not with data where one category contains points inside a circle, the other class includes points outside the line circle • The data points must be independent of each other. But in reality, the data points are influenced by each other
Linear Support Vector Regressor in Support Vector Machine (SVM)	• SVM is a good choice when we have no idea about the data • Works well with unstructured and semi-structured data such as text, images, and implants • The kernel trick is the real power of SVM. With a suitable Kernel function, any complex problem can be solved • Relatable to multidimensional data • The risk of overfitting is low in SVM	• Choosing "good" kernel functionality is not easy • Long training time for large data sets • Difficult to understand and interpret the final model, the weighted variables • Since the final model is not easily discernible, it is not possible to make minor corrections to the model
Random Forest	• Random forest is considered a precise and powerful method • The random forest does not suffer from the overfitting problem because it takes the average of all predictions, which cancels out biases • The algorithm can be used in both classification and regression • The relative importance of features can be seen to help select the features that contribute the most to the classification process	• Random forest is slow in predicting because it has many decision trees. Whenever it makes a prediction, all the trees in the forest have to expect the same input and then vote on it. This whole process takes time • Models are more challenging to understand than decision trees because with decision trees; we can easily make decisions by following a path in the tree
Decision Tree	• Easy to understand, accessible to model • No need to standardize features • It can be applied in both Classification and Regression • It is possible to model nonlinear relations • Can model interactions between different descriptive features	• If the continuous feature is used, the tree can become huge and less interpretable • Small changes in the data can lead to a completely different tree • If the number of features is relatively large, but the number of samples is small, it can lead to data mismatch
KNN	• Easy to deploy • Works well on fundamental recognition problems • Flexibility in choosing distance/distance • Flexible handling of multi-class cases • Practical if the training data is large enough	• Having difficulty in determining k matches • Changing k can change the class result of the data • The accuracy of KNN can be severely degraded with high size data since there is very little difference between nearest and farthest neighbors • KNN can be skewed class distribution

Fig. 2 Advantages and limitations of supervised learning algorithms in machine learning

duplicates, outliers, etc.), this study also describes the data. Still, it leaves the raw data as it is for calculation. Then proceed to visualize the data through the matrix. iv) Model selection and training. v) Test and refine the model. vi) Monitor and maintain the system when the model is stable.

Figure 3 shows the typical experiment and cross-section of the short circular CFST column, the data set in this study is also performed in the same way. Then, performing ML through regression analysis with six proposed algorithms is as described in the literature section. In addition, there is a K-fold cross-validation (K-fold CV) technique in the training model to increase the stability of the prediction models to

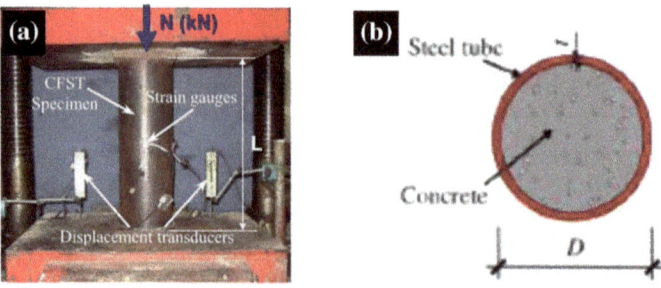

Fig. 3 **a** CFST column experiment [7]; **b** Cross-section of CFST Columns.

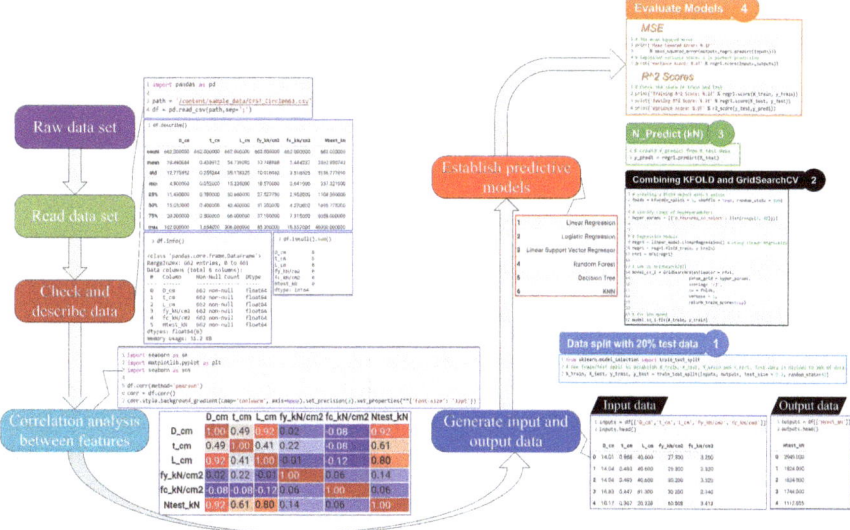

Fig. 4 The process of setting up a prediction model using machine learning

avoid overfitting and underfitting. Figure 4 illustrates the prediction procedure using machine learning algorithms.

3.1 K-fold Cross-Validation (K-fold CV)

K-fold CV helps us evaluate a model more fully and accurately when we have a small data set. Then we decide whether the model is suitable for the data, the current problem, or not to give the following action. As shown in Fig. 5 on the train ride, the

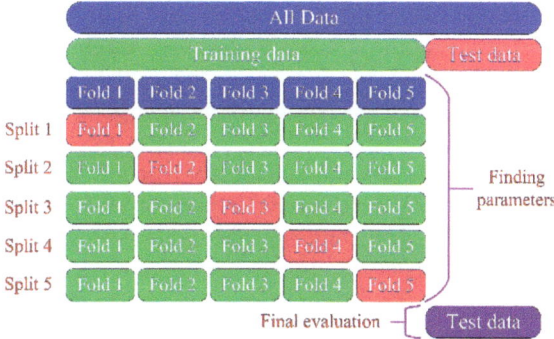

Fig. 5 How to K-fold CV works

test data part will be kept separate and reserved for the final evaluation step to check the "reaction" of the model when encountering completely unseen data.

The training data part will be randomly divided into K parts (K is an integer, choose 5 or 10, etc.). Then training the model K times; each time, the train will choose 1 part as validation data and K-1 as training data. The final model evaluation result will be the average of the evaluation results of K training times. That is why we evaluate more objectively and accurately.

3.2 Mean Squared Error (MSE)

The MSE criterion is utilized to compare the reliability of each model. In statistics, the estimator's mean square error (MSE) (of the unobserved quantity estimation procedure) measures the mean square of the mistakes – that is, the mean squared difference between the estimated value and observed value. MSE is a function of risk, corresponding to the expected value of squared error loss. The MSE is almost always positive (but not zero) due to randomness, or the estimator doesn't consider information that could produce a more accurate estimate.

$$MSE = \frac{1}{n} \sum_{i=1}^{n} (y_i - \bar{y}_i)^2 \tag{1}$$

where y_i is the independent variable, y_i is the estimated value.

4 Results and discussion

From setting up the model to predict the ultimate compressive strength of short circular CFST column from the raw data set of 663 specimens, the results between the two training models and the test models are obtained as shown in Fig. 6. Figure 6 shows the predicted results of the first seven specimens for the ultimate strength of CFST columns using the mentioned machine learning models. It can be seen that the Random Forest, Decision Tree, and KNN algorithms give good prediction results for this problem.

Assessing which predictive model is suitable for practical applications is carried out effectively when the sample data size is big enough (663 specimens for this study). The MSE standard is often used for comparing the results. Using the model reliability measured by MSE makes model selection easier because the smaller value of the MSE gives a better prediction. The MSE values of the prediction models are shown in Fig. 7. But to be more clear and intuitive about the effectiveness of the models in the correlation between input and output data, between predicted and actual compressive strengths (N_pred and N_test), the variance scores and the R_square (R^2) statistical

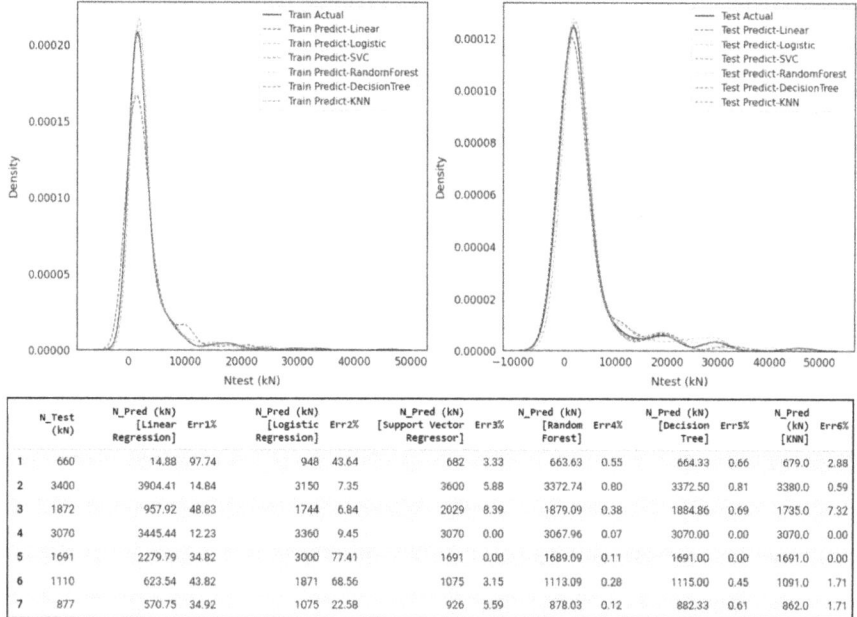

	N_Test (kN)	N_Pred (kN) [Linear Regression]	Err1%	N_Pred (kN) [Logistic Regression]	Err2%	N_Pred (kN) [Support Vector Regressor]	Err3%	N_Pred (kN) [Random Forest]	Err4%	N_Pred (kN) [Decision Tree]	Err5%	N_Pred (kN) [KNN]	Err6%
1	660	14.88	97.74	948	43.64	682	3.33	663.63	0.55	664.33	0.66	679.0	2.88
2	3400	3904.41	14.84	3150	7.35	3600	5.88	3372.74	0.80	3372.50	0.81	3380.0	0.59
3	1872	957.92	48.83	1744	6.84	2029	8.39	1879.09	0.38	1884.86	0.69	1735.0	7.32
4	3070	3445.44	12.23	3360	9.45	3070	0.00	3067.96	0.07	3070.00	0.00	3070.0	0.00
5	1691	2279.79	34.82	3000	77.41	1691	0.00	1689.09	0.11	1691.00	0.00	1691.0	0.00
6	1110	623.54	43.82	1871	68.56	1075	3.15	1113.09	0.28	1115.00	0.45	1091.0	1.71
7	877	570.75	34.92	1075	22.58	926	5.59	878.03	0.12	882.33	0.61	862.0	1.71

Fig. 6 The ultimate compressive strength of the short circular column CFST

	Models	MSE	Variance Scores Inputs|Output	R^2 Scores N_test|N_pred
1	Linear Regression	2389001.48	0.91	0.93
2	Logistic Regression	5893374.88	0.18	0.82
3	Linear Support Vector Regressor	64841.46	0.50	0.99
4	Random Forest	154699.68	0.99	0.99
5	Decision Tree	20273.83	1.00	1.00
6	KNN	45109.80	1.00	1.00

Fig. 7 The MSE and R_square values using different ML models in predicting the ultimate compressive strength of CFST columns

measure are used, respectively. Looking at Fig. 7, with the variance and R^2, it can be seen that the Random Forest, Decision Tree, and KNN models are efficient for this problem.

5 Conclusion

The ultimate compressive strength of short circular CFST columns is predicted by developing codes using machine learning models. The proposed formulas in the current design standards give reliable results, but these standards have limitations. Advanced analytical methods using the FEM simulation by commercial software is complicated, expensive, and takes time for modeling and verifying the accuracy. A developing ML algorithm can simplify code that can find answers faster with measurable accuracy. On the other hand, when there are changes to a large data set, ML can adapt well to that new data. In the future, with the big data of experiments, new formulas can be proposed for design standards in the civil engineering field using machine learning.

Acknowledgements The authors gratefully acknowledge the financial support provided by the Scientific Research Fund of Ho Chi Minh City Open University, Vietnam.

References

1. ACI318-14 (2014) Building code requirements for structural concrete and commentary. American Concrete Institute, America.
2. Eurocode 4 (2004) Design of composite steel and concrete structures - Part 1.1: General rules and rules for buildings. EN1994-1, Europe.
3. Pham D-D, Nguyen P-C (2021) Behavior analysis and design of concrete-filled steel circular-tube short columns subjected to axial compression. arXiv preprint arXiv:.06488.
4. Nguyen P-C, Pham D-D, Tran T-T, Nghia-Nguyen T (2021) Modified numerical modeling of axially loaded concrete-filled steel circular-tube columns. Eng Technol Appl Sci Res 11(3):7094–7099
5. Jayalekshmi S, Jegadesh JS, Goel A (2018) Empirical approach for determining axial strength of circular concrete filled steel tubular columns. J Inst Eng: Ser A 99(2):257–268
6. Zarringol M, Thai H-T, Thai S, Patel V (2020) Application of ANN to the design of CFST columns. Structures 28:2203–2220
7. Ren Q-X, Zhou K, Hou C, Tao Z, Han L-H (2018) Dune sand concrete-filled steel tubular (CFST) stub columns under axial compression: experiments. Thin-walled Struct 124:291–302

Matrix Strength Effects on the Tensile Resistance of Strain Hardening Fiber-Reinforced Concrete with High Strength Steel Fibers

Ngoc Thanh Tran and Ngoc Minh Phuong To

Abstract This paper aims to evaluate the influence of matrix strength on the tensile resistance of strain hardening fiber-reinforced concretes (SHFRCs) by performing direct tensile test. Two types of high strength steel fibers, twisted (T) and smooth (S) fibers, were reinforced at the volume content of 1.5% in different matrices, normal strength concrete (C1 with compressive strength of 28 MPa), high strength concrete (C2 with compressive strength of 84 MPa), and ultra-high strength concrete (C3 with compressive strength of 180 MPa). The test results indicated that T fibers in all matrices produced strain hardening behavior under direct tension whereas S fibers in C1 generated strain softening behavior. The tensile resistance of twisted fibers enhanced rapidly with the enhancement of matrix strength from 28 to 84 MPa, but increased slowly as the matrix strength enhanced from 84 to 180 MPa, while S fibers showed significant improvements in tensile resistance as the matrix strength enhanced. T fibers generated much better tensile resistance in C1 and C2 but slightly higher tensile resistance in C3 than S fibers. A SHFRC with post cracking strength of 11.5 MPa and strain capacity of 0.66% was obtained by using only 1.5% fiber volume content.

Keywords Ultra-high strength concrete · High strength steel fibers · Strain hardening behavior · Tensile resistance · Matrix strength

1 Introduction

The requirement for strong and tough building materials has recently been raised in order to enhance the resistance of civil structure [5]. One of favorable building materials for improving resistance of civil structure is strain hardening fiber-reinforced concrete (SHFRC) owing to its very high tensile strength, ductility and energy absorption capacity. However, recent research on SHFRC demonstrated that a large amount

N. T. Tran (✉) · N. M. P. To
Institute of Civil Engineering, Ho Chi Minh City University of Transport, Ho Chi Minh, Vietnam
e-mail: ngocthanh.tran@ut.edu.vn

of steel fibers (higher than 2.0% in volume) has been required to produce strain hardening behavior [6]. The high volume content of fiber results in the enhancement of cost and the decrease of workability of SHFRC and further limit the application of this material in practical structure [4]. Thus, the volume content of fiber should be minimized to promote the application of SHFRCs.

To achieve strain hardening behavior of SHFRC with optimum fiber volume content, the matrix-fiber interface bond properties play a significant role [3]. By utilizing high strength deformed steel fibers (HSDSF) with the additional mechanical bond, several SHFRCs with small amount of fiber but high tensile performance, have been established. Kim et al. [2] indicated that among four fiber types including twisted, hooked, spectra, and PVA fibers in matrix strength of 56 MPa, twisted fiber exhibited the best performance in term of load carrying capacity, energy absorption capacity and multiple cracking behavior whereas the PVA fibers showed the worst performance. In the same manner, Kim et al. [1] also concluded that deformed twisted fibers produced significant better performance than hooked fibers in high strength matrix of 84 MPa. However, the effectiveness of deformed steel fibers in ultra-high strength matrix (more than 150 MPa) showed the contrary tendencies. Wille et al. [9] demonstrated that deformed twisted fiber in ultra-high-strength concrete produced better performance in term of tensile strength and ductility than hooked and smooth fibers, ultra-high-strength concrete with 1.5% deformed steel fibers could generate a tensile strength of 13 MPa and a maximum strain capacity of 0.6% On the contrary, Yoo et al. [8] investigated that deformed steel fibers including twisted and hooked fibers produced poorer tensile performance than smooth fiber because of the severe matrix damage from excessive mechanical anchorage and fiber congestion of deformed steel fibers. Based on the above research, the effectiveness of high strength steel fibers (HSSF) in various matrix strength resulted in different tendencies. Hence, it is important to understand the influences of matrix strength on tensile resistance of SHFRCs with HSSF.

Thus, in this study, we proposed an experimental program to evaluate influence of matrix strength on the tensile resistance of SHFRCs with HSDSF in comparison to smooth steel fibers. The outcomes of this research are expected to investigate the suitable matrix composition and strength for HSSF in SHFRCs. Furthermore, the low cost SHFRCs with high tensile resistance can be developed.

2 Experimental program

In order to evaluate the influences of matrix strength on the tensile resistance of SHFRCs, an experimental program was prepared as shown in Fig. 1. Total six test series were tested. Two types of steel fiber including twisted (T) fibers and smooth (S) fibers were reinforced to three types of matrices including normal concrete (C1), high strength concrete (C2) and ultra-high strength concrete (C3) with different compressive strengths (28, 84 and 180 MPa). The dosages of fiber used was 1.5% by volume and each series had at least three specimens.

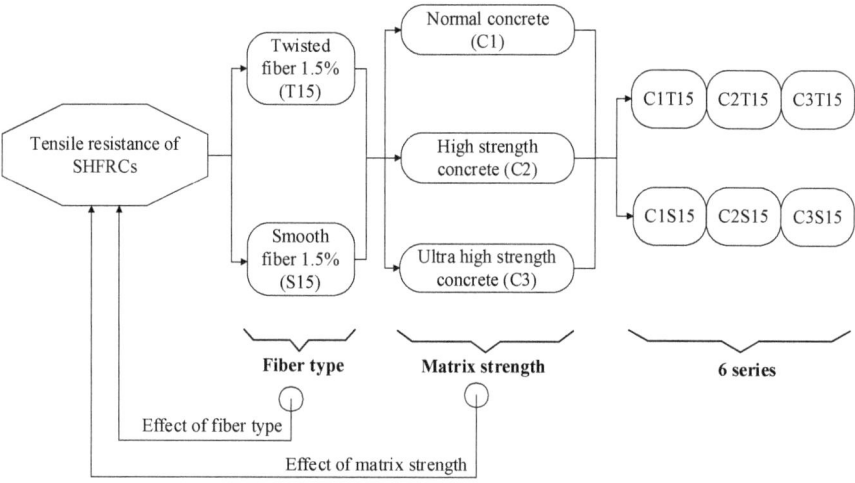

Fig. 1 Detail of experimental program

2.1 Materials and specimen preparation

Table 1 presents the composition of all matrices and their compressive strength. A wide range of matrix strengths from 28 to 180 MPa was evaluated. The water over cement ratios (W/C) of C1, C2, C3 are 0.65, 0.26 and 0.2, respectively, while their compressive strength regarding to cylinder specimens (D100 × 200) were 28, 84 and 180 MPa. Table 2 provides fiber properties: T fiber is deformed fiber using triangular section while S fiber is non-deformed fiber using circular section.

Table 1 Composition of matrices by ratio and compressive strength

Matrix	Cement	Fly ash	Sand	Silica fume	Silica powder	Super plasticizer	Water	Compressive strength at 14 days age (MPa)
C1	0.70	0.30	3.50	–	–	0.01	0.65	28
C2	0.80	0.20	1.00	0.07	–	0.04	0.26	84
C3	1.00	–	1.10	0.25	0.30	0.07	0.2	180

Table 2 Properties of high strength steel fibers

Fiber Type	Diameter (mm)	Length (mm)	Density (g/cm³)	Tensile strength (MPa)	Elastic modulus (GPa)
Twisted (T)	0.3	30	7.90	2760	200
Smooth (S)	0.3	30	7.90	2580	200

All the matrices were prepared by using a Hobart mixer. All dry materials except water, super-plasticizer and fibers in each matrix were first mixed within 5 min. Then, water was injected and mixed about 6 min. Next, super-plasticizer was slowly put and further mixed. When the mixture exhibited the sufficient workability and viscosity for fiber dispersion, fibers were slowly added into the mixture by using hand and continuously mixed for 2 min. After mixing, the mixture containing fibers was placed in molds and the casting direction was parallel to the length of the specimen. Then, a plastic sheet was used to wrap all the specimens at room temperature for two days. After the specimens were demolded, various water tanks with different temperatures were prepared to cure specimens. Specimens of C1 and C2 were put in water tank at 25°C during 14 days while specimens of C3 were immersed in water tank at 90°C during 3 days. At the age of 14 days, the direct tensile tests were set up to test all specimens.

2.2   Test setup and procedure

Figure 2. shows specimen geometry and direct tensile test set-up. Although various test methods such as splitting tests and bending tests have been applied to investigate tensile resistance of SHFRCs, the direct tensile test conducted with a clear force path has been reported to be the best method by which to accurately determine the tensile resistance of SHFRCs [7]. The specimens had dog-bone shape with the cross section of 25×50 mm^2. The gauge length was 100 mm and the boundary condition was hinge to hinge. A universal testing machine (UTM) was used to conduct the direct tensile test. The displacement velocity was controlled at 1 mm/min while the data

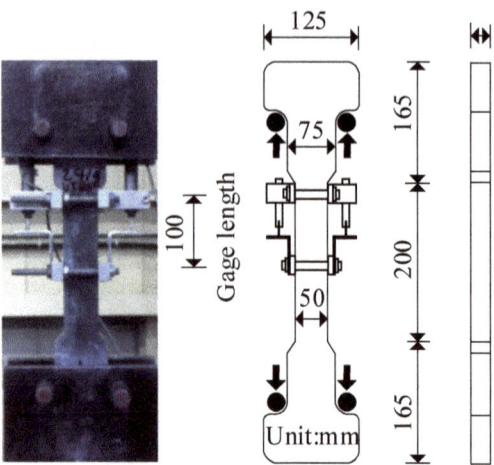

Fig. 2 Specimen geometry and test set up

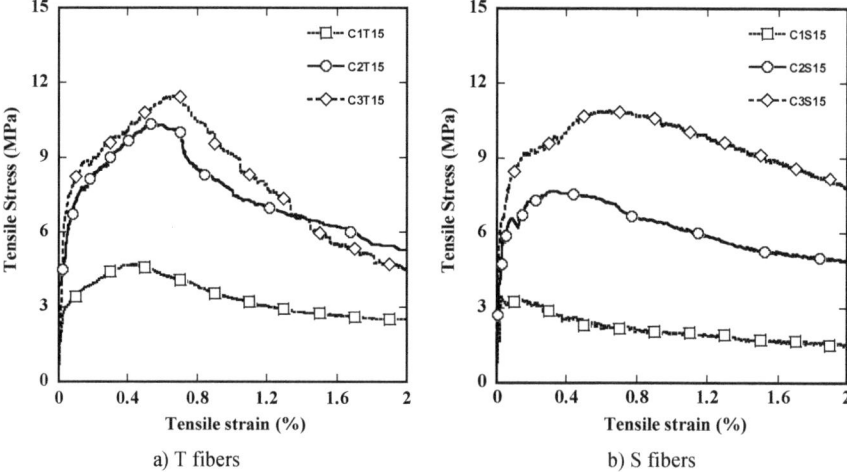

a) T fibers   b) S fibers

Fig. 3 Tensile stress and strain curves of SHFRCs **a** T fibers **b** S fibers

acquisition frequency was maintained at 1 Hz. A load cell was used to record the tensile stress while two linear variable differential transformers (LVDTs) were used to obtain the displacement. The direction of applied load was parallel to the length of the specimen.

3   Results and discussion

3.1   Tensile response of SHFRCs

The tensile stress and strain relationships of SHFRCs are illustrated in Fig. 3. T fibers showed strain hardening behavior in all matrices. On the contrary, S fibers exhibited strain softening behavior in C1 although they also produced strain hardening behavior in C2 and C3. The tensile behavior of SHFRCs was strongly dependent on matrix strength on fiber type. The higher matrix strength resulted in better performance and T fibers produced better performance than S fibers in the same matrix. The detail of matrix strength and fiber type effect will be discussed in detail in next parts.

3.2   Influence of matrix strength on the tensile resistance of SHFRCs

The tensile resistance of SHFRCs was estimated through following parameters: the post cracking strength represented to ultimate tensile strength, is the peak strength of tensile stress and strain curves and the strain capacity represented to ductility, is the strain at the peak strength.

Figure 4 shows influences of matrix strength on the tensile resistance including post cracking strength and strain capacity of SHFRCs. As the matrix strength increased the post cracking strength and the strain capacity of SHFRCs increased. However, the tensile resistance of T fibers increased rapidly with the enhancement of matrix strength from 28 to 84 MPa, but increased slowly as matrix strength enhanced from 84 to 180 MPa, while smooth fibers showed significant improvements in tensile resistance when the matrix strength enhanced. In detail, the post cracking strength and the strain capacity of T fibers increased 114% and 34%, respectively with the enhancement of matrix strength 28 to 84 MPa, but they increased 15% and 20%, respectively with the enhancement of matrix strength from 84 to 180 MPa. In addition, the post cracking strength and the strain capacity of S fibers increased 120% and 1451%, respectively with the enhancement of matrix strength from 28 to 84 MPa, and they increased 42% and 90%, respectively with the enhancement of matrix strength from 84 to 180 MPa. The enhancement of matrix strength produced the improvements in the adhesive-chemical and friction bond strength between fiber and surrounding matrix and further led to the increase in tensile resistance of SHFRCs [3]. Furthermore, the higher strength matrix containing finer aggregates, lower water-to-binder ratio and more cementitious material produced the lower porosity of the interfacial zone between fiber and matrix and the greater autogenous shrinkage [10]. On the other hand, a SHFRC with post cracking strength of 11.5 MPa and strain capacity of 0.66% was obtained by using only 1.5% fiber volume content.

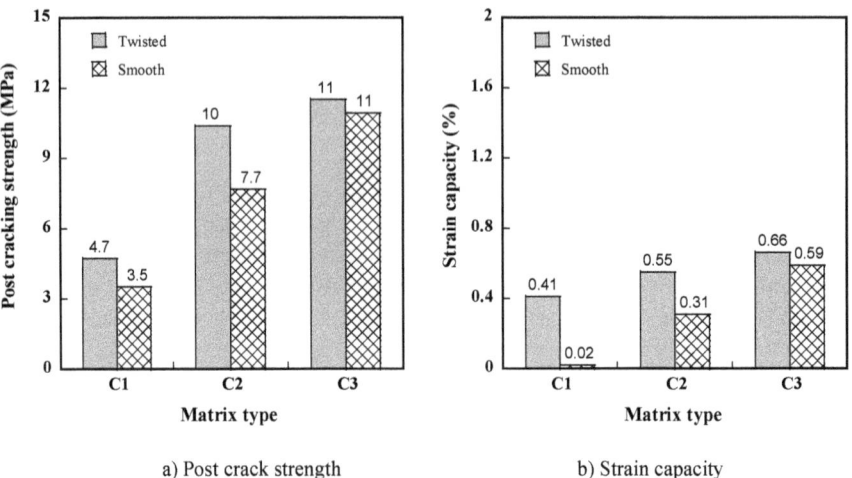

a) Post crack strength   b) Strain capacity

Fig. 4 Effect of matrix strength on the tensile resistance of SHFRCs **a** Post crack strength **b** strain capacity

3.3 Influence of fiber type on the tensile resistance of SHFRCs

The influence of fiber type on the tensile resistance of SHFRCs are also illustrated in Fig. 4. T fibers generated much better tensile resistance in C1 and C2 but slightly higher tensile resistance in C3 than S fibers. The post cracking strength of T fibers in C1, C2 and C3 was 34%, 30%, and 5% higher than those of S fibers, respectively while the strain capacity of T fibers was 1950%, 77%, and 12% higher than those of S fibers, respectively. T fibers exhibited better tensile resistance than S fibers because they utilized the mechanical bond strength owing to untwisted torsional moment resistance during pullout process [3]. However, the additional mechanical bond resistance of T fibers in C3 led to the unexpected severe damage of the surrounding matrix when fibers were pulled out and resulted in the slightly different in tensile resistance between twisted and smooth fibers.

4 Conclusion

An experimental program was conducted to evaluate the matrix strength influences on the tensile resistance of SHFRCs. Two types of high strength steel fibers and three types of matrix were considered. The following conclusions can be withdrawn:

- T fibers exhibited strain hardening behavior in all matrices. On the contrary, S fibers exhibited strain softening behavior in C1 although they also produced strain hardening behavior in C2 and C3.
- The tensile resistance of SHFRCs showed improvements when the matrix strength enhanced. However, the tensile resistance of T fibers increased rapidly with the enhancement of matrix strength from 28 to 84 MPa, but increased slowly with the enhancement of matrix strength from 84 to 180 MPa, while S fibers showed significant improvements in tensile resistance as the matrix strength enhanced.
- T fibers generated much better tensile resistance in C1 and C2 but slightly higher tensile resistance in C3 than S fibers.
- A SHFRC with post cracking strength of 11.5 MPa and strain capacity of 0.66% was obtained by using only 1.5% fiber volume content.

References

1. Kim DJ, El-Tawil S, Naaman AE (2009) Rate dependent tensile behavior of high performance fiber reinforced cementitious composites. Mater Struct 42:399–414
2. Kim DJ, El-Tawil S, Naaman AE (2008) Comparative flexural behavior of four fiber reinforced cementitious composites. Cem Concr Com 30:917–928

3. Ngo TT, Tran NT, Kim DJ, Pham TC (2021) Effects of corrosion level and inhibitor on pullout behavior of deformed steel fiber embedded in high performance concrete. Constr Build Mater 280:1–13
4. Tran NT, Tran TK, Jeon JK, Park JK, Kim DJ (2016) Fracture energy of ultra-high-performance fiber-reinforced concretes at high strain rates. Cem Concr Res 79:169–184
5. Tran NT, Tran TK, Kim DJ (2015) High rate response of ultra-high-performance fiber-reinforced concretes under direct tension. Cem Concr Res 69:72–87
6. Tran TK, Tran NT, Kim DJ (2021) Enhancing impact resistance of hybrid ultra-high-performance fiber-reinforced concretes through strategic use of polyamide fibers. Constr Build Mater 271:1–16
7. Tran TK, Tran NT, Nguyen DL, Kim DJ, Park JK, Ngo TT (2021) Dynamic fracture toughness of ultra-high-performance fiber-reinforced concrete under impact tensile loading. Struct Concr 22:1845–1860
8. Yoo DY, Kim S, Kim JJ, Chun B (2019) An experimental study on pullout and tensile behavior of ultra-high-performance concrete reinforced with various steel fibers. Constr Build Mater 206:46–61
9. Wille K, Kim DJ, Naaman AE (2011) Strain-hardening UHP-FRC with low fiber contents. Mater Struct 44:583–598
10. Wu Z, Shi C, Khayat KH (2019) Investigation of mechanical properties and shrinkage of ultra-high performance concrete: Influence of steel fiber content and shape. Compos B Eng 174:1–12

Numerical Investigation and Comparative Study of Aerodynamic Characteristics in Rectangular, Corner Cut and Chamfered Sections by Using LES

Daichi Tanimoto and Hiroshi Katsuchi

Abstract In recent years, the length of bridges increases, and the main towers are becoming taller, thus it is necessary to improve the wind resistant performance of their main towers. Currently, corner-cut and chamfered cross-sections are used as static countermeasure of vibration in towers. However, there were some cases of vortex excitation due to the wind direction and cross-sectional shape when the main tower of a real cable-stayed bridge was examined. Therefore, it is required to examine the effective cross-section by understanding the flow and how corner-cut controls the separated flow. In this study, a comparative study of rectangular, corner-cut, and chamfered cross sections was conducted by using computational fluid dynamics analysis. It was confirmed that the drag coefficient and the variation components of lift coefficients were smaller in the corner-cut and chamfered cross sections than in the rectangular cross sections. It was also confirmed that the width of the separated flow was smaller in the corner-cut and chamfered cross section than in the rectangular cross section, and the pressure loss at the top surface and in the wake was also smaller. In addition, there was no significant difference in the aerodynamic characteristics between the corner cut and chamfered cross sections.

Keywords Corner cut · Chamfered · Aerodynamic characteristics · Tower · CFD

1 Introduction

Most of the main towers of long span bridges are designed on the basis of rectangular cross-sections with a side ratio of 0.5 to 1.0. As shown by Nakaguchi et al. [1], the drag and pressure coefficients of rectangular cross-sections reach their maximum values at a side ratio of 0.5 to 0.7, and the cross-section is known to be aerodynamically unstable. Okajima et al. [2] concluded that the rectangular bluff body with corner-cut is superior to the rectangular bluff body in terms of wind resistance. Therefore, the corner-cut cross-section is used for the main towers of long-span bridges as a static

D. Tanimoto · H. Katsuchi (✉)
Structural Engineering, Yokohama National University, Yokohama, Japan
e-mail: katsuchi@ynu.ac.jp

© The Author(s), under exclusive license to Springer Nature Singapore Pte Ltd. 2023
J. N. Reddy et al. (eds.), *ICSCEA 2021*, Lecture Notes in Civil Engineering 268,
https://doi.org/10.1007/978-981-19-3303-5_81

countermeasure of vibration. However, an out-of-plane bending response of 12 cm has been observed in the Hakucho Suspension Bridge in Hokkaido, Japan, at a wind speed of about 5 m/s (when the vibration control system was stopped) under construction [3]. In the Great Bridge in Denmark, out-of-plane bending motion of 40 cm was observed at a wind speed of 23 m/s during under construction [4], indicating that the main towers are shaken by wind. In addition, experiments on the cross-section of A real cable-stayed bridge have confirmed that vortex excitation occurs depending on the wind direction and cross-sectional shape. Furthermore, the instability of vehicles passing through the leeward side of the main towers is an issue. Thus, it is necessary to understand the detailed flow around the corner cut section and to clarify the pressure distribution around the section. In this study, the aerodynamic force coefficients, pressures, and flow velocities of chamfered, corner-cut, and rectangular cross sections were compared using computational fluid dynamics.

2   Numerical Method

2.1   Algorithms

The turbulence model is the LES model, which spatially filters the Navier–Stokes equations to separate the GS (Grid Scale) component, which can be solved on a grid, from the SGS (Sub Grid Scale) component, which cannot be solved on a grid. While the static SGS model uses model coefficient, the dynamic SGS model determine the model coefficient dynamically. It has many desirable features such as correct asymptotic behavior near a solid wall and in laminar flow, capability of dealing with energy backscatter, etc. In addition, compared to the basic Smagorinsky model, the one-equation SGS model overcomes the deficiency of the local balance assumption between the SGC energy production and dissipation adopted in algebraic eddy viscosity models that occur in high Reynolds number bands, and can be successfully used in high Reynolds analysis [5]. Therefore, the Dynamic One Equation Model was used in this study. The Dynamic One Equation Model developed by Huang et al. [5] is shown below: The Navier–Stokes equation is filtered by assuming eddy viscosity to obtain Eq. (1)

$$\frac{\partial \rho \overline{u_i}}{\partial t} + \frac{\partial \left(\rho \overline{u_i u_j}\right)}{\partial x_j} = -\frac{\partial}{\partial x_j}\left(\overline{p} + \tfrac{2}{3}\rho k_{SGS}\right) + \frac{\partial}{\partial x_j}\left[2\rho(\nu_{SGS} + \nu)\overline{S}_{ij}\right] \qquad (1)$$

where u_i is the GS component of velocity, ν is the kinetic viscosity of fluid, $\overline{p}$ represents the GS component of pressure, ρ is the fluid density, ν_{SGS} is the SGS eddy viscosity and is given by

$$\nu_{SGS} = C_\nu \Delta_\nu \sqrt{k_{SGS}} \qquad (2)$$

where C_v denotes a constant instead of a dynamically determined parameter, Δ_v represents the characteristic length, and k_{SGS} is the SGS kinetic energy. Δ_v is calculated as

$$\Delta_v = \frac{\overline{\Delta}}{1 + C_k \frac{\overline{\Delta}^2 \overline{S}^2}{k_{SGS}}} \tag{3}$$

This equation is used to meet the appropriate asymptotic behaviour to the wall. In which C_k denotes a model constant, and $\overline{S}$ is defined as $\overline{S} = \sqrt{2\overline{S}_{ij}\overline{S}_{ij}}$ Besides, k_{SGS} is obtained from a transportation equation as follows

$$\frac{\partial k_{SGS}}{\partial t} + \overline{u}_j \frac{\partial k_{SGS}}{\partial x_i} = -\tau_{ij}\overline{S}_{ij} - C_\varepsilon \frac{k_{sgs}^{3/2}}{\overline{\Delta}} + \frac{\partial}{\partial x_i}\left[(C_d \Delta_v \sqrt{k_{SGS}} + v) \frac{\partial k_{SGS}}{\partial x_i} \right] - \varepsilon_w \tag{4}$$

The last term ε_w is an additional dissipation term given by

$$\varepsilon_w = 2v \frac{\partial \sqrt{k_{SGS}}}{\partial x_j} \frac{\partial \sqrt{k_{SGS}}}{\partial x_j} \tag{5}$$

The production term $-\tau_{ij}\overline{S}_{ij}$ is constructed as follows:

$$-\tau_{ij}\overline{S}_{ij} = \left((C_w^* \overline{\Delta})^2 \frac{\left(S_{ij}^d S_{ij}^d\right)^{3/2}}{\left(\overline{S}_{ij}\overline{S}_{ij}\right)^{5/2} + \left(S_{ij}^d S_{ij}^d\right)^{5/4}} \right) |\overline{S}|^2 - \frac{2}{3} k_{SGS} \delta_{ij} \overline{S}_{ij} \tag{6}$$

where C_w^* is modified model constant and S_{ij}^d denotes the traceless symmetric part of the square of the velocity gradient tensor. Using the above model, the present study was conducted using the Pimple algorithm, an unsteady solver for incompressible flows, with density: $\rho = 1.205 \text{ kg/m}^3$ and kinematic viscosity: $v = 1.511 \times 10^{-5} \text{ m}^2/\text{s}$. The drag coefficient, Strouhal number, and pressure coefficient used to verify the accuracy of the model and other factors are defined as follows.

$$C_D = \frac{F_D}{0.5\rho U_\infty^2 S} \tag{7}$$

where C_D is drag force coefficient, and F_D denotes the drag force, S is the projected area, U_∞ represents the velocity at infinity.

$$C_P = \frac{p - p_\infty}{0.5\rho U_\infty^2} \tag{8}$$

where C_P is the pressure coefficient, and p_∞ represents the pressure at infinity.

$$S_t = \frac{fD}{U} \tag{9}$$

where f is the frequency and D denotes the reference length.

2.2 Analysis Conditions

According to the experiments of Okajima et al. [2], the jump Reynolds of the cross-section with edge length ratio (B/D) of 0.66 and corner cut ratio of 0.115 (h/D) and 0.180 (b/D) was confirmed to change the aerodynamic characteristics significantly at Re = 8000.Therefore, in this study, we set Re = 10,000 and conducted the analysis. The object was set to D = 0.06 m and B = 0.0396 m in order to achieve the same conditions as in the previous experiments (Fig. 1). The other conditions are shown in Table 1.

The model and mesh were created using the academic version of ANSYS. In this study, a three-dimensional model was created, and the dimensions were determined to keep the blockage ratio below 5%. In numerical fluid analysis, the axial length may affect the analysis results depending on the Reynolds number and the target cross section. The results of verifying the influence of the axial direction for a rectangular cross section with a side ratio of 0.66 are shown in Figs. 2 and 3.

From these results, we conclude that the axial length of 0.096 (m) in a rectangular cross section is sufficiently convergent. In addition, when using LES for turbulence modeling, it is generally necessary to set y^+ defined in Eq. (10) to be less than 1, and

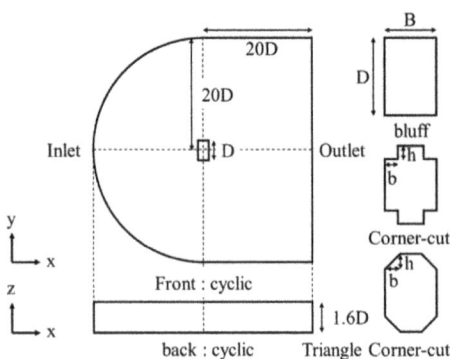

Fig.1 Measurement and Boundary Conditions

Table 1 Specification

Conditions	(Unit)	Conditions	(Unit)
Inlet(x,y)	(2.5, 0) (m/s)	y^+	1
Air density	1.205 (kg/m^3)	Analysis	Implicit method
Viscosity coefficient	1.821E−05 (Pa*s)	Time	5.0 (s)
Re number	1.0×10^4	Time step	CoNumber < 2
Turbulence intensity	1%	B/D	0.66
Temperature	25°C	b/D	0.0115
Wind tunnel	1.21×10^{-1} (m^2)	h/D	0.018

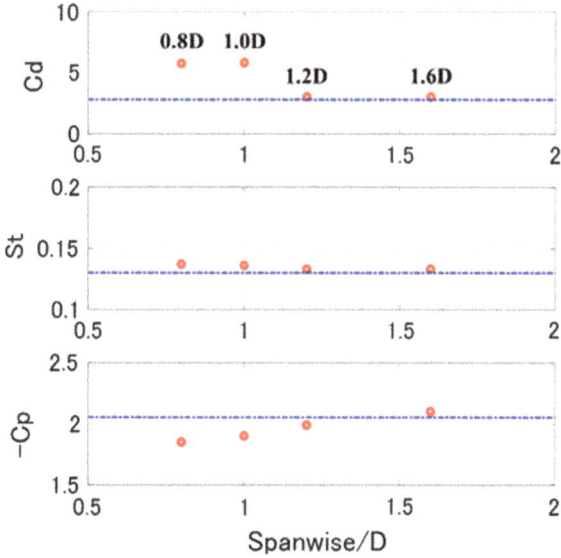

Fig.2 Convergence of each value in rectangular cross section in axial direction

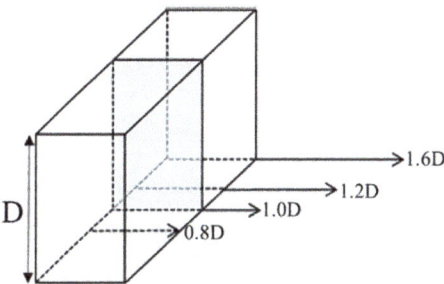

Fig.3 Schematic of the model

the mesh size was determined based on these results.

$$y^+ \equiv \frac{y}{\nu}\sqrt{\frac{\tau_w}{\rho}} \tag{10}$$

where y is the distance from wall and ν denotes kinematic viscosity, τ_w is wall shear stress given by

$$\tau_w = \rho\nu\left(\frac{\partial u}{\partial y}\right)_y \tag{11}$$

3 Numerical Results

3.1 Static Forces and Validation

The accuracy of the model is verified by comparing the results of previous experiments and analyses. The drag coefficient, the Strouhal number, and the pressure coefficient measured at the center of the leeward face are compared with the experiments of Nakaguchi et al. [1] and Nishimura et al. [6] in a rectangular cross section. Although the drag coefficient is overestimated by the analytical results, it is within the maximum error of about 5%. The Strouhal number was overestimated by the analysis, but the error was less than 7%. The pressure coefficient is underestimated by the analysis, but the maximum error is less than 8%. Therefore, the reproducibility of this analysis for rectangular cross sections was confirmed. For the corner-cut section, the results were compared with the experimental results of Okajima et al. [2] using the Strouhal number and the pressure coefficient on the upper side. In the experiment, the pressure was measured at points P1 to P4, and the largest error was observed at P4. The Strouhal number was 0.23 in the experiment, and two Strouhal numbers of 0.179/0.21 were confirmed in this analysis. These results indicate that the bluff body with corner-cut is not sufficient, but some reproducibility was confirmed (Fig. 4, Table 2).

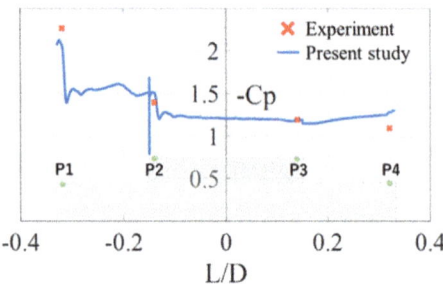

Fig.4 Validation of Corner-Cut model for Cp

Table 2 Validation of Rectangular model

	Re	Cd	St	-Cpd
Present study	10,000	2.84	0.138	1.89
Nakaguchi	20,000	2.8	0.13	2.05
Nishimura	20,000	2.7	0.13	1.85

Table 3 Comparison of Drag and fluctuation component of Lift Coefficient

	Rectangular	Corner-cut	Chamfered
Cd	2.68	1.88	1.86
Cl rms	1.504	1.027	1.057
Velocity in the Wake [u/U]	0.350	0.490	0.500

3.2 Comparison of Drag, Lift Coefficient and Velocity

The flow velocity at a point 2.5D away from the center of the leeward face is shown in Table 3, as well as the drag coefficient and the variable components of lift coefficient for rectangular, corner-cut and chamfered cross sections. The drag coefficient showed a similar trend in the corner-cut and chamfered sections and was about 30% smaller than that in the rectangular section. The variation of the lift coefficient was about 32% smaller in the corner-cut cross section and about 30% smaller in the chamfered cross section than in the rectangular cross section. In addition, the flow velocity in the wake was 40% larger in the corner-cut cross section and 43% larger in the chamfered cross section than in the rectangular cross section. This indicates that the velocity loss in the wake of the corner-cut and chamfered sections is smaller than that of the rectangular section.

3.3 Velocity at Leading and Trailing Edge

Figure 5 is an overview of the flow velocity measurement points set up at the leading and trailing edges and the results measured at each cross section. A group of measuring points is as shown in Fig. 5. It should be noted that the position of the first separation point is different between the rectangular section and the corner cut and chamfer section. With the bottom of the leading edge of the corner-cut as a reference, the corner-cut and chamfered sections showed similar velocity distributions, and the separation width which is defined the distance from the bottom of the leading edge of the corner cut to the maximum velocity point was reduced by about 50% compared

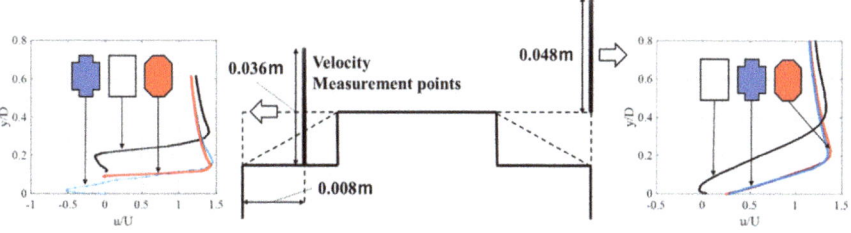

Fig. 5 Velocity at leading and trailing edge

to the rectangular section. If the separation width is defined as the distance from the first separation point to the maximum velocity point, the separation width is reduced by about 23% in the corner-cut cross section and by about 30% in the chamfered cross section compared to the rectangular cross section. At the trailing edge, a group of measurement points is shown in Fig. 5. The separation width of the trailing edge which is defined as the distance from the corner of the rectangular cross section to the maximum flow velocity point decreased by about 52% in the corner-cut cross section and by about 49% in the chamfered cross section. This is because the position of the first separation point was changed by installing a corner cut or chamfer at the leading edge, and the distance between the separated flow and the top surface was reduced.

3.4 Pressure Coefficients at Top Surface

Pressure measurement points were placed along the top surface of each rectangular, corner-cut and chamfered cross section (Fig. 6). The pressure coefficients obtained by time-averaging the measured pressures and applying them in Eq. (8). The pressure coefficients for the rectangular section and the corner cut are similar at the leading edge. The reason for this is that the relationship between the separated flow and the wall surface in the rectangular and corner sections is similar. The pressure coefficient of the rectangular section is linear at the front edge, while that of the corner-cut section is unstable. The pressure coefficients of the corner-cut and chamfered sections are larger when $x/D = -0.15$. This trend is also observed at the first separation point, indicating that the separated flow occurs at this point too, which is the second separation point. The pressure was measured up to a distance of 0.1 m from the center of the leeward face. Figure 7 shows the pressure coefficients obtained by time-averaging the measured pressures and applying them in Eq. (8). At the point where the pressure coefficient is maximum, the pressure coefficient is about 12%

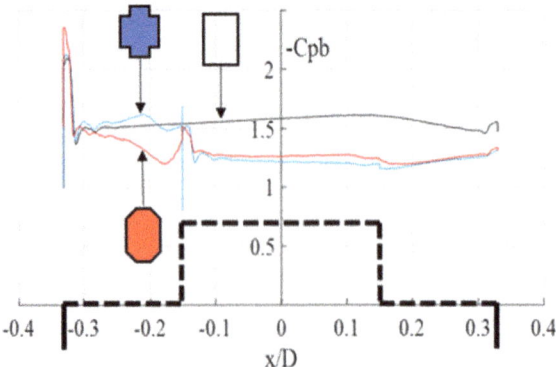

Fig. 6 Pressure Coefficients at the Top Surface

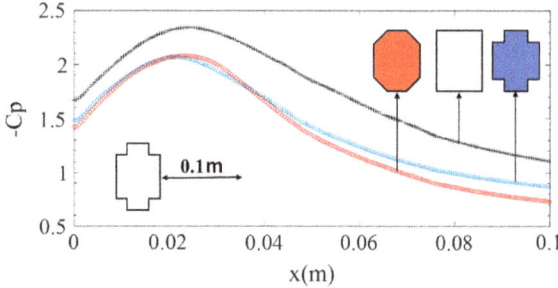

Fig. 7 Pressure coefficient at the Wake for Bodies

lower for the corner-cut cross section and about 11% lower for the chamfered cross section compared to the rectangular cross section. In other words, the pressure loss in the wake is smaller in the corner-cut and chamfered cross sections than in the rectangular cross section.

4 Conclusion

The aerodynamic characteristics of rectangular, corner-cut and chamfered cross sections were compared using CFD. The following is a summary:

(1) There is no significant difference in the drag coefficient and the variation components of lift coefficients, surrounding pressure coefficients and flow velocity distributions between the corner-cut and chamfered cross-sections, and the aerodynamic characteristics are likely to change by the same mechanism.

(2) Installing a corner cut or chamfer at the leading edge changes the position of the first separation point and the separation width decreases in the corner cut and chamfered cross section compared to the rectangular cross section. As a result, the distance between the separated flow and the top surface decreases, and the pressure coefficient at the top surface becomes smaller than that of the rectangular cross section.

(3) The pressure and velocity losses in the wake are reduced due to the smaller separation width.

(4) Thus, the pressure difference between the windward face and leeward face is reduced in the corner-cut and chamfered cross sections compared to the rectangular cross section, and the drag coefficient becomes smaller.

As future work, it is necessary to investigate the reason why the separation width of the corner-cut and chamfered models decreases when the separation width is defined as the distance between the first separation point and the maximum velocity point at the leading edge. Besides, A model with an inclined angle to the wind at each cross-section should be analyzed to study the effect.

References

1. Nakaguchi H, Hashimoto H, Muto S (1968) An experimental study on aerodynamic drag of rectangular cylinders. J Jpn Soc Aeronaut Eng 16(168):1–5
2. Okajima A, Ueno H, Abe A (1991) Influence of Reynolds number on flow and aerodynamic characteristics of bluff bodies with rectangular section of cut corners. J Wind Eng Ind Aerodyn 49:1–13
3. Siringoringo MD, Fujino Y (2012) Observed along-wind vibration of a suspension bridge tower. J Wind Eng Ind Aerodyn 103:107–121
4. Larose GL, Zasso A, Melelli S, Casanova D (1998) Field measurements of the wind-induced response of a 254 m high free-standing bridge pylon. J Wind Eng Ind Aerodyn 74–76:891–902
5. Huang S, Li QS (2010) A new dynamic one-equation subgrid-scale model for large eddy simulations. Int J Numer Meth Eng 81:835–865
6. Nishimura H (2002) A study of the wind force on several rectangular prisms. J Appl Mech 6:689–698

Post-buckling Analysis of Circular Functionally Graded Microplates Based on Isogeometric Analysis

Son Thai, Dieu T. T. Do, Vu Xuan Nguyen, and Qui X. Lieu

Abstract This paper presents a novel numerical model for investigating the post-buckling response of circular Functionally Graded (FG) microplates. The size-dependent effect, which is observed experimentally for small-scale structures, is captured based on the Modified Strain Gradient Theory (MSGT). The Third-order Shear Deformation Theory (TSDT) proposed by Reddy is used to represent the kinematic relations of the plates and capture the shear deformation effect, while effective material properties varing in the thickness direction of the plates follow the mixture rule. The governing equations of the post-buckling problems are derived by using the principle of virtual work and then are discretized following the framework of Isogeometric analysis (IGA) by using Non-Uniform Rational B-Splines (NURBS) basis function to satisfy the C^2-continuity requirement. Various numerical examples are conducted to verify the accuracy of the proposed numerical model and study the effects of size parameters and material distributions on the post-buckling responses of circular microplates.

Keywords Microplates · Functionally graded materials · Post-buckling · Geometrical nonlinearity · Isogeometric analysis

1 Introduction

In recent years, the use of Functionally graded (FG) microbeams and microplates as components in advanced technologies has become popular and attract considerable

S. Thai (✉) · V. X. Nguyen · Q. X. Lieu
Faculty of Civil Engineering, Ho Chi Minh City University of Technology (HCMUT), 268 Ly Thuong Kiet Street, District 10, Ho Chi Minh City, Vietnam
e-mail: son.thai@hcmut.edu.vn

Vietnam National University Ho Chi Minh City, Linh Trung Ward, Thu Duc City, Ho Chi Minh City, Vietnam

D. T. T. Do
Duy Tan Research Institute for Computational Engineering (DTRice), Duy Tan University, Da Nang 550000, Vietnam

© The Author(s), under exclusive license to Springer Nature Singapore Pte Ltd. 2023 893
J. N. Reddy et al. (eds.), *ICSCEA 2021*, Lecture Notes in Civil Engineering 268,
https://doi.org/10.1007/978-981-19-3303-5_82

attention from various researchers [1]. For small-scale structures, their structural responses under mechanical effects are remarkably influenced by the so-called size effect, which was observed and verified in experimental studies [2]. It has been claimed that utilization of the classical elasticity is not adequate to account for the size effect in those small-scales structures. Consequently, various non-classical continuum theories have been proposed to account for this mechanical phenomenon of structures of small sizes. The common point of those theories is the adoption of the so-called length-scale parameters, which are introduced in additional terms of the constitutive equations or the strain energy formula. Amongst those theories, the MSGT proposed by Lam et al. [2] has been widely adopted by a great number of studies to investigate the mechanical response of microbeams and microplates. In the MSGT, three length-scale parameters, which are related to the dilatation gradient tensor, deviatoric gradient tensor, and rotation gradient tensor, are introduced to capture the size-dependent effects. Recent studies on the mechanical behavior of FG microplates using the MSGT have been conducted based on analytical approach [3], finite-strip approach [4], and IGA approach [5–7]. It is noted that the analytical approaches are only suitable for the plate problems with simples geometries and boundary conditions, while other numerical methods must fulfill the requirements of higher-continuity of interpolation functions when the MSGT is employed. To eliminate the drawback of the traditional finite element method on geometrically modeling circular plates and fulfill the requirements of higher continuity of inter-polation functions, the IGA approach [8, 9] is threrefore adopted in this study to investigate the post-buckling response of circular FG microplates. The fundamental concept of IGA is to incorporate the tools of Computer-Aided Design (CAD) with the numerical methods, where the NURBS basis functions inherited from CAD are employed to construct smooth and high-order continuity interpolations.

This study aims to develop a numerical model to study the post-buckling response of FG microplates by using MSGT and IGA approach. Governing equations of geometrically nonlinear problems are derived in accordance to the principle of virtual work, then the NURBS basic functions are used to construct the global nonlinear system equations. The arc-length method is adopted to solve the nonlinear equation to trace the post-buckling paths. Various numerical examples are also conducted to verify the accuracy and investigate the influence of the size-dependent effect.

2 Theoretical Formulations

According to the MSGT [2], the virtual strain energy is expressed as a summation of four distinct components of classical strain tensor and high-order gradient tensors as follow:

$$\delta U = \int\limits_{V} \left(\sigma_{ij}\delta\varepsilon_{ij} + p_i\delta\varsigma_i + \tau^{(1)}_{ijk}\delta\eta^{(1)}_{ijk} + m^s_{ij}\delta\chi^s_{ij} \right) dV \tag{1}$$

in which ε_{ij} is the classical strain tensor based on von Karman's assumption, ς_i is the dilatation gradient tensor, $\eta_{ijk}^{(1)}$ is the deviatoric gradient tensor, and χ_{ij}^s is the symmetric part of rotation gradient tensor. Those terms are given as follows

$$\varepsilon_{ij} = \left(u_{i,j} + u_{j,i} + u_{m,i}u_{m,j}\right)/2 \tag{2}$$

$$\varsigma_i = \varepsilon_{mm,i} \tag{3}$$

$$\chi_{ij}^s = \left(e_{imn}u_{n,mj} + e_{jmn}u_{n,mi}\right)/4 \tag{4}$$

$$\eta_{ijk}^{(1)} = \eta_{ijk}^s - \left(\delta_{ij}\eta_{mmk}^s + \delta_{jk}\eta_{mml}^s + \delta_{ki}\eta_{mmj}^s\right)/5 \tag{5}$$

$$\eta_{ijk}^s = \left(u_{i,jk} + u_{j,ki} + u_{k,ij}\right)/3 \tag{6}$$

where u_i denote the components of the displacement vector, δ_{ij} and e_{imn} represent the Knocker delta and permutation symbols, respectively. The classical stress and high-order stresses work-conjugating to the deformation measures are expressed by

$$\sigma_{ij} = 2\mu\varepsilon_{ij} + \lambda\varepsilon_{kk}\delta_{ij} \tag{7}$$

$$p_i = 2\mu l_0^2 \varsigma_i \tag{8}$$

$$\tau_{ijk}^{(1)} = 2\mu l_1^2 \eta_{ijk}^{(1)} \tag{9}$$

$$m_{ij}^s = 2\mu l_2^2 \chi_{ij}^s \tag{10}$$

in which λ and μ denote the Lamé constants, l_0, l_1 and l_2 represents the length scale parameters.

The material properties vary in the thickness direction of FG microplates according to the mixture rule [10], which are expressed as follows

$$P(z) = (P_c - P_m)(z/2 + 0.5)^n + P_m \tag{11}$$

where $P(z)$ could be a typical material property, i.e. Young's modulus $E(z)$, Poisson's ratio $v(z)$. P_c and P_m are the properties of ceramic and metal materials, respectively, and n denotes the gradient index used to describe the profile of material gradation through the plates' thickness. By adopting the TSDT of Reddy [11], the displacement field is given by

$$
\begin{cases}
u_1 = u + f(z)\theta_x - g(z)w_{,x} \\
u_2 = v + f(z)\theta_y - g(z)w_{,y} \\
u_3 = w
\end{cases}
\qquad
\begin{cases}
f(z) = z - 4z^3/3h^2 \\
g(z) = 4z^3/3h^2
\end{cases}
\tag{12}
$$

where (u, v, w) and (θ_x, θ_y) are the displacements and rotations of an arbitrary point in the mid-plane of the plate, and h is the thickness of the plates.

By substituting Eqs. (2)–(6) to Eqs. (7)–(10) and taking integral over thickness h, the constitutive relations are expressed in terms of stress resultants forms as follows

$$
\hat{\sigma} = \hat{\mathbf{D}}_\varepsilon \{\hat{\varepsilon} + \hat{\varepsilon}_{nl}\}
\tag{13}
$$

$$
\hat{\mathbf{p}} = \hat{\mathbf{D}}_\varsigma \{\hat{\varsigma} + \hat{\varsigma}_{nl}\}
\tag{14}
$$

$$
\hat{\tau} = \hat{\mathbf{D}}_\eta (\hat{\eta} + \hat{\eta}_{nl})
\tag{15}
$$

$$
\hat{\mathbf{m}} = \hat{\mathbf{D}}_\chi \hat{\chi}
\tag{16}
$$

in which $\hat{\sigma}$, $\hat{\mathbf{p}}$, $\hat{\tau}$ and $\hat{\mathbf{m}}$ are stress resultants corresponding to classical stress tensor and high-order stress tensors. $\hat{\varepsilon}$, $\hat{\varsigma}$, $\hat{\eta}$, and $\hat{\chi}$ denote the linear parts of classical strain tensor and strain gradient tensors, while $\hat{\varepsilon}_{nl}$, $\hat{\varsigma}_{nl}$, and $\hat{\eta}_{nl}$ are those related to nonlinear terms. Details of those can be found in [5–7]. It is noted that the matrix $\mathbf{\Gamma}_\eta$ is mistyped in those studies and it should be

$$
diag(\mathbf{\Gamma}_\eta) = \{1\ 1\ 1\ 3\ 3\ 6\ 3\ 3\ 3\}
\tag{17}
$$

By utilizing the small strain assumption and the principle of virtual work, the governing equation of the post-buckling problem can be written with respect to the initial configuration as follows

$$
\int_\Omega \delta\left(\hat{\varepsilon} + \frac{1}{2}\varepsilon_{nl}\right)^T \hat{\mathbf{D}}_\varepsilon \left(\hat{\varepsilon} + \frac{1}{2}\varepsilon_{nl}\right) + \int_\Omega \delta(\hat{\varsigma} + \varsigma_{nl})^T \hat{\mathbf{D}}_\varsigma \left(\hat{\varsigma} + \varsigma_{nl}\right) d\Omega
$$
$$
+ \int_\Omega \delta(\hat{\eta} + \eta_{nl})^T \hat{\mathbf{D}}_\eta \mathbf{\Gamma}_\eta (\hat{\eta} + \eta_{nl}) d\Omega + \int_\Omega \delta\hat{\chi}^T \hat{\mathbf{D}}_\chi \mathbf{\Gamma}_\chi \hat{\chi} d\Omega = \int_S \delta\mathbf{u}^T \hat{\mathbf{t}} dS
\tag{18}
$$

where $\hat{\mathbf{t}}$ denotes the traction force applied in the boundary S.

3 NURBS-Based Discretizations

In this study, the NURBS basis functions are used to interpolate the dependent variables and the geometries of circular plates, the displacement variables $\bar{\mathbf{u}} =$

$\{ u \ v \ \theta_x \ \theta_y \ w \}^T$ are approximated by

$$\bar{\mathbf{u}} = \sum_{c}^{m \times n} R_c(\xi, \eta) \mathbf{d}_c \tag{19}$$

in which R_c is the NURBS basis function associated with control c, $m \times n$ represents the number of control points of an element, and $\mathbf{d}_c = \{ u_c \ v_c \ \theta_{xc} \ \theta_{yc} \ w_c \}^T$ is the corresponding vector of degree of freedoms of a control point. The global nonlinear system equations for post-buckling problems is

$$\left(\mathbf{K}_\varepsilon + \mathbf{K}_\varsigma + \mathbf{K}_\eta + \mathbf{K}_\chi \right) \mathbf{d} = \mathbf{f}_m \tag{20}$$

The details of $\mathbf{K}_\varepsilon$, $\mathbf{K}_\varsigma$, $\mathbf{K}_\eta$, and $\mathbf{K}_\chi$ are presented in [5, 7]. It was observed that a C^2-continuity requirement of interpolations must be met to construct those stiffness matrices. By using NURBS basic functions, this demand is accommodated efficiently [5, 6]. The post-buckling paths (load-dsiplacement relations) are traced by adopting the Arc-length technique with imperfection, where the deformed geometry of the first mode obtained in linear buckling analysis is employed as a small initial imperfection with an amplitude of $h/1000$ and the criteria for converged solutions is 0.1%.

4   Numerical Examples

To verify the accuracy of the proposed numerical model, a post-buckling problem of a circular plate investigated previously in [12] and [13] is revisited. The circular plate, as illustrated in Fig. 1a, has a radius of $r = 3$ m and a thickness of $h = 0.0625$ m. The plate is clamped movable ($\theta_x = \theta_y = w = 0$ at the circumference) and made from homogeneous material with $E = 3 \times 10^7$ kN/m^2, $v = 0.3$. The results of post-buckling paths obtained from this study and those from [12] and [13] are presented in Fig. 1b, in which the maximum deflection w/h ratio occurring at the center of the plate is plotted versus the compressive load parameter ratio t/t_{cr}. It can be seen that a good agreement between the result obtained from the present study and those from [12] and [13] is achieved. A slight difference between the present path and that given in [13] could be attributed to the neglection of the shear deformation effect. The computed linear buckling load is $t_{cr} = 1094.02$ kN, while the exact is 1094.01 kN. It is noted that a mesh convergence study was conducted and the mesh size of the plate is 8×8 with cubic basic functions yields good solutions with the lowest computational time, hence this mesh size is used for the remaining numerical examples.

To study the influences of size-dependent effect and material variation through the plate thickness, circular microplates with a radius of 10 μm are employed. The plates are assumed to be made from two material phrases: Al ($E_m = 70$ GPa, $v_m =$

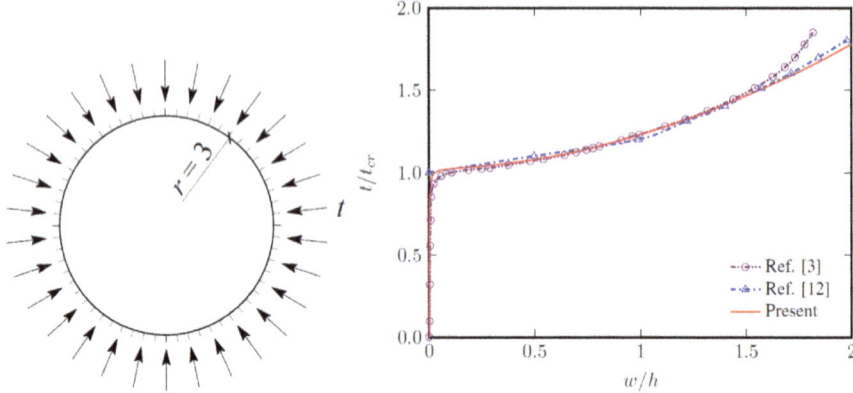

Fig. 1 a Circular plate with clamped movable boundary and **b** Load deflection curve at the center of the plate

0.3) and Al_2O_3 ($E_c = 380$ GPa, $\nu_c = 0.3$) with the bottom surface being Al-rich and the top surface being Al_2O_3-rich.

In Fig. 2, the influence of the material gradient index (n) on the post-buckling paths is presented. The plate in this example has a radius of $r = 10$ μm, $h = r/100$, the length scale parameters are assumed to be $l_0 = l_1 = l_2 = 0.1\ h$. Overall, it is seen that the buckling load ($\bar{t} = tr^2 / E_m h^3$) decreases with the increase of n. This is due to the fact that the elevation of n gives rise to metallic volume in the plate, consequently, the stiffness of the plates is reduced. Moreover, the buckling load of clamped movable plates is much higher than that of simply supported movable plates ($w = 0$ at the circumference). Another important point should be pointed out that the bifurcation buckling is only observed for clamped plates and simply supported plates with homogeneous material distributions ($n = 0$, $n = \infty$). It is noted that the

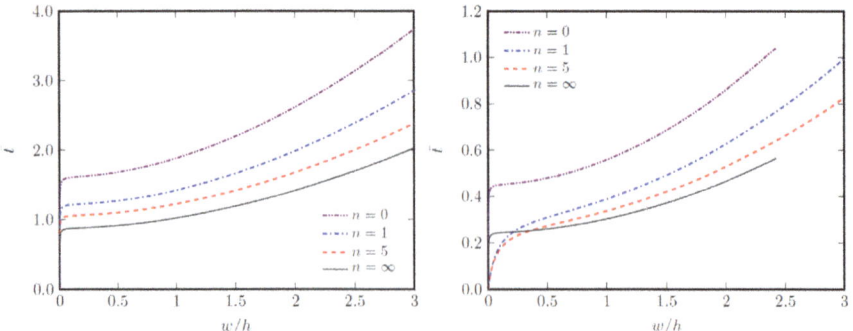

Fig. 2 Influence of material gradient index n **a** Clamped movable plate, **b** Simply supported movable plate

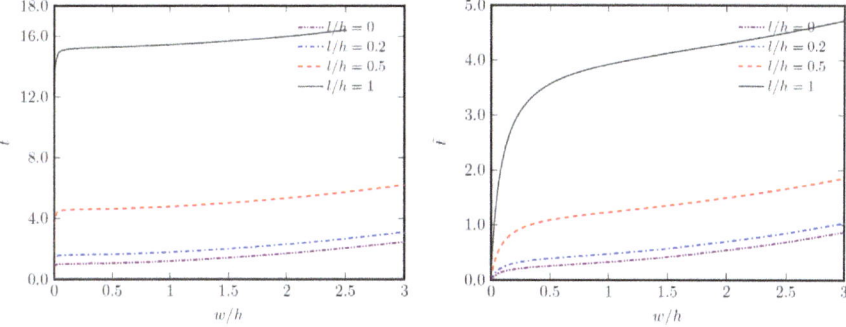

Fig. 3 Influence of length scale parameters **a** Clamped movable plate, **b** Simply supported movable plate

load is applied at the middle plane of the plates and nonhomogeneous distributions of materials could induce the eccentric bending immediately when the load is applied.

The influence of the length scale parameter is depicted in Fig. 3. The employed data for this example is $r = 10\ \mu m$, $h = r/100$ and $n = 2$. All three length-scale parameters are assumed to be equal ($l_0 = l_1 = l_2 = l$). As can be seen from the figures, the value of length scale parameters has a significant influence on the buckling and post-buckling behavior of the plates. As the length scale increases and approaches the thickness of the plate, the buckling loads increase considerably. Therefore, it can be concluded that the stiffness of the plates increases with the rise of length scale parameters relative to the plates' dimensions.

In the last example, the influence of plates' thickness ratios is investigated. A simply supported movable microplate with $r = 10\ \mu m$, $l_0 = l_1 = l_2 = 0.2\ h$, and $n = 2$. As presented in Fig. 4, the shear deformation effect can be neglected in case the

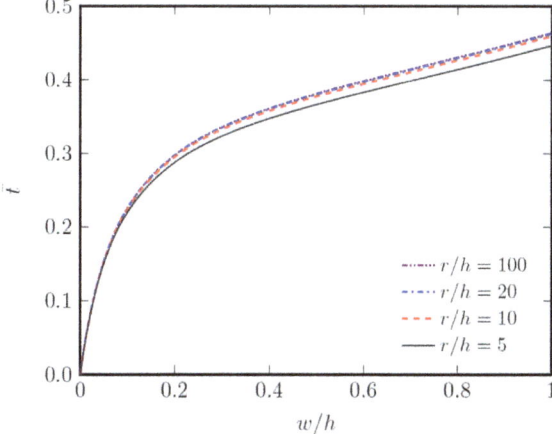

Fig. 4 Influence of shear deformation effect

plate is relatively thin ($r/h = 100 \sim r/h = 20$), however, when $r/h = 10$ and $r/h = 5$, the influence of the shear deformation effect is more considerable and it should be taken into account when analyzing the behavior of the plates.

5   Conclusions

In this study, the post-buckling behavior of FG microplates is successfully investigated. By using the MSGT, the size-dependent effect in microplates is captured. The displacement field of plates is based on TSDF and the governing equation is derived based on the principle of virtual energy. The IGA approach is used to discretize the governing equation, which requires C^2-continuity interpolation. Verification and parametric studies are conducted to prove the accuracy of the present approach and study the influence of material and length-scale parameters. The size effect captured by MSGT with three length scale parameters has a pronounced effect on the buckling load of microplates, especially when the plates' thickness is close to the length-scale parameters. In addition, it was found that the clamped boundary condition is capable of producing the bifurcation buckling phenomenon, while eccentric bending is observed in simply supported plates with nonhomogeneous materials.

Acknowledgements We acknowledge the support of time and facilities from Ho Chi Minh City Univerisity of Technology (HCMUT), VNH-HCM for this study.

References

1. Thai H-T, Kim S-E (2015) A review of theories for the modeling and analysis of functionally graded plates and shells. Compos Struct 128:70–86
2. Lam DCC, Yang F, Chong ACM, Wang J, Tong P (2003) Experiments and theory in strain gradient elasticity. J Mech Phys Solids 51(8):1477–1508
3. Zhang B, He Y, Liu D, Lei J, Shen L, Wang L (2015) A size-dependent third-order shear deformable plate model incorporating strain gradient effects for mechanical analysis of functionally graded circular/annular microplates. Compos B Eng 79:553–580
4. Ashoori Movassagh A, Mahmoodi MJ (2013) A micro-scale modeling of Kirchhoff plate based on modified strain-gradient elasticity theory. Eur J Mech A Solids 40:50–59
5. Thai S, Thai H-T, Vo TP, Nguyen-Xuan H (2017) Nonlinear static and transient isogeometric analysis of functionally graded microplates based on the modified strain gradient theory. Eng Struct 153:598–612
6. Thai S, Thai H-T, Vo TP, Patel VI (2017) Size-dependant behaviour of functionally graded microplates based on the modified strain gradient elasticity theory and isogeometric analysis. Comput Struct 190:219–241
7. Thai S, Thai H-T, Vo TP, Reddy JN (2017) Post-buckling of functionally graded microplates under mechanical and thermal loads using isogeomertic analysis. Eng Struct 150:905–917
8. Hughes TJR, Cottrell JA, Bazilevs Y (2005) Isogeometric analysis: CAD, finite elements, NURBS, exact geometry and mesh refinement. Comput Methods Appl Mech Eng 194(39–41):4135–4195

9. Nguyen VP, Anitescu C, Bordas SPA, Rabczuk T (2015) Isogeometric analysis: an overview and computer implementation aspects. Math Comput Simul 117:89–116
10. Reddy JN (2000) Analysis of functionally graded plates. Int J Numer Meth Eng 47(1–3):663–684
11. Reddy JN (1984) A simple higher-order theory for laminated composite plates. J Appl Mech 51(4):745–752
12. Thompson JMT, Hunt GW (1973) A general theory of elastic stability. Wiley, London
13. Katsikadelis JT, Babouskos N (2007) Post-buckling analysis of plates. A bem based mesh-less variational solution. 8th HSTAM International Congress on Mechanics. 8th HSTAM International Congress on Mechanics, Patras.

Stochastic Vibration Responses of Laminated Composite Beams Based on a Quasi-3D Theory

Xuan-Bach Bui, Trung-Kien Nguyen, T. Truong-Phong Nguyen, and Van-Trien Nguyen

Abstract Stochastic vibration responses of laminated composite beams based on a quasi-3D shear deformation theory are proposed in this paper. The mechanical properties of constituent materials are assumed to be uncertain, thus the free vibration responses can be modeled as random variables. A very large number of simulations is performed for propagating the overall uncertainty in the material properties to vibration behaviours by Monte Carlo simulation method. The higher-order shear deformation beam theory with nonlinear variations of both axial and transverse displacements is used and a trigonometric-series solution is developed to solve characteristic equations of motions. Novel numerical results are obtained to investigate the effects of uncertain material properties on the natural frequencies of the laminated composite beams.

Keywords Stochastic responses · Laminated composite beams · Vibration

1 Introduction

Thanks to the performance in high strength- and stiffness-to-weight ratios, multi-layered composite materials under beam structures have been used in many engineering fields such as mechanical engineering, aerospace engineering, construction, etc. Practically however, the mechanical performance of composite materials can be inconsistent, that probably results in the fabrication process or other unexpected factors. It is therefore necessary to account for these uncertainties in the behaviour analysis of laminated composite beams. Large applications of laminated composite beams led to the development of computational theories and methods with different approaches, only some representative references are herein cited [1–9]. For stochastic analysis, several methods have been used to model and propagate the uncertainty in stochastic computational simulations. Monte Carlo simulation method is known as the most straightforward and intuitive one which simply run the computational

X.-B. Bui (✉) · T.-K. Nguyen · T. T.-P. Nguyen · V.-T. Nguyen
Faculty of Civil Engineering, Ho Chi Minh City University of Technology and Education, Ho Chi Minh City, Vietnam
e-mail: bachbx@hcmut.edu.vn

© The Author(s), under exclusive license to Springer Nature Singapore Pte Ltd. 2023
J. N. Reddy et al. (eds.), *ICSCEA 2021*, Lecture Notes in Civil Engineering 268,
https://doi.org/10.1007/978-981-19-3303-5_83

model as many times as the accuracy required [10–12]. Nonetheless, when the physical model is complicated, the Monte Carlo method demands much computing time and infeasible to obtain desired sample outputs. In other word, stochastic numerical methods based on polynomial chaos expansion that speeds up the computing process while maintaining the accuracy have attracted considerable attention in predicting stochastic responses of laminated composite plates [13–16]. Though the polynomial chaos expansion requires a priori less computational cost, this approach appears to be complicated to implement and program. A literature review showed that there were still gaps in current research on stochastic behaviours of laminated composite beams and therefore, it motivates the investigation in this paper.

The objective of this paper is to propose stochastic vibration behaviours of laminated composite beams considering uncertainties in the material properties. It is based on a higher-order shear deformation theory which accounts for a higher-order variation of both axial and transverse displacements. The uncertainty of material properties are described through a probability distribution. This uncertainty will be propagated through the Ritz-method-based quasi-3D beam model to obtain the statistics of the outputs. The Monte Carlo simulation method will be used to propagate the uncertainty of material properties. Numerical results are presented to investigate the effects of uncertain material properties on the natural frequencies of laminated composite beams.

2 Theoretical Formulation

Consider a laminated composite beam with length L and rectangular cross-section $b \times h$ as shown in Fig. 1. It is made of n plies of orthotropic materials in different fibre angles with respect to the x-axis.

2.1 Displacements, Strains and Stresses

The displacement field of the present theory is given by:

$$u(x, z, t) = u_0(x, t) - z w_{0,x} + f(z)\theta_0(x, t) \tag{1a}$$

$$w(x, z, t) = w_0(x, t) + g(z)w_{z0}(x, t) \tag{1b}$$

where $u_0, w_0, w_{z0}, \theta_0$ are four variables to be determined; $g(z) = f'(z)$ where $f(z) = \frac{5z}{4} - \frac{5z^3}{3h^2}$ is the nonlinear shear function satisfying the condition $f'(z = \pm h/2) = 0$; the comma in subscript is used to indicate the differentiation of coordinate that follows.

The non-zero strains related to the displacements in Eq. (1) are given by:

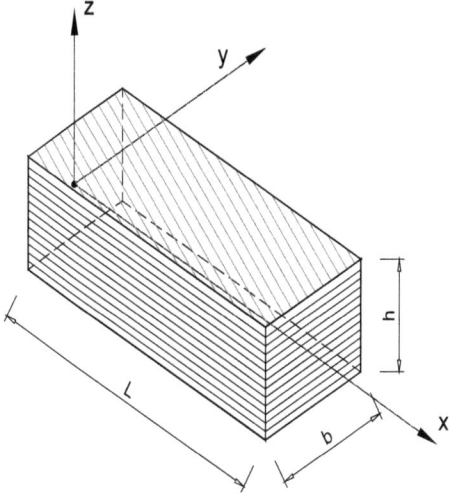

Fig. 1 Geometry of a laminated composite beam

$$\varepsilon_x(x, z) = u_{,x} = u_{0,x} - z w_{0,xx} + f\theta_{0,x} \tag{2a}$$

$$\varepsilon_z(x, z) = g''(z) w_{0z} \tag{2b}$$

$$\gamma_{xz}(x, z) = g(z)\big(w_{z0,x} + \theta_0\big) \tag{2c}$$

Moreover, the assumption of a plane stress state in (x, z)-plan leads to the stress–strain relation as follows:

$$\left\{\begin{array}{c} \sigma_x \\ \sigma_z \\ \sigma_{xz} \end{array}\right\} = \left(\begin{array}{ccc} Q'_{11} & Q'_{13} & 0 \\ Q'_{13} & Q'_{33} & 0 \\ 0 & 0 & Q'_{55} \end{array}\right) \left\{\begin{array}{c} \varepsilon_x \\ \varepsilon_z \\ \gamma_{xz} \end{array}\right\} \tag{3}$$

where $Q'_{11}, Q'_{13}, Q'_{33}, Q'_{55}$ are reduced stiffness constants of materials. These coefficients are related to the stiffness components of materials in global coordinates C'_{ij} as follows [9]:

$$Q'_{11} = C'_{11} + \frac{C'^2_{16}C'_{22} - 2C'_{12}C'_{16}C'_{26} + C'^2_{12}C'_{66}}{C'^2_{26} - C'_{22}C'_{66}} \tag{4a}$$

$$Q'_{13} = C'_{13} + \frac{C'_{16}C'_{22}C'_{36} + C'_{12}C'_{23}C'_{66} - C'_{16}C'_{23}C'_{26} - C'_{12}C'_{26}C'_{36}}{C'^2_{26} - C'_{22}C'_{66}} \tag{4b}$$

$$Q'_{33} = C'_{33} + \frac{C'^2_{36}C'_{22} - 2C'_{23}C'_{26}C'_{36} + C'^2_{23}C'_{66}}{C'^2_{26} - C'_{22}C'_{66}} \tag{4c}$$

$$Q'_{55} = C'_{55} - \frac{C'^2_{45}}{C'_{44}} \tag{4d}$$

2.2 Energy Formulation

The total energy of the laminated composite beams is composed of the strain energy Π_S and kinetic energy Π_K. The strain energy of the laminated composite beams is given by:

$$\Pi_S = \frac{1}{2} \int_V (\sigma_x \varepsilon_x + \sigma_z \varepsilon_z + \sigma_{xz} \gamma_{xz}) dV$$

$$= \frac{1}{2} \int_0^L \left[A_{11} u_{0,x}^2 + D_{11} w_{0,xx}^2 + H_{11}^s \theta_{0,x}^2 - 2B_{11} u_{0,x} w_{0,xx} + 2B_{11}^s u_{0,x} \theta_{0,x} - 2D_{11}^s w_{0,xx} \theta_{0,x} \right. \tag{5}$$

$$\left. + N_{33}^s w_{0z}^2 + 2T_{13}^s u_{0,x} w_{z0} - 2M_{13}^s w_{0,xx} w_{z0} + 2E_{13}^s \theta_{0,x} w_{z0} + A_{55}^s (\theta_0 + w_{z0,x})^2 \right] dx$$

where the stiffness components of laminated composite beams are defined as follows:

$$\left(A_{ij}, B_{ij}, D_{ij}, B_{ij}^s, D_{ij}^s, H_{ij}^s \right) = \int_{-h/2}^{h/2} (1, z, z^2, f, zf, f^2) Q'_{ij} b \, dz \tag{6a}$$

$$\left(A_{ij}^s, E_{ij}^s, T_{ij}^s, M_{ij}^s, N_{ij}^s \right) = \int_{-h/2}^{h/2} (g^2, fg_{,z}, g_{,z}, zg_{,z}, g_{,z}^2) Q'_{ij} b \, dz \tag{6b}$$

The kinetic energy of the laminated composite beams is given by:

$$\Pi_K = \frac{1}{2} \int_V \rho(z) \left(\dot{u}^2 + \dot{w}^2 \right) dV$$

$$= \frac{1}{2} \int_0^L \left[I_0 \dot{u}_0^2 + I_2 \dot{w}_{0,x}^2 + K_2 \dot{\theta}_0^2 - 2I_1 \dot{u}_0 \dot{w}_{0,x} + 2J_1 \dot{u}_0 \dot{\theta}_0 - 2J_2 \dot{w}_{0,x} \dot{\theta}_0 + I_0 \dot{w}_0^2 + 2L_1 \dot{w}_0 \dot{w}_{z0} + L_2 \dot{w}_{z0}^2 \right] dx \tag{7}$$

where the superscript dot is used to indicate the differentiation of the variable with the time t; ρ is the mass density; $(I_0, I_1, I_2, J_1, J_2, K_2, L_1, L_2)$ are terms of inertia defined as follows:

$$(I_0, I_1, I_2, J_1, J_2, K_2, L_1, L_2) = \int_{-h/2}^{h/2} \rho\left(1, z, z^2, f, zf, f^2, g, g^2\right) b \, dz \qquad (8)$$

By combining Eqs. (5) and (7), the total energy of the laminated composite beams is given by:

$$
\begin{aligned}
\Pi = \frac{1}{2} \int_0^L & \left[A_{11} u_{0,x}^2 + D_{11} w_{0,xx}^2 + H_{11}^s \theta_{0,x}^2 - 2B_{11} u_{0,x} w_{0,xx} + 2B_{11}^s u_{0,x} \theta_{0,x} - 2D_{11}^s w_{0,xx} \theta_{0,x} \right. \\
& \left. + N_{33}^s w_{0z}^2 + 2T_{13}^s u_{0,x} w_{z0} - 2M_{13}^s w_{0,xx} w_{z0} + 2E_{13}^s \theta_{0,x} w_{z0} + A_{55}^s \left(\theta_0 + w_{z0,x}\right)^2 \right] dx \\
- \frac{1}{2} \int_0^L & \left[I_0 \dot{u}_0^2 + I_2 \dot{w}_{0,x}^2 + K_2 \dot{\theta}_0^2 - 2I_1 \dot{u}_0 \dot{w}_{0,x} + 2J_1 \dot{u}_0 \dot{\theta}_0 - 2J_2 \dot{w}_{0,x} \dot{\theta}_0 + I_0 \dot{w}_0^2 + 2L_1 \dot{w}_0 \dot{w}_{z0} + L_2 w_{z0}^2 \right] dx
\end{aligned}
$$
$$(9)$$

2.3 Trigonometric-Series Solutions

The solution field u_0, w_0, θ_0, w_{z0} can be approximated under series of shape functions and associated variables as follows:

$$\{u_0(x,t), \theta_0(x,t)\} = \sum_{j=1}^{m} \psi_j(x)\{u_j(t), \theta_j(t)\} \qquad (10a)$$

$$\{w_0(x,t), w_{z0}(x,t)\} = \sum_{j=1}^{m} \varphi_j(x)\{w_j(t), w_{zj}(t)\} \qquad (10b)$$

where u_j, w_j, θ_j, w_{zj} are variables to be determined; $\psi_j(x)$, $\varphi_j(x)$ are shape functions. The approximations in Eq. (10) are known as Ritz's one in which it is noted that the accuracy of this approach depends on the construction of shape functions. These functions should be continuous, complete and orthogonal. In the present study, the functions of approximation $\psi_j(x)$, $\varphi_j(x)$ are selected under trigonometric ones that satisfy kinematic boundary conditions. Three typical boundary conditions (simply-supported: S-S, clamped-clamped: C-C, clamped-free: C-F) are considered as Table 1.

Substituting Eq. (10) into Eq. (9) and using Lagrange's equations lead to:

$$Kp + M\ddot{p} = F(t) \qquad (11)$$

where $p = \begin{bmatrix} u_0 & w_0 & w_{z0} & theta_0 \end{bmatrix}^T$; K and M are stiffness and mass matrix, respectively, which are given by:

Table 1 Kinematic boundary conditions and trigonometric shape functions

BCs	$x = 0$	$x = L$	$\psi_j(x)$	$\varphi_j(x)$
S-S	$w_0 = 0,\ w_{z,0} = 0$	$w_0 = 0,\ w_{z,0} = 0$	$\cos\frac{j\pi x}{L}$	$\sin\frac{j\pi x}{L}$
C-F	$u_0 = 0,\ w_0 = 0,\ w_{0,x} = 0, \theta_0 = 0,\ w_{z,0} = 0$		$\sin\frac{(2j-1)\pi}{2L}x$	$1 - \cos\frac{(2j-1)\pi}{2L}x$
C-C	$u_0 = 0,\ w_0 = 0,\ w_{0,x} = 0, \theta_0 = 0,\ w_{z,0} = 0$	$u_0 = 0,\ w_0 = 0,\ w_{0,x} = 0, \theta_0 = 0,\ w_{z,0} = 0$	$\sin\frac{2j\pi x}{L}$	$\sin^2\frac{j\pi x}{L}$

$$K = \begin{bmatrix} K^{11} & K^{12} & K^{13} & K^{14} \\ ^T K^{12} & K^{22} & K^{23} & K^{24} \\ ^T K^{13} & ^T K^{23} & K^{33} & K^{34} \\ ^T K^{14} & ^T K^{24} & ^T K^{34} & K^{44} \end{bmatrix}, \quad M = \begin{bmatrix} M^{11} & M^{12} & M^{13} & 0 \\ ^T M^{12} & M^{22} & M^{23} & M^{24} \\ ^T M^{13} & ^T M^{23} & M^{33} & 0 \\ 0 & ^T M^{24} & 0 & M^{44} \end{bmatrix} \quad (12)$$

The components of stiffness and mass matrix are determined as follows:

$$K_{ij}^{11} = A_{11} \int_0^L \psi_{i,x}\psi_{i,x}dx, \ K_{ij}^{12} = -B_{11} \int_0^L \psi_{i,x}\varphi_{i,xx}dx, \ K_{ij}^{13} = B_{11}^s \int_0^L \psi_{i,x}\psi_{j,x}dx, \ K_{ij}^{14} = T_{13}^s \int_0^L \psi_{i,x}\varphi_j dx$$

$$(13a)$$

$$K_{ij}^{22} = D_{11} \int_0^L \varphi_{i,xx}\varphi_{j,xx}dx, \ K_{ij}^{23} = -D_{11}^s \int_0^L \varphi_{i,xx}\psi_{j,x}dx, \ K_{ij}^{24} = -M_{13}^s \int_0^L \varphi_{i,xx}\varphi_j dx \quad (13b)$$

$$K_{ij}^{33} = H_{11}^s \int_0^L \psi_{i,x}\psi_{j,x}dx + A_{55}^s \int_0^L \psi_i \psi_j dx, \ K_{ij}^{34} = E_{13}^s \int_0^L \psi_{i,x}\varphi_j dx + A_{55}^s \int_0^L \psi_i \varphi_{j,x}dx \quad (13c)$$

$$K_{ij}^{44} = N_{33}^s \int_0^L \varphi_i \varphi_j dx + A_{55}^s \int_0^L \varphi_{i,x}\varphi_{j,x}dx \quad (13d)$$

$$M_{ij}^{11} = I_0 \int_0^L \psi_i \psi_j dx, \ M_{ij}^{12} = -I_1 \int_0^L \psi_i \varphi_{j,x}dx, \ M_{ij}^{13} = J_1 \int_0^L \psi_i \psi_j dx \quad (13e)$$

$$M_{ij}^{22} = I_0 \int_0^L \varphi_i \varphi_j dx + I_2 \int_0^L \varphi_{i,x}\varphi_{j,x}dx, \ M_{ij}^{23} = -J_2 \int_0^L \varphi_{i,x}\psi_j dx, \ M_{ij}^{24} = L_1 \int_0^L \varphi_i \varphi_j dx \quad (13f)$$

$$M_{ij}^{33} = K_2 \int_0^L \psi_i \psi_j dx, \ M_{ij}^{44} = L_2 \int_0^L \psi_i \psi_j dx \quad (13g)$$

It is worth noticing that the free vibration responses can be derived by expressing $p(t) = pe^{i\omega t}$ and solving the characteristic equation $(K - \omega^2 M)p = 0$ in which ω is natural frequencies of the laminated composite beams; $i^2 = -1$ is imaginary part.

2.4 Monte Carlo Simulation

The material properties of laminated composite beams are supposed to be random according to a required distribution. In order to propagate the variability in material properties to the vibration responses of laminated composite beams, Monte Carlo simulation method will be used. This technique requires a generation of random numbers set from material properties and then these are used to obtain the vibration responses and its statistics. The following statistics of the responses of the laminated composite beams are used for computations:

$$E[X] = \int_{-\infty}^{\infty} x f(x) dx \tag{14a}$$

$$SD = \sqrt{\sigma} = \sqrt{\frac{\sum_{i=1}^{n} (x_i - \mu)^2}{n - 1}} \tag{14b}$$

where $E[X]$ is expectation of the variable set X; x is the value in the sample space; $f(x)$ is probability density function (PDF); σ is standard deviation of the set of random numbers; n is number of samples. In addition to the expectation and variance, higher-order statistics such as the skewness $\tilde{\mu}_3$ and kurtosis $Kurt$, confidence interval CI are also measured as follows:

$$\tilde{\mu}_3 = \frac{\sum_i^N (x_i - \mu)^3}{(n - 1) \times \sigma^3} \tag{15a}$$

$$Kurt = \frac{E(x - \mu)^4}{\sigma^4} \tag{15b}$$

$$CI = \mu \pm z \frac{SD}{\sqrt{n}} \tag{15c}$$

where z is the required confidence interval (%).

3 Numerical Examples

A number of numerical examples are performed in this section to investigate the accuracy and efficiency of the present theory. The laminated composite beams are composed of orthotropic material layers of the same thickness. The means, standard-to-mean ratio (COV) and distribution of material properties are given in Table 2. The effects of material properties uncertainty on the vibration behaviours of laminated composite beams are observed with different lay-ups and boundary conditions.

As a first example, in order to study the convergence, laminated composite beams with mean material properties given in Table 2 are considered. The results are computed with three boundary conditions (BC) S-S, C-C and C-F, and different fiber orientations, [0°/90°], [0°/90°/0°], [45°/−45°] and [45°/−45°/45°]. The fundamental frequencies are reported in Table 3 for various values of number of series m. The results obtained from Table 3 show that the solutions converge quickly for all responses and boundary conditions, the number of series $m = 10$ can be considered as the convergence point for the natural frequencies of the laminated composite beams, therefore this value will be used for the sequel computations.

As a second example, in order to investigate stochastic responses of the laminated composite beams, Monte Carlo simulation method with number of samples $N_s = 10^5$

Table 2 Material properties and geometry of the laminated composite beams

Properties	Mean [9]	COV	Distribution
E_1 (GPa)	120	0.1	Lognormal
E_2 (GPa)	3	0.1	Lognormal
$G_{12} = G_{13}$ (GPa)	$0.6E_2$	0.1	Lognormal
G_{23} (GPa)	$0.5E_2$	0.1	Lognormal
ν_{12}	0.25	0.1	Lognormal
ρ (kg/m^2)	1500	0.1	Lognormal
L (m)	0.381	–	–
h (m)	0.1905	–	–
b (m)	0.0254	–	–

Table 3 Convergence of the fundamental frequencies (Hz)

BC	Lay-ups	m						
		2	4	6	8	10	12	14
S-S	0°/90°	213.754	213.754	213.754	213.754	213.754	213.754	213.754
	45°/−45°	102.389	102.389	102.389	102.389	102.389	102.389	102.389
	0°/90°/0°	482.307	482.307	482.307	482.307	482.307	482.307	482.307
	−45°/45°/−45°	102.389	102.389	102.389	102.389	102.389	102.389	102.389
C-C	0°/90°	469.254	465.780	464.802	464.361	464.124	463.986	463.900
	45°/−45°	231.998	230.455	229.993	229.764	229.625	229.532	229.467
	0°/90°/0°	900.487	885.486	880.776	878.679	877.582	876.941	876.535
	−45°/45°/−45°	231.998	230.455	229.993	229.764	229.625	229.532	229.467
C-F	0°/90°	77.200	76.927	76.848	76.811	76.790	76.776	76.766
	45°/−45°	36.969	36.767	36.699	36.665	36.644	36.630	36.620
	0°/90°/0°	180.377	179.600	179.375	179.272	179.215	179.179	179.156
	−45°/45°/−45°	36.969	36.767	36.699	36.665	36.644	36.630	36.620

is performed in which six parameters in Table 2 are randomly varied according to the lognormal distribution. It is noted that the lognormal distribution is chosen instead of normal distribution to avoid negative values in material property input. Besides, the coefficient $COV = 0.1$ is applied for all parameters. Table 4 presents four statistical moments of simulation outputs computed for laminated composite beams with different lay-ups and boundary conditions. The mean μ, standard deviation σ, skewness $\tilde{\mu}_3$ and kurtosis $Kurt$ are calculated with four layer-ups 0°/90°, 45°/−45°, 0°/90°/0°, −45°/45°/−45°, and three boundary conditions S-S, C-F and C-C. The means and deterministic values obtained from the present theory are compared to those of Nguyen et al. [9] based on a quasi-3D deterministic beam model. It can be seen that there are excellent agreements between the deterministic models, no significant differences of the means of fundamental frequencies and those of Nguyen

Table 4 Fundamental frequencies (Hz) of laminated composite beams with arbitrary lay-ups and boundary conditions

BC	Lay-ups	Present				Present (deterministic)	Nguyen [9]
		μ	σ	Kurtosis	Skewness		
S–S	0°/90°	214.300	13.538	3.054	0.193	213.754	213.116
	45°/−45°	102.468	6.293	3.036	0.170	102.389	−
	0°/90°/0°	483.159	31.689	3.060	0.184	482.307	482.295
	−45°/45°/−45°	102.472	6.320	3.064	0.174	102.389	−
C–C	0°/90°	465.298	28.570	3.056	0.169	464.124	462.889
	45°/−45°	229.812	14.113	3.040	0.167	229.625	−
	0°/90°/0°	878.536	51.876	3.046	0.174	877.582	876.952
	−45°/45°/−45°	229.722	14.075	3.029	0.168	229.625	−
C–F	0°/90°	77.006	4.877	3.040	0.195	76.790	76.562
	45°/−45°	36.665	2.262	3.077	0.199	36.644	−
	0°/90°/0°	179.557	12.256	3.082	0.199	179.215	179.177
	−45°/45°/−45°	36.670	2.254	3.041	0.184	36.644	−

et al. [9] are found. Moreover, it is observed that the ratios of standard deviation and mean is about 6%. The kurtosis values are slightly higher than 3, which indicates that the present distribution of responses has rather heavier tail and more outliers than the normal distribution. The positive skewness in Table 4 means the data is right-skewed which is a characteristic of lognormal distribution (Fig. 2).

Furthermore, the effect of number of simulations on the accuracy of output distribution is demonstrated in Fig. 3. The curves are plotted for 0°/90°/0° laminated composite beams with the span-to-thickness ratio $L/h = 20$, C-C and C-F boundary conditions. The 80% and 99% confidence interval of the mean value of simulation

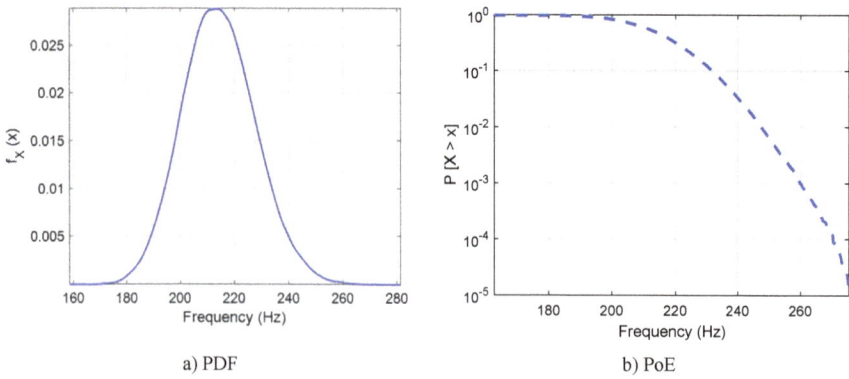

a) PDF   b) PoE

Fig. 2 Probability density function (PDF) and Probability of exceedance (PoE) of the fundamental frequency (Hz) for 0°/90° laminated composite beam ($L/h = 20$, S-S boundary condition)

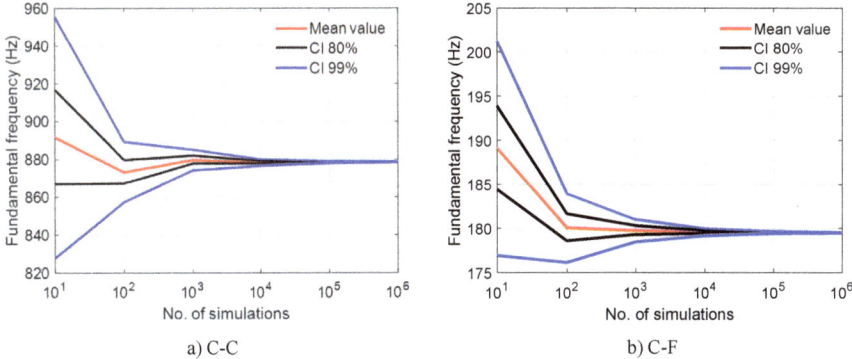

Fig. 3 Mean and confidence interval of the fundamental frequencies (Hz) of 0°/90°/0° laminated composite beam ($L/h = 20$, C-C and C-F boundary conditions)

outputs are shown for the number of simulations 10^1, 10^2, 10^3, 10^4, 10^5 and 10^6. It can be observed from these graphs that in the present beam model, the true value of mean fundamental frequencies can be achieved when the Monte Carlo simulation has 10^5 or more samples.

Figure 4 presents the probability of exceedance (PoE) of the fundamental frequencies for the composite beam with $-45°/45°/-45°$ ply composition and S-S boundary condition. In Fig. 4a, the PoE is plotted 10 times each of which has the number of samples $N_s = 10^5$. Noticingly, the tails of the plots past $P[X > x] = 10^{-3}$ have fluctuations. This is due to there are very few samples at the very small probability of occurrence. In Fig. 4b, the setup is similar to that of Fig. 4a but with the number of samples $N_s = 10^4$. The similar fluctuation is seen past the horizontal line where $P[X > x] = 10^{-2}$. Therefore, by plain observation, the outputs of Monte Carlo simulation is stable up to the point where $P[X > x] = 10^{-(\log(N_s)-2)}$.

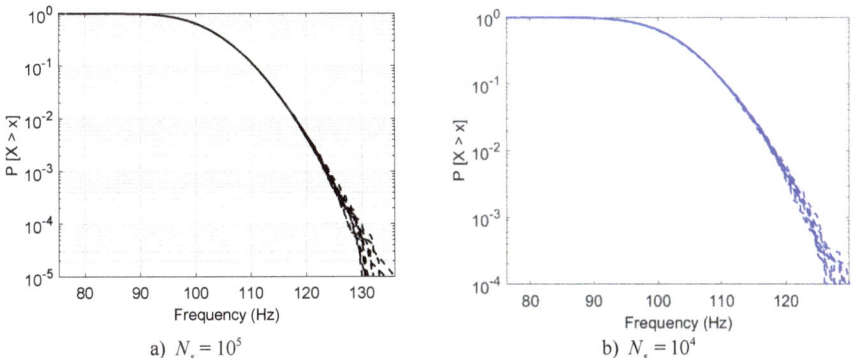

Fig. 4 Probability of exceedance (PoE) of the fundamental frequencies (Hz) of $-45°/45°/-45°$ laminated composite beam ($L/h = 20$, S-S boundary conditions)

4 Conclusions

This article presents stochastic vibration behaviours of laminated composite beams based on a quasi-3D theory. The stochastic mechanical properties of component materials are propagated into the vibrational responses of the composite beams with arbitrary lay-ups and boundary conditions. The beam model is based on the higher-order shear deformation theory with nonlinear formulation of axial and transverse displacements. A trigonometric-series solution is utilised to solve the equations of motion. A high number of Monte Carlo simulations is conducted to investigate the effects of stochastic uncertainties on the natural frequencies of the laminated composite beams. The outputs of these simulations are presented as probability density functions, probability of exceedance and several statistical moments. The numerical results obtained from this paper showed that the present model is simple and efficient in predicting stochastic vibration responses of the laminated composite beams. Novel results can be used as benchmarks for the future researches.

Acknowledgements This research is funded by Vietnam National Foundation for Science and Technology Development (NAFOSTED) Under Grant No. 107.02-2018.312.

References

1. Aydogdu M (2005) Vibration analysis of cross-ply laminated beams with general boundary conditions by Ritz method. Int J Mech Sci 47(11):1740–1755
2. Kant T, Marur SR, Rao GS (1997) Analytical solution to the dynamic analysis of laminated beams using higher order refined theory. Compos Struct 40(1):1–9
3. Khdeir A, Reddy J (1994) Free vibration of cross-ply laminated beams with arbitrary boundary conditions. Int J Eng Sci 32(12):1971–1980
4. Nguyen T-K et al (2017) Trigonometric-series solution for analysis of laminated composite beams. Compos Struct 160:142–151
5. Vo TP, Thai H-T (2012) Static behavior of composite beams using various refined shear deformation theories. Compos Struct 94(8):2513–2522
6. Nguyen N-D et al (2018) New Ritz-solution shape functions for analysis of thermo-mechanical buckling and vibration of laminated composite beams. Compos Struct 184:452–460
7. Jun L, Yuchen B, Peng H (2017) A dynamic stiffness method for analysis of thermal effect on vibration and buckling of a laminated composite beam. Arch Appl Mech 87(8):1295–1315
8. Khdeir AA, Reddy JN (1997) An exact solution for the bending of thin and thick cross-ply laminated beams. Compos Struct 37(2):195–203
9. Nguyen N-D et al (2018) Ritz-based analytical solutions for bending, buckling and vibration behavior of laminated composite beams. Int J Struct Stab Dyn 18(11):1850130
10. Nguyen HX et al (2017) Stochastic buckling behaviour of laminated composite structures with uncertain material properties. Aerosp Sci Technol 66:274–283
11. Li J et al (2016) Stochastic thermal buckling analysis of laminated plates using perturbation technique. Compos Struct 139:1–12
12. Grover N et al (2017) Influence of parametric uncertainties on the deflection statistics of general laminated composite and sandwich plates. Compos Struct 171:158–169

13. Peng X et al (2019) Uncertainty analysis of composite laminated plate with data-driven polynomial chaos expansion method under insufficient input data of uncertain parameters. Compos Struct 209:625–633
14. Chandra S et al (2019) Stochastic dynamic analysis of composite plate with random temperature increment. Compos Struct 226:111159
15. Chen N-Z, Guedes Soares C (2008) Spectral stochastic finite element analysis for laminated composite plates. Comput Methods Appl Mech Eng 197(51):4830–4839
16. Chakraborty S et al (2016) Stochastic free vibration analysis of laminated composite plates using polynomial correlated function expansion. Compos Struct 135:236–249

The Extension of Multi-layer Moving Plate Method (MMPM) for Analysis of Functionally Graded (FG) Sandwich Plate

Tan Ngoc Than Cao and Van Hai Luong

Abstract This paper presents an extension of Multi-layer Moving Plate Method (MMPM) for static and free vibration analyses of functionally graded (FG) sandwich plate resting on a viscoelastic foundation. The FG sandwich plate is composed of two parallel FG plates, in which upper plate is connected with lower plate by a viscoelastic layer and the lower plate rests on a viscoelastic foundation. The convergence and accuracy of this method are validated by comparing the results obtained from this study with other published results. Next, several numerical results are presented to examine the static and free vibration behaviors of FG sandwich plate with various volume fraction exponents, boundary conditions, geometric parameters, connected layer and foundation stiffness.

Keywords Functionally graded sandwich plate · Multi-layer moving plate method · Static and free vibration analyses

1 Introduction

Sandwich structures including two parallel beams/plates joined by a viscoelastic layer are very effective in reducing and controlling vibration response of flexible structures. Thus, the static and free vibration behaviors of sandwich beam/plates have attracted much researchers' attention in recent decades. Chonan and Sendai [1] studied the vibration responses of prestressed double strip-plate system with infinite length. The vibration analyses of two connected rectangular plates were presented in the studies of Kukla [2], Oniszczuk [3], Hedrih [4], De Rosa and Lippiello [5].

T. N. T. Cao (✉)
College of Technology, Can Tho University (CTU), Can Tho, Vietnam
e-mail: ctnthan@ctu.edu.vn

V. H. Luong
Faculty of Civil Engineering, Ho Chi Minh City University of Technology (HCMUT), 268 Ly Thuong Kiet Street, District 10, Ho Chi Minh City, Vietnam

Vietnam National University Ho Chi Minh City, Linh Trung Ward, Thu Duc District, Ho Chi Minh City, Vietnam

© The Author(s), under exclusive license to Springer Nature Singapore Pte Ltd. 2023
J. N. Reddy et al. (eds.), *ICSCEA 2021*, Lecture Notes in Civil Engineering 268,
https://doi.org/10.1007/978-981-19-3303-5_84

However, the static responses and free vibration behaviors of functionally graded (FG) sandwich plate, which consists of two connnected parallel FG plates, have not been studied yet. Therefore, more studies on this problem are essential to achieve deeper knowledge about the behavior of FG sandwich plate.

The Finite Element Method (FEM) is a widely-known powerful numerical method for solving a variety of complex problems, including responses of structures under moving loads. However, this method has some drawbacks. In fact, the position of load and vector of displacement have to be modified at each time step in the solution procedure owing to the change of position of moving load. In the effort to overcome these drawbacks, Koh et al. [6] proposed the Moving Element Method (MEM) to study the dynamic responses of train-track system. The extensions of MEM for analyzing static and dynamic behaviors of homogenous Mindlin plate, composite laminate plate and FG plate under moving loads were carried out in the studies of Luong et al. [7], Cao et al. [8] and Luong et al. [9], respectively. Recently, Cao et al. [10] successfully developed the Multi-layer Moving Plate Method (MMPM) to investigate the dynamic responses of two connected parallel plates under moving load.

The extension of MMPM to examine static responses and free vibration behaviors of FG sandwich plate is presented in this paper. First, the static and free vibration analyses of FG sandwich plates are considered. The calculated results and published results are compared to verify the accuracy of proposed method. Next, the effects of volume fraction exponents, boundary conditions, geometric parameters, connected layer and foundation stiffness on static and free vibration responses of FG sandwich plate are investigated.

2   Formulation and Methodology

2.1   Weak Form for the FG Sandwich Plate Resting on a Viscoelastic Foundation

A functionally graded (FG) sandwich plate is composed of two parallel FG plates connected by a viscoelastic layer and rests on a viscoelastic foundation as shown in Fig. 1. Upper plate is subjected to a moving load P which moves along the longitudinal middle line with velocity V and acceleration a. The connected layer and foundation are modeled by vertical spring stiffness k_{w1}, k_{w2} and damping c_1, c_2, respectively. Length L, width B and thickness $h_i (i = 1, 2)$ of upper and lower plates are the same. For convenience, $i = 1$, 2 stands for upper and lower plates, respectively. The FG plates have ceramic rich at top surface and metal rich at bottom surface. The variations of Young' modulus $E_1(z)$, $E_2(z)$ and density $\rho_1(z)$, $\rho_2(z)$ of FG plates are described by the simple power law distribution (Reddy [11]).

$$E_1(z) = (E_{c1} - E_{m1})V_{c1}(z) + E_{m1}; \quad E_2(z) = (E_{c2} - E_{m2})V_{c2}(z) + E_{m2} \quad (1)$$

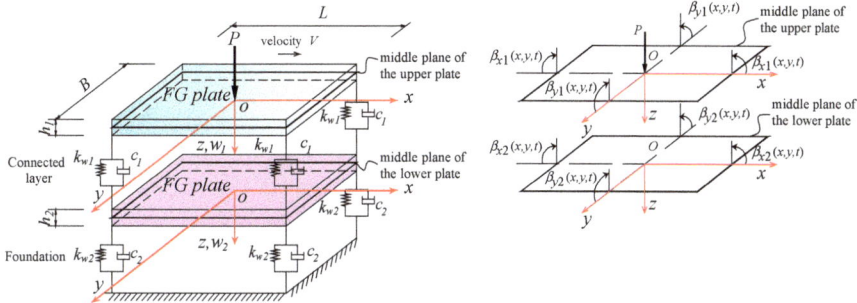

Fig. 1 Model of FG sandwich plate and positive directions of rotations β_{x1}, β_{y1} and β_{x2}, β_{y2}

$$\rho_1(z) = (\rho_{c1} - \rho_{m1})V_{c1}(z) + \rho_{m1}; \quad \rho_2(z) = (\rho_{c2} - \rho_{m2})V_{c2}(z) + \rho_{m2} \quad (2)$$

$$V_{c1}(z) = \left(\frac{z}{h_1} + \frac{1}{2}\right)^n; \quad V_{c2}(z) = \left(\frac{z}{h_2} + \frac{1}{2}\right)^n (0 \leq n \leq \infty) \quad (3)$$

where E_{c1}, E_{c2} and ρ_{c1}, ρ_{c2} are Young' modulus and density of ceramic, respectively; E_{m1}, E_{m2} and ρ_{m1}, ρ_{m2} are Young' modulus and density of metal, respectively; $V_{c1}(z)$ and $V_{c2}(z)$ are the volume fraction of ceramic; z is the thickness coordinate; and n is the volume fraction exponent.

The reference planes of two plates are chosen at middle planes that occupy the domains Ω_1, $\Omega_2 \subset R^2$. The vectors of three independent field variables $\mathbf{u}_1$, $\mathbf{u}_2$, curvature κ_1, κ_2 and shear strains γ_1, γ_2 of plates are respectively written as

$$\mathbf{u}_1 = \begin{bmatrix} w_1 & \beta_{x1} & \beta_{y1} \end{bmatrix}^T; \quad \mathbf{u}_2 = \begin{bmatrix} w_2 & \beta_{x2} & \beta_{y2} \end{bmatrix}^T \quad (4)$$

$$\kappa_1 = \mathbf{L}_d\boldsymbol{\beta}_1; \quad \kappa_2 = \mathbf{L}_d\boldsymbol{\beta}_2 \quad (5)$$

$$\gamma_1 = \nabla w_1 + \boldsymbol{\beta}_1; \quad \gamma_2 = \nabla w_2 + \boldsymbol{\beta}_2 \quad (6)$$

where w_1 and w_2 are the vertical deflections; $\boldsymbol{\beta}_1 = \begin{bmatrix} \beta_{x1} & \beta_{y1} \end{bmatrix}^T$ and $\boldsymbol{\beta}_2 = \begin{bmatrix} \beta_{x2} & \beta_{y2} \end{bmatrix}^T$ are the vector of rotations, in which β_{x1}, β_{y1} and β_{x2}, β_{y2} are the rotations of middle plane of the upper and lower plates around the y-axis and x-axis, respectively; $\nabla = \begin{bmatrix} \partial/\partial x & \partial/\partial y \end{bmatrix}^T$; and $\mathbf{L}_d$ is a differential operator matrix defined by

$$\mathbf{L}_d = \begin{bmatrix} \partial/\partial x & 0 \\ 0 & \partial/\partial y \\ \partial/\partial y & \partial/\partial x \end{bmatrix} \quad (7)$$

The D'Alembert–Lagrange equations of motion of upper and lower plates are respectively established as follow

$$
\begin{aligned}
&\int_{\Omega_1} \delta\boldsymbol{\kappa}_1^T \mathbf{D}_{b1} \boldsymbol{\kappa}_1 d\Omega_1 + \int_{\Omega_1} \delta\boldsymbol{\gamma}_1^T \mathbf{D}_{s1} \boldsymbol{\gamma}_1 d\Omega_1 \\
&+ \int_{\Omega_1} \delta\mathbf{u}_1^T \mathbf{m}_1 \ddot{\mathbf{u}}_1 d\Omega_1 + \int_{\Omega_1} \delta w_1^T k_{w1}(w_1 - w_2) d\Omega_1 + \int_{\Omega_1} \delta w_1^T c_1(\dot{w}_1 - \dot{w}_2) d\Omega_1 \\
&= \int_{\Omega_1} \delta\mathbf{u}_1^T \mathbf{b}_1(x, y, t) d\Omega_1
\end{aligned}
$$

$$(8)$$

$$
\begin{aligned}
&\int_{\Omega_2} \delta\boldsymbol{\kappa}_2^T \mathbf{D}_{b2} \boldsymbol{\kappa}_2 d\Omega_2 + \int_{\Omega_2} \delta\boldsymbol{\gamma}_2^T \mathbf{D}_{s2} \boldsymbol{\gamma}_2 d\Omega_2 \\
&+ \int_{\Omega_2} \delta\mathbf{u}_2^T \mathbf{m}_2 \ddot{\mathbf{u}}_2 d\Omega_2 - \int_{\Omega_2} \delta w_2^T k_{w1}(w_1 - w_2) d\Omega_2 \\
&- \int_{\Omega_2} \delta w_2^T c_1(\dot{w}_1 - \dot{w}_2) d\Omega_2 + \int_{\Omega_2} \delta w_2^T k_{w2} w_2 d\Omega_2 + \int_{\Omega_2} \delta w_2^T c_2 \dot{w}_2 d\Omega_2 \\
&= \int_{\Omega_2} \delta\mathbf{u}_2^T \mathbf{b}_2(x, y, t) d\Omega_2
\end{aligned}
$$

$$(9)$$

where single dot and double dots over a symbol denote velocity and acceleration, respectively; mass matrices $\mathbf{m}_1$, $\mathbf{m}_2$, flexural rigidity matrices $\mathbf{D}_{b1}$, $\mathbf{D}_{b2}$, and shear rigidity matrices $\mathbf{D}_{s1}$, $\mathbf{D}_{s2}$ are given as follow

$$
\mathbf{m}_1 = \int_{-h_1/2}^{h_1/2} \rho_1(z) \begin{bmatrix} 1 & 0 & 0 \\ 0 & z^2 & 0 \\ 0 & 0 & z^2 \end{bmatrix} dz \,; \quad
\mathbf{D}_{b1} = \int_{-h_1/2}^{h_1/2} z^2 \begin{bmatrix} Q_{11}^1 & Q_{12}^1 & 0 \\ Q_{12}^1 & Q_{22}^1 & 0 \\ 0 & 0 & Q_{66}^1 \end{bmatrix} dz \,;
$$

$$
\mathbf{D}_{s1} = \kappa_s \int_{-h_1/2}^{h_1/2} \begin{bmatrix} Q_{55}^1 & 0 \\ 0 & Q_{44}^1 \end{bmatrix} dz
$$

$$(10)$$

$$
\mathbf{m}_2 = \int_{-h_2/2}^{h_2/2} \rho_2(z) \begin{bmatrix} 1 & 0 & 0 \\ 0 & z^2 & 0 \\ 0 & 0 & z^2 \end{bmatrix} dz \,; \quad
\mathbf{D}_{b2} = \int_{-h_2/2}^{h_2/2} z^2 \begin{bmatrix} Q_{12}^1 & Q_{12}^2 & 0 \\ Q_{12}^2 & Q_{22}^2 & 0 \\ 0 & 0 & Q_{66}^2 \end{bmatrix} dz \,;
$$

$$
\mathbf{D}_{s2} = \kappa_s \int_{-h_2/2}^{h_2/2} \begin{bmatrix} Q_{55}^2 & 0 \\ 0 & Q_{44}^2 \end{bmatrix} dz
$$

$$(11)$$

in which $Q_{11}^1 = Q_{22}^1 = E_1(z)/(1 - \upsilon_1)(1 + \upsilon_1)$, $Q_{11}^2 = Q_{22}^2 = E_2(z)/(1 - \upsilon_2)(1 + \upsilon_2)$, $Q_{12}^1 = \upsilon_1 E_1(z)/(1 - \upsilon_1)(1 + \upsilon_1)$, $Q_{12}^2 = \upsilon_2 E_2(z)/(1 - \upsilon_2)(1 + \upsilon_2)$, $Q_{44}^1 = Q_{55}^1 = Q_{66}^1 = E_1(z)/2(1 + \upsilon_1)$, $Q_{44}^2 = Q_{55}^2 = Q_{66}^2 = E_2(z)/2(1 + \upsilon_2)$; the shear correction factor is chosen $\kappa_s = 5/6$; the Poisson' ration $\upsilon_1 = \upsilon_2$ are assumed to be constant.

The load vectors $\mathbf{b}_1$ and $\mathbf{b}_2$ in Eqs. (8) and (9) is given as

$$
\mathbf{b}_1(x, y, t) = \begin{bmatrix} P\delta(x - S)\delta(y - 0) & 0 & 0 \end{bmatrix}^T \,; \quad \mathbf{b}_2(x, y, t) = \begin{bmatrix} 0 & 0 & 0 \end{bmatrix}^T
$$

$$(12)$$

where S is the distance traveled by the load at any instant t; δ denotes the Dirac delta function.

2.2 Multi-Layer Moving Plate Method (MMPM)

The FG sandwich plate is discretized into a number of multi-layer moving plate elements as shown in Fig. 2. In the view of MMPM (Cao et al. [10]), a coordinates system (r, s) is attached to the moving load and for that reason it moves at the same velocity of the moving load. The correlation between moving coordinates system (r, s) and fixed coordinates systems (x, y) is expressed as

$$r = x - S; \quad s = y \tag{13}$$

Using the coordinate transformation, Eqs. (8) and (9) are rewritten in the coordinate system (r, s) as

$$
\begin{aligned}
&\int_{\Omega_1} \delta\boldsymbol{\kappa}_1^T \mathbf{D}_{b1}\boldsymbol{\kappa}_1 d\Omega_1 + \int_{\Omega_1} \delta\boldsymbol{\gamma}_1^T \mathbf{D}_{s1}\boldsymbol{\gamma}_1 d\Omega_1 \\
&+ \int_{\Omega_1} \delta\mathbf{u}_1^T \mathbf{m}_1 (V^2 \tfrac{\partial^2 \mathbf{u}_1}{\partial r^2} - 2V \tfrac{\partial^2 \mathbf{u}_1}{\partial r \partial t} - a \tfrac{\partial \mathbf{u}_1}{\partial r} + \tfrac{\partial^2 \mathbf{u}_1}{\partial t^2}) d\Omega_1 \\
&+ \int_{\Omega_1} \delta w_1^T k_{w1}(w_1 - w_2) d\Omega_1 \\
&- \int_{\Omega_1} \delta w_1^T c_1(V \tfrac{\partial w_1}{\partial r} - V \tfrac{\partial w_2}{\partial r}) d\Omega_1 \\
&+ \int_{\Omega_1} \delta w_1^T c_1(\tfrac{\partial w_1}{\partial t} - \tfrac{\partial w_2}{\partial t}) d\Omega_1 = \int_{\Omega_1} \delta\mathbf{u}_1^T \mathbf{b}_1(r, s, t) d\Omega_1
\end{aligned}
\tag{14}
$$

$$
\begin{aligned}
&\int_{\Omega_2} \delta\boldsymbol{\kappa}_2^T \mathbf{D}_{b2}\boldsymbol{\kappa}_2 d\Omega_2 + \int_{\Omega_2} \delta\boldsymbol{\gamma}_2^T \mathbf{D}_{s2}\boldsymbol{\gamma}_2 d\Omega_2 \\
&+ \int_{\Omega_2} \delta\mathbf{u}_2^T \mathbf{m}_2 (V^2 \tfrac{\partial^2 \mathbf{u}_2}{\partial r^2} - 2V \tfrac{\partial^2 \mathbf{u}_2}{\partial r \partial t} - a \tfrac{\partial \mathbf{u}_2}{\partial r} + \tfrac{\partial^2 \mathbf{u}_2}{\partial t^2}) d\Omega_2 \\
&- \int_{\Omega_2} \delta w_2^T k_{w1}(w_1 - w_2) d\Omega_2 + \int_{\Omega_2} \delta w_2^T c_1(V \tfrac{\partial w_1}{\partial r} - V \tfrac{\partial w_2}{\partial r}) d\Omega_2 \\
&- \int_{\Omega_2} \delta w_2^T c_1(\tfrac{\partial w_1}{\partial t} - \tfrac{\partial w_2}{\partial t}) d\Omega_2 + \int_{\Omega_2} \delta w_2^T k_{w2} w_2 d\Omega_2 \\
&+ \int_{\Omega_2} \delta w_2^T c_2(-V \tfrac{\partial w_2}{\partial r} + \tfrac{\partial w_2}{\partial t}) d\Omega_2 = \int_{\Omega_2} \delta\mathbf{u}_2^T \mathbf{b}_2(r, s, t) d\Omega_2
\end{aligned}
\tag{15}
$$

where V is velocity of load at any instant t.

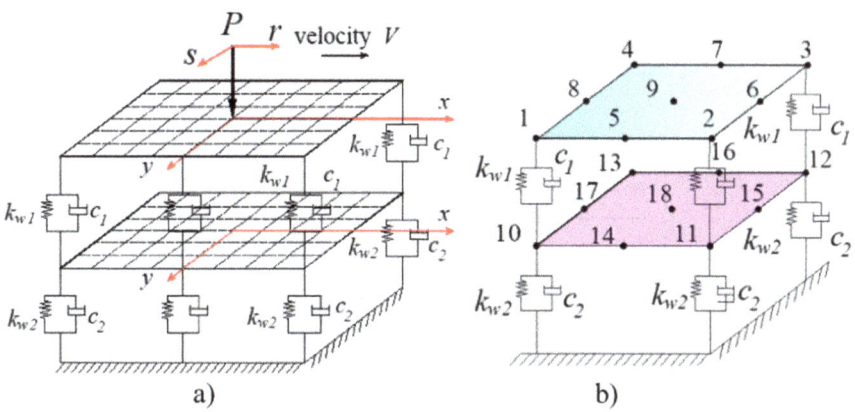

Fig. 2 **a** Discretization of FG sandwich plate into multi-layer moving plate elements; **b** A typical multi-layer moving plate element

Figure 2b shows a multi-layer moving plate element which is modeled by two layers of serendipity quadrilateral nine-node (Q9) elements. Each node has 3 DOFs, and this element has 18 nodes with 54 DOFs. The element displacement vector $\mathbf{d}^e$ is defined as

$$
\mathbf{d}^e = \left[\underbrace{w_1^1\ \beta_{r1}^1\ \beta_{s1}^1\ \dots\ w_1^9\ \beta_{r1}^9\ \beta_{s1}^9}_{\text{upper plate}}\ \underbrace{w_2^{10}\ \beta_{r2}^{10}\ \beta_{s2}^{10}\ \dots\ w_2^{18}\ \beta_{r2}^{18}\ \beta_{s2}^{18}}_{\text{lower plate}} \right]^T_{54\times1} \tag{16}
$$

The vectors of displacement, bending and shear strain of plates can be respectively expressed as

$$
\mathbf{u}_1 = \mathbf{N}_1\mathbf{d}^e;\ \ \mathbf{u}_2 = \mathbf{N}_2\mathbf{d}^e;\ \ w_1 = \mathbf{N}_{w1}\mathbf{d}^e;\ \ w_2 = \mathbf{N}_{w2}\mathbf{d}^e;\ \ \kappa_1 = \mathbf{B}_{b1}\mathbf{d}^e; \\
\kappa_2 = \mathbf{B}_{b2}\mathbf{d}^e;\ \ \gamma_1 = \mathbf{B}_{s1}\mathbf{d}^e;\ \ \gamma_2 = \mathbf{B}_{s2}\mathbf{d}^e \tag{17}
$$

where

$$
\mathbf{N}_1 = \left[\begin{array}{cccccccccccc} N_1 & 0 & 0 & \dots & N_9 & 0 & 0 & 0\,0\,0 & \dots & 0\,0\,0 \\ 0 & N_1 & 0 & \dots & 0 & N_9 & 0 & 0\,0\,0 & \dots & 0\,0\,0 \\ 0 & 0 & N_1 & \dots & 0 & 0 & N_9 & 0\,0\,0 & \dots & 0\,0\,0 \end{array} \right]_{3\times54} ;
$$
$$
\underbrace{\qquad\qquad}_{(3\times27)} \qquad \underbrace{\qquad\qquad}_{(3\times27)}
$$
$$
\mathbf{N}_2 = \left[\begin{array}{cccccccccccc} 0\,0\,0 & \dots & 0\,0\,0 & N_1 & 0 & 0 & \dots & N_9 & 0 & 0 \\ 0\,0\,0 & \dots & 0\,0\,0 & 0 & N_1 & 0 & \dots & 0 & N_9 & 0 \\ 0\,0\,0 & \dots & 0\,0\,0 & 0 & 0 & N_1 & \dots & 0 & 0 & N_9 \end{array} \right]_{3\times54}
$$
$$
\underbrace{\qquad\qquad}_{(3\times27)} \qquad \underbrace{\qquad\qquad}_{(3\times27)} \tag{18}
$$

$$
\mathbf{N}_{w1} = \left[\underbrace{N_1\,0\,0\ \dots\ N_9\,0\,0}_{(1\times27)}\ \underbrace{0\,0\,0\ \dots\ 0\,0\,0}_{(1\times27)} \right]_{1\times54} ;
$$
$$
\mathbf{N}_{w2} = \left[\underbrace{0\,0\,0\ \dots\ 0\,0\,0}_{(1\times27)}\ \underbrace{N_1\,0\,0\ \dots\ N_9\,0\,0}_{(1\times27)} \right]_{1\times54} \tag{19}
$$

$$
\mathbf{B}_{b1} = \left[\begin{array}{cccccccccc} 0 & N_{1,r} & 0 & \dots & 0 & N_{9,r} & 0 & 0\,0\,0\ \dots\ 0\,0\,0 \\ 0 & 0 & N_{1,s} & \dots & 0 & 0 & N_{9,s} & 0\,0\,0\ \dots\ 0\,0\,0 \\ 0 & N_{1,s} & N_{1,r} & \dots & 0 & N_{9,s} & N_{9,r} & 0\,0\,0\ \dots\ 0\,0\,0 \end{array} \right]_{3\times54} ;
$$
$$
\underbrace{\qquad\qquad}_{(3\times27)} \qquad \underbrace{\qquad\qquad}_{(3\times27)}
$$
$$
\mathbf{B}_{b2} = \left[\begin{array}{cccccccccc} 0\,0\,0\ \dots\ 0\,0\,0 & 0 & N_{1,r} & 0 & \dots & 0 & N_{9,r} & 0 \\ 0\,0\,0\ \dots\ 0\,0\,0 & 0 & 0 & N_{1,s} & \dots & 0 & 0 & N_{9,s} \\ 0\,0\,0\ \dots\ 0\,0\,0 & 0 & N_{1,s} & N_{1,r} & \dots & 0 & N_{9,s} & N_{9,r} \end{array} \right]_{3\times54}
$$
$$
\underbrace{\qquad\qquad}_{(3\times27)} \qquad \underbrace{\qquad\qquad}_{(3\times27)} \tag{20}
$$

$$\mathbf{B}_{s1} = \begin{bmatrix} \underbrace{N_{1,r}\ N_1\ 0\ \dots\ N_{9,r}\ N_9\ 0}_{(2\times27)}\ \underbrace{0\ 0\ 0\ \dots\ 0\ 0\ 0}_{(2\times27)} \\ N_{1,s}\ 0\ N_1\ \dots\ N_{9,s}\ 0\ N_9\ 0\ 0\ 0\ \dots\ 0\ 0\ 0 \end{bmatrix}_{2\times54} ;$$

$$\mathbf{B}_{s2} = \begin{bmatrix} \underbrace{0\ 0\ 0\ \dots\ 0\ 0\ 0}_{(2\times27)}\ \underbrace{N_{1,r}\ N_1\ 0\ \dots\ N_{9,r}\ N_9\ 0}_{(2\times27)} \\ 0\ 0\ 0\ \dots\ 0\ 0\ 0\ N_{1,s}\ 0\ N_1\ \dots\ N_{9,s}\ 0\ N_9 \end{bmatrix}_{2\times54} \tag{21}$$

Equations (14) and (15) are rewritten in short and familiar forms as

$$\mathbf{M}_1^e \ddot{\mathbf{d}}^e + \mathbf{C}_1^e \dot{\mathbf{d}}^e + \mathbf{C}_1^e \mathbf{d}^e = \mathbf{P}_1^e \tag{22}$$

$$\mathbf{M}_2^e \ddot{\mathbf{d}}^e + \mathbf{C}_2^e \dot{\mathbf{d}}^e + \mathbf{C}_2^e \mathbf{d}^e = \mathbf{P}_2^e \tag{23}$$

where mass matrices $\mathbf{M}_1^e$, $\mathbf{M}_2^e$, damping matrices $\mathbf{C}_1^e$, $\mathbf{C}_2^e$, stiffness matrices $\mathbf{K}_1^e$, $\mathbf{K}_2^e$, and load vector $\mathbf{P}_1^e$, $\mathbf{P}_2^e$ of element are respectively written as.

$$\mathbf{M}_1^e = \mathbf{m}_1 \int_{\Omega_{1e}} \mathbf{N}_1^T \mathbf{N}_1 d\Omega_{1e}; \quad \mathbf{M}_2^e = \mathbf{m}_2 \int_{\Omega_{2e}} \mathbf{N}_2^T \mathbf{N}_2 d\Omega_{2e} \tag{24}$$

$$\mathbf{C}_1^e = -2\mathbf{m}_1 V \int_{\Omega_{1e}} \mathbf{N}_1^T \mathbf{N}_{1,r} d\Omega_{1e} + c_1 \int_{\Omega_{1e}} \mathbf{N}_{w1}^T \mathbf{N}_{w1} d\Omega_{1e} - c_1 \int_{\Omega_{1e}} \mathbf{N}_{w1}^T \mathbf{N}_{w2} d\Omega_{1e} \tag{25}$$

$$\mathbf{C}_2^e = -2\mathbf{m}_2 V \int_{\Omega_{2e}} \mathbf{N}_2^T \mathbf{N}_{2,r} d\Omega_{2e} - c_1 \int_{\Omega_{2e}} \mathbf{N}_{w2}^T \mathbf{N}_{w1} d\Omega_{2e} + c_1 \int_{\Omega_{2e}} \mathbf{N}_{w2}^T \mathbf{N}_{w2} d\Omega_{2e} \\ + c_2 \int_{\Omega_{2e}} \mathbf{N}_{w2}^T \mathbf{N}_{w2} d\Omega_{2e} \tag{26}$$

$$\mathbf{K}_1^e = \int_{\Omega_{1e}} \mathbf{B}_{b1}^T \mathbf{D}_{b1} \mathbf{B}_{b1} d\Omega_{1e} + \int_{\Omega_{1e}} \mathbf{B}_{s1}^T \mathbf{D}_{s1} \mathbf{B}_{s1} d\Omega_{1e} + \mathbf{m}_1 V^2 \int_{\Omega_{1e}} \mathbf{N}_1^T \mathbf{N}_{1,rr} d\Omega_{1e} \\ - \mathbf{m}_1 a \int_{\Omega_{1e}} \mathbf{N}_1^T \mathbf{N}_{1,r} d\Omega_{1e} + k_{w1} \int_{\Omega_{1e}} \mathbf{N}_{w1}^T \mathbf{N}_{w1} d\Omega_{1e} - k_{w1} \int_{\Omega_{1e}} \mathbf{N}_{w1}^T \mathbf{N}_{w2} d\Omega_{1e} \\ - c_1 V \int_{\Omega_{1e}} \mathbf{N}_{w1}^T \mathbf{N}_{w1,r} d\Omega_{1e} + c_1 V \int_{\Omega_{1e}} \mathbf{N}_{w1}^T \mathbf{N}_{w2,r} d\Omega_{1e} \tag{27}$$

$$\mathbf{K}_2^e = \int_{\Omega_{2e}} \mathbf{B}_{b2}^T \mathbf{D}_{b2} \mathbf{B}_{b2} d\Omega_{2e} + \int_{\Omega_{2e}} \mathbf{B}_{s2}^T \mathbf{D}_{s2} \mathbf{B}_{s2} d\Omega_{2e} + \mathbf{m}_2 V^2 \int_{\Omega_{2e}} \mathbf{N}_2^T \mathbf{N}_{2,rr} d\Omega_{2e} \\ - \mathbf{m}_2 a \int_{\Omega_{2e}} \mathbf{N}_2^T \mathbf{N}_{2,r} d\Omega_{2e} - k_{w1} \int_{\Omega_{2e}} \mathbf{N}_{w2}^T \mathbf{N}_{w1} d\Omega_{2e} + k_{w1} \int_{\Omega_{2e}} \mathbf{N}_{w2}^T \mathbf{N}_{w2} d\Omega_{2e} \\ + c_1 V \int_{\Omega_{2e}} \mathbf{N}_{w2}^T \mathbf{N}_{w1,r} d\Omega_{2e} \\ - c_1 V \int_{\Omega_{2e}} \mathbf{N}_{w2}^T \mathbf{N}_{w2,r} d\Omega_{2e} \\ + k_{w2} \int_{\Omega_{2e}} \mathbf{N}_{w2}^T \mathbf{N}_{w2} d\Omega_{2e} - c_2 V \int_{\Omega_{2e}} \mathbf{N}_{w2}^T \mathbf{N}_{w2,r} d\Omega_{2e} \tag{28}$$

$$\mathbf{P}_1^e = \int_{\Omega_{1e}} \mathbf{N}_1^T \mathbf{b}_1(r, s, t) d\Omega_{1e}; \quad \mathbf{P}_2^e = \int_{\Omega_{2e}} \mathbf{N}_2^T \mathbf{b}_2(r, s, t) d\Omega_{2e} \tag{29}$$

in which, $(\cdot)_{,r}$ is the first partial derivatives and $(\cdot)_{,rr}$ is second partial derivatives with respect to r.

Finally, the mass, damping, stiffness matrices, and load vector of a multi-layer moving plate element are established as.

$$\mathbf{M}_e = \mathbf{M}_1^e + \mathbf{M}_2^e; \ \mathbf{C}_e = \mathbf{C}_1^e + \mathbf{C}_2^e; \ \mathbf{K}_e = \mathbf{K}_1^e + \mathbf{K}_2^e; \ \mathbf{P}_e = \mathbf{P}_1^e + \mathbf{P}_2^e \qquad (30)$$

The equation of motion of the FG sandwich plate is written as

$$\mathbf{M}\ddot{d} + C\dot{d} + Kd = P \qquad (31)$$

where $\ddot{\mathbf{d}}$, $\dot{\mathbf{d}}$, and $\mathbf{d}$ are global acceleration, global velocity, and global displacement vectors, respectively; $\mathbf{M}$ is global mass matrix; $\mathbf{C}$ is global damping matrix; $\mathbf{K}$ is global stiffness matrix; and $\mathbf{P}$ is the global load vector.

The static and free vibration are respectively calculated from Eq. (31) as

$$\mathbf{Kd} = \mathbf{P} \qquad (32)$$

$$(\mathbf{K} - \omega^2 \mathbf{M})\mathbf{d} = 0 \qquad (33)$$

in which ω is the natural frequency.

3 Numerical Results

3.1 Static Analysis

In this section, the geometric parameters of plate are defined as $L/B = 1$ and $L/h_1 = L/h_2 = 5$. The upper and lower FG plates comprise metal (aluminum, Al) and ceramic (zirconia, ZrO_2). The material paramerters are as follow: $E_{m1} = E_{m2} = 70\,\text{GPa}$, $E_{c1} = E_{c2} = 200\,\text{GPa}$, $\rho_{m1} = \rho_{m2} = 2707\,\text{kg/m}^3$, $\rho_{c1} = \rho_{c2} = 5700\,\text{kg/m}^3$, $\upsilon_1 = \upsilon_2 = 0.3$. The FG sandwich plate is subjected to a uniform load q. The non-dimensional center deflection $\overline{w}_i = 10^2 w_i\left(\frac{L}{2}, \frac{B}{2}\right) E_{mi} h_i^3/(12(1 - \upsilon_i^2)qL^4)$ $(i = 1, 2)$ is used to report the obtained results. The convergence of non-dimensional center deflection $\overline{w}_1$ of upper plate against the mesh density $N \times N$ for simply supported (SSSS) and clamped (CCCC) FG sandwich plate with various volume fraction exponent n is presented in Table 1. The stiffness and damping coefficients of connected layer and foundation are as follow: $k_{w1} = k_{w2} = 0\,\text{N/m}^3$ (free connection), $c_1 = c_2 = 0\,\text{Ns/m}^3$. The results of Nguyen-Xuan et al. [12] using an edge-based smoothed finite element method are also presented to give the solution's comparision. Table 1 shows that calculated results and those of Nguyen-Xuan et al. [12] agree excellently.

Table 1 Non-dimensional center deflection $\overline{w}_1$ against the mesh density $N \times N$ for simply supported (SSSS) and clamped (CCCC) FG sandwich plate under a uniform load q ($L/B = 1, L/h_1 = L/h_2 = 5$)

Boundary condition	Method	Meshing $N \times N$	Volume fraction exponent n			
			0	0.5	1	2
SSSS	MMPM	4×4	0.1700	0.2228	0.2518	0.2824
		8×8	0.1701	0.2231	0.2520	0.2824
		12×12	0.1701	0.2231	0.2521	0.2825
		16×16	0.1702	0.2231	0.2521	0.2825
	Nguyen-Xuan et al. [12]		0.1703	0.2232	0.2522	0.2827
CCCC		4×4	0.0787	0.1023	0.1165	0.1332
		8×8	0.0779	0.1013	0.1154	0.1318
		12×12	0.0777	0.1011	0.1151	0.1314
		16×16	0.0776	0.1010	0.1150	0.1313
	Nguyen-Xuan et al. [12]		0.0777	0.1012	0.1152	0.1313

Next, Table 2 shows the non-dimensional center deflection $\overline{w}_i$ ($i = 1, 2$) of simply supported FG sandwich plate with various connected layer and foundation stiffness. The stiffness parameter $k_{wi} = 0 \, \text{N/m}^3$ represents free connection and $k_{wi} = 1 \times 10^{16} \, \text{N/m}^3$ represents rigidity connection. It can be seen that that the sandwich plate separates into two independent plates when the stiffness parameters are $k_{w1} = k_{w2} = 0 \, \text{N/m}^3$. Thus, the center deflection values of upper plate reach those of single FG plate and the center deflection values of lower plate are zero. Furthermore, when the stiffness parameters are $k_{w1} = 1 \times 10^{16} \, \text{N/m}^3$, the system is composed two parallel FG plates with rigidity connection and the center deflection values are the smallest.

3.2 Free Vibration Analysis

In this section, the free vibration behaviors of FG sandwich plate are further investigated. The FG plate comprises metal (aluminum, Al) and ceramic (Aluminum oxide, Al_2O_3). The material characteristics and geometric paramaters of plate are as follow: $E_{c1} = E_{c2} = 380 \, \text{GPa}$, $E_{m1} = E_{m2} = 70 \, \text{GPa}$, $\rho_{c1} = \rho_{c2} = 3800 \, \text{kg/m}^3$, $\rho_{m1} = \rho_{m2} = 2707 \, \text{kg/m}^3$, $v_1 = v_2 = 0.3$, $L/B = 1$, $k_{w1} = k_{w2} = 0 \, \text{N/m}^3$ (free connection), $c_1 = c_2 = 0 \, \text{Ns/m}^3$. The convergence of first non-dimensional frequency $\overline{\omega} = \omega h_1 \sqrt{\rho_{c1}/E_{c1}}$ against the mesh density $N \times N$ of simply supported (SSSS) FG sandwich plate is illustrated in Table 3. It can be seen that the results obtained in this study are convergence and agree well with those of Nguyen-Xuan et al. [12].

Next, Table 4 presents the four lowest non-dimensional frequency $\overline{\omega} = \omega h_1 \sqrt{\rho_{c1}/E_{c1}}$ of simply supported FG sandwich plate with a changing connected

Table 2 Non-dimensional center deflection $\bar{w}_i$ ($i = 1, 2$) for simply supported FG sandwich plate under a uniform load q with various connected layer and foundation stiffness ($L/B = 1, L/h_1 = L/h_2 = 5$)

Foundation coefficients	Center deflection	Volume fraction exponent n			
		0	0.5	1	2
$k_{w1} = 0\,\text{N/m}^3\ c_1 =$ $0\,\text{Ns/m}^3$ $k_{w2} = 0\,\text{N/m}^3\ c_2 =$ $0\,\text{Ns/m}^3$	$\bar{w}_1$	0.1716	0.2250	0.2542	0.2851
	$\bar{w}_2$	0.0000	0.0000	0.0000	0.0000
$k_{w1} =$ $1 \times 10^8\,\text{N/m}^3\ c_1 =$ $0\,\text{Ns/m}^3$ $k_{w2} = 0\,\text{N/m}^3\ c_2 =$ $0\,\text{Ns/m}^3$	$\bar{w}_1$	0.1711	0.2242	0.2532	0.2838
	$\bar{w}_2$	3.80×10^{-4}	6.52×10^{-4}	8.32×10^{-4}	1.04×10^{-4}
$k_{w1} =$ $1 \times 10^{16}\,\text{N/m}^3\ c_1 =$ $0\,\text{Ns/m}^3$ $k_{w2} = 0\,\text{N/m}^3\ c_2 =$ $0\,\text{Ns/m}^3$	$\bar{w}_1$	0.0857	0.1124	0.1271	0.1424
	$\bar{w}_2$	0.0857	0.1124	0.1271	0.1424
$k_{w1} =$ $1 \times 10^8\,\text{N/m}^3\ c_1 =$ $0\,\text{Ns/m}^3$ $k_{w2} =$ $1 \times 10^{16}\,\text{N/m}^3\ c_2 =$ $0\,\text{Ns/m}^3$	$\bar{w}_1$	0.1711	0.2242	0.2533	0.2838
	$\bar{w}_2$	1.71×10^{-9}	2.42×10^{-9}	2.53×10^{-9}	2.83×10^{-9}

Table 3 Convergence of first non-dimensional frequency $\bar{\omega} = \omega h_1 \sqrt{\rho_{c1}/E_{c1}}$ of (SSSS) FG sandwich plate

Ratio L/h	Method	Meshing $N \times N$	Volume fraction exponent n				
			0	0.5	1	4	10
$L/h_1 = 5$ $L/h_2 = 5$	MMPM	4×4	0.2137	0.1858	0.1746	0.1156	0.1405
		8×8	0.2135	0.1856	0.1744	0.1154	0.1404
		12×12	0.2135	0.1856	0.1744	0.1154	0.1404
		16×16	0.2135	0.1856	0.1744	0.1154	0.1404
	Nguyen-Xuan et al. [12]		0.2121	0.1821	0.1635	0.1399	0.1327

layer stiffness. It can be seen that when the connected layer stiffness $k_{w1} = 0\,\text{N/m}^3$, the 1st order frequency and the 2nd order frequency are the same. The 3rd order and the 4th order have the same situation. It can be explained that as the stiffness of connected layer is zero, the FG sandwich plate separates into two independent and parallel FG plates. Consequently, the frequencies of two plates will alternate. However, when the stiffness coefficient of the connected layer is $k_{w1} = 1 \times 10^{16}\,\text{N/m}^3$,

Table 4 Non-dimensional frequency $\bar{\omega} = \omega h_1 \sqrt{\rho_{c1}/E_{c1}}$ of simply supported FG sandwich plate (Al/Al$_2$O$_3$) with various connected layer stiffness coefficients

Foundation coefficients	Mode number	Volume fraction exponent n			
		0	0.5	1	4
$k_{w1} = 0\,\text{N/m}^3\ c_1 = 0\,\text{Ns/m}^3$	1	0.2135	0.1856	0.1744	0.1554
$k_{w2} = 0\,\text{N/m}^3\ c_2 = 0\,\text{Ns/m}^3$	2	0.2135	0.1856	0.1744	0.1554
	3	0.4649	0.4052	0.3771	0.3316
	4	0.4649	0.4052	0.3771	0.3316
$k_{w1} = 1 \times 10^{16}\,\text{N/m}^3\ c_1 = 0\,\text{Ns/m}^3$	1	0.2135	0.1856	0.1744	0.1554
$k_{w2} = 0\,\text{N/m}^3\ c_2 = 0\,\text{Ns/m}^3$	2	0.4629	0.4052	0.3771	0.3316
	3	0.4744	0.4076	0.3823	0.3342
	4	0.6798	0.5883	0.5478	0.4754

the repetition frequency of the FG sandwich plate are eliminated. It can be reason that the rigid connected double-layer plate can be regarded as a single plate.

4 Conclusion

In this paper, the Multi-layer Moving Plate Method (MMPM) has been extended to investigate the static and free vibration behaviors of FG sandwich plate resting on a viscoelastic foundation. The formulations of multi-layer moving plate element mass, damping and stiffness matrices of the FG sandwich plate are derived. The accuracy of the method is verified by comparing calculated results with published results. Next, a parametric study is performed to examine the static and free vibration behaviors of FG sandwich plate with various volume fraction exponents, boundary conditions, geometric parameters, connected layer and foundation stiffness. Several conclusions can be withdrawn as follow.

(i) As the connected layer stiffness is zero, the FG sandwich plate separates into two independent and parallel FG plates. Thus, the deflection of the upper plate center reaches those of single FG plate and the deflection of lower plate center is zero. In addition, the natural frequencies of the two plates will alternate.

(ii) As the connected layer stiffness is rigidity, the FG sandwich plate is composed two parallel FG plates with rigidity connection. Thus, the deflection of plate is the smallest and the repetition frequency of the FG sandwich plate can be eliminated.

Acknowledgements We acknowledge the support of time and facilities from Ho Chi Minh University of Technology (HCMUT), VNU-HCM for this study.

References

1. Chonan S (1979) Moving load on initially stressed thick plates attached together by a flexible core. Ingenieur-Archiv 48:143–154
2. Kukla S (1999) Free vibration of a system of two elastically connected rectangular plates. J Sound Vib 225(1):29–39
3. Oniszczuk Z (2000) Free transverse vibrations of an elastically connected rectangular simply supported double-plate complex system. J Sound Vib 236(4):595–608
4. Hedrih K (2006) Transversal vibrations of double-plate systems. Acta Mech Sin 22(5):487–501
5. De Rosa MA, Lippiello M (2009) Free vibrations of simply supported double plate on two models of elastic soils. Int J Numer Anal Meth Geomech 33:331–353
6. Koh CG, Ong JSY, Chua DKH, Feng J (2003) Moving element method for train-track dynamics. Int J Numer Methods Eng 56:1549–1567
7. Luong VH, Cao TNT, Reddy JN, Ang KK, Tran MT, Dai J (2018) Static and dynamic analyses of Mindlin plates resting on viscoelastic foundation by using moving element method. Int J Struct Stab Dyn 18(11):1850131
8. Cao TNT, Luong VH, Vo HN, Nguyen XV, Bui VN, Tran MT (2018) A moving element method for the dynamic analysis of composite plate resting on a Pasternak foundation subjected to a moving load. Int J Comput Method 16(8):1850124 (1–19)
9. Luong VH, Cao TNT, Qui-X L, Nguyen XV (2020) Moving element method for dynamic analyses of functionally graded plates resting on pasternak foundation subjected to moving harmonic load. Int J Struct Stab Dyn 20(1):2050003 (25 pages)
10. Cao TNT, Reddy JN, Lieu QX, Nguyen XV, Luong VH (2019) A multi-layer moving plate method for dynamic analysis of viscoelastically connected double-plate systems subjected to moving loads. Adv Struct Eng 24(9):1798–1913
11. Reddy JN (2000) Analysis of functionally graded plates. Int J Numer Methods Eng 47:663–684
12. Nguyen-Xuan H, Tran LV, Nguyen-Thoi T, Vu-Do HC (2011) Analysis of functionally graded plates using an edge-based smoothed finite element method. Compos Struct 93:3019–3039

The Relationship Between the Speeds of Moving Load and the Dynamic Responses of Doubled-Plates Floating on the Shallow Water in Mekong Delta by Using IMEM Method

Ngoc Thuan Do, Xuan Vu Nguyen, Cong Huan Nguyen, Tran Nam Hai, Takayuki Suzuki, and Van Hai Luong

Abstract In this study, the responses of doubled-plates floating on the shallow water in the Mekong Delta under moving loads are considered. Particularly, the structures are modelled by two thin plates which are connected through Winkler-type elastic layers. Meanwhile, the linear shallow-water wave theory is adopted to model the water. In addition, a recently novel numerical method, which is namely Integrated Moving Element Method (IMEM), is employed to solve the coupled system of numerical equations in this study because of the advantages of this method in comparison to the traditional Finite Element Method (FEM). According to gained numerical results, the effects of the ratio between the thickness of layers as well as the stiffness of core on the Dynamic Amplification Factor (DAF) of the floating double-plates will be investigated.

Keywords IMEM · Moving load · Double-plate · FEM

N. T. Do
Ngo Quyen University, 229B Bach Dang Street, Phu Cuong Ward, Thu Dau Mot City, Binh Duong, Vietnam

N. T. Do · X. V. Nguyen · T. N. Hai · V. H. Luong (✉)
Faculty of Civil Engineering, Ho Chi Minh City University of Technology (HCMUT), 268 Ly Thuong Kiet Street, District 10, Ho Chi Minh City, Vietnam
e-mail: lvhai@hcmut.edu.vn

Vietnam National University Ho Chi Minh City, Linh Trung Ward, Thu Duc District, Ho Chi Minh City, Vietnam

C. H. Nguyen
Faculty of Civil Engineering, Sai Gon Technology University, 180 Cao Lo Street, Ward 4, District 8, Ho Chi Minh City, Vietnam

T. Suzuki
Department of Civil Engineering, Yokohama National University, 79-5 Tokiwadai, Hodogayaku, Yokohama 240-8501, Japan

© The Author(s), under exclusive license to Springer Nature Singapore Pte Ltd. 2023   929
J. N. Reddy et al. (eds.), *ICSCEA 2021*, Lecture Notes in Civil Engineering 268,
https://doi.org/10.1007/978-981-19-3303-5_85

1 Introduction

Nowadays, the demands for the ground to serve the economic development are extremely increasing. Especially, the increase of urbanization process and climate change, the countries that have longshore or island nations have implemented sea encroachment projects to solve housing for people and developing infrastructures. However, this method requires huge costs. To solve that issue, the experts suggested a highly feasible solution that is constructing a Very Large Floating Structures (VLFS). This method has been proved to be effective because it is simple to conduct, and we can customize this structure in various surface areas. That VLFS might be used to exploit energy, filter water, and make the area reserve. Besides, the VLFS is also used as a floating airport to solve traffic problems, which will save a large amount of land for economic development.

Since VLFS has a thickness that is much smaller than its length and width so elastic behavior is more important than rigid mass motion. The interaction between the fluid and structure that is taken into account by elastic deformation is called hydroelastic behavior. It plays a central role in the design of VLFS and has attracted the attention of many researchers around the world. This can be seen in the study of Watanabe et al. [1]. In addition, there are two-dimensional or three-dimensional theories that have been developed to analyze the hydroelastic behavior of a flexural floating structure, the problems were limited to the small amplitude linear wave theory. Different forms of these theories have been used for analytical methods [2, 3] and numerical methods [4, 5]. Besides, the Boundary Element Method—Finite Element Method (BEM–FEM)—a hybrid method—has been developed for hydroelastic behavior analysis. Regarding this method, Ismail [6] also applied the fundamental solution of Laplace's equation as a Green function to propose his method. In order to diversify the computational model, another form of the wave theory was applied—the linear shallow-water theory (tidal theory). Accordingly, the displacement and slope of the water surface do not assume to be small, and this theory is not a consequence of the linear wave theory. This theory made accurate predictions such as tides, currents in rivers or coastlines, etc. [7]. Some studies on the behavior of floating structures subjected to moving loads under the shallow water hypothesis can be mentioned such as Sturova [8] analyzed the behavior of an elastic circular plate caused by external loads using the combinatorial oscillation method. A time-domain method has been developed to analyze the hydroelastic behavior of a strip of heterogeneous thickness floating in the shallow water of variable depth by Sturova [9]. In order to develop a floating structure model, some research works proposed different types of floating structures, in which a two-layer composite beam system linked together through an elastic layer is a typical example [10]. The girder system is widely used in practice, such as floating floor of rail [11], sandwich beam using a bonding layer, and vibration damping system. Oniszczuk [12] investigated the free vibration of two parallel simply supported beams with multiple spans connected by Winkler-type elastic layers. Abu-Hilal [13] analyzed the dynamic behavior of this system under the impact of moving loads.

Under the impact of the liquid below, the plate structure around the contact position will be affected more than on the ground. In this paper, the use of the doubled-plates model will help to limit this problem, the dynamic pressure of the liquid acting on the upper plate will be significantly reduced because it has to go through the elastic layer. In this study, the IMEM method [14] was developed for the hydroelastic problem of infinitely wide soft-core doubled-plates floating on calm shallow water with a flat seabed. This method offers a more efficient and convenient approach to the dynamics problems with loads move in the fixed domain.

2 Theoretical Basis

2.1 Model of Floating Doubled-Plates

The plate structural model of two isotropic thin plates, which are parallel to each other and connected through a Winkler-type elastic layer, is shown in Fig. 1. The monitored domain of two plates having the same dimensions and having material characteristics for the upper and lower plates are respectively: thickness h_t and h_b, mass per area m_t and m_b, the elastic modulus of the material E_t and E_b, the Poison's ratio of the material v_t and v_b, and flexural rigidity of the plate D_t and D_b. The sandwich layer between two plates has a constant stiffness k_c. The water depth is H, and the moving load is $F(t)$.

The coordinate system $O_t x_t y_t z_t$, $O_b x_b y_b z_b$ of the upper and lower plate is chosen so that the plane $O_t x_t y_t$ and $O_b x_b y_b$ coincides with the neutral plane of each plate. The displacement in the z direction of a point on the neutral plane of two plates is w_t and w_b, respectively. According to thin plate theory, the displacement field at any point in the neutral surface of the plate: u_t, v_t, w_t and u_b, v_b, w_b in the x, y, and z directions is represented by the displacement field at the corresponding point on the neutral plane of the plate as follows:

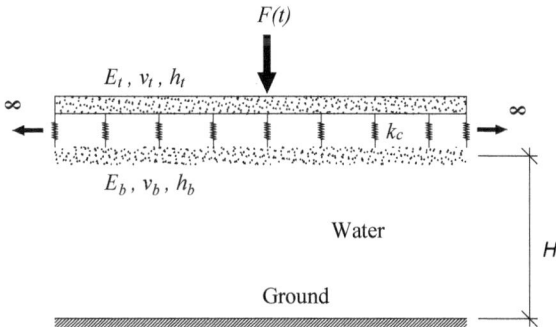

Fig. 1 Doubled-plates connected by a Winkler-type elastic layer

$$\begin{cases} u_t(x_t, y_t, z_t) = -z_t \dfrac{\partial w_t(x_t, y_t)}{\partial x_t} \\ v_t(x_t, y_t, z_t) = -z_t \dfrac{\partial w_t(x_t, y_t)}{\partial y_t} \ z_t = \left[-\dfrac{h_t}{2}, \dfrac{h_t}{2} \right] \\ w_t(x_t, y_t, z_t) = w_t(x_t, y_t) \\ u_b(x_b, y_b, z_b) = -z_b \dfrac{\partial w_b(x_b, y_b)}{\partial x_b} \\ v_b(x_b, y_b, z_b) = -z_b \dfrac{\partial w_b(x_b, y_b)}{\partial y_t} \ z_b = \left[-\dfrac{h_b}{2}, \dfrac{h_b}{2} \right] \\ w_b(x_b, y_b, z_b) = w_b(x_b, y_b) \end{cases} \tag{1}$$

The equation of motion of the isotropic two-plate system can be deduced:

$$m_t \frac{\partial^2 w_t}{\partial t^2} + a\nabla^4(w_t) + D_t\nabla^4(w_t) + k_c(w_t - w_b) = F(t)\delta(x - d)\delta(y) \tag{2}$$

$$m_b \frac{\partial^2 w_t}{\partial t^2} + a\nabla^4(w_b) + D_b\nabla^4(w_b) + k_c(-w_t + w_b) = p_z \tag{3}$$

2.2 Wave Theory

To describe the motion of fluid, the fluid is assumed to be incompressible, non-viscous, non-vortex, and has a density ρ. Therefore, a potential function satisfying Laplace's equation can be used to describe the motion of the fluid. Additionally, the pressure of the liquid satisfies the linear equation of Bernoulli. At the wet side, it is assumed that there is no gap between the structure and the liquid. Besides assuming that the seabed is flat and the liquid does not flow through the seabed, this means that the vertical velocity of the liquid at the seabed is zero. An assumption of non-perturbation is applied to the region far from the location of the load. The system of reduced equations is shown below

$$\frac{\partial w}{\partial t} + h\left(\frac{\partial^2}{\partial x^2} + \frac{\partial^2}{\partial y^2} \right)\phi = 0 \tag{4}$$

$$\frac{\partial \phi}{\partial x}; \frac{\partial \phi}{\partial y} = 0, x \to \infty, y \to \infty \tag{5}$$

$$p_z = -\rho \left. \frac{\partial \phi}{\partial t} \right|_{z=0} \tag{6}$$

2.3 The Moving Element Method for the Doubled-Plates on Shallow Water

In the IMEM, an orthogonal coordinate system moves along with the load whose origin is located at the load position as follows:

$$r = x - d; \; s = y; \; z = z \tag{7}$$

After converting the equations to the moving coordinate system, the differential equations and boundary conditions are rewritten in the moving coordinate system. Although the horizontal dimension of the plate is assumed to be infinite, in numerical simulation, a finite domain of the plate-water model is used. The boundary of the finite domain is taken far enough from the moving load so that displacement, rotation, velocity potential, velocity, moment, and shear force are close to zero.

The 4-node Hermite quadrilateral element has matrix of shape function $\mathbf{N}_t$ and $\mathbf{N}_b$ is used to model the behavior of the upper and lower plate. While the 4-node linear element with shape function φ is used to model the motion of the liquid, and the displacement vector has the form as Eq. (9).

$$
\begin{aligned}
\mathbf{w}_t^e &= \begin{bmatrix} w_{1t}, \varphi_{1t}^y, \varphi_{1t}^x, w_{2t}, \varphi_{2t}^y, \varphi_{2t}^x, w_{3t}, \varphi_{3t}^y, \varphi_{3t}^x, \\ w_{4t}, \varphi_{4t}^y, \varphi_{4t}^x \end{bmatrix} \\
\mathbf{w}_b^e &= \begin{bmatrix} w_{1b}, \varphi_{1b}^y, \varphi_{1b}^x, w_{2b}, \varphi_{2b}^y, \varphi_{2b}^x, w_{3b}, \varphi_{3b}^y, \varphi_{3b}^x, \\ w_{4b}, \varphi_{4b}^y, \varphi_{4b}^x \end{bmatrix} \\
\boldsymbol{\Phi}^e &= \begin{bmatrix} \phi_1 & \phi_2 & \phi_3 & \phi_4 \end{bmatrix}
\end{aligned}
\tag{8}
$$

The general equation of motion of the floating doubled-plate model in the moving coordinate system is rewritten as follows:

$$
\begin{aligned}
&\begin{bmatrix} \mathbf{M}_t & 0 & 0 \\ 0 & \mathbf{M}_b & 0 \\ 0 & 0 & 0 \end{bmatrix} \begin{bmatrix} \ddot{\mathbf{w}}_t \\ \ddot{\mathbf{w}}_b \\ \ddot{\boldsymbol{\Phi}} \end{bmatrix} + \begin{bmatrix} \mathbf{C}_t & 0 & 0 \\ 0 & \mathbf{C}_b & -\rho \mathbf{L}_2 \\ 0 & \mathbf{Q}_1 & 0 \end{bmatrix} \begin{bmatrix} \dot{\mathbf{w}}_t \\ \dot{\mathbf{w}}_b \\ \dot{\boldsymbol{\Phi}} \end{bmatrix} \\
&+ \begin{bmatrix} \mathbf{K}_t + \mathbf{K}_c^{tt} & -\mathbf{K}_c^{tb} & 0 \\ -\mathbf{K}_c^{tb} & \mathbf{K}_b + \mathbf{K}_c^{bb} & \rho \cdot d \cdot \mathbf{Ldr}_2 \\ 0 & -d \cdot \mathbf{Q}_2 & -\mathbf{H}_{sh} \end{bmatrix} \begin{bmatrix} \mathbf{w}_t \\ \mathbf{w}_b \\ \boldsymbol{\Phi} \end{bmatrix} + \begin{bmatrix} \mathbf{P} \\ 0 \\ 0 \end{bmatrix}
\end{aligned}
\tag{9}
$$

where the mass matrices $\mathbf{M}_t$ and $\mathbf{M}_b$, damping matrices $\mathbf{C}_t$ and $\mathbf{C}_b$, stiffness matrices $\mathbf{K}_t$ and $\mathbf{K}_b$ of the top and bottom plates respectively, with the connection matrix $\mathbf{K}_c$ is concatenated from the element matrices as follows

$$
\mathbf{K}_t^e = \int_{S_e} \left(\mathbf{B}_t^T \mathbf{D}_t \mathbf{B}_t \right) dS - m_t \frac{\partial d}{\partial t}^2 \int_{S_e} \mathbf{N}_{t,r}^T \mathbf{N}_{t,r} dS - m_t \frac{\partial^2 d}{\partial t^2} \int_{S_e} \mathbf{N}_t^T \mathbf{N}_{t,r} dS
$$

$$-\frac{\partial d}{\partial t}\int\limits_{S_e} c_{vd}\mathbf{N}_t^T\mathbf{N}_{t,r}dS - \alpha_t\int\limits_{S_e}\left(\mathbf{B}_t^T\mathbf{D}_t\mathbf{B}_{t,r}\right)dS \tag{10}$$

$$\mathbf{K}_b^e = \int\limits_{S_e}\left(\mathbf{B}_b^T\mathbf{D}_b\mathbf{B}_b\right)dS - m_b\frac{\partial d}{\partial t}^2\int\limits_{S_e}\mathbf{N}_{b,r}^T\mathbf{N}_{b,r}dS - m_b\frac{\partial^2 d}{\partial t^2}\int\limits_{S_e}\mathbf{N}_b^T\mathbf{N}_{b,r}dS$$

$$+\rho g\int\limits_{S_e}\mathbf{N}_b^T\mathbf{N}_b dS - \frac{\partial d}{\partial t}c_b\int\limits_{S_e}\mathbf{N}_b^T\mathbf{N}_{b,r}dS - \frac{\partial d}{\partial t}\int\limits_{S_e}c_{vd}\mathbf{N}_b^T\mathbf{N}_{b,r}dS \tag{11}$$

$$-\alpha_b\int\limits_{S_e}\left(\mathbf{B}_b^T\mathbf{D}_b\mathbf{B}_{b,r}\right)dS$$

$$\mathbf{M}_t^e = m_t\int\limits_{S_e}\mathbf{N}_t^T\mathbf{N}_t dS$$

$$\mathbf{M}_b^e = m_b\int\limits_{S_e}\mathbf{N}_b^T\mathbf{N}_b dS \tag{12}$$

$$\mathbf{C}_t^e = \alpha_t\int\limits_{S_e}\left(\mathbf{B}_t^T\mathbf{D}_t\mathbf{B}_t\right)dS + \int\limits_{S_e}c_{vd}\mathbf{N}_t^T\mathbf{N}_t dS - 2m_t\frac{\partial d}{\partial t}\int\limits_{S_e}\mathbf{N}_{t,r}^T\mathbf{N}_t dS$$

$$\mathbf{C}_b^e = c_b\int\limits_{S_e}\mathbf{N}_b^T\mathbf{N}_b dS + \alpha_b\int\limits_{S_e}\left(\mathbf{B}_b^T\mathbf{D}_b\mathbf{B}_b\right)dS + \int\limits_{S_e}c_{vd}\mathbf{N}_b^T\mathbf{N}_b dS \tag{13}$$

$$-2m_b\frac{\partial d}{\partial t}\int\limits_{S_e}\mathbf{N}_{b,r}^T\mathbf{N}_b dS$$

$$\mathbf{K}_c^{tt,e} = k_c\int\limits_{S_e}\mathbf{N}_t^T\mathbf{N}_t dS$$

$$\mathbf{K}_c^{tb,e} = k_c\int\limits_{S_e}\mathbf{N}_t^T\mathbf{N}_b dS \tag{14}$$

$$\mathbf{K}_c^{bb,e} = k_c\int\limits_{S_e}\mathbf{N}_b^T\mathbf{N}_b dS$$

The matrix $\mathbf{L}_2$ and $\mathbf{Ldr}_2$ play the role of converting the dynamic pressure of the liquid to the nodes on the plate element. Matrix $\mathbf{Q}_1$ and $\mathbf{Q}_2$ will convert from displacement vector of plate element to velocity vector at respective fluid elements. Finally, the element matrix $\mathbf{H}_{sh}$ represents the relationship between the velocity potential of two directions in the horizontal plane and the vertical motion of the water surface in shallow water. The above matrices are also concatenated from the element matrices as follows

$$\mathbf{L}_2^e = \int\limits_{S_e}\left(\mathbf{N}^T\varphi\right)dS; \; \mathbf{Ldr}_2^e = \int\limits_{S_e}\left(\mathbf{N}_{,r}^T\varphi\right)dS \tag{15}$$

$$\mathbf{Q}_1^e = \int_{Se} \varphi^T \mathbf{N} dS; \quad \mathbf{Q}_2^e = \int_{Se} \varphi^T \mathbf{N}_{,r} dS \tag{16}$$

$$\mathbf{H}_{sh}^e = \int_{Se} h(\varphi_{,r}\varphi_{,r} + \varphi_{,s}\varphi_{,s}) dS \tag{17}$$

3 Numerical Results

The survey model is a floating doubled-plates whose thickness of the bottom plate is $h_b = 0.17$ m and the elastic modulus for both layers is 500 MPa. To simulate the IMEM, the wet side of the investigated fluid region is a rectangular region of length 400 m and breadth 200 m with the center of the region at the point of loading. The load acting on the plate is simulated by a small vehicle whose weight 235 kg includes the driver, which acts as a load moving along the x-axis with velocities $v = 1.5$ m/s; $v = 2.5$ m/s; $v = 3.5$ m/s; $v = 4$ m/s; $v = 5.5$ m/s; $v = 7$ m/s; $v = 10$ m/s, respectively. The depth of the survey water area is static water with $H = 1$ m.

3.1 The Responses of Doubled-Plates Floating When the Ratio Between the Thickness of Layers and Load Velocity Changes

The thickness of the upper plate varies by: $h_t = 0.5h_b$, $h_t = h_b$, $h_t = 1.5h_b$, $h_t = 2h_b$, $h_t = 2.5h_b$. The stiffness of the core is $k_c = \rho g = 10$ kN/m³. Figure 2 shows the maximum vertical displacement of the upper plate in the x-axis direction when changing the thickness of the plate and load velocity.

The results in the figure show that there is a large distance between the displacement lines of the plate, but the distance between the lines tends to decrease and converge when increasing the thickness of the upper plate. This issue also reveals that when the plate thickness approaches the optimal value, the maximum displacement of the plate is almost unchanged.

In case the speed of the load changes, the change of displacement value is divided into 3 segments. The first part has the speed from $v = 1.5$ m/s to $v = 3.5$ m/s, the middle part has the speed from $v = 3.5$ m/s to $v = 7$ m/s, the last part has the speed $v = 7$ m/s to $v = 10$ m/s. The displacement in the middle segment depends significantly on the speed of the load, while the first and last sections have displacement almost independent of the speed of the moving load.

Figure 3 shows the maximum vertical displacement of the lower plate in the x direction. The results show that the distance between these displacement lines is

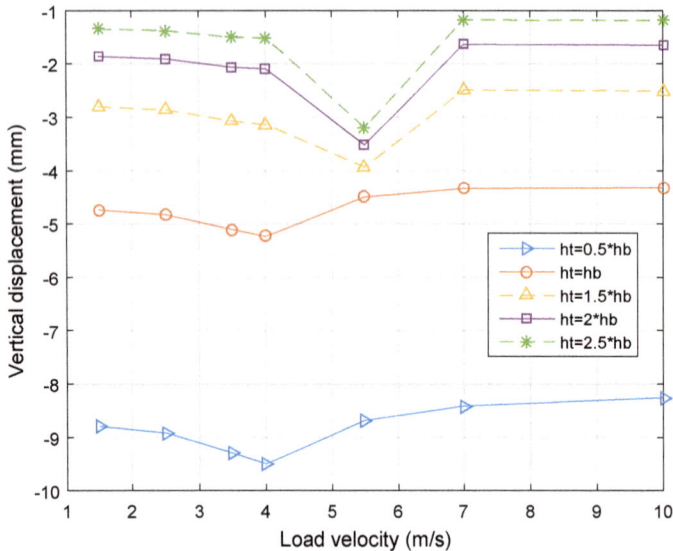

Fig. 2 Maximum vertical displacement of upper plate with the change of plate thickness and load velocity

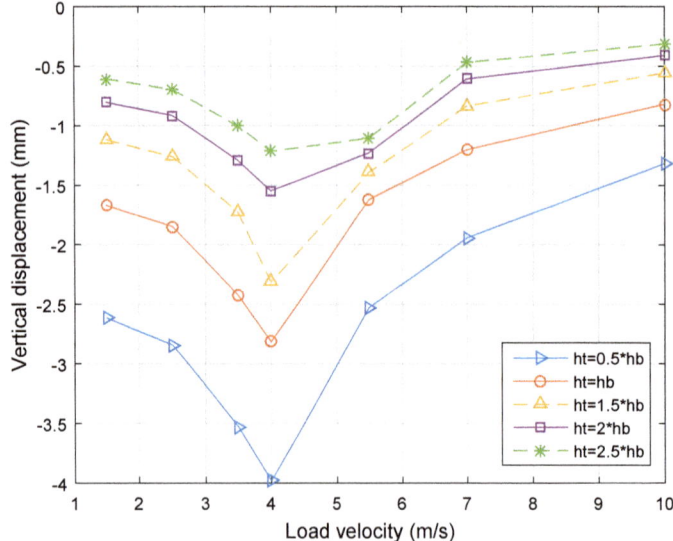

Fig. 3 Maximum vertical displacement of bottom plate with change of plate thickness and load velocity

quite large, and these lines also tend to converge gradually when the thickness of the upper plate is increased. And when the thickness has reached the optimal value, the displacements will not change. When the velocity of the load is changed, the displacement line is divided into two parts: for the speed is less than 4 m/s, the higher the speed, the higher the displacement; With a speed greater than 4 m/s, the higher the speed, the more the displacement tends to decrease. Displacement value reaches its maximum at $v = 4$ m/s.

Besides, when comparing the displacement of the upper plate with the lower plate, we find that the displacement lines of the lower plate have a steeper slope than the displacement lines of the upper plate. This means that the displacement of the lower plate is affected more than the upper plate when changing the velocity of the load. The reason is that when we use doubled-plates, the upper plate has been significantly reduced the influence of the bottom water surface because the lower plate and the connecting spring layer have absorbed some energy. In addition, the diagram also shows the general tendency that the displacement at the center of the upper plate is always much larger than that of the lower plate. But when gradually increasing the upper plate to a certain thickness, the displacement values of the two plates are almost equal.

From the displacement of the plate, the dynamic amplification factor (DAF) is obtained as shown in Fig. 5. The results show that when the upper plate has a thickness equal to or smaller than the lower plate, the DAF does not change significantly and does not depend much on the velocity of the load. But when the upper plate thickness is larger than the lower plate thickness, the DAF line is divided into distinct segments:

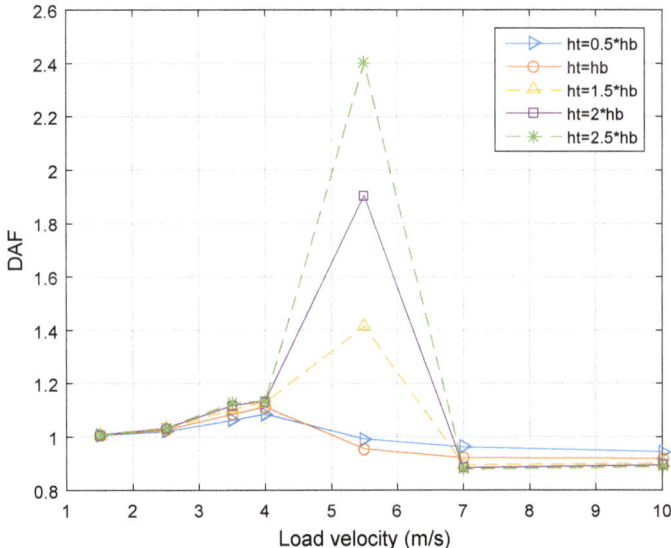

Fig. 4 The dynamic amplification factor of the upper plate with varying plate thickness and load velocity

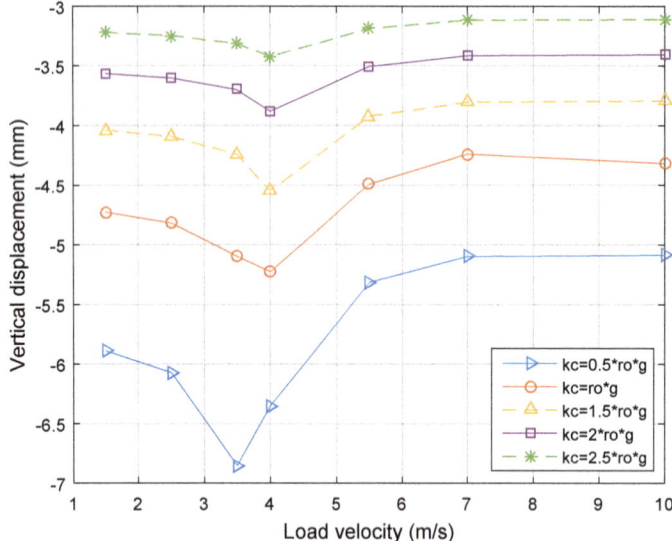

Fig. 5 Maximum vertical displacement of upper plate with varying the stiffness of the core and load velocity

the first segment with the velocity $v = 1.5$ m/s to $v = 4$ m/s; the middle segment has the velocity $v = 4$ m/s to $v = 7$ m/s, and the final segment has the velocity $v = 7$ m/s to $v = 10$ m/s. In which, the first section has the DAF > 1, which means the dynamic displacement is larger than the static displacement but the difference is not much, the influence of load velocity and plate thickness on the DAF is not obvious. The middle segment is a jump in the magnitude of the DAF, the maximum value of the DAF corresponding to the load velocity $v = 5.5$ m/s. The last segment has the DAF < 1, the dynamic displacement value is smaller than the static displacement and the difference is not large, the DAF has almost no change when changing the load velocity and plate thickness.

3.2 The Responses of Doubled-Plates Floating When the Stiffness of the Core and Load Velocities Changes

The plate thickness is $ht = hb = 0.17$ m. The stiffness of the core varies by: $k_c = 0.5 \, \rho g = 5$ kN/m³, $k_c = \rho g = 10$ kN/m³, $k_c = 15$ kN/m³, $k_c = 20$ kN/m³, $k_c = 25$ kN/m³.

The result of the maximum vertical displacement of the upper plate when changing the stiffness of the core and the load velocity is shown in Fig. 5.

The results show that: the distance between the displacement line is quite large when changing the stiffness of the core. However, these distances tend to get smaller

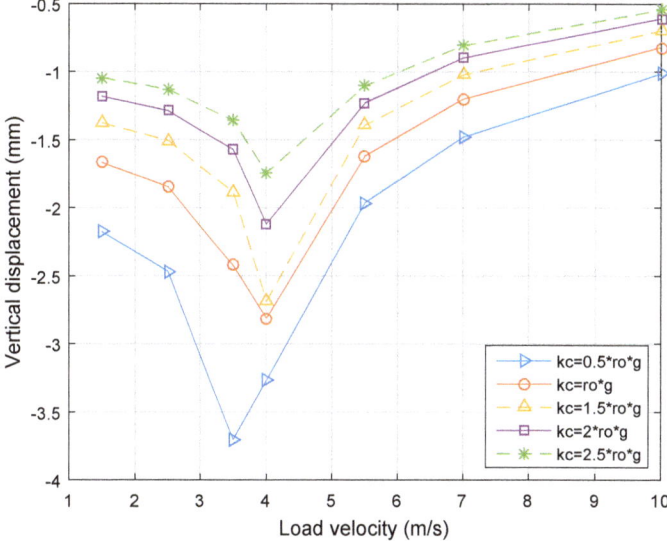

Fig. 6 Maximum vertical displacement of lower plate with varying the stiffness of the core and load velocity

and converge when increasing the stiffness of the core to an optimal value. On the other hand, when increasing the stiffness of the core, the displacement lines tend to gradually change to a straight line, which means a decrease in the dependence of the load velocity.

In the case when changing the load velocity, the maximum displacement of the plate has a rather large change in the range from $v = 2.5$ m/s to $v = 7$ m/s; But then $v > 7$ m/s, the displacement of the plate does not change much, which means it is less affected by the moving speed of the load.

Figure 6 shows the maximum vertical displacement of the lower plate. The results also show that the distance between these displacement lines is quite large, and these lines also tend to converge gradually when we increase the stiffness of the core to the optimal value. In case of changing the speed of the load, the trend of displacement is divided into two parts. When the speed is less than 4 m/s, the higher the speed, the higher the displacement; With a speed greater than 4 m/s, the higher the speed, the more the displacement tends to decrease. The maximum displacement value corresponds to the load speed at 4 m/s.

Figure 7 shows the DAF of the plate when changing the stiffness of the core and load velocity. The results show that the graph has a rather steep slope when the load velocity changes in the range from $v = 1.5$ m/s to $v = 7$ m/s, showing that the load velocity has a great influence on the DAF of the plate. However, when the speed is $v > 7$ m/s, the DAF is the equilibrium line, which means the DAF does not depend on the moving speed of the load. Additionally, when changing the stiffness of the core, there is also a clear effect on the DAF, and if we increase the stiffness of the

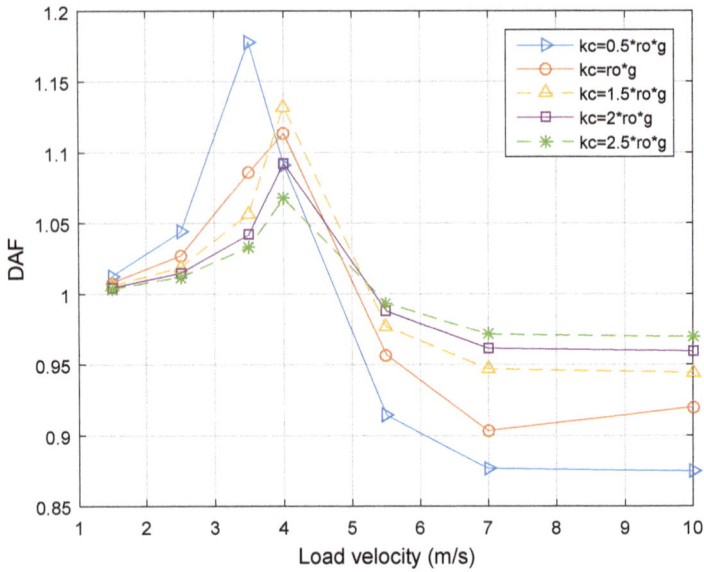

Fig. 7 The dynamic amplification factor of upper plate with varying the stiffness of the core and load velocity

core, the DAF will gradually converge and there will be no much change when the stiffness of the core to the optimal value.

4 Conclusions

The paper presents a doubled-plates model with an elastic core of the Winkler-type. This model is produced for the purpose of reducing the effect of water waves on vehicles moving above the floating structure. Besides, in order to increase the analysis efficiency for the moving load problem on shallow water, the IMEM has been developed. Several conclusions are drawn through numerical simulations are as follows:

- The multi-layer plate model is an effective model in reducing the dependence of the upper plate displacement on the vibration of the lower water surface since the energy of the vibration is absorbed by the lower plate and the stiffness of the core of the plate.
- For multi-layer structures, the design process should use the upper plate thicker than the lower plate, to balance the work between the two panels. It is the optimal thickness of the structure, which makes the most of the work efficiency of the material.

- The stiffness of the core should be more than twice the buoyancy stiffness, which will reduce the influence of water waves on the structure directly above, and the dependence on the speed of the load.

Acknowledgements This research is funded by Japan International Cooperation Agency Project for ASEAN University Network/Southeast Asia Engineering Education Development Network (JICA Project for AUN/SEED-Net) in the framework of Collaborative Education Program (CEP) under Program Contract No. HCMUT CEP 2101.

References

1. Watanabe E, Utsunomiya T, Wang CM (2004) Hydroelastic analysis of pontoon-type VLFS: a literature survey. Eng Struct 26:245–256. https://doi.org/10.1016/j.engstruct.2003.10.001
2. Sahoo T (2013) Mathematical techniques for wave interaction with flexible structures. CRC Press, Boca Raton
3. Kim JW, Webster WC (1998) The drag on an airplane taking off from a floating runway. J Mar Sci Technol 3(2):76–81. https://doi.org/10.1007/BF02492562
4. Fujikubo M, Yao T (2001) Structural modeling for global response analysis of VLFS. Mar Struct 14:295–310
5. Kashiwagi M (2004) Transient responses of a VLFS during landing and take off of an airplane. J Mar Sci Technol 9(1):14–23. https://doi.org/10.1007/s00773-003-0168-0
6. Ismail RES (2016) Time-domain three dimensional BE-FE method for transient response of floating structures under unsteady loads. Lat Am J Solids Struct 13(7):1340–1359. https://doi.org/10.1590/1679-78251688
7. Stoker JJ (1992) Water waves: the mathematical theory with applications. Wiley, Hoboken
8. Sturova IV (2003) The action of an unsteady external load on a circular elastic plate floating on shallow water. J Appl Math Mech 67(3):407–416. https://doi.org/10.1016/S0021-8928(03)90024-4
9. Sturova IV (2008) Effect of bottom topography on the unsteady behaviour of an elastic plate floating on shallow water. J Appl Math Mech 72(4):417–426. https://doi.org/10.1016/j.jappmathmech.2008.08.012
10. Vũ NX, Thân CTN, Việt BH, Hải LV (2019) Phương pháp phần tử chuyển động cho phân tích ứng xử tấm nổi nhiều lớp chịu tải trọng di chuyển trên vùng nước nông. Tạp chí Xây Dựng 12:118–122
11. Hussein MFM, Hunt HEM (2006) Modelling of floating-slab tracks with continuous slabs under oscillating moving loads. J Sound Vib 297(1–2):37–54. https://doi.org/10.1016/j.jsv.2006.03.026
12. Oniszczuk Z (2000) Free transverse vibrations of elastically connected simply supported double-beam complex system. J Sound Vib 232(2):387–403. https://doi.org/10.1006/jsvi.1999.2744
13. Abu-Hilal M (2006) Dynamic response of a double Euler-Bernoulli beam due to a moving constant load. J Sound Vib 297(3–5):477–491. https://doi.org/10.1016/j.jsv.2006.03.050
14. Reddy JN, Nguyen XV, Than Cao TN, Lieu QX, Luong VH (2020) An integrated moving element method (IMEM) for hydroelastic analysis of infinite floating Kirchhoff-Love plates under moving loads in a shallow water environment. Thin-Walled Struct 155:106934. https://doi.org/10.1016/j.tws.2020.106934

Transportation and Infrastructure Sessions

Determinants of Risky Riding Behaviors Among High School Students in Ho Chi Minh City, Vietnam

Thong Vo Manh, Long Nguyen Xuan, and Minh Chu Cong

Abstract This study examines the causal relationship between personality traits and latent factors in the Theory of Planned Behavior (TPB) in the Ho Chi Minh (HCM) City area. Four hundred nine students in grades 10, 11, and 12 have fully completed survey questionnaires to assess personality traits (anxiety, sensation seeking, anger, altruism, normlessness), latent factors (attitudes, subjective norms, perceived behavioral control, intentions), and risky riding behaviors. Research results show that sensation seeking directly impacts risky riding behaviors. In contrast, altruism has an indirect effect on these behaviors through attitudes and perceived behavioral control. The study also found that personality traits, latent factors, and risky riding behaviors did not differ between males and females. This study also revealed that personality traits play an important role in predicting risky riding behaviors in high school students, and the causal relationship can be explained through the TPB. In addition, a solution to improve students' attitudes and knowledge about traffic safety will be proposed from the results of this study.

Keywords Personality traits · Theory of Planned Behavior · Risky riding behaviors

T. V. Manh · L. N. Xuan (✉) · M. C. Cong
Faculty of Civil Engineering, Ho Chi Minh City University of Technology (HCMUT), 268 Ly Thuong Kiet Street, District 10, Ho Chi Minh City, Vietnam
e-mail: nxlong@hcmut.edu.vn

Vietnam National University Ho Chi Minh City, Linh Trung Ward, Thu Duc District, Ho Chi Minh City, Vietnam

1 Introduction

A road crash statistics in the first six months of 2019 [1] reported 640 traffic accidents with 265 deaths and 407 injures in Hanoi, 42 crashes with 33 deaths and 15 injures in Da Nang, 16 cases with 16 deaths in Hai Phong. Ho Chi Minh City (HCM City) had the highest number of accidents which is 1669 cases with 304 deaths and 1147 injures. It can be seen that the road traffic mortality rate in HCM City is 3.4 times higher than that of other cities (Fig. 1). Therefore, traffic accidents in HCM City are still very complicated, and this is the reason for choosing the city as a case study.

A survey with 4000 households in HCM City [2] showed that accidents were due to speeding (15.3%), reckless overtaking (13.4%), riding too close (9.2%), using a cell phone while riding (4.3%), running in the opposite direction (8.2%), drinking and riding (7.7%). Furthermore, high school students accounted for 70% of the total children traffic accidents in HCM City from 2010 to 2015 [3]. Therefore, traffic accidents among people under the age of 18 are a problem in HCM City.

According to the Theory of Planned Behavior (TPB), intention to perform a behavior can be used to predict individual behaviors in reality and the intention also was affect by attitude, subjective norm, and perceived behavioral control [4]. Therefore, the TPB factors were used to explain risky riding behaviors in many studies. Previous studies showed the effects of personality traits on risky behaviors. Steinbakk et al. [5] found that sensation-seeking, altruism, and normlessness affected indirectly preferred speed in work zone through attitudes towards speeding. In addition, risk perception had a negative effect on speeding in work zone for the country areas. Another study [6] showed that driving under the influence of using alcohol/substance and some personality traits such as anger and normlessness lead to risky driving behavior. However, in Viet Nam, studies that have focused on explaining the effect of personality traits to risky behaviors have been scarce. Some studies (e.g., [7]) are often only interested in the influence of latent factors on individual behaviors. Therefore, this study combined personality traits into the TPB, thereby creating a relationship between personality traits and risky riding behaviors of high school students in HCM City. The specific aim of the study is to identify latent factors

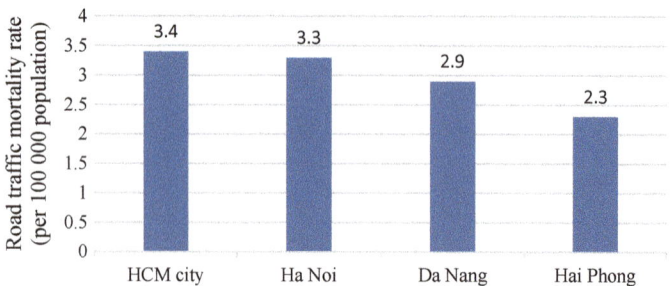

Fig. 1 Road traffic mortality rate of HCM City and other provinces in the first 6 months of 2019

Table 1 Main features of respondents ($N = 409$)

Demonstrate	Frequency	Percentage (%)
Gender		
Male	203	49.60
Female	206	50.40
Vehicle		
Motorcycle over 50 cc	129	31.50
Motorcycle 50 cc or below	164	40.10
Bicycle	93	22.70
Electric motorcycle	23	5.60
Accidents in 3 years		
Yes	111	27.10
No	298	72.90

(attitudes, subjective norms, perceived behavioral control, intention) in TPB and personality traits that lead to risky behaviors for the students in HCM City.

2 Methodology

2.1 Sample

High school students from Tran Cao Van High School, Go Vap District, HCM City, aged between 15 and 18 years, were selected for this study. A total of 500 questionnaires were distributed to the students based on convenience, and only students who were willing to engage in this survey answered the questionnaires. After incomplete questionnaires were deleted, 409 samples were obtained from the participants. The demographic information is presented in Table 1.

2.2 Questionnaire

In the questionnaire, *personality traits* were measured by 36 items in total, containing altruism (8 questions), anxiety (8 questions), sensation seeking (8 questions), anger (8 questions), normlessness (4 questions). All items use a five-point Likert-type scale, ranging from 1: Strongly disagree to 5: Strongly agree. *Risk perception* had two questions. Respondents were asked to judge the probability of them being involved in a traffic accident in the future. The first question was measured on a seven-point scale ranging from 1: Not probable at all to 7: Very probable. Then, respondents were asked how worried and concerned they were regarding being hurt in a traffic accident. The second question used a seven-point scale ranging from 1: Not worried at all to 7: Very worried. *Attitude towards traffic safety* was measured by 15 questions regarding three aspects of attitudes: rule obedience (9 questions), speeding (5 questions), fun riding (3 questions). All items were evaluated on a five-point Likert-type scale ranging from 1: Strongly agree to 5: Strongly disagree. *Subjective norm* had six questions. Items were scored ranging on a five-point Likert scale between 1: Strongly disagree and 5: Strongly agree. *Perceived behavioral control* was assessed by four items. The first two items were used to measure riding experience, and two questions were used to measure riding skills. The following two items were scored on a five-point Likert-type scale ranging from 1: Strongly agree to 5: Strongly disagree. *Intention* was measured by 6 questions related to 3 aspects: speeding (2 questions), alcohol using (2 questions), and rule violation (2 questions). The items were scored on a 5-point Likert scale ranging from 1 = Strongly agree to 5 = Strongly disagree. *Risky riding behaviors* consisted of 26 questions in this research, including speeding (5 questions), lane violations (5 questions), wrong direction (4 questions), overtaking (4 questions), not-giving-way (3 questions), lack of observation (5 questions). The scale ranged from 1: Very often to 5: Never.

2.3 Statistical Analysis

Figure 2 shows analysis steps to develop a Structural Equation Modeling (SEM) model. The Exploratory Factor Analysis (EFA) was performed to explore the underlying structure of measurement items in this study by abstracting factors from questionnaire responses. The goodness-of-fit indices satisfy the conventional requirements (Factor loading ≥ 0.3, Kaiser- Meyer-Olkin (KMO) > 0.05, Total variance extracted > 50%, Eigenvalues > 1), indicating a satisfactory model fit (Hair et al., 2009). Cronbach's alpha was used to estimate internal consistency, and its values greater than 0.6 are generally accepted as indicating a moderately reliable scale [8]. Based on the two-step approach in Multivariate Data Analysis [9], the study applied

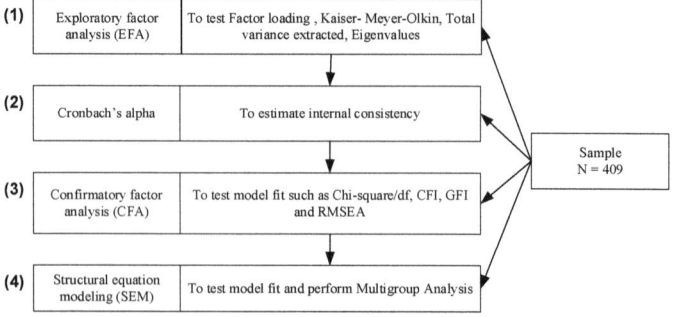

Fig. 2 The flowchart of analysis steps to develop the SEM model

the Confirmatory Factor Analysis (CFA) to validate reliabilities and goodness-of-fit for the exploratory factors. Thereafter, the structural equation modeling (SEM) was conducted to investigate the relations between personality traits, risk perception, latent factors in TPB, and risky riding behaviors. To check the Goodness-of-fit of the model, the Goodness-of-Fit Index (GFI) and Comparative Fit Index (CFI) are gradually approaching 1, indicating a better fit, the Square Error of Approximation (RMSEA) should be less than 0.08, and Chi-square/df should be less than 5 [9].

3 Results

The EFA model explains 52.3% of variance–covariance in the original matrix of 67 measurement items. After adopting the varimax rotation approach, principal factors abstracted from these items are personality traits, risk perception, latent factors. According to results from the CFA model, goodness-of-fit indices such as Chi-square/df (1.670), GFI (0.806), CFI (0.889), RMSEA (0.038) are suitable with the recommended values in Multivariate Data Analysis. GFI ranges from 0.8 to 0.9 can be accepted for analysis of Baumgartner and Homburg [10].

The SEM model is constructed to investigate the relationships between personality traits, risk perception, and latent factors. Figure 3 shows that the model indices fits the data well with Chi-square/df = 1.602, GFI = 0.800, CFI = 0.882, RMSEA = 0.038. For clarity, statistically significant paths (p-value < 0.01) are bold, and nonsignificant paths (p-value > 0.01) are dashed. Personality traits explain 4% of the variance in attitude, 6% in the subjective norm, 2% in perceived behavioral control and 20% in

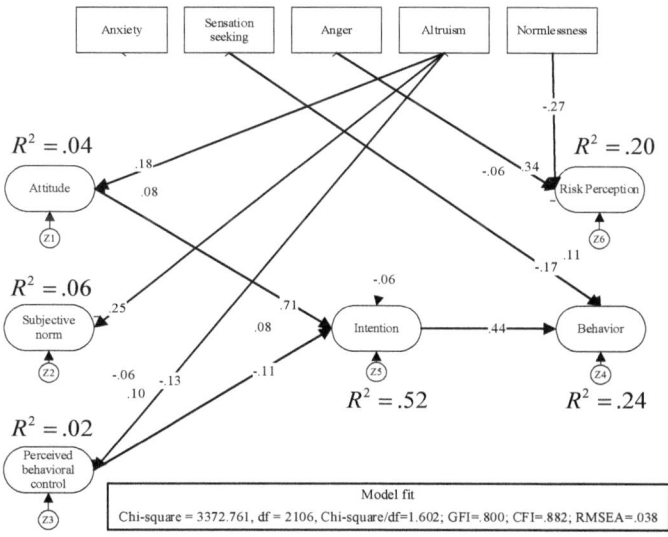

Fig. 3 The risky riding behaviour model

perceived risk. 52% of the variance in behavioral intention is explained by attitude and perceived behavioral control. It means that TPB factors strongly influence intention, and this finding is similar to the study of Elliott et al. [11]. Moreover, 24% of the variance in behaviors is explained by this model. Although the proportion of variance in this study was not so high, Elhoushy and El-Said [12] showed that when variance was 29.4%, TPB factors had a stronger influence on behavior.

This study analyzes the effects of latent factors on risky riding behavior. The direct effects are path coefficients between two specific structures. The indirect effects are the product of all coefficients along the path between two structures that involve intervening structures. The total consequences are defined to be the sum of direct effects and indirect effects. Table 2 shows that the indirect impact of attitude on risky riding behaviours through intention is the strongest among other factors. It means riders who believe that performing violations in traffic will put them at-risk situations are less likely to intend to make violations and low frequency. The indirect effect of subjective norm on risky riding behaviours implies that students who feel that advice regarding traffic safety from relatives is correct are less likely to participate in risky behaviours. In contrast, the indirect effect of perceived behavioural control shows that students who lack experiences and riding skills will perform risky behaviours more often than those who own more experiences and riding skills.

The study also analyzes the direct and indirect effects of personality traits on risky riding behaviors. As shown in Table 3, the indirect influence of altruism through latent factors and risk perception is most substantial among other personality traits. It means

Table 2 Effects of latent factors on risky riding behaviours

Latent factors	Direct effect	Indirect effect through intention	Total effect
Attitude	N/A*	0.312	0.312
Subjective norm	N/A	0.035	0.035
Perceived behavioural control	N/A	−0.048	−0.048

* N/A is represented when a path does not exist

Table 3 Effects of personality traits on risky riding behaviours

Personality trait	Direct effect	Indirect effect through risk perception and latent factors	Total effect
Anxiety	N/A	−0.001	−0.001
Sensation seeking	−0.17	0.003	−0.167
Anger	N/A	0.004	0.004
Altruism	0.11	0.045	0.155
Normlessness	N/A	−0.007	−0.007

Table 4 Fit indices for the invariant and non-variant model

Model	χ^2	df	RMSEA	CFI
Invariant	6401.774	4228	0.036	0.815
Non-invariant	6378.811	4212	0.036	0.816

that students with altruistic characteristics are more likely to not engage in risky behaviors. Sensation seeking characteristic has the most significant direct effects on behaviors. Students with the personality trait may perform risky behaviours. The indirect effect indicates that these students may have vehicle riding skills and experience. Thus it will help students decrease risky behaviours. The remaining three personality traits only have indirect effects. In particular, the indirect impact of anxiety and normlessness on the behaviors relates to increasing engagement in risk-taking behaviors. However, anger riders are less likely to perform risky riding behaviors than those with anxiety and normlessness.

A multi-group analysis was carried out to examine the differences in the structure of SEM model between males and females. The strength of the structural relations was constrained to be equal across gender. Table 4 shows that the difference between the invariant model (with constraints) and the non-invariant model (without constraints) is $\Delta\chi^2$ (Delta Chi-squared) = 22.963 and Δdf (Delta degrees of freedom) = 16, which means that the two models are the same. Therefore, its structural relationships have no difference between males and females.

4 Discussion and Implication

This study investigates the relationship between personality traits, latent factors in TPB toward risky riding behaviors. Riders with sensation seeking and altruism characteristics affected riding risky behaviors. Specifically, students with altruistic characteristics perceive the disadvantages of breaking traffic rules and have experience and skills in controlling vehicles. Therefore, they have less intention to engage in violation behaviors, which makes them less likely to perform risky riding behaviors. Sensation seeking riders are more likely to engage in risky behaviors. However, they also own experiences and skills of riding vehicles, which may help them to reduce risky behaviors. Moreover, there is no difference in the structural relationships of males and females in the study. It means gender does not affect the performance of risky behaviors. This result differed from Ian Glendon et al. [13] when Glendon indicated that young men are more likely to participate in accidents because they are often rated higher than young women in driving skills. The reason for these differences may be that this study focused on high school students whose ages ranged from 15 to 18 years old. Meanwhile, Glendon's study used young drivers aged 18–25 and old drivers aged 45–60. Therefore, the difference between the two studies can be that the psychology of adolescents is not similar to adulthood. In the "Road Safety for

Young People in Europe" book [14], it is said that "Knowledge without any practical experience of hazardous situations is not sufficient to avoid danger. And skills and knowledge are of no use if the road user prefers to adopt a risk behavior". Therefore, improving driving skills and knowledge of traffic regulations should be a priority in traffic safety education in schools [15].

The other finding shows that attitude strongly affects risky behaviors through intention and indirectly reduces the effects of personality traits. Therefore, educational methods that develop a positive attitude will also indirectly change the impact of personality on behaviors. In Vietnam, traffic safety programs are usually interested in propaganda regarding the consequences of a traffic accident on channels, print media, and network media. From 2011 to 2021, the government of Vietnam has been concerned about road safety education at schools and considers it a core subject. However, traffic accidents in Vietnam have no noticeable improvement. Therefore, corresponding countermeasures to change attitudes towards traffic safety should be considered. Content in classroom lectures should mention the disadvantages of performing risky behavior, accompanied by severe consequences affecting students themselves such as deaths, injury, and disablement. Moreover, the advantages of safety behaviors such as stop at the red light, wearing a helmet, pay attention before turning should be added in educational content.

The study has some limitations that should be considered. The limitation related to the risky riding behaviors consisted of many categories. However, items that use to measure latent factors only included some violations such as speeding, rules violation and using alcohol. Therefore, it could not completely explain the causes of risky riding behaviors. The effect of sensation-seeking on risky behavior was also not clearly explained in this study because sensation-seeking insignificantly affects behavior through TPB factors. Future studies should address the relationship between sensation seeking and risky behaviors among high school drivers, and items regarding latent variables should be added to various types of violations.

Acknowledgements This study is funded by Vietnam National University—Ho Chi Minh City (VNU-HCM) under grant number C2021-20-36. We acknowledge the support of time and facilities from Ho Chi Minh City University of Technology (HCMUT), VNU-HCM for this study.

References

1. Ministry of Transport (2019) Traffic Safety Bulletin. https://mt.gov.vn/vn/chuyen-muc/1013/ban-tin-an-toan-giao-thong.aspx
2. Minh CC, Long NX, Tu TT, Nathan H (2019) Assessment of motorcycle ownership, use, and potential changes due to transportation policies in Ho Chi Minh City, Vietnam. J Transp Eng A 145(12)
3. Vu AT, Man Nguyen DV (2018) Analysis of child-related road traffic accidents in Vietnam. IOP Conf Ser Earth Environ Sci 143:012074
4. Ajzen I (1991) The theory of planned behavior. Organ Behav Hum Decis Process 50(2):179–211

5. Steinbakk RT, Ulleberg P, Sagberg F, Fostervold KI (2019) Speed preferences in work zones: the combined effect of visible roadwork activity, personality traits, attitudes, risk perception and driving style. Transp Res Part F 62(1):390–405
6. Disassa A, Kebu H (2019) Psychosocial factors as predictors of risky driving behavior and accident involvement among drivers in Oromia Region, Ethiopia. Heliyon 5(6):e01876
7. Dinh DD, Vũ NH, McIlroy RC, Plant KA, Stanton NA (2020) Effect of attitudes towards traffic safety and risk perceptions on pedestrian behaviours in Vietnam. IATSS Res 44(3):238–247
8. Hinton PR, McMurray I, Brownlow C (2014) SPSS explainted. Routledge
9. Hair JF, Black WC, Babin BJ, Anderson RE (2009) Multivariate data analysis. Prentice Hall
10. Baumgartner H, Homburg C (1996) Applications of structural equation modeling in marketing and consumer research: a review. Int J Res Mark 13(2):139–161
11. Elliott MA, Thomson JA, Robertson K, Stephenson C, Wicks J (2013) Evidence that changes in social cognitions predict changes in self-reported driver behavior: causal analyses of two-wave panel data. Accid Anal Prev 50:905–916
12. Elhoushy S, El-Said OA (2020) Hotel managers' intentions towards female hiring: an application to the theory of planned behaviour. Tour Manag Perspect 36:100741
13. Ian Glendon A, Dorn L, Davies DR, Matthews G, Taylor RG (1996) Age and gender differences in perceived accident likelihood and driver competences. Risk Anal 16(6):755–762
14. European Commission DG Mobility and Transport (2013) Road safety for young people in Europe
15. Le LV, Chu MC, Nguyen LX (2020) The development of safe riding guidelines for young riders—a case study of Phu Yen, Vietnam. IATSS Res 45(2):226–233

Effect of Dowel Bar Distance in Jointed Concrete Pavement Based on ABAQUS Program

Manh Tuan Nguyen and Ngoc Tuong Vy Phan

Abstract Two dimensional finite element have been used for analysis concrete pavement for the past three decades to analyze rigid pavement response. In recent years with the development of a computer and an algorithm, the 3D finite element method (FEM) have been used as a powerful tool for a structure analysis, especially for the pavement analysis. This study concerns about the effect of dowel distance on the jointed concrete pavement such as the shear force along the dowel bar. The 3D FEM was established from Abaqus software. The 3D FEM was also verified in comparison with experiment data. The result of dowel distance could be used for further design in the jointed concrete pavement used in Vietnam.

Keywords Concrete pavement · Jointed concrete pavement · Finite element method · Abaqus · Dowel bar

1 Introduction

The increasing of distresses in Vietnam roads came from both the development of traffic volume, heavy trucks and weather. As a result, concrete pavement is a very good solution due to its duration, for example the concrete pavement was used in the intersections of Saigon East–West Highway due to rutting problem in 2013. But the studies about concrete pavement in Vietnam is not widely mentioned in comparison with asphalt concrete. This maybe comes from the price and the curing time of cement concrete in the construction as well as the maintenance stage.

The methods applied for the design of rigid or concrete pavements were based on the closed-form solutions obtained from the static analysis of infinitely long plates resting on elastic foundation for a long time ago as shown in the specification of

M. T. Nguyen (✉)
Faculty of Civil Engineering, Ho Chi Minh City University of Technology (HCMUT), Ho Chi Minh City, Vietnam
e-mail: nmanhtuan@hcmut.edu.vn

N. T. V. Phan
Urban Infrastructure Faculty, Mien Tay Construction University, Vinh Long City, Vietnam

© The Author(s), under exclusive license to Springer Nature Singapore Pte Ltd. 2023 955
J. N. Reddy et al. (eds.), *ICSCEA 2021*, Lecture Notes in Civil Engineering 268,
https://doi.org/10.1007/978-981-19-3303-5_87

22TCN 223-95 [1] or Huang [2]. Later, the limitations of closed-from solutions were largely overcome by the development of computers and efficient numerical technique such as Finite Element Method (FEM) in which the 2D model was used before the 3D. The 3D model with FEM provides more realistic simulations for the concrete pavements in comparison with the simple 2D models [3, 4]. Hence, a 3D concrete pavement structure is analyzed using the finite element code ABAQUS in this study.

There are many studies about the cement concrete around the world except for the dowel bar in the jointed concrete pavement. Sii et al. [5] developed a 3D FEM, which was not mentioned in the paper, for shear force and moment along the dowel bar. Davids et al. [6] used EverFE2.2 which is a 3D FEM for analysis of the jointed plain pavement in order to examine the effect of dowel locking and slab–base shear transfer on the pavement stresses due to the thermal gradients and the uniform slab shrinkage,

This paper shows the accuracy of the 3D model obtained from the general commercial FE-program ABAQUS and was verified by comparison with the experiment result derived from Siddique et al. [7]. The geometries and parameters in the modeling and experiment were selected as consistent as possible to ensure a high comparability of the 3D model. The paper also presents the effect of dowel distance into the shear force along the dowel bar.

2 Analysis of Concrete Pavement

2.1 Description of 3D Model and Model Verification

A 3D concrete pavement structure is analyzed using the finite element code from ABAQUS in this paper. The model consists three layers in which the top, middle, and bottom layer are concrete slab, cement treated base, and subgrade, respectively. The thickness and geometry of concrete slab and base layer are shown in Table 1. The dowel bars are located at the mid-depth of the concrete slab as shown in Fig. 1 and its dimension and interval presented in Table 1. The table also shows the material properties used in FEM. The elastic properties which are modulus of elasticity and Poisson's ratio are assigned for FEM model. All values for 2.1 part in the Table 1 are from Siddique et al. [7] in order to compare the deflection based the FEM model and experiment result based Siddique et al. [7]. Other values for 2.2 part are similar to Vietnam pavement design. The base layer and subgrade were converted into an equivalent subgrade which has the modulus of subgrade reaction as shown in Fig. 1.

The modulus of subgrade reaction (k) can be expressed using the following expression, from theory of elasticity solution for a rigid plate on a semi-infinite elastic soil medium subjected to a concentrated load (based on Huang [2]):

Table 1 Geometrical and material properties of the concrete pavement

Information		Unit	Value for 2.1 part	Value for 2.2 part
Concrete slab	Thickness	Mm	300	250
	Length	Mm	5000	4500
	Width	Mm	3700	3500
	Density	kg/m³	2400	
	Thermal expansion ratio	mm/mm/°C	1.1×10^{-5}	1.1×10^{-5}
Base	Thickness	Mm	101.6	
Dowel bar	Diameter	Mm	32	32
	Length	Mm	450	450
	Interval	mm	300	
	Elastic modulus	MPa	200,000	200,000
	Poisson'ratio		0.3	0.3
Material properties of pavement	Elastic modulus of concrete slab	MPa	28,960	28,000
	Poisson'ratio of concrete slab		0.15	0.15
	Elastic modulus of base layer	MPa	6500	350
	Poisson'ratio of base layer		0.15	0.4
	Elastic modulus of subgrade	MPa	270	50
	Poisson'ratio of subgrade		0.2	0.4

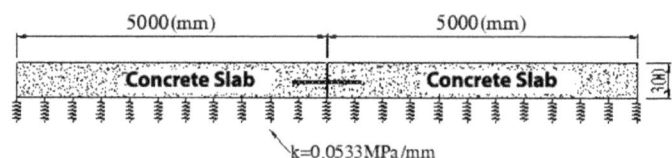

Fig. 1 The position of dowel bar in FEM model in vertical alignment

$$k = \frac{2E}{\pi\left(1 - \gamma^2\right)a} \qquad (1)$$

with E = Elastic modulus of a pavement layer; v = Poisson's ratio of subgrade and base layer = 0.4; a = radius of loading = 193 mm. The base layer has the finite thickness (H_1) and the subgrade thickness is semi-infinite. The elastic modulus E is the combination of elastic modulus of base (E_1) and elastic modulus of subgrade (E_o) which can be obtained based on 22TCN 211-06 [8] as follows:

- Calculate the diameter of loading at the bottom of concrete slab

$$D = D_o + H \tag{2}$$

where D_o is the load standard diameter which is 330mm; and $H = 300mm$ is the thickness of concrete slab;
- Determine the E which is elastic modulus on top of base layer in Fig. 3.3 from 22TCN 221-06 [8]: H_1/D and E_o/E_1  ratio $k' = E/E_1$

$$or \qquad E = k'E_1 \tag{3}$$

From Eqs. 1, 2, and 3, the calculated k is 0.0533 MPa/mm.

The ABAQUS/CAE (Version 6.10) FE package was used Hibbit and Sorenson [3, 4]. The C3D20RT (Continuum 3D-20 node thermally couple brick, triquadratic displacement, trillinear temperature, reduced integration) element is used for concrete slab in which the mesh size is $100 \times 100 \times 75$ mm. the B32 (2-node cubic beam) element is obtained for dowel bar. A spring element is assigned between dowel bar and concrete slab. The concrete slab is on top of subgrade which is the elastic foundation and has k value (termed the modulus of subgrade reaction). The 3D FE model before and after meshing can be seen in Fig. 2. Under the different temperature between top and bottom of concrete slab which is 20 degree Celcius as same as the

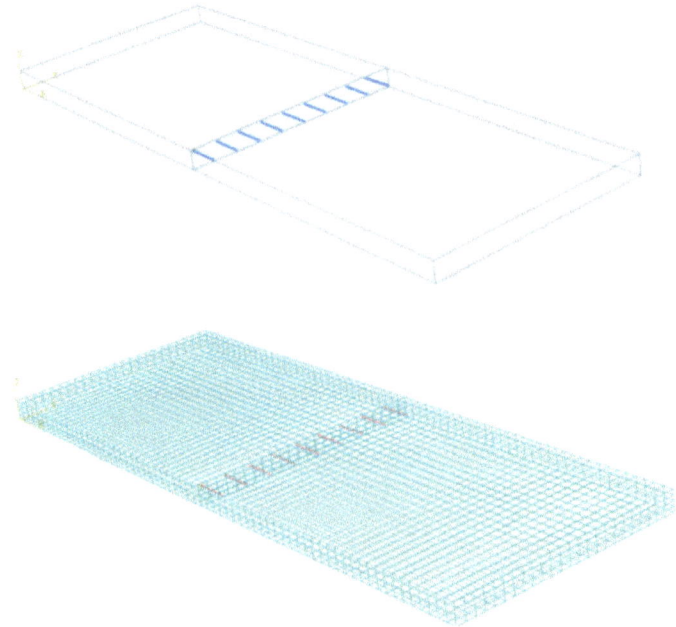

Fig. 2 Pavement modelling before and after meshing

value based Siddique et al. [7], the deflection and stress results can be obtained as shown in Figs. 3 and 4, respectively. The unit of deflection and stress are mm and MPa.

From Fig. 4, the maximum deflection is in the middle or centre of a concrete slab which is 0.6358 mm. From Siddique et al. [7], the measured deflection in slab centre is 0.63 mm. As the result, the 3D FEM can give a reasonable value in comparison with experiment. It is also concluded that the 3D FEM is reliable and good enough for further study. In terms of stress due to the different temperature between top and bottom of concrete slab, the maximum stress is in the center of slab in Fig. 5.

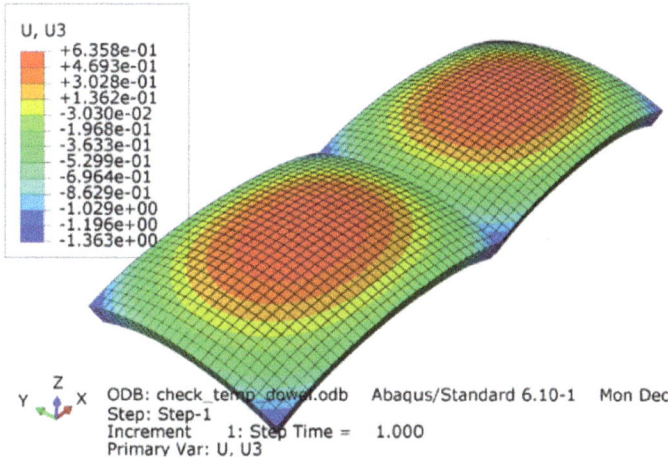

Fig. 3 Deflection of concrete slab

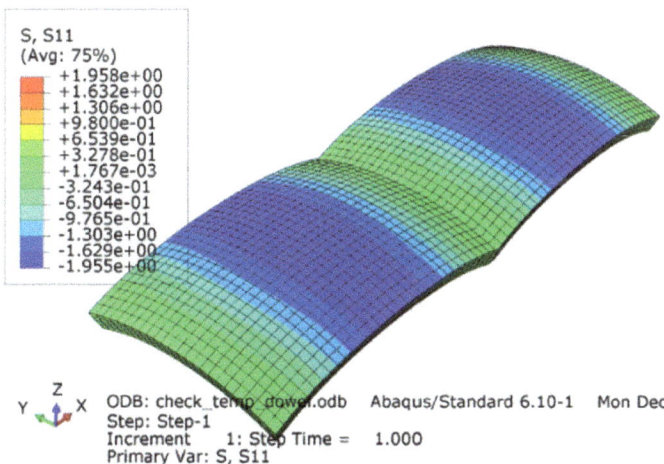

Fig. 4 Principal stress of concrete slab

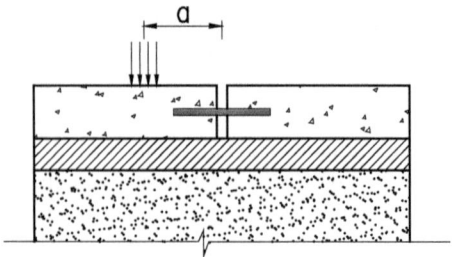

Fig. 5 Location of axle load

2.2 Effect of Dowel Distance

In order to study the dowel distance effect in dowel, 4 distances which are 225, 300, 450 and 600 mm were chosen. The above 3D FEM used in this parametric study with the different parameters as shown in Table 1. The axle load which is 100 kN; the length and width of contact wheel load, which is 50 kN, are 199 mm and 135 mm, respectively. In this study, the axle load location is in the middle of dowel bar or "a" value is zero as shown in Fig. 5. For the dowel accumulated width based on an edge of slab concrete, there are many maximum shear forces for many dowel bars as presented in Table 2. In particularly, the Fig. 6 also shows the shear force along the dowel bar no.3 (termed D3 in Table 2 for 300 mm distance case); and the maximum shear force in D3 is 3.682 kN which places in Table 2.

From maximum shear force along dowel bars in Table 2, the Fig. 7 can be obtained. The Fig. 7 shows that the shorter distance or the more dowel bars in a section has the lower maximum shear force. Certainly, the more shear force in dowel bar, the more quality of concrete slab material is needed.

Table 2 Maximum shear force in all dowel

Dowel no	Dowel distance (mm)							
	225		300		450		600	
	Accumulated width	Shear force (kN)	Accumulated width	Shear force (kN)	Accumulated width	Shear force (kN)	Accumulated width	Shear force (kN)
D1	225	3.321	300	3.442	225	3.586	300	3.745
D2	450	3.422	600	3.592	675	3.806	900	3.970
D3	675	3.53	900	3.682	1125	3.836	1500	3.778
D4	900	3.586	1200	3.582	1575	3.646	2100	3.970
D5	1125	3.548	1500	3.472	2025	3.646	2700	3.973
D6	1350	3.422	1800	3.422	2475	3.836	3300	3.742
D7	1575	3.353	2100	3.472	2925	3.806		
D8	1800	3.328	2400	3.582	3375	3.586		
D9	2025	3.353	2700	3.682				
D10	2250	3.422	3000	3.592				
D11	2475	3.548	3300	3.442				
D12	2700	3.585						
D13	2925	3.53						
D14	3150	3.422						
D15	3375	3.321						
Max		3.586		3.682		3.836		3.973

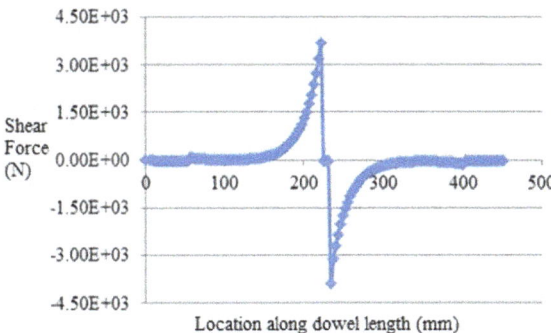

Fig. 6 Shear force along the dowel bar No3

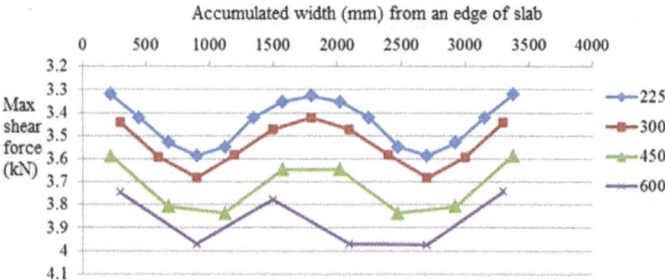

Fig. 7 Maximum shear force in each dowel bar

3 Conclusion

In this paper, the 3D finite element method based on ABAQUS was conducted and
verified in comparison with experiment result of Siddique et al. [7]. The 3D FEM
was used for checking the shear force along the dowel bar when the dowel distance
was changed. The following conclusions are from the obtained results:

- The 3D FEM from Abaqus is appropriate in terms of comparison with experiment.
  As the result, it can be used for further parametric studies for jointed concrete
  pavement;
- The 3D FEM was set to find out the shear force along the dowel bar. The shorter
  distance of dowel bar or the more dowel bars in a section has the lower maximum
  shear force. The shear force is useful in application the concrete type and finding
  the appropriate distance of dowel bar at a jointed pavement.

This 3D FEM could be used for investigation the effect of axle load, concrete slab
size, base layer type, and so on.

References

1. 22TCN 223-95 (1995) Rigid pavement for road—Design guide. Ministry of Transportation,
   Vietnam
2. Huang YH (2004) Pavement analysis and design, 2nd edn. Pearson Prentice Hall
3. Hibbit K, Sorenson I (2005) Abaqus—Getting Started with Abaqus—Version 6.10, Interactive
   edition
4. Hibbit K, Sorenson I (2005) Abaqus—Theory and User's Manual—Version 6.10
5. Sii HB, Chai GW, Staden RV et al (2014) Development of prediction model for doweled joint
   concrete pavement using three-dimensional finite element analysis. Appl Mech Mater 587–
   589:1047–1057
6. Davids WG, Wang Z, Turkiyyah G et al (2003) Three-dimensional finite element analysis of
   jointed plain concrete pavement with EverFE2.2. J Transp Res Board 1853:92–99

7. Siddique ZQ, Hossain M, Meggers M (2005) Temperature and curling measurements on concrete pavement. Proceedings of the 2005 Mid-Continent Transportation Research Symposium
8. 22TCN 211-06 (2006) Flexible pavement—Requirements and Design Guide. Ministry of Transportation, Vietnam

Field Assessment of Asphalt Concrete Incorporating Noise and Surface Friction

Ngoc Tram Hoang and Manh Tuan Nguyen

Abstract Pavement surface friction ensures pavement skid resistance when the vehicle tire contacts to the pavement surface, helping the vehicle to circulate safely. Friction evaluation has become an important tool in the evaluation of pavement surface quality. Macro-texture and micro-texture are two basic components when analyzing pavement surface friction. To assess friction in many asphalt concrete surface, the study used a British pendulum and sand patch test is conducted. In addition, the British pendulum will simulate the process of interaction between the vehicle tire and the pavement surface for measuring noise. Noise generated from testing will be measured by specialized equipment. The paper evaluates the pavement surface friction from the field experiments. The paper also shows the relationship between noise and surface friction.

Keywords Friction · Noise · Asphalt concrete · Hot mix asphalt · British pendulum

1 Introduction

When the economic development of a country is more and more increasing, there are needs of life without pollutions, especially noise and environment pollution, as well as safety on vehicles. Then, tire-pavement interaction noise and surface friction are two of a significant parameters for pavement management system in road system around the world. Although pavement noise is very important, the interaction between tires and pavement was less studied in Vietnam.

Noise is the subjects that researchers in Vietnam and the world concerned in their studies as follows:

- The studies about noise in Vietnam focus on the general evaluation of noise from many vehicles in an area such as Trinh and Nguyen [1] and Tran [2]. Trinh and

N. T. Hoang · M. T. Nguyen (✉)
Faculty of Civil Engineering, Ho Chi Minh City University of Technology (HCMUT), 268 Ly Thuong Kiet Street, District 10, Ho Chi Minh City, Vietnam
e-mail: nmanhtuan@hcmut.edu.vn

© The Author(s), under exclusive license to Springer Nature Singapore Pte Ltd. 2023 965
J. N. Reddy et al. (eds.), *ICSCEA 2021*, Lecture Notes in Civil Engineering 268,
https://doi.org/10.1007/978-981-19-3303-5_88

Nguyen [1] studied noise based sound meters in 9 locations of some streets in South of Hue city. The noise values are from 50.1 to 78.8 dB which are more than limit of QCVN 26:2010/BTNMT [3]. Tran [2] showed the monitoring data in 150 locations on 30 streets from the Department of Natural Resources and Environment of Hochiminh city. The results show that almost noises are over the limit.

- And for the world, there are many studies about noise under vehicles impact such as: Tan Li [4] who showed literature review of models on tire-pavement interaction in a laboratory as well as in streets; Paul R. Donavan [5] evaluated the different noise for different pavement types; De Chen et al. [6] concentrated on relationship between porous pavement and noise.

Pavement surface friction is one of the key elements required for ensuring highway safety. Macro-texture and micro-texture are two basic components when analyzing pavement surface friction. They depend on the physical properties of aggregates, mixture design, and construction quality. There are numerous methods and devices for measuring pavement surface friction. The British pendulum and sand patch test are well-known methods used in Vietnam and around the world for macro-texture and micro-texture. There were some significant studies about asphalt pavement friction in Vietnam as follows:

- The surface friction of open-graded friction course, which nominal maximum size of aggregate (NMAS) is 9.5 mm, was investigated by Nguyen [7];
- Another open-graded friction course in which NMAS is 12.5 mm was developed for better surface friction than one which has NMAS 9.5 mm [8].

For pavement surface friction, there are many studies [9, 10] about the surface friction by using many methods and correlation lab and field-testing, but there is not any relationship between micro-texture and macro-texture friction type.

In this study, micro-texture and macro-texture were investigated for roads in campuses of Hochiminh city University of Technology and a highway in District 9 by sand patch method and British pendulum device. Beside the friction, the field noises under British pendulum device were also measured by a sound meter in order to evaluate the different from two asphalt mixture types which are dense-graded and open-graded asphalt concrete.

2 Field Assessment

2.1 Project Information

Field experiments locate in two campuses of Ho Chi Minh city University of Technology, where are in Ho Chi Minh city and Binh Duong province, and in the highway of Long Thanh Dau Giay, district 9. The asphalt concretes of pavement surfaces in the campuses are dense-graded asphalt mixtures and another in the highway is open-graded friction mixture. Figure 1 and Table 1 shows the information of five roads in campuses and the highway for this study. There are 30 measured points for five roads from No. 1 to No. 5, and there are 12 measured points for highway due to traffic.

Road No.1	Road No.2	Road No.3
Road No.4	Road No.5	Highway HoChiMinh – Long Thanh – Dau Giay

Fig. 1 Pavement locations

Table 1 Test locations

Location	Road name	Surface type	Measured points
Ho Chi Minh city University of Technology—campus 1 (Ho Chi Minh city)	Road No. 1	Dense-graded asphalt concrete NMAS 12.5 mm	30
	Road No. 2		30
	Road No. 3		30
Ho Chi Minh city University of Technology—campus 2 (Binh Duong province)	Road No. 4	Dense-graded asphalt concrete NMAS 9.5 mm	30
	Road No. 5		30
Disctrict 9	Highway Ho Chi Minh–Long Thanh–Dau Giay	Open-graded asphalt concrete NMAS 12.5 mm	12

2.2 *Surface Friction*

Two types of test which are the sand patch method and the British pendulum method are used in this study. The sand patch test measures the macro-texture of surface and another measures the micro-texture of surface (Fig. 2).

The sand patch test method is used to assess the surface texture depth of a road surface which is the average depth of voids below the high points of the surface. This test is conducted based TCVN 8866-2011 [11] as follows: Sweep off all dust and other loose particles in the surface; Pour the contents of the cylinder which volume is 25 cm^3 onto the road surface into a small pile on the swept test site; Use the sand spreader to gently work the beads down into the surface voids, gently work the beads down into the surface voids in a circular spiral motion from the centre outwards until the diameter of the circle and the beads have completely filled the voids and the sand patch is leveled to the highest points on the surface; Measure the diameter of the circle at 4 evenly spaced diameters to the nearest 5 mm and record these measurements. Average the four readings to determine the average diameter (D in mm) of the circle. Then texture depth (h in mm) of each site test can be calculated as the following equation:

Sand patch method

British pendulum

Sound meter under British pendulum

Fig. 2 Measurement of surface friction and interaction noise

$$h = \frac{4V}{\pi D^2} \tag{1}$$

where cylinder volume $V = 25{,}000 \text{ mm}^3 = 25 \text{ cm}^3$. In this study, at each measured point, there is one value of texture depth.

British pendulum procedure from ASTM E303-93 [12] follows: Level the instrument accurately by turning leveling screws until the bubble is centered in the spirit level; Zero adjust for pendulum; Adjust the length of the contact path of the slider between 124 and 127 mm; Apply water for the test area; Execute swings five time and record the reading at each location or each measured point to have the SRT value. The measured SRT values were taken at about 30 degree Celsius and were converted into the standard temperature which is 20 degree Celsius by adding 2 units of SRT based British Standard 7976 [13].

2.3 Noise

The noise is measured by a well-known Japanese Rion SCA-405 sound meter which was qualified by Quatest3 (Quality Assurance and Testing Center 3) in Hochiminh city. The pavement interaction noise is created by the rubber slider of the British pendulum. The British pendulum simulates the process of interaction between the vehicle tire and the pavement surface to measure noise. The rubber slider in the British pendulum is used because of its convenience in transportation and the quality of rubber under the slider. At each location, before switching the pendulum to create the noise, the slider length of the contact path should be adjusted between 124 and 127 mm as same as the slider length when friction measurement. The Rion sound meter is placed in perpendicular with the slider also between 124 and 127 mm to record the noise. The noise in decibel (dB) is the maximum sound when the rubber slider of the pendulum contact the pavement surface in this study. For the interaction noise, there are five measured values at each measured points as same as British pendulum.

2.4 Test Results and Discussion

Table 2 presents the measured SRT values based on the British Pendulum and the texture depth based on the sand patch test for Road No. 2. The average STR values calculated from 5 measured times at each measured points. Similarly, the average noise is also the mean value from 5 measured times as same as SRT values. Finally, the average value of SRT, noise, and texture depth are calculated as shown in Table 2. Similarly, the procedure of data analysis for other roads and highway is the same as the Road No. 2. Other measured data of roads and highway are not shown because of the limitation of paper. Table 3 is the summarization of all roads and a highway.

Table 2 Measured values from British Pendulum, sand patch test and noise for Road No. 2

Measured point	SRT value (at 30 °C)					Average SRT for 5 measured times	Average noise (dB) for 5 measured times	Average texture depth (mm)
	1	2	3	4	5			
1	86.00	84.00	85.00	84.00	84.00	84.60	88.16	0.63
2	87.00	86.00	81.00	85.00	86.00	85.00	88.00	0.65
3	84.00	84.00	82.00	86.00	84.00	84.00	88.54	0.72
4	85.00	85.00	83.00	86.00	85.00	84.80	88.16	0.72
5	85.00	84.00	80.00	84.00	83.00	83.20	88.42	0.64
6	86.00	88.00	84.00	86.00	86.00	86.00	87.98	0.71
7	87.00	85.00	80.00	86.00	86.00	84.80	88.26	0.73
8	85.00	86.00	80.00	86.00	84.00	84.20	88.46	0.72
9	86.00	86.00	81.00	87.00	87.00	85.40	88.06	0.72
10	88.00	85.00	80.00	86.00	84.00	84.60	88.22	0.64
11	86.00	86.00	81.00	86.00	88.00	85.40	88.06	0.72
12	85.00	86.00	81.00	87.00	86.00	85.00	88.14	0.62
13	85.00	86.00	82.00	83.00	82.00	83.60	88.34	0.71
14	85.00	88.00	80.00	83.00	85.00	84.20	88.26	0.65
15	87.00	88.00	81.00	83.00	83.00	84.40	88.28	0.71
16	87.00	87.00	81.00	85.00	84.00	84.80	88.26	0.72
17	88.00	86.00	81.00	87.00	84.00	85.20	88.22	0.65
18	84.00	86.00	80.00	84.00	83.00	83.40	88.68	0.70
19	85.00	87.00	81.00	85.00	84.00	84.40	88.26	0.73
20	84.00	83.00	78.00	85.00	84.00	82.80	88.40	0.65
21	87.00	89.00	82.00	87.00	87.00	86.40	87.86	0.67
22	85.00	86.00	79.00	84.00	84.00	83.60	88.46	0.70
23	83.00	85.00	80.00	81.00	85.00	82.80	88.12	0.65
24	84.00	82.00	82.00	86.00	84.00	83.60	88.52	0.66
25	85.00	84.00	83.00	88.00	84.00	84.80	88.34	0.64
26	86.00	89.00	84.00	86.00	83.00	85.60	88.20	0.66
27	85.00	86.00	86.00	86.00	83.00	85.20	88.18	0.65
28	88.00	86.00	85.00	85.00	83.00	85.40	88.20	0.74
29	88.00	87.00	83.00	86.00	85.00	85.80	88.14	0.64
30	86.00	87.00	85.00	86.00	83.00	85.40	88.18	0.75
AVERAGE						**84.61**	**88.25**	**0.68**

Table 3 Average value of SRT, noise and texture depth for all road in this study

Location	SRT (at 30 °C)	SRT (at 20 °C)	Noise (dB)	Texture depth (mm)
Road No. 1	82.36	84.36	89.24	0.54
Road No. 2	84.61	86.61	88.25	0.68
Road No. 3	83.81	85.81	89.02	0.58
Road No. 4	81.02	83.02	89.87	0.47
Road No. 5	81.48	83.48	89.43	0.52
Highway	92.72	94.72	86.4	0.78

Based on TCVN 8860-2011 [11], if the texture depth is from 0.45 to 0.8 mm, the operating speed could be from 80 to 120 km/h. As the result, the pavement surface of all roads and highway are very good for high speed operation. The open-graded asphalt concrete gives more friction than dense-graded concrete, from 14.71 to 65.96% in terms of texture depth. The relationships can be expressed by many regression types, as shown in Table 4. In Table 4, the regression equations and the coefficients of determination (R^2) from 5 typical types are shown. The polynomial type could be chosen for two relationships. Figs. 3 and 4 show the relationship

Table 4 Regression equation for two relationships

No	Regression type	Relationship between noise and SRT		Relationship between texture depth and SRT	
		Equation	R^2	Equation	R^2
1	Linear	$y = -0.284x + 113.216$	0.970	$y = 0.025x - 1.542$	0.871
2	Logarithmic	$y = -25.306\ln(x) + 201.495$	0.973	$y = 2.214\ln(x) - 9.274$	0.880
3	Exponential	$y = 117.194e^{-0.003x}$	0.971	$y = 0.020e^{0.039x}$	0.831
4	Power	$y = 319.610x^{-0.288}$	0.974	$y = 0.000x^{3.515}$	0.842
5	Polynomial (2nd order)	$y = 0.010x^2 - 1.985x + 188.783$	**0.979**	$y = -0.003x^2 + 0.513x - 23.237$	**0.959**

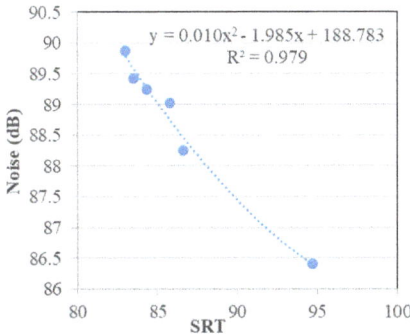

Fig. 3 Noise and SRT (at 20 °C) relationship

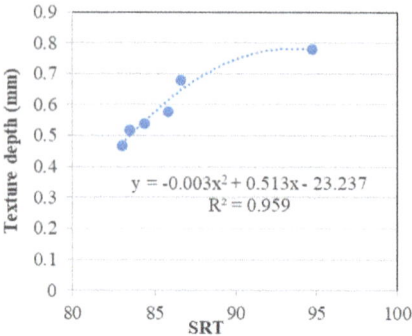

Fig. 4 Texture depth and SRT (at 20 °C) relationship

between noise and/or texture depth and SRT value at 20 °C for all roads and highway. As a result, the sand patch test can be used in replacement for the British pendulum device due to its cost. The interaction noise from British pendulum and pavement surface depends on surface friction from test data.

3 Remarkable Conclusion

In this study, the micro-texture based on British pendulum device and the macro-texture based on sand patch test were investigated for 5 roads in campuses of Hochiminh city University of Technology and a highway in District 9. The field noises under British pendulum device were also measured by a sound meter. The following conclusions are from test result:

- The surface friction of five roads and a highway are very good based on the sand patch test;
- There is a rather good relationship between the micro-texture and the macro-texture based on the polynomial relationship between texture depth and SRT value which has the coefficients of determination at 0.959.
- The interaction noise from the pavement surface also have a very good relation with friction.

References

1. Trịnh THC, Nguyen TNH (2012) Assessing the impact of noise from road traffic activities on people living along some roads in the south of Hue city. Sci J Hue Univ 73(4)
2. Tran DM (2018) A critical assessment of traffic noise pollution in Ho Chi Minh city, Vietnam. International conference on water resources and sustainability, China

3. QCVN 26:2010/BTNMT (2010) National technical regulation on noise. Ministry of Natural Resources and Environment, Hanoi
4. Tan L (2018) A state-of-the-art review of measurement techniques on tire–pavement interaction noise. Measurement 128:325–351
5. Paul RD (2006) Comparative measurements of tire/pavement noise in Europe and the United States. NITE Study
6. De C, Cheng L, Tingting W et al (2018) Prediction of tire–pavement noise of porous asphalt mixture based on mixture surface texture level and distributions. J Const Build Mat 173:801–810
7. Nguyen PM (2013) Research on determining reasonable material composition of high friction course asphalt pavement in Vietnam. PhD Thesis
8. Nguyen TB, Nguyen MT (2015) Initial application of high friction course asphalt concrete $D_{max} = 19mm$ in material conditions of Hochiminh city. J Transp 11:62–64
9. Miao Y, Zhanping Y, Guoxiong W et al (2020) Measurement and modeling of skid resistance of asphalt pavement: A review. J. Const Build Mat 260(10):119878
10. Shahin E, Cesare S, Giulio D, Riccardo L (2018) Recycling asphalt pavement and tire rubber: A full laboratory and field case study. J Const Build Mat 176:283–294
11. TCVN 8866-2011 (2011) Standard test method for measuring pavement macrotexture depth using a volumetric technique. Ministry of Science and Technology, Ha Noi
12. ASTM E303-93 (2013) Standard test method for measuring surface frictional properties using the British Pendulum tester. American Society for Testing and Materials, West Conshohocken, PA
13. BS 7976-1:2002 (2002) Pendulum testers—Part 1: Specification, British Standards

Hamburg Wheel Tracking Assessment of Hot Mix Asphalt Using RFCC

Anh-Thang Le

Abstract The paper presents research results on the Hamburg wheel tracking test (HWTT) of asphalt concrete using RFCC. RFCC (Residue Fluid Catalytic Cracking) is waste from the oil refinery. RFCC is used as filler powder in the hot mix asphalt. Tests for rutting resistance were performed following AASHTO T312:2004 and EN12697-33. Dense graded asphaltic concrete of the nominal maximum aggregate sizes of 12.5 mm with and without RFCC as filler powder was tested using HWTT. The rutting resistance properties of asphalt concrete using RFCC are compared with conventional asphalt concrete using limestone powder. Experimental results show that RFCC can help asphalt concrete improve its resistance to rutting.

Keywords RFCC · HMA · Filler powder · HWTT · Rutting resistance

1 Introduction

RFCC (Residue Fluid Catalytic Cracking) is waste from the oil refinery. This waste is currently redundant in a large amount. RFCC waste recycling is the main topic of the article. RFCC is fine in size and contains metallic oxidize components that can replace mineral powder in asphalt concrete. A reasonable amount of RFCC admixture into the aggregate can improve the Marshall properties of the asphalt mix. However, the performance of asphalt mixtures containing RFCC has not been tested much.

The increase in traffic volume and axle load of vehicles, which distresses in asphaltic layers, has shortened the life cycle of hot asphalt pavement (HMA). Rutting is primary distress in flexible pavements that occurs along the wheel path of heavily trafficked roads. Rutting depth is defined as the accumulated permanent deformation that remains after the removal of load. Asphalt concrete mix should be checked for rutting resistance because it is vital for preventive maintenance and rehabilitation strategies.

A.-T. Le (✉)
Faculty of Civil Engineering, Ho Chi Minh City University of Technology and Education (HCMUTE), 01 Vo Van Ngan Street, Thu Duc City, Ho Chi Minh City, Vietnam
e-mail: thangla@hcmute.edu.vn

© The Author(s), under exclusive license to Springer Nature Singapore Pte Ltd. 2023   975
J. N. Reddy et al. (eds.), *ICSCEA 2021*, Lecture Notes in Civil Engineering 268,
https://doi.org/10.1007/978-981-19-3303-5_89

The material requirements used for pavement structures are always high in rutting resistance. However, the asphalt concrete quality using industrial waste is typically not as good as the quality of traditional asphalt mixture. Therefore, materials using industrial waste always need to be tested with modern laboratory equipment to check their ability to meet the material requirements of rutting resistance.

This study checks the rutting resistance of hot mix asphalt (HMA) containing RFCC by the Hamburg wheel tracking test (HWTT). HWTT is a laboratory-controlled rut depth test that uses a loaded steel wheel to apply a moving load on compacted asphalt mixture specimens to simulate traffic load applied on asphalt pavements. For the comparison, dense-graded asphaltic concrete of the nominal maximum aggregate sizes of 12.5 mm with and without RFCC as filler powder was tested using HWTT. The article concentrates on evaluating and analyzing the rutting resistance of HMA with RFCC as filler under a moving load.

HWTD test outputs, the curve of the relationship between rut depth and the number of wheel passes is shown in Fig. 1. The curve obtains three parameters. They include post-compaction consolidation, creep slope, stripping slope, and stripping inflection point. Post-compaction consolidation is defined as deformation (mm) at 1,000 wheel passes. Creep slope is the inverse of deformation rate in the first steady-state region of the plot, between post compaction and stripping inflection point. Creep slope is the number of wheel passes required to create 1 mm of rut depth, rutting due to plastic flow. The stripping inflection point is the intersection of creep slope and stripping slope. After the stripping inflection point, the stripping slope is the inverse deformation rate, the second steady-state in the plot. The wheel passed number at the stripping inflection point and the stripping slope are usually considered because they are related to the moisture resistance of HMA [1].

Loaded Wheel Tester (LWT) has been used as a proof test for rutting resis-tance during the HMA mix design and Quality Control/Quality Assurance (QC/QA) process [3]. One common type of LWT used in the USA is the Asphalt Pavement

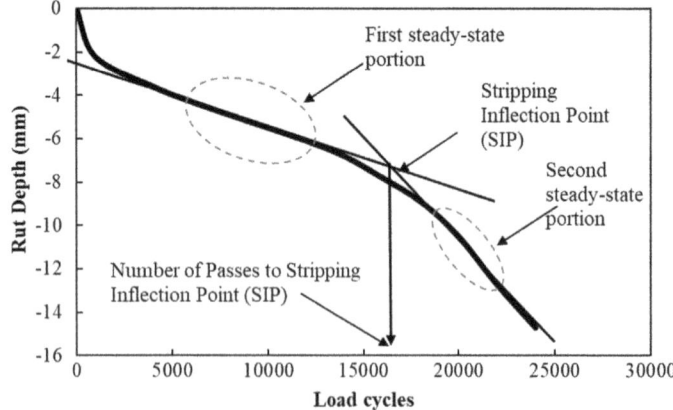

Fig. 1 Typical Hamburg wheel tracking device test results [1, 2]

Analyzer (APA). The APA, a complicated system, has been used to evaluate the rutting, fatigue, and moisture resistance of HMA mixtures. The Study of Xingwei Chen et al. [3] reports a strong correlation between the DS obtained from a flat solid rubber-wheeled (FLWT) and the Asphalt Pavement Analyzer (APA) rut depth for the dense-graded mixtures (25, 19, and 13-mm) designed with the Marshall mix design method, and a 13 mm SMA mixture. Besides, many researchers have reported that Dynamic stability is a characteristic of rutting depth development influenced by mixture gradation, asphalt, and testing temperatures. The criteria for DS were proposed in Decision No. 1677/QD-BGTVT [2] by the Ministry of transportation of Vietnam. Thus, Dynamic stability (DS) could be selected to characterize the rutting of hot-mix asphalt (HMA) in the study. Based on DS, the rut depth resistance of HMA with RFCC was discussed.

2 Dynamic Stability

Dynamic stability is evaluated based on the curve of the relationship between rut depth and the number of wheel passes. Dynamic stability is the number of load repetitions to generate 1-mm rutting during the last 15-min of one-hour testing. In this study, both equations proposed by Decision No. 1677/QD-BGTVT and Kim, Kwang-Woo were estimated for the curves of the experiment. The advantage of the equation proposed by Kim, Kwang-Woo is based on the shape of rutting depth curves and does not rely on correction factors.

Decision No. 1677/QD-BGTVT [2] estimated the dynamic stability based on the rutting depth data at 45 and 60 min and calculated as the following equation.

$$DS = \frac{(t_2 - t_1)}{(D_2 - D_1)} \times 50 \times C_1 \times C_2 \tag{1}$$

where D_1 is the rutting data (mm) corresponding to the time $t_1 = 45$ min, D_2 is the rutting data (mm) corresponding to the time $t_2 = 60$ min, C_1 is the test machine correction factor, and C_2 is the specimen coefficient.

According to Kim, Kwang-Woo et al. [4], the dynamic stability of the asphalt concrete sample at the experimental stages is calculated according to the following equation.

$$DS_0(cycle/mm) = \frac{600}{D_{3600} - D_{3000}} \tag{2}$$

$$DS_1(cycle/mm) = \frac{500}{D_{500}} \tag{3}$$

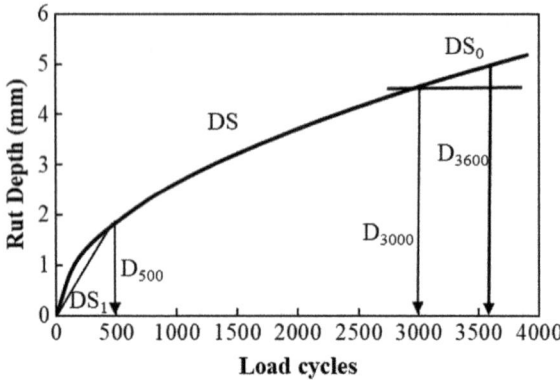

Fig. 2 The curve shows how to evaluate dynamic stabilities [4]

$$DS(cycle/mm) = \frac{1}{2}[DS_1 \times \frac{D_{3600}}{D_{500}} + DS_0] = \frac{1}{2}\left[\frac{500}{D_{500}} \times \frac{D_{3600}}{D_{500}} + \frac{600}{D_{3600} - D_{3000}}\right]$$

$$(4)$$

where DS_0, DS_1, and DS are the dynamic stability of the sample asphalt, which could be calculated following Fig. 2.

3 Experimental Program

3.1 Material

This study uses an asphalt binder, penetration grade of 60/70 distributed by Petrolimex Corporation in Vietnam, for all asphalt concrete specimens. Some essential characteristics of the asphalt binder are summarized in Table 1. It could be seen that asphalt properties are satisfied the current Vietnamese specification TCVN 7493-2005[5].

The aggregate used in this study is from Tan-Cang quarry, Dong-Nai Province. The chosen nominal maximum aggregate size (NMAS) is 12.5 mm. The aggregate gradation is selected because it is used widely in the flexible pavement in Vietnam. Materials including large aggregates, mineral powders, and RFCC satisfy TCVN8819-2011 [4]. Figure 3 shows the aggregate gradation curve used to design the HMA 12.5 mm, while the raw materials are presented in Fig. 4. Filler powder content in the asphalt concrete mix is 5% for both cases of the limestone powder (LS) and RFCC.

Table 1 Properties of asphalt binder 60/70

No	Properties	Values	Requirement
1	Penetration at 25 °C, 0.1 mm	70	60–70
2	The softening point, °C	49.3	46
3	Flashpoint, °C	318	min 232
4	Ductility at 25 °C, cm	>110	min 100
5	Adhesion with aggregate	Level 5	min level 3
6	Specific gravity, g/cm^3	1.037	1–1.05
7	Solubility, %	99.83	min 99
8	Retained penetration after thin film oven, %	74.96	min 54
9	Paraffin wax content, %	1.52	max 2.2

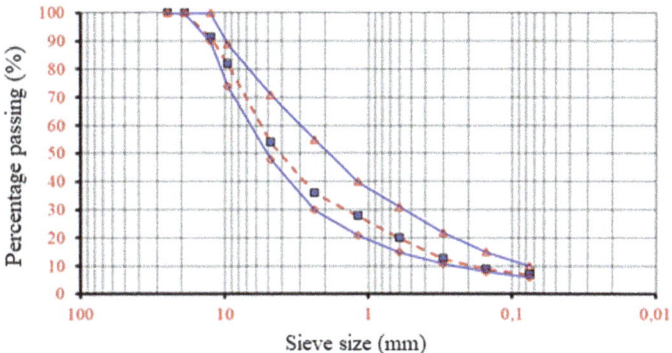

Fig. 3 Aggregate gradation of the HMA 12.5 mm

Fig. 4 Materials in this study: **a** Asphalt binder, **b** Aggregates, **c** Limestone, and **d** RFCC

3.2 Mix Design

The Marshall mix design method according to TCVN 8820:2011 [6] was used for determining the optimum asphalt content. The Marshall method of mix design is for dense-graded hot mix asphalt mixes and is popularly used everywhere.

The above aggregate gradation was used for the asphalt mixtures with different binder contents, 4.0, 4.5, 5.0, 5.5, and 6%. Marshall criterion was tested to determine the optimum binder content for the HMA 12.5 mm; Conventional mineral powder is limestone powder. Tables 2 and 3 show the Marshall test results, including mass density, residual air void, Marshall stability, Marshall flow, VMA, and VFA corresponding to different binder contents. Based on technical requirements, according to TCVN 8820-2011 [6], the chosen optimal asphalt content is 5.1% by weight of hot mix asphalt mixture in case of the conventional HMA 12.5 mm (Table 2). In HMA 12.5 mm with RFCC, the chosen optimal asphalt content is 5.0% by weight of the HMA mixture (Table 3).

Table 2 Marshall test results of conventional HMA 12.5 mm

Binder content (%)	Mass density (g/cm³)	Air void (%)	VFA (%)	VMA (%)	Marshall stability (kN)	Marshall flow (mm)
TCVN 8819-2011		3 ÷ 6	–	>14	>8	1.5 ÷ 4
4.0	2.337	7.49	55.06	16.66	9.36	1.8
4.5	2.369	5.55	65.28	15.99	11.51	2.4
5.0	2.373	4.82	70.40	16.27	14.08	3.0
5.5	2.374	4.33	74.06	16.67	12.64	3.6
6.0	2.359	4.36	75.27	17.63	11.12	3.9

Table 3 Marshall test results of HMA 12.5 mm with RFCC

Binder content (%)	Mass density (g/cm³)	Air void (%)	VFA (%)	VMA (%)	Marshall stability (kN)	Marshall flow (mm)
TCVN 8819-2011		3 ÷ 6	–	>14	>8	1.5 ÷ 4
4.0	2.527	7.40	55.39	16.58	19.07	2.8
4.5	2.508	5.72	64.56	16.14	15.78	3.0
5.0	2.493	4.10	78.79	15.64	14.90	3.1
5.5	2.482	2.16	85.40	14.79	13.33	3.4
6.0	2.467	0.57	96.04	14.37	11.23	5.5

3.3 Hamburg Wheel-Track Test

HWTT was based upon a device that utilizes rubber tires wheels. The HMA slab spec-imens had the size of $320 \times 260 \times 50$ mm. The device is operated by moving a pair of reciprocating steel wheels across the HMA slab specimen surface (Fig. 6) submerged in hot water, held at 50 °C. According to Decision No. 1617/QD-BGTVT, the test method uses method-A, developed based on AASHTO T312:2004 [7] and EN12697-33 [8]. Figure 5 shows the experiment equipment, which was at the laboratory in the University of Transport, the campus in Ho Chi Minh city.

The device can test a pair of specimens simultaneously, and specimens are compacted to (7 ± 2) percent air voids. The steel wheels have a diameter of 203 mm, a width of (47 ± 5)mm, and are capable of generating 50 ± 5 passes per minute. Each steel wheel weighs 702 N. The length of the slabs is 320 mm long by 260 mm wide by 50 mm thickness, as shown in Fig. 6. Linear variable differential transformers (LVDTs) measure rut depth or deformation at 11 points along the length of each specimen at an accuracy of 0.01 mm. The device automatically ends the test when

Fig. 5 Hamburg wheel-track test (HWTT) equipment

Fig. 6 Samples of HMA with RFCC after Hamburg wheel-track test

the preset number of wheel passes, considering 15,000 passes, is reached or a rut depth of 20 mm, whichever occurs first.

4 Result and Discussion

Figure 7 shows the rutting depth results of the conventional HMA samples during the Hamburg Wheel-Track test, compared to the HMA using RFCC. In the conventional HMA samples, 12.5 mm NMAS, the accumulated rutting depth was 8.5 mm after 15,000 loading passes. The accumulated rutting depth after 15,000 loading passes was 2.16 mm in the HMA samples with RFCC, 12.5 mm NMAS. According to Decision No. 1617/QD-BGTVT, the required final rutting depth must be less than 12.5 mm after 15,000 passes. Thus, both the conventional HMA and HMA with RFCC were satisfied the requirement of Decision No. 1617/QD-BGTVT of the Vietnam Ministry of Transport. Besides, the accumulated rutting depth of HMA with RFCC after 15,000 loading passes was four times higher than conventional HMA. It indicates that RFCC can help asphalt concrete improve its resistance to rutting.

This result shows that the activity of RFCC is high. It is known that RFCC has powerful hydrophilic properties. When mixed with asphalt and crushed stone aggregate, RFCC is responsible for filling the pores in the mixture particles and increasing the suction force of the binder with large aggregates. Besides, RFCC has pozzolanic characteristics [9], increasing the hardness and stability of the concrete and reducing the rate of HWTT rutting depth.

Table 4 summarizes the rutting depth, in mm, at the loading passes of 500, 3000, and 3600 times. Those values were substituted into Eqs. (2)–(4), calculating DS_0, DS_1, and DS. Table 5 presents DS and DS_1 values for comparison. Generally speaking, DS values from Eq. (1) were not much different from Eq. (4). The different error was about 10%. Besides, the DS increasing tendency of HMA with SFCC was maintained.

In the conventional HMA samples, DS and DS_1 were 1875 and 391 loading passes per millimeter, respectively. In addition, DS and DS_1 were 5357 and 877

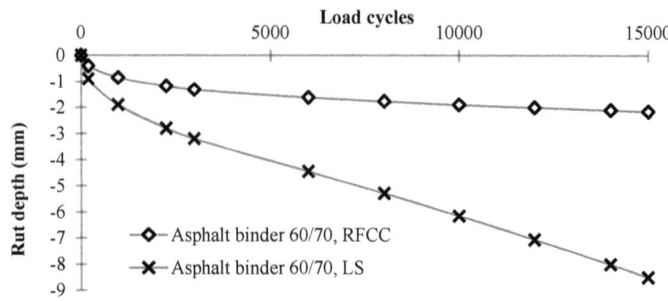

Fig. 7 Rutting depth curves of HMA with/without RFCC

Table 4 Dynamic Stability estimation

Mixture description	Rut depth (mm) at the passes of			DS_0	DS_1	Final rut depth
	500	3000	3600			(mm)
C12.5	1.28	3.2	3.45	2400	391	8.5
C12.5-RFCC	0.57	1.32	1.38	10,000	877	2.16

Table 5 Comparison of dynamic stability

Mixture description	DS_1	$DS^{(*)}$	DS	Percentage increase dynamic stability compared to conventional HMA (%)		
				DS_1 (%)	$DS^{(*)}$ (%)	DS (%)
C12.5	391	1726	1875	100	100	100
C12.5-RFCC	877	6062	5357	224	351	286

Note (*) DS was estimated by Eq. (4)

loading passes per millimeter, respectively, in the HMA samples with RFCC. DS_1 of HMA with RFCC was 2.24 times higher than that of conventional HMA. It indicates that HMA samples with RFCC have more stiffener and higher rutting resistance in post-compaction consolidation.

According to Decision No. 1617/QD-BGTVT, the required DS is higher than 1000 loading passes per millimeter for the conventional HMA [10]. In addition, the required DS is higher than 2800 loading passes per millimeter for the polymer HMA [11]. Thus, both the conventional HMA and HMA with RFCC were satisfied the requirement of Decision No. 1617/QD-BGTVT of the Vietnam Ministry of Transport. Besides, the rutting resistance of HMA with RFCC could be comparable to the polymer HMA, an technology of HMA for rutting depth resistance in a pavement.

5 Conclusion

In this paper, the optimal asphalt content of the control HMA, the conventional HMA with the limestone filler, and the HMA with RFCC were chosen using the Marshall method, in which the nominal maximum aggregate size is 12.5 mm. Then, these results were applied to prepare the samples of HWTT. HWTT had been done for both mixtures of the control HMA and HMA with RFCC. Based on the experiment results, the following conclusions are from the obtained results:

- HMA with RFCC were satisfied the requirement of Decision No. 1617/QD-BGTVT of the Vietnam Ministry of Transport;
- The accumulated rutting depth after 15,000 loading passes was four times higher than that of control HMA.

- The asphalt concrete mixtures with RFCC are better than conventional mixtures in terms of rutting resistance. HMA samples with RFCC have more stiffener and higher rutting resistance in post-compaction consolidation, based on DS_1.

References

1. Yildirim Y et al. (2007) Hamburg wheel-tracking database analysis. Texas Department of Transportation and Federal Highway Administration, FHWA/TX-05/0-1707-7
2. MOT (2014) Decision No. 1677/QD-BGTVT, in Technical regulation on wheel tracking depth testing method of asphalt concrete determined by the wheel tracking device. Ministry of Transportation, Vietnam
3. Chen X, Huang B, Xu Z (2007) Comparison between flat rubber wheeled loaded wheel tester and asphalt pavement analyzer. Road Mat Pave Des 8(3):595–604
4. Kim K-W, Doh Y-S (2006) Development of feasible dynamic stability in wheel tracking test for asphalt concrete mixtures. Int J High Eng 8(1):77–87
5. MOST (2005) TCVN 7493-2005, in Bitumen—Specifications. Ministry of Science and Technology, Vietnam
6. MOST (2011) TCVN 8820: 2011, in Standard Practice for Asphalt Concrete Mix Design Using Marshall Method. Ministry of Science and Technology, Vietnam
7. AASHTO (2004) AASHTO T 312: 2004, in standard method of test for preparing and determining the density of asphalt mixture specimens by means of the superpave gyratory compactor. American Association of State Highway and Transportation Officials
8. CEN (2019) EN 12697-33, in Bituminous mixtures. Test method. Specimen prepared by roller compactor. European committee for standardization
9. Torres Castellanos N, Torres Agredo J (2010) Using spent fluid catalytic cracking (FCC) catalyst as pozzolanic addition a review. Ingeniería e investigación 30(2):35–42
10. MOST (2011) TCVN 8819:2011, in Specification for Construction of Hot Mix Asphalt Concrete Pavement and Acceptance. Ministry of Science and Technology, Vietnam
11. MOT (2006) 22 TCN 356-06, in Construction process and acceptance of polymer asphalt concrete pavement. Ministry of Transportation, Vietnam

Laboratory Assessment of Recycled Polyethylene into Hot Mix Asphalt in Ho Chi Minh City

Ba Tu Vu and Manh Tuan Nguyen

Abstract In Ho Chi Minh city, the flexible pavement system is getting more and more damages come from variety of reasons such as heavy traffic loading and limitation of maintenance procedure. Many methods were introduced and one of the most effective solutions is polymer modified asphalt concrete instead of conventional asphalt concrete as well as polymer modified asphalt concrete is relatively expensive. As a result, recycled polyethylene from waste plastic bag into hot mix asphalt concrete mixtures to create polymer modified asphalt concrete with a reasonable cost is a solution. Recycled polyethylene used in this study is granular form which size is from 1 to 3 mm. The contents of polyethylene replaced asphalt binder are 6, 9, 12, 15, 18% by weight of asphalt binder. The paper shows the mix design of asphalt concrete with recycled polyethylene and evaluated tests including Marshall stability, indirect tensile strength, Cantabro durability and resilient modulus. The 12% could be the best content of PE.

Keywords Polymer · Polyethylene · Asphalt concrete · Hot mix asphalt · Green technology

1 Introduction

Nowadays, the increasing of distresses in Vietnam roads came from both the development of traffic volume, heavy trucks and bad maintenance strategy from a road management agency. There are many solutions which have been used and one of them is the application of polymer modified asphalt (PMA) in asphalt concrete for the surface course. The polymer modified asphalt binder increases not only the mechanical properties but also life of asphalt concrete [1] because polymers are used in order to reduce the permanent deformation, improve cohesive strength, and reduce low-temperature thermal cracking.

B. T. Vu · M. T. Nguyen (✉)
Faculty of Civil Engineering, Ho Chi Minh City University of Technology (HCMUT), Ho Chi Minh City, Vietnam
e-mail: nmanhtuan@hcmut.edu.vn

© The Author(s), under exclusive license to Springer Nature Singapore Pte Ltd. 2023
J. N. Reddy et al. (eds.), *ICSCEA 2021*, Lecture Notes in Civil Engineering 268,
https://doi.org/10.1007/978-981-19-3303-5_90

Besides, the development of industries and population make the amount of waste materials into the environment. The treatment of these wastes, especially non-decomposing waste, is an issue. Recycling waste into useful products is considered one of the sustainable solutions to this issue. Therefore, research and application of waste materials are increasingly around the world and in Vietnam. Many studies have been done to find effective applications of waste products, to meet the performance, stability, environmental issues, and the needs of efficiency and economy [2].

Polymer has been applied in asphalt concrete since years of 1980 in United States and Europe. Some significant polymers used to modify asphalt binder are SBS (Styrene–Butadiene–Styrene), SBR (Styrene-Butadiene Rubber), PB (Polybuatadiene Rubber), Elvaloy, EVA (Ethylene Vinyl Acetate), PE (Polyethylene) and others [3]. In Vietnam, PMA has not been used widely because the price of asphalt concrete used PMA is quite expensive in comparison with a conventional asphalt concrete which uses a conventional asphalt binder. As a result, a polymer which has a reasonable price could be a solution for this situation. Polyethylene (PE) may be that polymer.

Polyethylene (PE) is an elastomeric polymer. In Vietnam, a million of plastic waste are released into the environment every year. On the other hand, they cannot be decomposed after hundreds of years, causing serious pollution. Therefore, the reuse of PE from waste is a very important work because its contribution is not only the quality of asphalt, but also the environmental. Justo and Veeragavan [2] studied the application of plastic bags as a modifier in asphalt concrete which content is from 0 to 12% by weight of asphalt binder. The result showed that the 8% content is very good for a significant improving stability, strength, fatigue life of asphalt concrete. Awwad and Shabeed [4] used two types of PE including the low density PE and the high density PE to modify the asphalt binder. In their study, the optimal asphalt content was determined based on the Marshall method and seven PE contents which are 6, 8, 10, 12, 14, 16 and 18% by weight of optimal asphalt binder were evaluated. The research results showed that 12% of high density PE was the best choice because it increased the stability, decreased density, increased air void and voids of mineral aggregate. Punith và Veeraragavan [5] evaluated the effect of the low density PE from a plastic bag to modify the asphalt binder which penetration grade is 80/100. The low density PE contents were 2.5, 5.0, 7.5, and 10% by weight of asphalt binder. The results showed that the asphalt concrete had 5% of low density PE gave resilient modulus better than the conventional asphalt concrete about 28%, and increased tensile strength, moisture susceptibility. Moreover, the PE modified asphalt concrete had 2.5 times of life and the thickness was thinner in comparison with normal asphalt concrete. Ahmed et al. [6] evaluated the effect of the low and high density of PE polymer modified asphalt, which penetration grade is 40/50, on hot mix asphalt concrete based on Marshall stability, wheel tracking test. They found that the using of high density PE to modified asphalt mixtures gave better properties to asphalt mixtures from modified mixtures by the low density PE.

In Vietnam, there is not many studies which concern the PE in asphalt concrete. Nguyen QP et al. [7] applied plastic bags into asphalt concrete by used the dried method. They only evaluated the asphalt concrete by using the Marshall test.

This paper shows the evaluation of five PE contents, which are 6, 9, 12, 15, 18% by weight of asphalt binder, modified dense-graded asphalt concrete. The paper also focuses on mechanical properties of PE modified asphalt concrete including the Marshall stability, indirect tensile strength, resilient modulus, and Cantabro loss test based the Vietnam specifications.

2 Experimental Program

2.1 Material

In this study, one type of asphalt binder, which penetration grade is 60/70 from Shell in Singapore and distributed by Petrolimex Corporation in Vietnam, is used. Some basic properties of asphalt binder were checked and presented in Table 1. All the properties are satisfied current Vietnam specification TCVN 7493-2005 (Bitumen—specification) [8] which is similar from ASTM D for asphalt binder or bitumen.

The aggregate used in this study is from BMT hot mix asphalt plant in Long An province. The chosen nominal maximum aggregate size is 12.5 mm because this type is used widely in flexible pavement in Vietnam. All the coarse and fine aggregate used in this work were sieved and recombined in the proper proportions to meet the surface course gradation in Table 2 as required by TCVN 8819-2011 [9].

In order to evaluate the effect of Polyethylene (PE) on asphalt concrete, one type of PE, which is the high density PE and from recycled plastic bags, is used. The size of PE is from 1 to 3 mm as shown in Fig. 1. The specific gravity of PE is 0.95 g/cm3. In this study, the chosen contents of PE are 6, 9, 12, 15, 18% by weight of asphalt binder based on literature review in the introduction part.

Table 1 Properties of asphalt binder 60/70

No	Properties	Results	Requirement from TCVN 7493-2005
1	Penetration at 25 degree Celcius, 0.1 mm	70	60–70
2	Softening point from ring and ball test, degree Celcius	49.3	46
3	Flash point, degree Celcius	318	min 232
4	Ductility at 25 degree Celcius, cm	> 110	min 100
5	Adhesion with aggregate	Level 5	min level 3
6	Specific gravity, g/cm3	1.037	1–1.05
7	Solubility, %	99.83	min 99
8	Retained penetration after thin film oven, %	74.96	min 54
9	Paraffin wax content, %	1.52	max 2.2

Table 2 Aggregate gradation

Sieve size (mm)		Passing percent	
	Used in this study	Bottom limit from TCVN 8819:2011	Top limit from TCVN 8819:2011
19	100	100	100
12.5	96	90	100
9.5	88	74	89
4.75	59,5	48	71
2.36	37,5	30	55
1.18	23,4	21	40
0.6	16,25	15	31
0.3	11,25	11	22
0.15	8	8	15
0.075	7	6	10

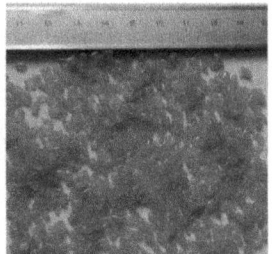

Fig. 1 Materials in this study (from left to right: Asphalt binder, aggregates and PE)

2.2 Mix Design

The Marshall method or TCVN 8820:2011 [10] is used for determining the optimal asphalt content. The Marshall method of mix design is for dense-graded hot mix asphalt mixes, and is used almost everywhere in the world. Firstly, the aggregate and asphalt binder were chosen based on their requirements from the specification. Secondly, the gradation was chosen based on Table 2 which is for dense-graded asphalt concrete. Thirdly, the optimal asphalt content without PE or control mixture was determined based on relationships between five asphalt binder contents and Marshall stability, Marshall flow, density, void in mineral aggregate and air void. The chosen optimal asphalt content is 5.44% by weight of hot mix asphalt mixture.

After the mix design of control mixture, the PE contents, which are 6, 9, 12, 15, and 18% by weight of asphalt binder in the mixture, were added to control mixture.

2.3 Mechanical Properties of Asphalt Concrete

In order to evaluate the effect of PE content on asphalt concrete, the Marshall stability, indirect tensile strength, resilient modulus and Cantabro loss test were conducted.

Marshall Stability was conducted according to TCVN 8820:2011 specification [10]. Before the Marshall stability test, all specimens were kept in water at 60 °C for 40 min and then loaded with a constant compression rate of 50 mm/min. The maximum load applied to the specimen is the Marshall stability. And the Marshall flow is the displacement value in accordance with the Marshall stability.

The indirect tensile strength (IDT) was obtained by a compressive load based on TCVN 8862-2011 [11]. The specimens were also prepared by the Marshall compactor. The compressive load increased continuously and evenly with displacement rate (50 mm/min) regulations until damage. The tensile strength of the material is calculated from the load which creates the damage of specimen (P) as follows:

$$R = 2P/\pi HD \qquad (1)$$

where D and H are the specimen diameter and height, respectively (mm).

Before the resilient test, all specimens were prepared based on the 22TCN 211-06 specification [12] and put in the chamber at 30 °C. The resilient modulus tests were conducted by using the 0.5 MPa of static loading pressure (p) into asphalt concrete specimens. The resilient modulus (E) of asphalt concrete is determined from resilient displacement (L) as follows:

$$E = pH/L \qquad (2)$$

where H is also the specimen height (mm).

The Cantabro abrasion loss test was conducted based on TCVN 11,415:2016 [13]. The specimen was put into the drum of Los Angeles machine without steel ball for 300 cycles. The Cantabro abrasion loss (CL) is determined from the weight before and after testing in Eq. (3). The requirement for Cantabro loss test is that the CL should be less than 20%.

$$CL = 100(A - B)/B \qquad (3)$$

where A and B are the weight of specimen before and after testing, respectively (Fig. 2).

3 Result and Discussion

In this study, effect of PE contents on Marshall stability, flow, indirect tensile strength, resilient modulus, and Cantabro loss are shown in Figs. 3 4, 5, and 6. In Fig. 4, the

Fig. 2 Set-up for Marshall stability, indirect tensile strength, resilient modulus, and Cantabro loss test (from left to right)

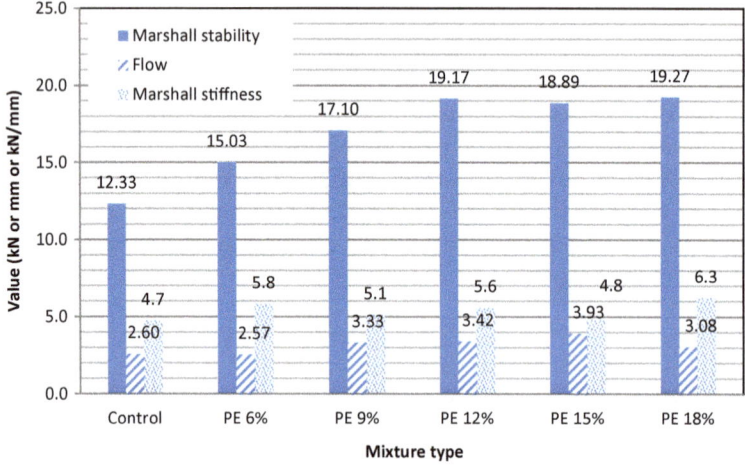

Fig. 3 Marshall stability (kN), flow (mm), and stiffness (kN/mm) results of all mixtures

flows of all mixtures are in the limit range of TCVN 8819:2011 which is from 2 to 4 mm. In the stability, the PE modified asphalt concrete or PE mixture is better than conventional asphalt concrete or control mixture. When PE replaced asphalt binder from 8 to 18%, the stability increased from 21 to 56%. Another evaluation in the Marshall test is the Marshall stiffness which is the ratio of Marshall stability to flow. The mixture exhibits a higher Marshall stiffness which means that the mixture has more resistance to shear stresses, and permanent deformation. According to Fig. 3, the increasing of the PE content in the mixture causes the increasing of mixture stiffness. In Fig. 4, the indirect tensile strengths of all mixtures are shown. The PE in asphalt mixtures increased the tensile strength but there are not too much different from the control mixture except the mixture that has 18% PE. In Fig. 5, the resilient modulus values of all mixtures are shown. For the resilient modulus, the PE modified

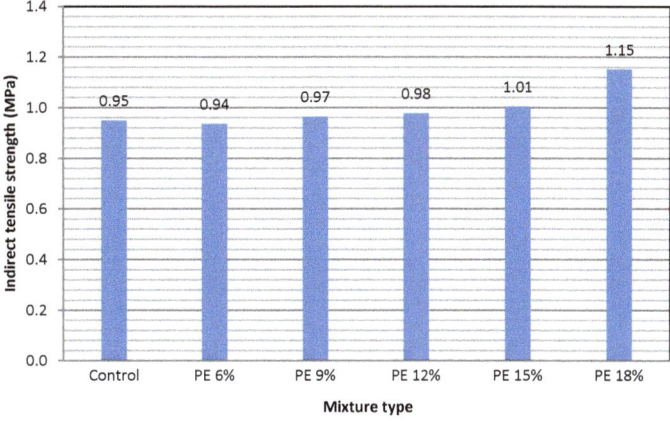

Fig. 4 Indirect tensile strength results of all mixtures

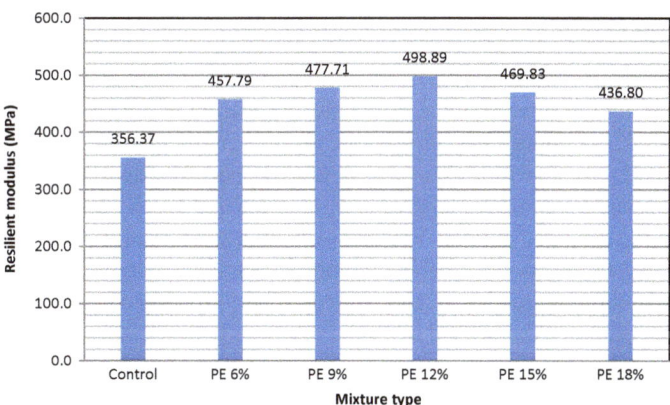

Fig. 5 Resilient modulus results of all mixtures

asphalt concrete is better than the control mixture. When PE replaced asphalt binder from 8 to 18%, the stability increased from 20 to 40%. As a result, the thickness of asphalt concrete used PE can be applied thinner in comparison to conventional asphalt concrete or the price of flexible pavement for road construction can be decreased because the thickness of pavement could be determined from resilient modulus [12]. In Fig. 6, all Cantabro loss values of all mixtures are presented. These values are less than the limit value of 20% or all mixtures could be applied for the surface course. The results also show that the mixture used bigger content of PE is more abrasion than other mixtures.

Finally, the best content of PE could be 12% for PE modified asphalt concrete in terms of Marshall stability, indirect tensile strength, and resilient modulus.

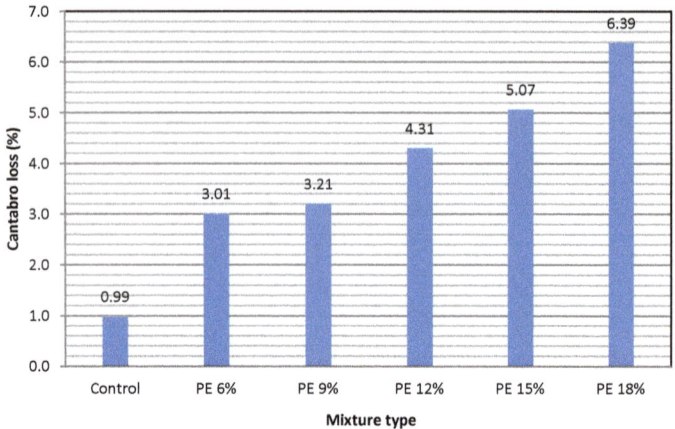

Fig. 6 Cantabro abrasion loss results of all mixtures

4 Conclusion

In this paper, the literature review about the application of PE in asphalt concrete is presented. The optimal asphalt content of the control mixture, which nominal maximum aggregate size is 12.5 mm, designed by using Marshall method is 5.44%. PE modified asphalt concrete mixtures were conducted by the dried method. The contents of PE replaced asphalt binder are 6, 9, 12, 15, and 18%. All mixtures were evaluated based on the Marshall stability, Marshall stiffness, indirect tensile strength, resilient modulus, and Cantabro loss. The following conclusions are from the obtained results:

- The content of PE used in asphalt concrete is less than 18% based on the literature review;
- The asphalt concrete mixtures used PE are better than conventional mixture in terms of Marshall stability, Marshall stiffness, indirect tensile strength, and resilient modulus. However, the PE reduces the abrasion property of asphalt concrete;
- The pavement structure used PE modified asphalt concrete can reduce the thickness in comparing with control asphalt concrete because higher resilient modulus.

References

1. Deef-Allah EMM, Mohamady A (2014) Performance evaluation of polymer modified asphalt mixtures. Int J ICT Aided Arch Civil Eng 1(2):35–52

2. Justo CEG, Veeraragavan A (2003) Utilization of waste plastic bags in bituminous mix for improved performance of roads. Banglore University, Bengaluru
3. Yildirim Y (2007) Polymer modified asphalt binders. J Constr Build Mat 21(1):66–72
4. Awwad MT, Shbeeb L (2007) The use of polyethylene in hot asphalt mixtures. Am J Appl Sci 4(6):390–396
5. Punith VS, Veeraragavan A (2010) Evaluation of reclaimed polyethylene modified asphalt pavements. J Test Eval 38(5):541–548
6. Ahmed NY, Sundus A, Al-Harbi M (2014) Effect of density of the polyethylene polymer on the asphalt mixtures. J Babylon Univ Eng Sci 22(4):674–774
7. Nguyen QP, Nguyen HQ, Le TA (2018) Applied study waste plastics based dried method on improvement of Marshall stability for asphalt concrete. Vietnam J Transp 1+2:54–58
8. TCVN 7493-2005 (2005) Bitumen—specifications. Ministry of Science and Technology, Vietnam
9. TCVN 8819:2011 (2011) Specification for construction of hot mix asphalt concrete pavement and acceptance. Ministry of Science and Technology, Vietnam
10. TCVN 8820:2011 (2011) Standard practice for asphalt concrete mix design using Marshall method. Ministry of Science and Technology, Vietnam
11. TCVN 8862:2011 (2011) Standard test method for splitting tensile strength of aggregate material bonded by adhesive binders. Ministry of Science and Technology, Vietnam
12. 22TCN 211-06 (2006) Flexible pavement—Requirements and design guide. Ministry of Transportation, Vietnam
13. TCVN 11415:2016 (2016) Asphalt concrete—Determination method of Cantabro abrasion loss. Ministry of Science and Technology, Vietnam

Water Resources Session

A Comparison Study of Water Pipe Failure Prediction Models

Thi Minh Lanh Pham and Quang Truong Nguyen

Abstract Water leakages have been a major problem for water supply companies, one of the main causes of this problem is pipe failure in the water supply network. The risk prediction models of the pipe failure are also constantly being improved to determine the location of these leaks accurately and quickly. The statistical model is the Logistic Regression model (LR) used for failure prediction in groups of pipes. Machine Learning approaches, particularly the Decision Tree model (DT) and the Artificial Neural Network model (ANN) are compared in predicting individual pipe failure. In this paper, we proposed applying these three models to predict pipe failure for the DMA17 water supply network in ward 17, Go Vap District, Ho Minh city. Using Area Under the Curve (AUC) value to evaluate the modeling results, comparing this value showed that the ANN was the most suitable for water pipe failure prediction.

Keywords Pipe failure · Predicting model · Water supply network · Artificial neural network · Regression logistic model

1 Introduction

During the operation of water supply networks, a certain amount of water loss always occurs. There are many causes of water loss, mainly technical reasons such as leaks in the network of water supply, construction mistakes, broken pipes due to excavation,…. Water loss is mainly leaking on the pipeline because the water supply pipes are buried underground, so it is difficult to find leaks. This is a challenge for the water company in the world as well as in Vietnam. With the aim of early detection and elimination of water leakage risk on the water supply network, three models of pipe

T. M. L. Pham (✉)
Department of Urban Infrastructure Engineering, Ho Chi Minh City Architecture University, Ho Chi Minh City, Vietnam
e-mail: lanh.phamthiminh@uah.edu.vn

Q. T. Nguyen
Faculty of Civil Engineering, Ho Chi Minh City University of Technology, VNU-HCM, Ho Chi Minh City, Vietnam

© The Author(s), under exclusive license to Springer Nature Singapore Pte Ltd. 2023
J. N. Reddy et al. (eds.), *ICSCEA 2021*, Lecture Notes in Civil Engineering 268,
https://doi.org/10.1007/978-981-19-3303-5_91

failure prediction were proposed by Lanh P. T. M, including the Logistic Regression model [1], Decision Tree model [2], and Artificial Neural Network model [3]. The three models are applied for predicting the possibility of pipe failure on the DMA17 Trung An water supply network in Go Vap District, Ho Chi Minh city. Based on the forecasting results, the results of the three models will be evaluated by the Area Under the Curve (AUC) value. In this study, some conclusions of the predictive performance of each model are discussed Material and methods.

1.1 Data Set

According to Lanh P. T. M et al. [2], the data set employed for the pipe failure predicting model on the water supply network requires 8 variables: physical characteristics (Diameter—D, Length—L, Material—Mat), Age (A) and the feature of the inside working environment (Road Diameter—RD). Also, quantitative properties for network characteristics include road code which features the outside working environment (Road code—R), the number of junctions per pipeline (Number junction—N0), and the number of prior pipe failures (Prior).

The study area is the water supply network of DMA17 (Fig. 1) (DMA-Area meter district) in ward 17, Go Vap district, managed by Trung Water Supply Company, Saigon Water Corporation. Data is collected from the GIS management and Technical management department of the company in which year from 2014–2020.

DMA17 water supply network includes diameter pipes of 25 mm, diameters main pipes from 50 to 250 mm, and 1683 customers meter. According to statistics from 2014 to July 2020, the network continuously expanded as shown in Table 1, up to

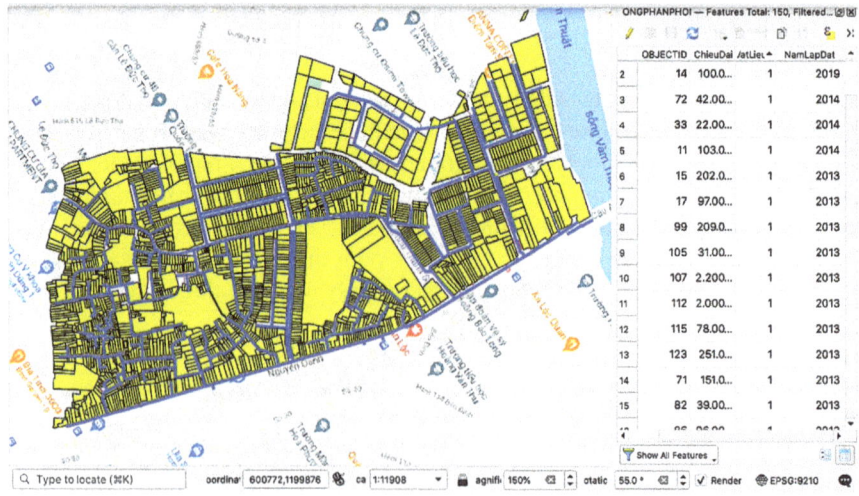

Fig. 1 GIS data of DMA17, Trung An, Go Vap district (year from 2014–2020)

Table 1 Total pipes length over years

Mat	Year						
	2014	2015	2016	2017	2018	2019	7/2020
uPVC (m)	216.38	508.77	618.56	704.96	999.72	1087.56	1244.92
PE (m)	620.73	1396.38	1518.75	1569.55	1688.49	1730.72	2153.05
HDPE (m)	10.50	10.50	11.59	11.59	11.59	11.59	18.72
Total (m)	**847.61**	**1915.65**	**2148.90**	**2286.10**	**2699.80**	**2829.87**	**3416.68**

the time of inspection, the total pipe length is 3416.68 m. Branch pipes have three types of materials: uPVC, PE, and HDPE with length rates is respectively 36.4, 63, and 0.6%.

1.2   Three Models of Pipe Failure Prediction

According to previous studies, in conjunction with the water supply network conditions in Vietnam, the following three predictions of pipe failure models has been built as follows:

Logistic regression model (LR): In general, LR is the basic model in the statistical field according to author T. Wengström, pipe failure was evaluated by failure rate per 1 km so the prediction model is exactly 70% [4]. To improve the efficiency of the predictive model, the authors Lanh. P.T.M et al. directly assessed the probability of pipe failure in binary [1], the result demonstrated that LR predicted well. In this study, LR was used for pipe failure prediction based on the 8 independent variables and one the dependent variable form 0|1.

  Decision tree model (DT): The decision tree used classification or regression tree algorithm into predictive modeling where the relationship between features and outcome is nonlinear. The tree starts at the root node that will be split into child nodes by Gini indexes, the algorithm continues split until a stop criterion is reached [5]. The advantage of DT is to be able to remove outliers from the system, each of which will be isolated at individual nodes. The DT is called a non-parametric model, so when used there are no constraints between the variables as well as for each variable and data set with null values.

  Artificial Neural Network model (ANN): Based on the idea of simulating the human brain, an Artificial Neural Network consists of many neurons connected to process information through algorithms. Some studies example H. Al-barqawi et al. (2008), the ANN approach was employed to determine the efficiency of the water distribution system as well as the rate of pipe failure during the operation [6]. In addition, some authors proposed an Analytic Hierarchy Process analysis into the ANN to increase model performance [7]. The data are classified into different levels by certain criteria, then the weight of each factor and aggregate analysis are evaluated to give the average value of each factor's contribution to the pipe failure. With the development of the current technology, the ANN model was increasingly more used in research. The feature of the neural network in this study includes eight variables of the input layer, one pipe failure of the output layer, with a Sigmoidal activation function, and one hidden layer. The Feed-forward network structure has a simple algorithm and has high efficiency, the variables in the predictive pipe failure model are independent so this structure will be applied in this study.

1.3 Methodology

LR is one of the traditional models in statistical models. On otherwise, the DT model and ANN model are the machine learning models. The performance measures were recently used to assess the statistical model accuracy: R square, Root Mean Square Error, and the machine learning models: Receiver Operating characteristic Curve (ROC). However, ROC was used to assess the statistical model performance too, so in this study, three models were measured accuracy by ROC.

The name Receiver Operating characteristic Curve (ROC) came from signal detection theory developed during World War II for analysis of radar images. Until the 1950s, the signal detection theory was recognized as useful for interpreting medical test results [8] and recent years have seen an increase in the use of ROC in machine learning. The model is evaluated by two basic evaluation measures, including specificity and sensitivity. The model of predicting pipe failure is assumed with two parameters (pipe nonfailure - Negative and pipe failure - Positive). The results of the predictive model will be divided into 4 groups as shown in Table 2, its name is the confusion matrix.

The Type II error has larger consequences due to the pipe failure but the model has been failed to detect, it is not only to cause water loss but also the risk of contamination intrusion into the pipe. A Type I error has an effect on the testing cost of the pipe but does not affect quality water and has less serious consequences. Thus, when the prediction results of the model have these two types of error, a performance measure is used to evaluate these results. The accuracy of the predictive model is

Table 2 Confusion matrix

Actual	Predicted	
	Positive	Negative
Positive	True Positives[1] (TP)	False Negatives[2] (FN)
Negative	False Positives[1] (FP)	True Negatives[2] (TN)

– False Positive[1] (FP) the equivalent of making a Type I error means that the pipe is working perfectly fine but the model predicts that the pipe is in failure.
– False Negatives[2](FN) the equivalent of making a Type II error means the pipe failure but the model predicts that the pipe is not in failure.

evaluated by two statistical parameters: the sensitivity and specificity parameter, they are estimated as:

$$\text{Sensitivity} = \text{TPR} = \frac{\text{TP}}{\sum \text{Positive}} = \frac{\text{TP}}{\text{TP} + \text{FN}} \tag{1}$$

$\rightarrow$

$$\text{FNR} = \frac{\text{FN}}{\text{TP} + \text{FN}} = \frac{\text{FN}}{\sum \text{Positive}} \tag{2}$$

$$\text{Specificity} = \text{TNR} = \frac{\text{TN}}{\sum \text{Negative}} = \frac{\text{TN}}{\text{TN} + \text{FP}} = 1 - \frac{\text{FP}}{\text{TN} + \text{FP}} \tag{3}$$

$\rightarrow$

$$\text{FPR} = \frac{\text{FP}}{\text{TN} + \text{FP}} = \frac{\text{FP}}{\sum \text{Negative}} = 1 - \text{Specificity} \tag{4}$$

where $\sum$ Positive   - Total positive
$\sum$ Negative   - Total negative
TPR   - True Positives Rate (F=1)
TNR   - True Negatives Rate (F=0)
FPR   - False Positives Rate type I
FNR   - False Negatives Rate type II.

ROC is a curve connecting points with the coordinate of FPR = 1—Specificity (if the specificity increases, the FPR decreases) and the intersection is the value of Sensitivity [8]. Another advantage of using the ROC is a single measure called the Area Under the Curve (AUC) index. As the name indicates, it is an area under the curve calculated in the ROC space by the trapezoidal rule, which is adding up all trapezoids under the curve. Although the theoretical range of the AUC index is between 0 and 1, the actual index ranges from 0.5 for accuracy of a chance to 1.0 for perfect accuracy [9]. For example, AUC = 0.85 indicates that the model predicts the probability of pipe failure is more accurate than pipe nonfailure is 85%, so the higher the AUC value, the predictive model is more accurate.

RStudio software. Developed in 1996 by Ross Ihaka and Robert Gentleman [10] R software is a new statistical analysis language to replace commercial statistical software such as SPSS, SAS, Stata, Statistica. R language quickly responded to statisticians and participated in building support tools for software, after more than 10 years R has become a popular statistical language for researchers. R-Studio is a software developed from R software with a more user-friendly interface, allowing high-level programming by branch structure (if / else), and iterative structure (for). With the advantages of this software, the author chose RStudio to simulate three models, including LR, DT, and ANN, thereby predicting the possibility of pipe failure and comparing with the data set.

2   Results

2.1   Statistical Analysis

Using software R analyzed the correlation between the pipe failure and 8 variables in the DMA17 water supply network as shown in Table 3.

2.2   Modeling Results

Input data. Statistics data in the DMA17 water supply network are 2597 columns divided in a 7: 3 scale equivalent to 1817: 780. Four data sets as shown in Table 4 when input the model have the following size matrix:

- Training data has 8 variables in a matrix [8 × 1817]
- Target data has 1 variable resulting in a matrix [1 × 1817]
- Testing data is [8 × 780] matrix
- Validation data is a matrix [1 × 780] used to verify the results of the predictive model.

Discussion. The testing data set are two binary values 0|1, where 0 – pipe is nonfailure and 1- pipe failure, compare these values with the predicted model results as boxplots shown in Fig. 2, in which the graph is 0, 1 respectively with the group of pipes nonfailure and pipes failure.

The results of three models LR (1), DT (2), ANN (3) all show that the algebraic values of the possibility F value range from 0 to 1 (shown in Fig. 2). Comparison of three graphs shows that the results of model (1) have the widest dispersion and model (3) predicts the closest value to reality. However, this is only a graphical overview, a study by using AUC standards is needed to be conducted to confirm which model is really good.

By using the plot command in the RStudio software, the ROC is defined in Fig. 3 The ROC is a curve representing the sensitivity and specificity of the result, a good model has the largest value of the area under the ROC (AUC). The graph shows that model (3) has the highest value however the difference between DT and ANN is not much.

The LR gives that the dependency coefficients of the F variable and the other variables and these coefficients are fixed, so when giving the testing data set, AUC is only 0.897. To overcome this disadvantage, the DT does not use the coefficient, the dependent variable is directly classified based on the information in the data set, so

Table 3 The statistical analysis of data set

No	Type	Variable	Value	Affect the pipe (F = 1)
1	Physical characteristics	Age (A)	3 months to 12 years	Pipe failure mainly appears on material type uPVC and PE with ages from 5 to 7 years old and from 11 to 13 years old
2		The number of junction (N0)	2 to 100 junctions	
3		Material (Mat)	uPVC, HDPE, PE	
4		Diameter (D)	25 mm	Brand pipes were only one kind of 25 mm diameter and failure on the short pipes (length < 8 m)
5		Length (L)	0.1 d´ên 12 m	
6	Characteristic of the outside environment	Road code (R)	Road code from 2 to 129	The observation pipe was the pipe connecting the main pipe to the customer meters. It means that they were working with low pressure and low external load, so the pipe failure due to the load was not evident in this data and the failure positions were on the different roads
7	Prior pipe failure	Prior	0 to 1	Prior pipe failure data shows that the number of failures the first time is higher than the second time and mainly concentrated on the pipes (RD) is 100 mm diameter
8	Characteristic of the inside environment	Diameter pipeline (RD)	50, 100, 150, 200 mm	

the result will be higher than the LR (AUC = 0.962). However, the results of the DT will depend on the characteristics of the training data, if it has similar characteristics with validation data, the results will be accurate, otherwise, the model efficiency will not be high. ANN is not only data classification but also adjusted functions by including bias parameters to correct the results closest to the target data, so the model efficiency is the highest (AUC = 0.973).

Table 4 Exemplary four data set

		Training data					1818	Testing data				779
		1	2	3	4	...	1818	1	2	3	...	779
1	**A**	2.02	7.84	11.47	5.58	...	6.81	2.01	0.04	11.47	...	5.58
2	**N0**	113	113	22	22	...	13	22	6	22	...	22
3	**Mat**	1	1	3	3	...	3	1	3	3	...	3
4	**D**	25	25	25	25	...	25	25	25	25	...	25
5	**L**	2.0	1.3	1.8	1.3	...	2.8	1.6	2.7	2.3	...	1.3
6	**R**	91	91	28	28	...	31	28	72	28	...	28
7	**Prior**	1	0	0	0	...	0	1	0	0	...	0
8	**RD**	200	200	100	100	...	100	100	100	100	...	100
		Target data						**Validation data**				
1	**F**	0	0	1	1	...	1	0	0	1	...	1

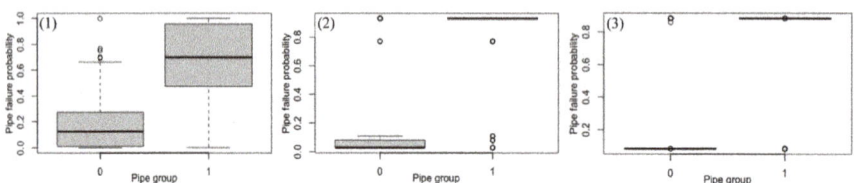

Fig. 2 Three results of water pipe failure prediction model
(1) Logistic Regression model; (2) Decision Tree model; (3) Artificial Neural Network model

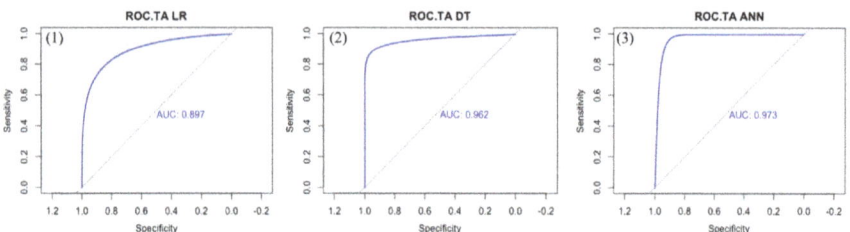

Fig. 3 ROC of pipe failure prediction model in Trung An water supply network
(1) Logistic Regression model; (2) Decision Tree model; (3) Artificial Neural Network model

3 Conclusion

The study has proposed three models to evaluate the possibility of pipe failure in the water supply network including LR, DT, and ANN. According to the results of comparing models by using the AUC index, it is all shows that the ANN model has the best performance. The result also shows that the ANN was used for failure prediction based on 8 variables which are the operating characteristics of the water supply network. Using future work will include the application of the modeling approach to more datasets.

Acknowledgements This research is funded by Vietnam National University Ho Chi Minh City (VNU-HCM) under grant number C2020-20-19

References

1. Minh LPT, Ha HP, Thi AV, Van HT, Thu T, Quang TN, Le Dinh H (2017) Application of logistic regression model to predict pipe failure on the water distribution system. J Water Resour Sci Technol 39:127–140
2. Minh LPT, Thi AV, Van HP, Ha, (2018) Proposing a predictive model of pipe failure on the water supply network. J Water Resour Environ Eng 60:3–9
3. Minh LPT, Dac BH, Quang TN (2020) Application of artificial neural networks to predict pipe failure in the water supply network. J Water Resour Environ Eng 72:93–100
4. Wengström T (1993) Comparative analysis of pipe break rates. Chalmers University Technology, pp 39
5. Wei-Yin Loh (2011) WIREs Data mining Knowl Discov. Wiley, pp 14–23
6. Al-barqawi H, Zayed T (2008) Infrastructure management: Integrated AHP/ANN model to evaluate municipal water mains performance. J Infrastructure Syst 305–318
7. Achim D, Ghotb F, McManus KJ (2007) Prediction of water pipe asset life using neural networks. J Infrastructure Syst 13(1):26–30
8. Kelly HZ, Aiyi L, Andriy IB, Licila O-M, Howard ER (2012) Statistical evaluation of diagnostic performance topics in ROC analysis. Chapman& Hall/CRC Biostatistics Series, pp 3–12
9. Fawcett T (2006) An introduction to ROC analysis. Pattern Recogn Lett 27(8):861–874
10. Ihaka R, Gentleman R (1996) R: A language for data analysis and graphics. J Comput Graph Stat 5(3):299–314

A Method for Calculating Unsteady Seepage at Riverbank

Giang Nguyen Mong, Hong Tran Thi My, Hoa Nguyen Thi Thanh, and Giang Le Song

Abstract The paper presents a method to calculate the unstable seepage at riverbank. The Laplace's equation in saturated zone is solved by the finite element method. Time derivative terms are discretized using implicit finite difference schema. The method considers three special factors of seepage including the change in time of water level of river; the variation in time of the phreatic line; and the variation of discharge face on river slopes. In response to the changes in time of the phreatic line, the computation mesh is modified at each step. The computation grid modification is performed using transformation. Although the implicit finite difference schema is used for discretization of time derivative terms, it is not required to know old pressure head at the nodes in new location. The method has been verified through a number of problems performed by other authors, including steady and unsteady seepage through rectangular earth dams, seepage flows through dams with a periodic boundary condition. The calculation results show that the method has a good accuracy.

Keywords Unsteady seepage · Riverbank · Finite element method

1 Introduction

Seepage is a problem applied in many engineering disciplines such as irrigation, geology, environment… One of these problems is seepage on steep banks like riverbanks, slopes of earth dams… for erosion studies. The presence of phreatic line and seepage surface on the slope and their variation over time make the calculation particularly difficult.

To calculate seepage flow, one can use the analytic method [1, 3, 8, 11, 12]. However, for complex problems, the calculation can only be done by numerical

G. N. Mong · H. T. T. My · H. N. T. Thanh · G. Le Song (✉)
Faculty of Civil Engineering, University of Technology (HCMUT), Vietnam National University HCMC, 268 Ly Thuong Kiet Street, District 10, Ho Chi Minh City, Vietnam
e-mail: lsgiang@hcmut.edu.vn

H. N. T. Thanh
Ho Chi Minh University of Natural Resources and Environment, Ho Chi Minh City, Vietnam

J. N. Reddy et al. (eds.), *ICSCEA 2021*, Lecture Notes in Civil Engineering 268,
https://doi.org/10.1007/978-981-19-3303-5_92

methods. The finite element [4–6, 9] and finite difference [7, 10] are the most widely used methods. Among these, the methods were studied by Dou et al. [4], Fu et al. [6], and GEO-SLOPE International, Ltd. [13] is based on the finite element method that can be applied to calculate seepage flows at slope banks. Using the boundary element method Bazyar and Talebi [2] have also developed a method capable of similar ability.

It can be seen that all the above authors solved the governing equation in both saturated and unsaturated regions. In this way, the accuracy of the phreatic surface is highly dependent on the mesh resolution. This can lead to a large number of elements and make it difficult to compute. This paper presents another method capable of calculating seepage flow at the slope but solving the governing equation only in the saturated region. Eliminate the unsaturated region, not only the calculated domain is reduced, but also the difficulties related to the nonlinearity of the governing equation in this region no longer exist. These will save computation time. The method has been compared with ones of other authors and gave similar results.

2 Method

Figure 1 shows a schematic diagram of the 2-dimensional porous media. The saturated aquifer is the Ω region enclosed in the closed perimeter ABCDEF. EF is the phreatic line and DE is the leakage surface. In region Ω, the seepage is described by equation [16]:

$$\frac{\partial}{\partial x}\left(k\frac{\partial H}{\partial x}\right) + \frac{\partial}{\partial z}\left(k\frac{\partial H}{\partial z}\right) = 0 \tag{1}$$

With boundary conditions:

$$\text{ABC:} \quad u_n = -k\left[\frac{\partial H}{\partial x}n_x + \frac{\partial H}{\partial z}n_z\right] = 0 \tag{2}$$

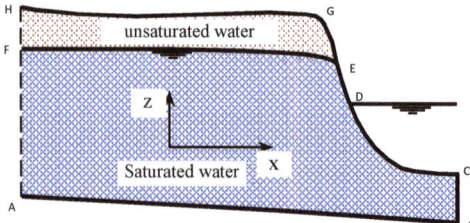

Fig. 1 Diagram of seepage flow

$$\text{AF:} \quad H = f_1(t) \tag{3}$$

$$\text{CD:} \quad H = f_2(t) \tag{4}$$

$$\text{DE:} \quad H = z \tag{5}$$

$$\text{EF:} \quad H = \eta \tag{6a}$$

$$S_y \frac{\partial \eta}{\partial t} - w + u \frac{\partial \eta}{\partial x} = R_g \tag{6b}$$

In these equations: H – the pore water pressure head ($H = z + p/\gamma$); k - the coefficient of permeability; η - the elevation of the phreatic face; u, w – the two component of seepage velocity ($u_i = -k\partial H/\partial x_i$); R_g – the replenish groundwater (above saturated surface); S_y – the specific yield.

Equation (1) is solved using the finite element method. The solution of the pressure head H will be found in the form of an approximation function:

$$\hat{H} = \sum_{i=1}^{m} \phi_i N_i = \mathbf{N}^T.\mathbf{H} \tag{7}$$

In these function $\mathbf{N}^T = [N_1, N_2, ..., N_m]$ - vector of interpolation function; $\mathbf{H}^T = [H_1, H_2, ..., H_m]$ - vector of pressure head at the nodes; m - the number of nodes. By substituting (7) into (1) and using Galerkin formulation [16] the equation for H_i is obtained:

$$\int_{\Omega} \left[\frac{\partial}{\partial x} \left(k \frac{\partial \hat{H}}{\partial x} \right) + \frac{\partial}{\partial z} \left(k \frac{\partial \hat{H}}{\partial z} \right) \right] N_i d\Omega = 0 \tag{8}$$

Equation (8) is then transformed to the weak form using the Gauss-Ostrogradsky formula:

$$\int_\Omega k \left[\frac{\partial \hat{H}}{\partial x} \frac{\partial N_i}{\partial x} + \frac{\partial \hat{H}}{\partial z} \frac{\partial N_i}{\partial z} \right] d\Omega - \oint_\Gamma k \frac{\partial \hat{H}}{\partial n} N_i d\Gamma = 0 \qquad (9)$$

In fact, because pressure head is known on the boundaries AF, CD and DE (except for point E), and $\partial H / \partial n = 0$ on the boundary ABC, the second integral of (9) is calculated on the boundary EF and part of DE near point E only. Using (6b), (9) obtain new form:

$$\int_\Omega k \left[\frac{\partial \hat{H}}{\partial x} \frac{\partial N_i}{\partial x} + \frac{\partial \hat{H}}{\partial z} \frac{\partial N_i}{\partial z} \right] d\Omega - \int_{EF} \left(S_y \frac{\partial \hat{H}}{\partial t} - R_g \right) N_i dx$$

$$- \int_{DE} k \left(\frac{\partial \hat{H}}{\partial x} n_x + \frac{\partial \hat{H}}{\partial z} n_z \right) N_i d\Gamma = 0 \qquad (10)$$

Substituting (7) into (10) and let i have value from 1 to m, (10) gives the system of m equations with m unkhown values of H at m nodes:

$$\left[\int_\Omega \left(\frac{\partial \mathbf{N}}{\partial x} \frac{\partial \mathbf{N}^T}{\partial x} + \frac{\partial \mathbf{N}}{\partial z} \frac{\partial \mathbf{N}^T}{\partial z} \right) d\Omega - \int_{DE} k\mathbf{N} \left(\frac{\partial \mathbf{N}^T}{\partial x} n_x + \frac{\partial \mathbf{N}^T}{\partial z} n_z \right) d\Gamma \right]$$

$$\mathbf{H} - \int_{EF} S_y \mathbf{N}.\mathbf{N}^T dx \frac{\partial \mathbf{H}}{\partial t} = - \int_{EF} R_g \mathbf{N} dx \qquad (11)$$

The Ω region is divided into Ne triangular elements (the elements with area Ω_e) in which the vertex of the triangles coinside with the computational nodes:

$$\Omega = \sum_{Ne} \Omega_e \qquad (12)$$

Choosing the interpolation function on each element as a linear function, performing integrals, (11) leads to the equation:

$$-\mathbf{M}.\frac{\partial \mathbf{H}}{\partial t} + \mathbf{K}.\mathbf{H} = -\mathbf{P} \qquad (13)$$

With matrices M, K and vector P generated from integrals. M and K are sparse matrices with many zero elements. So these matrices are stored in band matrices and the nodes are optimally numbered to reduce the band size. (13) will be discretized in time:

$$\left(\mathbf{K} - \frac{1}{\Delta t}\mathbf{M}\right).\mathbf{H}^{n+1} = -\left(\mathbf{P} + \frac{1}{\Delta t}\mathbf{M}.\mathbf{H}^{n}\right) \tag{14}$$

Or:

$$\mathbf{A}.\mathbf{H}^{n+1} = \mathbf{B} \tag{15}$$

where:

$$\mathbf{A} = \mathbf{K} - \frac{1}{\Delta t}\mathbf{M} \text{ and } \mathbf{B} = -\left(\mathbf{P} + \frac{1}{\Delta t}\mathbf{M}.\mathbf{H}^{n}\right) \tag{16}$$

By solving (15), pressure head $\mathbf{H}$ at the time step $n + 1$ is obtained. From (6a), the phreatic line is determined also. It is noted that due to the change of phreatic line, the computational mesh may need to be re-created after each time step. But (14) uses only old value of $\mathbf{H}$ on phreatic line, so there is not the need to determine the old value of $\mathbf{H}$ at new position of nodes.

3 Test Calculations

3.1 Steady Seepage Through a Rectangular Earth Dam

The first test was performed for a steady seepage through a rectangular earth dam. This test is also used by many authors to evaluate their methods such as Dou et al. [4] or Kazemzadeh-Parsi and Daneshmand [14]. Figure 3 shows a schematic diagram of the problem. The dam is 5 m wide and is placed on an impervious surface. AB and EF are the infiltration head boundaries with values of 10 m and 2 m, respectively. On CE there can be a seepage face and D is the free drainage point. The phreatic line BD will be determined in the calculation (Fig. 2).

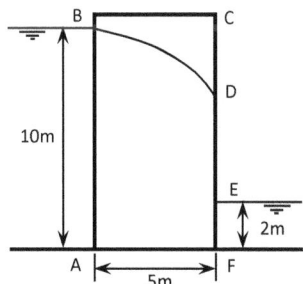

Fig. 2 Diagram of water seepage through a rectangular earth dam

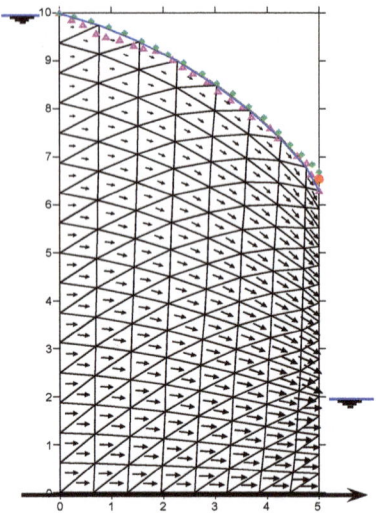

Fig. 3 Grid, velocity field and phreatic surface at steady state
(triangular symbol: calculation by Kazemzadeh-Parsi and Daneshmand [14]; rhombus symbol:
calculation of Dou et al. [4]; circular symbol: solution of Lee and Leap [15])

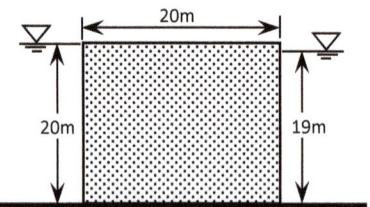

Fig. 4 Diagram of seepage calculation through rectangular earthen dam

The dam is divided into 256 elements with 153 nodes. Grid after the convergence to the stable solution is shown in Fig. 4. The average area of the elements is 0.17m^2. For comparison, the figure also introduces the saturation results of Dou et al. [4], Kazemzadeh-Parsi and Daneshmand [14], and Lee and Leap [15]. The calculated phreatic line quite coincides with the reference results.

Calculation results also show that the seepage flow through the dam is $9.61\ \text{m}^3/\text{day}$, which shows quite consistent with the calculated results of Dou et al. [4] which is $9.59\ \text{m}^3/\text{day}$ and with analytical solution which is $9.60\ \text{m}^3/\text{day}$.

3.2 Transient Seepage Through a Rectangular Earth Dam

The diagram of the problem is presented in Fig. 4. The permeability coefficient of dam material is k = 1.0 m/s. Initially, the water levels upstream and downstream

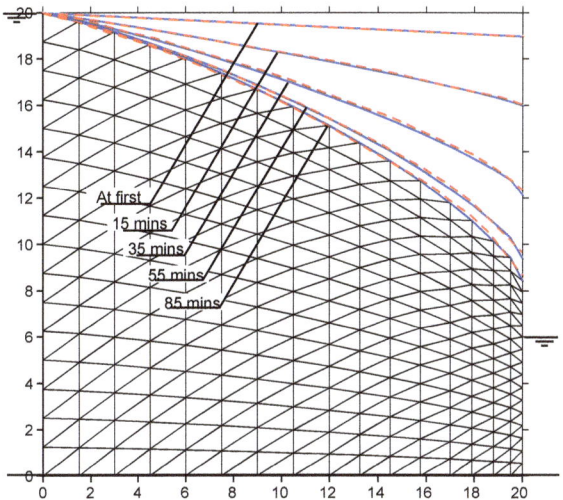

Fig. 5 Grid at 85 min and phreatic lines at the times
(Continuous line: present calculation; Broken line: calculation of Bazyar and Talebi [2])

of the dam were 20 m and 19 m respectively. The groundwater level in the dam decreased linearly from the upstream surface to the downstream surface. By the time, the downstream water level gradually decreased from 19 to 6 m in 65 min and then remained constant. This problem was also calculated by Bazyar and Talebi [2] by the boundary element method.

The dam is divided into 512 elements with an average area of $0.61 m^2$. Figure 5 shows the grid at the 85th minute. On the figure the phreatic lines at some time are also shown in comparing with results of Bazyar and Talebi [2]. The calculation results are also very consistent with the results calculated by the author of the discussion.

3.3 Seepage in Unconfined Aquifer with a Periodic Boundary Condition

The diagram of the problem is presented in Fig. 6. The dam is 48 m long, waterproof at -18.96 m deep. The permeability coefficient of dam material is k = $1.67.10^{-4}$ m/s and the specific yield is S_y = 0.025. Initially, the water level on both sides of the dam is at 0 m. Then while the water level at the downstream surface is maintained, the upstream water level fluctuates in a sinusoidal pattern:

$H(0, t) = A_0 + B_0 \sin \omega t$ (With $A_0 = 0$, $B_0 = 4$ m and with T = 24 h)

This problem has been solved by Fu and Jin [6] by both the analytic and finite element methods. Figure 7 is shown as the result of oscillation of the saturated surface

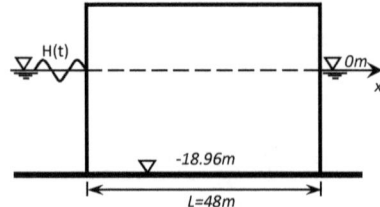

Fig. 6 Calculation diagram

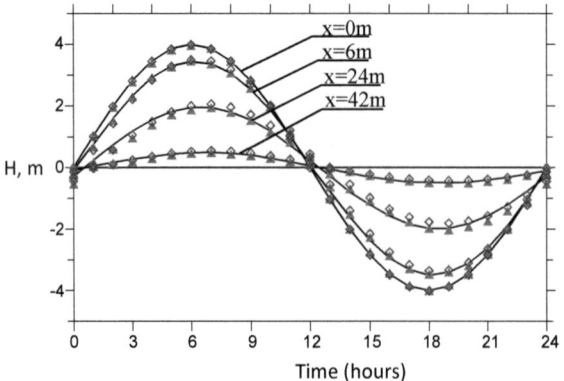

Fig. 7 Oscillation of the phreatic surface at 4 positions (Continuous line: Fu and Jin's analytic solution [6], rhombus symbol: present calculation, triangular symbol: Fu and Jin's calculation [6])

at 4 positions, including comparing the solution of Fu and Jin [6]. Note that while Fu and Jin solve the seepage equation for both the saturated and unsaturated zones, the method in this paper only solves the seepage equation in the saturated zone. Therefore, the pattern of seepage in the unsaturated zone, as well as the interaction of this zone with the saturated zone in the two methods are different. However, the results calculated by the 2 methods are similar and quite consistent with the analytical solution.

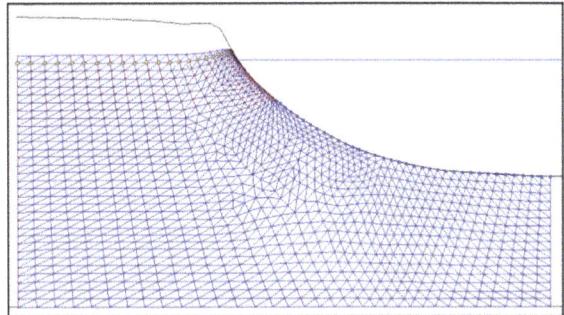

Fig. 8 Mesh of seepage in river bank

3.4 Seepage at the River Bank

The mesh of the seepage case at the riverbank is shown in Fig. 8. The river bed is 20 mdeep. The permeability coefficient of the bank soil k = 1.05 m/day and the specific yield is $S_y = 0.15$.

The river water level fluctuates according to the sinusoidal law with a period of 12-h and an amplitude of 1.5 m. The water level difference between the base and the top is 3.0 m, similar to the tidal fluctuation in rivers in the South of Vietnam. The time step is calculated as dt = 0.1 h. Although the program is stable when running with larger time steps (tried running with dt = 0.1 h) but to ensure accuracy 0.1 h time step was used.

Since this is a problem of testing the possibility of the method, the data are hypothetical. However, the parameters are roughly equivalent to the actual conditions in rivers in the South.

Figure 9 presents four instantaneous states of seepage at the riverbank corresponding to different water levels and states in one cycle of oscillation. The calculation results show that when the river water level drops, the groundwater at the river bank cannot fall in time, resulting in a large difference of water level in the river bank and in the river. This situation will weaken the stability and increase the possibility of riverbank erosion, as mensioned by Toan Duong Thi and Duc Do Minh [17].

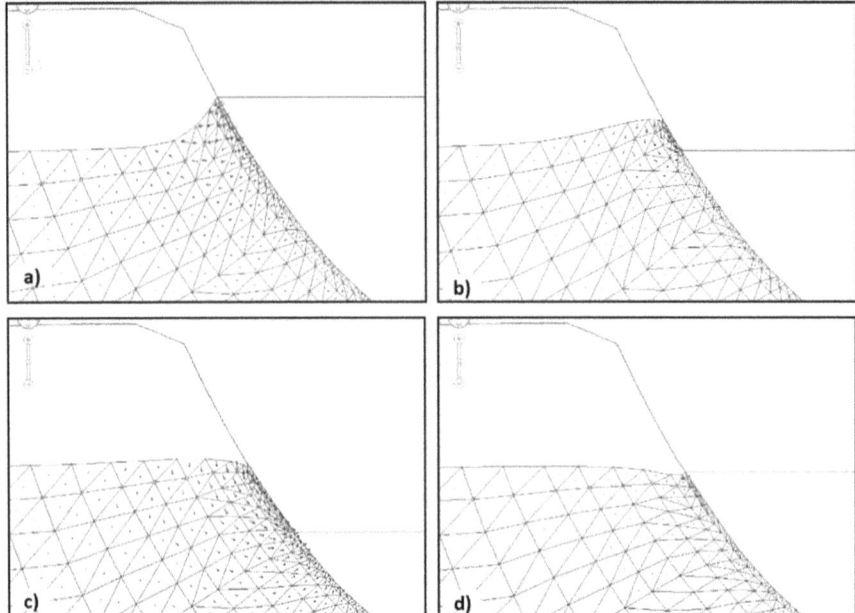

Fig. 9 Instant infiltration **a** When the river water level is highest; **b** When the river water level falls to the average water level; **c** When the water level falls to the lowest level; **d** When the river water level rises to the average water level

4   Conclusions

The study has presented the use of the finite element method to solve the seepage equation in the saturation zone together with unsteady boundary conditions at the water level boundary and at the phreatic surface. The method also considers the presence of a seepage surface on the slope. This makes it possible for the method to calculate seepage flow at a slope such as a riverbank or at an earth dam roof, where the flow structure is quite complex. The calculation results of the 3 test problems show that the method has quite a good accuracy. The calculation results for the test problem atc the river bank is also reasonable.

References

[1] Bansal RK (2017) Unsteady seepage flow over sloping beds in response to multiple localized recharge. Appl Water Sci 7:777–786
[2] Csoma. (2001) The analytic element method for groundWater flow modelling. Periodica Polytechnica Ser Civ Eng 39(1):1–22

[3] Dou Z, Wu J, Zhang H, Huang K (2017) The solution of unconfined water seepage problem in saturated-unsaturated soil using Bathe algorithm and Signorini condition. IOP Conf Ser Earth Environ Sci 69:012170

[4] Du EX, Sun JH (2012) Finite element analysis of unsteady seepage flow through the earth dam of Huangbizhuang reservoir. Adv Mater Res 594–597:1892–1896

[5] Fu J, Jin S (2009) A study on unsteady seepage flow through dam. J Hydrodyn 21(4):499–504

[6] Gupta CS, Bruch JC Jr, Comincioli V (1989) Three-dimensional unsteady seepage through an earth dam with accretion. Eng Comput 3:2–10

[7] Hong N, Xing L, Sheng-nan N, Zhang W (2015) Analytical study of unsteady nested groundwater flow systems. Math Probl Eng Article ID 284181

[8] Kalaidzidou-Paiku N, Karamouzis D, Moraitis D (1997) A finite element model for the unsteady groundwater flow over sloping beds. Water Resour Manage 69:69–81

9. Langevin CD, Hughes JD, Banta ER, Niswonger RG, Panday S, Provost AM (2017) Documentation for the MODFLOW 6 Groundwater Flow Model. Techniques and Methods 6-A55. https://doi.org/10.3133/tm6A55

[10] Nguyen VU, Raudkivi AJ (1983) Analytical solution for transient twodimensional unconfined groundwater flow. Hydrol Sci 28(2):209–219

[11] Xin Y, Zh, Zhou, Li M, Ch, Zhuang (2020) Analytical solutions for unsteady groundwater flow in an unconfined aquifer under complex boundary conditions. Water 12(2):75

[12] GEO-SLOPE International, Ltd. (2012). Seepage modeling with SEEP/W an engineering methodology. July 2012 Edition

[13] Kazemzadeh-Parsi MJ, Daneshmand F (2011) Unconfined seepage analysis in earth dams using smoothed fixed grid finite element method. Int J Numer Anal Meth Geomech 36(6):780–797

[14] Lee KK, Leap DI (1997) Simulation of a free-surface and seepage face using boundary-fitted coordinate system method. J Hydrol 196(1/4):297–309

[15] Zienkiewicz OC, Morgan K (1983) Finite element and approximation. Wiley 196(1/4):297–309

[16] Harr ME (1962) Groundwater and Seepage. McGraw-Hill

[17] Duong Thi T, Duc DM (2019) Riverbank stability assessment under river water level changes and hydraulic erosion. Water. 11:2598 https://doi.org/10.3390/w11122598

An Approach of Seepage Analysis Through Earth Dams Considering the Uncertainties of Soil Hydraulic Conductivity

Thi Tuyet Giang Vo and Vo Trong Nguyen

Abstract This study focuses on an approach for analyzing seepage mechanism through earth dams whose soil hydraulic conductivities are considered as random variables. The approach employs finite element analysis integrated into a popular software for seepage consideration and an easy-to-implement algorithm for the random field generation. Hence, a Monte Carlo method is used for the further analysis. Based on the analysis of the outcomes, there are three aspects that need to be warned when using this method. Firstly, the correct algorithm of the random number generation leads the relevant observed frequency distributions to normal-distribution shapes. Secondly, the refinement of mesh can lead to more accurate results in terms of numerical issues but that can change the scale of fluctuation of random field. Finally, a number of 500 samples for a Monte Carlo simulation is acceptable.

Keywords Seepage · Embankments · Finite-element modelling · Uncertainty

1 Introduction

Seepage analysis through earth-fill dams plays an important role when considering the stability of these works. As pointed out by ICOLD (1994) and [10], a high percentage of embankment dam failures were caused by the seepage mechanism. Empirical, analytical, and mathematical models for seepage analysis have been built and used for decades in designing the associated projects. But as discussed by [8], such approaches have made contributions to the literature but using deterministic parameters such as hydraulic conductivities needs cautions and they suggested using spatially random fields.

T. T. G. Vo (✉) · V. T. Nguyen
Faculty of Civil Engineering, Ho Chi Minh City University of Technology (HCMUT), 268 Ly Thuong Kiet Street, District 10, Ho Chi Minh City, Vietnam
e-mail: tuyetgiang.vo@hcmut.edu.vn

Vietnam National University Ho Chi Minh City, Linh Trung Ward, Thu Duc District, Ho Chi Minh City, Vietnam

© The Author(s), under exclusive license to Springer Nature Singapore Pte Ltd. 2023   1019
J. N. Reddy et al. (eds.), *ICSCEA 2021*, Lecture Notes in Civil Engineering 268,
https://doi.org/10.1007/978-981-19-3303-5_93

Using probabilistic treatment in seepage analysis through embankment dams is a trending study in recent decades. For examples, probabilistic techniques have been used including the soil consolidation analysis [3], the bearing capacity assessment of footings [14], the determination of failure probability of slopes (Genevois and Romeo 2003) and the seepage analysis (Calamak and Yanmaz 2016, 2018). In these studies, the uncertainty of geotechnical properties such as hydraulic conductivity, undrained shear strength, and the others were considered. Some further studies in this problem can be listed [1, 2, 8, 11, 15, 18–21, 24].

Recently, Calamak and Yanmaz (2016, 2018) analysed the seepage within embankment dams. They employed finite element method integrated into SEEP/W incorporating the uncertainty of materials of the dams and using a Monte Carlo simulation (MCS). This approach has advantages such as employing the popular commercial software for seepage analysis and using a technique for generating random numbers that is simple compared to other techniques. However, there are some aspects that need to be discussed when using this approach and this is the purpose of this paper. More specifically, the technique for checking the acceptability of the algorithm of random number generation is discussed. Next, although finer meshes lead to more accurate in the numerical manner, this meshing can result in the change of basics of random fields. Finally, the minimum number of samples for a Monte Carlo simulation is also analysed.

2 Theoretical Description

In this study, seepage analysis is in steady state, so the governing equation is as follows (Geo-Slope International Ltd 2012)

$$\frac{\partial}{\partial x}\left(k_x \frac{\partial H}{\partial x}\right) + \frac{\partial}{\partial y}\left(k_y \frac{\partial H}{\partial y}\right) + Q = 0 \tag{1}$$

where, H is the total head; k_x and k_y are the hydraulic conductivities in x and y directions (since the problem is considered in the isotropic environment so $k_x = k_y$); Q is the applied boundary flux.

Finite element method is applied to solve the problem through SEEP/W. The meshes within SEEP/W are provided with quadrilaterals only or triangles only or the combination of them. Besides, when calculating the numerical integrations, the software offers 4 and 9 Gauss points for the quadrilaterals and 1 and 3 for the triangles.

Regarding the hydraulic conductivities, since the analysis generates free surfaces within dams so there are the unsaturated zones existing. There are several models for estimating these conductivities including unsaturation. This study uses the model suggested by [22] due to the popularity of this model. Then

$$K(h) = K_s K_r(h) \tag{2}$$

where, K_s is the hydraulic conductivity at saturation, $K_r(h)$ is the normalized form of unsaturated hydraulic conductivity and h is the pressure head.

Within the saturated zones, h is greater than 0 so that $K_r(h) = 1$ and within the unsaturated zones, $K_r(h)$ is smaller than 1. $K_r(h)$ can be calculated using the parameters α and n. Treating the uncertainty of soils, besides K_s, two parameters α and n be spatially variable. However, since the layers of unsaturation above the unsaturated zones are comparatively small so in this study α and n are considered as deterministic ones. Generating random variables is applied using their probability density functions (PDF) defined with the means and the coefficients of variation (COV). Hydraulic conductivity of soils also follows a lognormal distribution, which is shown in several studies (see Calamak and Yanmaz 2016, 2018). Therefore, generating this parameter can be implemented using the following equations (K hereafter denotes the saturated hydraulic conductivity)

$$\sigma_{\ln K}^2 = \ln \left(1 + \frac{\sigma_K^2}{\mu_K^2} \right) \tag{3}$$

$$\mu_{\ln K} = \ln \mu_K - \frac{1}{2}\sigma_{\ln K}^2 \tag{4}$$

$$K = \exp(\mu_{\ln K} + \sigma_{\ln K} r) \tag{5}$$

where, r is the random number which can be generated using the following equation [4]

$$r = (-2 \ln u_1)^{1/2} \sin(2\pi u_2) \tag{6}$$

or

$$r = (-2 \ln u_1)^{1/2} \cos(2\pi u_2) \tag{7}$$

where, u_1 and u_2 are independent random variables from the same uniform density function on the interval of (0,1). The generated hydraulic conductivities then are assigned to respective Gauss points. The implementation of these equations into SEEP/W is conducted by coding in C# and using the provided tools to be integrated into the Add-In directory of SEEP/W. The outcomes are collected using Monte Carlo simulation (MCS) meaning that a number of repetitive simulations are conducted, and the statistics techniques are employed to get inference of the data.

3 The Further Discussion Using this Approach

This section is to discuss further aspects when using this approach. This study employs a dam whose geometry is displayed as in Fig. 1. The geometry employed is

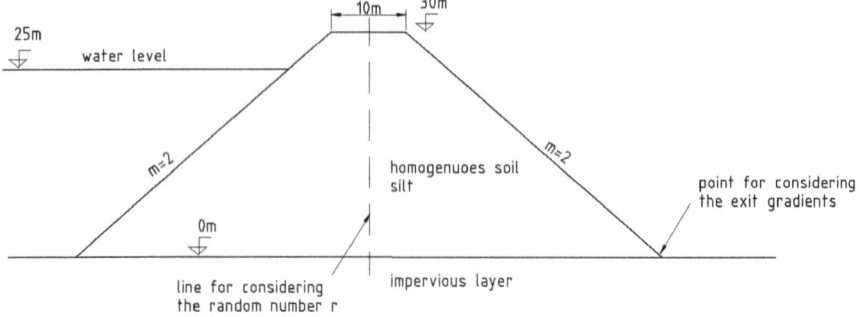

Fig. 1 The geometry of dam used for the analysis

Table 1 Parameters for the analysis

Mean of hydraulic conductivity	1×10^{-6} (m/s)
COV of hydraulic conductivity	2
α (van Genuchten model)	0,02
n (van Genuchten model)	1,40

simple, excluding items in dams such as filters and drains. This study uses the mesh of triangles and 3 Gauss-point elements for the analysis. The average element size is 2,5 m that is generated automatically by SEEP/W, based on the optimization of the given geometry of dam. The outcomes for the further analysis are based on the total flow rates through the dam and the exit gradients at the toe of the downstream face of the dam (Fig. 1). These are two of the important outcomes resulted from any seepage analysis through earth dams. The soil parameters are shown in the Table 1. These values are based on a study of [7].

3.1 The Generation of Random Numbers

The random number r plays an important role in this approach as they represent the random fields of hydraulic conductivity within soil bodies. However, since using the existing software (i.e., SEEP/W), the users must write a code to generate the random numbers which is compiled into the Add-In directory of SEEP/W. That should be cautious since the distribution of these numbers should be in a normal-distribution form. Regarding this matter, in this study, based on the coding written, the authors can extract the random numbers and show them in Fig. 2 based on a realization. The distribution has obeyed the normal form, ensuring the acceptability of the algorithm. The data based on the default mesh size of 2,5 m for element sizes. Hence, there are 2.289 Gauss points, corresponding to 2.289 random numbers.

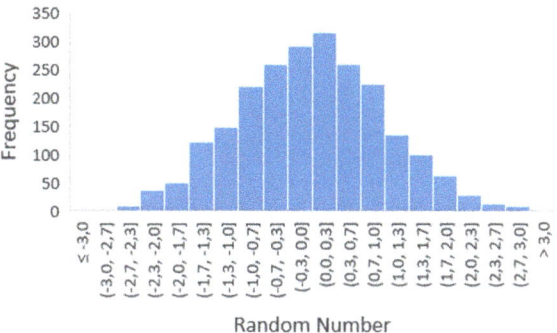

Fig. 2 A distribution of the random number r of a realization

In case the mesh is not fine enough the mesh should be revised to get finer one so that the number of random numbers are large enough to get the acceptable distribution. This technique is just used for checking the accuracy of the random number generation algorithm since the finer mesh can lead to the change of input of the scale of fluctuation of random field as discussed below.

3.2 The Mesh refinement

As pointed out by [16], in finite element analysis, the accuracy of the solution is judged based on the refinement of mesh. The analysis can use h-refinement or p-refinement for the issue. It is clear that the finer meshes, the more accurate the outcomes. However, using this approach, the finer meshes lead to the issue of scale of fluctuation of random variables. Figure 3 shows the distribution of the random numbers along the prescribed line (Fig. 1). The figure shows that the finer meshes lead to the more rapid changes of the random numbers so that effecting the scale of fluctuation.

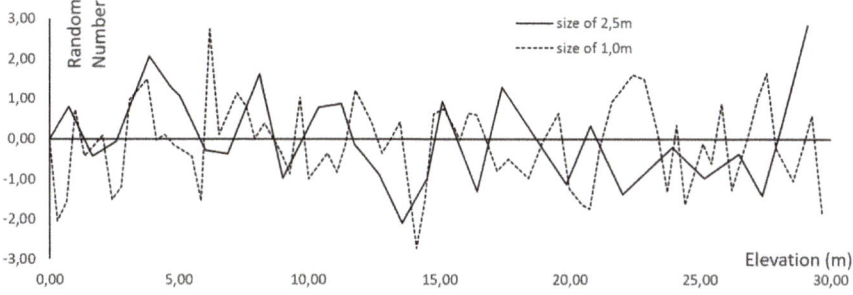

Fig. 3 Random numbers of two different sizes of mesh of a realization

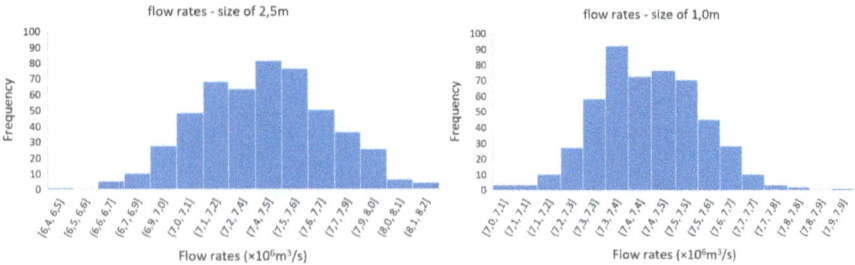

Fig. 4 The distributions of flow rates of two different sizes of mesh of a realization

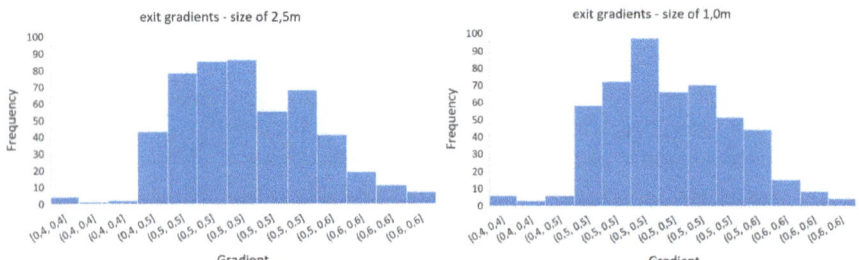

Fig. 5 The distributions of exit gradients of two different sizes of mesh of a realization

Table 2 Statistical parameters of two different sizes of mesh

Mesh sizes	Mean (flow rates) ($\times 10^6 \text{m}^3/\text{s}$)	SD (flow rates) ($\times 10^6 \text{m}^3/\text{s}$)	COV (flow rates)	Mean (exit gradients)	SD (exit gradients)	COV (exit gradients)
2,5 m	7,401	0,300	0,041	0,509	0,038	0,074
1,0 m	7,420	0,126	0,017	0,505	0,036	0,072

Furthermore, the study conducted two sets of MCS regarding the element size of 2,5 m and the finer one of 1,0 m. The results are shown in Fig. 4, 5 and Table 2. While the results related to exit gradients show insignificantly different in terms of the means and the COVs, the ones of flow rates show the significant difference in COV. Therefore, the refinement of mesh leads to the more accurate in terms of numerical issues but changing the outcomes in terms of uncertainty. This is a matter that needs to be warned in using this approach.

3.3 Number of Samples

Using this approach may lead to the problem of time consuming since the number of samples in Monte Carlo method should be large enough to get the accurate statistic

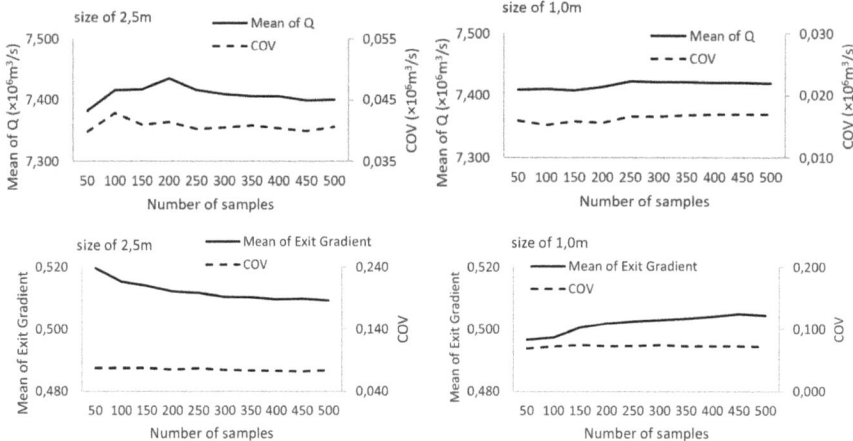

Fig. 6 Number of samples

results. For example, based on Wikipedia (2021), number of samples to approximate the value of π in a simulation is around 30.000. It is a massive task doing with SEEP/W.

This study conducts the relation between the numbers of simulations and the means and COVs. The results show in Fig. 6, stating that the trends change rapidly in the range less than 400. Beyond this range, the values become relatively stable. Based on the above analysis, the number of 500 can be used. However, it is recommended to use the larger one if possible.

4 Conclusions

The analysis of seepage through earth dams is an important task when considering the failures of dams caused by the seepage flows. However, analyses using deterministic parameters related to seepage, such as hydraulic conductivity may lead to inaccuracy in the outcomes. This study focuses on an approach used recently by some researchers. This approach has advantages compared to others such as employing finite element analysis integrated into SEEP/W, a popular software for seepage analysis, and using relatively easy-to-implement algorithm for generating random numbers within the interested areas. However, based on the seepage analysis through a homogenous earth dam, there are some points that need to be warned as follows:

- Based on the outcome of a realization, the algorithm of random number generation applied within this study can be reliable since the frequency distribution fit well the normal-distribution shape.

- Based on the outcomes of the 2-m-element-size and 1-m-element-size realizations, that needs to be concluded that the refinement of mesh can leads to more accurate results in terms of numerical issues but that can change the scale of fluctuation of random field.
- The number of samples for a MCS must be at least 500 to get the acceptable results.

Acknowledgements This research is funded by Vietnam National University Ho Chi Minh City (VNU-HCM) under grant number C2020-20-21.

References

1. Ahmed AA (2009) Stochastic analysis of free surface flow through earth dams. Comput Geotech 36(7):1186–1190
2. Ahmed AA (2013) Stochastic analysis of seepage under hydraulic structures resting on anisotropic heterogeneous soils. J Geotech Geoenviron Eng 139(6):1001–1004
3. Bari M, Shahin M, Nikraz H (2013) Probabilistic analysis of soil consolidation via prefabricated vertical drains. Int J Geomech https://doi.org/10.1061/(ASCE)GM.1943-5622.0000244, 877–881
4. Box GEP, Muller ME (1958) A note on the generation of random normal deviates. Ann Math Stat 29(2):610–611
5. Calamak M, Yanmaz AM (2018) Assessment of core-filter configuration performance of rock-fill dams under uncertainties. Int J Geomech 18(4):1532–3641
6. Calamak M, Yanmaz AM (2016) Uncertainty quantification of transient unsaturated seepage through embankment dams. Int J Geomech https://doi.org/10.1061/(ASCE)GM.1943-5622.0000823
7. Carsel RF, Parrish RS (1988) Developing joint probability distributions of soil water retention charasteristics. Water Resour Res 24(5):755–769
8. Fenton G, Griffiths D (1996) Statistics of free surface flow through stochastic earth dam. J Geotech Eng https://doi.org/10.1061/(ASCE)0733-9410(1996)122:6(427), 427–436
9. Fenton G, Vanmarcke EH (1990) Simulation of random fields via local average subdivision. J Eng Mech 116(8):1733–1749
10. Foster M, Fell R, Spannagle M (2000) The statistics of embankment dam failures and accidents. Can Geotech J 37(5):1000–1024
11. Freeze RA (1975) A stochastic-conceptual analysis of one-dimensional groundwater flow in nonuniform homogeneous media. Water Resour Res 11(5):725–741
12. Genevois R, Romeo R (2003) Probability of failure occurrence and recurrence in rock slopes stability analysis. Int J Geomech 10.1061 /(ASCE)1532–3641(2003)3:1(34), 34–42
13. Geo-Slope International Ltd. (2012) Seepage modelling with SEEP/W. Calgary, Canada
14. Griffiths D, Fenton G, Manoharan N (2006) Undrained bearing capacity of two-strip footings on spatially random soil." Int J Geomech https://doi.org/10.1061/(ASCE)1532-3641(2006)6:6(421) 421–427
15. Gutjahr AL, Gelhar LW (1981) Stochastic models of subsurface flow: Infinite versus finite domains and stationarity. Water Resour Res 17(2):337–350
16. Hutton DV (2004) Fundamentals of finite element analysis, 1st edn. McGraw-Hill, New York
17. ICOLD (1994) Embankment dams granular filters and drains, Bulletin 95, CIGB ICOLD
18. Le TMH, Gallipoli D, Sanchez M, Wheeler SJ (2012) Stochastic analysis of unsaturated seepage through randomly heterogeneous earth embankments. Int J Numer Anal Methods Geomech 36(8):1056–1076

19. Lin GF, Chen CM (2004) Stochastic analysis of spatial variability in unconfined groundwater flow. Stochastic Environ Res Risk Assess 18(2):100–108
20. Mantoglou A, Gelhar LW (1987) Stochastic modeling of largescale transient unsaturated flow systems. Water Resour Res 23(1):37–46
21. Tartakovsky DM (1999) Stochastic modeling of heterogeneous phreatic aquifers. Water Resour Res 35(12):3941–3945
22. van Genuchten MT (1980) A closed-form equation for predicting the hydraulic conductivity of unsaturated soils. Soil Sci Soc Am J 44:892–898
23. Wikipedia: Monte Carlo Method (2021) https://en.wikipedia.org/wiki/Monte_Carlo_method
24. Zhang D (1999) Nonstationary stochastic analysis of transient unsaturated flow in randomly heterogeneous media. Water Resour Res 35(4):1127–1141

Estimation of Sustainable Aquifer Yields in the Saigon River Basin

Tran Thanh Long, Sucharit Koontanakulvong, and Phu Nhat Truyen

Abstract Since the 1990s, under the pressure of socio-economic growth in Ho Chi Minh City and nearby provinces, groundwater becomes an essential resource for domestic and industrial purposes. At the same time, the surface water supply utilities still struggle to satisfy the total water demand in the developing region. The conjunctive use of surface and groundwater is an essential strategy of water supply management that has to be considered coupling with aquifer yield and surface water capability within a basin. The study aims to determine optimal intensity pumping and estimate aquifers yield within sustainable drawdown criteria. The optimal pumping intensity was analyzed from maximum existing pumping per square kilometer in four aquifers with no negative impacts on Saigon River Basin's groundwater system. The study estimated the sustainable aquifers yield through increasing pumpage until the lowest drawdown of aquifers meets the drawdown criteria. Here, the drawdown criteria considered the groundwater pumping laws of Vietnam and the hydraulic gradient at the salinity interface. Regarding drawdown criteria and hydraulic gradient at salinity interface criteria, the drawdown constraints are –20 m. MSL for aquifer 2 and –30 m. MSL for aquifer 3 and aquifer 4. The simulation inputs utilized the historical 20-year climate data. According to the pumping scheme in the existing area, the optimal intensity pumping for aquifer 2, aquifer 3, and aquifer four are found to be 2000 m^3/day/km^2, 3500 m^3/day/km^2, and 4000 m^3/day/km^2, respectively. Based on the optimal pumping intensity, the sustainable aquifers yield in Saigon River

T. T. Long (✉)
Faculty of Civil Engineering, Hochiminh City University of Technology, 268 Ly Thuong Kiet Street, District 10, Ho Chi Minh City, Vietnam
e-mail: ttlong@hcmut.edu.vn

T. T. Long · P. N. Truyen
Vietnam National University Ho Chi Minh City, Linh Trung Ward, Thu Duc District, Ho Chi Minh City, Vietnam

S. Koontanakulvong
Department of Water Resources Engineering, Faculty of Engineering, Chulalongkorn University, 254 Phayathai Road, Pathumwan, Bangkok 10330, Thailand

P. N. Truyen
Faculty of Geology and Petroleum Engineering, Hochiminh City University of Technology, 268 Ly Thuong Kiet Street, District 10, Ho Chi Minh City, Vietnam

Basin are estimated to be at the rate of 1.9 MCM/day, which includes 57,699 m^3/day from aquifer 1, 465,233 m^3/day from aquifer 2, 455,151 m^3/day from aquifer 3, and 925,836 m^3/day from aquifer 4.

Keywords Optimal pumping intensity · Sustainable aquifers yield · Saigon River Basin

1 Introduction

Under the scientific and technological revolution, the urban growth rates acceler-ated rapidly in developing countries where water resource scarcity is facing [1]. The conjunctive use of surface and groundwater is an essential strategy of water supply management that couple aquifer yield and surface water capability within a basin to meet the increasing water demand [2, –4]. Under the pressure of socio-economic growth in Ho Chi Minh City and nearby provinces, groundwater becomes an essential resource for domestic and industrial purposes. At the same time, the surface water supply utilities still struggle to satisfy the total water demand in the new urbanized area. However, since groundwater is free resources and easy to exploit, the ground-water has been violated excessively, leading to a severe decline in groundwater levels in some areas [5, 6]. To preserve groundwater resources in the Saigon River, there are numerous publication investigated the potential yield of the aquifers in the Saigon River basin, such as Chân and Kỳ [6], Khai [7], Tuan and Koontanakulvong [8], DWPRIS [9]. However, most studies focused on the potential groundwater resources and aquifers sustainable yield and without optimal pumping intensity in the Saigon river Basin. Hence, it is difficult for decision-making on the pumping control in the study area.

More intensive groundwater resources planning for sustainable socio-economic development, the study proposed a method to investigate optimum pumping intensity and aquifers sustainable yield intensively within sustainable drawdown criteria under historical climate. The optimal pumping intensity was analyzed from maximum existing pumping per square kilometer in four aquifers with no negative impact on the groundwater system in Saigon River Basin. The sustainable aquifers yield was estimated by increasing pumpage until the lowest drawdown of aquifers meets the drawdown criteria. The drawdown criteria considered the groundwater pumping laws of Vietnam and the hydraulic gradient at the salinity interface.

2 Study Area

The study area stretches from latitude 10.320 E to 11.201 E and from longitude 106.215 N to 107.024 N with an area of 6,640 km^2. It covers all Ho Chi Minh City areas and some districts of Dong Nai, Binh Phuoc, Binh Duong, Long An, and Tay

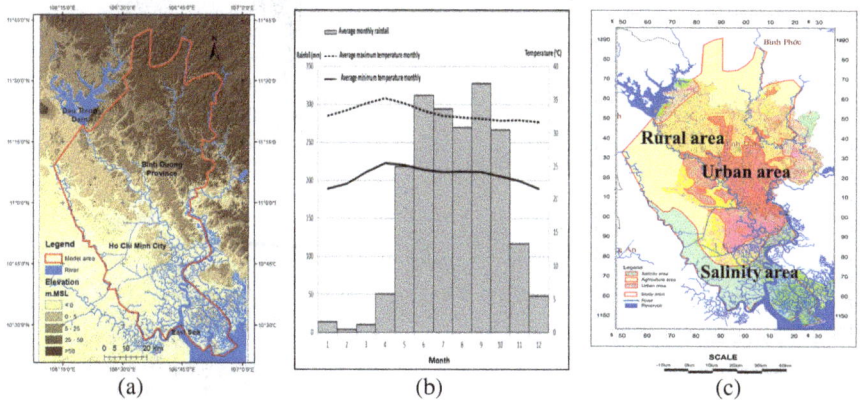

Fig. 1 Study area **a** topography; **b** rainfall and temperature); **c** Land use (red: urban area, yellow: rural area, green: salinity area)]

Ninh Province (Fig. 1). The area has a tropical climate, specifically a tropical wet and dry climate, with an average humidity of 75%. The climate year divides into two distinct seasons. Mean annual rainfall is at 1,612 mm, and mean annual temperature is at 27 °C. Terraced plain mainly characterizes the topography of the area with elevation varies from 0 to 70 MSL. Regarding the hydrogeology map, 3 aquifers interacted with the river system and distributing from top to bottom respectively as follows: upper-Pleistocene (qp$_3$), Upper Middle Pleistocene (qp$_{2-3}$), Lower Pleistocene (qp$_1$), and 1 aquifer disconnect with river system: Holocene (qh). Piezometer head of upper-Pleistocene (qp3), Upper Middle Pleistocene (qp$_{2-3}$), Lower Pleistocene (qp$_1$) oscillate following the fluctuation of rainfall and river stage. Under increasing abstraction rapidly, groundwater levels in all aquifers are declining with an annual rate of 0.04 m in the upper Pleistocene aquifer and 0.9 m in the lower Pleistocene aquifer. Currently, the abstraction is estimated at 800,000 m^3/day and occupies 34% of the water supply.[10, 11]

3 Methodology

Groundwater-flow model is used to simulate aquifer response, in terms of the head (groundwater level) and fluxes into and out of an aquifer, to natural and human-induced stresses; The governing equation represents in the three-dimensional movement of groundwater is

$$\frac{\delta}{\delta x}\left[K_{xx}\frac{\delta h}{\delta x}\right] + \frac{\delta}{\delta y}\left[K_{yy}\frac{\delta h}{\delta y}\right] + \frac{\delta}{\delta z}\left[K_{zz}\frac{\delta h}{\delta z}\right] + W = S_s\frac{\delta h}{\delta t} \tag{1}$$

where

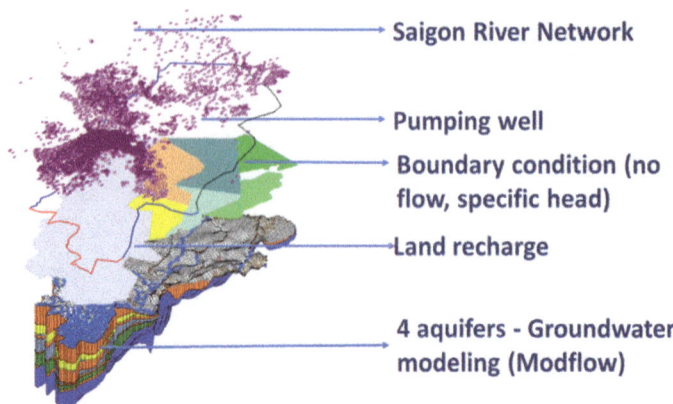

Fig. 2 Conceptual groundwater modeling in Saigon River basin

K_{xx}, K_{yy}, and K_{zz} are the values of hydraulic conductivity along the x, y, and z coordinate axes, and h is the potentiometric head (hydraulic head).

W is a volumetric flux per unit volume representing sources and/or sinks of water, where negative values are water extractions and positive values are injections/recharge. It is a function of space and time (i.e. $W = W(x, y, z, t)$).

Ss is the specific storage of the porous material and function of space.

t is time.

The conceptual groundwater modeling includes the Saigon river network, pumping well, boundary conditions, land recharge, aquifers properties. The river recharge was calculated between different piezometric heads between rivers and aquifers. The pumping well data were utilized from the survey of DWPRIS [9, 11, 12]. The boundary conditions were set as specific flow and general head, which follow the river and seawater level. Land recharge is calculated by multiplying rainfall with the land recharge coefficient [13]. The aquifers' properties were followed from the hydrogeology map of DWPRIS [9].

4   Results and Discussion

4.1   Calibration/Verification Groundwater Model

The groundwater modeling simulated groundwater system in the area during the period 1995–2017. The calibration is in the period 1995–2006, and verification is between 2007 and 2017. Results of calibration and verification indicate that the computed and observed are correspondence. The R^2 is 0.74 – 0.8. The ME is from 0.29 to 0.71. Likewise, SD is from 0.34 to 0.99. Figure 3 shows a sample of calibration and verification piezometric heads at well Q202-aquifer 2.

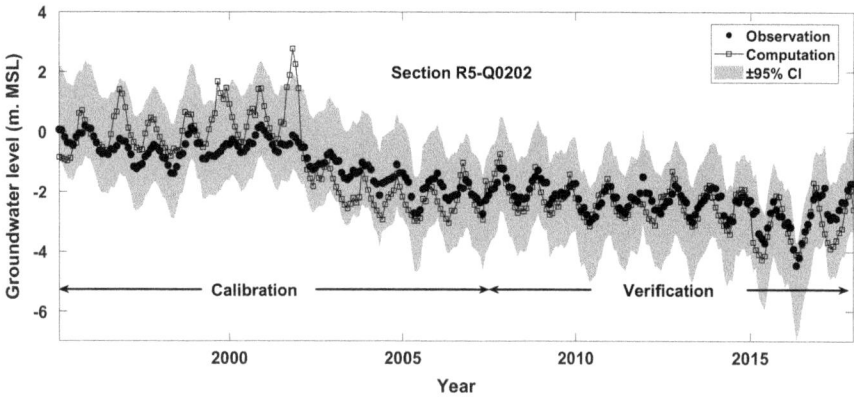

Fig. 3 Calibration and verification results from well Q202 aquifer 2

Table 1 Calibration and verification of groundwater model

	Calibration					Verification				
	Max	Min	Mean	SD	R^2	Max	Min	Mean	SD	R^2
Aquifer 1	2.12	0.04	0.61	0.67	0.76	2.16	0.03	0.71	0.63	0.78
Aquifer 2	3.53	0.02	0.53	0.99	0.8	3.17	0.01	0.73	0.81	0.79
Aquifer 3	1.29	0.02	0.29	0.42	0.81	1.1	0.07	0.32	0.34	0.83
Aquifer 4	1.7	0.01	0.39	0.69	0.78	1.88	0.01	0.29	0.67	0.74

4.2 Optimal Pumping Intensity for Saigon River Basin

This part describes the proposed optimal pumping intensity for the Saigon River Basin. Since the abstraction in aquifer 1 is low, the study focuses on aquifer 2, aquifer 3, aquifer 4. According to thousands of pumping well in the Saigon River Basin, the study analyzed the groundwater pumping intensity in each kilometer block for three aquifers. The groundwater pumping intensity of 3 aquifers in the Saigon River Basin varies from 0–5000 m^3/day/km^2 (see Fig. 4). The pumping intensity in three aquifers in the rural area is below 500 m^3/day/km^2. It can be explained that the population is not high in the rural area of the Saigon River Basin. However, in the urban area with a high population, the pumping intensity of aquifer 2 and aquifer 4 is over 4000 m^3/day/km^2, which causes the piezometric heads of both aquifers decreasing down below 30 m. MSL. Aquifer 3 has low abstraction because the thickness of aquifer 3 is too thin.

To estimate appropriate pumping intensity, the pumping intensity of 3 aquifers was compared with groundwater drawdown in Saigon River Basin. To avoid increasing salt intrusion from growing groundwater pumping, the appropriate intensity of 3 aquifers was selected base on drawdown criteria. The piezometric heads of aquifer 1 and aquifer 2 are not allowed to be below –20 m. MSL. While the piezometric

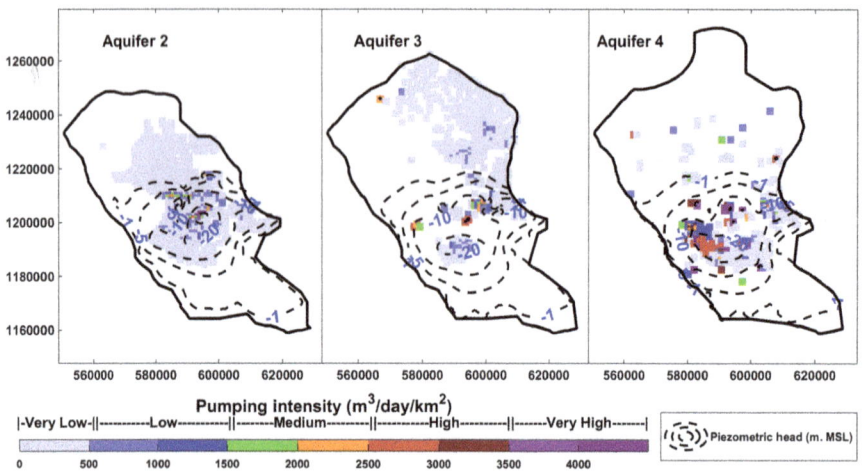

Fig. 4 Pumping intensity and piezometric heads in Saigon River in May 2016

heads of aquifer 3 and aquifer 4 are not allowed to be below −30 m [14]. Hence, the proposed sustainable pumping intensity for aquifer 2, aquifer 3 and aquifer 4 were 2500 m³/day/km², 3500 m³/day/km² and 4000 m³/day/km², respectively (Fig. 5).

The sustainable groundwater yield downstream was then estimated by the "trial error" method. Concerning drawdown criteria, appropriate intensity pumping, and 20-year climate data, the sustainable pumping yield in the existing urban area is 0.88 MCM/day. The sustainable pumping yield in the new area is 1.02 MCM/day (Fig. 6).

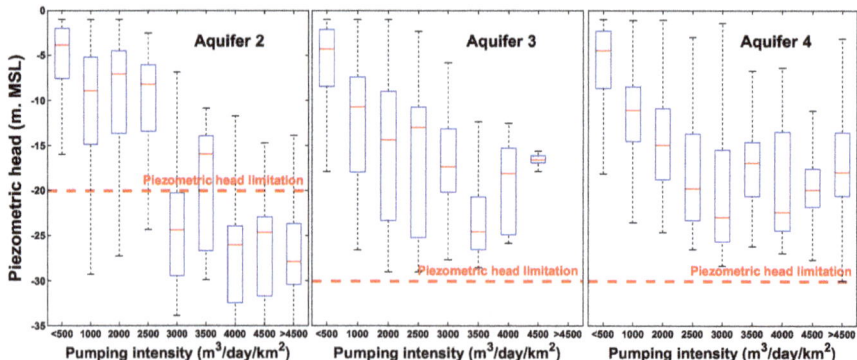

Fig. 5 The piezometric head of 3 aquifers under different pumping intensity

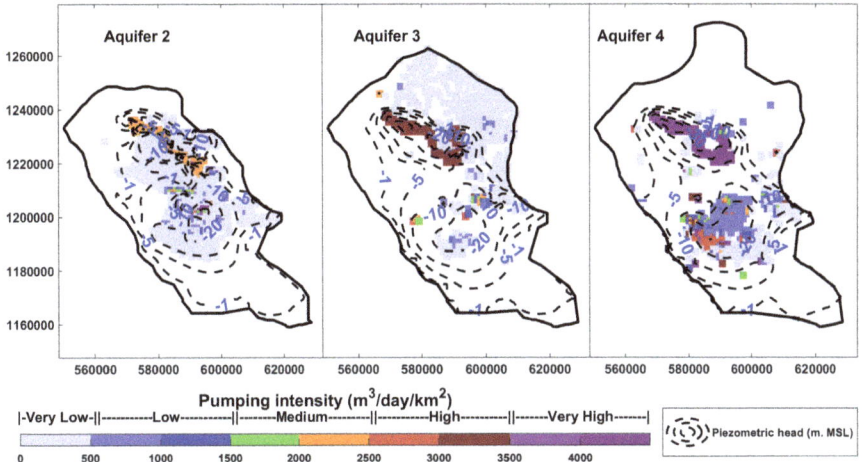

Fig. 6 Optimal pumping intensity and piezometric heads in Saigon River under 20 year historical climate

5 Conclusions and Discussions

According to the pumping in the existing area, the proposed pumping intensity was recognized for aquifer 1, aquifer 2, and aquifer 4 as 2000 m^3/day/km^2, 3500 m^3/day km^2, and 4000 m^3/day/km^2, respectively. With the drawdown criteria, appropriate pumping intensity, and 20-year climate data, the sustainable pumping yield in the existing urban area is 0.88 MCM/day. The sustainable pumping yield in the new area is 1.02 MCM/day. Hence, the total sustainable pumping yield in the existing area and the new area is 1.9 MCM/day. The optimal intensity pumping can be modified for further water resource planning efficiency and sustainability in the Saigon River. The approach in this study can be applied to estimate sustainable pumping yield in groundwater hotspot such as urban area.

Acknowledgements We would like to acknowledge the support time and facilities from Ho Chi Minh City University of Technology (HCMUT), VNU-HCM, and the Water Resources System Research Unit of the Faculty of Engineering, Chulalongkorn University for this study. The authors also thank the staff at Division for Water Resources Planning and Investigation for the South of Vietnam, Southern Regional Hydrometeorology Center Department of Resources and Environmental, Center for Nuclear Techniques in Ho Chi Minh City for data collection.

References

1. Vairavamoorthy K, Gorantiwar SD, Pathirana A (2008) Managing urban water supplies in developing countries—Climate change and water scarcity scenarios. Phys Chem Earth, Parts A/B/C 33(5):330–339

2. Peralta R and Shulstad R (2004) Optimization modeling for groundwater and conjunctive use water policy development. Utah State University, Logan
3. Emch PG, Yeh WW (1998) Management model for conjunctive use of coastal surface water and groundwater. J Water Resour Plan Manag 124(3):129–139
4. Bejranonda W, Koontanakulvong S, Koch M, Suthidhummajit C (2007) Groundwater modeling for conjunctive use patterns investigation in the upper Central Plain of Thailand. 2007:161–174
5. Le Vo P (2007) Urbanization and water management in Ho Chi Minh City, Vietnam-issues, challenges, and perspectives. GeoJournal 70(1):75–89
6. Chân NĐ, Kỳ NV (2010) Nguồn hình thành trữ lượng nước dưới đất vùng lưu vực sông Sài Gòn. VIETNAM J Earth Sci 32(2):156–164
7. Khai HQ and Koontanakulvong S (2014) Impact of Climate Change on groundwater recharge in Ho Chi Minh City Area, Chulalongkorn University
8. Tuan PV, Koontanakulvong S (2018) Groundwater and river Interaction parameter estimation in Saigon river. Vietnam Eng J 22(1):257–267
9. Division for Water Resources Planning and Investigation For The South Of Viet Nam (DWPRIS) (2019) Impact assessment of the exploitation of underground water to protection aquifers. Hochiminh city
10. Nga N (2006) State of groundwater management in HCM city the 5th Research meeting on the sustainable water management policy. Hochiminh
11. Division for Water Resources Planning and Investigation for The South of Viet Nam (DWPRIS) (2016) Results of local real survey determining current situation Of water resources under the exploitation and use; problems of landscape, landscape, loss, loss of water responsible by exploiting underground water, Division for Water Resources Planning And Investigation For The South Of Viet Nam
12. Division for Water Resources Planning and Investigation for The South Of Viet Nam (DWPRIS) (2019) Planning groundwater in Hochiminh city, Hochiminh
13. Long TT, Koontanakulvong S (2019) Deep percolation characteristics via soil moisture sensor approach In Saigon river basin, Vietnam. Int J Civ Eng Technol 10(03):10
14. Ho Chi Minh City People's Committee (2007) Decision 69/2007/QD-UBND of Ho Chi Minh City People's Committee promulgating the Plan for reducing underground water exploitation and filling underground wells in Ho Chi Minh City, in Decision 69/2007/QD-UBND, HCMCPs. Committee, Editor, Hochiminh Vietnam

Microplastic Removal Time in Saigon River

Tuan Dang Pham, Minh Huy Nguyen, and Thu Ha Nguyen

Abstract Microplastics have been a serious problem for the aquatic life because they can convey toxic additives to the food web networks where human is at the highest level. However, the harm is supposed to be less if microplastics are out of the water body, which can be defined as floating at the water surface or burying at the water bottom. In this paper, we processed the microscope images of microplastics collected in Sai Gon River reported in a previous sampling campaign using MATLAB image processing application, and predicted removal time of different microplastic types using an appropriate velocity formula. We found that the removal time of fiber MPs in Saigon River can range between 12 to 130 h and that of flat fragment MPs is about 2 to 35 h. The removal time of one-dimensional MPs is mostly controlled by their diameter while two-dimensional MPs are affected by all of the geometrical factors. One-dimensional MPs in Saigon River tend to stay longer in the water column body, therefore, may have more negative impacts on the aquatic health than two-dimensional MPs.

Keywords Microplastics · Removal time · Settling velocity · Geometrical factors · Saigon River

1 Introduction

The term microplastics (MPs) is used to describe a heterogeneous class of synthetic particles ranging in size from few microns to 5 mm. They are most originally from large pieces of plastic and degraded into small pieces during their time in aquatic environments [3]. Nowadays, MPs appear in every ocean, river, or any aquatic environments, from the surface to the bottom ([5, 7, 9]. One the most concerned problems of MPs is the removal time, which is the detention period of MPs within the water column. During this particular period, MPs can be mistaken as food, and thus the removal time becomes the most vulnerable time for aquatic creatures. In other

T. D. Pham · M. H. Nguyen · T. H. Nguyen (✉)
Faculty of Civil Engineering, Ho Chi Minh City University of Technology, VNU-HCM, Ho Chi Minh City, Vietnam
e-mail: thuhatnn@hcmut.edu.vn

© The Author(s), under exclusive license to Springer Nature Singapore Pte Ltd. 2023   1037
J. N. Reddy et al. (eds.), *ICSCEA 2021*, Lecture Notes in Civil Engineering 268,
https://doi.org/10.1007/978-981-19-3303-5_95

words, the longer the removal time, the more harmful MPs can be to the food web. And obviously, the removal time is controlled by the terminal velocity (in addition to the hydrodynamic conditions such as turbulence, currents, and tides). Therefore, good understanding of MP terminal velocity behavior is important to predict MP distributions for remediation planning.

Terminal velocity is the constant vertical speed that an object attains when all the forces exerting on it balance. When it moves downward, the terminal velocity is called settling velocity. The terminal velocity is controlled by MP density, geometrical properties like size, shapes, and other interference like ambient physical, chemical, and biological effects [12].

An MP particle density depends on 3 elements: (i) density of the source plastic materials, (ii) additives and the manufacturing process, (iii) the history of the MP residence in aquatic environment. The two first elements are the initial characteristics when it was produced from the factory, but in the natural aquatic environment, MP could be changed or combined with other components. Because of that, density of MPs is usually unknown [8].

MPs are varied in shapes, which can affect their movement characteristics through the water column. It can be categorized into 3 types of shapes. The first type is one-dimensional (1-D) MPs, which includes threadlike polymers, like ropes, fishing lines, fibers or anything that has one direction much larger than the others. 92% of 1-D samples collected in Sai Gon River are synthetic fibers (70% Polyester, 9% PET and minor fractions of PE or PP, PP-PE copolymer, rayon, PP vistalon, viscose, acrylic with density about 1.32 g/cm^3) (Lahens et al. 2017). The second type is two-dimensional (2-D) MPs, such as films, flakes, flat fragments or things that have two directions much larger the other. 40% of 2-D samples collected in Sai Gon River made by LDPE (0.9 g/cm^3) (Lahens et al. 2017). The third type is three-dimensional (3-D) MPs, including fragments, pellets, spherules or anything which has three dimensions not much larger than each other.

This paper aims to review some available formulas that can be used to calculate MP terminal velocity. Next, we apply an appropriate formula to predict MP terminal velocity and removal time in Saigon River, Vietnam using the experimental data from previous sampling campaign [11]. Only 1-D and 2-D MPs are analyzed in this paper due to the limit of 3-D MP images.

2   The terminal velocity

The formulas to predict settling velocity are explicitly established for natural sediment [1, 2, 4]. These formulas are employed to also predict MP settling velocity [6, 14]. Herein, we review two formulas that are either popular or convenient to apply to predict MP settling velocity in Saigon River.

2.1 Dietrich's Formula (1982)

One of the most popular settling velocity formula is from [2]. In this formula, all four important factors (size, density, shape and roundness) are considered and combined into three empirical formula R_1, R_2 and R_3 which has made this formula more practical to use in calculation but still having acceptable precision in form of

$$v_* = R_3 10^{R_1 + R_2},$$

(1)

where v_* is the dimensionless settling velocity; $R_3 = f$ (roundness) is the formula for predicting the ratio of the settling velocity of an angular particle to that of a well-rounded particle; $R_1 = f$ (density, size) is the formula for predicting the settling velocity of sphere; $R_2 = f$ (shape) is the formula for predicting the ratio of the settling velocity of a non-spherical, well-rounded particle to be settling velocity of a sphere with the same dimensionless diameter D_*. The shape in R_2 is represented using the Corey shape factor CFS $= c/\sqrt{a \times b} \geq 0.15$, where a, b and c are the longest, intermediate and shortest axes of the particle, respectively. The Dietrich's formula (1) has a compulsory condition that CFS must not be less than 0.15. This problem is a limitation for the cases of 1-D and 2-D MPs which have CFS < 0.15.

2.2 Nguyen et al.'s Formula (2020) [13]

$$v = 2\frac{-3\mu P + \sqrt{(3\mu p)^2 + A\rho_f g(\rho - \rho_f)V}}{A\rho_f} > 0, \text{ sinking for } \rho > \rho_f,$$

(2)

$$v = 2\frac{3\mu P - \sqrt{(3\mu P)^2 - A\rho_f g(\rho - \rho_f)V}}{A\rho_f} < 0, \text{ rising for } \rho < \rho_f,$$

(3)

where g is the gravitational acceleration; V, A and P are the aggregate solid volume, projected area and outer perimeter, respectively; ρ, ρ_f and μ are the aggregate density, fluid density and fluid viscosity; and v is the terminal velocity.

This formula is obtained by solving the force balance exerting on a moving particle in still water including gravitational, buoyancy, viscosity, and impact forces. It requires only physical properties of MPs without any empirical parameters. These physical properties can be straightforwardly retrieved from the MP images, therefore, is convenient for MP velocity prediction. Moreover, this formula is applicable for both MPs heavier and lighter than water. For this reason, this formula is chosen for all the analysis in this paper.

2.3 Image Processing Using MATLAB

All microplastic microscope images used in this paper were published in [11] and kindly provided by the authors. The microscope images were first processed using the Color Thresholder application in MATLAB to isolate MPs from the background. Next, the geometrical properties including area, perimeter and length were calculated in pixel and metric units. Finally, the formulas (2) and (3) above are used to determine the terminal velocity.

In the Color Thresholder in application, the HSV mode was chosen, in which the Hue, Saturation and Value channels were adjusted to separate the MP pixels from the background pixels (Fig. 1). Firstly, Hues are color appearance parameters of a color including three ranges of primary colors (red, blue, green) and three ranges of secondary colors (orange, green and violet). With a chosen range of Hue, the initial picture will only be displayed in this range of color, any other colors will become black. Depending on the image color, the Hue must be set in the color range of MPs or out of the background pixels' range. Secondly, Saturation is the purity of the color, which show how much an original color is combined with white. The less white is added, the higher the saturation is. Because the MPs are not only be distinguished from the background by colors but also by saturation, adjusting the saturation thresholding can increase the isolation effect. Finally, the Value is the maximum intensity of a pixel that is helpful in removing noise. Now the initial background pixels became black and the MP pixels were kept. For further analyses, the images were converted into binary mode, in which the MP pixels are 1 (white) and the background pixels are 0 (black) (Fig. 2).

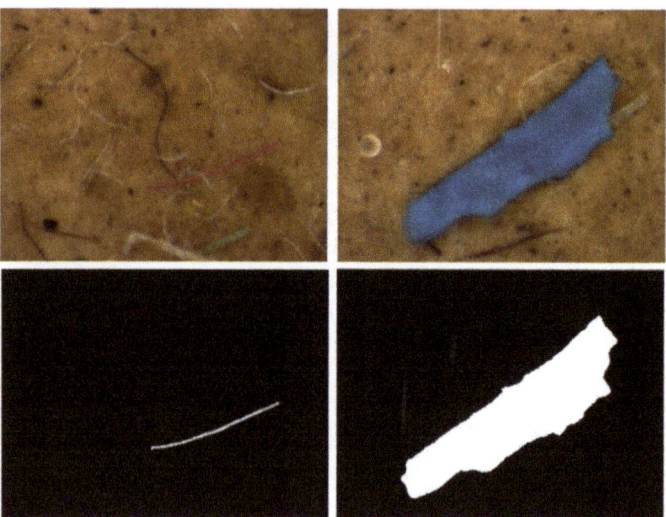

Fig. 1 Example microscope images of MPs in Sai Gon River. First row: original images; second row: processed binary images; left: 1-D MPs; right: 2-D MPs

The color thresholding process above may leave some holes within the MP images due to some interference on MP surfaces. These holes were automatically filled to produce completed MP images using a MATLAB function. Next, MP area was retrieved as the total number of MP white pixels and the perimeter was measured as the total boundary pixels. At this stage, these parameters were in pixels, therefore, they were converted into μm, where 1 pixel equals 3 μm, 1 square pixel equals 9 μm^2 (based on the microscope resolution). Only the MP volume could not be determined from the image, it requires assumptions and calculations (see Sect. 4).

3 Data analysis

We realized that when three geometrical factors (P, V and A) were transformed into the P/A and V/A ratios, the velocity formula would only depend on two factors and became more manageable to study. Besides, the V/A ratio would be calculated instead of the unknown volume V. Therefore, Eq. (2) and (3) are transformed into

$$v = 2\frac{\mp 3\mu P/A \pm \sqrt{\left(3\mu P/A\right)^2 \pm \rho_f g\left(\rho - \rho_f\right)V/A}}{\rho_f} \tag{4}$$

While the P/A ratio was immediately calculated because P and A can be measured through image processing, V/A ratio needs some further steps to calculate.

For 1-D MPs, $A = L \times D$ and $P = 2(L + D)$, solving for L and D, then

$$\frac{V}{A} = \frac{\frac{\pi D^2}{4} \times L}{L \times D} = \frac{\pi D}{4}, \tag{5}$$

where L and D are 1-D MP length and diameter, respectively.

For 2-D MPs,

$$V/A = \text{Thickness}. \tag{6}$$

We assumed the thickness in range of 10–40 μm, based on materials, colors and the relevant report [9].

4   Results and discussion

4.1   1-D Particle

1-D MP density is about 1.32 g/cm^3 larger than water density so the behavior of MPs is sinking. The MP minimum and maximum diameters are 9 μm and 30 μm, respectively. As mentioned above, the *V/A* ratio of the 1-D MPs equals to $\pi D/4$, therefore, instead of analyzing *V/A* ratio, we analyze relationship between diameter and terminal velocity as well as the removal time. Figure 3a and b illustrate that terminal velocity increases dramatically with increasing diameter and obviously the removal time is inversely proportional to the diameter as the result. For the largest diameter of 30 μm, terminal velocity peaks at 3.4 × 10^{-4} m/s and the removal time hits a low of about 12 h.

The trends in Fig. 3c and d are opposite to those in Fig. 3a and b, the terminal velocity significantly decreases when the *P/A* ratio increases and the removal time thus sharply rises. When the minimum *P/A* ratio is 2.25 × 10^5 m^{-1}, the terminal velocity reaches a low of 3.2 × 10^{-5} m/s, corresponding to the removal time of MPs reaching a high of nearly 130 h. The plots illustrate that the terminal velocity is controlled strictly by only one of the two ratios (i.e., *P/A*, *V/A*), that can explain why the trend of four plots are only steadily upward or downward, with further analysis we finally conclude that terminal velocity of our 1-D MP samples mostly depend

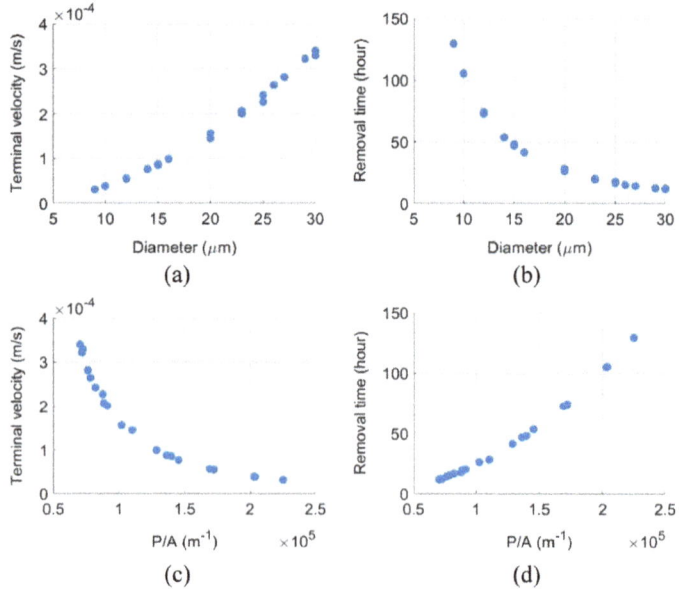

Fig. 3 1-D MP analyses. **a** Terminal velocity and **b** removal time as functions of diameter; **c** terminal velocity and (d) removal time as functions of P/A ratio

on diameter. Another minority distribution is from the length, if the MPs has longer length, it will have the faster terminal velocity and shorter removal time. As we can say, the 1-D MPs with large diameter and great length will be less harmful to the aquatic environment than the others.

4.2   2-D Particle

2-D MPs in Saigon River mostly have density of about 0.9 g/cm^3, lighter than water density so the behavior of MPs is floating and rising. These MP minimum thickness is 10 μm and maximum is 40 μm. There are also two plots of P/A and V/A ratios, where the V/A ratio of 2-D MPs represents thickness. In some aspects, the diameter of 1-D MPs could be considered as thickness, therefore, the behavior of 2-D MPs perhaps related with thickness like the 1-D MPs. The trend of terminal velocity of 2-D MPs versus thickness is also upward (Fig. 4a), similarly to the one in 1-D MPs (Fig. 3a). Compare the Fig. 4a to the Fig. 3a, they do have similar trend, the trend of the terminal velocity is still upward. Nevertheless, in this case we notice that there are some points despite having the same thickness, the terminal velocity is different. While the 2-D MP P/A ratio is independent of the V/A ratio, both P/A and V/A ratios of the 1-D MPs are linked to diameter, thus Fig. 4a is more scattered than Fig. 3a. The maximum terminal velocity is 1.4×10^{-3} m/s, minimum removal time is 2 h and 56 min when the thickness is 40 μm.

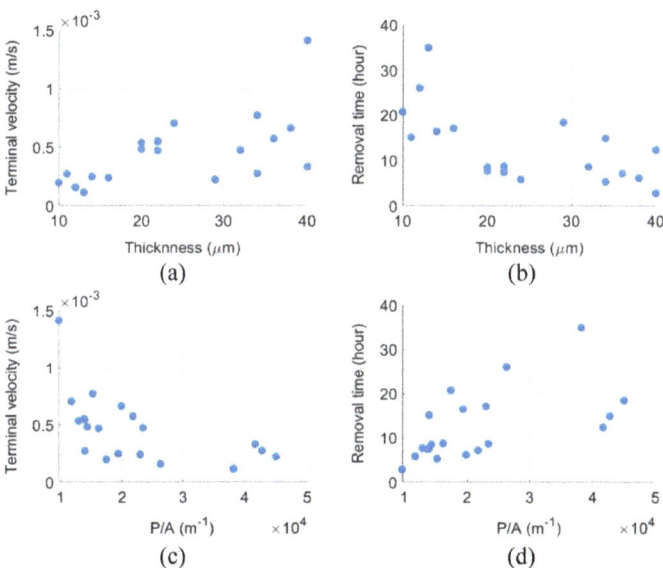

Fig. 4. 2-D MP analyses (a) Terminal velocity and (b) removal time as functions of diameter; (c) terminal velocity and (d) removal time as functions of P/A ratio

The general trend of P/A versus terminal velocity plot in Fig. 4c is downward; although similar to Fig. 3c, the data does not follow the trend completely. The minimum terminal velocity is 1.18×10^{-4} mm/s, and the maximum removal time is 35 h when the P/A ratio is 3.8×10^4 m^{-1} (smaller than the maximum ratio 4.5×10^4 m^{-1}). In the final analysis, the data and plots show that terminal velocity of 2-D MPs is affected by both P/A ratio and thickness (V/A ratio), when the P/A ratio falls and thickness rises, the terminal velocity will be faster and the removal time thus will be short and manifestly less dangerous to the surrounding.

The 1-D MP settling velocity calculated herein is slightly lower than the experimental results [15] while our 2-D rising velocity is smaller than that reported by [14] but much larger than that in Ref. [10]. The difference of our calculation from other experimental results can be explained by the dissimilarity in MP materials, size, shapes, and even the settling medium. Moreover, the formulas (2) and (3) used in our analyses are yet to considered the effects of artificial MPs in hydrodynamics like settling orientations of long 1-D MPs or the hydrophobicity of MP surfaces reducing drag force. Therefore, implementations of formula (2) and (3) is needed for better precise prediction of MP movement.

5 Conclusion

From the our analysis, MPs with large P/A and small V/A ratios have long removal time, and thus would have more chance to interact with aquatic creatures. We found that 1-D MPs would take more time to be removed from the water column than 2-D MPs (i.e., 12–130 h compare to 2–35 h, respectively). Therefore, 1-D MPs is likely to be more harmful to the environment than the 2-D MPs. The removal time of 1-D MPs is mostly controlled by their diameter while 2-D MPs are affected by both P/A and V/A.

Acknowledgements T.H.N is supported by the project JEAI PLASTIC of the The French National Research Institute for Sustainable Development (IRD). We acknowledge the support of time and facilities from HCMUT, VNU-HCM for this study.

References

1. Stokes GG (1951) On the effect of the internal friction of fluids on the motion of pendulums, vol 9. Pitt Press, Cambridge
2. Dietrich WE (1982) Settling velocity of natural particles. Water Resour Res 18(6):1615–1626
3. Andrady AL (2011) Microplastics in the marine environment. Mar Pollut Bull 62(8):1596–1605
4. Maggi F (2013) The settling velocity of mineral, biomineral, and biological particles and aggregates in water. J Geophys Res Ocean 118(4):2118–2132
5. Van Cauwenberghe L, Vanreusel A, Mees J, Janssen CR (2013) Microplastic pollution in deep-sea sediments. Environ Pollut 182:495–499

6. Khatmullina L, Isachenko I (2017) Settling velocity of microplastic particles of regular shapes. Mar Pollut Bull 114(2):871–880
7. Lebreton LC, Van der Zwet J, Damsteeg JW, Slat B, Andrady A, Reisser J (2017) River plastic emissions to the world's oceans. Nat Commun 8:15611
8. Chubarenko I, Esiukova E, Bagaev A, Isachenko I, Demchenko N, Zobkov M, Efimova I, Bagaeva M, Khatmullina L (2018) Behavior of microplastics in coastal zones. In: Zeng E (ed) Microplastic contamination in aquatic environments. Elsevier, Amsterdam, The Netherlands, pp 175–223
9. Lahens L, Strady E, Kieu-Le TC, Dris R, Boukerma K, Rinnert E, Gasperi J, Tassin B (2018) Macroplastic and microplastic contamination assessment of a tropical river (Saigon River, Vietnam) transversed by a developing megacity. Environ Pollut 236:661–671
10. Waldschläger K, Schüttrumpf H (2019) Effects of particle properties on the settling and rise velocities of microplastics in freshwater under laboratory conditions. Environ Sci Technol 53(4):1958–1966
11. Strady E, Kieu-Le TC, Gasperi J, Tassin B (2020) Temporal dynamic of anthropogenic fibers in a tropical river-estuarine system. Environ Pollut 259:113897
12. Nguyen TH (2020) Flocculation dynamics of cell-associated suspended particulate matter. Doctoral dissertation, University of Sydney
13. Nguyen TH, Tang FHM, Maggi F (2020) Sinking of microbial-associated microplastics in natural waters. PLoS ONE 15:1–20
14. Van Melkebeke M, Janssen C, De Meester S (2020) Characteristics and sinking behavior of typical microplastics including the potential effect of biofouling: implications for remediation. Environ Sci Technol 54(14):8668–8680
15. Khatmulina L, Chubarenko I (2021) Thin synthetic fibers sinking in still and convectively mixing water: laboratory experiments and projection to oceanic environment. Environ Pollut 288: 117714

On the Flood Reduction Effect of Reservoirs in the Vu Gia—Thu Bon Basin in October 2020 Flood

Hoa Nguyen Thi Thanh, Hong Tran Thi My, Thao Nguyen Thi Thach, and Giang Le Song

Abstract In October of 2020 on the Vu Gia—Thu Bon river basin, there were two large floods. Due to resulting heavy losses, there have been doubts about the effects of upstream reservoirs in flood prevention and control. The aim of this study was to re-evaluate the role of those reservoirs. The numerical modeling method was used for this analysis. An integrated 1D/2D model was developed using the flow in the rivers was as 1D and the flow on the floodplains as 2D. There are two computational scenarios to be considered. In scenario considering the presence of the reservoirs, the discharge imposed at the upper model boundary is the actual one. This discharge is the result of operation of the reservoirs. In the scenario considering no reservoirs, the discharge imposed at the model upper boundary is calculated from the actual discharge to the reservoirs, with the regulation of the reservoirs neglected. The simulation results of these two flood scenarios clearly show that the reservoirs have significantly reduced floods downstream.

Keywords Vu Gia—Thu Bon delta · Reservoirs · Flood · Integrated 1D–2D model

1 Introduction

Over the past years, there have been consecutive large storms in the central region of Vietnam. Specifically, in October 2020 there were four storms, with two of them landing in Quang Nam and Quang Ngai provinces (storms No.6 and No.9). Heavy rain caused flooding in the lower basin of Thu Bon—Vu Gia river.

On the upstream portion of Thu Bon—Vu Gia river, there are ten primary hydropower plants in operation or under construction (according to Decision No. 1537/QD-TTg dated September 7, 2015). In addition, there are 36 small and medium

H. N. T. Thanh (✉) · H. T. T. My · T. N. T. Thach · G. Le Song
Faculty of Civil Engineering, University of Technology (HCMUT), Vietnam National University HCMC, 268 Ly Thuong Kiet Street, District 10, Ho Chi Minh City, Vietnam
e-mail: ntthoa.sdh19@hcmut.edu.vn; ntthoa@hcmunre.edu.vn

H. N. T. Thanh
Ho Chi Minh University of Natural Resources and Environmen, Ho Chi Minh City, Vietnam

© The Author(s), under exclusive license to Springer Nature Singapore Pte Ltd. 2023   1047
J. N. Reddy et al. (eds.), *ICSCEA 2021*, Lecture Notes in Civil Engineering 268,
https://doi.org/10.1007/978-981-19-3303-5_96

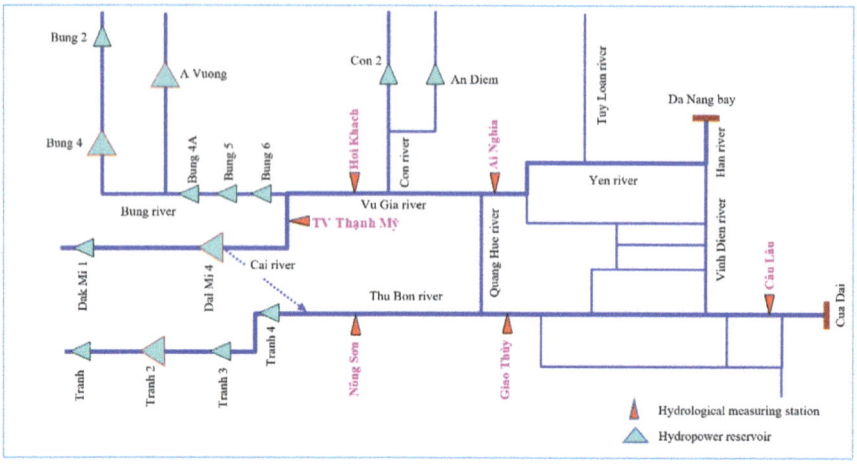

Fig. 1 Diagram of hydroelectric steps on the Vu Gia—Thu Bon river

hydropower plants, half of which are already in operation (Fig. 1). Most of these hydropower plants are not involved in flood control, except for four plants namely Song Tranh 2, Dak Mi 4, Song Bung 4 and A Vuong. The operation of these four reservoirs should impact on floods downstream, which has been the subject of multiple differeing opinions. The paper arms are to present an integrated 1D2D model capable of calculating flood flows on the Vu Gia—Thu Bon floodplain in detail and accuracy and using the model to clarify the real effects of these four reservoirs in flood control in the downstream area.

2 Method

The numerical modeling method was used in this study. The study is based on the 1D2D3D integrated model of the Vu Gia—Thu Bon (Fig. 2) established by Tran Thi My Hong et al. [3]. The model was built using F28 software [1]

Flows in rivers and in culverts crossing the road are considered as one-dimensional and solved using the Saint–Venant equation [1]:

$$\frac{\partial A}{\partial t} + \frac{\partial Q}{\partial s} = q_l \tag{1}$$

$$\frac{\partial Q}{\partial t} + \frac{\partial}{\partial s}\left(\frac{Q^2}{A}\right) + gA\frac{\partial \eta}{\partial s} + gA\frac{|Q|Q}{K^2} - u_l q_l = 0 \tag{2}$$

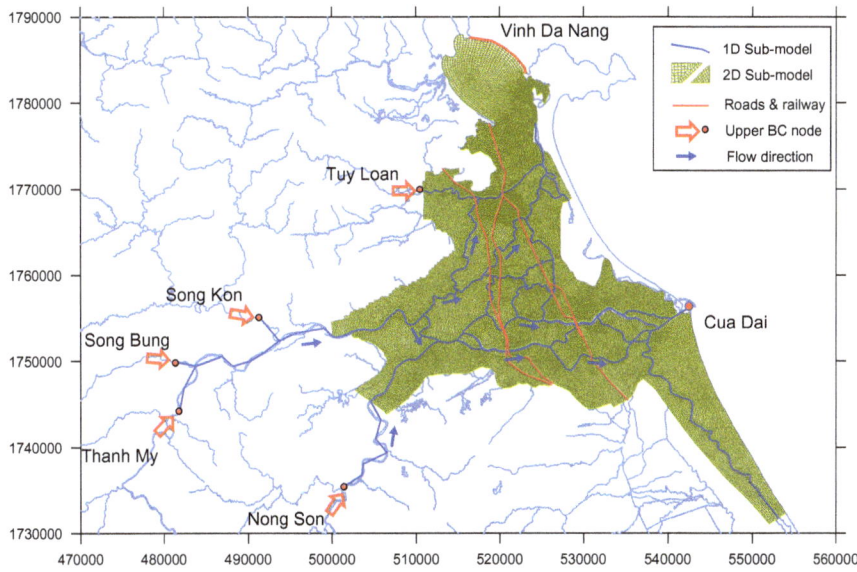

Fig. 2 The computational mesh of the Vu Gia—Thu Bon delta

In these equations: η—the water level; Q, A and K—the flow rate, the cross section area and the discharge module for 1D flow; q_l and u_l—the lateral inflow along the river and its axial component of velocity.

The 2D flow over the floodplain and in Da Nang Bay is solved from the shallow water equations [1]. In conservative form, they are written:

$$\frac{\partial \eta}{\partial t} + \nabla \cdot \mathbf{q} = q_v \tag{3}$$

$$\frac{\partial \mathbf{q}}{\partial t} + \nabla \cdot \mathbf{F}(\mathbf{q}) = \mathbf{b}(\mathbf{q}) \tag{4}$$

In these equations: $\mathbf{q} = \left[q_x, q_y\right]^T = D\mathbf{U}$—the vector of flow rate per unit width of 2D flow; $\mathbf{U} = \left[u_x, u_y\right]^T$—the depth-averaged velocity vector; D—the water depth; ∇—the differential operator; $\mathbf{F}(\mathbf{q})$—the flux vector of flow rate; and $\mathbf{b}(\mathbf{q})$—the vector of external forces. The flux vector, $\mathbf{F}(\mathbf{q})$, and external forces vector, $\mathbf{b}(\mathbf{q})$, have the forms:

$$\mathbf{F}(\mathbf{q}) = \begin{bmatrix} q_x\mathbf{U} - A_H D\partial\mathbf{U}/\partial x \\ q_y\mathbf{U} - A_H D\partial\mathbf{U}/\partial y \end{bmatrix} \tag{5}$$

$$\mathbf{b}(\mathbf{q}) = \begin{bmatrix} -gD\partial\eta/\partial x - (\tau_{bx} - \tau_{wx})/\rho + fq_y + u_a q_v \\ -gD\partial\eta/\partial y - (\tau_{by} - \tau_{wy})/\rho - fq_x + v_a q_v \end{bmatrix} \tag{6}$$

where f—Coriolis parameter; (τ_{wx}, τ_{wy})—the shear stress on water surface due to the wind; (τ_{bx}, τ_{by})—the bottom shear stress; A_H—the eddy viscosity; q_v and u_a, v_a - lateral inflow per unit surface area and its velocity components.

Equations (1–4) are solved using the finite volume method [1]. The model has fiev boundaries in upstream. The discharge imposed at the nodes Nong Son (Thu Bon river) and Thanh My (Vu Gia river) are measured data of the hydrological station located there. Discharge at the Bung river node is the total of discharge of Song Bung 4 and A Vuong reservoirs and discharge of the catchment part behind Song Bung 4 and A Vuong to the boundary node. The discharge of this catchment part is calculated:

$$Discharge_{After\,Song\,Bung4} = S_{After\,Song\,Bung4r} \frac{Discharge_{Song\,Bung4} + Discharge_{AVuong}}{S_{Song\,Bung4} + S_{AVuong}}$$

(7)

The boundary discharge at the nodes of Con and Tuy Loan rivers are calculated using the hydrological HMS model developed by Tran Thi My Hong et al. [2]. At 2D nodes in Da Nang Bay and at 1D node at Cua Dai estuary (Hoi An) water-level calculated from the tidal constants are imposed. Also, rainfall on 2D elements is imposed based on observed data in this basin [5].

3 Calculations and Analysis

3.1 Simulation of October 2020's Flood in the Vu Gia—Thu Bon low Basin

In October 2020, there are five floods on the Vu Gia—Thu Bon basin. One of two biggest floods occurred at the beginning of October (10–12/10/2020) and other one at the end of October (27–28/10/2020). These two floods were the result of storms No. 6 and No. 9 that flowed directly into the Quang Nam—Quang Ngai area. Total rainfall at some rain gauge stations in upper Vu Gia—Thu Bon basin during the floods is presented in Table 1 below.

The discharge to the upstream nodes in October 2020 is shown in Fig. 3. During the early October flood, the peak discharge in Nong Son was 6,530 m³/s and in Thanh

Table 1 Total precipitation at upstream stations during floods

Flood	Hydrological measuring station (m³/s)					Average rainfall per day
	Nong Son	Thanh My	Tra My	Hiep Duc	Tien Phuoc	
10–12 Oct. Flood	734.0	483.0	561.0	527.0	529.0	188.9
27–28 Oct. Flood	106.0	109.0	209.0	138.0	126.0	68.8

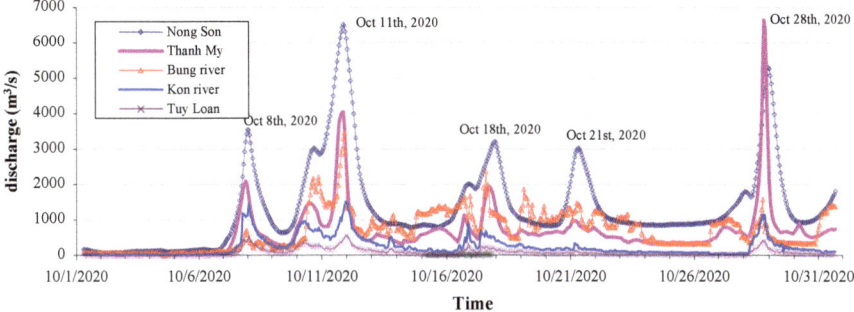

Fig. 3 Discharge flows to the boundary nodes of the downstream Vu Gia—Thu Bon

My was 4,090 m³/s. In the late October flood, the peak discharge in Nong Son was 5,300 m³/s and in Thanh My was 6,800 m³/s.

Figure 4 shows the calculated water level at the hydrological stations downstream of Vu Gia—Thu Bon. In general, the calculated results are quite consistent with the observed data. This comparison shows that the model is reliable.

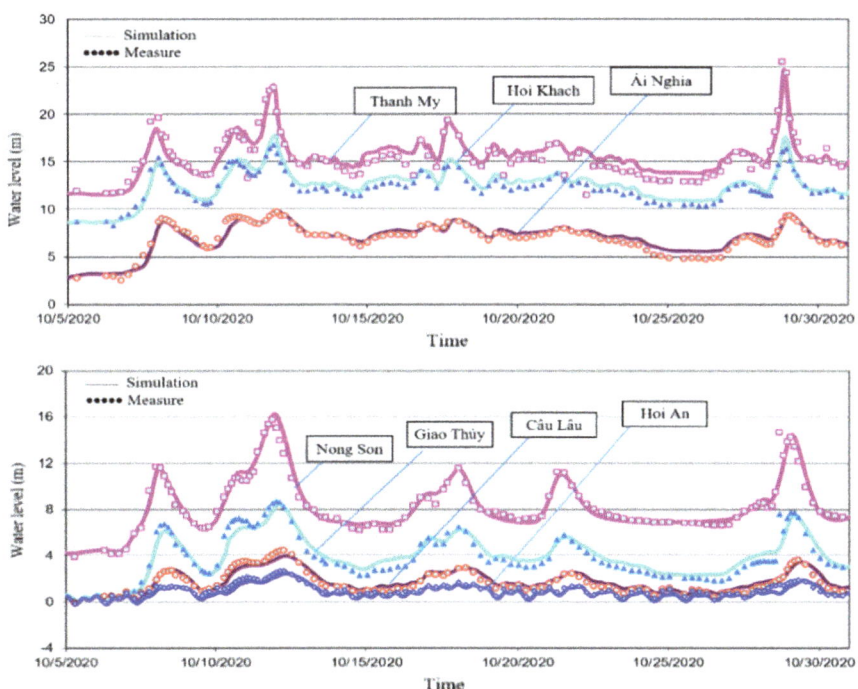

Fig. 4 Comparing calculated and measured water levels in the downstream Vu Gia—Thu Bon

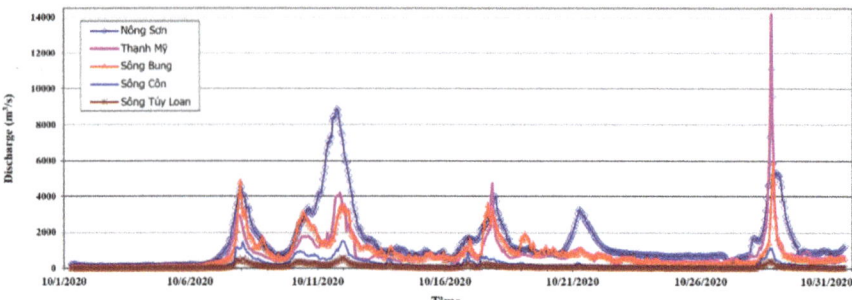

Fig. 5 Discharge to the boundary nodes of the Vu Gia—Thu Bon in the case absence of four reservoirs

3.2 Regulatory Efficiency of four Reservoirs in October 2020 Floods

The main impact of hydropower reservoirs on the flood in downstream area is to change the flood flow. In this study, other effects such as the change of vegetation cover on water accumulation time were ignored. To assess the impact of hydropower reservoirs, flood simulation were made with the assumption that 4 reservoirs Song Tranh 2, Dak Mi 4, Song Bung 4, and A Vuong do not regulate floods. This means that the downstream area will receive all rainwater from the upstream during the flood.

The water discharge to the upstream boundary nodes is evaluated from the actual discharge plus the difference due to regulation by the reservoirs. The discharge data is also referenced from the website of Quang Nam Provincial Disaster Prevention and Search and Rescue Command [5]. Figure 5 shows the discharge to the upstream boundary nodes of the model in the case of absence of four hydropower reservoirs.

The main impact of hydropower reservoirs on flooding in downstream areas is to change the flood flow. In this study, other effects such as the change of vegetation cover on water accumulation time were ignored. To assess the impact of hydropower reservoirs, flood simulations were made with the assumption that four reservoirs— Song Tranh 2, Dak Mi 4, Song Bung 4, and A Vuong do not regulate floods. This means that downstream areas will receive all rainwater from the upstream flow during the flood. The water discharge to the upstream boundary nodes is evaluated based on the actual discharge plus the difference due to regulation by the reservoirs. The discharge data is also referenced from the website of Quang Nam Provincial Disaster Prevention and Search and Rescue Command [5]. Figure 5 shows the discharge to the upstream boundary nodes for the model considering the absence of the four hydropower reservoirs.

Under assumed conditions that there is no reservoir, for the flood in early October, the peak discharge in Nong Son increased from 6,530 m³/s to 8,900 m³/s and in Thanh My from 4,090 m³/s increased to 4,110 m³/s. The peak discharge at Song Bung also increased from 3,490 m³/s to 3,530 m³/s. For the late October flood, the

Table 2 Comparison of flood discharge to the boundary nodes in the case of presence and absence of reservoirs

Flood	Actual discharge (m³/s)			Discharge without reservoirs (m³/s)		
	Nong Son	Thanh My	Bung river	Nong Son	Thanh My	Bung river
Early October flood	6.530	4.090	3.490	8.900	4.110	3.530
Late October flood	5.300	6.800	1.590	11.120	13.400	5.960

peak discharge in Nong Son increased from 5,300 m³/s to 11,120 m³/s and in Thanh My increased from 6,800 m³/s to 13,400 m³/s. The peak discharge at Song Bung also increased from 1,590 m³/s to 5,960 m³/s. Comparing flood maximum discharge to the upper nodes with and without reservoirs is presented in the Table 2 below.

Figure 6 compares the calculated water levels at downstream hydrological stations in the two cases with and without the four reservoirs. This figure shows that during the early October flood on the Vu Gia branch, the reservoirs did not reduce the peak flood water level, but did cause the flood to recede faster. On the Thu Bon branch, the reservoirs have reduced both the peak flood water level and the flood duration. During the late October flood, the reservoirs proved their role in flood regulation in

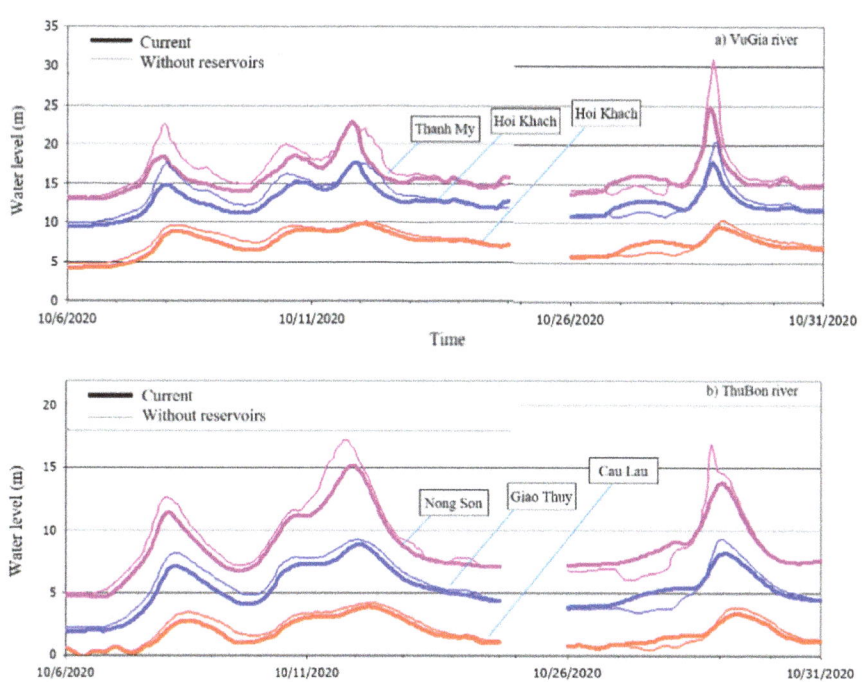

Fig. 6 Water level at downstream Vu Gia—Thu Bon in the cases of presence and absence of flood regulation of four reservoirs

both Vu Gia and Thu Bon branches. Flood peak water level and flood duration were both decreased significantly.

Comparison of the actual flood map with the flood map in the absence of flood regulation for the four reservoirs is shown in Figs. 7 and 8. The figure clearly shows that the lower Vu Gia—Thu Bon river is inundated more deeply and the flooded area is also larger in the case where flooding is not regulated by the four upstream reservoirs.

The comparison of inundation levels between current case and without 4 reservoirs is presented in Table 3 below:

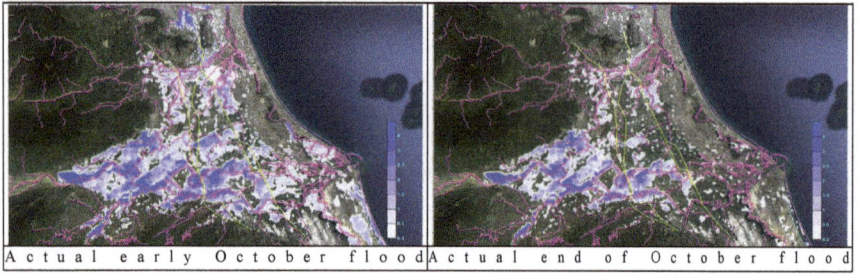

Fig. 7 Innundation map of the VG-TB lower basin during October 2020—current case

Fig. 8 Innundation map of the VG-TB lower basin during October 2020— without 4 reservoirs

Table 3 Inundation area at flood peak (ha)

Depth (m)	The beginning of October (10–12/10/2020)			The end of October (27–28/10/2020)		
	With 4 reservoirs (ha)	Without 4 reservoirs (ha)	Difference (%)	With 4 reservoirs (ha)	Without 4 reservoirs (ha)	Difference (%)
≥ 0.5	26,717	32,277	20.8	17,587	25,739	46.4
≥ 1.0	18,119	23,755	31.1	9,728	17,134	76.1
≥ 2.0	6,800	9,619	41.5	3,806	7,108	86.7
≥ 4.0	1,186	1,720	45.0	724	1,260	74.1

Table 4 Nash–Sutcliffe coefficient

Hydrological stations	Nong Son	Thanh My	Hoi Khach	Ai Nghia	Giao Thuy	Cau Lau
Nash–Sutcliffe coefficient	0,93	0,82	0,87	0,90	0,90	0,90

4 Conclusions

The paper presented simulations of two floods in October 2020 on the lower Vu Gia— Thu Bon river. Through two scenarios of the model, the calculated results against the reality are consistent, with the Nash–Sutcliffe coefficients at specific positions as follows (Table 4):

The calculations show that in the scenario where four hydropower plants Song Tranh 2, Dak Mi 4, Song Bung 4 and A Vuong do not participate in flood regulation, the flow to upstream nodes will significantly increase and will increase inundation in the downstream of Vu Gia—Thu Bon rivers, in terms of both inundation depth and flooded area. Thus, the upstream hydropower reservoir has performed well the function of flood prevention for downstream areas.

Acknowledgements The authors would like to thank the staff of Nong Son and Thanh My hydrology stations for sharing the October 2020 flood data of Vu Gia—Thu Bon river.

References

1. Giang Le Song (2011) Development of an integrated software for calculation of urban flood flow, Report B2007-20-13TĐ
2. Le Song Giang, Cong Hoai Huyng, Quang Truong Nguyen, Ngoc Minh Nguyen, My Hong Tran Thi (2014) Integrated 1D–2D model for flood simulation in the Vu Gia-Thu Bon delta, National Hydro-mechanics conference 2014. Sci Technol Develop J 17(4):43–50
3. Hong Tran Thi My, Thao Nguyen Thi Thach, Giang Le Song (2019). Impact assessment of the construction works on the flow of the Vu Gia - Thu Bon river system. In: 22nd National hydro-mechanics conference
4. http://avuong.com/thuydienavuong/. Accessed 03 Nov 2020
5. http://pctt.quangnam.vn/index.php/sa-liau/2017-05-22-06-39-31. Accessed 03 Nov 2020

Uncertainty-Based Seepage Analysis Through Different Types of Earth Dams

Thi Tuyet Giang Vo and Vo Trong Nguyen

Abstract This study presents the seepage analysis through three popular types of earth-fill dams whose hydraulic conductivities are considered as spatially random. This study focuses on evaluating two important parameters of seepage which are the flow rate and gradient. These outcomes then are compared with ones resulted from the deterministic calculations. Hence, within all three types of dams, there is a probability of 100% in which the flow rates calculated from the random method are greater than ones from the deterministic method. As a meanwhile, there is a probability of about 50% in which the gradients calculated from the random method are greater than ones from the deterministic method.

Keywords Seepage · Embankments · Finite-element modelling · Uncertainty

1 Introduction

Seepage analysis is always employed for the stability investigation of earth-fill dams [8]. As discussed by [7], approaches using deterministic parameters such as hydraulic conductivities need cautions and they suggested using spatially random fields when dealing with the issues.

Employing the randomness of soil parameters in seepage analysis through earth dams is a trending study in recent decades. For examples, probabilistic techniques have been used including the soil consolidation analysis [2], the bearing capacity assessment of footings [12], the determination of failure probability of slopes [10] and the seepage analysis [4, 5]. In these studies, the uncertainty of geotechnical properties such as hydraulic conductivity, undrained shear strength, and the others were considered. Some further studies in this problem can be listed [1, 7, 9, 14–16, 18, 20].

T. T. G. Vo (✉) · V. T. Nguyen
Faculty of Civil Engineering, Ho Chi Minh City University of Technology (HCMUT), Ly Thuong Kiet Street, District 10, Ho Chi Minh City, Vietnam
e-mail: tuyetgiang.vo@hcmut.edu.vn

Vietnam National University Ho Chi Minh City, Linh Trung Ward, Thu Duc District, Ho Chi Minh City, Vietnam

© The Author(s), under exclusive license to Springer Nature Singapore Pte Ltd. 2023
J. N. Reddy et al. (eds.), *ICSCEA 2021*, Lecture Notes in Civil Engineering 268,
https://doi.org/10.1007/978-981-19-3303-5_97

The current study employs an approach recently applied in Calamak and Yanmaz [4, 5] for analysing the seepage within embankment dams. It employs finite element method integrated into SEEP/W, a popular software for seepage investigation, incorporating the simple algorithm for the random field generation and using Monte Carlo simulations (MCS) for the further analysis. This study uses several types of earth dams to examine the effect of different geometries of dams into the outcomes. More specifically, this study focuses on evaluating two important parameters of seepage which are the flow rate and gradient. These outcomes then are compared with ones resulted from the determisnistic calculations.

2 Theoretical Description

The seepage analysis is in a steady state, so the governing equation is as follows [11]

$$\frac{\partial}{\partial x}\left(k_x \frac{\partial H}{\partial x}\right) + \frac{\partial}{\partial y}\left(k_y \frac{\partial H}{\partial y}\right) + Q = 0 \tag{1}$$

where, H is the total head; k_x and k_y are the hydraulic conductivities in x and y directions (since the problem is considered in isotropic environment so $k_x = k_y$); Q is the applied boundary flux.

Finite element method is applied to solve the problem. The meshes within SEEP/W are provided with quadrilaterals only or triangles only or the combination of them. Besides, when calculating the numerical integrations, the software offers 4 and 9 Gauss points for quadrilaterals and 1 and 3 for triangles.

Regarding the hydraulic conductivities, since the analysis geometry generates free surfaces within dams so there are the unsaturated zones existing. There are several models for estimating these conductivities including the unsaturation. The current study uses the model suggested by [19],

$$K(h) = K_s K_r(h) \tag{2}$$

where, K_s is the hydraulic conductivity at saturation, $K_r(h)$ is the normalized form of unsaturated hydraulic conductivity and h is the pressure head. Within the saturated zones, h is greater than 0 so that $K_r(h) = 1$ and within the unsaturated zones, $K_r(h)$ is smaller than 1. $K_r(h)$ can be calculated using the parameters α and n. Treating the uncertainty of soils, besides K_s, two parameters α and n be spatially variable. However, since the layers of unsaturation above the unsaturated zones are comparatively small so in this study α and n are considered as deterministic ones. Generating random variables is applied using their probability density functions (PDF) defined with the means and the coefficients of variation (COV). Hydraulic conductivities of soils also follow the lognormal distribution, which is shown in several studies (see [4, 5]). Therefore, generating this parameter can be implemented using the following equations (K hereafter denotes the saturated hydraulic conductivity)

$$\sigma_{\ln K}^2 = \ln\left(1 + \frac{\sigma_K^2}{\mu_K^2}\right) \tag{3}$$

$$\mu_{\ln K} = \ln \mu_K - \frac{1}{2}\sigma_{\ln K}^2 \tag{4}$$

$$K = \exp(\mu_{\ln K} + \sigma_{\ln K} r) \tag{5}$$

where, r is the random number which can be generated using the following equation [3]

$$r = (-2\ln u_1)^{1/2}\sin(2\pi u_2) \text{ or } r = (-2\ln u_1)^{1/2}\cos(2\pi u_2) \tag{6}$$

where, u_1 and u_2 are independent random variables from the same uniform density function on the interval of $(0,1)$. The generated hydraulic conductivities then are assigned to respective Gauss points. The implementation of these equations into SEEP/W is conducted by coding in C# and using the provided tools to be integrated into the Add-In directory of SEEP/W. The outcomes are collected using Monte Carlo Simulation (MCS) meaning that a number of repetitive simulations are conducted, and statistics techniques are employed to get the inference of the data.

3   Application into Different Types of Dams

In a companion study of the Authors, using this approach can lead to some concerns as follows:

- The distributions of random numbers r should obey normal-form ones within the meshes that are fine enough.
- The refinement of meshes leads to the more accurate results in terms of numerical issues [13] but changing the scale of fluctuation of the problem. If the analysis considers the scale of fluctuation, other methods of generation of random field should be used (e.g., [6].
- The number of samples in MCS should be large enough to get the accurate results. However, it is a massive task doing with SEEP/W. The number of 500 is suggested due to the acceptable results but it is recommended to do with the larger numbers.

However, this approach has the advantages that are the employment of finite element methods integrated into SEEP/W, a popular software in analyzing the seepage flow, and a technique for generating the random fields is simple compared to other existing techniques so that if the concerns above can be accepted, this

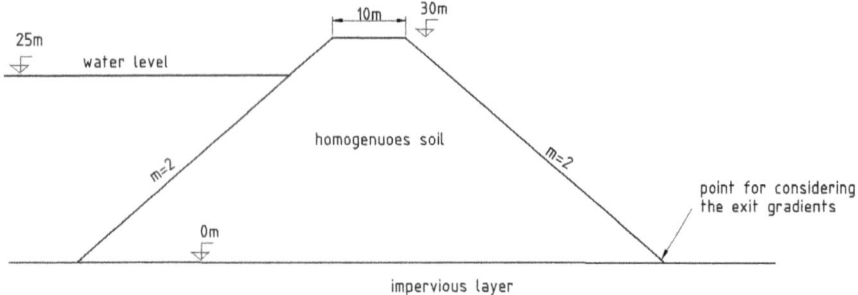

Fig. 1 The geometry of homogeneous dam

approach can be employed for the seepage analysis through earth dams considering the uncertainty in hydraulic conductivity.

In this study, the approach is employed for the analysis regarding the different types of earth dams. More specifically, beside the homogenous dams (Fig. 1), the ones installed with central-core and inclined-core [17] are used (Figs. 2 and 3). The geometries are simple, excluding necessary items in dams such as filters and drains. This study uses the mesh of triangles and 3 Gauss-point elements for the analysis. The element size of 2.5 m is automatically generated by the software. The outcomes for the further analysis are based on the total flow rates through the dam and the gradients (see Figs. 1, 2 and 3 for the interested points of gradients). These are the important outcomes resulted from any seepage analysis through earth dams. The soil parameters are shown in the Table 1. Since this study only investigates the impact of types of dams when using this approach so the other parameters are fixed such as COV, α and n. The readers can get these values based on the existing studies somewhere.

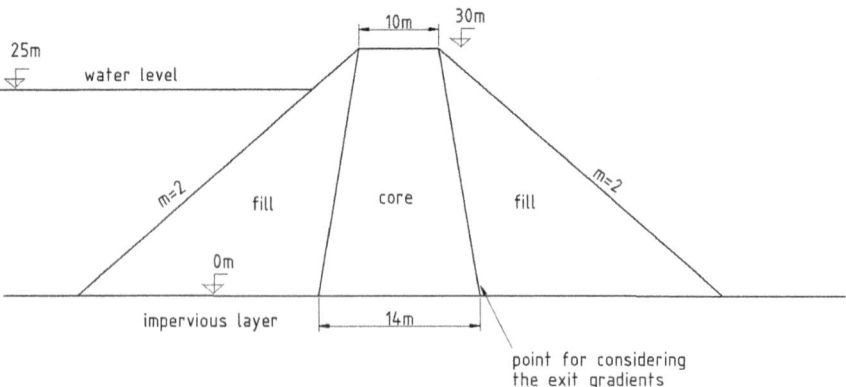

Fig. 2 The geometry of central-core dam

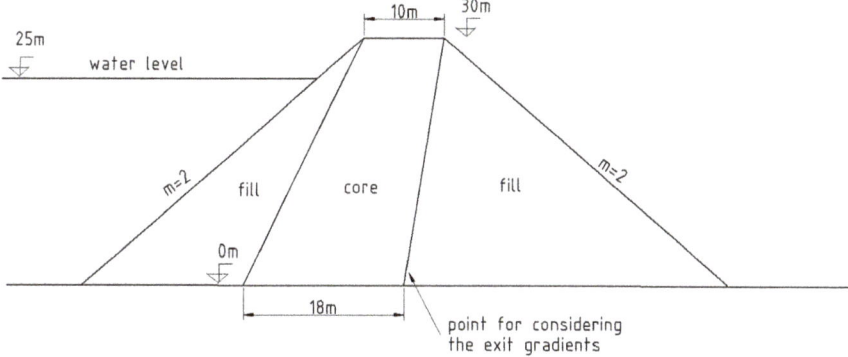

Fig. 3 The geometry of inclined-core dam

Table 1 Parameters for the analysis

Parameters	Fill soil	Core soil
Mean of hydraulic conductivity	1×10^{-6} (m/s)	1×10^{-8} (m/s)
COV of hydraulic conductivity	2	2
α (van Genuchten model)	0,02	0,02
n (van Genuchten model)	1,40	1,40

3.1 Random Numbers Generated Within the Parts

The random number r plays an important role in this approach as they represent the random fields of hydraulic conductivity within soil bodies. The generations of random numbers are examined, represented in Figs. 4, 5 and 6. The distributions of random numbers within the whole bodies of all types (homogenous, central-core and inclined-core) show that the mesh refinements are large enough so that they obey the normal-form ones.

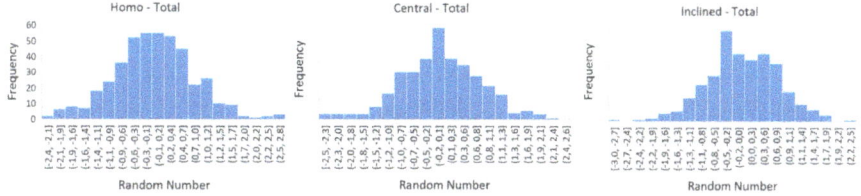

Fig. 4 The distributions of the random numbers r in the whole bodies of three dams

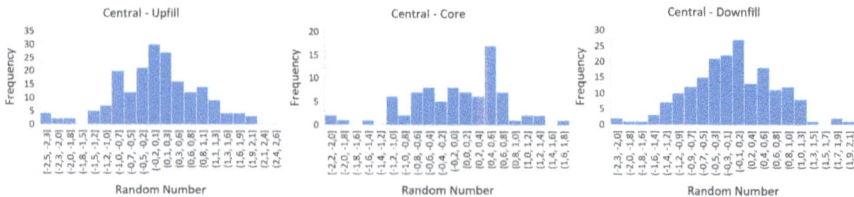

Fig. 5 The distributions of the random numbers r within parts of the central-core dam

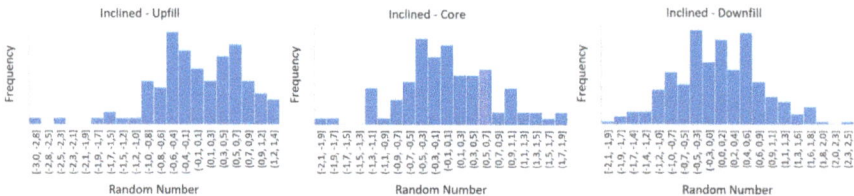

Fig. 6 The distributions of the random numbers r within parts of the inclined-core dam

However, more quantitatively, these frequency distributions are analysed using the Chi-square Test. The results show that they fit well with the relevant normal distributions. Therefore, the algorithm for random number generation is reliable.

3.2 Monte Carlo Simulations

This section presents the outcomes after the MCSs of the types of earth dams in Figs. 7 and 8, Tables 2 and 3. The results from the deterministic parameters are also shown for the comparison. There are some points that need to be discussed as follows:

- The Table 3 shows that there are the possibility of 100% the flow rates calculated from the random method are greater than ones from the deterministic method and the possibility of about 50% the gradients calculated from the random method are greater than ones from the deterministic method. This observation is particularly important, revealing that it is cautious when just using deterministic methods for designing.
- The COVs of the total flows of central-core and inclined-core dams are much greater than one of the homogenous dams. And it can be concluded that the more complicated geometries of dams, the more variable the results.
- Using the Chi-squared Test, the frequency distributions of flow rates fit well the relevant normal distributions. However, those of gradients do not fit well the relevant normal distributions and having the right-skewed trend.

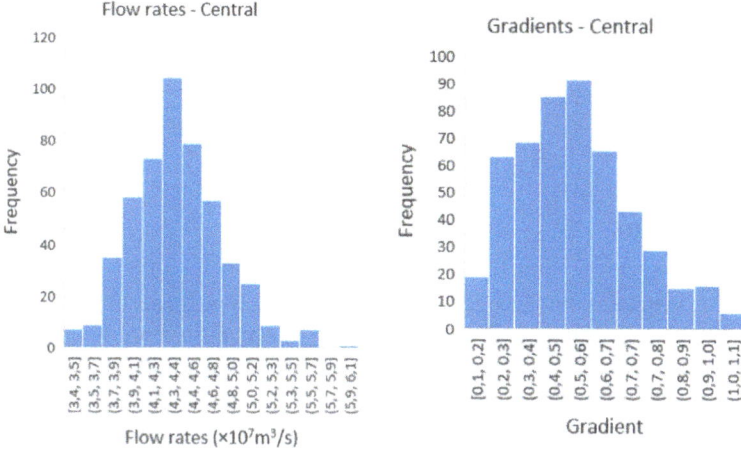

Fig. 7 The outcome of MCS of the central-core dam

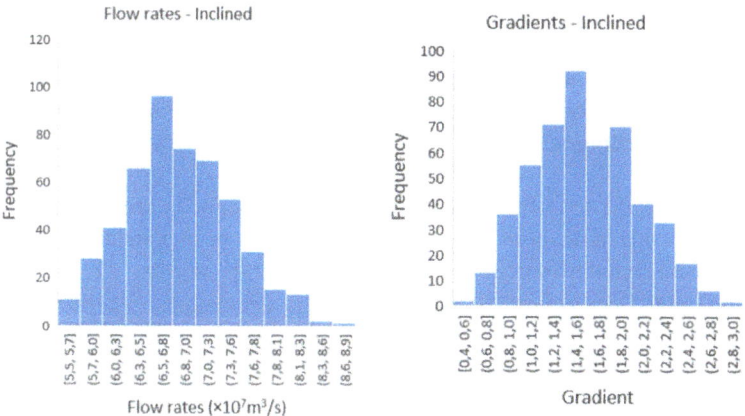

Fig. 8 The outcome of MCS of the inclined-core dam

Table 2 Statistical parameters of different types of dams

Types	Flow rates				Gradients			
	Mean ($\times 10^7\,\mathrm{m^3/s}$)	SD ($\times 10^7\,\mathrm{m^3/s}$)	COV	Deter. ($\times 10^7\,\mathrm{m^3/s}$)	Mean	SD	COV	Deter
Homogeneous	74,005	3,005	0,041	41,560	0,509	0,038	0,074	0,508
Central-core	4,400	0,415	0,094	2,458	0,507	0,206	0,406	0,542
Inclined-core	6,860	0,597	0,087	3,797	1,548	0,464	0,300	1,558

Table 3 The possibility that the results from the random method greater than those from the deterministic method

Types	Flow rate (%)	Gradient (%)
Homogeneous	100	51
Central-core	100	43
Inclined-core	100	49

4 Conclusions

This study investigates the seepage through the different types of earth-fill dams using the uncertainty of hydraulic conductivity of soils. The analysis employs an approach that has advantages such as using finite element analysis by SEEP/W, a popular software for seepage analysis, and incorporating a simple algorithm for generating random fields. The results show that there are some points that need to be cautious when only using the deterministic parameters when dealing with seepage analysis through earth dams as follows:

- The analysis shows that there are the possibility of 100% the flow rates calculated from the random method are greater than ones from the deterministic method and the possibility of about 50% the gradients calculated from the random method are greater than ones from the deterministic method. This observation is particularly important, revealing that it is cautious when just using deterministic methods for designing.
- The COVs of the total flows of central-core and inclined-core dams are much greater than one of the homogenous dams. And it can be concluded that the more complicated geometries of dams, the more variable the results.

Acknowledgements This research is funded by Vietnam National University Ho Chi Minh City (VNU-HCM) under grant number C2020-20-21.

References

1. Ahmed AA (2009) Stochastic analysis of free surface flow through earth dams. Comput Geotech 36(7):1186–1190
2. Bari M, Shahin M, Nikraz H (2013) Probabilistic analysis of soil consolidation via prefabricated vertical drains. Int J Geomech 877–881. doi: 10.1061/(ASCE)GM.1943-5622.0000244
3. Box GEP, Muller ME (1958) A note on the generation of random normal deviates. Ann Math Stat 29(2):610–611
4. Calamak M, Yanmaz AM (2018) Assessment of core-filter configuration performance of rock-fill dams under uncertainties. Int J Geomech 18(4):1532–3641
5. Calamak M, Yanmaz AM (2016) Uncertainty quantification of transient unsaturated seepage through embankment dams. Int J Geomech. doi:10.1061/(ASCE)GM.1943-5622.0000823
6. Fenton G, Vanmarcke EH (1990) Simulation of random fields via local average subdivision. J Eng Mech 116(8):1733–1749

7. Fenton G, Griffiths D (1996) Statistics of free surface flow through stochastic earth dam. J Geotech Eng 427–436. doi:10.1061/(ASCE)0733-9410(1996)122:6(427)
8. Foster M, Fell R, Spannagle M (2000) The statistics of embankment dam failures and accidents. Can Geotech J 37(5):1000–1024
9. Freeze RA (1975) A stochastic-conceptual analysis of one-dimensional groundwater flow in nonuniform homogeneous media. Water Resour Res 11(5):725–741
10. Genevois R, Romeo R (2003) Probability of failure occurrence and recurrence in rock slopes stability analysis. Int J Geomech 34–42. doi:10.1061/(ASCE)1532-3641(2003)3:1(34)
11. Geo-Slope International Ltd. (2012) Seepage modelling with SEEP/W. Calgary, Canada
12. Griffiths D, Fenton G, Manoharan N (2006) Undrained bearing capacity of two-strip footings on spatially random soil. Int J Geomech 421–427. doi:10.1061/(ASCE)1532-3641(2006)6:6(421)
13. Hutton DV (2004) Fundamentals of finite element analysis, 1st edn. McGraw-Hill, New York
14. Le TMH, Gallipoli D, Sanchez M, Wheeler SJ (2012) Stochastic analysis of unsaturated seepage through randomly heterogeneous earth embankments. Int J Numer Anal Methods Geomech 36(8):1056–1076
15. Lin GF, Chen CM (2004) Stochastic analysis of spatial variability in unconfined groundwater flow. Stochastic Environ Res Risk Assess 18(2):100–108
16. Mantoglou A, Gelhar LW (1987) Stochastic modeling of largescale transient unsaturated flow systems. Water Resour Res 23(1):37–46
17. Novak P, Moffat AIB, Nalurri C, Narayanan R (2007) Hydraulic Structures, 4th edn. Taylor and Francis, Abingdon
18. Tartakovsky DM (1999) Stochastic modeling of heterogeneous phreatic aquifers. Water Resour Res 35(12):3941–3945
19. van Genuchten MT (1980) A closed-form equation for predicting the hydraulic conductivity of unsaturated soils. Soil Sci Soc Am J 44:892–898
20. Zhang D (1999) Nonstationary stochastic analysis of transient unsaturated flow in randomly heterogeneous media. Water Resour Res 35(4):1127–1141

9 789811 933059